Applications of Metal-Organic Framework Composites
Exploring the Versatility of MOFs

Applications of Metal-Organic Framework Composites

Exploring the Versatility of MOFs

Edited by

Hafezeh Nabipour
Department of Chemical and Biochemical Engineering, University of Western Ontario, London, ON, Canada

Sohrab Rohani
Department of Chemical and Biochemical Engineering, University of Western Ontario, London, ON, Canada

ELSEVIER

Elsevier
Radarweg 29, PO Box 211, 1000 AE Amsterdam, Netherlands
125 London Wall, London EC2Y 5AS, United Kingdom
50 Hampshire Street, 5th Floor, Cambridge, MA 02139, United States

Copyright © 2025 Elsevier Inc. All rights reserved, including those for text and data mining, AI training, and similar technologies.

For accessibility purposes, images in electronic versions of this book are accompanied by alt text descriptions provided by Elsevier. For more information, see https://www.elsevier.com/about/accessibility.

Books and Journals published by Elsevier comply with applicable product safety requirements. For any product safety concerns or queries, please contact our authorised representative, Elsevier B.V., at productsafety@elsevier.com.

Publisher's note: Elsevier takes a neutral position with respect to territorial disputes or jurisdictional claims in its published content, including in maps and institutional affiliations.

No part of this publication may be reproduced or transmitted in any form or by any means, electronic or mechanical, including photocopying, recording, or any information storage and retrieval system, without permission in writing from the publisher. Details on how to seek permission, further information about the Publisher's permissions policies and our arrangements with organizations such as the Copyright Clearance Center and the Copyright Licensing Agency, can be found at our website: www.elsevier.com/permissions.

This book and the individual contributions contained in it are protected under copyright by the Publisher (other than as may be noted herein).

Notices

Knowledge and best practice in this field are constantly changing. As new research and experience broaden our understanding, changes in research methods, professional practices, or medical treatment may become necessary.

Practitioners and researchers must always rely on their own experience and knowledge in evaluating and using any information, methods, compounds, or experiments described herein. In using such information or methods they should be mindful of their own safety and the safety of others, including parties for whom they have a professional responsibility.

To the fullest extent of the law, neither the Publisher nor the authors, contributors, or editors, assume any liability for any injury and/or damage to persons or property as a matter of products liability, negligence or otherwise, or from any use or operation of any methods, products, instructions, or ideas contained in the material herein.

ISBN: 978-0-443-26743-7

For Information on all Elsevier publications
visit our website at https://www.elsevier.com/books-and-journals

Publisher: Candice Janco
Acquisitions Editor: Rafael Guilherme Trombaco
Editorial Project Manager: Andrea Dulberger
Production Project Manager: Paul Prasad Chandramohan
Cover Designer: Greg Harris

Typeset by MPS Limited, Chennai, India

Contents

List of contributors xvii
About the editors xxi
Preface xxiii
Acknowledgments xxv

1. **Introduction**
   Roghayyeh Ghasemzadeh and Kamran Akhbari

   Introduction 1
   Metal-organic frameworks 2
   Types of metal-organic frameworks 3
   Structures of metal-organic frameworks 9
   Properties of metal-organic frameworks 17
   Synthetic processes of metal-organic frameworks 26
   Activation of metal-organic frameworks 37
   Characterization of metal-organic frameworks 42
   Conclusion 46
   References 46

2. **Metal-organic framework composites**
   Megha Prajapati and Chhaya Ravi Kant

   Emergence and properties of metal-organic frameworks 61
   Structural and functional regulation of metal-organic
   frameworks and their composites 61
   Characterization of metal-organic frameworks and their composites 63
   Structural and morphological characterization 63
   Surface area and porosity characterization 64
   Chemical and compositional analysis 64
   Metal-organic framework polymer composites: synthesis and
   applications 65
   In situ polymerization 65
   Integrating prefabricated polymer ligands 66
   Grafting polymers onto metal-organic framework ligands post synthesis 67
   Adding preformed metal-organic frameworks to preformed polymers 67
   Crystallizing metal-organic frameworks around presynthesized
   polymers 68
   Metal-organic framework polymer composites for various applications 68

Metal-organic framework@carbon composites 71
In situ synthesis techniques 72
Ex situ synthesis techniques 73
Applications of metal-organic framework@carbon composites 73
Metal-organic framework@metal nanoparticles 78
Synthesis of metal-organic framework@metal nanoparticle composites 78
Solvent free gas phase loading technique 78
Solid grinding technique 79
Liquid impregnation method 79
Template synthesis technique 79
Applications of metal-organic framework@metal nanoparticle composites 80
Metal-organic framework@MXene composites 82
Synthesis of metal-organic framework@MXene composites 83
Applications of metal-organic framework@MXene composites 85
Conclusions and future prospects 85
References 87

3. **Metal-organic framework composites for supercapacitors**
   Shavita Salora, Jyoti Jangra, Shubham Kumar Patial and Suman Singh

   Introduction 97
   Fundamental of metal-organic frameworks 101
   Conventional synthesis techniques for metal-organic frameworks 102
   Advanced design strategies for enhanced performance 103
   Postsynthetic modification in metal-organic frameworks 104
   Presynthetic modification in metal-organic frameworks 105
   Topology of metal-organic frameworks 105
   Tailoring metal-organic frameworks for specific supercapacitor applications 106
   Metal-organic framework-based composite materials 110
   Rationale for combining metal-organic frameworks with other materials 110
   Types of metal-organic framework composites for supercapacitors 111
   Applications of metal-organic frameworks and metal-organic framework composites in supercapacitors 114
   Pure metal-organic framework electrodes for supercapacitors 114
   Metal-organic framework composites as electrode materials 115
   Conclusion 122
   References 123

4. **Metal-organic framework composites for carbon capture**
   Ying Liu and Zongbi Bao

   Introduction 135
   Metal-organic framework composites for CO_2 capture 137
   Shaped metal-organic framework composites 137
   Structural metal-organic framework-monolith composites 141

Amine-functionalized metal-organic framework composites	144
Metal-organic framework and carbon composites	151
Metal-organic framework and ionic liquid composites	162
Other metal-organic framework composites	167
Summary and perspective	169
References	169

5. **Metal-organic framework composites in fuel cells**
   Sreejitha Raj, Akhila Raman, Vikas Rajan and Appukuttan Saritha

Introduction	179
Metal-organic frameworks in fuel cells	180
Metal-organic frameworks and metal-organic framework composites as precursors for electrocatalysts	181
Metal-organic framework composites as electrocatalyst support materials	192
Metal-organic framework composites in fuel cell electrolytes	198
Functionalization of metal-organic framework composites in fuel cell electrolytes	199
Conclusion and future work	202
References	202

6. **Metal-organic framework composites for solar cells**
   Arife Gencer Imer

Introduction	211
Photovoltaic technology	212
Fundamentals of photovoltaic	213
Metal-organic framework composites	214
Synthesis of metal-organic framework composites	215
Application of metal-organic framework composites in photovoltatic devices	216
Metal-organic framework composites as working electrodes	217
Metal-organic framework composites in perovskite solar cells	228
Metal-organic framework composites in organic solar cells	236
Conclusion	237
Acknowledgment	238
References	238

7. **Metal-organic framework composites for electromagnetic interference shielding**
   Majed Amini, Aliakbar Isari, Seyyed Alireza Hashemi and Mohammad Arjmand

Introduction	251
Metal-organic frameworks, their composites, and their shielding mechanisms	253

Metal-organic framework shielding composites 258
Metal-organic framework composites with intrinsically conductive polymers 259
Metal-organic frameworks composites with MXenes 261
Metal-organic framework-based carbonaceous composites 264
Porous metal-organic framework-based shielding constructs 266
Metal-organic framework composite challenges and future prospects 270
References 271

8. Metal-organic framework composites for water splitting

Amir Kazemi and Faranak Manteghi

Introduction 277
Fundamentals of water splitting 279
Electrocatalytic water splitting: principles and mechanisms 279
Photocatalytic water splitting: principles and mechanisms 282
Merit of metal-organic frameworks for water splitting 282
Extremely high porosity combined with facile access to active sites 282
Optimizing metal nodes' performance 283
Impact of ligand engineering 283
Challenges and strategies water-splitting over metal-organic framework–based electrocatalysts 284
Metal-organic framework composites 284
Metal-organic framework composites with conductive materials 284
Metal-organic framework composites with porous materials 291
Other metal-organic framework composites 299
Conclusion and perspectives 304
References 307

9. Metal-organic framework composites for food applications

P.S. Sharanyakanth and Mahendran Radhakrishnan

Introduction 311
Metal-organic framework composites in food applications 311
Nanoparticle composites 311
Graphene oxide composites 312
Silica composites 313
Organic polymer composites 313
Quantum dot composites 314
Polyoxometalate composites 314
Activated carbon composites 315
Thin film on substrates 315
Aluminum composites 316
Heterostructures/hybrid composites 317

Enzyme composites	317
Composites of molecular species	318
Metal-organic framework cellulose composites	318
Applications of metal-organic framework composites in food systems	319
Active food contact materials	319
Antimicrobial nanocarriers	320
Regulated discharge of nano-systems for active compounds	320
Food packaging substance nanofillers	321
Food nanoreactors	321
Food material sensors	321
Immobilizers and stabilizers for enzymes and active compounds	322
Food cleaning	324
Conclusions	325
References	326

10. Metal-organic framework composites for adsorption of volatile organic compounds
Alireza Davoodi and Kamran Akhbari

Introduction	337
Volatile organic compound definitions	339
Variety and hazardous of volatile organic compounds	339
Less noticed source of volatile organic compounds	341
Strategies for reduction and elimination of volatile organic compounds	341
Suitable metal-organic frameworks for volatile organic compounds removal	342
Adsorption of volatile organic compounds by the metal-organic frameworks	343
Metal-organic framework composites for volatile organic compounds removal	343
Graphene, graphite, and their derivatives metal-organic framework composites	345
Metal, semimetal, and their salt metal-organic framework composites	348
Fiber and polymer metal-organic framework composites	349
Ordered mesoporous carbon, carbon nanotube, activated carbon, and another metal-organic framework composites	351
Adsorption in metal-organic framework composites	354
Parameter of volatile organic compounds adsorption	355
Thermodynamic parameters	355
Dynamic sorption methods	356
Disadvantages and possible costs	357
Metal-organic framework composites recycle	358
Conclusions and prospects	358
Conclusions	358

Prospects 359
AI disclosure 359
References 360

11. Metal-organic framework composites for antibacterial applications
Mehdi Ghaffari, Nazanin Habibi and Mohammad Reza Saeb

Introduction 373
Generic features of metal-organic frameworks 375
Surface engineering of metal-organic frameworks for biomedical applications 377
Bulk engineering of metal-organic frameworks for biomedical applications 377
Classification of metal-organic framework composites based on antibacterial properties 380
Fundamentals of antibacterial metal-organic frameworks and metal-organic framework composites 380
Mechanisms of bacteriostatic agents 381
Bactericidal mechanism 383
Metal-organic framework composites in antibacterial applications 387
Fundamentals 387
Silver-based metal-organic frameworks 388
Copper-based metal-organic frameworks 395
Iron-based metal-organic frameworks 396
Zeolitic imidazolate framework-based metal-organic frameworks 396
Zirconium-based metal-organic frameworks 397
Challenges with antibacterial metal-organic framework composites 398
Conclusion and future direction 399
References 400

12. Metal-organic framework composites for biomedical applications
Maryam Poostchi, Justyna Kucińska-Lipka, Mohsen Khodadadi Yazdi and Mohammad Reza Saeb

Introduction 411
Application of metal-organic frameworks in biomedical engineering 413
Application of metal-organic framework composites in biomedical engineering 416
Concluding remarks and future direction 429
References 430

13. Metal-organic framework composites for catalysis
Ramin Ebrahimi and Kamran Akhbari

Introduction	439
Metal-organic framework composites for catalysis	442
Carbon/metal-organic framework composites	442
Metal nanoparticle/metal-organic framework composites	449
Preparation methods for metal nanoparticle/metal-organic framework composites	450
Catalytic performances of metal nanoparticle/metal-organic framework composites	451
Common catalytic performances	451
Metal oxide nanoparticle/metal-organic framework composites	462
Preparation methods of metal oxide nanoparticle/metal-organic framework composites	463
Catalytic performance of metal oxide nanoparticle/metal-organic framework composites	464
Common catalytic performances	464
Photocatalysis	466
Polymer/metal-organic framework composites	471
Preparation methods of polymer/metal-organic framework composites	471
Catalytic performances of polymer/metal-organic framework composites	472
Common catalytic performances	472
Photocatalysis	480
Metal-organic framework/metal-organic framework composites	481
Preparation methods of metal-organic framework/metal-organic framework composites	481
Electrocatalysis	486
Photocatalysis	486
Polyoxometalate/metal-organic framework composites	488
Preparation methods of polyoxometalate/metal-organic framework composites	489
Catalytic performances of polyoxometalate/metal-organic framework composites	489
Common catalytic performances	489
Photocatalysis	496
Metal-organic framework/silica composites	498
Preparation methods of metal-organic framework/silica composites	498
Catalytic performances of metal-organic framework/silica composites	499
Common catalytic performances	499
Photocatalysis	502
Zeolite/metal-organic framework composites	502
Preparation methods of zeolite/metal-organic framework composites	503
Catalytic performances of zeolite/metal-organic framework composites	504

Common catalytic performances	504
Photocatalysis	508
Conclusion	511
AI disclosure	512
References	512

14. Metal-organic framework composites as electrocatalyst for carbon dioxide conversion
Khatereh Roohi, Jasper Coppen, Arjan Mol and Peyman Taheri

Introduction	539
Importance of the electrochemical carbon dioxide reduction reaction	539
Metal-organic framework composites for carbon dioxide reduction reaction	541
Scopes and objectives	542
Electrocatalytic mechanism of carbon dioxide reduction reaction	542
Electrochatalyst materials for carbon dioxide reduction reaction	544
Mechanistic study of carbon dioxide reduction reaction	545
Metal-organic framework composites as electrocatalysts	546
Recent advances on metal-organic framework composites for electrocatalytic carbon dioxide reduction reaction	547
Pristine metal-organic frameworks	547
Metal-organic frameworks as supports	548
Metal-organic frameworks on conductive substrates	552
Metal-organic frameworks on porous materials	553
Metal-organic framework-polymer/polyoxometalate composites	554
Metal-organic framework/MXene composites	556
Conclusion and outlook	556
References	558

15. Metal-organic framework composites for catalytic desulfurization and denitrogenation of fuels
Peiwen Wu

Background	567
Metal-organic framework composites for oxidative desulfurization of fuel oils	571
Metal-organic framework composites for oxidative desulfurization with H_2O_2 as the oxidant	571
Metal-organic framework composites for photo-catalytic oxidative desulfurization of fuel oils	580
Metal-organic framework composites-derived catalysts for oxidative desulfurization of fuel oils	585
Metal-organic framework composites-based porous ionic liquids for oxidative desulfurization of fuel oils	589
Metal-organic framework composites for oxidative denitrogenation of fuel oils	591

Metal-organic framework composites for oxidative denitrogenation of fuel oils using H_2O_2 as the oxidant	592
Metal-organic framework composites for oxidative denitrogenation of fuel oils using O_2 as the oxidant	595
Metal-organic framework composites-derived catalysts for oxidative denitrogenation of fuel oils	595
Conclusions and outlook	597
AI disclosure	600
References	600

16. Metal-organic framework composites for environmental remediation
Muhammad Altaf Nazir and Sami Ullah

Introduction to wastewater treatment	609
Industrial waste and water pollution	609
Current state of wastewater treatment	610
Metal-organic frameworks	611
Zeolitic imidazolate frameworks	612
Carboxylate frameworks	613
Nanoporous carbons derived from metal-organic frameworks	613
Adsorption mechanism	615
Influential factors on adsorption	616
Photocatalytic degradation mechanism	618
Factors influencing the photocatalytic degradation	620
Adsorption of pollutants	621
Adsorptive removal of organic dye by metal-organic framework composites	621
Adsorptive removal of pharmaceutical compounds by metal-organic framework composites	624
Adsorptive removal of heavy metals by metal-organic framework composites	626
Oil water separation by metal-organic framework composites	629
Degradation of pollutants	631
Degradation of organic dye by metal-organic framework composites	631
Degradation of pharmaceutical compounds by metal-organic framework composites	635
Degradation of heavy metals by metal-organic framework composites	637
Degradation of oil spills by metal-organic framework composites	639
Conclusion and perspective	641
References	643

17. Metal-organic framework composites for sensing applications
Ali Aghakhani

Introduction to sensors	657
Metal-organic frameworks-based optical sensors	658

Vapochromism sensors based on metal-organic frameworks	658
Luminescence sensors-based on metal-organic framework composites	665
Dielectric sensor based on metal-organic framework composites	673
Mechanical sensors based on metal-organic framework composites	684
Electrical sensor based on metal-organic framework composites	688
Conclusions	697
AI disclosure	698
References	698

18. Metal-organic framework composites for enhancing fire safety of polymers
Hafezeh Nabipour and Sohrab Rohani

Introduction	707
Flame retardants and mechanism of flame retardancy	709
Gas phase mechanism	709
Solid phase mechanism	710
Metal-organic framework composites as flame retardants	711
Flame retardant mechanism of metal-organic framework composites	712
Metal-organic framework composites applied in fire retardancy of epoxy resin	713
Metal-organic framework composites applied in fire retardancy of polyurethanes	717
Metal-organic framework composites applied in fire retardancy of polylactic acid	722
Metal-organic framework composites applied in fire retardancy of unsaturated polyester resin	725
Metal-organic framework composites applied in fire retardancy of polyurethane elastomer	727
Metal-organic framework composites applied in fire retardancy of polystyrene	728
Metal-organic framework composites applied in fire retardancy of polyurea	729
Metal-organic framework composites applied in fire retardancy of other polymers	730
Challenges and current limitations	735
Conclusion	736
References	737

19. Metal-organic framework composites as sensors for gases and volatile compounds
Alaa Bedair, Reda M. Abdelhameed, Marcello Locatelli and Fotouh R. Mansour

Introduction	745

Characterization of metal-organic framework composites	747
Metal-organic framework-based graphene/graphene derivatives composites for gas sensing	748
Metal-organic framework-based carbon nanotube composites for gas sensing	751
Metal-organic framework-based metal and metal oxide composites for gas sensing	755
Metal-organic framework-on-metal-organic framework composites for gas sensing	760
Perspectives	761
Conclusion	762
References	763

20. Metal-organic framework composites for immobilized enzyme applications
Muhammad Rezki, Seiya Tsujimura and K.S. Shalini Devi

Introduction	773
Preparation of enzyme-immobilized metal-organic framework composites	775
Effect of metal-organic framework hydrophobicity on enzyme immobilization	776
Effect of buffers	778
Effect of ligands	779
Effect of metals	780
Metal-organic framework composites for improving enzyme immobilization	781
Enhancement of immobilized enzyme activity using metal-organic framework composites	781
Improvement of immobilized enzyme stability in metal-organic framework composites	784
Recent fabrication and application of enzyme-immobilized metal-organic framework composites	785
Applications of enzyme-immobilized metal-organic framework composites	787
Biosensors	787
Enzyme-immobilized metal-organic framework composites for the environment	788
Metal-organic framework–enzyme composites for cancer therapy	789
Metal-organic framework–enzyme composites for wound healing-related applications	790
Metal-organic framework–enzyme composites in food and industrial applications	791
Metal-organic framework–enzyme composites for environmental application	792
Conclusion and future perspectives	792
References	793

21. Metal-organic framework composites: industrial applications
Fatemeh Shahrab and Kamran Akhbari

Introduction	803
Prominent industrial methods for metal-organic framework composites synthesis	809
Industrial applications	820
Chemical purification	822
Gases storage	830
Catalysis	837
Batteries and supercapacitors	841
Sensor and biosensor	845
Drug delivery	863
Metal-organic framework-based nanocomposites for bioimaging	873
Atmospheric water-harvesting	875
Food packaging	878
Conclusion	884
References	885

22. Perspectives and challenges of application of metal-organic framework composites
Hafezeh Nabipour and Sohrab Rohani

Introduction	915
Preparation of metal-organic framework composites	917
Metal-organic framework composites and their applications	918
Carbon-based metal-organic framework composites	918
Metal-based material@metal-organic framework composites	919
Metal oxide/metal-organic framework composites	919
Metal sulfide/metal-organic framework composites	919
Polymer@metal-organic framework composites	920
Polyoxometalate@metal-organic framework composites	920
Metal-organic framework@metal-organic framework composites	920
Perspectives on metal-organic framework composites	921
Challenges in the application of metal-organic framework composites	924
Future directions	926
Conclusion	926
References	927

Index 931

List of contributors

Reda M. Abdelhameed Applied Organic Chemistry Department, Chemical Industries Research Institute, National Research Centre, Giza, Egypt

Ali Aghakhani Department of Food Science, Engineering and Technology, Faculty of Agriculture, College of Agriculture & Natural Resources University of Tehran, Alborz, Karaj, Iran

Kamran Akhbari School of Chemistry, College of Science, University of Tehran, Tehran, Iran

Majed Amini Nanomaterials and Polymer Nanocomposites Laboratory, School of Engineering, The University of British Columbia, Kelowna, BC, Canada

Mohammad Arjmand Nanomaterials and Polymer Nanocomposites Laboratory, School of Engineering, The University of British Columbia, Kelowna, BC, Canada

Zongbi Bao Key Laboratory of Biomass Chemical Engineering of Ministry of Education, College of Chemical and Biological Engineering, Zhejiang University, Hangzhou, P.R. China

Alaa Bedair Department of Analytical Chemistry, Faculty of Pharmacy, University of Sadat City, Sadat City, Monufia, Egypt

Jasper Coppen Department of Materials Science and Engineering, Faculty of Mechanical Engineering, Delft University of Technology, Mekelweg, Delft, The Netherlands

Alireza Davoodi School of Chemistry, College of Science, University of Tehran, Tehran, Iran

K.S. Shalini Devi Graduate School of Pure and Applied Science, University of Tsukuba, Tsukuba, Ibaraki, Japan; Department of Material Sciences, Institute of Pure and Applied Sciences, University of Tsukuba, Tsukuba, Ibaraki, Japan

Ramin Ebrahimi School of Chemistry, College of Science, University of Tehran, Tehran, Iran

Arife Gencer Imer Physics Department, Van Yuzuncu Yil University, Tuşba, Van, Turkey

Mehdi Ghaffari Department of Polymer Engineering, Faculty of Engineering, Golestan University, Gorgan, Iran

Roghayyeh Ghasemzadeh School of Chemistry, College of Science, University of Tehran, Tehran, Iran

Nazanin Habibi Department of Polymer Engineering, Faculty of Engineering, Golestan University, Gorgan, Iran

Seyyed Alireza Hashemi Nanomaterials and Polymer Nanocomposites Laboratory, School of Engineering, The University of British Columbia, Kelowna, BC, Canada

Aliakbar Isari Nanomaterials and Polymer Nanocomposites Laboratory, School of Engineering, The University of British Columbia, Kelowna, BC, Canada

Jyoti Jangra Academy of Scientific and Innovative Research (AcSIR), Ghaziabad, Uttar Pradesh, India; Materials Science and Sensor Applications, CSIR-Central Scientific Instruments Organisation, Chandigarh, India

Amir Kazemi Research Laboratory of Inorganic Chemistry and Environment, Department of Chemistry, Iran University of Science and Technology, Tehran, Iran; Department of Chemical and Biochemical Engineering, Western University, London, ON, Canada

Justyna Kucińska-Lipka Department of Polymer Technology, Faculty of Chemistry, Gdańsk University of Technology, Gdańsk, Poland

Ying Liu Key Laboratory of Biomass Chemical Engineering of Ministry of Education, College of Chemical and Biological Engineering, Zhejiang University, Hangzhou, P.R. China

Marcello Locatelli Department of Science, University "G. d'Annunzio" of Chieti-Pescara, Chieti, Italy

Fotouh R. Mansour Medicinal Chemistry Department, Faculty of Pharmacy, King Salman International University (KSIU), Ras Sudr, South Sinai, Egypt; Department of Pharmaceutical Analytical Chemistry, Faculty of Pharmacy, Tanta University, Tanta, Egypt

Faranak Manteghi Research Laboratory of Inorganic Chemistry and Environment, Department of Chemistry, Iran University of Science and Technology, Tehran, Iran

Arjan Mol Department of Materials Science and Engineering, Faculty of Mechanical Engineering, Delft University of Technology, Mekelweg, Delft, The Netherlands

Hafezeh Nabipour Department of Chemical and Biochemical Engineering, University of Western Ontario, London, ON, Canada

Muhammad Altaf Nazir Institute of Chemistry, The Islamia University of Bahawalpur, Bahawalpur, Pakistan

Shubham Kumar Patial Academy of Scientific and Innovative Research (AcSIR), Ghaziabad, Uttar Pradesh, India; Materials Science and Sensor Applications, CSIR-Central Scientific Instruments Organisation, Chandigarh, India; School of Science, STEM College, RMIT University, Melbourne, VIC, Australia

Maryam Poostchi Department of Polymer Technology, Faculty of Chemistry, Gdańsk University of Technology, Gdańsk, Poland

Megha Prajapati Department of Applied Sciences and Humanities, Indira Gandhi Delhi Technical University for Women, Delhi, India; Electronics Materials Lab, College of Science and Engineering, James Cook University, Townsville, QLD, Australia

List of contributors

Mahendran Radhakrishnan Centre of Excellence in Nonthermal Processing, National Institute of Food Technology, Entrepreneurship and Management, Thanjavur (NIFTEM-T, formerly Indian Institute of Food Processing Technology, IIFPT), Thanjavur, Tamil Nadu, India

Sreejitha Raj Department of Mechanical Engineering, Amrita Vishwa Vidyapeetham, Kollam, Kerala, India

Vikas Rajan Department of Mechanical Engineering, Amrita Vishwa Vidyapeetham, Kollam, Kerala, India

Akhila Raman Department of Chemistry, Amrita Vishwa Vidyapeetham, Kollam, Kerala, India

Chhaya Ravi Kant Department of Applied Sciences and Humanities, Indira Gandhi Delhi Technical University for Women, Delhi, India

Muhammad Rezki Graduate School of Pure and Applied Science, University of Tsukuba, Tsukuba, Ibaraki, Japan

Sohrab Rohani Department of Chemical and Biochemical Engineering, University of Western Ontario, London, ON, Canada

Khatereh Roohi Department of Materials Science and Engineering, Faculty of Mechanical Engineering, Delft University of Technology, Mekelweg, Delft, The Netherlands

Mohammad Reza Saeb Department of Pharmaceutical Chemistry, Medical University of Gdańsk, Gdańsk, Poland

Shavita Salora Academy of Scientific and Innovative Research (AcSIR), Ghaziabad, Uttar Pradesh, India; Materials Science and Sensor Applications, CSIR-Central Scientific Instruments Organisation, Chandigarh, India

Appukuttan Saritha Department of Chemistry, Amrita Vishwa Vidyapeetham, Kollam, Kerala, India

Fatemeh Shahrab School of Chemistry, College of Science, University of Tehran, Tehran, Iran

P.S. Sharanyakanth Centre of Excellence in Nonthermal Processing, National Institute of Food Technology, Entrepreneurship and Management, Thanjavur (NIFTEM-T, formerly Indian Institute of Food Processing Technology, IIFPT), Thanjavur, Tamil Nadu, India

Suman Singh Academy of Scientific and Innovative Research (AcSIR), Ghaziabad, Uttar Pradesh, India; Materials Science and Sensor Applications, CSIR-Central Scientific Instruments Organisation, Chandigarh, India

Peyman Taheri Department of Materials Science and Engineering, Faculty of Mechanical Engineering, Delft University of Technology, Mekelweg, Delft, The Netherlands

Seiya Tsujimura Department of Material Sciences, Institute of Pure and Applied Sciences, University of Tsukuba, Tsukuba, Ibaraki, Japan

Sami Ullah Institute of Chemistry, The Islamia University of Bahawalpur, Bahawalpur, Pakistan

Peiwen Wu School of Chemistry and Chemical Engineering, Jiangsu University, Zhenjiang, Jiangsu Province, P.R. China

Mohsen Khodadadi Yazdi Division of Electrochemistry and Surface Physical Chemistry, Faculty of Applied Physics and Mathematics, Gdańsk University of Technology, Gdańsk, Poland; Advanced Materials Center, Gdańsk University of Technology, Gdańsk, Poland

About the editors

Hafezeh Nabipour is a researcher in the Department of Chemical and Biochemical Engineering at the University of Western Ontario. She previously served as an Associate Professor at the University of Science and Technology of China (USTC) for 3 years. She has authored over 80 ISI-indexed articles in peer-reviewed international journals. Her research focuses on synthesizing innovative metal-organic frameworks, designing novel bio-based polymers, and developing advanced complexes for applications in drug delivery, fire safety, and carbon dioxide capture.

Sohrab Rohani is a Full Professor in the Department of Chemical and Biochemical Engineering at the University of Western Ontario, where he also served as the past chair. His research focuses on solid-state chemistry and pharmaceutical control, the application of artificial intelligence in pharmaceutical crystallization, and the synthesis and applications of metal-organic frameworks. He has authored over 400 refereed journal publications.

Preface

The field of metal-organic frameworks (MOFs) has rapidly evolved, establishing itself as a fundamental aspect of modern materials science. MOF composites, in particular, have drawn significant attention from researchers worldwide due to their versatile structural properties, diverse functionality, and extensive range of applications. In response to the growing needs for a comprehensive resource, *Applications of Metal-Organic Framework Composites: Exploring the Versatility of MOFs* was created to serve as an authoritative guide for professionals and researchers in chemistry, materials science, and engineering.

This book bridges foundational principles with the latest research, offering valuable insights into the synthesis, functionalization, and characterization of MOF composites. Various chapters explore MOF applications in critical areas such as supercapacitors, fuel cells, solar cells, electromagnetic interference shielding, water splitting, food safety, adsorption of volatile organic compounds, sensors, immobilization of enzymes, carbon capture, catalysis, energy storage, environmental remediation, fire safety in polymers, CO_2 conversion, and biomedical technologies. Through reviews and expert perspectives, this book aims to provide a well-rounded understanding of MOF composites and inspire innovation.

Intended for researchers, educators, and students, this volume is designed for clarity and ease of navigation, enabling readers to easily access the latest developments and practical approaches. The chapters, written by experts from various fields, address current challenges, limitations, and future possibilities in this dynamic domain.

We hope this book serves as an essential reference to advance the science and applications of MOF composites, empowering readers to make meaningful contributions to this evolving field. We trust that it will inspire fresh ideas, foster collaborations, and drive breakthroughs in material design and real-world applications.

Hafezeh Nabipour
Sohrab Rohani

Acknowledgments

We would like to express our sincere gratitude to all the contributors for generously sharing their knowledge and expertise, making *Applications of Metal-Organic Framework Composites: Exploring the Versatility of MOFs* a comprehensive and invaluable resource. Our heartfelt thanks also go to the Faculty of Engineering at Western University and the Natural Sciences and Engineering Research Council of Canada (NSERC) for their support. Additionally, we are deeply appreciative of Elsevier for their steadfast collaboration and assistance in bringing this book to completion.

Chapter 1

Introduction

Roghayyeh Ghasemzadeh, and Kamran Akhbari
School of Chemistry, College of Science, University of Tehran, Tehran, Iran

Introduction

Porous materials play an important role in modern chemistry, materials science, and daily life. Although they have recently received scientific attention, their history dates back to antiquity. A wide range of applications has been explored for them, including catalysis, adsorption/separation, biomedicine, energy, and environmental technologies (Cai et al., 2021; Day et al., 2021). Porous materials are classified into microporous, mesoporous, and macroporous materials based on the size of their pores. According to the International Union of Pure and Applied Chemistry (IUPAC) notation, microporous materials have pore diameters of less than 2 nm, mesoporous materials between 2 and 50 nm, and macroporous materials have pore diameters greater than 50 nm (Lakshmi et al., 2022). Porous materials include activated charcoal, natural zeolites, synthetic zeolites, ceramics, mesoporous silica, activated carbon, metal oxides, polymers, microporous phosphates, porous metal, polymer foams, porous glass, covalent organic frameworks (COFs), and hydrogen-bonded organic frameworks (HOFs) usually either entirely organic or inorganic (Cai et al., 2021; Day et al., 2021; Huo, 2011; Lakshmi et al., 2022; Liu & Liu, 2020). They are often amorphous porous solids with undefined structures and irregular pores that hinder understanding structure–property relationships. For example, zeolites are solids with a crystalline structure, periodic structure, and intrinsic acidity that are favored for use in industrial adsorption and catalysis. Unfortunately, the limited kinds of pore structures and the difficulty in controlling acidic site distribution make it difficult to achieve rational control (Cai et al., 2021). In the late 1980s, it became evident that coordination polymers and coordination complexes can be highly crystalline. Most of the early work on 2D and 3D crystalline coordination polymers was done by Richard Robson. Later, Susumu Kitagawa advanced the field by designing porous hybrid inorganic–organic materials during the 1980s and 1990s. After the development of stable and permanently

2 Applications of Metal-Organic Framework Composites

porous metal-organic frameworks (MOFs) by Omar Yaghi in the late 1990s, porous coordination materials became popular (Day et al., 2021).

Metal-organic frameworks

MOFs, also known as porous coordination polymers (PCPs), are an emerging class of organic–inorganic hybrid and crystalline porous materials that are constructed by metal ions/clusters and organic ligands (Fig. 1.1). As a result of their structural and functional tunability, MOFs are one of the hottest fields in chemistry and materials science (Cai et al., 2021; Ghasemzadeh & Akhbari, 2023a, 2023b, 2023c, 2024; Wei, 2014; Xu & Kitagawa, 2018).

The framework structure, pore environment, and functionality of MOFs can be precisely controlled by carefully choosing desirable metal ions and organic ligands, rationally designing topological structures, and postsynthesis modification. Organic–inorganic hybrid structures merge the beneficial characteristics of organic and inorganic components while also endowing them with exceptional properties. Due to their high surface area and porosity, they provide abundant and easily accessible functional sites, which facilitate the adsorption and enrichment of substrate molecules. Having uniform pore sizes

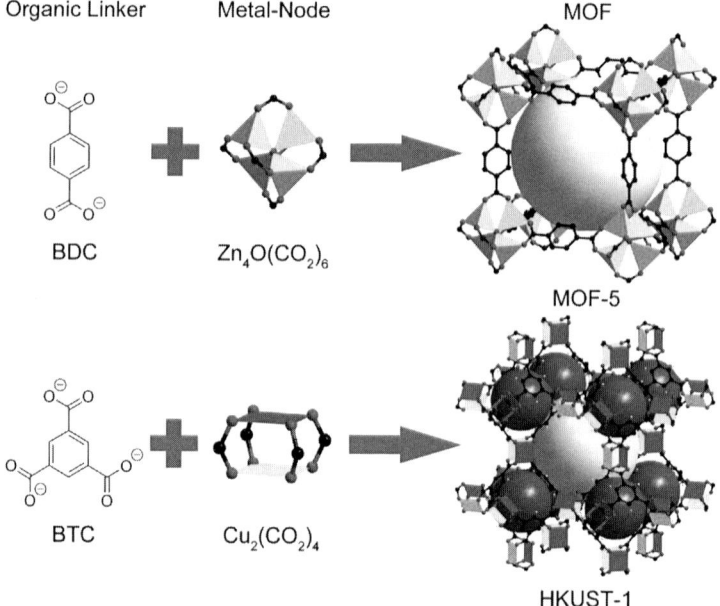

FIGURE 1.1 Schematic representation of the synthesis of MOF-5 and HKUST-1 using various metal ions and organic ligands. Free spaces in the framework are represented by yellow and blue spheres (Wei, 2014).

and shapes allows MOFs to react with substrates and products of different sizes and shapes, resulting in size-selective permeability. MOFs possess tunable pore wall environments at the atomic/molecular level, whether de novo or postsynthesis, which enable substrate and product recognition as well as substrate-active site interactions to be adjusted (Cai et al., 2021). Since MOFs are compatible with a wide range of materials, including quantum dots (QDs), nanoparticles (NPs), polyoxometalates (POMs), molecular species, silica, semiconductors, biomolecules, polymers, carbon material, dyes, metal ion/ligand immobilization, textile fibers, mixed-metal/linker, COFs, and other MOFs coupling, they provide an ideal platform in which functional species can be incorporated, resulting in a wide variety of MOFs composites (Cai et al., 2021; Chen & Xu, 2019; Ghasemzadeh & Akhbari, 2023a, 2023b, 2023c, 2024; Ghasemzadeh et al., 2018; Ma et al., 2020; Wang et al., 2020; Wei, 2014).

Types of metal-organic frameworks

MOFs are a fascinating research topic in modern materials chemistry and engineering that has contributed immensely to science's frontiers (Yadav et al., 2021). MOFs as functional materials have exploded in popularity due to their diverse applications and rapid advancements in design and synthesis. As a result, there are staggering numbers of MOFs with different compositions and structures, making classification increasingly difficult. Several methods have been developed to categorize them, tailored to certain needs and applications (Cao et al., 2022). There are many different types of MOFs, which can be classified in several ways. In general, they can be classified according to their dimensions and pore size (Cao et al., 2022; Zhang et al., 2022).

Classification of metal-organic frameworks-based on connection dimensions

MOFs are classified according to Yaghi's suggestion based on the dimensionality of the infinite [.node-linker-node.] connections and the spatial extension of the voids. It has been suggested by Yaghi that MOFs can be classified into four categories: zero-dimensional (0D), one-dimensional (1D), two-dimensional (2D), and three-dimensional (3D) (Jeremias, 2014).

0D metal-organic frameworks
The dimensions of 0D MOFs are finite in every direction. 0D MOFs are discrete complexes with one or more voids (Fig. 1.2A) (Jeremias, 2014).

1D metal-organic frameworks
1D MOFs can extend in one direction indefinitely. Stacking of 1D chains generates voids that cause porosity. 1D pores in 1D MOFs are typically channels (Fig. 1.2B) (Jeremias, 2014).

4 Applications of Metal-Organic Framework Composites

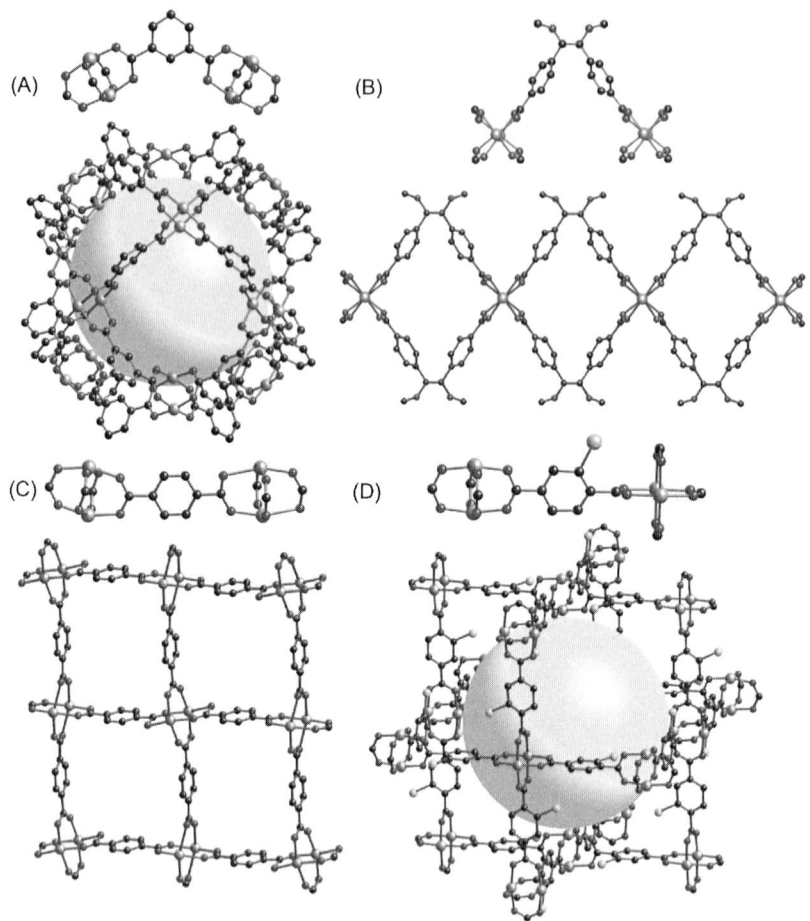

FIGURE 1.2 Examples for 0D, 1D, 2D, and 3D MOFs based on Yaghi's suggestion. (A) MOF-1 (0D), (B) MOF-222 (1D), (C) MOF-2 (2D), and (D) MOF-101 (3D) (Yaghi et al., 2003).

2D metal-organic frameworks

2D MOFs consist of stacked or infinite layers and have porosity caused by the channels generated by the pillared windows. Layer-like voids are also found in 2D MOFs (Fig. 1.2C) (Jeremias, 2014).

3D metal-organic frameworks

3D MOFs are composed of infinite coordination networks that extend spatially in all three dimensions. A 3D MOF's porosity is determined by the geometry of its node/linker coordination. As the building blocks are assembled, isolated, inaccessible 0D pores or cage-like 3D pores (intersecting channel networks) form (Fig. 1.2D) (Jeremias, 2014).

Classification of metal-organic frameworks-based on inorganic subnetworks dimensions

Depending on the chemical conditions, it can gradually increase the dimensionality of the inorganic subnetwork of MOFs, from 0D coordination polymers to 3D. In 2008, Férey suggested considering the dimensionality of inorganic subnetworks instead of the dimensionality of the [...node-linker-node...] network. This concept can also be used to categorize complex structures of several MIL-type MOFs, which may have inorganic 1D chains, 2D grids, or 3D networks (Fig. 1.3) (Férey, 2008; Yaghi et al., 2003).

Classification of metal-organic frameworks-based on morphological dimensions

MOFs can be considered a subclass of coordination polymers that can be extended into 1D, 2D, and 3D, using the terminology adopted by IUPAC in 2013 (Raptopoulou, 2021). MOFs can be classified based on their morphological dimensions into 1D, 2D, and 3D structures (Fig. 1.4) (Abdelhamid, 2023).

One-dimensional metal-organic frameworks with 1D morphology

The 1D structure is characterized by being oriented in a particular direction. The 1D MOF has similar morphological properties to the 1D structure, including unique geometric and electronic properties. The characteristics of MOF material and 1D structures are combined in 1D MOF material. The properties of this material include adjustable metal centers, ample active sites, and good flexibility, as well as a large surface area, controllable crystal structure, and porous structure. 1D MOFs are classified as nanowires,

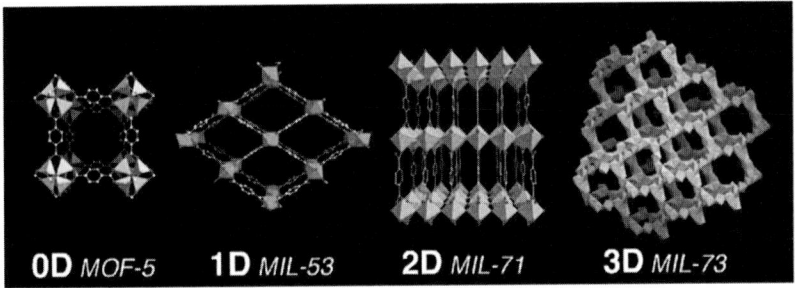

FIGURE 1.3 Examples of metal-organic frameworks with various inorganic sub-networks dimensions: MOF-5 (0D), MIL-53 (1D), MIL-71 (2D), MIL-73 (3D) (Férey, 2008).

6 Applications of Metal-Organic Framework Composites

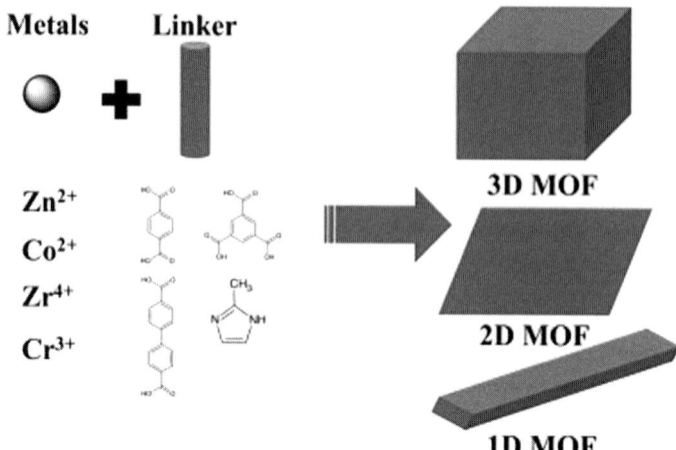

FIGURE 1.4 Examples of metal-organic frameworks with various morphological dimensions: 1D, 2D, and 3D (Abdelhamid, 2023).

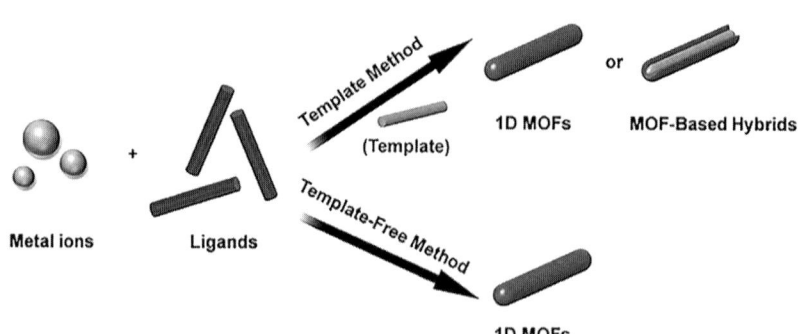

FIGURE 1.5 Schematic for preparing 1D MOFs using templates and templates-free methods (Su et al., 2023).

nanotubes, nanorods, and nanoribbons. As yet, 1D MOF research is in its infancy and focuses primarily on developing new synthetic methods. There are two types of synthesis methods for 1D MOFs, template and template-free, as shown in Fig. 1.5 (Pan et al., 2023; Su et al., 2023).

Two-dimensional metal-organic frameworks with 2D morphology

The discovery of graphene has led to an increase in research interest in 2D nanomaterials. 2D materials fall under the category of nanomaterials that have

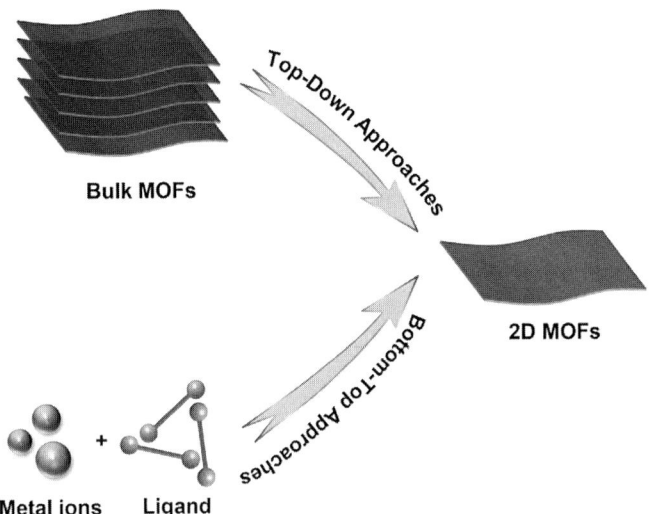

FIGURE 1.6 A diagram of the creation of 2D metal-organic frameworks using both top–down and bottom–top strategies (Su et al., 2023).

sheet-like structures and transverse dimensions greater than 100 nm. Research into 2D MOF nanosheets is growing because of their unique properties, such as their ultrathin thickness and large surface area with highly accessible active sites. The synthesis of 2D MOF materials has been divided into two main categories: top–down and bottom–top strategies (Fig. 1.6) (Hu, Mei, et al., 2019; Su et al., 2023; Zhao et al., 2018).

Three-dimensional metal-organic frameworks with 3D morphology

In 3D MOFs, the morphology of the structure is unique in three dimensions (Fig. 1.7). 3D MOFs exhibit fantastic morphologies, large specific surface areas, and large porosities compared to low-dimensional MOF materials (Su et al., 2023).

Classification of metal-organic frameworks-based on pore size

MOFs are unique crystal structures with multiple geometries, structures, component types, and pore sizes that find applications across a wide range of industries. A MOF with high porosity, specific pore size, and unparalleled chemical and thermal stability can be achieved through careful engineering of synthesis steps. MOFs' porosity determines their ability in various applications. Therefore, categorizing MOFs based on the size of their pores and knowing their advantages and limitations can provide insight into selecting the right MOF. According to IUPAC definition, MOFs are divided into

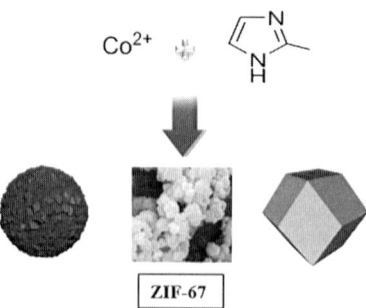

FIGURE 1.7 Schematics for synthesizing 3D ZIF-67 with 3D morphology (Bibi et al., 2021).

micropores, mesopores, and macropores based on pore size. Pore sizes of microporous MOFs are less than 2 nm, mesoporous MOFs are between 2 and 50 nm, and macroporous MOFs are greater than 50 nm (Zhang et al., 2022).

Microporous metal-organic frameworks

Microporous MOFs have pores smaller than 2 nm, which are the first category of MOFs. The small pores of these materials are ideal for targeting absorption and separation of small molecules such as gases, storing energy, and converting it in solar cells and electric batteries. Because of their small pores, microporous MOFs are limited in various applications due to their incapacity to penetrate and spread large molecules into their active sites. As a result, researchers have focused on the design of MOFs with larger pores due to their highly porous structures (Zhang et al., 2022).

Mesoporous metal-organic frameworks

Mesoporous MOFs have pores between 2 and 50 nm, providing an ideal space for molecules to enter the interior of its cavity. Thus, they are more effective in a wide range of applications, including drug delivery, catalysts, and dye absorption (Zhang et al., 2022).

Macroporous metal-organic frameworks

Lastly, macroporous MOFs with pore sizes greater than 50 nm fall into this category of MOFs. Because of their large internal spaces, this MOF category provides suitable sites for transferring large molecules such as proteins and enzymes. Moreover, macroporous MOFs have been shown to offer enhanced catalytic activity, greater efficiency, and more suitable sites for polymerization reactions due to their large space (Zhang et al., 2022).

Structures of metal-organic frameworks

The first reports of MOF structures date back to the 1960s by Tomic and others, but research on MOFs has greatly increased since the 1990s, especially with the discovery of MOF-based porous materials by Yaghi and his group. The basic structure of MOFs is composed of metals or metal clusters (also called secondary building units, SBUs) linked by organic ligands. As a result of the combination of organic and inorganic building blocks, there are a variety of variations in pore size, shape, and structure, as well as a number of opportunities for functionalization. Combining organic and inorganic building blocks provides many variations in pore size, shape, and structure, along with many possibilities for functionalization. In addition to SBU geometry, the shape and size of organic ligands control the structure of MOFs. Fig. 1.8 illustrates some representative metal nodes as SBUs and organic linkers to obtain diverse MOF crystalline structures (Jiao et al., 2019; Meena et al., 2023).

MOFs are organic–inorganic hybrid solids composed of organic linkers and metallic clusters that have a uniform, infinite framework structure (Li et al., 2012). Over 100,000 structures have been reported in the MOFs subset of the Cambridge Structural Database (CSD) in the last few decades, and MOF-related publications have steadily increased (Zhang et al., 2020). The lack of a broadly accepted definition during the development of this new type of hybrid material led to several parallel names being used. PCP seems to have been the most widely adopted among them.

Metal-organic frameworks structures based on configuration

Several types of MOFs are described in the configuration source, such as isoreticular metal-organic frameworks (IRMOFs), porous coordination networks (PCNs), zeolitic imidazolate frameworks (ZIFs), zeolite-like metal-organic framework (ZMOFs), metal-azolate frameworks (MAFs), and metal-biomolecule frameworks (MBioFs) (Al Obeidli et al., 2022; Huo et al., 2021; Jiao et al., 2019; Kampouraki et al., 2019; Li et al., 2024; Meena et al., 2023; Yusuf et al., 2022).

Isoreticular metal-organic frameworks

A series of MOFs known as IRMOFs own similar network topologies. IRMOF-1 (MOF-5) was first synthesized by Omar M. Yaghi's group (Fig. 1.9). IRMOFs-n (n=1–8, 10, 12, 14, 16) were synthesized using $[Zn_4O]^{6+}$ SBU and aromatic carboxylates with similar pcu topology (Mai & Liu, 2019; Yusuf et al., 2022).

Porous coordination networks

PCNs are stereo-octahedral materials with a hole-cage-hole topology, and they belong to the class of MOFs with a three-dimensional structure. The PCNs include PCN-222 (Fig. 1.10), PCN-333, PCN-224, and PCN-57 (Mai & Liu, 2019).

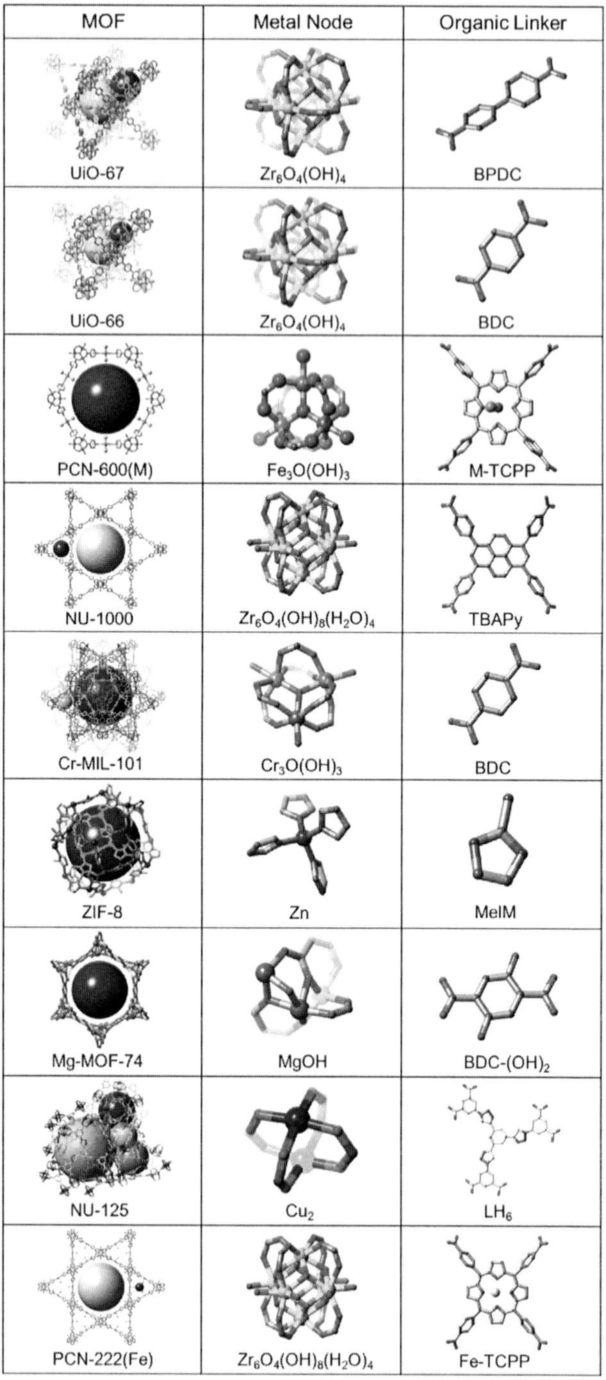

MOF	Metal Node	Organic Linker
UiO-67	$Zr_6O_4(OH)_4$	BPDC
UiO-66	$Zr_6O_4(OH)_4$	BDC
PCN-600(M)	$Fe_3O(OH)_3$	M-TCPP
NU-1000	$Zr_6O_4(OH)_8(H_2O)_4$	TBAPy
Cr-MIL-101	$Cr_3O(OH)_3$	BDC
ZIF-8	Zn	MeIM
Mg-MOF-74	MgOH	$BDC-(OH)_2$
NU-125	Cu_2	LH_6
PCN-222(Fe)	$Zr_6O_4(OH)_8(H_2O)_4$	Fe-TCPP

FIGURE 1.8 Representations of various structures of metal-organic frameworks are based on the shape, size of organic ligands, and geometry of SBUs (Howarth et al., 2017).

Zinc metal center (secondary building unit) Terephthalic acid IRMOF-1 unit cell IRMOF-1 framework

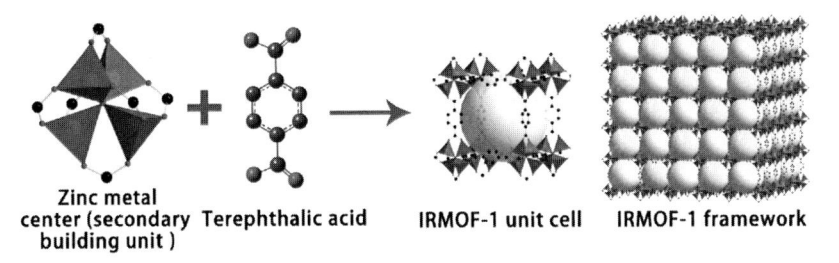

FIGURE 1.9 Illustrations of isoreticular metal-organic framework-1's secondary building units, organic linker, unit cell, and framework (Huo et al., 2021).

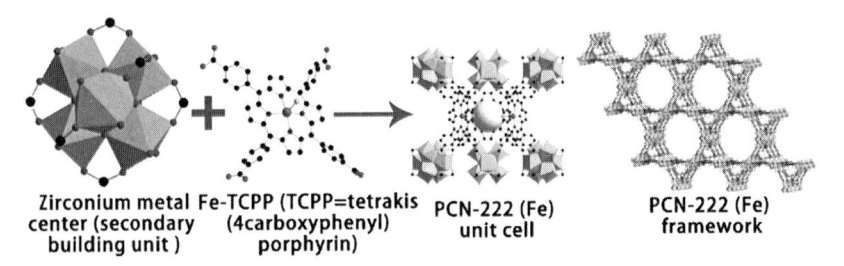

Zirconium metal center (secondary building unit) Fe-TCPP (TCPP=tetrakis (4carboxyphenyl) porphyrin) PCN-222 (Fe) unit cell PCN-222 (Fe) framework

FIGURE 1.10 Illustrations of porous coordination network-222 (Fe)'s secondary building unit, organic linker, unit cell, and framework (Huo et al., 2021).

Zeolitic imidazolate frameworks

ZIFs are composed of tetrahedrally-coordinated transition metal ions (e.g., Fe, Co, Cu, Zn) connected by imidazolate linkers. ZIFs exhibit zeolite-like topologies due to their metal-imidazole-metal angle that is similar to the 145 degrees Si–O–Si angle in zeolites. ZIFs include ZIF-8 (Fig. 1.11), ZIF-90, ZIF-L, ZIF-71, ZIF-67, and ZIF-7, as well as other ZIFs (Bhattacharjee et al., 2014; Molina et al., 2024; Yusuf et al., 2022).

Zeolite-like metal-organic framework

ZMOFs are a subset of MOFs with topologies and features similar to traditional inorganic zeolites. ZMOFs' unique pore systems and cage-based cavities with modular intra- and/or extra-framework components enable them to tailor pore size, shape, and properties to specific applications. A series of indium ZMOFs with rho (CPM-519-RHO), gis (CPM-520-GIS), and abw (CPM-521-ABW) topologies have been demonstrated with oxalic acid as the organic ligand (Fig. 1.12) (Eddaoudi et al., 2015; Liu & Liu, 2021).

12 Applications of Metal-Organic Framework Composites

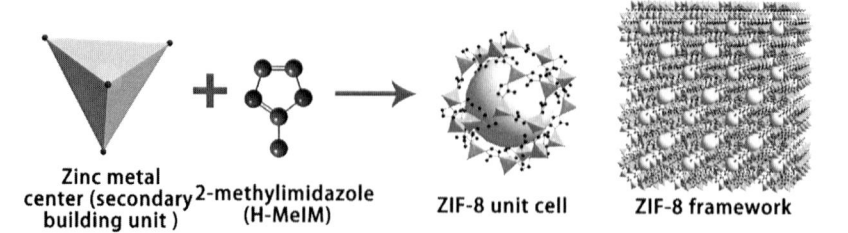

FIGURE 1.11 Illustrations of zeolitic imidazolate framework-8's secondary building unit, organic linker, unit cell, and framework (Huo et al., 2021).

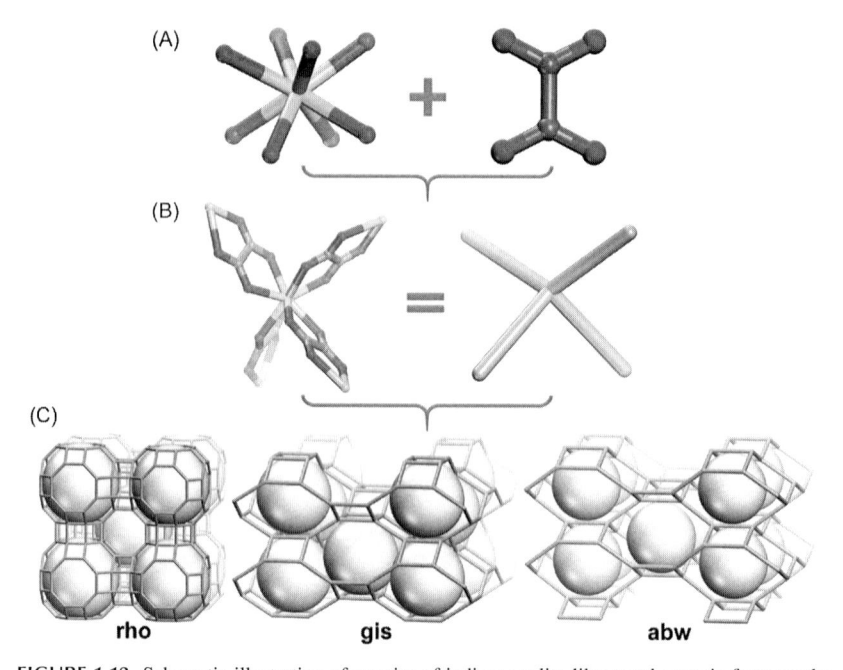

FIGURE 1.12 Schematic illustration of a series of indium zeolite-like metal-organic frameworks using oxalic acid as a ligand with zeolitic topologies like rho, gis and abw (Liu and Liu, 2021).

Metal-azolate frameworks

MAFs, a subclass of MOFs, are composed of metal nodes and azolate linkers (Fig. 1.13). Azoles, the five-membered aromatic nitrogen heterocycles, including imidazole (Him), pyrazole (Hpz), 1,2,4-triazole (Htz), 1,2,3-triazole (v-trizole, Hvtz), and tetrazole (Httz), can also be deprotonated to form the corresponding azolate (or azolide) anions. Due to azolate ligands' strong

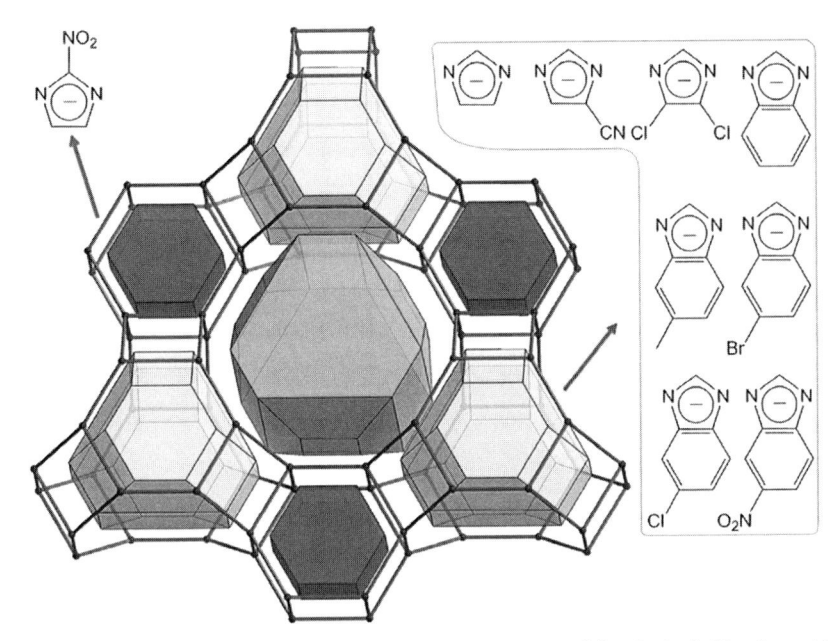

FIGURE 1.13 Zinc(II) imidazolate frameworks constructed by mixing 2-nitroimidazolate with other imidazolates (Zhang et al., 2012).

coordination ability in bridging metal ions, metal azolate frameworks have emerged as a promising new kind of MOF (Zhang et al., 2012).

Metal-biomolecule frameworks

A variety of biomolecules have been developed by nature that have the potential to bridge metal ions. MBioFs have been designed using these molecules as excellent biocompatible building blocks. The MBioFs are made up of metal ions linked to organic molecules derived from biological sources such as amino acids, proteins, peptides, nucleic acids, porphyrins, and saccharides. There are also numerous acids and bases naturally present in humans, animals, or plants as either free acids or carboxylate salts. It is clear that some of these molecules are candidates for MBioFs linkers because they have more than one carboxylate group (Fig. 1.14) (Imaz et al., 2011; Wang, Wang, et al., 2020).

Metal-organic frameworks structures based on university names

Researchers have also used an acronym containing the names of the universities where the materials were prepared. MOFs structures include Materials Institute Lavoisier (MIL) and University of Oslo (UiO), Hong Kong University of Science and Technology (HKUST), Instituto de Tecnología Química Metal-Organic

14 Applications of Metal-Organic Framework Composites

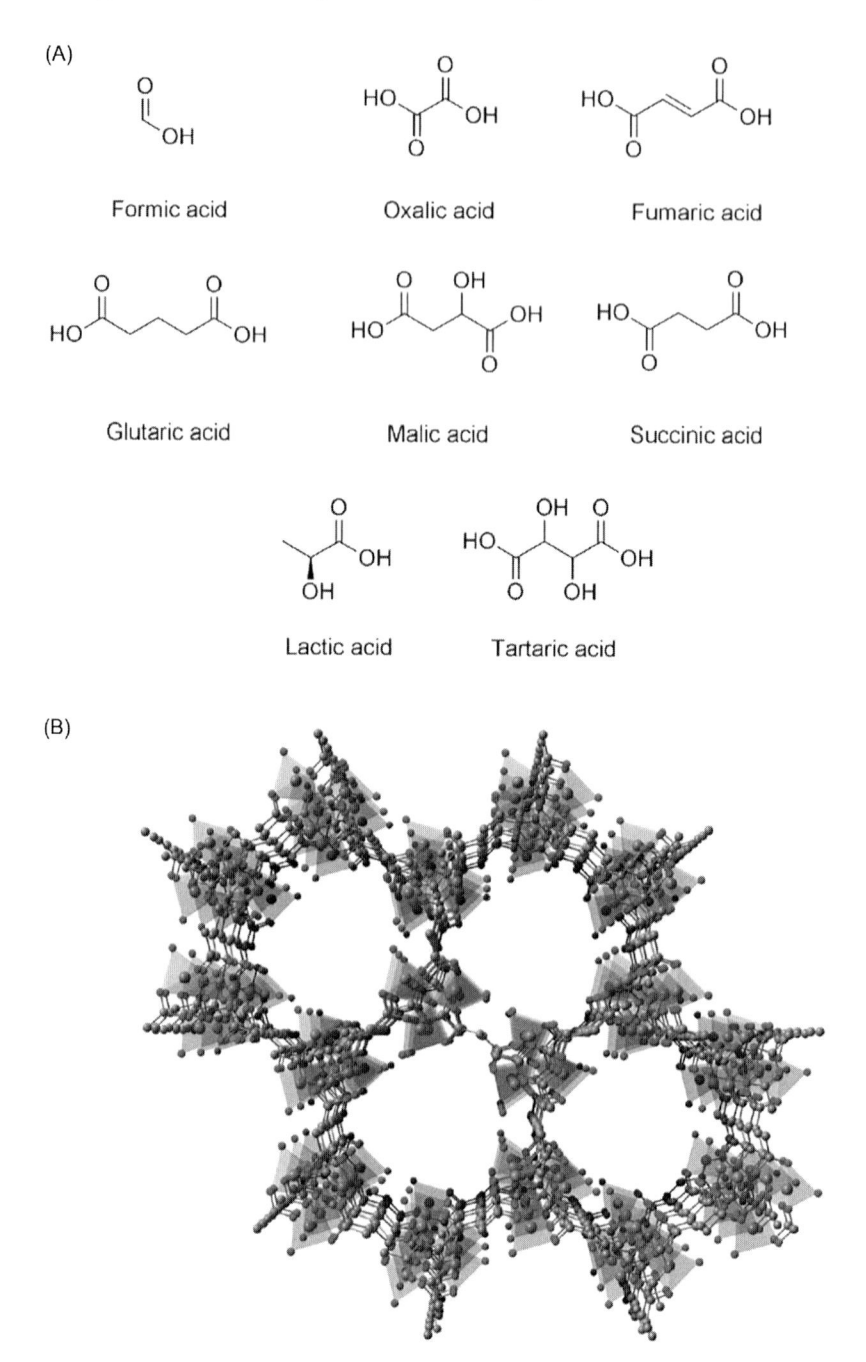

FIGURE 1.14 (A) Typical dicarboxylate ligands found naturally in humans, animals, and plants. (B) A schematic representation of the porous structure [Fe$_3$O(MeOH)$_3$(fum)$_3$(CO$_2$CH$_3$)].4.5MeOH (Imaz et al., 2011).

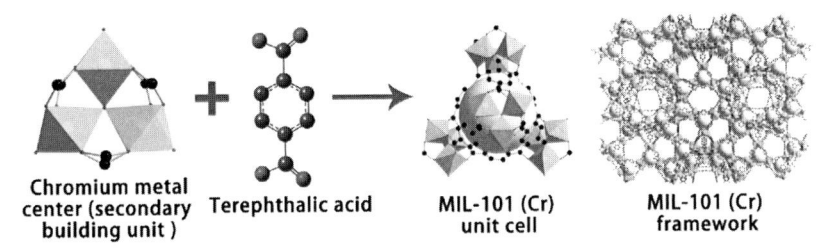

FIGURE 1.15 Illustrations of Materials Institute Lavoisier-101(Cr)'s secondary building unit, organic linker, unit cell, and framework (Huo et al., 2021).

Framework (ITQMOF), Seoul National University (SNU), Jilin University China (JUC), Cambridge University KRICT (CUK), Pohang University of Science and Technology (POST), Northwestern University (NU), Dresden University of Technology (DUT), University of Nottingham (NOTT), Christian-Albrechts-University (CAU) (Huo et al., 2021; Li et al., 2024; Li et al., 2012; Meena et al., 2023; Yusuf et al., 2022), Tarbiat Modares University (TMU) (Bigdeli et al., 2020), and Materials from University of Tehran (MUT) (Karimi Alavijeh et al., 2019) among others. A number of MOF structures have been investigated, including MIL, UiO, and MUT.

Materials Institute Lavoisier

MILs are synthesized using different elements and organic compounds containing two carboxylic functional groups. Upon outward incitement, MIL were capable of changing their pore size arrangement freely. There are several MILs, including MIL-101, MIL-100, MIL-53, MIL-88, MIL-125, etc. (Yusuf et al., 2022). Fig. 1.15 shows a schematic of structure MIL-101(Cr) (Huo et al., 2021).

University of Oslo

The UiO series of MOFs contains Zr^{4+} and dicarboxylic acid ligands, thereby creating a three-dimensional porous structure. There are many advantages to UiO, including their uniform size, large specific surface area, high stability, and active clusters of Zr–O. Despite their different ligand lengths, UiO-64, UiO-66, UiO-67, UiO-68, UiO-69, and various derivatives possess the same reticular structure. The change of the ligand does not affect the thermal stability of UiO (Ru et al., 2021). A schematic of the UiO-66 structure is shown in Fig. 1.16 (Huo et al., 2021).

Materials from University of Tehran

MUT are MOFs that contain main group and transition metal ions and dicarboxylic acid ligands. There are several different types of MUT, including

16 Applications of Metal-Organic Framework Composites

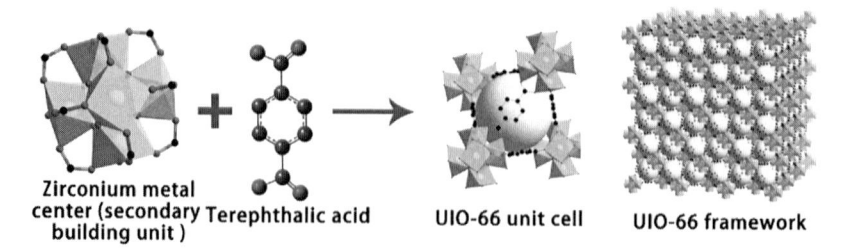

FIGURE 1.16 Illustrations of University of Oslo-66's secondary building unit, organic linker, unit cell, and framework (Huo et al., 2021).

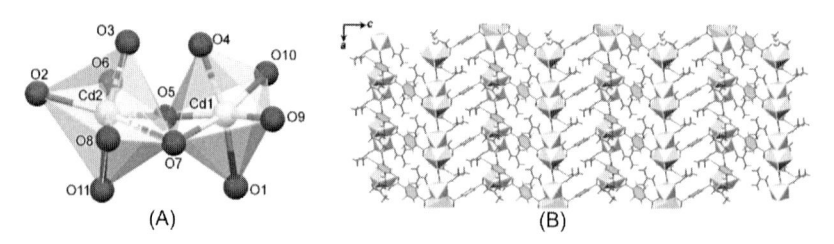

FIGURE 1.17 (A) Coordinate geometry of crystallographically independent cadmium centers within the framework of MUT-11. (B) A representation of the MUT-11 framework, showing the three-dimensional (3D) network in the crystallographic ac-plane, with polyhedral SBUs (Parsaei et al., 2023).

MUT-1, MUT-2, MUT-3, MUT-4, MUT-5, MUT-6, MUT-7, MUT-8, MUT-11 (Fig. 1.17), MUT-16 and etc. (Ghasemzadeh et al., 2024; Karimi Alavijeh et al., 2019, 2021, 2022; Parsaei, Akhbari & Tylianakis, Froudakis, et al., 2022; Parsaei, Akhbari & Tylianakis, & Froudakis, 2023; Parsaei, Akhbari, & Kawata, 2023; Parsaei, Akhbari, & White, 2022; Salimi et al., 2022).

Metal-organic frameworks structure-based on metal-organic framework's network

It's common to use the empirical formula for MOFs that display metals, ligands, and their stoichiometry in almost all published papers. Also, O'Keeffe and his colleagues proposed a systematic terminology to classify known structures according to the structure of a MOF's network. A 3D MOF network with a given geometrical linkage topology can be identified by using a three-letter symbol, such as 'dia,' 'cub,' or its extension, 'pcu-a,' 'cub-d,' etc. Using this approach can aid in the description and understanding of structures and also create a blueprint for the design of new MOFs (Li et al., 2012).

Properties of metal-organic frameworks

The emergence of MOFs has been one of the most important events in the chemistry and materials science communities over the past two decades. MOFs are a new class of porous materials that have opened up new research possibilities. MOFs have intriguing properties such as porosity, tunable pore size, customizable structures, flexibility, stability, magnetism, optical and chemical properties, open metal sites, electrical conductivity, proton conductivity, and host-guest interactions (Aggarwal et al., 2022; Ahmadi et al., 2022; Alhumaimess, 2020; Brunner & Rauche, 2020; Li et al., 2016; Ren et al., 2013). Due to their fascinating properties, MOFs have applications in numerous fields, including supercapacitors (Choi et al., 2014; Du et al., 2018; Otun et al., 2024; Wang, Xun, et al., 2020), carbon capture (Ansone-Bertina et al., 2022; Ding et al., 2019; Simmons et al., 2011; Tao & Xu, 2024), fuel cells (Li & Gao, 2022; Ponnada et al., 2021; Ren et al., 2013; Wen et al., 2020), solar cells (Heo et al., 2020; Maza et al., 2016; Nevruzoglu et al., 2016; Ryu et al., 2018), sensing (Achmann et al., 2009; Kumar et al., 2015; Lei et al., 2014; Olorunyomi et al., 2021), gas storage/separation (Akhbari & Morsali, 2013a, 2015; Jiang et al., 2022; Li & Wang, Fan, et al., 2022), drug delivery (Karimi Alavijeh & Akhbari, 2020, 2022, 2023, 2024; Parsaei & Akhbari, 2022a, 2022b, 2022c, 2023, 2024), antibacterial agents (Arshadi Edlo & Akhbari, 2023, 2024; Davoodi et al., 2023; Kalati & Akhbari, 2021, 2022; Karimi Alavijeh et al., 2018; Khattami Kermanshahi & Akhbari, 2024; Mohammadi Amidi & Akhbari, 2024; Mohammadi Amidi et al., 2023; Soltani & Akhbari, 2022a, 2022b; Soltani et al., 2022a), as templates (Ghasemzadeh & Akhbari, 2023a, 2023b, 2023c, 2024; Guo et al., 2019; Hu, Masoomi, et al., 2019; Mirzadeh & Akhbari, 2016), catalysis (Bavykina et al., 2020; Hu et al., 2018; Obeso et al., 2023; Salimi et al., 2023; Yang & Gates, 2019), electrocatalysis (Li et al., 2023; Shahbazi Farahani et al., 2022; Shen & Kung, 2023; Zheng & Lee, 2021), and photocatalysis (Bedia et al., 2019; Ghasemzadeh & Akhbari, 2023a, 2023b, 2023c, 2024; Ghasemzadeh et al., 2024; Qiu et al., 2018, 2024; Wang, Gao, et al., 2020).

Porosity of metal-organic frameworks

The porosity of MOFs, or the ratio of void volume to MOF total volume, is one of its most important properties. Due to their high porosity and specific surface area, MOFs are suitable for a variety of uses, including the encapsulation of biomolecules, gas storage, separation, catalysis, and sensing (Ahmadi et al., 2022). The porosities of MOFs can be systematically tuned through careful selection of the molecular building blocks. Additionally, predesigning or postsynthetic methods can introduce a range of functional sites/groups into organic linkers, metal ions/clusters, or pore spaces (Li et al., 2016). To control MOF porosity, it's common to alter the dimensions or chemical functionalities of organic linkers while maintaining the same underlying MOF topology

18 Applications of Metal-Organic Framework Composites

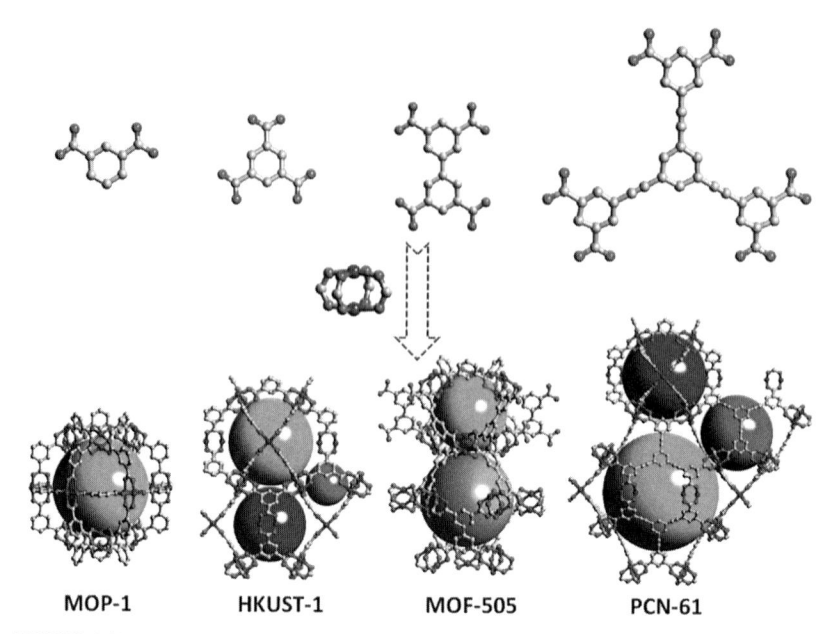

FIGURE 1.18 Changing in porosity of MOP-1, HKUST-1, MOF-505, and PCN-61 with different types of organic linkers (He et al., 2014).

(Zhang et al., 2020). Generally, high porosity MOFs have longer linkers, while low porosity MOFs usually have shorter linkers (Shen, 2023). The porosity of MOP-1, HKUST-1, MOF-505, and PCN-61 rises when organic linkers with increasing length are incorporated, as demonstrated in Fig. 1.18 (He et al., 2014).

Tunable pore size of metal-organic frameworks

The pore size in MOFs can be tuned by choosing the right building block molecules, changing the functional groups on the organic ligands or metal ions, or modifying presynthetic and postsynthetic modifications. Usually, the organic linker plays a key role in forming porous MOFs and tailoring the pore shapes and sizes. Adding functional groups to organic linkers (–OH, –Cl, –Br, and –NH$_2$) can alter their properties, including their hydrophilicity. The pore size and interactions between frameworks and guest molecules will be further modified by the various functional groups, which will make MOFs a suitable material for gas selectivity and separation. As shown in Fig. 1.19, the pore sizes of the RMOF-*n* series can be adjusted by utilizing various organic linkers with different structures and geometries to coordinate with metal nodes (Ahmadi et al., 2022; Shen, 2023).

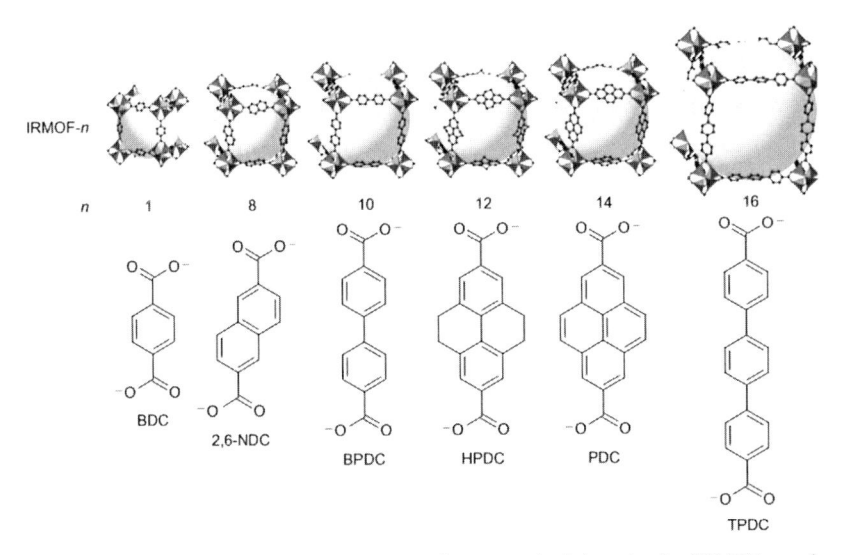

FIGURE 1.19 Altering the pore size using various organic linkers in the IRMOF-*n* series (Tsivadze et al., 2019).

Customizable structure of metal-organic frameworks

Since MOFs contain both inorganic and organic units, starting materials can be chosen more freely. MOFs can be constructed with different interactions between metal ions and organic ligands due to their flexible coordination chemistry. Theoretically, an infinite number of possible frameworks can be constructed. Depending on the organic linkers' length, geometry, and functional groups, the shape, size, and interior decoration of pores could be altered (Alhumaimess, 2020). With the right choice of metallic centers and specific ligands and an organized synthetic approach, crystal structure and topology are controlled. A schematic representation of MOF structures is shown in Fig. 1.20 to illustrate their impressive structural variability (Rocío-Bautista et al., 2019).

Flexibility of metal-organic frameworks

MOFs can be flexible yet robust enough to maintain their structure. A change in external conditions, such as temperature, pressure, chemical medium, etc., can cause movement in the framework. A MOF's structure can be adjusted to accommodate gases or liquids injected. Also, host-guest interactions and host behavior affect the response. It is possible to shrink or expand pores or channels reversibly, without breaking bonds, when gases or liquids fill or evacuate them. The dynamic or "breathing effect" refers to this type of response. In dynamic effects, the linker and/or metal clusters have to be flexible in their coordination to change the bond angles slightly without breaking the bonds (Alhumaimess, 2020). Fig. 1.21 illustrates the breathing

20 Applications of Metal-Organic Framework Composites

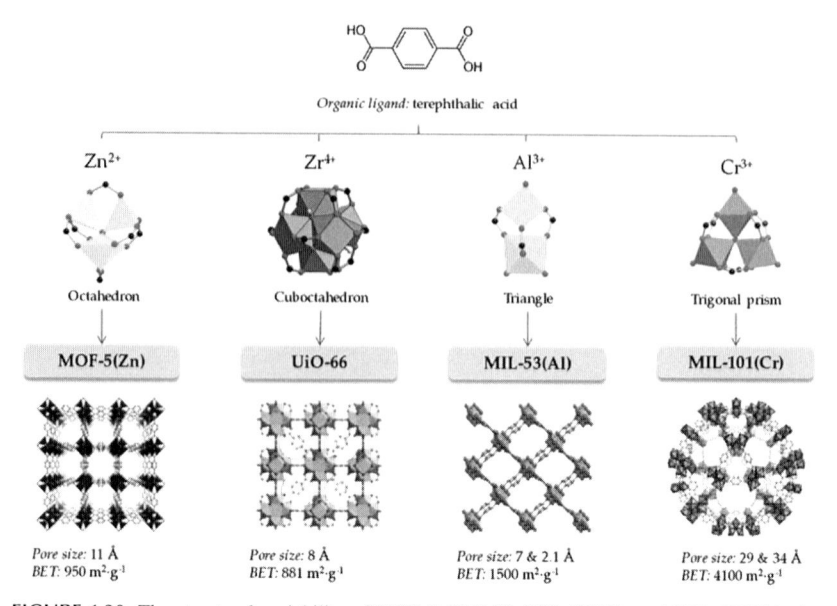

FIGURE 1.20 The structural variability of MOF-5, UiO-66, MIL-53(Al), and MIL-101(Cr), due to their flexible coordination chemistry between different metal ions and terephthalic acid (Rocío-Bautista et al., 2019).

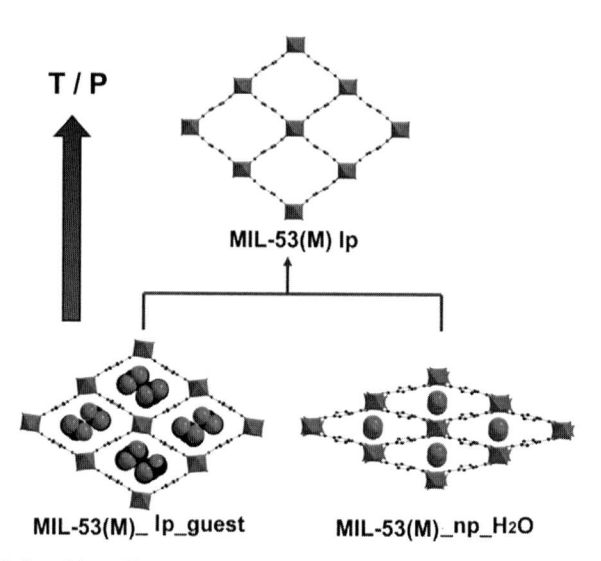

FIGURE 1.21 Breathing effects of Materials Institute Lavoisier-53(M) with temperature or pressure (*lp*, Large pore; *np*, narrow pore) (Wu et al., 2019).

effects of MIL-53(M) with temperature and pressure. Upon removing trapped molecules from Cr- and Al-based MIL-53, it causes the large pore (lp) to change directly from either the loaded narrow pore (np_guest) or the loaded large pore (lp_guest) forms to the empty large pore (lp) (Wu et al., 2019).

Stability of metal-organic frameworks

The stability of MOFs is crucial for a variety of practical applications. MOFs usually suffer from chemical, thermal, mechanical, and electrochemical stabilities (Baumann et al., 2019; Jiao et al., 2019). Different factors affect a MOF's stability, such as the metal ions and ligands used in synthesis, the geometry of metal-ligand coordination, the rigidity of ligands, and hydrophobic groups (Annamalai et al., 2022).

Chemical stability of metal-organic frameworks

MOFs' chemical stability can be described as their ability to maintain their long-range ordered structures in water and certain chemical environments, like acidic and alkaline conditions. The strength of metal-ligand bonds determines the stability of MOFs (Wen et al., 2021). Based on the hard and soft acids and bases (HSAB) theory, metal nodes and deprotonated organic linkers can be considered Lewis acid-base pairs, and their coordination bonds can be estimated. Those coordination bonds between hard acid metal ions such as Zr^{4+} and Cr^{3+} with smaller ionic radii and higher oxidation states (Fig. 1.22) and hard bases (the oxygen atoms in carboxylate linkers) are much stronger than those between hard acids and soft bases (neutral pyridine linkers). Conversely, soft acid metal ions with larger ionic radii and lower oxidation states (Fig. 1.22), such as Zn^{2+} and Ni^{2+}, can form stable frameworks with charged soft base linkers, such as azolates. In contrast, pairing hard acids and soft bases, or soft acids and hard bases, usually results in less stable MOFs (Zhang et al., 2020).

The HSAB theory can be used to design MOF materials that achieve the stability required by their intended applications; MOFs with matching

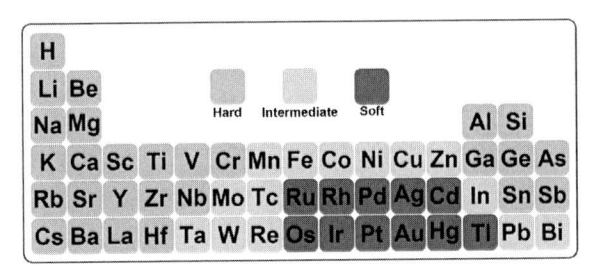

FIGURE 1.22 The classification of hard, soft, and intermediate acids on the periodic table (Zhang et al., 2020).

metal-organic linker pairs tend to be more stable (Fig. 1.23) (Wen et al., 2021; Zhang et al., 2020).

In addition to the metal-linker bonds, the design and modification of linkers can also stabilize the final frameworks. In MOFs, hydrophobic ligands can prevent water from contacting metal-linker bonds, which leads to water stability. Commonly hydrophobic functional groups such as fluoro, alkyl, and aromatic groups can be incorporated in MOFs through isoreticular substitution, de novo design, or postsynthesis modifications. A substitution of hydrophobic linkers with isoreticular linkers can improve the stability of some parent MOFs in the presence of water (Fig. 1.24) (Zhang et al., 2020).

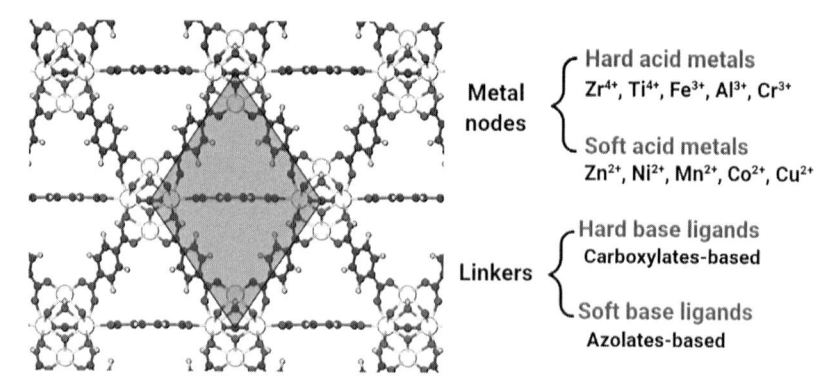

FIGURE 1.23 Structures of University of Oslo-66 and common hard and soft acids and bases (Wen et al., 2021).

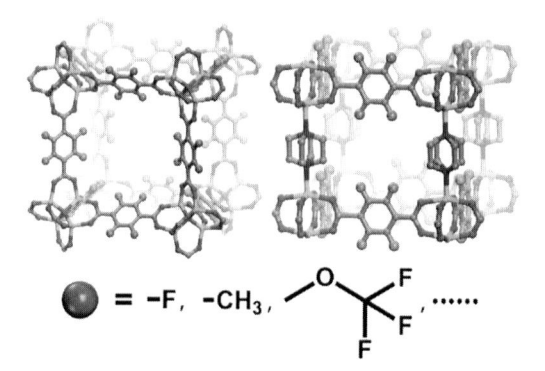

FIGURE 1.24 Schematic of the construction of hydrophobic metal-organic frameworks by introducing hydrophobic groups onto organic linkers using reticular chemistry (Zhang et al., 2020).

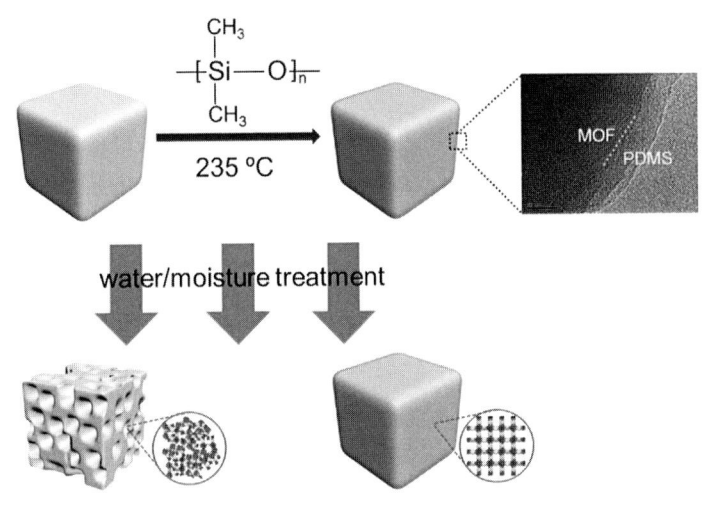

FIGURE 1.25 Illustration of polydimethylsiloxane coating on the surface of metal-organic frameworks and improvement of moisture/water resistance (Zhang et al., 2014).

However, increasing hydrophobicity by functionalizing ligands results in porosity shrinkage, while surface coatings do not. In order to improve moisture/water stability while maintaining their intrinsic properties (high surface area, pore texture, and crystalline structure), a facile polydimethylsiloxane (PDMS) coating treatment is used to form a hydrophobic layer on the surface of MOFs. PDMS-coated MOFs exhibit high hydrophobic behavior and excellent moisture/water tolerance as well as good porosity retention (Fig. 1.25) (Wen et al., 2021; Zhang et al., 2014).

Thermal stability of metal-organic frameworks

The thermal stability of MOFs is also important for their applications requiring elevated temperatures, such as high-temperature gas phase reactions. The node-linker bond strength and the number of linkers connected to each metal node determine thermal stability. MOFs with high thermal stability are often synthesized from metal ions with high valence, such as Ln^{3+}, Al^{3+}, Zr^{4+}, and Ti^{4+} (Jiao et al., 2019).

Mechanical stability of metal-organic frameworks

From an engineering perspective, MOFs' mechanical stability under vacuum or pressure is another key factor for their industry and practical applications. Despite their extraordinary porosity and high surface area, MOFs are inherently less mechanically stable due to their extraordinary porosity. Mechanical loading can result in phase changes, partial pore collapse, or even amorphization of a material. Often, grinding and ball milling reduce

MOF crystallinity and negatively affect its performance. MOF mechanical degradation is caused by capillary forces exerted by the removal of strongly coupled, pore-filling solvent molecules, especially water. As a result, it can be easy to confuse this with hydrolytic degradation. Moreover, the functional groups on the organic ligands may either enhance mechanical stability by forming nonbonded secondary networks or weaken MOFs by destabilizing bonded networks. As a consequence, it is crucial to design, test, and optimize functional groups that can synergistically match the secondary network with the bonding network of MOFs (Jiao et al., 2019).

Electrochemical stability of metal-organic frameworks
The electrochemical stability of MOFs can be enhanced by selecting appropriate synthetic parameters. MOFs with redox-inactive nodes and shorter, more rigid linkers exhibit greater thermal and chemical stability. However, high porosity and flexibility provide superior ion storage and transport. Multimetallic frameworks, flexible linkers, modulated host-guest interactions, and crystal size manipulations can all be used to tune mechanical properties. Electrodes that experience dendrite formation may need structural rigidity to prevent dangerous short-circuiting. Interestingly, MOF malleability may also be useful in devices undergoing volume expansion or contraction, where structural integrity is most critical. Further, stimuli-responsive flexible MOFs could be used as a safety feature, where a temperature, voltage, or mechanical signal activates a shut-off mechanism to protect the device and its users (Baumann et al., 2019).

Open metal sites in metal-organic frameworks
Metal sites in MOFs have a significant impact on their properties. Metals in MOFs often act as Lewis acids and coordinate with labile solvent molecules or counterions during synthesis. The activation of MOFs causes the weakly bonded species to be removed, resulting in open metal sites (OMS) or coordinatively unsaturated sites (CUS) or open coordination sites (OCS) (Fig. 1.26). The use of MOFs with open metal sites has been proposed for many different applications, including catalysis, gas storage and capture, separation, and sensing (Denysenko et al., 2014; Kökçam-Demir et al., 2020; Ren et al., 2013).

Chemical properties of metal-organic frameworks
Chemical properties of MOFs, such as the chemical composition and chemistry of the molecular building units, can affect their structural rigidity and flexibility. As a result of various chemical interactions within MOFs, such as coordination bonds and hydrogen bonds, MOFs can exhibit various chemical properties. Postsynthetic modification (PSM) of MOFs can regulate their chemical properties, such as pore reactivity (Ahmadi et al., 2022).

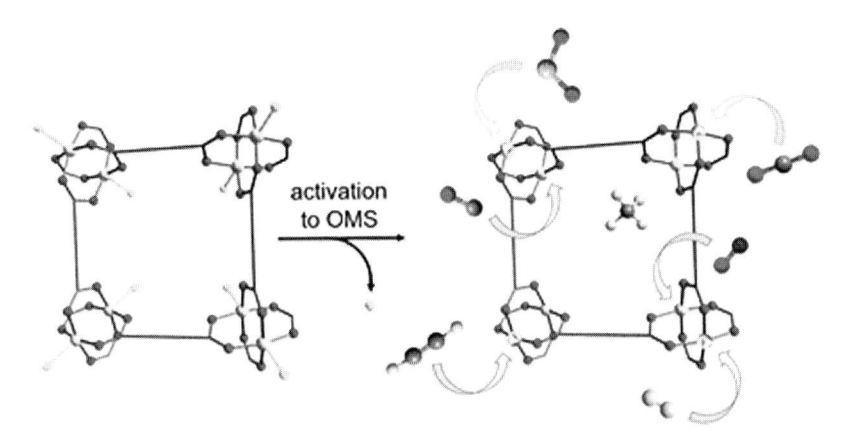

FIGURE 1.26 The generation of open metal sites on metal atoms in metal-organic frameworks through activation and its applications (Kökçam-Demir et al., 2020).

Optical properties of metal-organic frameworks

A characteristic of MOFs is their optical properties, which are mainly observed in lanthanide MOFs (Ln-MOFs). Incorporating a guest molecule into the pores of MOFs can change their optical properties. The optical properties of MOFs can lead to their use in optical applications (Ahmadi et al., 2022).

Magnetic properties of metal-organic frameworks

The magnetic properties of MOFs are influenced by the intrinsic nature of metal ions and the interactions between organic ligands and metal ions. Magnetism in MOFs can be achieved by incorporating paramagnetic metals or open-shell organic ligands in their structure. MOFs with magnetic properties are made of paramagnetic metals, especially the first-row transition metals, such as V, Cr, Mn, Fe, Co, Ni, and Cu. Paramagnetic metal ions and organic radicals cooperate to generate the magnetic properties of MOFs (Ahmadi et al., 2022).

Electrical conductivity of metal-organic frameworks

Most MOFs are electrical insulators and don't offer good conjugation pathways for charge transport or conductivity along with permanent porosity. Due to their low electrical conductivity, MOFs cannot be used in many other desirable applications, such as electrocatalysts and electrodes for fuel cells, capacitors, sensors, etc. A crystalline MOF with promising conductivity properties and/or charge mobility can be fabricated with a wise choice of electroactive metals and organic components. The construction of conductive 2D and 3D MOFs with high charge mobility and/or electrical conductivity has made great progress in the past six years despite a challenging task (Li et al., 2016).

Proton conductivity of metal-organic frameworks

MOFs are promising candidates for the development of porous proton-conducting materials, due to their tunability, functionality, and excellent stability. The high proton conductivity of MOFs can be achieved either by increasing the amount of proton carrier molecules or by changing the pore environment to improve the proton conduction pathway. Proton conductivity can be significantly enhanced by introducing guest molecules inside MOF pores that act as proton carriers or proton-conducting media due to their hydrogen bonding network. In fuel cells, MOFs may be used as proton conductors (Li et al., 2016; Sharma et al., 2023).

Host–guest interactions of metal-organic frameworks

MOFs offer inherent superiority as hosts due to their component diversity, structural designability, and tunable porosities. The resulting host–guest MOFs have a variety of properties and applications, and they can be modulated easily by tuning the host, the guest, and the host–guest interactions. The host-guest synergy may also result in new features, such as electron/energy transfer, redox reaction, or other processes, which can result in emergent properties distinct from the characteristics of individual hosts or guests. As a result, host–guest MOFs have gained considerable attention and have become hot research subjects (Chang, 2023).

Synthetic processes of metal-organic frameworks

The porous structure, morphological characteristics, and crystalline nature of MOFs are influenced by experimental conditions during MOF synthesis. In this regard, selecting a synthesis method that controls the physiochemical properties of the ingredients is crucial. In addition, it is essential to consider the economic and environmental aspects of large-scale synthesis. MOFs can be generated using various synthetic methods based on the resulting frameworks and properties (Mahjoob et al., 2023). These methods include hydrothermal, solvothermal, reflux, ionothermal, room temperature synthesis, diffusion, slow evaporation, electrochemical, microwave, sonochemical, mechanochemical, microfluidics, spray-drying, template, and PSM techniques (Annamalai et al., 2022; Dey et al., 2014; Li et al., 2024; Mahjoob et al., 2023; Noori & Akhbari, 2017; Yusuf et al., 2022).

Hydrothermal method

The hydrothermal method involves the use of organic ligands and metal salts in a water medium to synthesize MOF single crystals. Using water as a solvent, this method is known as hydrothermal. The hydrothermal method enabled the creation of MOFs with varying sizes, morphologies, and lattice parameters. In general, hydrothermal synthesis of MOFs involves combining ionic species

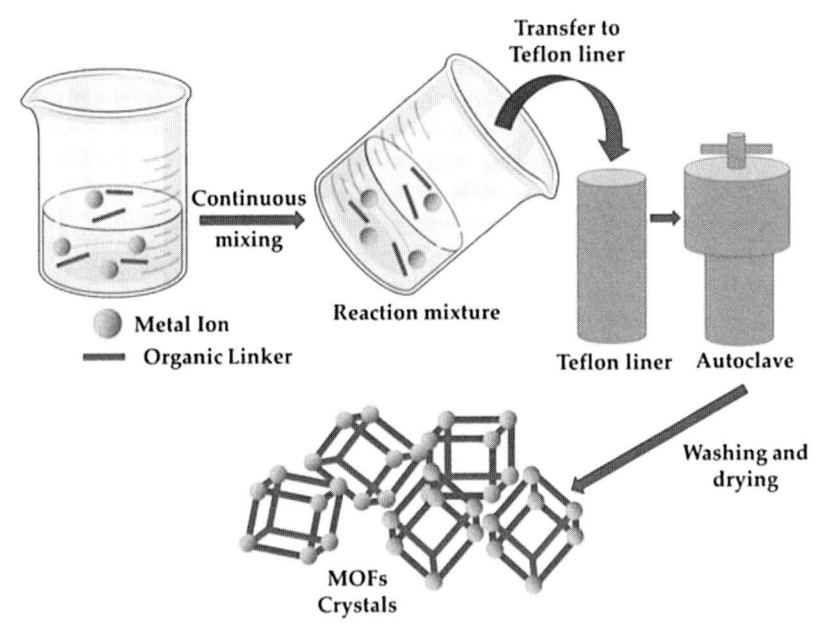

FIGURE 1.27 A schematic illustration of the hydrothermal method for synthesis of metal-organic frameworks (Gatou et al., 2023).

and organic ligands in a water medium and holding the combination at extreme temperatures (greater than 99°C) in an autoclave for long periods of time (Fig. 1.27). An autoclave is used to create a temperature difference inside a growth chamber to promote crystal growth by supplying nutrients via water. Hydrothermal synthesis allows compound crystallization at high vapor pressures and temperatures. The advantages of hydrothermal synthesis are simplicity and safety. It has been used to generate high-performance MOFs and their composites with controllable size and structure. It is possible that during this process, the reagents undergo unpredicted chemical transformations to provide new ligands in situ. The disadvantages of this technique include the need for expensive autoclaves, the high temperature and pressure, the inability to observe crystal growth, and the longer synthesis time (Gatou et al., 2023; Li et al., 2024; Mahjoob et al., 2023).

Solvothermal method

MOFs are commonly prepared using the solvothermal method. This method is preferred because it is simple, convenient, crystallinity-enhancing, and high-yielding. Metal salts and organic ligands are stirred in protic or aprotic organic solvents. Examples of aprotic solvents include N,N-diethylformamide (DEF), N,N-dimethylformamide (DMF), N-methyl-2-pyrrolidone (NMP), dimethyl

28 Applications of Metal-Organic Framework Composites

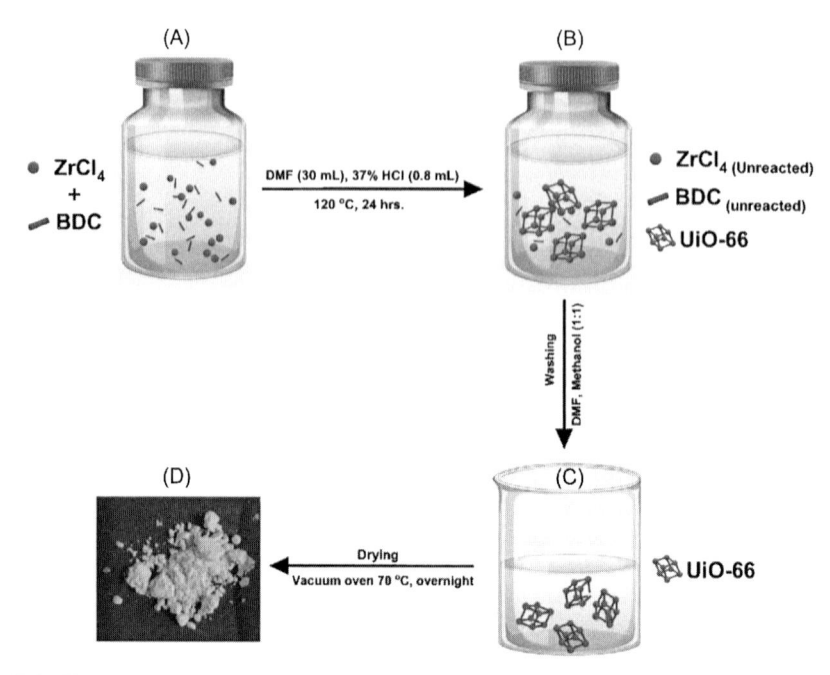

FIGURE 1.28 Mixing metal ion and ligand in DMF (A), synthesizing UiO-66 (B), washing UiO-66 to remove unreacted precursors, and drying UiO-66 (D) (Kobaisy et al., 2023).

sulfoxide (DMSO), N,N-dimethylacetamide (DMA), acetonitrile, and toluene. Among the protic solvents are methanol, ethanol, and mixed solvents. Mixtures of solvents can be used to avoid problems associated with the distinct solubility of the initial components. This mixture is poured into a closed vessel under elevated pressure and temperature for several hours or even a day. Low-temperature reactions are carried out in glass vials (Fig. 1.28), and high-temperature reactions (>400K) are carried out in stainless steel autoclaves lined with Teflon. To achieve higher pressure, the closed vessel is heated at a temperature higher than the boiling point of the aprotic or protic solvent (Yusuf et al., 2022). Solvothermal synthesis is also a time-consuming synthesis process (the growth of crystals occurs within hours or days), primarily used to make MOFs with the least amount of solvent. Using this technique, crystalline MOFs are produced by cooperating organic ligands with isolated metal ions, which allows their controlled synthesis. The solvothermal method brings together the benefits of both hydrothermal and sol-gel methods, allowing for precise control over the size, shape, and crystallinity of MOFs by altering temperature, solvent, precursor, and reaction duration. Solvothermal synthesis has the advantages of being simple and safe, easy to operate, having crystallinity, and excellent yields. In this process, new ligands formed in situ can generate new types of MOF structures that cannot be generated under mild reaction conditions when the reagents undergo an unprecedented chemical transformation. Costly autoclaves, the observation of growing crystals, time-consuming processes that can take days or weeks, and the use of hazardous solvents in the reaction mixture are the major drawbacks of this technique (Li et al., 2024; Salimi et al., 2024a).

Reflux method

A closed system is commonly used to prepare MOF materials via hydrothermal/solvothermal methods. As a result, hydrothermal/solvothermal techniques are largely unobservable when it comes to monitoring, controlling, and optimizing their processes. The reflux method is more convenient, faster, and greener than hydrothermal/solvothermal methods. The reflux method offers several advantages, including a simple synthesis protocol, environmental friendliness, controllability of surface morphologies, and cost savings (Fig. 1.29) (Lei et al., 2022; Liang et al., 2018; Mahjoob et al., 2023).

Ionothermal method

Ionothermal synthesis is the use of ionic liquids instead of water/organic solvents for the synthesis of MOFs (Fig. 1.30). Using ionic liquids as solvents in chemical synthesis has a number of advantages, including excellent solvating properties, recyclability, and high thermal stability. In order to overcome solvothermal synthesis' drawbacks, including insoluble inorganic precursors and hydrogen bond interference, ionothermal synthesis has become a viable alternative. A low-volatility ionic liquid is used as both a solvent and a structure-directing agent to

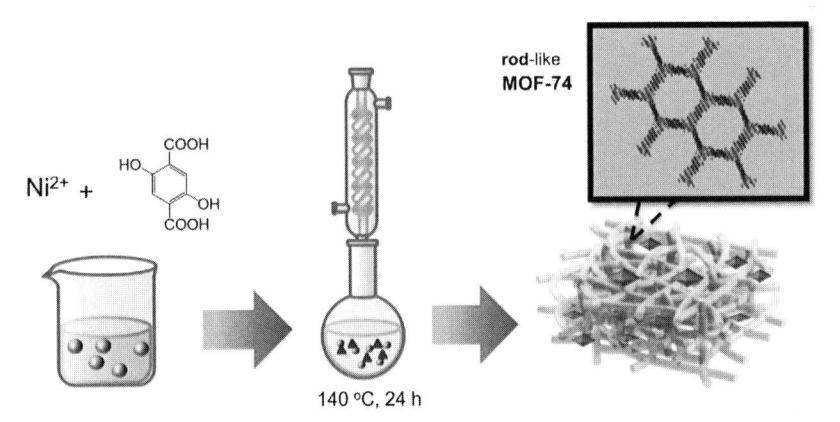

FIGURE 1.29 A schematic illustration of the reflux method for synthesis of metal-organic framework-74(Ni) (Lei et al., 2022).

30 Applications of Metal-Organic Framework Composites

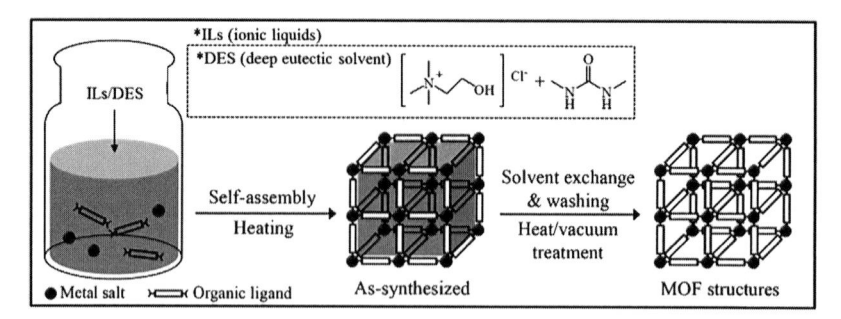

FIGURE 1.30 A schematic illustration of the ionothermal method for synthesis of metal-organic frameworks (Lee et al., 2013).

dissolve the precursors efficiently and promote the synthesis of MOFs with enhanced properties. As a result of their low vapor pressure, high solubility in organic molecules, high thermal stability, wide liquid range, and nonflammability, ionic liquids are an environmentally friendly reagent that can be used in the synthesis of MOFs as well as other materials such as zeolites and chalcogenides. Additionally, the careful selection of counterions and templates from ionic liquids promises to provide a wide variety of reaction environments for the generation of MOFs. The disadvantages of ionothermal method include the inseparable cations that reside in the pores of MOFs from the solvents (Anumah et al., 2019; Li et al., 2024).

Room temperature synthesis method

In the room temperature synthesis method, no heating is necessary for the production or crystallization of MOFs from the reaction of organic ligands with isolated metal ions. MOFs are crystallized under mild conditions while the solvent used in the reaction mixture is kept at room temperature. During the reaction, bases such as sodium hydroxide and triethylamine are used to deprotonate the organic ligands, leading to the formation of MOF. MOFs are commonly prepared using this method because of its simplicity, good yield, and low energy consumption. The process takes hours to days, and some of the reactions must be conducted under hazardous conditions using DMF (solvent), hydrofluoric acid (acid modulator), and so on (Fig. 1.31) (Habtemariam et al., 2022; Li et al., 2024).

Diffusion method

The diffusion method is commonly used to synthesize MOFs as single crystals or nanosized crystals at room temperature. There are three main types of diffusion: gel diffusion, liquid phase diffusion, and gas phase diffusion. In gel diffusion, certain materials (such as organic ligands dispersed in a gel

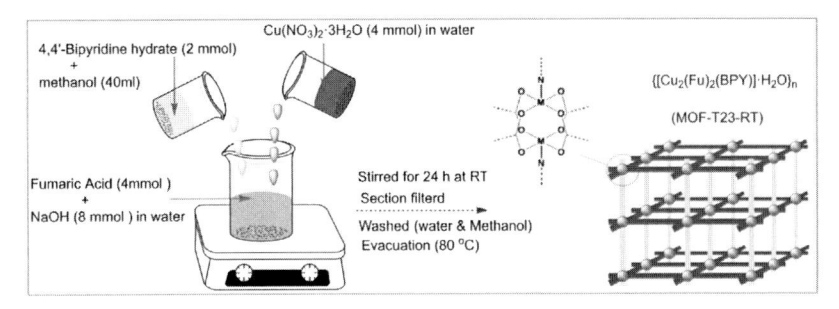

FIGURE 1.31 A schematic illustration for synthesis of metal-organic framework-T23-RT via the room temperature synthesis method (Habtemariam et al., 2022).

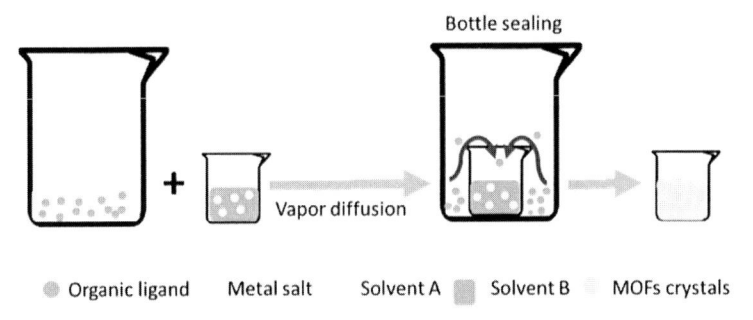

FIGURE 1.32 A schematic illustration of the gas phase diffusion method for synthesis of metal-organic frameworks (Bian et al., 2018).

substance) are mixed into another material (such as a solution of metal ions) for a period of time in order to form MOF crystals. Liquid phase diffusion involves dissolving metal salts and ligands in solvents of different densities, placing both solutions on the contact surface, and precipitating crystals through diffusion. The gas-phase diffusion process involves dissolving metal salts and organic ligands in appropriate solvents and allowing the volatile solvents to diffuse into the solution to achieve saturation and precipitate crystals (Fig. 1.32). Diffusion-based MOF synthesis is often carried out in mild conditions, but it is time-consuming (Akhbari & Morsali, 2013; Annamalai et al., 2022; Bian et al., 2018; Moghadam et al., 2018; Wang et al., 2023).

Slow evaporation method

The slow evaporation method is a traditional method for preparing MOFs that does not require external energy sources. The slow evaporation process involves slowly evaporating a solution of the starting materials at a fixed temperature, usually room temperature, to concentrate the solution (Fig. 1.33).

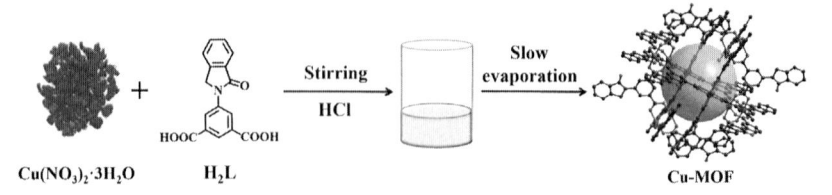

FIGURE 1.33 A schematic illustration of the slow evaporation method for synthesis of Cu-metal-organic framework (Ansari et al., 2019).

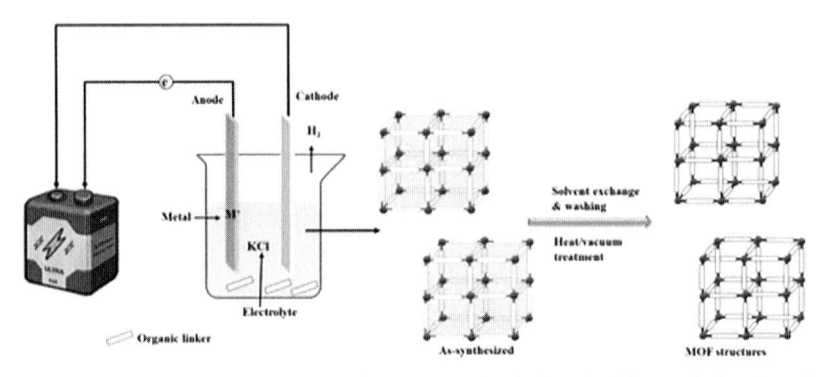

FIGURE 1.34 A schematic illustration of the electrochemical method for synthesis of metal-organic frameworks (Yusuf et al., 2022).

It is possible to disperse reagents in a mixture of solvents, increasing their solubility and accelerating the process by allowing low-boiling solvents to evaporate more rapidly. Despite its advantages as a room-temperature process, this method still takes longer than other conventional methods (Akhbari et al., 2007, 2009, 2011; Dey et al., 2014).

Electrochemical method

The electrochemical formation of MOFs can be performed at room temperature and pressure to avoid the risks and energy associated with a high-temperature and high-pressure process. The metal component of MOFs is further exposed to anodic oxidation in the presence of an electrical current during the metal electrochemical fabrication, resulting in a fast and cost-effective process. Additionally, this strategy has several benefits, including excellent yield, reduced energy consumption, simple operation, environmental friendliness, and no need for special tools. An electrochemical characteristic, including supporting electrolyte and applied voltage, plays a crucial role in determining the morphological characteristics (size, structure, and distribution of particles) and the production of the response (Fig. 1.34) (Mahjoob et al., 2023).

Microwave method

MOFs can be synthesized using microwave methods, which use microwave energy between 300 and 300,000 MHz. The fabrication of MOFs using this approach is energy-efficient and environmentally friendly. The microwave synthesis provides the necessary heat by interacting between the moving charges in the polar solution and the microwave radiation. The use of this method ensures a constant temperature rise throughout the reaction, which leads to the development of crystals in MOF fabrication. By utilizing microwave heating, crystal size and structure can be controlled more effectively, reaction times are reduced, and production efficiency, selectivity, and purity can be improved. Therefore, MOFs can be synthesized using microwaves with great efficiency (Fig. 1.35) (Akhbari & Morsali, 2015; Gatou et al., 2023).

Sonochemical method

The sonochemical synthesis, or ultrasound-assisted technique (20–10 MHz), is a rapid and environmentally friendly method of preparing MOF crystals. This technique accelerates homogeneous nucleation and smaller particle sizes while reducing the crystallization time compared to solvothermal methods. When high-energy ultrasound is interacting with liquid, acoustic cavitation (the process of creating, growing, and collapsing a bubble under changing pressure) occurs, followed by high temperatures of 5000–25,000K, high pressures, and rapid heating and cooling (Fig. 1.36). Due to its limited heating depth and low yield, this technique is not suitable for large-scale MOF synthesis (Li et al., 2024; Mirzadeh et al., 2017; Salimi et al., 2024b; Shahangi Shirazi & Akhbari, 2016; Usefi et al., 2019).

Mechanochemical method

The mechanochemical synthesis of MOFs offers a lot of potential for both environmental and economic benefits. In this process, the intermolecular bonds were broken mechanically by grinding the metal salts and organic ligands together in a ball mill or mortar (Fig. 1.37). This technique produces high

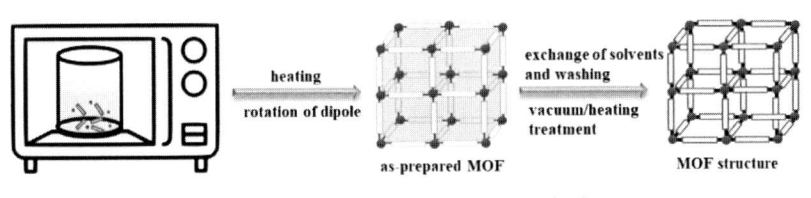

FIGURE 1.35 A schematic illustration of the microwave method for synthesis of metal-organic frameworks (Yusuf et al., 2022).

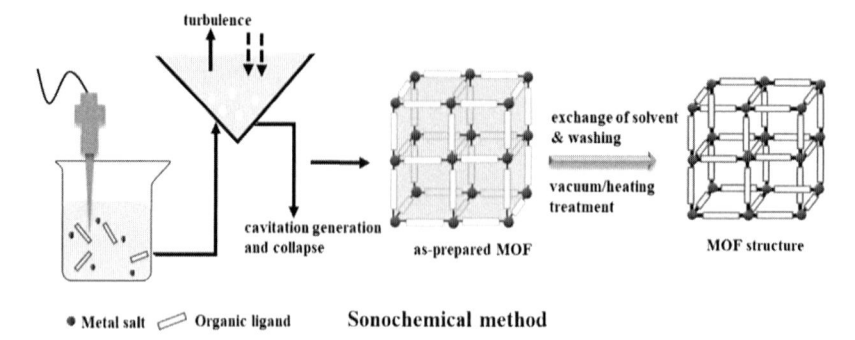

FIGURE 1.36 A schematic illustration of the sonochemical method for synthesis of metal-organic frameworks (Yusuf et al., 2022).

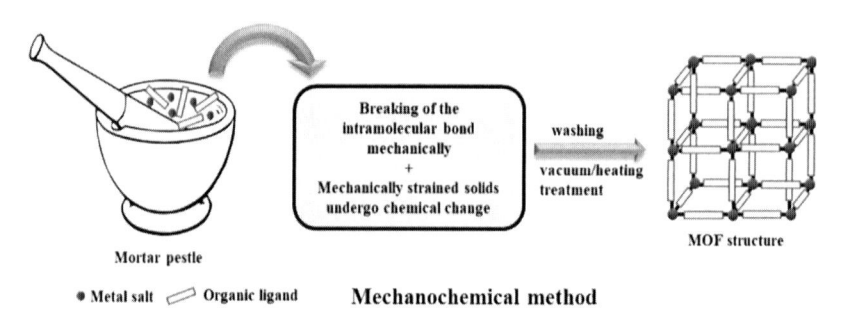

FIGURE 1.37 A schematic illustration of the mechanochemical method for synthesis of metal-organic frameworks (Yusuf et al., 2022).

yields of products at room temperature by eliminating solvents, producing safe by-products, and speeding up reaction times. However, adding a minimum quantity of solvent stimulates mechanochemical reactions by increasing the molecular mobility of reactants (Akhbari & Morsali, 2013b; Li et al., 2024; Moeinian, Akhbari, Boonmak, et al., 2016; Moeinian, Akhbari, Kawata, et al., 2016).

Microfluidics method

The microfluidics method is a common method for preparing nanoparticles and has recently been utilized to synthesize MOFs. The microfluidics method uses microchips to control fluid flows at a microscale and nanoscale. With this method, MOFs can be continuously produced by controlling precise parameters during synthesis, enabling intensification, versatility, and scalability. MOFs are synthesized using a conventional solvothermal approach in

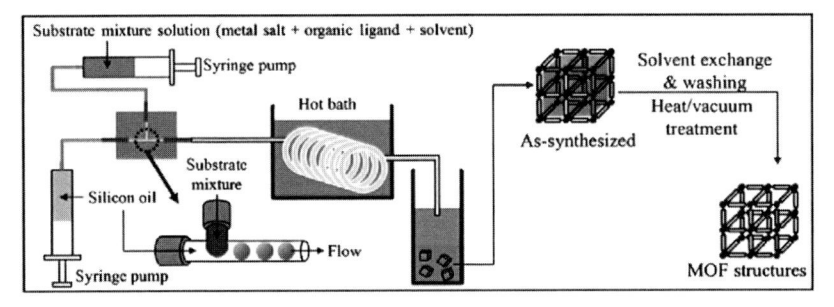

FIGURE 1.38 A schematic illustration of the microfluidic method for synthesis of metal-organic frameworks (Lee et al., 2013).

72 hours, but on a microfluidic chip in 1 minutes. Several advantages of microfluidic-assisted synthesis include morphological control, fine particle size, good reproducibility, and the lack of large areas for work and high equipment costs. In comparison with conventional approaches, this approach also offers defect-free hollow fiber membranes based on MOFs that conserve reactants. Although this process shows several advantages over others, MOF preparation at high temperatures has some drawbacks. In addition to requiring a large area and high wattage, this technique is costly, high maintenance, and releases harmful waste or by-products (Fig. 1.38) (Li et al., 2024; Raptopoulou, 2021).

Spray-drying method

The spray-drying method involves atomizing a solution of MOF precursors into a spray of microdroplets using a two-liquid nozzle. In this process, one or more solutions are injected into the inner nozzle at a specific speed (feed rate). Compressed air or nitrogen is discharged from the outer nozzle at another constant speed (dispersion rate). Then, each atomized droplet is heated at a specific temperature through a gas and evaporates. As the solvent evaporates, the precursor's concentration at the surface rises until large MOF particles are formed (Fig. 1.39) (Yusuf et al., 2022).

Template method

New MOFs can be synthesized using template molecules, which are difficult to obtain by conventional methods. A wide range of organic molecules have been used as template molecules, including organic solvents, organic amines, carboxylic acids, N-heterocyclic aromatic compounds, ionic liquids, surfactants, and other organic molecules. These organic compounds affect MOF synthesis and crystallization in different ways. Other molecules that can function as templates are coordination compounds, polyoxometalates, block copolymers, MOFs, polystyrene spheres, graphene oxide, and rarely

36 Applications of Metal-Organic Framework Composites

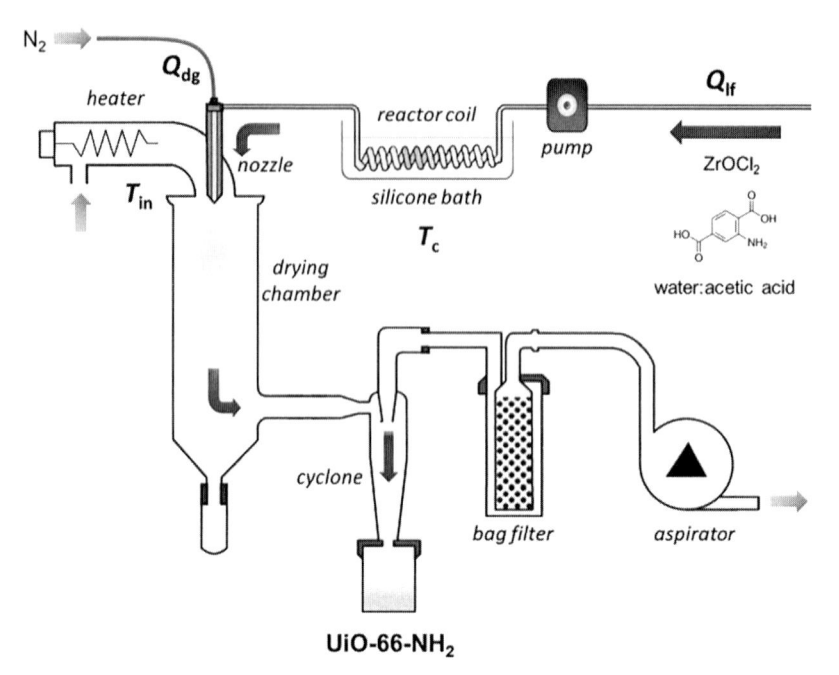

FIGURE 1.39 A schematic illustration of the spray-drying method for synthesis of metal-organic frameworks (Avci-Camur et al., 2018).

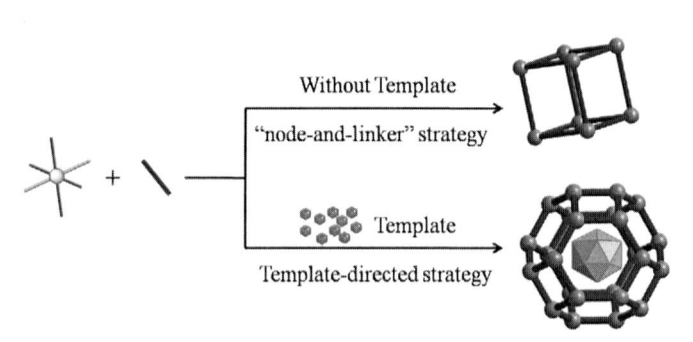

FIGURE 1.40 A schematic illustration of the construction of metal-organic frameworks without a template and with a template (Guo et al., 2019).

biomacromolecules. For the preparation of hierarchical porous materials, template synthesis generates mesoporous and microporous channels that can host large molecules like proteins and enzymes. The most common synthetic method for hierarchical MOFs uses reticular chemistry to obtain MOFs of the same topology with variable pores (Fig. 1.40) (Raptopoulou, 2021).

Postsynthetic modification method

PSM involves the introduction of functional groups after the MOFs have been synthesized. This method has been widely used to prepare isostructural MOFs with a variety of chemical and physical properties. There are four types of PSM, as illustrated in Fig. 1.41. These include covalent PSM, which is made by covalent modification of organic linkers; dative PSM, which involves the coordination of a metal center to a linker with an empty coordination site; inorganic PSM, which is made by altering SBUs; and ionic PSM, produced by exchanging a counter-ion in a cationic or anionic MOF. Generally, a reactive group is required on the linker of a MOF for covalent or dative PSM. In MOFs, this functional group is often called a "tag," which refers to a group or functionality that is stable and nonstructure-defining during MOF synthesis but can be modified postsynthesis (Bedia et al., 2019; Li et al., 2024; Raptopoulou, 2021).

Activation of metal-organic frameworks

Synthesis of MOFs has experienced significant developments and progress over the past few decades, while other important aspects have received less attention. Purification and activation of MOFs are also important aspects that require attention. A fundamental understanding of MOFs' intrinsic properties is crucial to their application. Usually, as-synthesized MOFs contain guest molecules (solvents, unreacted linkers, clusters, and modulators), so it is necessary to remove these guest species through an activation process in order to gain access to their pore space for application (Fig. 1.42). It is also essential to select activation methods that do not collapse the framework structurally, causing

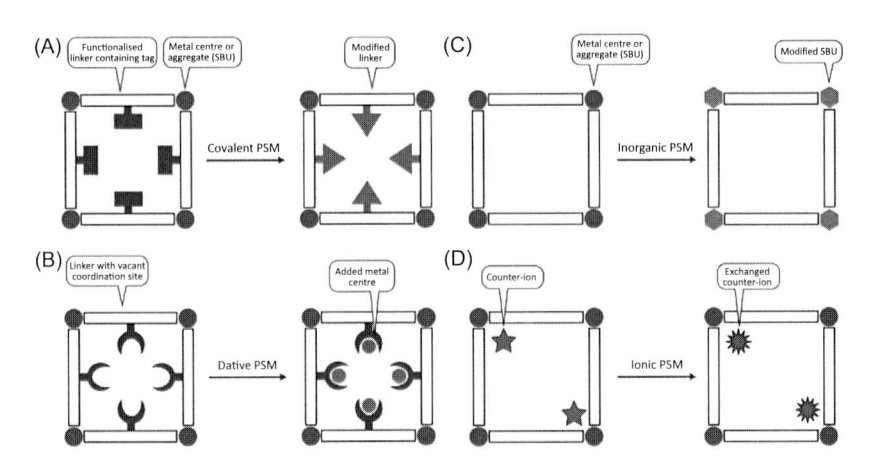

FIGURE 1.41 Methods for postsynthetic modification of metal-organic frameworks (iXamena and Gascon, 2013).

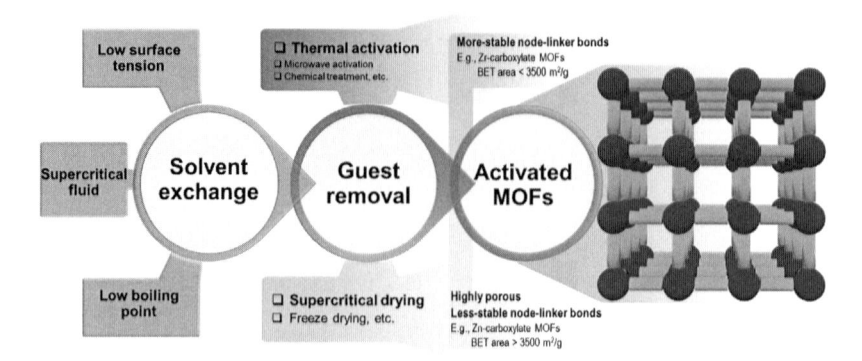

FIGURE 1.42 Overview of the workflow and strategies of metal-organic framework activation (Zhang et al., 2020).

partial or total loss of porosity. Various methods can be used to activate MOFs, including heating, solvent exchange, supercritical carbon dioxide (ScCO$_2$), freeze-drying, microwaves, photothermal, and chemical (acid treatment) (Akeremale et al., 2023; Al Amery et al., 2020; Zhang et al., 2020).

Heating activation

In the activation method, the solvent is evaporated from the pores and the residual precursors are sublimated by the heating process in an air environment. It is preferable to identify the decomposition temperature of a MOF from its thermogravimetric analysis (TGA) profile before performing this method. The MOF can then be heated at a temperature less than half that of decomposition. Using this method, methanol and ethanol can be completely removed from MOFs. To enhance the efficiency of this method, it is recommended to use inert gas to homogenize the powder during the heating process. Otherwise, this method has poor performance since it leads to partial damage to the activated MOF and may also leave some residual reaction components inside the pores. The second process is heating and vacuuming. This process requires three conditions. Firstly, the MOF powder has to be dried at low temperature to remove moisture. Second, heating should take place at a temperature that is determined by the MOF's TGA profile. In the third step, evacuation should be performed at a very low pressure during the heating process (Fig. 1.43) (Al Amery et al., 2020).

Solvent-exchange activation

In this technique, a solvent with a high boiling point is switched to one with a lower boiling point, gently activated under vacuum. Many MOFs are synthesized in solvents, especially those with high boiling points such as DMF, DEF, and DMSO. In order to activate MOFs, they are usually activated

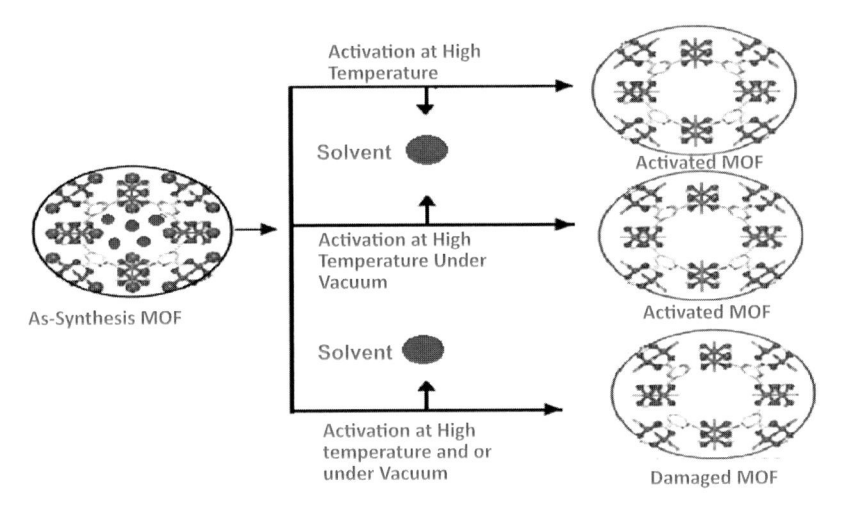

FIGURE 1.43 A schematic illustration of the heating and vacuum activation method for metal-organic frameworks (Al Amery et al., 2020).

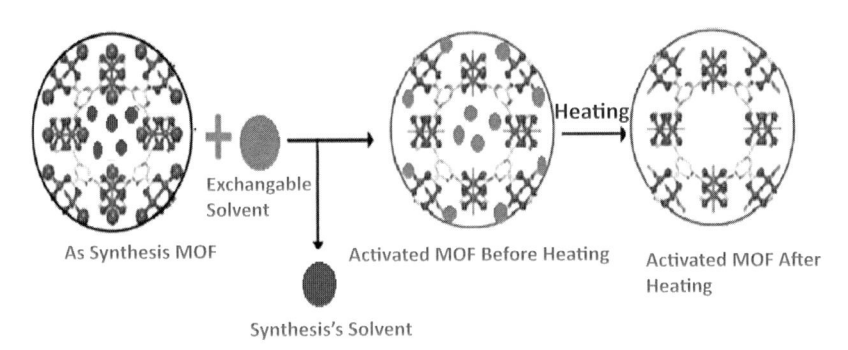

FIGURE 1.44 A schematic illustration of the solvent exchange activation method for metal-organic frameworks (Al Amery et al., 2020).

at elevated temperatures after exchanging with low-boiling-point solvents like acetone, methanol, and dichloromethane. Since fluids with lower boiling points usually possess weaker intermolecular interactions, surface tension and capillary forces are minimized during activation (Fig. 1.44) (Akeremale et al., 2023; Zhang et al., 2020)

Supercritical carbon dioxide activation

To activate supercritical CO_2, the crystals are soaked in ethanol for three days to exchange all guest solvents with ethanol. Every 24 hours, the soaking

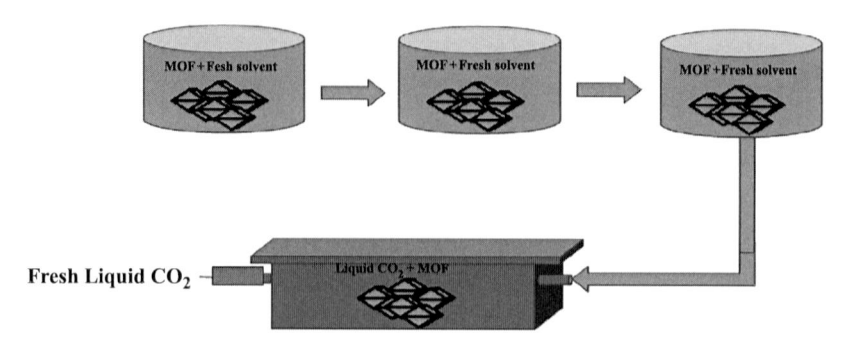

FIGURE 1.45 A schematic illustration of the supercritical CO_2 activation method for metal-organic frameworks (Al Amery et al., 2020).

solution is replaced with new ethanol. The ethanol-exchanged samples were soaked in liquid CO_2 for over 6 hours in a supercritical dryer, purged every hour with fresh liquid CO_2. It is necessary to maintain a temperature between –5°C and +5°C during this period. The temperature is then raised to 40°C (which is above the critical point of CO_2 [31°C]) for 30 minutes. After 18 hours, the supercritical CO_2 is slowly released (Fig. 1.45). In comparison to solvent exchange activation and conventional activation, which both have the potential to collapse frameworks, $ScCO_2$ activation produces a milder outcome. In this sense, $ScCO_2$ is a suitable medium for MOF activation in that the guest solvent molecules do not undergo a liquid-to-gas phase shift and instead travel straight through a supercritical phase, preventing the drawbacks associated with liquid-to-gaseous phase switching (Akeremale et al., 2023; Al Amery et al., 2020).

Freeze-drying activation

The process of freeze-drying is widely used in food processing, biological applications, and activating porous materials. As part of this procedure, the sample is exchanged with a solvent with a high freezing point compatible with the MOF (e.g., benzene with a freezing point of 5.5°C) and then cooled down to freeze the solvents. Sublimation can be used to remove the frozen solvents under vacuum below their freezing point, which eliminates the surface tension otherwise present during liquid–gas transformation (Zhang et al., 2020).

Microwave activation

Microwave heating can accelerate chemical reactions through electromagnetic radiation, which makes it useful in various chemical processes. The microwave method uses wave penetration to transfer heat energy directly to guest molecules, unlike conventional heat energy, which is transferred through thermal diffusion (Fig. 1.46) (Zhang et al., 2020).

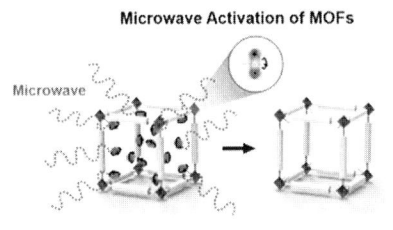

FIGURE 1.46 A schematic illustration of the microwave activation method for metal-organic frameworks (Lee et al., 2019).

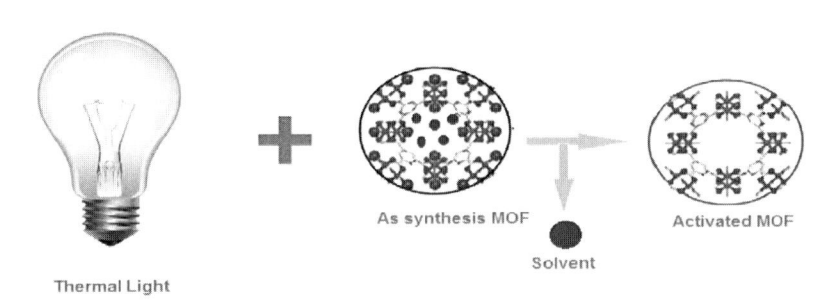

FIGURE 1.47 A schematic illustration of the photothermal activation method for metal-organic frameworks (Al Amery et al., 2020).

Photothermal activation

A photothermal activation method uses photo-irradiation to heat up MOFs in one step. It is called the "photothermal effect." Light can be converted into heat to increase the local temperature of a MOF while minimizing heat loss and diffusion. After irradiation, the localized light-to-heat conversion in the MOF structures causes the guest molecules to be rapidly removed, reducing the activation time considerably. This activation method requires MOFs that are photothermally active (Fig. 1.47) (Al Amery et al., 2020).

Chemical activation

MOFs can also be made active through chemical processing (acid treatment). Each of the aforementioned activation methods focuses on neutral species guest solvent molecules. However, when the bounded solvent molecules are charged (ionic) or have a high boiling point, the coordination site of an evacuated material cannot be cleaned by simple heating. It is because these molecules are nonvolatile (for ions) and low-volatile. Thus, chemical treatment may be the most effective method if solvent molecules, such as ionic species or benzoic acid, were coordinated in the framework during synthesis (sometimes they also function as modulators) (Akeremale et al., 2023).

Characterization of metal-organic frameworks

The resulting MOFs that are prepared using different methodologies should be further characterized using diverse physicochemical techniques to determine their properties. The methods such as X-ray diffraction (XRD) (single-crystal X-ray diffraction (SCXRD) and powder X-ray diffraction (PXRD)), physisorption isotherm, Fourier-transform infrared (FT-IR) spectroscopy, TGA, scanning electron microscopy (SEM), transmission electron microscopy (TEM), energy-dispersive X-ray (EDX) spectroscopy, inductively coupled plasma (ICP) spectroscopy, nuclear magnetic resonance (NMR) spectroscopy, and solid-state nuclear magnetic resonance (SS-NMR) spectroscopy listed below are more helpful for characterizing MOFs (Bedia et al., 2019).

X-ray diffraction

XRD is mainly used to identify the fingerprint of crystalline materials and their morphology. MOFs are porous materials composed of highly ordered crystals. XRD is a more valuable method for identifying MOFs than other techniques. The use of XRD analysis is preferable for characterization of unknown materials because it can provide important information about their structures. MOFs are polymeric and engineering materials that possess a high degree of crystallinity, resulting in each MOF having its distinct structure. MOFs are polymeric and engineering materials with a high degree of crystallinity, resulting in each MOF having its unique structure. Therefore, XRD analysis should be performed as the first characterization to identify synthesized materials and their purity before performing further investigations. XRD analysis is used to determine a new crystal structure and confirm existing crystal structures via SCXRD and PXRD (Abid et al., 2023).

Single-crystal X-ray diffraction

SCXRD is a powerful, nondestructive tool for elucidating atomically-precise snapshots of MOF structures and providing global structural information such as framework topology and chemical functionality. The new crystalline structure of an unknown MOF can be determined and refined using SCXRD data. However, SCXRD requires a perfect crystal having a large size (>100 μm). Single crystal XRD is not capable of analyzing certain large crystals because they may not be of high quality and not suitable for single crystal XRD. To solve this issue, their structure is computationally resolved using the data of PXRD (Abid et al., 2023; Albalad et al., 2021).

Powder X-ray diffraction

The use of PXRD has become common to determine the structural parameters and crystallinity of MOFs. MOF structural identification can be accomplished using a simulated pattern generated by a single crystal X-ray or by comparing the synthesized MOF's diffractogram with a previously published one in the literature or via computational modeling. This method can be used to determine the crystalline structure, recognize various polymorphic forms, discern between crystalline and amorphous materials, and calculate the percentage of crystallinity. The crystallographic properties, including unit cell size, lattice parameters, and crystallite size, can be ascertained after the crystalline structure of the MOF has been identified (Bedia et al., 2019).

Physisorption isotherm

The porosity of MOFs can be evaluated according to the adsorption isotherms of nitrogen (N_2) or argon (Ar) at their boiling points, at 77 K and 87 K, respectively. Adsorption data is usually arranged as "quantity of gas adsorbed" plotted against "relative pressure," which is known as a physisorption isotherm. IUPAC classified physisorption isotherms into six classic types (i.e., from type I to type VI) (Fig. 1.48). The majority of rigid MOF physisorption isotherms are based on either type I (microporous) or type IV (mesoporous), although

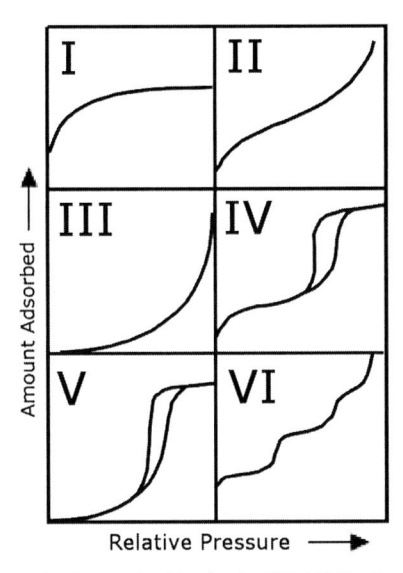

FIGURE 1.48 Physisorption isotherms classification by IUPAC (I=microporous, II=macroporous, III=nonporous or microporous, IV=mesoporous, V=microporous/mesoporous, VI=nonporous) (Yoganathan et al., 2010).

numerous flexible MOF isotherms show geometries that differ from the six typical varieties. These isotherms can provide information on the MOFs' surface area, pore volume, and pore size distribution (Ahmadi et al., 2022; Zhang et al., 2020).

Fourier-transform infrared spectroscopy

The most common method for identifying surface functional groups that are important to MOF performance is FT-IR spectroscopy. FT-IR is employed during the first phases of MOF synthesis to identify impurities and leftover reactants on both the outer surfaces and inner pores. Additionally, it serves as a valuable tool for detecting unreacted organic linkers and coordinated solvents within MOF structures (Abid et al., 2023).

Thermogravimetric analysis

The mass loss of a material in a controlled atmosphere is measured by TGA. This method can be utilized to determine the MOF's thermal stability and estimate its solvent-accessible pore volume. Different carrier gases (N_2, air, or O_2) used for the TGA determine the MOF decomposition. The TGA curve represents the weight loss of the MOF (%) as a function of temperature. TGA plot usually shows a decrease in mass between 40°C and 110°C, which is attributed to the elimination of solvent molecules. There is a second decrease in mass between 300°C and 600°C, which is generally thought to be the result of the decomposition of the MOF framework, leading to the formation of metal oxides. It is important to consider that this weight loss does not necessarily result from a structural alteration. Therefore, it is essential to supplement this investigation with a variable temperature PXRD examination to verify any changes in the structure (Bedia et al., 2019).

Scanning electron microscopy

One helpful technique for measuring the various MOF features, such as crystal size, shape, and elemental composition, is SEM. Because most MOFs are insulators, charging effects and other image abnormalities can hinder or even totally prevent the capture of high-quality SEM images. Coating the sample with a conducting material (such as gold or osmium) to reduce charge buildup from the electron gun is the most popular way to resolve these problems. Researchers can use a field emission scanning electron microscope (FESEM) to get around this problem. This apparatus improves the spatial resolution when operating at low potentials thanks to a field emission gun that shoots incredibly concentrated electron beams. This property even prevents the electron beam from damaging certain MOFs that are sensitive, hence lessening the charging effect on insulating material (Bedia et al., 2019; Howarth et al., 2017).

Transmission electron microscopy

TEM is commonly used to analyze grain size, particle size, and crystallographic data, including plane indices and dislocations. Various software applications can be utilized for the analysis of microscopy pictures, such as ImageJ and CellProfiler. Using these techniques, one can ascertain the size of particles by measuring various particles and generating a histogram, as well as calculate the distance between crystallographic planes. This technique is highly advantageous for characterizing MOFs that have been modified through the addition of NPs, as the acquired pictures yield information regarding the size and distribution of these NPs (Bedia et al., 2019).

Energy dispersive X-ray spectroscopy

The SEM and TEM microscopies can be combined with EDX spectroscopy to determine the quantitative and qualitative elemental composition of MOFs. EDX analysis of the chemical composition resulted in mapping results for every element that can be used to determine the homogeneity of the synthesized MOF (Bedia et al., 2019).

Inductively coupled plasma spectroscopy

Confirming purity or elemental ratios in MOF samples can be done by using inductively coupled plasma optical emission spectroscopy (ICP-OES) and inductively coupled plasma mass spectroscopy (ICP-MS). ICP-MS is capable of detecting very low concentrations (as low as ppt [parts per trillion]) for most elements compared to ICP-OES (as low as ppb [parts per billion]). The lower detection limits for ICP-MS are crucial when analyzing limited sample quantities (Howarth et al., 2017).

Nuclear magnetic resonance spectroscopy

NMR spectroscopy can be applied to determine MOF purity, linker ratios, and the modulator that remains, as well as the absence of solvent after activation of MOFs (Howarth et al., 2017).

Solid-state nuclear magnetic resonance spectroscopy

In particular, SS-NMR spectroscopy can be utilized for MOF characterization to examine the local chemical environment inside an MOF. This method is highly valuable for determining the identification or chemical state of a particular functional group within a MOF, such as revealing the oxidation state of a phosphorus atom within a MOF. SS-NMR can also be utilized to acquire insights into the supramolecular interactions and kinetics of small molecule docking within a MOF (Howarth et al., 2017).

Conclusion

MOFs have emerged as a promising class of crystalline porous materials with a wide range of applications due to their unique properties and customizable structures. MOFs consist of metal ions or metal clusters interconnected by organic ligands, allowing for a diverse range of pore sizes, shapes, and structures. The ability to fine-tune these parameters through careful selection of components and postsynthesis modifications enhances their versatility and functionality. The development of MOFs represents a significant breakthrough in material science, opening new avenues for research and application. Their customizable nature, combined with a rich diversity of potential structures and functionalities, positions them at the forefront of innovative solutions in various scientific domains. As research progresses, the understanding and utilization of MOFs will likely continue to expand, paving the way for advanced applications in energy, environmental science, medicine, and beyond.

References

Abdelhamid, H. N. (2023). Metal-organic frameworks (MOFs) as a unique theranostic nanoplatforms for therapy and imaging. *Inorganic Nanosystems: Theranostic Nanosystems, 2*, 323–350. https://doi.org/10.1016/B978-0-323-85784-0.00006-6, https://www.sciencedirect.com/book/9780323857840.

Abid, H. R., Azhar, M. R., Iglauer, S., Rada, Z. H., Al-Yaseri, A., & Keshavarz, A. (2023). Physiochemical characterization of metal organic framework materials: A mini review. *Heliyon, 10*(1), e23840. https://doi.org/10.1016/j.heliyon.2023.e23840.

Achmann, S., Hagen, G., Kita, J., Malkowsky, I. M., Kiener, C., & Moos, R. (2009). Metal-organic frameworks for sensing applications in the gas phase. *Sensors, 9*(3), 1574–1589. https://doi.org/10.3390/s90301574, http://www.mdpi.com/1424-8220/9/3/1574/pdf.

Aggarwal, V., Solanki, S., & Malhotra, B. D. (2022). Applications of metal-organic framework-based bioelectrodes. *Chemical Science, 13*(30), 8727–8743. https://doi.org/10.1039/d2sc03441g, http://pubs.rsc.org/en/journals/journal/sc.

Ahmadi, M., Ebrahimnia, M., Shahbazi, M. A., Keçili, R., & Ghorbani-Bidkorbeh, F. (2022). Microporous metal–organic frameworks: Synthesis and applications. *Journal of Industrial and Engineering Chemistry, 115*, 1–11. https://doi.org/10.1016/j.jiec.2022.07.047, http://www.sciencedirect.com/science/journal/1226086X.

Akeremale, O. K., Ore, O. T., Bayode, A. A., Badamasi, H., Adedeji Olusola, J., & Durodola, S. S. (2023). Synthesis, characterization, and activation of metal organic frameworks (MOFs) for the removal of emerging organic contaminants through the adsorption-oriented process: A review. *Results in Chemistry, 5*, 100866. https://doi.org/10.1016/j.rechem.2023.100866, http://www.journals.elsevier.com/results-in-chemistry.

Akhbari, K., & Morsali, A. (2013a). Modulating methane storage in anionic nano-porous MOF materials via post-synthetic cation exchange process. *Dalton Transactions, 42*(14), 4786–4789. https://doi.org/10.1039/c3dt32846e.

Akhbari, K., & Morsali, A. (2013b). Solid- and solution-state structural transformations in flexible lead(ii) supramolecular polymers. *CrystEngComm, 15*(44), 8915–8918. https://doi.org/10.1039/c3ce41171k.

Akhbari, K., & Morsali, A. (2013c). Solid-state structural transformations of two AgI supramolecular polymorphs to another polymer upon absorption of HNO_3 vapors. *Inorganic Chemistry, 52*(6), 2787–2789. https://doi.org/10.1021/ic3008415.

Akhbari, K., & Morsali, A. (2015). Needle-like hematite nano-structure prepared by directed thermolysis of MIL-53 nano-structure with enhanced methane storage capacity. *Materials Letters, 141*, 315–318. https://doi.org/10.1016/j.matlet.2014.11.110, http://www.journals.elsevier.com/materials-letters/.

Akhbari, K., Morsali, A., Hunter, A. D., & Zeller, M. (2007). $[Tl_2(\mu\text{-Htdp})_2(\mu\text{-}H_2O)]n$ (H_2tdp=4,4′-thiodiphenol) – A one-dimensional thallium(I) coordination polymer with a large tetranuclear metallacycle: Thermal, emission and structural studies. *Inorganic Chemistry Communications, 10*(2), 178–182. https://doi.org/10.1016/j.inoche.2006.10.014.

Akhbari, K., Alizadeh, K., Morsali, A., & Zeller, M. (2009). A new two-dimensional thallium(I) coordination polymer with 4-hydroxybenzylidene-4-aminobenzoate: Thermal, structural, solution and solvatochromic studies. *Inorganica Chimica Acta, 362*(8), 2589–2594. https://doi.org/10.1016/j.ica.2008.11.028, http://www.journals.elsevier.com/inorganica-chimica-acta/.

Akhbari, K., Hemmati, M., & Morsali, A. (2011). Fabrication of silver nanoparticles and 3D interpenetrated coordination polymer nanorods from the same initial reagents. *Journal of Inorganic and Organometallic Polymers and Materials, 21*(2), 352–359. https://doi.org/10.1007/s10904-010-9444-8.

Al Amery, N., Abid, H. R., Al-Saadi, S., Wang, S., & Liu, S. (2020). Facile directions for synthesis, modification and activation of MOFs. *Materials Today Chemistry, 17*, 100343. https://doi.org/10.1016/j.mtchem.2020.100343.

Al Obeidli, A., Ben Salah, H., Al Murisi, M., & Sabouni, R. (2022). Recent advancements in MOFs synthesis and their green applications. *International Journal of Hydrogen Energy, 47*(4), 2561–2593. https://doi.org/10.1016/j.ijhydene.2021.10.180, http://www.journals.elsevier.com/international-journal-of-hydrogen-energy/.

Albalad, J., Sumby, C. J., Maspoch, D., & Doonan, C. J. (2021). Elucidating pore chemistry within metal-organic frameworksviasingle crystal X-ray diffraction; from fundamental understanding to application. *CrystEngComm, 23*(11), 2185–2195. https://doi.org/10.1039/d1ce00067e, http://pubs.rsc.org/en/journals/journal/ce.

Alhumaimess, M. S. (2020). Metal–organic frameworks and their catalytic applications. *Journal of Saudi Chemical Society, 24*(6), 461–473. https://doi.org/10.1016/j.jscs.2020.04.002, http://www.sciencedirect.com/science/journal/13196103.

Annamalai, J., Murugan, P., Ganapathy, D., Nallaswamy, D., Atchudan, R., Arya, S., Khosla, A., Barathi, S., & Sundramoorthy, A. K. (2022). Synthesis of various dimensional metal organic frameworks (MOFs) and their hybrid composites for emerging applications – A review. *Chemosphere, 298*, 134184. https://doi.org/10.1016/j.chemosphere.2022.134184.

Ansari, S. N., Kumar, P., Gupta, A. K., Mathur, P., & Mobin, S. M. (2019). Catalytic CO_2 fixation over a robust lactam-functionalized Cu(II) metal organic framework. *Inorganic Chemistry, 58*(15), 9723–9732. https://doi.org/10.1021/acs.inorgchem.9b00684, http://pubs.acs.org/journal/inocaj.

Ansone-Bertina, L., Ozols, V., Arbidans, L., Dobkevica, L., Sarsuns, K., Vanags, E., & Klavins, M. (2022). Metal–organic frameworks (MOFs) containing adsorbents for carbon capture. *Energies, 15*, 3473. https://doi.org/10.3390/en15093473.

Anumah, A., Louis, H., Hamzat, A. T., Amusan, O. O., Pigweh, A. I., Akakuru, O. U., Adeleye, A. T., & Magu, T. O. (2019). Metal-organic frameworks (MOFs): Recent advances in synthetic methodologies and some applications. *Chemical Methodologies, 3*, 283–305.

Arshadi Edlo, A., & Akhbari, K. (2023). Modulating the antibacterial activity of a CuO@HKUST-1 nanocomposite by optimizing its synthesis procedure. *New Journal of Chemistry, 47*(45), 20770–20776. https://doi.org/10.1039/d3nj03914e, http://pubs.rsc.org/en/journals/journal/nj.

Arshadi Edlo, A., & Akhbari, K. (2024). Modulated antibacterial activity in ZnO@MIL-53(Fe) and CuO@MIL-53(Fe) nanocomposites prepared by simple thermal treatment process. *Applied Organometallic Chemistry, 38*(2), e7326. https://doi.org/10.1002/aoc.7326, http://onlinelibrary.wiley.com/journal/10.1002/(ISSN)1099-0739.

Avci-Camur, C., Troyano, J., Pérez-Carvajal, J., Legrand, A., Farrusseng, D., Imaz, I., & Maspoch, D. (2018). Aqueous production of spherical Zr-MOF beads: Via continuous-flow spray-drying. *Green Chemistry, 20*(4), 873–878. https://doi.org/10.1039/c7gc03132g, http://pubs.rsc.org/en/journals/journal/gc.

Baumann, A. E., Burns, D. A., Liu, B., & Thoi, V. S. (2019). Metal-organic framework functionalization and design strategies for advanced electrochemical energy storage devices. *Communications Chemistry, 2*(1), 86. https://doi.org/10.1038/s42004-019-0184-6, https://www.nature.com/articles/s42004-019-0184-6.

Bavykina, A., Kolobov, N., Khan, I. S., Bau, J. A., Ramirez, A., & Gascon, J. (2020). Metal-organic frameworks in heterogeneous catalysis: Recent progress, new trends, and future perspectives. *Chemical Reviews, 120*(16), 8468–8535. https://doi.org/10.1021/acs.chemrev.9b00685, http://pubs.acs.org/journal/chreay.

Bedia, J., Muelas-Ramos, V., Peñas-Garzón, M., Gómez-Avilés, A., Rodríguez, J. J., & Belver, C. (2019). A review on the synthesis and characterization of metal organic frameworks for photocatalytic water purification. *Catalysts, 9*(1), 52. https://doi.org/10.3390/catal9010052, https://www.mdpi.com/2073-4344/9/1/52/pdf.

Bhattacharjee, S., Jang, M. S., Kwon, H. J., & Ahn, W. S. (2014). Zeolitic imidazolate frameworks: Synthesis, functionalization, and catalytic/adsorption applications. *Catalysis Surveys from Asia, 18*(4), 101–127. https://doi.org/10.1007/s10563-014-9169-8, http://www.sprigerlink.com.

Bian, Y., Xiong, N., & Zhu, G. (2018). Technology for the remediation of water pollution: A review on the fabrication of metal organic frameworks. *Processes, 6*(8), 122. https://doi.org/10.3390/pr6080122.

Bibi, S., Pervaiz, E., & Ali, M. (2021). Synthesis and applications of metal oxide derivatives of ZIF-67: A mini-review. *Chemical Papers, 75*(6), 2253–2275. https://doi.org/10.1007/s11696-020-01473-y, http://www.springer.com/11696.

Bigdeli, F., Rouhani, F., Morsali, A., & Ramazani, A. (2020). Ultrasonic-assisted synthesis of the nanostructures of a Co(II) metal organic framework as a highly sensitive fluorescence probe of phenol derivatives. *Ultrasonics Sonochemistry, 62*, 104862. https://doi.org/10.1016/j.ultsonch.2019.104862.

Brunner, E., & Rauche, M. (2020). Solid-state NMR spectroscopy: An advancing tool to analyse the structure and properties of metal-organic frameworks. *Chemical Science, 11*(17), 4297–4304. https://doi.org/10.1039/d0sc00735h, http://pubs.rsc.org/en/journals/journal/sc.

Cai, G., Yan, P., Zhang, L., Zhou, H. C., & Jiang, H. L. (2021). Metal-organic framework-based hierarchically porous materials: Synthesis and applications. *Chemical Reviews, 121*(20), 12278–12326. https://doi.org/10.1021/acs.chemrev.1c00243, http://pubs.acs.org/journal/chreay.

Cao, Z., Momen, R., Tao, S., Xiong, D., Song, Z., Xiao, X., Deng, W., Hou, H., Yasar, S., & Altin, S. (2022). Metal–organic framework materials for electrochemical supercapacitors. *Nano-micro letters, 14*, 181.

Chang, Z. (2023). Recent progress in host–guest metal–organic frameworks: Construction and emergent properties. *Coordination Chemistry Reviews, 476*, 214921.

Chen, L., & Xu, Q. (2019). Metal-organic framework composites for catalysis. *Matter, 1*(1), 57–89. https://doi.org/10.1016/j.matt.2019.05.018, http://www.cell.com/matter.

Choi, K. M., Jeong, H. M., Park, J. H., Zhang, Y. B., Kang, J. K., & Yaghi, O. M. (2014). Supercapacitors of nanocrystalline metal-organic frameworks. *ACS Nano, 8*(7), 7451–7457. https://doi.org/10.1021/nn5027092, http://pubs.acs.org/journal/ancac3.

Davoodi, A., Akhbari, K., & Alirezvani, M. (2023). Prolonged release of silver and iodine from ZIF-7 carrier with great antibacterial activity. *CrystEngComm, 25*(27), 3931–3942. https://doi.org/10.1039/d3ce00529a, http://pubs.rsc.org/en/journals/journal/ce.

Day, G. S., Drake, H. F., Zhou, H. C., & Ryder, M. R. (2021). Evolution of porous materials from ancient remedies to modern frameworks. *Communications Chemistry, 4*(1), 114. https://doi.org/10.1038/s42004-021-00549-4, https://www.nature.com/articles/s42004-021-00549-4.

Denysenko, D., Grzywa, M., Jelic, J., Reuter, K., & Volkmer, D. (2014). Scorpionate-type coordination in MFU-4l metal-organic frameworks: Small-molecule binding and activation upon the thermally activated formation of open metal sites. *Angewandte Chemie – International Edition, 53*(23), 5832–5836. https://doi.org/10.1002/anie.201310004, http://onlinelibrary.wiley.com/journal/10.1002/(ISSN)1521-3773.

Dey, C., Kundu, T., Biswal, B. P., Mallick, A., & Banerjee, R. (2014). Crystalline metal-Organic frameworks (MOFs): Synthesis, structure and function. *International Union of Crystallography, India Acta Crystallographica Section B: Structural Science, Crystal Engineering and Materials, 70*(1), 3–10. https://doi.org/10.1107/S2052520613029557, http://journals.iucr.org/b/services/authorbdy.html.

Ding, M., Flaig, R. W., Jiang, H. L., & Yaghi, O. M. (2019). Carbon capture and conversion using metal-organic frameworks and MOF-based materials. *Chemical Society Reviews, 48*(10), 2783–2828. https://doi.org/10.1039/c8cs00829a, http://pubs.rsc.org/en/journals/journal/cs.

Du, W., Bai, Y. L., Xu, J., Zhao, H., Zhang, L., Li, X., & Zhang, J. (2018). Advanced metal-organic frameworks (MOFs) and their derived electrode materials for supercapacitors. *Journal of Power Sources, 402*, 281–295. https://doi.org/10.1016/j.jpowsour.2018.09.023, https://www.journals.elsevier.com/journal-of-power-sources.

Eddaoudi, M., Sava, D. F., Eubank, J. F., Adil, K., & Guillerm, V. (2015). Zeolite-like metal-organic frameworks (ZMOFs): Design, synthesis, and properties. *Chemical Society Reviews, 44*(1), 228–249. https://doi.org/10.1039/c4cs00230j, http://pubs.rsc.org/en/journals/journal/cs.

Férey, G. (2008). Hybrid porous solids: Past, present, future. *Chemical Society Reviews, 37*(1), 191–214. https://doi.org/10.1039/b618320b.

Gatou, M. A., Vagena, I. A., Lagopati, N., Pippa, N., Gazouli, M., & Pavlatou, E. A. (2023). Functional MOF-based materials for environmental and biomedical applications: A critical review. *Nanomaterials, 13*(15), 2224. https://doi.org/10.3390/nano13152224, http://www.mdpi.com/journal/nanomaterials.

Ghasemzadeh, R., & Akhbari, K. (2023a). Embedding of copper(i) oxide quantum dots in MOF-801 for the photocatalytic degradation of acid yellow 23 under visible light. *Journal of Chemistry, 47*(33), 15760–15770. https://doi.org/10.1039/d3nj01395b, http://pubs.rsc.org/en/journals/journal/nj.

Ghasemzadeh, R., & Akhbari, K. (2023b). Heterostructured Ag@MOF-801/MIL-88A(Fe) nanocomposite as a biocompatible photocatalyst for degradation of reactive black 5 under visible light. *Inorganic Chemistry, 62*(43), 17818–17829. https://doi.org/10.1021/acs.inorgchem.3c02616, http://pubs.acs.org/journal/inocaj.

Ghasemzadeh, R., & Akhbari, K. (2023c). Band gap engineering of MOF-801 via loading of γ-Fe$_2$O$_3$ quantum dots inside it as a visible light-responsive photocatalyst for degradation of acid orange 7. *Crystal Growth & Design, 23*(9), 6359–6368. https://doi.org/10.1021/acs.cgd.3c00272.

Ghasemzadeh, R., & Akhbari, K. (2024). Templated synthesis of ZnO quantum dots via double solvents method inside MOF-801 as emerging photocatalyst for photodegradation of Acid Blue 25 under UV light. *Journal of Photochemistry and Photobiology A: Chemistry, 448*, 115306. https://doi.org/10.1016/j.jphotochem.2023.115306.

Ghasemzadeh, R., Armanmehr, M. H., Abedi, M., Fateh, D. S., & Bahreini, Z. (2018). Phosphine-free synthesis and characterization of type-II ZnSe/CdS core-shell quantum dots. *Journal of Molecular Structure, 1151*, 106–111. https://doi.org/10.1016/j.molstruc.2017.09.012, https://www.journals.elsevier.com/journal-of-molecular-structure.

Ghasemzadeh, R., Akhbari, K., & Kawata, S. (2024). Ag@MUT-16 nanocomposite as a Fenton-like and plasmonic photocatalyst for degradation of quinoline yellow under visible light. *Dalton transactions*. Iran: Royal Society of Chemistry. http://pubs.rsc.org/en/journals/journal/dt, https://doi.org/10.1039/d4dt00322e.

Guo, X., Geng, S., Zhuo, M., Chen, Y., Zaworotko, M. J., Cheng, P., & Zhang, Z. (2019). The utility of the template effect in metal-organic frameworks. *Coordination Chemistry Reviews, 391*, 44–68. https://doi.org/10.1016/j.ccr.2019.04.003, http://www.journals.elsevier.com/coordination-chemistry-reviews/.

Habtemariam, T. H., Raju, V. J. T., & Chebude, Y. (2022). Room temperature synthesis of pillared-layer metal-organic frameworks (MOFs). *RSC Advances, 12*(50), 32652–32658. https://doi.org/10.1039/d2ra05878b, http://pubs.rsc.org/en/journals/journal/ra.

He, Y., Li, B., O'Keeffe, M., & Chen, B. (2014). Multifunctional metal-organic frameworks constructed from meta-benzenedicarboxylate units. *Chemical Society Reviews, 43*(16), 5618–5656. https://doi.org/10.1039/c4cs00041b, http://pubs.rsc.org/en/journals/journal/cs.

Heo, D. Y., Do, H. H., Ahn, S. H., & Kim, S. Y. (2020). Metal-organic framework materials for perovskite solar cells. *Polymers, 12*(9), 2061. https://doi.org/10.3390/POLYM12092061, https://res.mdpi.com/d_attachment/polymers/polymers-12-02061/article_deploy/polymers-12-02061.pdf.

Howarth, A. J., Peters, A. W., Vermeulen, N. A., Wang, T. C., Hupp, J. T., & Farha, O. K. (2017). Best practices for the synthesis, activation, and characterization of metal−organic frameworks. *Chemistry of Materials, 29*(1), 26–39. https://doi.org/10.1021/acs.chemmater.6b02626, http://pubs.acs.org/journal/cmatex.

Hu, M. L., Safarifard, V., Doustkhah, E., Rostamnia, S., Morsali, A., Nouruzi, N., Beheshti, S., & Akhbari, K. (2018). Taking organic reactions over metal-organic frameworks as heterogeneous catalysis. *Microporous and Mesoporous Materials, 256*, 111–127. https://doi.org/10.1016/j.micromeso.2017.07.057, https://www.sciencedirect.com/science/article/abs/pii/S1387181117305292?via%3Dihub.

Hu, M. L., Masoomi, M. Y., & Morsali, A. (2019). Template strategies with MOFs. *Coordination Chemistry Reviews, 387*, 415–435. https://doi.org/10.1016/j.ccr.2019.02.021, http://www.journals.elsevier.com/coordination-chemistry-reviews/.

Hu, T., Mei, X., Wang, Y., Weng, X., Liang, R., & Wei, M. (2019). Two-dimensional nanomaterials: Fascinating materials in biomedical field. *Science Bulletin, 64*(22), 1707–1727. https://doi.org/10.1016/j.scib.2019.09.021, http://link.springer.com/journal/11434.

Huo, Q. (2011). Synthetic chemistry of the inorganic ordered porous materials. *Modern Inorganic Synthetic Chemistry*, 339–373. https://doi.org/10.1016/B978-0-444–53599-3.10016-2, http://www.sciencedirect.com/science/book/9780444535993.

Huo, Yp, Liu, S., Gao, Zx, Ning, Ba, & Wang, Y. (2021). State-of-the-art progress of switch fluorescence biosensors based on metal-organic frameworks and nucleic acids. *Microchimica Acta, 188*(5), 168. https://doi.org/10.1007/s00604-021-04827-9, http://www.springer/at/mca.

Imaz, I., Rubio-Martínez, M., An, J., Solé-Font, I., Rosi, N. L., & Maspoch, D. (2011). Metal-biomolecule frameworks (MBioFs). *Chemical Communications, 47*(26), 7287–7302. https://doi.org/10.1039/c1cc11202c, http://pubs.rsc.org/en/journals/journal/cc.

iXamena, F. X. L., & Gascon, J. (2013). Metal organic frameworks as heterogeneous catalysts. *Royal Society of Chemistry*, 446. https://doi.org/10.1039/9781849737586-FP001.

Jeremias, F. (2014). *Synthesis and characterization of metal-organic frameworks for heat transformation applications*, 17 (Unpublished content). https://www.researchgate.net/publication/272825099_Synthesis_and_Characterization_of_Metal-Organic_Frameworks_for_Heat_Transformation_Applications?channel=doi&linkId=54f0621d0cf2495330e64c8d&showFulltext=true.

Jiang, C., Wang, X., Ouyang, Y., Lu, K., Jiang, W., Xu, H., Wei, X., Wang, Z., Dai, F., & Sun, D. (2022). Recent advances in metal-organic frameworks for gas adsorption/separation. *Nanoscale Advances, 4*(9), 2077–2089. https://doi.org/10.1039/d2na00061j, pubs.rsc.org/en/journals/journalissues/na?_ga2.190536939.1555337663.1552312502-1364180372.1550481316#!issueidna001002&typecurrent&issnonline2516-0230.

Jiao, L., Seow, J. Y. R., Skinner, W. S., Wang, Z. U., & Jiang, H. L. (2019). Metal–organic frameworks: Structures and functional applications. *Materials Today, 27*, 43–68. https://doi.org/10.1016/j.mattod.2018.10.038, http://www.journals.elsevier.com/materials-today/.

Kalati, M., & Akhbari, K. (2021). Optimizing the metal ion release and antibacterial activity of ZnO@ZIF-8 by modulating its synthesis method. *New Journal of Chemistry, 45*(48), 22924–22931. https://doi.org/10.1039/d1nj04534b, http://pubs.rsc.org/en/journals/journal/nj.

Kalati, M., & Akhbari, K. (2022). Copper(II) nitrate and copper(II) oxide loading on ZIF-8; synthesis, characterization and antibacterial activity. *Journal of Porous Materials, 29*(6), 1909–1917. https://doi.org/10.1007/s10934-022-01302-5, https://www.springer.com/journal/10934.

Kampouraki, Z. C., Giannakoudakis, D. A., Nair, V., Hosseini-Bandegharaei, A., Colmenares, J. C., & Deliyanni, E. A. (2019). Metal organic frameworks as desulfurization adsorbents of DBT and 4,6-DMDBT from fuels. *Molecules (Basel, Switzerland), 24*(24), 4525. https://doi.org/10.3390/molecules24244525, https://www.mdpi.com/1420-3049/24/24/4525/pdf.

Karimi Alavijeh, R., & Akhbari, K. (2020). Biocompatible MIL-101(Fe) as a smart carrier with high loading potential and sustained release of curcumin. *Inorganic Chemistry, 59*(6), 3570–3578. https://doi.org/10.1021/acs.inorgchem.9b02756, http://pubs.acs.org/journal/inocaj.

Karimi Alavijeh, R., & Akhbari, K. (2022). Improvement of curcumin loading into a nanoporous functionalized poor hydrolytic stable metal-organic framework for high anticancer activity against human gastric cancer AGS cells. *Colloids and Surfaces B: Biointerfaces, 212*, 112340. https://doi.org/10.1016/j.colsurfb.2022.112340.

Karimi Alavijeh, R., & Akhbari, K. (2023). Improved cytotoxicity and induced apoptosis in HeLa cells by co-loading vitamin E succinate and curcumin in nano-MIL-88B-NH2. *ChemBioChem, 24*(20), e202300415. https://doi.org/10.1002/cbic.202300415, http://onlinelibrary.wiley.com/journal/10.1002/(ISSN)1439-7633.

Karimi Alavijeh, R., & Akhbari, K. (2024). Cancer therapy by nano MIL-n series of metal-organic frameworks. *Coordination Chemistry Reviews, 503*, 215643. https://doi.org/10.1016/j.ccr.2023.215643.

Karimi Alavijeh, R., Beheshti, S., Akhbari, K., & Morsali, A. (2018). Investigation of reasons for metal–organic framework's antibacterial activities. *Polyhedron, 156*, 257–278. https://doi.org/10.1016/j.poly.2018.09.028, http://www.journals.elsevier.com/polyhedron/.

Karimi Alavijeh, R., Akhbari, K., & White, J. (2019). Solid–liquid conversion and carbon dioxide storage in a calcium-based metal–organic framework with micro- and nanoporous channels. *Crystal Growth & Design. 19*(12), 7290–7297. https://doi.org/10.1021/acs.cgd.9b01174.

Karimi Alavijeh, R., Akhbari, K., Tylianakis, E., Froudakis, G. E., & White, J. M. (2021). Twofold homointerpenetrated metal–organic framework with the potential for anticancer drug loading using computational simulations. *Crystal Growth & Design, 21*(11), 6402–6410. https://doi.org/10.1021/acs.cgd.1c00868.

Karimi Alavijeh, R., Akhbari, K., Bernini, M. C., García Blanco, A. A., & White, J. M. (2022). Design of calcium-based metal–organic frameworks by the solvent effect and computational investigation of their potential as drug carriers. *Crystal Growth & Design, 22*(5), 3154–3162. https://doi.org/10.1021/acs.cgd.2c00032.

Khattami Kermanshahi, P., & Akhbari, K. (2024). The antibacterial activity of three zeolitic-imidazolate frameworks and zinc oxide nanoparticles derived from them. *RSC Advances, 14*(8), 5601–5608. https://doi.org/10.1039/d4ra00447g, http://pubs.rsc.org/en/journals/journal/ra.

Kobaisy, A. M., Elkady, M. F., Abdel-Moneim, A. A., & El-Khouly, M. E. (2023). Surface-decorated porphyrinic zirconium-based metal-organic frameworks (MOFs) using post-synthetic self-assembly for photodegradation of methyl orange dye. *RSC Advances, 13*(33), 23050–23060. https://doi.org/10.1039/d3ra02656f, http://pubs.rsc.org/en/journals/journal/ra.

Kumar, P., Deep, A., & Kim, K. H. (2015). Metal organic frameworks for sensing applications. *TrAC – Trends in Analytical Chemistry, 73*, 39–53. https://www.sciencedirect.com/science/article/abs/pii/S0165993615001090?via%3Dihub, https://doi.org/10.1016/j.trac.2015.04.009, 73.

Kökçam-Demir, Ü., Goldman, A., Esrafili, L., Gharib, M., Morsali, A., Weingart, O., & Janiak, C. (2020). Coordinatively unsaturated metal sites (open metal sites) in metal-organic frameworks: Design and applications. *Chemical Society Reviews, 49*(9), 2751–2798. https://doi.org/10.1039/c9cs00609e, http://pubs.rsc.org/en/journals/journal/cs.

Lakshmi, D. S., Radha, K. S., Roberto, C.-M., & Marek, T. (2022). Emerging trends in porogens toward material fabrication: Recent progresses and challenges. *Polymers, 14*(23), 5209. https://doi.org/10.3390/polym14235209.

Lee, E. J., Bae, J., Choi, K. M., & Jeong, N. C. (2019). Exploiting microwave chemistry for activation of metal-organic frameworks. *ACS Applied Materials and Interfaces, 11*(38), 34989–34996. https://doi.org/10.1021/acsami.9b12201, http://pubs.acs.org/journal/aamick.

Lei, J., Qian, R., Ling, P., Cui, L., & Ju, H. (2014). Design and sensing applications of metal-organic framework composites. *TrAC – Trends in Analytical Chemistry, 58*, 71–78. https://doi.org/10.1016/j.trac.2014.02.012, https://www.sciencedirect.com/science/article/abs/pii/S0165993614000557?via%3Dihub.

Lee, Y. R., Kim, J., & Ahn, W. S. (2013). Synthesis of metal-organic frameworks: A mini review. *Korean Journal of Chemical Engineering, 30*(9), 1667–1680. https://doi.org/10.1007/s11814-013-0140-6.

Lei, L., Cheng, Y., Chen, C., Kosari, M., Jiang, Z., & He, C. (2022). Taming structure and modulating carbon dioxide (CO_2) adsorption isosteric heat of nickel-based metal organic framework (MOF-74(Ni)) for remarkable CO_2 capture. *Journal of Colloid and Interface Science, 612*, 132–145. https://doi.org/10.1016/j.jcis.2021.12.163, http://www.elsevier.com/inca/publications/store/6/2/2/8/6/1/index.htt.

Li, B., Wen, H. M., Cui, Y., Zhou, W., Qian, G., & Chen, B. (2016). Emerging multifunctional metal–organic framework materials. *Advanced Materials, 28*(40), 8819–8860. https://doi.org/10.1002/adma.201601133, http://www3.interscience.wiley.com/journal/119030556/issue.

Li, C., Zhang, H., Liu, M., Lang, F. F., Pang, J., & Bu, X. H. (2023). Recent progress in metal–organic frameworks (MOFs) for electrocatalysis. *Industrial Chemistry & Materials, 1*(1), 9–38. https://doi.org/10.1039/d2im00063f.

Li, D., Yadav, A., Zhou, H., Roy, K., Thanasekaran, P., & Lee, C. (2024). Advances and applications of metal-organic frameworks (MOFs) in emerging technologies: A comprehensive review. *Global Challenges, 8*(2), 2300244. https://doi.org/10.1002/gch2.202300244, https://onlinelibrary.wiley.com/journal/20566646.

Li, J. R., Sculley, J., & Zhou, H. C. (2012). Metal-organic frameworks for separations. *Chemical Reviews, 112*(2), 869–932. https://doi.org/10.1021/cr200190s.

Li, X. M., & Gao, J. (2022). Recent advances of metal–organic frameworks-based proton exchange membranes in fuel cell applications. *SusMat, 2*(5), 504–534. https://doi.org/10.1002/sus2.88, https://onlinelibrary.wiley.com/journal/26924552.

Li, Y., Wang, Y., Fan, W., & Sun, D. (2022). Flexible metal-organic frameworks for gas storage and separation. *Dalton Transactions, 51*(12), 4608–4618. https://doi.org/10.1039/d1dt03842g, http://pubs.rsc.org/en/journals/journal/dt.

Liang, Y., Huang, H., Kou, L., Li, F., Lü, J., & Cao, H. L. (2018). Synthesis of metal–organic framework materials by reflux: A faster and greener pathway to achieve super-hydrophobicity and photocatalytic application. *Crystal Growth & Design, 18*(11), 6609–6616. https://doi.org/10.1021/acs.cgd.8b00854.

Liu, T., & Liu, G. (2020). Porous organic materials offer vast future opportunities. *Nature Communications, 11*(1), 4984. https://doi.org/10.1038/s41467-020-15911-8, http://www.nature.com/ncomms/index.html.

Liu, X., & Liu, Y. (2021). Recent progress in the design and synthesis of zeolite-like metal-organic frameworks (ZMOFs). *Dalton Transactions, 50*(10), 3450–3458. https://doi.org/10.1039/d0dt04338a, http://pubs.rsc.org/en/journals/journal/dt.

Ma, K., Idrees, K. B., Son, F. A., Maldonado, R., Wasson, M. C., Zhang, X., Wang, X., Shehayeb, E., Merhi, A., Kaafarani, B. R., Islamoglu, T., Xin, J. H., & Farha, O. K. (2020). Fiber composites of metal-organic frameworks. *Chemistry of Materials, 32*(17), 7120–7140. https://doi.org/10.1021/acs.chemmater.0c02379, http://pubs.acs.org/journal/cmatex.

Mahjoob, M. K. M., Akbarizadeh, M., Hasheman, P., Rattanapan, N., Moradi-Gholami, A., Amin, H. I. M., Jalil, A. T., & Saleh, M. M. (2023). Recent advances in metal-organic frameworks synthesis and characterization with a focus on electrochemical determination of biological and food compounds, and investigation of their antibacterial performance. *Chinese Journal of Analytical Chemistry, 51*(8), 100286. https://doi.org/10.1016/j.cjac.2023.100286, https://www.journals.elsevier.com/chinese-journal-of-analytical-chemistry.

Mai, Z., & Liu, D. (2019). Synthesis and applications of isoreticular metal–organic frameworks IRMOFs- n (n=1, 3, 6, 8). *Crystal Growth & Design, 19*(12), 7439–7462. https://doi.org/10.1021/acs.cgd.9b00879.

Maza, W. A., Haring, A. J., Ahrenholtz, S. R., Epley, C. C., Lin, S. Y., & Morris, A. J. (2016). Ruthenium(ii)-polypyridyl zirconium(iv) metal–organic frameworks as a new class of sensitized solar cells. *Chemical Science, 7*(1), 719–727. https://doi.org/10.1039/C5SC01565K.

Meena, K., Dixit, S., & Tripathi, B. (2023). Metal-organic frameworks (MOFs) as solid state hydrogen storage system: A critical review. *Energy and Environment Focus, 7*(1), 1–16. https://doi.org/10.1166/eef.2023.1271.

Mirzadeh, E., & Akhbari, K. (2016). Synthesis of nanomaterials with desirable morphologies from metal-organic frameworks for various applications. *CrystEngComm, 18*(39), 7410–7424. https://doi.org/10.1039/c6ce01076h, http://www.rsc.org/publishing/journals/CE/article.asp?type=CurrentIssue.

Mirzadeh, E., Akhbari, K., Phuruangrat, A., & Costantino, F. (2017). A survey on the effects of ultrasonic irradiation, reaction time and concentration of initial reagents on formation of kinetically or thermodynamically stable copper(I) metal-organic nanomaterials. *Ultrasonics Sonochemistry, 35*, 382–388. https://doi.org/10.1016/j.ultsonch.2016.10.016, https://www.sciencedirect.com/science/article/abs/pii/S1350417716303558?via%3Dihub.

Moeinian, M., Akhbari, K., Boonmak, J., & Youngme, S. (2016). Similar to what occurs in biological systems; irreversible replacement of potassium with thallium in coordination polymer nanostructures. *Polyhedron, 118*, 6–11. https://doi.org/10.1016/j.poly.2016.07.039, http://www.journals.elsevier.com/polyhedron/.

Moeinian, M., Akhbari, K., Kawata, S., & Ishikawa, R. (2016). Solid state conversion of a double helix thallium(i) coordination polymer to a corrugated tape silver(i) polymer. *RSC Advances, 6*(85), 82447–82449. https://doi.org/10.1039/c6ra09423f, http://pubs.rsc.org/en/journals/journalissues.

Moghadam, Z., Akhbari, K., & Phuruangrat, A. (2018). Irreversible conversion of nanoporous lead (II) metal–organic framework to a nonporous coordination polymer upon thermal treatment. *Polyhedron, 156*, 48–53. https://doi.org/10.1016/j.poly.2018.09.009, http://www.journals.elsevier.com/polyhedron/.

Mohammadi Amidi, D., & Akhbari, K. (2024). Iodine-loaded ZIF-7-coated cotton substrates show sustained iodine release as effective antibacterial textiles. *New Journal of Chemistry, 48*(5), 2016–2027. https://doi.org/10.1039/d3nj05198f.

Mohammadi Amidi, D., Akhbari, K., & Soltani, S. (2023). Loading of ZIF-67 on silk with sustained release of iodine as biocompatible antibacterial fibers. *Applied Organometallic Chemistry, 37*(1), e6913. https://doi.org/10.1002/aoc.6913, http://onlinelibrary.wiley.com/journal/10.1002/(ISSN)1099-0739.

Molina, M. A., Rodríguez-Campa, J., Flores-Borrell, R., Blanco, R. M., & Sánchez-Sánchez, M. (2024). Sustainable synthesis of zeolitic imidazolate frameworks at room temperature in water with exact Zn/linker stoichiometry. *Nanomaterials, 14*(4), 348. https://doi.org/10.3390/nano14040348, http://www.mdpi.com/journal/nanomaterials.

Nevruzoglu, V., Demir, S., Karaca, G., Tomakin, M., Bilgin, N., & Yilmaz, F. (2016). Improving the stability of solar cells using metal-organic frameworks. *Journal of Materials Chemistry A, 4*(20), 7930–7935. https://doi.org/10.1039/c6ta02609e, http://pubs.rsc.org/en/journals/journalissues/ta.

Noori, Y., & Akhbari, K. (2017). Post-synthetic ion-exchange process in nanoporous metal-organic frameworks; an effective way for modulating their structures and properties. *RSC Advances, 7*(4), 1782–1808. https://doi.org/10.1039/c6ra24958b, http://pubs.rsc.org/en/journals/journalissues.

Obeso, J. L., Flores, J. G., Flores, C. V., Huxley, M. T., de los Reyes, J. A., Peralta, R. A., Ibarra, I. A., & Leyva, C. (2023). MOF-based catalysts: Insights into the chemical transformation of greenhouse and toxic gases. *Chemical Communications, 59*(68), 10226–10242. https://doi.org/10.1039/d3cc03148a, http://pubs.rsc.org/en/journals/journal/cc.

Olorunyomi, J. F., Geh, S. T., Caruso, R. A., & Doherty, C. M. (2021). Metal-organic frameworks for chemical sensing devices. *Materials Horizons, 8*(9), 2387–2419. https://doi.org/10.1039/d1mh00609f, http://pubs.rsc.org/en/journals/journal/mh.

Otun, K. O., Mukhtar, A., Nafiu, S. A., Bello, I. T., & Abdulsalam, J. (2024). Incorporation of redox-activity into metal-organic frameworks for enhanced supercapacitors: A review. *Journal of Energy Storage, 84*. https://doi.org/10.1016/j.est.2024.110673. https://www.sciencedirect.com/science/journal/2352152X.

Pan, X., Zhu, Q., Yu, K., Yan, M., Luo, W., Tsang, S. C. E., & Mai, L. (2023). One-dimensional metal-organic frameworks: Synthesis, structure and application in electrocatalysis. *Next Materials, 1*(1), 100010. https://doi.org/10.1016/j.nxmate.2023.100010.

Parsaei, M., & Akhbari, K. (2022a). MOF-801 as a nanoporous water-based carrier system for in situ encapsulation and sustained release of 5-FU for effective cancer therapy. *Inorganic Chemistry, 61*(15), 5912–5925. https://doi.org/10.1021/acs.inorgchem.2c00380, http://pubs.acs.org/journal/inocaj.

Parsaei, M., & Akhbari, K. (2022b). Smart multifunctional UiO-66 metal-organic framework nanoparticles with outstanding drug-loading/release potential for the targeted delivery of quercetin. *Inorganic Chemistry, 61*(37), 14528–14543. https://doi.org/10.1021/acs.inorgchem.2c00743, http://pubs.acs.org/journal/inocaj.

Parsaei, M., & Akhbari, K. (2022c). Synthesis and application of MOF-808 decorated with folic acid-conjugated chitosan as a strong nanocarrier for the targeted drug delivery of quercetin. *Inorganic Chemistry, 61*(48), 19354–19368. https://doi.org/10.1021/acs.inorgchem.2c03138, http://pubs.acs.org/journal/inocaj.

Parsaei, M., & Akhbari, K. (2023). Magnetic UiO-66-NH$_2$ core-shell nanohybrid as a promising carrier for quercetin targeted delivery toward human breast cancer cells. *ACS Omega, 8*(44), 41321–41338. https://doi.org/10.1021/acsomega.3c04863, http://pubs.acs.org/journal/acsodf.

Parsaei, M., Akhbari, K., & White, J. (2022). Modulating carbon dioxide storage by facile synthesis of nanoporous pillared-layered metal-organic framework with different synthetic routes. *Inorganic Chemistry, 61*(9), 3893–3902. https://doi.org/10.1021/acs.inorgchem.1c03414, http://pubs.acs.org/journal/inocaj.

Parsaei, M., Akhbari, K., Tylianakis, E., Froudakis, G. E., White, J. M., & Kawata, S. (2022). Computational study of two three-dimensional Co(II)-based metal–organic frameworks as quercetin anticancer drug carriers. *Crystal Growth & Design, 22*(12), 7221–7233. https://doi.org/10.1021/acs.cgd.2c00900.

Parsaei, M., Akhbari, K., Tylianakis, E., & Froudakis, G. E. (2023). Computational simulation of a three-dimensional Mg-based metal–organic framework as nanoporous anticancer drug carrier. *Crystal Growth & Design, 23*(11), 8396–8406. https://doi.org/10.1021/acs.cgd.3c01058.

Parsaei, M., Akhbari, K., & Kawata, S. (2023). Computational simulation of CO$_2$/CH$_4$ separation on a three-dimensional Cd-based metal–organic framework. *Crystal Growth & Design, 23*(8), 5705–5718. https://doi.org/10.1021/acs.cgd.3c00366.

Parsaei, M., Akhbari, K., Tylianakis, E., & Froudakis, G. E. (2024). Effects of fluorinated functionalization of linker on quercetin encapsulation, release and hela cell cytotoxicity of Cu-based MOFs as smart pH-stimuli nanocarriers. *Chemistry – A European Journal, 30*(1). https://doi.org/10.1002/chem.202301630. http://onlinelibrary.wiley.com/journal/10.1002/(ISSN)1521-3765.

Ponnada, S., Kiai, M. S., Gorle, D. B., Nowduri, A., & Sharma, R. K. (2021). Insight into the role and strategies of metal-organic frameworks in direct methanol fuel cells: A review. *Energy and Fuels, 35*(19), 15265–15284. https://doi.org/10.1021/acs.energyfuels.1c02010, http://pubs.acs.org/journal/enfuem.

Qiu, J., Zhang, X., Feng, Y., Zhang, X., Wang, H., & Yao, J. (2018). Modified metal-organic frameworks as photocatalysts. *Applied Catalysis B: Environmental, 231*, 317–342. https://doi.org/10.1016/j.apcatb.2018.03.039, https://www.sciencedirect.com/science/article/abs/pii/S0926337318302388?via%3Dihub.

Qiu, J., Dai, D., & Yao, J. (2024). Tailoring metal–organic frameworks for photocatalytic H$_2$O$_2$ production. *Coordination Chemistry Reviews, 501*(36), 215597.

Raptopoulou, C. P. (2021). Metal-organic frameworks: Synthetic methods and potential applications. *Materials, 14*(2), 1–32. https://doi.org/10.3390/ma14020310, https://www.mdpi.com/1996-1944/14/2/310/pdf.

Ren, Y., Chia, G. H., & Gao, Z. (2013). Metal-organic frameworks in fuel cell technologies. *Nano Today, 8*(6), 577–597. https://doi.org/10.1016/j.nantod.2013.11.004, http://www.elsevier.com/wps/find/journaldescription.cws_home/706735/description#description.

Rocío-Bautista, P., Taima-Mancera, I., Pasán, J., & Pino, V. (2019). Metal-organic frameworks in green analytical chemistry. *Separations, 6*(3), 33. https://doi.org/10.3390/separations6030033.

Ru, J., Wang, X., Wang, F., Cui, X., Du, X., & Lu, X. (2021). UiO series of metal-organic frameworks composites as advanced sorbents for the removal of heavy metal ions: Synthesis, applications and adsorption mechanism. *Ecotoxicology and Environmental Safety, 208*, 111577. https://doi.org/10.1016/j.ecoenv.2020.111577.

Ryu, U. J., Jee, S., Park, J. S., Han, I. K., Lee, J. H., Park, M., & Choi, K. M. (2018). Nanocrystalline titanium metal-organic frameworks for highly efficient and flexible perovskite solar cells. *ACS Nano, 12*(5), 4968–4975. https://doi.org/10.1021/acsnano.8b02079, http://pubs.acs.org/journal/ancac3.

Salimi, S., Akhbari, K., Farnia, S. M. F., & White, J. M. (2022). Multiple construction of a hierarchical nanoporous manganese(II)-based metal–organic framework with active sites for regulating N_2 and CO_2 Trapping. *Crystal Growth & Design. 22*(3), 1654–1664. https://doi.org/10.1021/acs.cgd.1c01183.

Salimi, S., Farnia, S. M. F., Akhbari, K., & Tavasoli, A. (2023). Engineered catalyst based on MIL-68(Al) with high stability for hydrogenation of carbon dioxide and carbon monoxide at low temperature. *Inorganic Chemistry, 62*(43), 17588–17601. https://doi.org/10.1021/acs.inorgchem.3c01094, http://pubs.acs.org/journal/inocaj.

Salimi, S., Akhbari, K., Farnia, S. M. F., Tylianakis, E., Froudakis, G. E., & White, J. M. (2024a). Solvent-directed construction of a nanoporous metal-organic framework with potential in selective adsorption and separation of gas mixtures studied by grand canonical Monte Carlo simulations. *ChemPlusChem, 89*(1), e202300455. https://doi.org/10.1002/cplu.202300455, http://onlinelibrary.wiley.com/journal/10.1002/(ISSN)2192-6506.

Salimi, S., Akhbari, K., Farnia, S. M. F., Tylianakis, E., Froudakis, G. E., & White, J. M. (2024b). Nanoporous metal–organic framework based on furan-2,5-dicarboxylic acid with high potential in selective adsorption and separation of gas mixtures. *Crystal Growth & Design, 24*(10), 4220–4231. https://doi.org/10.1021/acs.cgd.4c00349.

Shahangi Shirazi, F., & Akhbari, K. (2016). Sonochemical procedures; The main synthetic method for synthesis of coinage metal ion supramolecular polymer nano structures. *Ultrasonics Sonochemistry, 31*, 51–61. https://doi.org/10.1016/j.ultsonch.2015.12.003, https://www.sciencedirect.com/science/article/pii/S1350417715300961?via%3Dihub.

Shahbazi Farahani, F., Rahmanifar, M. S., Noori, A., El-Kady, M. F., Hassani, N., Neek-Amal, M., Kaner, R. B., & Mousavi, M. F. (2022). Trilayer metal-organic frameworks as multifunctional electrocatalysts for energy conversion and storage applications. *Journal of the American Chemical Society, 144*(8), 3411–3428. https://doi.org/10.1021/jacs.1c10963, http://pubs.acs.org/journal/jacsat.

Sharma, A., Lim, J., & Lah, M. S. (2023). Strategies for designing metal–organic frameworks with superprotonic conductivity. *Coordination Chemistry Reviews, 479*, 214995. https://doi.org/10.1016/j.ccr.2022.214995.

Shen, C. H., & Kung, C. W. (2023). Stable and electrochemically "inactive" metal–organic frameworks for electrocatalysis. *ChemElectroChem, 10*(23), e202300375. https://doi.org/10.1002/celc.202300375, http://onlinelibrary.wiley.com/journal/10.1002/(ISSN)2196-0216.

Shen, Y. (2023). Investigating the structure of metal-organic frameworks and the behaviors of adsorbed guest molecules in MOFs via solid-state NMR spectroscopy, *Electronic thesis and dissertation repository*. 9139. https://ir.lib.uwo.ca/etd/9139/.

Simmons, J. M., Wu, H., Zhou, W., & Yildirim, T. (2011). Carbon capture in metal-organic frameworks – A comparative study. *Energy and Environmental Science, 4*(6), 2177–2185. https://doi.org/10.1039/c0ee00700e.

Soltani, S., & Akhbari, K. (2022a). Embedding an extraordinary amount of gemifloxacin antibiotic in ZIF-8 framework with one-step synthesis and measurement of its H_2O_2-sensitive release and potency against infectious bacteria. *Journal of Chemistry, 46*(40), 19432–19441. https://doi.org/10.1039/d2nj02981b, http://pubs.rsc.org/en/journals/journal/nj.

Soltani, S., & Akhbari, K. (2022b). Facile and single-step entrapment of chloramphenicol in ZIF-8 and evaluation of its performance in killing infectious bacteria with high loading content and controlled release of the drug. *CrystEngComm, 24*(10), 1934–1941. https://doi.org/10.1039/d1ce01593a.

Soltani, S., Akhbari, K., & Phuruangrat, A. (2022a). Incorporation of silver nanoparticles on Cu-BTC metal–organic framework under the influence of reaction conditions and investigation of their antibacterial activity. *Applied Organometallic Chemistry, 36*(6), e6634. https://doi.org/10.1002/aoc.6634, http://onlinelibrary.wiley.com/journal/10.1002/(ISSN)1099-0739.

Soltani, S., Akhbari, K., & Phuruangrat, A. (2022b). Improved antibacterial activity by incorporation of silver sulfadiazine on nanoporous Cu-BTC metal-organic-framework. *Inorganica Chimica Acta, 543*, 121182. https://doi.org/10.1016/j.ica.2022.121182.

Su, Y., Yuan, G., Hu, J., Feng, W., Zeng, Q., Liu, Y., & Pang, H. (2023). Recent progress in strategies for preparation of metal-organic frameworks and their hybrids with different dimensions. *Chemical Synthesis, 3*(1). https://doi.org/10.20517/cs.2022.24. https://www.oaepublish.com/articles/cs.2022.24.

Tao, Y. R., & Xu, H. J. (2024). A critical review on potential applications of metal-organic frameworks (MOFs) in adsorptive carbon capture technologies. *Applied Thermal Engineering, 236*, 121504. https://doi.org/10.1016/j.applthermaleng.2023.121504.

Tsivadze, A. Y., Aksyutin, O. E., Ishkov, A. G., Knyazeva, M. K., Solovtsova, O. V., Men'Shchikov, I. E., Fomkin, A. A., Shkolin, A. V., Khozina, E. V., & Grachev, V. A. (2019). Metal-organic framework structures: Adsorbents for natural gas storage. *Russian Chemical Reviews, 88*(9), 925–978. https://doi.org/10.1070/RCR4873.

Usefi, S., Akhbari, K., & White, J. (2019). Sonochemical synthesis, structural characterizations and antibacterial activities of biocompatible copper(II)coordination polymer nanostructures. *Journal of Solid State Chemistry, 276*, 61–67. https://doi.org/10.1016/j.jssc.2019.04.016, http://www.elsevier.com/inca/publications/store/6/2/2/8/9/8/index.htt.

Wang, H. S., Wang, Y. H., & Ding, Y. (2020). Development of biological metal-organic frameworks designed for biomedical applications: From bio-sensing/bio-imaging to disease treatment. *Nanoscale Advances, 2*(9), 3788–3797. https://doi.org/10.1039/d0na00557f, http://pubs.rsc.org/en/journals/journalissues/na?_ga2.190536939.1555337663.1552312502-1364180372.1550481316#!issueidna001002&typecurrent&issnonline2516-0230.

Wang, K. B., Xun, Q., & Zhang, Q. (2020). Recent progress in metal-organic frameworks as active materials for supercapacitors. *EnergyChem, 2*(1), 100025. https://doi.org/10.1016/j.enchem.2019.100025, https://www.sciencedirect.com/science/article/abs/pii/S2589778019300284?via%3Dihub.

Wang, Q., Gao, Q., Al-Enizi, A. M., Nafady, A., & Ma, S. (2020). Recent advances in MOF-based photocatalysis: Environmental remediation under visible light. *Inorganic Chemistry Frontiers, 7*(2), 300–339. https://doi.org/10.1039/c9qi01120j, http://pubs.rsc.org/en/journals/journal/qi.

Wang, S. J., Zhang, Z. Y., Tan, Y., Liang, K. X., & Zhang, S. H. (2023). Review on the characteristics of existing hydrogen energy storage technologies. *Energy Sources, Part A: Recovery, Utilization, and Environmental Effects, 45*(1), 985–1006. https://doi.org/10.1080/15567036.2023.2175938.

Wei, Z. (2014). Unpublished content New design and synthetic strategies of metal-organic frameworks. (Unpublished content). https://core.ac.uk/download/pdf/79648478.pdf.

Wen, X., Zhang, Q., & Guan, J. (2020). Applications of metal–organic framework-derived materials in fuel cells and metal-air batteries. *Coordination Chemistry Reviews, 409*, 213214. https://doi.org/10.1016/j.ccr.2020.213214.

Wen, Y., Zhang, P., Sharma, V. K., Ma, X., & Zhou, H. C. (2021). Metal-organic frameworks for environmental applications. *Cell Reports Physical Science, 2*(2), 100348. https://doi.org/10.1016/j.xcrp.2021.100348, https://www.cell.com/cell-reports-physical-science/aims.

Wu, L., Chaplais, G., Xue, M., Qiu, S., Patarin, J., Simon-Masseron, A., & Chen, H. (2019). New functionalized MIL-53(In) solids: Syntheses, characterization, sorption, and structural flexibility. *RSC Advances, 9*(4), 1918–1928. https://doi.org/10.1039/C8RA08522F, http://pubs.rsc.org/en/journals/journal/ra.

Xu, Q., & Kitagawa, H. (2018). MOFs: new useful materials–A special issue in honor of Prof. Susumu Kitagawa. *Advanced Materials, 30*(37), e1803613. https://doi.org/10.1002/adma.201803613.

Yadav, S., Dixit, R., Sharma, S., Dutta, S., Solanki, K., & Sharma, R. K. (2021). Magnetic metal-organic framework composites: Structurally advanced catalytic materials for organic transformations. *Materials Advances, 2*(7), 2153–2187. https://doi.org/10.1039/d0ma00982b, www.rsc.org/journals-books-databases/about-journals/materials-advances/.

Yaghi, O. M., O'Keeffe, M., Ockwig, N. W., Chae, H. K., Eddaoudi, M., & Kim, J. (2003). Reticular synthesis and the design of new materials. *Nature, 423*(6941), 705–714. https://doi.org/10.1038/nature01650.

Yang, D., & Gates, B. C. (2019). Catalysis by metal organic frameworks: Perspective and suggestions for future research. *ACS Catalysis, 9*(3), 1779–1798. https://doi.org/10.1021/acscatal.8b04515, http://pubs.acs.org/page/accacs/about.html.

Yoganathan, R. B., Mammucari, R., & Foster, N. R. (2010). A green method for processing polymers using dense gas technology. *Materials. 3*(5), 3188–3203. https://doi.org/10.3390/ma3053188, http://www.mdpi.com/1996-1944/3/5/3188/pdf, Canada.

Yusuf, V. F., Malek, N. I., & Kailasa, S. K. (2022). Review on metal-organic framework classification, synthetic approaches, and influencing factors: Applications in energy, drug delivery, and wastewater treatment. *ACS Omega, 7*(49), 44507–44531. https://doi.org/10.1021/acsomega.2c05310, pubs.acs.org/journal/acsodf.

Zhang, J. P., Zhang, Y. B., Lin, J. B., & Chen, X. M. (2012). Metal azolate frameworks: From crystal engineering to functional materials. *Chemical Reviews, 112*(2), 1001–1033. https://doi.org/10.1021/cr200139g.

Zhang, W., Hu, Y., Ge, J., Jiang, H. L., & Yu, S. H. (2014). A facile and general coating approach to moisture/water-resistant metal-organic frameworks with intact porosity. *Journal of the American Chemical Society, 136*(49), 16978–16981. https://doi.org/10.1021/ja509960n, http://pubs.acs.org/journal/jacsat.

Zhang, W., Taheri-Ledari, R., Saeidirad, M., Qazi, F. S., Kashtiaray, A., Ganjali, F., Tian, Y., & Maleki, A. (2022). Regulation of porosity in MOFs: A review on tunable scaffolds and related effects and advances in different applications. *Journal of Environmental Chemical Engineering, 10*(6), 108836. https://doi.org/10.1016/j.jece.2022.108836.

Zhang, X., Chen, Z., Liu, X., Hanna, S. L., Wang, X., Taheri-Ledari, R., Maleki, A., Li, P., & Farha, O. K. (2020). A historical overview of the activation and porosity of metal-organic frameworks. *Chemical Society Reviews, 49*(20), 7406–7427. https://doi.org/10.1039/d0cs00997k, http://pubs.rsc.org/en/journals/journal/cs.

Zhao, M., Huang, Y., Peng, Y., Huang, Z., Ma, Q., & Zhang, H. (2018). Two-dimensional metal-organic framework nanosheets: Synthesis and applications. *Chemical Society Reviews, 47*(16), 6267–6295. https://doi.org/10.1039/c8cs00268a, http://pubs.rsc.org/en/journals/journal/cs.

Zheng, W., & Lee, L. Y. S. (2021). Metal-organic frameworks for electrocatalysis: Catalyst or precatalyst? *ACS Energy Letters, 6*(8), 2838–2843. https://doi.org/10.1021/acsenergylett.1c01350, http://pubs.acs.org/journal/aelccp.

Chapter 2

Metal-organic framework composites

Megha Prajapati[1,2], and Chhaya Ravi Kant[1]
[1]*Department of Applied Sciences and Humanities, Indira Gandhi Delhi Technical University for Women, Delhi, India,* [2]*Electronics Materials Lab, College of Science and Engineering, James Cook University, Townsville, QLD, Australia*

Emergence and properties of metal-organic frameworks

Reticular chemistry enacts the predesign of geometry-guided periodic, stable frameworks like MOFs by the union of organic and inorganic molecular building blocks using facile, directed chemical reactions. MOFs are considered crystalline porous materials consisting of primary components such as inorganic clusters/organic linkers and metal ions known as secondary building units (SBUs) coupled to form a networked structure via adjustable coordination bonding (Masoomi et al., 2019; Prajapati, Singh, et al., 2023). Since their discovery in the early 1990s, MOFs have sparked significant research interest across various domains. This is due to their exceptionally ordered porosity, remarkably high specific surface areas (ranging from 1000 to 10,000 m^2/g), tunable pore sizes and characteristics, flexible architectures, and ease of preparation and modification. As a novel and intriguing subclass of crystalline porous inorganic–organic hybrids, precise design and construction of various MOF architectures are crucial (Chai et al., 2021). In recent decades, there have been many opportunities for reticular design, which have led to the creation of a wide range of MOF (> 100,000) structures with distinctive characteristics. In terms of practicability concerns, BASF, Toyota, and Ford Global Technologies produce MOFs on a tonne scale; they are accessible at Strem Chemical, Sigma-Aldrich, and MOF technologies (Wang & Astruc, 2020; Yap et al., 2017). A study of the World Intellectual Property Organization (WIPO) database showed that the number of MOF-related patents released by researchers worldwide is continuously increasing. This suggests that MOFs are making significant progress toward commercial use (Yap et al., 2017).

Structural and functional regulation of metal-organic frameworks and their composites

The major approaches for regulating the structural characteristics and functionalities of MOFs are as follows: (1) regulating the metallic ions and linkers for MOFs; (2) designing materials derived from MOF templates; (3) controlling the morphologies of MOFs; and (4) building MOF composites. MOFs with a variety of metal nodes, ligands, and their numerous combinations offer a vast domain for structural and compositional regulation (Chai et al., 2021). However, in their pure state, MOFs have several shortcomings, including low chemical stability, which limits their application potential. Thus, the structured synthesis of MOFs with added functionalities is essential to meet practical applications. MOF templates serve as skeletons or precursors for deriving materials such as metal hydroxides, metal oxides, metal phosphides, and carbon-based materials. This process enhances their properties, including large specific surface area, high porosity, and substantial active sites, which are crucial for achieving targeted functionalities across multiple applications. For instance, Prajapati et al. (2023, 2024) synthesized NiCo and CoMn layered double hydroxides (LDH) using Co-ZIF as a template and employed them as anode in supercapacitor applications. Moreover, the characteristics and structure of these MOF-derived materials can be modified by selecting the parent MOF and employing specific design strategies during the conversion process. With a wide range of applications, MOFs with different topologies and dimensionalities, including 1D wires (Zou et al., 2018), 2D sheets, 3D hierarchical structures (Hong et al., 2019), and zero-dimensional (0D) particles (Min et al., 2019; Wang et al., 2018), have been created.

Furthermore, ordered crystalline MOFs with high porosity, adaptability, and flexibility have been combined with various functional materials possessing unique catalytic, magnetic, and optical properties to create MOF composites. These composites inherit advantageous properties from both constituents, offering novel physical and chemical characteristics and enhanced performance that surpasses what each component can achieve individually. Thus far, active species such as metal nanoparticles (MNPs), polyoxometalates (POMs), metal oxides/phosphides/sulfides, carbon quantum dots (CQDs) (Kumari, Prajapati & Ravi Kant, 2024), polymers (Yildirim Kalyon et al., 2022), graphene (Karimzadeh et al., 2019), biomolecules, carbon nanotubes (CNTs) (Zhang et al., 2018), have been efficaciously encapsulated into MOF composites/hybrids with intriguing usage for varied applications, including energy storage/conversion, heterogeneous catalysis, chemical sensors, bio-applications, drug delivery, magnetic materials, optical devices, and many other numerous technologies (Liu et al., 2021). Usually, the properties of MOF composites depend on the composition and architecture of the constituent MOF. These characteristics should be thoroughly studied using various characterization techniques to tailor the composites for specific applications.

Characterization of metal-organic frameworks and their composites

The morphology, structure, and composition of MOF composites can be analyzed using different techniques such as scanning electron microscopy, energy-dispersive X-ray spectrometry, transmission electron microscopy, powder X-ray diffraction, Fourier-transform infrared spectroscopy, thermogravimetric analysis, and Brunauer-Emmett-Teller, which have been briefly outlined below.

Structural and morphological characterization

X-ray diffraction (XRD) is employed to investigate the crystallographic structure, crystal size, different polymorphic forms, and the presence of defects in as-synthesized MOF composites. Structural identification is achieved by comparing the diffractogram of the synthesized MOF composites with previously reported data in the literature. Alternatively, computational modeling techniques such as density functional theory (DFT) can be utilized to provide theoretical insights into electronic behaviors such as band gap and density of states. Once the crystalline structure of the MOF is identified, it becomes feasible to determine various crystallographic parameters, including unit cell dimensions, lattice parameters, and crystallite size (Nath & Das, 2021). These parameters are crucial for identifying and characterizing the crystal phase, confirming material purity, and detecting structural distortions or phase changes. This allows researchers to better predict and manipulate the properties of MOFs and their composites for desired applications in areas such as catalysis, gas storage, separation, and drug delivery.

Scanning electron microscopy (SEM) provides insights into surface morphology, texture, particle shape, size, and component distribution within composites. It effectively detects microstructural changes occurring during composite synthesis and verifies the preservation of the MOF framework structure. Due to the typically insulating nature of MOF samples, coating their surfaces with a conductive material, typically gold, is often necessary for SEM analysis. Researchers employ field emission SEM (FE-SEM) to achieve high-resolution images, utilizing a field emission gun that produces a tightly focused electron beam at low voltages (0.02–5 kV), thereby enhancing spatial resolution (Orasugh et al., 2020).

Transmission electron microscopy (TEM)/high-resolution TEM provides detailed morphological information, interfacial interactions, and crystallinity (via selected area electron diffraction) of MOF composites. These methods are crucial for characterizing MOF@NP composites, facilitating the determination of nanoparticle size, distribution, and morphology within or on the MOF structure. Recently, high-resolution TEM (HR-TEM) has been preferred over conventional TEM due to its superior spatial resolution. HR-TEM enables direct imaging of individual atoms and precise analysis of atomic arrangements. It is highly effective in identifying and characterizing defects such as dislocations, vacancies, and grain boundaries, which significantly influence a material's mechanical, electronic, and chemical properties (Radhika et al., 2020).

Surface area and porosity characterization

Surface area and porosity are crucial quantitative metrics that dictate the performance of MOF composites in various applications such as gas adsorption, sensing, and catalysis. The Brunauer-Emmett-Teller (BET) method is essential for characterizing specific surface area by measuring nitrogen adsorption–desorption isotherms at 77 K, typically within a relative pressure range of 0.05–0.3. This technique evaluates both external and pore areas to determine the composite's specific surface area in m²/g. The Barrett-Joyner-Halenda (BJH) method complements BET by analyzing nitrogen adsorption on the composite's surface at the boiling point of liquid nitrogen, generating an adsorption isotherm. This approach allows for the evaluation of pore size distribution and specific pore volume. Using BJH, the precise specific surface area can be calculated independently of the external area through particle size estimation. Different types of adsorption isotherms (Type I, II, IV, etc.) indicate distinct pore structures such as microporous, mesoporous, or macroporous, which are critical for performance in various applications. Micropores (< 2 nm) increase surface area and are crucial for gas adsorption and catalysis. Mesopores (2–50 nm) balance surface area and pore volume, enhancing molecular transport, making them suitable for catalysis and drug delivery. Macropores (> 50 nm) facilitate fluid movement and provide structural support, which is important for filtration and lightweight construction. Tailoring the pore size distribution optimizes a material's surface area, transport properties, and mechanical strength to meet specific application requirements effectively (Bardestani et al., 2019; Kumari et al., 2023).

Chemical and compositional analysis

Energy dispersive X-ray spectroscopy (EDX or EDS) is commonly coupled with TEM and SEM to provide valuable insights into the distribution and composition of elements within samples. This analysis is crucial for confirming the chemical composition, identifying the presence of metals, organic linkers, and other elements within the composite, and ensuring the correct stoichiometry of both the MOF and the incorporated materials (Kumari et al., 2024).

X-ray photoelectron spectroscopy (XPS) offers detailed information on the elemental composition and chemical states of elements within MOF composites, including surface chemistry and oxidation states. By analyzing shifts in binding energies or changes in peak intensities, XPS can elucidate the nature of bonding and interaction mechanisms at the interface. This information is crucial for understanding the electronic properties of the composite, which are relevant for applications in sensing and electronic devices (Kumari, Prajapati & Kant, 2024; Prajapati et al., 2024).

Fourier-transform infrared spectroscopy (FT-IR) is employed to identify functional groups and monitor the coordination between metal nodes and organic ligands in composites. Comparing FT-IR spectra before and after composite formation helps identify new bonds or interactions, as well as changes in existing bonds such as hydrogen bonding, coordination bonds, or van der Waals interactions. This is achieved by observing variations in peak intensity, shifts in peak positions, or the appearance/disappearance of specific peaks (Kumari et al., 2022; Prajapati, Ravi Kant, et al., 2023).

Raman spectroscopy complements the FT-IR technique by providing more detailed information on the vibrational modes and chemical bonds present at the surface. This is useful for understanding the bonding and structure of MOFs with composites (Kumari et al., 2023).

Metal-organic framework polymer composites: synthesis and applications

The chapter further advances the classification of various MOF composites, shedding light on their synthesis techniques, characteristic properties, and integration for high-performance applications.

MOF-polymer composites represent a noteworthy class that has recently attracted significant attention from researchers due to their enhanced functionalities. MOFs are inherently crystalline, brittle, and porous solids, in stark contrast to the flexibility and processability of polymers. These distinct properties drive the hybridization of porous MOFs and flexible polymers to create numerous advanced, highly functional porous composites. Such composites retain the desirable properties of both materials and offer advantages over traditional single-component polymer films. Importantly, studies have shown that MOFs influence the structure of polymers, while polymers can modulate MOF growth and characteristics. MOF-polymer composites exhibit compelling attributes such as enhanced access to active sites and improved molecular transport of MOFs, making them suitable for applications in sensing, air purification, therapeutic delivery, and biomedical applications.

Five broad techniques for synthesizing MOF-polymer composites have been identified, and their schematic representations are depicted in Fig. 2.1 (Yang et al., 2021).

In situ polymerization

Monomers in both liquid and gas phases permeate into the MOF pores, facilitated by heat, light, direct interactions, or chemical initiators (Santos et al., 2015). This technique results in linear polymer chains stabilized by

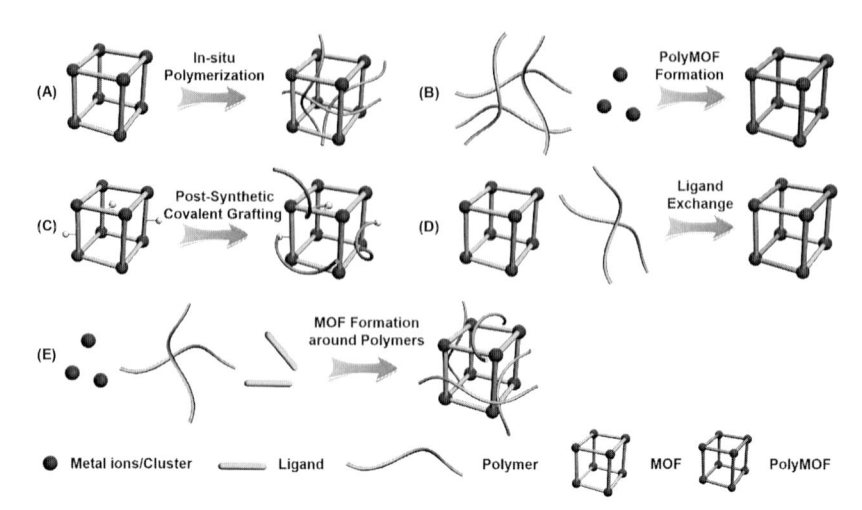

FIGURE 2.1 Schematic illustration of various techniques for preparing metal-organic framework@polymer composites: (A–E) in situ polymerization; formation of metal-organic frameworks from ligands; framework growth around polymers; post-synthetic covalent grafting; ligand exchange; and metal-organic framework formation around polymers, respectively. (Yang et al., 2021).

strong intermolecular interactions with the MOF walls. Many vinyl polymers, including polyacrylonitrile (Uemura et al., 2015), poly(vinyl acetate) (Hwang et al., 2017), and poly(methyl methacrylate) (Uemura et al., 2008), have successfully been incorporated into the MOFs. The moderate reaction conditions of these polymerization reactions are crucial for thermally and chemically unstable MOFs. The ease of application to practically any framework makes this process widely acceptable (Gamage et al., 2016). Researchers have also employed in situ polymerization on sacrificial MOF templates to create porous 3-D polymer networks with three-dimensional pore topologies. Free-standing 3-D porous polymers can be created by removing the MOF skeleton through an etching process (Sundriyal et al., 2018), although controlling pore blocking and ensuring uniform diffusivity of monomers across the MOF template can be challenging (Zhang et al., 2015) (Fig. 2.1A).

Integrating prefabricated polymer ligands

The second synthesis approach integrates prefabricated polymeric ligands during the synthesis and crystallization of MOFs, yielding polyMOFs. These materials combine the processability and chemical stability of polymers with the regularity, crystallinity, and porosity of MOF structures. Various polyMOFs, such as UiO and IRMOFs, have been extensively studied (Zhang et al., 2015). Employing polymeric ligands offers the additional advantage of easily altering the polymer molecular weight, structure, composition, and loading inside the MOF. However, it has been demonstrated that while MOFs alone can form small and rigid crystalline structures with well-organized ligands, incorporating polymeric building blocks can be challenging entropically and kinetically. This can lead to structural defects and chain intertwining (Zhang et al., 2016, 2017) (Fig. 2.1B).

Grafting polymers onto metal-organic framework ligands post synthesis

Grafting polymers onto MOF ligands postsynthesis is the third method gaining popularity for synthesizing MOF-polymer composites. This method utilizes ligands with reactive functionalities, such as $-CHO$, $-NH_2$, $-N_3$, $-OH$, $-Br$, and $C=C$ bonds, to prepare functionalized MOFs beforehand (Tanabe & Cohen, 2011). Next, conventional organic synthesis methods are used to graft polymeric units onto the MOF structures. For example, He et al. created grafted polymers and NH_2-MIL-101 (Al) (He et al., 2019) that change from hydrophilic to hydrophobic upon heating, demonstrating controlled regulation of MOF crystal aggregation in solution. Additionally, trifluoroacetic acid was reacted with UiO-66-NH_2@PtBMA to produce UiO-66-NH_2@PMAA (PMAA=polymethacrylic acid) (Hou et al., 2016). Additionally, layer-by-layer sequential grafting enables better control over polymer thickness and density without compromising the surface area of MOFs. While postsynthesis grafting offers several advantages, it also presents limitations. For instance, the use of reactive ligands may restrict the manipulation of MOF structures to achieve the desired density and thickness (Fig. 2.1C).

Adding preformed metal-organic frameworks to preformed polymers

Another approach for forming composites is by adding preformed MOFs to preformed polymers. This method is beneficial in cases where direct polymerization of monomer ligands into MOFs is impractical or incompatible. Additionally, this technique has been employed to incorporate biopolymers into MOFs, enabling an increase in the pore size of MOFs while preserving their well-defined structure. Peng et al. (2018), using this strategy, introduced single-stranded DNA (ssDNA) into an isoreticular series from Ni(-dodbc)$_2$-II to Ni(dodbc)$_2$-(2,5-dioxidoterephthalate ligand) with varied bond lengths. One major advantage of this approach over in situ methods is the ability to control the loading of preformed polymers within the MOF (Peng et al., 2018). Moreover, this method is often simpler to implement compared to other design strategies, involving soaking MOFs in solvents or polymer melts or exchanging ligands after the synthesis process. Additionally, it offers a better understanding of the bulk polymer structure–property relationships in MOF@polymer composites. However, polymers may obstruct pores and hinder functional accessibility in MOF@polymer composites. Polymer diffusion and its homogeneous distribution can be challenging, depending on the structure and crystallite size of the MOFs; MOFs with 1-D channels facilitate swift diffusion, whereas those with cages may impede this process (Kitao et al., 2015) (Fig. 2.1D).

Crystallizing metal-organic frameworks around presynthesized polymers

Lastly, we explore how MOFs can crystallize around presynthesized polymers, which either become trapped inside by a ship-in-a-bottle effect or bond to the internal MOF surface. Some of the previously mentioned issues, such as clogging of MOF pores and nonuniform polymer distribution, can be alleviated using this technique. Lyu *et al.* inserted polyvinylidene fluoride (PVDF)-capped protein molecules directly into the MOF formation to embed them into ZIF-8 (Lyu et al., 2014). The ZIF-8 MOF forms a composite structure that encapsulates the polymer-capped protein, crystallizing around zinc metal nodes linked by 2-methylimidazole (2-MIM) linkers. It is presumed that cross-linking sites between alginate polymeric chains and the MOF facilitate strong interactions (Fig. 2.1E).

A wide range of approaches for producing MOF/polymer composites offers great versatility, incorporating various polymeric units into MOFs. MOF@polymer composites have attracted attention for numerous applications in host-guest chemistry across diverse domains, as detailed in the following section.

Metal-organic framework polymer composites for various applications

Researchers worldwide are customizing the properties of MOF@polymer composites to achieve outstanding internal surface area, tunable pore size, and volume, while maintaining the regular structure of MOFs. These composites find applications in various fields such as water purification, energy storage and conversion, gas separation, biomedical applications, sensing, and more.

In energy storage applications, MOFs exhibit high porosity and specific surface areas. However, their low conductivity poses a challenge in achieving optimal electrochemical performance (Sundriyal et al., 2018). Intercalating MOFs with conducting polymers (CPs) such as polyethylene dioxythiophene (PEDOT), polypyrrole (Ppy), and polyaniline (PANI) has been introduced to mitigate these drawbacks. These CPs are synthesized easily and demonstrate utility in supercapacitor devices due to their high pseudocapacitance via a faradic charge storage mechanism (Guo et al., 2016). For instance, to fabricate a binder-free, self-standing electrode for supercapacitors, Chang *et al.* synthesized Ni-MOF nanosheet@PANI supported on Ni foam (Cheng et al., 2019). PANI, a common and popular CP, offers very high theoretical specific capacitance, low cost, and not only ensures strong mechanical adhesion but also promotes growth on Ni-MOF, enhancing conductivity. Ni-MOF@PANI/NF, used as an anode in an asymmetric supercapacitor (SC) device, demonstrates high energy density (45.6 at 850.0 W/kg) and excellent cycling stability with capacitance retention of 81.6% over 10,000 discharge–charge cycles (Cheng et al., 2019). Similarly, Wang, Li, et al. (2020) synthesized Ni-MOF@Ppy composite using a facile method, fabricating it as an anode, wherein Ppy enhances electron transfer kinetics through the Ni-MOF. Also, Salunkhe et al. (2016) reported Zn MOF-derived carbon@PANI composites synthesized via a hydrothermal method and employed them in supercapacitor applications. PANI nanorods grow uniformly on the surface of nanoporous carbon due to the abundance of nucleation sites. Interestingly, the nanoporous carbon retains its original polyhedral structure, potentially preventing the particle aggregation commonly observed in other studies involving carbon-PANI composites.

Furthermore, a plethora of studies chronicle the implementation of MOFs in biomedical applications (Gandara-Loe et al., 2020; Wang et al., 2016). Pristine MOFs offer high porosity and tunability but exhibit challenges such as poor loading, uncontrolled release of drugs or biomolecules, and low stability in physiological environments. To enhance their performance in areas such as drug delivery, photothermal cancer therapy, bioimaging, and others, researchers are exploring the development of new porous platforms using MOF@polymers for biomedical applications. Wang et al. (2016), for instance, synthesized UiO-66 MOFs and coated them with PANI to form UiO-66@PANI composite, which has a pore size range appropriate for cellular absorption and presents excellent stability. Its efficacy was tested by injecting UiO-66@PANI into phosphate-buffered saline (PBS) medium in mice afflicted with colon cancer. The mice treated daily with UiO-66@PANI and exposed to near-infrared radiation (NIR) for 10 days showed a significant up to 93% suppression in tumor size. In contrast, the mice in the control experiments that received daily treatment but were not exposed to radiation showed better recovery (Wang et al., 2016).

Drug delivery and biosensing have been identified as other potential application domains of MOFs@polymers composites. Huang et al. (2017) synthesized PEDOT nanotubes (PEDOT NTs) covered with porphyrin-based MOF nanocrystals (MOF-525) to develop a reliable biosensing system with enhanced selectivity and sensitivity for dopamine (DA). In these composites, PEDOT NTs act as an electrocatalytic surface, while MOF-525 serves as a charge collector, facilitating rapid ion/electron transport. The particles aggregate to form an interpenetrating conductive network, increasing the number of electrochemically active sites. This synergy leverages MOF-525 charge transport channels, demonstrated practically in the detection of DA produced by living PC12 cells (Huang et al., 2017).

During the past few years, MOF@polymers have been widely utilized in membrane-based liquid phase separation, especially for the decontamination of different pollutants such as toxic and radioactive metal ions (Li et al., 2018), aromatic pollutants (Lian & Yan, 2016), heavy metal pollutants (Kobielska et al., 2018), and dyes. MOF@polymer composites prove to be potential materials for removing these contaminants. Supervisory interactions, including hydrogen bonds, van der Waals forces, and π–π stacking between the polymers and MOFs, enhance their sorption properties. This review provides insights into the application of MOF membranes for water regeneration and wastewater treatment (Li et al., 2020). Reports indicate that thin-film nanocomposite membranes, formed by embedding MOFs on a polyimide substrate, exhibit higher solvent permeability. This improvement is attributed to increased hydrophilicity and porosity (Gnanasekaran et al., 2019). The composites increase the free volume of membranes, thereby enhancing the availability of active porous sites for the separation of gases and liquids.

A polyamide (PA) film supported by a plastic sheet, such as polyvinylidene fluoride (PVDF) or PSU, is widely favored for water filtration. Zirehpour and colleagues applied an interfacial polymerization (IP) process to form a PA layer on a polyethersulfone (PES) substrate. They enhanced the membrane by incorporating MOF nanocrystals through ultrasonic irradiation, resulting in a thin-film composite membrane (TCM) that increased pure water penetrability by 129.0% (Zirehpour et al., 2017). The presence of an organic ligand in the MOF crystal structure enhances compatibility between the organic polymer and MOF by facilitating the formation of covalent bonds, in addition to noncovalent interactions such as hydrogen bonds. In a similar vein, Gnanasekaran et al. (2019) created polymeric membranes integrated with MOF-5 using the phase inversion approach with PES, PVDF, and cellulose acetate (CA). The composites exhibited enhanced hydrophilicity with a 0.5 weight percent loading of MOF-5, resulting in improvements in permeability of 70%, 80.58%, and 46.47% for PVDF, PES, and CA membranes, respectively.

Fu et al. (2017) synthesized UiO-66@chitosan with strong mechanical stability using an ice-templating process, providing high adsorption capacity for pollutants in water treatment. Johari et al. (2021) synthesized MIL-100 implanted in a PES membrane, demonstrating notable removal of dyes (reactive red [RR120], reactive orange [RO16], and methylene blue [MB]) with over 98.5% rejection achieved. In contrast to the pristine PES membrane, the hydrophobicity of the MIL-100@PES membrane reduced water permeability and increased fouling tendency. Moreover, the hybrid membrane still achieved > 92% chemical oxygen demand (COD) removal and nearly 100% total suspended solids (TSS) rejection, promising for treating wastewater from textiles.

MOF@polymers find applications in ion detection fields such as glucose, DNA, and numerous harmful pesticides and antibiotics (Kobielska et al., 2018). Wang et al. (2017) synthesized UiO-66-NH$_2$ to detect cadmium ions, leveraging its strong chemical stability, abundance of amine groups, and water stability. They mixed it with PANI to enhance the conductivity of the composite material through accelerated electron-ion diffusion. UiO-66-NH$_2$ provides structural support to PANI, offering a large surface area and numerous adsorption sites. In contrast, PANI expands conduction routes, facilitating electron flow between the electrode surface and solution. The UiO-66-NH$_2$@PANI-modified electrode demonstrated superior performance with the lowest detection limit (LOD) of 0.3 μg/L for cadmium ions, showcasing improved stability, repeatability, and a larger linear sensing range (Wang et al., 2017). Similar to this, Li, Cai, et al. (2021) synthesized MOF-867@Ppy composites using interfacial polymerization and postmodification techniques to create a homogeneous, free-standing thin film for cadmium ion electrochemical sensing. The hybrid film exhibited robust structural integrity and high sensitivity, achieving a LOD of 0.29 μg/L and excellent selectivity for cadmium ions (Li, Liu, et al., 2021).

Moreover, MOF@polymer composites exhibit changes in color, resistance, luminescence, dimensions (contraction/expansion), mass, or other responses to specific stimuli, enabling the detection of trace foreign elements and ions. In order to identify a particular spectrum of pesticides and antibiotics in contaminated water, Dutta et al. (2023) synthesized iMOF-14C@PVDF (iMOF=ionic MOF; C=cationic), which demonstrated high selective photoluminescence (PL) quenching. The composite showed sensitivity to various nitro-functionalized toxins, including pesticides such as chlorpyrifos (CHPS) and nitrofuran antibiotics like nitrofurazone (NFZ) and nitrofurantoin (NFT). It exhibited a PL turn-off response to trace levels of NFZ and NFT, achieving detection limits as low as 20 and 100 ppb, respectively. The ability to design MOF@polymer composites with high structural tunability allows for precisely defined pore sizes and shapes that exhibit robust responses to trace amounts of specific species in diverse environments. This characteristic makes them highly sought-after active materials in pollution sensors (Dutta et al., 2023).

Metal-organic framework@carbon composites

There is also a growing research interest in the multidisciplinary field of MOF@carbonaceous materials, encompassing MOF@CNT (carbon nanotube), MOF@CF (carbon fiber), MOF@fullerene, MOF@GO (graphite or graphene oxide), MOF@rGO (reduced graphene oxide), MOF@CQD (carbon quantum dot), MOF@PC (porous carbon), and many others (Liu et al., 2016). In addition to improving electrical conductivity, the controlled integration of MOFs with carbon-based materials reveals a novel pore structure that offers higher stability.

MOF@carbonaceous composites can be designed using both in situ and ex situ synthesis techniques as elaborated below.

In situ synthesis techniques

The carbon substrate suspension is mixed with the MOF precursor solution using methods such as one-pot synthesis, seeded growth, and layer-by-layer assembly. Various in situ techniques are employed in these processes:

One-pot synthesis approach

In this approach, carbon-related materials are introduced in situ into well-dissolved MOF precursor solutions during the MOF synthesis solvothermal reactions. Lin et al. (2014) encapsulated branched poly-(ethylenimine)-capped carbon quantum dots (BPEI-CQDs) into the ZIF-8 via pot synthesis. However, creating a defect-free, dense membrane in a one-pot synthesis can be challenging. Furthermore, heterogeneous crystallization of MOFs on the surface of carbonaceous substrates lacks functional groups.

Seeded growth method

Seeded growth is the preferred two-step in situ growth process to create defect-free, continuous MOF@carbon membranes with efficient orientation control (Yao & Wang, 2014). Firstly, MOF crystals are synthesized and placed as seeds onto substrates, leading to oriented MOF crystallization and growth. The seeding process is crucial and can be carried out using a variety of methods, including heating, wiping, spin coating, and rubbing dip coating. A novel hybrid MOF@bucky-paper was reported by Gholidoust et al. (2019) adopting surface-seeding of Mg-MOF-74 crystals on CNT mats, also known as bucky-papers (BP). This resulted in improved MOF@CNT interfacial interaction, stronger dispersive forces, and less agglomeration plasma leading to better functionalities. With this method, new opportunities for designing nano- and macro-MOF adsorbents and membranes using graphitic materials as templates are being opened up (Gholidoust et al., 2019).

Layer-by-layer assembly

Researchers have also used the in situ layer-by-layer assembly technique, also known as liquid phase epitaxy, to integrate MOFs with carbon-related material like carbon fibers. This method exercises a stringent control over homogeneity, structure penetration, and thickness of MOFs with manipulation in the number of growth cycles. Wu et al. (2019) synthesized multilayer Cu-MOF/MWCNT films on modified glassy carbon electrodes (GCE) using electrodeposition of MWCNTs onto the GCE substrate followed by Cu-MOF crystallization to form a Cu-MOF/MWCNT composite layer.

The facile in situ fabrication techniques lead to stable, homogeneous multilayer films with large molecules, a contact surface area loaded with reactive sites. However, lack of proper control in in situ techniques may result in excessive carbon deposition that may hinder the coordination reactions required for MOF synthesis or at times even lead to the collapse of the MOF-carbon composite framework structure.

Ex situ synthesis techniques

Better control is provided by ex situ synthesis techniques, wherein presynthesized MOFs are combined with carbon-based materials through self-assembly, mechanical methods, and direct mixing. Different ex situ approaches are classified as follows:

Direct mixing and mechanical process

In the direct mixing approach, the textural (pore size and volume) and crystallinity features of the MOF@carbon matrix are modified through mechanical procedures like compression and ball milling followed by thermal treatments. However, mechanical pressure adversely affects the surface area and pore volume, deteriorating the adsorption capacity of the material and is thus not preferred for practical applications (Choi et al., 2014).

Self-assembly technique

Self-assembly is a process wherein preexisting disorganized components experience strong local interactions to realign, and self-assembly reduces its free energy, forming an organized structure or pattern, but without any external forces. A highly stable Co MOF@CNTs hybrid was fabricated by Fang et al. (2016) using a self-assembly method wherein CNTs were encapsulated into the Co-MOF polyhedron to form a hierarchical structure. Co (II) and nitrogen atoms of Co-MOFs showed a strong coordination with CNTs, resulting in a robust, stable structure suitable for harsh catalytic reactions (Fang et al., 2016).

Applications of metal-organic framework@carbon composites

Carbon, being a versatile material with immensely rich properties, renders MOF@carbon composites to have diverse applications across varied fields like energy storage, sensing, biomedical, wastewater treatment, and many more.

Carbon materials, such as CNTs, graphene oxide (GO), reduced graphene oxide (rGO), activated carbon (AC), carbon aerogels, and hydrogels, have a proven track record of enhancing the electrical properties of MOFs while maintaining an appropriate pore size distribution along with chemical, thermal, and mechanical characteristics. These composites have emerged as the preferred choice for effective energy storage applications, with graphene and its derivatives, such as GO and rGO, being the most referenced. They showcase high electrical conductivity, superior thermal and chemical stabilities, as well as a vast surface area. For example, Rahmanifar et al. (2018) synthesized bimetallic NiCo-MOF@rGO nanocomposite via a facile one-pot procedure, with rGO and the MOF framework laying a synergistic impact to act as a support and nanoscale current collector in the nanocomposite. In addition, the asymmetric NiCo-MOF@rGO//AC device exhibits good cyclability (91.6% capacitance retention after 6000 charge–discharge cycles (Fig. 2.2A)), a high specific capacitance of 181.4 F/g at 1 A/g, and an exceptional specific power of up to 42.5 kW/kg at 50 A/g. Furthermore, these devices (Fig. 2.2B) upon being charged operated a blue LED for 25 minutes, a green LED for 25 minutes, and a red LED for over one hundred and forty minutes, demonstrating their potential for real-world energy storage and demonstrating devices' potential for real-world applications in energy storage as shown in (Fig. 2.2C–F) (Rahmanifar et al., 2018).

Xu et al. (2020) assembled bimetal NiCo MOF on carbon cloth (CC) and utilized it as an anode for a supercapacitor. Directly grown NiCo MOF on the surface of conductive carbon fiber, demonstrates faster electron transport kinetics during Faradic redox reactions. Owing to the distinct nanostructure and morphology of NiCo MOF/CC, the synthesized electrode material exhibits an areal capacity of 1.01 C/cm^2 at 2 mA/cm^2. Fabricating a hybrid supercapacitor device based on NiCo MOF/CC as an anode and rGO/CNT as a cathode exhibited further performance improvement with a projected areal capacitance of 846 mF/cm^2 at 1 mA/cm^2 (177.7 F/g at 0.21 A/g) (Xu et al., 2020). Moreover, NiCo MOF@CQDs nanocomposite was successfully synthesized by Kumari, Prajapati, and Ravi Kant (2024) using a one-step solvothermal process. NiCo-MOF's distinct structure and thin sheets offer

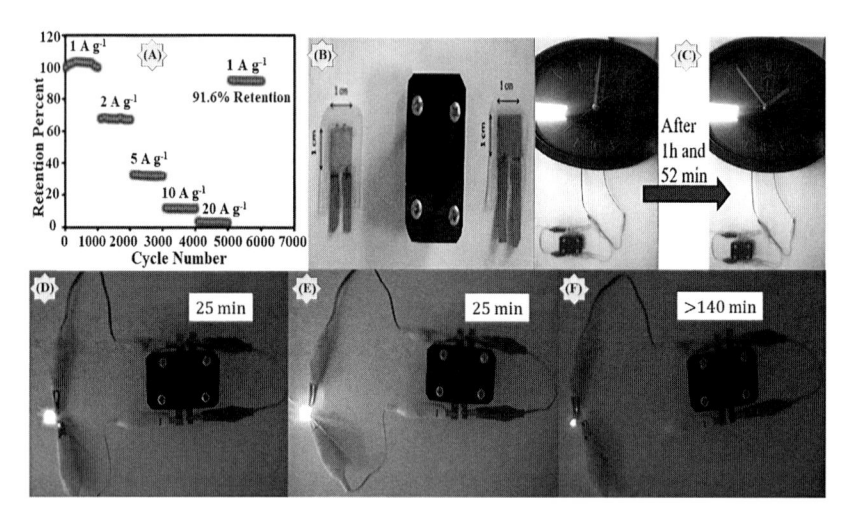

FIGURE 2.2 (A) Illustration of long-term cycling performance; (B–F) Demonstration of the practical applicability of the fabricated asymmetric NiCo-MOF@rGO//AC device by lighting various LEDs. (Rahmanifar et al., 2018).

plenty of electroactive sites for electrolyte diffusion, while CQDs, which are mesoporous spheres with a radius range of 20–50 nm, are fabricated from biowaste (waste edible soybean oil), which is an economically advantageous aspect of green synthesis. Consequently, the in situ integration of CQDs in NiCo MOF compensates for the rapid electro-kinetics and allows for rapid ion migration and efficient redox reaction. At a current density of 0.5 A/g, the combined effect of CQDs and NiCo MOF nanosheets produced a superior specific capacitance of 1063.02 F/g (Kumari, Prajapati & Ravi Kant, 2024).

Research is evolving at a fast pace, driven by the need for innovative biomedical solutions able to address challenges in diagnostics, therapy, and regenerative medicine. The versatility and tunability of MOFs and carbon composites offer exciting opportunities for future biomedical technologies. Karimzadeh et al. (2019) for instance, synthesized carboxymethylcellulose/MOF-5/GO bionanocomposite (CMC/MOF-5/GO) via the solvothermal method and loaded it with the drug tetracycline (TC) for controlled drug release. The TC-loaded CMC/MOF-5/GO showed better long-term stability and stronger antibacterial activity in contrast to TC-loaded GO (Karimzadeh et al., 2019). Likewise, Yao et al. (2019) synthesized Mn carbonyl-modified Fe (III)-based nano MOFs (MIL-100) coated with PEGylated carbon NPs. The composites loaded with doxorubicin (DOX) act as theragnostic nanoplatforms for CO-DOX combination therapy that responds to NIR light. When an 808 nm laser is used to irradiate the magnetic carbon core, it exhibits a photothermal action allowing on-demand release of DOX and CO. Consequently, MCM@PEG-CO-DOX NPs effectively destroyed the tumor when combined with photothermal therapy (PTT). Furthermore, tumor dual-mode imaging using photoacoustic imaging and magnetic resonance imaging (PAI and MRI) was also made possible by the synthesized MCM@PEG-CO-DOX NPs (Yao et al., 2019).

Sarabaegi et al. (2022) synthesized a NiCo-MOF@CF (carbon nanofiber) via a facile electrospinning method. This study effectively fabricates a label-free impedimetric aptasensor (Apt) employing NiCo-MOF@CF to detect *Helicobacter pylori*. The Apt/NiCo-MOF@CF/GCE operates in serum blood with an acceptable recovery rate, a low LOD, and a broad linear range (Sarabaegi et al., 2022). Pooresmaeil et al. (2021) synthesized chitosan-coated Zn-MOF hybrid microspheres with graphene oxide (CS/Zn-MOF@GO) as an anticancer drug carrier as shown in (Fig. 2.3). Additionally, the results of the MTT assay and cellular uptake, respectively, demonstrated the capacity for internalization of cells with a cell viability of 83% and did not significantly cytotoxicate MDA MB 231 as human breast cancer cells. Overall, these composites set forth promising applications for anticancer drug delivery, and more research into their drug combination with different medications for tumor drug delivery therapies may be conducted in the future.

MOFs exhibit strong adsorption of pollutants; however, by altering their architecture and physicochemical properties, their adsorption capability can be further enhanced. CNTs, AC, and GO are regarded as promising carbon

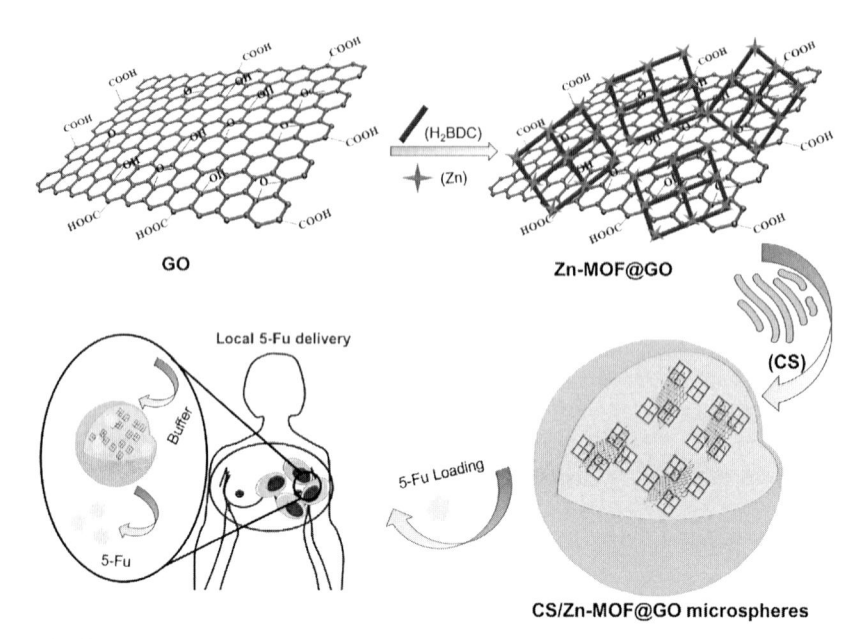

FIGURE 2.3 Schematic representation outlines the process of synthesizing CS/Zn-MOF@GO microspheres and the subsequent local delivery of 5-FU through the 5-FU@CS/Zn-MOF@GO microspheres (Pooresmaeil et al., 2021).

precursors for composites because of their many benefits, which include low cost, nontoxicity, tunable surface chemistry, and ease of functionalization with diverse materials. Controlling the physicochemical characteristics of MOFs during their synthesis, such as their size, shape, and morphology, is another benefit of the GO and CNTs platforms. A large number of research reports are available to corroborate the performance of MOF@carbonaceous materials in the area of wastewater treatment application.

For example, Yang et al. (2020) synthesized a highly stable membrane of Sm (samarium) MOF@GO skeleton. The Sm-MOF prevented snapping of GO layers in aqueous solutions and further increased its layer spacing, assisting improved permeance (26 L/m^2.h.bar) and rejection (> 91%) to the organic dye Rhodamine B (RhB) (Yang et al., 2020). Samy et al. (2021) mixed MOF-808 and CNTs in a variable ratio and studied both the oxidation and degradation pathways for the photocatalytic degradation of diazinon and carbamazepine (CB). The reactor was used to treat real liquid pharmaceutical and agrochemical wastes, resulting in a 93% and 76% decrease in total organic carbon (TOC), respectively. According to cost-estimation research, the total cost of a large-scale photocatalytic system, including operational and amortization expenses, was 2.523 \$/m^3, which is quite reasonable. Bi et al. (2021) reported a very effective hexavalent uranium, U(VI) adsorbent, PCN-22@GO-COOH by growing a mesoporous Zr-MOF (PCN-222) on the carboxylated graphene oxide (GO-COOH) using ultrasonication assisted stirring as shown in (Fig. 2.4). Due to its excellent hydrophilicity, large surface-to-volume ratio, and other excellent physicochemical properties, the PCN-222@GO-COOH demonstrated potential use in selective, quick, and highly efficient extraction of U(VI) from the wastewater (Bi et al., 2021).

Activated carbon holds significant potential for fabricating composites with MOFs due to its unique laminar structure and surface properties (Kumari et al., 2023). However, its strong hydrophilic properties can cause it to dissolve readily in water or other polar solvents. The concept that MOF@AC composites increase dispersion forces and create additional void spaces was widely acknowledged (Abd El Salam, 2023). Abd El Salam (2023) adopted a green approach for the synthesis of aluminum-based MOF@nanoporous activated carbon (Al MOFs@AC) from lemon peel waste as an affordable biosorbent to remove adsorptive removal of Eriochrome Black T (EBT) dye. Since EBT was negatively charged and Al-MOF and Al-MOF@AC were positively charged in an acidic environment, electrostatic attractions, including hydrogen bonding, π–π stacking, and electrostatic contact between the functional groups of EBT, such as sulfonate, nitro, and amino groups, and the electron cloud at the metal node, may arise. Moreover, succinate linkers in MOFs have the ability to enhance the adsorbate-adsorbent contact by strengthening the van der Waals interaction between the hydrocarbon chains of the linker and the aromatic rings of EBT. As a result, the

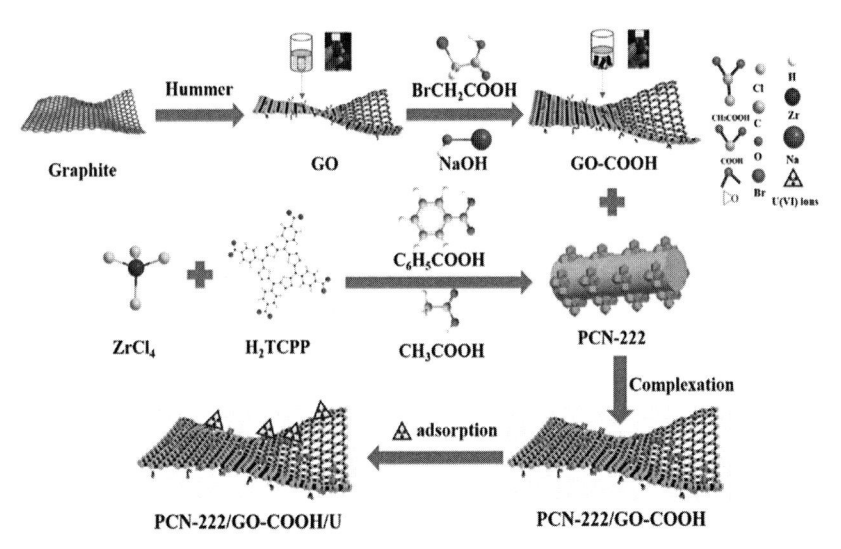

FIGURE 2.4 Schematic depiction of the synthesis of PCN-222@GO-COOH (Bi et al., 2021).

synthesized nanoporous Al-MOF@AC was efficient in removing 96% of EBT (Abd El Salam, 2023).

MOFs, with their highly porous structure and presence of Lewis basic/acidic or open metal functional sites with varying affinities toward guest molecules, act as viable hosts for the reversible adsorption and release of guest molecules. Further, the tunable pore dimensions and diverse optical characteristics of MOFs may give reasonable control over their sensing attributes. However, there are still several obstacles that prevent single-phase MOFs from being used as chemical sensors. Two factors that contribute to their low luminescence quantum yield are the first, their poor electrocatalytic capabilities and intrinsic poor conductivity, and the second, the weak metal-ligand charge transfer. As a result, MOF@carbon composites are formed that have increased conductivity or high luminescence activities, either as a result of the combined impact of the two components or boosted qualities of the integrated carbonaceous materials (Liu et al., 2016).

Metal-organic framework@metal nanoparticles

Considered to be the most promising materials, MNPs have emerged as one of the major research materials in recent years. Favorable chemical activity, ultrahigh surface area, and high intensity are prominent characteristics of MNPs. However, NPs pose restrictions due to certain inevitable issues like intrinsic instability brought on by their high surface-area-to-volume ratio. MOFs have a pervasive porous structure that can be pervaded by MNPs to create a more stable material. Moreover, the pores in MOF provide support to MNPs, preventing particle aggregation.

Synthesis of metal-organic framework@metal nanoparticle composites

MNPs in MOFs can be immobilized using one of two primary methods. The first and most common method, referred to as the "ship in bottle" approach, entails embedding MNPs in a MOF matrix. To introduce the metal precursors, different techniques like liquid impregnation technique, chemical vapor deposition (CVD), and solid grinding are used to impregnate them. The metal precursors are then reduced to metal atoms. Another strategy, sometimes referred to as the "bottle around the ship" or template synthesis approach, starts with the individual synthesis of MNPs and then adds appropriate chemicals to build the MOF around the MNPs (Chen et al., 2019; Hermes et al., 2005).

Solvent free gas phase loading technique

One type of gas-phase loading and solvent-free technique is CVD, encasing metal precursors in porous MOF frameworks and then reducing them under thermal or photochemical conditions. Avoiding the contention between the metal precursors and solvent, this approach typically works with more volatile metal precursors, supports good thermal stability, and in particular allows high loading levels (Hermes et al., 2005).

Solid grinding technique

The gas phase approach being tedious, a simpler and more efficient approach, like the solid grinding method, is on the card as an efficient solvent-free way to incorporate MNPs into porous materials. This method involves mixing a volatile organometallic precursor with the host MOF; the sublimated vapor of the volatile precursor penetrates the voids of the MOF, and with further exposure to H_2 gas at a low temperatures, the MNPs are formed, embedded within the MOF (Ishida et al., 2008). Li et al. (2019) adopted a similar strategy and reported an encapsulation method that uses mechanochemistry to trap MNPs inside MOF matrices. Using a facile ball mill, the metal precursor-supported MNPs hybrids were sacrificed in situ to create the MOF@MNPs composites (Li et al., 2019). Numerous intriguing MOF@MNPs composites with good crystallinity and no visible degradation have been fabricated. Though the solid-state grinding process is reasonably simple to operate, its controllability and homogeneity are not at par. Consequently, various research teams are investigating the liquid impregnation technique.

Liquid impregnation method

The traditional approach of loading MNPs onto a carrier in solution-phase is the liquid impregnation method. The basic idea behind this method is the pressure differential between the exterior and interior of the cavities, which pulls the active components inside and has been successfully used to fabricate Pd, Au, Ag, and Cu NPs in various MOF configurations (Jiang et al., 2011; Yang et al., 2009). Zhao et al. (2012) produced Ni@MOF-5 and analyzed its catalytic activity using crotonaldehyde as a probe molecule; Ni@MOF-5 demonstrated outstanding catalytic activity for the hydrogenation of C=C bonds under moderate circumstances. Crotonaldehyde had a conversion of over 90.0% at 1000 °C, and butyraldehyde had a selectivity of over 98.0% after 40 minutes (Zhao et al., 2012).

Template synthesis technique

The template synthesis, or "bottle around the ship" approach, refers to the process of building MOFs around the as-prepared metallic NPs using various synthesis procedures. Presynthesized metals, such as metal ions and organic linkers, are often added to the MOF precursor solution. The main advantage of this approach over the "ship in the bottle" method is the ability to carefully control the NPs' size, shape, and composition during the presynthesis phase.

MOF topologies and shapes can be controlled in the interim by varying their distinctive reaction parameters. Additionally, this technique can successfully prevent the reducing agent from damaging the MOF during the reduction of the metal precursors, as well as the aggregation of NPs on the outer surface of MOFs. It is important to note that in this approach, the metallic NPs serve as seeds or nucleation centers for the MOF framework construction. Furthermore, the binder, such as polyvinylpyrrolidone (PVP), on the surface of created metallic NPs is essential to anchor the heterogeneous development of MOF materials (Duan et al., 2020). Kuo et al. (2012) wrapped the Pd NPs in Cu_2O before forming the core-shell ZIF-8@Pd NPs using a self-sacrificing template technique. Hydrogen ions are emitted during ZIF-8 crystal formation, which is facilitated by the presence of Cu_2O. The Cu_2O layer was completely removed by these hydrogen ions through a corrosive action, resulting in yolk-shell ZIF-8@Pd NPs octahedral nanostructures, with cavities 230 nm in diameter and a 100 nm thick outer ZIF-8 layer. High catalytic activity was demonstrated by these materials in the hydrogenation of cyclohexene (Kuo et al., 2012).

MOF@MNPs composites, depending on the type of metal and MOF frameworks chosen, have been widely used in a range of applications by leveraging the vast surface area and size selectivity of MOFs, in addition to the controlled growth of MNPs within MOF matrices.

Applications of metal-organic framework@metal nanoparticle composites

Researchers are actively exploring various combinations and fabrication techniques to optimize the performance of MOF@NPs composites for wastewater treatment (Rehan et al., 2024). These composites hold promise in addressing challenges associated with water pollution by efficiently removing harmful substances from wastewater streams. The combination of NPs with MOFs often leads to synergistic effects, where the properties of the individual components complement each other (Chen et al., 2020; Niu et al., 2020). For example, NPs can provide catalytic activity or improve the transport properties within the composite, thereby enhancing the overall performance in pollutant removal (Sharma et al., 2023). Moreover, the MOF-based composites can exhibit improved stability and durability when NPs are integrated into their structure. NPs can also provide mechanical support and enhance the overall structural integrity of the composite, making it more resistant to physical and chemical stresses during wastewater treatment operations (Lu et al., 2012). Additionally, the presence of NPs can facilitate the regeneration of MOF composites, extending their lifespan and reducing operational costs.

A cost-effective method was described by Rehan et al. (2024), which combined the catalytic trait of Ag NPs anchored on the of surface viscose fibers (VF) and MIL-125-NH_2 that further enhances the photocatalytic capabilities. Ag NPs display improved photocatalytic behavior under visible light irradiation. It was noted that VF-Ag-MOF exhibited superior sonocatalytic and sonophotocatalytic activity in the process of breaking down sulfa drugs. Moreover, the improved qualities of composites, which could be justified by photogenerated electrons in the conduction band of MIL-125-NH$_2$ were expected to move rapidly to the valence band of the Ag NPs and then to the conduction band of the Ag NPs because of the formation of a Schottky barrier at the junction between the Ag NPs and MIL-125-NH$_2$. In VF-Ag-MIL-125-NH$_2$ composites, the significant electron transport process known as Z-scheme heterojunction effectively suppresses electron/hole pair recombination. Moreover, the composites' photocatalytic activity was increased when there was an adequate distance between the photogenerated charge carriers (electrons and holes) and prevented their recombination (Rehan et al., 2024).

Mercury (Hg) is one of the most dangerous heavy metals and poses a serious risk to both human health and the environment because of its high toxicity, rapid pace of spread, and high rate of bioaccumulation. In this regard, Wang, Xu, et al. (2020) synthesized an Ag NPs@COF-LZU1 composite, utilizing an easy one-step "postsynthesis" approach and investigating the adsorption capability for Hg ions from acidic solutions for the first time. In terms of its high adsorption capacity, ultrahigh Ag atom utilization (150%), quick adsorption kinetics, high selectivity and stability, and reusability, the composites proved to be a capable adsorbent for Hg (II) ions. Additionally, the DFT calculations reported by researchers demonstrated that the COF material served as both an electron donor and a supporting matrix during the Ag-Hg nanoalloy formation (Wang, Xu, et al., 2020).

Also, Hu et al. (2021) synthesized N-doped porous carbon rods formed from MOFs that included FeCo NPs using the Co-doped Fe-MOFs as templates and a simple carbonization procedure to form FeCo NPCs and demonstrated their application as a suspension catalyst in hetero-EF to remove and degrade TC (Hu et al., 2021).

Research on MOF@MNPs composite-based sensing applications has revealed a wide range and complexity of synergistic effects between MOFs and MNPs. Different synergistic processes can be employed to describe the interactions of the MOF@MNPs composite utilized in the application of sensing. Firstly, MNPs act as active sites that are regulated by MOFs. This is the most facile and typical synergistic process; it results in MOF@MNPs composites with superior catalytic activity, which are widely used in surface-enhanced Raman scattering (SERS), electrochemical, and colorimetric sensors, among other applications. In addition, MNPs function as active sites, and size selectivity is determined by porous MOFs. Target molecules smaller than the MOF pores can reach the encapsulated MNPs (active sites) in MOF@MNPs composites with core-shell architectures, while larger molecules cannot pass through the MOFs. MOF@MNPs-based sensors show antiinterference and enhanced selectivity performance because of the MOFs' molecular sieving activity. As host matrices, MOFs in particular can provide a platform for the incorporation of MNPs to create MOF@MNPs composites, which ensures that the MOF pore is accessible for both reactants and products in addition to making the metal NPs well-dispersed.

Plenty of NPs, such as Ag, Au, and Pt, are utilized in detection and sensing applications because of their distinctive structures and ease of aggregation (Bagheri et al., 2018; Ma et al., 2019).

Ma et al. (2019) synthesized Pt NPs-supported Zr(IV)-based MOF (UiO-66-NH$_2$) nanocomposites (Pt@UiO-66-NH$_2$) with a large surface area through a facile reduction method. Pt NPs were anchored onto UiO-66-NH$_2$, leveraging the amino groups of UiO-66-NH$_2$ for strong interfacial contact and highly dispersed Pt NPs. These UiO-66-NH$_2$@Pt NPs nanocomposites are crucial for reducing the use of noble metals while providing abundant adsorption sites and an ultrahigh surface area, facilitating the immobilization of acetylcholinesterase (AChE) biosensors for detecting organophosphorus pesticides (OPs). The resulting biosensor is effective for identifying residual OPs in the environment (Ma et al., 2019).

Also, an electrochemical apt based on 2D Au NCs@MOF-521 nanosheets has been effectively fabricated employing a one-pot method (Fig. 2.5A) for the sensing of cocaine by Su et al. (2017), as illustrated in Fig. 2.5C. The oligonucleotide and the Au NCs@MOF-521 framework exhibited covalent bonding interactions and significant π–π stacking, enabling the subsequent attachment of the cocaine apt onto the electrode surface modified by the nanocomposite (Fig. 2.5B). The apt offers high electrochemical activity, high specific surface area, and biocompatibility, and traces of cocaine were detected with LOD values, within a linear range of 0.001–1.0 ng/mL, as determined by differential pulse voltammetry (DPV) (Su et al., 2017).

Also, the MOF-5@Au NPs electrode-based electrochemical DNA sensor was successfully fabricated by Yang et al. (2016). High electrocatalytic activity was demonstrated by Au NPs with significant specific surface areas during the electro-oxidation of hydrazine. The low electron transfer power and electron tunneling that DNA-modified Au NPs present can be overcome by the distinctive structural properties of MOF-5. The electrocatalytic activity of the MOF-5@Au NPs composites was significantly increased, and the DNA detection limit was improved by the distinct porous structure of MOF-5. Au NPs and MOFs together significantly increased the signal strength and limit (Yang et al., 2016).

Metal-organic framework@MXene composites

A growing family of 2D materials of transition metal nitrides and carbides known as MXenes is attracting a lot of attention in research because of their strong redox activity, profuse hydrophilic surface, and superior electrical conductivity. MXenes can be fabricated by selective etching of the *A* elements (where *A*=Al) from the MAX phases (layered ternary nitrides and carbides, such as Ti$_3$AlC$_2$, Ta$_4$AlC$_3$, and Ti$_2$AlC). Due to the presence of a number of

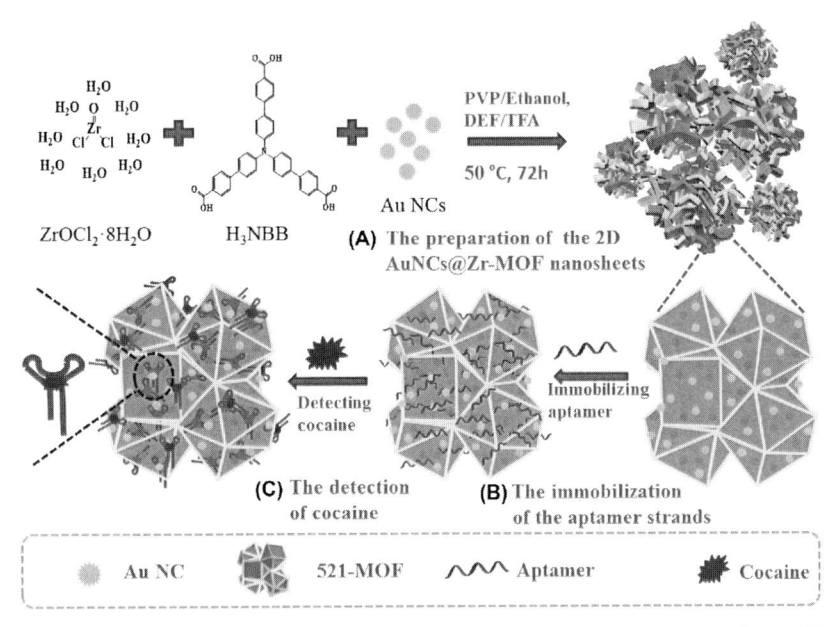

FIGURE 2.5 (A) Schematic of the synthesis process; (B) Immobilization of 2D Au NCs@Zr-MOF; (C) Cocaine detection using 2D Au NCs@Zr-MOF (Su et al., 2017).

functional groups (like –O, –OH, and –F) with charged negativity, it is possible to create a networked structure with other components, enabling fast ion-electron transport via electrostatic interaction (Wang et al., 2022; Xue et al., 2017). Furthermore, adaptable layer spacing in MXenes provides ample space for embedding various intercalators, which is advantageous for building sandwich structures with improved lamellar accumulation. MXenes are therefore the perfect materials to form composites with MOFs to deliver optimal performance.

Synthesis of metal-organic framework@MXene composites

MOF@MXene composites are commonly synthesized via two techniques, that is, in situ and ex situ.

The in situ growth technique deals with the direct synthesis of MOFs on MXene surfaces or into their interlayer gaps, offering advantages of simple operation and consistent morphology in the final products. To synthesize the composites, MXenes are added to solutions containing well-dissolved metal ions and organic ligands. This process should be carried out using the same synthesis technique as MOFs. MOF@MXene hybrids with desired characteristics can be produced by meticulously controlling reaction parameters, including precursor concentrations and reaction control parameters. Direct integration of the two elements enables close interaction and mutually reinforcing effects between MXenes and MOFs. The composite interface has a rather high bonding degree as a result of spontaneous development. At the same time, it simplifies and eliminates the tedious multistep procedure. Since in situ production of composites frequently needs consistent and controlled pressure and temperature, hydrothermal/solvothermal is the most prevalent synthesis technique used by researchers (Aldhaher et al., 2024; Zhuang et al., 2023). The hydrothermal method facilitates crystal growth and rapid nucleation, leading to well-characterized hybrid structures. Exercising a control on the reaction parameters (e.g., reaction temperature, reaction time, solvents, and pH), the size and morphology of as-prepared materials can be regulated. Li, Liu, et al. (2021) synthesized MIL-NH$_2$ on the surface of Ti$_3$C$_2$ MXene nanosheets via the hydrothermal method and reported coordination of Ti with the NH$_2$ group in the composites, enabling close interfacial contact between Ti$_3$C$_2$ and MIL-NH$_2$ as well as a different pathway for electron transfer from MIL-NH$_2$ to Ti$_3$C$_2$. Consequently, the Ti$_3$C$_2$@MIL-NH$_2$ composites exhibit superior photocatalytic H$_2$ evolution activity (Li, Liu, et al., 2021).

While in situ synthesis shows significant potential for synthesized MOF@MXene composites, the diversity of MOF@MXene hybrids is still limited by the instability and easy oxidation of MXenes during the hybridization process. Ex situ synthesis, wherein the as-synthesized MOF and MXene are integrated using methods like self-assembly, is a significant alternative synthesis strategy to prevent damage to the chemical and structural characteristics of MXene (Wang et al., 2020). The use of presynthesized MOFs offers the advantage that the chemical reactions during MOF synthesis do not adversely impact the composite structure, resulting in versatile composites with enhanced functionalities. Self-assembly techniques are fast-paced and user-friendly and rely on the inherent qualities of MOF and MXene components to spontaneously assemble into ordered structures. Self-assembly can be promulgated by electrostatic interactions, noncovalent interactions, π–π stacking, van der Waals forces, hydrogen bonding, and other forces, which results in the spontaneous formation of a certain ordered structure. This method has a number of benefits, including scalability, simplicity, and capacity to create complex patterns. Obtaining composites with consistent morphologies and sizes requires careful adjustment of the precursor, solvent, pH, temperature, and other synthesis variables. For instance, Liu et al. (2020) synthesized three-dimensional porous NiCo-MOF@Ti$_3$C$_2$ composites using the hydrogen bonding-induced self-assembly method. Initially, NiCo-MOF was synthesized via coordination interaction of bimetallic ions (Ni^{2+} and Co^{2+}) with benzene-1,4-dicarboxylate (BDC) linker at room temperature. In this process, six O atoms participate in octahedral coordination of Ni and Co atoms to form 2D bimetal layers parted by BDC ligand molecules, procreating an abundance of –COOH groups on the nanosheets of MOF. Subsequently, by etching Ti$_3$AlC$_2$ with hydrofluoric acid (HF) solution, alkalizing, and then exfoliating using sonication, multilayered Ti$_3$C$_2$ MXene with increased interlayer spacing is produced. The surface of the Ti$_3$C$_2$ nanosheets is heavily anchored by terminal groups (–OH, –O, and –F). Finally, the Ti$_3$C$_2$ nanosheet solution was added to the NiCo-MOF, resulting in an interconnected porous structure with MXene nanosheets and interlayer hydrogen bonds of MOF. As fabricated, 3D NiCo MOF@Ti$_3$C$_2$ porous composites are utilized as anode materials in lithium-ion batteries. The composite exhibits wider interlayer spacing and higher accessible surface area for electrochemical reactions, and rapid Li$^+$ infiltration into electrodes resulted in higher electrochemical performance (Liu et al., 2020).

Applications of metal-organic framework@MXene composites

MOF@MXene composites offer several advantages, including enhanced structural stability, high porosity, good electroconductivity, and numerous active sites, effectively mitigating the drawbacks of both individual components. The low electron conductivity and chemical stability issues of MOFs are effectively countered by the addition of MXene. Conversely, MOFs implanted within MXene can address the challenges of severe oxidative degradation and self-stacking observed in MXene alone. Numerous studies have demonstrated that MOF@MXene composites outperform pure MXene and pristine MOFs in terms of electrochemical properties, such as specific capacities, Tafel slope, cycling stability, and rate performance. This superiority is attributed to their well-designed structure, abundant active sites, strong interfacial interactions, and conductive channels, all crucial factors for enhancing electrochemical performance (Aldhaher et al., 2024). MOF@MXene composites also find numerous applications in other domains such as water purification, controlled drug release, and ion detection (Do et al., 2021; Liu et al., 2022).

However, further details have been curtailed in this chapter due to word limit restrictions.

Conclusions and future prospects

The primary goal of this chapter is to provide a comprehensive review of significant advancements in MOF composites, which have evolved significantly over the last few decades, offering deeper insights into MOF composite interactions and underlying chemistry. MOFs themselves are esteemed materials with notable porosity and properties suitable for a myriad of applications, with countless MOF frameworks currently in use. However, challenges such as inadequate chemical and mechanical stability, along with poor conductivity, hinder their full potential. These limitations are addressed by burgeoning research on MOF-based composites. The chapter elucidates research on various functional materials added to different MOFs through simple yet highly effective techniques to enhance MOF performance beyond the state-of-the-art in several significant applications. Considering these favorable characteristics, this review highlights key research contributions to major developments in designing MOF composites with different functional materials like polymers, carbonaceous materials, MXenes, and MNPs, and their applications primarily in energy storage—especially supercapacitors—sensing, wastewater treatment, and biomedical fields. We discuss critical findings in these fields that are intriguing to readers.

The introduction of polymers into MOFs can improve and positively impact critical features such as ease of synthesis, processability, MOF dispersion and assembly through controlled interactions with solvents, biocompatibility, and ion kinetics and transport enhancement. For instance, integrating conductive polymers with MOFs creates hierarchical porous structures with enhanced ion transport pathways, leading to improved energy storage performance. Researchers have enhanced the electrochemical parameters of supercapacitor electrodes by incorporating MOFs into polymer matrices like PEDOT or PANI. MOF polymers and their composites represent versatile materials with significant potential for addressing challenges associated with wastewater treatment, selectively adsorbing pollutants such as heavy metal ions and organic dyes to mitigate water pollution and ensure access to clean water resources.

Combining MOFs with carbon-based materials yields various novel capabilities such as enhanced stability, template effects, and structural changes essential for practical requirements and enhancing MOFs' potential applications. This enhancement is substantiated by GO and rGO, renowned for their excellent electrical conductivity and mechanical strength, which contribute to the overall robustness and durability of MOF composites. Through controlled synthesis processes, researchers can tailor the properties of these composites for specific applications such as selective molecule adsorption and advanced sensing. Overall, the synergistic effects between MOFs and carbon materials lead to the development of versatile materials with superior performance in energy storage, gas sensing, environmental remediation, and beyond.

The hybridization of MOFs with MXenes presents intriguing opportunities across various applications, particularly in energy storage and conversion, biomedical applications, and sensing. These composites leverage MOFs' high surface area, porosity, and tunable properties alongside MXenes' excellent mechanical strength and conductivity, offering synergistic effects for enhanced performance. MOF-MXene composites hold potential for sensing and detection applications by enhancing selectivity and sensitivity toward target analytes through functionalizing MOF pores with MXene nanosheets or incorporating MXene NPs into MOF matrices. Moreover, MOFs' high surface areas and tunable pore sizes make them suitable for encapsulating and delivering therapeutic agents such as drugs or biomolecules. By incorporating MXenes into MOF structures, researchers can enhance the stability and biocompatibility of delivery systems while offering opportunities for targeted drug release and controlled drug delivery.

MOFs combined with NPs provide a multifunctional platform with diverse applications, particularly in sensing and wastewater treatment. In sensing, MOF-MNP hybrids offer enhanced sensitivity and selectivity crucial for detecting pollutants and analytes in environmental monitoring and biomedical diagnostics. For example, MOF-based sensors incorporating noble metal NPs like gold or silver exhibit exceptional sensitivity to toxic gases or volatile organic compounds (VOCs) due to synergistic effects between MOF pores and metal NPs, ensuring rapid and accurate detection of environmental pollutants for timely intervention and mitigation strategies. Additionally, in wastewater treatment, MOF-MNP composites demonstrate remarkable catalytic activity for degrading pollutants and removing heavy metal ions. For instance, MOF-supported metal NPs such as palladium or platinum catalyze the degradation of organic contaminants through advanced oxidation processes in water, while MOF-based adsorbents loaded with metal NPs efficiently capture and remove toxic heavy metal ions like cadmium.

In conclusion, this chapter outlines recent advances in synthesizing MOF composites and their applications. Further exploration using simulation studies and artificial intelligence could expedite parameter optimization through predictive analysis, improving resource management. This chapter provides insight into developing novel synthesis strategies for MOF composites with enhanced functionalities and encourages researchers to create composite structures inspired by MOFs, advancing toward technologically advanced, environmentally sustainable, economically viable, and socially desirable applications.

References

Abd El Salam, H. M. (2023). Bio-sustainable alternatives synthesis of nanoporous activated carbon @Al-MOF for the adsorption of hazardous organic dyes from wastewater. *Water, Air, & Soil Pollution, 234*(9), 567. https://doi.org/10.1007/s11270-023-06572-6.

Aldhaher, A., Rabiee, N., & Iravani, S. (2024). Exploring the synergistic potential of MXene-MOF hybrid composites: A perspective on synthesis, properties, and applications. *Hybrid Advances, 5*, 100131. https://doi.org/10.1016/j.hybadv.2023.100131.

Bagheri, N., Khataee, A., Hassanzadeh, J., & Habibi, B. (2018). Visual detection of peroxide-based explosives using novel mimetic Ag nanoparticle/ZnMOF nanocomposite. *Journal of Hazardous Materials, 360*, 233–242. https://doi.org/10.1016/j.jhazmat.2018.08.013, http://www.elsevier.com/locate/jhazmat.

Bardestani, R., Patience, G. S., & Kaliaguine, S. (2019). Experimental methods in chemical engineering: Specific surface area and pore size distribution measurements—BET, BJH, and DFT. *Canadian Journal of Chemical Engineering, 97*(11), 2781–2791. https://doi.org/10.1002/cjce.23632, http://onlinelibrary.wiley.com/journal/10.1002/(ISSN)1939-019X.

Bi, C., Zhang, C., Ma, F., Zhang, X., Yang, M., Nian, J., Liu, L., Dong, H., Zhu, L., Wang, Q., Guo, S., & Lv, Q. (2021). Growth of a mesoporous Zr-MOF on functionalized graphene oxide as an efficient adsorbent for recovering uranium (VI) from wastewater. *Microporous and Mesoporous Materials, 323*, 111223. https://doi.org/10.1016/j.micromeso.2021.111223.

Chai, L., Pan, J., Hu, Y., Qian, J., & Hong, M. (2021). Rational design and growth of MOF-on-MOF heterostructures. *Small (Weinheim an der Bergstrasse, Germany), 17*(36), 2100607. https://doi.org/10.1002/smll.202100607, http://onlinelibrary.wiley.com/journal/10.1002/(ISSN)1613-6829.

Chen, F., Shen, K., Chen, J., Yang, X., Cui, J., & Li, Y. (2019). General immobilization of ultrafine alloyed nanoparticles within metal-organic frameworks with high loadings for advanced synergetic catalysis. *ACS Central Science, 5*(1), 176–185. https://doi.org/10.1021/acscentsci.8b00805, http://pubs.acs.org/journal/acscii.

Chen, S. S., Hu, C., Liu, C.-H., Chen, Y.-H., Ahamad, T., Alshehri, S. M., Huang, P.-H., & Wu, K. C.-W. (2020). De Novo synthesis of platinum-nanoparticle-encapsulated UiO-66-NH$_2$ for photocatalytic thin film fabrication with enhanced performance of phenol degradation. *Journal of Hazardous Materials, 397*, 122431. https://doi.org/10.1016/j.jhazmat.2020.122431.

Cheng, Q., Tao, K., Han, X., Yang, Y., Yang, Z., Ma, Q., & Han, L. (2019). Ultrathin Ni-MOF nanosheet arrays grown on polyaniline decorated Ni foam as an advanced electrode for asymmetric supercapacitors with high energy density. *Dalton Transactions, 48*(13), 4119–4123. https://doi.org/10.1039/c9dt00386j, http://pubs.rsc.org/en/journals/journal/dt.

Choi, K. M., Jeong, H. M., Park, J. H., Zhang, Y. B., Kang, J. K., & Yaghi, O. M. (2014). Supercapacitors of nanocrystalline metal-organic frameworks. *ACS Nano, 8*(7), 7451–7457. https://doi.org/10.1021/nn5027092, http://pubs.acs.org/journal/ancac3.

Do, H. H., Cho, J. H., Han, S. M., Ahn, S. H., & Kim, S. Y. (2021). Metal–organic-framework- and mxene-based taste sensors and glucose detection. *Sensors (Basel, Switzerland), 21*(21), 7423. https://doi.org/10.3390/s21217423, https://www.mdpi.com/1424-8220/21/21/7423/pdf.

Duan, M., Jiang, L., Zeng, G., Wang, D., Tang, W., Liang, J., Wang, H., He, D., Liu, Z., & Tang, L. (2020). Bimetallic nanoparticles/metal-organic frameworks: Synthesis, applications and challenges. *Applied Materials Today, 19*, 100564. https://doi.org/10.1016/j.apmt.2020.100564, http://www.journals.elsevier.com/applied-materials-today/.

Dutta, S., Mandal, W., Desai, A. V., Fajal, S., Dam, G. K., Mukherjee, S., & Ghosh, S. K. (2023). A luminescent cationic MOF and its polymer composite membrane elicit selective sensing of antibiotics and pesticides in water. *Molecular Systems Design & Engineering, 8*(12), 1483–1491. https://doi.org/10.1039/d3me00008g.

Fang, Y., Li, X., Li, F., Lin, X., Tian, M., Long, X., An, X., Fu, Y., Jin, J., & Ma, J. (2016). Self-assembly of cobalt-centered metal organic framework and multiwalled carbon nanotubes hybrids as a highly active and corrosion-resistant bifunctional oxygen catalyst. *Journal of Power Sources, 326*, 50–59. https://doi.org/10.1016/j.jpowsour.2016.06.114.

Fu, Q., Wen, L., Zhang, L., Chen, X., Pun, D., Ahmed, A., Yang, Y., & Zhang, H. (2017). Preparation of ice-templated MOF-polymer composite monoliths and their application for wastewater treatment with high capacity and easy recycling. *ACS Applied Materials and Interfaces, 9*(39), 33979–33988. https://doi.org/10.1021/acsami.7b10872, http://pubs.acs.org/journal/aamick.

Gamage, N. D. H., McDonald, K. A., & Matzger, A. J. (2016). MOF-5-polystyrene: Direct production from monomer, improved hydrolytic stability, and unique guest adsorption. *Angewandte Chemie - International Edition, 55*(39), 12099–12103. https://doi.org/10.1002/anie.201606926, http://onlinelibrary.wiley.com/journal/10.1002/(ISSN)1521-3773.

Gandara-Loe, J., Souza, B. E., Missyul, A., Giraldo, G., Tan, J.-C., & Silvestre-Albero, J. (2020). MOF-based polymeric nanocomposite films as potential materials for drug delivery devices in ocular therapeutics. *ACS Applied Materials & Interfaces, 12*(27), 30189–30197. https://doi.org/10.1021/acsami.0c07517.

Gholidoust, A., Maina, J. W., Merenda, A., Schütz, J. A., Kong, L., Hashisho, Z., & Dumée, L. F. (2019). CO$_2$ sponge from plasma enhanced seeded growth of metal organic frameworks across carbon nanotube bucky-papers. *Separation and Purification Technology, 209*, 571–579. https://doi.org/10.1016/j.seppur.2018.07.085, http://www.journals.elsevier.com/separation-and-purification-technology/.

Gnanasekaran, G., Balaguru, S., Arthanareeswaran, G., & Das, D. B. (2019). Removal of hazardous material from wastewater by using metal organic framework (MOF) embedded polymeric membranes. *Separation Science and Technology (Philadelphia), 54*(3), 434–446. https://doi.org/10.1080/01496395.2018.1508232, http://www.tandf.co.uk/journals/titles/01496395.asp.

Guo, S. N., Zhu, Y., Yan, Y. Y., Min, Y. L., Fan, J. C., Xu, Q. J., & Yun, H. (2016). Metal-organic framework)-polyaniline sandwich structure composites as novel hybrid electrode materials for high-performance supercapacitor. *Journal of Power Sources, 316*, 176–182. https://doi.org/10.1016/j.jpowsour.2016.03.040.

He, S., Wang, H., Zhang, C., Zhang, S., Yu, Y., Lee, Y., & Li, T. (2019). A generalizable method for the construction of MOF@polymer functional composites through surface-initiated atom transfer radical polymerization. *Chemical Science, 10*(6), 1816–1822. https://doi.org/10.1039/c8sc03520b, http://pubs.rsc.org/en/journals/journal/sc.

Hermes, S., Schröter, M. K., Schmid, R., Khodeir, L., Muhler, M., Tissler, A., Fischer, R. W., & Fischer, R. A. (2005). Metal@MOF: Loading of highly porous coordination polymers host lattices by metal organic chemical vapor deposition. *Angewandte Chemie - International Edition, 44*(38), 6237–6241. https://doi.org/10.1002/anie.200462515.

Hong, H., Liu, J., Huang, H., Atangana Etogo, C., Yang, X., Guan, B., & Zhang, L. (2019). Ordered macro-microporous metal-organic framework single crystals and their derivatives for rechargeable aluminum-ion batteries. *Journal of the American Chemical Society, 141*(37), 14764–14771. https://doi.org/10.1021/jacs.9b06957, http://pubs.acs.org/journal/jacsat.

Hou, L., Wang, L., Zhang, N., Xie, Z., & Dong, D. (2016). Polymer brushes on metal-organic frameworks by UV-induced photopolymerization. *Polymer Chemistry, 7*(37), 5828–5834. https://doi.org/10.1039/c6py01008c, http://pubs.rsc.org/en/Journals/JournalIssues/PY.

Hu, T., Deng, F., Feng, H., Zhang, J., Shao, B., Feng, C., Tang, W., & Tang, L. (2021). Fe/Co bimetallic nanoparticles embedded in MOF-derived nitrogen-doped porous carbon rods as efficient heterogeneous electro-Fenton catalysts for degradation of organic pollutants. *Applied Materials Today, 24*, 101161. https://doi.org/10.1016/j.apmt.2021.101161.

Huang, T. Y., Kung, C. W., Liao, Y. T., Kao, S. Y., Cheng, M., Chang, T. H., Henzie, J., Alamri, H. R., Alothman, Z. A., Yamauchi, Y., Ho, K. C., & Wu, K. C. W. (2017). Enhanced charge collection in MOF-525–PEDOT nanotube composites enable highly sensitive biosensing. *Advanced Science, 4*(11), 1700261. https://doi.org/10.1002/advs.201700261, http://onlinelibrary.wiley.com/journal/10.1002/(ISSN)2198-3844.

Hwang, J., Lee, H. C., Antonietti, M., & Schmidt, B. V. K. J. (2017). Free radical and RAFT polymerization of vinyl esters in metal-organic-frameworks. *Polymer Chemistry, 8*(40), 6204–6208. https://doi.org/10.1039/c7py01607g, http://pubs.rsc.org/en/Journals/JournalIssues/PY.

Ishida, T., Nagaoka, M., Akita, T., & Haruta, M. (2008). Deposition of gold clusters on porous coordination polymers by solid grinding and their catalytic activity in aerobic oxidation of alcohols. *Chemistry – A European Journal, 14*(28), 8456–8460. https://doi.org/10.1002/chem.200800980, http://www3.interscience.wiley.com/cgi-bin/fulltext/121388974/PDFSTART, Japan.

Jiang, H. L., Akita, T., Ishida, T., Haruta, M., & Xu, Q. (2011). Synergistic catalysis of Au@Ag core-shell nanoparticles stabilized on metal-organic framework. *Journal of the American Chemical Society, 133*(5), 1304–1306. https://doi.org/10.1021/ja1099006.

Johari, N. A., Yusof, N., Lau, W. J., Abdullah, N., Salleh, W. N. W., Jaafar, J., Aziz, F., & Ismail, A. F. (2021). Polyethersulfone ultrafiltration membrane incorporated with ferric-based metal-organic framework for textile wastewater treatment. *Separation and Purification Technology, 270*, 118819. https://doi.org/10.1016/j.seppur.2021.118819.

Karimzadeh, Z., Javanbakht, S., & Namazi, H. (2019). Carboxymethylcellulose/MOF-5/graphene oxide bio-nanocomposite as antibacterial drug nanocarrier agent. *BioImpacts: BI, 9*(1), 5–13. https://doi.org/10.15171/bi.2019.02, http://bi.tbzmed.ac.ir/.

Kitao, T., Bracco, S., Comotti, A., Sozzani, P., Naito, M., Seki, S., Uemura, T., & Kitagawa, S. (2015). Confinement of single polysilane chains in coordination nanospaces. *Journal of the American Chemical Society, 137*(15), 5231–5238. https://doi.org/10.1021/jacs.5b02215, http://pubs.acs.org/journal/jacsat.

Kobielska, P. A., Howarth, A. J., Farha, O. K., & Nayak, S. (2018). Metal–organic frameworks for heavy metal removal from water. *Coordination Chemistry Reviews, 358*, 92–107. https://doi.org/10.1016/j.ccr.2017.12.010, http://www.journals.elsevier.com/coordination-chemistry-reviews/.

Kumari, D., Prajapati, M., & Ravi Kant, C. (2024). Highly efficient non-enzymatic electrochemical glucose biosensor based on copper metal organic framework coated on graphite sheet. *ECS Journal of Solid State Science and Technology, 13*(4), 047007. https://doi.org/10.1149/2162-8777/ad3fe5, https://iopscience.iop.org/journal/2162-8777.

Kumari, R., Kharangarh, P. R., Singh, V., Jha, R., & Ravi Kant, C. (2023). Sequential processing of nitrogen-rich, biowaste-derived carbon quantum dots combined with strontium cobaltite for enhanced supercapacitive performance. *Journal of Alloys and Compounds, 969*, 172256. https://doi.org/10.1016/j.jallcom.2023.172256, https://www.journals.elsevier.com/journal-of-alloys-and-compounds.

Kumari, R., Kumar Sharma, S., Singh, V., & Ravi Kant, C. (2022). Facile, two-step synthesis of activated carbon soot from used soybean oil and waste engine oil for supercapacitor electrodes. *Materials Today: Proceedings, 67*, 483–489. https://doi.org/10.1016/j.matpr.2022.07.253, https://www.sciencedirect.com/science/journal/22147853.

Kumari, R., Prajapati, M., & Kant, C. R. (2024). X-ray irradiation-induced enhancement of supercapacitive properties of bio-derived activated carbon. *Journal of Electronic Materials, 53*, 4985–4996. https://doi.org/10.1007/s11664-024-11048-2, https://www.springer.com/journal/11664.

Kumari, R., Prajapati, M., & Ravi Kant, C. (2024). NiCo MOF@carbon quantum dots anode with soot derived activated carbon cathode: An efficient asymmetric configuration for sustainable high-performance supercapacitors. *Advanced Sustainable Systems, 8*(10), 2400109. https://doi.org/10.1002/adsu.202400109, http://www.advsustainsys.com.

Kumari, R., Singh, V., & Ravi Kant, C. (2023). Enhanced performance of activated carbon-based supercapacitor derived from waste soybean oil with coffee ground additives. *Materials Chemistry and Physics, 305*, 127882. https://doi.org/10.1016/j.matchemphys.2023.127882.

Kuo, C. H., Tang, Y., Chou, L. Y., Sneed, B. T., Brodsky, C. N., Zhao, Z., & Tsung, C. K. (2012). Yolk-shell nanocrystal@ZIF-8 nanostructures for gas-phase heterogeneous catalysis with selectivity control. *Journal of the American Chemical Society, 134*(35), 14345–14348. https://doi.org/10.1021/ja306869j.

Li, J., Wang, X., Zhao, G., Chen, C., Chai, Z., Alsaedi, A., Hayat, T., & Wang, X. (2018). Metal-organic framework-based materials: Superior adsorbents for the capture of toxic and radioactive metal ions. *Chemical Society Reviews, 47*(7), 2322–2356. https://doi.org/10.1039/c7cs00543a, http://pubs.rsc.org/en/journals/journal/cs.

Li, J., Wang, H., Yuan, X., Zhang, J., & Wei Chew, J. (2020). Metal-organic framework membranes for wastewater treatment and water regeneration. *Coordination Chemistry Reviews, 404*, 213116. https://doi.org/10.1016/j.ccr.2019.213116.

Li, X., Zhang, Z., Xiao, W., Deng, S., Chen, C., & Zhang, N. (2019). Mechanochemistry-assisted encapsulation of metal nanoparticles in MOF matrices: Via a sacrificial strategy. *Journal of*

Materials Chemistry A, 7(24), 14504–14509. https://doi.org/10.1039/c9ta03578h, http://pubs.rsc.org/en/journals/journal/ta.

Li, Y., Cai, Y., Shao, K., Chen, Y., & Wang, D. (2021). A free-standing poly-MOF film fabricated by post-modification and interfacial polymerization: A novel platform for Cd^{2+} electrochemical sensors. *Microporous and Mesoporous Materials, 323*, 111200. https://doi.org/10.1016/j.micromeso.2021.111200.

Li, Y., Liu, Y., Wang, Z., Wang, P., Zheng, Z., Cheng, H., Dai, Y., & Huang, B. (2021). In-situ growth of Ti_3C_2@MIL-NH_2 composite for highly enhanced photocatalytic H_2 evolution. *Chemical Engineering Journal, 411*, 128446. https://doi.org/10.1016/j.cej.2021.128446.

Lian, X., & Yan, B. (2016). A lanthanide metal-organic framework (MOF-76) for adsorbing dyes and fluorescence detecting aromatic pollutants. *RSC Advances, 6*(14), 11570–11576. https://doi.org/10.1039/c5ra23681a, http://pubs.rsc.org/en/journals/journal/ra.

Lin, X., Gao, G., Zheng, L., Chi, Y., & Chen, G. (2014). Encapsulation of strongly fluorescent carbon quantum dots in metal-organic frameworks for enhancing chemical sensing. *Analytical Chemistry, 86*(2), 1223–1228. https://doi.org/10.1021/ac403536a.

Liu, C., Wang, J., Wan, J., & Yu, C. (2021). MOF-on-MOF hybrids: Synthesis and applications. *Coordination Chemistry Reviews, 432*, 213743. https://doi.org/10.1016/j.ccr.2020.213743.

Liu, M., Wang, J., Song, P., Ji, J., & Wang, Q. (2022). Metal-organic frameworks-derived In_2O_3 microtubes/$Ti_3C_2T_x$ MXene composites for NH_3 detection at room temperature. *Sensors and Actuators B: Chemical, 361*, 131755. https://doi.org/10.1016/j.snb.2022.131755.

Liu, X. W., Sun, T. J., Hu, J. L., & Wang, S. D. (2016). Composites of metal-organic frameworks and carbon-based materials: Preparations, functionalities and applications. *Journal of Materials Chemistry A, 4*(10), 3584–3616. https://doi.org/10.1039/c5ta09924b, http://pubs.rsc.org/en/journals/journal/ta.

Liu, Y., He, Y., Vargun, E., Plachy, T., Saha, P., & Cheng, Q. (2020). 3D porous Ti_3C_2 MXene/NiCo-MOF composites for enhanced lithium storage. *Nanomaterials, 10*(4), 695. https://doi.org/10.3390/nano10040695, https://www.mdpi.com/2079-4991/10/4/695/pdf.

Lu, G., Li, S., Guo, Z., Farha, O. K., Hauser, B. G., Qi, X., Wang, Y., Wang, X., Han, S., Liu, X., Duchene, J. S., Zhang, H., Zhang, Q., Chen, X., Ma, J., Loo, S. C. J., Wei, W. D., Yang, Y., Hupp, J. T., & Huo, F. (2012). Imparting functionality to a metal-organic framework material by controlled nanoparticle encapsulation. *Nature Chemistry, 4*(4), 310–316. https://doi.org/10.1038/nchem.1272.

Lyu, F., Zhang, Y., Zare, R. N., Ge, J., & Liu, Z. (2014). One-pot synthesis of protein-embedded metal-organic frameworks with enhanced biological activities. *Nano Letters, 14*(10), 5761–5765. https://doi.org/10.1021/nl5026419, http://pubs.acs.org/journal/nalefd.

Ma, L., He, Y., Wang, Y., Wang, Y., Li, R., Huang, Z., Jiang, Y., & Gao, J. (2019). Nanocomposites of Pt nanoparticles anchored on UiO66-NH_2 as carriers to construct acetylcholinesterase biosensors for organophosphorus pesticide detection. *Electrochimica acta, 318*, 525–533. https://doi.org/10.1016/j.electacta.2019.06.110.

Masoomi, M. Y., Morsali, A., Dhakshinamoorthy, A., & Garcia, H. (2019). Mixed-metal MOFs: Unique opportunities in metal–organic framework (MOF) functionality and design. *Angewandte Chemie, 131*(43), 15330–15347. https://doi.org/10.1002/ange.201902229.

Min, H., Wang, J., Qi, Y., Zhang, Y., Han, X., Xu, Y., Xu, J., Li, Y., Chen, L., Cheng, K., Liu, G., Yang, N., Li, Y., & Nie, G. (2019). Biomimetic metal–organic framework nanoparticles for cooperative combination of antiangiogenesis and photodynamic therapy for enhanced efficacy. *Advanced Materials, 31*(15), e1808200. https://doi.org/10.1002/adma.201808200, http://onlinelibrary.wiley.com/journal/10.1002/(ISSN)1521-4095.

Nath, D., & Das, R. (2021). Experimental (XRD) and theoretical (DFT) analysis for understanding the influence of SHI irradiation on the stacking fault energy in CdSe nanocrystals. *Journal of Alloys and Compounds, 879*, 160456. https://doi.org/10.1016/j.jallcom.2021.160456.

Niu, H., Liu, Y., Mao, B., Xin, N., Jia, H., & Shi, W. (2020). In-situ embedding MOFs-derived copper sulfide polyhedrons in carbon nanotube networks for hybrid supercapacitor with superior energy density. *Electrochimica Acta, 329*, 135130. https://doi.org/10.1016/j.electacta.2019.135130.

Orasugh, J. T., Ghosh, S. K., & Chattopadhyay, D. (2020). Nanofiber-reinforced biocomposites. *Nanofiber-reinforced biocomposites Fiber-Reinforced Nanocomposites: Fundamentals and Applications,* 199–233. https://doi.org/10.1016/B978-0-12-819904-6.00010-4, https://www.sciencedirect.com/book/9780128199046.

Peng, S., Bie, B., Sun, Y., Liu, M., Cong, H., Zhou, W., Xia, Y., Tang, H., Deng, H., & Zhou, X. (2018). Metal-organic frameworks for precise inclusion of single-stranded DNA and transfection in immune cells. *Nature Communications, 9*(1), 1293. https://doi.org/10.1038/s41467-018-03650-w, http://www.nature.com/ncomms/index.html.

Pooresmaeil, M., Asl, E. A., & Namazi, H. (2021). A new pH-sensitive CS/Zn-MOF@GO ternary hybrid compound as a biofriendly and implantable platform for prolonged 5-fluorouracil delivery to human breast cancer cells. *Journal of Alloys and Compounds, 885*, 160992. https://doi.org/10.1016/j.jallcom.2021.160992, https://www.journals.elsevier.com/journal-of-alloys-and-compounds.

Prajapati, M., Ravi Kant, C., Allende, S., & Jacob, M. V. (2023). Metal organic framework derived NiCo layered double hydroxide anode aggregated with biomass derived reduced graphene oxide cathode: A hybrid device configuration for supercapattery applications. *Journal of Energy Storage, 73*, 109264. https://doi.org/10.1016/j.est.2023.109264, http://www.journals.elsevier.com/journal-of-energy-storage/.

Prajapati, M., Singh, V., Jacob, M. V., & Ravi Kant, C. (2023). Recent advancement in metal-organic frameworks and composites for high-performance supercapatteries. *Renewable and Sustainable Energy Reviews, 183*, 113509. https://doi.org/10.1016/j.rser.2023.113509, https://www.journals.elsevier.com/renewable-and-sustainable-energy-reviews.

Prajapati, M., Ravi Kant, C., & Jacob, M. V. (2024). Binder free cobalt manganese layered double hydroxide anode conjugated with bioderived rGO cathode for sustainable, high-performance asymmetric supercapacitors. *Journal of Electroanalytical Chemistry, 961*, 118242. https://doi.org/10.1016/j.jelechem.2024.118242.

Radhika, M. G., Gopalakrishna, B., Chaitra, K., Gopalakri Bhatta, L. K., Venkatesh, K., Sudha Kamath, M. K., & Kathyayini, N. (2020). Electrochemical studies on Ni, Co & Ni/Co-MOFs for high-performance hybrid supercapacitors. *Materials Research Express, 7*(5), 054003. https://doi.org/10.1088/2053-1591/ab8d5d.

Rahmanifar, M. S., Hesari, H., Noori, A., Masoomi, M. Y., Morsali, A., & Mousavi, M. F. (2018). A dual Ni/Co-MOF-reduced graphene oxide nanocomposite as a high performance supercapacitor electrode material. *Electrochimica Acta, 275*, 76–86. https://doi.org/10.1016/j.electacta.2018.04.130, http://www.journals.elsevier.com/electrochimica-acta/.

Rehan, M., Montaser, A. S., El-Shahat, M., & Abdelhameed, R. M. (2024). Decoration of viscose fibers with silver nanoparticle-based titanium-organic framework for use in environmental applications. *Environmental Science and Pollution Research, 31*(9), 13185–13206. https://doi.org/10.1007/s11356-024-31858-5, https://www.springer.com/journal/11356.

Salunkhe, R. R., Tang, J., Kobayashi, N., Kim, J., Ide, Y., Tominaka, S., Kim, J. H., & Yamauchi, Y. (2016). Ultrahigh performance supercapacitors utilizing core-shell nanoarchitectures from a metal-organic framework-derived nanoporous carbon and a conducting polymer. *Chemical Science, 7*(9), 5704–5713. https://doi.org/10.1039/c6sc01429a, http://pubs.rsc.org/en/Journals/JournalIssues/SC.

Samy, M., Ibrahim, M. G., Fujii, M., Diab, K. E., ElKady, M., & Gar Alalm, M. (2021). CNTs/MOF-808 painted plates for extended treatment of pharmaceutical and agrochemical wastewaters in a novel photocatalytic reactor. *Chemical Engineering Journal, 406*, 127152. https://doi.org/10.1016/j.cej.2020.127152, http://www.elsevier.com/inca/publications/store/6/0/1/2/7/3/index.htt.

Santos, V. P., Wezendonk, T. A., Jaén, J. J. D., Dugulan, A. I., Nasalevich, M. A., Islam, H. U., Chojecki, A., Sartipi, S., Sun, X., Hakeem, A. A., Koeken, A. C. J., Ruitenbeek, M., Davidian, T., Meima, G. R., Sankar, G., Kapteijn, F., Makkee, M., & Gascon, J. (2015). Metal organic framework-mediated synthesis of highly active and stable Fischer-Tropsch catalysts. *Nature Communications, 6*, 6451. https://doi.org/10.1038/ncomms7451, http://www.nature.com/ncomms/index.html.

Sarabaegi, M., Roushani, M., Hosseini, H., Saedi, Z., & Lemraski, E. G. (2022). A novel ultrasensitive biosensor based on NiCo-MOF nanostructure and confined to flexible carbon nanofibers with high-surface skeleton to rapidly detect Helicobacter pylori. *Materials Science in Semiconductor Processing, 139*, 106351. https://doi.org/10.1016/j.mssp.2021.106351, https://www.journals.elsevier.com/materials-science-in-semiconductor-processing.

Sharma, I., Kaur, J., Poonia, G., Mehta, S. K., & Kataria, R. (2023). Nanoscale designing of metal organic framework moieties as efficient tools for environmental decontamination. *Nanoscale Advances, 5*(15), 3782–3802. https://doi.org/10.1039/d3na00169e, http://pubs.rsc.org/en/journals/journalissues/na?_ga2.190536939.1555337663.1552312502-1364180372.1550481316#!issueidna001002&typecurrent&issnonline2516-0230.

Su, F., Zhang, S., Ji, H., Zhao, H., Tian, J. Y., Liu, C. S., Zhang, Z., Fang, S., Zhu, X., & Du, M. (2017). Two-dimensional zirconium-based metal-organic framework nanosheet composites embedded with Au nanoclusters: A highly sensitive electrochemical aptasensor toward detecting cocaine. *ACS Sensors, 2*(7), 998–1005. https://doi.org/10.1021/acssensors.7b00268, http://pubs.acs.org/journal/ascefj.

Sundriyal, S., Kaur, H., Bhardwaj, S. K., Mishra, S., Kim, K. H., & Deep, A. (2018). Metal-organic frameworks and their composites as efficient electrodes for supercapacitor applications. *Coordination Chemistry Reviews, 369*, 15–38. https://doi.org/10.1016/j.ccr.2018.04.018, http://www.journals.elsevier.com/coordination-chemistry-reviews/.

Tanabe, K. K., & Cohen, S. M. (2011). Postsynthetic modification of metal–organic frameworks—A progress report. *Chemical Society reviews, 40*(2), 498–519. https://doi.org/10.1039/C0CS00031K.

Uemura, T., Ono, Y., Kitagawa, K., & Kitagawa, S. (2008). Radical polymerization of vinyl monomers in porous coordination polymers: Nanochannel size effects on reactivity, molecular weight, and stereostructure. *Macromolecules, 41*(1), 87–94. https://doi.org/10.1021/ma7022217.

Uemura, T., Kaseda, T., Sasaki, Y., Inukai, M., Toriyama, T., Takahara, A., Jinnai, H., & Kitagawa, S. (2015). Mixing of immiscible polymers using nanoporous coordination templates. *Nature Communications, 6*(1), 7473. https://doi.org/10.1038/ncomms8473.

Wang, B., Li, W., Liu, Z., Duan, Y., Zhao, B., Wang, Y., & Liu, J. (2020). Incorporating Ni-MOF structure with polypyrrole: Enhanced capacitive behavior as electrode material for supercapacitor. *RSC Advances, 10*(21), 12129–12134. https://doi.org/10.1039/c9ra10467d, http://pubs.rsc.org/en/journals/journal/ra.

Wang, L., Xu, H., Qiu, Y., Liu, X., Huang, W., Yan, N., & Qu, Z. (2020). Utilization of Ag nanoparticles anchored in covalent organic frameworks for mercury removal from acidic waste water. *Journal of Hazardous Materials, 389*, 121824. https://doi.org/10.1016/j.jhazmat.2019.121824.

Wang, Q., & Astruc, D. (2020). State of the art and prospects in metal-organic framework (MOF)-based and MOF-derived nanocatalysis. *Chemical Reviews, 120*(2), 1438–1511. https://doi.org/10.1021/acs.chemrev.9b00223, http://pubs.acs.org/journal/chreay.

Wang, S., McGuirk, C. M., d'Aquino, A., Mason, J. A., & Mirkin, C. A. (2018). Metal–organic framework nanoparticles. *Advanced Materials, 30*(37). https://doi.org/10.1002/adma.201800202, http://onlinelibrary.wiley.com/journal/10.1002/(ISSN)1521-4095.

Wang, W., Wang, L., Li, Y., Liu, S., Xie, Z., & Jing, X. (2016). Nanoscale polymer metal–organic framework hybrids for effective photothermal therapy of colon cancers. *Advanced Materials, 28*(42), 9320–9325. https://doi.org/10.1002/adma.201602997, http://www3.interscience.wiley.com/journal/119030556/issue.

Wang, Y., Liu, Y., Wang, C., Liu, H., Zhang, J., Lin, J., Fan, J., Ding, T., Ryu, J. E., & Guo, Z. (2020). Significantly enhanced ultrathin NiCo-based MOF nanosheet electrodes hybrided with $Ti_3C_2T_x$ MXene for high performance asymmetric supercapacitors. *Engineered Science, 9*, 50–59. https://doi.org/10.30919/es8d903, http://www.espublisher.com/journals/articledetails/210/.

Wang, Y., Wang, L., Huang, W., Zhang, T., Hu, X., Perman, J. A., & Ma, S. (2017). A metal-organic framework and conducting polymer based electrochemical sensor for high performance cadmium ion detection. *Journal of Materials Chemistry A, 5*(18), 8385–8393. https://doi.org/10.1039/c7ta01066d, http://pubs.rsc.org/en/journals/journalissues/ta.

Wang, Y., Xiong, Y., Huang, Q., Bi, Z., Zhang, Z., Guo, Z., Wang, X., & Mei, T. (2022). A bifunctional VS_2-Ti_3C_2 heterostructure electrocatalyst for boosting polysulfide redox in high performance lithium-sulfur batteries. *Journal of Materials Chemistry A, 10*(36), 18866–18876. https://doi.org/10.1039/d2ta05604f, http://pubs.rsc.org/en/journals/journal/ta.

Wu, L., Lu, Z., & Ye, J. (2019). Enzyme-free glucose sensor based on layer-by-layer electrodeposition of multilayer films of multi-walled carbon nanotubes and Cu-based metal framework modified glassy carbon electrode. *Biosensors and Bioelectronics, 135*, 45–49. https://doi.org/10.1016/j.bios.2019.03.064, http://www.elsevier.com/locate/bios.

Xu, S., Liu, R., Shi, X., Ma, Y., Hong, M., Chen, X., Wang, T., Li, F., Hu, N., & Yang, Z. (2020). A dual CoNi MOF nanosheet/nanotube assembled on carbon cloth for high performance hybrid supercapacitors. *Electrochimica Acta, 342*, 136124. https://doi.org/10.1016/j.electacta.2020.136124.

Xue, Q., Pei, Z., Huang, Y., Zhu, M., Tang, Z., Li, H., Huang, Y., Li, N., Zhang, H., & Zhi, C. (2017). Mn_3O_4 nanoparticles on layer-structured Ti_3C_2 MXene towards the oxygen reduction reaction and zinc–air batteries. *Journal of Materials Chemistry A, 5*(39), 20818–20823. https://doi.org/10.1039/c7ta04532h.

Yang, G., Zhang, D., Zhu, G., Zhou, T., Song, M., Qu, L., Xiong, K., & Li, H. (2020). A Sm-MOF/GO nanocomposite membrane for efficient organic dye removal from wastewater. *RSC Advances, 10*(14), 8540–8547. https://doi.org/10.1039/d0ra01110j, http://pubs.rsc.org/en/journals/journal/ra.

Yang, H., Han, L., Liu, J., Li, Y., Zhang, D., Liu, X., & Liang, Z. (2016). Highly sensitive electrochemical biosensor assembled by Au nanoparticle /MOF-5 composite electrode for DNA detection. *International Journal of Electrochemical Science, 14*(6), 5491–5507. https://doi.org/10.20964/2019.06.49, http://www.electrochemsci.org/papers/vol14/140605491.pdf.

Yang, Q., Xu, Q., Liu, B., Zhong, C., & Berend, S. (2009). Molecular simulation of CO_2/H_2 mixture separation in metal-organic frameworks: Effect of catenation and electrostatic interactions. *Chinese Journal of Chemical Engineering, 17*(5), 781–790. https://doi.org/10.1016/S1004-9541(08)60277-3.

Yang, S., Karve, V. V., Justin, A., Kochetygov, I., Espín, J., Asgari, M., Trukhina, O., Sun, D. T., Peng, L., & Queen, W. L. (2021). Enhancing MOF performance through the introduction of polymer guests. *Coordination chemistry reviews, 427*, 213525. https://doi.org/10.1016/j.ccr.2020.213525.

Yao, J., Liu, Y., Wang, J., Jiang, Q., She, D., Guo, H., Sun, N., Pang, Z., Deng, C., Yang, W., & Shen, S. (2019). On-demand CO release for amplification of chemotherapy by MOF functionalized magnetic carbon nanoparticles with NIR irradiation. *Biomaterials, 195*, 51–62. https://doi.org/10.1016/j.biomaterials.2018.12.029, http://www.journals.elsevier.com/biomaterials/.

Yao, J., & Wang, H. (2014). Zeolitic imidazolate framework composite membranes and thin films: Synthesis and applications. *Chemical Society reviews, 43*(13), 4470–4493. https://doi.org/10.1039/C3CS60480B.

Yap, M. H., Fow, K. L., & Chen, G. Z. (2017). Synthesis and applications of MOF-derived porous nanostructures. *Green Energy and Environment, 2*(3), 218–245. https://doi.org/10.1016/j.gee.2017.05.003, http://www.keaipublishing.com/en/journals/green-energy-and-environment/.

Yildirim Kalyon, H., Gencten, M., Gorduk, S., & Sahin, Y. (2022). Novel composite materials consisting of polypyrrole and metal organic frameworks for supercapacitor applications. *Journal of Energy Storage, 48*, 103699. https://doi.org/10.1016/j.est.2021.103699.

Zhang, X., Xu, Y., & Ye, B. (2018). An efficient electrochemical glucose sensor based on porous nickel-based metal organic framework/carbon nanotubes composite (Ni-MOF/CNTs). *Journal of Alloys and Compounds, 767*, 651–656. https://doi.org/10.1016/j.jallcom.2018.07.175.

Zhang, Y., Feng, X., Li, H., Chen, Y., Zhao, J., Wang, S., Wang, L., & Wang, B. (2015). Photoinduced postsynthetic polymerization of a metal-organic framework toward a flexible stand-alone membrane. *Angewandte Chemie - International Edition, 54*(14), 4259–4263. https://doi.org/10.1002/anie.201500207, http://onlinelibrary.wiley.com/journal/10.1002/(ISSN)1521-3773.

Zhang, Y. Y., Wei, Y., Tang, X. Y., & Shi, M. (2017). Dual-role of PtCl2 catalysis in the intramolecular cyclization of (hetero)aryl-allenes for the facile construction of substituted 2,3-dihydropyrroles and polyheterocyclic skeletons. *Chemical Communications, 53*(44), 5966–5969. https://doi.org/10.1039/c7cc01684k, http://pubs.rsc.org/en/journals/journal/cc.

Zhang, Z., Nguyen, H. T. H., Miller, S. A., & Cohen, S. M. (2015). PolyMOFs: A class of interconvertible polymer-metal-organic-framework hybrid materials. *Angewandte Chemie - International Edition, 54*(21), 6152–6157. https://doi.org/10.1002/anie.201502733, http://onlinelibrary.wiley.com/journal/10.1002/(ISSN)1521-3773.

Zhang, Z., Nguyen, H. T. H., Miller, S. A., Ploskonka, A. M., Decoste, J. B., & Cohen, S. M. (2016). Polymer-metal-organic frameworks (polyMOFs) as water tolerant materials for selective carbon dioxide separations. *Journal of the American Chemical Society, 138*(3), 920–925. https://doi.org/10.1021/jacs.5b11034, http://pubs.acs.org/journal/jacsat.

Zhao, H., Song, H., & Chou, L. (2012). Nickel nanoparticles supported on MOF-5: Synthesis and catalytic hydrogenation properties. *Inorganic Chemistry Communications, 15*, 261–265. https://doi.org/10.1016/j.inoche.2011.10.040.

Zhuang, X., Zhang, S., Tang, Y., Yu, F., Li, Z., & Pang, H. (2023). Recent progress of MOF/MXene-based composites: Synthesis, functionality and application. *Coordination Chemistry Reviews, 490*, 215208. https://doi.org/10.1016/j.ccr.2023.215208.

Zirehpour, A., Rahimpour, A., & Ulbricht, M. (2017). Nano-sized metal organic framework to improve the structural properties and desalination performance of thin film composite forward osmosis membrane. *Journal of Membrane Science, 531*, 59–67. https://doi.org/10.1016/j.memsci.2017.02.049, http://www.elsevier.com/locate/memsci.

Zou, L., Hou, C. C., Liu, Z., Pang, H., & Xu, Q. (2018). Superlong single-crystal metal-organic framework nanotubes. *Journal of the American Chemical Society, 140*(45), 15393–15401. https://doi.org/10.1021/jacs.8b09092, http://pubs.acs.org/journal/jacsat.

Chapter 3

Metal-organic framework composites for supercapacitors

Shavita Salora[1,2,*], Jyoti Jangra[1,2,*], Shubham Kumar Patial[1,2,3,*], and Suman Singh[1,2]

[1]Academy of Scientific and Innovative Research (AcSIR), Ghaziabad, Uttar Pradesh, India, [2]Materials Science and Sensor Applications, CSIR-Central Scientific Instruments Organisation, Chandigarh, India, [3]School of Science, STEM College, RMIT University, Melbourne, VIC, Australia

Introduction

The global energy demand is projected to double in approximately two decades. Energy significantly influences our social and economic development, shaping our lives. With the reduction of fossil fuels and changes in the environment, energy production and storage applications are receiving worldwide attention (Yang et al., 2022). A main worry regarding energy harvesting from fossil fuels is that it is not going to last for a longer period. Researchers and industry have been compelled to find an alternative solution to deal with this issue. In recent decades, there has been a significant increase in research on renewable energy sources like solar energy, geothermal energy, wind energy, biofuels, and more. Additionally, there has been notable research interest in electrochemical energy storage devices such as supercapacitors, rechargeable batteries, and so on (Azcárate et al., 2017; Holze, 2009; McKone et al., 2017).

Metal-organic frameworks (MOFs) are a novel class of porous materials with a hybrid organic/inorganic nature that enables the combination of properties from both organic and inorganic porous materials. MOF porosity, as well as pore sizes, exceeds that of zeolites. MOFs are coordination polymers. MOFs are coordination polymers formed by connecting metal ions with organic linkers. In the past decade, these materials have attracted the attention of researchers because of their significant properties. MOFs with a

---

* These authors contributed equally to this chapter.

crystalline nature generally possess extremely high surface areas, typically 1000–10,000 m^2/g, and tunable pore sizes/characteristics, which were measured by Brunauer–Emmett–Teller (BET) (Arnold et al., 2007). As reported in the literature, the first successful demonstration was carried out for MOF-5. Crystallites from this material could be firmly and selectively anchored on modified gold substrates (Hermes et al., 2005). The potential of this approach was initially demonstrated in a microporous manganese(II) formate framework by Arnold *et al.* on porous substrates (Arnold et al., 2007). In the literature, greater than 20,000 MOFs are reported. HKUST-1, Fe-BTC, MIL-53, and ZIF-8 are the most currently utilized MOFs. MOFs have several applications, including gas storage, energy conversion, chemical sensing, drug delivery, proton conductivity, catalysis, etc. (Furukawa et al., 2013; Yaghi et al., 2003). Early studies revealed that during pyrolysis, MOFs collapse under high-temperature conditions, damaging their original morphologies and resulting in featureless bulk materials (2017) (Dang et al., 2017; Jiang et al., 2011; Yang et al., 2015). Various nanostructures are derived from MOFs with different morphologies by selectively choosing precursors for MOF synthesis and control strategies for pyrolysis. Other transition metals can be used to replace metals in MOF structures, and more heterometallic MOFs can also be synthesized (Furukawa et al., 2010; Wang & Cohen, 2009). Thermal stability and water stability are crucial for using MOFs in catalysis. In the starting period of MOF research in catalysis, structures like MOF-538 and MOF-17739, known for their impressive surface areas and gas storage capabilities, were also employed as catalysts (Chae et al., 2004; Rosi et al., 2003). However, their application in catalysis was hindered by limited thermal stability and sensitivity to moisture. Accordingly, various nanostructures derived from MOFs with different morphologies have been prepared, such as 0D polyhedral, hollow, and core-shell structures; 1D rods, hollow rods, and tubes; 2D sheets, ribbons, and sandwiches; and 3D arrays, frameworks, and honeycomb-like structures (Kaneti et al., 2017).

MOF degradation can occur through two primary mechanisms: the breaking of metal-ligand bonds and the formation of more stable products compared to the original MOFs (Antwi-Baah & Liu, 2018; Low et al., 2009). Therefore, the intrinsic structures of MOFs (internal factors)—such as the charge density of metal ions, connection numbers of metal ions/clusters, basicity, configuration, and the hydrophobicity of ligands—are crucial for the chemical stability of MOFs. After taking these things into account, building MOFs with good stability is made possible by the careful design of linkers and the choice of metal nodes (Lu et al., 2014).

Energy storage devices play a crucial role in modern technology by addressing the increasing need for efficient, reliable, and sustainable energy management. These devices store energy in various forms and release it when needed, helping to balance energy supply and demand. The fundamental principles of energy storage cover a wide range of technologies, each with unique mechanisms and applications. One of the most prevalent forms of energy storage is electrochemical storage, prominently represented by batteries. Supercapacitors, or ultracapacitors, represent another form of electrochemical storage but differ from batteries in their energy storage mechanism.

The initial idea for a supercapacitor (SC) was formed on the concept of the electric doublelayer. It exists at the interface between a conductor and the electrolyte solution it contacts. Hermann von Helmholtz first put forward this theory, which was later expanded upon by Gouy, Chapman, Grahame, and Stern. The electric double-layer theory forms the basis of electrochemistry. It is used to examine the electrochemical operations at the interface between the charged electrode material and the electrolyte. Since the theory was first proposed, this knowledge has led to the development of various electrochemical theories and technologies, such as electrochemical supercapacitors, fuel cells, and batteries. Manufacturing and construction details help us to classify the different types of supercapacitors. Supercapacitors are available in various shapes, such as flat, cylindrical, or rectangular (Şahin et al., 2022; Zhang & Pan, 2015). The operation principle of supercapacitors is based on energy storage. Supercapacitors are divided into three main groups depending on the energy storage method. Based on their energy storage mechanism, SCs can be categorized as electric double-layer capacitors (EDLCs) and pseudo capacitors (PCs). In EDLC, charge storage occurs between the electrolyte and electrodes. PCs involve fast and reversible Faradaic redox reactions to increase the capacitance of the SC. Current versus voltage curves Fig. 3.1 serves to classify different charge storage modes.

The life expectancy of SC is much longer than that of other charge storage devices, and since there is no charge transfer between the electrolyte and electrode in a non-Faradaic process, no chemical reaction is involved. Due to the chemical nature of SC, it can store charges longer than capacitors and batteries. However, because the liquid electrolytes can evaporate over time, SC with polymer electrodes has a shorter lifespan. The temperature and the applied voltage influence SC evaporation, which plays a critical role in its lifetime (Rajan & Rahman, 2014).

Choosing the correct values and ratings of supercapacitors is very important when replacing the batteries in the system. It is important to connect capacitance and energy in terms of watts per hour. Battery charging and discharging time are calculated based on the ampere-hour rating. An equation relating ampere-hour and capacitance is given below (Yang et al., 2022). Where Ah is ampere hour, F is Faraday, V_{min} and V_{max} are terminating voltage levels (Manikandan, 2015).

$$Ah = \{(V_{min} + V_{max})/2\}*\{F/3600\}$$

MOF-based materials, for example, carbon, metal oxides, and metal sulfides, inherit porous structures and large surface areas, providing adequate active sites (Aijaz et al., 2017; Dong et al., 2023; Liu et al., 2018).

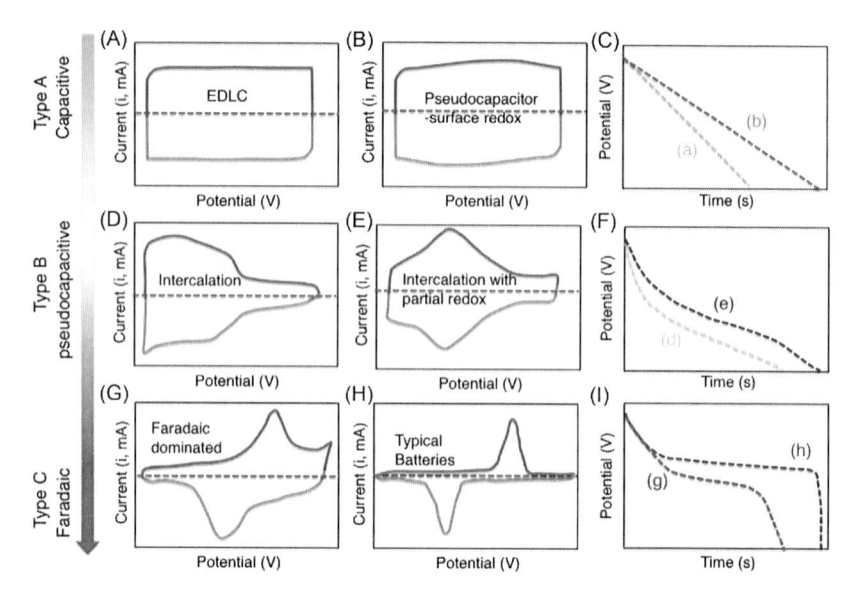

FIGURE 3.1 (A, B, D, E, G, and H) Cyclic voltammetry (CV) and (C, F, and I) Galvanostatic charge discharge (GCD) curves of different materials for energy-storage (Gogotsi & Penner, 2018).

Furthermore, due to their wide range of favorable electrochemical properties, they are suitable for various applications, including batteries, supercapacitors, photocatalysis, and electrocatalysis. Metal sulfides generated from MOFs have a lot of applications in electrocatalysis and energy storage because of their electrochemical characteristics (Abazari et al., 2019; Moradi et al., 2019; Oveisi et al., 2019; Radwan et al., 2021; Shi et al., 2004). However, MOFs can exhibit both electrochemical double-layer capacitance similar to carbon materials and pseudocapacitance similar to metal oxides/hydroxides due to their porous features, ordered crystalline structure, and variable-valence transition metal ions. Le Quoc Bao et al. found that combining reduced graphene oxide (rGO) with conducting polyaniline (PANI) and Zn-MOF has significantly improved performance. This is due to enhanced electronic conductivity, higher pseudocapacitance behavior, and the porous morphology of the created structures, enabling efficient ion transfer. The composites containing Zn-MOF appear to have better stability due to their crystalline structure, which is maintained during long charge/discharge processes (Quoc Bao et al., 2021).

This chapter provides an in-depth discussion of various synthesis methods, including conventional and advanced strategies. Finally, the chapter will detail the use of MOFs and MOF-based composite materials in supercapacitor applications.

Fundamental of metal-organic frameworks

Due to their unique functional and structural characteristics, MOFs are increasingly recognized as a significant class of porous materials. These frameworks are constructed from metal ions or metal ion clusters connected by organic linkers. Choosing primary building units (PBUs) is crucial in determining MOFs' final structure and properties. Additionally, various synthesis methods and parameters, such as temperature, reaction time, pressure, pH, and solvent, influence the characteristics of the resulting MOFs. Several synthetic techniques can create MOFs, including slow diffusion (Chen et al., 2005), hydrothermal/solvothermal methods (Shen et al., 2013), electrochemical synthesis (Van Assche et al., 2012), mechanochemical approaches (Pichon et al., 2006), microwave-assisted heating (Jhung et al., 2003), and ultrasound (Jung et al., 2010), as represented in Fig. 3.2. Each technique can yield distinct structures and properties of MOFs based on the specific synthesis conditions applied.

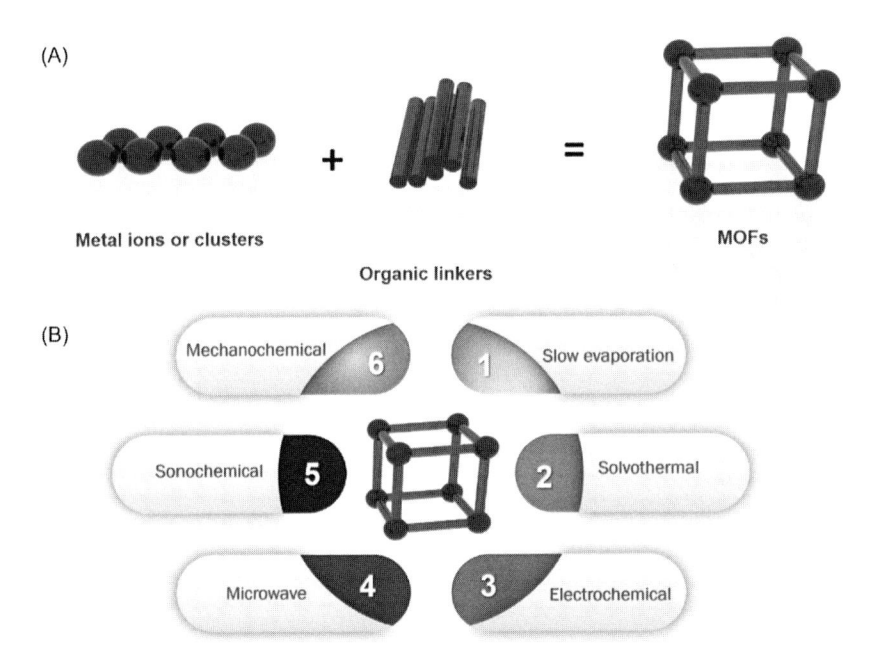

FIGURE 3.2 (A) Illustration of the fundamental components and architecture of MOFs. (B) Common techniques employed for synthesizing MOFs in their initial form (Radwan et al., 2021). *MOFs*, Metal-organic frameworks.

Conventional synthesis techniques for metal-organic frameworks

Diffusion method

The diffusion method involves gradually introducing different species to interact. One approach is solvent liquid diffusion, where two layers of solvents with different densities are formed: a precipitant solvent and a solvent containing the product, separated by a layer of solvent. Crystal growth occurs at the interface as the precipitant solvent slowly diffuses into the other layer. Another approach involves using physical barriers, such as vials of different sizes, to diffuse the reactants slowly. Gels can also be used as crystallization and diffusion media to slow diffusion and prevent bulk material precipitation. This method is particularly effective in obtaining single crystals suitable for X-ray diffraction analysis, especially when the products are not highly soluble (Abazari et al., 2019; Moradi et al., 2019; Oveisi et al., 2019).

Hydro/solvo thermal method

The hydrothermal/solvothermal methods use self-assembly from soluble precursors. Initially used for synthesizing zeolites, this technique has been adapted for MOF synthesis. In a sealed autoclave under autogenous pressure, temperatures range from 80°C to 260°C, influenced by the cooling rate at the end of the reaction. While this method often requires long reaction times, sometimes several days for solvothermal and hydrothermal techniques and several weeks for the diffusion method, it is highly effective for synthesizing MOFs (Shi et al., 2004).

Microwave method

Microwave-assisted synthesis generates small metal and oxide particles by heating the solution with microwaves, creating nanosized metal crystals for an hour or more. Although not commonly used for producing crystalline MOFs, this technique offers rapid synthesis and control over particle shape and size. The microwave method, which can create uniform seeding conditions similar to a single X-ray analysis, allows for faster synthesis cycles and precise control over crystal morphology (Morsali et al., 2015; Tompsett et al., 2006; Zhang et al., 2008).

Electrochemical method

The electrochemical synthesis produces MOF powders on an industrial scale, avoiding anions like nitrates from metal salts and allowing for lower reaction temperatures and rapid synthesis compared to solvothermal methods. This method limits bulk crystallization by producing metal ions in situ near the support surface, reducing the risk of crystal accumulation during membrane synthesis. The lower temperatures also reduce the risk of thermally induced cracking during cooling. The electrochemical method allows for fine-tuning synthesis parameters by adjusting voltage or applying specific signals, making it more versatile compared to solvothermal techniques (Mueller et al., 2006; Wu et al., 2008).

Mechanochemical method

Mechanochemical synthesis uses mechanical force to induce chemical reactions and physical phenomena. This method, which involves mechanically breaking intramolecular bonds, has a long history in synthetic chemistry and has been applied to multicomponent reactions, inorganic solid-state chemistry, polymer science, and organic synthesis. Mechanochemical synthesis is environmentally friendly, often avoiding using organic solvents and operating at ambient temperature. Reactions can yield products with tiny components and quantitative yields in short durations (10–16 minutes). This method can use metal oxides as starting materials, producing only water as a by-product and employing metal acetates or carbonates to obtain well-crystallized compounds. Adding small amounts of solvents, known as liquid-assisted grinding (LAG), can accelerate reactions and influence structure-directing properties. Ion- and liquid-assisted grinding (ILAG) has been shown to selectively produce pillared-layered MOFs, demonstrating the structural directing influence of ions and solvents (Beldon et al., 2010; Hoskins & Robson, 1989).

Sonochemistry method

Sonochemistry involves using high-energy ultrasound (20 kHz–10 MHz) to induce chemical reactions. Ultrasound, a cyclic mechanical vibration, can create cavitation and microjets that clean, activate, and corrode surfaces, dispersing agglomerations of smaller particles. Due to extreme conditions and shear solid forces, chemical reactions can occur in the cavity, interface, or bulk media. These conditions can generate radicals, break bonds, and form molecules excitedly. Ultrasonication enhances the dissolution of starting compounds. Sonochemistry is widely used for synthesizing organic and nanomaterials. In MOF science, sonochemical synthesis aims to be quick, environmentally friendly, energy-efficient, user-friendly, and applicable at ambient temperature. The rapid reactions facilitated by sonochemistry are suitable for scaling up MOF production (Bang & Suslick, 2010).

Various techniques have been developed for synthesizing MOFs, leveraging various available components and process parameters, and creating thousands of different MOF materials.

Advanced design strategies for enhanced performance

Pore window and pore size optimization

In MOFs, the pore window and size are dictated by the choice of organic and inorganic building components. MOF pore structures exhibit significant diversity in size and shape, including tubular pores, slits, cylinders, or spheres, forming in zero-dimensional (0D), one-dimensional (1D), two-dimensional (2D), or three-dimensional (3D) configurations. The diameters of these pores can span from angstroms (Å) to nanometers (nm), with internal surface areas ranging from a few hundred μ/g to as much as 7800 m²/g (Lin et al., 2020). Research has shown that precise control over pore size can significantly enhance the selectivity. The challenge lies in designing and synthesizing MOFs with finely tuned pore apertures. Various studies have explored the creation of MOFs with specific pore sizes and shapes that enhance properties, demonstrating the potential of these materials in practical applications (Chen et al., 2007; Pham & Space, 2020).

For instance, Li et al. highlighted the importance of pore shapes and sizes; they showed that MOFs could be optimized selectively for various applications by exploiting the size-exclusion effect of the pores (Li et al., 2013). Similarly, Qin and colleagues developed a Cd-MOF with pores that preferentially adsorb CO_2 over methane due to the size-exclusive impacts (Qin et al., 2014). These studies underscore the role of pore size and surface properties in achieving selective properties. Further advancements include the design of MOFs that undergo structural changes in response to external stimuli, such as hydration or pressure variations, altering their pore sizes and enhancing their properties. Flexible MOFs exhibiting such "breathing" phenomena can adjust their pore sizes, improving adsorption performance under different conditions (Aaron & Tsouris, 2005).

Postsynthetic modification in metal-organic frameworks

Postsynthetic modification (PsSM) is a technique used to enhance the properties of MOFs after their initial synthesis. PsSM is considered a flexible tool to achieve desired structural and functional modifications, which are often challenging through direct synthesis methods. This approach allows the introduction of functional groups, modification and exchange of organic linkers, and cation exchange within the MOF structure, significantly improving their conductivity and selectivity (Asghar, Iqbal, Noor, et al., 2020). One notable example of PsSM is the modification of IRMOF-3 with acetic anhydride, which incorporates amide groups throughout the MOF network. This modification enhances the attraction between target molecules and amine groups, increasing the adsorption capacity (Wang & Cohen, 2007). Similarly, amine-functionalized MOFs, such as MOF-177, have improved selectivity at varying temperatures compared to their nonfunctionalized counterparts (Gaikwad et al., 2021).

PsSM techniques have been employed to modify various MOFs, enhancing conductivity and selectivity. For instance, Mg-MOF-74 was modified with hydrazine, resulting in higher ion flow values without compromising pore size expansion (Choi et al., 2012). A solvent-assisted ligand incorporation (SALI) strategy was developed to introduce fluorinated chains into the MOF framework, enhancing attractive forces between the quadrupole moment and the C-F dipoles (Hu et al., 2016). PsSM has enhanced MOFs' thermal and water stability while increasing their conductivity. For example, UiO-66(Hf) MOFs modified with various functional groups showed significant improvements in conductivity. Another study modified Cr-MIL-101-SO$_3$H with tris(2-aminoethyl) amine (TAEA), resulting in higher adsorption capacities and selectivity (Fujiwara et al., 1995).

Presynthetic modification in metal-organic frameworks

Presynthetic modification (PrSM) involves the functionalization of organic linkers before the synthesis of MOFs. This approach aims to decorate the pores with functional groups that enhance CO_2 affinity without significantly increasing regeneration energies. Presynthetic modifications can substantially influence the overall structure and properties of MOFs (Asghar, Iqbal, Aftab, et al., 2020).

Functional groups such as hydroxyl, carboxyl, and amino groups are commonly used for grafting onto MOF ligands. Solvothermal techniques have shown potential for presynthetic modifications, producing high-performance MOFs with enhanced supercapacitive properties (Mihaylov et al., 2015). For example, incorporating cellulose with metal ions resulted in cellulose-modified MOFs with better performance than their pristine counterparts. Deng and coworkers demonstrated the incorporation of various functional groups into MOFs using presynthetic modification, producing novel structures with improved selectivity (Moghaddam et al., 2018). Another study involved the presynthetic functionalization of Zr-MOF, which enhanced energy-storing applications. The presynthetic modification offers a strategic approach to designing MOFs with enhanced gas adsorption properties by carefully selecting and incorporating functional groups into the organic linkers before synthesis.

Topology of metal-organic frameworks

The topology of MOFs, defined by the structural network formed by metal nodes and linkers, plays a crucial role in the selective capture and separation of ions. Designing MOFs with specific topologies, such as large cavities for gas storage and small, well-defined pore windows, can optimize their performance for targeted supercapacitive applications. Strategies for designing selective MOFs based on topology include choosing appropriate organic linkers and metal cluster secondary building units (SBUs) (Guillerm et al., 2014). For instance, Yu et al. synthesized a multicage MOF termed NUM-3, which exhibited good selective capture of ions compared to others due to its unique topology (Yu et al., 2017).

Similarly, developing MOF-74 isomers with different topologies has demonstrated the importance of pore structure in environmental applications. For example, UTSA-74, an isomer of Zn-MOF-74, showed enhanced CO_2 binding due to two open metal active sites with different coordination geometries (Luo et al., 2016). Furthermore, NbOFFIVE-1Ni, a 3D microporous framework, demonstrated effective pore sieving due to its specific topology, which includes microporous channels and open metal sites for metal ion binding (Lin et al., 2020).

Tailoring metal-organic frameworks for specific supercapacitor applications

In the following sections, we explore the design and control of MOF-derived electrocatalysts across different scales, including active sites, interfaces, pore structures, and morphologies. We emphasize the influence of the structures and compositions of MOF-based precursors and the conversion processes on the creation of MOF-derived materials with specific characteristics at various scales (Fig. 3.3).

Design of active sites

Experimental and theoretical investigations into electrochemical processes in well-defined single-crystalline or polycrystalline systems have advanced our knowledge of the relationships between structure, composition, and activity. These findings inform the design of active sites with high intrinsic activity. Key properties or descriptors, like the adsorption-free energy of reactive intermediates, effectively capture these relationships (De Luna et al., 2019). Utilizing descriptors is a powerful method for connecting atomiclevel and molecularlevel electrochemical processes to overall activity. This approach helps in designing and creating active sites with high intrinsic activity. By carefully selecting MOF precursors and controlling the conversion conditions, it is possible to fabricate active sites or phases such as metals, metal compounds, carbon, and atomically dispersed metal sites (ADMSs) with desirable descriptor values. Furthermore, the structural and compositional diversity, along with the tunability of MOF-derived materials, allows for precise adjustment of the electronic structures of active sites. This can be achieved by introducing additional metal elements, heteroatom doping, and reducing particle size (Yang et al., 2021).

Adding metal elements to form bimetallic alloys or metal compounds has significantly enhanced electrochemical performance compared to monometallic counterparts due to optimized electronic structures, increased electrical conductivity, and improved stability. Bimetallic MOFs have been used as precursors to fabricate these alloys or compounds, benefiting from the uniform distribution of metal elements in the MOF precursors. Secondary metal

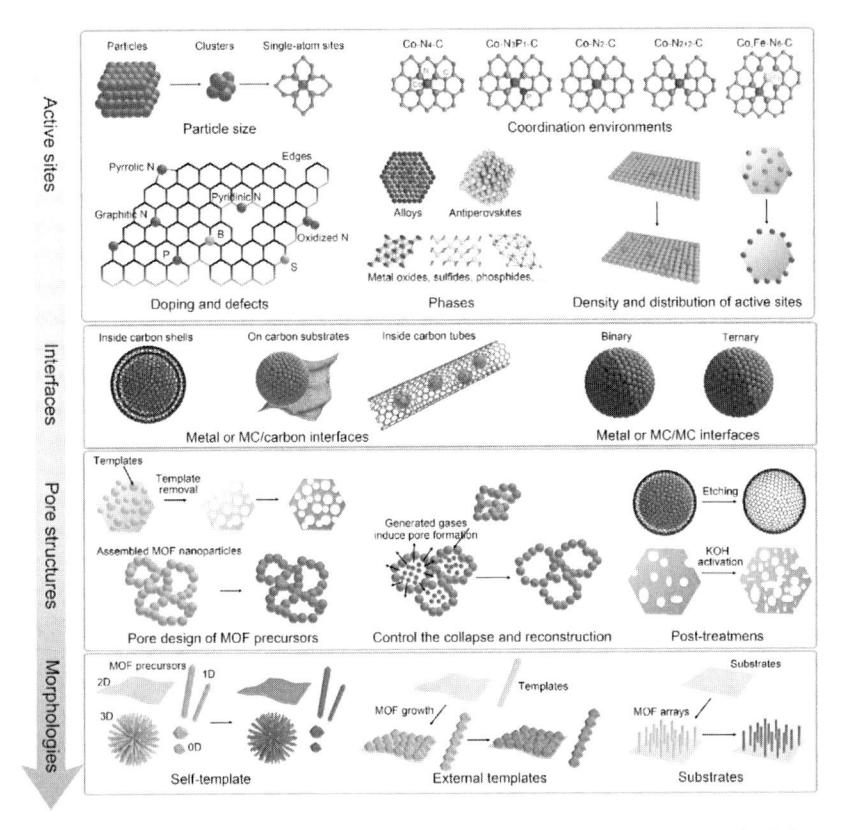

FIGURE 3.3 Schematic representations of the multiscale design of materials derived from metal-organic frameworks (Zhang et al., 2018).

elements can also be introduced as guests into MOF pores, as demonstrated by doping Ni, Mn, and Fe into CoP using ZIF-67 with Ni(acac)$_2$, Mn(acac)$_2$, and Fe(C$_5$H$_5$)$_2$ as precursors (Pan et al., 2019).

In MOF-derived materials, carbon components serve as active sites for electrocatalysis in addition to metal components. Introducing heteroatoms such as nitrogen, sulfur, boron, and phosphorus into the carbon matrix alters the electronic structure of neighboring carbon atoms, thereby creating numerous active sites for the adsorption and activation of reactants and intermediates (Liu, 2017). Organic ligands in MOFs facilitate the doping of carbon components with heteroatoms in MOF-derived materials. For instance, nitrogen-doped carbon with a high nitrogen content can be produced by pyrolysing ZIF-8, which is constructed using 2-methylimidazolate ligands. Additionally, introducing other guest molecules into MOF pores can provide sources for multiple heteroatom dopings, resulting in carbon doped with elements such as nitrogen, sulfur, and phosphorus (Li et al., 2014). Reducing the particle size of active phases is a widely used approach to enhance an electrocatalyst's electrochemical performance, as it increases the number of low-coordinated sites on the surface. For materials derived from MOFs, methods to reduce particle size include incorporating additional metal elements that can be easily removed through pyrolysis or chemical etching, optimizing pyrolytic processes, and using substrates such as graphene oxides (GO) and ZIF-8 to disperse MOF precursors and prevent aggregation during pyrolysis (Liang, 2017). These strategies can produce smaller particle sizes and abundant active sites.

Interface engineering

At a larger scale, interfaces between different components in an electrocatalyst play a crucial role. Constructing interfaces or heterostructures in hybrid electrocatalysts has proven efficient in achieving superior activity, selectivity, and stability due to synergistic effects between components (Zhao et al., 2018). These effects include charge redistribution, fast electron transport, stabilization of active components, and provision of distinct sites for adsorption and activation of reactive molecules. These effects work together to enhance the overall electrochemical performance of hybrid electrocatalysts.

Interfaces in MOF-derived materials can be finely tuned by carefully designing MOF precursors and controlling conversion conditions. The organic–inorganic hybrid nature of MOFs allows for the facile construction of metal/metal compound-carbon interfaces through pyrolysis in an inert atmosphere (Liang et al., 2019). MOF precursors containing transition metals can catalyze the formation of graphitic carbon at high temperatures, leading to metal@carbon structures with enhanced electrochemical activity and stability. External carbon sources such as graphene and carbon nanotubes (CNTs) have also been utilized to create metal compound-carbon interfaces. For example, the pyrolysis and phosphorylation of ZIF-67 with added guests like urea and boric acid produce CoP nanoparticles encapsulated by boron- and nitrogen-co-doped CNT heterostructures. Furthermore, interfaces between metal/metal compounds and metal compounds have been developed using precursors with multiple metal components in MOF-derived materials (Tabassum et al., 2017).

Pore structure

Pores in porous materials are categorized based on their size: micropores (<2 nm), mesopores (2–50 nm), and macropores (>50 nm). In electrocatalysts, a hierarchical porous structure is highly beneficial. Micropores accommodate active M–N–C sites, while mesopores and macropores transport O_2 and H^+ to these active sites. Additionally, they offer pathways for water drainage and diffusion of reactants (Chen et al., 2020; He et al., 2020).

Direct pyrolysis of MOF crystals results in hierarchical pore structures as the well-defined pore structures of MOFs undergo collapse and reconstruction during pyrolysis. Controlling these processes is challenging. Strategies to control the pore structures of MOF-derived materials include tuning MOF precursor pore structures, controlling MOF collapse and reconstruction during pyrolysis, and creating pores through posttreatment (He et al., 2020). For instance, hard-template methods using 3D-ordered polystyrene spheres to create mesopores/macropores in MOF precursors and assembling MOF nanocrystals can generate hierarchical porous carbon. Introducing components like triazole and dicyandiamide into MOF precursors can induce abundant pore formation during pyrolysis. Additionally, coating MOF nanocrystals with mesoporous silica layers can induce mesopore formation. Posttreatments, such as chemical etching and secondary thermal activation, can further tune the pore structure of MOF-derived materials (Amali et al., 2014).

Morphologies

Electrocatalyst morphology significantly influences mass transport to and from active sites. Controlled fabrication of electrocatalysts with 0D, 1D, 2D, and 3D morphologies has advanced in recent decades (Guan et al., 2019). Nanoparticles (0D) enable the study of structure-composition-performance relationships at the nanoscale but may face issues like considerable contact resistance and low chemical stability. Constructing 1D (e.g., nanowires, nanorods, nanotubes) or 2D (e.g., nanosheets, nanoflakes) morphologies overcomes some issues of 0D nanoparticles due to higher electron mobility and structural robustness. Recent research shows that assembling 0D, 1D, and 2D structures into hierarchical 3D morphologies (e.g., foams, arrays, and interconnected networks) improves electrocatalytic performance by providing interconnected conductive networks for electron transport and open channels for mass transport (Liu, Zhang, et al., 2020). These 3D morphologies host a high density of active sites on their open surfaces, ensuring accessibility to reactants and electrolytes and achieving high active site utilization efficiency. The hierarchical 3D morphology benefits from forming desirable three-phase interfaces (gas, liquid electrolyte, and solid electrocatalyst) for gas-involving reactions like the oxygen evolution reaction (OER) and oxygen reduction reaction (ORR).

Materials derived from MOFs exhibit various morphologies, such as 0D nanoparticles, 1D nanowires and nanorods, 2D nanosheets and nanoribbons, and 3D networks, foams, and arrays. These diverse structures are achieved using various MOF precursors and treatment techniques. The morphology of MOF-derived materials can be adjusted through self-templated or externally templated methods. In self-templated methods, MOFs are designed with specific morphologies, which are then maintained during pyrolysis (Liu, Zhang, et al., 2020). External templates like CNTs, graphene, and metal oxide arrays support MOFs and influence the nucleation and growth of MOF crystals, optimizing the electrochemical performance of their derivatives. The structural and compositional versatility of MOF precursors and various treatment methods allow for the multiscale design of MOF-derived materials. This approach optimizes active sites, interfaces, pore structures, and morphologies, enhancing the overall electrochemical performance of electrocatalysts for specific applications. Similar multiscale design strategies can be applied to covalent organic frameworks (COFs). COFs, used as precursors for carbon-based electrocatalysts, differ from MOFs because they do not contain metal sources, resulting in metal-free porous carbon with heteroatom doping after pyrolysis. However, integrating metal sources into COFs can yield COF-derived materials with improved electrocatalytic properties (Yu et al., 2018).

Metal-organic framework-based composite materials

Poor electrical conductivity has posed a significant challenge in improving the electrochemical performance of many MOFs. This issue can be addressed by integrating MOFs with other conductive substances such as carbon materials, conducting polymers, and metals. Carbon materials such as graphene, carbon fibers, carbon cloth, and CNTs are commonly incorporated into MOFs through solvothermal or hydrothermal techniques. This involves combining metal ions, ligands, and solvents with carbon materials, enabling the MOFs to grow directly on the carbon substrate under carefully controlled conditions (Feng et al., 2021). Similarly, comparable methods can integrate metals like nickel foam with MOFs. Electrodeposition is another technique frequently employed to grow MOFs on carbon materials and nickel foam surfaces uniformly (Kazemi et al., 2018).

Integrating MOFs with conducting polymers can be categorized into two approaches: growing MOFs on conducting polymer substrates or incorporating conducting polymers into MOFs. The latter involves chemical or electrochemical synthesis in a system containing a polymeric monomer, MOFs, an oxidant, and a solvent. New methods for creating metal oxide@MOF composites have also been developed, such as the in situ self-transformation route to fabricate MnO_x@MOF composites using MOF-Mn hexacyanoferrate hydrate nanocubes (Zhang et al., 2016). MOF-based ternary composites can be prepared through multistep processes, like the two-step hydrothermal method used to grow Ni-based MOFs on carbon cloth with Co_3O_4 (Zhang et al., 2018).

Rationale for combining metal-organic frameworks with other materials

Combining MOFs with other functional materials aims to leverage the strengths of each component to overcome the inherent limitations of MOFs. Carbon materials provide excellent electrical conductivity and structural stability, making them ideal partners for MOFs in composite materials. The integration process, such as solvothermal and hydrothermal reactions or electrodeposition, ensures uniform growth and optimal interface contact, enhancing overall performance.

Conducting polymers can be integrated with MOFs by growing MOFs on polymer substrates or embedding polymers within MOFs. This integration improves the composite's conductivity and mechanical properties. Chemical or electrochemical synthesis methods facilitate incorporating conducting polymers into MOFs, broadening the range of possible applications.

Moreover, novel techniques for preparing MOF-derived materials, including microwave-assisted heating, CO_2 laser engraving, and alkaline hydrolysis, offer more efficient and controllable synthesis routes. These methods enable the production of carbon, metal oxides, and other derivatives from MOFs while maintaining their porous structure (Van Lam et al., 2020). For example, microwave-assisted heating has been used to fabricate MOF-5-derived carbon materials, while CO_2 laser engraving transforms MOF-74(Ni) into Ni nanoparticles/porous carbon, offering rapid processing and uniform particle distribution. Hybridizing MOFs with conductive materials like carbon, polymers, and metals significantly enhances their electrochemical performance and structural stability. These composite materials are synthesized through various innovative techniques that ensure uniform growth and optimal functionality, paving the way for their application in advanced electrode materials and beyond (Van Lam et al., 2020).

Types of metal-organic framework composites for supercapacitors

These properties make MOFs promising candidates for various applications, including supercapacitors. In supercapacitors, MOFs are often integrated with other materials to form composites that enhance their electrochemical performance, stability, and suitability for specific applications. This section explores different types of MOF composites used in supercapacitor electrodes, focusing on their compositions and electrochemical properties.

Metal-organic frameworks/conductive polymer composite electrodes

Conductive polymers (CPs) are a class of organic polymers characterized by a matrix of conjugated bonds that enable electrical conductivity. Their function as electrodes in supercapacitors is primarily based on the Faraday storage mechanism. CPs exhibit diverse potential applications in electrical energy storage due to their low manufacturing cost, straightforward synthesis, good stability, reversible Faradaic redox capabilities, and high pseudocapacitance. Extensive research focuses on integrating CPs with MOFs to develop composite materials with superior electrochemical performance suitable for flexible symmetrical supercapacitors (FSSCs). Notably, poly(3,4-ethylenedioxythiophene) (PEDOT), polypyrrole (PPy), and polyaniline (PANI) are widely studied for their high conductivity and promising properties in these applications (Hao et al., 2016; Miao et al., 2016).

Metal-organic frameworks@polyaniline

MOFs combined with conductive polymers such as PANI offer synergistic effects that improve the overall electrochemical performance of supercapacitors. PANI, known for its high conductivity and reversible Faradaic redox capabilities, enhances the charge storage capacity of MOFs by providing additional pseudocapacitance. For instance, Wang et al. developed a flexible hybrid conductive electrode (PANI-ZIF-67-CC) by depositing PANI on cobalt-based MOF crystals (ZIF-67). This composite electrode showed significantly improved capacitance and stability, making it suitable for flexible supercapacitor applications (Liu et al., 2019; Wang et al., 2015).

Metal-organic frameworks@polypyrrole

PPy, another conductive polymer, has been integrated with MOFs to exploit its high conductivity, excellent elasticity, and flexibility. Yamauchi et al. demonstrated a strategy where PPy tubes were used as a conductive interconnecting framework to support the ZIF-67. This composite structure maintained the porous nature of MOFs while improving mechanical stability and electron/ion diffusion pathways. The resulting supercapacitor electrodes exhibited enhanced specific capacitance and mechanical robustness, crucial for wearable and flexible electronics (Xu et al., 2017; Yan et al., 2021).

Metal-organic frameworks@poly(3,4-ethylenedioxythiophene)

PEDOT, known for its high charge storage capacity and electroactive potential window, has been combined with MOFs to enhance supercapacitors' flexibility and electrochemical performance. Fu et al. developed flexible porous electrodes by electrochemically depositing PEDOT on MOF/GO composites. This approach not only improved the specific capacitance of the electrodes but also enhanced their mechanical flexibility and cycling stability. The synergistic effects between PEDOT and MOFs create multiple pathways for efficient charge storage and ion transport, making them ideal for high-performance supercapacitor applications (Fu et al., 2016; Zhao et al., 2017).

Metal-organic framework/carbon-based material composite electrodes

Carbon-based electrode materials are highly favored for energy storage devices due to their exceptional chemical and thermal stability, flexibility, high electrical conductivity, and lightweight nature. Typically employed in EDLC electrodes, these materials store charge electrostatically at the electrode-electrolyte interface (Chen et al., 2017). EDLCs offer high-rate capability and excellent cycle stability, maintaining 95%–100% of their initial capacitance even after 1000–10,000 cycles. However, the energy density of pure carbon-based electrodes is relatively low because of the physical charge storage limitations of EDLCs. To meet the increasing demand for energy storage devices with superior electrochemical performance, there is an urgent need to integrate other pseudocapacitive materials with carbon-based materials. Currently, the development of composite materials combining MOFs with carbon-based materials like activated carbon (AC), graphene, and CNTs has shown promise in enhancing charge transfer characteristics. This approach represents a viable strategy for improving the electrochemical properties of supercapacitor electrodes.

Metal-organic frameworks@carbon nanotubes composites
CNTs are often incorporated into MOFs to enhance their electrical conductivity and structural stability. MOF/CNT composites offer a high surface area and improved charge storage capacity, which is crucial for high-performance supercapacitors. For example, Zhang et al. synthesized MOF/CNT electrodes that exhibited enhanced specific capacitance and excellent cycling stability due to the synergistic effects between MOFs and CNTs (Zhang et al., 2017).

Metal-organic frameworks@graphene nanocomposites
With its high electrical conductivity and mechanical strength, graphene has been combined with MOFs to develop nanocomposite electrodes for supercapacitors. These electrodes offer enhanced charge storage capacity, rapid ion transport, and superior mechanical flexibility. Liu et al. reported on MOF@graphene nanocomposites that achieved high specific capacitance and long-term cycling stability, making them suitable for applications in energy storage devices requiring high power densities and long cycle life (El-Kady et al., 2016).

Metal-organic frameworks/metal oxide composite electrodes
MOFs combined with metal oxides leverage both materials' unique properties to enhance supercapacitor electrodes' performance. Metal oxides such as manganese oxide (MnO_x), cobalt oxide (Co_3O_4), and nickel oxide (NiO) provide additional redox-active sites and pseudocapacitance, thereby increasing the overall energy density of the supercapacitors. For instance, Zhang et al. developed MOF/MnO_x composite electrodes with improved capacitance retention and cycling stability, highlighting their potential for practical supercapacitor applications (Li et al., 2020).

Metal-organic framework/other material composites
Metal-organic framework/carbon-based metal composites
In addition to CNTs and graphene, MOFs have been combined with other carbon-based materials, such as carbon fibers, MXenes, and carbon black, to enhance supercapacitor electrodes' conductivity and mechanical stability. These composites offer tailored properties such as high specific surface area, excellent electrical conductivity, and improved charge storage

capacity, making them versatile for various energy storage applications (Zhao et al., 2019).

Metal-organic framework/inorganic nanoparticle composites

MOFs integrated with inorganic nanoparticles (e.g., metal nanoparticles, metal sulfides) provide additional functionalities such as catalytic activity and enhanced electrochemical performance. These composites exhibit improved specific capacitance, stability, and electrochemical activity, making them suitable for advanced supercapacitor applications requiring high efficiency and durability (Zhang et al., 2021).

MOF composites hold tremendous potential for advancing supercapacitor technology by addressing key challenges such as low electrical conductivity, poor mechanical stability, and limited charge storage capacity. By integrating MOFs with conductive polymers, carbon-based materials, metal oxides, and other nanomaterials, researchers can tailor the properties of supercapacitor electrodes to meet specific application requirements in energy storage, portable electronics, and flexible devices.

Applications of metal-organic frameworks and metal-organic framework composites in supercapacitors

Pure metal-organic framework electrodes for supercapacitors

Yaghi and Li were the first to propose the definition of MOFs in 1995 (Yaghi & Li, 1995). Since then, thousands of MOFs have been explored as electrode material for energy storage and other applications. Electrolyte ion diffusion into the material's pores is the most required need for an efficient supercapacitor. Hence, there is a need to design the pore size of the MOFs. The pristine MOFs, without any modification and functionalization in the morphology and structure, have been widely studied because of their abundant raw material and facile synthesis. Díaz et al. (2012) used the MOF-based material Co-MOF for supercapacitor application for the first time, which was synthesized by partial substitution of Zn by Co, which was then employed as electrode material and exhibited EDLC behavior, and their electrochemical performance accelerated the development of the pure MOFs as electrode material despite their low electrical conductivity. Zeolitic imidazolate frameworks (ZIFs), pristine MOFs favored as electrode material, and typical ZIF-8 and ZIF-67 have been employed as electrode material, in which ZIF-8 is composed of zinc metal ions and 2-methylimidazole, which provide high surface area, thermal stability, and facile synthesis, whereas ZIF-67 is composed of cobalt ions and 2-methylimidazole (Xiao et al., 2020). However, the ZIF MOFs deliver low electrical conductivity, which restricts their application in supercapacitors. Hence, their derivative has also been fabricated by pyrolysis or deposition, which can exhibit excellent electrochemical performance. Zhang and coworkers synthesized NiCo-MOF by the one-step solvothermal method, which showed a unique three-dimensional spherical structure and an excellent specific capacitance of 639.8 F/g at a current density of 1 A/g. Other work done by Xiao et al. (2020) synthesized $Ni_3(HITP)_2$ by electrophoretic deposition method and used it in the fabrication of a commercial device-based symmetrical supercapacitor device, which exhibited the areal specific capacitance of 15.69 mF/cm^2 with the potential window of 1 V and exhibited enhanced cyclic stability. Silver-based MOFs were also synthesized by Shalini and others in 2022 (Shalini et al., 2022), which served as efficient electrode materials with enhanced electrochemical performance. They exhibited conductive networks, surface activity, and high electrical conductivity. The assembled device demonstrated a specific capacitance of 289.4 C/g at a current density of 1 A/g, with an energy density of 48.69 Wh/kg and a power density of 608.73 W/kg, respectively. Similarly, Ma et al. (2021) synthesized a 3D cluster of copper MOF Cu-atrz-BDC, which showed a higher specific capacitance of 5525 F/g at a current density of 1 A/g and capacitive retention, maintaining 886 F/g even after 1000 cycles in a three-electrode assembly. The assembled asymmetrical supercapacitor devices, fabricated with Cu-atrz-BDC as the positive electrode and rGO as the negative electrode, displayed good electrochemical performance with enhanced cyclic stability, fast ion charge transportation, and lower internal resistance but showed poorer energy and power density compared to the silver-based MOF. Various other pristine MOFs, such as Zr-MOF, UiO-66, Ce-MOF, etc., have been synthesized, usually by solvothermal or hydrothermal methods, and have shown improved pseudocapacitive properties, along with high volumetric and areal capacitance. Wang and colleagues (Ma et al., 2021; Wang, 2020) synthesized 2D nanosheet MOF composites with exfoliated graphene sheets, which displayed remarkable cyclic stability and higher areal capacitance of 18.9 mF/cm^2. As MOFs are composed of secondary building units that play a crucial role in their structure, influencing varying porosity and pore size, their connectivity enhances both structural stability and porosity. Two primary factors influencing MOF stability are the strength of the coordination bonds and the rigidity of the framework structure, determined by the valence metal center with higher coordination numbers. Pristine MOFs have proven promising as electrode materials for supercapacitors. Therefore, further exploration of MOFs exhibiting good electrical conductivity, morphology, and higher porosity is encouraged.

Metal-organic framework composites as electrode materials

Conductive polymer composites

MOFs comprise organic linkers and metal ion clusters with tunable structure and adjustable porosity but exhibit lower energy density and electrical conductivity. Hence, to resolve the problems related to the pristine MOFs, various MOFs with enhanced conductivity are synthesized by their derivatives or their composites, providing higher electrical conductivity and structural stability (Zhong et al., 2020). There has been extensive literature reported on the composite of MOFs with carbon-based materials such as graphene, CNTs, GO, etc.

Conductive polymers area promising material with remarkable electrical conductivity, chemical stability, and high surface area, but due to their agglomeration, their theoretical capacity is very low, as agglomeration leads to the reduction of active sites (Jayaramulu et al., 2021; Shen et al., 2021). Conductive polymers are generally prepared by chemical oxidation methods, electrochemical polymerization or photo-induced polymerization methods. Previous studies were done on PANI, PPy, or PEDOT. Recent research done by Tian et al.(2020) synthesized the composite of MOF with PANI, MOF-808@PANI material, via a solvothermal method, which exhibited higher capacitive retention. The composite was synthesized by first activating the MOFs, followed by ultrasonication and vacuum drying. The activated MOF and PANI composite showed better electrochemical performance than the pure MOF with a capacitive retention of 99.7% even after 10,000 cycles. Cao et al. (2022) synthesized the UiO-66/PANI composite in which PANI molecular chains were deposited on the pores of UiO-66, which formed an interpenetrating network and wasused as electrode material, which showed the capacitance of 1015 F/g at the current density of 1 A/g. The synthesized material was used for the development of a symmetrical flexible device, which exhibited a specific capacitance of 647 F/g at a current density of 1 A/g, as shown in Fig. 3.4. Cyclic voltammetry curves of NiCo-MOF-74(1)/T-PPy-m samples at a scan rate of 20 mV/s are shown in Fig. 3.4(A), and Fig. 3.4(B) shows the galvanostatic charge discharge curve. Electrochemical impedance spectroscopy spectra areshown in Fig. 3.4(D) and galvanostatic charge–discharge curves at different current densities (Fig. 3.4(E)), cyclic stability curves (Fig. 3.4(F)). Similarly, Liu, Wang, et al. (2020) reported the composite of PANI with MIL-101, which exhibited a capacitance of 1197 F/g. Zheng (2022) synthesized 3D networked MOF and PPy composite by heating PPy, which served as the support for the growth of MOF on its surface. The three-electrode assembly was used for performing the electrochemical characterization, which showed a specific capacitance of 597.6 F/g and an areal capacitance of 2.33 F/cm^2. Shi et al. (2022) synthesized the highly conductive and stable Ni-MOF/PEDOT composite via in situ chemical vapor polymerization method and then used it as electrode material and delivered a specific capacitance of 1401 F/g with a capacitive retention of 80% even after 1000 cycles. Zhang et al. (2019) fabricated POAP/Zn MOF (POAP: poly orthoaminophenol), and due to their porous structure and electrochemically active centers, their electrochemical performance has been enhanced and showed the capacitance of 580 F/g at 1 A/g.

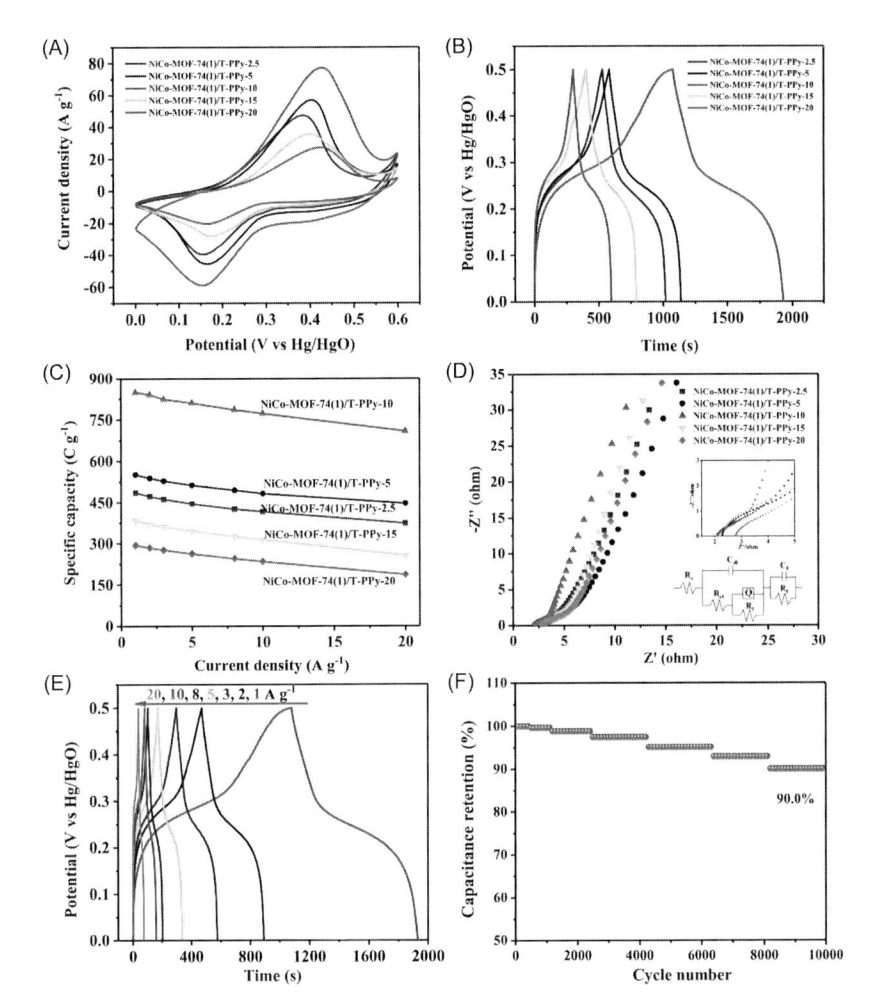

FIGURE 3.4 Electrochemical analysis of NiCo-MOF/T-PPy-m. (A) Cyclic voltammetry curves at scan rate of 20 mV/s; (B) Galvanostatic charge discharge curves comparison at 1 A/g; (C) Graph between specific capacity and current density; (D) Electrochemical impedance spectroscopy curve in the frequency range of 0.01 10^5 Hz (the inset is the fitted equivalent circuit); (E) Galvanostatic charge discharge curves, and (F) cyclic performances at 10 A/g of NiCo-MOF/T-PPy-10 (Cao et al., 2022).

Metal oxide and hydroxide composites

Transition metal oxides or hydroxides show high theoretical specific capacitance and also offer high pseudocapacitance, hence gaining the researchers' attention. In recent research conducted by Hussain et al., 2022 the nanopolyhedron-structured CeO_2@ZIF-8 composite synthesized by the wet chemical method was reported. When used for supercapacitor applications in a three-electrode assembly with 3 M KOH electrolyte, it showed a specific capacitance of 424.5 F/g. When employed in an asymmetrical supercapacitor device, it exhibited a capacitance of 89 F/g, with an energy density of 31.3 Wh/kg and a power density of 800 W/kg, respectively. Li et al.(2019) reported the composite of NiO/Ni-MOF nanostructures, which showed a capacitance of 1176.6 F/g and a specific capacity of 163.4 mAh/g. The synthesized material was then used to assemble the asymmetrical supercapacitor device with activated carbon and was able to achieve an energy density of 31.3 Wh/kg and a power density of 374.2 W/kg with a cyclic stability of 88.7% after 2000 cycles. The composite of MOFs with bimetallic metal oxide was also synthesized because bimetallic MOFs can provide more metal ion centers, which facilitate redox reactions. Shayeh and Salari (2020) fabricated the composite of bimetallic oxide with MOF, in which $MnCo_2O_4$ was first prepared via the solvothermal synthesis method and utilized the bimetallic MnCo-MOF as a precursor. The $MnCo_2O_4$/Ni-MOF composite was synthesized by growing the Ni-MOF on the $MnCo_2O_4$. The electrochemical study was performed, which showed the specific capacitance of the composite was 957.11 F/g at a current density of 1 A/g. The composite was then used as electrode material in an asymmetrical supercapacitor device and showed an energy density of 35.6 Wh/kg with a power density of 749.91 W/kg. These improved electrochemical performances are due to their unique nanoflower structure with small balls, which can offer higher surface area and provide abundant active sites.

Transition metal hydroxide has also been explored as a composite material with MOFs, and their synergistic effects exhibited enhanced electrochemical performance and electrical conductivity (Cai et al., 2018). Jiang and coworkers synthesized a composite of MOF with metal hydroxides, Fe-MOF@Ni(OH)$_2$ composite by in situ physical growth, and the synthesized composite showed enhanced electrochemical performance with high rate capability. The specific capacitance of 188 mAh/g was achieved at the current density of 1 A/g. Further, the asymmetrical supercapacitor device was also assembled, which delivered an energy density of 67.1 Wh/kg and capacitance retention of 100% after 4000 cycles. Xia et al. (2021) synthesized a multicomponent structure NiCoFeLDH@ZIF-67 composite, which delivers a specific capacitance of 1202.08 F/g at the current density of 0.5 A/g. Zheng, Zhou, et al. (2022) synthesized vertically oriented Ni-MOF@Co(OH)$_2$ flakes and delivered the specific capacitance of 1448 F/g at the current density of 1448 F/g. As the 2D nanosheets facilitate the transportation of electrolyte ions during the intercalation or de-intercalation process, a higher energy density of 45.7 Wh/kg was achieved when used in a hybrid supercapacitor device. MOFs and metal oxide or metal hydroxide composites have shown various advantages, but still, the issue related to the volume expansion, which deforms their structure, is a major concern related to such materials, and hence, there is a need to solve the issues by controlling the reaction condition.

Metal oxides or hydroxides exhibit higher surface area, yet they have a lower number of active sites, which reduces their specific capacity (Iqbal et al., 2020). Various research has been performed on metal oxide-based electrodes, but due to poor electrical conductivity and cyclic stability, their commercial applicability is obstructed. Phosphite-based materials have also been explored owing to their electrical conductivity and fast ion transportation, and they exhibit both metallic and semiconductor behavior with cyclic stability (Chhetri et al., 2022). Alam and Iqbal (2021) showed that cobalt phosphate nanoflakes deposited on Ni foam exhibited battery-like behavior with a specific capacitance of 1990 F/g.

Li et al. (2022) fabricated 2D–3D NiZnCoP electrodes by directly coating them on carbon cloth, providing a large number of active sites. These electrodes exhibited a specific capacitance of 2816 F/g at a current density of 1 A/g, attributed to the unique interfacial contacts and fast diffusion of electrolyte ions. The asymmetrical supercapacitor device assembled delivered an energy density of 62.5 Wh/kg with a capacitive retention of 89.5% after the 10,000 cycles. Xu (2021) reported the interconnected nanoparticles of Ni_2P dispersed on the amorphous carbon complex and used as electrode material, which exhibited a specific capacity of 3631 mC/cm^2, and the hybrid supercapacitor delivered the energy density of 44 Wh/kg at the power density of 800 W/kg. Similarly, Zhao et al. (2021) synthesized high-energy-density electrode material CoP/Mo-NiCoP nanoplates with a specific capacity of 892.6 C/g at 1 A/g with a rate capability of 78.9% at a current density of 20 F/g, as shown in Fig. 3.5. Fig. 3.5(A) shows the schematic of the fabrication of the supercapacitor device, while Fig. 3.5(B–H) displays the electrochemical performance of the assembled device. Cyclic voltammetry and galvanostatic charge–discharge curves are also illustrated, and the Ragone plot is shown in Fig. 3.5(I).

Carbon based composites

CNTs, 1-D material, are widely used as electrode material due to their high electrical conductivity, chemical stability, and higher surface area, and their unique hollow structure produces short pathways for the diffusion of ions and faster reaction. MOF@CNT composite demonstrates high specific surface area and good electrical conductivity (Nishihara et al., 2009).

Tan et al. (2017) synthesized MOF and CNT composite. When employed in supercapacitor applications, the material demonstrated a specific capacitance of 75.1 F/g at a current density of 1 A/g, attributed to its suitable porosity and electrical conductivity. You et al. (2016) nucleated prussian blue (PB) cubic nanoparticles on CNTs to form composite material (PB/CNT) by using the hydrothermal method, and the synthesized composite delivered improved electrical conductivity. Wang et al. (2019) synthesized conductive Zr-MOF and CNT nanocomposites decorated by Mn, which exhibit high electrical

120 Applications of Metal-Organic Framework Composites

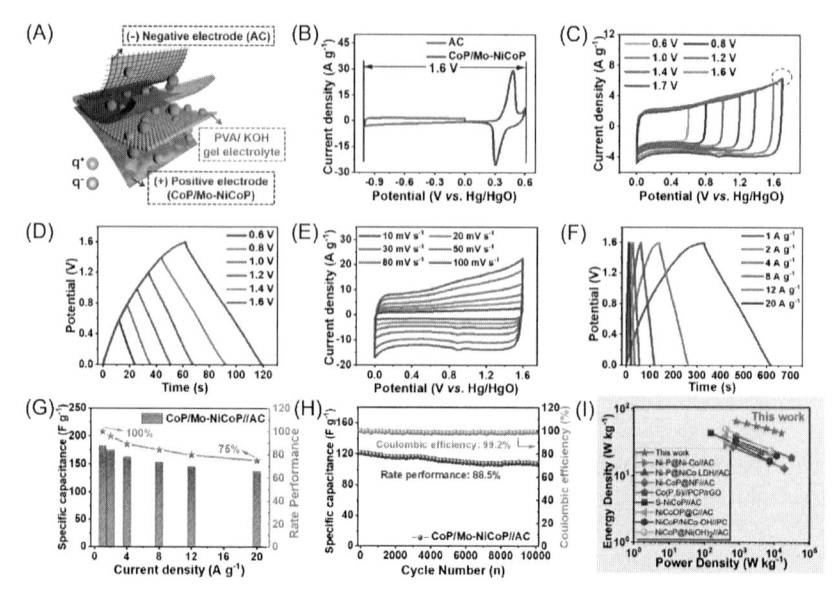

FIGURE 3.5 Electrochemical analysis of NiCo-MOF/T-PPy-m. (A) Cyclic voltammetry curves at scan rate of 20mV/s; (B) Galvanostatic charge discharge curves comparison at 1A/g; (C) Graph between specific capacity and current density; (D) Electrochemical impedance spectroscopy curve in the frequency range of 0.01105 Hz (the inset is the fitted equivalent circuit); (E) Galvanostatic charge discharge curves, and (F) cyclic performances at 10A/g of NiCo-MOF/T-PPy-10. (G) Specific capacity variations with the current density. (H) Cycling performance and corresponding Coulombic efficiency at 4 A/g for 10,000 cycles. (I) Ragone plots of the present and previously reported asymmetric SCs (Zhao et al., 2021).

conductivity because of CNT bridges interconnecting multiple MOF nanocrystals and enhanced redox active sites because of Mn sites supported by the Zr-MOF. Tang et al. (2022) synthesized Ni-HCF@CNF with high density, uniform distribution, and good stability and deposited it on carbon nanofiber support. The unique structure of composite Ni-HCF@CNF, the electrode showed a high specific capacitance of 507 F/g at a current density of 1 A/g and a capacitive retention of 67% at a current density of 10 A/g. Han et al. (2022) synthesized facile and efficient cathodic electrodeposition of Co-MOF, which was coated on flexible carbon fiber cloth and displayed an areal capacitance of 1784 mF/cm^2 at the current density of 1 mA/cm^2, outstanding cycling stability, and rate capacity.

Xiao et al. (2019) prepared 2D Ni-MOF composites by hydrothermal process and LEG (liquid-exfoliation graphene)-based composites, which are more electrically conductive. Zhong et al. (2020) synthesized a composite of Ni-MOF with rGO (Ni-MOF/rGO) via simple solvothermal and calcination methods. The optimized Ni-MOF/rGO composite exhibited a specific capacity of 954 F/g at a current density of 1 A/g and maintained 720 F/g at 5 A/g. In the

TABLE 3.1 Comparison of different metal-organic framework-based composites.

Electrode material	Electrolyte	Specific capacitance	Energy density (Wh/kg)	Power density (W/kg)	Cyclic stability	Reference
ZIF-67@PANI	1 M KOH	512 F/g	71.1	504.72	92.3% (9000 cycles)	Liu et al. (2021)
Ni-MOF@PEDOT	–	1401 F/g	40.6	450	80.6% (1000 cycles)	Shi et al. (2022)
Graphene@ZIF-67/PANI-NT	1 M Na_2SO_4	20,416.70 mF/g	1.814	134	75% (3000 cycles)	Ramandi and Entezari (2022)
CeO_2@ZIF-8	–	132 F/g	18.3	1000	90% (5000 cycles)	Rabani et al. (2021)
NiO@Ni-MOF	–	163.4 mA/g	31.3	374.2	88.7% (2000 cycles)	Wang et al. (2021)
Fe-MOF@Ni(OH)$_2$	–	188.75 F/g	67.1	800	100% (4000 cycles)	Jiang et al. (2021)
Ni-MOF@Co(OH)$_2$	–	1448 F/g	45.7	22,400	87.3% (8000 cycles)	Shi et al. (2021)
Cu-MOF/rGO	1 M Na2SO4	867.09 F/g	30.56	0.6	131.65% (5000 cycle)	Krishnan et al. (2022)
Ni-MOF/CNT	6 M KOH	1765 F/g	36.6	480	–	Wen et al. (2015)
Ni-MOF/PANI/NF	2 M KOH	3626.4 mF/cm^2	41.2	375	79.1% (10000) cycles	Liu, Wang, et al. (2020)
PPNF@Co–Ni MOF	3 M KOH	1096.2 F/g	93.6	1600	85.5% (10,000) cycle	Tian et al. (2020)
Cu-CAT-NWAs/PPy	3 M KCl	485 mF/cm^2	468	1.1	87% (5000 cycle)	Hou et al., 2020
NiCo-MOF@PNTs	2 M KOH	1109 F/g	41.2	375	79.1% (10,000 cycle)	Liu, Wang, et al.(2020)

case of graphene nanoribbons (GNR) or GO, Wang et al. (2020) fabricated UiO-66-GNR and UiO-66-GO nanocomposites, where redox-active sites were modified within the MOF structure, enhancing their pseudocapacitive behavior. Krishnan et al. (2022) anchored N-rich Cu-MOF by ultrasonic treatment, with channels for ion transport and high surface area on rGO to obtain Cu-MOF/rGO composites, and the synthesized composite demonstrated the capacity retention of 131.65% after 50,000 cycles.

Shen et al. (2021) synthesized Ce-MOF-808-CNT nanocomposites via a solvothermal process by adjusting the ratios between Ce-MOF and CNT, and these nanocrystals provide redox active sites, which significantly improved the capacitance of the material. In addition, there are some other composites of MOF and carbon that have been reported by researchers and show good electrochemical performance on supercapacitors (Shen et al., 2021; Wang et al., 2022). Table 3.1 summarizes the performance of various composites of MOFs in supercapacitor applications.

Conclusion

MOFs and their composites have been used for different energy storage applications like batteries (Li-ion, Na-ion, K-ion, Zn-ion) and supercapacitors. MOFs have a high surface area, pore size, and changeable pore structure. MOFs can have different organic linkers and metal nodes, and we can use them according to the requirements of our applications. All these properties help them to be best suited for different energy harvesting applications. Different synthesis strategies for MOFs and MOF composites have been discussed in this paper. Recent progress of these materials in various applications has also been reviewed. These applications include supercapacitors, Li-ion batteries, Li-S batteries, and Li-O_2 batteries. MOFs can be used as both the anode and cathode materials for Li-ion batteries. When MOFs are used as anode materials, their electrochemical performance is affected by morphologies, particle sizes, and solvent molecules. As cathode materials, MOFs with redox-active sites in both organic linkers and metal nodes are preferred because they can store more Li ions per the MOF formula.

Significant progress has been made in the research of energy storage devices utilizing MOF/functional materials composites. Particularly in the integration of different functional materials and MOFs, such as MOF/polymer composites and MOF/MXene composites for the advancements in batteries and supercapacitors. The complementary effects of MOFs and functional materials lead to much better electrochemical energy storage (EES) devices and electrochemical characteristics than conventional EES devices. It also has some practical challenges. As MOFs/MOF composites have the potential to become active electrode materials due to their manageable chemical composition and porous structure, the cycle stability and capacity of MOFs/MOF composite cathodes are not optimal. MOFs can change into an amorphous phase or break down irreversibly when charging and discharging. In addition, MOFs might disintegrate and their framework collapse in aqueous, acidic, and alkali environments. MOFs, due to their poor conductivity, are limited in their applications and development. Cost is a major issue when using MOFs/MOF composite electrodes. The use of expensive ligands and complex cationic technologies has posed numerous challenges for large-scale production.

In conclusion, MOF composites play a crucial role in advancing the field of EES by integrating functional materials that enhance the electrochemical properties of pure MOFs. Despite existing challenges, research on MOF composites has progressed in recent years with the development of novel functional materials. We believe this chapter aids in understanding the application of MOF composites as promising materials in energy storage and serves as a valuable reference for future studies

References

Aaron, D., & Tsouris, C. (2005). Separation of CO_2 from flue gas: A review. *Separation Science and Technology, 40*(1–3), 321–348. https://doi.org/10.1081/SS-200042244.

Abazari, R., Mahjoub, A. R., & Shariati, J. (2019). Synthesis of a nanostructured pillar MOF with high adsorption capacity towards antibiotics pollutants from aqueous solution. *Journal of Hazardous Materials, 366*, 439–451. https://doi.org/10.1016/j.jhazmat.2018.12.030, www.elsevier.com/locate/jhazmat.

Aijaz, A., Masa, J., Rösler, C., Xia, W., Weide, P., Fischer, R. A., Schuhmann, W., & Muhler, M. (2017). Metal–organic framework derived carbon nanotube grafted cobalt/carbon polyhedra grown on nickel foam: An efficient 3D electrode for full water splitting. *ChemElectroChem, 4*(1), 188–193. https://doi.org/10.1002/celc.201600452.

Alam, S., & Iqbal, M. Z. (2021). Nickel-manganese phosphate: An efficient battery-grade electrode for supercapattery devices. *Ceramics International, 47*(8), 11220–11230. https://doi.org/10.1016/j.ceramint.2020.12.247, https://www.journals.elsevier.com/ceramics-international.

Amali, A. J., Sun, J. K., & Xu, Q. (2014). From assembled metal-organic framework nanoparticles to hierarchically porous carbon for electrochemical energy storage. *Chemical Communications, 50*(13), 1519–1522. https://doi.org/10.1039/c3cc48112c.

Antwi-Baah, R., & Liu, H. (2018). Recent hydrophobic metal-organic frameworks and their applications. *Materials, 11*(11), 2250.

Arnold, M., Kortunov, P., Jones, D. J., Nedellec, Y., Kärger, J., & Caro, J. (2007). Oriented crystallisation on supports and anisotropic mass transport of the metal-organic framework manganese formate. *European Journal of Inorganic Chemistry, 2007*(1), 60–64. https://doi.org/10.1002/ejic.200600698.

Asghar, A., Iqbal, N., Aftab, L., Noor, T., Kariuki, B. M., Kidwell, L., & Easun, T. L. (2020). Ethylenediamine loading into a manganese-based metal–organic framework enhances water stability and carbon dioxide uptake of the framework. *Royal Society Open Science, 7*(3), 191934. https://doi.org/10.1098/rsos.191934.

Asghar, A., Iqbal, N., Noor, T., & Khan, J. (2020). Ethylendiamine (EDA) loading on MOF-5 for enhanced carbon dioxide capture applications. *IOP Conference Series: Earth and Environmental Science, 471*(1), 012009. https://doi.org/10.1088/1755-1315/471/1/012009.

Van Assche, T. R. C., Desmet, G., Ameloot, R., De Vos, D. E., Terryn, H., & Denayer, J. F. M. (2012). Electrochemical synthesis of thin HKUST-1 layers on copper mesh. *Microporous and Mesoporous Materials, 158*, 209–213. https://doi.org/10.1016/j.micromeso.2012.03.029.

Azcárate, C., Mallor, F., & Mateo, P. (2017). Tactical and operational management of wind energy systems with storage using a probabilistic forecast of the energy resource. *Renewable Energy, 102*, 445–456. https://doi.org/10.1016/j.renene.2016.10.064, http://www.journals.elsevier.com/renewable-and-sustainable-energy-reviews/.

Bang, J. H., & Suslick, K. S. (2010). Applications of ultrasound to the synthesis of nanostructured materials. *Advanced Materials, 22*(10), 1039–1059. https://doi.org/10.1002/adma.200904093, http://www3.interscience.wiley.com/cgi-bin/fulltext/123269284/PDFSTART.

Beldon, P. J., Fábián, L., Stein, R. S., Thirumurugan, A., Cheetham, A. K., & Friščić, T. (2010). Rapid room-temperature synthesis of zeolitic imidazolate frameworks by using mechanochemistry. *Angewandte Chemie – International Edition, 49*(50), 9640–9643. https://doi.org/10.1002/anie.201005547.

Cai, C., Zou, Y., Xiang, C., Chu, H., Qiu, S., Sui, Q., Xu, F., Sun, L., & Shah, A. (2018). Broccoli-like porous carbon nitride from ZIF-8 and melamine for high performance supercapacitors. *Applied Surface Science, 440*, 47–54. https://doi.org/10.1016/j.apsusc.2017.12.242.

Cao, Y., Wu, N., Yang, F., Yang, M., Zhang, T., Guo, H., & Yang, Wu (2022). Interpenetrating network structures assembled by "string of candied haws"-like PPY nanotube-interweaved NiCo-MOF-74 polyhedrons for high-performance supercapacitors. *Colloids and Surfaces A: Physicochemical and Engineering Aspects, 646*, 128954. https://doi.org/10.1016/j.colsurfa.2022.128954.

Chae, H. K., Siberio-Pérez, D. Y., Kim, J., Go, Y. B., Eddaoudi, M., Matzger, A. J., O'Keeffe, M., & Yaghi, O. M. (2004). A route to high surface area, porosity and inclusion of large molecules in crystals. *Nature, 427*(6974), 523–527. https://doi.org/10.1038/nature02311.

Chen, B., Ma, S., Zapata, F., Fronczek, F. R., Lobkovsky, E. B., & Zhou, H.-C. (2007). Rationally designed micropores within a metal−organic framework for selective sorption of gas molecules. *Inorganic Chemistry, 46*(4), 1233–1236. https://doi.org/10.1021/ic0616434.

Chen, H., Liang, X., Liu, Y., Ai, X., Asefa, T., & Zou, X. (2020). Active site engineering in porous electrocatalysts. *Advanced Materials, 32*(44), 2002435. https://doi.org/10.1002/adma.202002435.

Chen, X., Paul, R., & Dai, L. (2017). Carbon-based supercapacitors for efficient energy storage. *National Science Review, 4*(3), 453–489. https://doi.org/10.1093/nsr/nwx009, http://nsr.oxfordjournals.org/.

Chen, X. Y., Zhao, B., Shi, W., Xia, J., Cheng, P., Liao, D. Z., Yan, S. P., & Jiang, Z. H. (2005). Microporous metal-organic frameworks built on a Ln3 cluster as a six-connecting node. *Chemistry of Materials, 17*(11), 2866–2874. https://doi.org/10.1021/cm050526o.

Chhetri, K., Kim, T., Acharya, D., Muthurasu, A., Dahal, B., Bhattarai, R. M., Lohani, P. C., Pathak, I., Ji, S., Ko, T. H., & Kim, H. Y. (2022). Hollow carbon nanofibers with inside-outside decoration of bi-metallic MOF derived Ni-Fe phosphides as electrode materials for asymmetric supercapacitors. *Chemical Engineering Journal, 450*, 138363. https://doi.org/10.1016/j.cej.2022.138363.

Choi, S., Watanabe, T., Bae, T.-H., Sholl, D. S., & Jones, C. W. (2012). Modification of the Mg/DOBDC MOF with amines to enhance CO_2 adsorption from ultradilute gases. *The Journal of Physical Chemistry Letters, 3*(9), 1136–1141. https://doi.org/10.1021/jz300328j.

Dang, S., Zhu, Q.-L., & Xu, Q. (2017). Nanomaterials derived from metal–organic frameworks. *Nature Reviews Materials, 3*(1), 1–14.

Dong, A., Lin, Y., Guo, Y., Chen, D., Wang, X., Ge, Y., Li, Q., & Qian, J. (2023). Immobilization of iron phthalocyanine on MOF-derived N-doped carbon for promoting oxygen reduction in zinc-air battery. *Journal of Colloid and Interface Science, 650*, 2056–2064. https://doi.org/10.1016/j.jcis.2023.06.043.

De Luna, P., Hahn, C., Higgins, D., Jaffer, S. A., Jaramillo, T. F., & Sargent, E. H. (2019). What would it take for renewably powered electrosynthesis to displace petrochemical processes? *Science (New York, N.Y.), 364*(6438), eaav3506.

Díaz, R., Orcajo, M. G., Botas, J. A., Calleja, G., & Palma, J. (2012). Co8-MOF-5 as electrode for supercapacitors. *Materials Letters, 68*, 126–128. https://doi.org/10.1016/j.matlet.2011.10.046.

El-Kady, M. F., Shao, Y., & Kaner, R. B. (2016). Graphene for batteries, supercapacitors and beyond. *Nature Reviews Materials, 1*(7), 16033.

Feng, J., Liu, L., & Meng, Q. (2021). Enhanced electrochemical and capacitive deionization performance of metal organic framework/holey graphene composite electrodes. *Journal of Colloid and Interface Science, 582*, 447–458. https://doi.org/10.1016/j.jcis.2020.08.091, http://www.elsevier.com/inca/publications/store/6/2/2/8/6/1/index.htt.

Fu, D., Li, H., Zhang, X.-M., Han, G., Zhou, H., & Chang, Y. (2016). Flexible solid-state supercapacitor fabricated by metal-organic framework/graphene oxide hybrid interconnected with PEDOT. *Materials Chemistry and Physics, 179*, 166–173. https://doi.org/10.1016/j.matchemphys.2016.05.024.

Fujiwara, M., Ando, H., Tanaka, M., & Souma, Y. (1995). Hydrogenation of carbon dioxide over CuZn-chromate/zeolite composite catalyst: The effects of reaction behavior of alkenes on hydrocarbon synthesis. *Applied Catalysis A, General, 130*(1), 105–116. https://doi.org/10.1016/0926-860X(95)00108-5.

Furukawa, H., Ko, N., Go, Y. B., Aratani, N., Choi, S. B., Choi, E., Yazaydin, A.Ö., Snurr, R. Q., O'Keeffe, M., Kim, J., & Yaghi, O. M. (2010). Ultrahigh porosity in metal-organic frameworks. *Science (New York, N.Y.), 329*(5990), 424–428. https://doi.org/10.1126/science.1192160.

Furukawa, H., Cordova, K. E., O'Keeffe, M., & Yaghi, O. M. (2013). The chemistry and applications of metal-organic frameworks. *Science (New York, N.Y.), 341*(6149). https://doi.org/10.1126/science.1230444.

Gaikwad, S., Kim, Y., Gaikwad, R., & Han, S. (2021). Enhanced CO_2 capture capacity of amine-functionalized MOF-177 metal organic framework. *Journal of Environmental Chemical Engineering, 9*(4), 105523. https://doi.org/10.1016/j.jece.2021.105523.

Gogotsi, Y., & Penner, R. M. (2018). Energy storage in nanomaterials – Capacitive, pseudocapacitive, or battery-like? *ACS Nano, 12*(3), 2081–2083. https://doi.org/10.1021/acsnano.8b01914, http://pubs.acs.org/journal/ancac3.

Guan, A., Yang, C., Quan, Y., Shen, H., Cao, N., Li, T., Ji, Y., & Zheng, G. (2019). One-dimensional nanomaterial electrocatalysts for CO_2 fixation. *Chemistry - An Asian Journal, 14*(22), 3969–3980. https://doi.org/10.1002/asia.201900819, http://onlinelibrary.wiley.com/journal/10.1002/(ISSN)1861-471X.

Guillerm, V., Kim, D., Eubank, J. F., Luebke, R., Liu, X., Adil, K., Lah, M. S., & Eddaoudi, M. (2014). A supermolecular building approach for the design and construction of metal-organic frameworks. *Chemical Society Reviews, 43*(16), 6141–6172. https://doi.org/10.1039/c4cs00135d, http://pubs.rsc.org/en/journals/journal/cs.

Han, Y., Cui, J., Yu, Y., Chao, Y., Li, D., Wang, C., & Wallace, G. G. (2022). Efficient metal-oriented electrodeposition of a Co-based metal-organic framework with superior capacitive performance. *ChemSusChem, 15*(14), 2200644. https://doi.org/10.1002/cssc.202200644, http://onlinelibrary.wiley.com/journal/10.1002/(ISSN)1864-564X.

Hao, C., Yang, B., Wen, F., Xiang, J., Li, L., Wang, W., Zeng, Z., Xu, B., Zhao, Z., Liu, Z., & Tian, Y. (2016). Flexible all-solid-state supercapacitors based on liquid-exfoliated black-phosphorus nanoflakes. *Advanced Materials, 28*(16), 3194–3201. https://doi.org/10.1002/adma.201505730.

He, Y., Liu, S., Priest, C., Shi, Q., & Wu, G. (2020). Atomically dispersed metal-nitrogen-carbon catalysts for fuel cells: Advances in catalyst design, electrode performance, and durability improvement. *Chemical Society Reviews, 49*(11), 3484–3524. https://doi.org/10.1039/c9cs00903e, http://pubs.rsc.org/en/journals/journal/cs.

Hermes, S., Schröter, M. K., Schmid, R., Khodeir, L., Muhler, M., Tissler, A., Fischer, R. W., & Fischer, R. A. (2005). Metal@MOF: Loading of highly porous coordination polymers host lattices by metal organic chemical vapor deposition. *Angewandte Chemie International Edition, 44*(38), 6237–6241. https://doi.org/10.1002/anie.200462515.

Holze, R. (2009). *Supercapacitors–Materials, systems, and applications*. Springer.

Hoskins, B. F., & Robson, R. (1989). Infinite polymeric frameworks consisting of three dimensionally linked rod-like segments. *Journal of the American Chemical Society, 111*(15), 5962–5964. https://doi.org/10.1021/ja00197a079.

Hou, R., Miao, M., Wang, Q., Yue, T., Liu, H., Park, S., Qi, K., & Xia, B. Y. (2020). Integrated conductive hybrid architecture of metal–organic framework nanowire array on polypyrrole membrane for all-solid-state flexible supercapacitors. *Advanced Energy Materials, 10*(1), 1901892.

Hu, Z., Nalaparaju, A., Peng, Y., Jiang, J., & Zhao, D. (2016). Modulated hydrothermal synthesis of UiO-66(Hf)-type metal-organic frameworks for optimal carbon dioxide separation. *Inorganic Chemistry, 55*(3), 1134–1141. https://doi.org/10.1021/acs.inorgchem.5b02312, http://pubs.acs.org/journal/inocaj.

Hussain, I., Iqbal, S., Hussain, T., Cheung, W. L., Khan, S. A., Zhou, J., Ahmad, M., Khan, S. A., Lamiel, C., Imran, M., AlFantazi, A., & Zhang, K. (2022). Zn–Co-MOF on solution-free CuO nanowires for flexible hybrid energy storage devices. *Materials Today Physics, 23*, 100655.

Iqbal, J., Numan, A., Omaish Ansari, M., Jagadish, P. R., Jafer, R., Bashir, S., Sharifah, M., Ramesh, K., & Ramesh, S. (2020). Facile synthesis of ternary nanocomposite of polypyrrole incorporated with cobalt oxide and silver nanoparticles for high performance supercapattery. *Electrochimica Acta, 348*, 136313. https://doi.org/10.1016/j.electacta.2020.136313.

Jayaramulu, K., Horn, M., Schneemann, A., Saini, H., Bakandritsos, A., Ranc, V., Petr, M., Stavila, V., Narayana, C., Scheibe, B., Kment, Š., Otyepka, M., Motta, N., Dubal, D., Zbořil, R., & Fischer, R. A. (2021). Covalent graphene-MOF hybrids for high-performance asymmetric supercapacitors. *Advanced Materials, 33*(4), 2004560. https://doi.org/10.1002/adma.202004560, http://onlinelibrary.wiley.com/journal/10.1002/(ISSN)1521-4095.

Jhung, S. H., Chang, J. S., Hwang, J. S., & Park, S. E. (2003). Selective formation of SAPO-5 and SAPO-34 molecular sieves with microwave irradiation and hydrothermal heating. *Microporous and Mesoporous Materials, 64*(1–3), 33–39. https://doi.org/10.1016/S1387-1811(03)00501-8, www.elsevier.com/inca/publications/store/6/0/0/7/6/0.

Jiang, H. L., Liu, B., Lan, Y. Q., Kuratani, K., Akita, T., Shioyama, H., Zong, F., & Xu, Q. (2011). From metal-organic framework to nanoporous carbon: Toward a very high surface area and hydrogen uptake. *Journal of the American Chemical Society, 133*(31), 11854–11857. https://doi.org/10.1021/ja203184k.

Jiang, S., Li, S., Xu, Y., Liu, Z., Weng, S., Lin, M., Xu, Y., Jiao, Y., & Chen, J. (2021). An iron based organic framework coated with nickel hydroxide for energy storage, conversion and detection. *Journal of Colloid and Interface Science, 600*, 150–160. https://doi.org/10.1016/j.jcis.2021.05.014.

Jung, D. W., Yang, D. A., Kim, J., Kim, J., & Ahn, W. S. (2010). Facile synthesis of MOF-177 by a sonochemical method using 1-methyl-2-pyrrolidinone as a solvent. *Dalton Transactions, 39*(11), 2883–2887. https://doi.org/10.1039/b925088c.

Kaneti, Y. V., Tang, J., Salunkhe, R. R., Jiang, X., Yu, A., Wu, K. C. W., & Yamauchi, Y. (2017). Nanoarchitectured design of porous materials and nanocomposites from metal-organic frameworks. *Advanced Materials, 29*(12), 1604898. https://doi.org/10.1002/adma.201604898, http://onlinelibrary.wiley.com/journal/10.1002/(ISSN)1521-4095.

Kazemi, S. H., Hosseinzadeh, B., Kazemi, H., Kiani, M. A., & Hajati, S. (2018). Facile synthesis of mixed metal-organic frameworks: Electrode materials for supercapacitors with excellent areal capacitance and operational stability. *ACS Applied Materials and Interfaces, 10*(27), 23063–23073. https://doi.org/10.1021/acsami.8b04502, http://pubs.acs.org/journal/aamick.

Krishnan, S., Gupta, A. K., Singh, M. K., Guha, N., & Rai, D. K. (2022). Nitrogen-rich Cu-MOF decorated on reduced graphene oxide nanosheets for hybrid supercapacitor applications with enhanced cycling stability. *Chemical Engineering Journal, 435*, 135042. https://doi.org/10.1016/j.cej.2022.135042.

Van Lam, D., Sohail, M., Kim, J.-H., Lee, H. J., Han, S. O., Shin, J., Kim, D., Kim, H., & Lee, S.-M. (2020). Laser synthesis of MOF-derived Ni@carbon for high-performance pseudocapacitors. *ACS Applied Materials & Interfaces, 12*(35), 39154–39162. https://doi.org/10.1021/acsami.0c10235.

Li, C., Wang, J., Yan, Y., Huo, P., & Wang, X. (2022). MOF-derived NiZnCo-P nano-array for asymmetric supercapacitor. *Chemical Engineering Journal, 446*, 137108. https://doi.org/10.1016/j.cej.2022.137108.

Li, G., Cai, H., Li, X., Zhang, J., Zhang, D., Yang, Y., & Xiong, J. (2019). Construction of hierarchical NiCo$_2$O$_4$@Ni-MOF hybrid arrays on carbon cloth as superior battery-type electrodes for flexible solid-state hybrid supercapacitors. *ACS Applied Materials & Interfaces, 11*(41), 37675–37684. https://doi.org/10.1021/acsami.9b11994.

Li, J. R., Yu, J., Lu, W., Sun, L. B., Sculley, J., Balbuena, P. B., & Zhou, H. C. (2013). Porous materials with pre-designed single-molecule traps for CO_2 selective adsorption. *Nature Communications, 4*, 1538. https://doi.org/10.1038/ncomms2552.

Li, J. S., Li, S. L., Tang, Y. J., Li, K., Zhou, L., Kong, N., Lan, Y. Q., Bao, J. C., & Dai, Z. H. (2014). Heteroatoms ternary-doped porous carbons derived from MOFs as metal-free electrocatalysts for oxygen reduction reaction. *Scientific Reports, 4*, 5130. https://doi.org/10.1038/srep05130, www.nature.com/srep/index.html.

Li, Z., Ge, X., Li, C., Dong, S., Tang, R., Wang, C., Zhang, Z., & Yin, L. (2020). Rational microstructure design on metal–organic framework composites for better electrochemical performances: Design principle, synthetic strategy, and promotion mechanism. *Small Methods, 4*(3), 1900756. https://doi.org/10.1002/smtd.201900756.

Liang, R., Hu, A., Li, M., Ran, Z., Shu, C., & Long, J. (2019). Cobalt encapsulated within porous MOF-derived nitrogen-doped carbon as an efficient bifunctional electrocatalyst for aprotic lithium-oxygen battery. *Journal of Alloys and Compounds, 810*, 151877. https://doi.org/10.1016/j.jallcom.2019.151877, https://www.journals.elsevier.com/journal-of-alloys-and-compounds.

Liang, Z. B. (2017). Edge-abundant porous Fe_3O_4 nanoparticles docking in nitrogen-rich graphene aerogel as efficient and durable electrocatalyst for oxygen reduction. *ChemElectroChem, 4*(10), 2442–2447.

Lin, R. B., Xiang, S., Zhou, W., & Chen, B. (2020). Microporous metal-organic framework materials for gas separation. *Chem, 6*(2), 337–363. https://doi.org/10.1016/j.chempr.2019.10.012, http://www.cell.com/chem/home.

Liu, C., Huang, X., Wang, J., Song, H., Yang, Y., Liu, Y., Li, J., Wang, L., & Yu, C. (2018). Hollow mesoporous carbon nanocubes: Rigid-interface-induced outward contraction of metal-organic frameworks. *Advanced Functional Materials, 28*(6), 1705253. https://doi.org/10.1002/adfm.201705253.

Liu, J., Zhang, H., Qiu, M., Peng, Z., Leung, M. K. H., Lin, W.-F., & Xuan, J. (2020). A review of non-precious metal single atom confined nanomaterials in different structural dimensions (1D–3D) as highly active oxygen redox reaction electrocatalysts. *Journal of Materials Chemistry A, 8*(5), 2222–2245. https://doi.org/10.1039/c9ta11852g.

Liu, J. L. (2017). Design strategies toward advanced MOF-derived electrocatalysts for energy-conversion reactions. *Advanced Energy Materials, 7*(23), 1700518.

Liu, P., Yan, J., Guang, Z., Huang, Y., Li, X., & Huang, W. (2019). Recent advancements of polyaniline-based nanocomposites for supercapacitors. *Journal of Power Sources, 424*, 108–130. https://doi.org/10.1016/j.jpowsour.2019.03.094.

Liu, P.-Y., Zhao, J.-J., Dong, Z.-P., Liu, Z.-L., & Wang, Y.-Q. (2021). Interwoving polyaniline and a metal-organic framework grown in situ for enhanced supercapacitor behavior. *Journal of Alloys and Compounds, 854*, 157181. https://doi.org/10.1016/j.jallcom.2020.157181.

Liu, Y., Wang, Y., Chen, Y., Wang, C., & Guo, L. (2020). NiCo-MOF nanosheets wrapping polypyrrole nanotubes for high-performance supercapacitors. *Applied Surface Science, 507*, 145089. https://doi.org/10.1016/j.apsusc.2019.145089.

Low, J. J., Benin, A. I., Jakubczak, P., Abrahamian, J. F., Faheem, S. A., & Willis, R. R. (2009). Virtual high throughput screening confirmed experimentally: Porous coordination polymer hydration. *Journal of the American Chemical Society, 131*(43), 15834–15842. https://doi.org/10.1021/ja9061344, http://pubs.acs.org/doi/pdfplus/10.1021/ja9061344.

Lu, W., Wei, Z., Gu, Z. Y., Liu, T. F., Park, J., Park, J., Tian, J., Zhang, M., Zhang, Q., Gentle, T., Bosch, M., & Zhou, H. C. (2014). Tuning the structure and function of metal-organic frameworks via linker design. *Chemical Society Reviews, 43*(16), 5561–5593. https://doi.org/10.1039/c4cs00003j, http://pubs.rsc.org/en/journals/journal/cs.

Luo, F., Yan, C., Dang, L., Krishna, R., Zhou, W., Wu, H., Dong, X., Han, Y., Hu, T.-L., O'Keeffe, M., Wang, L., Luo, M., Lin, R.-B., & Chen, B. (2016). UTSA-74: A MOF-74 isomer with two accessible binding sites per metal center for highly selective gas separation. *Journal of the American Chemical Society, 138*(17), 5678–5684. https://doi.org/10.1021/jacs.6b02030.

Ma, Y., Gao, G., Su, H., Rong, H., Lai, L., & Liu, Q. (2021). A Cu_4 cluster-based MOF as a supercapacitor electrode material with ultrahigh capacitance. *Ionics, 27*(4), 1699–1707. https://doi.org/10.1007/s11581-021-03954-w.

Manikandan, J. (2015). Rapid smart phone charging using super capacitor. *International Journal of Advanced Research in Electrical, Electronics and instrumentation Engineering, 4*, 2175–2180.

McKone, J. R., DiSalvo, F. J., & Abruña, H. D. (2017). Solar energy conversion, storage, and release using an integrated solar-driven redox flow battery. *Journal of Materials Chemistry A, 5*(11), 5362–5372. https://doi.org/10.1039/c7ta00555e, http://pubs.rsc.org/en/journals/journalissues/ta.

Miao, F., Shao, C., Li, X., Wang, K., & Liu, Y. (2016). Flexible solid-state supercapacitors based on freestanding nitrogen-doped porous carbon nanofibers derived from electrospun polyacrylonitrile@polyaniline nanofibers. *Journal of Materials Chemistry A, 4*(11), 4180–4187. https://doi.org/10.1039/C6TA00015K.

Mihaylov, M., Andonova, S., Chakarova, K., Vimont, A., Ivanova, E., Drenchev, N., & Hadjiivanov, K. (2015). An advanced approach for measuring acidity of hydroxyls in confined space: A FTIR study of low-temperature CO and $^{15}N_2$ adsorption on MOF samples from the MIL-53(Al) series. *Physical Chemistry Chemical Physics, 17*(37), 24304–24314.

Moghaddam, Z. S., Kaykhaii, M., Khajeh, M., & Oveisi, A. R. (2018). Synthesis of UiO-66-OH zirconium metal-organic framework and its application for selective extraction and trace determination of thorium in water samples by spectrophotometry. *Spectrochimica Acta – Part A: Molecular and Biomolecular Spectroscopy, 194*, 76–82. https://doi.org/10.1016/j.saa.2018.01.010.

Moradi, E., Rahimi, R., & Safarifard, V. (2019). Sonochemically synthesized microporous metal–organic framework representing unique selectivity for detection of Fe^{3+} ions. *Polyhedron, 159*, 251–258. https://doi.org/10.1016/j.poly.2018.11.062, http://www.journals.elsevier.com/polyhedron/.

Morsali, A., Monfared, H. H., Morsali, A., & Janiak, C. (2015). Ultrasonic irradiation assisted syntheses of one-dimensional di(azido)-dipyridylamine Cu(II) coordination polymer nanoparticles. *Ultrasonics Sonochemistry, 23*, 208–211. https://doi.org/10.1016/j.ultsonch.2014.06.005.

Mueller, U., Schubert, M., Teich, F., Puetter, H., Schierle-Arndt, K., & Pastré, J. (2006). Metal-organic frameworks – Prospective industrial applications. *Journal of Materials Chemistry, 16*(7), 626–636. https://doi.org/10.1039/b511962f.

Nishihara, H., Itoi, H., Kogure, T., Hou, P. X., Touhara, H., Okino, F., & Kyotani, T. (2009). Investigation of the ion storage/transfer behavior in an electrical double-layer capacitor by using ordered microporous carbons as model materials. *Chemistry – A European Journal, 15*(21), 5355–5363. https://doi.org/10.1002/chem.200802406.

Oveisi, M., Alinia Asli, M., & Mahmoodi, N. M. (2019). Carbon nanotube based metal-organic framework nanocomposites: Synthesis and their photocatalytic activity for decolorization of colored wastewater. *Inorganica Chimica Acta, 487*, 169–176. https://doi.org/10.1016/j.ica.2018.12.021, http://www.journals.elsevier.com/inorganica-chimica-acta/.

Pan, Y., Sun, K., Lin, Y., Cao, X., Cheng, Y., Liu, S., Zeng, L., Cheong, W. C., Zhao, D., Wu, K., Liu, Z., Liu, Y., Wang, D., Peng, Q., Chen, C., & Li, Y. (2019). Electronic structure and d-band center control engineering over M-doped CoP (M = Ni, Mn, Fe) hollow polyhedron frames for boosting hydrogen production. *Nano Energy, 56*, 411–419.

Pham, T., & Space, B. (2020). Insights into the gas adsorption mechanisms in metal–organic frameworks from classical molecular simulations. *Topics in Current Chemistry, 378*, 215–279. 10.1007/978-3-030-47340-2_7.

Pichon, A., Lazuen-Garay, A., & James, S. L. (2006). Solvent-free synthesis of a microporous metal-organic framework. *CrystEngComm, 8*(3), 211–214. https://doi.org/10.1039/b513750k.

Qin, L., Ju, Z.-M., Wang, Z.-J., Meng, F.-D., Zheng, H.-G., & Chen, J.-X. (2014). Interpenetrated metal–organic framework with selective gas adsorption and luminescent properties. *Crystal Growth & Design, 14*(6), 2742–2746. https://doi.org/10.1021/cg500269h.

Quoc Bao, L., Nguyen, T.-H., Fei, H., Sapurina, I., Ngwabebhoh, F. A., Bubulinca, C., Munster, L., Bergerová, E. D., Lengalova, A., Jiang, H., Trong Dao, T., Bugarova, N., Omastova, M., Kazantseva, N. E., & Saha, P. (2021). Electrochemical performance of composites made of rGO with Zn-MOF and PANI as electrodes for supercapacitors. *Electrochimica Acta, 367*, 137563. https://doi.org/10.1016/j.electacta.2020.137563.

Rabani, I., Karuppasamy, K., Vikraman, D., ul haq, Z., Kim, H.-S., & Seo, Y.-S. (2021). Hierarchical structured nano-polyhedrons of CeO_2@ZIF-8 composite for high performance supercapacitor applications. *Journal of Alloys and Compounds, 875*, 160074. https://doi.org/10.1016/j.jallcom.2021.160074.

Radwan, A., Jin, H., He, D., & Mu, S. (2021). Design engineering, synthesis protocols, and energy applications of MOF-derived electrocatalysts. *Nano-Micro Letters, 13*(1), 132. https://doi.org/10.1007/s40820-021-00656-w.

Rajan, R. S., & Rahman, M. M. (2014). Lifetime analysis of super capacitor for many power electronics applications. *IOSR Journal of Electrical and Electronics Engineering, 9*(1), 55–58. https://doi.org/10.9790/1676–09145558.

Ramandi, S., & Entezari, M. H. (2022). Design of new, efficient, and suitable electrode material through interconnection of ZIF-67 by polyaniline nanotube on graphene flakes for supercapacitors. *Journal of Power Sources, 538*, 231588. https://doi.org/10.1016/j.jpowsour.2022.231588.

Rosi, N. L., Eckert, J., Eddaoudi, M., Vodak, D. T., Kim, J., O'Keeffe, M., & Yaghi, O. M. (2003). Hydrogen storage in microporous metal-organic frameworks. *Science (New York, N.Y.), 300*(5622), 1127–1129. https://doi.org/10.1126/science.1083440.

Şahin, M. E., Blaabjerg, F., & Sangwongwanich, A. (2022). A comprehensive review on supercapacitor applications and developments. *Energies, 15*, 2022.

Shalini, S. S., Balamurugan, R., Velmathi, S., & Bose, A. C. (2022). Systematic investigation on the electrochemical performance of pristine silver metal-organic framework as the efficient electrode material for supercapacitor application. *Energy and Fuels, 36*(13), 7104–7114. https://doi.org/10.1021/acs.energyfuels.2c01034, http://pubs.acs.org/journal/enfuem.

Shayeh, J. S., & Salari, H. (2020). Dendritic fibrous nano metal organic framework: A magnetic core-shell structure as high performance material for electrochemical capacitors. *Journal of Energy Storage, 32*, 101734. https://doi.org/10.1016/j.est.2020.101734.

Shen, C. H., Chuang, C. H., Gu, Y. J., Ho, W. H., Song, Y. D., Chen, Y. C., Wang, Y. C., & Kung, C. W. (2021). Cerium-based metal-organic framework nanocrystals interconnected by carbon nanotubes for boosting electrochemical capacitor performance. *ACS Applied Materials and Interfaces, 13*(14), 16418–16426. https://doi.org/10.1021/acsami.1c02038, http://pubs.acs.org/journal/aamick.

Shen, L., Wu, W., Liang, R., Lin, R., & Wu, L. (2013). Highly dispersed palladium nanoparticles anchored on UiO-66(NH$_2$) metal-organic framework as a reusable and dual functional visible-light-driven photocatalyst. *Nanoscale, 5*(19), 9374–9382. https://doi.org/10.1039/c3nr03153e.

Shi, L., Yang, W., Zha, X., Zeng, Q., Tu, D., Li, Y., Yang, Y., Xu, J., & Chen, F. (2022). In situ deposition of conducting polymer on metal organic frameworks for high performance hybrid supercapacitor electrode materials. *Journal of Energy Storage, 52*, 104729. https://doi.org/10.1016/j.est.2022.104729.

Shi, X., Deng, T., & Zhu, G. (2021). Vertically oriented Ni-MOF@Co(OH)$_2$ flakes towards enhanced hybrid supercapacitior performance. *Journal of Colloid and Interface Science, 593*, 214–221. https://doi.org/10.1016/j.jcis.2021.02.096, http://www.elsevier.com/inca/publications/store/6/2/2/8/6/1/index.htt.

Shi, X., Zhu, G., Qiu, S., Huang, K., Yu, J., & Xu, R. (2004). Zn$_2$[(S)-O$_3$PCH$_2$NHC$_4$H$_7$CO$_2$]$_2$: A homochiral 3D zinc phosphonate with helical channels. *Angewandte Chemie International Edition, 43*(47), 6482–6485. https://doi.org/10.1002/anie.200460724.

Tabassum, H., Guo, W., Meng, W., Mahmood, A., Zhao, R., Wang, Q., & Zou, R. (2017). Metal-organic frameworks derived cobalt phosphide architecture encapsulated into B/N Co-doped graphene nanotubes for all pH value electrochemical hydrogen evolution. *Advanced Energy Materials, 7*(9), 2017.

Tan, B., Wu, Z. F., & Xie, Z. L. (2017). Fine decoration of carbon nanotubes with metal organic frameworks for enhanced performance in supercapacitance and oxygen reduction reaction.

Science Bulletin, 62(16), 1132–1141. https://doi.org/10.1016/j.scib.2017.08.011, http://link.springer.com/journal/11434.

Tang, Z., Shang, X., Hu, B., Nie, P., Shi, W., Yang, J., & Liu, J. (2022). Fabrication of various metal hexacyanoferrates@CNF through acid-regulation for high-performance supercapacitor with superior stability. *Carbon, 187*, 47–55. https://doi.org/10.1016/j.carbon.2021.10.076.

Tian, D., Song, N., Zhong, M., Lu, X., & Wang, C. (2020). Bimetallic MOF nanosheets decorated on electrospun nanofibers for high-performance asymmetric supercapacitors. *ACS Applied Materials & Interfaces, 12*(1), 1280–1291. https://doi.org/10.1021/acsami.9b16420.

Tompsett, G. A., Conner, W. C., & Yngvesson, K. S. (2006). Microwave synthesis of nanoporous materials. *Chemphyschem: A European Journal of Chemical Physics and Physical Chemistry, 7*(2), 296–319. https://doi.org/10.1002/cphc.200500449, http://onlinelibrary.wiley.com/journal/10.1002/(ISSN)1439-7641.

Wang, G., Yan, Z., Wang, N., Xiang, M., & Xu, Z. (2021). NiO/Ni metal–organic framework nanostructures for asymmetric supercapacitors. *ACS Applied Nano Materials, 4*(9), 9034–9043. https://doi.org/10.1021/acsanm.1c01628.

Wang, L., Feng, X., Ren, L., Piao, Q., Zhong, J., Wang, Y., Li, H., Chen, Y., & Wang, B. (2015). Flexible solid-state supercapacitor based on a metal–organic framework interwoven by electrochemically-deposited PANI. *Journal of the American Chemical Society, 137*(15), 4920–4923. https://doi.org/10.1021/jacs.5b01613.

Wang, M. (2020). Phthalocyanine-based 2D conjugated metal-organic framework nanosheets for high-performance micro-supercapacitors. *Advanced Functional Materials, 30*(30), 2020.

Wang, Y.-S., Chen, Y.-C., Li, J.-H., & Kung, C.-W. (2019). Toward metal–organic-framework-based supercapacitors: Room-temperature synthesis of electrically conducting MOF-based nanocomposites decorated with redox-active manganese. *European Journal of Inorganic Chemistry, 2019*(26), 3036–3044. https://doi.org/10.1002/ejic.201900584.

Wang, Y., Chen, N., Liu, Y., Zhou, X., Pu, B., Qing, Y., Zhang, M., Jiang, X., Huang, J., Tang, Q., Zhou, B., & Yang, W. (2022). MXene/graphdiyne nanotube composite films for free-standing and flexible solid-state supercapacitor. *Chemical Engineering Journal, 450*, 138398. https://doi.org/10.1016/j.cej.2022.138398.

Wang, Y. S., Liao, J. L., Li, Y. S., Chen, Y. C., Li, J. H., Ho, W. H., Chiang, W. H., & Kung, C. W. (2020). Zirconium-based metal-organic framework nanocomposites containing dimensionally distinct nanocarbons for pseudocapacitors. *ACS Applied Nano Materials, 3*(2), 1448–1456. https://doi.org/10.1021/acsanm.9b02297, https://pubs.acs.org/journal/aanmf6.

Wang, Z., & Cohen, S. M. (2007). Postsynthetic covalent modification of a neutral metal-organic framework. *Journal of the American Chemical Society, 129*(41), 12368–12369. https://doi.org/10.1021/ja074366o.

Wang, Z., & Cohen, S. M. (2009). Postsynthetic modification of metal–organic frameworks. *Chemical Society Reviews, 38*(5), 1315–1329. https://doi.org/10.1039/b802258p.

Wen, P., Gong, P., Sun, J., Wang, J., & Yang, S. (2015). Design and synthesis of Ni-MOF/CNT composites and rGO/carbon nitride composites for an asymmetric supercapacitor with high energy and power density. *Journal of Materials Chemistry A, 3*(26), 13874–13883. https://doi.org/10.1039/C5TA02461G.

Wu, Y., Kobayashi, A., Halder, G. J., Peterson, V. K., Chapman, K. W., Lock, N., Southon, P. D., & Kepert, C. J. (2008). Negative thermal expansion in the metal–organic framework material Cu 3 (1,3,5-benzenetricarboxylate) 2. *Angewandte Chemie International Edition, 47*(46), 8929–8932. https://doi.org/10.1002/anie.200803925.

Xia, F., Kang, S., Xue, F., & Bu, X. (2021). Simple reductive synthesis of a novel mixed-lanthanide metal–organic framework with excellent cycling ability as a binder-free supercapacitor electrode. *Materials Letters, 282*, 128715. https://doi.org/10.1016/j.matlet. 2020.128715.

Xiao, X., Zou, L., Pang, H., & Xu, Q. (2020). Synthesis of micro/nanoscaled metal-organic frameworks and their direct electrochemical applications. *Chemical Society Reviews, 49*(1), 301–331. https://doi.org/10.1039/c7cs00614d, http://pubs.rsc.org/en/journals/journal/cs.

Xiao, Y., Wei, W., Zhang, M., Jiao, S., Shi, Y., & Ding, S. (2019). Facile surface properties engineering of high-quality graphene: Toward advanced Ni-MOF heterostructures for high-performance supercapacitor electrode. *ACS Applied Energy Materials, 2*(3), 2169–2177. https://doi.org/10.1021/acsaem.8b02201.

Xu, F. (2021). Coral–like Ni_2P@C derived from metal–organic frameworks with superior electrochemical performance for hybrid supercapacitors. *Electrochimica Acta, 380*, 2021.

Xu, X., Tang, J., Qian, H., Hou, S., Bando, Y., Hossain, M. S. A., Pan, L., & Yamauchi, Y. (2017). Three-dimensional networked metal–organic frameworks with conductive polypyrrole tubes for flexible supercapacitors. *ACS Applied Materials & Interfaces, 9*(44), 38737–38744. https://doi.org/10.1021/acsami.7b09944.

Yaghi, O. M., & Li, H. (1995). Hydrothermal synthesis of a metal-organic framework containing large rectangular channels. *Journal of the American Chemical Society, 117*(41), 10401–10402. https://doi.org/10.1021/ja00146a033.

Yaghi, O. M., O'Keeffe, M., Ockwig, N. W., Chae, H. K., Eddaoudi, M., & Kim, J. (2003). Reticular synthesis and the design of new materials. *Nature, 423*(6941), 705–714. https://doi.org/10.1038/nature01650.

Yan, J., Huang, Y., Liu, X., Zhao, X. X., Li, T., Zhao, Y., & Liu, P. (2021). Polypyrrole-based composite materials for electromagnetic wave absorption. *Polymer Reviews, 61*(3), 646–687. https://doi.org/10.1080/15583724.2020.1870490.

Yang, M., Jiao, L., Dong, H., Zhou, L., Teng, C., Yan, D., Ye, T.-N., Chen, X., Liu, Y., & Jiang, H.-L. (2021). Conversion of bimetallic MOF to Ru-doped Cu electrocatalysts for efficient hydrogen evolution in alkaline media. *Science Bulletin, 66*(3), 257–264. https://doi.org/10.1016/j.scib.2020.06.036.

Yang, Y., Lun, Z., Xia, G., Zheng, F., He, M., & Chen, Q. (2015). Non-precious alloy encapsulated in nitrogen-doped graphene layers derived from MOFs as an active and durable hydrogen evolution reaction catalyst. *Energy and Environmental Science, 8*(12), 3563–3571. https://doi.org/10.1039/c5ee02460a, http://www.rsc.org/Publishing/Journals/EE/About.asp.

Yang, Y., Han, Y., Jiang, W., Zhang, Y., Xu, Y., & Ahmed, A. M. (2022). Application of the supercapacitor for energy storage in China: Role and strategy. *Applied Sciences, 12*(1), 354. https://doi.org/10.3390/app12010354.

You, Y., Yao, H. R., Xin, S., Yin, Y. X., Zuo, T. T., Yang, C. P., Guo, Y. G., Cui, Y., Wan, L. J., & Goodenough, J. B. (2016). Subzero-temperature cathode for a sodium-ion battery. *Advanced Materials, 28*(33), 7243–7248. https://doi.org/10.1002/adma.201600846.

Yu, C., Wang, Y., Cui, J., Yu, D., Zhang, X., Shu, X., Zhang, J., Zhang, Y., Vajtai, R., Ajayan, P. M., & Wu, Y. (2018). MOF-74 derived porous hybrid metal oxide hollow nanowires for high-performance electrochemical energy storage. *Journal of Materials Chemistry A, 6*(18), 8396–8404. https://doi.org/10.1039/c8ta01426d, http://pubs.rsc.org/en/journals/journal/ta.

Yu, M. H., Zhang, P., Feng, R., Yao, Z. Q., Yu, Y. C., Hu, T. L., & Bu, X. H. (2017). Construction of a multi-cage-based MOF with a unique network for efficient CO_2 capture. *ACS Applied Materials and Interfaces, 9*(31), 26177–26183. https://doi.org/10.1021/acsami.7b06491, http://pubs.acs.org/journal/aamick.

Zhang, B., Wang, J., Su, X., Duan, H., Cai, H., Wang, J., Yang, S., & Huo, S. (2017). Enhanced microwave absorption properties of epoxy composites containing graphene decorated with core–shell Fe_3O_4@polypyrrole nanoparticles. *Journal of Materials Science: Materials in Electronics, 28*(16), 12122–12131. https://doi.org/10.1007/s10854-017-7026-z.

Zhang, F., Ma, J., & Yao, H. (2019). Ultrathin Ni-MOF nanosheet coated $NiCo_2O_4$ nanowire arrays as a high-performance binder-free electrode for flexible hybrid supercapacitors. *Ceramics International, 45*(18), 24279–24287. https://doi.org/10.1016/j.ceramint.2019.08.140, https://www.journals.elsevier.com/ceramics-international.

Zhang, L., Zhang, Y., Huang, S., Yuan, Y., Li, H., Jin, Z., Wu, J., Liao, Q., Hu, L., Lu, J., Ruan, S., & Zeng, Y. (2018). Co_3O_4/Ni-based MOFs on carbon cloth for flexible alkaline battery-supercapacitor hybrid devices and near-infrared photocatalytic hydrogen evolution. *Electrochimica Acta, 281*, 189–197.

Zhang, F., Zhang, J., Song, J., You, Y., Jin, X., & Ma, J. (2021). Anchoring Ni-MOF nanosheet on carbon cloth by zeolite imidazole framework derived ribbonlike Co_3O_4 as integrated composite cathodes for advanced hybrid supercapacitors. *Ceramics International, 47*(10), 14001–14008. https://doi.org/10.1016/j.ceramint.2021.01.269.

Zhang, S., & Pan, N. (2015). Supercapacitors performance evaluation. *Advanced Energy Materials, 5*(6).

Zhang, W. X., Yang, Y. Y., Zai, S. B., Seik, W. N., & Chen, X. M. (2008). Syntheses, structures and magnetic properties of dinuclear copper(II)-lanthanide(III) complexes bridged by 2-hydroxymethyl-1- methylimidazole. *European Journal of Inorganic Chemistry,* (5), 679–685. https://doi.org/10.1002/ejic.200701041, http://www3.interscience.wiley.com/cgi-bin/fulltext/117876612/PDFSTART.

Zhang, Y. Z., Cheng, T., Wang, Y., Lai, W. Y., Pang, H., & Huang, W. (2016). A simple approach to boost capacitance: Flexible supercapacitors based on manganese oxides@MOFs via chemically induced in situ self-transformation. *Advanced Materials, 28*(26), 5242–5248. https://doi.org/10.1002/adma.201600319, http://www3.interscience.wiley.com/journal/119030556/issue.

Zhao, D., Zhang, Q., Chen, W., Yi, X., Liu, S., Wang, Q., Liu, Y., Li, J., Li, X., & Yu, H. (2017). Highly flexible and conductive cellulose-mediated PEDOT:PSS/MWCNT Composite films for supercapacitor electrodes. *ACS Applied Materials & Interfaces, 9*(15), 13213–13222. https://doi.org/10.1021/acsami.7b01852.

Zhao, G., Rui, K., Dou, S. X., & Sun, W. (2018). Heterostructures for electrochemical hydrogen evolution reaction: A review. *Advanced Functional Materials, 28*(43), 1803291. https://doi.org/10.1002/adfm.201803291, http://onlinelibrary.wiley.com/journal/10.1002/(ISSN)1616-3028.

Zhao, W., Peng, J., Wang, W., Jin, B., Chen, T., Liu, S., Zhao, Q., & Huang, W. (2019). Interlayer hydrogen-bonded metal porphyrin frameworks/MXene hybrid film with high capacitance for flexible all-solid-state supercapacitors. *Small (Weinheim an der Bergstrasse, Germany), 15*(18), 1901351. https://doi.org/10.1002/smll.201901351, http://onlinelibrary.wiley.com/journal/10.1002/(ISSN)1613-6829.

Zhao, Y., Dong, H., Yu, J., Chen, R., Liu, Q., Liu, J., Li, R., Wang, X., & Wang, J. (2021). Binder-free metal-organic frameworks-derived CoP/Mo-doped NiCoP nanoplates for high-performance quasi-solid-state supercapacitors. *Electrochimica Acta, 390*, 138840. https://doi.org/10.1016/j.electacta.2021.138840.

Zheng, L. (2022). Unique core-shell $Co_2(OH)_2CO_3$@MOF nanoarrays with remarkably improved cycling life for high performance pseudocapacitors. *Electrochimica Acta, 412*, 2022.

Zheng, S., Zhou, H., Xue, H., Braunstein, P., & Pang, H. (2022). Pillared-layer Ni-MOF nanosheets anchored on Ti_3C_2 MXene for enhanced electrochemical energy storage. *Journal of Colloid and Interface Science, 614*, 130–137. https://doi.org/10.1016/j.jcis.2022.01.094.

Zhong, Y., Cao, X., Ying, L., Cui, L., Barrow, C., Yang, W., & Liu, J. (2020). Homogeneous nickel metal-organic framework microspheres on reduced graphene oxide as novel electrode material for supercapacitors with outstanding performance. *Journal of Colloid and Interface Science, 561*, 265–274. https://doi.org/10.1016/j.jcis.2019.10.023.

Chapter 4

Metal-organic framework composites for carbon capture

Ying Liu, and Zongbi Bao
Key Laboratory of Biomass Chemical Engineering of Ministry of Education, College of Chemical and Biological Engineering, Zhejiang University, Hangzhou, P.R. China

Introduction

The rising atmospheric concentration of carbon dioxide (CO_2), a major greenhouse gas, has been identified as the primary contributor to global climate change (Shakun et al., 2012). Since the mid-20th century, annual global CO_2 emissions have increased dramatically, reaching a staggering 37.55 billion tons in 2023 (see Fig. 4.1A). The main sources of anthropogenic CO_2 emissions (see Fig. 4.1B) include the burning of fossil fuels, industrial processes (e.g., cement and steel production), and land-use changes (e.g., deforestation) (Feng et al., 2015; Rackley, 2017). Consequently, atmospheric CO_2 levels have risen from less than 320 ppm in the 1960s to 421 ppm in 2023 (see Fig. 4.2), accounting for approximately 60% of the observed global warming. To mitigate the devastating impacts of climate change, the Paris Agreement aims to limit the global average temperature increase to well below 2 K and pursue efforts to restrict it to 1.5 K. Achieving this ambitious target necessitates the urgent deployment of carbon capture, utilization, and storage (CCUS) technologies, with carbon capture being the critical initial step.

Existing carbon capture technologies include amine absorption, adsorption, membrane separation, and cryogenic separation (Haring, 2007). Amine scrubbing, which utilizes alkanolamine-based absorbents such as monoethanolamine (MEA) and triethanolamine (TEA), is the most extensively employed industrial carbon capture method (Rochelle, 2009). However, amine-based absorbents have several drawbacks, including high heat capacity, high regeneration energy requirements, and corrosive properties. As an alternative, pressure-swing adsorption (PSA) or temperature-swing adsorption (TSA) using highly efficient carbon capture adsorbents has emerged as a promising technology for effective, energy-saving, and long-term stable carbon capture in industrial settings (Yang, 1988). The primary challenges

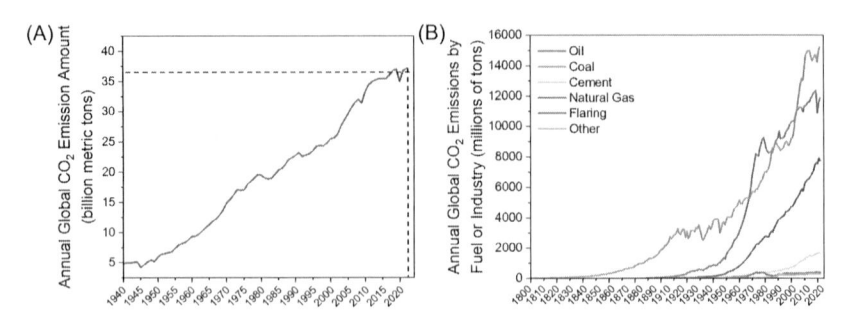

FIGURE 4.1 (A) Annual global CO_2 emission amount since 1940; (B) annual global CO_2 emissions by fuel or industry from 1800 to 2022.
Data from www.statista.com.

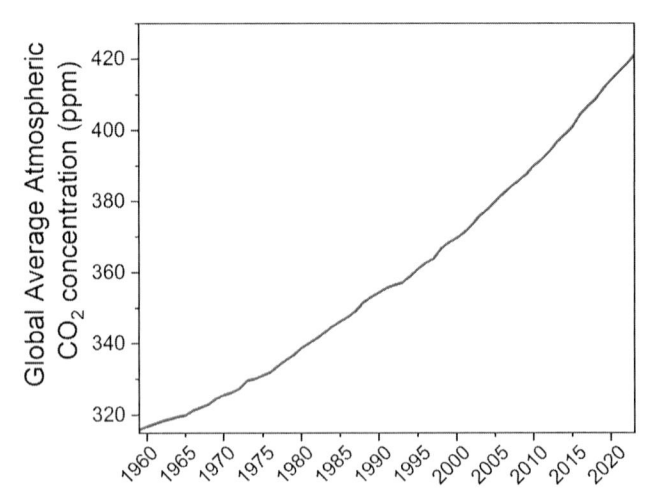

FIGURE 4.2 Average CO_2 levels in the atmosphere worldwide from 1959 to 2023.
Data from www.statista.com.

for carbon capture materials are their renewability, regeneration energy consumption, and the ability to maintain high performance under realistic operating conditions, which ultimately govern the efficiency and economic viability of the carbon capture process.

Metal-organic frameworks (MOFs), a class of crystalline porous materials consisting of metal ions or clusters coordinated to organic ligands (Glover & M, 2019; Kaskel, 2016), have attracted considerable attention for their potential applications in gas storage and separation, catalysis, sensing, and energy (Altintas et al., 2022; Chuah et al., 2022; Langmi et al., 2014; Ma et al., 2022; Vodyashkin et al., 2023; Wang et al., 2016; Wang, Chen, et al., 2023;

Wang, Huang, et al., 2023; Zhang et al., 2021). MOFs with multifunctional groups, such as open metal sites (OMSs) and Lewis basic sites (LBSs), exhibit strong interactions with CO_2 due to its large quadrupole moment and polarizability, making them promising candidates for carbon capture (Bose et al., 2023). However, MOFs still face several challenges, including poor water stability, limited diffusion in microporous channels, and difficulties in scaling up and shaping (Boyd et al., 2019; Hu et al., 2022; Hu, Hu et al., 2023; Hu, Jiang et al., 2023; Li et al., 2013; Mukherjee et al., 2019; Zhu et al., 2023). To overcome these limitations, MOF composites have been developed by rationally combining MOFs with other materials, such as organic/inorganic binders, structural monoliths, alkyl amines, carbon-based porous materials, and ionic liquids. These composites often demonstrate enhanced structural stability, mechanical strength, and improved separation efficiency through thermodynamic-kinetic synergistic effects, showing great potential for industrial carbon capture applications.

Metal-organic framework composites for CO_2 capture

Shaped metal-organic framework composites

The majority of MOF materials are synthesized in powder form, which can lead to various issues during practical applications, such as dust pollution, increased pressure drops, pipeline clogging, and difficulties in transportation, loading, and unloading. These challenges hinder the large-scale and commercial development of MOFs and limit their potential in industrial settings. To facilitate the industrial application of MOFs, it is essential to process and shape them into suitable forms. Shaped MOF materials not only circumvent the aforementioned issues but also facilitate reuse and recycling, further expanding their range of applications (Cortés & Rojas Macías, 2021).

MOFs, unlike organic polymers, are nonthermoplastic and insoluble in solvents, rendering traditional solvent or melt processing techniques unsuitable for their structural processing. Currently, the shaping methods for MOFs can be categorized into two main approaches: direct shaping methods via in situ synthesis and indirect shaping methods via postsynthesis, with the latter further divided into mechanical extrusion, compounding, granulation, and other techniques. The choice of shaping method depends on the specific application requirements. Once shaped, MOFs must retain their porosity and functionality while withstanding the wear and tear caused by industrial operations to prevent disintegration back into powder form. Fig. 4.3 shows shaped MOFs prepared by several shaping methods (Abramova et al., 2022; Bo et al., 2020; Chen et al., 2016; Garai et al., 2016; Lee et al., 2020; Valekar et al., 2017).

Wet granulation is the simplest, most efficient, and most widely used method for shaping MOFs (Lee et al., 2016). The process involves mixing MOF powder and a binder under solvent moistening conditions, followed by

138 Applications of Metal-Organic Framework Composites

FIGURE 4.3 Examples of shaped metal-organic frameworks (MOFs) prepared using several shaping methods. (A) Granulated MIL-100(Fe); (B) granulated MIL-101(Cr); (C) shaped HKUST-1; (D) UiO-66 extrudates; (E) ZIF-8 foam; (F) Mg-MOF-74 foam; (G) MOF-coated paper; (H) ZIF-8-coated film; (I) UTSA-16(Co)-coated monolith; (J) 3D print MOF monolith; and (K) UiO-66 beads (Abramova et al., 2022; Bo et al., 2020; Chen et al., 2016; Garai et al., 2016; Cortés & Rojas Macías, 2021; Valekar et al., 2017; Zhang et al., 2011).

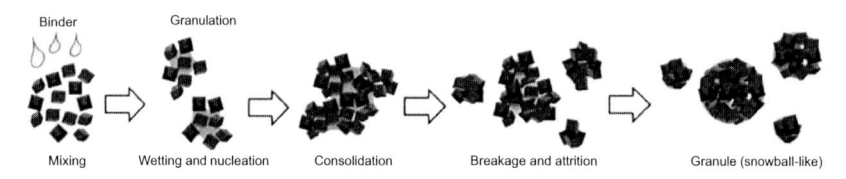

FIGURE 4.4 Schematic illustration of wet granulation process (Lee et al., 2016).

nucleation, consolidation, breakage, attrition, and granule formation, allowing the powder particles to adhere to the binder and form mechanically stable particles (see Fig. 4.4). The most critical factors to consider when selecting a binder are its stability and compatibility with the original powder. For shaping inorganic materials, inorganic binders such as alumina, silica sol, or clay are commonly used, and the final shaped materials are obtained through high-temperature calcination. However, this shaping method is not suitable for MOFs, and organic binders, including cellulose derivatives (hydroxyethyl cellulose, hydroxypropyl cellulose, hydroxypropyl methylcellulose) and polymers (polyethylene glycol, polyvinyl alcohol, polyacrylamide, polyethylene oxide, povidone), are more frequently employed.

MIL-101(Cr), a MOF with outstanding CO_2 adsorption performance and stability, has been extensively studied in the literature for its potential in carbon capture applications (Liu, Ning, et al., 2013; Zhang et al., 2011). Researchers have demonstrated that increasing the pressure from 30 to 500 bar at room temperature results in a proportional increase in CO_2 adsorption capacity, ranging from 22.9 to 40.0 mmol/g. Hong et al. (2015) developed a method to prepare MIL-101(Cr) monoliths using extrusion techniques and bentonite clay as a binder. The process involves mixing MIL-101(Cr) powder, water, and binder to form a paste, which is then extruded into blocks using specialized machinery. The material is dried and sintered at 423 K for approximately 33 hours to obtain the final structured MOF composite. The equilibrium CO_2 adsorption capacity of the purified MIL-101(Cr) monoliths with 40 wt.% binder, calculated from breakthrough experiments, reached 1.95 mmol/g at 2 bar, which is considered acceptable for industrial applications.

Another promising MOF for CO_2 capture is the zinc-based Calgary Framework 20 (CALF-20), which has garnered significant attention from researchers. CALF-20 is synthesized by reacting zinc ions, 1,2,4-triazolate and oxalate to form a robust framework with 3D pore channels. The aperture sizes viewed from three different directions are 2.7 Å×2.9 Å, 1.9 Å×3.1 Å, and 2.7 Å×3.0 Å, respectively, making it suitable for CO_2 capture (Lin et al., 2021). At 293 K and 1.2 bar, CALF-20 exhibits a CO_2 capacity of 4.07 mmol/g, with a relatively high adsorption enthalpy of −39 kJ/mol and a considerable selectivity of 230 for a CO_2/N_2 (10/90, v/v) mixture. A CALF-20-polysulfone composite (20 wt.% polysulfone) with a particle size between 1 and 3 mm was prepared using a simple slurry under ambient conditions, followed by solvent exchange. The CO_2 capture performance of the structured CALF-20 composites was evaluated under both dry and humid conditions. For industrial applications, carbon capture material must be able to absorb CO_2 from postcombustion flue gases containing water vapor and acid gases at temperature above 373 K and withstand the stresses associated with the regeneration process, which involves temperature, pressure, and vacuum changes. CALF-20 demonstrates excellent stability, water resistance and considerable resistance to wet acidic gases, even when exposed to flue gases for extended periods. Moreover, CALF-20 can adsorb CO_2 at relative humidity levels exceeding 40%, and the presence of CO_2 has been shown to inhibit water adsorption by CALF-20. The inexpensive and easily available feedstocks, high yields (>90%), simple batch synthesis process, and time space yield of up to 550 kg/(m^3.day) make CALF-20 a highly promising material for CO_2 capture applications.

Zhang's group reported a rigid MOF called Zn-ox-mtz (ox=oxalic acid, mtz=3-methyl-1H-1,2,4-triazole) with 1D channels and narrow pore windows (3.5 Å) as molecular sieve adsorbents to realize CO_2 capture (Hu et al., 2024). At 298 K and 1 bar, Zn-ox-mtz absorbed more than 3.43 mmol/g CO_2 while adsorbing negligible amounts of CH_4 and N_2. In comparison, its isomer ZU-301 achieved only about half the CO_2 uptake. The IAST (ideal adsorbed solution theory) selectivity of Zn-ox-mtz for CO_2/N_2 (15/85, v/v) at ambient temperature and pressure exceeded 1×10^6, the highest reported for MOFs, and reached up to 2.7×10^5 for CO_2/CH_4 (50/50, v/v). Hydroxypropyl cellulose (HPC) was used as a binder, and a binder to powder mass ratio of 1:15 was sufficient to achieve granulation with high MOF loading (Fig. 4.5A). The saturated CO_2 adsorption capacity at atmospheric pressure of the formed Zn-

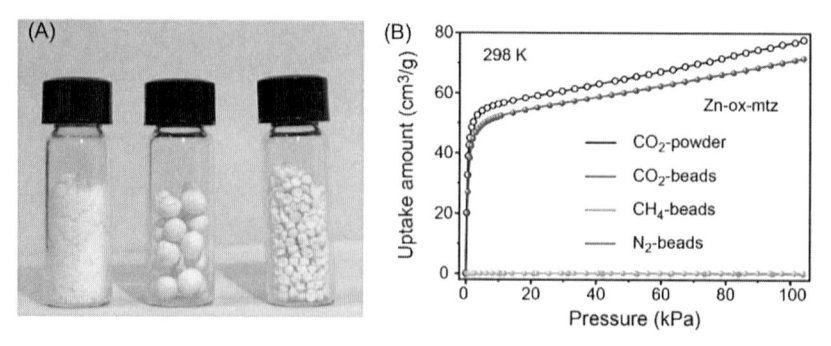

FIGURE 4.5 (A) Photo of Zn-ox-mtz powder, spherical beads, and cylindrical beads. (B) The single-component adsorption isotherms of CO_2, CH_4, and N_2 on Zn-ox-mtz powder and beads (Hu et al., 2024).

ox-mtz was 3.20 mmol/g, which maintained the good sequestration of N_2 and CH_4 (Fig. 4.5B).

In the production of ethylene (C_2H_4), the composition of the hydrotreating products is complex, and the dehydrogenation conversion rate is only 50%–60%, resulting in crude C_2H_4 product containing many impurities, including CO_2, acetylene (C_2H_2), and ethane (C_2H_6). These impurities have similar structures and properties to C_2H_4, making them difficult to remove. To obtain high purity (>99.9%) C_2H_4, multistage processes such as chemical adsorption, catalytic hydrogenation, and cryogenic distillation are employed. However, this purification process is associated with substantial equipment costs and energy consumption. Chen and coworkers developed a promising solid adsorbent (Zn-fa-atz(2)) to produce high purity C_2H_4 by physical adsorption (Yang et al., 2024). Single-component isotherms at 298 K demonstrated that the order of adsorption capacity of Zn-fa-atz(2) at 14, 25, and 33 kPa is $C_2H_6 > C_2H_2 > CO_2 > C_2H_4$, $C_2H_2 \approx CO_2 > C_2H_6 > C_2H_4$, and $CO_2 \approx C_2H_2 > C_2H_6 > C_2H_4$, respectively. Fixed bed breakthrough results confirmed that high purity C_2H_4 (>99.9%) can be obtained directly from CO_2/C_2H_2/C_2H_4/C_2H_6 mixtures with the strongest reservation of CO_2. Composite spherical particles of Zn-fa-atz(2) with a particle size of around 2.5 mm were prepared by using polyether sulfone (PES) as organic binder with 20 wt.% addition (Fig. 4.6A). SEM (scanning electron microscope) images showed that Zn-fa-atz(2) were perfectly embedded in internal polymer matrix (Fig. 4.6B and C). BET (Brunauer–Emmett–Teller) results indicated that the composite retained most of the microporous pores, and adsorption kinetics experiments showed that the diffusion of gas molecules within the composite was not significantly hindered. Breakthrough experiments also demonstrated the preferential capture of CO_2 and the one-step purification of C_2H_4 from a four-component mixture (Fig. 4.6D).

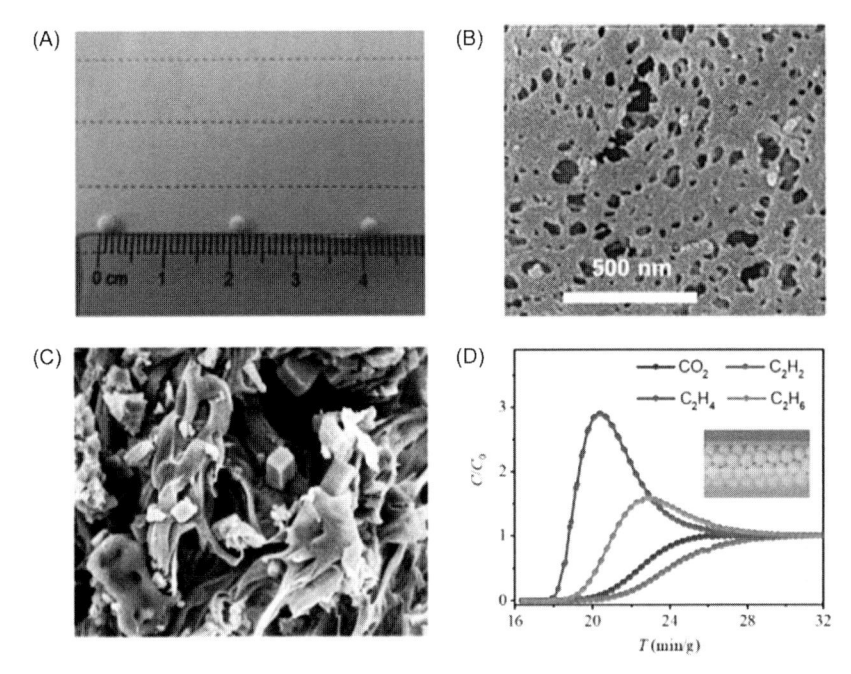

FIGURE 4.6 (A) Photograph of Zn-fa-atz(2)/PES beads; (B and C) SEM images of the bead surface and cross-section; and (D) fixed-bed column breakthrough curves for Zn-fa-atz(2)/PES beads used to separate an equimolar mixture of $CO_2/C_2H_2/C_2H_4/C_2H_6$ (Yang et al., 2024).

Structural metal-organic framework-monolith composites

Although shaping MOFs through granulation has proven to be an effective strategy for structuring adsorbents, issues such as high pressure drop, limited mass and heat transfer, and weight loss due to particle abrasion can be detrimental in the long term. These challenges are expected to be addressed by directly coating MOF powders onto structured monoliths. Due to their homogeneous channels, structured MOF-monolith composites can overcome the problems of pressure drop and particle wear that are common in conventional packed bed systems. Especially at high gas velocities, the thin-walled structure provides better mass and heat transfer properties compared to fixed and fluidized beds.

Rezaei and coworkers reported a simple method to concentrate the crystals of MOF-74(Ni) and UTSA-16(Co) on the surface of monolithic structures, achieving a MOF loading up to 76%–80% (see Fig. 4.7A) (Lawson et al., 2017). The fabrication process of the MOF-monolith composite adsorbent (MOF-MCA) is straightforward, and the carbon capture capability of the MOFs was retained. The adsorption capacities of the prepared MOF-74(Ni)

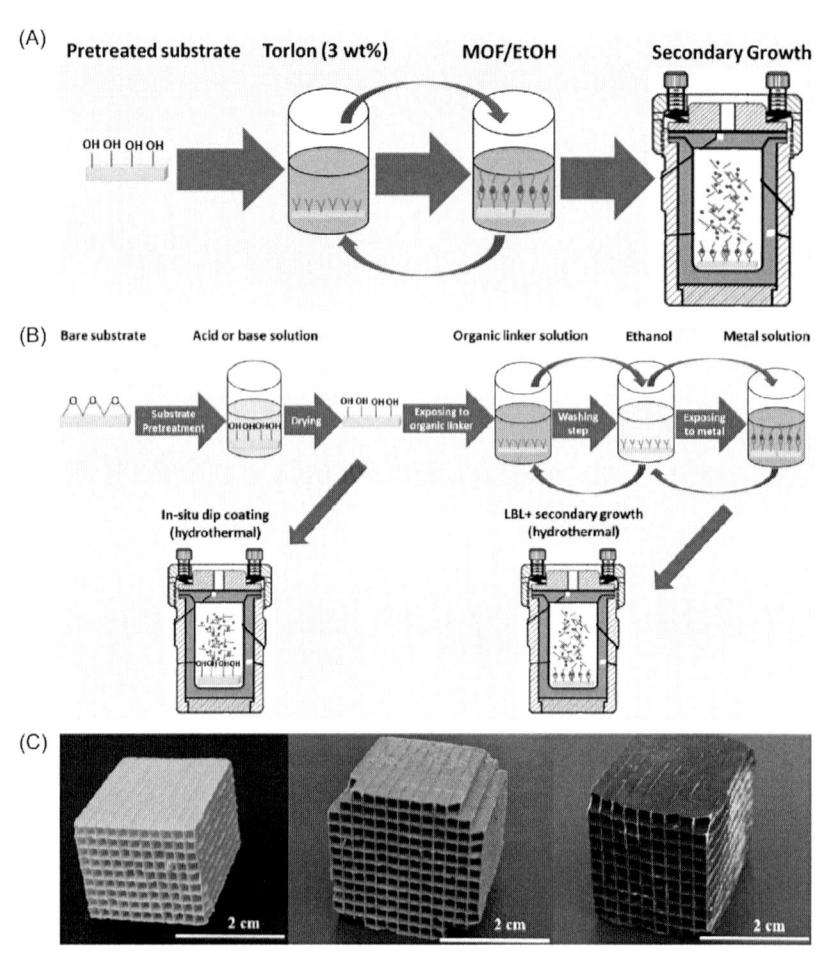

FIGURE 4.7 (A) Schematic of polymer phase separation procedure used for preparing MOF-monolith composite adsorbent (MOF-MCA); (B) schematic of layer-by-layer assembly+secondary growth and in situ dip coating (ISDC) techniques used for preparing MOF-coated monoliths; and (C) picture of bare (left) and coated cordierite monoliths with MOF-74(Ni) (middle) and UTSA-16(Co) (right) (Lawson et al., 2017; Rezaei et al., 2017).

monolith and UTSA-16(Co) monolith for CO_2 were 2.5 and 1.6 mmol/g, respectively, showing an improvement compared to the adsorption capacity of the pristine MOFs. However, considering the MOF loading ratio on the composites, the CO_2 adsorption capacity of the composites still lags behind that of the original powders, possibly due to difficulties in fully activating the pores in the loaded MOF films. More experiments and studies are needed to enhance and improve this aspect to achieve higher capacity and better kinetics. They also investigated the immobilization of two MOFs, MOF-74(Ni) and

UTSA-16(Co), on commercial cordierite monolith and systematically evaluated their adsorption properties in carbon capture (see Fig. 4.7B and C) (Rezaei et al., 2017).

Various bottom-up growth techniques, such as layer-by-layer (LBL) assembly and in situ dip coating (ISDC), have been employed to control the nucleation and growth of MOF crystals on monolithic substrates, optimizing the loading, thickness, and adsorption properties of the MOF films (Chen et al., 2023; Ellis et al., 2021). The choice of a suitable coating method depends mainly on the type of MOF material. The results showed that LBL secondary growth is suitable for MOF-74(Ni) films to achieve ~52 wt.% MOF loading, while ISDC is a promising method to achieve ~55 wt.% MOF loading for UTSA-16(Co). The MOF-coated monolithic materials exhibited relatively moderate CO_2 adsorption capacity and fast adsorption kinetics compared to the original powders. Adsorption isotherms revealed that the bare support did not exhibit any CO_2 uptake, while the MOF-coated monoliths exhibited partial CO_2 uptake at the same partial pressure. Compared to the original powders, the CO_2 capacities of the composites at ambient conditions were 28% (1.7 mmol/g) and 29% (1.1 mmol/g) of those of the original powders, respectively. Despite a mass loading of over 50 wt.% for both monoliths, the capacities were lower than expected due to incomplete activation of the coated substrates, as deduced from the porosity data. The dynamic adsorption properties of the powders MOF-coated monoliths were evaluated for CO_2/N_2 (10/90, v/v) mixtures. Although the composites had a lower dynamic CO_2 capacity than the original powders, the composites had a shorter mass transfer region as reflected in the penetration curves, indicating that CO_2 has less mass transfer resistance in the composites. This suggests that the structured MOF composites offer improved mass transfer characteristics compared to the original powder materials, which is beneficial for industrial gas separation processes.

Traditional structuring techniques, such as epitaxial growth, dip coating, spin coating, and polymer immobilization, for fabricating MOF-coated monolithic substrates have limitations, including low MOF loading per weight of coated monolith, pore blockage, and channel restriction, which are detrimental to practical applications. Recent advances in additive manufacturing (3D printing) have made internal doping extremely realistic, practical, and efficient, increasing flexibility when designing structures with customized geometries. In 2017 Rezaei's group used 3D printing to fabricate MOF-74(Ni) and UTSA-16(Co) monoliths with MOF loading up to 80% and 85%, respectively (Fig. 4.8A), to characterize their physical and structural properties and evaluate their adsorption performance for removing CO_2 from air (Thakkar et al., 2017). When exposed to 5000 ppm CO_2 at 298 K, the CO_2 capacity on 3D-print MOF74(Ni) and UTSA-16(Co) monoliths was 1.35 and 1.31 mmol/g, respectively, which was 79% and 87% of the uptake of MOF analogues under the same conditions. The CO_2 uptake of 3D-print MOF-74(Ni) and UTSA-16(Co) monoliths at 1.1 bar was 4.0 and 3.0 mmol/g,

144 Applications of Metal-Organic Framework Composites

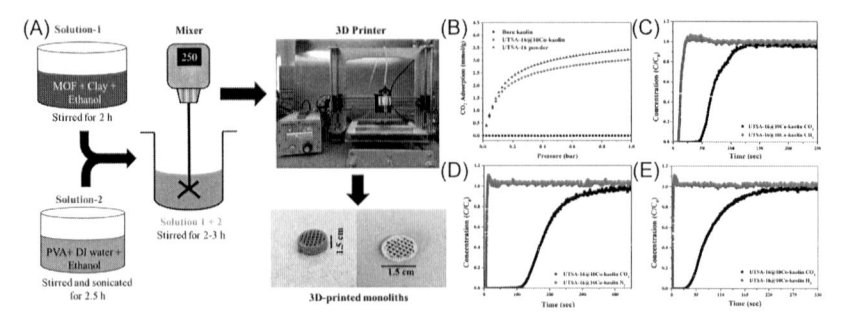

FIGURE 4.8 (A) Schematic of the 3D-printing procedure used to prepare metal-organic framework monoliths; (B) CO2 adsorption isotherms for 3D-printed bare kaolin, UTSA-16@10Co-kaolin, and UTSA-16 powder at 298 K; and (C–E) breakthrough profiles for CO_2/CH_4, CO_2/N_2, and CO_2/H_2 using 3D-printed UTSA-16@10Co-kaolin monolith at 298 K and 1 bar (Lawson et al., 2018; Thakkar et al., 2017).

respectively, which was slightly lower than that of the powder (4.7 mmol/g for MOF-74(Ni) and 3.5 mmol/g for UTSA-16(Co)). The as-synthesized 3D-print MOF-monoliths retain the physical properties and mechanical integrity of MOF powders and not only have good cyclic stability, but also exhibit relatively fast adsorption kinetics.

Furthermore, they investigated the feasibility of preparing other MOF (MOF-74(Ni), UTSA-16(Co), MIL-101(Cr), and HKUST-1) monoliths using 3D printing techniques, and the monolithic support materials that were considered for 3D printing included zeolite 13X, mesoporous silica, bentonite clay, and kaolin (Lawson, Al-Naddaf et al., 2018). The experimental results showed that the 3D-print UTSA-16@10Co-kaolin monolith had the best CO_2 capture performance. Compared to previously developed 3D-print UTSA-16 monoliths, UTSA-16@10Co-kaolin monoliths have not only higher MOF loading but also higher compressive stresses, indicating their robust structure. The CO_2 adsorption isotherms obtained at 298 K are shown in Fig. 4.8B. The 3D-print bare kaolin showed no significant CO_2 uptake, but UTSA-16@10Co-kaolin adsorbed considerable CO_2 (3.1 mmol/g), which is 89% of the uptake of the UTSA-16 powder. Breakthrough experiments confirmed that the 3D-print UTSA-16@10Co-kaolin monoliths can preferentially capture CO_2 from CO_2/CH_4, CO_2/N_2, and CO_2/H_2 mixtures (Fig. 4.8C–E) (Lawson et al., 2018).

Amine-functionalized metal-organic framework composites

Postsynthetic functionalization of MOFs with alkyl amines has been the most successful approach to enhance CO_2 capture performance. MOFs with open metal sites are the most common choice for the insertion of aminoalkanes, as the high density of these sites provides suitable locations for amine incorporation.

MIL-101(Cr), with its large pore volume, high BET surface area (2600–4500 m^2/g), high density of open metallic chromium sites (3 mmol/g), and good stability and moisture resistance, is an ideal candidate for loading various amines to produce efficient carbon capture adsorbents. In 2013 Chen's group reported a series of polyethyleneimine (PEI)-incorporated MIL-101 adsorbents with different PEI loadings (Lin et al., 2013). Although the surface areas and pore volumes of amine-functionalized MIL-101 decreased significantly, all MOF composites showed increased CO_2 capacity at low pressure. PEI is able to form strong interactions with CO_2 due to its high amine density and the accessible primary amine sites at the ends of the chain. With 100 wt.% PEI loading, the CO_2 capacity of the MOF composites at 0.15 bar reached 4.2 mmol/g at 298 K and 3.4 mmol/g at 323 K, together with fast adsorption kinetics (adsorption equilibrium could be rapidly reached in 5 minutes). More importantly, an extremely high CO_2 selectivity for CO_2/N_2 (15/85, v/v) mixtures was observed in these MOF composites. The CO_2/N_2 selectivity was 770 and 1200 at 298 K and 323 K, respectively. They also found that MOF particle size and PEI molecular weight have an important influence on CO_2 capture capacity (Lin et al., 2014). On the one hand, the smaller the MIL-101(Cr) particles, the easier PEI is loaded into the pores, and the smaller the surface area/pore volume ratio. This results in lower CO_2 capacity but better CO_2/N_2 selectivity. On the other hand, the lower molecular weight linear PEI can easily diffuse into the internal pores, effectively preventing N_2 from adsorbing. Compared to MIL-101(Cr), the PEI/MIL-101 composite has a higher selective CO_2 capacity (Fig. 4.9A–D). The reduction of the MOF crystal size can significantly improve the CO_2/N_2 selectivity, while the loading of branched PEI can improve the CO_2 capacity at high temperatures (Fig. 4.9E and F).

Walton, Sholl, and Jones, et al. firstly prepared the MIL-101(Cr)-TREN composite (TREN=tris(2-amino ethyl)), which exhibited high CO_2 capacity (2.8 mmol/g) (Darunte et al., 2016). However, in temperature swing adsorption cycles at 400 ppm CO_2 partial pressure, the MIL-101(Cr)-TREN composite showed an obvious loss of amine. This was due to the evaporation of the weakly adsorbed TREN under desorption conditions. The MIL-101(Cr)-PEI-800 containing less volatile PEI has good cycle stability. The MIL-101(Cr)-PEI-800 composite with a loading of 1.0–1.1 mmol of PEI/g of MOF maintained a good balance between the CO_2 capacity and the CO_2 adsorption kinetics.

Rezaei and coworkers prepared the PEI-MIL-101 monolith and the TEPA-MIL-101 monolith (TEPA=tetraethylenepentamine) by means of pre- and postfunctionalization approaches using 3D printing technology (Fig. 4.10) (Lawson et al., 2019). The composites showed higher CO_2 adsorption capacities at low pressures than the pristine MOFs (Darunte et al., 2016; Hong et al., 2009). At 3000 ppm and 298 K, the CO_2 capacities of the preimpregnated MOF monoliths (the contents of TEPA and PEI are 3.5 mmol/g and 5.5 mmol/g) were 1.6 and 1.4 mmol/g, respectively. However, the composite still suffers

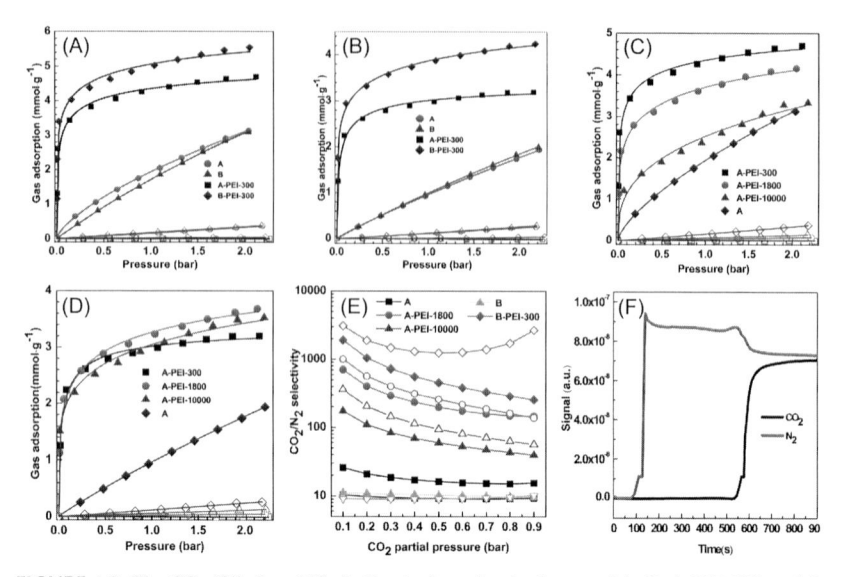

FIGURE 4.9 The CO_2 (filled) and N_2 (hollow) adsorption isotherms of A, B, A-PEI-300, and B-PEI-300 at (A) 298 K and (B) 323 K. The CO_2 (filled) and N_2 (hollow) adsorption isotherms of MIL-101(Cr) before and after loading PEI at (C) 298 K and (D) 323 K. Lines are the fitted isotherms, A and B represent MIL-101(Cr) with particle size of ~150 nm and ~250 nm. 300, 1000, 1800 represent the molecular weight of PEI. (E) The selectivity for the CO_2/N_2 mixture at a total pressure of 1 bar and temperatures of 298 K (filled) and 323 K (hollow), respectively. (F) Breakthrough curves of B-PEI-300 with an equimolar CO_2/N_2 mixture at 298 K (Lin et al., 2014).

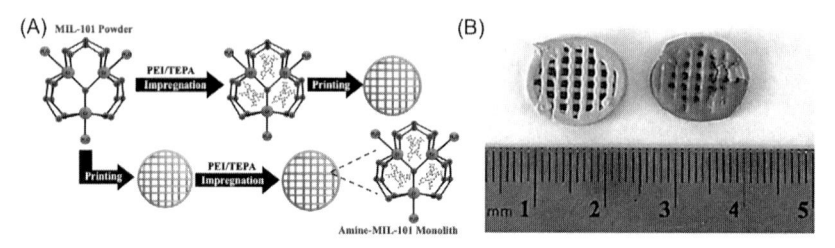

FIGURE 4.10 (A) Schematic of the pre- and postimpregnated MIL-101 monoliths formation processes. (B) Photo of 3D-print bare (left) and amine-functionalized (right) MIL-101 monoliths (Lawson et al., 2019).

from slow adsorption kinetics, with the key being a reduction in the overall wall thickness of the monoliths, thereby increasing the accessibility of the amine to the CO_2 molecules.

Developing adsorbents capable of capturing carbon under humid conditions is important for treating flue gases in practical industries. Xia and Li prepared the PEI-embedded ZIF-8 composite (PEI@ZIF-8), which demonstrated water

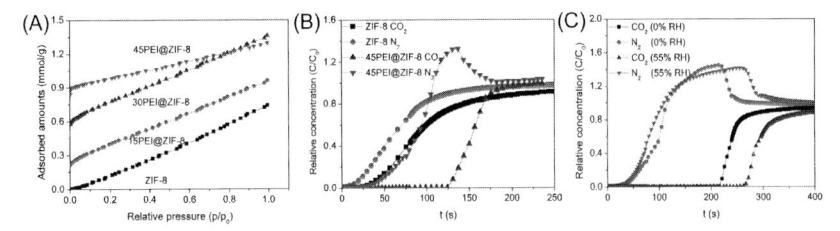

FIGURE 4.11 (A) The CO_2 adsorption isotherms of PEI@ZIF-8 composites with different PEI loading at 298 K. (B) Breakthrough curves of CO_2/N_2 (50/50, v/v) mixture on ZIF-8 and 45PEI@ZIF-8 at 303 K. (C) Breakthrough curves of CO_2/N_2 (50/50, v/v) mixture on 45PEI@ZIF-8 at 338 K and under different relative humidity (Xian et al., 2015).

vapor enhanced carbon capture performance (Xian et al., 2015). Under dry condition, the CO_2 capacity and CO_2/N_2 selectivity of 45PEI@ZIF-8 (PEI loading is 45 wt.%) calculated from breakthrough experiments are 0.95 mmol/g and 25.4, respectively (see Fig. 4.11). However, the CO_2 capacity and CO_2/N_2 selectivity can be improved to be 1.99 mmol/g and 89.3 under humid condition with 55% relative humidity (RH). The synergistic mechanism of water vapor and PEI is responsible for this enhancement. According to DFT calculation, under dry conditions, two amine groups in PEIs can react with a single CO_2 molecule, whereas in humid conditions, the presence of water vapor leads to the formation of bicarbonate, and a single bicarbonate molecule can react with a single amine group in PEIs. This encourages more CO_2 molecules to be adsorbed on the PEI@ZIF-8 composites, highlighting the potential of amine-functionalized MOFs for efficient CO_2 capture under realistic conditions.

The high density of open metal sites in M_2(dobpdc) (M=Mg, Mn, Fe, Co, Ni, Zn; dobpdc=4,4′-dioxidobiphenyl-3,3′-dicarboxylate) can be functionalized with diamines, alcohol amines, and alkoxy alkylamines, with the potential to further enhance CO_2 capture capability (Kim et al., 2020; McDonald et al., 2012, 2015; Siegelman et al., 2019). The mmen-M_2(dobpdc) (M=Mg and Mn, mmen=dimethyl ethylene diamine) was selected for further growth on a honeycomb cordierite monolith that is wash-coated with α-alumina for the construction of the monolith-supported contactor for the capture of CO_2 from simulated flue gas with a CO_2 concentration of 10%. The MOF composites adsorbed 2.37 mmol/g and 2.88 mmol/g at 10% and 100% CO_2 partial pressure, respectively (Darunte et al., 2017). The performance remained stable over multiple cycles and humid conditions. Hong and coworkers proposed a scalable MOF microbead synthesis method (see Fig. 4.12A) (Choe et al., 2021). After mixing MOF powder with alumina sol, the spray drying method can be used to produce MOF/Al microbeads with a particle size distribution of 30–70 μm, which have high mechanical strength (wear index <0.6%). The microbeads are functionalized with een (N-ethylethylenediamine) and coated with long

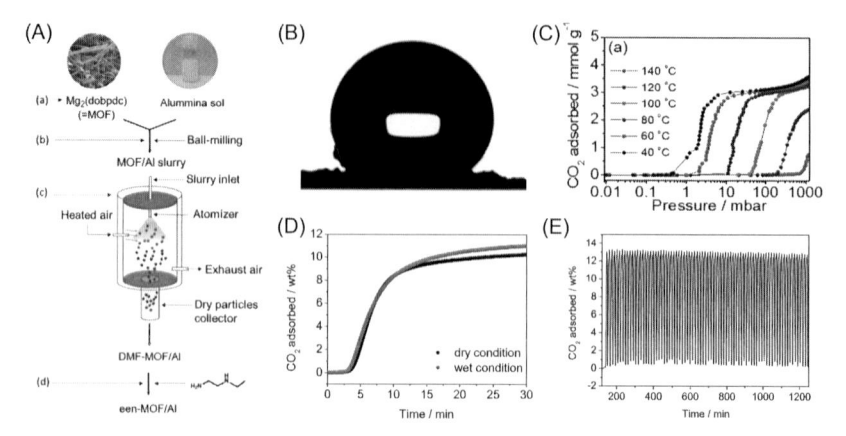

148 Applications of Metal-Organic Framework Composites

FIGURE 4.12 (A) Schematic of the spray-drying process of MOF/Al microbeads. (B) Contact angle of een-MOF/Al-Si. (C) Adsorption isotherms of CO_2 for een-MOF/Al-Si at various temperatures. (D) CO_2 adsorption capacities of een-MOF/Al-Si under dry and wet conditions; and (E) temperature-swing adsorption (TSA) cycles of een-MOF/Al-Si over 100 cycles (Choe et al., 2021).

alkyl chain silanes. This results in a composite material (een-MOF/Al-Si) with improved CO_2 capture capacity and high hydrophobicity (see Fig. 4.12B and C). For the separation of a CO_2/N_2 mixture with a CO_2 content of 15%, the selectivity was up to 53,182 at 313 K and decreased at higher temperatures. Its working capacity does not have a significant decrease even after 100 cycles of adsorption and desorption (see Fig. 4.12E). The een-MOF/Al-Si is promising for use as a practical adsorbent in industrial CO_2 capture due to its large working capacity, fast adsorption kinetics, good reusability, excellent long-term performance, and improved structural stability under humid conditions.

Although the diamine-Mg_2(dobpdc) exhibit excellent CO_2 capture performance, the adsorption amount of CO_2 is limited to one CO_2 molecule per diamine, and only a small amount of additional CO_2 is adsorbed by physical adsorption. Long and coworkers found that the introduction of pip2 (pip2=1-(2-aminoethyl piperidine)) into Mg_2(dobpdc) has significantly increased the CO_2 adsorption capacity to approximately 1.5 CO_2 molecules per diamine (Fig. 4.13A and B) (Zhu et al., 2024). FT-IR (Fourier transform infrared spectrometer) and NMR (nuclear magnetic resonance) spectroscopy results show that chemical and physical adsorptions occur simultaneously in pip2-Mg_2(dobpdc) (Fig. 4.13C and D). On the one hand, CO_2 reacts with pip2 to form ammonium carbamate chains; on the other hand, physically adsorbed CO_2 molecules occupy the space between adjacent ammonium carbamate chains along the skeleton pores, and there is a stable intermolecular interaction between the ammonium carbamate chains. Binder-free pellets of the materials

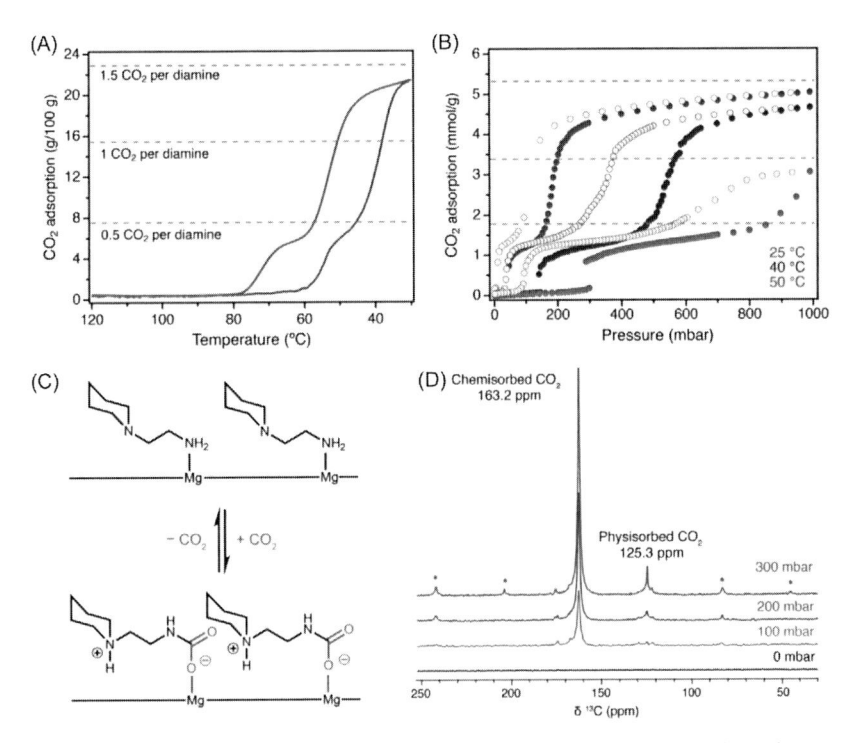

FIGURE 4.13 (A) Adsorption (blue) and desorption (red) isobars pip2-Mg$_2$(dobpdc) under pure CO$_2$ (~1 bar), as measured by thermogravimetric analysis. (B) Pure CO$_2$ adsorption isotherms for pip2-Mg$_2$(dobpdc) obtained at different temperatures. (C) Illustration of the proposed mechanism for CO$_2$ chemisorption in pip2-Mg$_2$(dobpdc) based on solid-state NMR data. (D) Solid-state MAS ^{13}C NMR spectra for a sample of pip2-Mg$_2$(dobpdc) dosed at different ^{13}CO$_2$ pressures (Zhu et al., 2024).

were also prepared. The pip2-Mg$_2$(dobpdc) was granulated and sieved to obtain 20–40 mesh particles for fixed bed breakthrough experiments. The results showed good cyclicity and stability, and a high dynamic CO$_2$ capacity of 5.1 mmol/g.

To address the problem of reduced CO$_2$ capture performance of diamine-functionalized Mg$_2$(dobpdc) under humid conditions due to diamine loss, a hydrophobic carbonate compound (tert-butyl decarbonate, Boc) was used to protect the een-Mg$_2$(dobpdc) (Fig. 4.14A–C) (Choe et al., 2024). Boc effectively prevents water from entering the framework by reacting rapidly with the diamine to form dense secondary and tertiary hydrophobic amines. Even under simulated flue gas conditions with a humidity of 10%, the Boc-protected MOF still showed an excellent CO$_2$ capture capability. It was confirmed that the Boc protection led to an improvement in the hydrophobicity of the framework according to the CO$_2$ adsorption isobars (15% partial

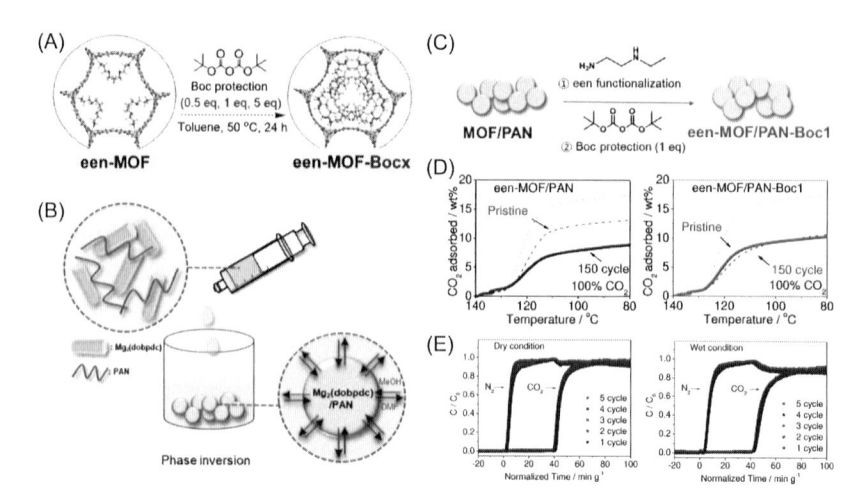

FIGURE 4.14 (A) Scheme of Boc protection for een-MOF. (B) Scheme of the Mg$_2$(dobpdc)/PAN composite beads preparation. (C) Preparation of een-MOF/PAN-Boc1. (D) Pure CO$_2$ isobars of een-MOF/PAN and een-MOF/PAN-Boc1 before and after humid cycle tests. (E) Cyclic breakthrough tests for een-MOF/PAN-Boc1 under dry (15% CO$_2$ and 85% N$_2$) and wet conditions (15% CO$_2$ and 75% N$_2$ with 10% H$_2$O) (Choe et al., 2024).

pressure) under dry and 10% humidity conditions (Fig. 4.14D). In the temperature range 353–363 K, wet isobars and dry isotherms of een-MOF-Boc1 and een-MOF-Boc5 overlap, indicating that the presence of Boc effectively resists coadsorption of H$_2$O and achieves selective CO$_2$ capture under these conditions. Adsorption–desorption experiments confirm that the material achieves better cycling performance even under 10% wet conditions due to the introduction of the bulky hydrophobic Boc group. After ball milling, the original powder was mixed with PAN (polyacrylonitrile) in DMF solution to give Mg$_2$(dobpdc)/PAN mixture, which was then added dropwise to methanol to form composite microspheres. The formed composite retained porosity, and the BET surface area (2913 m^2/g) was slightly lower than that of the original powder (3383 m^2/g). The introduction of the Boc groups can maintain the crystallinity, CO$_2$ adsorption capacity, and een content of the composite, thus significantly improving the reusability and stability of the material. The breakthrough experiments further verified the adsorption and separation performance of the material (Fig. 4.9E), suggesting that een-MOF/PAN-Boc1 is a potential adsorbent that can efficiently capture CO$_2$ from humid flue gases.

Pellet-based systems suffer from high pressure drops and extended mass transfer zones, resulting in an inefficient utilization of the adsorbent. In contrast, hollow fiber sorbents provide local heat management by flowing water in the fiber hole, offer low pressure drop, shorten the mass transfer zone, enable rapid temperature change and, if a waterproof layer is installed on the hole side, enable effective heat integration. Using stable metal oxide precursors, Koh and coworkers prepared the diamine-Mg$_2$(dobpdc)/PEI and tetraamine-Mg$_2$(dobpdc)/PEI hollow fibers by the "dry-jet, wet-quench" method, which exhibited dynamic CO_2 capacities of 0.70 and 1.99 mmol/g, respectively, at 308 K and 400 ppm CO_2 partial pressure (Lee et al., 2021). A MOF hollow fiber composite with 70% MOF content (2-ampd-Mg$_2$(dobpdc)/PES, 2-ampd=2-(aminomethyl)piperidine, PES=poly(ether sulfone)) can be prepared by adding Mg$_2$(dobpdc) to PES for direct spinning and then introducing 2-ampd during the solvent exchange process after spinning (Quan et al., 2022). A unique two-steps adsorption isotherm was generated due to the steric hindrance of 2-ampd on Mg$_2$(dobpdc). The CO_2 capacity at 40 mbar is approximately 2.0 mmol CO_2/g fiber and 3.0 mmol CO_2/g MOF. The two-step adsorption also leads to a "shock–wave–shock" shaped breakthrough curve. Compared to the granular fixed bed breakthrough curve, the hollow fiber breakthrough curve is steeper, indicating a higher mass transfer rate.

Metal-organic framework and carbon composites

Although various types of MOFs have been widely used for adsorption, gas molecules cannot completely fill the large voids inside the MOFs due to the weak interaction between the framework walls and the adsorbents. Moreover, the practical application of MOFs for carbon capture is limited by poor hydrothermal stability and insufficient interaction between gas molecules and MOFs. Carbon-based porous materials, such as activated carbon (AC), carbon nanotubes (CNTs), graphene oxide (GO), and cellulose, exhibit excellent water resistance and structural stability. Combining MOFs with these carbon-based porous materials can effectively enhance the structural stability of MOF materials and provide abundant mesopores and macropores, which serve as excellent diffusion channels for gas molecules, greatly improving gas adsorption and separation performance.

Competitive adsorption of H_2O affects the performance of activated carbon (AC) in capturing CO_2 from flue gas. Hydrophobic ZIF-8 can help to mitigate this issue. Since the surface of carbon materials does not offer enough active sites for the growth of ZIF-8, polydopamine (PDA), which is adhesive and rich in active groups (-OH and -NH$_2$), is used as a binder between the original AC and the hydrophobic ZIF-8 (Ji et al., 2023). The composite (AC@ZIF-8) with long-lasting high surface area and hydrophobicity can be obtained by further growth of ZIF-8 on PDA-coated AC. The CO_2 capacity of the composite was 41.69 cm^3/g, only 13.61% lower than that of AC under dry conditions, but much higher than that of ZIF-8 (16.72 cm^3/g). The -OH and -NH$_2$ groups in PDA are a high affinity for CO_2. The hydrophobic ZIF-8 layer effectively prevents water molecules from contacting the AC inside, and under high humidity, the CO_2 capacity of the composite only dropped by 16.74%.

Porous carbon materials have the advantages of a large specific surface area, a large pore volume, a high degree of chemical inertness, low cost, and good mechanical stability. It has been shown that the CO_2 adsorption capacity at normal pressure is maximized by the micropores (< 1 nm) in carbon materials. In addition, the presence of nitrogen content in carbon materials also determines their basicity, which is also conducive to CO_2 adsorption. Jaroniec and coworkers prepared the composites of Cu-BTC, ordered mesoporous nonactivated carbon (OMC), mesoporous activated carbon (AC), and nitrogen-containing microporous carbon (NC) as CO_2 adsorbents (Liu et al., 2019). The formation of additional micropores in the heterogeneous interface region greatly increased the surface area and porosity of the composites. As a result of the additional micropores, the CO_2 absorption capacity of the composite is greatly enhanced compared to the parent Cu-BTC and carbon materials. At 298 K, NC-Cu-BTC demonstrated a higher CO_2 uptake (4.51 mmol/g) and CO_2/N_2 selectivity (18.9) than that of Cu-BTC (CO_2 uptake of 4.07 mmol/g, and CO_2/N_2 selectivity of 16.8).

The nitrogen-doped porous hierarchical carbon monolith (HCM) has a polar surface and abundant micropores, which can provide a microenvironment for growing MOF grains to maximize volumetric CO_2 adsorption capacity. Cu-BTC was synthesized in situ using HCM as a matrix and HCM-$Cu_3(BTC)_2$ composites were obtained by a stepwise impregnation crystallization method (see Fig. 4.15A) (Qian et al., 2012). The SEM images clearly show the growth of MOF crystals within the macropores of the HCM matrix, and the sponge-like skeleton of the HCM does not change after MOF growth (Fig. 4.15B–E). HCM-$Cu_3(BTC)_2$ composite has a CO_2 capacity of 22.7 cm^3/cm^3, almost double that of the original HCM (Fig. 4.15F). HCM plays a major role in capturing CO_2 at low relative pressures, whereas $Cu_3(BTC)_2$ plays a major role at relatively high pressures. Under ambient condition, the composite with the best CO_2 capture performance, HCM-$Cu_3(BTC)_2$-3, exhibits a selectivity of 18.4 for CO_2/N_2 (Fig. 4.15G). The breakthrough experiment also verified the composite's ability to capture CO_2 from the CO_2/N_2 mixture, and the separation factor estimated from the breakthrough curve was 67–100. In addition, purging with inert gas at room temperature can achieve rapid release of CO_2.

Hollow fibers have proven to be excellent gas–solid contactors in adsorptive separation processes, mainly due to their unique combination of high surface area and packed volume porosity. Hollow polymeric fibers have many advantages for the incorporation of inorganic compounds and the synthesis of composite materials. However, they suffer from chemical instability and swelling problems under various hydrothermal MOF synthesis conditions. In contrast, carbon hollow fibers (CHF) have high porosity and structural integrity under MOF synthesis conditions. First, the hollow carbon

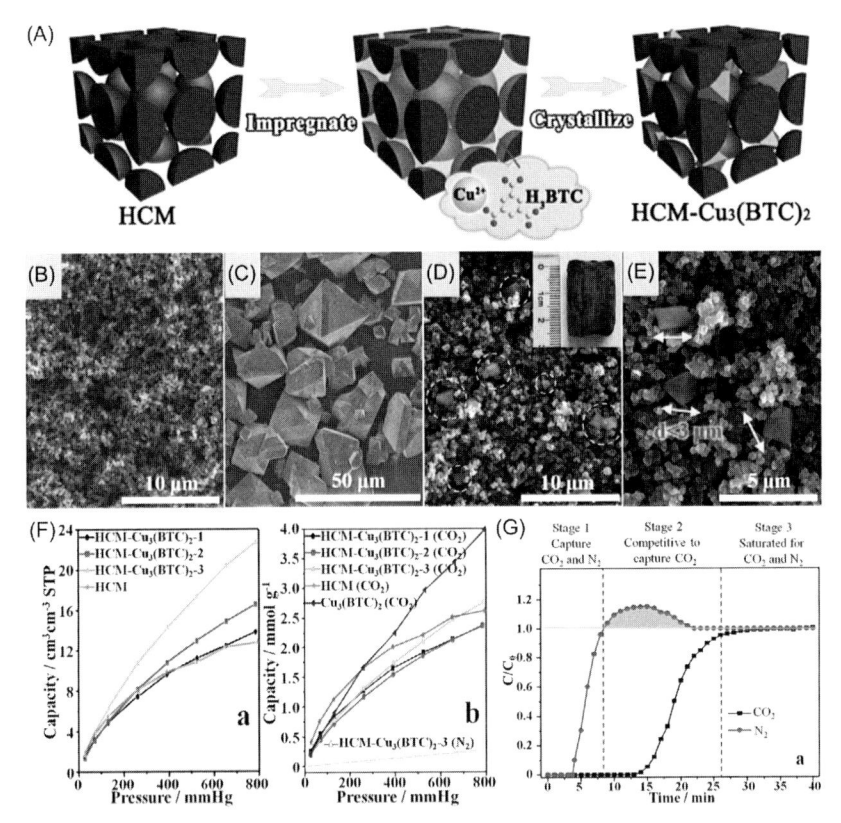

FIGURE 4.15 (A) Scheme procedure to fabricate HCM-Cu$_3$(BTC)$_2$ composites; SEM micrographs of (B) HCM, (C) Cu$_3$(BTC)$_2$, and (D and E) HCM-Cu$_3$(BTC)$_2$-3. (F) The CO$_2$ and adsorption isotherms on the volumetric and gravimetric basis. (G) Breakthrough curve of HCM-Cu$_3$(BTC)$_2$-3 using a stream of CO$_2$/N$_2$ (16/84, v/v) at 298 K (Qian et al., 2012).

fibers were functionalized to increase the surface hydroxyl groups, and then MOFs (MOF-74 and UTSA-16) were grown into the pores and outer surface of the hollow fibers by dip coating or LBL techniques. Finally, the composites with 37%–38% MOF loading were prepared (Lawson et al., 2018). At 298 K and 1 bar, the CO$_2$ capacity of CHF-MOF-74 and CHF-UTSA-16 was 2.0 and 1.2 mmol/g, respectively. The results of adsorption response time indicated that the MOF composites exhibit better kinetics and reach equilibrium faster than the parent MOF powders.

CNTs have excellent mechanical strength, hydrophobicity, and adsorption properties. The separation performance and mechanical stability of MOF composites are expected to be enhanced if CNTs are properly incorporated into MOFs. Anbia and colleagues prepared the MWCNT@MIL-101 hybrid composite (MWCNT=multiwall carbon nanotubes), which had a larger micropore volume, better CO_2 capacity, and higher thermal stability than the original MOF (Anbia & Hoseini, 2012). Another successful example of combining MOFs with CNTs is the CNT@Cu-BTC composite, which has significantly improved CO_2 and CH_4 capture capabilities. At 298 K and 18 bar, the CO_2 capacity of the CNT@Cu-BTC was 595 mg/g, twice as high as that of the original MOF (295 mg/g) (Xiang, Hu, et al., 2011). Cao et al. studied the carbon capture performance and CO_2/CH_4 separation ability of the hybrid composite CNT@$Cu_3(BTC)_2$ (Xiang, Peng, et al., 2011). The CO_2 selectivity of the composite was improved by the ultramicroporous structure generated at the interface of MOF crystals grown on the surface of CNTs. Zhu and coworkers synthesized a series of ZIF-8 incorporated and hydroxyl functionalized CNT composites with different CNTs loadings (Yang et al., 2014). The introduction of the mesopores of the CNTs into the microporous system of the ZIF-8 is expected to improve the CO_2 adsorption capacity and the CO_2/N_2 selectivity, and to realize the preparation of bulk materials with hierarchical structures. It has been confirmed that the ZIF-8 particles in the composite have a similar crystal structure and morphology to pure ZIF-8, but the thermal stability of the composite is significantly improved. There is a synergistic effect between the ZIF-8 and the CNTs. Incorporating CNTs can provide more nucleating sites and promote nucleation and crystallization of ZIF-8.

GO is an important precursor to graphene. It has a 2D-layered conjugated structure and is abundant in oxygen-containing functional groups, including hydroxyl and epoxy groups, mostly distributed in the graphene layer, and carbonate groups at the edge of the layer. The advantages of dense atomic arrangement, rich oxygen functionalities, unique structure, and supramolecular properties make GO a promising growth matrix for composite materials, which not only increases porosity and dispersing forces, but also reduces inner pore space to enhance adsorption.

Adding aminated graphite oxide (AGO) to MOF-5 resulted in a significant increase in porosity as new pores are created at the graphite/MOF interface (Zhao et al., 2013). Compared to the original MOF, the CO_2 capacity and moisture resistance of the MOF composite were improved. The combination of GO and the NbO-type copper tetracarboxylate framework can generate a new composite (MOF-505@GO) with higher CO_2 capacity and CO_2/CH_4 and CO_2/N_2 selectivity (Chen, Lv et al., 2017). This is due to the new micropores and unsaturated metal sites formed in the composite and the enhanced surface dispersion forces.

Maji and Rao et al. were the first to grow and fix nanoscale MOFs on the surface of GO, resulting in a hybrid nanocomposite (ZIF-8@GO) (Kumar et al., 2013). GO acts as a structure-directing agent, and its content is able to regulate the size and morphology of the ZIF-8 nanocrystals (from hexagonal to spherical), thus influencing the morphology and porosity of the composites at the nanoscale. Compared to ZIF-8 (27 wt.%) and GO (33 wt.%), the CO_2 capacity of the composite material was significantly improved (49 wt.% for ZG-4, 72 wt.% for ZG-20). This capacity improvement resulted from the cumulative effect of ZIF-8 and GO, and because GO can provide specific CO_2 interactions through different polar functional groups.

UiO-66 is a classic Zr-based MOF with high chemical and thermal stability. The composite obtained by its combination with GO has a higher BET surface area and better CO_2 capacity (Cao et al., 2015). However, the CO_2 capacity decreases as the amount of GO incorporation increases. This may be due to the fact that as the concentration of GO increases, its degree of filling in the pores of the MOF increases, resulting in the reduction of the pore windows and the blocking of the pores.

Improved carbon capture and hydrogen storage performance was also observed for the hybrid MOF composites based on Cu-BTC and GO layers. At 273 K and 1 bar, the CO_2 capacity increased from 6.39 mmol/g to 8.26 mmol/g (Liu, Sun, et al., 2013). Xia, Li and coworkers have synthesized a composite from Cu-BTC and GO using the solvothermal method (Huang et al., 2014). The composite has a higher BET surface area and pore volume compared to the original MOF. Adsorption experiments show that GrO@Cu-BTC not only has a high CO_2 capacity (8.19 mmol/g), but also maintains a high CO_2/CH_4 selectivity (14) (Fig. 4.16A and B). The temperature-programmed desorption

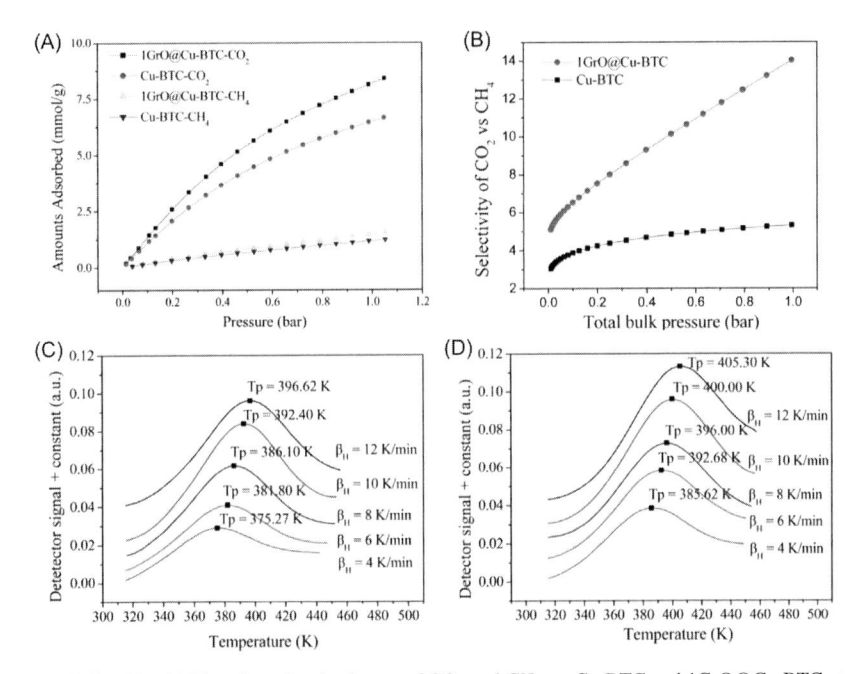

FIGURE 4.16 (A) The adsorption isotherms of CO_2 and CH_4 on Cu-BTC and 1GrO@Cu-BTC at 273 K; (B) IAST selectivities of CO_2/CH_4 (50/50, v/v) mixtures on Cu-BTC and 1GrO@Cu-BTC at 273 K; temperature-programmed desorption (TPD) spectra of CO_2 on (C) Cu-BTC and (D) 1GrO@Cu-BTC at different heating rates (Huang et al., 2014).

(TPD) results show that the desorption activation energy of CO_2 on 1GrO@Cu-BTC (68.6 kJ/mol) is higher than that on Cu-BTC (56.7 kJ/mol), indicating a stronger interaction between the composite material and CO_2 (Fig. 4.16C and D). They also proposed an ultrafast synthesis method to rapidly synthesize the composite of GO and HKUST-1 (GrO@HKUST-1) within 1 minute at room temperature (Xu et al., 2016). The BET surface area and pore volume of GrO@HKUST-1 are higher than those of the parent HKUST-1. At 273 K and 1 bar, the CO_2 capacity and CO_2/N_2 selectivity were 9.02 mmol/g (32% improvement over HKUST-1) and 186 (1.8-folds of HKUST-1), respectively.

Highly crystalline and defect-free MOFs are usually accompanied by an additional resistance to the diffusion of guest molecules and a limited accessibility to the internal active sites. Defect engineering is a versatile approach to unleashing the potential value of the specific surface area, active sites, and porosity of MOFs. GO was used as a modulator to promote defect formation in MOF by changing the crystallization time, nucleation, and crystal growing process (Fig. 4.17) (Niu et al., 2023). The structural characterization results showed that the MOF in the composite material contained a large number of cluster defects, which were stabilized by GO. This resulted not only in a higher defect frequency, but also in a higher thermal stability of the

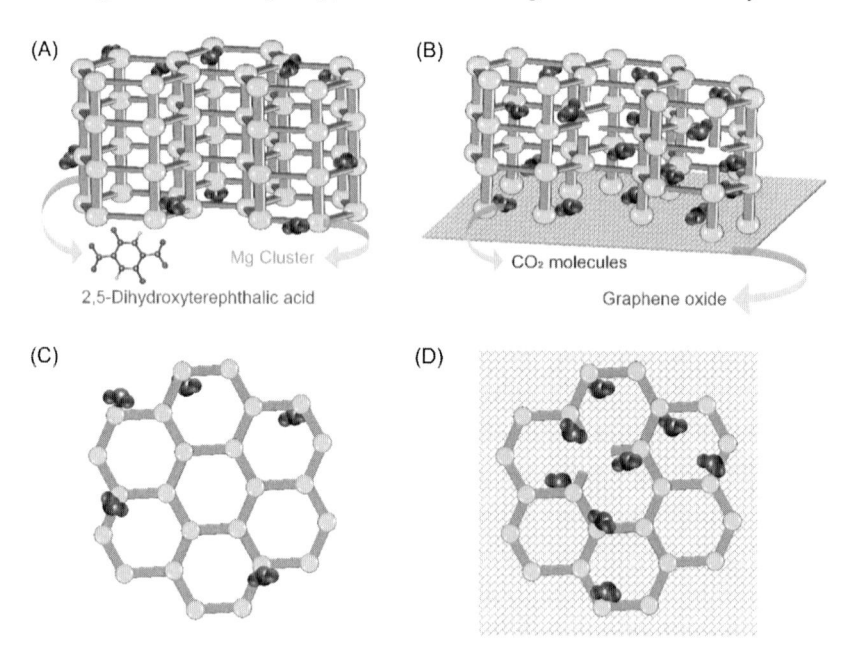

FIGURE 4.17 Schematic illustration of (A, B) Mg-MOF-74 and (C, D) MOF@GO composite with CO_2 adsorbed on the framework (Niu et al., 2023).

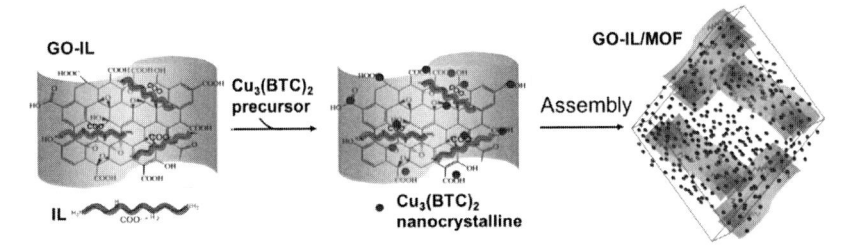

FIGURE 4.18 Formation mechanism of the GO-IL/MOF composite (Bian et al., 2014).

composite material. Compared to the original Mg-MOF-74, the surface area and total pore volume of MOF@GO composite increased by 18% and 15%, respectively. The diffusion resistance was also reduced, which means that gases can more easily approach the Mg^{2+} sites inside the MOF. At 0.1 bar and 298 K, the CO_2 adsorption of the MOF composite reached 6.06 and 9.17 mmol/g, respectively, which are 19.29% and 16.37% higher than that of the original Mg-MOF-74. The CO_2/N_2 selectivity is about 17.36% higher than that of the original Mg-MOF-74 (Fig 4.18A–D).

The GO-IL@MOF ternary composite (IL=Ionic Liquid) wase synthesized by self-assembly by Bian et al. (2014) (see Fig. 4.18). Amine or imidazole cations were first adsorbed on the surface of GO to prevent GO deformation, and then a large number of active sites were provided by ionic liquid-assisted growth to adsorb Cu^{2+}. A layer of nanoscale $Cu_3(BTC)_2$ particles can be uniformly grown on the surface of GO by reasonably controlling the synthesis conditions. Finally, the ternary composites (GO-Tac@MOF, GO-TBF4@MOF, and GO-BBF4@MOF) were formed. Among them, GO-TAc@MOF-60 showed the best CO_2 adsorption performance, with the CO_2 adsorption amount reaching 5.62 mmol/g at 298 K and 1 bar, and the CO_2/N_2 selectivity reaching 20.81. Moreover, the MOF composite has good CO_2/N_2 dynamic separation performance and cyclic stability. The IL-assisted strategy proposed in this work can be widely used for the preparation of multifunctional MOF composites.

Spherical polyacrylonitrile beads (MIL-101(Cr)@GO/PAN, PAN= polyacrylonitrile matrix) with grain sizes of 2–3 mm and good mechanical properties were produced by combining water-resistant MIL-101(Cr) with GO (Fig. 4.19A and B) (Gebremariam et al., 2023). This composite has a hierarchical pore structure and MOF loading up to 80%. During the in situ growth of MIL-101(Cr), the GO acts as a structural support for the nucleation of the MOF on the surface of the GO. There are two types of pores in the composite, one is the micropores of the MOF and the addition of GO creates new mesopores. The CO_2 adsorption kinetics of the composites were faster than those of the MOFs, suggesting that the additional porous channels facilitated the

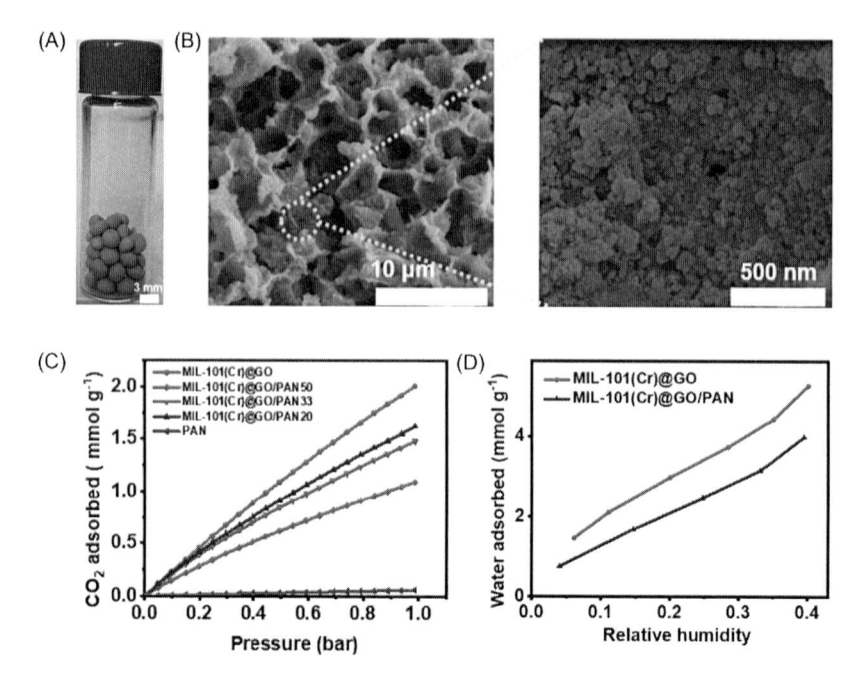

FIGURE 4.19 (A) Photograph and (B) cross-sectional SEM images of MIL-101(Cr)@GO/PAN20; (C) CO$_2$ adsorption isotherms of MIL-101(Cr)@GO/PAN beads (per total material mass) compared to MIL-101(Cr)@GO powder at 298 K; and (D) water vapor adsorption isotherm of MIL-101(Cr)@GO/PAN20 beads at 298 K and up to 40% RH, compared with MIL-101(Cr)@GO (Gebremariam et al., 2023).

transport of CO$_2$ within the composite structure and made it easier to contact the active sites in the MOFs. The CO$_2$ capacity of MIL-101(Cr)@GO/PAN20 is 2.53 mmol/g at 273 K and 1 bar, which is about 80% of that of MIL-101(Cr)@GO powder (Fig. 4.19C). The CO$_2$/N$_2$ selectivity of MIL-101(Cr)@GO/PAN20 is 28.6 and 15.5 for CO$_2$/N$_2$ mixed gas containing 15% CO$_2$ at 0.1 bar and 1 bar, respectively. The excellent hydrophobicity of MIL-101(Cr)@GO/PAN20 makes it promising for capturing CO$_2$ from humid flue gas (Fig. 4.19D).

The incorporation of amine functionalities into the pores of MOFs provides polar sites that can be highly interactive with CO$_2$. However, the stiffness and brittleness of the MOF structure dictates its existence in powder form, resulting in poor processability, weak binding affinity, and slow diffusion kinetics, limiting its industrial application in CO$_2$ capture. Cellulose-modified MOFs have been shown to overcome the above shortcomings and significantly improve CO$_2$ absorption capacity. The world's most abundant renewable biomass, cellulose, has the advantages of large surface area, high aspect ratio, diverse surface chemistry, excellent mechanical strength, and biocompatibility. Benefiting from the high density of hydrogen bonds in the cellulose backbone, cellulose can provide a good matrix for oriented MOF attachment, avoiding particle aggregation and improving MOF processability/recyclability for carbon capture.

Bamboo powder cellulose (BM), with its high strength, high stability, and natural channel structure, can anchor NH_2-MIL-101 nanoparticles by means of hydrogen bonding to form the amine-functionalized cellulose-based MOF composite (NH_2-MIL-101@BM) (see Fig. 4.20A) (Shi et al., 2024). The physical sorption synergistic effect and chemisorption synergistic effect between NH_2-MIL-101 and BM can not only enhance the mass transfer of CO_2 by reducing the diffusion resistance of the hierarchical pore structure, but also provide stronger adsorption sites for CO_2 by increasing the adsorption energy of open metal site-CO_2 or amine site-CO_2. At 20 wt.% BM doping, the MOF composite (NH_2-MIL-101@20%BM) has an excellent initial CO_2 adsorption capacity of 13.4 mmol/g at 298 K and good cycle stability after 15 cycles (see Fig. 4.20B and C).

Valencia and Abdelhamid (2019) presented a new synthetic strategy for the preparation of nanocellulose leaf-like zeolitic imidazolate framework (ZIF-L)

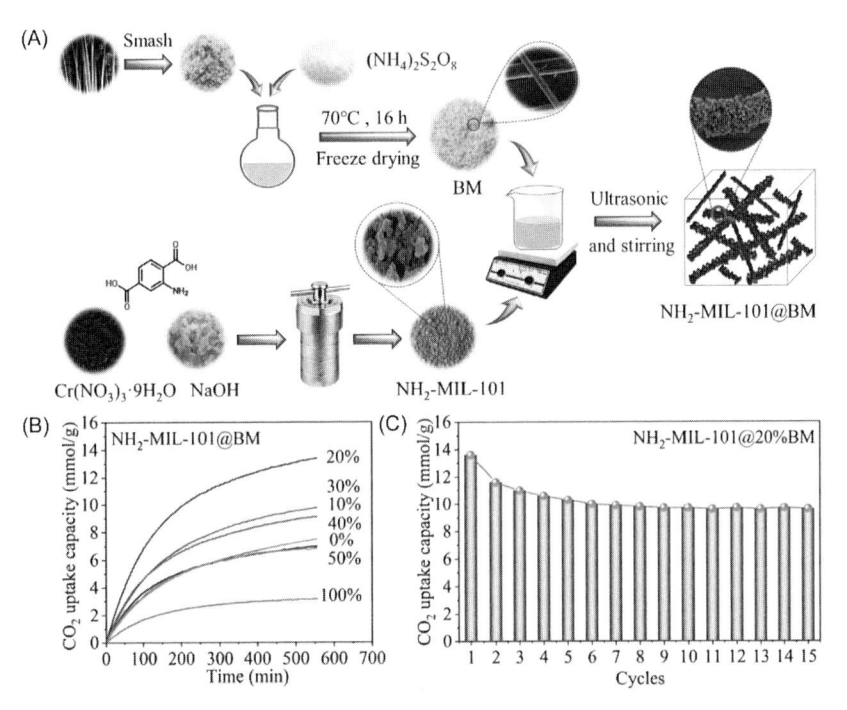

FIGURE 4.20 (A) Illustration of the synthesis process for NH2-MIL-101@BM composite sorbents; (B) CO_2 uptake of NH2-MIL-101@BM composites with varying BM doping levels; and (C) CO_2 uptake capacity of NH_2-MIL-101@20%BM over 15 cycles (Shi et al., 2024).

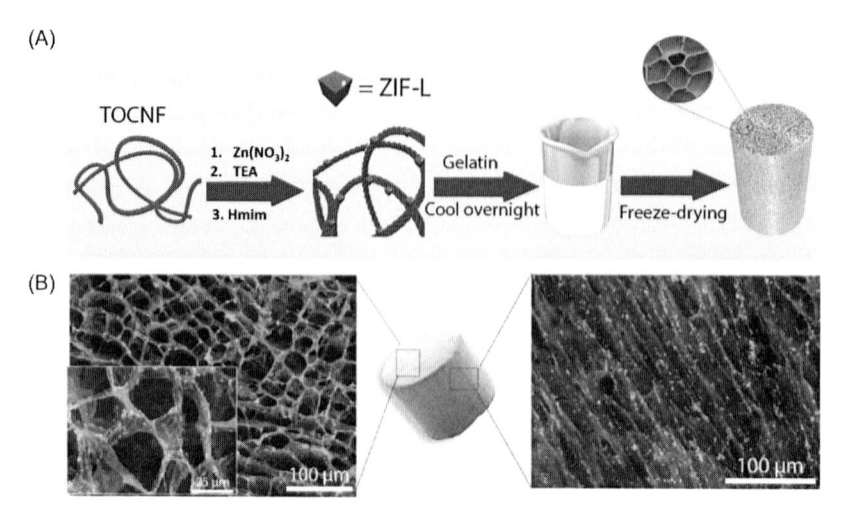

160 Applications of Metal-Organic Framework Composites

FIGURE 4.21 (A) Schematic illustration of the fabrication process of ZIF-L@TOCNF foams. (B) The microstructure of freeze-dry hybrid foams (Valencia & Abdelhamid, 2019).

foams using water as a solvent at room temperature (Fig. 4.21). The method is facile, fast, and eco-friendly. At low ZIF-L loading (21%–50%), the composite foams have good CO_2 adsorption capacity and high CO_2 selectivity. The synthesis process includes in situ MOF growth, gelatin matrix incorporation, and freeze-drying. The foam composite has an ultra-light weight (the density is only 19.18–37.4 kg/m; Feng et al., 2015), and its hierarchical porous structure can enhance the diffusion of the CO_2 molecules and accelerate the adsorption and regeneration kinetics. The synergistic effect between nanocellulose and ZIF-L improves the carbon capture capability of the hybrid foam. The hierarchical porous structure of the foam composite promotes rapid mass transfer of CO_2 to the adsorbent through mesopores and macropores, and strong interaction between CO_2 and ZIF-L in micropores. Multi adsorption–desorption cycles at room temperature confirmed reusability of hybrid foam composite as CO_2 adsorbent.

Abdelhamid and Mathew prepared the binder-free 3D-printed cellulose-ZIF-8 composites and investigated the adsorption performance of the composites. The whole process involved gel formation and in situ growth of MOF crystals (Fig. 4.22A) (Abdelhamid et al., 2023). The TMEPO (2,2,6,6-tetramethylpiperidine-1-oxyl radical) cellulose nanofibril (TOCNF) was selected as the matrix to form the TOCNF/Hmim and TOCNF/ZnO nanosheets. It was then used as the ligand and metal source for the in situ growth of ZIF-8, respectively. The 3D print CelloZIF_ZnO composite showed a better adsorption performance with a CO_2 capacity of 0.63 mmol/g at 1 bar, corresponding to the performance of pure ZIF-8 (0.85 mmol/g) (Fig. 4.22B and C). Additionally, the composites were also

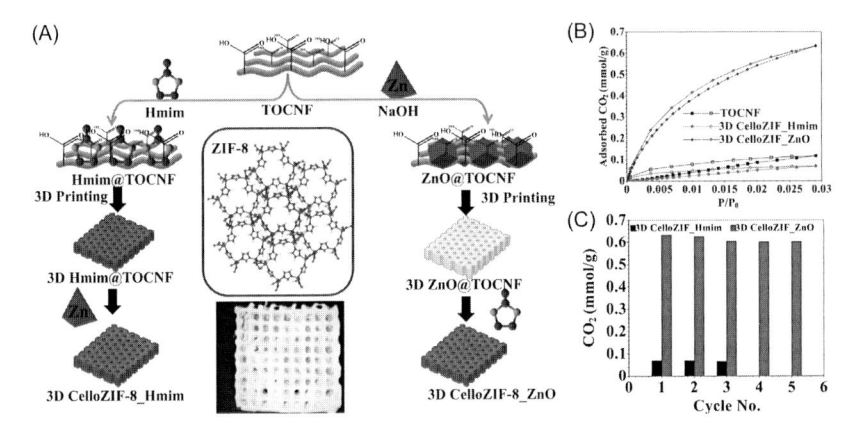

FIGURE 4.22 (A) Schematic representation for 3D printing of CelloZIF-8 composites. (B) The CO_2 adsorption isotherms of TOCNF and 3D print CelloZIF composites. (C) The CO_2 adsorption recyclability of 3D print CelloZIF composites (Abdelhamid et al., 2023).

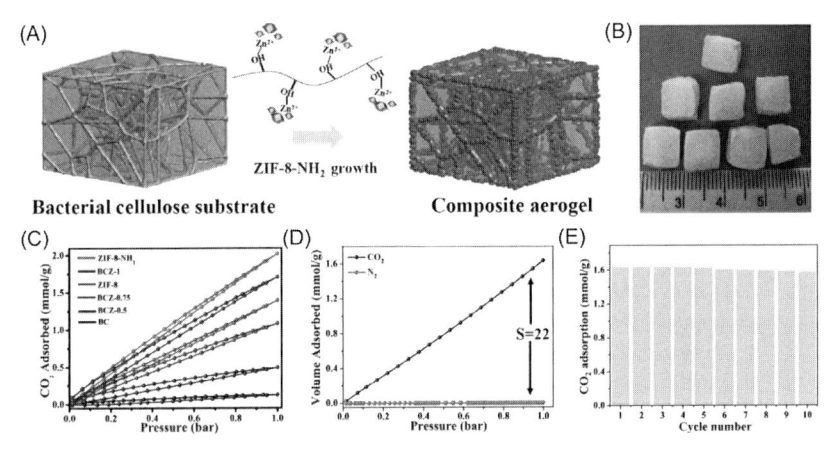

FIGURE 4.23 (A) Schematic illustration of the fabrication of ZIF-8-NH_2@BC foams; (B) photograph of ZIF-8-NH_2@BC foams; (C) CO_2 adsorption isotherms of ZIF crystals and composite foams at 273 K; (D) single-component adsorption isotherms and calculated CO_2/N_2 selectivity for BCZ-1 at 298 K; and (E) CO_2 adsorption–desorption cycles of BCZ-1 (Ma et al., 2021).

efficient at adsorbing organic dyes and removing heavy metal ions, showing excellent catalytic performance for dye degradation, with potential for customized and commercial applications.

Bacterial cellulose (BC), which is ubiquitous in nature, sustainable and biodegradable, has been used as a substrate for the growth of ZIF and as a matrix for the support and dispersion of ZIF particles (Fig. 4.23A and B) (Ma et al., 2021).

Using an in situ growth method, amino-functionalized ZIF-8 (ZIF-8-NH$_2$) was prepared in the BC substrate. The ZIF crystals are uniformly wrapped around the cellulose fibers, and chelation between the zinc ions and hydroxyl groups gives the composite greater interfacial affinity and compatibility. The prepared ZIF-8-NH$_2$@BC foam has a high CO$_2$ capacity (1.63 mmol/g) at 273 K and CO$_2$/N$_2$ selectivity (22) at 298 K (Fig. 4.23C and D). It can also be fully regenerated at 353 K, and the CO$_2$ capacity of BCZ-1 showed only a slight decrease after 10 cycles of adsorption and desorption (Fig. 4.23E). This hybrid foam also has the benefits of low density, flexibility, and mechanical strength. The addition of ZIF-8-NH$_2$ did not significantly affect BC substrate structural integrity and mechanical properties. The BCZ-1 composite had a high level of compressive strength and retained more than 90% of the compressive strain after 100 cycles.

Jones' group used 3D printing technology to produce the composites composed of cellulose acetate (CA) and adsorbent particles MIL-101(Cr) (Wang et al., 2024). These composites feature interpenetrated macroporous polymeric scaffolds with uniformly distributed MOF microcrystals, and the adsorbent loading can reach 70 wt.%. Then, the branched PEI was successfully incorporated to afford the CA/MIL-101(Cr)/PEI composite. The composite can adsorb up to 1.05 mmol/g CO$_2$ at 253 K and 400 ppm, has a working capacity of 0.95 mmol/g and good cycle stability, making it suitable for direct air capture at lower temperatures.

Metal-organic framework and ionic liquid composites

Ionic liquids (ILs) are a class of molten salts with melting points below 100°C, composed entirely of ions. They have unique properties such as negligible vapor pressure, high thermal stability, and tunable physicochemical properties. ILs have been widely used in gas separation processes due to their high CO$_2$ solubility and selectivity. However, the practical application of ILs in gas separation is hindered by their high viscosity, which leads to slow mass transfer and diffusion rates. Incorporating ILs into porous materials, such as MOFs, can overcome these limitations while preserving the advantageous properties of both components. The combination of ILs and MOFs can be achieved through various methods, including impregnation, encapsulation, and ionothermal synthesis (Pandya et al., 2024; Ullah et al., 2022).

A commonly used method is to incorporate ILs into the pores of the MOF. This requires that the hydrophilicity and hydrophobicity of the two composites are consistent, and that the pore size of the MOF is large enough to accommodate the ILs. In addition, the structural integrity of the MOF during synthesis after the introduction of ILs and the long-term stability of the selected ionic liquid and MOF must be taken into account. Using simulation calculations, Jiang and coworkers evaluated the CO$_2$ capture performance of the IRMOF-1 composite ([BMIM][PF$_6$]/IRMOF-1, [BMIM][PF$_6$]=1-n-butyl-3-methylimidazolium hexafluorophosphate) after incorporation of IL

(Chen et al., 2011). The results showed that due to the confinement effect of the nanoporous IRMOF-1 framework, the IL pairs in the composites are more ordered than those in the bulk phase. After adding ILs to the MOF, PF_6^- occupied metal cluster corners, becoming strong ionic adsorption sites for CO_2. With increasing IL ratio in composites, ions began to occupy the open pores and were more uniformly distributed, thus localizing CO_2 molecules more uniformly. At a mass ratio of IL to MOF of 1.5, the composite has a CO_2/N_2 (15/85, v/v) selectivity of about 70 at 300 K and 1 bar, making it a potential adsorbent for efficient flue gas carbon capture.

$Cu_3(BTC)_2$ (also termed as HKUST-1) with open metal sites and internal pore cages is also a good candidate for incorporating ILs. Keskin and Uzun incorporated 1-butyl-3-methylimidazolium tetrafluoroborate ([BMIM][BF$_4$]) into $Cu_3(BTC)_2$ to enhance the carbon capture capability and separation selectivity of MOF (Sezginel et al., 2016). The IL initially exists inside the MOF and the composite material is still in a dry powder state; when the loading amount is higher than 30 wt.%, the sample surface becomes sticky, indicating that the pores inside the MOF are filled and the IL starts to exist on the outer surface of the MOF. The experimental results show that the addition of [BMIM][BF$_4$] has a significant effect on the separation performance of the MOFs for carbon capture and gas storage.

More recent research has focused on the combination of ILs with ZIF-8, which has a sodalite (SOD) cage structure. When the hydrophobic ionic liquid ([BMIM][PF$_6$]) was incorporated into the hydrophobic ZIF-8, the specific surface area (415 m²/g (Rackley, 2017)) and pore volume (0.2197 cm³/g (Feng et al., 2015)) of the resulting composite were reduced compared to the original MOF (1208 m²/g (Rackley, 2017), 0.6332 cm³/g (Feng et al., 2015)), but the ZIF-8 crystal structure remained unchanged after IL incorporation (Kinik et al., 2016). The incorporation of IL creates new and more potent adsorption sites for CO_2 at low pressure. The interaction between the IL and the MOF improves the ability of the MOF to capture and separate CO_2 by at least twofold. The IL ([BMIM][BF$_4$]) was added to ZIF-8 at different loadings to investigate the effect of IL incorporation on MOF gas absorption and separation properties (Koyuturk et al., 2017). At low pressure, CO_2 uptake increased to a value higher than that of pristine ZIF-8, while CH_4 uptake was lower than the ZIF-8 under the same conditions. As a result, the gas separation performance of the ZIF-8 and IL composite was significantly improved.

The cage size within ZIF-8 is approximately 1.12 nm, connected by a small window of 3.4 Å. The small window size is expected to separate CO_2/N_2 by screening, as it lies between the kinetic diameters of CO_2 (3.3 Å) and N_2 (3.6 Å). However, the rotational flexibility of the ligand in ZIF-8 reduces this screening effect and the material ultimately shows moderate CO_2 selectivity. To further improve the carbon capture capacity of ZIF-8, the cut-off size needs to be further reduced. The ionic liquid [BMIM][Tf$_2$N] (1-butyl-3-methylimidazolium bis(trifluoromethyl-sulfonyl) imide) was confined in the SOD cage of ZIF-8 by in situ ion thermal synthesis (Ban et al., 2015). The effective cage size of ZIF-8 was tuned between CO_2 and N_2. The mixed matrix membranes prepared from the IL-modified ZIF-8 composite showed remarkable permeability and selectivity for the CO_2/N_2 and CO_2/CH_4 separations, which exceeded the upper limit of the polymer membranes. Significant performance improvements were also observed in the composites formed by ZIF-8 and the IL ([BMIM][BF$_4$]), with CO_2/CH_4 and CO_2/N_2 selectivities increasing by 3.3-fold and 4.5-fold, respectively, compared to the original MOF.

Although the presence of ILs in the interior of the MOF cage improves the selectivity of the MOF for various gas separations, this improvement is limited by the fact that the ILs occupies a large part of the pore space. However, when ILs are deposited or coated on the outer surface of MOFs, the pore space of the MOFs can still be used for gas adsorption.

The core-shell IL/MOF composite was prepared using the hydrophilic IL ([HEMIM][DCA], 1-(2-hydroxyethyl)-3-methylimidazolium dicyanamide) and the hydrophobic ZIF-8 by wet impregnation at stoichiometric IL loading of 40 wt.%, using acetone as solvent (Zeeshan et al., 2018). It was confirmed that the IL was deposited as a shell layer covering the outer surface of the ZIF-8 framework by a series of characterizations including FT-IR, XPS (X-ray photoelectron spectroscopy), and TEM (transmission electron microscope). The composite adsorbs 5.7 times as much CO_2 as pure MOF at 1 mbar and 298 K, and its CO_2/CH_4 selectivity is 45 times that of pure MOF. The optimal CO_2/CH_4 selectivity is greater than 110 at 1 mbar. It gradually decreases to about 11 when the pressure is increased to 1 bar. The corresponding selectivities of the pristine ZIF-8 at these pressures were only 2.4 and 2.5, respectively. At 1 bar and 298 K, the composite's selectivity for CO_2/CH_4 (50/50, v/v) and CO_2/CH_4 (5/95, v/v) mixtures is 30 and 70. The results of the conductor-like screening model for realistic solvents (COSMO-RS) calculations showed that CH_4 is poorly soluble in [HEMIM][DCA], while CO_2 has a solubility that is more than one order of magnitude higher. It appears that the outer layer of IL acts as a sieve, allowing only the CO_2 to pass into the pore cages of the MOF.

Using triethylene tetramine lactate, [TRIEN]L, an amino-rich functionalized IL with high affinity for CO_2, a series of core-shell composite adsorbents (named BUCT-CX) were obtained by impregnation through rational adjustment of the ratio of IL to MOF (Yang et al., 2023). The obtained BUCT-C19 exhibits a high Henry selectivity of 246.5 for CO_2/C_2H_2 at 298 K. Breakthrough experiments show that BUCT-C19 can be used to directly produce high purity (≥99.9%) C_2H_2 from CO_2/C_2H_2 mixtures without any further desorption process. The ionic liquid's amino groups can selectively recognize CO_2, which is more soluble than C_2H_2 in the IL shell. Once CO_2 diffuses into the outer IL, the MOF inside the composite can further provide abundant adsorption space, enhancing the material's screening effect on CO_2/C_2H_2 and achieving CO_2 capture. The DFT calculations show that the binding energy between IL and CO_2 is greater than that between C_2H_2, and the MD simulation results also confirm that the number of CO_2 molecules in both the IL layer and the MOF core is much greater than that of C_2H_2. This solution-based separation strategy can effectively avoid having to precisely control the internal structure and properties of the MOF, has the potential for large scale production, and is expected to meet the real industrial needs for CO_2/C_2H_2 separation. The [TRIEN]L@MIL-101(Cr) (BUCT-C20) composite with MIL-101(Cr) as the core also showed the ability to efficiently capture CO_2 from CO_2/C_2H_2, indicating that the design method of such composites is scalable.

Conventional imidazolyl ILs usually have high solubility for CO_2, but this process requires high pressure and the viscous nature result in insufficient adsorption capacity at the gas–liquid interface, making it difficult for further industrialization. Encapsulating ILs in MOF nanocages is an effective solution. Ban and Yang assembled ionic liquid molecular layers (ILMLs) with a thickness of 1.7–18 nm on the outer surface of the MOF-808 (Fig. 4.24A–E) (Zhao et al., 2022). At low pressure, IL itself has weak adsorption effect on CO_2, but MOF-808 with 2.8 nm thick ILMLs shows good adsorption effect on CO_2 (Fig. 4.24F). The CO_2 capacity is 3 mmol/g at 298 K and 1 bar, and it can reach 5.19 mmol/g at 273 K, which is more than twice as high as that of MOF-808 under the same conditions. The composites showed a high selectivity (372 and 478) for CO_2/N_2 mixtures with a CO_2 concentration of 15% and 5%, respectively (Fig. 4.24G and H). The CO_2 capacity per gram of ILMLs was 1000 times higher than that of the bulk liquid. The composite's CO_2 capacity was as high as 14.4 mmol/g at 20 bar (Fig. 4.24I). Breakthrough experiments indicated that composite can dynamically separate CO_2/CH_4 (50/50, v/v) and CO_2/N_2 (15/85, v/v) mixtures (Fig. 4.24J and K). A CH_4 product of greater than 99.6% purity can be obtained directly before CO_2 elution. The IL-coated MOF-808 composite has a higher dynamic saturated adsorption capacity of CO_2 than pure MOF. It also has good water resistance, thermal stability, cyclic stability, and rapid desorption.

Mesoporous poly(ionic liquids) (MPIL), as a solid material with a well-developed pore structure, have a higher CO_2 adsorption capacity and faster adsorption/desorption rate than conventional ILs, and also have excellent design and processing capabilities. A hydrophobic mesoporous ionic copolymer (PMAC) was prepared by radical copolymerization of amino-functionalized imidazolium-IL monomer and divinylbenzene (DVB). Then ZIF-8 was in situ grown on the surface of PMAC to generate the novel CO_2 adsorbent (ZPMAC) (Guo et al., 2018). There is a coordination effect between the zinc ions in ZIF-8 and the nitrogen atoms on the amino-alkyl chain in PMAC. This changes the pore structure of PMAC, increasing the amount of CO_2 adsorbed and also accelerating the rate of CO_2 adsorption. The CO_2 capacity of PMAC is 0.51 mmol/g at 298 K and 1 bar. After ZIF-8 growth, the CO_2 capacity of the composite increased by 56% (0.8 mmol/g). The CO_2 capacity of the

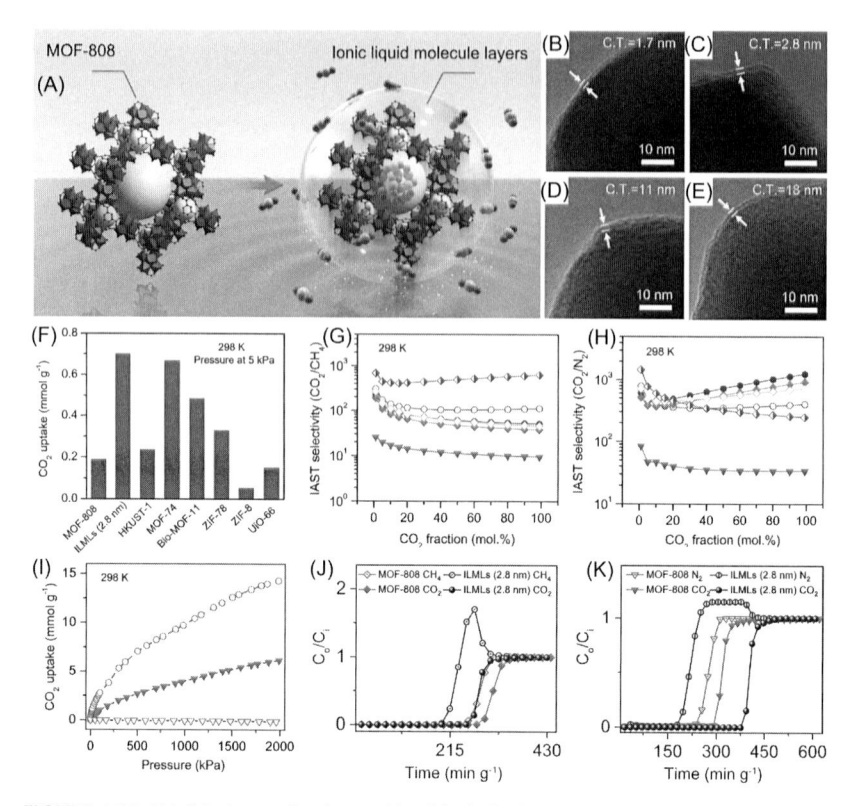

FIGURE 4.24 (A) Solution-mediated assembly of ionic liquid molecular layers (ILMLs) on MOF-808; (B–E) ILMLs observed under TEM; (F) comparison of low-pressure CO_2 uptake with other MOFs from the literature; predicted IAST selectivity of (G) CO_2/N_2 and (H) CO_2/CH_4 for MOF-808 (*solid triangle*), and composites with 1.7 nm ILMLs (*solid circle*), 2.8 nm ILMLs (*hollow circle*), 5.2 nm ILMLs (*solid diamond*), 11 nm ILMLs (*hollow diamond*), and 18 nm ILMLs (*hollow/solid diamond*); (I) high-pressure adsorption data (up to 20 bar) for bulk ionic liquid, MOF-808, and MOF-808 with 2.8 nm ILMLs; and breakthrough curves for the composite with 2.8 nm ILMLs for (J) CO_2/CH_4 (50/50, v/v) and (K) $CO_2/N_2/H_2O$ (14.5/84.5/3.0, v/v/v) at 298 K and 1 bar (Zhao et al., 2022).

composite containing less than 11% ZIF-8 is slightly lower than that of pure ZIF-8 (0.84 mmol/g). The performance of the composite is even better than that of pure ZIF-8 under lower pressure conditions. Its CO_2 capacity at 10 mbar is 0.04 mmol/g, which is about 6.9 times that of ZIF-8. This is because ZPMAC combines the rich amino groups of PMAC with the microporous structure of ZIF-8. Under dilute conditions, PMAC's rich amino groups significantly promote CO_2 adsorption. Additionally, ZPMAC displayed a faster CO_2 adsorption kinetics than PMAC and ZIF-8, because its hierarchical porous structure provides a continuous network of micro- and mesopores that can rapidly transport CO_2 molecules and increase the rate of adsorption and diffusion of CO_2.

Other metal-organic framework composites

Apart from the aforementioned MOF composites, various other types of MOF-based composites have been developed for CO_2 capture applications, aiming to enhance the adsorption capacity, selectivity, and stability of the materials. These composites include MOF-polymer composites, MOF-metal oxide composites, and MOF-MOF composites (Zhu et al., 2016; Chen, Li et al., 2017; Wan et al., 2023; Xin et al., 2016). The siliceous mesocellular foam (MCF) has a continuous 3D pore structure consisting of uniform spherical pores with diameters of 16–42 nm connecting by regular windows. The composite generated from MCF and MOF combine the advantages of both two materials, with the mesopores of MCF promoting gas transfer and the micropores of MOF being beneficial for gas adsorption and separation. By assembling HKUST-1 on MCF, a series of laminated composites with different ratios of micropores and mesopores can be synthesized, where the volume ratio of micropores to mesopores can be adjusted by changing the amount of MOF used (Xin et al., 2016). A fixed bed breakthrough experiment was performed at 303 K and 1 bar with a mixture of CO_2/N_2 (10/90, v/v), and it was calculated that the CO_2 capacity of the composite could reach 1.4 mmol/g, which was higher than that of MOF and MCF and the mechanical mixture of MOF and MCF. The CO_2 capacity of the composite can be further increased to 1.68 mmol/g after hydrothermal treatment. The -OH groups provided by MCF and the water molecules remaining in the pores after hydrothermal treatment are responsible for the CO_2 adsorption improvement. Furthermore, the composite exhibited good durability and better adsorption kinetics than the pure MOF. MCF functionalized with hydroxyl, carboxyl, and amino groups can also be combined with HKUST-1 in the one-pot process (Zhu et al., 2016). Amino-functionalized MCF provides the best match of both coordinative strength and surface density of surface functionalities, which minimizes the structural distortion of the MOF, giving the resulting composite (HK#MCF-NH$_2$) optimal textural properties and CO_2 adsorption performance. The CO_2 capacity of HK#MCF-NH$_2$ is 3.89 mmol/g, 16.1% higher than pure MOF and can be maintained after 25 adsorption–desorption cycles.

MOF@MOF composites and their derivatives have been widely used in catalysis, gas storage, and other fields, but there is still a paucity of literature on the practical industrial applications of MOF@MOF composites with a hydrophobic structure and excellent carbon capture performance. Mg-MOF-74 has excellent CO_2 capture performance (~8 mmol/g) under dry conditions. However, its CO_2 capture performance is significantly impaired under humid conditions, with an adsorption capacity of only about 16% of that under dry conditions. The reason for this is that water molecules, which are in coexistence with CO_2, are in strong competition with the CO_2 for adsorption. Using Mg-MOF-74 as functional core and hydrophobic ZIF-8

168 Applications of Metal-Organic Framework Composites

as protective shell and waterproof layer, a new core-shell MOF@MOF composite was prepared by epitaxial growth method (Wan et al., 2023). Various characterizations including SEM, HAADF-STEM (high-angle annular dark-field scanning transmission electron microscopy), EDS (energy dispersive spectrometer), PXRD (powder X-ray diffraction), FT-IR, and XPS confirmed the successful synthesis of the core-shell composite with an average shell thickness of 28 nm. The water contact angle results show that the Mg-MOF-74@ZIF-8 composite has moderate hydrophobicity, which is a great improvement compared to the unmodified Mg-MOF-74. The hydrophobicity of the composite becomes stronger as the assembled cycle increases. The results of fixed column breakthrough experiments under humid conditions showed that the dynamic CO_2 capture capacity of MOF-74@ZIF-8-3 could be 3.56 mmol/g.

Zhao's group fabricated the novel self-supporting MOFs foams (Li et al., 2024). The boron nitride nanosheets (BNNS) and MOFs (HKUST-1, MIL-100(Fe), ZIF-8) crystals acted as "rebar" and "aggregate," respectively, while PEI acted as an adhesive to create a self-supporting, thermally conductive MOFs@BNNS-PEI foams (Fig. 4.25). This new "self-supporting foam" strategy avoids using large amounts of polymer and maintains high MOF loading (>70%). On the other hand, the foam composite's high specific surface area and abundant nitrogen-containing micropores favor CO_2 adsorption, while PEI can reduce the interfacial thermal resistance between BNNS and MOF, thus overcoming the problems of poor internal heat transfer and limited CO_2 diffusion in MOFs. The CO_2 capacity of MOFs@BNNS-PEI foams increased more than 35%–42% over pure MOF, and the CO_2 desorption rate increased at least fivefold.

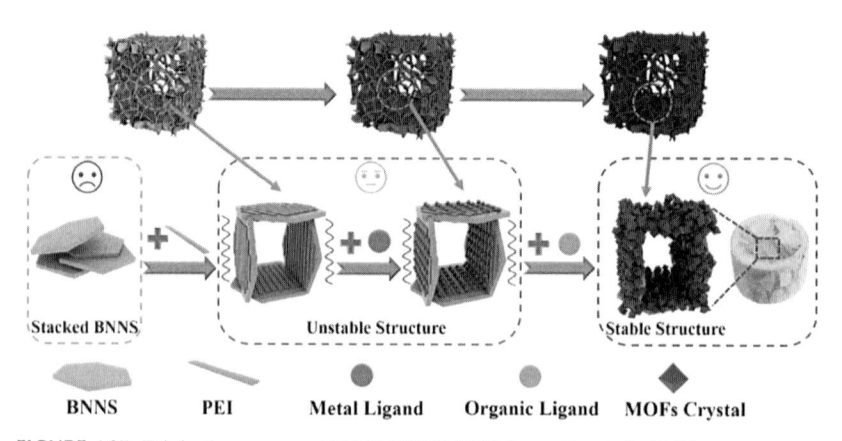

FIGURE 4.25 Fabrication process of MOFs@BNNS-PEI foams (Li et al., 2024).

Summary and perspective

In summary, MOF composites obtained by combining MOFs with inexpensive and readily available components such as inorganic/organic binders, structured monoliths, carbon-based porous materials, and ionic liquids can significantly enhance the carbon capture performance and robustness of MOFs. Shaped MOF composites can effectively overcome the issues of high pressure drop and powder clogging in fixed beds. Structured MOF-monolith composites prepared by impregnation, in situ growth, or 3D printing are expected to replace conventional packed bed systems, fundamentally addressing particle abrasion and mass and heat transfer limitations under high gas velocity conditions. Amine functionalization can improve the carbon capture capacity of MOF composites through physical and chemical adsorption mechanisms, but requires sufficient space and sites in the treated MOFs. Porous carbon materials are ideal candidates for increasing the surface area and pore volume of MOF composites, enhancing CO_2 affinity, and improving gas diffusion and mass transfer. Assembling ionic liquids inside or outside the pores of MOFs can promote CO_2 capture by acting as strong CO_2 anchors, modifying pore structures, or exploiting differences in gas solubility for molecular sieving.

Despite the significant progress, there is still much room for improving the CO_2 adsorption capacity and selectivity of MOF composites. The fabrication processes of MOF composites are complicated, and fine control over assembly, impregnation, coating, and in situ growth steps is crucial. Designing appropriate production devices and technological routes to ensure product repeatability and scalability will be the focus of future research. In addition, the long-term stability, reusability, and cost-effectiveness of MOF composites under realistic industrial conditions need to be carefully evaluated. With continuous research efforts, MOF composites are expected to make significant contributions to the development of efficient carbon capture technologies and the realization of a sustainable low-carbon future.

References

Abdelhamid, H. N., Sultan, S., & Mathew, A. P. (2023). Binder-free three-dimensional (3D) printing of cellulose-ZIF8 (CelloZIF-8) for water treatment and carbon dioxide (CO2) adsorption. *Chemical Engineering Journal, 468*, 143567. https://doi.org/10.1016/j.cej.2023.143567.

Abramova, A., Couzon, N., Leloire, M., Nerisson, P., Cantrel, L., Royer, S., Loiseau, T., Volkringer, C., & Dhainaut, J. (2022). Extrusion-spheronization of UiO-66 and UiO-66_NH2 into robust-shaped solids and their use for gaseous molecular iodine, xenon, and krypton adsorption. *ACS Applied Materials and Interfaces, 14*(8), 10669–10680. https://doi.org/10.1021/acsami.1c21380, http://pubs.acs.org/journal/aamick.

Altintas, C., Erucar, I., & Keskin, S. (2022). MOF/COF hybrids as next generation materials for energy and biomedical applications. *CrystEngComm, 24*(42), 7360–7371. https://doi.org/10.1039/d2ce01296k.

Anbia, M., & Hoseini, V. (2012). Development of MWCNT@MIL-101 hybrid composite with enhanced adsorption capacity for carbon dioxide. *Chemical Engineering Journal, 191*, 326–330. https://doi.org/10.1016/j.cej.2012.03.025.

Ban, Y., Li, Z., Li, Y., Peng, Y., Jin, H., Jiao, W., Guo, A., Wang, P., Yang, Q., Zhong, C., & Yang, W. (2015). Confinement of ionic liquids in nanocages: Tailoring the molecular sieving properties of ZIF-8 for membrane-based CO2 capture. *Angewandte Chemie International Edition, 54*(51), 15483–15487. https://doi.org/10.1002/anie.201505508.

Bian, Z., Zhu, X., Jin, T., Gao, J., Hu, J., & Liu, H. (2014). Ionic liquid-assisted growth of Cu3(BTC)2 nanocrystals on graphene oxide sheets: Towards both high capacity and high rate for CO2 adsorption. *Microporous and Mesoporous Materials, 200*, 159–164. https://doi.org/10.1016/j.micromeso.2014.08.012.

Bo, R., Taheri, M., Liu, B., Ricco, R., Chen, H., Amenitsch, H., Fusco, Z., Tsuzuki, T., Yu, G., Ameloot, R., Falcaro, P., & Tricoli, A. (2020). Hierarchical metal-organic framework films with controllable meso/macroporosity. *Advancement of Science, 7*(24). https://doi.org/10.1002/advs.2020023682002.

Bose, S., Sengupta, D., Rayder, T. M., Wang, X., Kirlikovali, K. O., Sekizkardes, A. K., Islamoglu, T., & Farha, O. K. (2023). Challenges and opportunities: Metal–organic frameworks for direct air capture. *Advanced Functional Materials, 34*(43), 2307478.

Boyd, P. G., Chidambaram, A., García-Díez, E., Ireland, C. P., Daff, T. D., Bounds, R., Gładysiak, A., Schouwink, P., Moosavi, S. M., Maroto-Valer, M. M., Reimer, J. A., Navarro, J. A. R., Woo, T. K., Garcia, S., Stylianou, K. C., & Smit, B. (2019). Data-driven design of metal–organic frameworks for wet flue gas CO2 capture. *Nature, 576*(7786), 253–256. https://doi.org/10.1038/s41586-019-1798-7, http://www.nature.com/nature/index.html.

Cao, Y., Zhao, Y., Lv, Z., Song, F., & Zhong, Q. (2015). Preparation and enhanced CO2 adsorption capacity of UiO-66/graphene oxide composites. *Journal of Industrial and Engineering Chemistry, 27*, 102–107. https://doi.org/10.1016/j.jiec.2014.12.021.

Chen, C., Li, B., Zhou, L., Xia, Z., Feng, N., Ding, J., Wang, L., Wan, H., & Guan, G. (2017). Synthesis of hierarchically structured hybrid materials by controlled self-assembly of metal–organic framework with mesoporous silica for CO2 adsorption. *ACS Applied Materials and Interfaces, 9*(27), 23060–23071. https://doi.org/10.1021/acsami.7b08117.

Chen, D.-H., Gliemann, H., & Woll, C. (2023). Layer-by-layer assembly of metal-organic framework thin films: Fabrication and advanced applications. *Chemical Physics Reviews, 4*(1). https://doi.org/10.1063/5.0135019.

Chen, Y., Hu, Z., Gupta, K. M., & Jiang, J. (2011). Ionic liquid/metal-organic framework composite for CO2 capture: A computational investigation. *Journal of Physical Chemistry C, 115*(44), 21736–21742. https://doi.org/10.1021/jp208361p.

Chen, Y., Huang, X., Zhang, S., Li, S., Cao, S., Pei, X., Zhou, J., Feng, X., & Wang, B. (2016). Shaping of metal–organic frameworks: From fluid to shaped bodies and robust foams. *Journal of the American Chemical Society, 138*(34), 10810–10813. https://doi.org/10.1021/jacs.6b06959.

Chen, Y., Lv, D., Wu, J., Xiao, J., Xi, H., Xia, Q., & Li, Z. (2017). A new MOF-505@GO composite with high selectivity for CO2/CH4 and CO 2/N2 separation. *Chemical Engineering Journal, 308*, 1065–1072. https://doi.org/10.1016/j.cej.2016.09.138.

Choe, J. H., Kim, H., Yun, H., Kang, M., Park, S., Yu, S., & Hong, C. S. (2024). Boc protection for diamine-appended MOF adsorbents to enhance CO2 recyclability under realistic humid conditions. *Journal of the American Chemical Society, 146*(1), 646–659. https://doi.org/10.1021/jacs.3c10475, http://pubs.acs.org/journal/jacsat.

Choe, J. H., Park, J. R., Chae, Y. S., Kim, D. W., Choi, D. S., Kim, H., Kang, M., Seo, H., Park, Y.-K., & Hong, C. S. (2021). Shaping and silane coating of a diamine-grafted metal-organic framework for improved CO2 capture. *Communications Materials, 2*. https://doi.org/10.1038/s43246-020-00109-8.

Chuah, C. Y., Lee, H., & Bae, T.-H. (2022). Recent advances of nanoporous adsorbents for light hydrocarbon (C1 – C3) separation. *Chemical Engineering Journal, 430*. https://doi.org/10.1016/j.cej.2021.132654.

Cortés, P. H., & Rojas Macías, S. R. (Eds.). (2021). *Metal-organic frameworks in biomedical and environmental field*. Springer.

Darunte, L. A., Oetomo, A. D., Walton, K. S., Sholl, D. S., & Jones, C. W. (2016). Direct air capture of CO2 using amine functionalized MIL-101(Cr). *ACS Sustainable Chemistry and Engineering, 4*(10), 5761–5768. https://doi.org/10.1021/acssuschemeng.6b01692, http://pubs.acs.org/journal/ascecg.

Darunte, L. A., Terada, Y., Murdock, C. R., Walton, K. S., Sholl, D. S., & Jones, C. W. (2017). Monolith-supported amine-functionalized Mg2(dobpdc) adsorbents for CO2 capture. *ACS Applied Materials and Interfaces, 9*(20), 17042–17050. https://doi.org/10.1021/acsami.7b02035, http://pubs.acs.org/journal/aamick.

Ellis, J. E., Crawford, S. E., & Kim, K.-J. (2021). Metal–organic framework thin films as versatile chemical sensing materials. *Materials Advances, 2*, 6169–6196. https://doi.org/10.1039/D1MA00535A.

Feng, K., Davis, S. J., Sun, L., & Hubacek, K. (2015). Drivers of the US CO2 emissions 1997-2013. *Nature Communications, 6*, 7714. https://doi.org/10.1038/ncomms8714, http://wwwnature.com/ncomms/index.html.

Garai, B., Mallick, A., & Banerjee, R. (2016). Photochromic metal–organic frameworks for inkless and erasable printing. *Chemical Science, 7*(3), 2195–2200. https://doi.org/10.1039/c5sc04450b.

Gebremariam, S. K., Mathai Varghese, A., Reddy, K. S. K., Fowad AlWahedi, Y., Dumée, L. F., & Karanikolos, G. N. (2023). Polymer-aided microstructuring of moisture-stable GO-hybridized MOFs for carbon dioxide capture. *Chemical Engineering Journal, 473*, 145569. https://doi.org/10.1016/j.cej.2023.145286, wwwelsevier.com/inca/publications/store/6/0/1/2/7/3/index.htt.

Glover, T. G., & M, B. (2019). Gas adsorption in metal-organic frameworks. *Fundamentals and Applications, 2019*.

Guo, Q., Chen, C., Zhou, L., Li, X., Li, Z., Yuan, D., Ding, J., Wan, H., & Guan, G. (2018). Design of ZIF-8/ion copolymer hierarchically porous material: Coordination effect on the adsorption and diffusion for carbon dioxide. *Microporous and Mesoporous Materials, 261*, 79–87. https://doi.org/10.1016/j.micromeso.2017.11.007.

Haring, H.-W. (2007). *Industrial gases processing*, 217–238. 2007.

Hong, D.-Y., Lee, K. Y., Serre, C., Ferey, G., & Chang, J.-S. (2009). Porous chromium terephthalate MIL-101 with coordinatively unsaturated sites: Surface functionalization, encapsulation, sorption and catalysis. *Advanced Functional Materials, 19*(10), 1537–1552. https://doi.org/10.1002/adfm.200801130.

Hong, W. Y., Perera, S. P., & Burrows, A. D. (2015). Manufacturing of metal-organic framework monoliths and their application in CO2 adsorption. *Microporous and Mesoporous Materials, 214*, 149–155. https://doi.org/10.1016/j.micromeso.2015.05.014, www.elsevier.com/inca/publications/store/6/0/0/7/6/0.

Hu, P., Hu, J., Zhu, M., Xiong, C., Krishna, R., Zhao, D., & Ji, H. (2023). Induced-fit-identification in a rigid metal-organic framework for ppm-level CO2 removal and ultra-pure CO enrichment. *Angewandte Chemie - International Edition, 62*(40), e202305944. https://doi.org/10.1002/anie.202305944, http://onlinelibrarywiley.com/journal/10.1002/(ISSN)1521-3773.

Hu, P., Liu, H., Wang, H., Zhou, J., Wang, Y., & Ji, H. (2022). Synergic morphology engineering and pore functionality within a metal–organic framework for trace CO2 capture. *Journal of Materials Chemistry A, 10*(2), 881–890. https://doi.org/10.1039/d1ta09974d.

Hu, Y., Chen, Y., Yang, W., Hu, J., Li, X., Wang, L., & Zhang, Y. (2024). Efficient carbon dioxide capture from flue gas and natural gas by a robust metal–organic framework with record selectivity and excellent granulation performance. *Separation and Purification Technology, 343*, 127099. https://doi.org/10.1016/j.seppur.2024.127099.

Hu, Y., Jiang, Y., Li, J., Wang, L., Steiner, M., Neumann, R. F., Luan, B., & Zhang, Y. (2023). New-generation anion-pillared metal–organic frameworks with customized cages for highly efficient CO2 capture. *Advanced Functional Materials, 33*(14), 2213915. https://doi.org/10.1002/adfm.202213915.

Huang, W., Zhou, X., Xia, Q., Peng, J., Wang, H., & Li, Z. (2014). Preparation and adsorption performance of GrO@Cu-BTC for separation of CO2/CH4. *Industrial and Engineering Chemistry Research, 53*(27), 11176–11184. https://doi.org/10.1021/ie501040s.

Ji, Y., Liu, X., Li, H., Jiao, X., Yu, X., & Zhang, Y. (2023). Hydrophobic ZIF-8 covered active carbon for CO2 capture from humid gas. *Journal of Industrial and Engineering Chemistry, 121*, 331–337. https://doi.org/10.1016/j.jiec.2023.01.036.

Kaskel, S. (2016). *The chemistry of metal-organic frameworks: Synthesis, characterization and applications*. Wiley VCH.

Kim, E. J., Siegelman, R. L., Jiang, H. Z. H., Forse, A. C., Lee, J. H., Martell, J. D., Milner, P. J., Falkowski, J. M., Neaton, J. B., Reimer, J. A., Weston, S. C., & Long, J. R. (2020). Cooperative carbon capture and steam regeneration with tetraamine-appended metal-organic frameworks. *Science, 369*(6502), 392–396. https://doi.org/10.1126/science.abb3876, https://science.sciencemag.org/content/369/6502/392/tab-pdf.

Kinik, F. P., Altintas, C., Balci, V., Koyuturk, B., Uzun, A., & Keskin, S. (2016). [BMIM][PF6] Incorporation doubles CO2 selectivity of ZIF-8: Elucidation of interactions and their consequences on performance. *ACS Applied Materials and Interfaces, 8*(45), 30992–31005. https://doi.org/10.1021/acsami.6b11087, http://pubs.acs.org/journal/aamick.

Koyuturk, B., Altintas, C., Kinik, F. P., Keskin, S., & Uzun, A. (2017). Improving gas separation performance of ZIF-8 by [BMIM][BF4] incorporation: Interactions and their consequences on performance. *Journal of Physical Chemistry C, 121*(19), 10370–10381. https://doi.org/10.1021/acs.jpcc.7b00848, http://pubs.acs.org/journal/jpccck.

Kumar, R., Jayaramulu, K., Maji, T. K., & Rao, C. N. R. (2013). Hybrid nanocomposites of ZIF-8 with graphene oxide exhibiting tunable morphology, significant CO2 uptake and other novel properties. *Chemical Communications, 49*(43), 4947–4949. https://doi.org/10.1039/c3cc00136a.

Langmi, H. W., Ren, J., North, B., Mathe, M., & Bessarabov, D. (2014). Hydrogen storage in metal-organic frameworks: A review. *Electrochimica Acta, 128*, 368–392. https://doi.org/10.1016/j.electacta.2013.10.190.

Lawson, S., Al-Naddaf, Q., Krishnamurthy, A., Amour, M. S., Griffin, C., Rownaghi, A. A., Knox, J. C., & Rezaei, F. (2018). UTSA-16 growth within 3D-printed Co-kaolin monoliths with high selectivity for CO2/CH4, CO2/N2, and CO2/H2 separation. *ACS Applied Materials and Interfaces, 10*(22), 19076–19086. https://doi.org/10.1021/acsami.8b05192, http://pubs.acs.org/journal/aamick.

Lawson, S., Griffin, C., Rapp, K., Rownaghi, A. A., & Rezaei, F. (2019). Amine-functionalized MIL-101 monoliths for CO2 removal from enclosed environments. *Energy and Fuels, 33*(3), 2399–2407. https://doi.org/10.1021/acs.energyfuels.8b04508, http://pubs.acs.org/journal/enfuem.

Lawson, S., Hajari, A., Rownaghi, A. A., & Rezaei, F. (2017). MOF immobilization on the surface of polymer-cordierite composite monoliths through in-situ crystal growth. *Separation and*

Purification Technology, 183, 173–180. https://doi.org/10.1016/j.seppur.2017.03.072, http://www.journals.elsevier.com/separation-and-purification-technology/.

Lawson, S., Rownaghi, A. A., & Rezaei, F. (2018). Carbon hollow fiber-supported metal–organic framework composites for gas adsorption. *Energy Technology, 6*(4), 694–701. https://doi.org/10.1002/ente.201700657, http://onlinelibrary.wiley.com/journal/10.1002/(ISSN)2194-4296.

Lee, D. W., Didriksen, T., Olsbye, U., Blom, R., & Grande, C. A. (2020). Shaping of metal-organic framework UiO-66 using alginates: Effect of operation variables. *Separation and Purification Technology, 235.* https://doi.org/10.1016/j.seppur.2019.116182, http://www.journals.elsevier.com/separation-and-purification-technology/.

Lee, U.-H., Valekar, A. H., Hwang, Y. K., & Chang, J.-S. (2016). *Granulation and shaping of metal-organic frameworks.* Wiley, 551–572. https://doi.org/10.1002/9783527693078.ch18.

Lee, Y. H., Kwon, Y., Kim, C., Hwang, Y. E., Choi, M., Park, Y., Jamal, A., & Koh, D. Y. (2021). Controlled synthesis of metal-organic frameworks in scalable open-porous contactor for maximizing carbon capture efficiency. *JACS Au, 1*(8), 1198–1207. https://doi.org/10.1021/jacsau.1c00068, https://pubs.acs.org/page/jaaucr/about.html.

Li, J.-R., Yu, J., Lu, W., Sun, L.-B., Sculley, J., Balbuena, P. B., & Zhou, H.-C. (2013). Porous materials with pre-designed single-molecule traps for CO2 selective adsorption. *Nature Communications, 4.* https://doi.org/10.1038/ncomms2552.

Li, S., Yu, X., Liu, L., Zhou, L., Jia, Z., Chen, J., Zhao, Z., & Zhao, Z. (2024). Thermal-conductive MOFs@BN self-supporting foams for synchronously boosting CO2 adsorption/desorption. *Advanced Functional Materials.* https://doi.org/10.1002/adfm.202401955, http://onlinelibrary.wiley.com/journal/10.1002/(ISSN)1616-3028.

Lin, J. B., Nguyen, T. T. T., Vaidhyanathan, R., Burner, J., Taylor, J. M., Durekova, H., Akhtar, F., Mah, R. K., Ghaffari-Nik, O., Marx, S., Fylstra, N., Iremonger, S. S., Dawson, K. W., Sarkar, P., Hovington, P., Rajendran, A., Woo, T. K., & Shimizu, G. K. H. (2021). A scalable metal-organic framework as a durable physisorbent for carbon dioxide capture. *Science, 374*(6574), 1464–1469. https://doi.org/10.1126/science.abi7281, https://www.science.org/doi/10.1126/science.abi7281.

Lin, Y., Lin, H., Wang, H., Suo, Y., Li, B., Kong, C., & Chen, L. (2014). Enhanced selective CO2 adsorption on polyamine/MIL-101(Cr) composites. *Journal of Materials Chemistry A, 2*(35), 14658–14665. https://doi.org/10.1039/c4ta01174k, http://pubs.rsc.org/en/journals/journalissues/ta.

Lin, Y., Yan, Q., Kong, C., & Chen, L. (2013). Polyethyleneimine incorporated metal-organic frameworks adsorbent for highly selective CO2 capture. *Scientific Reports, 3.* https://doi.org/10.1038/srep01859.

Liu, Q., Ning, L., Zheng, S., Tao, M., Shi, Y., & He, Y. (2013). Adsorption of carbon dioxide by MIL-101(Cr): Regeneration conditions and influence of flue gas contaminants. *Scientific Reports, 3.* https://doi.org/10.1038/srep02916.

Liu, S., Sun, L., Xu, F., Zhang, J., Jiao, C., Li, F., Li, Z., Wang, S., Wang, Z., Jiang, X., Zhou, H., Yang, L., & Schick, C. (2013). Nanosized Cu-MOFs induced by graphene oxide and enhanced gas storage capacity. *Energy and Environmental Science, 6*(3), 818–823. https://doi.org/10.1039/c3ee23421e.

Liu, Y., Ghimire, P., & Jaroniec, M. (2019). Copper benzene-1,3,5-tricarboxylate (Cu-BTC) metal-organic framework (MOF) and porous carbon composites as efficient carbon dioxide adsorbents. *Journal of Colloid and Interface Science, 535,* 122–132. https://doi.org/10.1016/j.jcis.2018.09.086.

Ma, H., Wang, Z., Zhang, X.-F., Ding, M., & Yao, J. (2021). In situ growth of amino-functionalized ZIF-8 on bacterial cellulose foams for enhanced CO2 adsorption. *Carbohydrate Polymers, 270,* 118376. https://doi.org/10.1016/j.carbpol.2021.118376.

Ma, X., Kang, J., Wu, Y., Pang, C., Li, S., Li, J., Xiong, Y., Luo, J., Wang, M., & Xu, Z. (2022). Recent advances in metal/covalent organic framework-based materials for photoelectrochemical sensing applications. *Trends in Analytical Chemistry, 157.* https://doi.org/10.1016/j.trac.2022.116793.

McDonald, T. M., Lee, W. R., Mason, J. A., Wiers, B. M., Hong, C. S., & Long, J. R. (2012). Capture of carbon dioxide from air and flue gas in the alkylamine-appended metal-organic framework mmen-Mg 2(dobpdc). *Journal of the American Chemical Society, 134*(16), 7056–7065. https://doi.org/10.1021/ja300034j.

McDonald, T. M., Mason, J. A., Kong, X., Bloch, E. D., Gygi, D., Dani, A., Crocellà, V., Giordanino, F., Odoh, S. O., Drisdell, W. S., Vlaisavljevich, B., Dzubak, A. L., Poloni, R., Schnell, S. K., Planas, N., Lee, K., Pascal, T., Wan, L. F., Prendergast, D., ... Long, J. R. (2015). Cooperative insertion of CO2 in diamine-appended metal-organic frameworks. *Nature, 519*(7543), 303–308. https://doi.org/10.1038/nature14327, http://www.nature.com/nature/index.html.

Mukherjee, S., Sikdar, N., O'Nolan, D., Franz, D. M., Gascon, V., Kumar, A., Scott, H. S., Madden, D. G., Kruger, P. E., Space, B., & Zaworokto, M. J. (2019). Trace CO2 capture by an ultramicroporous physisorbent with low water affinity. *Science Advances, 5*(11). https://doi.org/10.1126/sciadv.aax9171.

Niu, J., Li, H., Tao, L., Fan, Q., Liu, W., & Tan, M. C. (2023). Defect engineering of low-coordinated metal-organic frameworks (MOFs) for improved CO2 access and capture. *ACS Applied Materials and Interfaces, 15*(26), 31664–31674. https://doi.org/10.1021/acsami.3c06183, http://pubs.acs.org/journal/aamick.

Pandya, I., El Seoud, O. A., Assiri, M. A., Kumar Kailasa, S., & Malek, N. I. (2024). Ionic liquid/metal organic framework composites as a new class of materials for CO2 capture: Present scenario and future perspective. *Journal of Molecular Liquids, 395.* https://doi.org/10.1016/j.molliq.2023.123907.

Qian, D., Lei, C., Hao, G. P., Li, W. C., & Lu, A. H. (2012). Synthesis of hierarchical porous carbon monoliths with incorporated metal-organic frameworks for enhancing volumetric based CO2 capture capability. *ACS Applied Materials and Interfaces, 4*(11), 6125–6132. https://doi.org/10.1021/am301772k.

Quan, W., Holmes, H. E., Zhang, F., Hamlett, B. L., Finn, M. G., Abney, C. W., Kapelewski, M. T., Weston, S. C., Lively, R. P., & Koros, W. J. (2022). Scalable formation of diamine-appended metal-organic framework hollow fiber sorbents for postcombustion CO2 capture. *JACS Au, 2*(6), 1350–1358. https://doi.org/10.1021/jacsau.2c00029, https://pubs.acs.org/page/jaaucr/about.html.

Rackley, S. A. (2017). *Carbon capture and storage carbon capture and storage.* Elsevier, 1–677. https://www.sciencedirect.com/book/9780128120415, https://doi.org/10.1016/C2015-0-01587-8.

Rezaei, F., Lawson, S., Hosseini, H., Thakkar, H., Hajari, A., Monjezi, S., & Rownaghi, A. A. (2017). MOF-74 and UTSA-16 film growth on monolithic structures and their CO2 adsorption performance. *Chemical Engineering Journal, 313,* 1346–1353. https://doi.org/10.1016/j.cej.2016.11.058, www.elsevier.com/inca/publications/store/6/0/1/2/7/3/index.htt.

Rochelle, G. T. (2009). Amine scrubbing for CO2 capture. *Science, 325*(5948), 1652–1654. https://doi.org/10.1126/science.1176731.

Sezginel, K. B., Keskin, S., & Uzun, A. (2016). Tuning the gas separation performance of CuBTC by ionic liquid incorporation. *Langmuir, 32*(4), 1139–1147. https://doi.org/10.1021/acs.langmuir.5b04123, http://pubs.acs.org/journal/langd5.

Shakun, J. D., Clark, P. U., He, F., Marcott, S. A., Mix, A. C., Liu, Z., Otto-Bliesner, B., Schmittner, A., & Bard, E. (2012). Global warming preceded by increasing carbon dioxide concentrations during the last deglaciation. *Nature, 484*(7392), 49–54. https://doi.org/10.1038/nature10915.

Shi, L., Hu, T., Xie, R., Wang, H., Li, J., Li, S., Liu, Y., Zhi, Y., Yao, K., & Shan, S. (2024). Dual synergistic effect of the amine-functionalized MIL-101@cellulose sorbents for enhanced CO2 capture at ambient temperature. *Chemical Engineering Journal, 481*, 148566. https://doi.org/10.1016/j.cej.2024.148566.

Siegelman, R. L., Milner, P. J., Forse, A. C., Lee, J. H., Colwell, K. A., Neaton, J. B., Reimer, J. A., Weston, S. C., & Long, J. R. (2019). Water enables efficient CO2 capture from natural gas flue emissions in an oxidation-resistant diamine-appended metal-organic framework. *Journal of the American Chemical Society, 141*(33), 13171–13186. https://doi.org/10.1021/jacs.9b05567, http://pubs.acs.org/journal/jacsat.

Thakkar, H., Eastman, S., Al-Naddaf, Q., Rownaghi, A. A., & Rezaei, F. (2017). 3D-printed metal-organic framework monoliths for gas adsorption processes. *ACS Applied Materials and Interfaces, 9*(41), 35908–35916. https://doi.org/10.1021/acsami.7b11626, http://pubs.acs.org/journal/aamick.

Ullah, A., Shah, M. U. H., Ahmed, J., Younas, M., & Othman, M. H. D. (2022). *Ionic liquids and metal-organic frameworks as advanced environmental materials for CO2 capture. Handbook of energy materials*. Singapore: Springer, 1–29. https://doi.org/10.1007/978-981-16-4480-1_84-1.

Valekar, A. H., Lee, S. G., Cho, K. H., Lee, U. H., Lee, J. S., Yoon, J. W., Hwang, Y. K., Cho, S. J., & Chang, J. S. (2017). Shaping of porous metal-organic framework granules using mesoporous ρ-alumina as a binder. *RSC Advances, 7*(88), 55767–55777. https://doi.org/10.1039/c7ra11764g, http://pubs.rsc.org/en/journals/journalissues.

Valencia, L., & Abdelhamid, H. N. (2019). Nanocellulose leaf-like zeolitic imidazolate framework (ZIF-L) foams for selective capture of carbon dioxide. *Carbohydrate Polymers, 213*, 338–345. https://doi.org/10.1016/j.carbpol.2019.03.011, http://www.elsevier.com/wps/find/journaldescription.cws_home/405871/description#description.

Vodyashkin, A. A., Sergorodceva, A. V., Kezimana, P., & Stanishevskiy, Y. M. (2023). Metal-organic framework (MOF)—a universal material for biomedicine. *International Journal of Molecular Sciences, 24*. https://doi.org/10.3390/ijms24097819.

Wan, Y., Kong, D., Xiong, F., Qiu, T., Gao, S., Zhang, Q., Miao, Y., Qin, M., Wu, S., Wang, Y., Zhong, R., & Zou, R. (2023). Enhancing hydrophobicity via core–shell metal organic frameworks for high-humidity flue gas CO2 capture. *Chinese Journal of Chemical Engineering, 61*, 82–89. https://doi.org/10.1016/j.cjche.2023.03.002.

Wang, L., Han, Y., Feng, X., Zhou, J., Qi, P., & Wang, B. (2016). Metal–organic frameworks for energy storage: Batteries and supercapacitors. *Coordination Chemistry Reviews, 307*, 361–381. https://doi.org/10.1016/j.ccr.2015.09.002.

Wang, L., Huang, H., Zhang, X., Zhao, H., Li, F., & Gu, Y. (2023). Designed metal-organic frameworks with potential for multi-component hydrocarbon separation. *Coordination Chemistry Reviews, 484*, 215111. https://doi.org/10.1016/j.ccr.2023.215111.

Wang, W., Chen, D., Li, F., Xiao, X., & Xu, Q. (2023). Metal-organic framework-based materials as platforms for energy applications. *Chem, 10*(1), 86–133.

Wang, Y., Rim, G., Song, M. G., Holmes, H. E., Jones, C. W., & Lively, R. P. (2024). Cold temperature direct air CO2 capture with amine-loaded metal-organic framework monoliths. *ACS Applied Materials and Interfaces, 16*(1), 1404–1415. https://doi.org/10.1021/acsami.3c13528, http://pubs.acs.org/journal/aamick.

Xian, S., Xu, F., Ma, C., Wu, Y., Xia, Q., Wang, H., & Li, Z. (2015). Vapor-enhanced CO2 adsorption mechanism of composite PEI@ZIF-8 modified by polyethyleneimine for CO2/N2 separation. *Chemical Engineering Journal, 280*, 363–369. https://doi.org/10.1016/j.cej.2015.06.042.

Xiang, Z., Hu, Z., Cao, D., Yang, W., Lu, J., Han, B., & Wang, W. (2011). Metal–organic frameworks with incorporated carbon nanotubes: Improving carbon dioxide and methane storage capacities by lithium doping. *Angewandte Chemie International Edition, 50*(2), 491–494. https://doi.org/10.1002/anie.201004537.

Xiang, Z., Peng, X., Cheng, X., Li, X., & Cao, D. (2011). CNT@Cu3(BTC)2 and metal–organic frameworks for separation of CO2/CH4 mixture. *The Journal of Physical Chemistry C, 115*(40), 19864–19871. https://doi.org/10.1021/jp206959k.

Xin, C., Jiao, X., Yin, Y., Zhan, H., Li, H., Li, L., Zhao, N., Xiao, F., & Wei, W. (2016). Enhanced CO2 adsorption capacity and hydrothermal stability of HKUST-1 via introduction of siliceous mesocellular foams (MCFs). *Industrial and Engineering Chemistry Research, 55*(29), 7950–7957. https://doi.org/10.1021/acs.iecr.5b04022, http://pubs.acs.org/journal/iecred.

Xu, F., Yu, Y., Yan, J., Xia, Q., Wang, H., Li, J., & Li, Z. (2016). Ultrafast room temperature synthesis of GrO@HKUST-1 composites with high CO2 adsorption capacity and CO2/N2 adsorption selectivity. *Chemical Engineering Journal, 303*, 231–237. https://doi.org/10.1016/j.cej.2016.05.143.

Yang, J., Tong, M., Han, G., Chang, M., Yan, T., Ying, Y., Yang, Q., & Liu, D. (2023). Solubility-boosted molecular sieving-based separation for purification of acetylene in core–shell IL@MOF composites. *Advanced Functional Materials, 33*(15). https://doi.org/10.1002/adfm.202213743.

Yang, R., Wang, Y., Cao, J.-W., Ye, Z.-M., Pham, T., Forrest, K. A., Krishna, R., Chen, H., Li, L., Ling, B.-K., Zhang, T., Gao, T., Jiang, X., Xu, X.-O., Ye, Q.-H., & Chen, K.-J. (2024). Hydrogen bond unlocking-driven pore structure control for shifting multi-component gas separation function. *Nature Communications, 15*. https://doi.org/10.1038/s41467-024-45081-w.

Yang, R. T. (1988). Gas separation by adsorption processes. *Chemical Engineering Science, 43*(4), 985. https://doi.org/10.1016/0009-2509(88)80096-4.

Yang, Y., Ge, L., Rudolph, V., & Zhu, Z. (2014). In situ synthesis of zeolitic imidazolate frameworks/carbon nanotube composites with enhanced CO2 adsorption. *Dalton Transactions, 43*(19), 7028–7036. https://doi.org/10.1039/c3dt53191k.

Zeeshan, M., Nozari, V., Yagci, M. B., Islk, T., Unal, U., Ortalan, V., Keskin, S., & Uzun, A. (2018). Core-shell type ionic liquid/metal organic framework composite: An exceptionally high CO2/CH4 selectivity. *Journal of the American Chemical Society, 140*(32), 10113–10116. https://doi.org/10.1021/jacs.8b05802, http://pubs.acs.org/journal/jacsat.

Zhang, Z., Huang, S., Xian, S., Xi, H., & Li, Z. (2011). Adsorption equilibrium and kinetics of CO2 on chromium terephthalate MIL-101. *Energy and Fuels, 25*(2), 835–842. https://doi.org/10.1021/ef101548g.

Zhang, Z., Peh, S. B., Kang, C., Chai, K., & Zhao, D. (2021). Metal-organic frameworks for C6–C8 hydrocarbon separations. *EnergyChem, 3*. https://doi.org/10.1016/j.enchem.2021.100057.

Zhao, M., Ban, Y., & Yang, W. (2022). Assembly of ionic liquid molecule layers on metal–organic framework-808 for CO2 capture. *Chemical Engineering Journal, 439*. https://doi.org/10.1016/j.cej.2022.135650.

Zhao, Y., Ding, H., & Zhong, Q. (2013). Synthesis and characterization of MOF-aminated graphite oxide composites for CO2 capture. *Applied Surface Science, 284*, 138–144. https://doi.org/10.1016/j.apsusc.2013.07.068.

Zhu, C., Zhang, Z., Wang, B., Chen, Y., Wang, H., Chen, X., Zhang, H., Sun, N., Wei, W., & Sun, Y. (2016). Synthesis of HKUST-1#MCF compositing materials for CO2 adsorption. *Microporous and Mesoporous Materials, 226*, 476–481. https://doi.org/10.1016/j.micromeso.2016.02.029, www.elsevier.com/inca/publications/store/6/0/0/7/6/0.

Zhu, Z., Parker, S. T., Forse, A. C., Lee, J. H., Siegelman, R. L., Milner, P. J., Tsai, H., Ye, M., Xiong, S., Paley, M. V., Uliana, A. A., Oktawiec, J., Dinakar, B., Didas, S. A., Meihaus, K. R., Reimer, J. A., Neaton, J. B., & Long, J. R. (2023). Cooperative carbon dioxide capture in diamine-appended magnesium-olsalazine frameworks. *Journal of the American Chemical Society, 145*(31), 17151–17163. https://doi.org/10.1021/jacs.3c03870, http://pubs.acs.org/journal/jacsat.

Zhu, Z., Tsai, H., Parker, S. T., Lee, J. H., Yabuuchi, Y., Jiang, H. Z. H., Wang, Y., Xiong, S., Forse, A. C., Dinakar, B., Huang, A., Dun, C., Milner, P. J., Smith, A., Guimarães Martins, P., Meihaus, K. R., Urban, J. J., Reimer, J. A., Neaton, J. B., & Long, J. R. (2024). High-capacity, cooperative CO2 capture in a diamine-appended metal-organic framework through a combined chemisorptive and physisorptive mechanism. *Journal of the American Chemical Society, 146*(9), 6072–6083. https://doi.org/10.1021/jacs.3c13381, http://pubs.acs.org/journal/jacsat.

Chapter 5

Metal-organic framework composites in fuel cells

Sreejitha Raj[1], Akhila Raman[2], Vikas Rajan[1], and Appukuttan Saritha[2]
[1]*Department of Mechanical Engineering, Amrita Vishwa Vidyapeetham, Kollam, Kerala, India;*
[2]*Department of Chemistry, Amrita Vishwa Vidyapeetham, Kollam, Kerala, India*

Introduction

Power generation using nonrenewable energy sources necessitates a shift toward green energy technology due to the environmental impact and the well-being of living beings (Zhong et al., 2008; Zou et al., 2016). Major energy sources for power production include nuclear energy (18%), fossil fuels (60%), and renewable energy (21%). Among these, fossil fuels—comprising coal, natural gas, and petroleum products—are depleting at a faster rate compared to other resources (Debe, 2012). As a result, renewable energy resources are regarded as a viable alternative. In the pursuit of clean and sustainable energy solutions, Sir William Grove's fuel cell concept can be considered an ideal substitute. The fundamental principle of a fuel cell is the electrochemical reaction that converts the chemical energy of a carbon-free fuel directly into electricity without undergoing any combustion (exothermic oxidation). Fuel cells are promising technology for stationary power production, as well as for industrial and transportation applications, due to their flexible fuel diversity and operating efficiency of 60% compared with conventional engines.

Fuel cells consist of two major components: the electrode (anode and cathode) and the electrolyte. According to the United States Department of Energy, fuel cells are classified based on the type of electrolyte used, such as proton exchange membrane fuel cell (PEMFC) (Shao et al., 2007), direct methanol fuel cell (DMFC) (Jing et al., 2020), phosphoric acid fuel cell (PAFC), molten carbonate fuel cell (MCFC), alkaline fuel cell (AFC), solid oxide fuel cell (SOFC), and microbial fuel cell (MFC). The key half-cell reactions at the anode and cathode of a fuel cell are hydrogen oxidation reactions (HOR) and oxygen reduction reactions (ORR), respectively. ORR is considered the crucial reaction in a fuel cell, necessitating the development of superior catalysts to enhance electrocatalytic efficiency. However, a significant

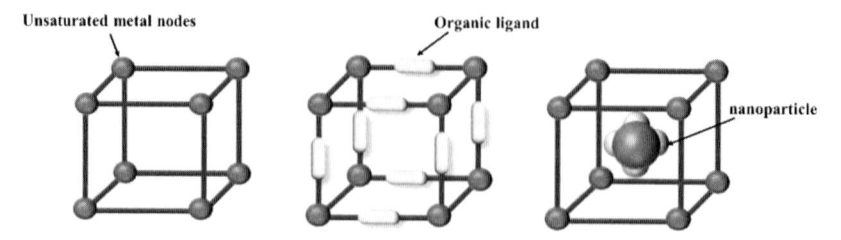

FIGURE 5.1 Active catalytic sites on various basic constituents of metal-organic framework materials (Zhang et al., 2024).

limitation in manufacturing is the cost and durability of the electrocatalyst (Cui & Yu, 2013; Nie et al., 2015). Platinum (Pt), gold, and other precious metals are commonly used catalysts with exceptional electrocatalytic features; yet, due to their high cost and limited availability, they are not economically viable (Zhe-qin et al., 2021).

Currently, fuel cells employ MOFs—an organic–inorganic hybrid catalyst with a porous structure composed of organic ligands and metal-containing nodes—to expedite electrochemical reactions (Oar-Arteta et al., 2017; Williams, 2017). MOFs have garnered significant interest and thorough investigation by experts in the chemical and engineering fields due to their homogeneous pore distribution, large specific surface area, variable chemical characteristics, and structural diversity compared to commercial porous materials. In 1995, Professor Omar Yaghi and Michael O'Keeffe introduced the new crystal chemistry of porous coordination compounds known as MOFs. To date, around 10,000 distinct MOF materials have been reported, with many yet to be discovered. The emergence of MOFs has significantly influenced researchers, leading to widespread applications across various industrial and scientific sectors, including gas adsorption/separation technology, catalysis, drug delivery, fuel storage, batteries, supercapacitors, and fuel cells (Bin Wu & Lou, 2017; Gascon et al., 2014; Wang et al., 2014; Wang et al., 2017).

The presence of additional catalytic active sites, as shown in Fig. 5.1, along with unsaturated metal nodes and organic ligands, and the ability to serve as a host for encapsulation, are the major structural characteristics that make MOFs distinct from other electrocatalysts (Zhang et al., 2024).

Metal-organic frameworks in fuel cells

In the current energy landscape, the increasing reliance on fossil fuels has led to unavoidable side effects. To achieve reliable and nonhazardous energy solutions, hydrogen-driven fuel cells offer a potential alternative to fossil fuels, enabling a carbon-free transformation of chemical energy directly into electric power (Liu et al., 2008). Hydrogen-driven fuel cells hold a significant position in future energy strategy, with potential for implementation across various sectors such as transportation, industry, and residential applications. However, commercialization is hindered by issues related to the performance, cost, and durability of electrodes and electrolyte membranes in fuel cells. The limited availability of precious metals like platinum and palladium, which are necessary as electrocatalysts, further hampers market penetration due to their high cost. Additionally, the expenses and complexity associated with electrolyte manufacturing impact commercial viability. Moreover, the overall durability of fuel cell systems is affected by cyclic operation. Therefore, the use of nonnoble metals for electrodes and electrolytes, such as MOFs (Williams, 2017; Wu et al., 2015), shows promise in reducing costs and improving the performance of fuel cells.

In recent decades, the chemistry of MOFs has undergone substantial development due to their ability to modify surface area and structural topology, making them adaptable to different compositions based on specific requirements (Oar-Arteta et al., 2017). From energy production to separation, storage, and utilization, MOFs play a tremendous and commendable role in addressing challenges faced by conventional fuel cell systems. They are primarily used in fuel cell electrodes as catalysts and catalyst support materials, replacing expensive noble metal catalysts. In electrolytes, MOFs combine with the polymer matrix to enhance ionic conduction.

This chapter delves into the multifaceted roles of MOFs and MOF composites in fuel cell technology, exploring their use as electrolytes, catalyst supports, and precursors for electrocatalysts. By highlighting the unique structural characteristics of MOFs, such as additional catalytic sites and unsaturated metal nodes, this discussion unveils how these materials have revolutionized the development and performance of fuel cells, paving the way for clean energy generation.

Metal-organic frameworks and metal-organic framework composites as precursors for electrocatalysts

The integration of MOFs with polymers, metal oxides, or even carbon materials results in the formation of MOF composites. These tailored materials, with their unprecedented performance and properties—including high mechanical strength, stability, and functionality—enhance their application in diverse fields such as energy, biomedical, and environmental sciences for storage, catalysis, separation, and sensing purposes. Due to their cost-effectiveness and acceptable electrocatalytic properties, MOF-derived hybrid materials with transition metals (Fe, Co, Ni) doped with heteroatoms (B, N, S) over carbon materials (carbon nanotube [CNT], graphene oxide [GO], reduced graphene oxide [rGO]) are considered potential candidates for electrocatalysis, offering an alternative to platinum group metal (PGM) catalysts in fuel cell operations.

In 2011, Proietti et al. (2011) reported a breakthrough in using MOFs as precursors for the synthesis of metal-N-C catalysts for fuel cells. They used Zeolite Imidazole Framework-8 (ZIF-8) as a non-PGM catalyst precursor to produce an Fe-N-C catalyst for PEMFCs, achieving a power density of nearly 0.75 W/cm² at 0.6 V, thanks to the Zn (II) core structure and Fe (II) shell (Barkholtz et al., 2016). However, the formation of Fe^{2+} or Fe^{3+} during the process can produce hydroxyl species that combine with the polymer membrane and ionomer, reducing the stability and lifespan of PEMFCs (Zhang et al., 2018a). Additionally, over time, the organic ionomer experienced significant deterioration due to the Fenton reaction, which occurs when Fe and peroxide are present in the electrode, producing free radicals. Later, Ma et al. (2011). developed a more durable catalyst with more active sites by replacing Fe with Co and using ZIF-67 as the precursor. By controlling the crystal size of the Co atom between 200 and 40 nm, they reported increased ORR activity with a continuous positive shift of nearly 0.8 V, representing enhanced active sites for ORR. However, when the particle size was further reduced to 20 nm, ORR activity dropped significantly due to the agglomeration of catalyst material, which inhibits mass transfer. It was inferred that when catalyst material operates under vigorous conditions, it may corrode at a quicker pace, triggering structural instability (Wang et al., 2018) and causing a decline in performance. Furthermore, Armel et al. (2015). studied the effect of ZIF-8 crystal morphology by varying the molar ratio of 2-Melm to Zn (II) from 40 to 140 via a ligand-excess method during the O_2 electro-reduction process. They observed that catalysts with a 40 molar ratio had a crystal size ranging from 1200 to 1800 nm, while molar ratios of 60, 80, 100, and 140 resulted in crystal sizes of 250–700, 160–400, 100–280, and 80–200 nm, respectively.

Currently, researchers are introducing heteroatom-based catalysts, particularly nitrogen, which, when doped with carbon-based materials, showcase enhanced catalytic activity and stability. For example, Zhang et al. (2014) fabricated nitrogen-doped porous carbon material by carbonizing a ZIF-7/glucose composite, achieving a half-wave potential ($E1/2$) of 0.7 V compared to the reversible hydrogen electrode (RHE).

Considering an iron-based MOF, Material Institute Lavoisier-101 (MIL-101(Fe)), as a source for Fe^{3+} (Li et al., 2015) was used as a precursor in the production of nitrogen-doped Fe/Fe₃C@graphitic layer/CNT hybrid, demonstrating outstanding electrocatalytic activity for ORR. Similarly, Fe–N coordination sites encapsulated inside ZIF-8 and embedded within the carbon nanofiber (CNF) as a 3D conductive network (Fe-Nx)-ZIF-8/CNF) (Miao et al., 2022) exhibited an $E1/2$ of 0.875 V (vs RHE). In contrast to large pore size (2–3 nm) and enhanced surface area, MIL-101-NH₂ is used as a host for generating highly porous graphitic carbon electrocatalysts. Zhu et al. (2017) prepared the catalyst by incorporating abundant Fe and N into the pores of the MOF structure, using dicyandiamide and $FeCl_3$ as sources for N and Fe, respectively, with their encapsulation restricted to the pores by blocking the outer surface of the MOF using a double solvent method followed by pyrolysis for synthesis. The ORR activity of the Fe/N-GPC catalyst was assessed with an onset potential value of 0.85 V and an $E1/2$ value of 0.63 V compared with RHE.

Despite the structural deformations that Fe- and Zn-doped ZIFs may experience during manufacture and operation, they continue to attract researchers' interest as PGM-free catalysts for fuel cell applications (da Silva Freitas et al., 2022). One of the crucial factors that influence catalytic performance, structural alterations, and the overall efficiency of MOF-derived materials is their pyrolysis temperature. MOF nanocomposites pyrolyze to provide highly porous, graphitized, MOF-derived nanomaterials doped with N or P, which are increasingly employed as effective electrocatalysts (Wang & Astruc, 2019). By adjusting the pyrolysis temperature, the characteristics of the material can be customized for different applications.

Deng et al. (2017) investigated the impact of the quantity of iron added during the synthesis process, particularly in relation to the pyrolysis temperature. They hypothesized that Fe could influence both the catalyst's crystal size and the performance of the fuel cell at a temperature of 900°C. Out of their samples, as shown in Fig. 5.2. Fe-ZIF-8/0.84 showed a current density of -6.48 mA/cm^2 at a voltage of 0.7 V, a higher value than the Pt/C catalyst.

Later, Tong et al. (2018) prepared an electrocatalyst using a combination of MIL-101(Fe) and polypyrrole (PPy) to examine the influence of pyrolysis temperature on the MOF. Their findings revealed that MIL-101/PPy pyrolyzed

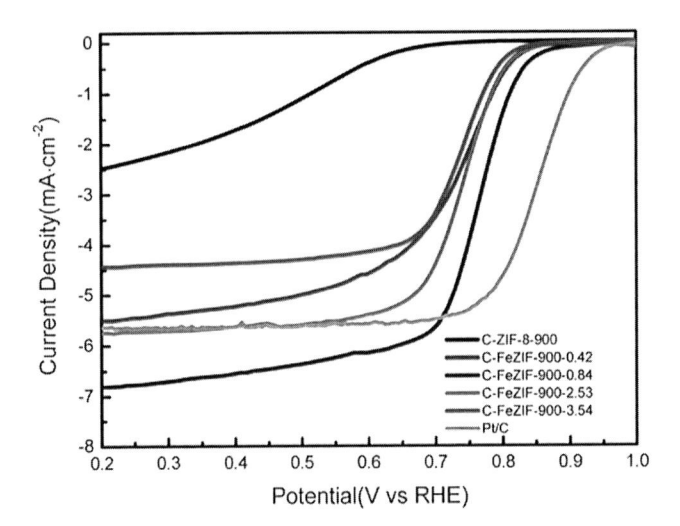

FIGURE 5.2 Linear Sweep Voltammetry (LSV) oxygen reduction reaction curves in O$_2$-saturated 0.1 M HClO$_4$ of C-ZIF-8–900, C-FeZIF-900-0.42, C-FeZIF-900-0.84, C-FeZIF-900-2.53, C-FeZIF-900-3.54, and 20 wt.% Pt/C (Deng et al., 2017).

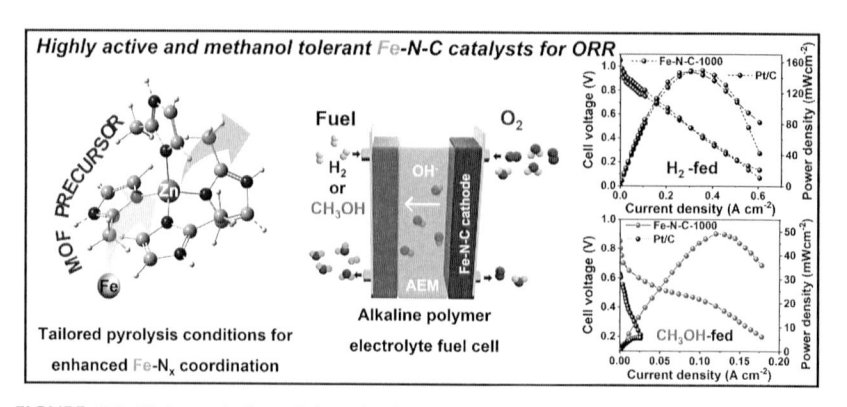

FIGURE 5.3 High catalytic activity of tailored Fe-N$_x$-C actives sites from metal-organic framework template toward oxygen reduction reaction in alkaline environment (da Silva Freitas et al., 2023).

at 900°C had a high nitrogen concentration, particularly 46.1%, which endowed the catalyst with a high density of active sites and strong cyclic durability. da Silva Freitas et al. (2023) investigated the influence of pyrolysis temperature on the performance of catalysts derived from bimetallic-type MOFs, as shown in Fig. 5.3. They synthesized the Fe-N-C catalyst by substituting Fe for Zn in ZIF-8, followed by pyrolysis at 700°C, 900°C, and 1000°C. The sample Fe-N-C-1000 exhibited exceptional ORR activity with an open circuit voltage (OCV) of 0.98 and maintained acidic tolerance even after 30,000 cycles. This supports the results reported by Liu, Zhao, et al. (2016), where they tailored an Fe-N-C catalyst from ZIF under different pyrolysis temperatures, with FeNC-20–1000 showing superior performance with a half-wave potential (HWP) of 0.77 V. In a different approach, Xiao, Yang, et al. (2020) replaced Fe with Mn and prepared an Mn-N-C electrocatalyst derived from Mn(II) with large nitrogen linkers. This catalyst demonstrated a half-wave potential ($E1/2$) of 0.81 V versus RHE and exhibited sustained stability even after 10,000 cycles.

Similarly, Xiao, Yuan, et al. (2020) integrated Mn$_3$O$_4$ onto an N-doped C composite prepared from a bimetallic Mn/Zn-MOF composite, achieving an efficiency with a half-wave potential ($E1/2$) of 0.81 V.

In 2017, a group of researchers (Armel et al., 2017) investigated whether Zn-based ZIF structures exhibit better ORR activity. To explore this, they used different ligands, including imidazolate (Im), benzimidazolate (bzim), 2-methylimidazolate (mim), and 2-ethylimidazolate (eim), synthesizing nine ZIF structures such as Zn(Im)(mim) (ZIF-61), Zn(eim)$_2$, Zn(Im)$_2$ (ZIF-7), Zn (Im)$_2$ (ZIF-4), Zn(eim)$_2$ (ZIF-14), Zn(mim)$_2$ (ZIF-8), and Zn(bzim)$_2$ (ZIF-11). Among all nine catalysts, ZIF-8 (Zn(mim)$_2$) demonstrated the best catalytic properties, featuring a suitable cavity size and the highest specific pore volume. Wang et al. (2019) sought to understand the correlation between ORR activity and crystal structure by synthesizing a set of catalysts with various imidazolate ligands that had undergone structural modification via a ball-milling process. They observed that reacting ZnO with mim, followed by treatment with Tris (1,10-phenanthroline) iron (II) perchlorate (TPIBP) and subsequent ball milling and pyrolysis, resulted in the catalyst Zn(mim)$_2$TPIBP, which exhibited superior ORR activity with an E1/2 of nearly 0.93 V compared with RHE. Luo, Han, et al. (2019) synthesized Fe/S@N/C-X hybrid electrocatalysts from ZIF-8, where the amount of Fe was controlled by adjusting the proportion of Fe and Zn precursors in the catalyst, resulting in the preparation of three batches of catalysts at 0.33, 0.25, and 0.5 ratios. Among them, Fe/S@N/C-0.5 shows the highest half-wave potential of 0.87 V. Replacing Fe with Co, along with phosphorus as a heteroatom source supported on CNT, was prepared by Shah et al. (2021), as illustrated in Fig. 5.4. Due to their large surface area, CNTs offer enhanced flexibility in transporting electrons and ions across the channel, resulting in an increased number of active sites for reactions. The sample P-Co-CNT exhibits a half-cell reaction potential of 0.88 V, which is superior to that of the more expensive Pt/C catalyst.

Recently, nonheteroatom catalysts have gained attention, such as those prepared from combinations of ZIF-8 and multiwalled carbon nanotubes (MWCNT) via pyrolysis at 900°C (Kumar et al., 2024). These catalysts can achieve a power density of 1557 mW/m² and a half-wave potential of 0.71 V versus RHE. When preparing catalysts for Direct Methanol Fuel Cells (DMFCs), it is crucial to consider the potential for methanol crossover, as methanol can permeate from the anode to the cathode through the membrane and impair the function of the cathode catalyst. Therefore, the ORR activity in the presence of methanol oxidation should be thoroughly analyzed. Chao and coworkers investigated the methanol oxidation performance of S, N-containing manganese MOF (Mn^{2+}[(Tdc)(4,4'-Bpy)]$_n$) with various morphologies, including 0D nanoparticles, 1D nanorods, and 3D bulk structures over carbon.

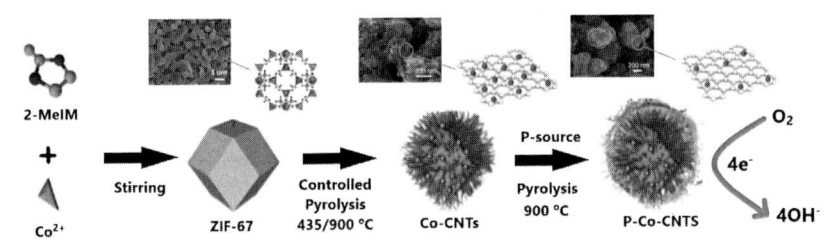

FIGURE 5.4 Diagrammatic representation of P-Co-carbon nanotubes nanostructure synthesis process (Shen et al., 2021).

They found that the nanorod structure displayed the highest ORR activity (Chao et al., 2020).

Given ZIF-8's promising properties in the energy sector, researchers are exploring its integration with various metal complexes. Barkholtz et al. (2015) present results for three iron additive combinations, as shown in Fig. 5.5. They paired ZIF-8 with Fe(II) acetate to form Zn(mim)$_2$(Fe(Ac)$_2$), with tris-1,10-phenanthroline Fe(II) perchlorate to create Zn(mim)$_2$(TPI), and with a mixture of Fe(Ac)$_2$ and 1,10-phenanthroline (Phen) to produce Zn(mim)$_2$(Fe(Ac)$_2$(Phen))$_6$. Among these, the inclusion of TPI demonstrated superior performance in terms of polarization curves compared to the other combinations.

In ORR, nitrogen-doped carbon materials encased in metal nanoparticles derived from MOFs control the interfacial charge transfer between Metal-N and Carbon-N, resulting in more active sites. However, MOFs often experience structural deterioration during pyrolysis, which can lead to the formation of uncontrolled bulk phases of additional metal particles and a significant reduction in electrochemical performance. Therefore, it is crucial to functionalize the surface of MOFs by integrating active components through surface engineering. Gao et al. (2017) report that an additional nitrogen source obtained from polyaniline (PAni) is combined with ML-101(Fe) to prepare a nitrogen–carbon (NC) hybrid catalyst, NC@Fe$_3$O$_4$. Their work highlights the remarkable electrocatalytic properties of this hybrid catalyst and provides

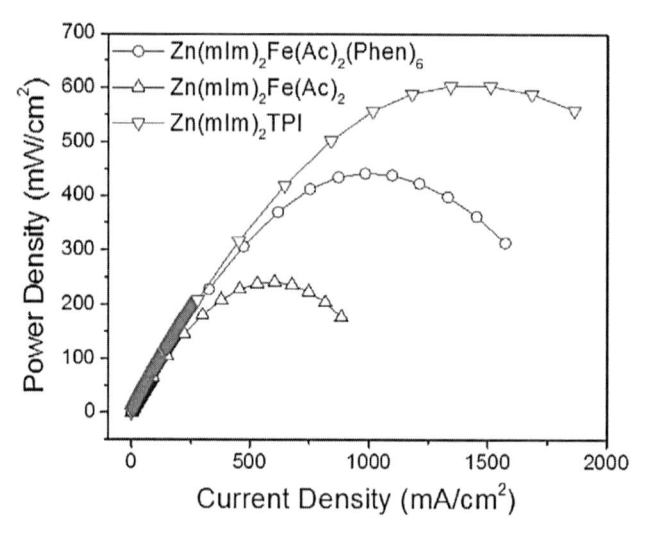

FIGURE 5.5 Power density curves corresponding to polarization curves of Zn(mIm)$_2$Fe(Ac)$_2$=241.5 mW/cm^2, Zn(mIm)$_2$Fe(Ac)$_2$(Phen)$_6$=441.9 mW/cm^2, and Zn(mIm)$_2$TPI=603.3 mW/cm^2 (Barkholtz et al., 2015).

valuable insights into the functionalization of MOFs for enhanced ORR activity, as shown in Fig. 5.6.

Including N-rich molecules in the precursors and breaking them down to release gases during pyrolysis helps create more abundant pores, which in turn facilitate the anchoring of more metal atoms. Specifically, incorporating these N-rich molecules into the precursors and their subsequent breakdown to release gases during pyrolysis will generate additional pores that serve to anchor more metal atoms (Chung et al., 2017). In contrast, Haixia Su et al. (2018) prepared N-doped Fe_3O_4/Fe_3N from ML-101(Fe), using PAni as an additional nitrogen source. The entire synthesis procedure is illustrated in Fig. 5.7, showing the process of mixing the MOF precursor with PAni, followed by initial N_2 treatment, leaching, and final N_2 treatment.

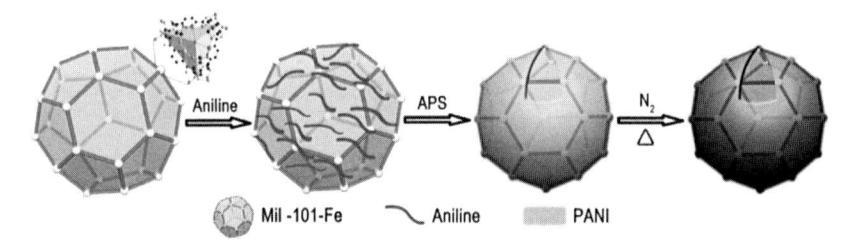

FIGURE 5.6 Schematic illustration of NC@Fe_3O_4 synthesis (Gao et al., 2017).

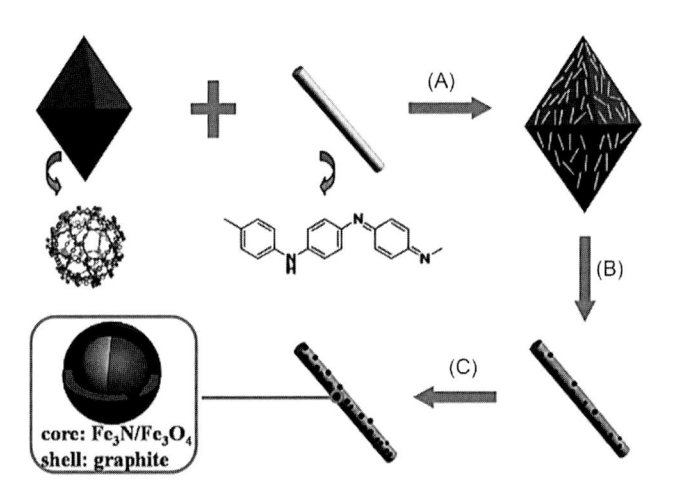

FIGURE 5.7 Preparation procedure: (A) mixing of ML-101(Fe) with PAni, (B) heat treatment in a nitrogen atmosphere, and (C) acid leaching followed by further heat treatment (Su et al., 2018).

The synthesis results in an N-doped MOF combined with graphene to form a core-shell structure. This structure exhibits ORR activity with a half-wave potential of 0.916 V. As depicted in Fig. 5.8, the typical core-shell structure is advantageous for ORR activity because it prevents the dissolution of Fe_3O_4/Fe_3N, even in strong alkaline media, due to the presence of graphitic carbon.

When functionalizing the surface with the eco-friendly biopolymer polydopamine (PDA) for the synthesis of a Co-coordinated nitrogen-doped carbonaceous nanosheet catalyst (Zhang et al., 2018b), the Co-MOF/PDA is prepared. PDA acts as a coating layer, as shown in Fig. 5.9, over the Co-MOF during pyrolysis, preventing the nanoparticles from agglomeration. The resulting catalyst demonstrates enhanced stability and exhibits a half-wave potential ($E1/2$) of 0.819 V, which is higher than that of the traditional Pt/C catalyst.

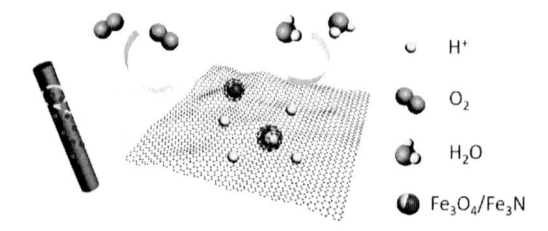

FIGURE 5.8 Working mechanism of ML-101(Fe) with PAni (Su et al., 2018).

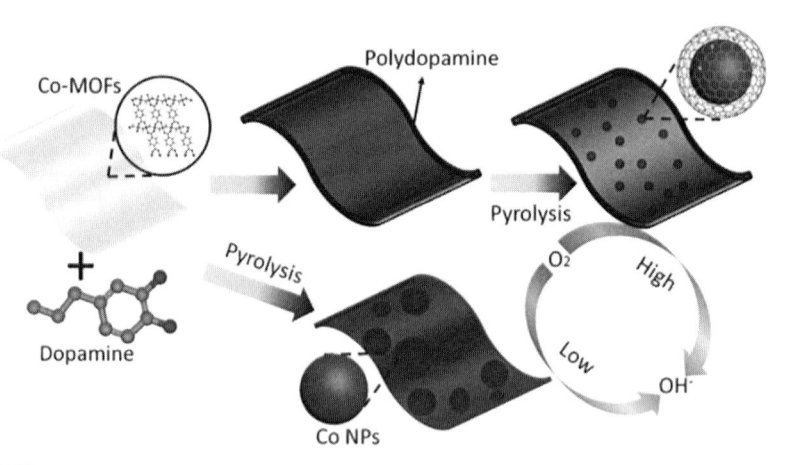

FIGURE 5.9 Schematic representation of Co coordinated N doped carbonaceous nanosheet catalyst for the synthesis of polydopamine (Zhang et al., 2018b).

As an additional nitrogen source, the Fe@ZIF-8 catalyst is synthesized with a urea coating, which decomposes during calcination to release ammonia that acts as a nitrogen source. Ammonia helps increase the nitrogen content in the catalyst. Furthermore, the addition of nitrogen-rich melamine during the synthesis of Cu single-atom nanosheets from a Cu/Zn bimetallic MOF leads to the creation of a highly porous structure with numerous active sites. This catalyst exhibits improved mass transport and demonstrates an impressive ORR performance with a half-wave potential of 0.9 V with respect to RHE (Lu et al., 2022). In contrast, incorporating a nitrogen–phosphorous–sulfur ternary doping into porous carbon using MOF-5 for ORR results in a limiting current density value of 11.6 mA/cm² at 0.6 V (Li et al., 2014). Additionally, core-shell-type catalysts prepared from MOFs have shown efficient performance in both acidic and alkaline media. For instance, Huang et al. (2019) synthesized a core-shell type catalyst that demonstrated effective performance in both alkaline and acidic media. Herein, as shown in Fig. 5.10, Fe(OH)$_3$@ZIF-8 serves as the core structure with carbon forming the shell, resulting in a larger Brunauer-Emmet-Teller surface area of approximately 1021 m²/g and an $E1/2$ of 0.88 V compared to RHE. Building on this concept, Zhang et al. (2020) used nitrogen-doped carbon as the shell structure anchored by Fe/Fe$_3$C as the core, which was then decorated on graphite. The resulting catalyst, Fe/Fe$_3$C@NC-G, exhibited efficient ORR activity with an $E1/2$ of 0.88 V.

During ORR, the H$^+$ ions obtained from the oxidation of H$_2$ at the anode combine with O$_2$ at the cathode to form water droplets in the presence of an electrocatalyst. Since ORR is considered a slow process that prevents achieving maximum efficiency, the cathode catalyst must be more stable and active compared to the anode catalyst. The most widely used electrocatalysts synthesized from MOFs to form MOF-hybrid electrocatalysts are listed in Table 5.1.

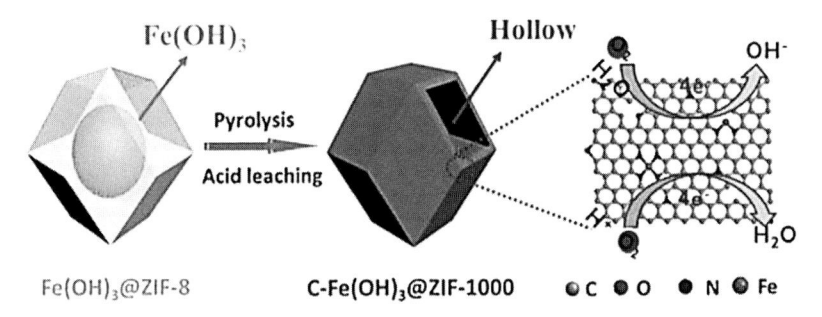

FIGURE 5.10 Procedure of preparing C-Fe(OH)$_3$@ZIF-8 in a core shell structure (Huang et al., 2019).

TABLE 5.1 Recent studies on metal-organic framework hybrids as precursors.

Sl. no.	Original MOF	Electrocatalyst	Power density (mW/m²)	Current density (mA/cm²)	Half wave potential (V)	Type of fuel cell	References
1	ZIF-8	Fe-NC	–	–	0.831	AFC	Hou et al. (2020)
2	ZIF-67	Cu-Co-NC	1008	–	0.14	MFC	Wang, Wei, et al. (2020)
3	ZIF-67	Co-NC-Gr	–	5.6	0.83	DMFC	Gao et al. (2020)
4	ZIF-67	Co-NC/rGO	–	5.6	0.82	PEMFC	Gao et al. (2021)
5	ZIF-8	Fe-NC	–	–	0.834	PEMFC	Zhang et al. (2022)
6	ZIF-67	Co/Fe-NC	1831	20.4	−0.3	MFC	Liang et al. (2024)
7	ZIF-8	Carboxylate/ Fe-NC	1330	–	–	PEMFC	Li et al. (2021)
8	ZIF-8	Fe-BN-C	–	6.154	0.859	AFC	Ma et al. (2024)
9	ZIF-8	Co/Zn-NC@PPy	–	4.99	0.977	AFC	Seyed Bagheri et al. (2022)
10	ZIF-8	Co/Zn-NC@PPy	–	5.48	0.5	PEMFC	Seyed Bagheri et al. (2022)
11	ZIF-8	Co-N-C	920	–	0.8	PEMFC	Wang, Zhang, et al. (2020)
12	MIL-101	Fe- FeN@N-C	–	6.04	0.813	AFC	Feng et al. (2020)
13	ZIF-67	Co@NPC-MWCNT	432	–	0.79	DEFC	Liu et al. (2022)
14	ZIF-8	Fe-N-CNT/C	–	–	0.84	PEMFC	Shen et al. (2021)

15	ZIF-8	Fe-NC-H	171.6	-5.82	0.865	MCFC	Zhang et al. (2023)
16	ZIF-8	Fe-NC	129.59	10	0.9	AFC	Ren et al. (2023)
17	ZIF-67	Co-Fe-NC	755	5.06	0.79	PEMFC	Zhu et al. (2023a)
18	ZIF-67	Co-N-C	–	5.33	0.82	AFC	Xuan et al. (2021)
19	ZIF-67	Cu/Co-NC	–	–	0.85	AFC	Parkash (2020a)
20	ZIF-67	Co-N-C/CNT	–	6.04	0.87	AEMFC	Tang et al. (2024)
21	UiO-66	NH$_2$-UiO-66 (Zr/Ni)	800	–	–	MFC	Noori et al. (2022)
22	ZIF-67	N-Co@C/CNT	–	5.6	0.83	AFC	Gao, Liu, et al. (2021)
23	ZIF-8	CoO/Co-N-C	96	–	0.84	AFC	Yang, Wang, et al. (2022)
24	ZIF-8	Co-Zn-N-C	1324.5	–	0.48	MFC	Ding et al. (2023)
25	ZIF-8	Fe-NC	241	–	0.5	PEMFC	Jafari et al. (2021)
26	ZIF-8	NC@CoFe	–	7.99	0.85	AFM	Guo et al. (2021)
27	ZIF-67	Co-N-PC@CNT	2479	27.83	–	MFC	Zhong, Liang, et al. (2021)
28	UiO-66	Ag/Fe-N-C	1261	5.85	0.76	MFC	Zhong et al. (2021)
29	ZIF-67	Ni/Co-NC	4335.6	6.24	–	MFC	Mo et al. (2021)

Metal-organic framework composites as electrocatalyst support materials

The performance of a fuel cell depends on the mass and charge transfer sites, or the active site density, of the electrodes. The active site density relies on the porosity of the supporting substrate, as these pores serve as hosts for catalytic sites to facilitate chemical reactions (Furukawa et al., 2010). Meanwhile, the porous nature of MOF composites is favored in the field of electrocatalysts. Work by Shui et al. (2015) highlights a nonprecious catalyst obtained by electrospinning a polymer solution of Tris-1,10-phenanthroline iron(II) perchlorate (TPI) and ZIF-8. This catalyst contains densely dispersed active catalytic sites, with a reported volumetric activity of nearly 450 A/cm³ at 0.8 V. To ensure effective charge and mass transfer, the catalyst should have a sufficient amount of micropores (pore diameter < 2 nm) and macropores (pore diameter > 50 nm), respectively, as shown in Fig. 5.11.

In Yuan et al. (2013) and Zhao et al. (2014), carbon is used as a catalyst support due to its sufficient microporous surfaces. However, the agglomeration of these surfaces can lead to the formation of large clusters or mesopore surfaces with a pore diameter of 2–50 nm. Mesopores have a larger volume-to-surface area ratio compared to micropores, which increases mass transfer resistance, adds more volume to the catalyst, and lowers catalyst density. Therefore, achieving a higher density of active sites requires a catalyst with a high density. Additionally, a single-atom Fe catalyst supported over a carbon substrate derived from MOF-5 was reported (Khan et al., 2016), appearing as a 3D cubic structure of Zn_4O (1,4-benzenedicarboxylate)$_3$. The synthesized C-MOF-5 produced a higher amount of FeN_x active sites with 2.35 wt.% Fe content and 2751 m²/g of specific surface area. The prepared Fe-SAC-MOF-5 delivers an $E1/2$ of 0.83 V compared with RHE in 0.5 M H_2SO_4 medium. The exceptional physicochemical properties of the carbon support derived from

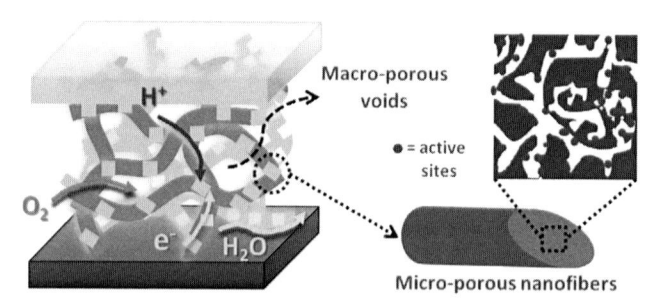

FIGURE 5.11 Schematic representation of macropore–micropore morphology and charge/mass transfer within a nanofibrous network catalyst, Fe/N/FC, at the fuel cell cathode (Shui et al., 2015).

MOF generate extensive electrochemically active surfaces, resulting in higher fuel cell performance.

In the late 2000s, the use of nonplatinum group metals (non-PGMs) became more appealing due to their easy availability and low cost. However, one of the biggest obstacles scientists encounter with non-PGM catalysts is metal dissolution, which arises from the continuous cyclic process, as explained in Fig. 5.12 (Xie et al., 2022). Additionally, non-PGM catalysts often suffer from low turnover frequency, leading to a decrease in ORR activity, as shown in Fig. 5.13 (Zhu et al., 2023b).

Hence, it may be necessary to use a combination of MOF and PGM nanoparticles or precious metals to substantially increase fuel cell performance. Among precious metals, platinum, silver, gold, palladium, and ruthenium are commonly used as electrocatalysts in fuel cells. In 2010, Nadeem et al. (2020) successfully synthesized a MOF-derived PGM catalyst by incorporating Co-ZIF or ZIF-67 into a carbon catalyst, demonstrating promising ORR performance. Instead of using a pristine Co catalyst, Diaz Duran (Khan et al., 2014) doped Co with nitrogen and prepared three different types of cathode catalysts: ZIF-67, CoNic, and $Co(CO_2)_2Pz$, all coated over a carbon support. Among these, $Co(CO_2)_2Pz$ exhibited an excellent open circuit

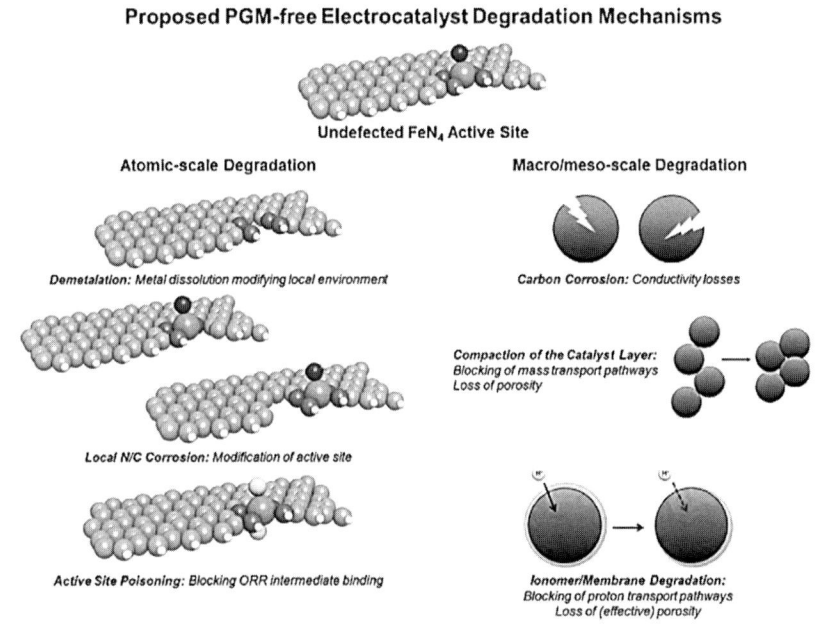

FIGURE 5.12 Proposed PGM-free electrocatalyst degradation mechanisms in atomic and macro/meso scale foe FeN_4 catalyst (Martinez et al., 2018).

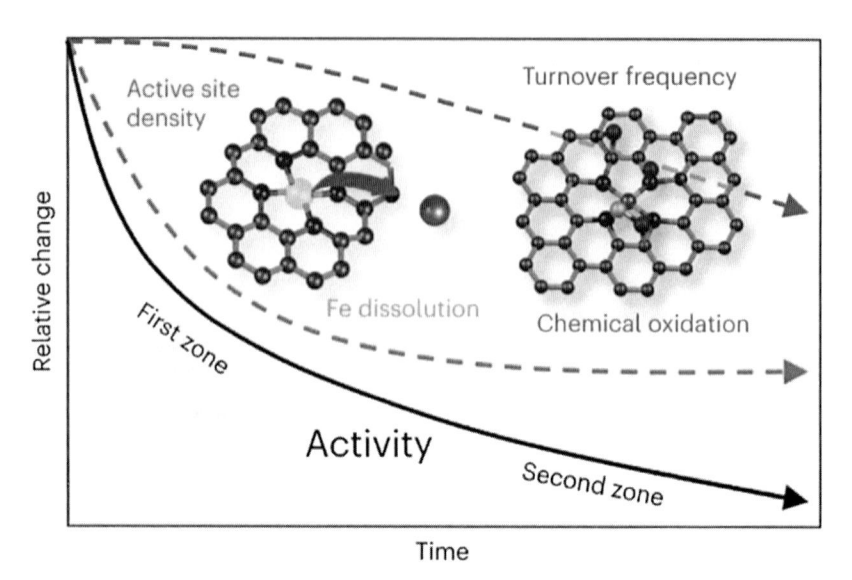

FIGURE 5.13 Morphological changes seen in catalyst with respect to time (Bae et al., 2023).

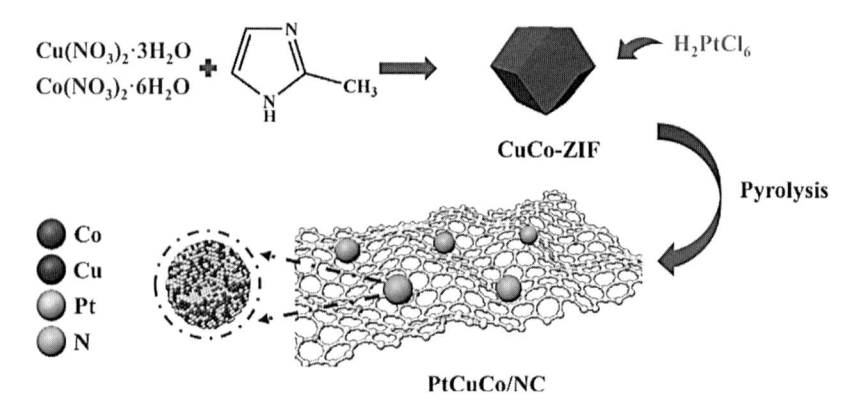

FIGURE 5.14 Production method of PtCuCo/NC from CuCo-ZIF incorporated by Cu element (Zhu et al., 2023b).

voltage (0.86 V) and a better power density of approximately 259 mW/cm², comparable to Pt/C.

Recently, researchers have reported numerous bimetallic (Cu and Co) MOF-based support materials using ZIF (CuCo-ZIF) along with Pt as an electrocatalyst for fuel cell applications, as represented in Fig. 5.14 (Khan et al., 2021). It is the well-organized and regular arrangement of metal centers that allows MOFs to act as both catalysts and hosts for encapsulating other metals. The ordered cell structure and 3D architectures offer maximum volumetric density, which is essential for a catalyst. The length of organic linkers can be adjusted to fine-tune the size of pores in MOFs, enabling optimal gas and liquid diffusion. Following annealing at 700°C, the process produces highly doped nitrogen in porous carbon (CuCo-NC). It was concluded that the proposed catalyst can exhibit excellent catalytic activity even after 10,000 cycles, with a power density of 642.86 mW/cm². Meanwhile, Tang et al. (2016) prepared a bimetallic electrocatalyst via thermal treatment for fuel cell applications. Here, a layered MOF material, $(C_8H_{10}CdO_7)_n \cdot 4H_2O$, is heat-treated at 900°C, resulting in a porous carbon substrate (PC900) with a BET surface area of 877 m²/g. Upon this substrate, bimetallic nanoparticles (Pt-Ni) with 7 wt.% Pt are encapsulated, forming Pt-Ni@PC900, which exhibited an $E1/2$ of 0.91 V, a higher value than the commercial 20 wt.% Pt catalyst used for fuel cell applications.

Parallel to this, Bae et al. (2023) successfully prepared porous carbon, as shown in Fig. 5.15, through the carbonization of MOF-5 (Zn_4O (1,4-benzene-dicarboxylate)$_3$) at 900°C. This carbon served as a support material for PtFe nanoparticles in a direct ethanol fuel cell (DEFC), achieving a power density of 121 W/cm², which is significantly higher compared to Pt on Vulcan carbon.

Meanwhile, a Pt/Ni bimetallic electrocatalyst supported on highly porous carbon, produced by carbonizing MOF-5 at 950°C for an extended period, demonstrated exceptional reliability and stability, even after 500 cyclic durability tests under continuous oxygen supply (Martinez et al., 2018). Subsequently, an investigation into the performance of various bimetallic catalyst nanoparticles, such as PtM (M=Fe, Co, Ni, Cu, Zn), supported over porous carbon derived from the carbonization of MOF-5 at 950°C for methanol oxidation reaction (MOR), revealed that PtCu/PC-950 and PtNi/PC-950 exhibited significantly higher specific mass activities compared to PtZn/PC-950, PtFe/PC-950, Pt/PC-950, and Pt/C (20%) (Díaz-Duran et al., 2019; Goenaga et al., 2010).

Another notable combination used in fuel cell applications involves MOF and graphene-based composites. The concentration of graphene affects the crystallization of MOF and enhances the electrocatalytic capabilities when an optimal amount of graphene is introduced during MOF synthesis. Due to weak van der Waals interactions and the formation of π–π bonds, rGO cannot be used directly in fuel cells, as the restacking of graphene sheets during the reaction limits its effectiveness. This restacking issue is addressed by coating a MOF layer on top of the rGO, allowing for controlled interactions and improved performance. As such, Yoo and Kim (2022) synthesized an electrocatalyst by combining Pt nanoparticles supported over rGO doped with MOF (MIL-101 (Fe)), as represented in Fig. 5.16. This electrocatalyst exhibited an ordered porous structure, facilitating the efficient transportation of fuel and end products.

Pt-rGO/Fe-MIL-100 achieves the highest electrochemical surface area (ECSA) of approximately 261.83 m²/kg, outperforming commercial Pt/C

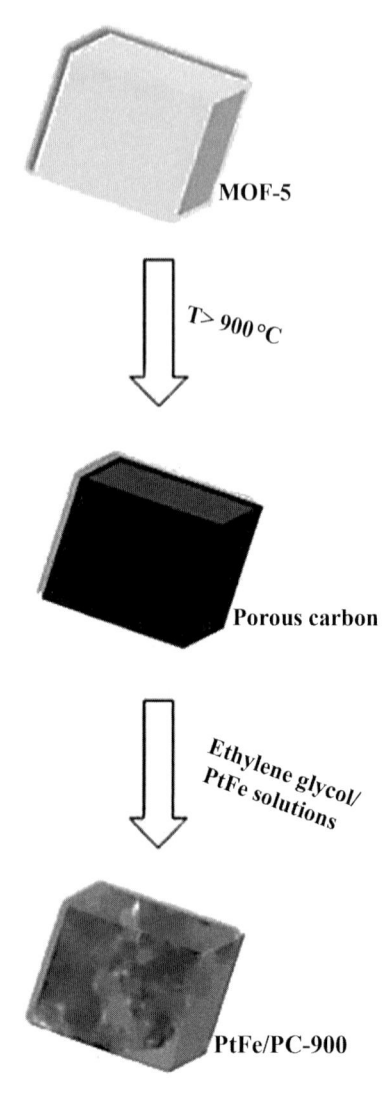

FIGURE 5.15 Schematic representation of preparing porous carbon and PtFe/PC-900 catalyst (Khan et al., 2014).

catalysts. The role of rGO in bimetallic MOFs, specifically from nickel oxide (NiO) and copper oxide (CuO), was investigated by Noor et al. (2020), showing that rGO enhances overall electrocatalytic activity for fuel cell reactions. The prepared rGO-NiO/CuO sample exhibited a current density of 437.28 mA/cm² at 0.9 V. During an oxygen reduction reaction, efficient propagation of electrons, protons, and fuel is crucial, necessitating adequate pores for active sites and

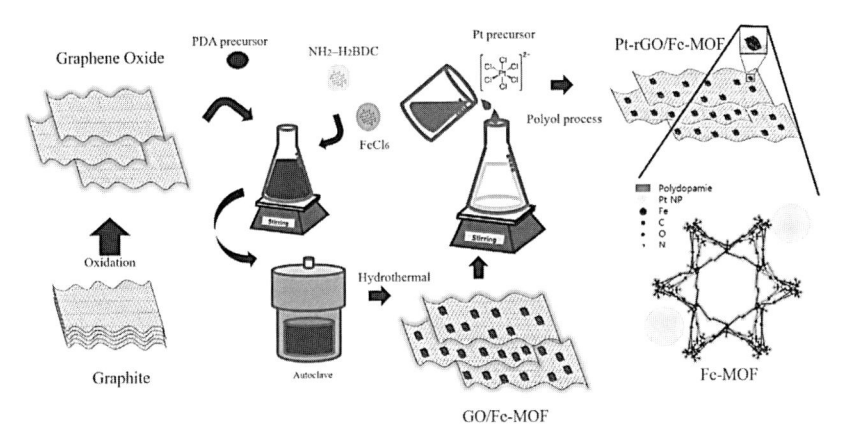

FIGURE 5.16 Schematic representation of Pt-rGO/Fe-MOF composite synthesis (Yoo and Kim, 2022).

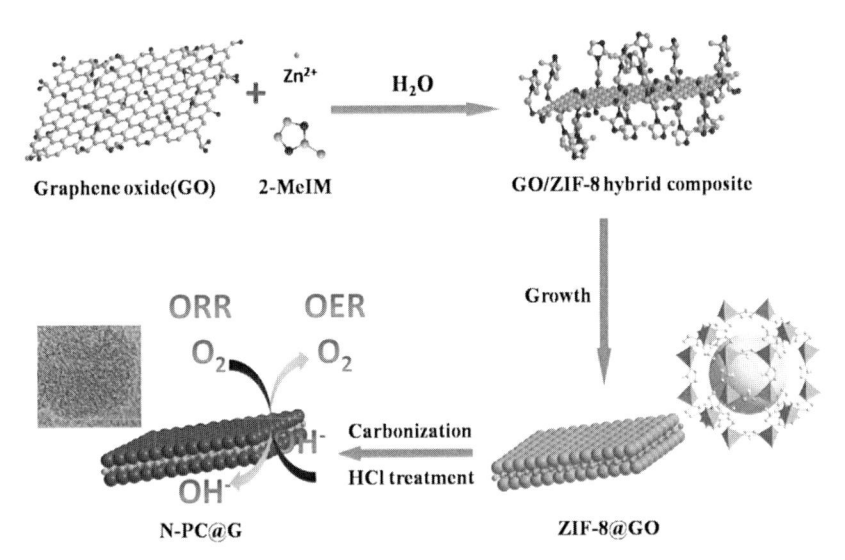

FIGURE 5.17 Diagrammatic representations of the nitrogen-doped porous carbon@graphene synthesis process with sandwich structure (Liu et al., 2016).

proper interconnection channels to ensure uninterrupted transport of these components. Conventional catalysts often suffer from morphological and structural damage, which necessitates in situ preparation of catalysts to create unique structural variations. In 2016, Liu et al. (2016b) synthesized a nitrogen-doped porous carbon over graphene catalyst in a sandwich-like structure by pyrolyzing ZIF-8@GO, as shown in Fig. 5.17. From the figure, the insitu fabrication of ZIF-8 on both sides of the graphene sheet, followed by etching to remove metallic zinc, results in the formation of the N-PC@G electrocatalyst. According to the ORR activity report, catalysts with 0.02 g of GO (N-PC@G-0.02) demonstrated better catalytic activity with a higher catalytic surface area of approximately 1094.3 m²/g. Copper-based MOFs are also gaining popularity as ORR catalysts in fuel cells. MOF-74, a copper-based MOF, was synthesized and encapsulated with Pt nanoparticles. Upon pyrolysis at 900°C, a homogenized hybrid functional material, Pt-Cu-PC-900, was produced. This electrocatalyst showed superior ORR activity compared to Pt/C catalysts (Parkash, 2020b; Parkash et al., 2022). Conversely, Cu-Pt can also be used as a support material for synthesizing nitrogen-doped PC material (Wang et al., 2021a). The resulting electrocatalyst exhibited a Pt activity 4.4 times higher than that of Pt/C. It was noted that using Cu as a dopant for Pt improves the d-band center of Pt atoms and helps in reducing the ORR overpotential.

Metal-organic framework composites in fuel cell electrolytes

Electrolytes play a crucial role in fuel cells by promoting ionic movement across the anode and cathode. Therefore, they must possess excellent ionic conductivity, good chemical and electrical stability, and high mechanical strength to ensure consistent fuel cell performance. However, the high cost and complex manufacturing processes of conventional electrolytes, like Nafion, impede their use in large-scale applications. This has spurred the search for alternative electrolyte substrates, such as MOF-based materials. MOFs, being versatile and reconfigurable materials, are strongly recommended for use as electrolyte membranes in fuel cells (Ramaswamy et al., 2014). To advance fuel cell technology, it is essential to develop electrodes and electrolytes tailored to specific applications. Consequently, these materials must have optimal structures and compositions to address performance challenges. Studies have shown that the coordination structure of MOFs plays a significant role as a proton carrier, leading to new strategies for enhancing fuel cell performance. These strategies include loading existing membrane materials, synthesizing novel electrolyte materials, or functionalizing MOFs and integrating them into polymer matrices (Meng et al., 2017). The polymer matrix provides flexibility and processability, while the MOF component enhances proton conductivity, even under harsh and acidic conditions. Together, the polymer and MOF create a robust network of hydrogen bonds that facilitates excellent proton conductivity.

Various polymer matrices, such as poly(vinyl alcohol) (PVA@PA@Zn-MOF-4) (Zhou et al., 2023), poly(ether-ether-ketone) (MOF-C-SO$_3$H@SPEEK) (Huang et al., 2021), poly(styrene-ethylene-butylene-styrene) (Gao et al., 2019), sulfonated polyimide (SPI-Fe-@MOF) (Yang et al., 2022), poly(benzimidazole) (ZIF-8@PBI) (Devrim & Colpan, 2024), and poly(vinylidene fluoride) (Zr-MOF@PVDF) (Uğur Nigiz & Akel, 2024), can be combined with different MOFs as fillers to fabricate various types of membranes for fuel cells. The network-like structure of

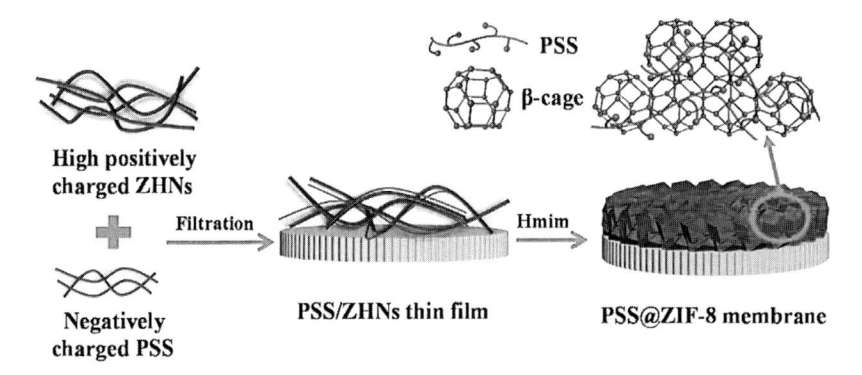

FIGURE 5.18 Schematic of the synthesis and the structure of the PSS@ZIF-8 membrane (Cai et al., 2019).

MOFs enables them to encapsulate particles, a strategy adopted by Cai et al. (2019). to encapsulate poly(4-styrene sulfonate) (PSS) into the nanochannels of ZIF-8, resulting in the PSS@ZIF-8 polymer composite membrane as shown in Fig. 5.18. The resultant membrane outperforms the Nafion membrane in terms of proton conductivity, measuring 2.59×10^{-1} S/cm at 80°C.

Even the pores of MOFs can be used for encapsulating nonpolymer materials, such as integrating MOFs with organic or inorganic compounds. Luo, Ren, et al. (2019) synthesized a high proton-conducting membrane by encapsulating the pores of MOF-808 with imidazole molecules. Im@MOF-808 achieved a proton conductivity of 3.45×10^{-2} S/cm at 338K and exhibited great durability at high temperatures. The $Zr_6O_5(OH)_3$ in MOF-808 has three hydroxyl groups per cluster that can ionize to generate more protons, resulting in strong proton conduction even at room temperature. Generally, thin membrane perfluorosulfonic acid (PFSA) is used to lower Ohmic losses or its associated membrane resistance, and recently, the short side chain (SSC) PFSA is gaining more interest as an electrolyte membrane in fuel cells regardless of their temperature. SSC PFSA, such as Aquivion, is favored for its robust conductivity even at minimal relative humidity and good resilience to chemical and electrochemical deterioration. Paul et al. (2020) studied the effect of this SSC when combined with MOF as an electrolyte in PEMFC. Their research showed that Aquivion SSC modified with Zn-based MOF (MOF-1) performed with a proton conductivity of 3.49×10^{-2} S/cm.

Functionalization of metal-organic framework composites in fuel cell electrolytes

In addition to the direct integration of MOFs into a polymer matrix, functionalizing MOF materials with functional groups like –COOH, –SO$_3$H, and –NH$_2$

before integration, or performing a postsynthesis modification, is another strategy to enhance membrane activity. Through functionalization, more hydrophilic groups are introduced, creating a uniform hydrogen bond network throughout the matrix that facilitates proton transfer. The most commonly reported functionalization involves adding sulfonic groups (–SO$_3$H), which serve as proton carriers by generating more hydrogen bond networks, thereby boosting ionic conductivity (Yang et al., 2017). Recently, Ryu et al. (2022) synthesized a dual sulfonated membrane using polysulfone (sPSF) and controllably sulfonated two MOF composites (sMOF) for a PEMFC. As depicted in Fig. 5.19, the prepared sPSF matrix is combined with MIL-101(Cr)-SO$_3$H and UiO-66(Zr)-SO$_3$H, resulting in an sPSF/MOF membrane with enhanced sulfonated sites that effectively manage the transportation of both ions and water molecules. It can be elucidated that using MIL-101(Cr)-SO$_3$H as the MOF filler with a 3 wt.% loading onto sPSF demonstrates an exceptional proton conductivity of 0.18 S/cm compared to UiO-66(Zr)- SO$_3$H.

Similarly, Roshanravan et al. (2021) developed a composite material for PEM microbial fuel cells by replacing MIL-101(Cr) with MIL-100(Fe) as a filler in an sPSF matrix, resulting in the sPSF/MIL-100(Fe)-7 membrane. This membrane exhibits an ionic conductivity of 2.55 mS/cm, significantly higher compared to the bare sPSF membrane with an ionic conductivity of 0.90 mS/cm.

Furthermore, Moorthy and Deivanayagam (2024) fabricated a poly(2,5-benzimidazole) (ABPBI) membrane loaded with sulfonated Co-based MOF (sCo-MOF), as depicted in Fig. 5.20, to enhance the performance of PEMFC. They prepared membranes with different proportions of sCo-MOF, and the 4 wt.% loaded membrane showed superior water uptake (23.25%) and high ion-exchange capacity (2.89 meq./g). Additionally, unit cell tests revealed that the membrane

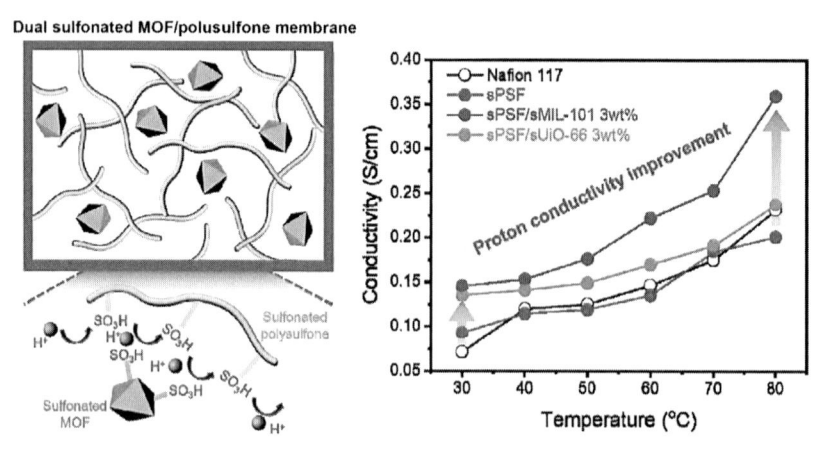

FIGURE 5.19 Mechanism of sulfonated metal-organic framework/polysulfone composite membrane and their conductivity versus temperature graph (Ryu et al., 2022).

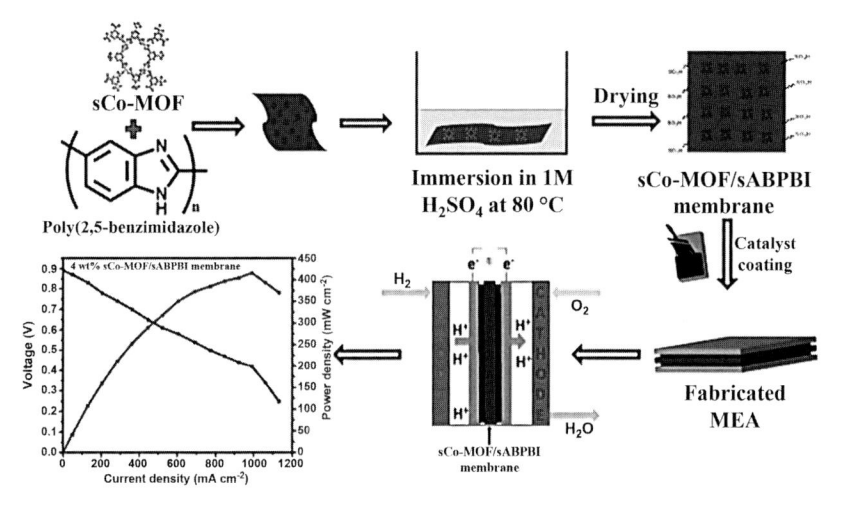

FIGURE 5.20 Scheme for the preparation of the sCo-MOF/sABPBI composite membrane, its incorporation into a fuel cell, and the VI characteristics of 4 wt.% sCo-MOF (Moorthy and Deivanayagam, 2024).

with 4 wt.% sCo-MOF demonstrated a good power density (415.8 mW/cm²) at 80°C, which was three times higher than that of the pristine ABPBI membrane.

Another composite membrane was prepared by anchoring MIL-101-NH$_2$ loaded with imidazole onto the surface of poly(arylene ether ketone sulfone) by Meng et al. (2022), which reported an exceptional proton conductivity value of 0.082 S/cm at 90°C. Similarly, Wang et al. (2021b) doped MOF-carrying guest molecules into a Nafion matrix to enhance membrane performance. High ionic conductivity Zn-MOF filler enriched with NH$_3$ molecules was incorporated into the Nafion polymer matrix, achieving a conductivity of 2.13×10^{-2} S/cm, which is 5.47 times higher than that of the bare Nafion membrane.

Concurrently, direct introduction of MOFs into the polymer matrix can lead to agglomeration and sedimentation, resulting in an uneven surface that is brittle and fragile, significantly reducing the MOF's activity in proton conduction for PEMs. An efficient approach to enhance the proper dispersion of MOFs and create continuous proton channels in the membrane is the in situ growth of MOFs within the polymer matrix. Sun et al. (2017a, 2017b) reported an in situ synthesized membrane where S-UiO-66 and ZIF-8 were grown on graphene oxide and carbon nanotubes, followed by doping into the polymer matrix. This method prevented MOF agglomeration and improved ionic conductivity. Similarly, Cai et al. (2021) studied the performance of a sulfonated MOF-808 grown in graphene oxide and doped into a poly(ether-ether ketone) polymer matrix. The prepared membrane exhibited high proton conductivity (0.196 S/cm) at 70°C with 90% humidity.

Conclusion and future work

Consumption and exploitation of energy are inextricably linked to an economy's rapid growth, necessitating the development of green energy technologies. Modern methods for energy production emphasize increasing the use of renewable resources, with fuel cells emerging as key energy-converting devices. However, the high cost of electrodes and electrolytes hampers their commercialization, highlighting the need for cost-effective raw materials in fuel cell systems. MOFs have demonstrated significant potential in fuel cell applications due to their unique structure and properties. This chapter reviews recent advances in MOF applications in fuel cells, focusing on key challenges such as improving electrodes and electrolytes. In fuel cells, the half-cell reactions—HOR at the anode and ORR at the cathode—require stable and effective electrocatalysts. MOF-derived materials are increasingly gaining attention as electrocatalysts. Initially, MOFs were used as precursors, which can be transformed into nanostructured catalysts through various physical and chemical treatments. MOFs, containing metals like Fe, Cr, Co, and Ni, have led to catalysts such as Fe-NC/ZIF-8, Co-NC/ZIF-67, and Cu-Co-NC/ZIF-67. The application of MOFs, from single-metal to bimetallic and multimetal combinations, shows promising potential for green energy sectors. Despite issues like structural degradation and efficiency loss during synthesis and use, these challenges can be mitigated through functionalization. Functionalization often involves integrating nitrogen with the MOF precursor, leading to various catalytic events. MOFs are not only used as precursors for non precious group metal catalysts but also as supports for precious group metal (PGM) catalysts, increasing active sites available for chemical reactions. The large specific surface area of MOFs facilitates better water management by accommodating substantial water molecules, thus modifying the proton exchange membrane of the fuel cell. MOFs can be restructured in various forms, including integrating them into existing membrane materials, synthesizing novel electrolytes, or functionalizing them and incorporating them into polymer matrices. However, challenges remain in the commercialization of fuel cells, particularly concerning stability and durability at elevated temperatures. Continuous research and development are essential to overcome these obstacles and achieve widespread commercialization. While the application of MOFs in fuel cells has seen progress, their composites are not yet extensively utilized, representing a promising area for future research.

References

Armel, V., Hannauer, J., & Jaouen, F. (2015). Effect of zif-8 crystal size on the O_2 electro-reduction performance of pyrolyzed Fe–N–C catalysts. *Catalysts, 5*(3), 1333–1351.

Armel, V., Hindocha, S., Salles, F., Bennett, S., Jones, D., & Jaouen, F. (2017). Structural descriptors of zeolitic–imidazolate frameworks are keys to the activity of Fe–N–C catalysts. *Journal of the American Chemical Society, 139*(1), 453–464.

Bae, G., et al. (2023). Unravelling the complex causality behind Fe–N–C degradation in fuel cells. *Nature Catalysis, 6*(12), 1140–1150.

Barkholtz, H. M., Chong, L., Kaiser, Z. B., Xu, T., & Liu, D.-J. (2015). Highly active non-PGM catalysts prepared from metal organic frameworks. *Catalysts, 5*(2), 955–965.

Barkholtz, H. M., Chong, L., Kaiser, Z. B., Xu, T., & Liu, D.-J. (2016). Enhanced performance of non-PGM catalysts in air operated PEM-fuel cells. *International Journal of Hydrogen Energy, 41*(47), 22598–22604.

Cai, Y. Y., et al. (2019). Achieving efficient proton conduction in a MOF-based proton exchange membrane through an encapsulation strategy. *Journal of Membrane Science, 590*, 117277.

Cai, Y. Y., Wang, J. J., Cai, Z. H., Zhang, Q. G., Zhu, A. M., & Liu, Q. L. (2021). Enhanced performance of sulfonated poly (ether ether ketone) hybrid membranes by introducing sulfated MOF-808/graphene oxide composites. *ACS Applied Energy Materials, 4*(9), 9664–9672.

Chao, S., Xia, Q., Wang, Y., Li, W., & Chen, W. (2020). Pristine S, N-containing Mn-based metal organic framework nanorods enable efficient oxygen reduction electrocatalysis. *Dalton Transactions, 49*(14), 4336–4342.

Chung, H. T., et al. (2017). Direct atomic-level insight into the active sites of a high-performance PGM-free ORR catalyst. *Science (New York, N.Y.), 357*(6350), 479–484.

Cui, C.-H., & Yu, S.-H. (2013). Engineering interface and surface of noble metal nanoparticle nanotubes toward enhanced catalytic activity for fuel cell applications. *Accounts of Chemical Research, 46*(7), 1427–1437.

da Silva Freitas, W., et al. (2023). Tailoring MOF structure via iron decoration to enhance ORR in alkaline polymer electrolyte membrane fuel cells. *Chemical Engineering Journal, 465*, 142987.

Debe, M. K. (2012). Electrocatalyst approaches and challenges for automotive fuel cells. *Nature, 486*(7401), 43–51.

Deng, Y., et al. (2017). Well-defined ZIF-derived Fe–N codoped carbon nanoframes as efficient oxygen reduction catalysts. *ACS Applied Materials & Interfaces, 9*(11), 9699–9709.

Devrim, Y., & Colpan, C. O. (2024). Assessment of polybenzimidazole/MOF composite membranes for the improvement of high-temperature PEM fuel cell performance. *International Journal of Hydrogen Energy, 58*, 470–478. https://doi.org/10.1016/j.ijhydene.2024.01.184.

Ding, F., et al. (2023). Bimetallic zeolite imidazolium framework derived multiphase Co/HNC as pH-universal catalysts with efficient oxygen reduction performance for microbial fuel cells. *Electrochimica Acta, 438*, 141548. https://doi.org/10.1016/j.electacta.2022.141548.

Díaz-Duran, A. K., Viva, F. A., & Roncaroli, F. (2019). High durability fuel cell cathodes obtained from cobalt metal organic frameworks. *Electrochimica Acta, 320*, 134623.

Feng, C., et al. (2020). 2-Methylimidazole as a nitrogen source assisted synthesis of a nano-rod-shaped Fe/FeN@N-C catalyst with plentiful FeN active sites and enhanced ORR activity. *Applied Surface Science, 533*, 147481. https://doi.org/10.1016/j.apsusc.2020.147481.

Furukawa, H., et al. (2010). Ultrahigh porosity in metal-organic frameworks. *Science (New York, N.Y.), 329*(5990), 424–428.

Gao, H., et al. (2020). Co/NC-Gr composite derived from ZIF-67: Effects of preparation method on the structure and electrocatalytic performance for oxygen reduction reaction. *International Journal of Hydrogen Energy, 45*(7), 4403–4416.

Gao, H., et al. (2021). MOF-derived N-doped carbon coated Co/RGO composites with enhanced electrocatalytic activity for oxygen reduction reaction. *Inorganic Chemistry Communications, 123*, 108330.

Gao, H., Liu, Y., Ma, Y., Meng, E., & Zhang, Y. (2021). Synthesis of N-doped Co@C/CNT materials based on ZIF-67 and their electrocatalytic performance for oxygen reduction. *Ionics, 27*(6), 2561–2569. https://doi.org/10.1007/s11581-021-04031-y.

Gao, S., et al. (2017). N-doped-carbon-coated Fe_3O_4 from metal-organic framework as efficient electrocatalyst for ORR. *Nano Energy, 40*, 462–470.

Gao, X., et al. (2019). Enhanced water transport in AEMs based on poly (styrene–ethylene–butylene–styrene) triblock copolymer for high fuel cell performance. *Polymer Chemistry, 10*(15), 1894–1903. https://doi.org/10.1039/C8PY01618F.

Gascon, J., Corma, A., Kapteijn, F., & Llabres i Xamena, F. X. (2014). Metal organic framework catalysis: Quo vadis? *Acs Catalysis, 4*(2), 361–378.

Goenaga, G., Ma, S., Yuan, S., & Liu, D.-J. (2010). New approaches to non-PGM electrocatalysts using porous framework materials. *Ecs Transactions, 33*(1), 579.

Guo, Z., et al. (2021). Core-shell structured metal organic framework materials derived cobalt/iron–nitrogen Co-doped carbon electrocatalysts for efficient oxygen reduction. *International Journal of Hydrogen Energy, 46*(14), 9341–9350. https://doi.org/10.1016/j.ijhydene.2020.11.210.

Hou, B., Wang, C. C., Tang, R., Zhang, Q., & Cui, X. (2020). High performance Fe–N–C oxygen reduction electrocatalysts by solid-phase preparation of metal–organic frameworks. *Materials Research Express, 7*(2), 25506.

Huang, H., Ma, Y., Jiang, Z., & Jiang, Z.-J. (2021). Spindle-like MOFs-derived porous carbon filled sulfonated poly (ether ether ketone): A high performance proton exchange membrane for direct methanol fuel cells. *Journal of Membrane Science, 636*, 119585. https://doi.org/10.1016/j.memsci.2021.119585.

Huang, J.-W., Cheng, Q.-Q., Huang, Y.-C., Yao, H.-C., Zhu, H.-B., & Yang, H. (2019). Highly efficient Fe–N–C electrocatalyst for oxygen reduction derived from core–shell-structured Fe (OH)$_3$@ zeolitic imidazolate framework. *ACS Applied Energy Materials, 2*(5), 3194–3203.

Jafari, M., Gharibi, H., & Parnian, M. J. (2021). Metal organic framework derived iron-nitrogen doped porous carbon support decorated with cobalt and iron as efficient nanocatalyst toward oxygen reduction reaction. *Journal of Power Sources, 499*, 229956. https://doi.org/10.1016/j.jpowsour.2021.229956.

Jing, F., Sun, R., Wang, S., Sun, H., & Sun, G. (2020). Effect of the anode structure on the stability of a direct methanol fuel cell. *Energy & Fuels, 34*(3), 3850–3857.

Khan, I. A., Badshah, A., Haider, N., Ullah, S., Anjum, D. H., & Nadeem, M. A. (2014). Porous carbon as electrode material in direct ethanol fuel cells (DEFCs) synthesized by the direct carbonization of MOF-5. *Journal of Solid State Electrochemistry, 18*, 1545–1555.

Khan, I. A., Badshah, A., Shah, F. U., Assiri, M. A., & Nadeem, M. A. (2021). Zinc-coordination polymer-derived porous carbon-supported stable PtM electrocatalysts for methanol oxidation reaction. *ACS Omega, 6*(10), 6780–6790.

Khan, I. A., Qian, Y., Badshah, A., Nadeem, M. A., & Zhao, D. (2016). Highly porous carbon derived from MOF-5 as a support of ORR electrocatalysts for fuel cells. *ACS Applied Materials & Interfaces, 8*(27), 17268–17275.

Kumar, R., et al. (2024). Catalyzing oxygen reduction by morphologically engineered ZIF-derived carbon composite catalysts in dual-chamber microbial fuel cells. *Journal of Environmental Chemical Engineering, 12*(2), 112242.

Li, J.-S., et al. (2014). Heteroatoms ternary-doped porous carbons derived from MOFs as metal-free electrocatalysts for oxygen reduction reaction. *Scientific Reports, 4*(1), 5130.

Li, J.-S., et al. (2015). Nitrogen-doped Fe/Fe$_3$C@ graphitic layer/carbon nanotube hybrids derived from MOFs: Efficient bifunctional electrocatalysts for ORR and OER. *Chemical Communications, 51*(13), 2710–2713.

Li, Y., et al. (2021). A general carboxylate-assisted approach to boost the ORR performance of ZIF-derived Fe/N/C catalysts for proton exchange membrane fuel cells. *Advanced Functional Materials, 31*(15), 2009645. https://doi.org/10.1002/adfm.202009645.

Liang, B., Su, M., Zhao, Z., & Liang, S. (2024). Iron-involved zeolitic imidazolate framework-67 derived Co/Fe-NC as enhanced ORR catalyst in air–cathode microbial fuel cell. *Journal of Electroanalytical Chemistry, 962*, 118260. https://doi.org/10.1016/j.jelechem.2024.118260.

Liu, B., Shioyama, H., Akita, T., & Xu, Q. (2008). Metal-organic framework as a template for porous carbon synthesis. *Journal of the American Chemical Society, 130*(16), 5390–5391.

Liu, S., et al. (2016). Metal-organic framework derived nitrogen-doped porous carbon@graphene sandwich-like structured composites as bifunctional electrocatalysts for oxygen reduction and evolution reactions. *Carbon, 106*, 74–83.

Liu, T., Zhao, P., Hua, X., Luo, W., Chen, S., & Cheng, G. (2016). An Fe–N–C hybrid electrocatalyst derived from a bimetal–organic framework for efficient oxygen reduction. *Journal of Materials Chemistry A, 4*(29), 11357–11364.

Liu, Z., et al. (2022). ZIF-67-derived Co nanoparticles embedded in N-doped porous carbon composite interconnected by MWCNTs as highly efficient ORR electrocatalysts for a flexible direct formate fuel cell. *Chemical Engineering Journal, 432*, 134192. https://doi.org/10.1016/j.cej.2021.134192.

Lu, F., et al. (2022). Cu-N4 single atoms derived from metal-organic frameworks with trapped nitrogen-rich molecules and their use as efficient electrocatalysts for oxygen reduction reaction. *Chemical Engineering Journal, 431*, 133242.

Luo, H.-B., Ren, Q., Wang, P., Zhang, J., Wang, L., & Ren, X.-M. (2019). High proton conductivity achieved by encapsulation of imidazole molecules into proton-conducting MOF-808. *ACS Applied Materials & Interfaces, 11*(9), 9164–9171.

Luo, X., Han, W., Ren, H., & Zhuang, Q. (2019). Metallic organic framework-derived Fe, N, S co-doped carbon as a robust catalyst for the oxygen reduction reaction in microbial fuel cells. *Energies, 12*(20), 3846.

Ma, J., et al. (2024). Preparation of Fe-BN-C catalysts derived from ZIF-8 and their performance in the oxygen reduction reaction. *RSC Advances, 14*(7), 4607–4613. https://doi.org/10.1039/D3RA07188J.

Ma, S., Goenaga, G. A., Call, A. V., & Liu, D. (2011). Cobalt imidazolate framework as precursor for oxygen reduction reaction electrocatalysts. *Chemistry–A European Journal, 17*(7), 2063.

Martinez, U., Babu, S. K., Holby, E. F., & Zelenay, P. (2018). Durability challenges and perspective in the development of PGM-free electrocatalysts for the oxygen reduction reaction. *Current Opinion in Electrochemistry, 9*, 224–232.

Meng, L., Zhang, Z., Ju, M., Xu, J., & Wang, Z. (2022). Enhancing proton conductivity of proton exchange membranes via anchoring imidazole-loaded MIL-101-NH2 onto sulfonated poly (arylene ether ketone sulfone) by chemical bonding. *International Journal of Energy Research, 46*(15), 23480–23492.

Meng, X., Wang, H.-N., Song, S.-Y., & Zhang, H.-J. (2017). Proton-conducting crystalline porous materials. *Chemical Society Reviews, 46*(2), 464–480.

Miao, W., et al. (2022). Single atom Fe-based catalyst derived from hierarchical (Fe, N)-ZIF-8/CNFs for high-efficient ORR activity. *Materials Chemistry Frontiers, 6*(21), 3213–3224.

Mo, R., et al. (2021). Highly efficient PtCo nanoparticles on Co–N–C nanorods with hierarchical pore structure for oxygen reduction reaction. *International Journal of Hydrogen Energy, 46*(29), 15991–16002.

Moorthy, S., & Deivanayagam, P. (2024). Sulfonated cobalt metal–organic framework embedded mixed matrix membrane towards fuel-cell applications. *ACS Applied Materials & Interfaces, 16*(12), 14712–14721.

Nadeem, M., et al. (2020). Pt-Ni@PC900 hybrid derived from layered-structure Cd-MOF for fuel cell ORR activity. *ACS Omega, 5*(5), 2123–2132.

Nie, Y., Li, L., & Wei, Z. (2015). Recent advancements in Pt and Pt-free catalysts for oxygen reduction reaction. *Chemical Society Reviews, 44*(8), 2168–2201.

Noor, T., et al. (2020). Nanocomposites of NiO/CuO based MOF with rGO: An efficient and robust electrocatalyst for methanol oxidation reaction in DMFC. *Nanomaterials, 10*(8), 1601.

Noori, M. T., Ezugwu, C. I., Wang, Y., & Min, B. (2022). Robust bimetallic metal-organic framework cathode catalyst to boost oxygen reduction reaction in microbial fuel cell. *Journal of Power Sources, 547*, 231947. https://doi.org/10.1016/j.jpowsour.2022.231947.

Oar-Arteta, L., Wezendonk, T., Sun, X., Kapteijn, F., & Gascon, J. (2017). Metal organic frameworks as precursors for the manufacture of advanced catalytic materials. *Materials Chemistry Frontiers, 1*(9), 1709–1745.

Parkash, A. (2020a). CTAB-capped copper nanoparticles coated on N doped carbon layer and encapsulated in ZIF-67: A highly-efficient ORR catalyst. *Journal of Porous Materials, 27*(5), 1377–1387. https://doi.org/10.1007/s10934-020-00913-0.

Parkash, A. (2020b). Pt nanoparticles anchored on Cu-MOF-74: An efficient and durable ultra-low Pt electrocatalyst toward oxygen reduction reaction. *ECS Journal of Solid State Science and Technology, 9*(6), 65021.

Parkash, A., Islam, M., Qureshi, K. M., & Arain, A. M. (2022). MOF-74 derived carbon-stabilized Pt/Cu-PC-900 nanoparticles: Ultra-low Pt content and improved electrocatalytic activity. *ECS Journal of Solid State Science and Technology, 11*(9), 91015.

Paul, S., Choi, S.-J., & Kim, H. J. (2020). Enhanced proton conductivity of a Zn (II)-based MOF/aquivion composite membrane for PEMFC applications. *Energy & Fuels, 34*(8), 10067–10077.

Proietti, E., et al. (2011). Iron-based cathode catalyst with enhanced power density in polymer electrolyte membrane fuel cells. *Nature Communications, 2*(1), 416. https://doi.org/10.1038/ncomms1427.

Ramaswamy, P., Wong, N. E., & Shimizu, G. K. H. (2014). MOFs as proton conductors–challenges and opportunities. *Chemical Society Reviews, 43*(16), 5913–5932.

Ren, J., Shi, Z., & Huang, Y. (2023). Zeolitic-imidazolate-framework-derived Fe-NC catalysts towards efficient oxygen reduction reaction. *International Journal of Hydrogen Energy, 48*(33), 12333–12341. https://doi.org/10.1016/j.ijhydene.2022.12.099.

Roshanravan, B., Younesi, H., Abdollahi, M., Rahimnejad, M., & Pyo, S.-H. (2021). Application of proton-conducting sulfonated polysulfone incorporated MIL-100 (Fe) composite materials for polymer-electrolyte membrane microbial fuel cells. *Journal of Cleaner Production, 300*, 126963.

Ryu, G. Y., et al. (2022). Dual-sulfonated MOF/polysulfone composite membranes boosting performance for proton exchange membrane fuel cells. *European Polymer Journal, 180*, 111601.

Seyed Bagheri, S. M., Gharibi, H., & Zhiani, M. (2022). Introduction of a new active and stable cathode catalyst based on bimetal-organic frameworks/PPy-sheet for alkaline direct ethanol fuel cell. *International Journal of Hydrogen Energy, 47*(56), 23552–23569. https://doi.org/10.1016/j.ijhydene.2022.05.142.

Shah, S. S., et al. (2021). Nanostructure engineering of metal–organic derived frameworks: Cobalt phosphide embedded in carbon nanotubes as an efficient ORR catalyst. *Molecules (Basel, Switzerland), 26*(21), 6672. https://doi.org/10.3390/molecules26216672.

Shao, Y., Yin, G., Wang, Z., & Gao, Y. (2007). Proton exchange membrane fuel cell from low temperature to high temperature: Material challenges. *Journal of Power Sources, 167*(2), 235–242.

Shen, S., et al. (2021). The development of a highly durable Fe-N-C electrocatalyst with favorable carbon nanotube structures for the oxygen reduction in PEMFCs. *Journal of Electrochemical Energy Conversion and Storage, 19*(1), 010905. https://doi.org/10.1115/1.4050725.

Shui, J., Chen, C., Grabstanowicz, L., Zhao, D., & Liu, D.-J. (2015). Highly efficient nonprecious metal catalyst prepared with metal–organic framework in a continuous carbon nanofibrous network. *Proceedings of the National Academy of Sciences, 112*(34), 10629–10634.

da Silva Freitas, W., et al. (2022). Metal-organic-framework-derived electrocatalysts for alkaline polymer electrolyte fuel cells. *Journal of Power Sources, 550*, 232135.

Su, H., et al. (2018). Metal–organic frameworks-derived core–shell Fe_3O_4/Fe_3N@ graphite carbon nanocomposites as excellent non-precious metal electrocatalyst for oxygen reduction. *Dalton Transactions, 47*(46), 16567–16577.

Sun, H., Tang, B., & Wu, P. (2017a). Rational design of S-UiO-66@GO hybrid nanosheets for proton exchange membranes with significantly enhanced transport performance. *ACS Applied Materials & Interfaces, 9*(31), 26077–26087.

Sun, H., Tang, B., & Wu, P. (2017b). Two-dimensional zeolitic imidazolate framework/carbon nanotube hybrid networks modified proton exchange membranes for improving transport properties. *ACS Applied Materials & Interfaces, 9*(40), 35075–35085.

Tang, H., et al. (2016). Metal–organic-framework-derived dual metal-and nitrogen-doped carbon as efficient and robust oxygen reduction reaction catalysts for microbial fuel cells. *Advanced Science, 3*(2), 1500265.

Tang, M., et al. (2024). An improvement on the electrocatalytic performance of ZIF-67 byin situself-growing CNTs on surface. *Nanotechnology, 35*(23), 235601. https://doi.org/10.1088/1361-6528/ad2f73.

Tong, J., et al. (2018). Composite of hierarchically porous N-doped carbon/carbon nanotube with greatly improved catalytic performance for oxygen reduction reaction. *ACS Sustainable Chemistry & Engineering, 6*(7), 8383–8391.

Uğur Nigiz, F., & Akel, M. (2024). Fabrication of Zr MOF-doped polyvinylidene fluoride membranes and testing in H-type microbial fuel cell. *International Journal of Hydrogen Energy, 54*, 1264–1272. https://doi.org/10.1016/j.ijhydene.2023.12.094.

Wang, C., et al. (2021a). Highly dispersed Cu atoms in MOF-derived N-doped porous carbon inducing Pt loads for superior oxygen reduction and hydrogen evolution. *Chemical Engineering Journal, 426*, 130749.

Wang, H., et al. (2021b). Proton conduction of nafion hybrid membranes promoted by NH_3-modified Zn-MOF with host–guest collaborative hydrogen bonds for H_2/O_2 fuel cell applications. *ACS Applied Materials & Interfaces, 13*(6), 7485–7497.

Wang, H., et al. (2019). Impacts of imidazolate ligand on performance of zeolitic-imidazolate framework-derived oxygen reduction catalysts. *ACS Energy Letters, 4*(10), 2500–2507.

Wang, H., Wei, L., Liu, J., & Shen, J. (2020). Hollow bimetal ZIFs derived Cu/Co/N co-coordinated ORR electrocatalyst for microbial fuel cells. *International Journal of Hydrogen Energy, 45*(7), 4481–4489.

Wang, H., Zhu, Q.-L., Zou, R., & Xu, Q. (2017). Metal-organic frameworks for energy applications. *Chem, 2*(1), 52–80.

Wang, Q., & Astruc, D. (2019). State of the art and prospects in metal–organic framework (MOF)-based and MOF-derived nanocatalysis. *Chemical Reviews, 120*(2), 1438–1511.

Wang, R., Zhang, P., Wang, Y., Wang, Y., Zaghib, K., & Zhou, Z. (2020). ZIF-derived Co–N–C ORR catalyst with high performance in proton exchange membrane fuel cells. *Progress in Natural Science: Materials International, 30*(6), 855–860. https://doi.org/10.1016/j.pnsc.2020.09.010.

Wang, X., et al. (2014). MOF derived catalysts for electrochemical oxygen reduction. *Journal of Materials Chemistry A, 2*(34), 14064–14070.

Wang, X. X., et al. (2018). Nitrogen-coordinated single cobalt atom catalysts for oxygen reduction in proton exchange membrane fuel cells. *Advanced Materials, 30*(11), 1706758.

Williams, K. (2017). Metal-organic frameworks and MOF-derived carbon materials for fuel cell applications. *Digital Commons*, 7452.

Wu, B., Lin, X., Ge, L., Wu, L., & Xu, T. (2015). A novel route for preparing highly proton conductive membrane materials with metal-organic frameworks (MOFs). *Chemical Communications, 3*(207890), 10715–10722.

Bin Wu, H., & Lou, X. W. (2017). Metal-organic frameworks and their derived materials for electrochemical energy storage and conversion: Promises and challenges. *Science Advances, 3*(12), eaap9252.

Xiao, L., Yang, J.-M., Huang, G.-Y., Zhao, Y., & Zhu, H.-B. (2020). Construction of efficient Mn-NC oxygen reduction electrocatalyst from a Mn (II)-based MOF with N-rich organic linker. *Inorganic Chemistry Communications, 118*, 107982.

Xiao, L., Yuan, J., & Zhu, H. (2020). Facile synthesis of MOF-derived Mn_3O_4@ N-doped carbon with efficient oxygen reduction. *Zeitschrift für anorganische und allgemeine Chemie, 646*(17), 1426–1431.

Xie, X., Shang, L., Xiong, X., Shi, R., & Zhang, T. (2022). Fe single-atom catalysts on MOF-5 derived carbon for efficient oxygen reduction reaction in proton exchange membrane fuel cells. *Advanced Energy Materials, 12*(3), 2102688.

Xuan, J.-P., Huang, N.-B., Zhang, J.-J., Dong, W.-J., Yang, L., & Wang, B. (2021). Fabricating Co–N–C catalysts based on ZIF-67 for oxygen reduction reaction in alkaline electrolyte. *Journal of Solid State Chemistry, 294*, 121788. https://doi.org/10.1016/j.jssc.2020.121788.

Yang, F., et al. (2017). A flexible metal–organic framework with a high density of sulfonic acid sites for proton conduction. *Nature Energy, 2*(11), 877–883.

Yang, J., et al. (2022). Hybrid proton exchange membrane used in fuel cell with amino-functionalized metal–organic framework in sulfonated polyimide to construct efficient ion transport channel. *Advanced Composites and Hybrid Materials, 5*(2), 834–842. https://doi.org/10.1007/s42114-022-00469-4.

Yang, J.-M., Wang, K.-A., & Zhu, H.-B. (2022). KOH-promoted in-situ construction of zeolitic imidazolate framework-derived CoO/Co-N-C hybrids jointly boosting oxygen reduction reaction. *Journal of Alloys and Compounds, 912*, 165198. https://doi.org/10.1016/j.jallcom.2022.165198.

Yoo, P. K., & Kim, S. (2022). Preparation and electrochemical activity of platinum catalyst-supported graphene and Fe-based metal-organic framework composite electrodes for fuel cells. *Journal of Industrial and Engineering Chemistry, 105*, 259–267.

Yuan, S., et al. (2013). A highly active and support-free oxygen reduction catalyst prepared from ultrahigh-surface-area porous polyporphyrin. *Angewandte Chemie, 32*(125), 8507–8511.

Zhang, D., et al. (2018a). Isolated Fe and Co dual active sites on nitrogen-doped carbon for a highly efficient oxygen reduction reaction. *Chemical Communications, 54*(34), 4274–4277.

Zhang, Y., et al. (2018b). Well-defined cobalt catalyst with N-doped carbon layers enwrapping: The correlation between surface atomic structure and electrocatalytic property. *Small (Weinheim an der Bergstrasse, Germany), 14*(6), 1702074.

Zhang, L.-L., Huang, G.-Y., Wang, K.-A., Shi, J.-Y., & Zhu, H.-B. (2023). Hollow polyhedral Fe-NC nanostructures derived from ZIF-8 highlighting markable morphology effect on electrochemical oxygen reduction and carbon dioxide reduction. *Journal of Alloys and Compounds, 936*, 168341.

Zhang, P., Sun, F., Xiang, Z., Shen, Z., Yun, J., & Cao, D. (2014). ZIF-derived in situ nitrogen-doped porous carbons as efficient metal-free electrocatalysts for oxygen reduction reaction. *Energy & Environmental Science, 7*(1), 442–450.

Zhang, P.-Y., et al. (2022). General carbon-supporting strategy to boost the oxygen reduction activity of zeolitic-imidazolate-framework-derived Fe/N/carbon catalysts in proton exchange membrane fuel cells. *ACS Applied Materials & Interfaces, 14*(27), 30724–30734. https://doi.org/10.1021/acsami.2c04786.

Zhang, Y., et al. (2020). Fabrication of core-shell nanohybrid derived from iron-based metal-organic framework grappled on nitrogen-doped graphene for oxygen reduction reaction. *Chemical Engineering Journal, 401*, 126001.

Zhang, Y., Yu, X., Hou, Y., Liu, C., Xie, G., & Chen, X. (2024). Current research status of MOF materials for catalysis applications. *Molecular Catalysis, 555*, 113851.

Zhao, D., et al. (2014). Electrocatalysts: Highly efficient non-precious metal electrocatalysts prepared from one-pot synthesized zeolitic imidazolate frameworks. *Advanced Materials, 26*(7), 1092.

Zhe-qin, C., Xiao-cong, Z., Yong-min, X., Jia-ming, L., Zhi-feng, X., & Rui-xiang, W. (2021). A high-performance nitrogen-rich ZIF-8-derived Fe-NC electrocatalyst for the oxygen reduction reaction. *Journal of Alloys and Compounds, 884*, 160980.

Zhong, C.-J., et al. (2008). Fuel cell technology: Nano-engineered multimetallic catalysts. *Energy & Environmental Science, 1*(4), 454–466.

Zhong, K., et al. (2021). Enhanced oxygen reduction upon Ag/Fe co-doped UiO-66-NH$_2$-derived porous carbon as bacteriostatic catalysts in microbial fuel cells. *Carbon, 183*, 62–75. https://doi.org/10.1016/j.carbon.2021.06.070.

Zhong, M., Liang, B., Fang, D., Li, K., & Lv, C. (2021). Leaf-like carbon frameworks dotted with carbon nanotubes and cobalt nanoparticles as robust catalyst for oxygen reduction in microbial fuel cell. *Journal of Power Sources, 482*, 229042. https://doi.org/10.1016/j.jpowsour.2020.229042.

Zhou, Y., et al. (2023). Facilitating the proton conductivity of polyvinyl alcohol based proton exchange membrane by phytic acid encapsulated Zn-azolate MOF. *Process Safety and Environmental Protection, 172*, 48–56. https://doi.org/10.1016/j.psep.2023.01.072.

Zhu, W., et al. (2023a). Defect-engineered ZIF-derived non-Pt cathode catalyst at 1.5 mg cm^{-2} loading for proton exchange membrane fuel cells. *Small (Weinheim an der Bergstrasse, Germany), 19*(43), 2302090. https://doi.org/10.1002/smll.202302090.

Zhu, H., et al. (2023b). Bimetallic ZIF-based PtCuCo/NC electrocatalyst Pt supported with an N-doped porous carbon for oxygen reduction reaction in PEM fuel cells. *ACS Applied Energy Materials, 6*(3), 1575–1584.

Zhu, Q.-L., Xia, W., Zheng, L.-R., Zou, R., Liu, Z., & Xu, Q. (2017). Atomically dispersed Fe/N-doped hierarchical carbon architectures derived from a metal–organic framework composite for extremely efficient electrocatalysis. *ACS Energy Letters, 2*(2), 504–511.

Zou, C., Zhao, Q., Zhang, G., & Xiong, B. (2016). Energy revolution: From a fossil energy era to a new energy era. *Natural Gas Industry B, 3*(1), 1–11.

Chapter 6

Metal-organic framework composites for solar cells

Arife Gencer Imer
Physics Department, Van Yuzuncu Yil University, Tuşba, Van, Turkey

Introduction

The rapid advancement of solar cell technology has created a demand for innovative materials that can enhance the efficiency and durability of photovoltaic (PV) devices (Adilbekova et al., 2020; Ahmadian-Yazdi et al., 2020; Atli & Yildiz, 2022; Atli, 2023; Bhattarai et al., 2022; Campagnol et al., 2013). Among the emerging materials, metal-organic frameworks (MOFs) and their composites have garnered significant attention due to their versatile properties, such as tunable porosity, high surface area, and structural diversity. These unique characteristics make MOF composites particularly attractive for applications in solar cells, where they can be tailored to optimize light absorption, charge transport, exciton recombination, and overall device performance (Atli & Yildiz, 2022; Atli, 2023; Ban et al., 2013; Bella et al., 2013).

While numerous studies have explored the use of MOF composites in solar cells, identifying highly efficient and ultra-stable MOFs remains crucial for contributing to the reduction of global warming through renewable energy technologies (Chen et al., 2018; Chi et al., 2015; Chiba et al., 2006; Do & Kim, 2023; Dong et al., 2022; Dong, Rao, et al., 2017; Dou et al., 2016; Eastham et al., 2018; Heo et al., 2020; Holliday et al., 2016). This chapter delves into the fundamental aspects of MOFs, including their structure, functionalization, and the mechanisms by which they enhance the efficiency of various types of solar cells. The chapter begins with a brief summary of PV technology and the commonly used methods for synthesizing MOF composites, including hydrothermal, solvothermal, and sol–gel techniques. It then highlights the various roles that MOFs and their composites play in dye-sensitized solar cells (DSSCs), such as their use in photoanodes and counter electrodes (CEs). Additionally, the review examines the incorporation of MOF composites in perovskite layers (PLs), interfacial layers (ILs), and charge transport materials in electron and hole transport layers (ETL/HTL) in perovskite solar cells (PSCs). The application of

MOFs in organic solar cells (OSCs) is also addressed. Finally, the review discusses current challenges and future opportunities for the advancement of MOFs and MOF composites in solar cell applications.

Photovoltaic technology

PV technology, a key player in renewable energy, converts sunlight directly into electricity, offering an environmentally friendly and cost-effective energy solution. Solar cells, which are the primary units of PV technology, come in two primary categories: (1) wafer-based and (2) thin-film based. Wafer-based solar cells are also known first-generation solar cells. They are typically made from crystalline silicon (c-Si) or III–V multijunctions like gallium arsenide (GaAs), and are known for their high efficiency but are costly to manufacture due to the need for highly purified materials. For instance, c-Si cells have achieved a record efficiency of 26.7%, nearing the theoretical limit of 29% (Yoshikawa et al., 2017), while GaAs cells, with a higher efficiency of 27.6% (Dimroth et al., 2014), are primarily used in space applications due to their high cost (Kayes et al., 2011).

To reduce the costs, research has shifted toward second-generation solar cells, based on thin-film layers, which require fewer materials and are more versatile in the application. These cells are fabricated by depositing thin layers of PV material on a substrate, making them cheaper and more adaptable than wafer-based cells. Commercialized thin-film technologies include amorphous silicon (a-Si), cadmium telluride (CdTe), and copper indium gallium selenide (CIGS). Despite their lower efficiencies around 10% for a-Si, 18% for commercial CdTe modules, and 17.4% for CIGS modules, these technologies allow for more efficient space utilization (Atli, 2023). However, the environmental concerns related to toxic elements like cadmium and tellurium in CdTe, and indium and selenium in CIGS, pose significant challenges.

Emerging thin-film technologies (known as third-generation solar cells), such as DSSCs, OSCs, and PCSs, are still in the research phase but show promise due to their lower manufacturing costs and potential for high efficiency (Atli & Yildiz, 2022; Atli et al., 2021; Hui et al., 2021; Jung et al., 2019; Kocak & Yildiz, 2021; Li, Chen, et al., 2022; O'Regan & Grätzel, 1991; Ozel et al., 2024; Wang, Xue, et al., 2019; Wang, Zhou, et al., 2019; Yildiz et al., 2021; Yoo et al., 2021). DSSCs, in particular, have been a focal point since their introduction in 1991 due to their ease of fabrication, environmental friendliness, and suitability for low-light conditions. Despite their relatively lower efficiency (around 13%), DSSCs are highly regarded for applications where sunlight is diffuse, such as indoor environments.

The development of PV technology has been marked by several milestones; from the discovery of the PV effect by Becquerel in 1839 to the first efficient silicon solar cell developed by Bell Laboratories in the 1950s, the introduction of DSSCs by Gratzel in 1991 (O'Regan & Grätzel, 1991), and

25.7% power conversion efficiency (PCE) of PSCs by Kojima in 2009 (Kojima et al., 2009). Each advancement has brought PV technology closer to meeting the growing energy needs of society, with ongoing research focused on improving efficiency, stability, and scalability of solar cells. In this chapter, we explore emerging thin-film technologies in other words third-generation solar cells that incorporate MOFs.

Fundamentals of photovoltaic

A solar cell transforms sunlight into electrical energy, enabling the production of electricity. Unlike nuclear and coal-based power plants, solar cells do not release hazardous waste and they are environmentally friendly. The operation of solar cells is based on the PV effect, where the exposure of the cell's surface to the sunlight generates voltage and current. This effect occurs when carriers, such as electrons or holes, within a material become excited, leading to the production of output power. Traditional PV cells are made from semiconductor materials, which allows them to absorb light, resulting in creating the photocurrent owing to an applied voltage. This external voltage arises from the junction of p-type and n-type semiconductors, creating an electric field due to band bending at their interface, caused by the difference in Fermi levels of both materials. As a result, the electrons excited by light move from the p-type to the n-type semiconductor, generating an electric current that flows through the circuit, and it can be used as energy.

The efficiency of a solar cell is determined by several key factors: open-circuit voltage (V_{OC}) represents the potential difference at the terminals of a PV when it is exposed to light, and the circuit is open, meaning no current flows. The semiconductor materials with larger bandgaps typically yield higher V_{OC}, which is influenced by the semiconductor's bandgap. A smaller bandgap is beneficial for absorbing a wider range of light wavelengths, but it can lead to a reduction in V_{OC}, necessitating the use of materials with an optimal bandgap, ideally around 1.4 eV, to maximize both V_{OC}. On the other hand, the short-circuit current density (J_{SC}) refers to the current density generated when the cell is illuminated and the circuit is shorted. J_{SC} is influenced by the intensity and spectrum of the incident light, as well as how efficiently the generated electrons and holes are collected without recombination losses, which can occur inside the cell or at interfaces (Chen et al., 2023; Uğur et al., 2022). The fill factor (FF) measures how closely the J–V curve of the solar cell approaches a rectangular shape under illumination. The PCE (η) is defined as the ratio of the output power, production of V_{max} and J_{max}, to the incident power per unit area. FF and η can be calculated by following equations:

$$FF = \frac{V_{max} \times J_{max}}{V_{oc} \times J_{sc}} \tag{6.1}$$

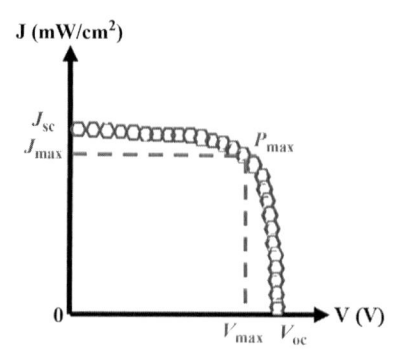

FIGURE 6.1 The current density–voltage characteristics of solar cell (Atli, 2023).

$$\eta = \frac{V_{OC} \times J_{SC} \times FF}{P_{in}} \times 100 \qquad (6.2)$$

herein, P_{in} refers to the incident solar power.

The typical J–V curves (Fig. 6.1) serve as the primary method for characterizing the solar cells. From a typical J–V curve, as previously mentioned key PV parameters J_{SC}, V_{OC}, FF, and overall η can be determined. The J_{SC} is indicated by the intercept of the ordinate at zero applied bias in the J–V curve. The V_{OC} corresponds to the abscissa intercept at open-circuit conditions in the J–V curve. The FF of a device, as given in Eq. (6.1) is calculated by dividing the maximum power output. Therefore the PCE (η) of a solar cell is determined by the ratio of the maximum electrical power output to P_{in}, as indicated in Eq. (6.2).

Metal-organic framework composites

Porous materials like foams, carbonaceous, porous metals, zeolites, and MOFs are extensively utilized across different kinds of disciplines due to their superior properties, particularly in their high specific surface area and the efficient moving of carriers through their pore networks. MOFs are porous, organic–inorganic hybrid compounds in which metal ions are coordinated with organic ligands, forming a three-dimensional structure (Fig. 6.2).

In MOFs, the inorganic components typically include different kinds of metals and organic linkers, combined with architecture of high surface area with porous structure, which allows for the transport of molecules or solvents. MOFs enable the customization of pore size and properties, making them versatile for various applications (Lin et al., 2017; Rosi et al., 2003; Silva Filho et al., 2020; Sun et al., 2013; Wang et al., 2014).

Due to their structural versatility, MOFs and MOF-derived materials denoted as MOF composites hold promise for the PV research, and there has been growing

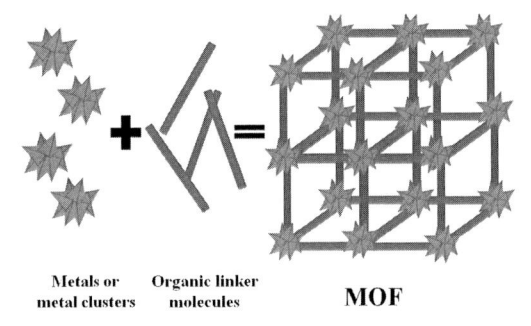

Metals or metal clusters   **Organic linker molecules**   **MOF**

FIGURE 6.2 Representative illustration of metals, linkers and MOF structure (Heo et al., 2020).

interest in employing MOFs for energy applications. By carefully selecting appropriate metal ions and ligands, MOF composites can achieve enhanced stability in various solution environments, which strengthens the coordination bonds within their frameworks. As a result, solar cells that incorporate MOF composite materials, such as organic–inorganic PSCs, DSSCs, and OSCs, benefit from improved efficiency and stability. This is due to the unique physical, optical, and electrical properties of MOF composites.

The use of MOF composites in solar cells presents multiple benefits. These include providing a highly porous and stable framework that facilitates the incorporation of guest compounds, offering excellent solution processability, allowing easy combination with other materials through techniques such as slot or spin coating to create microporous films, enabling precise adjustment of optoelectronic properties by modifying the metal salts and organic elements within the MOF and allowing nanoscale control over the BET surface area and porosity to improve their performance in solar cell applications.

In particular, there is growing interest in using MOF composites to improve the efficiency, long-term stability, and overall performance of PSCs, DSSCs, and OSCs devices. MOF composites act as a microporous scaffold, guiding the growth of anode materials, which enhances the PCE of emerging solar cells.

Synthesis of metal-organic framework composites

The majority of MOF and MOF composite syntheses are carried out in a liquid phase. Typically, the metal salts and organic ligands are either mixed after being prepared separately in a solvent, or they are directly inserted into the organic solvent together. Because the selection of solvent significantly influences the structural characteristics of the resulting MOF composite molecules, the synthesis method, and the type of solvent should be chosen based on the specific features desired in the final product.

The hydrothermal process is a widely used technique for synthesizing MOF and MOF composite materials for solar cell applications. In this method, the reaction takes place in a sealed vessel at elevated temperatures and pressures, allowing the formation of MOF composites with various sizes and structures by adjusting these parameters (Ma et al., 2015). While the hydrothermal method is effective in producing MOFs and their composites, it faces challenges, such as low yield and high cost. Additionally, the detailed reaction mechanisms involved in the formation of MOFs and their composites through this method are not yet fully understood.

The solvothermal technique is another prominent method for synthesizing MOFs and their composites. It is particularly valued for its high yield and precise control over the morphology of the resulting materials. The choice of solvent plays a critical role in the solvothermal process, with commonly used solvents including dimethyl formamide, ethanol, and methanol. These organic solvents can significantly influence the structure and size of MOFs, tailoring them for specific applications. However, the use of organic solvents poses environmental and health concerns due to their toxicity and difficulty in recycling. This underscores the need for developing safer and more environmentally friendly solvents for the solvothermal synthesis of MOFs (Ban et al., 2013; Pachfule et al., 2011).

The sol–gel method has recently gained popularity for producing inorganic nanomaterials, including MOF composites. This approach involves ion coprecipitation, leading to the formation of desired structures. The sol–gel method offers advantages over the hydrothermal process, such as the ability to create MOF composites with desired features by adjusting the reaction time and temperature. Moreover, MOF materials produced via the sol–gel method can serve as precursors for the development of MOF composite materials with enhanced properties (Sumida et al., 2017; Yu et al., 1995).

In addition to these aforementioned methods, MOF composites can be synthesized using alternative strategies, such as layer by layer deposition technique, electrodeposition techniques, microwave-assisted synthesis, and ultrasound methods. These methods offer diverse approaches to MOF composite synthesis, each with its unique advantages, and are being explored to expand the range of MOF derivative materials available for various applications, including in solar cells (Campagnol et al., 2013; Lv et al., 2017; Ni & Masel, 2006; Remya & Kurian, 2019; Son et al., 2008).

Application of metal-organic framework composites in photovoltaic devices

This section provides an overview of MOF composites application in solar cells, with a particular focus on recent studies utilizing MOF composites in the emerging thin film solar cells, which are considered potential successors to silicon-based PV devices.

Metal-organic framework composites as working electrodes

A typical DSSC features an electrolyte positioned between two transparent conductive electrodes as shown in Fig. 6.3. One of the electrodes is the photoanode/working electrode (WE), formed by a dye-loaded semiconductor, generally nanocrystalline TiO_2. Other one contains the photocathode/CE, usually thermally deposited platinum (Pt) (Boschloo & Hagfeldt, 2009; Grätzel, 2005; Hagfeldt et al., 2010).

For DSSCs, the PV mechanism involves three primary steps: light absorption, charge generation, and charge transport. Initially, the dye molecules absorb the light and become excited, transitioning to a higher energy state. The excited dye molecules inject electrons into the conduction band (CB) of the TiO_2 semiconductor. This process leaves the dye molecules in an oxidized state, while the electrons move through the TiO_2 network toward the photoanode, and are then carried through an external circuit to reach the CE, generating an electric current. The oxidized dye molecules are regenerated by electrons from redox reactions in the electrolyte, typically iodide/triiodide (I^-/I_3^-), completing the circuit (Grätzel, 2005).

The ability to control architectures, adjust pore sizes, achieve high surface areas, and utilize the distinctive thermal properties of MOF materials makes them ideal candidates for exploring their use in various components of the DSSC. Consequently, this section will examine the application of MOF composites in each DSSC components individually, with regard to the recent studies.

Metal-organic framework composites as working electrode

In a typical DSSC, the WE consists of a wide band gap semiconductor material. The semiconductor materials used in DSSCs must exhibit several key characteristics: a large surface area, strong loading capacity for dye molecules,

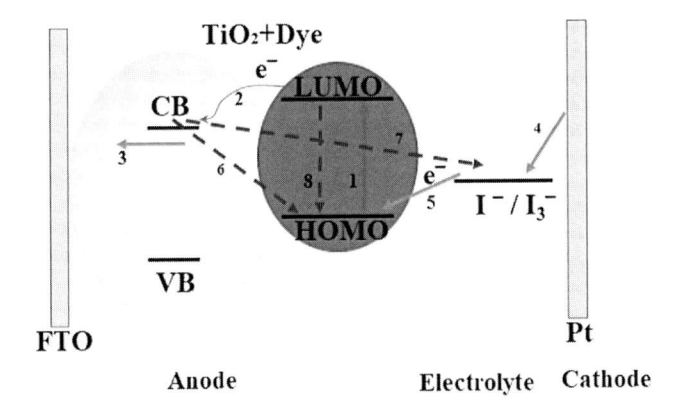

FIGURE 6.3 The schematic illustration of dye-sensitized solar cell (Uğur et al., 2022).

appropriate CB/valence band (VB) potentials, and high charge carrier mobility. Various materials have been explored as photoanodes in DSSCs, including TiO₂ (Nowotny et al., 2008; Park et al., 2000; Wang et al., 2006), ZnO (Fan et al., 2013), Nb₂O₅ (Ou et al., 2012), CeO₂ (Corma et al., 2004), Zn₂SnO₄ (Tan et al., 2007), SrTiO₃ (Daeneke et al., 2012), and BaSnO₃ (Shin et al., 2013). Among these, TiO₂ remains the most widely used semiconductor material in DSSCs (Kakiage et al., 2015). However, the rapid recombination of excitons in TiO₂ has led researchers to explore alternative photoanode materials for solar cell applications (Kaya et al., 2025).

MOF composites offer potential as photosensitizers for converting sunlight into electricity. Li, Pang, et al. (2011) reported the use of a zeolite imidazolate framework-8 (ZIF-8) in DSSCs for the first time. A thin layer (2 nm) of ZIF-8 was deposited on top of the TiO₂ layer, improving V_{OC} due to the material's properties, such as its high surface area and small pore size. Increasing the thickness of the ZIF-8 layer-enhanced V_{OC} by inhibiting interfacial charge recombination, although it also led to a decrease in J_{SC} and 5.34% efficiency.

In a follow-up study, a posttreatment method was proposed to mitigate the negative effects on J_{SC} and PCE (Li et al., 2014b). In this approach, ZIF-8 was added after dye sensitization, which helped maintain efficient photocurrent output despite the challenges posed by thicker MOF layers.

The combination of ZIF-8 with TiO₂ was used as photoanode, improving the efficiency of the DSSC from 7.75% to 9.42%, due to ZIF-8's effective prevention of electron–hole recombination (Gu et al., 2017).

Uğur et al., 2022 studied the surface modification of the TiO₂ layer by embedding nano-MIL-101 onto a commercial photoanode Fig. 6.4. A comparison of the J–V characteristics between devices with a bare WE and those with a nano-MIL-101 embedded WE showed that the nano-MIL-101-based device achieved a higher efficiency of 0.828%, while the commercial WE-based device reached an efficiency of 0.468%. The improved PV parameters are attributed to the enhanced dye-loading capacity of the nano-MIL-101-based device.

In recent study, the strategy was applied by incorporating, reduced-graphene oxide (RGO) with ZIF-8 and UiO-66, which improved charge transfer (He et al., 2020). The energy conversion efficiency was 7.33% for ZIF-8-RGO/TiO₂ and 7.67% for UiO-66-RGO/TiO₂. Later, same research group (He & Wang, 2021) combined a 3D graphane-based network (3DGN) with ZIF-8 to further increase the PCE of DSSC device. The optimal sample of ZIF-8/3DGN/TiO₂ achieved an efficiency of 8.77%, attributed to the better charge transfer in device due to an excellent electrical conductivity of the 3D graphene network.

Beyond ZIF-8, MOF composites based on copper (Cu), nickel (Ni), and cobalt (Co) have also been used in the fabrication of photoanodes. For instance, a TiO₂–Ni-MOF composite hybrid structure was developed and obtained 30% improvement in performance compared to bare TiO₂ photoanode (Ramasubbu et al., 2019). This enhancement is attributed to the Ni-MOF composite role in altering the absorption features and reducing the exciton loss.

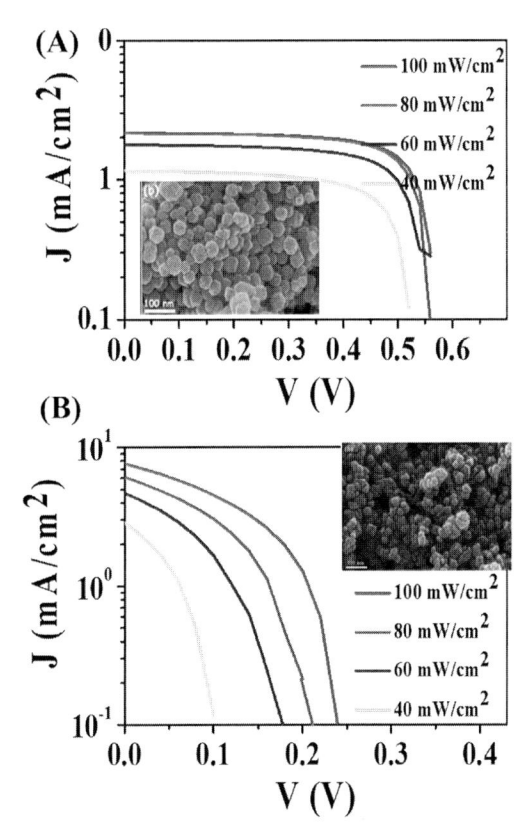

FIGURE 6.4 The current density–voltage plots of (A) nano-MIL-101 modified and (B) standard dye-sensitized solar cell under different illuminations, inset: the FESEM image of related working electrode (Uğur, Imer, Kaya, et al., 2022).

Cu-MOF/TiO$_2$, Ni-MOF/TiO$_2$, and pure TiO$_2$-based WEs were further investigated and compared to the solar cell efficiencies of various composites (Kumar et al., 2020). The device combined with Ni-MOF/TiO$_2$ WE exhibited the shortest diffusion time and lowest charge transfer resistance, leading to the highest energy conversion efficiency among the materials tested.

In another study, Al$_2$(BDC)$_3$, (MIL-53-Al) MOFs were explored to identify the charge-separated states using their photochemical properties (Lopez et al., 2011). These states were stabilized by the inclusion of guest molecules, leading to improved photocurrent and V_{OC} when applied in DSSCs.

TiO$_2$ was further derived from MOFs by converting MIL-125(Ti) into hierarchical TiO$_2$ (her-TiO$_2$) in anatase phase (Chi et al., 2015). This material served as a scattering layer, increasing J_{SC} and overall device performance by enhancing dye loading and surface area. Later, MIL-125-derived TiO$_2$ nanoparticles were used as a mesoporous (mp) layer in DSSCs resulting in efficiencies exceeding 7% owing to bigger electron lifetime and improved PCE (Dou et al., 2016).

Lee, Lim, et al. (2015) explored different Co based MOF composites (Co-NDC: cobalt-2,6-naphthalenedicarboxylic acid and Co-BDC: cobalt-benzene-dicarboxylate) doped with I_2 for use as WE in DSSCs. Specifically, the results showed the 1.12% of PCE for I_2@Co-NDC/TiO_2 and 0.96% of PCE for I_2@Co-BDC/TiO_2. The enhanced PCE was associated with the strong interaction between dopant with π-bonds of the MOF.

TiO_2 paste was derived from a MOF structure by Uğur et al., 2022, doped MIL-101 into TiO_2 and used it for the WE in DSSCs Fig. 6.5. The doped WE with MIL-101-based DSSC showed better performance, achieving an efficiency of 8.687% compared to 4.689% for the device without MOF doping.

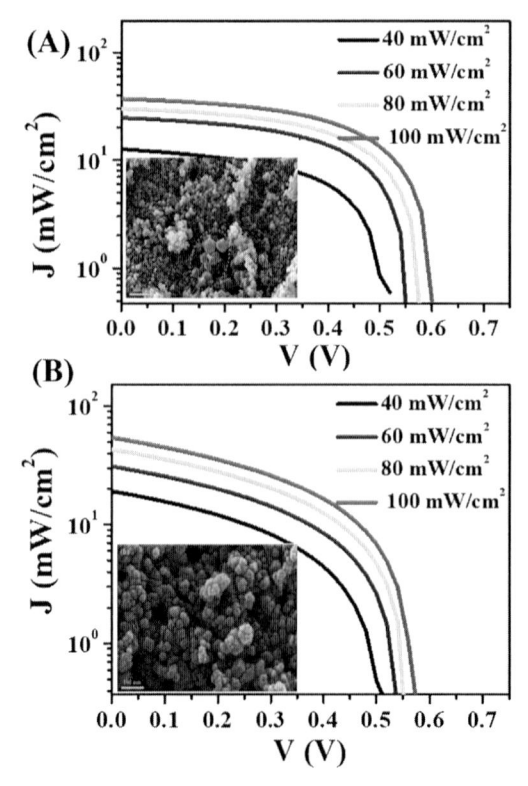

FIGURE 6.5 The current density–voltage plots of (A) nano-MIL-101 modified and (B) standart dye-sensitized solar cell under different illuminations, inset: The FESEM image of related working electrode (Uğur, Imer, & Gülcan, 2022).

This improvement was due to the enhanced dye-adsorbing capability of the MIL-101 structure, which provided a high surface area.

Except from TiO_2, MOFs have also been investigated to use common metal oxides like ZnO as WE for DSSCs. For instance, synthesized ZnO from a zinc-based MOF through calcination, achieved efficiencies of 0.14% and 0.15% in liquid-state DSSCs (Kundu et al., 2012). Similarly, MOF-5-modified ZnO was performed as WE for scattering layer further enhanced by ZIF-8 deposition, which improved light-harvesting efficiency (Li et al., 2014a).

In another study, Fe-polyoxometalates (POM@MOF(Fe)) derived ZnO layer was used as WE in DSSCs (Zhang et al., 2018). This modification increased PCE by 28%, due to enhanced light absorption with the presence of POM@MOF(Fe).

Metal-organic framework composites as counter electrodes

The CE functions as the contact in a DSSC, facilitating the re-entry of charge carriers from the external load to complete the circuit. Typically, the CE has catalyst layer deposited on fluorine doped tin oxide (FTO) substrate, and it should be compatible with the specific redox mediator. The primary role of the CE is to promote the rapid induction of the redox mediator, thereby it must have high electrocatalytic performance with minimal overpotential (Thomas et al., 2014).

In recent years, various materials, such as carbon (Wu et al., 2011; Zhu et al., 2011), transition metals (Li, Song, et al., 2011; Sun et al., 2014), and metal alloys (Peng et al., 2009) have been evaluated as potential CE materials for DSSCs. Among these, Pt has emerged as the most commonly used material due to its excellent catalytic properties and stability in different conditions (Thomas et al., 2014), even though its high cost significantly increases the overall cost of solar cells.

MOFs, with their uniformly distributed metal active sites, have emerged as promising alternatives for CEs in DSSCs. For example, MOF-525 combined with a conductive polymer and applied this mixture to carbon cloth by Chen et al. (2017), and they achieved higher performance of 8.91% with MOF-525 CE than that of Pt CE. The potential of ZIF-8 was further demonstrated in CEs by Zhao et al., who reported that incorporating ZIF-8 into the CE led to higher PCE compared to DSSCs without MOFs (Ahmed et al., 2018).

CE was developed by blending a copper-MOF-derived with poly(3,4-ethylenedioxythiophene) (PEDOT) (Yang et al., 2021). This resulted in an impressive 9.45% of PCE and notable stability, attributed to the strong adhesion on FTO substrate. Similarly, Zn-TCPP, (TCPP: tetrakis(4-carboxyphenyl)porphyrin) nanolayers embedded Pt nanoparticles were employed to create a promising CE for DSSCs, achieving a solar cell efficiency of 5.48% (Tian et al., 2020).

Additionally, MOFs have increasingly been used to derive carbon-based materials for CE applications in DSSCs. The process for creating these porous carbonaceous materials is relatively simple and advantageous, as their tunnel structure facilitates redox electrolyte (RE) diffusion and their large BET improves the electrocatalytic performance. The first introduced carbonaceous materials derived from ZIF-8 in DSSCs have achieved an efficiency of 7.32%, comparable to that of Pt electrodes (Sun et al., 2016).

Building on this, the synthesized nitrogen-doped porous graphitic carbon embedded with surface-oxidized CoO species was introduced using ZIF-67 as a self-template (Jing et al., 2016). This method was faster and more cost-effective than using separate templates. The carbonized material exhibited improved catalytic activity and stability due to the embedded metal particles, with the highest efficiency (7.92%) observed for ZIF-67 calcined at 850°C.

CoS_2 embedded in carbon nanocages were developed using ZIF-67 as a template (Cui et al., 2016). The resulting composite offered superior catalytic activity, enhanced electrical conductivity, and prevented CoS_2 aggregation, resulting in an outstanding efficiency of 8.2% in DSSCs, surpassing that of Pt-based devices.

Later, the synthesized ZnSe embedded in N-doped carbon cubes derived from ZIF-67 was constructed as CE in DSSC, achieving a higher energy conversion efficiency (8.69%) than Pt-based DSSCs (Jiang et al., 2018). This increase in the efficiency was ascribed to enhanced electron transfer and lower resistance in the carbon matrix.

Recently, ZIF-67 composite was also used to synthesize a graphene with cobalt oxide and tungsten carbide (Co_3O_4-WC-CN/rGO), which demonstrated superior catalytic activity and achieved an energy conversion efficiency of 7.38%, outperforming Pt-based CEs. This success was attributed to the uniform distribution of Co_3O_4 and WC within the graphene matrix and the material's high catalytic activity (Chen et al., 2018).

Despite these advancements, the application of MOF composites as CEs in DSSCs remains limited due to their poor conductivity. As a result, significant research efforts have been focused on using MOF composites as sacrificial agents to fabricate hybrid structures that serve as more efficient CEs in DSSCs.

Metal-organic framework composites as sensitizer dyes

In DSSCs, the sensitizer's function is to absorb incoming light. Upon absorbing sunlight, the sensitizer-dye becomes photoexcited, and then subsequently injects the photogenerated charges into the CB of TiO_2 in WE. After oxidation process, the dye is restored by the electrolyte. For effective charge injection/regeneration process in DSSCs, the dye must have several criteria. It should absorb solar light across the entire visible and ultra violet (UV) spectrum, ideally extending into the near-infrared region. Also, its lowest unoccupied molecular orbital (LUMO) level should be at a more reducing potential than the conduction level of TiO_2 material to ensure effective charge injection. Finally, its highest occupied molecular orbital (HOMO) level should be at a more oxidized than the RE to facilitate efficient dye regeneration (Grätzel, 2005). Ruthenium (Ru)-based sensitizers are most favorable with their stability and photochemical activity, which is the focus of attention since the inception of DSSCs (Kohle et al., 1997). A significant milestone in an advancement of sensitizer dye was marked by the investigation of the black dye (N3) in 1993 (Nazeeruddin et al., 1997). The black dye exhibited strong absorption, with an absorption edge near 800 nm. Later, same research group (Nazeeruddin et al., 1997) developed N719, which is an enhanced version of N3. Despite the superior photochemical performance of Ru-derived sensitizers in DSSCs, researches have turned their attention to more abundant and cost-effective alternatives, such as zinc-based sensitizers (Mathew et al., 2014) and other metal-based complexes sensitizers (Bessho et al., 2008; Gencer Imer et al., 2018; Linfoot et al., 2011). While high enough PCE has been obtained with the alternative sensitizer-dyes in DSSCs, challenges persist, such as significant desorption when incorporated into devices, variability in binding modes and interactions, and dye molecule aggregation, all of which undermine device performance.

To address these limitations, MOFs have been explored as sensitizers in DSSCs, due to their high photon responsivity and tunable architecture. MOF-199 was deposited on an mp-TiO$_2$ (mesoporous titanium oxide) layer as a sensitizer by Lee group (Lee, Shinde, et al., 2014), and they reported poor conductivity with MOF-199 sensitizer. The conductivity was enhanced by doping the neutral coordination polymer MOF-199, with I$_2$, due to the intermolecular interactions with the π-electrons, resulting in an efficiency of 0.26% for MOF-sensitized DSSCs.

The same research group modified this system by incorporating multi-walled carbon nanotubes (MWCNTs) into the TiO$_2$ layer, on which MOF-199 was deposited (Lee, Shin, et al., 2014). This modification led to a 60% improvement in overall efficiency compared to devices without MWCNTs. Additionally, the group successfully applied a Ru-MOF to promote efficient electron flow, and device efficiency increased from 0.46% to 1.22%, after doping with I$_2$.

In a separate approach, MOF-based photoanodes were developed by incorporating graphene into a Eu-MOF (benzene-1,3,5-tricarboxylates) (Kaur, Kim, et al., 2017). Graphene's superior conductivity enabled the Eu-MOF, which had low conductivity on its own, to form a composite with significantly improved conductivity. This composite was used as a sensitizer on top of TiO$_2$. The PCE of DSSCs using the Eu-MOF composite was demonstrated an energy conversion efficiency of 2.2%, due to the improved photocurrent generation.

In another study, the impact of two different synthesis methods for MOF-derived sensitizer dye was explored on DSSC efficiency (Maza et al., 2016). The results indicated that the overall performance was influenced by the synthesis method. The one-pot method resulted in better dye distribution and higher efficiency, while the postsynthetic route led to nonuniform dye distribution, likely due to diffusion limitations of Ru(bpy)$_2$Cl$_2$, hindered dye penetration into the MOF structure. Furthermore, the encapsulation of sensitizer within the pores of MOF structure blocked the redox couple's penetration, restricting dye regeneration and resulting in lower J_{SC} and V_{OC}.

To investigate the true contribution of MOF composites without additional guest molecules, Spoerke et al. (2017) constructed DSSCs using a pillared porphyrin framework. The authors ensured that solvent and unreacted linker molecules were excluded from the framework. A thin layer of TiO$_2$ was deposited using atomic layer deposition, and the MOF was placed on top. Although low efficiencies were observed due to limited surface area, the MOF-incorporated device performed better than the bare TiO$_2$ device, indicating the pure MOF contributed to enhanced performance. The electrochemical impedance spectroscopy (EIS) measurements showed a significant reduction in interfacial charge transfer resistance after the addition of the MOF, further supporting the improved efficiency of the device.

Metal-organic framework composites in redox elelctrolytes

In DSSCs, the electrolyte has an important function in determining both the stability and efficiency of the device. Typically, it consists of a redox couple and a solvent. Its primary function is to regenerate the oxidized dye adsorbed WE with charge carriers by transportation of them from the CE. For an electrolyte to be efficient, it must facilitate rapid regeneration process, calm down exciton recombination, and low absorption in the wavelength range of 400–700 nm. Achieving the aforementioned requirements depends on the high solubility and ionic mobility in the chosen solvent. Additionally, dye regeneration should occur with minimal overpotential (Hamann & Ondersma, 2011; Wang & Hu, 2012). The triiodide/iodide (I^{-3}/I^{-}) redox couple have been commonly used in DSSCs, achieving an efficiency of 11% when paired with the N749 dye as a sensitizer (Chiba et al., 2006). Its redox kinetic mechanism is favorable: I^{-} offers a fast dye regeneration rate, while the slow reduction of I^{-3} at the CE enhances carrier collection and minimizes e-hole recombinations. Devices using I^{-3}/I^{-} as electrolyte have demonstrated good stability under different conditions, making this redox couple a strong candidate for commercial applications. However, despite its widespread use, researchers have sought alternatives due to certain drawbacks, such as its absorption edge in UV region (which competes with the sensitizer) and its corrosive effects on collector metals, such as silver and copper in PV modules (Boschloo & Hagfeldt, 2009; Grätzel, 2001).

Additives like heterocyclic compounds and tert-butylpyridine (tBP) have been used in DSSCs to enhance performance of device by influencing the redox potential and recombination kinetics, thereby improving PV parameters (Nazeeruddin et al., 1993). To further boost performance, additives like deoxycholic acids and phosphonic acids have been employed, with chenodeoxycholic acid (cheno) proving particularly effective in significantly increasing the current generated by the device (Daeneke et al., 2011; Linfoot et al., 2011; Xie & Hamann, 2013).

For liquid-state DSSCs, various organic solvents have been tested, with acetonitrile emerging as the preferred choice due to its high solubility of salt components, low viscosity, and excellent chemical stability, leading to higher efficiency (Hagfeldt et al., 2010; Nazeeruddin et al., 1993, 1997; O'Regan & Grätzel, 1991).

Additionally, MOF composites have been utilized in quasisolid-state (QSS) DSSCs as a mediator for the redox couple, further enhancing device performance. To use MOF composites as the mediator, specific types of metal-organic gels (MOGs) have been developed by incorporating metal into organic gel systems. MOGs are primarily assembled from metal ions and bridging organic ligands, closely resembling the preparation of MOFs. The MOGs exhibit characteristics specified by metal that can be finely tuned by adjusting the materials involved. Due to their porous structure, MOGs can house electrolyte components while maintaining the features of electrolytes. Additionally, MOGs effectively diffuse, promoting the strong contact of the mp layer of WE and electrolyte.

A magnesium-based MOF composite (Mg-MOF) was first used as an electrolyte medium, and incorporated into a mixture of polyethylene glycol methacrylate and polyethylene glycol diacrylate (80:20 weight ratio). A polymer matrix was formed, as exposure to UV light, and then activated by sucking it in an electrolyte with I^{-3}/I^{-}. Increasing the amount of Mg-MOF content enhanced the electron lifetime, reduced the recombination, improved the long-term stability of the device, and achieved PCE of 4.8% (Bella et al., 2013).

In another study, Al^{+3} and 1,3,5-benzenetricarboxylate (H_3BTC) MOG were utilized as the medium for a polymer gel electrolyte. The device integrated with MOG-based electrolyte reached PCE of 8.69%, while traditional device demonstrated PCE of 9.13%, due to high recombination, supported by dark current, resulting in lower V_{oc} and overall PCE of MOG-based DSSC (Fan et al., 2014).

Recently, another Al-based MOG was used as a redox medium in QSS-DSSCs by the Su group (Dong, Rao, et al., 2017) In this matrix, the desired MOG was synthesized using the Al^{+3} and tBP, tBP acting as both a gelator and an active additive to improve device performance. The optimized DSSC achieved an efficiency of 8.25%, which was related to longer carrier lifetime with higher tBP concentration, due to reduced electron recombination from tBP adsorption onto the surface. Table 6.1 summarizes the solar cell parameters of the DSSC devices based on MOF composites in different layers to clearly express the impact of MOF composites on the performance of DSSCs.

To summarize the impact of MOF composites on DSSCs, the PV parameters and PCE have shown significant improvements due to the large

TABLE 6.1 The photovoltaic parameters of the dye-sensitized solar cell devices based on metal-organic framework composites in different components.

Position	MOF composites	J_{SC} (mA/cm^2)	V_{OC} (mV)	FF (%)	PCE (%)	Reference
WE	ZIF-8	10.28	753	0.69	5.34	Li, Pang, et al. (2011)
WE	ZIF-8	10.89	789	0.74	6.35	Li et al. (2014b)
WE	ZIF-8	14.39	897	0.73	9.42	Gu et al. (2017)
WE	ZIF-8-3DGN	20.9	681	0.62	8.77	He and Wang (2021)
WE	Cu-MOF	1.70	320	0.24	0.13	Kumar et al. (2020)
WE	UiO-66-RGO	18.6	678	0.61	8.77	He et al. (2020)
WE	Ni-MOF	27.32	624	0.52	8.85	Ramasubbu et al. (2019)
WE	Co-NDC	2.56	630	0.63	1.12	Lee et al. (2015)
WE	MIL-53-Al	0.0036	361	0.40	—	Lopez et al. (2011)
WE	Zn-MOF	—	—	—	0.15	Kundu et al. (2012)
WE	MOF-5	8.13	663	0.68	3.67	Li et al. (2014a)
WE	MIL-125(Ti)	19.1	660	0.55	7.1	Chi et al. (2015)
WE	MIL-125(Ti)	13.99	768	0.67	7.20	Dou et al. (2016)
CE	Cu-MOF-PEDOT	13.56	770	0.68	7.32	Yang et al. (2021)
CE	MOF-525	16.14	800	0.70	8.91	Chen et al. (2017)
CE	ZIF-8-PEDOT:PSS	11.46	852	0.70	7.56	Ahmed et al. (2018)
CE	ZIF-8	13.56	770	0.68	7.32	Sun et al. (2016)

CE	ZIF-67	13.29	740	0.80	7.92	Jing et al. (2016)
CE	ZIF-67	16.9	730	0.66	8.20	Cui et al. (2016)
CE	ZIF-67	12.27	740	0.66	6.02	Tan et al. (2018)
CE	Zn-TCPP	12.95	690	0.61	5.48	Tian et al. (2020)
Sensitizer	MOF-199	1.25	490	0.43	0.26	Lee, Shinde, et al. (2014)
Sensitizer	Ru-MOF	2.56	630	0.63	1.22	Lee, Kim, et al. (2014)
Sensitizer	Eu-MOF	20.0	449	0.46	2.3	Kaur, Sharma, et al. (2017)
Sensitizer	Pd porphyrin Zn-SURMOF	0.71	700	0.65	0.45	Liu et al. (2015)
Sensitizer	Zn(II)porphyrin Zn-SURMOF 2	0.023	267	–	0.017	Liu et al. (2016)
Sensitizer	RuDCBPY-ZrMOF	0.564	482	0.47	0.125	Maza et al. (2016)
Sensitizer	RuDCBPY-ZrMOF	8.50	521	0.48	0.0023	Spoerke et al. (2017)
Sensitizer	Cu-MOF	1.95	480	0.51	0.46	Lee, Shin, et al. (2014)
Electrolyte medium	Mg-MOF	12.6	690	0.55	4.80	Bella et al. (2013)
Electrolyte medium	Al-MOF	17.1	734	0.68	8.49	Fan et al. (2014)
Electrolyte medium	Al-tBP-MOG	17.08	704	0.70	8.25	Dong, Rao, et al. (2017)

CE, Counter electrode; *Co-NDC*, cobalt-2,6-naphthalenedicarboxylic acid; *DCBPY*, 2,2′-bipyridine-4,4′-dicarboxylic acid; *DGN*, 3,3′-diaminoguanidine; *FF*, fill factor; *Mg-MOF*, magnesium-based metal-organic framework composite; *MOG*, metal-organic gel; *MOF*, metal-organic framework; *NDC*, naphthalene-2,6-dicarboxylic acid; *PEDOT:PSS*, poly(3,4-etilendioksitiyofen) polistiren sülfonat; *PCE*, power conversion efficiency; *tBP*, tert-butylpyridine; *TCPP*, tetra(4-carboxyphenyl)porphyrin; *WE*, working electrode; *ZIF-8*, zeolite imidazolate framework-8; *3DGN*, 3D graphane-based network.

BET and porosity of MOF composites. MOF and their composites have been utilized in various roles, including as modifiers and dopant materials for the WE or CE, as sensitizers, and as redox media. While the specific MOF materials and their applications may vary, they consistently deliver similar outcomes. The contributions of MOF derivatives to DSSCs can be highlighted as follows: (1) they enhance the illuminated area in the TiO₂ film, (2) they improve interfacial carrier transport, leading to longer electron lifetimes, (3) they promote dye adsorption into the photoanode, and (4) they increase light harvesting and the overall performance of the devices. These findings offer valuable insights for the future design of more efficient DSSC devices.

Metal-organic framework composites in perovskite solar cells

PSC device typically consists of five components: the PL is in between electron/hole transport layer (ETL/HTL), FTO substrate is contacted with ETL, and metal electrode on the top of HTL (Fig. 6.6). The working principle of PSCs is divided into three stages: exciton generation by photon absorption, separation of electron–hole pairs, and transportation of separated charges. Specifically, the PL absorbs photons from sunlight to generate electrons in the CB and holes in the VB of perovskite material. Electrons transport to the ETL, and then to FTO anode, while holes move to the HTL, and then reach the metal cathode (Bhattarai et al., 2022). This movement of charges generates an electric current, which can be harnessed for power.

PSCs have emerged as promising alternatives to inorganic PVs owing to their excellent optical absorption, cost-effectiveness, and rapid charge transfer. Recent advancements in PSC longevity have been achieved through compositional engineering, particularly by substituting unstable organic components with inorganic ions, which significantly enhances stability. This section

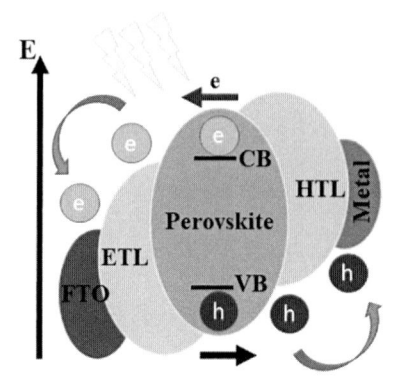

FIGURE 6.6 The schematic illustration of perovskite solar cell device (Do & Kim, 2023).

examines the role of MOF composites as functional additives in the PL, ETL, HTL, and interface in PSC.

Metal-organic framework composites as/in perovskite layer

MOFs have been extensively studied for their ability to improve the performance and stability of PL in PSCs. Due to their porous structures, MOF composites act as scaffolds that promote the growth of PLs and improve the overall crystallization process.

Zr-based porphyrin MOF (MOF-525) nanocrystals were incorporated into a perovskite precursor, creating MOF-525/perovskite heterojunction thin films. The uniform pore structure of the MOF-525 nanocrystals facilitated an improved arrangement of perovskite crystallites, resulting in a PCE of 14.5% (Chang et al., 2015).

Two types of Zr-based MOFs, (UiO-66 and MOF-808) were introduced into perovskite film to form hybrid layers, deposited onto NiOx HTLs in inverted PSCs. These MOFs effectively passivated surface defects and reduced moisture penetration, achieving a high PCE of 18.01% along with enhanced stability under ambient conditions (Lee et al., 2019).

Li et al. investigated Indium-based MOFs, such as [In$_2$(phen)$_3$Cl$_6$]·CH$_3$CN· 2H$_2$O]; labeled as (In2), which were incorporated into lead-based PSCs. In2 improved perovskite crystallization, reduced trap states, and increased the PCE from 15.41% to 17.15% (Li, Xia, et al., 2019). Another In-based MOF composite, [In$_{12}$O(OH)$_{16}$(H$_2$O)$_5$(BTC)$_6$]n; (In-BTC), further enhanced film quality and decreased structural defects, achieving a PCE of 20.87% (Zhou, Qiu, Fan, Zhang, et al., 2020).

A Zn-based MOF, ({[(Me$_2$ NH$_2$)$_3$ (SO$_4$)]$_2$[Zn$_2$ (ox)$_3$]}$_n$, ZnL), was utilized to passivate PLs, improving device stability and achieving a PCE of 21.39%. Another Zn-based MOF, (Zn-TTB:1-(triazol-1-ly)-4-tetrazol-5-ylmethyl) benzene, promoted the growth of enlarged, cross-linked perovskite grains, resulting in a high PCE of 23.14% while maintaining 90.1% of the initial efficiency after 300 h of continuous illumination (Li, Qiu, et al., 2022).

A POM-based MOF composites, CoW12@MIL-101(Cr), was integrated into MOF structures to simultaneously eliminate Pb° and passivate Pb^{+2} in PSCs. This approach improved both device performance and stability, with excellent retention of efficiency after heating at 85°C and prolonged storage under ambient conditions (Wang, Zhang, Lin, et al., 2021).

In addition, Cr-MOF composites have been incorporated into inorganic CsPbI$_2$Br PSCs, where a π-conjugated aromatic structure-based terpyridyl chromium MOF (Cr-MOF) was utilized. This MOF facilitated carrier transport and prevented the decomposition of the PL, resulting in a PCE of 17.02% and enhanced stability for up to 1000 h under ambient conditions (Yuan et al., 2021).

These studies collectively demonstrate that integrating MOFs with perovskites increases the durability of the PL. The porous structure of MOFs diminishes barrier of perovskites, thereby promoting the crystallizations. Furthermore, ligands like carbonyl and amino groups serve as additives in the perovskite precursor to aid alignment, leading to enhanced film quality. This, in turn, significantly improves both the performance and durability of PSCs.

Metal-organic framework composites in the electron transport layer

In addition to being integrated into the PLs of PSCs, MOF composites have also been investigated as layers of ETL and HTL, or inserted within these layers as an interfaial layer. For effective ETLs in PSCs, it is essential that they possess high mobility and well-align energy levels with other layers in device. Large surface areas and low defect densities are also key factors in improving PCE. TiO_2 is predominantly used as the ETL in PSCs due to its low cost and good structural durability (Bai et al., 2014; Chen & Mao, 2007; Wei et al., 2017). However, a wide bandgap of commercially available TiO_2 (3.3 eV in UV range) enables electron excitation and injection, it results in inefficient electron transport. To address this issue, reducing the bandgap has been proposed, and metal synthesized using solvothermal technique by Nguyen and Bark (2020). The incorporation of Co atoms introduced defects into the TiO_2 lattice, reducing the bandgap and inducing lattice distortion. Ti in (Ti)-MOF was oxidized to to TiO_2 by annealing, and the resulting 1 wt.% Co-doped TiO_2 MOF exhibited higher porosity than un-doped TiO_2, leading to improved PV performance. While the PCE of the undoped TiO_2 based PSCs was 12.32%, Co doping increased the PCE to 15.75%. The resistance value of R_{CT} and R_{REC} were significantly reduced with Co doping, facilitating charge transport and moderating exciton recombination. These enhancements were attributed to surface modification with Co doping.

In some studies, MOF composites were combined with TiO_2, while in others a MOF composite layer was deposited on the TiO_2, acting as a transport layer. For instance, ZIF-8 was inserted between the TiO_2 and PLs, and the resulting PSC device performance analysis were compared based on reaction time. The presence of ZIF-8 increased the PCE from 9.6% to 12.0%, with the optimal immersion time for ZIF-8 coating determined to be 2 minutes. The transfer rate of photoexcited charges and the absorption feature of the PL were enhanced with the presence of ZIF-8 MOF layer (Chung et al., 2018).

In another study, (Ti)-MOF denoted as MIL-125 is employed as an ETL in the PSC by depositing into the TiO_2 layer. The high CB edge of MOF-125 facilitated the transfer of photoexcited electrons by quantum tunneling. Additionally, the mp structure of the MOF-125 enlarged interfacial contact with the perovskite, and improved crystallization, thereby suppressing recombination of interface charges. As a result, the PV parameters of the device fabricated with TiO_2-MOF-125 ETL were significantly enhanced (Vinogradov et al., 2014).

Another MOF structure of MIL-125 was incorporated into a porous hier-TiO$_2$ nanostructures by sintering (Hou et al., 2017). The MIL-125 inherited hier-TiO$_2$ nanostructures providing larger spaces for perovskite growth compared to bare TiO$_2$. The resulting PSC demonstrated an enhancement in PCE from 6.4% to 16.6%, related to better stability.

Metal-organic framework composites in the hole transport layer

The performance of PSCs is heavily influenced by the HTL, which plays a vital role in preventing electron transfer to the cathode, thereby reducing the recombination of electron–hole pairs. The effectiveness and long-term performance of PSCs largely are related to the thermal stability, photochemical durability, and conductivity of the HTL. Spiro-OMeTAD, (2,2′,7,7′-Tetrakis[N,N-di(4-metoksifenil)amino]-9,9′-spirobifluoren), is one of the most commonly used hole transport material is Spiro-OMeTAD, and was initially able to achieve PCE of up to 9.7%. Because of low mobility of hole, the additives, such as lithium bis(trifluoromethanesulfonyl)imide (Li-TFSI) and tetra-tert-butylpyridine (TBP) should be utilized to increase its performance. Additionally, various MOFs have been integrated into HTLs to further enhance conductivity, stability, and oxidation properties in PSCs (Hua et al., 2016; Seo et al., 2018).

One notable example is the use of polyoxometalate with MOF structure (POM@MOF), specifically POM@Cu-BTC, as a dopant in the HTL. This MOF regulated the oxidation process of spiro-OMeTAD and improved its durability, increasing the PCE from 20.21% to 21.44% (Dong et al., 2019). Furthermore, the device retained approximately 90% of its initial efficiency after prolonged storage. Similarly, indium (In)-based and zinc (Zn)-based MOFs have been used as chemical dopants for spiro-OMeTAD. For example, the indium-based MOF (In2) enhanced light absorption and improved the PCE from 12.8% to 15.8%. Another indium-based MOF, [In$_{0.5}$K(3-qlc)Cl$_{1.5}$(H$_2$O)$_{0.5}$]$_{2n}$ (In10), when incorporated into spiro-OMeTAD, resulted in a 20% increase in PCE compared to cells without In10 MOF, although excessive amounts caused surface inhomogeneity and degradation of performance (Li, Wang, et al., 2019; Li, Xia, et al., 2019).

Zinc-based MOF was introduced for its function in controlling the spiro-OMeTAD's oxidation, acting as protective layer to prevent moisture penetration. These materials demonstrated a PCE of 20.64%, with 90% retention after 30 days of storage (Wang, Zhang, Yang, et al., 2021). In addition, polynuclear MOFs like Tb-TZB and Tb-FTZB (H$_2$TZB=4-(1H-tetrazol-5-yl)benzoic acid, and H$_2$FTZB=2-fluoro-4-(1H-tetrazol-5-yl)benzoic acid) were incorporated into HTLs, achieving a PCE of 21.31% and displaying good stability under extended storage conditions (Zhang et al., 2022).

Li-TFSI, a commonly used additive to improve hole mobility in spiro-OMeTAD, has the disadvantage of being hygroscopic, which accelerates perovskite degradation. To overcome this limitation, a Li-TFSI@NH$_2$-MIL-101 MOF was developed, offering tenfold higher stability and retaining over 85% of its initial PCE after 3600 h of storage (Wang, Zhang, Yang, et al., 2022). Additionally, the thermally stable In-Pyia MOF was introduced as a replacement for TBP, resulting in a PCE increase of 19.47%. The device maintained 80.9% of its initial efficiency after prolonged heat exposure and ambient conditions (Zhou, Qiu, Fan, Ye, et al., 2020).

Apart from the application of 3D MOF derivatives in PCS, two-dimensional (2D) MOFs have also emerged as an effective strategy for optimizing the performance of HTL due to their high electrical conductivity. For instance, 2D Cu-benzenehexathiol MOFs exhibited excellent conductivity and transmittance, making them potential substitutes for indium tin oxide (Jin et al., 2017). Additionally, 2D Pb-MOF hexagonal sheets were synthesized in the literature by Huang et al. (2019), and they mixed synthesized Pb-MOF with spiro-OMeTAD, which resulted in improved surface roughness, increased hydrophobicity, and enhanced stability. After 9 days, the device retained 54% of its initial efficiency, compared to 28% without the Pb-MOF.

Another material, 2D graphitic N-rich porous carbon (NPC), was developed for an additive material in HTL. It improved quality of the film by decrease the aggregation of lithium salts, and minimizing density of defects, facilitating faster hole transportation (Zhou, Qiu, Fan, Wang, et al., 2020).

Overall, the incorporation of MOF composites into HTLs significantly improved the performance of PSCs, leading to higher PCEs and enhanced long-term durability under various environmental conditions.

Metal-organic framework composites as the interface layer

A well-designed interlayer (IL) with appropriate energy levels and chemical bonding has enhanced electron–hole transport and reduced recombination. Interfacial engineering has proven to be an effective strategy for improving the stability of PSCs. Recent studies have incorporated highly stable MOF composites as ILs in PSCs to control crystallite size, improve perovskite film quality, and inhibit charge recombination, thereby enhancing overall device performance. MOFs are not only easy to synthesize but also offer tunable bandgaps, making them valuable components in PSCs.

ZIF-8 MOF structure was the first to be utilized as an IL on the top of mp-TiO$_2$. When applied to the mp-TiO$_2$ film, the ZIF-8 layer supported crystal growth, which behaves as an additional scaffold, during the beginning of perovskite film preparation (Chang et al., 2015). By optimizing the amount of ZIF-8, the grain size was reduced, and the film surface became smoother. As a result, a more efficient light-harvesting layer was formed on the mp-TiO$_2$ film.

Shen et al. (2018) employed a thin ZIF-8 interlayer on the top of mp-TiO$_2$. ZIF-8, being both chemically robust and thermally stable, provided a scaffold that regulated perovskite crystallization through cross-linking. This IL significantly improved the crystallinity and particle size of the perovskite material. EIS revealed that PSCs based on mp-TiO$_2$/ZIF-8 exhibited lower recombination rates and higher charge transfer efficiency compared to PSCs based on bare mp-TiO$_2$. The device's PCE increased from 14.75% to 16.99% with the addition of ZIF-8 (Shen et al., 2018). ZIF-8 was also utilized on the mp-TiO$_2$ film, where it functioned as an additional photon-absorbing layer in the UV region, thereby enhancing device performance (Chung et al., 2018). Similarly, the Eslamian group replaced the mp-TiO$_2$ layer with ZIF-8 as an IL, achieving a PCE of 16.8% (Ahmadian-Yazdi et al., 2020). The presence of the ZIF-8 interlayer significantly increased perovskite grain size, which reduced the trapped charges by defects. Moreover, the band bending at the interface created an energy barrier that protected electron–hole pairs from surface defect traps.

Recently, Pb-based MOF-525 was utilized as a scaffold at the TiO$_2$/perovskite interface, facilitating the high crystallization quality of methyl ammonium lead iodide (MAPbI$_3$) PL (Liu et al., 2021). This setup enhanced charge separation at the ETL/perovskite interlayer by resulting in the lower density of defects. The PSC device integrated with Pb-MOF-interelayer achieved a better PCE of 20.87%, compared to 16.85% for the PSC without the MOF (Liu et al., 2021).

Two different MOF structures of UiO-66 and MOF-808 were used as scaffolds at the interface of NiO$_x$/PLs. These scaffolds modulated the crystallization process, resulting in larger particle sizes in the PL. Additionally, space-charge-limited current tests indicated a lower trap-state density, leading to a PCE increase from 15.79% (without MOF) to 17.01% (Lee et al., 2019).

Beyond applications beneath the perovskite material, ZIF-8 added with methyl ammonium iodide (MAI) and formamidinium iodide (FAI) was also employed on top of the PL in ZIF-8@MAI- or ZIF-8@FAI-based PSCs (Li et al., 2021). The release of FAI/MAI compensated for defects in the perovskite film, while ZIF-8 provided moisture isolation. This approach achieved an efficiency of 19.13% and maintained over 93% of the initial PCE after 150 h of operation at maximum power input.

Recently, ZIF-8 was employed as a host matrix to encapsulate methylammonium chloride (MACl), forming an interlayer at the SnO$_2$/ perovskite interface. This strategy offered several advantages, including a reduction in SnO$_2$ oxygen vacancies, the formation of new bonds between ZIF-8 and unsaturated ions, and the supply of ions to reduce Pb vacancy defects in the perovskite film. Consequently, the PCS device assembled with MACl@ZIF-8 interlayer demonstrated high performance, achieving a PCE of 22.10%, thus opening new avenues for incorporating MOFs into PSCs (Jin et al., 2023). Table 6.2 summarizes the parameters and efficiencies of PSCs with MOF composites integrated into different layers to clearly illustrate the influence of MOF on solar cell performances.

To summarize the impact of MOF composites on PSCs, the PCE, and long-term performance have seen significant improvements due to the remarkable

TABLE 6.2 The photovoltaic parameters of the perovskite solar cell devices based on metal-organic frameworks in different layers.

Position	MOF composites	J_{SC} (mA/cm^2)	V_{OC} (V)	FF (%)	PCE (%)	Reference
Heterojunction	MOF-525	23.04	0.93	60.0	12.0	Chang et al. (2015)
Heterojunction	UiO-66	21.85	1.07	76.9	18.01	Lee et al. (2019)
Heterojunction	MOF-808	21.01	1.06	79.8	17.81	Lee et al. (2019)
Heterojunction	In2	23.18	1.04	71.0	17.15	Li, Xia, et al. (2019)
Heterojunction	In-BTC	22.99	1.10	77.0	19.63	Zhou, Qiu, Fan, Zhang, et al. (2020)
Heterojunction	ZnL	23.86	1.12	79.3	21.15	Li, Qiu, et al. (2022)
Heterojunction	CoW$_{12}$@MIL-101(Cr)	23.85	1.14	79.0	21.39	Wang, Zhang, Lin, et al. (2021)
Heterojunction	Zn-TTB	25.16	1.15	80.1	23.14	Wang, Zhang, Gai, et al. (2022); Wang, Zhang, Yang, et al. (2022)
Heterojunction	Cr-MOF	16.51	1.30	79.0	17.02	Yuan et al. (2021)
ETL	MIL-125	10.9	0.85	69.0	6.4	Vinogradov et al. (2014)
ETL	MIL-125	22.81	1.01	71.8	16.56	Hou et al. (2017)
ETL	nTi-MOF	23.18	1.08	75.5	18.94	Ryu et al. (2018)
ETL	Co-doped Ti MOF	24.08	1.03	64.9	15.75	Nguyen and Bark (2020)
ETL	ZIF-8	21.60	0.94	62.0	12.40	Chung et al. (2018)
ETL	ZIF-8 derived ZnO	22.10	1.11	73.9	18.10	Zhang et al. (2020)
ETL	ZIF-8 derived mpg-C	22.13	1.06	72.0	17.32	Zhang et al. (2019)
HTL	In2	21.03	1.01	74.0	15.8	Li et al. (2018)
HTL	POM@Cu-BTC	23.90	1.11	80.0	21.44	Dong et al. (2019)

HTL	In10	24.3	1.00	70.0	17.0	Li, Wang, et al. (2019)
HTL	Zn-based MOF	23.17	1.14	78.4	20.64	Li, Xia, et al. (2019)
HTL	Tb-FTZB	23.86	1.13	79.3	21.31	Wang, Zhang, Yang, et al. (2021)
HTL	NH$_2$-MIL-101	23.34	1.07	76.9	19.23	Wang, Zhang, Yang, et al. (2022)
HTL	In-Pyia	23.53	1.09	79.0	20.26	Zhou, Qiu, Fan, Ye, et al. (2020)
HTL	MOF-545	23.9	1.12	77.9	20.9	Dong et al. (2022)
HTL	2D Pb-MOF	19.57	1.00	67.3	13.17	Huang et al. (2019)
HTL	2D Graphite NPC	23.51	1.06	76.0	18.51	Zhou, Qiu, Fan, Wang, et al. (2020)
Interface	ZIF-8	22.82	1.02	73.0	16.99	Shen et al. (2018)
Interface	ZIF-8	19.8	0.97	62.0	12.0	Chung et al. (2018)
Interface	ZIF-8-10	21.8	1.23	59.0	16.8	Ahmadian-Yazdi et al. (2020)
Interface	MACl	24.07	1.16	79.1	22.10	Jin et al. (2023)
Interface	MOF-525	23.99	1.08	72.1	18.73	Liu et al. (2021)
Interface	UiO-66	20.25	1.07	78.5	17.01	Lee et al. (2019)
Interface	MOF-808	19.64	1.07	78.9	16.55	Lee et al. (2019)
Interface	ZIF-8	23.93	1.06	75.6	19.13	Li et al. (2021)
Interface	PCN-224 QDs	24.56	1.17	78.2	22.51	Liu et al. (2023)
Interface	ZrL3	22.58	1.20	81.3	22.02	Wu et al. (2020)

BTC, 1,3,5-Benzenetricarboxylic acid; FF, fill factor; MOF, metal-organic framework; NPC, N-rich porous carbon; PCE, power conversion efficiency; PCN-224 QDs, polymer coordination network encapsulated quantum dots.

structural and thermal stability provided by MOF structure. MOF composites have been utilized in various positions, including HTL, ETL, interlayers, and in PL. Although the specific MOF materials and their applications vary, they consistently yield similar outcomes. The contributions of MOF derivatives on the PV performance of PSCs can be expressed as follows: (1) MOF composites enhance the quality and crystallinity of perovskite films, (2) MOF composites boost carrier transport while reducing exciton recombination, and (3) MOF derivatives increase the overall stability of the devices. These findings offer valuable insights for the future design of PSCs aimed at achieving both high efficiency and long-term stability.

Metal-organic framework composites in organic solar cells

For OSCs, the process includes four key steps: absorption of sunlight absorption and generation of electron–hole pairs, diffusion of photogenerated charges, dissociation charges by applied field, and transportation of charges to the electrodes, as depicted in Marinova et al. (2017). When the incident sunlight reaches the p-type semiconductor material, electrons are excited from the HOMO to the LUMO by absorbing photon energy. These excited electrons then transfer to the LUMO level of the p-type semiconductor (acceptor) and move toward the cathode, while remaining holes migrate through the n-type semiconductor (donor) material to the anode of the OSC device. This flow of charges generates an electric current that is collected and used to power the external devices (Fig. 6.7).

OSCs offer notable advantages, including customizable structures, cost-effectiveness, large-scale production capabilities, flexibility, and the potential for translucent devices. However, their development is hampered by low PCE (Dong, Lv, et al., 2017; Eastham et al., 2018; Fan et al., 2018; Holliday et al., 2016; Huang et al., 2017; Lee, Le, et al., 2015; Li et al., 2017; Nguyen et al.,

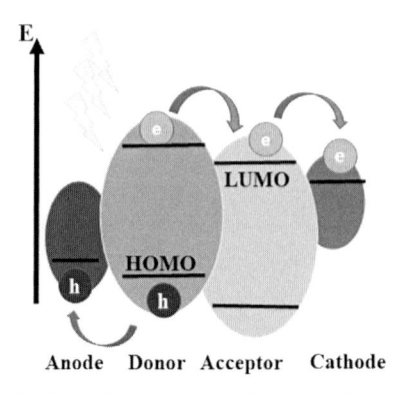

FIGURE 6.7 The schematic illustration of organic solar cell device (Do & Kim, 2023).

2015; Yu et al., 1995; Zhang et al., 2017). In OSCs, the interface layers have the critical position, such as the electron extraction layer (EEL) and the hole extraction layer. They should have essential properties of high conduction and good carrier transportation to function effectively.

Despite the increasing interest in dye-sensitized and PSCs, metal-MOFs have not been extensively explored in OSCs due to their generally low semiconductor properties. To date, only a limited number of studies have investigated the use of MOFs in OSCs. 2D materials, however, are particularly well-suited as additives to the IL, owing to their superior electrical and optical features and large BET areas (Adilbekova et al., 2020).

For example, polyethyleneimine ethoxylated (PEIE) was used to exfoliate Te-based MOF bulk into nanolayers, which were subsequently employed as the EEL in OSCs (Xing et al., 2018). This method increased the PCE from 9.91% to 10.39% for PEIE-derived OSC, attributed to the passivation of defects in the ZnO layer, and resulting in enhanced conductivity. Furthermore, the potential of MOFs in OSCs was demonstrated by Sasitharan et al. (2020), who utilized a mixture of Zn-based MOF nanolayers, (P3HT:poly(3-hex-ylthiophene-2,5diyl), and PCBM:[6,6]-phenyl-C61-butyric acid methyl ester) to form an active layer. This strategy enhanced the PCE from 2.67% (without MOF) to 5.2% (with MOF), likely due to the crucial role of the Zn-based MOF in structuring P3HT and preventing the agglomeration of PCBM.

Conclusion

In conclusion, MOF composites have emerged as a promising area of study in solar energy applications, driving extensive theoretical and experimental research to enhance their PV properties. They have shown potential for use as different components in DSSCs, offering notable efficiency improvements. Furthermore, MOF composites can be applied as interface modifiers and charge transport materials in PSCs. Recent studies have also explored their potential in OSCs, suggesting new avenues for innovation in solar technologies. Despite the rapid advancements in integrating MOF derivatives into various solar cell components, several key challenges remain to further improve solar cell efficiency.

One major challenge is the poor electron conductivity of MOF composites. Enhancing this requires novel designs and the synthesis of MOF derivatives with improved conductivity. Integrating MOF composites with highly conductive materials or developing inherently conductive MOF structures are potential solutions to this issue.

A deeper understanding of the relationship between MOF derivatives, their properties, and solar cell efficiency is essential. This would enable the development of more optimized MOF composites for PV applications. Additionally, further research on 2D MOF structure holds potential for solar cell advancement.

Growing MOF composites on conductive substrates remains a significant challenge due to the lack of chemical bonding. Developing new techniques to fabricate MOF thin films on conductive substrates is a priority for future work.

The conduction mechanism in MOF materials is still not fully understood. Therefore combined efforts involving simulations and experimental studies are needed to clarify the relationship between MOF composite structure and PCE, as well as to better understand their operation in solar devices.

Despite these challenges, MOF-based materials offer great potential for solar cells due to the diverse range of MOF composites available. Through compositional engineering, the bandgap (E_g) of MOF materials can be tailored to achieve optimal PCE.

Solar cell durability is crucial for industrial application. Future investigations should focus on testing MOF-based PSCs under extreme conditions, such as high humidity, to assess their long-term stability.

However, their application in OSC devices remains limited due to their optical properties. The development of photoluminescent MOF materials could yield promising results in this area.

Moreover, using advanced computational methods to identify the most suitable MOF omposites for solar cells could save time and reduce costs in experimental research. Finally, the excellent conductivity of conductive MOF composites and their remarkable stability show great promise for future PV applications.

Acknowledgment

The author would like to sincerely thank Mehmet İMER for his invaluable support.

References

Adilbekova, B., Lin, Y., Yengel, E., Faber, H., Harrison, G., Firdaus, Y., El-Labban, A., Anjum, D. H., Tung, V., & Anthopoulos, T. D. (2020). Liquid phase exfoliation of MoS_2 and WS_2 in aqueous ammonia and their application in highly efficient organic solar cells. *Journal of Materials Chemistry C, 8*(15), 5259–5264. https://doi.org/10.1039/d0tc00659a, http://pubs.rsc.org/en/journals/journal/tc.

Ahmadian-Yazdi, M. R., Gholampour, N., & Eslamian, M. (2020). Interface engineering by employing zeolitic imidazolate framework-8 (ZIF-8) as the only scaffold in the architecture of perovskite solar cells. *ACS Applied Energy Materials, 3*(4), 3134–3143. https://doi.org/10.1021/acsaem.9b02115, http://pubs.acs.org/journal/aaemcq.

Ahmed, A. S. A., Xiang, W., Saana Amiinu, I., & Zhao, X. (2018). Zeolitic-imidazolate-framework (ZIF-8)/PEDOT:PSS composite counter electrode for low cost and efficient dye-sensitized solar cells. *New Journal of Chemistry, 42*(21), 17303–17310. https://doi.org/10.1039/C8NJ03192D, http://pubs.rsc.org/en/journals/journal/nj.

Atli, A., Sutcu, I., Yildiz, Z. K., & Yildiz, A. (2021). Optimizing deposition parameters of DSSCs composed of blue TiO_2. *IEEE Journal of Photovoltaics, 11*(1), 118–123. https://doi.org/10.1109/jphotov.2020.3038602.

Atli, A., & Yildiz, A. (2022). Hybrid TiO$_2$ nanorods combined with a buffer layer for dye-sensitized solar cells. *Applied Solar Energy, 58*(3), 323–329. https://doi.org/10.3103/s0003701x22030045.

Bai, Y., Mora-Seró, I., De Angelis, F., Bisquert, J., & Wang, P. (2014). Titanium dioxide nanomaterials for photovoltaic applications. *Chemical Reviews, 114*(19), 10095–10130. https://doi.org/10.1021/cr400606n.

Atli, A. (2023). Fabrication and analysis of a dye sensitized solar cell: Hydrothermally grown hybrid photoanode and opaque Pt counter electrode. In Ph.D. thesis (p. 1), Department of Energy Systems Engineering, Yıldırım Beyazıt University.

Ban, Y., Li, Y., Liu, X., Peng, Y., & Yang, W. (2013). Solvothermal synthesis of mixed-ligand metal–organic framework ZIF-78 with controllable size and morphology. *Microporous and Mesoporous Materials, 173*, 29–36. https://doi.org/10.1016/j.micromeso.2013.01.031.

Bella, F., Bongiovanni, R., Kumar, R. S., Kulandainathan, M. A., & Stephan, A. M. (2013). Light cured networks containing metal organic frameworks as efficient and durable polymer electrolytes for dye-sensitized solar cells. *Journal of Materials Chemistry A, 1*(32), 9033–9036. https://doi.org/10.1039/c3ta12135f.

Bessho, T., Constable, E. C., Graetzel, M., Hernandez Redondo, A., Housecroft, C. E., Kylberg, W., Nazeeruddin, M. K., Neuburger, M., & Schaffner, S. (2008). An element of surprise – Efficient copper-functionalized dye-sensitized solar cells. *Royal Society of Chemistry, Switzerland Chemical Communications*, (32), 3717–3719. https://doi.org/10.1039/b808491b, http://pubs.rsc.org/en/journals/journal/cc.

Bhattarai, S., Mhamdi, A., Hossain, I., Raoui, Y., Pandey, R., Madan, J., Bouazizi, A., Maiti, M., Gogoi, D., & Sharma, A. (2022). A detailed review of perovskite solar cells: Introduction, working principle, modelling, fabrication techniques, future challenges. *Micro and Nanostructures, 172*, 207450. https://doi.org/10.1016/j.micrna.2022.207450.

Boschloo, G., & Hagfeldt, A. (2009). Characteristics of the iodide/triiodide redox mediator in dye-sensitized solar cells. *Accounts of Chemical Research, 42*(11), 1819–1826. https://doi.org/10.1021/ar900138m.

Campagnol, N., Van Assche, T., Boudewijns, T., Denayer, J., Binnemans, K., Vos, D. D., & Fransaer, J. (2013). High pressure, high temperature electrochemical synthesis of metal-organic frameworks: Films of MIL-100 (Fe) and HKUST-1 in different morphologies. *Journal of Materials Chemistry A, 1*(19), 5827–5830. https://doi.org/10.1039/c3ta10419b.

Chang, T. H., Kung, C. W., Chen, H. W., Huang, T. Y., Kao, S. Y., Lu, H. C., Lee, M. H., Boopathi, K. M., Chu, C. W., & Ho, K. C. (2015). Planar heterojunction perovskite solar cells incorporating metal-organic framework nanocrystals. *Advanced Materials, 27*(44), 7229–7235. https://doi.org/10.1002/adma.201502537, http://www3.interscience.wiley.com/journal/119030556/issue.

Chen, L., Chen, W., & Wang, E. (2018). Graphene with cobalt oxide and tungsten carbide as a low-cost counter electrode catalyst applied in Pt-free dye-sensitized solar cells. *Journal of Power Sources, 380*, 18–25. https://doi.org/10.1016/j.jpowsour.2017.11.057.

Chen, T. Y., Huang, Y. J., Li, C. T., Kung, C. W., Vittal, R., & Ho, K. C. (2017). Metal-organic framework/sulfonated polythiophene on carbon cloth as a flexible counter electrode for dye-sensitized solar cells. *Nano Energy, 32*, 19–27. https://doi.org/10.1016/j.nanoen.2016.12.019, http://www.journals.elsevier.com/nano-energy/.

Chen, X., & Mao, S. S. (2007). Titanium dioxide nanomaterials: Synthesis, properties, modifications and applications. *Chemical Reviews, 107*(7), 2891–2959. https://doi.org/10.1021/cr0500535.

Chen, B., Yang, Z., Jia, Q., Ball, R. J., Zhu, Y., & Xia, Y. (2023). Emerging applications of metal-organic frameworks and derivatives in solar cells: Recent advances and challenges. *Materials Science and Engineering: R: Reports, 152*, 100714. https://doi.org/10.1016/j.mser.2022.100714.

Chiba, Y., Islam, A., Watanabe, Y., Komiya, R., Koide, N., & Han, L. (2006). Dye-sensitized solar cells with conversion efficiency of 11.1%. *Japanese Journal of Applied Physics, 45*(7L), L638. https://doi.org/10.1143/jjap.45.l638.

Chi, W. S., Roh, D. K., Lee, C. S., & Kim, J. H. (2015). A shape- and morphology-controlled metal organic framework template for high-efficiency solid-state dye-sensitized solar cells. *Journal of Materials Chemistry A, 3*(43), 21599–21608. https://doi.org/10.1039/c5ta06731f, http://pubs.rsc.org/en/journals/journalissues/ta.

Chung, H. Y., Lin, C. H., Prabu, S., & Wang, H. W. (2018). Perovskite solar cells using TiO_2 layers coated with metal-organic framework material ZIF-8. *Journal of the Chinese Chemical Society, 65*(12), 1476–1481. https://doi.org/10.1002/jccs.201800173, http://onlinelibrary.wiley.com/journal/10.1002/(ISSN)2192-6549.

Corma, A., Atienzar, P., García, H., & Chane-Ching, J. Y. (2004). Hierarchically mesostructured doped CeO_2 with potential for solar-cell use. *Nature Materials, 3*(6), 394–397. https://doi.org/10.1038/nmat1129, http://www.nature.com/nmat/.

Cui, X., Xie, Z., & Wang, Y. (2016). Novel CoS_2 embedded carbon nanocages by direct sulfurizing metal–organic frameworks for dye-sensitized solar cells. *Nanoscale, 8*(23), 11984–11992. https://doi.org/10.1039/c6nr03052a.

Daeneke, T., Kwon, T. H., Holmes, A. B., Duffy, N. W., Bach, U., & Spiccia, L. (2011). High-efficiency dye-sensitized solar cells with ferrocene-based electrolytes. *Nature Chemistry, 3*(3), 211–215. https://doi.org/10.1038/nchem.966.

Daeneke, T., Uemura, Y., Duffy, N. W., Mozer, A. J., Koumura, N., Bach, U., & Spiccia, L. (2012). Aqueous dye-sensitized solar cell electrolytes based on the ferricyanide-ferrocyanide redox couple. *Advanced Materials, 24*(9), 1222–1225. https://doi.org/10.1002/adma.201104837.

Dimroth, F., Grave, M., Beutel, P., Fiedeler, U., Karcher, C., Tibbits, T. N. D., Oliva, E., Siefer, G., Schachtner, M., Wekkeli, A., Bett, A. W., Krause, R., Piccin, M., Blanc, N., Drazek, C., Guiot, E., Ghyselen, B., Salvetat, T., Tauzin, A., ... Schwarzburg, K. (2014). Wafer bonded four-junction GaInP/GaAs//GaInAsP/GaInAs concentrator solar cells with 44.7% efficiency. *Progress in Photovoltaics: Research and Applications, 22*(3), 277–282. https://doi.org/10.1002/pip.2475.

Do, H.H., & Kim, S.Y. (2023). Metal-organic frameworks based multifunctional materials for solar cells: A review. Symmetry, 15(10), 1830. https://doi.org/10.3390/sym15101830.

Dong, T., Lv, L., Feng, L., Xia, Y., Deng, W., Ye, P., Yang, B., Ding, S., Facchetti, A., & Dong, H. (2017). Noncovalent Se⋯O conformational locks for constructing high-performing optoelectronic conjugated polymers. *Advanced Materials, 29*, 1606025.

Dong, Y. J., Rao, H. S., Cao, Y., Chen, H. Y., Kuang, D. B., & Su, C. Y. (2017). In situ gelation of Al(III)-4-tert-butylpyridine based metal-organic gel electrolyte for efficient quasi-solid-state dye-sensitized solar cells. *Journal of Power Sources, 343*, 148–155. https://doi.org/10.1016/j.jpowsour.2017.01.051.

Dong, Y., Zhang, J., Yang, Y., Qiu, L., Xia, D., Lin, K., Wang, J., Fan, X., & Fan, R. (2019). Self-assembly of hybrid oxidant POM@Cu-BTC for enhanced efficiency and long-term stability of perovskite solar cells. *Angewandte Chemie International Edition, 58*(49), 17610–17615. https://doi.org/10.1002/anie.201909291.

Dong, Y., Zhang, J., Yang, Y., Wang, J., Hu, B., Wang, W., Cao, W., Gai, S., Xia, D., Lin, K., & Fan, R. (2022). Multifunctional nanostructured host-guest POM@MOF with lead sequestration capability induced stable and efficient perovskite solar cells. *Nano Energy, 97*, 107184. https://doi.org/10.1016/j.nanoen.2022.107184.

Dou, J., Li, Y., Xie, F., Ding, X., & Wei, M. (2016). Metal–organic framework derived hierarchical porous anatase TiO$_2$ as a photoanode for dye-sensitized solar cell. *Crystal Growth & Design, 16*(1), 121–125. https://doi.org/10.1021/acs.cgd.5b01003.

Eastham, N. D., Logsdon, J. L., Manley, E. F., Aldrich, T. J., Leonardi, M. J., Wang, G., Powers-Riggs, N. E., Young, R. M., Chen, L. X., Wasielewski, M. R., Melkonyan, F. S., Chang, R. P. H., & Marks, T. J. (2018). Hole-transfer dependence on blend morphology and energy level alignment in polymer: ITIC photovoltaic materials. *Advanced Materials, 30*(3). https://doi.org/10.1002/adma.201704263, http://onlinelibrary.wiley.com/journal/10.1002/(ISSN)1521-4095.

Fan, J., Hao, Y., Cabot, A., Johansson, E. M. J., Boschloo, G., & Hagfeldt, A. (2013). Cobalt(II/III) redox electrolyte in ZnO nanowire-based dye-sensitized solar cells. *ACS Applied Materials and Interfaces, 5*(6), 1902–1906. https://doi.org/10.1021/am400042s.

Fan, J., Li, L., Rao, H. S., Yang, Q. L., Zhang, J., Chen, H. Y., Chen, L., Kuang, D. B., & Su, C. Y. (2014). A novel metal-organic gel based electrolyte for efficient quasi-solid-state dye-sensitized solar cells. *Journal of Materials Chemistry A, 2*(37), 15406–15413. https://doi.org/10.1039/c4ta03120b, http://pubs.rsc.org/en/journals/journalissues/ta.

Fan, Q., Wang, Y., Zhang, M., Wu, B., Guo, X., Jiang, Y., Li, W., Guo, B., Ye, C., Su, W., Fang, J., Ou, X., Liu, F., Wei, Z., Sum, T. C., Russell, T. P., & Li, Y. (2018). High-performance as-cast nonfullerene polymer solar cells with thicker active layer and large area exceeding 11% power conversion efficiency. *Advanced Materials, 30*(6). https://doi.org/10.1002/adma.201704546, http://www3.interscience.wiley.com/journal/119030556/issue.

Gencer Imer, A., Syan, R. H. B., Gülcan, M., Ocak, Y. S., & Tombak, A. (2018). The novel pyridine based symmetrical Schiff base ligand and its transition metal complexes: synthesis, spectral definitions and application in dye sensitized solar cells (DSSCs). *Journal of Materials Science: Materials in Electronics, 29*(2), 898–905. https://doi.org/10.1007/s10854-017-7986-z, http://rd.springer.com/journal/10854.

Grätzel, M. (2001). Photoelectrochemical cells. *Nature, 414*(6861), 338–344. https://doi.org/10.1038/35104607.

Grätzel, M. (2005). Solar energy conversion by dye-sensitized photovoltaic cells. *Inorganic Chemistry, 44*(20), 6841–6851. https://doi.org/10.1021/ic0508371.

Gu, A., Xiang, W., Wang, T., Gu, S., & Zhao, X. (2017). Enhance photovoltaic performance of tris (2,2′-bipyridine) cobalt(II)/(III) based dye-sensitized solar cells via modifying TiO$_2$ surface with metal-organic frameworks. *Solar Energy, 147*, 126–132. https://doi.org/10.1016/j.solener.2017.03.045.

Hagfeldt, A., Boschloo, G., Sun, L., Kloo, L., & Pettersson, H. (2010). Dye-sensitized solar cells. *Chemical Reviews, 110*(11), 6595–6663. https://doi.org/10.1021/cr900356p.

Hamann, T. W., & Ondersma, J. W. (2011). Dye-sensitized solar cellredox shuttles. *Energy Environmental Science, 4*(2), 370–381. https://doi.org/10.1039/c0ee00251h.

Heo, D. Y., Do Ha, H., Ahn Sang, H., & Kim Soo, Y. (2020). Metal-organic framework materials for perovskite solar cells. *Polymers, 12*(9). https://doi.org/10.3390/polym12092061.

He, Y., & Wang, W. (2021). ZIF-8 and three-dimensional graphene network assisted DSSCs with high performances. *Journal of Solid State Chemistry, 296*, 121992. https://doi.org/10.1016/j.jssc.2021.121992.

He, Y., Zhang, Z., Wang, W., & Fu, L. (2020). Metal organic frameworks derived high-performance photoanodes for DSSCs. *Journal of Alloys and Compounds, 825*, 154089. https://doi.org/10.1016/j.jallcom.2020.154089.

Holliday, S., Ashraf, R. S., Wadsworth, A., Baran, D., Yousaf, S. A., Nielsen, C. B., Tan, C. H., Dimitrov, S. D., Shang, Z., Gasparini, N., Alamoudi, M., Laquai, F., Brabec, C. J., Salleo, A., Durrant, J. R., & McCulloch, I. (2016). High-efficiency and air-stable P3HT-based polymer solar cells with a new non-fullerene acceptor. *Nature Communications, 7*. https://doi.org/10.1038/ncomms11585, http://www.nature.com/ncomms/index.html.

Hou, X., Pan, L., Huang, S., Wei, O. Y., & Chen, X. (2017). Enhanced efficiency and stability of perovskite solar cells using porous hierarchical TiO_2 nanostructures of scattered distribution as scaffold. *Electrochimica Acta, 236*, 351–358. https://doi.org/10.1016/j.electacta.2017.03.192, http://www.journals.elsevier.com/electrochimica-acta/.

Huang, H., Yang, L., Facchetti, A., & Marks, T. J. (2017). Organic and polymeric semiconductors enhanced by noncovalent conformational locks. *Chemical Reviews, 117*(15), 10291–10318. https://doi.org/10.1021/acs.chemrev.7b00084, http://pubs.acs.org/journal/chreay.

Huang, L., Zhou, X., Wu, R., Shi, C., Xue, R., Zou, J., Xu, C., Zhao, J., & Zeng, W. (2019). Oriented haloing metal-organic framework providing high efficiency and high moisture-resistance for perovskite solar cells. *Journal of Power Sources, 433*, 226699. https://doi.org/10.1016/j.jpowsour.2019.226699.

Hua, Y., Xu, B., Liu, P., Chen, H., Tian, H., Cheng, M., Kloo, L., & Sun, L. (2016). High conductivity Ag-based metal organic complexes as dopant-free hole-transport materials for perovskite solar cells with high fill factors. *Chemical Science, 7*(4), 2633–2638. https://doi.org/10.1039/c5sc03569d.

Hui, W., Chao, L., Lu, H., Xia, F., Wei, Q., Su, Z., Niu, T., Tao, L., Du, B., Li, D., Wang, Y., Dong, H., Zuo, S., Li, B., Shi, W., Ran, X., Li, P., Zhang, H., Wu, Z., ... Huang, W. (2021). Stabilizing black-phase formamidinium perovskite formation at room temperature and high humidity. *Science, 371*(6536), 1359–1364. https://doi.org/10.1126/science.abf7652.

Jiang, X., Li, H., Li, S., Huang, S., Zhu, C., & Hou, L. (2018). Metal-organic framework-derived Ni–Co alloy@carbon microspheres as high-performance counter electrode catalysts for dye-sensitized solar cells. *Chemical Engineering Journal, 334*, 419–431. https://doi.org/10.1016/j.cej.2017.10.043.

Jing, H., Song, X., Ren, S., Shi, Y., An, Y., Yang, Y., Feng, M., Ma, S., & Hao, C. (2016). ZIF-67 derived nanostructures of Co/CoO and Co@N-doped graphitic carbon as counter electrode for highly efficient dye-sensitized solar cells. *Electrochimica Acta, 213*, 252–259. https://doi.org/10.1016/j.electacta.2016.07.129.

Jin, Z., Li, B., Xu, Y., Zhu, B., Ding, G., Wang, Y., Yang, J., Zhang, Q., & Rui, Y. (2023). Confinement of MACl guest in 2D ZIF-8 triggers interface and bulk passivation for efficient and UV-stable perovskite solar cells. *Journal of Materials Chemistry C, 11*(20), 6730–6740. https://doi.org/10.1039/d3tc00609c.

Jin, Z., Yan, J., Huang, X., Xu, W., Yang, S., Zhu, D., & Wang, J. (2017). Solution-processed transparent coordination polymer electrode for photovoltaic solar cells. *Nano Energy, 40*, 376–381. https://doi.org/10.1016/j.nanoen.2017.08.028.

Jung, E. H., Jeon, N. J., Park, E. Y., Moon, C. S., Shin, T. J., Yang, T. Y., Noh, J. H., & Seo, J. (2019). Efficient, stable and scalable perovskite solar cells using poly(3-hexylthiophene). *Nature, 567*(7749), 511–515. https://doi.org/10.1038/s41586-019-1036-3, http://www.nature.com/nature/index.html.

Kakiage, K., Aoyama, Y., Yano, T., Oya, K., Fujisawa, J.-ichi, & Hanaya, M. (2015). Highly-efficient dye-sensitized solar cells with collaborative sensitization by silyl-anchor and carboxy-anchor dyes. *Chemical Communications, 51*(88), 15894–15897. https://doi.org/10.1039/C5CC06759F.

Kaur, R., Kim, K. H., & Deep, A. (2017). A convenient electrolytic assembly of graphene-MOF composite thin film and its photoanodic application. *Applied Surface Science, 396*, 1303–1309. https://doi.org/10.1016/j.apsusc.2016.11.150, http://www.journals.elsevier.com/applied-surface-science/.

Kaur, R., Sharma, A. L., Kim, K. H., & Deep, A. (2017). A novel CdTe/Eu-MOF photoanode for application in quantum dot-sensitized solar cell to improve power conversion efficiency. *Journal of Industrial and Engineering Chemistry, 53*, 77–81. https://doi.org/10.1016/j.jiec.2017.04.002, http://www.sciencedirect.com/science/journal/1226086X.

Kayes, B. M., Nie, H., Twist R., Spruytte, S. G., Reinhardt, F., Kizilyalli, I. C., & Higashi, G. S. (2011) 27.6% Conversion efficiency, a new record for single-junction solar cells under 1 sun illumination. *Conference record of the IEEE photovoltaic specialists conference* (pp. 000004–000008). http://doi.org/10.1109/PVSC.2011.6185831.

Kaya, E., Gencer Imer, A., & Gülcan, M. (2025). Metal organic framework (MOF-5) and graphene oxide (GO) derived photoanodes for an efficient dye-sensitized solar cells. *Journal of Power Sources, 626*, 235811–235821. https://doi.org/10.1016/j.jpowsour.2024.235811.

Kocak, Y., & Yildiz, A. (2021). Carminic acid extracted from cochineal insect as photosensitizer for dye-sensitized solar cells. *International Journal of Energy Research, 45*(11), 16901–16907. https://doi.org/10.1002/er.6883.

Kohle, O., Grätzel, M., Meyer, A. F., & Meyer, T. B. (1997). The photovoltaic stability of bis (isothiocyanato)ruthenium(II)-bis-2,2′-bipyridine-4,4′-dicarboxylic acid and related sensitizers. *Advanced Materials, 9*(11), 904–906. https://doi.org/10.1002/adma.19970091111, http://www3.interscience.wiley.com/journal/119030556/issue.

Kojima, A., Teshima, K., Shirai, Y., & Miyasaka, T. (2009). Organometal halide perovskites as visible-light sensitizers for photovoltaic cells. *Journal of the American Chemical Society, 131*(17), 6050–6051. https://doi.org/10.1021/ja809598r.

Kumar, P. R., Ramasubbu, V., Sahaya Shajan, X., & Mothi, E. M. (2020). Porphyrin-sensitized quasi-solid solar cells with MOF composited titania aerogel photoanodes. *Materials Today Energy, 18*, 100511. https://doi.org/10.1016/j.mtener.2020.100511.

Kundu, T., Sahoo, S. C., & Banerjee, R. (2012). Solid-state thermolysis of anion induced metal-organic frameworks to ZnO microparticles with predefined morphologies: Facile synthesis and solar cell studies. *Crystal Growth and Design, 12*(5), 2572–2578. https://doi.org/10.1021/cg300174f.

Lee, C. C., Chen, C. I., Liao, Y. T., Wu, K. C. W., & Chueh, C. C. (2019). Enhancing efficiency and stability of photovoltaic cells by using perovskite/Zr-MOF heterojunction including bilayer and hybrid structures. *Taiwan Advanced Science, 6*(5). https://doi.org/10.1002/advs.201801715, http://onlinelibrary.wiley.com/journal/10.1002/(ISSN)2198-3844.

Lee, D. Y., Kim, E. K., Shin, C. Y., Shinde, D. V., Lee, W., Shrestha, N. K., Lee, J. K., & Han, S. H. (2014). Layer-by-layer deposition and photovoltaic property of Ru-based metal-organic frameworks. *RSC Advances, 4*(23), 12037–12042. https://doi.org/10.1039/c4ra00397g.

Lee, C. Y., Le, Q. V., Kim, C., & Kim, S. Y. (2015). Use of silane-functionalized graphene oxide in organic photovoltaic cells and organic light-emitting diodes. *Physical Chemistry Chemical Physics, 17*(14), 9369–9374. https://doi.org/10.1039/c5cp00507h, http://www.rsc.org/Publishing/Journals/CP/index.asp.

Lee, D. Y., Lim, I., Shin, C. Y., Patil, S. A., Lee, W., Shrestha, N. K., Lee, J. K., & Han, S. H. (2015). Facile interfacial charge transfer across hole doped cobalt-based MOFs/TiO$_2$ nanohybrids making MOFs light harvesting active layers in solar cells. *Journal of Materials Chemistry A, 3*(45), 22669–22676. https://doi.org/10.1039/c5ta07180a, http://pubs.rsc.org/en/journals/journalissues/ta.

Lee, D. Y., Shinde, D. V., Yoon, S. J., Cho, K. N., Lee, W., Shrestha, N. K., & Han, S. H. (2014). Cu-based metal-organic frameworks for photovoltaic application. *Journal of Physical Chemistry C, 118*(30), 16328–16334. https://doi.org/10.1021/jp4079663, http://pubs.acs.org/journal/jpccck.

Lee, D. Y., Shin, C. Y., Yoon, S. J., Lee, H. Y., Lee, W., Shrestha, N. K., Lee, J. K., & Han, S. H. (2014). Enhanced photovoltaic performance of Cu-based metal-organic frameworks sensitized solar cell by addition of carbon nanotubes. *Scientific Reports, 4*. https://doi.org/10.1038/srep03930.

Linfoot, C. L., Richardson, P., McCall, K. L., Durrant, J. R., Morandeira, A., & Robertson, N. (2011). A nickel-complex sensitiser for dye-sensitised solar cells. *Solar Energy, 85*(6), 1195–1203. https://doi.org/10.1016/j.solener.2011.02.023.

Lin, Y., Kong, C., Zhang, Q., & Chen, L. (2017). Metal-organic frameworks for carbon dioxide capture and methane storage. *Advanced Energy Materials, 7*(4). https://doi.org/10.1002/aenm.201601296, http://onlinelibrary.wiley.com/journal/10.1002/(ISSN)1614-6840.

Liu, Y., Liu, T., Guo, X., Hou, M., Yuan, Y., Shi, S., Wang, H., Zhang, R.-Z., Galiotis, C., & Wang, N. (2023). Porphyrinic metal–organic framework quantum dots for stable n–i–p perovskite solar cells. *Advanced Functional Materials, 33*, 2210028. https://doi.org/10.1002/adfm.202210028.

Liu, C. K., Wu, K. H., Lu, Y. A., Hsiao, L. Y., Lai, K. W., Chu, C. W., & Ho, K. C. (2021). Introducing postmetalation metal–organic framework to control perovskite crystal growth for efficient perovskite solar cells. *ACS Applied Materials and Interfaces, 13*(50), 60125–60134. https://doi.org/10.1021/acsami.1c22144, http://pubs.acs.org/journal/aamick.

Liu, J., Zhou, W., Liu, J., Fujimori, Y., Higashino, T., Imahori, H., Jiang, X., Zhao, J., Sakurai, T., Hattori, Y., Matsuda, W., Seki, S., Garlapati, S. K., Dasgupta, S., Redel, E., Sun, L., & Wöll, C. (2016). A new class of epitaxial porphyrin metal-organic framework thin films with extremely high photocarrier generation efficiency: Promising materials for all-solid-state solar cells. *Journal of Materials Chemistry A, 4*(33), 12739–12747. https://doi.org/10.1039/c6ta04898f, http://pubs.rsc.org/en/journals/journal/ta.

Liu, J., Zhou, W., Liu, J., Howard, I., Kilibarda, G., Schlabach, S., Coupry, D., Addicoat, M., Yoneda, S., Tsutsui, Y., Sakurai, T., Seki, S., Wang, Z., Lindemann, P., Redel, E., Heine, T., & Wöll, C. (2015). Photoinduced charge-carrier generation in epitaxial MOF thin films: High efficiency as a result of an indirect electronic band gap? *Angewandte Chemie International Edition, 54*(25), 7441–7445. https://doi.org/10.1002/anie.201501862.

Li, Y., Che, Z., Sun, X., Dou, J., & Wei, M. (2014a). Metal–organic framework derived hierarchical ZnO parallelepipeds as an efficient scattering layer in dye-sensitized solar cells. *Chemical Communications, 50*(68), 9769–9772. https://doi.org/10.1039/c4cc03352c.

Li, Y., Chen, C., Sun, X., Dou, J., & Wei, M. (2014b). Metal–organic frameworks at interfaces in dye-sensitized solar cells. *ChemSusChem, 7*(9), 2469–2472. https://doi.org/10.1002/cssc.201402143.

Li, Y., Chen, Z., Yu, B., Tan, S., Cui, Y., Wu, H., Luo, Y., Shi, J., Li, D., & Meng, Q. (2022). Efficient, stable formamidinium-cesium perovskite solar cells and minimodules enabled by crystallization regulation. *Joule, 6*(3), 676–689. https://doi.org/10.1016/j.joule.2022.02.003.

Li, C., Guo, S., Chen, J., Cheng, Z., Zhu, M., Zhang, J., Xiang, S., & Zhang, Z. (2021). Mitigation of vacancy with ammonium salt-trapped ZIF-8 capsules for stable perovskite solar cells through simultaneous compensation and loss inhibition. *Nanoscale Advances, 3*(12), 3554–3562. https://doi.org/10.1039/d1na00173f.

Li, Y., Pang, A., Wang, C., & Wei, M. (2011). Metal-organic frameworks: Promising materials for improving the open circuit voltage of dye-sensitized solar cells. *Journal of Materials Chemistry, 21*(43), 17259–17264. https://doi.org/10.1039/c1jm12754c.

Li, C., Qiu, J., Zhu, M., Cheng, Z., Zhang, J., Xiang, S., Zhang, X., & Zhang, Z. (2022). Multifunctional anionic metal-organic frameworks enhancing stability of perovskite solar cells. *Chemical Engineering Journal, 433*, 133587. https://doi.org/10.1016/j.cej.2021.133587.

Li, G. R., Song, J., Pan, G. L., & Gao, X. P. (2011). Highly Pt-like electrocatalytic activity of transition metal nitrides for dye-sensitized solar cells. *Energy and Environmental Science, 4*(5), 1680–1683. https://doi.org/10.1039/c1ee01105g.

Li, M., Wang, J., Jiang, A., Xia, D., Du, X., Dong, Y., Wang, P., Fan, R., & Yang, Y. (2019). Metal organic framework doped Spiro-OMeTAD with increased conductivity for improving perovskite solar cell performance. *Solar Energy, 188*, 380–385. https://doi.org/10.1016/j.solener.2019.05.078.

Li, M., Xia, D., Jiang, A., Du, X., Fan, X., Qiu, L., Wang, P., Fan, R., & Yang, Y. (2019). Enhanced crystallization and optimized morphology of perovskites through doping an indium-based metal–organic assembly: achieving significant solar cell efficiency enhancements. *Energy Technology, 7*(5). https://doi.org/10.1002/ente.201900027.

Li, M., Xia, D., Yang, Y., Du, X., Dong, G., Jiang, A., & Fan, R. (2018). Doping of [In$_2$(phen)$_3$Cl$_6$]·CH$_3$CN·2H$_2$O indium-based metal–organic framework into hole transport layer for enhancing perovskite solar cell efficiencies. *Advanced Energy Materials, 8*, 1702052.

Li, S., Ye, L., Zhao, W., Liu, X., Zhu, J., Ade, H., & Hou, J. (2017). Design of a new small-molecule electron acceptor enables efficient polymer solar cells with high fill factor. *Advanced Materials, 29*(46). https://doi.org/10.1002/adma.201704051, http://www3.interscience.wiley.com/journal/119030556/issue.

Lopez, H. A., Dhakshinamoorthy, A., Ferrer, B., Atienzar, P., Alvaro, M., & Garcia, H. (2011). Photochemical response of commercial MOFs: Al$_2$(BDC)$_3$ and its use as active material in photovoltaic devices. *Journal of Physical Chemistry C, 115*(45), 22200–22206. https://doi.org/10.1021/jp206919m.

Lv, D., Chen, Y., Li, Y., Shi, R., Wu, H., Sun, X., Xiao, J., Xi, H., Xia, Q., & Li, Z. (2017). Efficient mechanochemical synthesis of MOF-5 for linear alkanes adsorption. *Journal of Chemical & Engineering Data, 62*(7), 2030–2036. https://doi.org/10.1021/acs.jced.7b00049.

Marinova, N., Valero, S., & Delgado, J. L. (2017). Organic and perovskite solar cells: Working principles, materials and interfaces. *Journal of Colloid and Interface Science, 488*, 373–389. https://doi.org/10.1016/j.jcis.2016.11.021, http://www.elsevier.com/inca/publications/store/6/2/2/8/6/1/index.htt.

Mathew, S., Yella, A., Gao, P., Humphry-Baker, R., Curchod, B. F. E., Ashari-Astani, N., Tavernelli, I., Rothlisberger, U., Nazeeruddin, M. K., & Grätzel, M. (2014). Dye-sensitized solar cells with 13% efficiency achieved through the molecular engineering of porphyrin sensitizers. *Nature Chemistry, 6*(3), 242–247. https://doi.org/10.1038/nchem.1861.

Maza, W. A., Haring, A. J., Ahrenholtz, S. R., Epley, C. C., Lin, S. Y., & Morris, A. J. (2016). Ruthenium(ii)-polypyridyl zirconium(iv) metal–organic frameworks as a new class of sensitized solar cells. *Chemical Science, 7*(1), 719–727. https://doi.org/10.1039/C5SC01565K.

Ma, W., Yang, G., Jiang, K., Carpenter, J. H., Wu, Y., Meng, X., McAfee, T., Zhao, J., Zhu, C., Wang, C., Ade, H., & Yan, H. (2015). Influence of processing parameters and molecular weight on the morphology and properties of high-performance PffBT4T-2OD:PC$_{71}$BM organic solar cells. *Advanced Energy Materials, 5*(23). https://doi.org/10.1002/aenm.201501400, http://onlinelibrary.wiley.com/journal/10.1002/(ISSN)1614-6840.

Nazeeruddin, M. K., Kay, A., Rodicio, I., Humphry-Baker, R., Mueller, E., Liska, P., Vlachopoulos, N., & Graetzel, M. (1993). Conversion of light to electricity by cis-X2bis (2,2′-bipyridyl-4,4′-dicarboxylate)ruthenium(II) charge-transfer sensitizers (X=Cl-, Br-, I-,

CN-, and SCN-) on nanocrystalline titanium dioxide electrodes. *Journal of the American Chemical Society, 115*(14), 6382–6390. https://doi.org/10.1021/ja00067a063.

Nazeeruddin, M. K., Péchy, P., & Grätzel, M. (1997). Efficient panchromatic sensitization of nanocrystalline TiO_2 films by a black dye based on a trithiocyanato–ruthenium complex. *Chemical Communications, 18*, 1705–1706. https://doi.org/10.1039/a703277c.

Nguyen, T. M. H., & Bark, C. W. (2020). Synthesis of cobalt-doped TiO_2 based on metal-organic frameworks as an effective electron transport material in perovskite solar cells. *ACS Omega, 5*(5), 2280–2286. https://doi.org/10.1021/acsomega.9b03507, http://pubs.acs.org/journal/acsodf.

Nguyen, T. P., Van Le, Q., Choi, K. S., Oh, J. H., Kim, Y. G., Lee, S. M., Chang, S. T., Cho, Y. H., Choi, S., Kim, T. Y., & Kim, S. Y. (2015). MoS_2 nanosheets exfoliated by sonication and their application in organic photovoltaic cells. *Science of Advanced Materials, 7*(4), 700–705. https://doi.org/10.1166/sam.2015.1891, http://docserver.ingentaconnect.com/deliver/connect/asp/19472935/v7n4/s13.pdf?.

Ni, Z., & Masel, R. I. (2006). Rapid production of metal-organic frameworks via microwave-assisted solvothermal synthesis. *Journal of the American Chemical Society, 128*(38), 12394–12395. https://doi.org/10.1021/ja0635231.

Nowotny, M. K., Sheppard, L. R., Bak, T., & Nowotny, J. (2008). Defect chemistry of titanium dioxide. Application of defect engineering in processing of TiO_2-based photocatalysts. *Journal of Physical Chemistry C, 112*(14), 5275–5300. https://doi.org/10.1021/jp077275m.

O'Regan, B., & Grätzel, M. (1991). A low-cost, high-efficiency solar cell based on dye-sensitized colloidal TiO_2 films. *Nature, 353*(6346), 737–740. https://doi.org/10.1038/353737a0.

Ou, J. Z., Rani, R. A., Ham, M. H., Field, M. R., Zhang, Y., Zheng, H., Reece, P., Zhuiykov, S., Sriram, S., Bhaskaran, M., Kaner, R. B., & Kalantar-Zadeh, K. (2012). Elevated temperature anodized Nb_2O_5: A photoanode material with exceptionally large photoconversion efficiencies. *ACS Nano, 6*(5), 4045–4053. https://doi.org/10.1021/nn300408p, http://pubs.acs.org/journal/ancac3.

Ozel, K., Atilgan, A., & Yildiz, A. (2024). Multi-layered blocking layers for dye sensitized solar cells. *Journal of Photochemistry and Photobiology A: Chemistry, 448*, 115297. https://doi.org/10.1016/j.jphotochem.2023.115297.

Pachfule, P., Das, R., Poddar, P., & Banerjee, R. (2011). Solvothermal synthesis, structure, and properties of metal organic framework isomers derived from a partially fluorinated link. *Crystal Growth & Design, 11*(4), 1215–1222. https://doi.org/10.1021/cg101414x.

Park, N.-G., van de Lagemaat, J., & Frank, A. J. (2000). Comparison of dye-sensitized rutile- and anatase-based TiO_2 solar cells. *The Journal of Physical Chemistry. B, 104*(38), 8989–8994. https://doi.org/10.1021/jp9943651.

Peng, S., Shi, J., Pei, J., Liang, Y., Cheng, F., Liang, J., & Chen, J. (2009). $Ni_{1-x}Pt_x$ (x=0–0.08) films as the photocathode of dye-sensitized solar cells with high efficiency. *Nano Research, 2*(6), 484–492. https://doi.org/10.1007/s12274-009-9044-5.

Ramasubbu, V., Kumar, P. R., Mothi, E. M., Karuppasamy, K., Kim, H. S., Maiyalagan, T., & Shajan, X. S. (2019). Highly interconnected porous TiO_2-Ni-MOF composite aerogel photoanodes for high power conversion efficiency in quasi-solid dye-sensitized solar cells. *Applied Surface Science, 496*. https://doi.org/10.1016/j.apsusc.2019.143760, http://www.journals.elsevier.com/applied-surface-science/.

Remya, V. R., & Kurian, M. (2019). Synthesis and catalytic applications of metal–organic frameworks: A review on recent literature. *International Nano Letters, 9*(1), 17–29. https://doi.org/10.1007/s40089-018-0255-1.

Rosi, N. L., Eckert, J., Eddaoudi, M., Vodak, D. T., Kim, J., O'Keeffe, M., & Yaghi, O. M. (2003). Hydrogen storage in microporous metal-organic frameworks. *Science, 300*(5622), 1127–1129,. https://doi.org/10.1126/science.1083440.

Ryu, U., Jee, S., Park, J.-S., Han, I.-K., Lee, J.-H., Park, M., & Choi, K. M. (2018). Nanocrystalline titanium metal–organic frameworks for highly efficient and flexible perovskite solar cells. *ACS nano, 12*(5), 4968–4975.

Sasitharan, K., Bossanyi, D. G., Vaenas, N., Parnell, A. J., Clark, J., Iraqi, A., Lidzey, D. G., & Foster, J. A. (2020). Metal-organic framework nanosheets for enhanced performance of organic photovoltaic cells. *Journal of Materials Chemistry A, 8*(12), 6067–6075. https://doi.org/10.1039/c9ta12313j, http://pubs.rsc.org/en/journals/journal/ta.

Seo, J. Y., Kim, H. S., Akin, S., Stojanovic, M., Simon, E., Fleischer, M., Hagfeldt, A., Zakeeruddin, S. M., & Grätzel, M. (2018). Novel p-dopant toward highly efficient and stable perovskite solar cells. *Energy and Environmental Science, 11*(10), 2985–2992. https://doi.org/10.1039/c8ee01500g, http://pubs.rsc.org/en/journals/journal/ee.

Shen, D., Pang, A., Li, Y., Dou, J., & Wei, M. (2018). Metal–organic frameworks at interfaces of hybrid perovskite solar cells for enhanced photovoltaic properties. *Chemical Communications, 54*(10), 1253–1256. https://doi.org/10.1039/C7CC09452C.

Shin, S. S., Kim, J. S., Suk, J. H., Lee, K. D., Kim, D. W., Park, J. H., Cho, I. S., Hong, K. S., & Kim, J. Y. (2013). Improved quantum efficiency of highly efficient perovskite $BaSnO_3$-based dye-sensitized solar cells. *ACS Nano, 7*(2), 1027–1035. https://doi.org/10.1021/nn305341x.

Silva Filho, J. C., Venancio, E. C., Silva, S. C., Takiishi, H., Martinez, L. G., & Antunes, R. A. (2020). A thermal method for obtention of 2 to 3 reduced graphene oxide layers from graphene oxide. *SN Applied Sciences, 2*(8). https://doi.org/10.1007/s42452-020-03241-9.

Son, W. J., Kim, J., Kim, J., & Ahn, W. S. (2008). Sonochemical synthesis of MOF-5. *Chemical Communications,* (47), 6336–6338. https://doi.org/10.1039/b814740j, http://pubs.rsc.org/en/journals/journal/cc.

Spoerke, E. D., Small, L. J., Foster, M. E., Wheeler, J., Ullman, A. M., Stavila, V., Rodriguez, M., & Allendorf, M. D. (2017). MOF-sensitized solar cells enabled by a pillared porphyrin framework. *Journal of Physical Chemistry C, 121*(9), 4816–4824. https://doi.org/10.1021/acs.jpcc.6b11251, http://pubs.acs.org/journal/jpccck.

Sumida, K., Liang, K., Reboul, J., Ibarra, I. A., Furukawa, S., & Falcaro, P. (2017). Sol-gel processing of metal-organic frameworks. *Chemistry of Materials, 29*(7), 2626–2645. https://doi.org/10.1021/acs.chemmater.6b03934, http://pubs.acs.org/journal/cmatex.

Sun, X., Dou, J., Xie, F., Li, Y., & Wei, M. (2014). One-step preparation of mirror-like NiS nanosheets on ITO for the efficient counter electrode of dye-sensitized solar cells. *Chemical Communications, 50*(69), 9869–9871. https://doi.org/10.1039/c4cc03798g.

Sun, X., Li, Y., Dou, J., Shen, D., & Wei, M. (2016). Metal-organic frameworks derived carbon as a high-efficiency counter electrode for dye-sensitized solar cells. *Journal of Power Sources, 322*, 93–98. https://doi.org/10.1016/j.jpowsour.2016.05.025.

Sun, C. Y., Qin, C., Wang, X. L., & Su, Z. M. (2013). Metal-organic frameworks as potential drug delivery systems. *Expert Opinion on Drug Delivery, 10*(1), 89–101. https://doi.org/10.1517/17425247.2013.741583.

Tan, B., Toman, E., Li, Y., & Wu, Y. (2007). Zinc Stannate (Zn_2SnO_4) dye-sensitized solar cells. *Journal of the American Chemical Society, 129*(14), 4162–4163. https://doi.org/10.1021/ja070804f.

Tan, Y. X., Wang, F., & Zhang, J. (2018). Design and synthesis of multifunctional metal-organic zeolites. *Chemical Society Reviews, 47*(6), 2130–2144. https://doi.org/10.1039/c7cs00782e, http://pubs.rsc.org/en/journals/journal/cs.

Thomas, S., Deepak, T. G., Anjusree, G. S., Arun, T. A., Nair, S. V., & Nair, A. S. (2014). A review on counter electrode materials in dye-sensitized solar cells. *Journal of Materials Chemistry A, 2*(13), 4474–4490. https://doi.org/10.1039/c3ta13374e.

Tian, Y. B., Wang, Y. Y., Chen, S. M., Gu, Z. G., & Zhang, J. (2020). Epitaxial growth of highly transparent metal-porphyrin framework thin films for efficient bifacial dye-sensitized solar cells. *ACS Applied Materials and Interfaces, 12*(1), 1078–1083. https://doi.org/10.1021/acsami.9b19022, http://pubs.acs.org/journal/aamick.

Uğur, A., Imer, A. G., & Gülcan, M. (2022). Enhancement in the photovoltaic efficiency of dye-sensitized solar cell by doping TiO_2 with MIL-101 MOF structure. *Materials Science in Semiconductor Processing, 150*, 106951. https://doi.org/10.1016/j.mssp.2022.106951.

Uğur, A., Imer, A. G., Kaya, E., Karataş, Y., & Gülcan, M. (2022). Improved efficiency in dye sensitized solar cell (DSSC) by nano -MIL-101(Cr) impregnated photoanode. *Zeitschrift für Naturforschung A. 77*(1), 93–104. https://doi.org/10.1515/zna-2021-0175.

Vinogradov, A. V., Zaake-Hertling, H., Hey-Hawkins, E., Agafonov, A. V., Seisenbaeva, G. A., Kessler, V. G., & Vinogradov, V. V. (2014). The first depleted heterojunction TiO_2 –MOF-based solar cell. *Chemical Communications, 50*(71), 10210–10213. https://doi.org/10.1039/C4CC01978D.

Wang, H., & Hu, Y. H. (2012). Graphene as a counter electrode material for dye-sensitized solar cells. *Energy and Environmental Science, 5*(8), 8182–8188. https://doi.org/10.1039/c2ee21905k, http://www.rsc.org/Publishing/Journals/EE/About.asp.

Wang, C. C., Li, J. R., Lv, X. L., Zhang, Y. Q., & Guo, G. (2014). Photocatalytic organic pollutants degradation in metal-organic frameworks. *Energy and Environmental Science, 7*(9), 2831–2867. https://doi.org/10.1039/c4ee01299b, http://www.rsc.org/Publishing/Journals/EE/About.asp.

Wang, Q., Ito, S., Grätzel, M., Fabregat-Santiago, F., Mora-Seró, I., Bisquert, J., Bessho, T., & Imai, H. (2006). Characteristics of high efficiency dye-sensitized solar cells. *The Journal of Physical Chemistry. B, 110*(50), 25210–25221. https://doi.org/10.1021/jp064256o.

Wang, R., Xue, J., Wang, K. L., Wang, Z. K., Luo, Y., Fenning, D., Xu, G., Nuryyeva, S., Huang, T., Zhao, Y., Yang, J. L., Zhu, J., Wang, M., Tan, S., Yavuz, I., Houk, K. N., & Yang, Y. (2019). Constructive molecular configurations for surface-defect passivation of perovskite photovoltaics. *Science, 366*(6472), 1509–1513. https://doi.org/10.1126/science.aay9698, http://science.sciencemag.org/content/366/6472/1509/tab-pdf.

Wang, J., Zhang, J., Gai, S., Wang, W., Dong, Y., Hu, B., Li, J., Lin, K., Xia, D., Fan, R., & Yang, Y. (2022). Self-organized small molecules in robust MOFs for high-performance perovskite solar cells with enhanced degradation activation energy. *Advanced Functional Materials, 32*, 2203898.

Wang, W., Zhang, J., Lin, K., Dong, Y., Wang, J., Hu, B., Li, J., Shi, Z., Hu, Y., Cao, W., Xia, D., Fan, R., & Yang, Y. (2021). Construction of polyoxometalate-based material for eliminating multiple Pb-based defects and enhancing thermal stability of perovskite solar cells. *Advanced Functional Materials, 31*(52). https://doi.org/10.1002/adfm.202105884.

Wang, J., Zhang, J., Yang, Y., Dong, Y., Wang, W., Hu, B., Li, J., Cao, W., Lin, K., Xia, D., & Fan, R. (2022). Li-TFSI endohedral metal-organic frameworks in stable perovskite solar cells for anti-deliquescent and restricting ion migration. *Chemical Engineering Journal, 429*, 132481. https://doi.org/10.1016/j.cej.2021.132481.

Wang, J., Zhang, J., Yang, Y., Gai, S., Dong, Y., Qiu, L., Xia, D., Fan, X., Wang, W., Hu, B., Cao, W., & Fan, R. (2021). New insight into the Lewis basic sites in metal–organic framework-doped hole transport materials for efficient and stable perovskite solar cells. *ACS Applied Materials & Interfaces, 13*(4), 5235–5244. https://doi.org/10.1021/acsami.0c19968.

Wang, L., Zhou, H., Hu, J., Huang, B., Sun, M., Dong, B., Zheng, G., Huang, Y., Chen, Y., Li, L., Xu, Z., Li, N., Liu, Z., Chen, Q., Sun, L.-D., & Yan, C.-H. (2019). A Eu^{3+}-Eu^{2+} ion redox shuttle imparts operational durability to Pb-I perovskite solar cells. *Science, 363*(6424), 265–270. https://doi.org/10.1126/science.aau5701.

Wei, D., Ji, J., Song, D., Li, M., Cui, P., Li, Y., Mbengue, J. M., Zhou, W., Ning, Z., & Park, N. G. (2017). A TiO_2 embedded structure for perovskite solar cells with anomalous grain growth and effective electron extraction. *Journal of Materials Chemistry A, 5*(4), 1406–1414. https://doi.org/10.1039/c6ta10418e, http://pubs.rsc.org/en/journals/journalissues/ta.

Wu, S. F., Li, Z., Li, M. Q., Diao, Y. X., Lin, F., Liu, T. T., Zhang, J., Tieu, P., Gao, W. P., Qi, F., Pan, X. Q., Xu, Z. T., Zhu, Z. L., & Jen, A. K. Y. (2020). 2D metal–organic framework for stable perovskite solar cells with minimized lead leakage. *Nature Nanotechnology, 15*, 934.

Wu, M., Lin, X., Wang, T., Qiu, J., & Ma, T. (2011). Low-cost dye-sensitized solar cell based on nine kinds of carbon counter electrodes. *Energy and Environmental Science, 4*(6), 2308–2315. https://doi.org/10.1039/c1ee01059j.

Xie, Y., & Hamann, T. W. (2013). Fast low-spin cobalt complex redox shuttles for dye-sensitized solar cells. *Journal of Physical Chemistry Letters, 4*(2), 328–332. https://doi.org/10.1021/jz301934e.

Xing, W., Ye, P., Lu, J., Wu, X., Chen, Y., Zhu, T., Peng, A., & Huang, H. (2018). Tellurophene-based metal-organic framework nanosheets for high-performance organic solar cells. *Journal of Power Sources, 401*, 13–19. https://doi.org/10.1016/j.jpowsour.2018.08.078.

Yang, A. N., Lin, J. T., & Li, C. T. (2021). Electroactive and sustainable Cu-MoF/PEDOT composite electrocatalysts for multiple redox mediators and for high-performance dye-sensitized solar cells. *ACS Applied Materials and Interfaces, 13*(7), 8435–8444. https://doi.org/10.1021/acsami.0c21542, http://pubs.acs.org/journal/aamick.

Yildiz, A., Chouki, T., Atli, A., Harb, M., Verbruggen, S. W., Ninakanti, R., & Emin, S. (2021). Efficient Iron phosphide catalyst as a counter electrode in dye-sensitized solar cells. *ACS Applied Energy Materials, 4*(10), 10618–10626. https://doi.org/10.1021/acsaem.1c01628, http://pubs.acs.org/journal/aaemcq.

Yoo, J. J., Seo, G., Chua, M. R., Park, T. G., Lu, Y., Rotermund, F., Kim, Y. K., Moon, C. S., Jeon, N. J., Correa-Baena, J. P., Bulović, V., Shin, S. S., Bawendi, M. G., & Seo, J. (2021). Efficient perovskite solar cells via improved carrier management. *Nature, 590*(7847), 587–593. https://doi.org/10.1038/s41586-021-03285-w, http://www.nature.com/nature/index.html.

Yoshikawa, K., Kawasaki, H., Yoshida, W., Irie, T., Konishi, K., Nakano, K., Uto, T., Adachi, D., Kanematsu, M., Uzu, H., & Yamamoto, K. (2017). Silicon heterojunction solar cell with interdigitated back contacts for a photoconversion efficiency over 26%. *Nature Energy, 2*(5). https://doi.org/10.1038/nenergy.2017.32.

Yuan, S., Xian, Y., Long, Y., Cabot, A., Li, W., & Fan, J. (2021). Chromium-based metal–organic framework as a-site cation in $CsPbI_2$ Br perovskite solar cells. *Advanced Functional Materials, 31*(51). https://doi.org/10.1002/adfm.202106233.

Yu, G., Gao, J., Hummelen, J. C., Wudl, F., & Heeger, A. J. (1995). Polymer photovoltaic cells: Enhanced efficiencies via a network of internal donor-acceptor heterojunctions. *Science, 270*(5243), 1789–1791. https://doi.org/10.1126/science.270.5243.1789.

Zhang, Y.-N., Li, B., Fu, L., Li, Q., & Yin, L. (2020). MOF-derived ZnO as electron transport layer for improving light harvesting and electron extraction efficiency in perovskite solar cells. *Electrochimica Acta, 330*, 135280.

Zhang, W., Li, W., He, X., Zhao, L., Chen, H., Zhang, L., Tian, P., Xin, Z., Fang, W., & Zhang, F. (2018). Dendritic Fe-based polyoxometalates @ metal–organic framework (MOFs) combined with ZnO as a novel photoanode in solar cells. *Journal of Materials Science: Materials in Electronics, 29*(2), 1623–1629. https://doi.org/10.1007/s10854-017-8073-1.

Zhang, J., Li, J., Yang, Y., Yang, C., Dong, Y., Lin, K., Xia, D., & Fan, R. (2022). Functionalized rare-earth metal cluster-based materials as additives for enhancing the efficiency of perovskite solar cells. *ACS Applied Energy Materials, 5*(11), 13318–13326. https://doi.org/10.1021/acsaem.2c01909.

Zhang, Z., Luo, X., Wang, B., & Zhang, J. B. (2019). Electron transport improvement of perovskite solar cells via a ZIF-8-derived porous carbon skeleto. *ACS Applied Energy Materials, 2*, 2760–2768.

Zhang, S., Zhang, J., Abdelsamie, M., Shi, Q., Zhang, Y., Parker, T. C., Jucov, E. V., Timofeeva, T. V., Amassian, A., Bazan, G. C., Blakey, S. B., Barlow, S., & Marder, S. R. (2017). Intermediate-sized conjugated donor molecules for organic solar cells: Comparison of benzodithiophene and benzobisthiazole-based cores. *Chemistry of Materials, 29*(18), 7880–7887. https://doi.org/10.1021/acs.chemmater.7b02665, http://pubs.acs.org/journal/cmatex.

Zhou, X., Qiu, L., Fan, R., Wang, A., Ye, H., Tian, C., Hao, S., & Yang, Y. (2020). Metal–organic framework-derived N-rich porous carbon as an auxiliary additive of hole transport layers for highly efficient and long-term stable perovskite solar cells. *Solar RRL, 4*(3). https://doi.org/10.1002/solr.201900380.

Zhou, X., Qiu, L., Fan, R., Ye, H., Tian, C., Hao, S., & Yang, Y. (2020). Toward high-efficiency and thermally-stable perovskite solar cells: A novel metal-organic framework with active pyridyl sites replacing 4-tert-butylpyridine. *Journal of Power Sources, 473*, 228556. https://doi.org/10.1016/j.jpowsour.2020.228556.

Zhou, X., Qiu, L., Fan, R., Zhang, J., Hao, S., & Yang, Y. (2020). Heterojunction incorporating perovskite and microporous metal–organic framework nanocrystals for efficient and stable solar cells. *Nano-Micro Letters, 12*(1). https://doi.org/10.1007/s40820-020-00417-1.

Zhu, G., Pan, L., Lu, T., Xu, T., & Sun, Z. (2011). Electrophoretic deposition of reduced graphene-carbon nanotubes composite films as counter electrodes of dye-sensitized solar cells. *Journal of Materials Chemistry, 21*(38), 14869–14875. https://doi.org/10.1039/c1jm12433a, http://www.rsc.org/Publishing/Journals/jm/index.asp.

Chapter 7

Metal-organic framework composites for electromagnetic interference shielding

Majed Amini*, Aliakbar Isari*, Seyyed Alireza Hashemi*, and Mohammad Arjmand

Nanomaterials and Polymer Nanocomposites Laboratory, School of Engineering, The University of British Columbia, Kelowna, BC, Canada

Introduction

Electromagnetic (EM) waves, a pillar of advanced technological developments, have revolutionized different sectors, including telecommunication, healthcare, military, and beyond. Despite the numerous advantages of EM waves, their widespread usage caused an interrupting type of pollution, namely EM interference (EMI). EMI can adversely affect the proper function of sensitive electronics and jeopardize human health and living entities (Hashemi et al., 2022; Isari, Ghaffarkhah, Hashemi, Wuttke, et al., 2024). Therefore, developing protective barriers against EM waves is crucial and increasingly in demand to address these challenges. To this end, various effective metallic and superconductive shielding systems based on conductive nanomaterials have been developed. However, these systems present a new challenge, i.e., the induced reflection from the surface of conductive shields, which can be as harmful as the incident EM waves (Hashemi et al., 2024). This happens as a result of impedance mismatch between the shield and free space (377 Ω), increasing the surface reflections (Isari, Ghaffarkhah, Hashemi, Wuttke, et al., 2024). Tackling this problem has garnered interest in the manufacturing of absorption-dominant shielding systems, where (1) surface reflections are minimized through impedance matching with free space, and (2) electromagnetic wave dissipation parameters are enhanced. The dissipating parameters of the electromagnetic waves involve internal scattering within

---

* These authors contributed equally to this book chapter.

the porosities of a shield with finite electrical conductivity, as well as conduction, dielectric, and magnetic losses. The combined effect of these factors has proven to be highly effective in dissipating the energy of EM waves as heat, facilitating the absorption-dominant shielding construct (Isari et al., 2024; Xu, Li, et al., 2023).

Moreover, creating an absorption-dominant shielding system requires a careful balance between EM wave dissipation parameters and impedance matching. In this context, metal-organic frameworks (MOFs), the rising stars of reticular chemistry, have emerged as promising candidates (Panahi-Sarmad et al., 2023; Xu, Li, et al., 2023). MOFs, as highly porous crystalline compounds with microporosities or mesoporosities, consist of metal clusters bonded to organic ligands through coordination chemistry (Li et al., 2016). The controllable nature of coordination chemistry has facilitated the development of over 20,000 MOFs, each exhibiting unique crystal structures, ultra-large specific surface areas, and tunable morphologies and compositions (Xu, Li, et al., 2023; Zhou et al., 2012). Notably, MOFs can be transformed into functional nanostructures through postprocessing or thermal annealing, resulting in metal oxides, carbonaceous compounds, metal sulfides, or hybrid composites (Guan et al., 2017; Jiao et al., 2019; Wang et al., 2018). Despite their promising characteristics, such as large surface area and tunable porosity, MOFs face several limitations when used alone for EMI shielding. These limitations include low electrical conductivity and poor mechanical stability. The low conductivity of MOFs hinders their potential to effectively attenuate electromagnetic waves. Additionally, the inherent brittleness of MOF crystals makes it challenging to fabricate flexible or large-scale shielding materials (Panahi-Sarmad et al., 2023; Xu, Li, et al., 2023).

To address these limitations, recent studies have focused on the development of hybridized MOF composites (Han et al., 2020; Huang et al., 2022; Lin et al., 2023; Zhang et al., 2017). These composites combine MOFs with other materials to create hybrid structures that often exhibit enhanced EMI shielding performance compared to their individual components. MOF-based composites can be engineered to improve electrical conductivity, mechanical strength, and electromagnetic wave absorption characteristics via material hybridization. One approach in MOF composite development is the incorporation of electrically conductive materials. For example, integrating graphene (Zhang et al., 2017) or MXene (Verma et al., 2023) into MOF structures can increase electrical conductivity by several orders of magnitude. Furthermore, MOF composites offer the opportunity to introduce magnetic components, enhancing magnetic loss mechanisms in EMI shielding (Chen et al., 2022; Guo et al., 2023). The integration of magnetic nanoparticles, such as Fe_3O_4 (Deng et al., 2020) or CoFe (Wang et al., 2020), into MOF structures can create materials that exploit both dielectric and magnetic losses for EM wave attenuation.

The synergistic effects in MOF composites often result in improved impedance matching with free space, a critical factor in reducing surface reflection and enhancing absorption-dominant shielding. This is exemplified in the work of Zhang et al. (2024), where a MOF/CNT composite exhibited a maximum reflection loss of −52.7 dB at 4.67 GHz at a thickness of 1.6 mm. Moreover, current research in MOF composites for EMI shielding focuses on optimizing synthesis methods, improving long-term stability, and developing scalable production techniques (Guo et al., 2023). Challenges include achieving uniform distribution of conductive or magnetic components within the MOF matrix and maintaining the desirable properties of MOFs, such as large surface area, during composite formation (Chen et al., 2022; Wang et al., 2020; Zhang et al., 2024). Addressing these challenges is crucial for the practical implementation of MOF composites in real-world EMI shielding applications. As research progresses, MOF composites are increasingly being explored for multifunctional applications. Beyond EMI shielding, these materials show potential in areas such as gas storage (Li et al., 2019), catalysis (Konnerth et al., 2020), and energy storage (Peng et al., 2022). This multifunctionality presents opportunities for developing integrated solutions that address multiple technological challenges simultaneously, particularly in industries where space and weight constraints are critical, such as aerospace and portable electronics.

This book chapter provides a brief exploration of the shielding mechanisms of MOF-based composites. It summarizes recent advancements in incorporating MOFs with structural components that enhance the integrity of MOF-based composites, along with conductivity boosters, to develop EMI shields particularly those with absorption-dominant mechanisms. The chapter concludes by outlining potential strategies to further enhance MOF-based shielding composites.

Metal-organic frameworks, their composites, and their shielding mechanisms

EMI shielding is a critical aspect of modern technological development and is essential for protecting sensitive electronics and human health from the harmful effects of electromagnetic radiation (Bagotia et al., 2018; Dai et al., 2022; Zhou et al., 2020). The interaction of incident EM waves with shielding systems includes several main mechanisms, namely reflection, absorption, transmission, and multiple reflections. Reflection occurs due to impedance mismatch between a conductive material and free space when an electromagnetic wave reaches its surface (Kamkar et al., 2021; Saini & Arora, 2012). The impedance mismatch between the material and free space causes the waves to reflect, preventing them from penetrating the material. Metallic substrates with high electrical conductivity, like copper, aluminum, and silver, usually lead to a high degree of reflection (Zhao et al., 2021). While reflection can block EM waves effectively from penetrating the shield's body, it can result in secondary EM pollution. This secondary pollution occurs when the reflected waves interfere with other nearby electronic devices, creating additional EM noise that can be as harmful as the incident waves. Therefore, a careful structural design is required to manage the reflection mechanism and avoid generating secondary EM wave pollution (Zhou et al., 2022).

Absorption is a key mechanism in EMI shielding, where the EM energy is dissipated within the shield. This process involves converting the EM energy into heat, effectively attenuating the interference. Absorption-dominant structures are particularly crucial because they efficiently dissipate absorbed electromagnetic energy, preventing it from being reflected and potentially contributing to secondary electromagnetic pollution (Gebrekrstos et al., 2022; Kamkar et al., 2021; Li et al., 2024). The EM absorption character of a material is chiefly influenced by the electrical, dielectric and magnetic properties, which determine how well the material can absorb and dissipate EM waves. Accordingly, materials with high electrical conductivity, dielectric permittivity and magnetic permeability are often more effective at absorption, making them appropriate candidates for EMI shielding applications.

In the absorption mechanism, several mechanisms play critical roles in MOF-based shields: dielectric, conduction, and magnetic losses (Panahi-Sarmad et al., 2023; Xu, Li, et al., 2023). Dielectric losses are key phenomena that dissipate EM waves through electric dipoles, leading to polarization loss. In response to an alternating EM field, charged particles tend to align or reorient themselves, creating electric dipoles. These reorientations dissipate EM waves through polarization loss. Ohmic losses generally involve two primary mechanisms: electron migration and electron hopping, both of which contribute to Ohmic (conduction) loss, serving as an additional means to effectively dissipate electromagnetic wave energy (Isari et al., 2024). Magnetic losses arise from the interaction between EM waves and the magnetic domains within the shield's body, which leads to the dissipation of EM wave energy into thermal energy (Wang et al., 2023). This mechanism is typically associated with phenomena such as magnetic hysteresis, magnetic resonance, and eddy currents. Magnetic resonance, occurring at specific frequencies, enhances the dissipation of energy, making materials with high magnetic permeability particularly effective for EMI shielding (Xu et al., 2024). Eddy currents, which are induced loops of electrical current within the material, contribute to energy dissipation through resistive losses, thereby reducing the intensity of the EM wave (Ismail & Azis, 2024; Wei et al., 2023).

Transmission refers to the portion of EM waves that pass through the shield's body (Srivastava & Manna, 2022; Verma et al., 2023). Effective EMI shielding aims to minimize transmission by maximizing absorption within the material. The goal is to attenuate the EM waves as much as possible before they can pass through the shield (Li et al., 2024). Hence, achieving this requires a careful balance between absorption and reflection mechanisms to ensure that the shielding material effectively blocks and dissipates the EM waves (Jia et al., 2022). Multiple reflections involve the back-and-forth reflection of EM waves within a shield. When EM waves penetrate the front interface of a thin shielding system and reach the opposite interface, they may undergo re-reflection or transmission. This repeated reflection between the surface and back interfaces gradually dissipates the energy of the EM waves. Another mechanism, known as internal scattering, is characteristic of porous shielding systems. It enables EM waves to reflect repeatedly within the internal porosities of the shield, further facilitating the dissipation of EM energy (Isari et al., 2024).

MOFs and their carbonized forms can enhance magnetic losses. For instance, the carbonization of MOFs that contain metal ions such as Fe, Co, or Ni can lead to the formation of magnetic nanoparticles. These nanoparticles enhance the shield's ability to dissipate EM waves by increasing magnetic losses, making them highly effective in applications where magnetic loss mechanisms are dominant (Thi et al., 2024). Conduction losses, on the other hand, occur when EM waves induce currents within a conductive material. As these currents flow, they encounter resistance, which converts part of the EM energy into thermal energy, effectively reducing the strength of the wave. Materials with high electrical conductivity, such as those derived from carbonized MOFs, are particularly effective at enhancing conduction losses. The porous structure of MOFs, coupled with their large surface area, enables extensive interaction with EM waves, promoting the conduction of induced currents and maximizing energy dissipation (Panahi-Sarmad et al., 2023; Zhang, Tian, et al., 2023).

MOFs are uniquely positioned to optimize both magnetic and conduction losses due to their highly customizable structures. By selecting appropriate metal ions and organic ligands and through post-synthetic modifications such as hybridization or carbonization, MOFs can be tailored to exhibit the desired balance of magnetic and conductive properties. This flexibility allows for the creation of EMI shielding structures that are not only highly effective but also tailored to specific operational frequencies and environments, offering superior performance compared to conventional materials (Shu et al., 2021; Xu, Li, et al., 2023).

When MOFs are incorporated into polymeric materials or hybrid composites, their EMI shielding performance is notably enhanced. The polymer matrix acts as a supportive framework, not only reinforcing the overall structure but also ensuring uniform dispersion of the MOFs throughout the composite. This even distribution is key to effective EMI shielding because it allows the MOFs to interact with EM waves across the entire material (Yao et al., 2015; Zhang, Zheng, et al., 2023). The polymer also provides flexibility, making the composite more adaptable for various applications, such as wearable technology and aerospace components. In these composites, the polymer's insulating properties, combined with the magnetic and/or conductive characteristics of the MOFs, create versatile materials with enhanced electromagnetic shielding performance (Zhu et al., 2022).

In MOF-based composites, the shielding mechanism is a blend of the inherent properties of the MOFs and the advantages offered by the composite material. MOFs contribute potential magnetic and conductive properties that aid in absorbing and dissipating EM waves, while the composite material introduces additional pathways for wave dissipation, including dielectric loss and internal scattering within the material's microstructure (Fei et al., 2019). When MOFs are carbonized in these composites, their conductive and magnetic properties are further enhanced, allowing for better EM wave absorption and reflection than traditional shielding materials (Li et al., 2022). This combination makes MOF composites highly effective, allowing customizable solutions for EMI shielding in a range of environments, offering superior performance tailored to specific frequencies and conditions.

Measuring the shielding effectiveness of a material requires a standardized approach. To this end, the vector network analyzer (VNA) method is a widely employed approach for assessing the EMI shielding performance of the materials. The recorded S-parameters (S_{11}, S_{22}, S_{12}, and S_{21}) from a VNA are the main drivers of the VNA method, and S_{ab} denotes transmission from port b to port a (Fig. 7.1). Note that the homogeneity of the shield results in S_{11}/S_{22} and S_{12}/S_{21} parameter ratios approaching 1. The S-parameters can be used to determine the EMI shielding coefficients (absorbance [A], transmittance [T], and reflectance [R]) based on the transmission-line theory approach. The R coefficient can be calculated as the ratio of the reflected EM wave power (P_R) to the power of the incident EM wave (P_I), as shown in Eq. (7.1). Similarly, the T coefficient is determined by the ratio of the transmitted wave power (P_T) to P_I. (7.3) provides the formula for calculating absorbance based on reflectance and transmittance (Iqbal et al., 2020; Kondawar & Modak, 2020; Wang, Ma, et al., 2021).

$$R = \frac{P_R}{P_I} = |S_{11}|^2 = |S_{22}|^2 \tag{7.1}$$

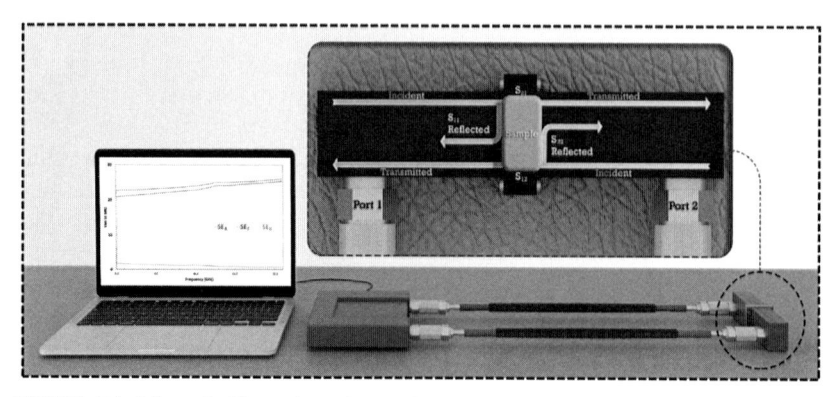

FIGURE 7.1 Schematic illustration of scattering parameters in the vector network analyzer method (Isari, Ghaffarkhah, Hashemi, Wuttke, et al., 2024).

$$T = \frac{P_T}{P_I} = \mid S_{12} \mid^2 = \mid S_{21} \mid^2 \tag{7.2}$$

$$1 = A + R + T \tag{7.3}$$

EMI shielding effectiveness (SE) is a parameter that can quantify a shield's ability to dissipate or reflect electromagnetic radiation. The higher a material's SE, the more efficiently it blocks EMI and reduces the transmission of EM waves. Total shielding effectiveness (SE_T) measures how well a shield can weaken or block incident EM waves. SE_T is defined as the logarithmic ratio of incident wave power to transmitted wave power, as expressed in the following equation (Wang, Tang, et al., 2021):

$$SE_T = 10 \log \frac{P_I}{P_T} = 10 \log \frac{1}{T} = 10 \log \frac{1}{S_{21}^2} \tag{7.4}$$

where P_T and P_I correspond to the transmitted and incident power of EM waves, respectively. Note that the value of SE_T, as demonstrated in Eq. (7.4), can be easily computed using the experimental scattering parameters obtained through VNA measurements. As a result, it serves as a vital point of reference for comparing the shielding efficiency of various structures. Moreover, the weight of the shield is a crucial reference point in the design of shielding structures (Amini et al., 2023). Accordingly, specific shielding effectiveness (SSE) over thickness (SSE/t) is introduced to assess the total shielding effectiveness of a shield while considering density (g/cm^3) and thickness of a shield (cm). The unit of this parameter is dB cm^2/g and can be calculated as follows (Isari, Ghaffarkhah, Hashemi, Yousefian, et al., 2024):

$$SSE/t = \frac{SE_T}{\rho \cdot t} \tag{7.5}$$

where t and ρ are indexed to the thickness and density of the shield, respectively. A higher SSE/t is preferred in lightweight shielding applications.

Structural design is crucial in optimizing EMI shielding performance by enhancing absorption mechanisms while suppressing transmission and reflection. The development of composite materials with specific architectures, such as 3D hierarchical structures and multiphase components, can boost shielding capabilities. For instance, gradient structure designs in layered and porous materials can facilitate an "absorb-reflect-reabsorb" process, further improving the material's effectiveness (Li et al., 2023; Sun et al., 2023; Xu, Hou, et al., 2023). This process involves the EM waves being absorbed, reflected, and then reabsorbed within the material, resulting in excellent EMI shielding performance with low surface reflection. This multistage attenuation process is particularly effective for reducing the secondary pollution associated with reflection-dominant shielding.

MOFs are emerging as game-changers in the field of EMI shielding (Zhu et al., 2022). Known for their unique structure and versatility, these materials introduce a novel approach to creating effective shields against unwanted EM waves. MOFs are engineered using a multiscale approach starting from molecular chemistry and microscale assembly to macroscale manufacturing. This comprehensive strategy allows for precise control over the material's properties, enhancing its ability to absorb and dissipate EM waves effectively (Panahi-Sarmad et al., 2023).

One of the interesting features of MOFs is their relatively high porosity, which boosts their ability to interact with and absorb EM waves. This means that MOFs don't just block interference—they actively absorb and distribute it, enhancing their shielding effectiveness. Furthermore, MOFs can be tailored with various chemical groups, enabling the fine-tuning of their dielectric and magnetic properties for specific applications. What sets MOFs apart is their flexibility in design and synthesis. This adaptability means that they can be tailored to meet the unique needs of various electronic systems, offering a level of customization that traditional materials can't match. Incorporating MOFs into various EMI shielding architectures leads to smarter, more efficient solutions that better absorb EM waves in different scenarios and applications. This represents a significant leap forward in the field, making MOFs an exciting development for the future of EMI shielding constructs (Fei et al., 2019; Ma et al., 2024; Thi et al., 2021).

Metal-organic framework shielding composites

The unique character of MOFs in dissipating the EM waves through the combined effect of dielectric, conduction, and/or magnetic losses has made them promising materials for EMI shielding (Bi et al., 2021; Chen et al., 2022). Recent studies have concentrated on embedding MOFs into shielding systems alongside other materials, serving as conductivity agents or integrity enhancers to address their limitations. This approach has resulted in the development of highly effective EMI shielding building blocks. This section reviews the latest developments in MOF-based composites for EMI shielding, highlighting the mechanisms that contribute to their effectiveness. The initial requirement of a proper shielding system is sufficient electrical conductivity and magnetic characteristics. MOFs typically fall short in these areas, particularly in terms of conductivity, making them less suitable for EMI shielding applications in their powdery or neat form (Bi et al., 2021; Chen et al., 2022; Deng et al., 2020; Guo et al., 2023). Combining MOFs with other functional materials, like conductive polymers, carbonaceous materials, and metallic nanoparticles, is a practical approach to improving the effectiveness of MOF-based EMI shields. Known as "MOF composites", these materials benefit not only from the individual properties of their components but also from unique synergistic effects, such as enhanced interfaces and magnetic–dielectric coupling (Huang et al., 2023; Sun et al., 2018). Recently, hybridizing MOFs with conductive materials like graphene, intrinsically conductive polymers, carbon nanotubes

(CNTs), and MXene has gained attention. These combinations have led to improved EMI shielding, introducing a new category of MOFs to the shielding science. The hybridization of MOFs with the aforementioned materials enhances EMI shielding by improving attenuation parameters, resulting in optimized absorption-dominant shielding systems (Panahi-Sarmad et al., 2023; Xu, Li, et al., 2023). In the following subsections, several of these practices are exemplified.

Metal-organic framework composites with intrinsically conductive polymers

Intrinsically conductive polymers are considered potential candidates for robust shielding systems thanks to their promising characteristics, including proper electrical conductivity, lightweight, cost-effectiveness, corrosion resistance, and flexibility. Combining these conductive polymers with MOFs enables the manufacturing of absorption-dominant EMI shielding materials across a broad frequency range. Among the most commonly used conductive polymers, polyaniline (PANI) and polypyrrole (PPy) can be mentioned, offering adjustable conductivity through variations in synthesis parameters (Bi et al., 2021; Wang et al., 2017). When these polymers are compounded with MOFs, they improve the EM characteristics of the composite material. This enhancement enables the production of absorbers that maintain low filler content and reduced thickness while achieving a broad effective absorption bandwidth (EAB). Research studies highlighted the excellent EM wave attenuation capability of hybrid composites made of MOFs and conductive polymers. For instance, Bi et al. (2021) developed a cobalt–zinc MOF-based composite with 2-methylimidazole as the ligand, which was subsequently carbonized at high temperatures and sequentially decorated with MoS_2 and PPy to form CoZn/C@MoS_2@PPy composites. Paraffin wax was added to the nanoparticles, yielding composites with different mass ratios of the as-synthesized compound. This structure, at a 30 wt.% mass ratio, displayed a minimum reflection loss (RL_{min}) of −49.18 dB. The remarkable absorption capabilities of these composites can be attributed to a combination of several synergistic factors, such as the presence of Zn and Co particles within the carbon matrix, along with the integration of MoS_2 and PPy, which introduces numerous interfacial polarizations (Fig. 7.2). Additionally, embedding various metallic particles within the carbon matrix induces dipole polarization, further contributing to increased dielectric loss. The graphitized carbon matrix and highly conductive PPy also play a crucial role in amplifying the conductive losses.

In another attempt (Wang et al., 2017), a hybrid composite consisting of PANI-coated MOF/Fe was synthesized. The fabrication process followed a two-step approach: (1) the production of MOF (Fe) particles via a hydrothermal route and (2) coating PANI onto the surface of these particles through in situ chemical polymerization. This resulted in a core-shell structure, where

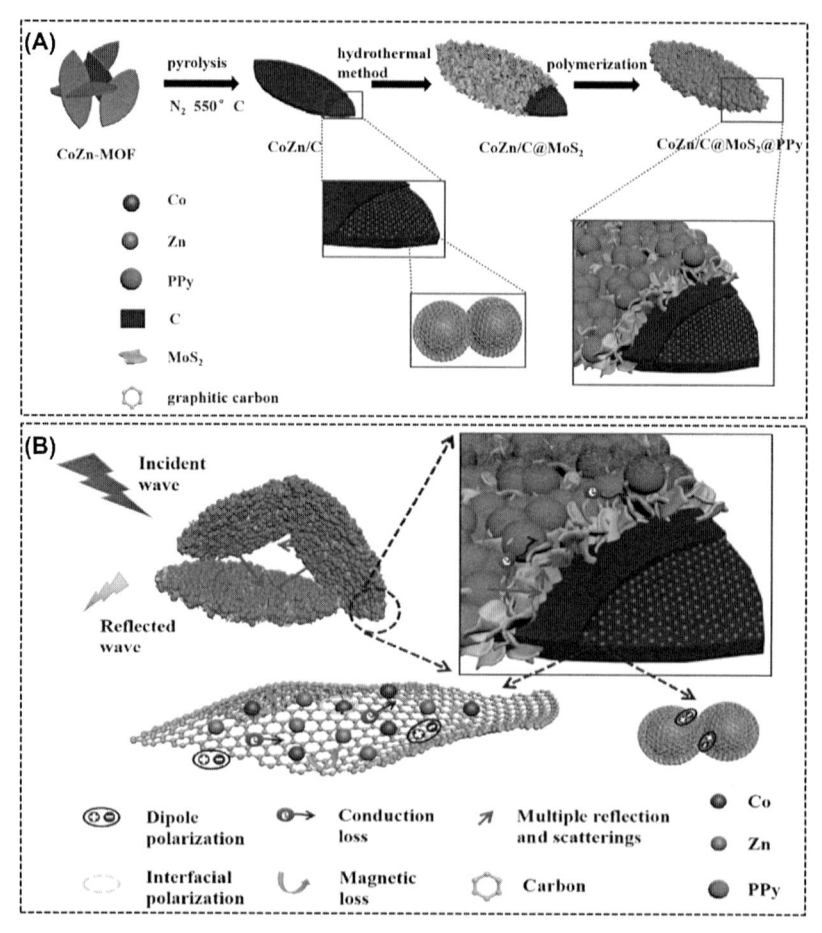

FIGURE 7.2 (A) Schematic illustration of CoZn/C@MoS$_2$@polypyrrole composite shield, and (B) its electromagnetic interference shielding mechanism (Bi et al., 2021).

the MOF particles were entirely encapsulated by PANI. Consequently, the composite exhibited improved microwave absorption properties compared to pure MOF (Fe). Specifically, the RL$_{max}$ was recorded at −41.4 dB (at 11.6 GHz), with an EAB below −10 dB spanning 5.5 GHz, covering a frequency range from 9.8 to 15.3 GHz. Remarkably, these absorption properties were achieved with a relatively thin material layer of just 2 mm. The unique structure (core-shell) facilitated increased interfacial polarization between the PANI and MOF (Fe), contributing to higher dielectric loss. Additionally, the porous nature of the MOF allowed for multiple internal reflections and scattering of incident EM waves, leading to more effective energy dissipation. The improved impedance matching of the resulting composite enabled greater penetration of microwaves into the material, reducing reflection and enhancing absorption. Furthermore, the conductive properties of PANI combined with the magnetic characteristics of the MOF (Fe) core contributed to both dielectric and magnetic losses, further improving the overall absorption efficiency (Wang et al., 2017). Similarly, a Co/C@PPy composite was synthesized (Sun et al., 2018) by growing PPy on ZIF-67 (cobalt-based zeolitic imidazolate frameworks) precursors using 2-methylimidazole as the ligand. In this regard, PPy was uniformly coated on the surface of the ZIF-67, improving the shielding properties of the resulting hybridized composite. The composite showed proper EM wave absorption properties with an RL_{min} of −44.76 dB at 17.32 GHz and an EAB of 6.56 GHz.

Huang et al. (2023) investigated the incorporation of MOFs-derived Bi_xSe_y@C sheets, which were synthesized through a solvothermal reaction followed by salinization pyrolysis treatment, into poly(3,4-ethylenedioxythiophene) polystyrene sulfonate (PEDOT:PSS) to create flexible films. The resulting Bi_xSe_y@C film, with an average size of approximately 7.32 μm, was dispersed in the PEDOT:PSS matrix. The composite films were manufactured using a vacuum filtration approach followed by a subsequent cold pressing, enabling free-standing composites with variable Bi_xSe_y@C content. The layered structure of these films contributed to their bi-functional capabilities. They found that when the nanoparticle content reached 70 wt.%, the resulting composite film exhibited an average SE_T value of 45.0 dB within the X-band frequency range (Huang et al., 2023). These examples highlighted the potential of MOF-conductive polymer composites in producing high-performance EMI shields with optimized thickness and density.

Metal-organic frameworks composites with MXenes

MXenes, a potent 2D material, have attracted considerable attention in shielding science due to their remarkable characteristics, including excellent water dispersibility, high electrical conductivity, customizable surface chemistry, ease of processing, and outstanding thermal stability (Guo et al., 2023). These properties make MXenes highly effective for EMI shielding. Nevertheless, their EMI shielding performance can be further enhanced by creating composites with other materials such as polymers, fibers, carbon derivatives, nanoferrite, and MOFs. Various studies have demonstrated the improved microwave absorption capabilities of MOF-MXene composites through different synthesis techniques and material combinations (Xu, Li, et al., 2023).

For instance, a CoFe-MIL-88A was synthesized, and a layer of MXene was decorated on the surface through electrostatic interactions (Chen et al., 2022). This composite was then carbonized at high temperatures, resulting in CoFe/C@TiO_2/C. Fig. 7.3(A) shows the morphological feature and fabrication process of the resulting shielding system. The resulting nanostructure achieved a RL_{min} of −20 dB with an absorption bandwidth of 6.1 GHz at a thickness of

2 mm. The sandwich-like structure of the composite facilitated multiple internal reflections of EM waves, while the numerous heterojunction surfaces enhanced interfacial polarization. Additionally, the presence of Fe and TiO_2 contributed to natural and exchange resonance, and the introduction of magnetic loss led to superior wave absorption performance (Chen et al., 2022). Fig. 7.3(B) showcases the shielding mechanism of the developed composites, proposing the combined effect of dielectric, conduction, and magnetic losses to dissipate the EM waves effectively.

To enhance impedance matching, researchers have been exploring ways to optimize porosity at various scales, allowing electromagnetic waves to better penetrate the shield and thereby promote absorption. For example, highly efficient EMI shielding material was developed by creating a hybrid carbon aerogel that integrated ZIF-67-derived Co/C nanoparticles and MXene into a cellulose-derived carbon matrix (Guo et al., 2023). The fabrication process involved a straightforward freeze–drying technique followed by thermal treatment, resulting in a hierarchically porous structure with numerous heterointerfaces. This hybrid aerogel demonstrated a remarkable EMI shielding effectiveness of 86.7 dB while maintaining a low density of just 85.6 mg/cm^3. The material's exceptional performance is attributed to the

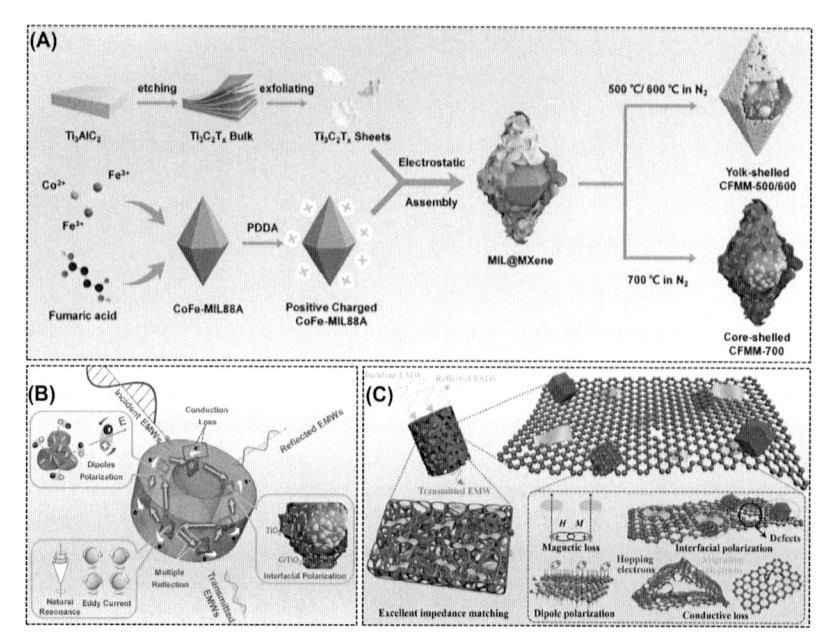

FIGURE 7.3 (A) Schematic demonstration of multiple-interface CoFe-MOF@Ti$_3$C$_2$T$_x$ MXene composite fabrication process (Chen et al., 2022) copyright? (B) Their electromagnetic waves attenuation mechanism (Chen et al., 2022) and (C) The shielding mechanism of hybrid ZIF-67/MXene/cellulose aerogel after carbonization (Guo et al., 2023).

synergistic effects of Ohmic and magnetic loss mechanisms, along with the presence of multiple heterointerfaces that enhance interfacial polarization losses. The microscale porosities of the aerogel also promote the multiple internal scattering, further dissipating the intruding EM waves. These features collectively contributed to the absorption-dominant shielding mechanism of the aerogel, where a portion of the incident waves is absorbed (Fig. 7.3C). The Co/C nanoparticles derived from ZIF-67 introduce magnetic loss into the system, while the MXene enhances conductive loss, both of which are critical in dissipating EM waves and enhancing the overall shielding effectiveness (Guo et al., 2023).

In another study, a Co/ZnO/$Ti_3C_2T_x$ (MOF/MXene) composite was prepared (Kong et al., 2020), and the impact of carbonization temperature on their EMI shielding properties was explored. The composite exhibited an RL_{min} of −44.22 dB. The perfect EMI shielding performance was attributed to multiple factors, including the unique structure that promoted microwave scattering and multiple internal reflections and dielectric and magnetic losses. The incorporation of ZnO nanoparticles enhanced conductive losses, while magnetic Co nanoparticles improved magnetic loss.

Further expanding on the multifunctional capabilities of MXene composites, Wu et al. (2021) fabricated MXene/CoNi/CNT composites at high temperatures using terephthalic acid. The resulting composite achieved an RL_{min} of −51.6 dB at 15.1 GHz with a thickness of 1.6 mm and an EAB of 4.5 GHz. The hybrid structure's excellent absorption properties were attributed to a combination of polarization loss, interlacing magnetic fields, and both conductive and magnetic losses. In another attempt, Deng et al. (2020) fabricated 1D heterostructures using MXene and CoNi-bimetal MOF, leading to the creation of 3D network structures intertwined with CNTs. This hybrid, with a matching thickness of 1.6 mm, achieved an RL_{min} of over −51.8 dB and an EAB of 6.5 GHz. The enhanced microwave absorption was attributed to the synergy between MXene, MOF, and CNTs, which resulted in improved impedance matching, promoted multiple internal reflections, and facilitated interfacial polarization. Moreover, Han et al., 2020 fabricated MXene nanosheet/MOF composites, specifically utilizing Co-MOF and Ni-MOF. The MXene@Co-MOF composite had an RL_{min} of −60.09 dB, while the MXene@Ni-MOF composite demonstrated an RL_{min} of -64.11 dB. These composites benefited from an accordion-like structure that disrupted current paths and improved charge accumulation, leading to enhanced microwave absorption properties (Han et al., 2020). Overall, these studies collectively illustrate the immense potential of MXene-based composites, particularly when combined with MOFs, for advanced EMI shielding applications. The synergy between MXene and other materials, coupled with structural innovations, enables impedance matching and effective EM wave attenuation across a broad range of frequencies.

Metal-organic framework-based carbonaceous composites

Conductive carbonaceous materials, such as graphene, carbon nanofibers (CNFs), and CNTs, are widely recognized as ideal templates for constructing lightweight microwave-absorbing materials due to their tunable surface chemistry, high conductivity, and large specific surface area. These materials also possess a large number of functional groups, such as hydroxyl and carboxyl on their surfaces, which can effectively absorb metal ions through electrostatic interactions, facilitating the in situ MOF formation. One notable study developed a MOF-derived composite upon the integration of reduced graphene oxide (rGO) and γ-Fe_2O_3 nanoparticles (Lin et al., 2023). The resulting composite was prepared through freeze–drying and subsequent thermal annealing to generate a potent shielding system with a hierarchical arrangement. The γ-Fe_2O_3 provided magnetic loss, while the rGO offered excellent electrical conductivity and dielectric loss, contributing to the overall EMI shielding effectiveness. At a low filler loading of 5 wt.% and density of 2.65 mg/cm^3, the composite achieved an RL_{min} of –60.5 dB and an EAB of 7.76 GHz, covering a broad frequency range. This demonstrates the potential of MOF-carbon composites to provide lightweight and efficient EMI shielding in advanced applications (Lin et al., 2023).

Ni-doped Fe_3O_4@C/rGO was found to exhibit good EMI shielding performance at low thicknesses (Huang et al., 2022). The composite's manufacturing was performed upon direct mixing of precursors, their gelation, and further freeze–drying to yield free-standing aerogels (Fig. 7.4A). The Fe_3O_4@C/rGO aerogel achieved an RL_{min} of –58.1 dB at 15.4 GHz with an EAB of 6.48 GHz at a thickness of 2.5 mm, while the Ni-doped version reached an RL_{min} of –46.2 dB at 14 GHz and an EAB of 7.92 GHz at 2.8 mm thickness. These results were obtained with ultralow filler contents, highlighting the efficiency of these composites in EMI shielding applications. These studies attribute the EMI shielding effectiveness to the synergistic effects of the hierarchically porous structure and heterointerface engineering. The 3D porous architecture of the aerogels allows for multiple reflections and scatterings of EM waves, enhancing impedance matching and wave absorption. Additionally, the combination of dielectric and magnetic losses from the MOF-derived components and the rGO matrix contributes to the high-performance EMI shielding observed in these materials (Fig. 7.4B) (Huang et al., 2022).

In another approach, Zhang et al. (2017) successfully synthesized Co/C-rGO composites. This composite exhibited a RL_{min} of –52 dB at 9.6 GHz and an EAB of 7.72 GHz. Similarly, Xu et al. (2020) developed a pomegranate-shaped CoNi@NC/rGO-600 nanocomposite by using CoNi-BTC/rGO as precursors. The resulting material showed outstanding microwave absorption performance, with a broad EAB of 6.7 GHz and an RL_{min} of –68.0 dB. In this comparative study, CoNi@NC-600, derived from CoNi-BTC without rGO,

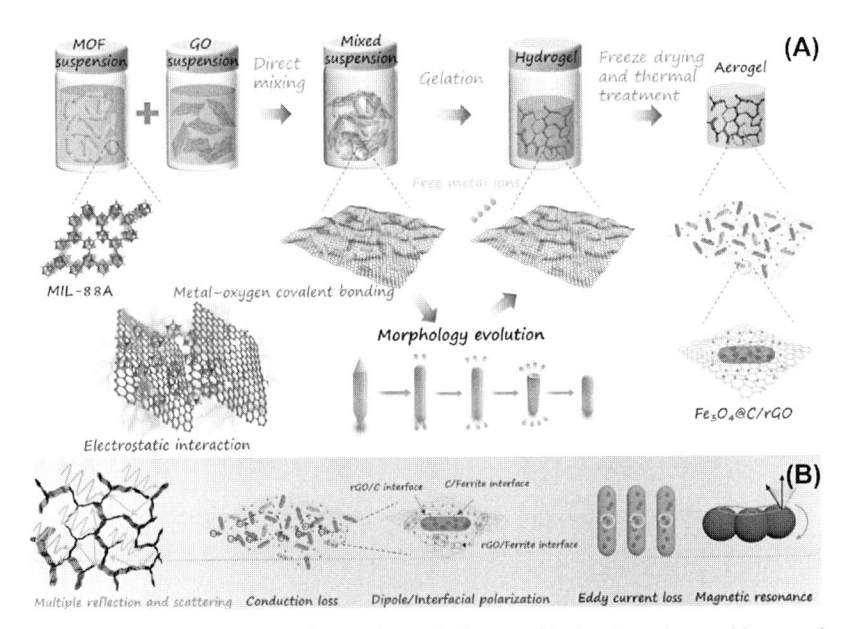

FIGURE 7.4 (A) A depiction of the metal-organic framework/reduced graphene oxide aerogels fabrication process, utilizing MOFs to trigger the gelation of graphene oxide directly and (B) Electromagnetic wave dissipation mechanism of the developed aerogel (Huang et al., 2022).

demonstrated inferior microwave absorption properties, highlighting the importance of rGO in providing conductive networks after thermal treatment.

Wang et al. (2020) investigated the hybrid effect of carbonaceous materials such as rGO and CNT along with MOF in developing the rGO/CNT-MOF nanocomposite shielding system. The resulting nanocomposites were prepared upon growing ZnCo-MOF on GO and CNT carriers, followed by adding Fe^{3+} into the pores of ZnCo-MOF. The rGO-CoFe@C composite achieved an RL_{min} of −36.08 dB and an EAB of 5.17 GHz. However, the CNT-CoFe@C composite showed superior performance, with an RL_{min} of −40 dB and an EAB of 5.62 GHz. The enhanced performance of CNT-CoFe@C was attributed to the high aspect ratio of CNTs and the formation of large π bonds from conjugated carbon atoms, which facilitated the creation of a three-dimensional (3D) conductive network (Wang et al., 2020).

3D materials have great potential for EMI shielding because of their microscale porosity and interconnected conductive networks. Song et al. (2021) developed ZnO/C@PG and ZnO/ZnFe$_2$O$_4$/C@PG composites by first growing MOF-5 and Fe(III)-MOF-5 on a 3D porous graphene (PG) framework. These were then subjected to high-temperature treatment. The resulting pyrolyzed materials retained a hollow structure on the nanoscale, which extended the EM waves' propagation path. The ZnO/ZnFe$_2$O$_4$/C@PG composite demonstrated greater magnetic loss compared to ZnO/C@PG, thanks to the incorporation of Fe(III). This resulted in an RL_{min} of –54.6 dB at 9.04 GHz and an effective absorption bandwidth of 5.36 GHz (Song et al., 2021).

These studies collectively illustrate the significant potential of MOF-carbon composites for EMI shielding. By combining the magnetic properties of MOF-derived particles with the conductive and dielectric characteristics of carbon materials, these composites offer a robust solution for attenuating EM waves across a wide frequency range. The synergy between the components, facilitated by advanced fabrication techniques, enables the design of lightweight, high-performance shielding materials suitable for various technological applications.

Porous metal-organic framework-based shielding constructs

In industries like aviation and aerospace, where the weight of constructs is a critical factor, the design of lightweight structures is essential. This has led to increased interest in porous materials such as foams, sponges, and aerogels, which meet the weight requirements while enhancing EMI shielding capabilities. The porous design increases the internal surface area, which helps absorb EM waves. A key challenge in designing porous structures for EMI shielding is achieving the right level of electrical conductivity. The material must have proper conductivity to allow EM waves to penetrate the shield, preventing them from reflecting back and causing additional interference. Once inside, the waves interact with multiple surfaces, undergoing scattering and energy dissipation as heat, effectively minimizing their intensity. However, if the material is too conductive, it can reflect too many waves at the surface, diminishing its ability to absorb them. Therefore, it is crucial to carefully control the conductivity to ensure the material is more effective at absorbing rather than reflecting EM waves (Isari et al., 2024).

Incorporating MOFs into porous structures is essential for enhancing their effectiveness in EMI shielding applications. While MOFs in their powder form have shown high EM wave absorbance potential, they often need to be processed into various constructs to be effective in practical applications. This transformation allows MOFs to better leverage their large surface area and tunable properties, making them more suitable for advanced applications. By creating porous structures, the material's ability to attenuate EM waves is greatly improved, leading to more efficient and effective EMI shielding (Panahi-Sarmad et al., 2023).

Processing MOFs into porous constructs is especially advantageous as it transforms them from simple powders into structured, functional materials with broader application potential. This approach not only improves the materials' conductivity and structural integrity but also enhances their suitability for EMI shielding in demanding environments. The ability to tailor the porosity, conductivity, and overall structure of these materials ensures they can meet the specific requirements of various applications, making them versatile and effective tools in combating electromagnetic interference (Panahi-Sarmad et al., 2023; Xu, Li, et al., 2023).

Yang Fei and colleagues developed lightweight Co/C@CNF aerogels and assessed their performance for EMI shielding applications (Fei et al., 2020). By utilizing a freeze-casting method followed by high-temperature carbonization, they developed a 3D interconnected network structure that exhibited high EMI shielding effectiveness. Specifically, the aerogel carbonized at 900 °C achieved a SE_T of 35.1 dB with a low density of 1.74 mg/cm^3, making it highly suitable for applications where weight and performance are critical. The superior EMI shielding performance of the Co/C@CNF aerogels is attributed to their unique structure, which effectively combines dielectric and magnetic loss mechanisms. The presence of Co/C nanoparticles within the carbonized framework enhances both dielectric and magnetic losses, resulting in an absorption-dominant EMI shielding mechanism (A > 0.5).

Lei and coworkers synthesized rGO/MXene/FeCoC nanocomposite aerogels and assessed their EMI shielding performance (Lei et al., 2024). The synthesis process involved coprecipitation, freeze–drying, and pyrolysis techniques, which were employed to construct a 3D porous structure comprising rGO, MXene, and FeCoC nanoparticles (Fig. 7.5A). The rGO/MXene/FeCoC aerogels achieved an outstanding RL_{min} of −61.4 dB at a matching thickness of 1.55 mm, with an EAB of 4.95 GHz. This remarkable EMI shielding performance is attributed to several key mechanisms, including the synergistic dielectric–magnetic losses, enhanced impedance matching, and the rich internal scattering of EM waves within the porous structure of aerogels. The MOF-derived FeCoC nanoparticles played a crucial role in improving the magnetic loss mechanism, contributing to the overall absorption-dominant shielding characteristic of the aerogel, as shown in Fig. 7.5(B).

Shao et al. synthesized graphene/polyimide/Co–N–C aerogels to study their EM wave absorption capabilities (Shao et al., 2024). The synthesis process for the production of these aerogels involved combining MOF-derived Co–N–C groups with GO and polyamic acid (PAA). This combination was processed through freeze–drying and thermal imidization, resulting in a lightweight, porous aerogel with robust mechanical properties. The PI/rGO/CNC aerogel demonstrated excellent EMI shielding performance, achieving an RL_{min} of −75 dB and a maximum EAB of 7.28 GHz. This high level of performance was largely due to the aerogel's multiple interfaces and the synergistic EM losses facilitated by the combined dielectric and magnetic components. The inclusion of MOF-derived Co–N–C species played a critical role in enhancing impedance matching and promoting both dielectric losses and magnetic losses. Fig. 7.6(A) illustrates the comprehensive EM wave absorption mechanism, indicating how the aerogel's structure supports various loss mechanisms, including conduction loss, magnetic loss, interfacial polarizations, and dipole polarizations.

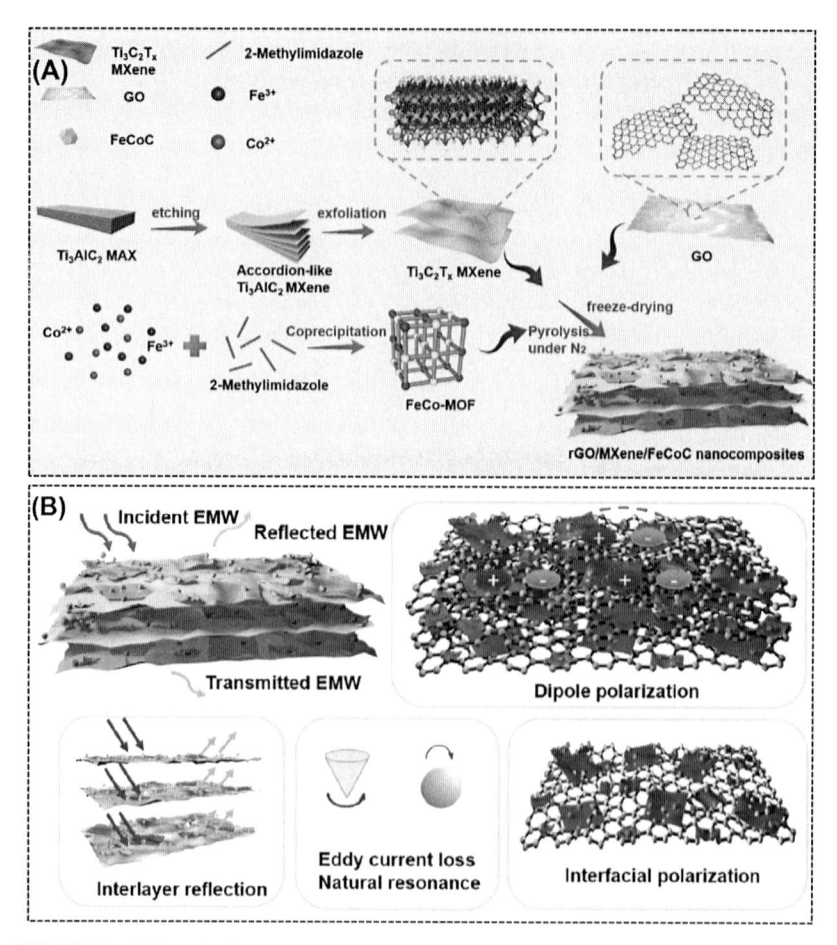

268 Applications of Metal-Organic Framework Composites

FIGURE 7.5 Schematic illustration of (A) reduced graphene oxide/MXene/FeCoC preparation process and their and (B) Electromagnetic wave absorption mechanisms (Lei et al., 2024).

It's important to focus on the multifunctionality of MOF-based porous constructs, as these materials can offer more than just a single function, like EMI shielding. By developing multifunctional materials, we can address multiple challenges in advanced technologies simultaneously, leading to more efficient and compact solutions. For example, a study by Sridhar et al. explored the use of MOF-derived Fe–N–C nanostructures for both sodium-ion batteries and EMI shielding applications (Sridhar et al., 2021). In this work, they developed a rapid microwave synthesis method to create 3D carbon nanostructures (3DCNS) from MOF precursors. These nanostructures, which include N-doped carbon nanotubes (N-CNTs) decorated on rGO, demonstrated potential in EMI shielding applications. The obtained experimental results indicated that the synthesized

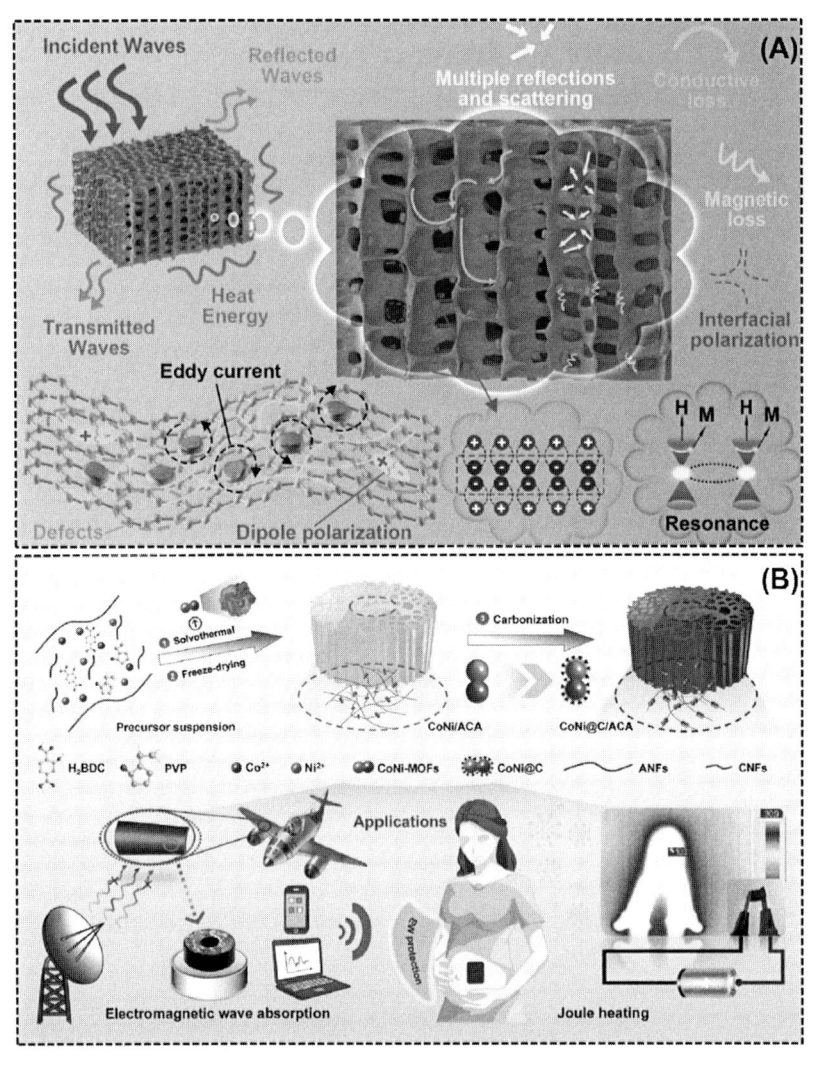

MOF composites for electromagnetic interference shielding **Chapter | 7** **269**

FIGURE 7.6 (A) Schematic illustration of microwave absorption mechanism of PI/rGO/CNC nanocomposite aerogels (Shao et al., 2024) and (B) Schematic illustration of CoNi@C/ACA aerogel preparation process and multifunctional applications (He et al., 2023).

MDCNT@CF (MOF-derived carbon nanotubes on carbon fibers) exhibited high microwave absorption performance across a frequency domain of 2–18 GHz. This high performance is attributed to the material's high porosity and extensive surface area, which improve the dissipation of EM waves.

Lightweight hybrid carbonaceous aerogels made of MOFs were manufactured for simultaneous EMI shielding and heat management applications

(He et al., 2023). In this regard, to manufacture the aerogels, bimetallic CoNi-MOFs were combined with aramid nanofibers (ANFs) and CNFs. Next, a freeze–drying process followed by carbonization was employed to form highly porous, lightweight constructs (Fig. 7.6B). This type of structure was specifically designed to maximize the scattering of EM waves within the material, enhancing its effectiveness at absorbing and dissipating EM waves. The CoNi@C/ACA aerogel demonstrated outstanding results in EMI shielding, with an RL_{min} of –66.57 dB and a wide absorption bandwidth of 6.3 GHz, even at a very low MOF loading of 1.8 wt.%. These results were achieved thanks to the material's capacity to balance and enhance both dielectric and magnetic losses effectively. This was made possible by the meticulously designed porous structure and the integration of CoNi@C particles.

Metal-organic framework composite challenges and future prospects

The increasing prevalence of EM waves in the environment, driven by the widespread use of telecommunication systems and advanced technological products, has posed a modern challenge: EM pollution. This form of pollution not only threatens the proper functioning of sensitive electronics but also jeopardizes the health of living organisms. Consequently, there is a growing need for the development of absorption-dominant shielding systems that can effectively attenuate EM wave energy while minimizing surface reflections caused by impedance mismatching.

MOF-based shielding composites have emerged as promising candidates for creating such absorption-dominant shields. Despite the inadequate electrical conductivity of MOFs, their carbonized forms or hybridization with other conductive nanomaterials have proven highly effective in enhancing EM wave dissipation through mechanisms such as conduction, dielectric, and magnetic losses. Furthermore, embedding MOFs in porous structures like aerogels can lead to multiple internal scatterings, which, in synergy with other EM wave attenuation phenomena, boost EM wave dissipation.

This chapter explored recent advancements in the fabrication of MOF-based shielding composites and porous aerogel compositions, providing a roadmap for future developments. The following prospects are proposed for further studies and the advancement of MOF-based shielding systems:

Intrinsically conductive MOF-based composite shields: A major limitation of MOFs in shielding applications is their poor conductivity, necessitating the enhancement of the conductive domain through carbonization. This challenge underscores the need to develop intrinsically conductive MOFs with improved electrical conductivity and EM wave dissipation properties.

Structural reinforcement: MOFs typically lack the structural integrity to form macroscopic constructs with custom sizes and features using neat powder. To overcome this, it is recommended that MOFs be hybridized with other materials that enhance both integrity and conductivity. This would enable the creation of mechanically robust MOF-based shielding systems with superior EM wave attenuation capabilities and minimized impedance mismatching. The structure could take the form of a hybrid composite or a porous building block, such as a cryogel or aerogel, which benefits from numerous interfaces and enhances internal scattering in conjunction with other EM wave dissipation factors.

Porous macroscopic MOF-based building blocks: Given the high potential of MOFs in dissipating EM waves, it is advisable to incorporate them into porous building blocks through scalable manufacturing methods. This could involve direct mixing of MOFs with precursors or in situ nucleation and growth of MOFs on supporting porous building blocks.

Impedance matching agent: MOF species can be considered potential candidates for EM wave attenuation, facilitating impedance matching with free space and minimizing surface reflections. This function could be utilized in multilayered structures, combining conductive and nonconductive layers to create robust EM wave absorption shielding systems.

References

Amini, M., Hosseini, H., Dutta, S., Wuttke, S., Kamkar, M., & Arjmand, M. (2023). Surfactant-mediated highly conductive cellulosic inks for high-resolution 3D printing of robust and structured electromagnetic interference shielding aerogels. *ACS Applied Materials & Interfaces, 15*(47), 54753–54765.

Bagotia, N., Choudhary, V., & Sharma, D. (2018). A review on the mechanical, electrical and EMI shielding properties of carbon nanotubes and graphene reinforced polycarbonate nanocomposites. *Polymers for Advanced Technologies, 29*(6), 1547–1567.

Bi, Y., Ma, M., Liu, Y., Tong, Z., Wang, R., Chung, K. L., Ma, A., Wu, G., Ma, Y., & He, C. (2021). Microwave absorption enhancement of 2-dimensional CoZn/C@MoS$_2$@PPy composites derived from metal-organic framework. *Journal of Colloid and Interface Science, 600*, 209–218.

Chen, F., Zhang, S., Ma, B., Xiong, Y., Luo, H., Cheng, Y., Li, X., Wang, X., & Gong, R. (2022). Bimetallic CoFe-MOF@ Ti$_3$C$_2$T$_x$ MXene derived composites for broadband microwave absorption. *Chemical Engineering Journal, 431*, 134007.

Dai, Y., Wu, X., Li, L., Zhang, Y., Deng, Z., Yu, Z. Z., & Zhang, H. B. (2022). 3D printing of resilient, lightweight and conductive MXene/reduced graphene oxide architectures for broadband electromagnetic interference shielding. *Journal of Materials Chemistry A, 10*(21), 11375–11385.

Deng, B., Xiang, Z., Xiong, J., Liu, Z., Yu, L., & Lu, W. (2020). Sandwich-like Fe&TiO 2@C nanocomposites derived from MXene/Fe-MOFs hybrids for electromagnetic absorption. *Nano-Micro Letters, 12*, 1–16.

Fei, Y., Liang, M., Chen, Y., & Zou, H. (2019). Sandwich-like magnetic graphene papers prepared with MOF-derived Fe$_3$O$_4$–C for absorption-dominated electromagnetic interference shielding. *Industrial & Engineering Chemistry Research, 59*(1), 154–165.

Fei, Y., Liang, M., Yan, L., Chen, Y., & Zou, H. (2020). Co/C@cellulose nanofiber aerogel derived from metal-organic frameworks for highly efficient electromagnetic interference shielding. *Chemical Engineering Journal, 392*, 124815.

Gebrekrstos, A., Orasugh, J. T., Muzata, T. S., & Ray, S. S. (2022). Cellulose-based sustainable composites: A review of systems for applications in EMI shielding and sensors. *Macromolecular Materials and Engineering, 307*(9), 2200185.

Guan, B. Y., Yu, X. Y., Wu, H. B., & Lou, X. W. (2017). Complex nanostructures from materials based on metal–organic frameworks for electrochemical energy storage and conversion. *Advanced Materials, 29*(47), 1703614.

Guo, Z., Ren, P., Yang, F., Wu, T., Zhang, L., Chen, Z., Huang, S., & Ren, F. (2023). MOF-derived Co/C and MXene co-decorated cellulose-derived hybrid carbon aerogel with a multi-interface architecture toward absorption-dominated ultra-efficient electromagnetic interference shielding. *ACS Applied Materials & Interfaces, 15*(5), 7308–7318.

Han, X., Huang, Y., Ding, L., Song, Y., Li, T., & Liu, P. (2020). $Ti_3C_2T_x$ MXene nanosheet/metal–organic framework composites for microwave absorption. *ACS Applied Nano Materials, 4*(1), 691–701.

Hashemi, S. A., Ghaffarkhah, A., Ahmadijokani, F., Yousefian, H., Mhatre, S. E., Sinelshchikova, A., Banvillet, G., Kamkar, M., Rojas, O. J., Wuttke, S., & Arjmand, M. (2024). Liquid-templated graphene aerogel electromagnetic traps. *Nanoscale, 16*(18), 8858–8867.

Hashemi, S. A., Ghaffarkhah, A., Hosseini, E., Bahrani, S., Najmi, P., Omidifar, N., Mousavi, S. M., Amini, M., Ghaedi, M., Ramakrishna, S., & Arjmand, M. (2022). Recent progress on hybrid fibrous electromagnetic shields: Key protectors of living species against electromagnetic radiation. *Matter, 5*(11), 3807–3868.

He, W., Zheng, J., Dong, W., Jiang, S., Lou, G., Zhang, L., Du, W., Li, Z., Li, X., & Chen, Y. (2023). Efficient electromagnetic wave absorption and Joule heating via ultra-light carbon composite aerogels derived from bimetal-organic frameworks. *Chemical Engineering Journal, 459*, 141677.

Huang, J., Qin, J., Meng, Q., Wang, L., Du, Y., & Shen, S. Z. (2023). Flexible free-standing BixSey@C/PEDOT:PSS thermoelectric composite film with high-performance electromagnetic interference shielding. *Applied Surface Science, 639*, 158162.

Huang, X., Wei, J., Zhang, Y., Qian, B., Jia, Q., Liu, J., Zhao, X., & Shao, G. (2022). Ultralight magnetic and dielectric aerogels achieved by metal–organic framework initiated gelation of graphene oxide for enhanced microwave absorption. *Nano-Micro Letters, 14*(1), 107.

Iqbal, A., Sambyal, P., & Koo, C. M. (2020). 2D MXenes for electromagnetic shielding: A review. *Advanced Functional Materials, 30*(47), 2000883.

Isari, A. A., Ghaffarkhah, A., Hashemi, S. A., Wuttke, S., & Arjmand, M. (2024). Structural design for EMI shielding: From underlying mechanisms to common pitfalls. *Advanced Materials, 36*(24), 2310683.

Isari, A. A., Ghaffarkhah, A., Hashemi, S. A., Yousefian, H., Rojas, O. J., & Arjmand, M. (2024). A journey from structured emulsion templates to multifunctional aerogels. *Advanced Functional Materials, 34*(44), 2402365.

Ismail, I., & Azis, R. S. (2024). A review of magnetic nanocomposites for EMI shielding: Synthesis, properties, and mechanisms. *Journal of Materials Science, 59*(13), 5293–5329.

Jia, X., Li, Y., Shen, B., & Zheng, W. (2022). Evaluation, fabrication and dynamic performance regulation of green EMI-shielding materials with low reflectivity: A review. *Composites Part B: Engineering, 233*, 109652.

Jiao, L., Seow, J. Y. R., Skinner, W. S., Wang, Z. U., & Jiang, H. L. (2019). Metal–organic frameworks: Structures and functional applications. *Materials Today, 27*, 43–68.

Kamkar, M., Ghaffarkhah, A., Hosseini, E., Amini, M., Ghaderi, S., & Arjmand, M. (2021). Multilayer polymeric nanocomposites for electromagnetic interference shielding: Fabrication, mechanisms, and prospects. *New Journal of Chemistry, 45*(46), 21488–21507.

Kondawar, S. B., & Modak, P. R. (2020). Theory of EMI shielding. *Materials for Potential EMI Shielding Applications, 9*–25.

Kong, M., Jia, Z., Wang, B., Dou, J., Liu, X., Dong, Y., Xu, B., & Wu, G. (2020). Construction of metal-organic framework derived Co/ZnO/Ti$_3$C$_2$T$_x$ composites for excellent microwave absorption. *Sustainable Materials and Technologies, 26*, e00219.

Konnerth, H., Matsagar, B. M., Chen, S. S., Prechtl, M. H., Shieh, F. K., & Wu, K. C. W. (2020). Metal-organic framework (MOF)-derived catalysts for fine chemical production. *Coordination Chemistry Reviews, 416*, 213319.

Lei, D., Liu, C., Dong, C., Wang, S., Zhang, P., Li, Y., Liu, J., Dong, Y., & Zhou, C. (2024). Reduced graphene oxide/MXene/FeCoC nanocomposite aerogels derived from metal–organic frameworks toward efficient microwave absorption. *ACS Applied Nano Materials, 7*(1), 230–242.

Li, B., Wen, H. M., Cui, Y., Zhou, W., Qian, G., & Chen, B. (2016). Emerging multifunctional metal–organic framework materials. *Advanced Materials, 28*(40), 8819–8860.

Li, H., Li, L., Lin, R.-B., Zhou, W., Zhang, Z., Xiang, S., & Chen, B. (2019). Porous metal-organic frameworks for gas storage and separation: Status and challenges. *EnergyChem, 1*(1), 100006.

Li, M., Feng, Y., & Wang, J. (2023). Asymmetric conductive structure design for stabilized composites with absorption dominated ultra-efficient electromagnetic interference shielding performance. *Composites Science and Technology, 236*, 110006.

Li, Q., Sun, Y., Li, G., Yang, X., & Zuo, X. (2022). Enhancing interfacial and electromagnetic interference shielding properties of carbon fiber composites via the hierarchical assembly of the MWNT/MOF interphase. *Langmuir: The ACS Journal of Surfaces and Colloids, 38*(46), 14277–14289.

Li, X., Li, K., Zhang, S., Zhang, J., Hu, X., Li, Y., & Liu, Y. (2024). Recent advances in mechanism, influencing parameters, and dopants of electrospun EMI shielding composites: A review. *Journal of Applied Polymer Science, 141*(2), e54788.

Lin, J., Qiao, J., Tian, H., Li, L., Liu, W., Wu, L., Liu, J., & Zeng, Z. (2023). Ultralight, hierarchical metal–organic framework derivative/graphene hybrid aerogel for electromagnetic wave absorption. *Advanced Composites and Hybrid Materials, 6*(5), 177.

Ma, X., Liu, S., Luo, H., Guo, H., Jiang, S., Duan, G., Zhang, G., Han, J., He, S., & Lu, W. (2024). MOF@ wood derived ultrathin carbon composite film for electromagnetic interference shielding with effective absorption and electrothermal management. *Advanced Functional Materials, 34*(4), 2310126.

Panahi-Sarmad, M., Samsami, S., Ghaffarkhah, A., Hashemi, S. A., Ghasemi, S., Amini, M., Wuttke, S., Rojas, O., Tam, K. C., & Jiang, F. (2023). MOF-based electromagnetic shields multiscale design: Nanoscale chemistry, microscale assembly, and macroscale manufacturing. *Advanced Functional Materials, 34*(43), 2304473.

Peng, Y., Xu, J., Xu, J., Ma, J., Bai, Y., Cao, S., Zhang, S., & Pang, H. (2022). Metal-organic framework (MOF) composites as promising materials for energy storage applications. *Advances in Colloid and Interface Science, 307*, 102732.

Saini, P., & Arora, M. (2012). Microwave absorption and EMI shielding behavior of nanocomposites based on intrinsically conducting polymers, graphene and carbon nanotubes. *New Polymers for Special Applications, 3*, 73–112.

Shao, S., Xing, S., Bi, K., Zhao, T., Wang, H., Tang, Y., Liu, J., & Wang, F. (2024). Fabrication of graphene/polyimide/Co-NC aerogel with reinforced electromagnetic losses and broadband absorption for highly efficient microwave absorption and thermal insulation. *Chemical Engineering Journal, 494*, 152976.

Shu, J. C., Cao, W. Q., & Cao, M. S. (2021). Diverse metal–organic framework architectures for electromagnetic absorbers and shielding. *Advanced Functional Materials, 31*(23), 2100470.

Song, S., Zhang, A., Chen, L., Jia, Q., Zhou, C., Liu, J., & Wang, X. (2021). A novel multi-cavity structured MOF derivative/porous graphene hybrid for high performance microwave absorption. *Carbon, 176*, 279–289.

Sridhar, V., Lee, I., & Park, H. (2021). Metal organic frameworks derived Fe-NC nanostructures as high-performance electrodes for sodium ion batteries and electromagnetic interference (EMI) shielding. *Molecules (Basel, Switzerland), 26*(4), 1018.

Srivastava, S. K., & Manna, K. (2022). Recent advancements in the electromagnetic interference shielding performance of nanostructured materials and their nanocomposites: A review. *Journal of Materials Chemistry A, 10*(14), 7431–7496.

Sun, Q. Y., Zhao, C., Qin, C., Li, F. M., Yu, H., & Wang, M. (2023). Multilayer three-dimensional woven silver nanowire networks for absorption-dominated electromagnetic interference shielding. *Advanced Materials Interfaces, 10*(11), 2202393.

Sun, X., Lv, X., Sui, M., Weng, X., Li, X., & Wang, J. (2018). Decorating MOF-derived nanoporous Co/C in chain-like polypyrrole (PPy) aerogel: A lightweight material with excellent electromagnetic absorption. *Materials, 11*(5), 781.

Thi, Q. V., Lee, Y., Cho, H. Y., Hong, J., Koo, C. M., Tung, N. T., & Sohn, D. (2021). Core-shell architecture of Ni-Co MOF wrapped by a heterogeneous FeBTC@PPy layer for high-performance EMI shielding. *Synthetic Metals, 281*, 116929.

Thi, Q. V., Tan, H. L., Tang, K. Y., Heng, J. Z. X., Loh, X. J., Ye, E., Sohn, D., & Truong, V. X. (2024). Innovation in hierarchical metal organic framework derivatives toward electromagnetic wave absorption. *Composite Structures, 345*, 118390.

Verma, R., Thakur, P., Chauhan, A., Jasrotia, R., & Thakur, A. (2023). A review on MXene and its' composites for electromagnetic interference (EMI) shielding applications. *Carbon, 208*, 170–190.

Wang, L., Ma, Z., Zhang, Y., Chen, L., Cao, D., & Gu, J. (2021). Polymer-based EMI shielding composites with 3D conductive networks: A mini-review. *SusMat, 1*(3), 413–431.

Wang, M., Tang, X. H., Cai, J. H., Wu, H., Shen, J. B., & Guo, S. Y. (2021). Construction, mechanism and prospective of conductive polymer composites with multiple interfaces for electromagnetic interference shielding: A review. *Carbon, 177*, 377–402.

Wang, S., McGuirk, C. M., d'Aquino, A., Mason, J. A., & Mirkin, C. A. (2018). Metal–organic framework nanoparticles. *Advanced Materials, 30*(37), 1800202.

Wang, X. X., Zheng, Q., Zheng, Y. J., & Cao, M. S. (2023). Green EMI shielding: Dielectric/magnetic "genes" and design philosophy. *Carbon, 206*, 124–141.

Wang, Y., Wang, H., Ye, J., Shi, L., & Feng, X. (2020). Magnetic CoFe alloy@C nanocomposites derived from ZnCo-MOF for electromagnetic wave absorption. *Chemical Engineering Journal, 383*, 123096.

Wang, Y., Zhang, W., Wu, X., Luo, C., Wang, Q., Li, J., & Hu, L. (2017). Conducting polymer coated metal-organic framework nanoparticles: Facile synthesis and enhanced electromagnetic absorption properties. *Synthetic Metals, 228*, 18–24.

Wei, Z., Cai, Y., Xie, Z., Meng, Y., Zhan, Y., Hu, X., & Xia, H. (2023). Green electromagnetic interference shielding films with unique and interconnected 3D magnetic/conductive interfaces. *Applied Surface Science, 636*, 157841.

Wu, F., Liu, Z., Wang, J., Shah, T., Liu, P., Zhang, Q., & Zhang, B. (2021). Template-free self-assembly of MXene and CoNi-bimetal MOF into intertwined one-dimensional heterostructure and its microwave absorbing properties. *Chemical Engineering Journal, 422*, 130591.

Xu, L., Si, R., Ni, Q., Chen, J., Zhang, J., & Ni, Q. Q. (2024). Synergistic magnetic/dielectric loss and layered structural design of Ni@carbon fiber/Ag@ graphene fiber/polydimethylsiloxane composite for high-absorption EMI shielding. *Carbon, 225*, 119155.

Xu, X., Li, D., Li, L., Yang, Z., Lei, Z., & Xu, Y. (2023). Architectural design and microstructural engineering of metal–organic framework-derived nanomaterials for electromagnetic wave absorption. *Small Structures, 4*(1), 2200219.

Xu, X., Ran, F., Fan, Z., Cheng, Z., Lv, T., Shao, L., & Liu, Y. (2020). Bimetallic metal–organic framework-derived pomegranate-like nanoclusters coupled with CoNi-doped graphene for strong wideband microwave absorption. *ACS Applied Materials & Interfaces, 12*(15), 17870–17880.

Xu, Y., Hou, M., Feng, Y., Li, M., Yang, S., & Wang, J. (2023). Heterogeneous asymmetric double-layer composite with ultra-low reflection for electromagnetic shielding. *Colloids and Surfaces A: Physicochemical and Engineering Aspects, 669*, 131545.

Yao, K., Gong, J., Tian, N., Lin, Y., Wen, X., Jiang, Z., Na, H., & Tang, T. (2015). Flammability properties and electromagnetic interference shielding of PVC/graphene composites containing Fe_3O_4 nanoparticles. *RSC Advances, 5*(40), 31910–31919.

Zhang, H., Zheng, X., Jiang, R., Liu, Z., Li, W., & Zhou, X. (2023). Research progress of functional composite electromagnetic shielding materials. *European Polymer Journal, 185*, 111825.

Zhang, J., Chen, J., Wang, X., Ma, Y., Li, Y., Yang, X., & Li, Y. (2024). Multi-electron transfer mechanism and band structures of CNTs in MOF-derived CNTs/Co hybrids for enhanced electromagnetic wave absorption property. *Materials Today Communications, 39*, 108936.

Zhang, K., Xie, A., Sun, M., Jiang, W., Wu, F., & Dong, W. (2017). Electromagnetic dissipation on the surface of metal organic framework (MOF)/reduced graphene oxide (RGO) hybrids. *Materials Chemistry and Physics, 199*, 340–347.

Zhang, X., Tian, X.-L., Qin, Y., Qiao, J., Pan, F., Wu, N., Wang, C., Zhao, S., Liu, W., & Cui, J. (2023). Conductive metal–organic frameworks with tunable dielectric properties for boosting electromagnetic wave absorption. *ACS Nano, 17*(13), 12510–12518.

Zhao, B., Hamidinejad, M., Wang, S., Bai, P., Che, R., Zhang, R., & Park, C. B. (2021). Advances in electromagnetic shielding properties of composite foams. *Journal of Materials Chemistry A, 9*(14), 8896–8949.

Zhou, H. C., Long, J. R., & Yaghi, O. M. (2012). Introduction to metal–organic frameworks. *ACS Publications*, 673–674.

Zhou, M., Gu, W., Wang, G., Zheng, J., Pei, C., Fan, F., & Ji, G. (2020). Sustainable wood-based composites for microwave absorption and electromagnetic interference shielding. *Journal of Materials Chemistry A, 8*(46), 24267–24283.

Zhou, S., Zhang, G., Nie, Z., Liu, H., Yu, H., Liu, Y., Bi, K., Geng, W., Duan, H., & Chou, X. (2022). Recent advances in 3D printed structures for electromagnetic wave absorbing and shielding. *Materials Chemistry Frontiers, 6*(13), 1736–1751.

Zhu, S. Q., Shu, J. C., & Cao, M. S. (2022). Novel MOF-derived 3D hierarchical needlelike array architecture with excellent EMI shielding, thermal insulation and supercapacitor performance. *Nanoscale, 14*(19), 7322–7331.

Chapter 8

Metal-organic framework composites for water splitting

Amir Kazemi[1,2], and Faranak Manteghi[1]
[1]*Research Laboratory of Inorganic Chemistry and Environment, Department of Chemistry, Iran University of Science and Technology, Tehran, Iran,* [2]*Department of Chemical and Biochemical Engineering, Western University, London, ON, Canada*

Introduction

As global energy demands continue to rise and environmental concerns intensify, the transition from nonrenewable fossil fuels to green and renewable energy sources has become imperative. Fossil fuels, while historically significant, contribute significantly to carbon dioxide (CO_2) emissions, exacerbating climate change and environmental degradation. In this context, hydrogen (H_2) has emerged as a promising clean energy carrier due to its high energy density (approximately 142 MJ/kg) and its carbon-free nature. Currently, industrial hydrogen production relies primarily on three methods: coal gasification, steam reforming, and water splitting. Coal gasification and steam reforming, which involve the reaction of nonrenewable fossil fuels with water, are energy-intensive processes that generate by-products such as carbon monoxide (CO) and CO_2. These by-products not only require further separation to obtain high-purity hydrogen but also exacerbate greenhouse gas emissions. Additionally, these methods operate under high temperatures and pressures, adding to their environmental impact (Kazemi et al., 2023; Shuai et al., 2024).

Water splitting involves two key half-reactions: the hydrogen evolution reaction (HER) at the cathode and the oxygen evolution reaction (OER) at the anode. The OER typically requires a higher overpotential than the HER, posing challenges for efficient and scalable hydrogen production. Thermodynamically, water splitting necessitates a Gibbs free energy change (ΔG) of around 237.2 kJ/mol, corresponding to a standard potential (ΔE) of 1.23 V versus a reversible hydrogen electrode (RHE). However, the practical application of this process is hindered by unfavorable thermodynamics and high overpotential requirements. Currently, platinum-based materials are recognized for their superior HER activity, while iridium (Ir), titanium (Ti), and ruthenium (Ru) compounds are effective for OER. However, the high cost, scarcity, and limited durability of these noble metals hinder their widespread applications. As a result, there is a critical need to develop highly active and durable alternatives. Significant progress has been made in this area, with earth-abundant metals, including phosphides, oxides, hydroxides, nitrides, borides, carbides, and chalcogenides, demonstrating promising performance for HER and OER. Another promising avenue for hydrogen production is solar-driven water splitting, utilizing sunlight as a virtually inexhaustible natural energy source. Numerous photocatalysts have been developed for this purpose, with titanium dioxide (TiO_2) being one of the most extensively studied due to its stability, low cost, and environmental friendliness (Gao, et al., 2024; Kazemi et al., 2024).

Metal-organic frameworks (MOFs) have emerged as a revolutionary class of materials with high porosity and tunable properties. Composed of organic ligands and metal ions linked by coordination bonds, MOFs exhibit excellent crystallinity, structural flexibility, and significant surface areas (1000 to 10,000 m²/g). MOFs have emerged as highly attractive materials for both electrocatalytic and photocatalytic water splitting due to their well-developed porous structures, high specific surface areas, and tunable properties. Unlike other porous materials, MOFs offer precise control over their structural and compositional characteristics, which can be optimized for enhanced catalytic performance. The versatility of MOFs in terms of their metal nodes and organic linkers allows for the customization of their electronic structure and bandgap, maximizing their efficacy as water-splitting catalysts. Recent developments have demonstrated that pristine MOFs can serve as effective bifunctional catalysts for both HER and OER. Additionally, combining MOFs with other materials, such as polyoxometalates, metal nanoparticles, and carbon nanotubes, enhances their electrical conductivity and catalytic performance. Despite the progress, existing reviews have yet to comprehensively address the full spectrum of MOF-based electrocatalysts, including synthesis methods, intrinsic activity enhancements, surface morphologies, MOF composites, catalytic activities, stability in water, and the underlying principles of HER and OER. MOF-based materials for water splitting can be categorized into three types: direct use of pristine MOFs, MOF-based composites, and MOF-derived compounds. While MOF-derived compounds may suffer from the loss of the ordered porous structure and the aggregation of metal sites, pristine MOFs and MOF-based composites continue to gain attention for their superior performance and stability (Kazemi, Pordsari, Tamtaji, Zainali, et al., 2025; Quan et al., 2024).

This chapter delves into the innovative applications and advancements in MOF composites for water splitting. It aims to provide a comprehensive overview of how MOFs and their composites can be optimized to enhance hydrogen production through both electrocatalytic and photocatalytic processes. Key areas of focus will include the synthesis methods for MOF-based composites, their structural and compositional adjustments to improve intrinsic activity, and the exploration of their surface morphologies. Additionally, this chapter will review various strategies for combining MOFs with other materials to boost their electrical conductivity and catalytic performance. The discussion will also cover the stability of MOF-based materials in water and the principles underlying their effectiveness in hydrogen evolution and oxygen evolution reactions. Through this examination, this chapter will highlight recent progress, current challenges, and future directions for leveraging MOF composites in sustainable water-splitting technologies.

Fundamentals of water splitting

Electrocatalytic water splitting: principles and mechanisms

In water splitting, the electrolysis process requires an external power source to drive the reaction. The fundamental principle of water splitting involves two half-reactions: the HER at the cathode and the OER at the anode, as shown in Fig. 8.1. When a potential difference is applied across the electrodes in an aqueous electrolyte, oxygen gas is generated at the anode and hydrogen gas at the cathode. Although the theoretical cell voltage required for water splitting is 1.23 V under standard conditions, a higher potential is often necessary due to serial resistance and activation barriers of the electrodes. This additional potential over the theoretical value is known as overpotential, which can be reduced by employing highly active HER and OER electrocatalysts (Hu et al., 2021; Liu et al., 2021).

The HER and OER occur in both acidic and alkaline solutions, each following distinct steps. In acidic solutions, HER involves the reduction of protons (H^+) to form hydrogen gas (H_2). In alkaline solutions, HER involves the reduction of water to generate hydrogen gas and hydroxide ions (OH^-). For OER, in acidic solutions, water is oxidized to form oxygen gas (O_2), protons (H^+), and electrons. In alkaline solutions, OER involves the oxidation of hydroxide ions to produce oxygen gas, water, and electrons. These reactions are crucial for processes like water splitting, as shown in Fig. 8.2.

The HER typically proceeds through two main steps, as shown in Table 8.1. First, an adsorbed hydrogen atom is formed from water in alkaline solutions or protons in acidic conditions, known as the Volmer reaction, which has a Tafel slope of 120 mV/dec. This is followed by the evolution of hydrogen through one of two distinct mechanisms.

In the Heyrovsky reaction, an adsorbed hydrogen atom reacts with an electron and a proton or water to produce molecular hydrogen, with a Tafel slope of 40 mV/dec. Alternatively, in the Tafel reaction, two adsorbed hydrogen atoms combine directly to complete the HER process, exhibiting a

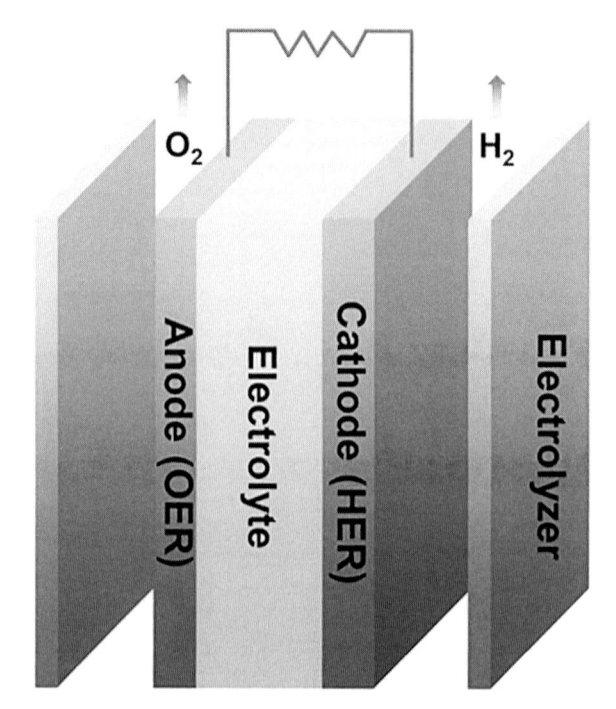

280 Applications of Metal-Organic Framework Composites

FIGURE 8.1. Schematic illustration of water electrolyzer for electrochemical water splitting (Hu et al., 2021).

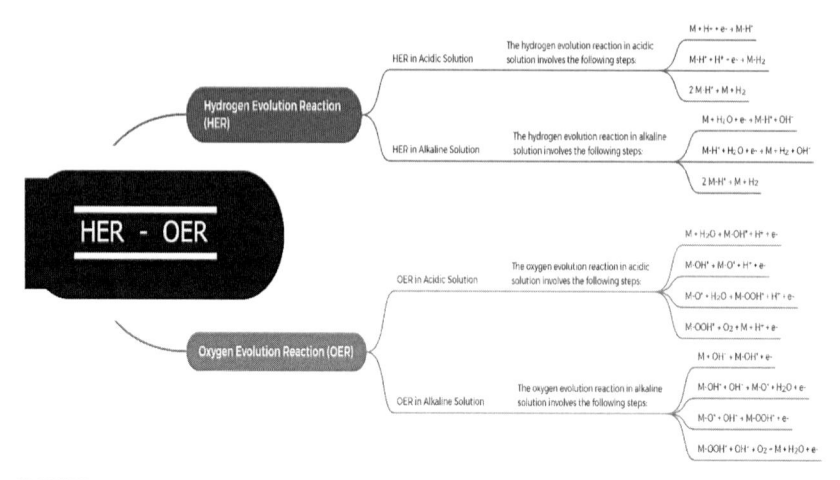

FIGURE 8.2 This diagram shows the steps of the hydrogen evolution reaction and oxygen evolution reaction in acidic and alkaline solutions, highlighting their roles in electrochemical processes.

TABLE 8.1 Hydrogen evolution reaction in acidic and alkaline media.

Acidic solution	Alkaline solution
$M + H^+ + e^- \rightarrow M\text{-}H^*$ Volmer	$M + H_2O + e^- \rightarrow M\text{-}H^* + OH^-$ Volmer
$M\text{-}H^* + H^+ + e^- \rightarrow M\text{-}H_2$ Heyrovsky	$M\text{-}H^* + H_2O + e^- \rightarrow M + H_2 + OH^-$ Heyrovsky
$2\,M\text{-}H^* \rightarrow M + H_2$ Tafel	$2\,M\text{-}H^* \rightarrow M + H_2$ Tafel

TABLE 8.2 Oxygen evolution reaction in acidic and alkaline media.

Acidic solution	Alkaline solution
$M + H_2O \rightarrow M\text{-}OH^* + H^+ + e^-$	$M + OH^- \rightarrow M\text{-}OH^* + e^-$
$M\text{-}OH^* \rightarrow M\text{-}O^* + H^+ + e^-$	$M\text{-}OH^* + OH^- \rightarrow M\text{-}O^* + H_2O + e^-$
$M\text{-}O^* + H_2O \rightarrow M\text{-}OOH^* + H^+ + e^-$	$M\text{-}O^* + OH^- \rightarrow M\text{-}OOH^* + e^-$
$M\text{-}OOH^* \rightarrow O_2 + M + H^+ + e^-$	$M\text{-}OOH^* + OH^- \rightarrow O_2 + M + H^+ + e^-$

Tafel slope of 30 mV/dec (Adegoke & Maxakato, 2022). The rate-determining step (RDS) of a particular HER process can be evaluated by analyzing the Tafel slope from the HER polarization curve. Notably, the efficiency of HER is influenced by the adsorption and desorption of hydrogen atoms on the electrocatalyst surface. A strong interaction between the hydrogen atom and the electrocatalyst can impede recombination, while a weak interaction can slow down the formation rate of adsorbed hydrogen atoms. Therefore the hydrogen adsorption-free energy (ΔG_{H*}) for an outstanding HER electrocatalyst should be neither too high nor too low (ideally close to zero). Due to its intricate process involving multiple intermediates, the OER is slower than the HER and requires a significantly larger overpotential to achieve a substantial current. Moreover, as shown in Table 8.2, the mechanism for OER is more complex than that of HER. The OER process involves the formation of intermediates such as OH^*, O^*, and OOH^* under both acidic and alkaline conditions. Similar to HER, the reaction kinetics for OER are significantly influenced by the bonding strength between active sites and oxygen-containing groups, which can be assessed through Tafel slopes. A Tafel slope of 120 mV/dec indicates that the rate-determining step (RDS) is the first electron transfer reaction in the formation of M–OH*. A slope of 60 mV/dec suggests that the RDS is the conversion of M–OH* to M–O*. If the

RDS is the third electron transfer reaction, a Tafel slope of 30 mV/dec is observed. Generally, a smaller Tafel value reflects faster reaction kinetics for the OER process (Kulkarni et al., 2023).

Photocatalytic water splitting: principles and mechanisms

Photocatalytic water splitting, driven by light irradiation, was first demonstrated by Fujishima and Honda (1972). Since then, this method has been extensively studied with various photocatalysts. The process generally involves three main steps as shown in Fig. 8.3: (1) the generation of electron–hole pairs in the conduction band (CB) and valence band (VB) upon photon absorption, provided the photon energy exceeds the bandgap of the photocatalyst; (2) the separation and migration of these photogenerated electrons and holes to the photocatalyst surface; and (3) the generation of hydrogen and oxygen through reactions between the electrons/holes and water molecules at the photocatalyst surface (Ahmad et al., 2015).

In theory, photocatalytic water splitting should occur with light around 1000 nm, due to a Gibbs free energy change of +237 kJ/mol (1.23 eV). However, most photocatalysts only respond to UV or visible light, with visible light being much more prevalent. This necessitates the development of photocatalysts that can be activated by visible light. Additionally, the CB of photocatalyst must be more negative than the H^+/H_2 redox potential (0.0 V vs. NHE), and the VB must be more positive than the O_2/H_2O redox potential (1.23 V vs NHE) to effectively drive the reaction.

Merit of metal-organic frameworks for water splitting

Extremely high porosity combined with facile access to active sites

MOFs have exceptionally high specific surface areas, due to their porous nature, which is essential for enhancing water-splitting activity. These porous structures can host and disperse highly active materials, boosting catalytic performance through synergistic effects between the guest materials and the

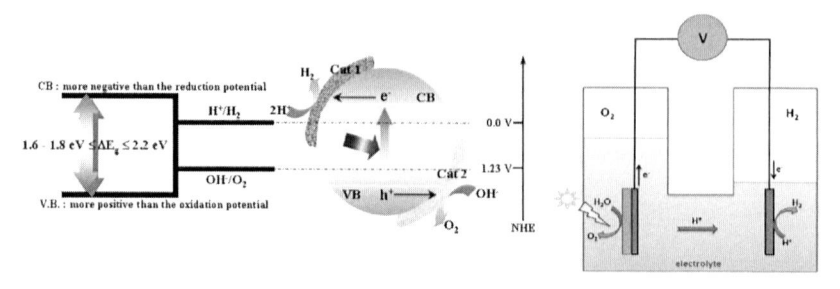

FIGURE 8.3 The principle of photocatalytic water splitting reactions (Ahmad et al., 2015).

MOFs. While incorporating active materials into MOF pores can reduce their specific surface area, the enhanced catalytic performance often outweighs this drawback. The porous nature also facilitates electrolyte penetration, increasing the accessibility of active sites and accelerating catalytic activity (Kazemi, Afshari, Baesmat, Bozorgnia, et al., 2024; Kazemi, Pordsari, Tamtaji, Manteghi, et al., 2025).

Optimizing metal nodes' performance

A wide variety of MOFs have been designed and fabricated, reflecting the diverse range of metal nodes and organic ligands available. The choice of metal nodes plays a crucial role in determining catalytic performance and can be strategically selected to optimize MOF activity. Isostructural MOFs with different metal species can exhibit significantly different performances for water splitting, particularly in electrocatalytic applications. For instance, variations in metal nodes within isostructural MOFs, such as those based on different metals, can lead to considerable differences in electrocatalytic activity. Additionally, bimetallic or multimetallic MOFs often demonstrate enhanced water-splitting activity due to synergistic effects between the metals. These effects can include an increased electrochemically active surface area, modified adsorption energy of intermediates, and improved electron transfer rates. The design of MOFs can also involve heteroatom doping, which can further enhance performance. For example, the incorporation of different metal ions or the use of specific metal ratios can improve catalytic activity by extending electron-hole recombination times and reducing carrier recombination rates. Overall, the metal nodes in MOFs are critical to their water-splitting performance, and the introduction of heteroatoms can significantly boost their catalytic efficiency compared to nondoped MOFs (Han et al., 2020; Kazemi, Afshari, Baesmat, Manteghi, et al., 2024).

Impact of ligand engineering

In MOF-based catalysts, the diversity of components, including various metal nodes and organic linkers, is a significant advantage. While the metal nodes critically influence water-splitting activity, the role of the organic ligands is also crucial. Organic ligands not only link metal nodes to form the porous structure but also affect the electronic properties of the metal nodes and light-harvesting efficiency, thereby regulating water-splitting activity. Functionalized ligands can modify the electronic density of the metal nodes and enhance light absorption, which influences the photocatalytic and electrocatalytic performance. Ligand engineering can improve photocatalytic activity by altering absorption properties or by affecting the electronic environment around the metal nodes. For instance, modifications to ligands can lead to enhanced stability and activity in water-splitting processes. Ligands that introduce π-electron-rich groups or create intraframework coupling pathways can significantly boost performance and stability. Additionally, the preservation of the porous structure during the catalytic process is essential for maintaining high performance. Overall, the strategic design and modification of organic ligands in MOFs are key factors in optimizing their water-splitting capabilities (Kazemi, Hosseini, et al., 2025; Kazemi, Pordsari, et al., 2024).

Challenges and strategies water-splitting over metal-organic framework–based electrocatalysts

The potential of MOFs as catalysts for water splitting has generated significant interest due to their unique compositional and structural characteristics. MOFs are regarded as highly promising electrocatalysts for water splitting because of these features, and notable advancements have been made in this area. However, the electrocatalytic performance of MOFs for water splitting remains limited, with only a few pure MOFs or MOF-based composites being directly used as electrocatalysts. The suboptimal performance of pristine MOFs and MOF-based materials can be attributed to their low conductivity, limited accessibility of active sites, and suboptimal adsorption energy for intermediates. To address these challenges, various strategies have been proposed to enhance the electrocatalytic activity of pure MOFs and MOF-based composites for water splitting (Torabi et al., 2025; Yang et al., 2020).

Metal-organic framework composites

Constructing composites has recently emerged as an effective strategy to develop materials with enhanced properties compared to their individual components. MOFs, with their large specific surface area, high porosity, versatile structures, various metal nodes, and functional groups, are excellent candidates for fabricating MOF-based composites to enhance water-splitting performance. Typically, the primary aim of creating MOF-based composites is to improve the inherent water-splitting activities of MOFs, which are promising as water-splitting electrocatalysts. Additionally, in some cases, MOFs within the composites can function as a supportive matrix (Liu et al., 2020).

Metal-organic framework composites with conductive materials

Combining MOFs with highly conductive materials is considered an effective strategy to further enhance electron transfer efficiency during the water-splitting process. Carbon-based materials, including graphene, acetylene black, carbon nanotubes, and carbon black, are frequently selected for the assembly of these composites due to their excellent conductivity and compatibility with MOFs.

Metal-organic framework/graphene composites

The use of MOF/graphene composites in water splitting is a promising area of research, aiming to enhance the efficiency and stability of the electrocatalytic process. These composites leverage the unique properties of MOFs, such as high surface area and tunable porosity, combined with the excellent electrical conductivity and mechanical strength of graphene. Graphene is a single layer of carbon atoms arranged in a two-dimensional honeycomb lattice. It is known for its excellent electrical conductivity, meaning it can efficiently transport electrons, which is important for driving electrochemical reactions like water splitting. Graphene is also mechanically strong, flexible and has a large surface area, making it an ideal material to support and enhance the properties of other materials like MOFs. When MOFs are combined with graphene to form a composite material, the resulting composite benefits from the high surface area and tunable porosity of the MOFs, which provide abundant active sites for the reactions to occur. At the same time, the graphene contributes its excellent electrical conductivity, which helps in the efficient transport of electrons during the reactions, and its mechanical strength, which enhances the stability of the composite. This combination of properties makes MOF/graphene composites highly effective for use in water splitting, where both efficiency and stability are crucial for practical applications. The integration of these properties not only enhances the individual strengths of MOFs and graphene but also allows for the development of advanced structures, such as freestanding 3D heterostructure films, which exhibit superior electrocatalytic performance (Gopi et al., 2022; Li et al., 2023). Yan et al. (2020) reported that a Ni-centered MOF/graphene oxide composite, when converted into a 3D heterostructure film, demonstrated excellent electrocatalytic activity. This composite showed low overpotentials of 95 and 260 mV for the HER and OER, respectively, to reach a current density of 10 mA/cm^2, which are critical steps in the water-splitting process. The improved performance of this material was attributed to the synergistic effects between N-doped carbon and nickel nanoparticles within the composite. These interactions enhance the efficiency of the catalytic process, making the material more effective in facilitating the required reactions for water splitting. The exceptional performance of the MOF/graphene oxide composite in water splitting is highlighted by its ability to combine the high surface area and tunable porosity of MOFs with the excellent electrical conductivity and mechanical strength of graphene oxide, which greatly enhances the efficiency and stability of the resulting electrocatalysts. Gopi et al. (2022) reported the development of a bifunctional electrocatalyst for water splitting, specifically designed for the HER and OER. This catalyst, which leverages a vanadium-doped bimetallic nickel-iron (Ni/Fe) nanoarray, is synthesized using a MOF supported on graphene oxide (GO). The use of MOF/GO composites is pivotal in this design, as it combines the high surface area and tunable porosity of MOFs with the excellent electrical conductivity and mechanical strength of graphene oxide. This integration enables the replacement of expensive noble metal-based catalysts with earth-abundant, nonprecious metals while maintaining high efficiency and stability. In their study, the authors prepared three different catalysts by varying the molar ratios of Ni to Fe, and they found that the catalyst with an equal molar ratio (1:1) of Ni to Fe exhibited the highest catalytic activity for both HER and OER. This specific catalyst demonstrated low overpotentials of 90 mV for HER and 210 mV for OER at a current density of 10 mA/cm².

Additionally, it achieved high current densities of 208 mA/cm² for HER in an acidic medium (0.5 M H_2SO_4) and 579 mA/cm² for OER in an alkaline medium (1 M KOH). The performance of this catalyst surpassed that of the other catalysts with different Ni/Fe ratios. The study also highlighted the crucial role of the MOF/GO composite in enhancing the catalytic performance. The synergistic effect of vanadium doping, combined with the improved corrosion resistance and increased catalytic efficiency imparted by the MOF/GO composite, significantly boosted the overall performance. Gas chromatography analysis confirmed the successful production of hydrogen (H_2) and oxygen (O_2) during water electrolysis, showcasing the effectiveness of the MOF/GO composite in achieving efficient water splitting. Based on the results shown in Fig. 8.4, it can be hoped that MOF/graphene composites provide a synergistic approach to water splitting, effectively utilizing the high surface area of MOFs and the conductivity of graphene to enhance catalytic performance.

Metal-organic framework/MXene composites

Metal-organic framework/MXene composites have emerged as promising materials for hydrogen generation through water splitting, offering an innovative solution to energy sustainability. MXenes, a class of two-dimensional (2D) materials with high electrical conductivity, large surface area, and rich surface chemistry, face challenges such as agglomeration and stability issues. MOFs, on the other hand, possess tunable porosity, high surface area, and adjustable bandgaps, making them ideal for photocatalysis. By integrating MXenes with MOFs, a synergistic effect is achieved, enhancing charge separation, reducing recombination, and improving overall photocatalytic efficiency. MXenes serve as excellent electron acceptors, while MOFs facilitate electron-hole pair generation upon light irradiation, leading to superior catalytic performance. The porous nature of MOFs prevents MXene agglomeration, ensuring long-term stability and efficient charge transport. Additionally, the functional groups on MXenes (—O, —OH, —F) act as active sites, further boosting catalytic activity. The tunable bandgap of MOFs allows optimization for specific light absorption, enhancing solar energy utilization in water splitting. Various synthesis methods, including hydrothermal, solvothermal, and vacuum-assisted filtration, enable precise control over the

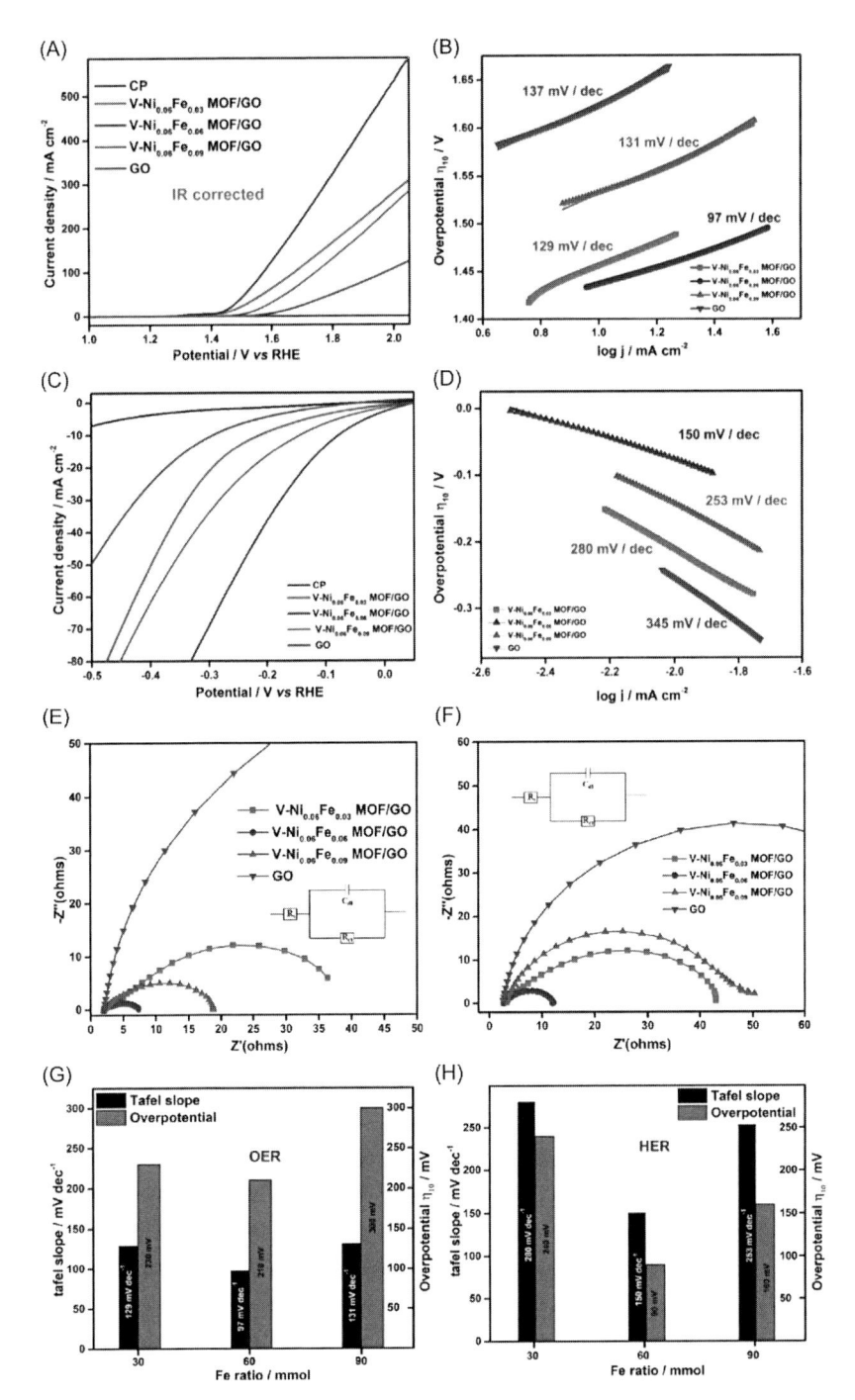

FIGURE 8.4 Electrocatalytic OER study (A) LSV curve, (B) Tafel slope, Electrocatalytic HER study (C) LSV curve (D) Tafel slope for HER, EIS in (E) 1 M KOH (OER), (F) 0.5 M H_2SO_4 (HER). Tafel slope and overpotential comparison for the synthesized catalyst with different ratios of Ni and Fe metal in (G) OER, and (H) HER. *EIS*, Electrochemical impedance spectra (Gopi et al., 2022).

composite's morphology and composition. The integration of MOFs and MXenes not only addresses their individual limitations but also enhances their durability and recyclability. This combination offers an eco-friendly and efficient alternative to traditional catalysts, making MXene-MOF hybrids highly competitive in sustainable hydrogen production. With tailored structural properties and improved performance, these materials hold great potential in advancing clean energy technologies and reducing environmental impact (Kumar et al., 2024; Zhang, Qian, et al., 2023). Zhao et al. (2017) reported significant results from the combination of two-dimensional MOFs with $Ti_3C_2T_x$ (MXene) nanosheets. The hybrid material was applied in the OER and achieved a current density of 10 mA/cm² at a potential of 1.64 V versus RHE and a Tafel slope of 48.2 mV/dec in 0.1 M KOH. These results outperformed the standard IrO_2-based catalysts and were comparable to or even better than previously reported state-of-the-art transition-metal-based catalysts. The study highlighted how the CoBDC layer provided a highly porous structure and large active surface area, while the $Ti_3C_2T_x$ nanosheets facilitated rapid charge and ion transfer, enhancing the overall electrocatalytic performance. In another study, Zong et al., 2021 reported the development of an ultrathin Ti_2NT_x MXene-wrapped MOFs-derived CoP composite, referred to as Ti_2NT_x@MOF-CoP, as a bifunctional catalyst with both low cost and high activity. This MOF/MXenes composite was synthesized by phosphating a precursor ZIF-67, which had been self-assembled with Ti_2NT_x MXene nanosheets. The composite demonstrated competitive hydrogen evolution performance across a wide pH range, with a notably low overpotential of 112 mV at a current density of 10 mA/cm² in an alkaline solution (pH=14). Furthermore, the Ti_2NT_x@MOF-CoP electrode showed strong performance in OER, with a low overpotential of 241 mV at 50 mA/cm². The study highlighted the MOF/MXenes composite's bifunctional activity in alkaline environments, where it showed impressive catalytic stability over 20 hours in a dual-electrode configuration for both HER and OER. Fig. 8.5 demonstrates that this MOF/MXenes composite has significant potential in hydrogen evolution and water oxidation, contributing to reduced carbon emissions and environmental protection. In summary, MOF/MXenes composites leverage the unique properties of both materials to address the challenges associated with water splitting, such as charge carrier recombination, catalyst stability, and efficient light absorption. These composites represent a promising approach to developing efficient and sustainable water-splitting technologies.

Metal-organic framework/metal nanoparticle composites

When MOFs combined with metal nanoparticles (MNPs), these MOF/MNP composites exhibit unique properties that make them highly effective for energy storage and heterogeneous catalysis. The porous nature of MOFs enables them to serve as ideal supports for MNPs, preventing the

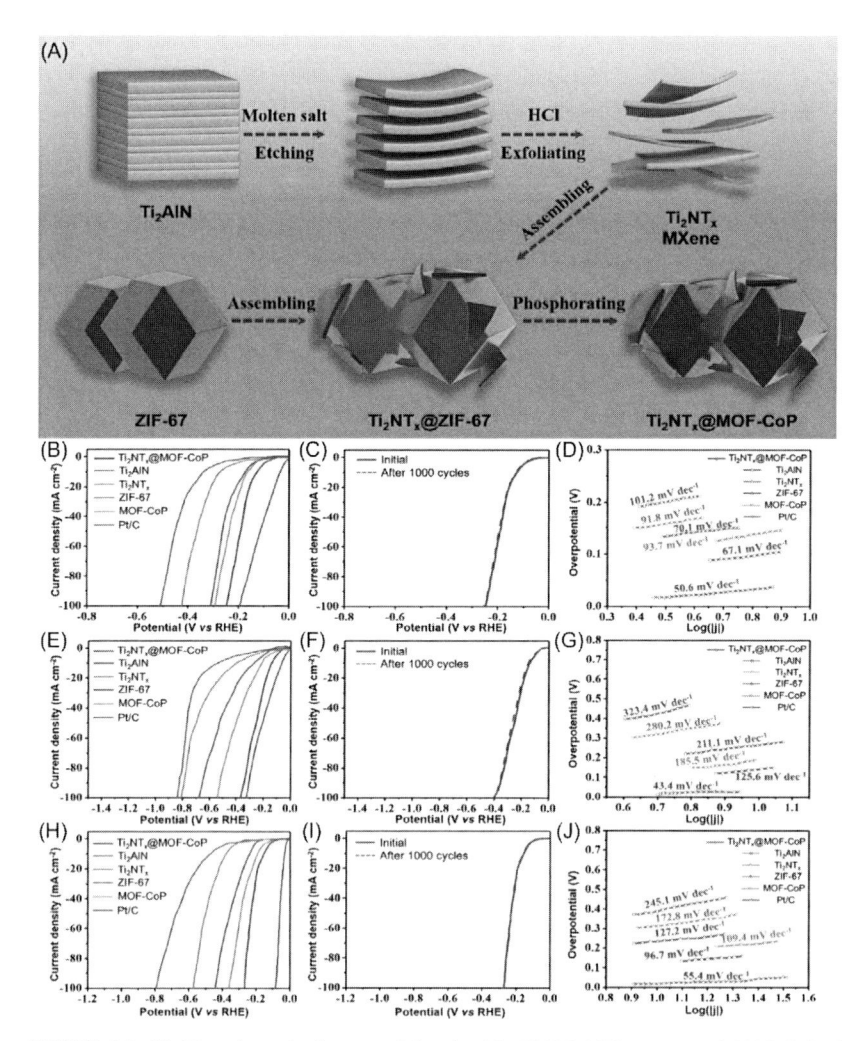

FIGURE 8.5 (A) The schematic diagram of the ultrathin Ti$_2$NT$_x$ MXene-wrapped MOF-derived CoP nanohybrids (Ti$_2$NT$_x$@MOF-CoP). The LSV curves of the Ti$_2$AlN, Ti$_2$NT$_x$ MXene, MOF-CoP and Ti$_2$NT$_x$@MOF-CoP electrodes for HER at 5 mV/s in (B) 1.0 M KOH, (E) 1.0 M PBS and (H) 0.5 M H$_2$SO$_4$; the LSV curves of the Ti$_2$NT$_x$@MOF-CoP electrodes before and after 1000 cycles in (C) 1.0 M KOH, (F) 1.0 M PBS and (I) 0.5 M H$_2$SO$_4$; the Tafel slope of the corresponding electrodes in (D) 1.0 M KOH, (G)1.0 M PBS, and (J) 0.5 M H$_2$SO$_4$ (Zong et al., 2021).

agglomeration of MNPs in confined spaces and thereby enhancing their catalytic performance. The ability of MOFs to stabilize and disperse nanoparticles uniformly also contributes to improved catalytic activity, making these composites particularly valuable in advanced catalytic processes such as water splitting. The application of MNPs in catalysis, however, faces several challenges, including a tendency to agglomerate, difficulties in recycling, and the complexity of nanoparticle extraction from reaction mixtures. To address these issues, MNPs are often deposited on or embedded within support materials, such as mesoporous silicates, metal oxides, zeolites, and carbon. Among these, MOFs stand out due to their regular porosity and the ability to tailor pore sizes and shapes to specific requirements. This characteristic not only enhances the accessibility of active sites but also improves the recovery and reusability of the nanoparticles, making MOF/MNP composites highly efficient and sustainable for catalytic applications (Jiao et al., 2018; Yang et al., 2017).

Noble metal-based MOF/MNP composites extend the concept of Pt-based composites to include other noble metals, such as gold (Au), palladium (Pd), rhodium (Rh), and iridium (Ir). These noble metals are known for their excellent catalytic properties, including high activity, selectivity, and resistance to oxidation, making them valuable for a variety of catalytic applications. When these metals are incorporated into MOFs as nanoparticles, similar benefits to those seen with Pt-based composites are observed. The MOF's porosity and high surface area provide a supportive environment that prevents nanoparticle agglomeration, enhances the accessibility of active sites, and stabilizes the nanoparticles during catalytic reactions. Additionally, the chemical environment within the MOF can be tailored to interact specifically with the noble MNPs, further enhancing catalytic performance. These composites are particularly useful in applications requiring high stability and activity under harsh conditions, such as in electrocatalysis for water splitting, carbon dioxide reduction, and fuel cell reactions. The ability to fine-tune the MOF structure and composition allows for the optimization of the catalytic properties of these noble MNPs, making MOF/MNP composites versatile and powerful catalysts in both industrial and environmental processes (Tao et al., 2016; Zheng et al., 2024). Pt-based MOF/MNP composites are hybrid materials that combine Pt nanoparticles with MOFs. These composites leverage the high electrocatalytic activity of Pt, one of the most efficient and widely used noble metals in catalysis, particularly for reactions like hydrogen evolution and oxygen reduction. When Pt nanoparticles are uniformly dispersed within the porous structure of MOFs, several benefits arise. The MOF's large surface area and tunable pore structure allow for optimal distribution and exposure of Pt active sites, which enhances catalytic efficiency. Additionally, the MOF matrix can prevent the agglomeration of Pt nanoparticles, maintaining their high surface area and catalytic activity over time. This uniform dispersion also leads to lower Pt loading while achieving the same or even higher catalytic performance, which is economically advantageous given the high cost of Pt. Moreover, the interaction between the Pt nanoparticles and the MOF's organic linkers or metal nodes can create synergistic effects, improving electron transfer and overall catalytic activity in processes such as water splitting (Liu et al., 2022; Ye et al., 2019). Yang et al. (2023) reported the development of an advanced electrocatalyst by anchoring well-dispersed Ir nanoparticles onto nickel MOF Ni-NDC nanosheets. The MOF/MNPs composite exhibited exceptional electrocatalytic performance for both the HER and OER, as well as overall water splitting across a wide pH range. The outstanding performance of this catalyst was attributed to the strong synergy between Ir and the MOF, facilitated by interfacial Ni–O–Ir bonds. Theoretical calculations further revealed that the charge redistribution induced by the Ni–O–Ir bridge optimized the adsorption of H_2O, OH*, and H*, which in turn accelerated the electrochemical kinetics for HER and OER. This study highlights the potential of MOF/MNP composites as highly effective bifunctional electrocatalysts for universal pH water splitting. Pt plays a crucial role in the MOF/Pt composites by acting as the active component responsible for catalyzing the HER. Its presence in the composite is key to achieving high performance. Li, Huang, et al. (2019) reported the development of a versatile synthesis platform using the MOF-808-EDTA-based metal ion trap method to prepare stable, uniformly distributed, and well-defined single-atom metal catalysts within a MOF architecture. Fig. 8.6 illustrates the approach to the synthesis of MOF/Pt composites. Specifically, single Pt^{2+} ions were efficiently captured by EDTA ligands postsynthetically modified and anchored at the Zr_6 cluster metal nodes of the MOF-808. Activation in a hydrogen atmosphere resulted in the formation of stable MOF-808-EDTA encapsulated single-atom Pt catalysts with well-defined atomic positions, confirmed through high-angle annular dark-field scanning transmission electron microscopy and extended X-ray absorption fine structure spectroscopy.

Theoretical density functional theory calculations further validated these experimental findings, providing detailed insights into the atomic-level structures of the catalysts. The Pt role in MOF/Pt composites was highlighted by the significant enhancement in photocatalytic hydrogen evolution activity. The synthesized MOF-808-EDTA encapsulated single-atom Pt catalyst exhibited a hydrogen generation rate three orders of magnitude higher than previously reported for MOF-based single-atom Pt catalysts. This improvement is attributed to the lower hydrogen binding free energy at the single-atom Pt sites within the MOF-808 architecture. The synthesis method was also successfully extended to other single-atom metal catalysts, suggesting it can serve as a general platform for creating stable, atomically dispersed bimetallic or tri-metallic alloy catalysts with tailored properties.

Metal-organic framework composites with porous materials

Combining MOFs with porous materials significantly enhances water-splitting efficiency by utilizing the unique properties of these materials. For example, MOF-polyoxometalates composites, MOF/covalent organic framework composites, and MOF/MOF composites leverage the high surface area of porous

292 Applications of Metal-Organic Framework Composites

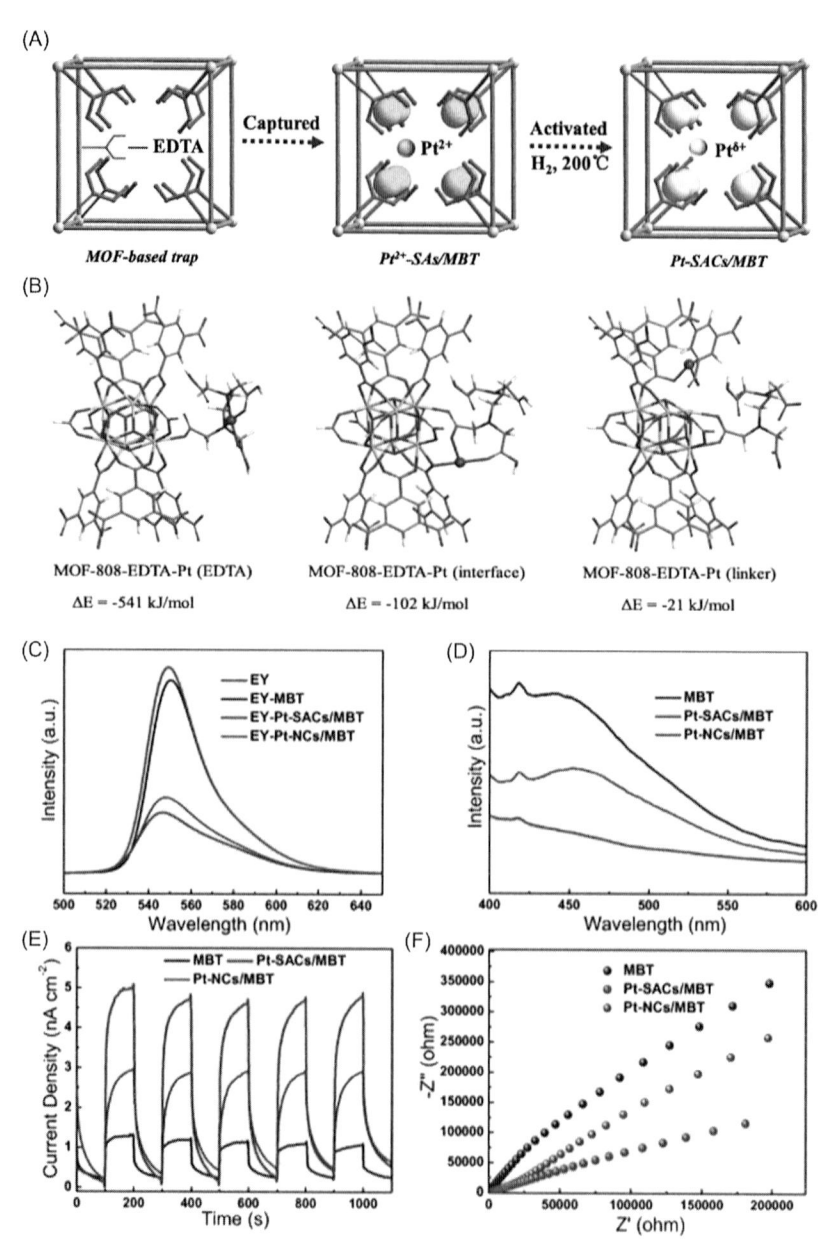

FIGURE 8.6 Schematic illustration of single-atom Pt catalysts encapsulated in MOF-808-EDTA via the single metal ion trap method. (B) Pt^{2+} metal ions incorporated into the Zr_6 cluster metal nodes of MOF-808. Carbon, oxygen, zirconium, nitrogen, platinum, and hydrogen atoms are depicted in black, red, green, blue, cyan, and white, respectively. (C) Steady-state PL spectra: 5 mg catalyst, 5 mg EY, 50 mL H_2O, excitation wavelength 480 nm. (D) Steady-state PL spectra of catalyst powders at an excitation wavelength of 325 nm. (E) Transient photocurrent responses and (F) EIS curves of the electrodes (Li, Huang, et al., 2019).

materials such as mesoporous silica or zeolites, which offers a greater number of active sites for catalysis, thus boosting overall catalytic activity. This porous structure improves the accessibility and diffusion of water molecules and other reactants to these active sites, resulting in more effective reactions. Additionally, it helps stabilize the MOF catalysts by preventing their agglomeration or degradation, ensuring reliable performance over time. The porous framework also facilitates the efficient removal and transport of gases, such as hydrogen and oxygen, produced during water splitting, reducing bubble accumulation and enhancing overall reaction efficiency. By integrating porous materials with MOFs, the resulting composites achieve a synergistic effect that optimizes the catalytic process and improves water-splitting performance.

Metal-organic framework/polyoxometalate composites

MOF–polyoxometalate (POM) composites combine MOFs with POMs to create advanced materials with enhanced catalytic properties. In these composites, the MOF acts as a porous support structure, offering a high surface area and a scaffold for incorporating POMs. POMs are metal oxide clusters known for their structural stability, redox activity, and electron transfer capabilities. When integrated into the MOF framework, POMs impart their unique properties, significantly improving the composite's catalytic performance. This combination enables more effective catalysis in various processes, including water splitting. The benefits of MOF/POM composites are substantial: the high surface area of the MOF, coupled with the redox activity of the POMs, markedly enhances catalytic activity. In addition, POMs can increase the structural stability of the MOF, preventing degradation and ensuring long-term functionality. The interaction between the MOF and POM also produces synergistic effects that optimize electron and charge transfer, further improving the efficiency of catalytic reactions. Thus MOF/POM composites represent a powerful strategy for developing efficient and durable catalysts for a wide range of chemical processes.

Among various support materials, MOFs have recently become a prominent choice as porous, crystalline, cage-like supports for incorporating POMs. The strategies of encapsulating and functionalizing these materials, whether pre- or postsynthetic, have garnered significant research interest due to their potential in designing and enhancing catalytic systems. Incorporating POMs into MOFs results in new materials known as POMOFs, which benefit from the combined advantages of both components. The reactivity and selectivity of POMs can be further refined by adjusting the MOF's functionality, allowing for precise tuning of catalytic properties. By selecting appropriate light-harvesting linkers for the MOFs and integrating redox-active POMs, the photocatalytic activity can be greatly enhanced through the efficient transfer of photogenerated charges from the MOFs to the POMs. Well-designed POMOF materials effectively bridge the gap between homogeneous and heterogeneous catalysis, offering significantly improved properties compared to either of the individual components (Lan et al., 2016; Li & Zhang, Li, et al., 2019). Zhang et al. (2015) reported the development of a charge-assisted self-assembly process to encapsulate a noble metal-free POM within a porous and phosphorescent MOF, specifically built from [Ru(bpy)$_3$]$^{2+}$-derived dicarboxylate ligands and Zr$_6$(μ_3-O)$_4$(μ_3-OH)$_4$ secondary building units. This innovative approach led to the formation of POMOF composites, where the hierarchical organization of photosensitizing and catalytic proton reduction components within the POM@MOF assembly enabled rapid multielectron transfer from the photoactive MOF framework to the encapsulated redox-active POMs upon photoexcitation. The results showed that the POM@UiO assemblies achieved hydrogen evolution rates of 699 and 193 μmol/h/g under visible light (>400 nm) with turnover numbers (TONs) of up to 79. The POM@UiO composites were effective for at least three reaction cycles with only a slight loss of activity and crystallinity. Further experiments demonstrated that the photocatalytic hydrogen evolution rate in a DMF/CH$_3$CN mixed solution reached 307 μmol/h in 14 hours and 540 μmol/h in 36 hours, which is 13 times higher than that of the homogeneous controls. These results underscore the significant enhancement in hydrogen evolution performance due to the unique structure and properties of the POMOF composites. In recent years, the development of nonprecious metal-based catalysts for electrochemical reactions, such as water splitting for hydrogen and oxygen production, has become a significant focus in materials science and electrochemistry. These catalysts offer a more accessible and cost-effective alternative to precious metals. A notable advancement in this field involves using MOFs as platforms for designing catalysts with unique properties. In this context, Zhang, Ran, et al. (2023) reported the development of a novel mesoporous CoMoO$_4$ catalyst with hollow tube structures (CoMoO$_4$ HTs), derived from POMOFs through a calcination process. The CoMoO$_4$ HTs catalyst was synthesized using a solvothermal method by introducing H$_3$PMo$_{12}$O$_{40}$ into the Co-BTC system. This approach utilized POMOF composites as an efficient preassembly platform. The CoMoO$_4$ HTs exhibited exceptional catalytic activity for both the OER and the HER in 1.0 M KOH, with low overpotentials of 210 mV and 75 mV at 10 mA/cm², respectively. Additionally, an overall water-splitting device employing CoMoO$_4$ HTs as both the anode and cathode required only a cell voltage of 1.57 V to achieve a current density of 10 mA/cm², demonstrating excellent electrochemical stability over 72 hours. The superior performance of the CoMoO$_4$ HTs is attributed to its high specific surface area and abundant ion transport channels resulting from its unique porous and hollow structure. As shown in Fig. 8.7, this work highlights the potential of POMOF composites as efficient electrocatalysts for overall water splitting.

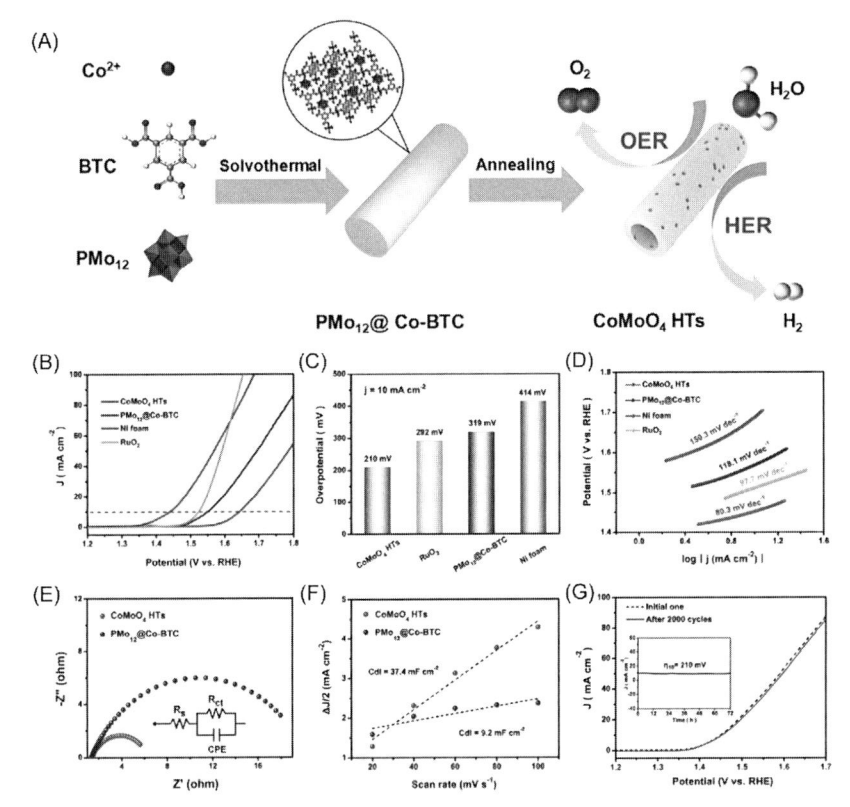

FIGURE 8.7 (A) Schematic illustration of the synthetic process for CoMoO$_4$ HTs, (B) LSV curves of CoMoO4 HTs, PMo12 @Co-BTC, Pt/C and Ni foam in 1.0 M KOH for the OER. (C) The comparison of the overpotentials at 10 mA/cm^2 of different electrocatalysts. (D) Tafel plots for CoMoO$_4$ HTs, PMo$_{12}$ @Co-BTC, Pt/C and Ni foam. (E) EIS Nyquist spectra (inset: a simplified equivalent circuit diagram) and (F) Cdl values (data were calculated from the CV curves for CoMoO$_4$ HTs and PMo$_{12}$ @Co-BTC. (G) LSV curves of CoMoO$_4$ HTs before and after 2000 CV cycles (inset: chronoamperometric measurement at 10 mA/cm^2 for CoMoO$_4$ HTs catalyst for 72 h) (Zhang, Ran, et al., 2023).

Metal-organic framework/covalent–organic framework composites

MOFs and covalent–organic frameworks (COFs) are two classes of porous crystalline materials known for their high surface areas, tunable pore sizes, and versatile chemical functionalities. MOFs are constructed from metal ions or clusters coordinated to organic ligands, while COFs are entirely composed of light elements (H, B, C, N, O) connected through strong covalent bonds. A MOF/COF composite integrates the distinct advantages of both MOFs and COFs into a single material. By combining these two frameworks, researchers can harness the metal-centered active sites of MOFs along with the robust covalent architecture and chemical stability of COFs. This synergy often leads to enhanced physical and chemical properties that are superior to those of the individual components. COF materials are a new class of photocatalysts known for their excellent hydrogen production capabilities, thanks to their porosity, large surface area, excellent light absorption, and adjustable band positions. Despite these advantages, the photocatalytic efficiency of pure COFs is often hindered by the rapid recombination of photogenerated electrons and holes. A proven strategy to mitigate this issue is the introduction of suitable species to form heterojunctions. MOFs another type of porous photocatalysts, share similar benefits with COFs but also offer diverse metal nodes that typically act as active sites in catalytic reactions. The combination of COFs and MOFs can effectively retain the porous nature of the original photocatalysts, which significantly enhances photocatalytic reactions by improving mass transfer. When applied to water splitting, the integration of COFs and MOFs into composites leverages the strengths of both materials. COFs contribute to high surface areas, excellent light-harvesting capabilities, and tunable electronic properties, which are crucial for efficient HER and OER. However, the fast recombination of electrons and holes in COFs can limit their photocatalytic performance. MOFs, on the other hand, introduce metal nodes that act as active catalytic sites, which can reduce the recombination rate and enhance overall photocatalytic efficiency (Wu et al., 2024; Yusran et al., 2023). Han et al. (2022) reported the development of a multifunctional MOF/COF heterostructured photocatalyst, specifically Cu-NH$_2$-MIL-125/TpPa-2-COF, designed for visible light–driven hydrogen production and the oxidation of amines to imines. By integrating MOFs and COFs, the researchers leveraged the covalently connected heterojunction and monodispersed Cu^{2+}/Cu$^+$ centers to significantly enhance photocatalytic performance, as shown in Fig. 8.8. The Cu ions, immobilized by –NH$_2$ groups within the MOF, served as effective active sites, facilitating the separation and migration of photogenerated charge carriers, thus prolonging the lifetime of excited electrons. The result was a hydrogen evolution rate of 9.21 mmol/g/h, which is markedly higher compared to systems without Cu coordination or pristine COFs, and superior to many other MOF and COF-based photocatalysts without Pt. This work demonstrates the effectiveness of combining MOFs and COFs in improving the efficiency and stability of photocatalysts for water splitting and other reactions. In conclusion, the combination of MOFs and COFs in the resulting composites offers significant advantages in photocatalytic applications. The integration facilitates more efficient charge separation and transfer during water splitting, thereby improving the overall efficiency of the process. Additionally, the porous structures inherited from both MOFs and COFs enhance the diffusion of reactants and products, further optimizing the performance of these composites in HER and OER processes. This synergy between MOFs and COFs not only improves photocatalytic activity but also ensures greater stability and durability in various catalytic reactions.

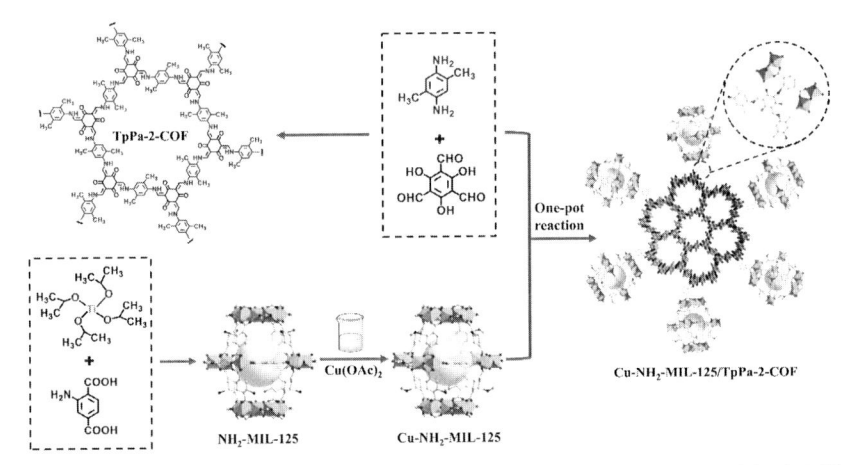

FIGURE 8.8 Schematic illustration of the synthesis of Cu-NH$_2$-MIL-125/TpPa-2-COF hybrid material (Han et al., 2022).

Metal-organic framework/metal-organic framework composites

Hybrid materials created by combining two or more different MOFs are known as MOF/MOF composites. The concept of MOF/MOF composites involves integrating different MOFs to create a new material that benefits from the properties of each individual framework. For example, one MOF might offer superior stability, while another provides enhanced catalytic activity. When combined, the composite can exhibit both of these advantages, leading to a material that is more versatile and effective than either of the original MOFs alone. These composites are particularly valuable because they allow for the customization of material properties. By carefully selecting and combining MOFs, researchers can design composites with specific characteristics tailored for particular applications. This has led to significant advancements in fields like energy storage, where MOF/MOF composites can be used to improve the efficiency and durability of batteries and supercapacitors. Additionally, these composites can offer improved selectivity and capacity in gas storage and separation, making them valuable for environmental and industrial processes. Overall, MOF/MOF composites represent a promising area of research, with the potential to create materials that combine the best features of different MOFs, leading to new and enhanced functionalities that can be applied in a wide range of scientific and industrial fields. Creating MOF/MOF composites involves carefully combining different MOFs to form a new material with enhanced or novel properties. The process of making these composites can vary depending on the desired characteristics of the final material, but several common methods are widely used in the field. One popular approach is the layer-by-layer assembly technique. In this method, different MOFs are sequentially grown on top of each other. The process begins by immersing a substrate in a solution containing the precursors for the first MOF, allowing it to crystallize and form a thin layer. After washing, the substrate is then immersed in a solution for the second MOF, which grows on top of the first. This method allows for precise control over the thickness and composition of each layer, making it possible to tailor the properties of the composite for specific applications, such as catalysis or gas separation. Another method used to create MOF/MOF composites is direct mixing and cocrystallization. In this process, the precursors for two or more MOFs are mixed together in a single solution, where they crystallize simultaneously to form a composite material. By carefully controlling factors such as temperature, pH, and concentration of the reactants, researchers can encourage the different MOFs to integrate at the molecular level. This method can produce composites with highly interconnected structures, leading to materials that exhibit synergistic properties where the combined performance of the MOFs is greater than the sum of their individual capabilities. Finally, postsynthetic modification is another technique used to create MOF/MOF composites. In this approach, a primary MOF is synthesized first, and then it undergoes a secondary modification to incorporate a second MOF. This can be done through processes like soaking the primary MOF in a solution containing the precursors for the second MOF, allowing it to grow within or on the surface of the existing structure. This method is particularly useful for creating composites with complex architectures and enhanced functionalities, such as increased stability or improved catalytic activity (Chai et al., 2021; Farooq et al., 2024; Kazemi, Afshari, et al., 2025).

The combination of different MOFs allows the composite to benefit from the catalytic strengths of each MOF. This can lead to higher efficiency in both HER and OER, improving the overall rate of water splitting. Since water splitting is a demanding process that requires catalysts to be stable under harsh conditions. MOF/MOF composites often exhibit enhanced stability compared to individual MOFs, making them more durable and reliable for long-term use in water-splitting applications. The interaction between different MOFs in a composite can lead to synergistic effects, where the combined properties of the MOFs produce a better catalytic performance than the sum of their individual contributions. This can result in lower energy requirements and higher hydrogen production rates. Melillo et al. (2023) reported the investigation of photocatalytic activity in various MOF-on-MOF composites for overall water splitting under simulated sunlight. They prepared composites with UiO-66 topologies, including UiO-66(Ce), UiO-66(Zr)-NH$_2$, UiO-66(Zr)-NH$_2$@UiO-66(Ce), and UiO-66(Ce)@UiO-66(Zr)-NH$_2$. The UiO-66(Zr)-NH$_2$@UiO-66(Ce) composite exhibited the most promising photocatalytic performance, achieving 708 μmol/g of H$_2$ and 320 μmol/g of O$_2$ after 22 hours of sunlight irradiation. This enhanced activity was attributed to its optimal energy band diagram and efficient photoinduced charge separation. The UiO-66(Zr)-NH$_2$@UiO-66(Ce) composite showed three times higher photocatalytic activity compared to the individual UiO-66(Zr)-NH$_2$ and UiO-66(Ce) solids. The results were competitive with existing MOF-based photocatalysts, and the study is expected to aid in the development of effective MOF-based photocatalysts for solar-driven water splitting. This study demonstrated that the UiO-66(Zr)-NH$_2$@UiO-66(Ce) MOF-on-MOF composite significantly outperforms individual MOFs in photocatalytic water splitting under simulated sunlight. This composite achieved superior hydrogen and oxygen production, highlighting its potential for efficient solar-driven water-splitting applications. MOF/MOF composites are very important in increasing water-splitting efficiency by integrating the strengths of different MOF components and optimizing their structural properties. This approach significantly improves the catalytic activity and promotes the potential of MOF/MOF composites for advanced water-splitting applications. Rui et al. (2018) reported the synthesis and evaluation of 2D MOF nanosheets decorated with Fe-MOF nanoparticles for water oxidation catalysis in an alkaline medium. They demonstrated that incorporating electrochemically inert Fe-MOF nanoparticles onto active Ni-MOF or NiCo-MOF nanosheets significantly enhanced catalytic activity. The hybrid Ni-MOF@Fe-MOF catalyst showed a reduced overpotential of 265 mV to achieve a current density of 10 mA/cm^2, with a ≈100 mV improvement due to the synergistic effects and 2D nanosheet morphology. The study also revealed that during the OER, NiO nanograins formed in situ, acting as active centers and contributing to the performance of the porous nanosheet catalysts. These findings offer new insights into MOF-based catalysts and pave the way for designing efficient MOF-derived nanostructures for electrocatalysis. MOF-on-MOF composites can exhibit improved properties such as increased stability, higher surface area, and enhanced catalytic activity. This is due to the synergistic effects that arise from combining different MOFs. Various techniques can be employed to create MOF-on-MOF composites, including layer-by-layer assembly, in situ growth, and postsynthetic modification. The method chosen can significantly affect the properties and performance of the final composite. Chen et al. (2022) reported the successful synthesis of a novel hybrid MOF composite, UiO-66-on-ZIF-67, denoted as MZU-Co$_x$Zr$_y$. This MOF/MOF composite demonstrated highly efficient and stable bifunctional performance for both the OER and HER in an alkaline medium, as shown in Fig. 8.9. Specifically, the MZU-Co$_{2.5}$Zr$_1$ composite showed remarkable OER performance with a low overpotential of 252 mV and an HER overpotential of 172 mV at 10 mA/cm^2. When used in a water-splitting device, it achieved low total potentials of 1.56 V at 10 mA/cm^2 and 1.59 V at 30 mA/cm^2, highlighting its excellent overall water-splitting performance. This work underscores the critical role of MOF/MOF composites in enhancing electrocatalytic efficiency and provides guidance for designing effective non-precious catalysts for energy conversion.

Other metal-organic framework composites

Metal-organic framework/carbon nanotube composites

MOF–carbon nanotube (CNT) composites have gained significant attention as advanced materials for water splitting, a crucial process for clean hydrogen

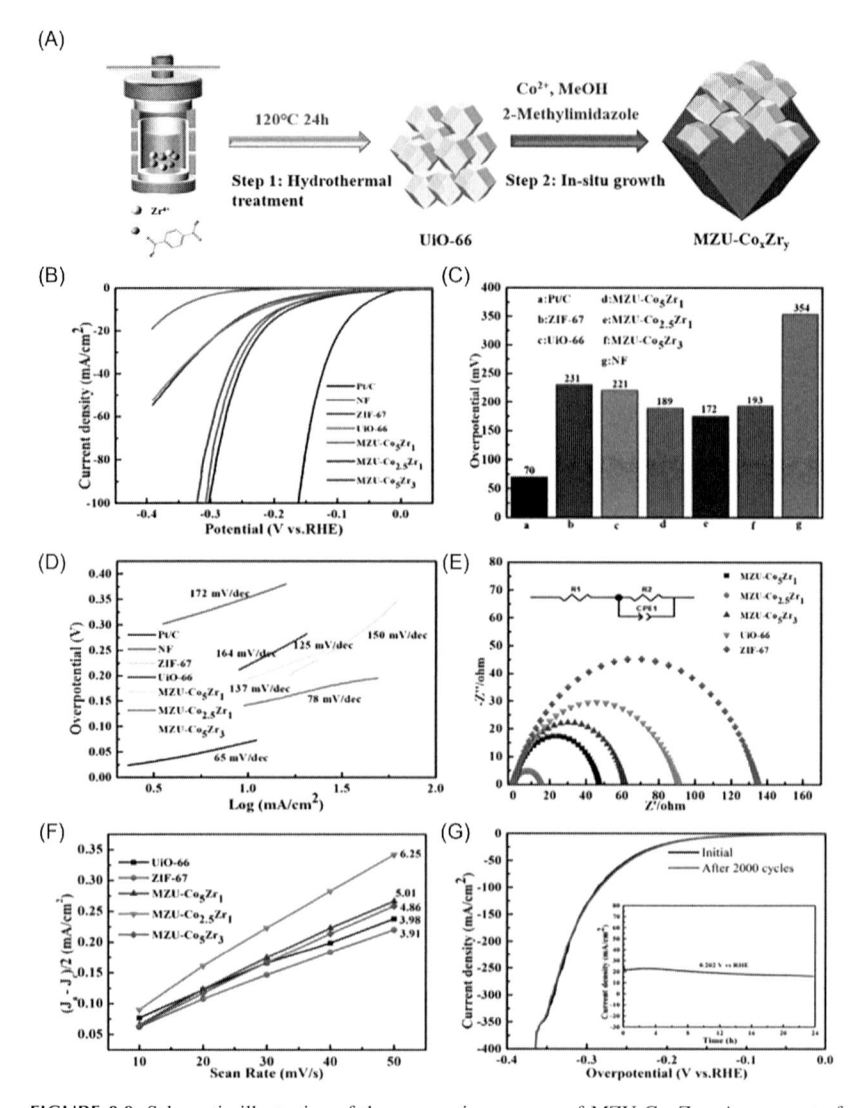

FIGURE 8.9 Schematic illustration of the preparation process of MZU-CoxZry. Assessment of HER performance of different catalysts in 1 mol/L KOH solution: (B) LSV polarization curves; (C) Overpotential required to achieve a current density of 10 mA/cm^2; (D) Tafel plots; (E) Electrochemical impedance spectroscopy (EIS); (F) Double-layer capacitance (Cdl) of UiO-66, ZIF-67, MZU-Co$_5$Zr$_1$, MZU-Co$_{2.5}$Zr$_1$, and MZU-Co$_5$Zr$_3$; (G) LSV curves of MZU-Co$_{2.5}$Zr$_1$ after the 1st and 2000th cycles (Chen et al., 2022).

production. MOF/CNT composites combine the exceptional electrical conductivity of CNTs with the high surface area, tunable porosity, and catalytic activity of MOFs, making them highly effective for this application. CNTs, composed of hexagonally arranged carbon atoms, provide a conductive network that facilitates rapid electron transfer, reducing charge recombination in water-splitting reactions. Their excellent mechanical stability further ensures long-term catalytic durability. Meanwhile, MOFs, with their high surface area and adjustable bandgap, serve as active catalytic sites for HER and OER. The synergy between CNTs and MOFs enhances charge separation, improves catalytic efficiency, and increases hydrogen production rates. In electrocatalytic water splitting, CNTs act as efficient charge transport channels, while MOFs provide abundant active sites for redox reactions. This combination minimizes energy loss and enhances overall catalytic performance. In photocatalytic applications, MOFs absorb solar energy, generating electron-hole pairs that are efficiently transferred by CNTs, leading to improved hydrogen generation under light irradiation. Furthermore, the porous structure of MOFs prevents CNT agglomeration, maintaining high catalytic efficiency over prolonged use. The integration of CNTs with MOFs also enhances catalyst stability, reducing degradation and improving recyclability. Various synthetic approaches, such as solvothermal and hydrothermal methods, allow precise control over the morphology and composition of MOF/CNT composites, optimizing their performance for water splitting. MOF/CNT composites represent a promising class of materials for sustainable hydrogen production, addressing key challenges in electrocatalysis and photocatalysis. Their ability to enhance charge transport, provide tunable catalytic properties, and improve stability makes them strong candidates for next-generation hydrogen energy technologies. (Chronopoulos et al., 2022; Zheng et al., 2019). Yaqoob et al. (2021) reported the electrocatalytic behavior of a cost-effective and efficient iron-nickel 2-aminoterephthalic acid MOF (FeNiNH$_2$BDC MOF) and its composites with 2–6 wt.% CNTs for the OER in alkaline media. Fig. 8.10 shows a detailed structural and morphological analysis of the solvothermally fabricated Fe–Ni-based MOF and its CNT composites using techniques. The synergetic effect of the bimetallic MOF and the conductive CNT support was found to enhance the catalytic performance of the MOF in water splitting. Specifically, the FeNiNH$_2$BDC MOF composite with 5 wt.% CNTs demonstrated a current density of 10 mA/cm² at an overpotential (η) of 0.22 V and an onset potential of 1.36 V versus RHE. These results are comparable to the performance of RuO$_2$, a state-of-the-art catalytic material for the OER in fuel cells. Moreover, the FeNiNH$_2$BDC-5 wt% CNTs composite exhibited a Tafel slope of 68.50 mV/dec and a turnover frequency (TOF) of 0.67 1/s, highlighting its potential as a promising alternative to more expensive catalysts based on Pt, Pd, Ir, and Ru for water splitting applications.

Metal-organic framework/metal sulfide composites

In the context of water splitting, MOF-metal sulfide composites have emerged as highly effective catalysts, particularly for the HER and the OER. Metal sulfides are a class of compounds composed of metal cations and sulfide anions. These materials are well-known for their semiconducting properties,

(A)

Metal salts+ Linker (1:1) +15ml DMF +2, 4, 5, 6 wt % CNTs

↓

Homogenous solution stirred for 1 hour followed by 2 hours' sonication after CNTs addition

↓

Resulting mixture poured into autoclave was heated overnight at 120 °C

↓

Solid brown product thoroughly washed with DMF and ethanol and vacuum dried at 75 °C

(B)

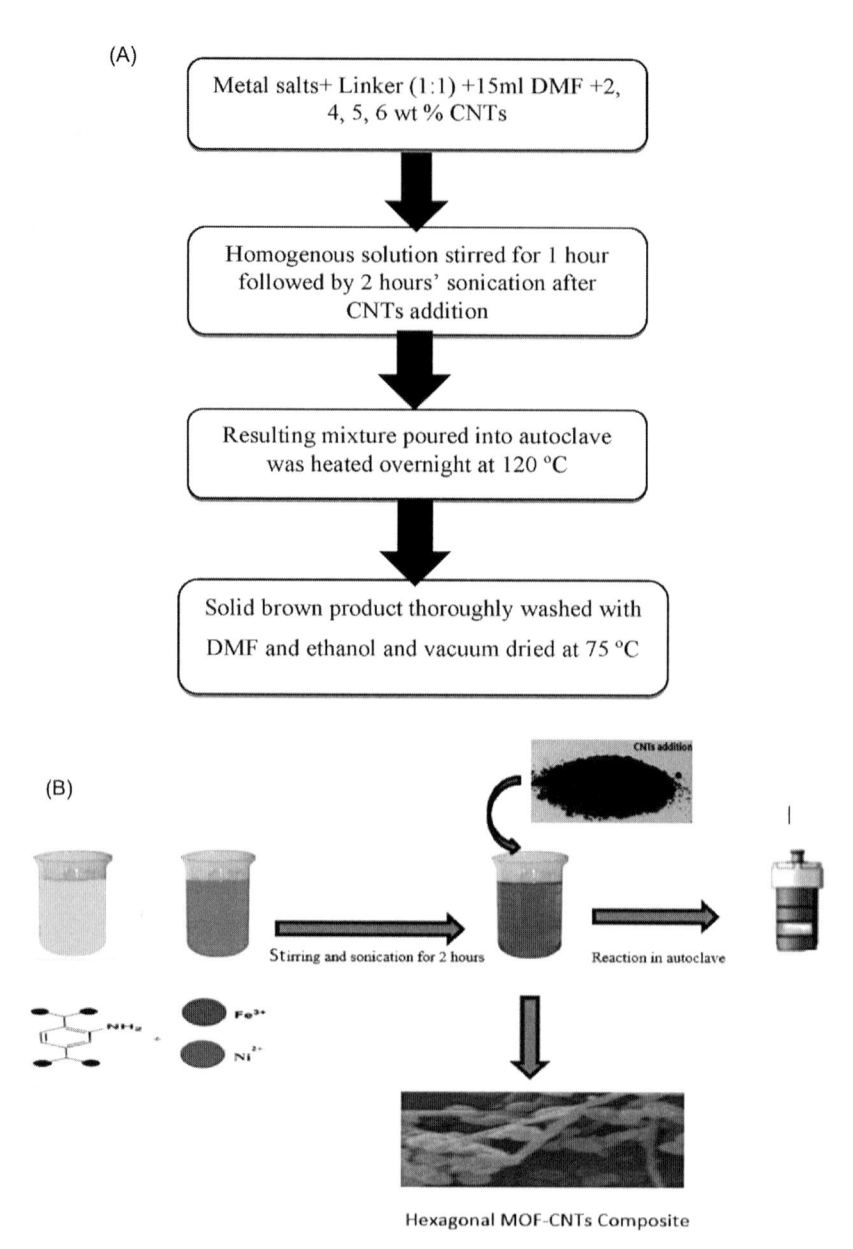

FIGURE 8.10 Stepwise fabrication process of FeNiNH$_2$BDC MOF and 2,4,5,6 wt.% CNTs composites. (B) Chart of synthetic scheme. *CNTs*, Carbon nanotubes (Yaqoob et al., 2021).

which make them highly suitable for applications in catalysis, energy storage, and electronic devices. Metal sulfides like molybdenum disulfide (MoS$_2$), nickel sulfide (NiS), and cobalt sulfide (CoS) have garnered particular interest in the field of catalysis due to their ability to facilitate redox reactions. They exhibit excellent catalytic activity, especially in reactions involving the transfer of electrons, such as the HER in water splitting. Additionally, metal sulfides possess a relatively high surface area and a range of tunable properties, such as bandgap and conductivity, which can be optimized for specific catalytic applications. MOF-metal sulfide composites are advanced materials that combine the structural advantages of MOFs with the catalytic properties of metal sulfides. In these composites, metal sulfides are either encapsulated within the porous structure of MOFs or formed on the surface of MOFs through various synthetic techniques. The combination of MOFs and metal sulfides creates a synergistic effect that enhances the overall performance of the material in catalytic applications. The porous nature of MOFs allows for the dispersion of metal sulfides, preventing agglomeration and maximizing the exposure of active sites. Moreover, the MOF matrix can stabilize the metal sulfides, protecting them from degradation and prolonging their catalytic life. The metal sulfides in the composite act as active sites for these reactions, facilitating the transfer of electrons necessary to split water into hydrogen and oxygen. The MOF component enhances the catalytic performance by providing a high surface area that increases the accessibility of reactants to the active sites. Additionally, the tunable structure of MOFs allows for the incorporation of various functional groups or metal centers that can further improve the efficiency of the catalytic process. MOF-metal sulfide composites are particularly advantageous in electrocatalytic water splitting. The metal sulfides provide excellent catalytic activity for the HER, while the MOF framework offers structural support and additional catalytic sites for the OER. This dual functionality is crucial for improving the overall efficiency of water splitting, as it ensures that both half-reactions proceed at an accelerated rate. Moreover, the integration of metal sulfides within the MOF matrix can enhance the stability of the catalyst, preventing the leaching of active components and ensuring long-term durability under operational conditions. Among the various electrocatalyst materials explored for sustainable and efficient hydrogen production, MoS_2 has garnered significant attention due to its distinctive two-dimensional structure, numerous catalytic active sites, and low hydrogen adsorption-free energy. However, the catalytic activity of MoS_2 is primarily derived from the S-Mo edges rather than its basal plane, and its poor conductivity significantly hampers its overall catalytic performance. Enhancing its activity for practical applications continues to be a major challenge (Hosseini & Safarifard, 2024; Van Nguyen et al., 2024). Liao et al. (2022) reported the synthesis of a highly efficient MoS_2-based electrocatalyst for the HER by constructing a composite system that enhances stability, reduces aggregation, and improves intrinsic conductivity. They synthesized a MoS_2 composite with an erbium-based MOF (Er-MOF/MoS_2) using a hydrothermal method. The Er-MOF/MoS_2 composite exhibited enhanced conductivity, more exposed active sites, and excellent HER activity in acidic media, with an overpotential of only 234 mV required to achieve 10 mA/cm^2 and a Tafel slope of 54.3 mV/dec. This work demonstrates a novel and effective strategy for preparing high-performance MoS$_2$-based electrocatalysts using a porous structure of rare earth MOF materials, as shown in Fig. 8.11.

In photocatalytic water splitting, MOF-metal sulfide composites also demonstrate significant potential. The metal sulfides can absorb visible light and generate electron-hole pairs, which are essential for driving the water-splitting reactions. The MOF component helps in the efficient separation and transfer of these charge carriers, reducing recombination losses and increasing the quantum efficiency of the process. This makes MOF-metal sulfide composites highly effective in converting solar energy into chemical energy, offering a promising approach to sustainable hydrogen production. Overall, while noble metals like Pt, Ti, Ru, and Ir are still some of the most efficient catalysts, nonnoble materials are rapidly catching up. Their lower cost, coupled with ongoing improvements in their catalytic performance, makes them attractive candidates for large-scale water-splitting applications (Ghamami et al., 2020; Li & Baek, 2019). Qiao et al. (2022) reported the development of an efficient photocatalyst for hydrogen production by optimizing the synthesis conditions for MoS$_2$ and MOF. They established that the optimal synthesis of MoS$_2$ involved a calcination temperature of 190°C and the addition of 1 wt.% polyethylene glycol surfactant. For the MOF synthesis, the best conditions included using copper nitrate as the copper precursor, a 30% ultrasonic amplitude, and a calcination temperature of 240°C. When 1 wt.% MOF was added to MoS$_2$, a flower-like structure with small particle size, uniform distribution, and large surface pores was formed. The resulting MOF/MoS$_2$ composite exhibited a modified unit with rough, porous octahedral structures that had a high specific surface area. This composite demonstrated a highly negative conduction band edge (−0.135 V), the smallest charge-transfer resistance (Rct=1.78 Ω), the largest photocurrent (11.1 mA/cm²), and low PL spectral peak intensity, along with excellent photocatalytic stability. These characteristics led to more active sites, enhanced electron transfer rates, and inhibited electron–hole recombination, ultimately achieving a maximum hydrogen production capacity of 626.3 μmol/g/h. Overall, MOF-metal sulfide composites represent a powerful class of materials for water-splitting applications. By combining the catalytic prowess of metal sulfides with the structural and functional versatility of MOFs, these composites offer enhanced efficiency, stability, and tunability in the water-splitting process. As research in this area continues, MOF-metal sulfide composites are poised to play a key role in the development of next-generation catalysts for clean energy technologies.

Conclusion and perspectives

This chapter has thoroughly examined the application of MOF composites as both effective electrocatalysts and photocatalysts for water splitting, offering a

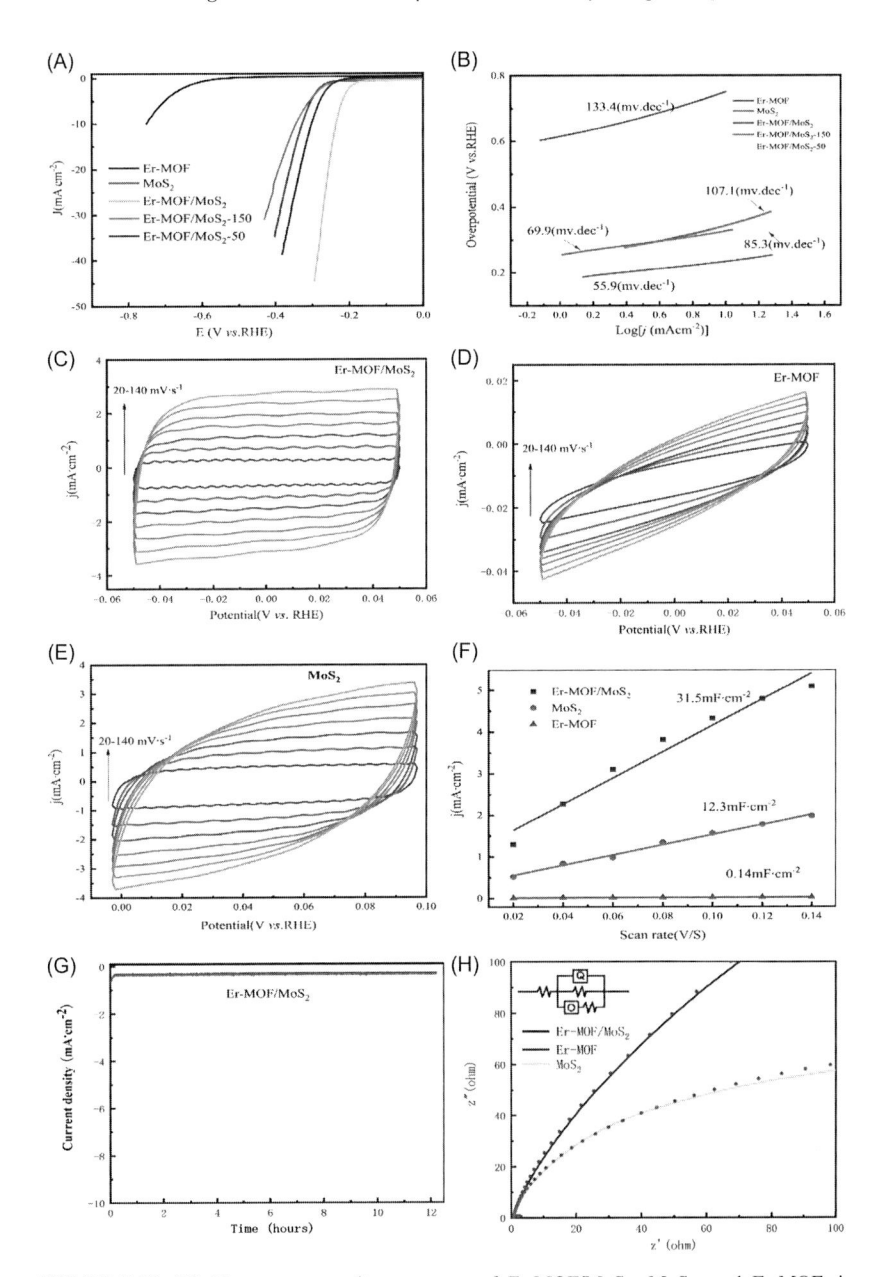

Metal-organic framework composites for water splitting **Chapter | 8** 305

FIGURE 8.11 (A) Linear sweep voltammograms of Er-MOF/MoS$_2$, MoS$_2$, and Er-MOF, in 0.5 M H$_2$SO$_4$ at a scan rate of 5 mV/s. (B) The corresponding Tafel plots for samples. (C–E) Cyclic voltammetry of Er-MOF, MoS$_2$ and Er-MOF/MoS$_2$ at different scan rate in 0.5 M H$_2$SO$_4$ in the non-faradaic potential region. (F) Capacitive current density as a function of scan rate from 20 to 140 mV/s. (G) Stability test for Er-MOF/MoS$_2$ at $\eta = 50$ mV. (H) EIS measure of Er-MOF, MoS$_2$ and Er-MOF/MoS$_2$, inset is corresponding equivalent circuit diagram consisting of an electrolyte resistance (R_s), a charge-transfer resistance (R_{ct}) (Liao et al., 2022).

comprehensive understanding of their potential and challenges. The discussion began by delving into the fundamentals of water splitting, highlighting the critical role that both electrocatalytic and photocatalytic processes play in the generation of clean energy. In the section on electrocatalytic water splitting, this chapter explored the underlying principles and mechanisms that drive HER and OER. These reactions are central to the process of splitting water into its constituent gases, and the chapter provided a detailed analysis of how MOFs can be optimized to enhance these reactions. The section on photocatalytic water splitting further expanded on the mechanisms that enable MOFs to harness solar energy to drive water splitting. This process not only offers a sustainable approach to hydrogen production but also leverages the unique structural features of MOFs to improve efficiency. The ability of MOFs to act as effective photocatalysts is closely tied to their tunable electronic properties and their capacity to facilitate charge separation and transfer. The chapter then moved to explore the merit of MOFs for water splitting, where several key advantages were identified. The extremely high porosity of MOFs, combined with their facile access to active sites, was emphasized as a major contributor to their catalytic performance. This high porosity allows for greater interaction between the catalytic sites and the reactants, enhancing the overall efficiency of the water-splitting process. The importance of optimizing metal nodes within MOFs was also discussed. By carefully selecting and positioning metal ions within the framework, it is possible to enhance the catalytic activity of the MOF. This section highlighted how different metal nodes can be tailored to improve the efficiency of HER and OER. Ligand engineering was another critical aspect covered in the chapter. This chapter detailed how modifying the ligands that connect the metal nodes within MOFs can lead to significant improvements in catalytic performance. Ligand modifications can influence the electronic properties of the MOF, as well as its stability and reactivity. The chapter did not shy away from addressing the challenges and strategies for improving water splitting over MOF-based electrocatalysts. It acknowledged the limitations of current MOF materials, particularly in terms of conductivity and stability, and offered insights into various strategies that could be employed to overcome these challenges. These strategies include the incorporation of conductive materials and the design of more stable MOF structures that can withstand the harsh conditions of water-splitting reactions.

In the section on Composite MOFs, this chapter explored the innovative approaches being taken to enhance the performance of MOFs by creating composites. The MOF/graphene composite was discussed for its ability to improve electrical conductivity, while the MOF/MXene composite was noted for its unique structural properties that can enhance both catalytic activity and stability. The chapter also covered MOF/MNP composites, which offer additional catalytic sites and improved electron transfer capabilities. This chapter then turned to composites with porous materials, such as MOF/COF composites and MOF/MOF composites. These materials were shown to provide synergistic effects that can lead to enhanced catalytic performance. The section highlighted how the combination of different porous materials can create a more efficient catalytic system. Finally, this chapter explored other MOF composites, such as MOF/CNT composites and MOF/metal sulfide composites. These materials offer unique properties that can further enhance the performance of MOFs in water-splitting applications. This chapter highlighted the versatility of MOFs and their composites, showing how they can be tailored to meet specific needs in the field of water splitting. This chapter has attempted to provide a comprehensive overview of the significant advancements in the use of MOF composites for water splitting. It has highlighted both the potential and the challenges associated with these materials and offered a clear direction for future research. With continued innovation and strategic development, MOF-based composites are poised to play a crucial role in the advancement of sustainable energy technologies, particularly in the area of water splitting.

References

Adegoke, K. A., & Maxakato, N. W. (2022). Empirical analysis and recent advances in metal-organic framework-derived electrocatalysts for oxygen reduction, hydrogen and oxygen evolution reactions. *Materials Chemistry and Physics, 289*, 126438.

Ahmad, H., Kamarudin, S., Minggu, L. J., & Kassim, M. (2015). Hydrogen from photo-catalytic water splitting process: A review. *Renewable and Sustainable Energy Reviews, 43*, 599–610.

Chai, L., Pan, J., Hu, Y., Qian, J., & Hong, M. (2021). Rational design and growth of MOF-on-MOF heterostructures. *Small (Weinheim an der Bergstrasse, Germany), 17*(36), 2100607.

Chen, P., Wang, M., Li, G., Jiang, H., Rezaeifard, A., Jafarpour, M., Wu, G., & Rao, B. (2022). Construction of ZIF-67-On-UiO-66 catalysts as a platform for efficient overall water splitting. *Inorganic Chemistry, 61*(46), 18424–18433.

Chronopoulos, D. D., Saini, H., Tantis, I., Zbořil, R., Jayaramulu, K., & Otyepka, M. (2022). Carbon nanotube based metal–organic framework hybrids from fundamentals toward applications. *Small (Weinheim an der Bergstrasse, Germany), 18*(4), 2104628.

Farooq, A., Nazir, M. S., ul Hassan, S., Akhtar, M. N., Hussain, M., Farooq, M., Aslam, A. A., Khan, A. A., & Ali, Z. (2024). Synergistic strategies in MOF on MOF photocatalysts: Review on exploring sustainable hydrogen generation from water splitting. *Nano-Structures & Nano-Objects, 39*, 101295.

Fujishima, A., & Honda, K. (1972). Electrochemical photolysis of water at a semiconductor electrode. *Nature, 238*(5358), 37–38.

Gao, G., Chen, X., Han, L., Zhu, G., Jia, J., Cabot, A., & Sun, Z. (2024). Advances in MOFs and their derivatives for non-noble metal electrocatalysts in water splitting. *Coordination Chemistry Reviews, 503*, 215639.

Ghamami, S., Kazemi Korayem, A., & Baqeri, N. (2020). Production and purification of titanium dioxide with titanium tetrachloride nanoparticles from eliminate concentrate of Kahnooj mine in Kerman. *Applied Chemistry Today, 15*(55), 189–206.

Gopi, S., Panda, A., Ramu, A., Theerthagiri, J., Kim, H., & Yun, K. (2022). Bifunctional electrocatalysts for water splitting from a bimetallic (V doped-NixFey) Metal–Organic framework MOF@ Graphene oxide composite. *International Journal of Hydrogen Energy, 47*(100), 42122–42135.

Han, Q., Dong, Y., Xu, C., Hu, Q., Dong, C., Liang, X., & Ding, Y. (2020). Immobilization of metal–organic framework MIL-100 (Fe) on the surface of BiVO4: A new platform for enhanced visible-light-driven water oxidation. *ACS applied materials & interfaces, 12*(9), 10410–10419.

Han, W., Shao, L.-H., Sun, X.-J., Liu, Y.-H., Zhang, F.-M., Wang, Y., Dong, P.-Y., & Zhang, G.-L. (2022). Constructing Cu ion sites in MOF/COF heterostructure for noble-metal-free photoredox catalysis. *Applied Catalysis B: Environmental, 317*, 121710.

Hosseini, S. M., & Safarifard, V. (2024). MoS2@ MOF composites: Design strategies and photocatalytic applications. *Materials Science in Semiconductor Processing, 169*, 107892.

Hu, E., Yao, Y., Cui, Y., & Qian, G. (2021). Strategies for the enhanced water splitting activity over metal–organic frameworks-based electrocatalysts and photocatalysts. *Materials Today Nano, 15*, 100124.

Jiao, L., Wang, Y., Jiang, H. L., & Xu, Q. (2018). Metal–organic frameworks as platforms for catalytic applications. *Advanced Materials, 30*(37), 1703663.

Kazemi, A., Afshari, M. H., Baesmat, H., Bozorgnia, B., Manteghi, F., Nabipour, H., Rohani, S., Aliabadi, H. A. M., Adibzadeh, S., & Saeb, M. R. (2024). Polydopamine-coated Zn-MOF-74 nanocarriers: Versatile drug delivery systems with enhanced biocompatibility and cancer therapeutic efficacy. *Journal of Inorganic and Organometallic Polymers and Materials*, 1–14.

Kazemi, A., Afshari, M. H., Baesmat, H., Keshavarz, S., Zeinali, F., Zahiri, S., Torabi, E., Manteghi, F., & Rohani, S. (2025). Room-temperature synthesis of pH-responsive MOF nanocarriers for targeted drug delivery in cancer therapy. *Journal of Polymers and the Environment*, 1–12.

Kazemi, A., Afshari, M. H., Baesmat, H., Manteghi, F., Nabipour, H., Rohani, S., & Saeb, M. R. (2024). Tunable Zn-MOF-74 nanocarriers coated with sodium alginate as versatile drug carriers. *Polymer Bulletin*, 1–17.

Kazemi, A., Hosseini, A. K., Pordsari, M. A., Tamtaji, M., Keshavarz, S., Manteghi, F., Tadjarodi, A., Ghaemi, A., Rohani, S., & Goddard, W. A. (2025). Enhancing CO_2 selectivity in MOFs through a dual-ligand strategy: Experimental and theoretical insights. *Journal of CO_2 Utilization, 93*, 103035.

Kazemi, A., Manteghi, F., & Tehrani, Z. (2024). Metal electrocatalysts for hydrogen production in water splitting. *ACS omega, 9*(7), 7310–7335.

Kazemi, A., Moghadaskhou, F., Pordsari, M. A., Manteghi, F., Tadjarodi, A., & Ghaemi, A. (2023). Enhanced CO2 capture potential of UiO-66-NH2 synthesized by sonochemical method: experimental findings and performance evaluation. *Scientific Reports, 13*(1), 19891.

Kazemi, A., Pordsari, M. A., Tamtaji, M., Afshari, M. H., Keshavarz, S., Zeinali, F., Baesmat, H., Zahiri, S., Manteghi, F., & Ghaemi, A. (2024). Unveiling the power of defect engineering in MOF-808 to enhance efficient carbon dioxide adsorption and separation by harnessing the potential of DFT analysis. *Chemical Engineering Journal, 494*, 153049.

Kazemi, A., Pordsari, M. A., Tamtaji, M., Manteghi, F., Ghaemi, A., Rohani, S., & Goddard, W. A., III (2025). Environmentally friendly synthesis and morphology engineering of mixed-metal MOF for outstanding CO2 capture efficiency. *Chemical Engineering Journal, 505*, 158951.

Kazemi, A., Pordsari, M. A., Tamtaji, M., Zainali, F., Keshavarz, S., Baesmat, H., Manteghi, F., Ghaemi, A., Rohani, S., & Goddard, W. A., III (2025). Eco-friendly synthesis and morphology control of MOF-74 for exceptional CO2 capture performance with DFT validation. *Separation and Purification Technology, 361*, 131328.

Kulkarni, R., Lingamdinne, L. P., Karri, R. R., Momin, Z. H., Koduru, J. R., & Chang, Y.-Y. (2023). Catalytic efficiency of LDH@ carbonaceous hybrid nanocomposites towards water splitting mechanism: Impact of plasma and its significance on HER and OER activity. *Coordination Chemistry Reviews, 497*, 215460.

Kumar, Y. A., Reddy, G. R., Ramachandran, T., Kulurumotlakatla, D. K., Abd-Rabboh, H. S., Hafez, A. A. A., Rao, S. S., & Joo, S. W. (2024). Supercharging the future: MOF-2D MXenes supercapacitors for sustainable energy storage. *Journal of Energy Storage, 80*, 110303.

Lan, Q., Zhang, Z. M., Qin, C., Wang, X. L., Li, Y. G., Tan, H. Q., & Wang, E. B. (2016). Highly dispersed polyoxometalate-doped porous Co3O4 water oxidation photocatalysts derived from POM@ MOF crystalline materials. *Chemistry–A European Journal, 22*(43), 15513–15520.

Li, C., & Baek, J.-B. (2019). Recent advances in noble metal (Pt, Ru, and Ir)-based electrocatalysts for efficient hydrogen evolution reaction. *ACS Omega, 5*(1), 31–40.

Li, G., Zhang, K., Li, C., Gao, R., Cheng, Y., Hou, L., & Wang, Y. (2019). Solvent-free method to encapsulate polyoxometalate into metal-organic frameworks as efficient and recyclable photocatalyst for harmful sulfamethazine degrading in water. *Applied Catalysis B: Environmental, 245*, 753–759.

Li, J., Huang, H., Liu, P., Song, X., Mei, D., Tang, Y., Wang, X., & Zhong, C. (2019). Metal-organic framework encapsulated single-atom Pt catalysts for efficient photocatalytic hydrogen evolution. *Journal of Catalysis, 375*, 351–360.

Li, Z., Guo, Y., Li, K., Wang, S., De Bonis, E., Cao, H., Mertens, S. F., & Teng, C. (2023). Shape control of bimetallic MOF/Graphene composites for efficient oxygen evolution reaction. *Journal of Electroanalytical Chemistry, 930*, 117144.

Liao, J., Xue, Z., Sun, H., Xue, F., Zhao, Z., Wang, X., Dong, W., Yang, D., & Nie, M. (2022). MoS2 supported on Er-MOF as efficient electrocatalysts for hydrogen evolution reaction. *Journal of Alloys and Compounds, 898*, 162991.

Liu, W., Cao, D., & Cheng, D. (2021). Review on synthesis and catalytic coupling mechanism of highly active electrocatalysts for water splitting. *Energy Technology, 9*(2), 2000855.

Liu, Y., Huang, D., Cheng, M., Liu, Z., Lai, C., Zhang, C., Zhou, C., Xiong, W., Qin, L., & Shao, B. (2020). Metal sulfide/MOF-based composites as visible-light-driven photocatalysts for enhanced hydrogen production from water splitting. *Coordination Chemistry Reviews, 409*, 213220.

Liu, Y., Yang, X., Guo, R., & Yao, Z. (2022). Engineering MOF-based nanocatalysts for boosting electrocatalytic water splitting. *International Journal of Hydrogen Energy, 47*(92), 39001–39017.

Melillo, A., Cabrero-Antonino, M., Ferrer, B., Dhakshinamoorthy, A., Baldoví, H. G., & Navalón, S. (2023). MOF-on-MOF composites with UiO-66-based materials as photocatalysts for the overall water splitting under sunlight irradiation. *Energy & Fuels, 37*(7), 5457–5468.

Qiao, Z., Wang, W., Liu, N., Huang, H.-T., Karuppasamy, L., Yang, H.-J., Liu, C.-H., & Wu, J. J. (2022). Synthesis of MOF/MoS2 composite photocatalysts with enhanced photocatalytic performance for hydrogen evolution from water splitting. *International Journal of Hydrogen Energy, 47*(96), 40755–40767.

Quan, L., Jiang, H., Mei, G., Sun, Y., & You, B. (2024). Bifunctional Electrocatalysts for Overall and Hybrid Water Splitting. *Chemical Reviews, 124*(7), 3694–3812.

Rui, K., Zhao, G., Chen, Y., Lin, Y., Zhou, Q., Chen, J., Zhu, J., Sun, W., Huang, W., & Dou, S. X. (2018). Hybrid 2D dual-metal–organic frameworks for enhanced water oxidation catalysis. *Advanced Functional Materials, 28*(26), 1801554.

Shuai, Q., Li, W.-L., Zhao, C., & Yu, J. (2024). Recent advances of computational simulations on carbon capture in MOFs. *Materials Today Communications,* 110050.

Tao, Z., Wang, T., Wang, X., Zheng, J., & Li, X. (2016). MOF-derived noble metal free catalysts for electrochemical water splitting. *ACS Applied Materials & Interfaces, 8*(51), 35390–35397.

Torabi, E., Kazemi, A., Tamtaji, M., Manteghi, F., Rohani, S., & Goddard, W. A. (2025). Sacrificial MOF-derived MnNi hydroxide for high energy storage supercapacitor electrodes via DFT-based quantum capacitance study. *Heliyon, 11*(1).

Van Nguyen, T., Tekalgne, M., Van Le, Q., Van Tran, C., Ahn, S. H., & Kim, S. Y. (2024). Recent progress and strategies of non-noble metal electrocatalysts based on MoS2/MOF for the hydrogen evolution reaction in water electrolysis: an overview. *Microstructures, 4*(4) N/A-N/A.

Wu, Y., Lv, H., & Wu, X. (2024). Design of two-dimensional porous covalent organic framework semiconductors for visible-light-driven overall water splitting: A theoretical perspective. *Chinese Journal of Structural Chemistry,* 100375.

Yan, L., Xu, Y., Chen, P., Zhang, S., Jiang, H., Yang, L., Wang, Y., Zhang, L., Shen, J., & Zhao, X. (2020). A freestanding 3D heterostructure film stitched by MOF-derived carbon nanotube microsphere superstructure and reduced graphene oxide sheets: a superior multifunctional electrode for overall water splitting and Zn–air batteries. *Advanced Materials, 32*(48), 2003313.

Yang, J., Shen, Y., Sun, Y., Xian, J., Long, Y., & Li, G. (2023). Ir nanoparticles anchored on metal-organic frameworks for efficient overall water splitting under pH-universal conditions. *Angewandte Chemie, 135*(17), e202302220.

Yang, M., Zhou, Y.-N., Cao, Y.-N., Tong, Z., Dong, B., & Chai, Y.-M. (2020). Advances and challenges of Fe-MOFs based materials as electrocatalysts for water splitting. *Applied Materials Today, 20,* 100692.

Yang, Q., Xu, Q., & Jiang, H.-L. (2017). Metal–organic frameworks meet metal nanoparticles: synergistic effect for enhanced catalysis. *Chemical Society Reviews, 46*(15), 4774–4808.

Yaqoob, L., Noor, T., Iqbal, N., Nasir, H., Zaman, N., & Talha, K. (2021). Electrochemical synergies of Fe–Ni bimetallic MOF CNTs catalyst for OER in water splitting. *Journal of Alloys and Compounds, 850,* 156583.

Ye, B., Jiang, R., Yu, Z., Hou, Y., Huang, J., Zhang, B., Huang, Y., Zhang, Y., & Zhang, R. (2019). Pt (1 1 1) quantum dot engineered Fe-MOF nanosheet arrays with porous core-shell as an electrocatalyst for efficient overall water splitting. *Journal of Catalysis, 380,* 307–317.

Yusran, Y., Zhao, J., Chen, F., & Fang, Q. (2023). Recent advances of covalent organic frameworks as water splitting electrocatalysts. *Organic Materials, 5*(04), 207–221.

Zhang, F., Qian, Y., Jin, Z., Fei, Z., Zhang, J., Mao, H., Kang, D. J., & Pang, H. (2023). Recent advances in MOFs/MXenes composites: Synthesis and their electrochemical energy applications. *Journal of Energy Storage, 72,* 108213.

Zhang, Z., Ran, J., Fan, E., Zhou, S., Chai, D.-F., Zhang, W., Zhao, M., & Dong, G. (2023). Mesoporous CoMoO4 hollow tubes derived from POMOFs as efficient electrocatalyst for overall water splitting. *Journal of Alloys and Compounds, 968,* 172169.

Zhang, Z.-M., Zhang, T., Wang, C., Lin, Z., Long, L.-S., & Lin, W. (2015). Photosensitizing metal–organic framework enabling visible-light-driven proton reduction by a Wells–Dawson-type polyoxometalate. *Journal of the American Chemical Society, 137*(9), 3197–3200.

Zhao, L., Dong, B., Li, S., Zhou, L., Lai, L., Wang, Z., Zhao, S., Han, M., Gao, K., & Lu, M. (2017). Interdiffusion reaction-assisted hybridization of two-dimensional metal–organic frameworks and Ti3C2T x nanosheets for electrocatalytic oxygen evolution. *ACS Nano, 11*(6), 5800–5807.

Zheng, F., Cao, S., Yang, Z., Sun, Y., Shen, Z., Wang, Y., & Pang, H. (2024). *Synthesis and catalytic application of MOF complexes containing noble metals. Energy & Fuels*.

Zheng, F., Zhang, Z., Zhang, C., & Chen, W. (2019). Advanced electrocatalysts based on metal–organic frameworks. *ACS Omega, 5*(6), 2495–2502.

Zong, H., Qi, R., Yu, K., & Zhu, Z. (2021). Ultrathin Ti2NTx MXene-wrapped MOF-derived CoP frameworks towards hydrogen evolution and water oxidation. *Electrochimica Acta, 393,* 139068.

Chapter 9

Metal-organic framework composites for food applications

P.S. Sharanyakanth, and Mahendran Radhakrishnan
Centre of Excellence in Nonthermal Processing, National Institute of Food Technology, Entrepreneurship and Management, Thanjavur (NIFTEM-T, formerly Indian Institute of Food Processing Technology, IIFPT), Thanjavur, Tamil Nadu, India

Introduction

Chemists from coordination and solid-state/zeolite chemistry conducted the initial investigation of metal-organic frameworks (MOFs) Recently, material scientists have also become increasingly interested in MOFs. Similar to how chemists modified MOFs postsynthesis through a sequence of chemical reactions to add functionalities, material scientists also draw knowledge from the creation of composite materials and often incorporate functional species—such as nanoparticles, organic dyes, nanofibers, small biomolecules, polymers, and nanowires—into MOFs through the composites they form. MOFs can be grown or deposited on a two-dimensional planar or curved substrate to form thin films, or they can be combined with other materials to make mixed structures to create MOF composites (Li & Huo, 2015). Given the expanding spectrum of MOFs and their composites, a thorough understanding of the topic is necessary. This chapter focuses mostly on the forms of MOF composites and their application in food systems, as summarized in Fig. 9.1.

Metal-organic framework composites in food applications

Nanoparticle composites

Metal nanoparticles' high chemical processes and specificities have drawn considerable attention. However, these particles tend to assemble and ignite due to their high surface energy and high surface-to-volume ratio. One way to efficiently control the accretion of metal nanoparticles in limited voids is to arrange them into porous materials like metal oxides, carbon, zeolites, and mesoporous silicates. As a novel category of porous substances, MOFs with huge surfaces and porosity are a new family of porous materials that can be used

Applications of Metal-Organic Framework Composites

MOFs	MOF Composite Types	Food Application
ZIF family IRMOF series MOF-5 MIL series HKUST-1	• Nanoparticle • Cellulose • Silica • Enzyme • Activated Carbon • Organic Polymer • Thin film substrate • Polyoxometalate • Graphene Oxide • Molecular species • Quantum Dot • Hybrid • Aluminium	➤ Food nanoreactors ➤ Food material sensors ➤ Active food contact material ➤ Food Cleaning ➤ Immobilizers and stabilizers for enzyme and active compounds ➤ Antimicrobial nanocarriers ➤ Regulated discharge of nano-system for active compound ➤ Nanofillers for food packaging substances

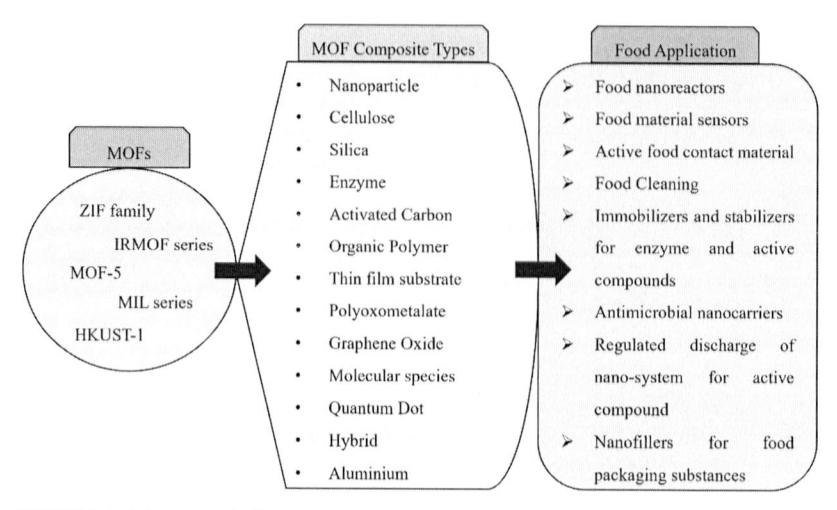

FIGURE 9.1 Meta–organic frameworks composites for food application.

as supports for metal nanoparticles. (Chen & Xu, 2019; Chen et al., 2018; Li et al., 2018; Yang et al., 2012). It has also been demonstrated that composites of MIL-100 (Fe) (MIL: Materials from Institute Lavoisier) and magnetic nanoparticles are rapidly and effortlessly absorbent for the extraction of acid dyes (Fan et al., 2018; Wang et al., 2016), which are harmful contaminants often found in food and beverages (Shar (Ma et al., 2020).

Through theoretical calculations and experimental evidence, Sr-MOF (strontium MOF) has been shown to have semiconducting properties (Usman et al., 2015). The MOF was discovered to have an electrical conductivity rate on the order of 106 S/cm using temperature-dependent current–voltage testing. An increase in temperature during the annealing process led to an exponential increase in the MOF's conductivity. The semiconducting transport actions of Sr-MOF were displayed by the Arrhenius conductivity plot in place of the thermally facilitated carriers and adjustable hopping. A study was published in which a composite of Sr-MOF, produced with a semiconductive organic ligand (1,4,5,8-naphthalene tetracarboxylic acid hydrate), was used to develop a solid-state direct white light-emitting diode illumination (Haider et al., 2016) that is used in water treatment and food processing (Prasad et al., 2020). The above-mentioned electroluminescent Sr-photoluminescent MOF's spectra verified the presence of distinct emission peaks that resulted in strontium's intermetallic electronic transitions, metallic energy state transitions, and the conversion of charges from metal to ligand.

Graphene oxide composites

Graphene oxide (GO) is particularly suitable for the production of MOF composites, and functional oxygen-containing graphene with chemical groups has recently attracted attention due to its exceptional properties, such as mechanical stability, large surface area, optical characteristics, and strong electrical properties (Li et al., 2015). Cobalt-based MOFs (CoMOFs) were simply electrochemically deposited on the reduced graphene oxide (rGO) electrode surface to create the three-dimensional, flexible, freestanding, and CoMOF composite (Topçu, 2020). Zirconium-based MOF (Zr-MOF) and rGO were utilized in an insitu method by Musyoka et al. (2017) to create a composite. Comparing this compound to Zr-MOF, it demonstrated higher hydrogen storage efficiency in storage applications. Besides, zeolitic imidazolate framework/graphene oxide (ZIF-67@GO) composites have effectively removed the carcinogenic dye malachite green from contaminated water. The adsorbent exhibited high removal efficiency and reusability, suggesting its potential as a promising material for addressing water pollution issues in aquaculture (Rehman Shah et al., 2021). Similarly, a magnetic composite (Fe_3O_4@SiO_2-GO-Ni-MOF) was developed for efficient epoxiconazole extraction from food. Combining graphene oxide, silica, and a nickel-based MOF, this material outperformed individual components. Optimized extraction and analysis enabled the detection of epoxiconazole in fruits and vegetables (Chen et al., 2020).

Silica composites

Porosity, stability, and hydrophilicity are just a few of the many nanoscale roles that silica particles and nanostructures can perform. These effective methods have sparked considerable interest in catalytic applications. MOF-silica composites are essentially divided into two categories: SiO_2@MOFs and MOFs@SiO_2. A MOF shell can grow into a prefabricated silica sphere, or dispersed silica particles can be injected into MOF pores or channels in the former method, whilst in the latter, silica is used as a covering layer that is created on the surface of MOF or as a tool to help MOF particles grow (Kou & Sun, 2018; Jo et al., 2011; Uemura et al., 2008). Silica-based nanomaterials, particularly mesoporous silica ($mSiO_2$:Si–O–Si framework), are extensively employed in enzyme immobilization due to their favorable properties such as high surface area, tunable porosity, stability, and biocompatibility. These characteristics render them promising candidates for enhancing enzyme performance in the food processing industry (Wu & Mu, 2022).

Organic polymer composites

MOF-polymer composites offer a promising platform for developing multifunctional materials with applications in various fields, including biomedicine. Integrating polymers with MOFs provides opportunities to combine the desirable properties of both materials, such as flexibility, biocompatibility, and porosity. By incorporating functional groups and targeting ligands, these composites can be designed for specific applications, such as drug delivery, imaging, and sensing. While the primary focus of this research has been on biomedical applications, the underlying principles and design strategies could be adapted for developing innovative materials with potential applications in the food industry (Kitao et al., 2017; Rowe et al., 2009). Xiong et al. (2019) investigated the potential of a Gd-based MOF (Gd-MOF) composite for removing phoxim, a pesticide commonly used in vegetable cultivation. The synthesized Gd-MOF composite exhibited a remarkable adsorption capacity for phoxim. The porous structure of the MOF composite and the interaction between Gd^{3+} ions and the phosphorous–oxygen bond in phoxim were identified as key factors contributing to the efficient removal of the pesticide. Moreover, the MOF demonstrated excellent recyclability, highlighting its potential for practical water treatment and food safety applications.

Quantum dot composites

Flexibility, microporosity, and high surface area of MOF composites with quantum dots allow for the creation of composite constituents with enhanced characteristics for a variety of uses, including energy storage, gas storage, sensing, and photocatalysis (Ramesh et al., 2019). MOF encapsulation of quantum dots (QDs) will increase MOF stability and control electron-hole part recombination rates. When QDs in various forms like oxide-, nitride-, chalcogenide-, and carbon-based compounds are incorporated into MOFs, the resulting composite materials exhibit improved qualities and potential uses (Aguilera-Sigalat & Bradshaw, 2016). For instance, the integration of luminescent carbon quantum dots (CQDs) with molecularly imprinted polymers (MIPs) using MOFs as a scaffold has created a novel sensing platform. The resulting CQDs@MOF@MIP composite demonstrates exceptional sensitivity and selectivity for quercetin detection in complex food matrices. This advanced material effectively converts chemical signals into measurable fluorescent outputs, showcasing its potential as a transformative analytical tool (Pan et al., 2020).

Polyoxometalate composites

Oxygen clusters with anionic metal in various additions, adjustable forms, and characteristics provide excellent prospects in numerous catalytic transformations, especially in reactions involving acids and oxidants. However, their limited stability and low specific area limit their implementation. One interesting method for stabilizing and improving polyoxometalates to increase their catalytic capabilities is to immobilize them in MOFs. Polyoxometalates can be employed as adaptable building blocks (bases, nodes, or prototypes in the cages) for developing polyoxometalate-based MOFs due to their compositional stability and structural robustness. Moreover, host–guest interactions can encapsulate polyoxometalates in MOF pores to create MOF polyoxometalate composites (Du et al., 2014; Wu et al., 2016).

Activated carbon composites

Kayal and Chakraborty (2018) demonstrated the potential of MOF fibers for CO_2 capture and methane recovery. Building on this, subsequent research has focused on MOF-activated carbon composites for removing organic contaminants from water. These composites, synthesized via hydrothermal methods, exhibited superior adsorption capacities compared to their components under various conditions. Mahmoodi et al. (2019) developed a green MOF composite using activated carbon derived from cucumber peel and MIL-101(Cr). Characterization studies confirmed the successful integration of MOF crystals onto the activated carbon surface. The composite's adsorption performance was effectively modeled using the surface reaction approach.

Hasanzadeh et al. (2020) developed a highly porous MIL-101(Cr)-activated carbon composite for efficient anionic dye removal from aqueous solutions. The composite exhibited superior adsorption kinetics compared to pure MIL-101(Cr), achieving 85% dye removal within 5 minutes under near-neutral pH conditions. These results highlight the potential of this material for rapid and effective wastewater treatment.

Thin film on substrates

The emergence of MOF-patterned thin films has ignited significant interest in their potential for nanotechnological applications. Their versatility across domains such as chemical sensing, catalysis, and membrane technology has spurred intensive research into fabricating MOF films with tailored porosity and chemical properties on various substrates (Allendorf et al., 2008; Liu et al., 2011). While this field is relatively nascent, rapid advancements have led to a growing body of literature exploring fabrication methods, characterization techniques, and potential applications (Ameloot et al., 2009; Bux et al., 2011; Hermes et al., 2005; Witters et al., 2013; Yoo & Jeong, 2008; Zhuang et al., 2011; Schoedel et al., 2010). Recent studies have also delved into the fundamental chemistry at the MOF-substrate interface, providing essential insights for future developments (Bétard & Fischer, 2012; Carné et al., 2011; Shekhah et al., 2011; Zacher et al., 2009; Bradshaw et al., 2012).

Several methods have been explored for depositing MOF thin films on substrates. Electrochemical and microwave-assisted approaches, while effective, are limited to conductive substrates (Ameloot et al., 2009; Rodenas et al., 2014; Yoo & Jeong, 2008). Direct growth techniques, including solvothermal and ambient synthesis, often result in polycrystalline films with varying degrees of homogeneity (Guo et al., 2009; Scherb et al., 2008). The layer-by-layer (LBL) method, or liquid phase epitaxy, was developed to address these limitations. This approach involves sequential deposition of organic linkers and metal salts, producing highly uniform and controllable MOF thin films, termed SURMOFs (surface-anchored, crystalline, and oriented metal-organic framework multilayers). The LBL method offers precise control over film thickness, crystallographic orientation, and interpenetration, surpassing the capabilities of traditional direct growth methods (Shekhah et al., 2009; So et al., 2013; Scherb et al., 2008).

Sappia et al. (2020) introduced a novel MOF composite thin film fabricated on a conductive PEDOT (poly(3,4-ethylene dioxythiophene)) substrate for sensitively detecting imazalil, a prevalent fungicide in the food industry. The developed sensor demonstrated exceptional performance in detecting trace amounts of imazalil, highlighting its potential for ensuring food safety and quality control. This technology offers a promising avenue for developing portable and cost-effective sensing platforms for pesticide residue monitoring in the agricultural and food processing sectors.

More recently, Makiura et al. (2010) and Motoyama et al. (2011) and associates have developed the LBL approach in conjunction with the Langmuir–Blodgett (LB) methodology to construct homogenous, highly ordered, and favorably concerned with MOF nanofilms onto nonmodified gold and silica surfaces. Metalloporphyrin building units make up the MOF films, and weak interactions like π stacking are used to stack the layers. With enough flexibility, this film growth technique can create a variety of well-ordered nanofilms with total control over structure growth in both in-plane and out-of-plane orientations concerning the substrate.

Aluminum composites

Alq_3 (tris(8-hydroxyquinoline)), a key component in electroluminescent materials, has shown promise in thin film applications due to its high efficiency and stability. To enhance Alq_3 properties, researchers have explored its incorporation into MOFs. This approach has led to a significant improvement in chromophore lifespan and emission yield by preventing Alq_3 aggregation. Additionally, modulating the Alq_3 content within the MOF composite resulted in a tunable blue emission, attributed to molecular interactions between Alq_3 and the MOF host. These findings underscore the potential of MOFs as effective matrices for tailoring the optical properties of embedded chromophores (Desai et al., 2007; Garbuzov et al., 1996; Yang et al., 2012).

A novel, ultra-sensitive electrochemical sensor was developed utilizing a Bi_2CuO_4@Al-MOF@UiO-67 nanocomposite. Combining the properties of MOFs and Bi_2CuO_4, this unique material architecture enabled simultaneous detection of Cd^{2+}, Cu^{2+}, Pb^{2+}, and Hg^{2+} ions with exceptional sensitivity and selectivity. The sensor demonstrated superior adsorption capacity and electrocatalytic activity towards these heavy metal ions. Importantly, the sensor exhibited remarkably low detection limits for all four analytes, facilitating their quantification in complex food matrices such as rice, sorghum, maize, milk, honey, and tea. This innovative approach holds significant promise for addressing food safety concerns related to heavy metal contamination (Zhang et al., 2024).

Heterostructures/hybrid composites

MOF-based materials have emerged as versatile platforms with applications spanning catalysis, environmental remediation, and energy storage. For instance, MOF composites incorporating tetranitro copper(II) phthalocyanine have shown promise as efficient catalysts, potentially replacing platinum-based alternatives in hydrogen production (Monama et al., 2019). Additionally, MOF/graphene composites have demonstrated exceptional adsorption capacities for pollutants, contributing to improved environmental sustainability. While these materials have not yet been extensively explored in the food industry, their unique properties suggest potential applications in areas such as food packaging, preservation, and safety (Alattar et al., 2018).

A Cu-MOF/graphene composite has been successfully employed to develop an electrochemical sensor for caffeine determination in tea and coffee samples. Compared to bare and graphene-modified electrodes, the composite exhibited enhanced electrocatalytic activity towards caffeine oxidation. The resulting sensor demonstrated good sensitivity, selectivity, and stability, making it a promising tool for caffeine quantification in food products (Venkadesh et al., 2021).

Integrating MOFs into electrochemical sensors has shown promise for sensitively detecting heavy metal contaminants in food and environmental samples. Wang et al. (2021) demonstrated this by developing a graphene aerogel (GA)-supported UiO-66-NH$_2$ MOF composite for the simultaneous detection of Cu^{2+}, Hg^{2+}, Pb^{2+}, and Cd^{2+} ions. The resulting electrochemical sensor exhibited excellent sensitivity and selectivity, enabling the detection of these metal ions at low nanomolar levels in vegetables. This approach holds the potential for developing advanced monitoring systems for food safety and environmental quality assessment.

Enzyme composites

Enzymes, while highly selective biocatalysts, suffer from limitations such as instability and susceptibility to deactivation. To address these challenges, researchers have explored the use of MOFs as enzyme immobilization matrices (Doonan et al., 2017; Lian et al., 2017; Drout et al., 2019). Beyond enzyme immobilization, MOF composites have shown potential for other applications, such as luminescence. For instance, the encapsulation of rhodamine within an adenine-based bio-MOF resulted in enhanced optical properties (Chen et al., 2018).

MOF-composite materials have shown promise as electrochemical sensors for nitrite detection in food samples. Saeb and Asadpour-Zeynali (2022) developed a core-shell ZIF-8@ZIF-67/Au NP composite for nitrite sensing, demonstrating enhanced sensitivity and selectivity compared to unmodified electrodes. The synergistic combination of MOFs and gold nanoparticles contributed to improved catalytic activity. Dong et al. (2017) explored using a Co-TCPP/ionic liquid/Cyt c biohybrid for nitrite detection in pickle juice, highlighting the potential of enzyme-based MOF composites for stable and sensitive sensing. These studies underscore the versatility of MOF-based materials for developing electrochemical sensors with practical applications in food safety monitoring.

Composites of molecular species

MOF composites offer a promising platform for stabilizing and enhancing the performance of molecular catalysts. Traditional catalysts, such as saline complexes, dyes, and porphyrins, suffer from limitations including oxidative degradation and self-aggregation. Encapsulating or integrating these molecules within MOF pores can significantly improve their stability and activity. The controlled pore environment of MOFs prevents catalyst degradation and facilitates interaction with substrates. This approach can revolutionize catalysis by providing a stable and efficient platform for various molecular (Kajiwara et al., 2016; Niu et al., 2018; Wu & Zhao, 2017).

A novel Eu-postfunctionalized bimetallic porphyrin MOF (PCN-221(Zr/Ce)@Eu-DPA-H$_4$btec) has been developed for the sensitive and selective detection of sulfonamide residues in food products. This fluorescent sensor exhibited excellent performance in detecting sulfamethazine, sulfamerazine, and sulfamethoxydiazine with low detection limits. The sensor was successfully applied to real food samples, demonstrating its practical utility. Moreover, the integration of the sensor into a fluorescent hydrogel enabled visual detection of sulfonamides, providing a rapid and user-friendly analytical tool (Jie et al., 2024).

Metal-organic framework cellulose composites

Cellulose, a versatile material with desirable properties such as biodegradability and sustainability, serves as an ideal substrate for MOF composite development (Cai et al., 2012; Demilecamps et al., 2015; Liao et al., 2016; Litschauer et al., 2011). Researchers have successfully integrated MOFs into cellulose aerogels to enhance their adsorption capabilities through a facile in situ growth method. The resulting composite materials exhibit remarkable adsorption capacity for heavy metal ions, such as Pb^{2+} and Cu^{2+}, demonstrating their potential for water treatment applications (Lei et al., 2019). Characterization studies confirmed the successful formation of MOFs within the cellulose matrix and their ability to maintain adsorptive properties after composite formation. These findings highlight the potential of MOF-cellulose composites as sustainable and effective adsorbents for environmental remediation.

A novel cellulose nanocrystal/metal-organic framework (CNC/MOF) composite has been developed as a high-performance stabilizer for Pickering emulsions in food packaging applications. The composite exhibited superior emulsion stabilization compared to pure MOFs, resulting in sized paper with enhanced hydrophobicity, gas barrier properties, and antimicrobial activity. The resulting packaging material effectively extended the shelf life of strawberries by preventing microbial growth and moisture loss. This study demonstrates the potential of CNC/MOF composites as multifunctional materials for advanced food packaging and other applications requiring water resistance and antimicrobial properties (Rui et al., 2024).

Applications of metal-organic framework composites in food systems

Active food contact materials

Various active materials that come in contact with food are studied to outspread the shelf life of food, for instance, ethanol emitters, flavor-absorbing/releasing systems, carbon dioxide absorbers/emitters, moisture absorbers, ethylene absorbers, temperature–time indicators, and oxygen scavengers (Bracone et al., 2016). Likewise, three unlike MOFs, Basolite C300, Zeolite Z13X, and Basolite A520, were used by Chopra et al. (2017) in the form of sachet ethylene (plant food hormone accountable for the aging and ripening of plant foods) absorbers to extend the life of the banana throughout its postharvest. The statistics on the storage of ethylene and its desorption show that MOFs might be a suitable means for bioactive compound delivery; the water vapor usage to attain the release of the volatile compounds is perfectly standardized (Ayala-Zavala et al., 2008). Chopra et al. (2017) identified that all the analyzed MOFs were effectual scavengers of ethylene.

To be beneficial in storage or food packaging, having a delivery system is necessary for 1-MCP (1-methylcyclopropene; plant regulatory to limit the ethylene production in stored vegetables and fruit) (Lee et al., 2006). However, delivery systems of 1-MCP depend on cyclodextrin encapsulation tailed by water vapor exposure for release (Sisler & Serek, 1997). Differences in this system have been projected with osmotically active materials and polymeric packaging to bind the amount of release from cyclodextrin (Lee et al., 2006; Macnish et al., 2004).

Despite its high 1-MCP loading capacity, Basolite C300 exhibited limitations in releasing the ethylene inhibitor. To address this, the development of MOF composites capable of efficient 1-MCP adsorption and controlled release is essential (Chopra et al., 2017). Incorporating MOFs into polymer matrices or other support structures could enhance their performance and durability. Research has shown that certain MOFs, such as zinc-tcbpe, calcium-4,4′-sulfonyldibenzoic acid, and copper-2,4,6-tris (3,5-dicarboxylphenylamino)-1,3,5-triazine, exhibit promising adsorption properties for 1-MCP. By optimizing MOF composition, structure, and integration with other materials, it is possible to develop MOF composites tailored for specific applications in the food industry, such as ethylene scavenging and controlled atmosphere packaging (Elangovan, 2010; Kuppler et al., 2009; Shekhah et al., 2011).

Antimicrobial nanocarriers

Reduced levels of harmful and spoilage microorganisms are directly linked to food safety and quality. MOFs have surfaced as carriers of antimicrobial substances in this context; these substances can reduce or inhibit the development of spoilage and pathogenic bacteria, that are major contributors to foodborne disease (Álvarez et al., 2017; Bracone et al., 2016; Gutiérrez et al., 2017, 2018). Accordingly, Wu et al. (2019) described antimicrobial activity against gram-negative *Escherichia coli* O157:H7 bacteria (4.4 log bacterial reduction) from Zn@MOFs loaded with volatile antimicrobial essential oil (thymol). Similarly, Duan et al., (2018) created and explained that Ag NP@HKUST-1@CF composites revealed antimicrobial activity against Gram-positive *Staphylococcus aureus* (99.41% inhibitory effect) because of Ag^+ ions immobilized on HKUST-1 (stated as MOF-199).

Regulated discharge of nano-systems for active compounds

The food industry is interested in controlled release systems because the active molecules they carry can be stabilized and released at a specific location, which is usually done to improve the active compounds' adsorption in the small intestine (Álvarez et al., 2017; Gutiérrez, 2018). For instance, Ma et al. (2020) used UiO-66-based carriers that are curcumin-loaded to achieve the controlled release of curcumin for 3 hours while the subject was gastrointestinal. Another groundbreaking study in this area was conducted by Lashkari et al. (2017). They used three different allyl isothiocyanate-loaded MOFs (HKUST-1, MOF-74(Zn), and RPM6-Zn) that were able to control the active substance's release because of differences in moisture: low moisture values reserved the active molecule release, while greater moisture values amplified the active compound release amount between 70% and 96% due to MOFs structural changes. When Wang et al. (2016) examined a MOF loaded with allyl isothiocyanate, they also observed a similar pattern: 27.3% load and custody for roughly two days.

Food packaging substance nanofillers

In general, MOFs are investigated as systems for the active material carrier (food nano-packaging) (Sharanyakanth & Radhakrishnan, 2020). However, MOFs have seen an intriguing surge lately when used as dynamic fillers in polymeric food packaging (Li, 2019). UiO-66 demonstrated exceptional water durability and could, at the very least, preserve the topological structure despite exposure to water for several months, thanks to the high valence of its Zr center (Liu, 2020). To make materials for apple cube coating, Zhao et al. (2020) prepared gelatin/chitosan-based films loaded with MOFs that contained capsaicin, an antibacterial compound found in red chilies. The materials were opaque and more water vapor resistant, and they also had antimicrobial activity against *E. coli*. Despite this, this field still needs a great deal more research because it hasn't been thoroughly investigated.

Food nanoreactors

Food nanoreactors are nanostructured materials with a high area-to-volume ratio. Increasing the rates at which raw materials are converted into products has been made possible by this characteristic (Samui et al., 2020). Despite this, there is little control over the synthesis of these materials, and they are not easily repeated. Because of this, a lot of research is being done in this area right now to find safe and affordable ways to produce these materials. Similarly, Navarro et al. (2019) created a MOF-based nanocomposite (protease@MIL-101(Al)–NH_2) loaded with protease (aspartic proteinase from *Aspergillus saitoi*, also called *Aspergillus phoenicis*), which can increase protein hydrolysates in comparison to the free enzyme, even at high pH values and temperatures (till 95°C).

Food material sensors

The primary use of electrochemical and fluorescent MOFs is as sensor devices to evaluate the safety and chemical composition of food (Liu & Yin, 2016). Functionalization and signal transmission of MOF composites as sensors are shown in Fig. 9.2. To detect patulin (mycotoxin) in samples of apple juice electrochemically (Hatamluyi et al., 2020) created and utilized Au@Cu-MOF/nitrogen-doped. The principal findings included a limit of detection (LOD) of 0.0007 pptv (pptv: parts per trillion by volume), a linear detection range (LDR) of 0.001–70.0 ppbv (ppbv: parts per billion by volume), recovery values ranging from 97.6% to 99.4% with relative standard deviations (RSDs) from 1.23% to 4.61%, and limits of quantification (LOQ) between 0.7 and 2.3 pptv. Besides, a novel electrochemical aptasensor for tobramycin detection in milk was developed using a CeO_2/CuO_x@mC nanocomposite derived from a bimetallic cerium/copper-based MOF. The high surface area and porous

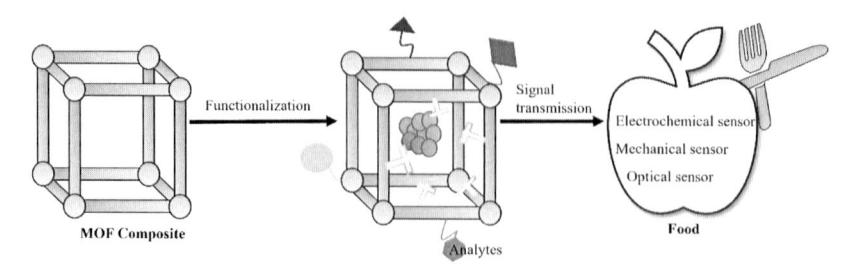

FIGURE 9.2 Functionalization and signal transmission of metal-organic framework composite as sensors in food.

structure of the MOF nanocomposite, combined with the electrochemical properties of the metal oxides, enhanced the sensor's sensitivity and selectivity, enabling the detection of tobramycin at ultralow levels. The aptasensor exhibited a wide linear range and a remarkably low LOD, demonstrating its potential for accurate and reliable food safety monitoring (Cheng et al., 2021).

The processes of surface-enhanced Raman scattering (SERS), phosphorescence, and surface plasmon resonance (SPR) can also be applied to MOF composites as optical sensors, and a few of them are listed in Table 9.1. It's also crucial to note that one major drawback of MOFs as sensors is that they require pricey equipment to provide a quantitative response, such as gas chromatography (GC), capillary electrophoresis (CE), and high-performance liquid chromatography (HPLC). In light of this, Tian et al. (2021) newly produced three distinct MOFs (zeolitic imidazolate framework-67 (ZIF-67), monoclonal antibodies-loaded MPF (MPFmAb), and metal–polydopamine framework (MPF)) to create a lateral flow assay (LFA) according to a strategy of competition for the measurable detection of tetracycline (TET, antibiotic) in samples of shrimp, milk, fish, and chicken by means of a camera. The results showed an LDR of 0.09–6 ppbv, an LOD of 0.045 ppbv, retrievals ranging from 91% to 114%, and RSDs < 4.7%.

Immobilizers and stabilizers for enzymes and active compounds

The food business is interested in active chemicals and enzymes because the former can serve as purposeful food additives and the latter has countless culinary uses (Gutiérrez, 2018). However, these compounds require stabilization due to their limited thermal and physicochemical stability. In light of this, peroxidase was stabilized on platinum NP-decorated 2D TCPP (Fe) (iron meso-tetra(4-carboxyphenyl) porphine chloride) and PCN-333 (Al) by Feng et al. (2015) and Chen et al. (2018), whereas protease and lipase were stabilized on ZIF-8 and MIL-101(Al)–NH$_2$, respectively, by Navarro et al. (2019) and Nadar and Rathod (2020). The reusable nature of these systems is a crucial feature that should be emphasized because, despite their possible high

TABLE 9.1 Applications of metal-organic framework (MOF) composites in sensors within the food system.

Sensing category	MOF composite	Food sample matrix	Target analytes	References
Electrochemical sensor	IRMOF-8	Honey	Chloramphenicol	Xiao et al. (2017)
	MIL-101(Cr)	Cucumber, Kidney bean	Omethoate or Methamidophos	Shi et al. (2017)
	Cu-MOF	Water	Nitrite	Yuan et al. (2016)
	Cu-MOF	Food packaging material	Bisphenol A	Da Silva et al. (2016)
	UiO-66-NH$_2$	Water	Cd^{2+}	Wang et al. (2017)
	Ni-MOF	Cow milk	Lactate	Manivel et al. (2018)
Fluorescent sensor	HKUST-1	Corn	Ochratoxin A	Hu et al. (2017)
	IRMOF-3	Water	*Staphylococcus arlettae*	Bhardwaj et al. (2016)
	UiO-66-NH$_2$	Drinking water	Hg^{2+}	Wu et al. (2016)
	MIL-53-NH$_2$(Fe)	Pastry cream	*Staphylococcus aureus*	Bhardwaj et al. (2017)
	ZIF-8	Water	Cu^{2+}	Lin et al. (2014)
	MIL-101	Milk powder	Pyrraline	Li et al. (2018)
Surface-enhanced Raman Scattering	MIL-101	Botanical dietary supplements	Fosinopril sodium enalaprilat	Zhang, Shi, et al. (2018)
	MIL-101	Vermicelli	Methenamine	Cai et al. (2018)

manufacturing costs, this feature makes them appealing. Conversely, the active ingredients meant for use as useful food additives are typically hydrophilic materials that need to be stabilized. As such, CD-based MOFs have effectively stabilized and transported various hydrophilic active compounds via hydrogen bonding (Moussa et al., 2016; Qiu et al., 2018). Examples of these compounds include catechin, which is antioxidant and anticancer; curcumin, which is an antarthritic, anticancer, antioxidant, antiinflammatory, and neuroprotective; glycyrrhizic acid, which has antiviral, antiinflammatory, and antimicrobial properties (Qiu, McClements, et al., 2019; Qiu, Wang, et al., 2019); resveratrol, which has antioxidant, antidepressant, anticancer, antimicrobial, and antiangiogenic properties (Qiu et al., 2018); and retinol palmitate, which is also known as an antioxidant and vitamin A palmitate (Zhang, Meng, et al., 2018). Achieving up to 22.54% (w/w) of encapsulation effectiveness by stabilizing and encapsulating menthol in CD-based MOFs, Hu et al. (2021) recently showed (Lv et al., 2017) that MOFs with sucralose-loaded CD-based exhibited a high encapsulation efficiency of 86.2% and thermal stability reaching up to 90°C.

Food cleaning

As porous materials, MOFs have been widely used for the removal, precontraction, and detection of various food impurities, including antibiotics, dyes, aromatic amines, herbicides, heavy metals, organophosphorus pesticides (OPPs), insecticides, mycotoxins, phytohormones, volatile organic compounds (VOCs), polycyclic aromatic hydrocarbons (PAHs), and herbicides using several extraction techniques, including magnetic solid-phase extraction (MSPE), dispersive solid-phase extraction (DSPE), microsolid-phase extraction (μ-SPE), matrix solid-phase dispersion extraction (MSPDE), stir-bar sorptive extraction, and others, SBSE (Hashemi et al., 2017; Rocío-Bautista et al., 2019; Villa et al., 2021; Zhang et al., 2019). Regardless of the extraction technique employed, MOFs have demonstrated the capacity to be employed in a broad range, often in the range of ppb (ppb: parts per billion), leading to low LOQ and LOD values. Furthermore, MOFs have demonstrated a strong ability to recover and low values of RSD when it comes to the extraction of food pollutants, making them accurate and efficient materials in this regard. In addition, MOFs can be used to refine fatty foods like sunflower, olive, and linseed oils. Among all the vegetable oils that were examined, Ti-MOF yielded the highest degree of impurity extraction when compared to Al-MOF and Zn-MOF, which showed similar yield behavior (Vlasova et al., 2016).

The most intriguing finding is that MOFs' superior adsorption capabilities and decreased toxicity allow them to be applied directly to clean food-based oils and beverages (Cirujano et al., 2017). Zhang et al. (2014) studied an adsorption material for drinking water defluorination by employing MOF MIL-96. The most noteworthy finding was that MIL-96 performed effectively at the maximal fluoride adsorption capacity of 31.69 mg/g and across a varied pH range (pH=3–10). In comparison to activated alumina or nano alumina, MIL-96 provided much higher defluorination efficiency and subsequent aluminum levels. Similarly, to remove fluoride from brick tea infusion, Ke et al. (2018) utilized a mixture of MOF-801 and calcium fumarate (CaFu) MOF as an adsorbent. The findings showed that MOF-801 could eliminate more than 80% of the fluoride present in brick tea infusion in the first five minutes. To purify unrefined vegetable oils, Vlasova et al. (2016) assessed the effectiveness of Al-MOF, Zn-MOF, and Ti-MOF. Ti-MOF demonstrated the highest efficacy in eliminating peroxide compounds and free fatty acids. With Ti-MOF, the greatest removal effectiveness of peroxide and free fatty acid components in vegetable oils was 64.8% and 93.4%, respectively. Furthermore, Ti-MOF is easily regenerable through ethanol washing, allowing for at least five recycling cycles. Using cysteine-functionalized UiO-66(NH_2), Liu et al. (2019) utilized a UiO-66(NH_2)@Au-Cys composite as an adsorbent to remove the mycotoxin patulin (PAT), which is present in apple juice. The high percentage of PAT removal (87%) from apple juice can be ascribed to the UiO-66(NH_2)@Au-Cys composite's surface, which has a large number of active sites with carboxyl, hydroxyl, and amine groups. Nutrients like vitamin C and phenol in apple juice showed no discernible loss, and the quantity of PAT in the juice after treatment with the composite could meet the WHO's suggested level of 50 µg/kg.

Conclusions

The preparation of MOF composites has been made possible by several factors, including the material's highly porous structure, chemical adaptability, and structural tolerance, as well as its low cost and availability for mass production. A viable and affordable method for creating novel materials with a combination of qualities that exceed the sum of their parts is through the design and synthesis of MOF composites. As described in this study, MOFs can be synthesized with a wide range of efficient materials, supplementing their porous structures with new functions such as metals, polymers, oxides, carbon, POMs, and more.

Although research into MOFs as platforms for creating composites is still in its early stages, rapid development is being achieved. When developing the composites, the structures and characteristics of the parent MOFs are the main factors to consider. It should be noted that most current research is concentrated on a select group of "star" MOFs, including the ZIF family, HKUST-1, the MIL series, MOF-5, and the extended IRMOF (IRMOF: isoreticular metal-organic framework) series. It is critically necessary to develop more viable large-scale synthesis techniques and alternative porous MOFs that are chemically and thermally stable. Meanwhile, the type of functional materials added to MOFs is crucial for influencing the composite properties and for constructing the composite structures. For instance, the surface functionalities of the substrates have demonstrated the importance of SAMs in nucleation, morphology, and characteristics. The type of interface that exists between the MOF and the secondary component also significantly influences composite performance. Thus, it is vital to understand the interactions at the interface to investigate structure–property correlations and to develop future generations of MOFs with superior performance. Additionally, a broader range of functional species should be incorporated into composite research. The integration of various guest moieties within a single MOF will create new opportunities for developing multifunctional materials.

MOF composites have gained recognition as crystalline porous coordination materials hybridized with intriguing nanotechnology-centered culinary applications. Due to their ability to undergo modification after synthesis and their large surface area of action, MOFs have primarily demonstrated definite food applications, such as sensors and extractors of various food impurities (including aromatic amines, antibiotics, heavy metals, dyes, herbicides, mycotoxins, insecticides, organophosphorus pesticides, phytohormones, and volatile organic compounds). Further research is needed to fully understand other food-related uses of MOFs, which include active food contact materials, antimicrobial carriers, controlled-release systems for active compounds, food nanoreactors, fillers for food packaging materials, and stabilizers and immobilizers for enzymes and active compounds.

The rapid expansion of MOF composites in recent years indicates a promising future for this novel class of functional materials, despite the numerous obstacles that still need to be overcome. With continued research on this fascinating topic, practical applications in several fields could one day be realized.

References

Aguilera-Sigalat, J., & Bradshaw, D. (2016). Synthesis and applications of metal-organic framework-quantum dot (QD at MOF) composites. *Coordination Chemistry Reviews, 307*, 267–291. https://doi.org/10.1016/j.ccr.2015.08.004, http://www.journals.elsevier.com/coordination-chemistry-reviews/.

Alattar, E. M., Elwasife, K. Y., Radwan, E. S., & Alagha, A. M. (2018). Effect of microwave treated water on the growth of corn (*Zea mays*) and pepper (*Capsicum annuum*) seedlings. *Romanian Journal of Biophysics, 28*, 10. https://www.rjb.ro/wp-content/uploads/03-Alattar-1.pdf.

Allendorf, M. D., Houk, R. J. T., Andruszkiewicz, L., Talin, A. A., Pikarsky, J., Choudhury, A., Gall, K. A., & Hesketh, P. J. (2008). Stress-induced chemical detection using flexible metal-organic frameworks. *Journal of the American Chemical Society, 130*(44), 14404–14405. https://doi.org/10.1021/ja805235k.

Álvarez, K., Famá, L., & J. Gutiérrez, T. (2017). Physicochemical, Antimicrobial and Mechanical Properties of Thermoplastic Materials Based on Biopolymers with Application in the Food Industry. In *Advances in Physicochemical Properties of Biopolymers (Part 1)*. Bentham Science Publishers, 358–400. https://doi.org/10.2174/9781681084534117010015.

Ameloot, R., Stappers, L., Fransaer, J., Alaerts, L., Sels, B. F., & De Vos, D. E. (2009). Patterned growth of metal-organic framework coatings by electrochemical synthesis. *Chemistry of*

Materials, 21(13), 2580–2582. https://doi.org/10.1021/cm900069fBelgium, http://pubs.acs.org/doi/pdfplus/10.1021/cm900069f.

Ayala-Zavala, J. F., Del-Toro-Sánchez, L., Alvarez-Parrilla, E., & González-Aguilar, G. A. (2008). High relative humidity in-package of fresh-cut fruits and vegetables: advantage or disadvantage considering microbiological problems and antimicrobial delivering systems? *Journal of Food Science, 73*(4), R41–R47.

Bétard, A., & Fischer, R. A. (2012). Metal-organic framework thin films: From fundamentals to applications. *Chemical Reviews, 112*(2), 1055–1083. https://doi.org/10.1021/cr200167v.

Bhardwaj, N., Bhardwaj, S. K., Mehta, J., Kim, K. H., & Deep, A. (2017). MOF-bacteriophage biosensor for highly sensitive and specific detection of *Staphylococcus aureus*. *ACS Applied Materials and Interfaces, 9*(39), 33589–33598. https://doi.org/10.1021/acsami.7b07818, http://pubs.acs.org/journal/aamick.

Bhardwaj, N., Bhardwaj, S. K., Mehta, J., Nayak, M. K., & Deep, A. (2016). Bacteriophage conjugated IRMOF-3 as a novel opto-sensor for: *S. arlettae*. *New Journal of Chemistry, 40*(9), 8068–8073. https://doi.org/10.1039/c6nj00899b, http://pubs.rsc.org/en/journals/journal/nj.

Bracone, M., Merino, D., Gonzalez, J. S., Alvarez, V. A., & Gutiérrez, T. J. (2016). Nanopackaging from natural fillers and biopolymers for the development of active and intelligent films. In *Natural polymers: Derivatives, blends and composites*. Argentina: Nova Science Publishers, Inc. Vol. I, 119–156. https://www.novapublishers.com/catalog/product_info.php?products_id=59405.

Bradshaw, D., Garai, A., & Huo, J. (2012). Metal–organic framework growth at functional interfaces: thin films and composites for diverse applications. *Chemical Society Reviews, 41*(6), 2344–2381. https://doi.org/10.1039/C1CS15276A.

Bux, H., Feldhoff, A., Cravillon, J., Wiebcke, M., Li, Y. S., & Caro, J. (2011). Oriented zeolitic imidazolate framework-8 membrane with sharp H_2/C_3H_8 molecular sieve separation. *Chemistry of Materials, 23*(8), 2262–2269. https://doi.org/10.1021/cm200555s.

Cai, J., Liu, S., Feng, J., Kimura, S., Wada, M., Kuga, S., & Zhang, L. (2012). Cellulose–silica nanocomposite aerogels by in situ formation of silica in cellulose gel. *Angewandte Chemie, 124*(9), 2118–2121. https://doi.org/10.1002/ange.201105730.

Cai, Y., Wu, Y., Xuan, T., Guo, X., Wen, Y., & Yang, H. (2018). Core-shell Au@metal-organic frameworks for promoting Raman detection sensitivity of methenamine. *ACS Applied Materials and Interfaces, 10*(18), 15412–15417. https://doi.org/10.1021/acsami.8b01765, http://pubs.acs.org/journal/aamick.

Carné, A., Carbonell, C., Imaz, I., & Maspoch, D. (2011). Nanoscale metal–organic materials. *Chemical Society Reviews, 40*(1), 291–305. https://doi.org/10.1039/c0cs00042f.

Chen, H., Qiu, Q., Sharif, S., Ying, S., Wang, Y., & Ying, Y. (2018). Solution-phase synthesis of platinum nanoparticle-decorated metal-organic framework hybrid nanomaterials as biomimetic nanoenzymes for biosensing applications. *ACS Applied Materials and Interfaces, 10*(28), 24108–24115. https://doi.org/10.1021/acsami.8b04737, http://pubs.acs.org/journal/aamick.

Chen, L., & Xu, Q. (2019). Metal-organic framework composites for catalysis. *Matter, 1*(1), 57–89. https://doi.org/10.1016/j.matt.2019.05.018.

Chen, S., Wan, S., Lan, Q., Zheng, Y., & Zhu, X. (2020). Magnetic graphene oxide-ultrathin nickel–organic framework composite for the extraction and determination of epoxiconazole in food samples. *RSC Advances, 10*(73), 44793–44797. https://doi.org/10.1039/d0ra08650a.

Cheng, W., Tang, X., Zhang, Y., Wu, D., & Yang, W. (2021). Applications of metal-organic framework (MOF)-based sensors for food safety: Enhancing mechanisms and recent advances. *Trends in Food Science and Technology, 112*, 268–282. https://doi.org/10.1016/j.

tifs.2021.04.004, http://www.elsevier.com/wps/find/journaldescription.cws_home/601278/description#description.

Chopra, S., Dhumal, S., Abeli, P., Beaudry, R., & Almenar, E. (2017). Metal-organic frameworks have utility in adsorption and release of ethylene and 1-methylcyclopropene in fresh produce packaging. *Postharvest Biology and Technology, 130*, 48–55. https://doi.org/10.1016/j.postharvbio.2017.04.001, www.elsevier.com/inca/publications/store/5/0/3/1/3/index.htt.

Cirujano, F. G., Luz, I., Soukri, M., Van Goethem, C., Vankelecom, I. F. J., Lail, M., & De Vos, D. E. (2017). Boosting the catalytic performance of metal–organic frameworks for steroid transformations by confinement within a mesoporous scaffold. *Angewandte Chemie, 129*(43), 13487–13491. https://doi.org/10.1002/ange.201706721.

Da Silva, C. T. P., Veregue, F. R., Aguiar, L. W., Meneguin, J. G., Moisés, M. P., Fávaro, S. L., Radovanovic, E., Girotto, E. M., & Rinaldi, A. W. (2016). AuNp@MOF composite as electrochemical material for determination of bisphenol A and its oxidation behavior study. *New Journal of Chemistry, 40*(10), 8872–8877. https://doi.org/10.1039/c6nj00936k, http://pubs.rsc.org/en/journals/journal/nj.

Demilecamps, A., Beauger, C., Hildenbrand, C., Rigacci, A., & Budtova, T. (2015). Cellulose-silica aerogels. *Carbohydrate Polymers, 122*, 293–300. https://doi.org/10.1016/j.carbpol.2015.01.022, http://www.elsevier.com/wps/find/journaldescription.cws_home/405871/description#description.

Desai, P., Shakya, P., Kreouzis, T., Gillin, W. P., Morley, N. A., & Gibbs, M. R. J. (2007). Magnetoresistance and efficiency measurements of Alq3-based OLEDs. *Physical Review B, 75*(9), 094423. https://doi.org/10.1103/physrevb.75.094423.

Dong, Z., Sun, Y., Chu, J., Zhang, X., & Deng, H. (2017). Multivariate metal-organic frameworks for dialing-in the binding and programming the release of drug molecules. *Journal of the American Chemical Society, 139*(40), 14209–14216. https://doi.org/10.1021/jacs.7b07392, http://pubs.acs.org/journal/jacsat.

Doonan, C., Riccò, R., Liang, K., Bradshaw, D., & Falcaro, P. (2017). Metal-organic frameworks at the biointerface: Synthetic strategies and applications. *Accounts of Chemical Research, 50*(6), 1423–1432. https://doi.org/10.1021/acs.accounts.7b00090, http://pubs.acs.org/journal/achre4.

Drout, R. J., Robison, L., & Farha, O. K. (2019). Catalytic applications of enzymes encapsulated in metal–organic frameworks. *Coordination Chemistry Reviews, 381*, 151–160. https://doi.org/10.1016/j.ccr.2018.11.009.

Du, D. Y., Qin, J. S., Li, S. L., Su, Z. M., & Lan, Y. Q. (2014). Recent advances in porous polyoxometalate-based metal-organic framework materials. *Chemical Society Reviews, 43*(13), 4615–4632. https://doi.org/10.1039/c3cs60404g, http://pubs.rsc.org/en/journals/journal/cs.

Duan, C., Meng, J., Wang, X., Meng, X., Sun, X., Xu, Y., Zhao, W., & Ni, Y. (2018). Synthesis of novel cellulose- based antibacterial composites of Ag nanoparticles@ metal-organic frameworks@carboxymethylated fibers. *Carbohydrate Polymers, 193*, 82–88. https://doi.org/10.1016/j.carbpol.2018.03.089.

Elangovan D. (2010). *Characterization of metal organic framework and polymer composites and the method of their preparation*. In Electronic Theses & Dissertations, ISBN 9781267053701.

Fan, J., Chen, D., Li, N., Xu, Q., Li, H., He, J., & Lu, J. (2018). Adsorption and biodegradation of dye in wastewater with Fe_3O_4@MIL-100 (Fe) core–shell bio-nanocomposites. *Chemosphere, 191*, 315–323. https://doi.org/10.1016/j.chemosphere.2017.10.042, www.elsevier.com/locate/chemosphere.

Feng, D., Liu, T. F., Su, J., Bosch, M., Wei, Z., Wan, W., Yuan, D., Chen, Y. P., Wang, X., Wang, K., Lian, X., Gu, Z. Y., Park, J., Zou, X., & Zhou, H. C. (2015). Stable metal-organic frameworks containing single-molecule traps for enzyme encapsulation. *Nature Communications, 6*, 5979. https://doi.org/10.1038/ncomms6979, http://www.nature.com/ncomms/index.html.

Garbuzov, D. Z., Bulović, V., Burrows, P. E., & Forrest, S. R. (1996). Photoluminescence efficiency and absorption of aluminum-tris-quinolate (Alq3) thin films. *Chemical Physics Letters, 249*(5-6), 433–437. https://doi.org/10.1016/0009-2614(95)01424-1.

Guo, H., Zhu, G., Hewitt, I. J., & Qiu, S. (2009). "Twin copper source" growth of metal-organic framework membrane: $Cu_3(BTC)_2$ with high permeability and selectivity for recycling H_2. *Journal of the American Chemical Society, 131*(5), 1646–1647. https://doi.org/10.1021/ja8074874China, http://pubs.acs.org/doi/pdfplus/10.1021/ja8074874.

Gutiérrez, T. J. (2018). Antibiofilm enzymes as an emerging technology for food quality and safety. In *Enzymes in food biotechnology: Production, applications, and future prospects*. Argentina: Elsevier, 321–342. http://www.sciencedirect.com/science/book/9780128132807, 10.1016/B978-0-12-813280-7.00019-0.

Gutiérrez, T. J., Ponce, A. G., & Alvarez, V. A. (2017). Nano-clays from natural and modified montmorillonite with and without added blueberry extract for active and intelligent food nanopackaging materials. *Materials Chemistry and Physics, 194*, 283–292. https://doi.org/10.1016/j.matchemphys.2017.03.052.

Haider, G., Usman, M., Chen, T. P., Perumal, P., Lu, K. L., & Chen, Y. F. (2016). Electrically driven white light emission from intrinsic metal-organic framework. *ACS Nano, 10*(9), 8366–8375. https://doi.org/10.1021/acsnano.6b03030, http://pubs.acs.org/journal/ancac3.

Hasanzadeh, M., Simchi, A., & Shahriyari Far, H. (2020). Nanoporous composites of activated carbon-metal organic frameworks for organic dye adsorption: Synthesis, adsorption mechanism and kinetics studies. *Journal of Industrial and Engineering Chemistry, 81*, 405–414. https://doi.org/10.1016/j.jiec.2019.09.031.

Hashemi, B., Zohrabi, P., Raza, N., & Kim, K. H. (2017). Metal-organic frameworks as advanced sorbents for the extraction and determination of pollutants from environmental, biological, and food media. *TrAC – Trends in Analytical Chemistry, 97*, 65–82. https://doi.org/10.1016/j.trac.2017.08.015, www.elsevier.com/locate/trac.

Hatamluyi, B., Rezayi, M., Beheshti, H. R., & Boroushaki, M. T. (2020). Ultra-sensitive molecularly imprinted electrochemical sensor for patulin detection based on a novel assembling strategy using Au@Cu-MOF/N-GQDs. *Sensors and Actuators, B: Chemical, 318*, 128219. https://doi.org/10.1016/j.snb.2020.128219, https://www.journals.elsevier.com/sensors-and-actuators-b-chemical.

Hermes, S., Schröder, F., Chelmowski, R., Wöll, C., & Fischer, R. A. (2005). Selective nucleation and growth of metal-organic open framework thin films on patterned COOH/CF3-terminated self-assembled monolayers on Au(111). *Journal of the American Chemical Society, 127*(40), 13744–13745. https://doi.org/10.1021/ja053523l.

Hu, S., Ouyang, W., Guo, L., Lin, Z., Jiang, X., Qiu, B., & Chen, G. (2017). Facile synthesis of $Fe_3O_4/g-C_3N_4$/HKUST-1 composites as a novel biosensor platform for ochratoxin A. *Biosensors and Bioelectronics, 92*, 718–723. https://doi.org/10.1016/j.bios.2016.10.006, www.elsevier.com/locate/bios.

Hu, Z., Li, S., Wang, S., Zhang, B., & Huang, Q. (2021). Encapsulation of menthol into cyclodextrin metal-organic frameworks: Preparation, structure characterization and evaluation of complexing capacity. *Food Chemistry, 338*, 127839. https://doi.org/10.1016/j.foodchem.2020.127839.

Jie, M., Lan, S., Zhu, B., Zhu, A., Yue, X., Xiang, Q., & Bai, Y. (2024). Europium functionalized porphyrin-based metal-organic framework heterostructure and hydrogel for visual ratiometric fluorescence sensing of sulfonamides in foods. *Food Chemistry, 458*, 140304. https://doi.org/10.1016/j.foodchem.2024.140304.

Jo, C., Lee, H. J., & Oh, M. (2011). One-Pot Synthesis of Silica@Coordination Polymer Core–Shell Microspheres with Controlled Shell Thickness. *Advanced Materials, 23*(15), 1716–1719. https://doi.org/10.1002/adma.201004208.

Kajiwara, T., Fujii, M., Tsujimoto, M., Kobayashi, K., Higuchi, M., Tanaka, K., & Kitagawa, S. (2016). Photochemical reduction of low concentrations of CO_2 in a porous coordination polymer with a ruthenium(II)-CO complex. *Angewandte Chemie - International Edition, 55*(8), 2697–2700. https://doi.org/10.1002/anie.201508941, http://onlinelibrary.wiley.com/journal/10.1002/(ISSN)1521-3773.

Kayal, S., & Chakraborty, A. (2018). Activated carbon (type Maxsorb-III) and MIL-101(Cr) metal organic framework based composite adsorbent for higher CH4 storage and CO_2 capture. *Chemical Engineering Journal, 334*, 780–788. https://doi.org/10.1016/j.cej.2017.10.080, www.elsevier.com/inca/publications/store/6/0/1/2/7/3/index.htt.

Ke, F., Peng, C., Zhang, T., Zhang, M., Zhou, C., Cai, H., Zhu, J., & Wan, X. (2018). Fumarate-based metal-organic frameworks as a new platform for highly selective removal of fluoride from brick tea. *Scientific Reports, 8*(1), 939. https://doi.org/10.1038/s41598-018-19277-2.

Kitao, T., Zhang, Y., Kitagawa, S., Wang, B., & Uemura, T. (2017). Hybridization of MOFs and polymers. *Chemical Society Reviews, 46*(11), 3108–3133. https://doi.org/10.1039/c7cs00041c, http://pubs.rsc.org/en/journals/journal/cs.

Kou, J., & Sun, L. B. (2018). Fabrication of metal-organic frameworks inside silica nanopores with significantly enhanced hydrostability and catalytic activity. *ACS Applied Materials and Interfaces, 10*(14), 12051–12059. https://doi.org/10.1021/acsami.8b01652, http://pubs.acs.org/journal/aamick.

Kuppler, R. J., Timmons, D. J., Fang, Q. R., Li, J. R., Makal, T. A., Young, M. D., Yuan, D., Zhao, D., Zhuang, W., & Zhou, H. C. (2009). Potential applications of metal-organic frameworks. *Coordination Chemistry Reviews, 253*(23-24), 3042–3066. https://doi.org/10.1016/j.ccr.2009.05.019.

Lashkari, E., Wang, H., Liu, L., Li, J., & Yam, K. (2017). Innovative application of metal-organic frameworks for encapsulation and controlled release of allyl isothiocyanate. *Food Chemistry, 221*, 926–935. https://doi.org/10.1016/j.foodchem.2016.11.072, www.elsevier.com/locate/foodchem.

Lee, Y. S., Beaudry, R., Kim, J. N., & Harte, B. R. (2006). Development of a 1-methylcyclopropene (1-MCP) sachet release system. *Journal of Food Science, 71*(1), C1–C6.

Lei, C., Gao, J., Ren, W., Xie, Y., Abdalkarim, S. Y. H., Wang, S., Ni, Q., & Yao, J. (2019). Fabrication of metal-organic frameworks@cellulose aerogels composite materials for removal of heavy metal ions in water. *Carbohydrate Polymers, 205*, 35–41. https://doi.org/10.1016/j.carbpol.2018.10.029, http://www.elsevier.com/wps/find/journaldescription.cws_home/405871/description#description.

Li, F., Jiang, X., Zhao, J., & Zhang, S. (2015). Graphene oxide: A promising nanomaterial for energy and environmental applications. *Nano Energy, 16*, 488–515. https://doi.org/10.1016/j.nanoen.2015.07.014, http://www.journals.elsevier.com/nano-energy/.

Li, G., Zhao, S., Zhang, Y., & Tang, Z. (2018). Metal-organic frameworks encapsulating active nanoparticles as emerging composites for catalysis: Recent progress and perspectives. *Advanced Materials, 30*(51), 1800702. https://doi.org/10.1002/adma.201800702, http://onlinelibrary.wiley.com/journal/10.1002/(ISSN)1521-4095.

Li, S., & Huo, F. (2015). Metal-organic framework composites: From fundamentals to applications. *Nanoscale, 7*(17), 7482–7501. https://doi.org/10.1039/c5nr00518c, http://pubs.rsc.org/en/journals/journal/nr.

Li, W. (2019). Metal–organic framework membranes: Production, modification, and applications. *Progress in Materials Science, 100*, 21–63. https://doi.org/10.1016/j.pmatsci.2018.09.003.

Lian, X., Fang, Y., Joseph, E., Wang, Q., Li, J., Banerjee, S., Lollar, C., Wang, X., & Zhou, H.-C. (2017). Enzyme–MOF (metal–organic framework) composites. *Chemical Society Reviews, 46*(11), 3386–3401. https://doi.org/10.1039/C7CS00058H.

Liao, Q., Su, X., Zhu, W., Hua, W., Qian, Z., Liu, L., & Yao, J. (2016). Flexible and durable cellulose aerogels for highly effective oil/water separation. *RSC Advances, 6*(68), 63773–63781. https://doi.org/10.1039/C6RA12356B.

Lin, X., Gao, G., Zheng, L., Chi, Y., & Chen, G. (2014). Encapsulation of strongly fluorescent carbon quantum dots in metal-organic frameworks for enhancing chemical sensing. *Analytical Chemistry, 86*(2), 1223–1228. https://doi.org/10.1021/ac403536a.

Litschauer, M., Neouze, M. A., Haimer, E., Henniges, U., Potthast, A., Rosenau, T., & Liebner, F. (2011). Silica modified cellulosic aerogels. *Cellulose, 18*(1), 143–149. https://doi.org/10.1007/s10570-010-9459-x, www.springer.com/journal/10570.

Liu, J., Sun, F., Zhang, F., Wang, Z., Zhang, R., Wang, C., & Qiu, S. (2011). In situ growth of continuous thin metal–organic framework film for capacitive humidity sensing. *Journal of Materials Chemistry, 21*(11), 3775. https://doi.org/10.1039/c0jm03123b.

Liu, M., Wang, J., Yang, Q., Hu, N., Zhang, W., Zhu, W., Wang, R., Suo, Y., & Wang, J. (2019). Patulin removal from apple juice using a novel cysteine-functionalized metal-organic framework adsorbent. *Food Chemistry, 270*, 1–9. https://doi.org/10.1016/j.foodchem.2018.07.072.

Liu, W., & Yin, X. B. (2016). Metal-organic frameworks for electrochemical applications. *TrAC – Trends in Analytical Chemistry, 75*, 86–96. https://doi.org/10.1016/j.trac.2015.07.011, www.elsevier.com/locate/trac.

Liu, X. (2020). Metal-organic framework UiO-66 membranes. *Frontiers of Chemical Science and Engineering, 14*(2), 216–232. https://doi.org/10.1007/s11705-019-1857-5.

Lv, N., Guo, T., Liu, B., Wang, C., Singh, V., Xu, X., Li, X., Chen, D., Gref, R., & Zhang, J. (2017). Improvement in thermal stability of sucralose by γ-cyclodextrin metal-organic frameworks. *Pharmaceutical Research, 34*(2), 269–278. https://doi.org/10.1007/s11095-016-2059-1.

Ma, P., Zhang, J., Liu, P., Wang, Q., Zhang, Y., Song, K., Li, R., & Shen, L. (2020). Computer-assisted design for stable and porous metal-organic framework (MOF) as a carrier for curcumin delivery. *LWT, 120*, 108949. https://doi.org/10.1016/j.lwt.2019.108949.

Macnish, A. J., Joyce, D. C., Irving, D. E., & Wearing, A. H. (2004). A simple sustained release device for the ethylene binding inhibitor 1-methylcyclopropene. *Postharvest Biology and Technology, 32*(3), 321–338. https://doi.org/10.1016/j.postharvbio.2003.12.003.

Mahmoodi, N. M., Taghizadeh, M., & Taghizadeh, A. (2019). Activated carbon/metal-organic framework composite as a bio-based novel green adsorbent: Preparation and mathematical pollutant removal modeling. *Journal of Molecular Liquids, 277*, 310–322. https://doi.org/10.1016/j.molliq.2018.12.050.

Makiura, R., Motoyama, S., Umemura, Y., Yamanaka, H., Sakata, O., & Kitagawa, H. (2010). Surface nano-architecture of a metal–organic framework. *Nature Materials, 9*(7), 565–571. https://doi.org/10.1038/nmat2769.

Manivel, P., Suryanarayanan, V., Nesakumar, N., Velayutham, D., Madasamy, K., Kathiresan, M., Kulandaisamy, A. J., & Rayappan, J. B. B. (2018). A novel electrochemical sensor based on a nickel-metal organic framework for efficient electrocatalytic oxidation and rapid detection of lactate. *New Journal of Chemistry, 42*(14), 11839–11846. https://doi.org/10.1039/C8NJ02118J, http://pubs.rsc.org/en/journals/journal/nj.

Monama, G. R., Hato, M. J., Ramohlola, K. E., Maponya, T. C., Mdluli, S. B., Molapo, K. M., Modibane, K. D., Iwuoha, E. I., Makgopa, K., & Teffu, M. D. (2019). Hierachiral 4-tetranitro copper(II)phthalocyanine based metal organic framework hybrid composite with improved electrocatalytic efficiency towards hydrogen evolution reaction. *Results in Physics, 15*, 102564. https://doi.org/10.1016/j.rinp.2019.102564, http://www.elsevier.com/wps/find/journaldescription.cws_home/725996/description#description.

Motoyama, S., Makiura, R., Sakata, O., & Kitagawa, H. (2011). Highly crystalline nanofilm by layering of porphyrin metal–organic framework sheets. *Journal of the American Chemical Society, 133*(15), 5640–5643. https://doi.org/10.1021/ja110720f.

Moussa, Z., Hmadeh, M., Abiad, M. G., Dib, O. H., & Patra, D. (2016). Encapsulation of curcumin in cyclodextrin-metal organic frameworks: Dissociation of loaded CD-MOFs enhances stability of curcumin. *Food Chemistry, 212*, 485–494. https://doi.org/10.1016/j.foodchem.2016.06.013, www.elsevier.com/locate/foodchem.

Musyoka, N. M., Ren, J., Langmi, H. W., North, B. C., Mathe, M., & Bessarabov, D. (2017). Synthesis of rGO/Zr-MOF composite for hydrogen storage application. *Journal of Alloys and Compounds, 724*, 450–455. https://doi.org/10.1016/j.jallcom.2017.07.040.

Nadar, S. S., & Rathod, V. K. (2020). Immobilization of proline activated lipase within metal organic framework (MOF). *International Journal of Biological Macromolecules, 152*, 1108–1112. https://doi.org/10.1016/j.ijbiomac.2019.10.199, www.elsevier.com/locate/ijbiomac.

Navarro, J., Almora Barrios, N., Lerma Berlanga, B., Ruiz-Pernía, J. J., Lorenz Fonfria, V. A., Tuñón, I., & Martí-Gastaldo, C. (2019). Translocation of enzymes into a mesoporous MOF for enhanced catalytic activity under extreme conditions. *Chemical Science, 10*(14), 4082–4088. https://doi.org/10.1039/c9sc00082h.

Niu, Z., Bhagya Gunatilleke, W. D. C., Sun, Q., Lan, P. C., Perman, J., Ma, J. G., Cheng, Y., Aguila, B., & Ma, S. (2018). Metal-organic framework anchored with a lewis pair as a new paradigm for catalysis. *Chemistry, 4*(11), 2587–2599. https://doi.org/10.1016/j.chempr.2018.08.018, http://www.cell.com/chem/home.

Pan, M., Xie, X., Liu, K., Yang, J., Hong, L., & Wang, S. (2020). Fluorescent carbon quantum dots—Synthesis, functionalization and sensing application in food analysis. *Nanomaterials, 10*(5), 930. https://doi.org/10.3390/nano10050930.

Prasad, A., Du, L., Zubair, M., Subedi, S., Ullah, A., & Roopesh, M. S. (2020). Applications of light-emitting diodes (LEDs) in food processing and water treatment. *Food Engineering Reviews, 12*(3), 268–289. https://doi.org/10.1007/s12393-020-09221-4, http://www.springer.com/life+sci/food+science/journal/12393.

Qiu, C., McClements, D. J., Jin, Z., Wang, C., Qin, Y., Xu, X., & Wang, J. (2019). Development of nanoscale bioactive delivery systems using sonication: Glycyrrhizic acid-loaded cyclodextrin metal-organic frameworks. *Journal of Colloid and Interface Science, 553*, 549–556. https://doi.org/10.1016/j.jcis.2019.06.064, http://www.elsevier.com/inca/publications/store/6/2/2/8/6/1/index.htt.

Qiu, C., Wang, J., Qin, Y., Xu, X., & Jin, Z. (2019). Characterization and mechanisms of novel emulsions and nanoemulsion gels stabilized by edible cyclodextrin-based metal-organic frameworks and glycyrrhizic acid. *Journal of Agricultural and Food Chemistry, 67*(1), 391–398. https://doi.org/10.1021/acs.jafc.8b03065, http://pubs.acs.org/journal/jafcau.

Qiu, C., Wang, J., Zhang, H., Qin, Y., Xu, X., & Jin, Z. (2018). Novel approach with controlled nucleation and growth for green synthesis of size-controlled cyclodextrin-based metal-organic frameworks based on short-chain starch nanoparticles. *Journal of Agricultural and Food Chemistry, 66*(37), 9785–9793. https://doi.org/10.1021/acs.jafc.8b03144, http://pubs.acs.org/journal/jafcau.

Ramesh, M., Muthukrishnan, M., Khan, A., & Azam, M. (2019). Metal-organic-framework-quantum dots (QD@ MOF) composites. In *Metal-organic framework composites*. Materials Research Foundations, 49–84 Vol. II.

Rehman Shah, H. U., Ahmad, K., Naseem, H. A., Parveen, S., Ashfaq, M., Rauf, A., & Aziz, T. (2021). Water stable graphene oxide metal-organic frameworks composite (ZIF-67@GO) for efficient removal of malachite green from water. *Food and Chemical Toxicology, 154*, 112312. https://doi.org/10.1016/j.fct.2021.112312, www.elsevier.com/locate/foodchemtox.

Rocío-Bautista, P., Taima-Mancera, I., Pasán, J., & Pino, V. (2019). Metal-organic frameworks in green analytical chemistry. *Separations, 6*(3), 33. https://doi.org/10.3390/separations6030033.

Rodenas, T., Van Dalen, M., García-Pérez, E., Serra-Crespo, P., Zornoza, B., Kapteijn, F., & Gascon, J. (2014). Visualizing MOF mixed matrix membranes at the nanoscale: Towards structure-performance relationships in CO_2/CH_4 separation over NH_2-MIL-53(Al)@PI. *Advanced Functional Materials, 24*(2), 249–256. https://doi.org/10.1002/adfm.201203462.

Rowe, M. D., Tham, D. H., Kraft, S. L., & Boyes, S. G. (2009). Polymer-modified gadolinium metal-organic framework nanoparticles used as multifunctional nanomedicines for the targeted imaging and treatment of cancer. *Biomacromolecules, 10*(4), 983–993. https://doi.org/10.1021/bm900043e, http://pubs.acs.org/doi/pdfplus/10.1021/bm900043e.

Rui, Z., Yu, D., & Zhang, F. (2024). Novel cellulose nanocrystal/metal-organic framework composites: Transforming ASA-sized cellulose paper for innovative food packaging solutions. *Industrial Crops and Products, 207*, 117771. https://doi.org/10.1016/j.indcrop.2023.117771.

Saeb, E., & Asadpour-Zeynali, K. (2022). A novel ZIF-8@ZIF-67/Au core–shell metal organic framework nanocomposite as a highly sensitive electrochemical sensor for nitrite determination. *Electrochimica Acta, 417*, 140278. https://doi.org/10.1016/j.electacta.2022.140278.

Samui, A., Happy, & Sahu, S. K. (2020). Integration of α-amylase into covalent organic framework for highly efficient biocatalyst. *Microporous and Mesoporous Materials, 291*, 109700. https://doi.org/10.1016/j.micromeso.2019.109700.

Sappia, L. D., Tuninetti, J. S., Ceolín, M., Knoll, W., Rafti, M., & Azzaroni, O. (2020). MOF@PEDOT composite films for impedimetric pesticide sensors. *Global Challenges, 4*(2), 1900076. https://doi.org/10.1002/gch2.201900076, https://onlinelibrary.wiley.com/journal/20566646.

Scherb, C., Schödel, A., & Bein, T. (2008). Directing the Structure of Metal–Organic Frameworks by Oriented Surface Growth on an Organic Monolayer. *Angewandte Chemie, 120*(31), 5861–5863. https://doi.org/10.1002/ange.200704034.

Schoedel, A., Scherb, C., & Bein, T. (2010). Oriented Nanoscale Films of Metal–Organic Frameworks By Room-Temperature Gel-Layer Synthesis. *Angewandte Chemie, 122*(40), 7383–7386. https://doi.org/10.1002/ange.201001684.

Sharanyakanth, P. S., & Radhakrishnan, M. (2020). Synthesis of metal-organic frameworks (MOFs) and its application in food packaging: A critical review. *Trends in Food Science & Technology, 104*, 102–116. https://doi.org/10.1016/j.tifs.2020.08.004.

Shekhah, O., Liu, J., Fischer, R. A., & Wöll, C. (2011). MOF thin films: Existing and future applications. *Chemical Society Reviews, 40*(2), 1081–1106. https://doi.org/10.1039/c0cs00147c.

Shekhah, O., Wang, H., Paradinas, M., Ocal, C., Schüpbach, B., Terfort, A., Zacher, D., Fischer, R. A., & Wöll, C. (2009). Controlling interpenetration in metal–organic frameworks by liquid-phase epitaxy. *Nature Materials, 8*(6), 481–484. https://doi.org/10.1038/nmat2445.

Shi, X., Lu, J., Yin, H., Qiao, X., & Xu, Z. (2017). A biomimetic sensor with signal enhancement of ferriferrous oxide-reduced graphene oxide nanocomposites for ultratrace levels quantification of methamidophos or omethoate in vegetables. *Food Analytical Methods, 10*(4), 910–920. https://doi.org/10.1007/s12161-016-0641-0, http://www.springer.com/life+sci/food+science/journal/12161.

Sisler, E. C., & Serek, M. (1997). Inhihitors of ethylene responses in plants at the receptor level: Recent developments. *Physiologia Plantarum, 100*(3), 577–582.

So, M. C., Jin, S., Son, H. J., Wiederrecht, G. P., Farha, O. K., & Hupp, J. T. (2013). Layer-by-layer fabrication of oriented porous thin films based on porphyrin-containing metal-organic frameworks. *Journal of the American Chemical Society, 135*(42), 15698–15701. https://doi.org/10.1021/ja4078705, http://pubs.acs.org/journal/jacsat.

Tian, Y., Bu, T., Zhang, M., Sun, X., Jia, P., Wang, Q., Liu, Y., Bai, F., Zhao, S., & Wang, L. (2021). Metal-polydopamine framework based lateral flow assay for high sensitive detection of tetracycline in food samples. *Food Chemistry, 339*, 127854. https://doi.org/10.1016/j.foodchem.2020.127854.

Topçu, E. (2020). Three-dimensional, free-standing, and flexible cobalt-based metal-organic frameworks/graphene composite paper: A novel electrochemical sensor for determination of resorcinol. *Materials Research Bulletin, 121*, 110629. https://doi.org/10.1016/j.materresbull.2019.110629.

Uemura, T., Hiramatsu, D., Yoshida, K., Isoda, S., & Kitagawa, S. (2008). Sol-gel synthesis of low-dimensional silica within coordination nanochannels. *Journal of the American Chemical Society, 130*(29), 9216–9217. https://doi.org/10.1021/ja8030906.

Usman, M., Mendiratta, S., Batjargal, S., Haider, G., Hayashi, M., Rao Gade, N., Chen, J. W., Chen, Y. F., & Lu, K. L. (2015). Semiconductor behavior of a three-dimensional strontium-based metal-organic framework. *ACS Applied Materials and Interfaces, 7*(41), 22767–22774. https://doi.org/10.1021/acsami.5b07228, http://pubs.acs.org/journal/aamick.

Venkadesh, A., Mathiyarasu, J., & Radhakrishnan, S. (2021). Voltammetric Sensing of Caffeine in Food Sample Using Cu-MOF and Graphene. *Electroanalysis, 33*(4), 1007–1013. https://doi.org/10.1002/elan.202060488.

Villa, C. C., Sánchez, L. T., Valencia, G. A., Ahmed, S., & Gutiérrez, T. J. (2021). Molecularly imprinted polymers for food applications: A review. *Trends in Food Science and Technology, 111*, 642–669. https://doi.org/10.1016/j.tifs.2021.03.003, http://www.elsevier.com/wps/find/journaldescription.cws_home/601278/description#description.

Vlasova, E. A., Yakimov, S. A., Naidenko, E. V., Kudrik, E. V., & Makarov, S. V. (2016). Application of metal-organic frameworks for purification of vegetable oils. *Food Chemistry, 190*, 103–109. https://doi.org/10.1016/j.foodchem.2015.05.078, www.elsevier.com/locate/foodchem.

Wang, H., Lashkari, E., Lim, H., Zheng, C., Emge, T. J., Gong, Q., Yam, K., & Li, J. (2016). The moisture-triggered controlled release of a natural food preservative from a microporous metal-organic framework. *Chemical Communications, 52*(10), 2129–2132. https://doi.org/10.1039/c5cc09634k, http://pubs.rsc.org/en/journals/journal/cc.

Wang, X., Xu, Y., Li, Y., Li, Y., Li, Z., Zhang, W., Zou, X., Shi, J., Huang, X., Liu, C., & Li, W. (2021). Rapid detection of cadmium ions in meat by a multi-walled carbon nanotubes enhanced metal-organic framework modified electrochemical sensor. *Food Chemistry, 357*, 129762. https://doi.org/10.1016/j.foodchem.2021.129762.

Wang, Y., Wang, L., Huang, W., Zhang, T., Hu, X., Perman, J. A., & Ma, S. (2017). A metal-organic framework and conducting polymer based electrochemical sensor for high performance cadmium ion detection. *Journal of Materials Chemistry A, 5*(18), 8385–8393. https://doi.org/10.1039/c7ta01066d, http://pubs.rsc.org/en/journals/journalissues/ta.

Witters, D., Vermeir, S., Puers, R., Sels, B. F., De Vos, D. E., Lammertyn, J., & Ameloot, R. (2013). Miniaturized layer-by-layer deposition of metal-organic framework coatings through digital microfluidics. *Chemistry of Materials, 25*(7), 1021–1023. https://doi.org/10.1021/cm400216m.

Wu, C.-D., & Zhao, M. (2017). Incorporation of molecular catalysts in metal–organic frameworks for highly efficient heterogeneous catalysis. *Advanced Materials, 29*(14), 1605446. https://doi.org/10.1002/adma.201605446.

Wu, H., & Mu, W. (2022). Application prospects and opportunities of inorganic nanomaterials for enzyme immobilization in the food-processing industry. *Current Opinion in Food Science, 47*, 100909. https://doi.org/10.1016/j.cofs.2022.100909.

Wu, L.-L., Wang, Z., Zhao, S.-N., Meng, X., Song, X.-Z., Feng, J., Song, S.-Y., & Zhang, H.-J. (2016). A metal–organic framework/DNA hybrid system as a novel fluorescent biosensor for mercury(II) ion detection. *Chemistry – A European Journal, 22*(2), 477–480. https://doi.org/10.1002/chem.201503335.

Wu, Y., Luo, Y., Zhou, B., Mei, L., Wang, Q., & Zhang, B. (2019). Porous metal-organic framework (MOF) carrier for incorporation of volatile antimicrobial essential oil. *Food Control, 98*, 174–178. https://doi.org/10.1016/j.foodcont.2018.11.011.

Xiao, L., Xu, R., Yuan, Q., & Wang, F. (2017). Highly sensitive electrochemical sensor for chloramphenicol based on MOF derived exfoliated porous carbon. *Talanta, 167*, 39–43. https://doi.org/10.1016/j.talanta.2017.01.078, https://www.journals.elsevier.com/talanta.

Xiong, T. A. N. G., Maolong, C. H. E. N., Zhou, X. U., & Yunhui, C. H. E. N. G. (2019). Preparation of gadolinium metal organic framework and its application for adsorption of phoxim. *Food and Machinery, 35*, 37–41.

Yang, G. S., Li, M. N., Li, S. L., Lan, Y. Q., He, W. W., Wang, X. L., Qin, J. S., & Su, Z. M. (2012). Controllable synthesis of microporous, nanotubular and mesocage-like metal-organic frameworks by adjusting the reactant ratio and modulated luminescence properties of Alq$_3$@MOF composites. *Journal of Materials Chemistry, 22*(34), 17947–17953. https://doi.org/10.1039/c2jm32990e.

Yoo, Y., & Jeong, H. K. (2008). Rapid fabrication of metal organic framework thin films using microwave-induced thermal deposition. *Chemical Communications*, (21), 2441–2443. https://doi.org/10.1039/b800061a, http://pubs.rsc.org/en/journals/journal/cc.

Yuan, B., Zhang, J., Zhang, R., Shi, H., Wang, N., Li, J., Ma, F., & Zhang, D. (2016). Cu-based metal–organic framework as a novel sensing platform for the enhanced electro-oxidation of nitrite. *Sensors and Actuators B: Chemical, 222*, 632–637. https://doi.org/10.1016/j.snb.2015.08.100.

Zacher, D., Shekhah, O., Wöll, C., & Fischer, R. A. (2009). Thin films of metal–organic frameworks. *Chemical Society Reviews, 38*(5), 1418–1429. https://doi.org/10.1039/b805038b.

Zhang, N., Yang, X., Yu, X., Jia, Y., Wang, J., Kong, L., Jin, Z., Sun, B., Luo, T., & Liu, J. (2014). Al-1,3,5-benzenetricarboxylic metal–organic frameworks: A promising adsorbent for defluoridation of water with pH insensitivity and low aluminum residual. *Chemical Engineering Journal, 252*, 220–229. https://doi.org/10.1016/j.cej.2014.04.090.

Zhang, Y., Xu, Y., Ma, Y., Luo, H., Hou, J., Hou, C., & Huo, D. (2024). Ultra-sensitive electrochemical sensors through self-assembled MOF composites for the simultaneous detection of multiple heavy metal ions in food samples. *Analytica Chimica Acta, 1289*, 342155. https://doi.org/10.1016/j.aca.2023.342155.

Zhang, Y., Li, G., Wu, D., Li, X., Yu, Y., Luo, P., Chen, J., Dai, C., & Wu, Y. (2019). Recent advances in emerging nanomaterials based food sample pretreatment methods for food safety screening. *TrAC Trends in Analytical Chemistry, 121*, 115669. https://doi.org/10.1016/j.trac.2019.115669.

Zhang, B. bin, Shi, Y., Chen, H., Zhu, Q. xia, Lu, F., & Li, Y. wei (2018). A separable surface-enhanced Raman scattering substrate modified with MIL-101 for detection of overlapping and invisible compounds after thin-layer chromatography development. *Analytica Chimica Acta, 97*, 35–43. https://doi.org/10.1016/j.aca.2017.10.006.

Zhao, J., Wei, F., Xu, W., & Han, X. (2020). Enhanced antibacterial performance of gelatin/chitosan film containing capsaicin loaded MOFs for food packaging. *Applied Surface Science, 510*, 145418. https://doi.org/10.1016/j.apsusc.2020.145418.

Zhang, G., Meng, F., Guo, Z., Guo, T., Peng, H., Xiao, J., Liu, B., et al. (2018). Enhanced stability of vitamin A palmitate microencapsulated by γ-cyclodextrin metal-organic frameworks. *Journal of Microencapsulation, 35*(3), 249–258. https://doi.org/10.1080/02652048.2018.1462417.

Zhuang, J. L., Ceglarek, D., Pethuraj, S., & Terfort, A. (2011). Rapid room-temperature synthesis of metal-organic framework HKUST-1 crystals in bulk and as oriented and patterned thin films. *Advanced Functional Materials, 21*(8), 1442–1447. https://doi.org/10.1002/adfm.201002529.

Chapter 10

Metal-organic framework composites for adsorption of volatile organic compounds

Alireza Davoodi, and Kamran Akhbari
School of Chemistry, College of Science, University of Tehran, Tehran, Iran

Introduction

The rapid progress of diverse industries has led to the release of numerous pollutants, including radioactive elements (Zou et al., 2016), metallic ions (Li et al., 2018), pesticides (Vymazal & Březinová, 2015), antibiotics (Martinez, 2009), dyes (Yagub et al., 2014), volatile organic compounds (VOCs) (McHale et al., 2012), and others, into the environment. This poses a significant risk to human health and the wellbeing of other organisms. VOCs are organic chemicals characterized by their relatively high vapor pressure at room temperature, typically exceeding 0.01 kPa at 20°C (Wang & Austin, 2006). These compounds vaporize at room temperature, rapidly transitioning states and becoming inhalable. Many VOCs commonly present in environments have been linked to both acute and chronic adverse health effects, such as sensory and skin irritation, headaches, breathing difficulties, heightened risk of asthma, and cancer (Irga et al., 2024), as well as significant physical and mental health concerns (Finewax et al., 2021). Certain naturally occurring VOCs come from biogenic sources in both terrestrial and marine environments, in addition to anthropogenic sources (Irga et al., 2024). Examples of VOCs detrimental effects on both human health and the environment include benzene, xylene, and toluene. Aldehydes (RCHO), aromatic compounds, and polycyclic aromatic hydrocarbons are among the common yet highly hazardous and carcinogenic VOCs (Kim et al., 2013). In the troposphere, certain VOCs undergo photochemical reactions with SO_2 and NO_x, leading to the formation of ozone (O_3) and photochemical pollution (Seinfeld & Pandis, 2016). Additionally, VOCs play a role in the creation of secondary organic aerosols (SOA), which in turn lead to environmental pollution (Fig. 10.1). Hence, it is imperative to significantly reduce the release

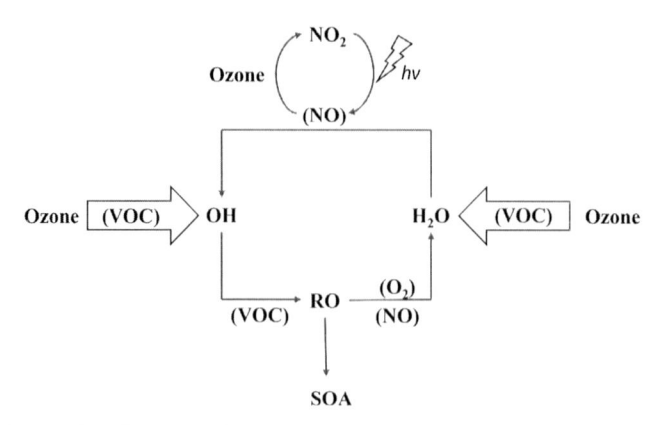

FIGURE 10.1 A simplified model for volatile organic compounds ozonolysis and radical cycling. (*SOA*, secondary organic aerosols; *VOC*, volatile organic compound).

of VOCs from various sources to prevent their entry (Kumar et al., 2021; Wang et al., 2017) into the environment. Various methods, including membrane separation, catalytic oxidation, absorption, adsorption, biodegradation, and other postprocessing management techniques, have been employed to reduce VOCs and prevent their release (Zhang et al., 2017). Among these methods, adsorption stands out as the most practical and cost efficient approach to lower and control VOC emissions into the environment (Wang, Zhang, et al., 2014). It should be noted that VOCs are not always harmful. In some cases, such as pollination and many other instances in nature, they have useful applications (Knudsen & Gershenzon, 2020). However, due to their increasing production and the numerous harm they cause to humans and nature, they have become a significant threat today.

There is an increasing need for adsorbents that are environmentally friendly and have high adsorption capacities for environmental pollutants (Pourebrahimi & Pirooz, 2022). Consequently, researchers have explored various materials, such as polysaccharide based adsorbents (Qi et al., 2021), natural clay (Orta et al., 2020), and metal-organic frameworks (MOFs) (Ghasemzadeh & Akhbari, 2023a, 2023b, 2024) for the decontamination of polluted environments. Among these, MOFs have garnered significant attention from researchers in environmental remediation (Wang et al., 2020; Wu et al., 2020). This is due to the high effectiveness of MOFs and their composites as adsorbents for environmental pollutants, owing to their exceptional porosity, pore architectures, functionality, and numerous active adsorption sites (Noori & Akhbari, 2017; Parsaei et al., 2023). MOFs can be readily constructed, designed, and fabricated from organic compounds (ligands) and metal ions under specific conditions and their recyclability is another notable advantage of MOFs (Akhbari & Morsali, 2013; Parsaei & Akhbari, 2022a). Incorporating MOFs can significantly enhance the adsorption selectivity of composite adsorbents, while the synergistic interaction between MOFs and functional materials can further improve adsorption properties. Additionally, researchers often fabricate multifunctional MOF composites by carefully integrating dual or multifunctional materials in a controllable manner (Elsaidi et al., 2018; Yang et al., 2021). Also, previous reports have shown that pure MOF adsorbents can be functionally modified by introducing functional groups onto their ligands (Alavijeh et al., 2019; Salimi et al., 2024). This modification allows for the creation of additional adsorption sites, thereby enhancing the adsorption capabilities of the MOF material. Consequently, MOF-composite adsorbents can also be combined with graphene oxide GO (Wei et al., 2020), polydopamine (–OH, –NH$_2$) (Yu et al., 2019) or chitosan (CS) nanofibers (–OH, –NH$_2$) (Niu et al., 2021), which take inspiration from the aforementioned approach. With these superior properties, the increasing utilization and development of MOF composites for the elimination of VOCs are clearly evident.

Volatile organic compound definitions

Different countries have their own definitions for VOCs. For instance, in Canada, Health Canada defines VOCs as organic compounds with boiling points typically ranging from 50°C to 250°C (Xu et al., 2016). The European Union's definition of VOCs includes any organic compound, with a vapor pressure of 0.01 kPa or more at 293.15 K (Burghardt & Pashkevich, 2018). In China, VOCs are recognized as substances originating from sources like automobiles and industrial production, possessing a saturated vapor pressure above 70.91 Pa under room temperature conditions (Gao et al., 2021). Each country's approach reflects its unique environmental challenges and regulatory priorities regarding VOCs.

Variety and hazardous of volatile organic compounds

Many chemicals fall within the category of VOCs. This section provides examples of some of the most prominent materials in this category. It is important to note that in the realm of chemical safety, particularly concerning VOCs, the immediately dangerous to life or health (IDLH) parameter holds significant importance (Table 10.1). Typically, VOCs are colorless, possess a notably low boiling point, and tend to exhibit high reactivity. Additionally, they frequently blend with noxious gases, underscoring the significance of understanding these characteristics (Singh et al., 2022). For instance, acetone as a VOC, poses health risks and inhaling it can lead to irritation of the eyes,

TABLE 10.1 Volatile organic compounds' properties and their corresponding immediately dangerous to life or health levels as proposed by National Institute for Occupational Safety and Health (NIOSH), (2016).

Compound	Formula	IDLH mg/m^3
Ethyl alcohol	C$_2$H$_5$OH	1900[b]
Methyl alcohol	CH$_3$OH	260[b]
Isopropyl alcohol	CH$_3$CHOHCH$_3$	980[b]
Phenol	C$_6$H$_5$OH	19[a]
Acetaldehyde	CH$_3$CHO	n.e
Formaldehyde	H$_2$CO	n.e
Benzaldehyde	C$_6$H$_5$CHO	NA
n-Hexane	C$_6$H$_{14}$	180[a]
Octane	C$_8$H$_{18}$	350[a]
Benzene	C$_6$H$_6$	3.2[a]
Toluene	C$_7$H$_8$	375[a]
Acetic acid	CH$_3$COOH	25[a]
Formic acid	HCOOH	9[a]
Ethyl acetate	CH$_3$COOC$_2$H$_5$	1400[a]
Chloroform	CHCl$_3$	9.78[b]
Acetone	(CH$_3$)$_2$CO	590[a]
Hydrogen peroxide	H$_2$O$_2$	1.4[a]
Ammonia	NH$_3$	18[a]
Dimethylamine	(CH$_3$)$_2$NH	18[a]
Hydrogen sulfide	H$_2$S	15[a]
Acetylene	(C$_2$H$_2$)	NA

ILDH, Immediately dangerous to life or health; *NA*, not available; *n.e*, none established (potential occupational carcinogenic).
[a] *8 h time weighted average (TWA).*
[b] *60 min short term exposure limit (STEL).*

nose, and throat. Ingesting acetone can cause headache, dizziness, and dermatitis (Aldehydes, 2007). Also, acetylene is highly dangerous due to its inherent instability, especially when subjected to liquefaction, pressurization, heating, or when mixed with air (Iftekhar Uddin et al., 2015; Lee et al., 2015). Benzene is notably worrisome due to its toxicity and carcinogenic nature.

Benzene is recognized as a carcinogen through all exposure routes and is a significant contributor to the development of leukemia and lymphomas (Ke et al., 2009). Common symptoms associated with formaldehyde exposure include irritation of the eyes, nose, and throat, typically occurring in ambient air. Furthermore, formaldehyde has been categorized as a human carcinogen due to its link to nasopharyngeal cancer, pulmonary damage, and likely leukemia (Castro-Hurtado et al., 2013). Methanol is another VOC that is extremely toxic, leading to acidosis and potential blindness. Methanol poisoning can cause symptoms, such as nausea, abdominal pain, headaches, blurred vision, shortness of breath, and dizziness (Patel et al., 2003). These descriptions cover some of the most well-known VOCs mentioned here. Additional VOCs are listed in Table 10.1, but it is important to note that the range of these substances extends far beyond those mentioned here. Many more VOCs exist beyond those included in this discussion.

Less noticed source of volatile organic compounds

As mentioned in previous sections, the sources of VOCs are primarily associated with the consumption of fossil fuels, such as burning gasoline, diesel, oil, and gas. However, VOC sources are not limited to these and extend further. Biogenic volatile organic compounds (bVOCs) are the VOCs released from biomass, including substances like methanol, acetone, and acetaldehyde. Soil plays a significant role in bVOC emissions due to various biotic and abiotic activities of organisms within it (Loreto & Fares, 2013). Also, plant species like sagebrush (Jaeger et al., 2016), bitter orange (González-Mas et al., 2019), corn (Shao et al., 2001), etc., can produce monoterpenes, monoterpenoid, ethyl acetate, rhizathalene, ethanol, aldehydes, ketones, acetone, and isoprene. Additionally, certain bacterial species, such as *Staphylococcus aureus* (Chen et al., 2018), *Lactobacillus helveticus* (Cuffia et al., 2018), and fungal species like *Phomopsis* (Santoso et al., 2020) and *Trametes versicolor* (Drilling & Dettner, 2009) have the capability to produce ethanol, ethyl acetate, acetaldehyde, naphthalene, sabinene, and various other VOCs (Daisy et al., 2002; Oro et al., 2018).

Strategies for reduction and elimination of volatile organic compounds

Considering the dangers of VOCs, it is very important to control them. To control these substances, one can use several techniques, some of which are discussed in this section. VOC destruction, recovery, and so on., are some techniques to control VOCs. For instance, thermal oxidation, catalytic oxidation, and biofiltration are subcategories of destruction methods. The thermal oxidation technology can attain a VOC elimination efficiency ranging from 95% to 99%, with the possibility of energy recuperation. However, additional treatment is required for combustion byproducts, particularly those containing halogenated VOCs (Khan & Kr Ghoshal, 2000). Catalytic oxidation systems, akin to thermal oxidizers, directly incinerate VOCs. They operate at lower temperatures. However, certain combustion byproducts have the potential to diminish catalytic activity (Parmar & Rao, 2008). In addition, techniques for VOC retrieval encompass condensation, adsorption, and membrane separation. Condensation is most effective for VOCs with boiling points exceeding 37.7°C and concentrations surpassing 5000 ppm (Volkamer et al., 2006). Additionally, VOCs are extracted from gas streams using adsorption, a process in which polluted air interacts with a liquid solvent and soluble VOCs transition into the liquid phase. An adsorber has the potential to achieve VOC elimination efficiencies of 90%–98% (William & Lead, 1997). Adsorption occurs when organic molecules are attracted to the surface and pores of the adsorbent through weak van der Waals forces. This type of adsorption is typically characterized by a low heat of adsorption and the establishment of a quickly reversible adsorption equilibrium. Activated carbon (AC)-based adsorption is a commonly used technique for reducing VOC emissions (Ruthven, 1984). MOF composites, with their numerous capabilities, can effectively trap VOCs using multiple methods (especially adsorption), which is the reason for the advantage of this category of chemicals in their use.

Suitable metal-organic frameworks for volatile organic compounds removal

The advancement of MOFs has been extensively pursued due to their versatility, ease of fabrication, tunable pore structure, and robustness (Karimi Alavijeh & Akhbari, 2020; Nakhaei et al., 2021; Parsaei & Akhbari, 2022c; Soltani & Akhbari, 2022a). A wide range of ligands and metal ions is available for synthesizing MOFs, resulting in diverse structural variations (Davoodi et al., 2023; Parsaei & Akhbari, 2022a). Nevertheless, a comprehensive understanding of the coordination and interaction between MOFs and VOCs is essential to elucidate their adsorption mechanisms (Qin et al., 2023; Siu et al., 2023). The adsorption capacity of MOFs stems from their porous structure, which provides a substantial surface area and significant pore volume (Karimi Alavijeh & Akhbari, 2024; Parsaei & Akhbari, 2022b; Parsaei et al., 2024). The expansive surface area furnishes ample active adsorption sites, while the deep pore volume enables the adsorbate to be effectively captured inside the MOF pores (Amidi & Akhbari, 2024; Arshadi Edlo & Akhbari, 2023; Kalati & Akhbari, 2021; Soltani & Akhbari, 2022b). While this chapter primarily delves into MOF composites, this section briefly examines the adsorption of VOCs by different types of materials, including MOFs, MILs, ZIFs, and so on, to facilitate further comparison of the findings.

Adsorption of volatile organic compounds by the metal-organic frameworks

In this section, examples of the adsorption of VOCs by MOFs are mentioned. For instance, the MIL series, standing for Materials of the Institute Lavoisier, comprises a range of MOFs, including MIL-53, MIL-47, MIL-100, and MIL-101 (Alivand et al., 2019). Variations among MILs primarily stem from the properties imparted by the different metals utilized during synthesis. Additionally, the choice of organic linker (ligand) varies across MILs; for instance, MIL-100 employs trimesic acid (1,3,5-benzenetricarboxylic acid = BTC), while MIL-101 utilizes terephthalic acid (benzene-1,4-dicarboxylic acid = BDC) in its synthesis (Hamon et al., 2009). But in the field of adsorption of VOCs, MIL-53 (Al or Cr) and MIL-47 (V) in the MIL series are renowned for their reversible adsorption characteristics, which are due to their breathing capability. Extensive research has been carried out on the reversible properties of MIL-53 and MIL-47 concerning hydrogen sulfide (H_2S) gas as a VOC under different pressures (Yot et al., 2012). Adsorption in MIL-53 and MIL-47 occurs via hydrogen bonding between the adsorbate and the carboxylic groups of the MILs (Alhamami et al., 2014). Similarly, about examples of MOFs, it can be noted that MOF-177 has been extensively synthesized from zinc as the metal and 4,4′,4″-benzene-1,3,5 triyl-trisbenzoic acid (H_3BTB) and employed for gas and VOC adsorption purposes (Chae et al., 2004). This material exhibits a high capacity for ammonia adsorption and the isoreticular MOF-3 (IRMOF-3) demonstrates highly effective performance in adsorbing chlorine gas (Britt et al., 2008). Another group of porous compounds that is valuable in this domain are known as Northeast Normal University (NENU) series. These MOFs, synthesized using metal ions and benzene-1,3,5-tribenzoate (BTB), exhibit excellent performance in adsorbing various VOCs, including benzene (Table 10.2). Another category of porous materials, known as ZIFs, displays remarkable adsorption characteristics and can significantly contribute to VOC removal. For instance, zeolitic imidazole framework-8 (ZIF-8) is one such example that has been extensively investigated (Qin et al., 2023). It should be noted that the aforementioned MOFs typically exhibit adsorption of various VOCs, with the highest reported adsorption among these VOCs as "Target VOC" in Table 10.2

Metal-organic framework composites for volatile organic compounds removal

In many instances, MOFs demonstrate a favorable performance in adsorbing VOCs, yet they may also exhibit drawbacks, such as limited adsorption rates and efficiency (El-Mehalmey et al., 2018; Furukawa et al., 2014). In addressing these issues, combining MOFs with other materials to create MOF composites presents a promising solution. Thus, besides offering the

TABLE 10.2 Investigated the adsorption performance of metal-organic frameworks against various volatile organic compounds.

MOF	Linker (ligand)	Metal ion	Target VOC	References
MIL-47	BDC	V	Hydrogen sulfide	Hamon et al. (2009) and Yot et al. (2012)
MIL-53	BDC	Cr	Hydrogen sulfide	Hamon et al. (2009)
MIL-53	BDC	Fe	Hydrogen sulfide	Hamon et al. (2009)
MIL-100	BTC	Fe	Phenol	Hamon et al. (2009)
MIL-100	BTC	Cr	Hydrogen sulfide	Hamon et al. (2009) and Liu et al. (2014)
MIL-101	BDC	Cr	Benzene	Alivand et al. (2019)
MOF-177	BTB	Zn	Ammonia	Britt et al. (2008) and Yang et al. (2013)
MOF-5	BDC	Zn	Benzene	Britt et al. (2008) and Yang et al. (2013)
MOF-74	DOBDC	Zn	Sulfure dioxide	Britt et al. (2008) and Yang et al. (2013)
IRMOF-3	NH_2-BDC	Zn	Chlorine	Britt et al. (2008) and Yang et al. (2013)
MOF-199	BTC	Cu	Tetrahydrothiophene	Britt et al. (2008) and Yang et al. (2013)
IRMOF-62	BDB	Zn	Benzene	Britt et al. (2008) and Yang et al. (2013)
NENU-511	BTB	Zn	Benzene	He et al. (2015)
NENU-512	BTB	Zn	Benzene	He et al. (2015)
NENU-513	BTB	Zn	Benzene	He et al. (2015)
NENU-514	BTB	Zn	Benzene	He et al. (2015)

BDC, Benzene-1,4-dicarboxylic acid; *BTB*, benzene-1,3,5-tribenzoate; *BTC*, 1,3,5-benzenetricarboxylic acid; *IRMOF*, isoreticular metal-organic framework; *MOF*, metal-organic frameworks; *NENU*, Northeast Normal University; *VOC*, volatile organic compound; H_4DOBDC, 2,5-Dihydroxyterephthalic acid.

potential for multifunctionality, composites serve as an effective structural substrate to improve adsorption and desorption efficiency. Immobilizing an MOF on a substrate or a substance on a MOF enhances the potential for various applications. Furthermore, despite the inherent differences between MOFs and other materials, the deliberate incorporation of multiple functional groups to create flexible composites with improved performance has proven to be a practical approach for achieving multifunctionality and enhancing effectiveness in various applications (Duan et al., 2020; Kalaj et al., 2020). Essentially, the creation of MOF composites enhances porosity, introduces additional functional groups, increases surface area, improves chemical and thermal stability, and facilitates recycling. This, in turn, promotes a more economical use of MOF composites. Moreover, the methods employed for synthesizing MOF composites are also pivotal in expanding their applications (Liu et al., 2023). Generally, MOF composites are synthesized through several approaches, each tailored to the specific type of composite (Cai et al., 2022). Hence, MOF composites are often regarded as promising candidates for adsorbent applications, owing to their inherent properties, such as diverse functionalities, robust stability, and ease of modification. A distinctive aspect of MOF composites lies in the integration of the capabilities of both MOFs and the secondary component, featuring various types of functional groups. Consequently, although individual components may not exhibit optimal adsorption performance, the resulting composite may demonstrate enhanced adsorption capabilities in the adsorption process, attributed to the synergistic effects of their chemical and physical properties (Aguilera-Sigalat & Bradshaw, 2016; Dhakshinamoorthy et al., 2018). In recent times, there has been a growing interest in MOF-based composites, as their adsorption properties can be augmented by the incorporation of other materials, such as mesoporous carbon (OMC), carbon nanotubes (CNTs) (Xiang et al., 2011), graphite oxide (GO) (Petit & Bandosz, 2011), metals (Meilikhov et al., 2010), and AC (Somayajulu Rallapalli et al., 2013).

Graphene, graphite, and their derivatives metal-organic framework composites

Before addressing this section, it is important to clarify certain points to eliminate potential ambiguities. In this section, which discusses composites containing graphene and graphite, the abbreviation GO is used for both materials. To maintain consistency with the original reported texts, the names of these materials have not been altered to distinguish between them. Therefore, to comply with copyright, the GO has been included unchanged, but its full name has been included to remove ambiguity. Bandosz and colleagues synthesized composites comprising MOFs (MOF-5, HKUST-1, and MIL-100(Fe)) and graphite oxide (GO), and subsequently evaluated their characteristics in adsorbing small gas molecules NH_3, H_2S, and NO_2. The study revealed that synergistic interactions involving porosity and chemical properties augmented the adsorption capacity for these hazardous gases (Petit & Bandosz, 2009, 2011, 2012). In another study by Sun et al., the adsorption of *n*-hexane by MIL-101 composite was investigated. This study clearly demonstrates that MIL-101 composites exhibit significantly better performance than pure MIL-101 in adsorbing this VOC. In the study, MIL-101(Cr)/GO composites (MIL-101@GO) were synthesized, where GO refers to graphite oxide. The composites were prepared with GO contents of 5% and 10% by weight relative to the starting material. The exceptional adsorption performance of MIL-101@GO-5% is attributed not only to its large surface area but also to the optimal amount of GO, which strengthens surface dispersive forces (Sun et al., 2014). Similarly, Dai et al. investigated the adsorption efficiency of toluene, comparing Cu-BTC with its composite form, Cu-BTC@GO (GO = graphene oxide). The composite exhibited a notably higher adsorption capacity of around 183 mg/g (nearly three times that of Cu-BTC (62.7 mg/g)) when 20% of the mass was replaced with GO (graphen oxide) (Dai et al., 2019). In another study, composite materials consisting of MOF-5 and GO (graphite oxide) in varying proportions were synthesized and assessed for their ability to remove ammonia under dry conditions. The findings reveal a synergistic impact leading to heightened ammonia adsorption in comparison to the individual components. This enhancement is attributed to increased dispersive forces within the composite's pore structure (from 56 to 82 mg/g) (Petit & Bandosz, 2010).

In some studies aimed at developing sensors or other applications for VOCs, researchers attempt to synthesize MOF composites. Ultimately, in addition to achieving their goal, they create a product with high adsorption capacity. For example, Tung et al. investigated graphene hybrid nanocomposites paired with MOFs including copper benzene-1,3,5-tricarboxylate (pG-Cu BTC) (pG=pristine graphene), zirconium 1,4-dicarboxybenzene (pG-UiO-66), and zinc salt of 2-methyl imidazole (pG-ZIF-8) to enhance the diagnostic efficiency of various VOCs, such as methanol, ethanol, chloroform, acetone, acetonitrile, and THF. The nanocomposite material exhibited improved sensing performance, with graphene serving as a highly conductive sensing element and MOFs with high surface area and adsorption capacity (Tung et al., 2020). In another study, an NH_3 sensor was developed based on a composite of reduced graphene oxide (rGO) coated with polypyrrole nanofibers and decorated with nanoscale Cu–BTC. The formation of Cu–BTC was facilitated by the composite, and Cu–BTC's excellent gas-adsorption properties can significantly enhance the ammonia adsorption on the PPy–rGO composite surface (Yin et al., 2018).

A solvent-free mechanochemical method is also suggested for the rapid synthesis of Cu-BTC@GO (GO = graphite oxide) composites, combining Cu-BTC with GO (graphite oxide). Their water stability and ability to adsorb toluene vapor were investigated, revealing that this method effectively

Metal-organic framework composites for adsorption **Chapter | 10** **347**

produces Cu-BTC@GO composites. The resulting Cu-BTC@GO composites exhibited a higher BET surface area and pore volume compared to the parent Cu-BTC, leading to a 47% increase in toluene uptake compared to Cu-BTC alone (from 6.2 to 9.1 mmol/g). This enhancement exceeded that of conventional ACs and zeolites. Additionally, the water stability of Cu-BTC@GO composites was significantly improved. Cu-BTC@GO composites were highly promising for practical applications in VOC adsorption due to their excellent water stability and high toluene adsorption capacity (Li et al., 2016).

In another study, a hybrid nanocomposite, UiO-67/graphene oxide (GO), was synthesized using a solvothermal method, leading to a substantial improvement in toluene adsorption. Among the tested composites, the one with 0.5% GO exhibited the highest adsorption capacity. This enhancement was attributed to the influence of GO, which strengthened $\pi\cdots\pi$ interactions, induced defect formation, and provided extra adsorption sites (Fig. 10.2) (Zhao et al., 2022).

The MOF composite GO@MIL-101 (GO = graphene oxide) displayed a larger surface area and pore volume compared to pristine MIL-101, with the crystal size of MIL-101 within the composite being smaller than that of the original MIL-101. As a result, GO@MIL-101 demonstrated enhanced adsorption capabilities for acetone in comparison to pristine MIL-101 (Zhou et al., 2014). Bandosz et al. synthesized two kinds of hybrid materials by combining MOFs with graphite oxide (GO). One hybrid was derived from zinc-containing MOF-5, while the other utilized copper-containing HKUST-1. The researchers

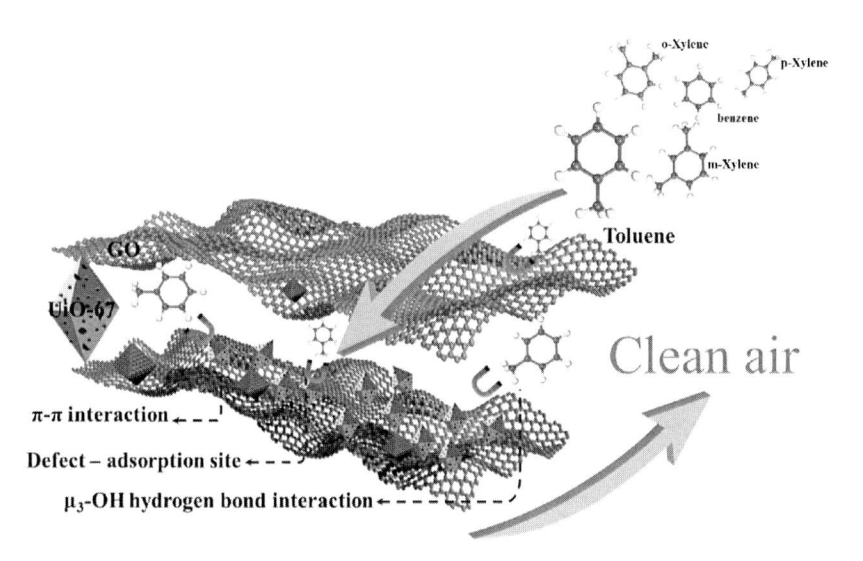

FIGURE 10.2 Feasible mechanisms for toluene gas adsorption in UiO-67/GO (Zhao et al., 2022).

observed that these composite materials retained similar crystalline structures and porosity to their respective parent MOFs. Furthermore, the ammonia adsorption capacity of these composites surpassed that of their parent MOFs. This enhancement was attributed to the formation of new pores between the two phases (MOF units and GO) within the hybrids (Bandosz & Petit, 2011).

Levasseur et al. produced composites consisting of HKUST-1 and GO (graphite oxide). They investigated the reactive adsorption of NO_2 under both dry and moist conditions and observed that, under dry conditions, the composites demonstrated a higher NO_2 breakthrough capacity compared to the individual components of MOF and GO (graphite oxide). The increased performance was attributed to the enhanced porosity of the composites and the reactive adsorption of NO_2 by the copper present in the HKUST-1 structure (Levasseur et al., 2010).

Metal, semimetal, and their salt metal-organic framework composites

In a different study investigating the composite's effectiveness in VOC degradation, TiO_2@UiO-66 demonstrated enhanced photocatalytic activity for converting toluene into CO_2, outperforming other UiO-based composites, especially in extended flow reactions. This superior performance was attributed to the synergistic effect between UiO-66 and TiO_2, which increased toluene adsorption (Zhang et al., 2021). Another TiO_2-based composite suitable for VOC adsorption and removal is the TiO_2@NH_2-MIL-125 photocatalyst. This material, synthesized via an in situ solvothermal method, features strong interfacial contact between NH_2-MIL-125 and TiO_2, making it highly effective for formaldehyde (HCHO) removal under UV irradiation. Compared to pure TiO_2 and NH_2-MIL-125, the composite demonstrated significantly improved adsorption and photocatalytic efficiency, attributed to the synergistic effects of NH_2-MIL-125's high adsorption capacity, TiO_2's fine dispersion, and enhanced interfacial charge transfer between the two components (Huang et al., 2019).

In another study, a series of composites consisting of SiO_2 incorporated into aluminum-based MOF (MIL-68) with varying SiO_2 loadings were synthesized using a simple and gentle compositing method for the efficient removal of aniline. The adsorption capabilities of aniline were compared among SiO_2, MIL-68(Al), a physical mixture of these two materials, and SiO_2@MIL-68(Al) composites. The results revealed a relatively high adsorption capacity for SiO_2@MIL-68(Al) toward aniline. Moreover, the composite exhibited rapid adsorption kinetics, along with excellent reusability. These findings suggest that SiO_2@MIL-68(Al) composites hold significant promise as effective adsorbents for aniline (Han et al., 2016). Also, Wang et al. assessed Fe_3O_4@SiO_2–MOF-177 as a favorable adsorbent for magnetic solid-phase extraction (MSPE) of phenols in environmental samples. As a result, the utilization of magnetic MOF materials as novel solid-phase extraction adsorbents has gained popularity due to their advantages, such as the exceptionally large surface area of MOFs and the convenient operation of MSPE (Wang, Lei, et al., 2014). A straightforward, highly sensitive, and durable technique utilizing magnetic MOFs composed of Fe_3O_4@SiO_2–MIL-101 as sorbents for MSPE was effectively created. This method enables the simultaneous detection of four pyrazole/pyrrole pesticides (Flusilazole, Fipronil, Chlorfenapyr, and Fenpyroximate) as VOCs in the environment (Ma et al., 2016).

Fiber and polymer metal-organic framework composites

For aniline removal, researchers developed a MOF-polyaniline (PANI) nanofiber array composite photothermal membrane. Designed for eliminating high-concentration VOCs from water, this membrane uses molecular sieving during solar-driven evaporation. The modified ZIF-8 layer, grown on the PANI nanofiber array, acts as a selective barrier, allowing water to evaporate while trapping VOC molecules. This system achieved both high VOC rejection and an impressive water evaporation rate across various VOC concentrations in water (Peng et al., 2022).

Topuz et al. successfully introduced hydrophobic properties to nanofibrous polyimide membranes by modifying them with $-CF_3$ groups while retaining their crystalline structure. They incorporated two Zr-based MOFs (pristine UiO-66 (1071 m²/g) and defective UiO-66 (1582 m²/g)) into the nanofibers, forming high-performance nanocomposites. These modified fibers effectively removed both polar formaldehyde and nonpolar VOCs like toluene, xylene, and mesitylene from the air. The integration of MOFs significantly enhanced the adsorption capacity of polyamide fibers, despite their initially low surface area (Topuz et al., 2022). In the realm of toluene adsorption, ZIF-67(Co) is uniformly attached to the macroporous surface of polyacrylates (PA) via interface coordination, resulting in a distinctive hierarchical porous structure. When compared to the pristine ZIF-67(Co), the toluene adsorption efficiency of ZIF-67(Co)@PA is enhanced by 2.75 times at 298 K (Fig. 10.3) (Wu et al., 2022).

Composite MOF foams were synthesized by directly growing UiO-66 on a polyurethane foam template. With optimal conditions, these composite materials retained the mtacrostructure and flexibility of the polyurethane foam, while exhibiting the microporosity, high surface area, and adsorption characteristics of UiO-66. The composite MOF foam, with its hierarchical porosity, exhibits a high adsorption capacity for benzene and n-hexane, retaining over 70% of the adsorption capacity of UiO-66 (Pinto et al., 2013).

Cui et al. utilized cellulose derivatives, including carboxymethyl cellulose (CMC) and polysaccharides like chitosan (CS) and sodium alginate (SA) (Fig. 10.4) as a gel-based platform for the growth of HKUST-1. The synthesized materials were then degassed via freeze-drying to create

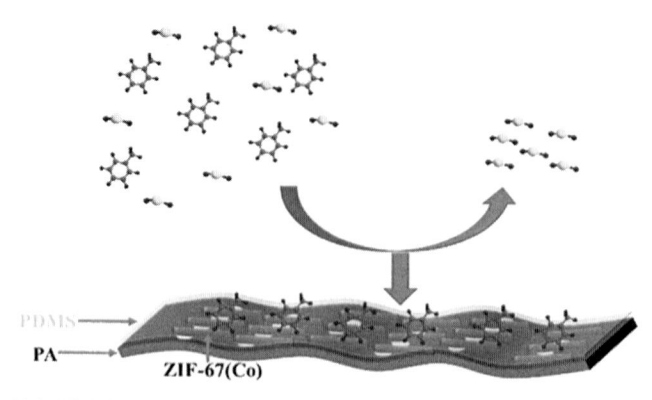

FIGURE 10.3 High-humidity-proof hierarchical porous P-ZIF-67(Co)-polymer composite materials by surface modification for highly efficient volatile organic compound adsorption (Wu et al., 2022).

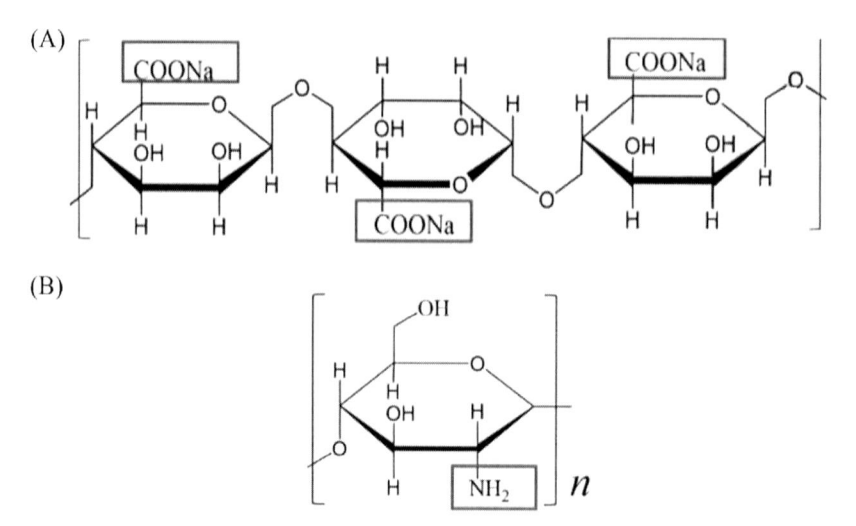

FIGURE 10.4 Molecular structures of polysaccharides (A) sodium alginate (SA) and (B) chitosan (CS) (Cui et al., 2019).

HKUST-1@cellulose foam. In VOC adsorption studies, this composite demonstrated superior toluene adsorption capacity at very low relative pressures, outperforming pristine HKUST-1 (Cui et al., 2019).

To enhance VOC adsorption under humid conditions, a hydrophobic barrier composed of short-chain polyethylene glycol (PEG) was introduced into MIL-101(Cr). The PEG chains effectively restricted water diffusion into the MOF while allowing toluene molecules to pass through. This modification, particularly in PEG5@MIL-101, notably increased both the adsorption capacity and diffusivity for toluene compared to the pristine MOF (Wang et al., 2021).

Agrawal and colleagues recently introduced a method for the in situ synthesis of ZIF-8 and ZIF-67 nanocrystals on carboxymethylated cotton (CM Cotton) fabric. VOC adsorption studies revealed that both ZIF-8@CM Cotton and ZIF-67@CM Cotton exhibit high adsorption capacities for various organic pollutants, including aniline, benzene, and styrene, even at an adsorption pressure of 1 torr. Moreover, these textiles can be easily regenerated by heating to 120°C without any loss of adsorption capacity (Jhinjer et al., 2021).

Ordered mesoporous carbon, carbon nanotube, activated carbon, and another metal-organic framework composites

Gao et al. enhanced the hydrophobic properties of MOFs by using ordered mesoporous carbon (OMC) as a porous template for the growth of MIL(Ti)-125-NH$_2$ (Ti-clusters). Their method involved using Ti-doped phenol-formaldehyde resin as the precursor for OMC synthesis, with TiO$_2$ etching during MOF formation creating a mesoporous structure. The MIL(Ti)-125-NH$_2$@OMC composite was tested for toluene adsorption, achieving a capacity of 0.1447 g/g under dynamic flow conditions at 80% relative humidity (1.5 times higher than pristine MIL-125-NH$_2$). This enhanced toluene uptake is attributed to faster mass diffusion through the mesoporous OMC channels and the material's inherent hydrophobicity (Fig. 10.5) (Gao et al., 2022).

A series of CNT@MIL-68(Al) composites with varying CNT loadings was synthesized, where CNT inclusion improved MIL-68(Al) dispersity and reduced particle size compared to pure MIL-68(Al). The optimal CNT content increased the BET surface area and the total volume of small micropores (< 16 Å), significantly enhancing phenol adsorption from aqueous solutions (Han et al., 2015).

Sheykhi et al. synthesized a composite material called MWCNT@MIL-53(Cr) (MWCNT=multiwalled carbon nanotube), and demonstrated that compared to pure MIL-53(Cr), this hybrid material shows enhanced CH$_4$ storage capacity at room temperature and superior thermal stability (Anbia et al., 2017).

Khoshakhlagh et al. used oxidized activated carbons (OACs) enriched with carboxyl (–COOH) groups as a template for the direct synthesis of Cu-BDC MOF. The resulting Cu-BDC@OAC composite achieved the highest toluene capture efficiency in dynamic breakthrough tests, showing approximately 12% and 50% higher efficiency than pure Cu-BDC and oxidized ACs, respectively (Khoshakhlagh et al., 2020). At the culmination of this section, all the composite materials discussed herein are succinctly summarized in Table 10.3.

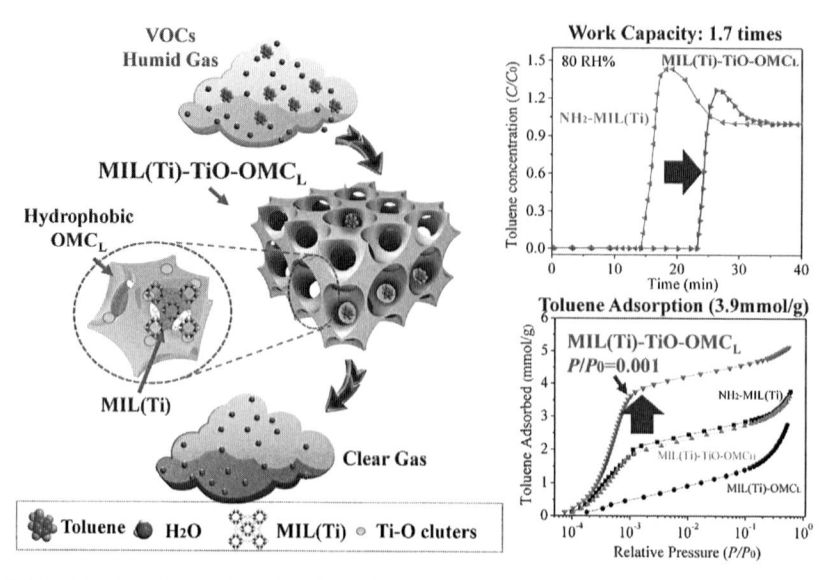

FIGURE 10.5 Toluene adsorption by ordered mesoporous carbon/metal-organic framework composite and quantitative investigation of its adsorption in MIL(Ti)-125-NH$_2$@OMC (Gao et al., 2022).

TABLE 10.3 Summary of metal-organic framework composite mentioned in this section "Metal-organic framework composites for volatile organic compounds removal."

MOF	Second part of MOF composite	Target VOC	References
MIL-100 (Fe)	Graphite oxide	Ammonia	Petit and Bandosz (2011)
MOF-5	Graphite oxide	Ammonia	Petit and Bandosz (2009)
HKUST-1	Graphite oxide	Ammonia, hydrogen sulfide, and nitrogen dioxide	Petit and Bandosz (2012)
MIL-101 (Cr)	Graphite oxide	n-hexane	Sun et al. (2014)
Cu-BTC	Graphen oxide	Toluene	Dai et al. (2019)
MOF-5	Graphite oxide	Ammonia	Petit and Bandosz (2010)

(*Continued*)

Cu-BTC, UiO-66, and ZIF-8	Pristine graphene	Methanol, ethanol, chloroform, acetone, acetonitrile, and tetrahydrofuran	Tung et al. (2020)
Cu-BTC	Reduced graphene oxide (rGO) and polypyrrole nanofibers	Ammonia	Yin et al. (2018)
Cu-BTC	Graphite oxide	Toluene	Li et al. (2016)
UiO-67	Graphen oxide	Toluene	Zhao et al. (2022)
MIL-101	Graphen oxide	Acetone	Zhou et al. (2014)
MOF-5	Graphite oxide	Ammonia	Bandosz and Petit (2011)
HKUST-1	Graphite oxide	Ammonia	Bandosz and Petit (2011)
HKUST-1	Graphite oxide	Nitrogen dioxide	Levasseur et al. (2010)
UiO-66	TiO_2	Toluene	Zhang et al. (2021)
NH_2-MIL-125	TiO_2	Formaldehyde	Huang et al. (2019)
MIL-68 (Al)	SiO_2	Aniline	Han et al. (2016)
SiO_2–MOF-178	Fe_3O_4	Phenol	Wang, Lei, et al. (2014)
SiO_2–MIL-101	Fe_3O_4	Flusilazole, fipronil, chlorfenapyr, and fenpyroximate	Ma et al. (2016)
ZIF-8	Poluanilinie	Anilline	Peng et al. (2022)
UiO-66	Polyamides	Formaldehyde, toluene, xylene, and mesitylene	Topuz et al. (2022)
ZIF-67 (Co)	Polyacrylates (PA)	Toluene	Wu et al. (2022)
UiO-66	Polyurethane (PU)	Benzene, n-hexane	Pinto et al. (2013)
HKUST-1	Carboxymethyl cellulose (CMC) and polysaccharides (SA and CS)[a]	Toluene	Cui et al. (2019)
MIL-101 (Cr)	Polyethylene glycol (PEG)	Toluene	Wang et al. (2021)

(Continued)

TABLE 10.3 (Continued)

MOF	Second part of MOF composite	Target VOC	References
ZIF-8, ZIF-67	Carboxymethylated cotton (CM Cotton)	Aniline, benzene, and styrene	Jhinjer et al. (2021)
NH_2-MIL-125	Ordered mesoporous carbon (OMC)	Toluene	Gao et al. (2022)
MIL-68 (Al)	Carbon nanotube (CNT)	Phenol	Han et al. (2015)
MIL-53 (Cr)	Multiwalled carbon nanotube (MWCNT)	Methane	Anbia et al. (2017)
Cu-BDC	Activated carbons (ACs)	Toluene	Khoshakhlagh et al. (2020)

BDC, Benzene-1,4-dicarboxylic acid; *BTC*, 1,3,5-benzenetricarboxylic acid; *MOF*, metal-organic framework; *VOC*, volatile organic compound; *ZIF*, zeolitic imidazole framework.
[a] Polysaccharides (SA and CS)*, as shown in Fig. 10.4.

Adsorption in metal-organic framework composites

Surface functionalization can be employed to adjust the quantity of active sites, which can adsorb specific pollutants, thereby establishing a relationship between structure and activity. Apart from adjusting metal nodes and linkers, enhancing adsorption sites can also be achieved by creating composites of MOFs with various materials, such as graphene, polymers, metal nanoparticles, quantum dots, natural enzymes, and other functional substances (Chen et al., 2020). Moreover, polymers have been predominantly utilized in the creation of MOF composites and have demonstrated encouraging results in adsorption applications. These composites can be formed either by integrating MOFs into polymer matrices or by growing MOFs directly onto polymer surfaces. Several polymers, including polyvinyl alcohol, polyvinylidene fluoride, PEG, polyethylenimine, polymethylacrylate, PANI, and polyimide, have been commonly utilized in the production of MOF-polymer composites (Kalaj et al., 2020). In contrast to their individual counterparts, MOF composites exhibit superior adsorption performance thanks to synergistic effects, and they also address the drawbacks associated with pristine MOFs, such as challenging recovery, cost concerns, and environmental toxicity (Yao et al., 2022). Similarly, the advantage of easy recovery through magnetic field systems is evident in MOF composites incorporating magnetic nanoparticles. While maintaining MOF stability in aqueous environments poses a significant challenge, recent advancements in crafting water-stable MOFs hold promise for expanding the applications of these remarkable materials (Wang et al., 2016). Various linkages, including electrostatic interactions, hydrophobic interactions, acid-base interactions, size-selective interactions, connections or stacking interactions, and hydrogen bonding,

TABLE 10.4 Comparison of toluene adsorption capacity for some metal-organic frameworks and metal-organic framework composites.

Sample	Surface area (m²/g)	Capacity (mmol/g)	References
MIL-101 (Fe)	337	1.07	Romero-Sáez et al. (2018)
MIL-101 (Cr)	2449	0.64	Subhan et al. (2019)
MOF-12	1500	0.19	Muhammad et al. (2011)
MOF-5	424	0.36	Romero-Sáez et al. (2018)
HKUST-1	1237	1.73	Cui et al. (2019)
HKUST-1@CMC	545	3.1	Cui et al. (2019)
HKUST-1@CMC-SA	491	2.14	Cui et al. (2019)
HKUST-1@CMC-CS	489	2.79	Cui et al. (2019)
Cu-BTC@GO (GO = graphen oxide)	512	10.74	Dai et al. (2019)
ZIF-67(Co)@PA	433	58.35	Wu et al. (2022)
MIL-125-NH$_2$@OMC	1069	9.8	Gao et al. (2022)
Cu-BDC@AC	712	12.9	Khoshakhlagh et al. (2020)

have been identified as occurring between the adsorption sites of MOF composite and pollutants (Hasan & Jhung, 2015; Zheng et al., 2022). To investigate more closely the adsorption of VOCs by MOF composites, Table 10.4 presents examples of toluene adsorption by several MOFs and various MOF composites, facilitating comparative analysis. It is evident that, in numerous instances, the surface area has decreased following the transformation from MOF to MOF composite. However, as discussed in this chapter, factors such as synergistic effects contribute to the increased adsorption of VOCs.

Parameter of volatile organic compounds adsorption

Thermodynamic parameters

The heat of adsorption can be estimated effectively using the Clausius–Clapeyron equation. The relationship between the retention volume

(V_g) and temperature (T) can be expressed mathematically to determine the heat of adsorption, denoted as ΔH_{ads} (Eq. 10.1), where R represents the ideal gas constant. Henry's law constant (H_{lc}), which quantifies the relationship between the partial pressure of a gas and its dissolved amount, is closely linked to the retention volume (V_g) (Padial et al., 2013).

$$AH_{ads} = -R\frac{\delta(\ln V_g)}{\delta(1/T)} \quad (10.1)$$

In adsorption, Henry's law constant (H_{lc}) is directly linked to the amount of gas adsorbed. The determination of H_{lc} can be expressed mathematically, as demonstrated in Eq. (10.2), where n_{ads} represents the quantity of gas adsorbed, m is the mass of the adsorbent utilized, and p denotes the outlet pressure of the gas.

$$H_{lc} = \frac{n_{ads}}{m \times p} \quad (10.2)$$

The heat of adsorption can also be estimated using a virial-type expansion that includes the virial coefficients A_i and B_i. Here, m and n represent the number of coefficients utilized to determine the isotherms Eq. (10.3) (Furukawa et al., 2007).

$$\ln P = \ln n_{ads} + \frac{1}{T}\sum_{i=0}^{m} A_i n^i{}_{ads} + \sum_{i=0}^{n} B_i n^i{}_{ads}$$
$$\Delta H_{ads} = -R \sum_{i=0}^{m} A_i n^i{}_{ads} \quad (10.3)$$

Dynamic sorption methods

The adsorption capacity of porous materials, such as MOF composites, to adsorb gas is primarily determined using dynamic sorption methods, as gas adsorption in real-world scenarios occurs dynamically. One commonly used method is static volumetric measurement, a semi-dynamic technique in which the adsorption process occurs in a closed chamber. At a specific pressure (P/P_0), the amount of gas adsorbed upon reaching saturation is recorded. Eq. (10.4) demonstrates that the amount of adsorbed gas can be calculated by assuming the gas component behaves as an ideal gas, where n represents the number of moles of gas, p is pressure, V is volume, T is temperature, and R is the ideal gas constant. By plotting the amount of gas adsorbed against the given pressure, the adsorption isotherm curve can be derived (Grande et al., 2013; Möller et al., 2017).

$$n_{\text{adsorbed}} = n_{\text{dose}} - n_{\text{free}} = \left(\frac{pV}{RT}\right)_{\text{dose}}$$
$$- (V_{\text{dose}} + V_{\text{chamber}})\left(\frac{p}{RT}\right)_{\text{chamber}} \quad (10.4)$$

Under forced-flow conditions, a dynamic method generates data that is displayed as a breakthrough curve, where adsorption time (the test duration) is plotted against the concentration of adsorbate in the effluent stream. The measurement is performed in an open system while maintaining a constant feed pressure. Controlling the gas flow rate is essential, as it determines the amount of VOCs introduced into the adsorption system. An increase in gas flow rate directly reduces the contact time between the VOCs and the MOF adsorbent, affecting adsorption efficiency. A series of equations derived from breakthrough data can be used to calculate the dynamic adsorption capacity (Bastos-Neto et al., 2011; Grant Glover et al., 2011). The concentration of VOCs over time (C_t, mg.min/m^3) is calculated by multiplying the concentration of the feed (C_{feed}, mg/m^3) by the length of time that the feed passed (t_f, min), as shown in Eq. (10.5) (Bastos-Neto et al., 2011).

$$Ct_{\text{feed}} = t_f C_{\text{feed}} \quad (10.5)$$

The subsequent eluted concentration of VOCs per unit time (C_{telution}, mg.min/m^3) conducted by the adsorbent until the flow stops can be calculated by integrating under the elution curve, as demonstrated by Eq. (10.6). This equation can also be applied for desorption purposes. Here, t_s represents the saturation time (min), t_n is the time at point n (min), and C_n is the concentration at time n (mg/m^3). It is important to note that there are several modified forms of these equations available for modeling the adsorption experiment (Grant Glover et al., 2011).

$$Ct_{\text{elution}} = \sum_{t=0}^{t_s} \frac{C_n + C_{n-1}}{2} \times (t_{n-1} - t_n) \quad (10.6)$$

Disadvantages and possible costs

While it is crucial to assess the safety of emerging materials, our understanding of the toxicity of MOFs composite remains limited. Only a handful of studies in biological contexts have explored the toxicity of MOF composites, making it a critical issue that warrants discussion prior to their use in environmental applications. Any potential toxicity associated with MOFs composite is likely linked to factors such as the metallic nodes, functional groups on the ligands, and the organic solvents utilized during MOF synthesis (Kumar et al., 2019).

Despite the numerous advantages stemming from the significant chemical and structural adjustability of MOFs, their manufacturing expenses pose a substantial obstacle to widespread commercial adoption. The production costs of MOFs composite primarily consist of raw materials, such as metal salts and organic ligands, along with processing expenses, which may involve non-recyclable organic solvents and activation-related costs. A technoeconomic analysis indicates that transitioning from solvothermal synthesis to aqueous synthesis can reduce the cost from over 35 $/kg to approximately 13 $/kg for large-scale MOF production. To facilitate industrial-scale utilization of MOFs, it is imperative to reduce production costs to below 10 $/kg (DeSantis et al., 2017).

Metal-organic framework composites recycle

Recent literature emphasizes the discovery of numerous MOF composites with impressive adsorption capabilities. However, adsorption can occasionally lead to irreversible collapse, significantly affecting the reusability of MOF composites. The regeneration of adsorbents is closely linked to operational expenses, underscoring the advantage of MOF composites with efficient regeneration processes. Moreover, enhancing their regenerative capacity could help offset the inherent costs associated with MOF composites (Kumar et al., 2019; Zhang et al., 2019). Furthermore, the cycle stability of MOF composites is a critical aspect that has been thoroughly investigated in the contexts of adsorption and catalysis (Petit, 2018). However, despite extensive study, degradation remains a possibility due to the inherent characteristics of MOF composites. Ideally, MOF composites would retain their physicochemical properties and be reusable after adsorption (Khutia et al., 2013). For recycling the MOF composites and exploring their adsorption and desorption cycles, it is essential to consider various factors. Numerous coordinating elements influence the interaction between the adsorbent and adsorbate (Majchrzak-Kucęba & Bukalak-Gaik, 2016). MOF composites can adsorb VOCs through various mechanisms that involve chemical bonding. Additionally, chemical interactions between VOCs and MOF composites can hinder the functions of MOFs (Hanikel, Prévot & Yaghi, 2020). Hence, the synthesis method and raw materials play crucial roles in achieving a stable product during the synthesis of MOFs and their composites. The challenges associated with separating MOF composite adsorbents can be addressed in various ways, such as transforming MOF particles into polymer matrix membranes, including electrospun nanofiber membranes, thin-film nanocomposite membranes, and mixed matrix membranes (Li et al., 2021; Lin et al., 2018).

Conclusions and prospects

Conclusions

In this chapter, some recent advancements in the development of MOF composites for VOC adsorption have been summarized. The reported works were categorized based on the types of MOF composites, and the adsorption effectiveness of MOF composites was compared with pristine MOFs and added ingredients. The significant adsorption properties of MOF composites have been highlighted in previous studies, attributed to their pore cavity structures, high surface area, and abundant functional sites. Building upon this foundation, the combination of suitable functional materials, such as graphene, graphite, membranes, polymers, metals, and semimetals with MOFs has led to synergistic adsorption performances. However, despite these advancements, limitations still exist for the industrial production of MOF composites adsorbents. Most MOF composite adsorbents remain impractical for real-world applications. As a result, greater focus should be placed on creating adsorbent materials with outstanding potential, and there should be strong confidence in the future progress of MOF composite adsorbents.

Prospects

Despite the significant progress made in the development of MOFs and MOF-based composites for VOC capture across different environments, there is still a long way to go in designing and synthesizing materials suitable for practical VOC treatment. Our approach to developing ideal materials for VOC treatment includes the following considerations:

1. Practical VOC treatment requires materials that can efficiently capture VOCs at trace levels even under moderate to high temperatures.
2. Effective water repellence is crucial for preventing a decrease in uptake under high humidity conditions.
3. The materials should be conducive to large-scale production at a reasonable cost.

While simultaneously meeting all these criteria poses challenges, material design can focus on optimizing pore size and enhancing π-interactions between VOCs and the surrounding adsorption cavity walls. Improving adsorption capacity at low pressures and moderate temperatures can significantly enhance the practical applicability of VOC capture materials. Moreover, it is essential to test and refine these materials in the presence of potentially interfering gases, such as CO, SO_2, H_2S, amines, and water vapor, which often accompany VOC emissions. It is hoped that the insights presented on the design and synthesis of MOFs and MOF-based composites will inspire researchers to develop more effective systems for capturing and decomposing VOCs in practical applications. By tackling the outlined challenges and continuously optimizing material properties, researchers can make substantial progress toward creating more efficient and reliable VOC adsorption solutions.

AI disclosure

While drafting this manuscript, the author(s) utilized GPT-4.o as a supporting tool to enhance readability and address potential grammatical or stylistic errors. It is important to emphasize that this AI tool was not used to create, modify, or analyze figures or scientific results. After leveraging this tool, the author(s) carefully reviewed and revised the content as necessary and assume full responsibility for the final version of the publication

References

Aguilera-Sigalat, J., & Bradshaw, D. (2016). Synthesis and applications of metal-organic framework-quantum dot (QD at MOF) composites. *Coordination Chemistry Reviews, 307*, 267–291. https://doi.org/10.1016/j.ccr.2015.08.004, http://www.journals.elsevier.com/coordination-chemistry-reviews/.

Akhbari, K., & Morsali, A. (2013). Modulating methane storage in anionic nano-porous MOF materials via post-synthetic cation exchange process. *Dalton Transactions, 42*(14), 4786–4789. https://doi.org/10.1039/c3dt32846e.

Alavijeh, R. K., Akhbari, K., & White, J. (2019). Solid–liquid conversion and carbon dioxide storage in a calcium-based metal–organic framework with micro- and nanoporous channels. *Crystal Growth & Design, 19*(12), 7290–7297. https://doi.org/10.1021/acs.cgd.9b01174.

Aldehydes (2007). *A comprehensive guide to the hazardous properties of chemical substances.* John Wiley & Sons, 160–192.

Alhamami, M., Doan, H., & Cheng, C. H. (2014). A review on breathing behaviors of metal-organic-frameworks (MOFs) for gas adsorption. *Materials, 7*(4), 3198–3250. https://doi.org/10.3390/ma7043198, http://www.mdpi.com/1996-1944/7/4/3198/pdf.

Alivand, M. S., Shafiei-Alavijeh, M., Tehrani, N. H. M. H., Ghasemy, E., Rashidi, A., & Fakhraie, S. (2019). Facile and high-yield synthesis of improved MIL-101(Cr) metal-organic framework with exceptional CO_2 and H_2S uptake; the impact of excess ligand-cluster. *Microporous and Mesoporous Materials, 279*, 153–164. https://doi.org/10.1016/j.micromeso.2018.12.033, http://www.elsevier.com/inca/publications/store/6/0/0/7/6/0.

Amidi, D. M., & Akhbari, K. (2024). Iodine-loaded ZIF-7-coated cotton substrates show sustained iodine release as effective antibacterial textiles. *New Journal of Chemistry, 48*(5), 2016–2027. https://doi.org/10.1039/d3nj05198f.

Anbia, M., Sheykhi, S., & Dehghan, R. (2017). MWCNT@MIL-53(Cr) nanoporous composite: Synthesis, characterization, and methane storage property. *Journal of Chemical and Petroleum Engineering, 51*(1), 21–26.

Arshadi Edlo, A., & Akhbari, K. (2023). Modulating the antibacterial activity of a CuO@HKUST-1 nanocomposite by optimizing its synthesis procedure. *New Journal of Chemistry, 47*(45), 20770–20776. https://doi.org/10.1039/d3nj03940e, http://pubs.rsc.org/en/journals/journal/nj.

Bandosz, T. J., & Petit, C. (2011). MOF/graphite oxide hybrid materials: Exploring the new concept of adsorbents and catalysts. *Adsorption, 17*(1), 5–16. https://doi.org/10.1007/s10450-010-9267-5.

Bastos-Neto, M., Moeller, A., Staudt, R., Böhm, J., & Gläser, R. (2011). Adsorption measurements of nitrogen and methane in hydrogen-rich mixtures at high pressures. *Industrial and Engineering Chemistry Research, 50*(17), 10211–10221. https://doi.org/10.1021/ie200652e.

Britt, D., Tranchemontagne, D., & Yaghi, O. M. (2008). Metal-organic frameworks with high capacity and selectivity for harmful gases. *Proceedings of the National Academy of Sciences, 105*(33), 11623–11627. https://doi.org/10.1073/pnas.0804900105.

Burghardt, T. E., & Pashkevich, A. (2018). Emissions of volatile organic compounds from road marking paints. *Atmospheric Environment, 193*, 153–157. https://doi.org/10.1016/j.atmosenv.2018.08.065, http://www.elsevier.com/locate/atmosenv.

Cai, J., Song, C., Gong, X., Zhang, J., Pei, J., & Chen, Z. (2022). Gradation of limestone-aggregate-based porous asphalt concrete under dynamic crushing test: Composition, fragmentation and stability. *Construction and Building Materials, 323*, 126532. https://doi.org/10.1016/j.conbuildmat.2022.126532.

Castro-Hurtado, I., Mandayo, G. G., & Castaño, E. (2013). Conductometric formaldehyde gas sensors. A review: From conventional films to nanostructured materials. *Thin Solid Films, 548*, 665–676. https://doi.org/10.1016/j.tsf.2013.04.083.

Chae, H. K., Siberio-Pérez, D. Y., Kim, J., Go, Y. B., Eddaoudi, M., Matzger, A. J., O'Keeffe, M., & Yaghi, O. M. (2004). A route to high surface area, porosity and inclusion of large molecules in crystals. *Nature, 427*(6974), 523–527. https://doi.org/10.1038/nature02311.

Chen, J., Tang, J. N., Hu, K. L., Zhao, Y. Y., & Tang, C. (2018). The production characteristics of volatile organic compounds and their relation to growth status of *Staphylococcus aureus* in milk environment. *Journal of Dairy Science, 101*(6), 4983–4991. https://doi.org/10.3168/jds.2017-13629.

Chen, L., Zhang, X., Cheng, X., Xie, Z., Kuang, Q., & Zheng, L. (2020). The function of metal-organic frameworks in the application of MOF-based composites. *Nanoscale Advances, 2*(7), 2628–2647. https://doi.org/10.1039/d0na00184h, http://pubs.rsc.org/en/journals/journalissues/na?_ga2.190536939.1555337663.1552312502-1364180372.1550481316#!issueidna001002&typecurrent&issnonline2516-0230.

Cuffia, F., Bergamini, C. V., Wolf, I. V., Hynes, E. R., & Perotti, M. C. (2018). Characterization of volatile compounds produced by *Lactobacillus helveticus* strains in a hard cheese model. *Food Science and Technology International, 24*(1), 67–77. https://doi.org/10.1177/1082013217728628, http://www.sagepub.com.

Cui, X., Sun, X., Liu, L., Huang, Q., Yang, H., Chen, C., Nie, S., Zhao, Z., & Zhao, Z. (2019). In-situ fabrication of cellulose foam HKUST-1 and surface modification with polysaccharides for enhanced selective adsorption of toluene and acidic dipeptides. *Chemical Engineering Journal. 369*, 898–907. https://doi.org/10.1016/j.cej.2019.03.129, https://www.sciencedirect.com/science/article/abs/pii/S1385894719305935.

Dai, Y., Li, M., Liu, F., Xue, M., Wang, Y., & Zhao, C. (2019). Graphene oxide wrapped copper-benzene-1,3,5-tricarboxylate metal organic framework as efficient absorbent for gaseous toluene under ambient conditions. *Environmental Science and Pollution Research, 26*(3), 2477–2491. https://doi.org/10.1007/s11356-018-3657-8, http://www.springerlink.com/content/0944-1344.

Daisy, B. H., Strobel, G. A., Castillo, U., Ezra, D., Sears, J., Weaver, D. K., & Runyon, J. B. (2002). Naphthalene, an insect repellent, is produced by *Muscodor vitigenus*, a novel endophytic fungus. *Microbiology, 148*(11), 3737–3741. https://doi.org/10.1099/00221287-148-11-3737, http://mic.sgmjournals.org.

Davoodi, A., Akhbari, K., & Alirezvani, M. (2023). Prolonged release of silver and iodine from ZIF-7 carrier with great antibacterial activity. *CrystEngComm, 25*(27), 3931–3942. https://doi.org/10.1039/d3ce00529a, http://pubs.rsc.org/en/journals/journal/ce.

DeSantis, D., Mason, J. A., James, B. D., Houchins, C., Long, J. R., & Veenstra, M. (2017). Techno-economic analysis of metal-organic frameworks for hydrogen and natural gas storage. *Energy and Fuels, 31*(2), 2024–2032. https://doi.org/10.1021/acs.energyfuels.6b02510, http://pubs.acs.org/journal/enfuem.

Dhakshinamoorthy, A., Asiri, A. M., Alvaro, M., & Garcia, H. (2018). Metal organic frameworks as catalysts in solvent-free or ionic liquid assisted conditions. *Green Chemistry, 20*(1), 86–107. https://doi.org/10.1039/c7gc02260c, http://pubs.rsc.org/en/journals/journal/gc.

Drilling, K., & Dettner, K. (2009). Electrophysiological responses of four fungivorous coleoptera to volatiles of trametes versicolor: Implications for host selection. *Chemoecology, 19*(2), 109–115. https://doi.org/10.1007/s00049-009-0015-9.

Duan, C., Ma, T., Wang, J., & Zhou, Y. (2020). Removal of heavy metals from aqueous solution using carbon-based adsorbents: A review. *Journal of Water Process Engineering, 37*, 101339. https://doi.org/10.1016/j.jwpe.2020.101339.

El-Mehalmey, W. A., Ibrahim, A. H., Abugable, A. A., Hassan, M. H., Haikal, R. R., Karakalos, S. G., Zaki, O., & Alkordi, M. H. (2018). Metal-organic framework@silica as a stationary phase sorbent for rapid and cost-effective removal of hexavalent chromium. *Journal of Materials Chemistry A, 6*(6), 2742–2751. https://doi.org/10.1039/c7ta08281a, http://pubs.rsc.org/en/journals/journal/ta.

Elsaidi, S. K., Sinnwell, M. A., Devaraj, A., Droubay, T. C., Nie, Z., Murugesan, V., McGrail, B. P., & Thallapally, P. K. (2018). Extraction of rare earth elements using magnetite@MOF composites. *Journal of Materials Chemistry A, 6*(38), 18438–18443. https://doi.org/10.1039/c8ta04750b, http://pubs.rsc.org/en/journals/journal/ta.

Finewax, Z., Pagonis, D., Claflin, M. S., Handschy, A. V., Brown, W. L., Jenks, O., Nault, B. A., Day, D. A., Lerner, B. M., Jimenez, J. L., Ziemann, P. J., & de Gouw, J. A. (2021). Quantification and source characterization of volatile organic compounds from exercising and application of chlorine-based cleaning products in a university athletic center. *Indoor Air, 31*(5), 1323–1339. https://doi.org/10.1111/ina.12781, http://onlinelibrary.wiley.com/journal/10.1111/(ISSN)1600-0668.

Furukawa, H., Gándara, F., Zhang, Y. B., Jiang, J., Queen, W. L., Hudson, M. R., & Yaghi, O. M. (2014). Water adsorption in porous metal-organic frameworks and related materials. *Journal of the American Chemical Society, 136*(11), 4369–4381. https://doi.org/10.1021/ja500330a, http://pubs.acs.org/journal/jacsat.

Furukawa, H., Miller, M. A., & Yaghi, O. M. (2007). Independent verification of the saturation hydrogen uptake in MOF-177 and establishment of a benchmark for hydrogen adsorption in metal-organic frameworks. *Journal of Materials Chemistry, 17*(30), 3197–3204. https://doi.org/10.1039/b703608f, http://www.rsc.org/Publishing/Journals/jm/index.asp.

Gao, Y., Peng, X., Zhang, Z., Zhang, W., Li, H., Chen, B., Li, S., Zhang, Y., & Chi, S. (2021). Ternary mixed-oxide synergy effects of nano TiO_2-FexOy-MOk (M=Mn, Ce, Co) on α-pinene catalytic oxidation process assisted by nonthermal plasma. *Materials Research Express, 8*(1), 015509. https://doi.org/10.1088/2053-1591/abdbf7.

Gao, Z., Wang, J., Muhammad, Y., Hu, P., Hu, Y., Chu, Z., Zhao, Z., & Zhao, Z. (2022). Hydrophobic shell structured NH_2-MIL(Ti)-125@mesoporous carbon composite via confined growth strategy for ultra-high selective adsorption of toluene under highly humid environment. *Chemical Engineering Journal, 432*, 134340. https://doi.org/10.1016/j.cej.2021.134340.

Ghasemzadeh, R., & Akhbari, K. (2023a). Heterostructured Ag@MOF-801/MIL-88A(Fe) nanocomposite as a biocompatible photocatalyst for degradation of reactive black 5 under visible light. *Inorganic Chemistry, 62*(43), 17818–17829. https://doi.org/10.1021/acs.inorgchem.3c02616, http://pubs.acs.org/journal/inocaj.

Ghasemzadeh, R., & Akhbari, K. (2023b). Band gap engineering of MOF-801 via Loading of γ-Fe_2O_3 quantum dots inside it as a visible light-responsive photocatalyst for degradation of acid orange 7. *Crystal Growth & Design, 23*(9), 6359–6368. https://doi.org/10.1021/acs.cgd. 3c00272.

Ghasemzadeh, R., & Akhbari, K. (2024). Templated synthesis of ZnO quantum dots via double solvents method inside MOF-801 as emerging photocatalyst for photodegradation of Acid Blue 25 under UV light. *Journal of Photochemistry and Photobiology A: Chemistry, 448*, 115306. https://doi.org/10.1016/j.jphotochem.2023.115306.

González-Mas, M. C., Rambla, J. L., López-Gresa, M. P., Blázquez, M. A., & Granell, A. (2019). Volatile compounds in citrus essential oils: A comprehensive review. *Frontiers in Plant Science, 10*. https://doi.org/10.3389/fpls.2019.00012, http://www.frontiersin.org/articles/10.3389/fpls.2019.00012/pdf.

Grande, C. A., Blom, R., Möller, A., & Möllmer, J. (2013). High-pressure separation of CH_4/CO_2 using activated carbon. *Chemical Engineering Science, 89*, 10–20. https://doi.org/10.1016/j.ces.2012.11.024, http://www.journals.elsevier.com/chemical-engineering-science/.

Grant Glover, T., Peterson, G. W., Schindler, B. J., Britt, D., & Yaghi, O. (2011). MOF-74 building unit has a direct impact on toxic gas adsorption. *Chemical Engineering Science, 66*(2), 163–170. https://doi.org/10.1016/j.ces.2010.10.002.

Hamon, L., Serre, C., Devic, T., Loiseau, T., Millange, F., Férey, G., & Weireld, G. D. (2009). Comparative study of hydrogen sulfide adsorption in the MIL-53(Al, Cr, Fe), MIL-47(V), MIL-100(Cr), and MIL-101(Cr) metal-organic frameworks at room temperature. *Journal of the American Chemical Society, 131*(25), 8775–8777. https://doi.org/10.1021/ja901587t, http://pubs.acs.org/doi/pdfplus/10.1021/ja901587t.

Han, T., Li, C., Guo, X., Huang, H., Liu, D., & Zhong, C. (2016). In-situ synthesis of SiO_2@MOF composites for high-efficiency removal of aniline from aqueous solution. *Applied Surface Science, 390*, 506–512. https://doi.org/10.1016/j.apsusc.2016.08.111, http://www.journals.elsevier.com/applied-surface-science/.

Han, T., Xiao, Y., Tong, M., Huang, H., Liu, D., Wang, L., & Zhong, C. (2015). Synthesis of CNT@MIL-68(Al) composites with improved adsorption capacity for phenol in aqueous solution. *Chemical Engineering Journal, 275*, 134–141. https://doi.org/10.1016/j.cej.2015.04.005, https://www.sciencedirect.com/journal/chemical-engineering-journal/vol/275/suppl/C.

Hanikel, N., Prévot, M. S., & Yaghi, O. M. (2020). MOF water harvesters. *Nature Nanotechnology, 15*(5), 348–355. https://doi.org/10.1038/s41565-020-0673-x, http://www.nature.com/nnano/index.html.

Hasan, Z., & Jhung, S. H. (2015). Removal of hazardous organics from water using metal-organic frameworks (MOFs): Plausible mechanisms for selective adsorptions. *Journal of Hazardous Materials, 283*, 329–339. https://doi.org/10.1016/j.jhazmat.2014.09.046, http://www.elsevier.com/locate/jhazmat.

He, W. W., Yang, G. S., Tang, Y. J., Li, S. L., Zhang, S. R., Su, Z. M., & Lan, Y. Q. (2015). Phenyl groups result in the highest benzene storage and most efficient desulfurization in a series of isostructural metal-organic frameworks. *Chemistry - A European Journal, 21*(27), 9784–9789. https://doi.org/10.1002/chem.201500815, http://onlinelibrary.wiley.com/journal/10.1002/(ISSN)1521-3765.

Huang, Q., Hu, Y., Pei, Y., Zhang, J., & Fu, M. (2019). In situ synthesis of TiO_2@NH_2-MIL-125 composites for use in combined adsorption and photocatalytic degradation of formaldehyde. *Applied Catalysis B: Environmental, 259*, 118106. https://doi.org/10.1016/j.apcatb.2019.118106.

Iftekhar Uddin, A. S. M., Phan, D. T., & Chung, G. S. (2015). Low temperature acetylene gas sensor based on Ag nanoparticles-loaded ZnO-reduced graphene oxide hybrid. *Sensors and Actuators, B: Chemical, 207*, 362–369. https://doi.org/10.1016/j.snb.2014.10.091.

Irga, P. J., Mullen, G., Fleck, R., Matheson, S., Wilkinson, S. J., & Torpy, F. R. (2024). Volatile organic compounds emitted by humans indoors– A review on the measurement, test conditions, and analysis techniques. *Building and Environment, 255*. https://doi.org/10.1016/j.buildenv.2024.111442, http://www.sciencedirect.com/science/journal/03601323.

Jaeger, D. M., Runyon, J. B., & Richardson, B. A. (2016). Signals of speciation: Volatile organic compounds resolve closely related sagebrush taxa, suggesting their importance in evolution. *The New Phytologist, 211*(4), 1393–1401. https://doi.org/10.1111/nph.13982, http://onlinelibrary.wiley.com/journal/10.1111/(ISSN)1469-8137.

Jhinjer, H. S., Singh, A., Bhattacharya, S., Jassal, M., & Agrawal, A. K. (2021). Metal-organic frameworks functionalized smart textiles for adsorptive removal of hazardous aromatic pollutants from ambient air. *Journal of Hazardous Materials, 411*. https://doi.org/10.1016/j.jhazmat.2021.125056, http://www.elsevier.com/locate/jhazmat.

Kalaj, M., Bentz, K. C., Ayala, S., Palomba, J. M., Barcus, K. S., Katayama, Y., & Cohen, S. M. (2020). MOF-polymer hybrid materials: From simple composites to tailored architectures. *Chemical Reviews, 120*(16), 8267–8302. https://doi.org/10.1021/acs.chemrev.9b00575, http://pubs.acs.org/journal/chreay.

Kalati, M., & Akhbari, K. (2021). Optimizing the metal ion release and antibacterial activity of ZnO@ZIF-8 by modulating its synthesis method. *New Journal of Chemistry, 45*(48), 22924–22931. https://doi.org/10.1039/d1nj04534b, http://pubs.rsc.org/en/journals/journal/nj.

Karimi Alavijeh, R., & Akhbari, K. (2020). Biocompatible MIL-101(Fe) as a smart carrier with high loading potential and sustained release of curcumin. *Inorganic Chemistry, 59*(6), 3570–3578. https://doi.org/10.1021/acs.inorgchem.9b02756, http://pubs.acs.org/journal/inocaj.

Karimi Alavijeh, R., & Akhbari, K. (2024). Cancer therapy by nano MIL-n series of metal-organic frameworks. *Coordination Chemistry Reviews, 503*, 215643. https://doi.org/10.1016/j.ccr.2023.215643.

Ke, M. T., Lee, M. T., Lee, C. Y., & Fu, L. M. (2009). A MEMS-based benzene gas sensor with a self-heating WO_3 sensing layer. *Sensors, 9*(4), 2895–2906. https://doi.org/10.3390/s90402895, http://www.mdpi.com/1424-8220/9/4/2895/pdf.

Khan, F. I., & Kr Ghoshal, A. (2000). Removal of volatile organic compounds from polluted air. *Journal of Loss Prevention in the Process Industries, 13*(6), 527–545. https://doi.org/10.1016/S0950-4230(00)00007-3, https://www.sciencedirect.com/journal/journal-of-loss-prevention-in-the-process-industries/vol/13/issue/6.

Khoshakhlagh, A. H., Golbabaei, F., Beygzadeh, M., Carrasco-Marín, F., & Shahtaheri, S. J. (2020). Toluene adsorption on porous Cu-BDC@OAC composite at various operating conditions: Optimization by response surface methodology. *RSC Advances, 10*(58), 35582–35596. https://doi.org/10.1039/d0ra06578a, http://pubs.rsc.org/en/journals/journal/ra.

Khutia, A., Rammelberg, H. U., Schmidt, T., Henninger, S., & Janiak, C. (2013). Water sorption cycle measurements on functionalized MIL-101Cr for heat transformation application. *Chemistry of Materials, 25*(5), 790–798. https://doi.org/10.1021/cm304055k.

Kim, K. H., Jahan, S. A., Kabir, E., & Brown, R. J. C. (2013). A review of airborne polycyclic aromatic hydrocarbons (PAHs) and their human health effects. *Environment International, 60*, 71–80. https://doi.org/10.1016/j.envint.2013.07.019, http://www.elsevier.com/locate/envint.

Knudsen, J. T., & Gershenzon, J. (2020). *The chemical diversity of floral scent*. Informa UK Limited, 57–78. 10.1201/9780429455612-5.

Kumar, P., Anand, B., Tsang, Y. F., Kim, K. H., Khullar, S., & Wang, B. (2019). Regeneration, degradation, and toxicity effect of MOFs: Opportunities and challenges. *Environmental Research, 176*. https://doi.org/10.1016/j.envres.2019.05.019, https://www.sciencedirect.com/journal/environmental-research/vol/176/suppl/C.

Kumar, V., Rickly, A. Lahib, C. M. F. Rosales, S. Dusanter, & P. Stevens (2021). Measurements of OH reactivity in a forest using the total OH loss rate method (TOHLM), AGU fall meeting abstracts, 2021, pp. A25M-1851.

Lee, C. S., Choi, J. H., & Park, Y. H. (2015). Development of metal-loaded mixed metal oxides gas sensors for the detection of lethal gases. *Journal of Industrial and Engineering Chemistry, 29*, 321–329. https://doi.org/10.1016/j.jiec.2014.10.048, http://www.sciencedirect.com/science/journal/1226086X.

Levasseur, B., Petit, C., & Bandosz, T. J. (2010). Reactive adsorption of NO_2 on copper-based metal-organic framework and graphite oxide/metal-organic framework composites. *ACS Applied Materials and Interfaces, 2*(12), 3606–3613. https://doi.org/10.1021/am100790v.

Li, J., Wang, X., Zhao, G., Chen, C., Chai, Z., Alsaedi, A., Hayat, T., & Wang, X. (2018). Metal-organic framework-based materials: Superior adsorbents for the capture of toxic and radioactive metal ions. *Chemical Society Reviews, 47*(7), 2322–2356. https://doi.org/10.1039/c7cs00543a, http://pubs.rsc.org/en/journals/journal/cs.

Li, Q., Xu, X., Guo, J., Hill, J. P., Xu, H., Xiang, L., Li, C., Yamauchi, Y., & Mai, Y. (2021). Two-dimensional MXene-polymer heterostructure with ordered in-plane mesochannels for high-performance capacitive deionization. *Angewandte Chemie, 133*(51), 26732–26738. https://doi.org/10.1002/ange.202111823.

Li, Y., Miao, J., Sun, X., Xiao, J., Li, Y., Wang, H., Xia, Q., & Li, Z. (2016). Mechanochemical synthesis of Cu-BTC@GO with enhanced water stability and toluene adsorption capacity. *Chemical Engineering Journal. 298*, 191–197. https://doi.org/10.1016/j.cej.2016.03.141 https://www.sciencedirect.com/science/article/abs/pii/S1385894716304053?via%3Dihub.

Lin, R., Villacorta Hernandez, B., Ge, L., & Zhu, Z. (2018). Metal organic framework based mixed matrix membranes: An overview on filler/polymer interfaces. *Journal of Materials Chemistry A, 6*(2), 293–312. https://doi.org/10.1039/c7ta07294e, http://pubs.rsc.org/en/journals/journal/ta.

Liu, B., Yang, F., Zou, Y., & Peng, Y. (2014). Adsorption of phenol and p-nitrophenol from aqueous solutions on metal-organic frameworks: Effect of hydrogen bonding. *Journal of Chemical and Engineering Data, 59*(5), 1476–1482. https://doi.org/10.1021/je4010239, http://pubs.acs.org/journal/jceaax.

Liu, K.-G., Bigdeli, F., Sharifzadeh, Z., Gholizadeh, S., & Morsali, A. (2023). Role of metal-organic framework composites in removal of inorganic toxic contaminants. *Journal of Cleaner Production, 404*, 136709. https://doi.org/10.1016/j.jclepro.2023.136709.

Loreto, F., & Fares, S. (2013). Biogenic volatile organic compounds and their impacts on biosphere-atmosphere interactions. *Developments in Environmental Science, 13*, 57–75. https://doi.org/10.1016/B978-0-08-098349-3.00004-9, http://www.elsevier.com/wps/find/bookdescription.cws_home/BS_DES/description#description.

Ma, J., Yao, Z., Hou, L., Lu, W., Yang, Q., Li, J., & Chen, L. (2016). Metal organic frameworks (MOFs) for magnetic solid-phase extraction of pyrazole/pyrrole pesticides in environmental water samples followed by HPLC-DAD determination. *Talanta, 161*, 686–692. https://doi.org/10.1016/j.talanta.2016.09.035, http://www.journals.elsevier.com/talanta.

Majchrzak-Kucęba, I., & Bukalak-Gaik, D. (2016). Regeneration performance of metal–organic frameworks: TG-vacuum tests. *Journal of Thermal Analysis and Calorimetry, 125*(3), 1461–1466. https://doi.org/10.1007/s10973-016-5624-2, http://www.springer.com/sgw/cda/frontpage/0,11855,1-40109-70-35752391-0,00.html.

Martinez, J. L. (2009). Environmental pollution by antibiotics and by antibiotic resistance determinants. *Environmental Pollution, 157*(11), 2893–2902. https://doi.org/10.1016/j.envpol.2009.05.051.

McHale, C. M., Zhang, L., & Smith, M. T. (2012). Current understanding of the mechanism of benzene-induced leukemia in humans: implications for risk assessment. *Carcinogenesis, 33*(2), 240–252. https://doi.org/10.1093/carcin/bgr297.

Meilikhov, M., Yusenko, K., Esken, D., Turner, S., Van Tendeloo, G., & Fischer, R. A. (2010). Metals@MOFs - Loading MOFs with metal nanoparticles for hybrid functions. *European Journal of Inorganic Chemistry,* (24), 3701–3714. https://doi.org/10.1002/ejic.201000473, http://onlinelibrary.wiley.com/doi/10.1002/ejic.201000473/pdf.

Möller, A., Eschrich, R., Reichenbach, C., Guderian, J., Lange, M., & Möllmer, J. (2017). Dynamic and equilibrium-based investigations of CO_2-removal from CH_4-rich gas mixtures on microporous adsorbents. *Adsorption, 23*(2-3), 197–209. https://doi.org/10.1007/s10450-016-9821-x.

Muhammad, Y., Lu, Y., Shen, C., & Li, C. (2011). Dibenzothiophene hydrodesulfurization over Ru promoted alumina based catalysts using in situ generated hydrogen. *Energy Conversion and Management, 52*(2), 1364–1370. https://doi.org/10.1016/j.enconman.2010.09.034.

Nakhaei, M., Akhbari, K., & Davoodi, A. (2021). Biocompatible MOF-808 as an iodophor antimicrobial agent with controlled and sustained release of iodine. *CrystEngComm, 23*(48), 8538–8545. https://doi.org/10.1039/d1ce00019e, http://pubs.rsc.org/en/journals/journal/ce.

National Institute for Occupational Safety and Health (NIOSH). (2016). An official website of the United States government, *NIOSH pocket guide to chemical hazards.* https://www.cdc.gov/niosh/npg/pgintrod.html.

Niu, C., Zhang, N., Hu, C., Zhang, C., Zhang, H., & Xing, Y. (2021). Preparation of a novel citric acid-crosslinked Zn-MOF/chitosan composite and application in adsorption of chromium(VI) and methyl orange from aqueous solution. *Carbohydrate Polymers, 258*, 117644. https://doi.org/10.1016/j.carbpol.2021.117644.

Noori, Y., & Akhbari, K. (2017). Post-synthetic ion-exchange process in nanoporous metal-organic frameworks; an effective way for modulating their structures and properties. *RSC Advances, 7*(4), 1782–1808. https://doi.org/10.1039/c6ra24958b, http://pubs.rsc.org/en/journals/journalissues.

Oro, L., Feliziani, E., Ciani, M., Romanazzi, G., & Comitini, F. (2018). Volatile organic compounds from *Wickerhamomyces anomalus, Metschnikowia pulcherrima* and *Saccharomyces cerevisiae* inhibit growth of decay causing fungi and control postharvest diseases of strawberries. *International Journal of Food Microbiology, 265*, 18–22. https://doi.org/10.1016/j.ijfoodmicro.2017.10.027, http://www.elsevier.com/locate/ijfoodmicro.

Orta, M.del M., Martín, J., Santos, J. L., Aparicio, I., Medina-Carrasco, S., & Alonso, E. (2020). Biopolymer-clay nanocomposites as novel and ecofriendly adsorbents for environmental remediation. *Applied Clay Science, 198*, 105838. https://doi.org/10.1016/j.clay.2020.105838.

Padial, N. M., Quartapelle Procopio, E., Montoro, C., López, E., Oltra, J. E., Colombo, V., Maspero, A., Masciocchi, N., Galli, S., Senkovska, I., Kaskel, S., Barea, E., & Navarro, J. A. R. (2013). Highly hydrophobic isoreticular porous metal-organic frameworks for the capture of harmful volatile organic compounds. *Angewandte Chemie - International Edition, 52*(32), 8290–8294. https://doi.org/10.1002/anie.201303484.

Parmar, G. R., & Rao, N. N. (2008). Emerging control technologies for volatile organic compounds. *Critical Reviews in Environmental Science and Technology, 39*(1), 41–78. https://doi.org/10.1080/10643380701413658.

Parsaei, M., & Akhbari, K. (2022a). MOF-801 as a nanoporous water-based carrier system for in situ encapsulation and sustained release of 5-FU for effective cancer therapy. *Inorganic Chemistry, 61*(15), 5912–5925. https://doi.org/10.1021/acs.inorgchem.2c00380, http://pubs.acs.org/journal/inocaj.

Parsaei, M., & Akhbari, K. (2022b). Smart multifunctional UiO-66 metal-organic framework nanoparticles with outstanding drug-loading/release potential for the targeted delivery of Quercetin. *Inorganic Chemistry, 61*(37), 14528–14543. https://doi.org/10.1021/acs.inorgchem.2c00743, http://pubs.acs.org/journal/inocaj.

Parsaei, M., & Akhbari, K. (2022c). Synthesis and application of MOF-808 decorated with folic acid-conjugated chitosan as a strong nanocarrier for the targeted drug delivery of Quercetin. *Inorganic Chemistry, 61*(48), 19354–19368. https://doi.org/10.1021/acs.inorgchem.2c03138, http://pubs.acs.org/journal/inocaj.

Parsaei, M., Akhbari, K., Tylianakis, E., & Froudakis, G. E. (2024). Effects of fluorinated functionalization of linker on Quercetin encapsulation, release and hela cell cytotoxicity of Cu-based MOFs as smart pH-stimuli nanocarriers. *Chemistry - A European Journal, 30*(1). https://doi.org/10.1002/chem.202301630, http://onlinelibrary.wiley.com/journal/10.1002/(ISSN)1521-3765.

Parsaei, M., Akhbari, K., Tylianakis, E., & Froudakis, G. E. (2023). Computational simulation of a three-dimensional Mg-based metal–organic framework as nanoporous anticancer drug carrier. *Crystal Growth & Design, 23*(11), 8396–8406. https://doi.org/10.1021/acs.cgd.3c01058.

Patel, N. G., Patel, P. D., & Vaishnav, V. S. (2003). Indium tin oxide (ITO) thin film gas sensor for detection of methanol at room temperature. *Sensors and Actuators B: Chemical, 96*(1-2), 180–189. https://doi.org/10.1016/s0925-4005(03)00524-0.

Peng, Y., Wei, X., Wang, Y., Li, W., Zhang, S., & Jin, J. (2022). Metal-organic framework composite photothermal membrane for removal of high-concentration volatile organic compounds from water via molecular sieving. *ACS Nano, 16*(5), 8329–8337. https://doi.org/10.1021/acsnano.2c02520, http://pubs.acs.org/journal/ancac3.

Petit, C. (2018). Present and future of MOF research in the field of adsorption and molecular separation. *Current Opinion in Chemical Engineering, 20*, 132–142. https://doi.org/10.1016/j.coche.2018.04.004, http://www.elsevier.com/wps/find/journaldescription.cws_home/725837/description#description.

Petit, C., & Bandosz, T. J. (2010). Enhanced adsorption of ammonia on metal-organic framework/graphite oxide composites: Analysis of surface interactions. *Advanced Functional Materials, 20*(1), 111–118. https://doi.org/10.1002/adfm.200900880, http://www3.interscience.wiley.com/cgi-bin/fulltext/122616692/PDFSTART.

Petit, C., & Bandosz, T. J. (2012). Exploring the coordination chemistry of MOF-graphite oxide composites and their applications as adsorbents. *Dalton Transactions, 41*(14), 4027–4035. https://doi.org/10.1039/c2dt12017h.

Petit, C., & Bandosz, T. J. (2009). MOF-graphite oxide nanocomposites: Surface characterization and evaluation as adsorbents of ammonia. *Journal of Materials Chemistry, 19*(36), 6521–6528. https://doi.org/10.1039/b908862h.

Petit, C., & Bandosz, T. J. (2011). Synthesis, characterization, and ammonia adsorption properties of mesoporous metal-organic framework (MIL(Fe))-graphite oxide composites: Exploring the limits of materials fabrication. *Advanced Functional Materials, 21*(11), 2108–2117. https://doi.org/10.1002/adfm.201002517.

Pinto, M. L., Dias, S., & Pires, J. (2013). Composite MOF foams: The example of UiO-66/polyurethane. *ACS Applied Materials and Interfaces, 5*(7), 2360–2363. https://doi.org/10.1021/am303089g.

Pourebrahimi, S., & Pirooz, M. (2022). Functionalized covalent triazine frameworks as promising platforms for environmental remediation: A review. *Cleaner Chemical Engineering, 2*, 100012. https://doi.org/10.1016/j.clce.2022.100012.

Qi, X., Tong, X., Pan, W., Zeng, Q., You, S., & Shen, J. (2021). Recent advances in polysaccharide-based adsorbents for wastewater treatment. *Journal of Cleaner Production, 315*, 128221. https://doi.org/10.1016/j.jclepro.2021.128221.

Qin, J., Yang, J., Huang, H., Fu, M., Ye, D., & Hu, Y. (2023). Tuning the hierarchical pore structure and the metal site in a metal-organic framework derivative to unravel the mechanism for the adsorption of different volatile organic compounds. *Environmental Science and Technology, 57*(41), 15703–15714. https://doi.org/10.1021/acs.est.3c03467, http://pubs.acs.org/journal/esthag.

Romero-Sáez, M., Dongil, A. B., Benito, N., Espinoza-González, R., Escalona, N., & Gracia, F. (2018). CO_2 methanation over nickel-ZrO_2 catalyst supported on carbon nanotubes: A comparison between two impregnation strategies. *Applied Catalysis B: Environmental, 237*, 817–825. https://doi.org/10.1016/j.apcatb.2018.06.045.

Ruthven, D. M. (1984). *Principles of adsorption and adsorption processes*. John Wiley & Sons.

Salimi, S., Akhbari, K., Morteza, S., Farnia, F., Tylianakis, E., Froudakis, G. E., & White, J. M. (2024). Nanoporous metal–organic framework based on furan-2,5-dicarboxylic acid with high potential in selective adsorption and separation of gas mixtures. *Crystal Growth & Design, 24*(10), 4220–4231. https://doi.org/10.1021/acs.cgd.4c00349.

Santoso, S. P., Angkawijaya, A. E., Bundjaja, V., Soetaredjo, F. E., & Ismadji, S. (2020). Metal-organic frameworks and their hybrid composites for adsorption of volatile organic compounds. In *Applications of metal-organic frameworks and their derived materials*. Wiley, 313–355. https://doi.org/10.1002/9781119651079.ch12, http://onlinelibrary.wiley.com/doi/book/10.1002/9781119651079.

Seinfeld, J. H., & Pandis, S. N. (2016). *Atmospheric chemistry and physics: From air pollution to climate change*. John Wiley & Sons, 181–184.

Shao, M., Czapiewski, K. V., Heiden, A. C., Kobel, K., Komenda, M., Koppmann, R., & Wildt, J. (2001). Volatile organic compound emissions from Scots pine: Mechanisms and description by algorithms. *Journal of Geophysical Research Atmospheres, 106*(17), 20483–20491. https://doi.org/10.1029/2000JD000248, http://onlinelibrary.wiley.com/journal/10.1002/(ISSN)2169-8996.

Singh, A., Kumar, K., Sikarwar, S., & Yadav, B. C. (2022). Highly sensitive and selective LPG sensor working below lowest explosion limit (LEL) at room temperature using as-fabricated indium doped SnO_2 thin film. *Materials Chemistry and Physics, 287*, 126275. https://doi.org/10.1016/j.matchemphys.2022.126275.

Siu, B., Chowdhury, A. R., Yan, Z., Humphrey, S. M., & Hutter, T. (2023). Selective adsorption of volatile organic compounds in metal-organic frameworks (MOFs). *Coordination Chemistry Reviews, 485*. https://doi.org/10.1016/j.ccr.2023.215119, http://www.journals.elsevier.com/coordination-chemistry-reviews/.

Soltani, S., & Akhbari, K. (2022a). Cu-BTC metal–organic framework as a biocompatible nanoporous carrier for chlorhexidine antibacterial agent. *JBIC Journal of Biological Inorganic Chemistry*, 1–7.

Soltani, S., & Akhbari, K. (2022b). Embedding an extraordinary amount of gemifloxacin antibiotic in ZIF-8 framework with one-step synthesis and measurement of its H_2O_2-sensitive release and potency against infectious bacteria. *New Journal of Chemistry, 46*(40), 19432–19441. https://doi.org/10.1039/d2nj02981b, http://pubs.rsc.org/en/journals/journal/nj.

Somayajulu Rallapalli, P. B., Raj, M. C., Patil, D. V., Prasanth, K. P., Somani, R. S., & Bajaj, H. C. (2013). Activated carbon@MIL-101(Cr): A potential metal-organic framework composite material for hydrogen storage. *International Journal of Energy Research, 37*(7), 746–753. https://doi.org/10.1002/er.1933.

Subhan, S., Ur Rahman, A., Yaseen, M., Ur Rashid, H., Ishaq, M., Sahibzada, M., & Tong, Z. (2019). Ultra-fast and highly efficient catalytic oxidative desulfurization of dibenzothiophene at ambient temperature over low Mn loaded Co-Mo/Al$_2$O$_3$ and Ni-Mo/Al$_2$O$_3$ catalysts using NaClO as oxidant. *Fuel, 237*, 793–805. https://doi.org/10.1016/j.fuel.2018.10.067, http://www.journals.elsevier.com/fuel/.

Sun, X., Xia, Q., Zhao, Z., Li, Y., & Li, Z. (2014). Synthesis and adsorption performance of MIL-101(Cr)/graphite oxide composites with high capacities of n-hexane. *Chemical Engineering Journal, 239*, 226–232. https://doi.org/10.1016/j.cej.2013.11.024, https://www.sciencedirect.com/journal/chemical-engineering-journal/vol/239/suppl/C.

Topuz, F., Abdulhamid, M. A., Hardian, R., Holtzl, T., & Szekely, G. (2022). Nanofibrous membranes comprising intrinsically microporous polyimides with embedded metal–organic frameworks for capturing volatile organic compounds. *Journal of Hazardous Materials, 424*, 127347. https://doi.org/10.1016/j.jhazmat.2021.127347.

Tung, T. T., Tran, M. T., Feller, J. F., Castro, M., Van Ngo, T., Hassan, K., Nine, M. J., & Losic, D. (2020). Graphene and metal organic frameworks (MOFs) hybridization for tunable chemoresistive sensors for detection of volatile organic compounds (VOCs) biomarkers. *Carbon, 159*, 333–344. https://doi.org/10.1016/j.carbon.2019.12.010, http://www.journals.elsevier.com/carbon/.

Volkamer, R., Jimenez, J. L., San Martini, F., Dzepina, K., Zhang, Q., Salcedo, D., Molina, L. T., Worsnop, D. R., & Molina, M. J. (2006). Secondary organic aerosol formation from anthropogenic air pollution: Rapid and higher than expected. *Geophysical Research Letters, 33*(17). https://doi.org/10.1029/2006GL026899, http://onlinelibrary.wiley.com/journal/10.1002/(ISSN)1944-8007/issues?year=2012.

Vymazal, J., & Březinová, T. (2015). The use of constructed wetlands for removal of pesticides from agricultural runoff and drainage: A review. *Environment International, 75*, 11–20. https://doi.org/10.1016/j.envint.2014.10.026, http://www.elsevier.com/locate/envint.

Wang, C., Liu, X., Keser Demir, N., Chen, J. P., & Li, K. (2016). Applications of water stable metal-organic frameworks. *Chemical Society Reviews, 45*(18), 5107–5134. https://doi.org/10.1039/c6cs00362a, http://pubs.rsc.org/en/journals/journal/cs.

Wang, D. K. W., & Austin, C. C. (2006). Determination of complex mixtures of volatile organic compounds in ambient air: An overview. *Analytical and Bioanalytical Chemistry, 386*, 1089–1098.

Wang, G. H., Lei, Y. Q., & Song, H. C. (2014). Evaluation of Fe$_3$O$_4$@SiO$_2$-MOF-177 as an advantageous adsorbent for magnetic solid-phase extraction of phenols in environmental water samples. *Analytical Methods, 6*(19), 7842–7847. https://doi.org/10.1039/c4ay00822g, http://pubs.rsc.org/en/journals/journal/ay.

Wang, J., Muhammad, Y., Gao, Z., Jalil Shah, S., Nie, S., Kuang, L., Zhao, Z., Qiao, Z., & Zhao, Z. (2021). Implanting polyethylene glycol into MIL-101(Cr) as hydrophobic barrier for enhancing toluene adsorption under highly humid environment. *Chemical Engineering Journal, 404*, 126562. https://doi.org/10.1016/j.cej.2020.126562.

Wang, Q., Gao, Q., Al-Enizi, A. M., Nafady, A., & Ma, S. (2020). Recent advances in MOF-based photocatalysis: Environmental remediation under visible light. *Inorganic Chemistry Frontiers, 7*(2), 300–339. https://doi.org/10.1039/c9qi01120j, http://pubs.rsc.org/en/journals/journal/qi.

Wang, S., Zhang, L., Long, C., & Li, A. (2014). Enhanced adsorption and desorption of VOCs vapor on novel micro-mesoporous polymeric adsorbents. *Journal of Colloid and Interface Science, 428*, 185–190. https://doi.org/10.1016/j.jcis.2014.04.055, https://www.sciencedirect.com/science/article/abs/pii/S0021979714002744?via%3Dihub.

Wang, T., Xue, L., Brimblecombe, P., Lam, Y. F., Li, L., & Zhang, L. (2017). Ozone pollution in China: A review of concentrations, meteorological influences, chemical precursors, and effects. *Science of the Total Environment, 575*, 1582–1596. https://doi.org/10.1016/j.scitotenv. 2016.10.081, http://www.elsevier.com/locate/scitotenv.

Wei, N., Zheng, X., Li, Q., Gong, C., Ou, H., & Li, Z. (2020). Construction of lanthanum modified MOFs graphene oxide composite membrane for high selective phosphorus recovery and water purification. *Journal of Colloid and Interface Science, 565*, 337–344. https://doi.org/10.1016/j. jcis.2020.01.031, https://www.sciencedirect.com/journal/journal-of-colloid-and-interface-science/vol/565/suppl/C.

William, J., & Lead, P. (1997). VOC control strategies in plant design, Chemical processing. *Project engineering annual, 44*.

Wu, H., Wang, P., Du, L., Jin, J., Mi, J., & Yun, J. (2022). Design of high-humidity-proof hierarchical porous P-ZIF-67(Co)-polymer composite materials by surface modification for highly efficient volatile organic compound adsorption. *Industrial and Engineering Chemistry Research, 61*(10), 3591–3600. https://doi.org/10.1021/acs.iecr.1c04434, http://pubs.acs.org/journal/iecred.

Wu, T., Liu, X., Liu, Y., Cheng, M., Liu, Z., Zeng, G., Shao, B., Liang, Q., Zhang, W., He, Q., & Zhang, W. (2020). Application of QD-MOF composites for photocatalysis: Energy production and environmental remediation. *Coordination Chemistry Reviews, 403*, 213097. https://doi.org/10.1016/j.ccr.2019.213097.

Xiang, Z., Peng, X., Cheng, X., Li, X., & Cao, D. (2011). CNT@Cu$_3$(BTC)$_2$ and metal–organic frameworks for separation of CO$_2$/CH$_4$ Mixture. *The Journal of Physical Chemistry C, 115*(40), 19864–19871. https://doi.org/10.1021/jp206959k.

Xu, J., Szyszkowicz, M., Jovic, B., Cakmak, S., Austin, C. C., & Zhu, J. (2016). Estimation of indoor and outdoor ratios of selected volatile organic compounds in Canada. *Atmospheric Environment, 141*, 523–531. https://doi.org/10.1016/j.atmosenv.2016.07.031, http://www.elsevier.com/locate/atmosenv.

Yagub, M. T., Sen, T. K., Afroze, S., & Ang, H. M. (2014). Dye and its removal from aqueous solution by adsorption: A review. *Advances in Colloid and Interface Science, 209*, 172–184. https://doi.org/10.1016/j.cis.2014.04.002.

Yang, K., Xue, F., Sun, Q., Yue, R., & Lin, D. (2013). Adsorption of volatile organic compounds by metal-organic frameworks MOF-177. *Journal of Environmental Chemical Engineering, 1*(4), 713–718. https://doi.org/10.1016/j.jece.2013.07.005.

Yang, S., Karve, V. V., Justin, A., Kochetygov, I., Espín, J., Asgari, M., Trukhina, O., Sun, D. T., Peng, L., & Queen, W. L. (2021). Enhancing MOF performance through the introduction of polymer guests. *Coordination Chemistry Reviews, 427*. https://doi.org/10.1016/j.ccr.2020. 213525, http://www.journals.elsevier.com/coordination-chemistry-reviews/.

Yao, Y., Wang, C., Na, J., Hossain, M. S. A., Yan, X., Zhang, H., Amin, M. A., Qi, J., Yamauchi, Y., & Li, J. (2022). Macroscopic MOF architectures: Effective strategies for practical application in water treatment. *Small (Weinheim an der Bergstrasse, Germany), 18*(8). https://doi.org/10.1002/smll.202104387, http://onlinelibrary.wiley.com/journal/10.1002/(ISSN)1613-6829.

Yin, Y., Zhang, H., Huang, P., Xiang, C., Zou, Y., Xu, F., & Sun, L. (2018). Inducement of nanoscale Cu–BTC on nanocomposite of PPy–rGO and its performance in ammonia sensing. *Materials Research Bulletin, 99*, 152–160. https://doi.org/10.1016/j.materresbull.2017.11.012, http://www.sciencedirect.com/science/journal/00255408.

Yot, P. G., Ma, Q., Haines, J., Yang, Q., Ghoufi, A., Devic, T., Serre, C., Dmitriev, V., Férey, G., Zhong, C., & Maurin, G. (2012). Large breathing of the MOF MIL-47(VIV) under mechanical pressure: A joint experimental-modelling exploration. *Chemical Science, 3*(4), 1100–1104. https://doi.org/10.1039/c2sc00745b, http://pubs.rsc.org/en/journals/journal/sc.

Yu, B., Ye, G., Chen, J., & Ma, S. (2019). Membrane-supported 1D MOF hollow superstructure array prepared by polydopamine-regulated contra-diffusion synthesis for uranium entrapment. *Environmental Pollution, 253*, 39–48. https://doi.org/10.1016/j.envpol.2019.06.114, http://www.journals.elsevier.com/environmental-pollution.

Zhang, J., Guo, Z., Yang, Z., Wang, J., Xie, J., Fu, M., & Hu, Y. (2021). TiO$_2$@UiO-66 composites with efficient adsorption and photocatalytic oxidation of VOCs: Investigation of synergistic effects and reaction mechanism. *ChemCatChem, 13*(2), 581–591. https://doi.org/10.1002/cctc.202001466, http://onlinelibrary.wiley.com/journal/10.1002/(ISSN)1867-3899.

Zhang, X., Gao, B., Creamer, A. E., Cao, C., & Li, Y. (2017). Adsorption of VOCs onto engineered carbon materials: A review. *Journal of Hazardous Materials, 338*, 102–123. https://doi.org/10.1016/j.jhazmat.2017.05.013, http://www.elsevier.com/locate/jhazmat.

Zhang, X., Lv, X., Shi, X., Yang, Y., & Yang, Y. (2019). Enhanced hydrophobic UiO-66 (University of Oslo 66) metal-organic framework with high capacity and selectivity for toluene capture from high humid air. *Journal of Colloid and Interface Science, 539*, 152–160. https://doi.org/10.1016/j.jcis.2018.12.056, https://www.sciencedirect.com/journal/journal-of-colloid-and-interface-science/vol/539/suppl/C.

Zhao, Q., Zhao, Z., Rao, R., Yang, Y., Ling, S., Bi, F., Shi, X., Xu, J., Lu, G., & Zhang, X. (2022). Universitetet i Oslo-67 (UiO-67)/graphite oxide composites with high capacities of toluene: Synthesis strategy and adsorption mechanism insight. *Journal of Colloid and Interface Science, 627*, 385–397. https://doi.org/10.1016/j.jcis.2022.07.059 https://www.sciencedirect.com/journal/journal-of-colloid-and-interface-science/vol/627/suppl/C.

Zheng, M., Chen, J., Zhang, L., Cheng, Y., Lu, C., Liu, Y., Singh, A., Trivedi, M., Kumar, A., & Liu, J. (2022). Metal organic frameworks as efficient adsorbents for drugs from wastewater. *Materials Today Communications, 31*, 103514. https://doi.org/10.1016/j.mtcomm.2022.103514.

Zhou, X., Huang, W., Shi, J., Zhao, Z., Xia, Q., Li, Y., Wang, H., & Li, Z. (2014). A novel MOF/graphene oxide composite GrO@MIL-101 with high adsorption capacity for acetone. *Journal of Materials Chemistry A, 2*(13), 4722–4730. https://doi.org/10.1039/c3ta15086k.

Zou, Y., Wang, X., Khan, A., Wang, P., Liu, Y., Alsaedi, A., Hayat, T., & Wang, X. (2016). Environmental remediation and application of nanoscale zero-valent iron and its composites for the removal of heavy metal ions: A review. *Environmental Science and Technology, 50*(14), 7290–7304. https://doi.org/10.1021/acs.est.6b01897, http://pubs.acs.org/journal/esthag.

Chapter 11

Metal-organic framework composites for antibacterial applications

Mehdi Ghaffari[1], Nazanin Habibi[1], and Mohammad Reza Saeb[2]
[1]Department of Polymer Engineering, Faculty of Engineering, Golestan University, Gorgan, Iran,
[2]Department of Pharmaceutical Chemistry, Medical University of Gdańsk, Gdańsk, Poland

Introduction

Bacterial diseases, such as skin, sinus, and ear infections and contaminations are among the main causes of morbidity and mortality in cirrhotic patients of human beings, likewise, animals are endangered by bacteria, and they also seriously threaten the environment. There are several infectious diseases caused by bacteria, such as strep throat, salmonella, and tuberculosis, which are well-known. There are also other types of bacterial diseases, such as cholera, diphtheria, bacterial meningitis, tetanus, Lyme disease, gonorrhea, and syphilis to be named. Typically, infectious bacteria are classified as Gram-positive or Gram-negative. To combat such diseases, antibacterial agents have been introduced, which attracted a great deal of attention in view of their ability to inhibit or retard bacterial growth or kill the bacteria. This specification made them one of the most promising medical aids and agents supporting human health. For example, since antibiotics have high bactericidal activity and low toxicity to mammalian cells, treatment of cells with antibiotics has always been the first choice to combat microbial infections (Cai & Liu, 2020). Therefore, the need for high-performance precursors resulting from nanotechnology could be considered as a major solution.

Antibacterial agents mostly originate from antibacterial bioactive materials. An antibacterial bioactive material can be defined as a material preserving the activity to kill the bacteria, suppress their growth, or inhibit their ability to reproduce. Natural antibacterial agents are frequently extracted from plants or the bacteria themselves (Liu et al., 2012). Natural antimicrobial peptides (AMPs) are the most common natural biopolymers available as cationic or anionic peptides. Additionally, apart from application and popularity of usage, antibacterial materials usually exhibit superior activities superior to their origins when engineered and used at nano- and micro-scale, which looks attractive for biomedical applications. The most commonly reported antibacterial nanomaterials include carbon nanotubes (Teixeira-Santos et al., 2021), MXenes (Seidi et al., 2023), graphene-based nanomaterials (Firouzjaei et al., 2018; Ji et al., 2016; Zou et al., 2016), nanoparticles (NPs) of metals (Chernousova & Epple, 2013), polymeric NPs (Rahimpour et al., 2018), and metal oxide NPs (Raghunath & Perumal, 2017). In general, selection of an antibacterial agent depends on the economy, its performance, and its availability, as well as the versatility in design and sustainability.

Metal-organic frameworks (MOFs), also known as porous coordination polymers, are a particular class of antibacterial materials, which has recently become known for their antibacterial activity (Seyedpour et al., 2020; Zirehpour et al., 2017). MOFs enjoy having active centers, which resembles the characteristics of metal/metal oxide NPs. Such active sites are stabilized through the formation of sufficiently strong chemical bonds in the main structure of MOFs (Wyszogrodzka et al., 2016). Due to abundant nanometer-size pores (even < 2 nm), MOFs have edge over many other types of nanomaterials, which can be seen from their ultra-high surface area (beyond a Langmuir surface area of 10,000 m^2/g), tunable and uniform porous structure, flexible functional metal sites and organic groups, and high designability. Moreover, the microstructure, composition, and functionality of MOFs can be rationally tuned by changing the optimized combination of organic and inorganic components, which relies on a remarkable antibacterial activity (Yan, Gopal, et al., 2022).

Despite the aforementioned promising features of MOFs, they suffer from a few weak points, such as poor chemical stability that limits their widespread use. Chemical modification of the surface of MOFs with functional groups or hybridization with functional nanomaterials making MOF composites are possible solutions to their low stability. MOF composites are advanced materials, which have gained significant attention for their potential antibacterial properties. There are some known mechanisms contributing to the antibacterial activity of MOF composites, which include but are not limited to the wider window of releasing metal ions, photocatalytic activity, and synergistic effects when incorporating them with both metal ions and organic antimicrobial agents within a single MOF structure. Moreover, the structural functionalization of MOFs can often enhance their overall antibacterial activity through synergistic interactions. Due to their versatility, MOF composites are taking a specific position in medical applications, water treatment, and packaging materials to inhibit bacterial growth and prevent infections (Li & Huo, 2015; Zhu & Xu, 2014). Nevertheless, antibacterial activity of MOF composites remains a challenge, as hybridization does not principally guarantee their superior antibacterial performance over the neat MOFs, making careful assessment essential.

This chapter provides insights into the evaluation of the antibacterial activity of MOF composites. In view of the possibility of engineering the microstructure of MOF composites, this chapter seeks to outline their potential as antibacterial nanomaterials by summarizing the related literature followed by categorizing and interpreting the latest advances in this field of research. Different types of MOFs are considered in this survey in view of their antibacterial activities and discussed based on their structural parameters, followed by consideration of more complex structures towards MOF composites. The latest advances in antibacterial performance of MOF composites are specifically tabulated and discussed. Eventually, challenging aspects of their application and future ahead of their development and commercialization are highlighted. The advent and progress in artificial intelligence (AI) and machine learning (ML) have been outlined as technological tools advancing and deepening the design and engineering of antibacterial agents, having future perspectives in scope.

Generic features of metal-organic frameworks

After over two decades of research and development, significant progress has been made in the synthesis and structural engineering of MOFs. Basically, MOFs are created by linking organic linkers with metal ion clusters or metal ions (Safaei et al., 2019). Various methods exist for synthesizing MOFs, which impact the physical properties of the resulting material, such as the crystalline structure, morphology, size, porosity, and surface area (Zhang, Yan, et al., 2023), which are illustrated through an evolutionary trend (Fig. 11.1).

Selecting an appropriate method for synthesizing MOFs is crucial for achieving the desired properties. For instance, hydrothermal synthesis stands out as one of the most widely used methods for the synthesis of MOFs. Recent experimental data suggests that the hydrothermal method results in high specific surface area, thermal stability, and crystallinity. Microwave-assisted synthesis offers a rapid nucleation process, leading to creation of MOFs with a uniform particle size distribution within a short and controllable reaction time. On the other hand, sonochemistry, which

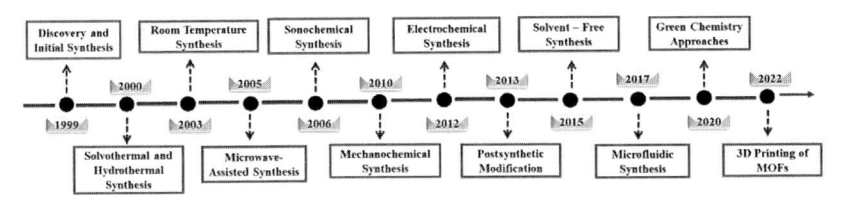

FIGURE 11.1 Timeline of evolution in the metal-organic framework synthesis.

relies on chemical reactions induced by the use of ultrasonic sounds, provides a fast, ecofriendly, cost-effective, user-friendly, and high-yield approach. Electrochemical synthesis is also characterized by its simplicity, rapidity, short reaction time, and lower energy consumption. Additionally, the mechanochemical method offers a clean and eco-friendly synthesis process benefiting from a minimized solvent volatilization. However, this method may exhibit lower crystallinity, such that the resulting structure can be easily compromised during the process (Lee et al., 2013; Safaei et al., 2019; Zhang, Peng, et al., 2022; Zhang, Yan, et al., 2023).

Due to diversity of synthesis methods, MOFs exhibit a range of unique properties, particularly depending on their surface chemistry and areas, making them suitable for various biomedical applications. Fig. 11.2 showcases examples of MOFs with distinct properties (Deria et al., 2014).

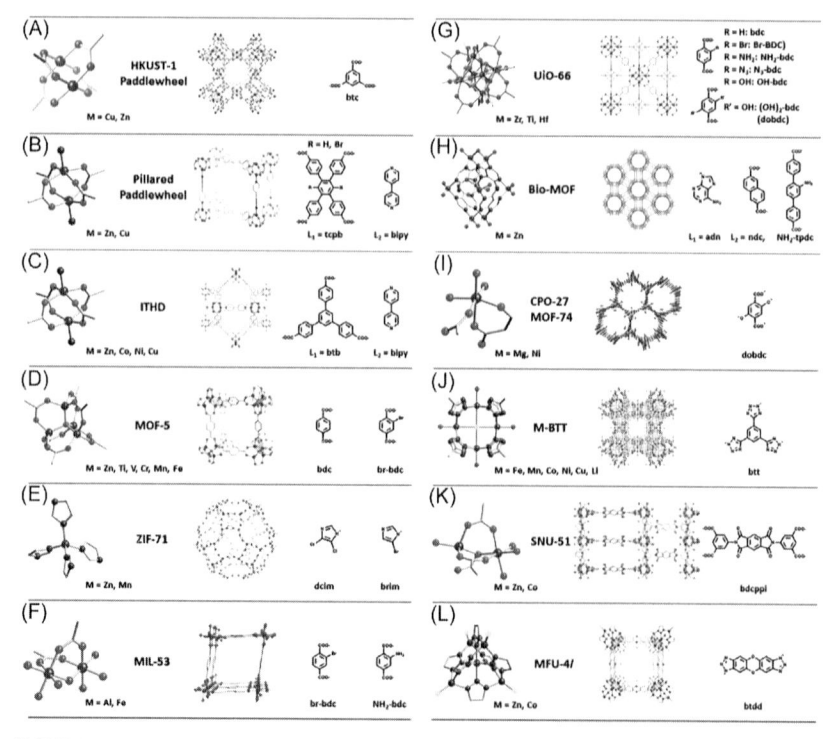

FIGURE 11.2 Possible (A to L) lattice structures (middle) and corresponding secondary building units (SBUs) (metal nodes (left), and organic linkers (right)) of some of the most commonly used metal-organic frameworks. Atom definition: blue—metal, red—oxygen, purple—nitrogen, gray—carbon, green—chlorine (Deria et al., 2014).

Surface engineering of metal-organic frameworks for biomedical applications

As discussed in the former section, large surface area and porosity are key generic features of MOFs, with surface areas ranging experimentally from 1000 m^2/g to almost a theoretical value of 10,000 m^2/g. These characteristics enable one for loading biomolecules and the encapsulation of various types of pharmaceuticals within MOFs (Lian et al., 2017). In addition, due to highly porous structure of MOFs, functional molecules can readily adsorb onto their surface (Bieniek et al., 2021; Velásquez-Hernández et al., 2021).

Although a synthesized MOF material may not principally reveal desired therapeutic properties, the functionality and chemical affinity of MOFs can be considered for biomedical applications, where surface modification plays a key role. This is particularly crucial in bio-applications, as the surface properties (such as surface area, charge, type of ligands, etc.) can influence factors like the blood circulation time, biodistribution, and the release of therapeutic agents. In fact, surface functionalization could render MOF NPs invisible to the immune system, especially the phagocytic system, holding a significant promise in drug delivery applications. This strategy aims to prolong the blood circulation times of therapeutic agents followed by an enhanced bioavailability. To rely on this function, appropriate ligands or antibodies (targeting cancer cells, for instance) could be attached to the surface of the MOF carrier. The primary objective of such therapy is to overcome biological barriers by concentrating the drug in specific target areas, thereby increasing its efficacy while minimizing adverse side effects (Seoane et al., 2015).

External surface modification of MOFs is commonly accomplished through coordination modulation and post-synthetic modification (Chen, Zhuang, et al., 2021; Wang, Mcguirk, et al., 2018). This modification could improve the colloidal stability by preventing agglomeration, enhance the loading capacity, and enable controlled payload release (Haddad et al., 2020; Horcajada et al., 2010). Table 11.1 summarizes recent studies focusing on the surface modification of MOFs.

Bulk engineering of metal-organic frameworks for biomedical applications

MOF-based composite materials have found applications in various biomedical fields, including cancer treatment, diabetes therapy, wound healing, brain treatment, etc.

Creating MOF-based composites by combining MOFs with other functional materials provides support for enhancing the surface area of MOFs, along with facilitates its separation and adsorption capacity. For instance, combining MOFs with an organic component can lead to optimizing relative cell viability, increasing bioactivity, and enhancing the bioavailability of the final nanocomposite (Hu & Zhao, 2015; Li & Huo, 2015).

TABLE 11.1 Recent investigations made into the surface engineering of MOFs.

MOF	Surface agent	Functionality	Refs.
MIL-MOFs	PEG	To control carrier interactions with the biological medium, improve MOFs "stealth" properties, and prolong carrier movement in the blood flow from a few minutes up to a few hours.	Horcajada et al. (2010)
ZIF-8	PVP	Long-term stability in a cell medium leads to higher cellular uptake. The presence of PVP could also help the escape of encapsulated proteins from endo-lysosomes and protect the activity of proteins throughout the delivery process.	Chen et al. (2018)
UiO	Nucleic acids	Higher cellular uptake, increased stability and effective entry into cells (without using viral transfection of cationic agents)	He et al. (2014) and Morris et al. (2014)
MIL-100(Fe), MIL-101(Cr)	Lipids	Improved colloidal stability, cellular uptake, and controlled release	Wuttke et al. (2015)
PCN-224	Folic acid	Targeted anticancer therapy	Park et al. (2016)
MIL-100(Fe)	CS	Improved chemical stability, colloidal stability, enhanced cellular uptake, and good biocompatibility	Hidalgo et al. (2017)

COF, covalent organic frameworks; *GO*, Graphene oxide; *MOF*, Metal-organic framework; *ZIF-8*, zeolitic imidazolate framework-8.

Various matrices can be used to integrate functional materials, such as NPs, bioentities, and polymers, resulting in composites with enhanced or novel properties compared to their parent frameworks (Liu, Liu, Du, et al., 2019; Liu, Liu, Huang, et al., 2019; Zheng et al., 2017).

By incorporating different functional NPs, like Au, Fe_3O_4, and MXene into MOF structures, the resulting nanocomposites can preserve the versatile crystalline and porous structures of MOFs while maintaining the unique biomimetic catalytic, optical-electrical, and magnetic properties of the NPs

(Pooresmaeil et al., 2021). Additionally, the synergistic effect of integrating these materials can lead to the emergence of new chemical and physical properties. Table 11.2 summarizes the outcomes of some investigations on the properties of MOF nanocomposites.

TABLE 11.2 A brief overview of some MOF nanocomposites applied in biomedical applications.

Matrix/ NPs	MOFs	Application	Function	Refs.
GO	Zn-MOF, Ag-MOF, MIL-Fe	Breast cancer, antibacterial dressing, and photothermal therapy	Improved drug delivery, enhanced drug adsorption values, and structure stability, alongside reducing cytotoxicity	Pooresmaeil et al. (2021), Lai et al. (2021), Meng et al. (2017)
MXene	MOF-5 ZIF-8 Zn-MOF	Drug delivery, gene delivery, biosensor, and wound healing	Achieved suitable drug payload, enhanced bioavailability, and controllable release	Wang, Sun, et al. (2021), Guo, Cheng, Liang, Zhang, Duan, et al. (2022)
COFs	MIL-88B (Fe) Co-MOF	Bacterial inhibition, and electrochemical aptasensor	Achieved antimicrobial efficacy, wound healing, and high stability	Zhang et al. (2021), Liu, Hu, et al. (2019)
Noble metal (Au)	ZIF-8, AL-MOF, MIL-101	Drug delivery, tumor therapy, antibacterial therapy, and cancer therapy	Controlled drug release, controlled system stability, decreased toxicity, and supported high drug loading capacity	Silva et al. (2019), Hu et al. (2020), Li, Shi, et al. (2021)
Fe$_3$O$_4$ NPs	UiO-66 MIL-100(Fe) ZIF-8	Magnetic resonance imaging, drug delivery, and cancer therapy	Achieved suitable drug loading capacity, improved catalytic performance, cell-killing effect, and biocompatibility	Zhao et al. (2016), Wang, Xu et al. (2018), Zhou et al. (2020)

COF, covalent organic frameworks; *GO*, Graphene oxide; *MOF*, Metal-organic framework; *ZIF-8*, zeolitic imidazolate framework-8GO.

Classification of metal-organic framework composites based on antibacterial properties

Fundamentals of antibacterial metal-organic frameworks and metal-organic framework composites

MOFs exhibit not merely high porosity but also a wide range of pore sizes, which can be controlled by selecting appropriate organic ligands. Actually, the control of pore size also known as porosity engineering alongside composition provides support for creation of MOFs with specific surface areas and tailored porosity (Mori and Takamizawa, 2000).

The customized MOFs with optimized composition, structure, and specific surface area can reveal desired antibacterial properties, which are attributed to the degradation of metal ions and organic ligands, as well as their synergistic actions (Wyszogrodzka et al., 2016). The type of metal present in the MOF structure plays a crucial role in determining the antibacterial activity, as these metals can store and in a controlled manner release ions, which is advantageous compared to metal/metal oxide systems. The release of metal ions is often considered the primary factor behind the antimicrobial effects (Liu et al., 2010; Moritz & Geszke-Moritz, 2013). For instance, metal ions can penetrate bacterial cell membranes, disrupting their structure.

MOF composites, either surface- or bulk-modified MOFs, can exhibit antibacterial properties undertaking different mechanisms, depending on their composition and structure, as recent studies have identified structures containing metal ions such as silver (Ag), iron (Fe), zinc (Zn), copper (Cu), magnesium (Mg), cobalt (Co), zirconium (Zr), and manganese (Mn) for their antibacterial and antimicrobial properties, each exhibiting unique mechanisms of action. For example, silver ions (Ag^+) can damage cell membranes, denature nucleic acids and enzymes, and interfere with electron transport, ultimately leading to bacterial inactivation (Miller et al., 2015).

Several types of MOF composites, such as silver-based MOFs or MOFs hybridized with polymers like polyvinyl alcohol (PVA) or PEG, have demonstrated excellent antibacterial activity against *Escherichia coli* and *Staphylococcus aureus* on the ground of the release of Ag^+ ions. Zeolitic imidazolate framework-8 (ZIF-8), when integrated into polymers, such as polyacrylonitrile or polylactic acid (PLA), has also found to be effective in killing bacteria through the release of Zn^{2+} ions and reactive oxygen species (ROS) generation. Additionally, Cr-based MOF known as MIL-101 incorporated into poly(methyl methacrylate) (PMMA) has shown potential in antibacterial applications benefiting from its large surface area and the porosity of MIL-101 that enhanced interaction with bacterial cells.

MOF-based hybrid nanocomposites have demonstrated antibacterial properties through direct contact or synergistic effects. There is a reason why some nanocomposites like MOFs hybridized with GO reveal antibacterial properties, which roots in their sharp edges leading to physical damage of bacteria upon contact (Pang et al., 2017). More complex hybridized structure were also studied, e.g., combination of Co-MOF with GO resulted in nanoporous GO/Co-MOF with enhanced antibacterial activity against pathogens like *E. coli* and *S. aureus*, along with an improved stability compared to Co-MOF alone (Hatamie et al., 2019).

Mechanisms of bacteriostatic agents

Antibacterial agents can be classified based on their mode of action as bacteriostatic (inhibiting bacterial growth) or bactericidal (killing bacteria).

Bacteriostatic agents typically function upon inhibiting protein synthesis or interfering with bacterial metabolic pathways. On the other hand, bacteriostatic agents prevent bacterial growth, so that their distinction from bactericidal agents can sometimes be blury, especially at high concentrations where bacteriostatic agents may exhibit bactericidal properties. (Zhang & Cheng, 2022).

Bacteriostatic antimicrobials are widely used because of their mechanism of action, which involve limiting bacterial growth by disrupting protein synthesis or DNA replication. Examples of bacteriostatic agents include tetracyclines, sulfonamides, spectinomycin, trimethoprim, chloramphenicol, macrolides, and lincosamides. These agents typically function in conjunction with the immune system to eliminate microorganisms from the body. Some bacteriostatic agents may also exhibit bactericidal effects at high concentrations. Table 11.3 briefly overviews the bacteriostatic antibacterial along with their function.

Photocatalytic bacteriostatic agents have emerged as a unique class of antimicrobials with benefits, such as environmental friendliness, high efficiency, and broad-spectrum bacteriostatic properties. Materials like TiO_2, ZnO, CeO_2, and g-C_3N_4, have been utilized in photocatalytic bacteriostasis, showing promising effects despite challenges like cost and toxicity (Kumar et al., 2014; Sun et al.,

TABLE 11.3 An overview of bacteriostatic antibacterial functions.

Bacteriostatic antibacterial	Function
Tetracyclines	Inhibits protein synthesis
Sulfonamides	Inhibit folate synthesis at initial stages
Spectinomycin	It binds to the 30S ribosomal subunit, thereby interrupting protein synthesis
Trimethoprim	It disturbs the tetrahydrofolate synthesis pathway
Chloramphenicol	Amphenicols work as protein synthesis inhibitors
Macrolides	Inhibits protein synthesis
Lincosamids	Inhibits protein synthesis

2017; Yang et al., 2018). More complex structures like CeO$_2$/GO composite were also applied for bacteriostatic function attributing to the two-dimensional planar structure of GO, which facilitates charge carrier transport.

The bacteriostatic properties of MOF composites are fueled by their ability to inhibit the growth and reproduction of bacteria without necessarily killing them, which are derived from the unique structural and chemical characteristics of MOFs in view of being tailored and functionalized for enhanced antimicrobial activity. Bacteriostatic MOF composites find applications in various fields, such as antibiotic carrier and osteogenesis promoter (Karakeçili et al., 2022), wound healing (Wang et al., 2023), biomedical implants (Yang et al., 2021), antibiotic therapy (Ghaffar et al., 2019), wound dressings (Wang, Zhou, et al., 2021; Zhang, Zhang, et al., 2023), etc.

For example, UiO-66 was selected as a carrier for the fosfomycin (FOS) antibiotic and incorporated in 3D CS scaffolds as an osteogenic differentiation promoter. Results showed that CS/UiO-66/FOS could significantly reduce the proliferation in *S. aureus* compared to that of CS scaffolds alone, which was ascribed to a bacteriostatic effect at pH 7.4 a bactericidal at pH 5.5 (Karakeçili et al., 2022).

In another MOF composite structure, ZIF-8@Ag, along with sulfonated poly(ether ether ketone) (SPEEK), formed a composite exhibiting superior bacteriostatic and bactericidal activities against *E. coli* and *S. aureus*, due to the antibacterial ability of Ag$^+$ and Zn^{2+} released from SP with ZIF-8 and Ag$^+$ (SPZA), as illustrated in Fig. 11.3 (Yang et al., 2021). Generally, Ag$^+$ and Zn^{2+} can catalyze the formation of ROS, such as hydrogen peroxide (H$_2$O$_2$), superoxide (O$_2^-$), and hydroxyl radicals (•OH). These ROS cause oxidative damage to cellular components, including DNA, proteins, and lipids leading to a considerable antibacterial activity.

In another research, a series of wound dressings were designed based on inclusion of Ag, synthesizing AgCu@MOF and loading it into PVA/CS wound

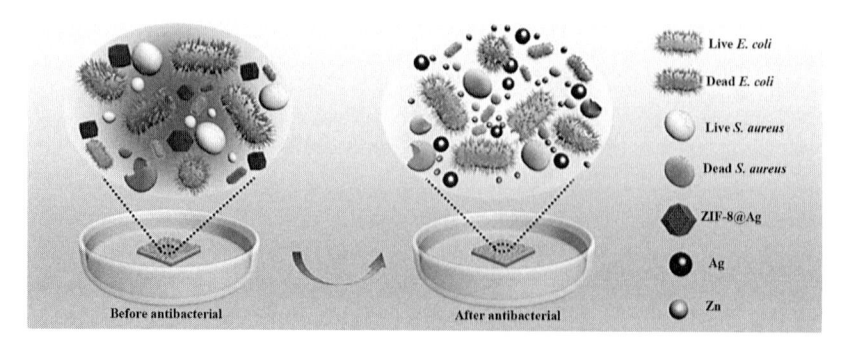

FIGURE 11.3 Antibacterial mechanism of Ag$^+$ ions-loaded ZIF-8 on sulfonated poly(ether ether ketone) (SP with ZIF-8 and Ag$^+$). *ZIF-8*, Zeolitic imidazolate framework-8 (Yang et al., 2021).

dressings. Generally, PCbM (PVA/CS/AgCu@MOF) indicated an excellent bacteriostatic and antibacterial activity, killed the bacteria on the wound, and also inhibited the reproduction of common bacteria (Zhang, Zhang, et al., 2023).

There is also evidence that MXene reveals bacteriostatic properties. Cu_2O/MXene composites have been shown to enhance bacteriostatic properties upon promoting ROS generation through improved electron–hole separation under photocatalysis. Similarly, $BiOI@CeO_2@Ti_3C_2$ MXene composites have demonstrated bacteriostatic efficiency against bacteria like *E. coli* and *S. aureus*, showcasing enhanced photocatalytic bacteriostatic abilities through a combined action mechanism (Mao et al., 2023; Wang et al., 2020). MOF/MXene-loaded PVA/CS wound dressing hydrogels recently showed an acceptable antimicrobial effect against *E. coli* and *S. aureus*as, preserving good electrical conductivity and mechanical strength (Zhang et al., 2024).

The development of polymers with antimicrobial properties is a crucial field of research with applications in various domains like healthcare, water purification, and food packaging. Polymers containing antimicrobial units, such as quaternary ammonium salts, guanidine, peptides, and antibiotics can effectively combat bacteria upon contact (Mao et al., 2023; Wang et al., 2020).

Studies have also shown promising results with polymer composites like tea tree oil-coated mesoporous silica, $Ag@TiO_2$-poly(*p*-dioxanone)-coated gauzes, and GO-modified starch/CS polymers, demonstrating bacteriostatic effects against a range of pathogens (Gao et al., 2020; Krystyjan et al., 2021; Liu et al., 2023).

In summary, the exploration of MOFs, MOF composites, bacteriostatic agents, and polymer composites with antimicrobial properties highlights diverse strategies with their complications and promises, pertinent to optimization. These structures can serve to combat bacterial infections and promote antimicrobial efficacy across various applications. Nevertheless, engineering of such MOF composite structures necessities design of experiments as well as the use of modeling and simulation supplementary tools in order to deepen our understanding of their antibacterial effects.

Bactericidal mechanism

To efficiently understand the antibacterial effects of MOF composites, there is a need for understanding the actions of metals and metal oxides, which contribute to MOF formation. The rise of antibiotic-resistant bacterial strains poses a significant threat to public health, necessitating the development of new bactericidal agents. The term "bactericidal" refers to substances capable of killing the bacteria. Certain antibacterials target bacterial cell walls or membranes to destroy bacteria. The effectiveness of bactericidal agents is influenced by the concentration and duration of exposure to the antibiotic, which includes disinfectants, antibiotics, and antiseptics (Hurdle et al., 2011).

There is a common agreement that bactericidal drugs exhibit potent antibacterial properties capable of direct killing of bacteria. In contrast, bacteriostatic antibiotics are believed to rely on phagocytic cells to clear bacteria, making them less effective in the absence of a robust immune response. This distinction has led to the recommendation that critically ill and immunosuppressed patients with bacterial infections should be treated with bactericidal antibiotics. Table 11.4 gives an overview of bactericidal antibacterials used in antibiotics.

For instance, polymyxins are cationic AMPs targeted at multidrug-resistant Gram-negative and Gram-positive bacterial infections. These peptides contain cationic L-α-γ-diaminobutyric acid residues, belonging to cationic AMPs. It is well known that the primary target of polymyxins against Gram-negative bacteria is the lipopolysaccharide component of the outer membrane. Upon breaching the barrier and reaching the cytoplasmic membrane, polymyxins disrupt membrane integrity (Yin et al., 2020). NPs have garnered attention as promising antimicrobial agents due to their exceptional antibacterial activity and large specific surface area. Moreover, bacteria exhibit low rates of resistance development toward metal and metal oxide NPs due to their complex antibacterial mechanisms (Pelgrift & Friedman, 2013). These NPs have been utilized as antimicrobial agents for detecting, diagnosing, and treating bacterial infections. The bactericidal mechanism of the metal and metal oxide nanocomposites involving the generation of ROS, including superoxide radical anions, hydrogen peroxide anions, and hydrogen peroxide. These species interact with the bacterial cell wall, causing damage to the membrane and hindering cell growth. This disruption leads to the leakage of internal cellular components, ultimately resulting in bacterial death (Pachaiappan et al., 2021). Additionally, the antibacterial efficacy of metal and metal oxide NPs is influenced by factors, such as size, shape, surface chemistry, and inherent structural components. Certain types of metal and metal oxide NPs, such as Ag, TiO_2, Au, CuO, Bi, Fe_2O_3, have been reported to possess bactericidal properties.

TABLE 11.4 An overview of bactericidal agents for antibiotic applications.

Bactericidal antibacterials	Function
β-Lactams antibiotic class (like penicillin and cephalosporins)	Inhibits the synthesis of the bacterial cell wall
Aminoglycosides	Inhibit protein synthesis
Quinolones and fluoroquinolones	Block the bacterial DNA replication
Glycopeptide (like vancomycin)	Inhibit cell wall synthesis
Polymyxins	Disrupt cell membrane
Metronidazole	Inhibit DNA synthesis
Rifamycin S	Inhibition of DNA-dependent RNA polymerase

The bactericidal mechanisms of MOF composites can be multifaceted, which is often affected by the specific design and composition of the MOF. These MOF composites serve in many fields, such as tissue engineering (Ansari-Asl et al., 2022), antimicrobial (Jasim Al-Khafaji et al., 2023), medical applications (Gwon et al., 2021), biomedical implants (Yang et al., 2021), wound dressing (Zhu et al., 2021), etc., with the bactericidal function.

Polymers can play a crucial role in this context. For example, Ansari-Asl et al. have coated a layer of CS onto Cu-MOF@polydimethylsiloxane (PDMS) sponge, and the results showed antibacterial and bactericidal properties of this MOF composite against *E. coli* and *S. aureus* due to the bactericidal properties of CS (Ansari-Asl et al., 2022). Also, Cu/dipicolinic acid-MOF/oxidized pectin/CS (Cu/DPA-MOF/OP/CS) hydrogel has been able to show inhibitory/fungicidal/bactericidal abilities in vitro conditions (Jasim Al-Khafaji et al., 2023). Polysiloxanes@Cu-MOF (PS@Cu-MOF) showed more than 80% bactericidal properties against *E. coli* and *S. aureus*, making it a promising candidate for implants, skin disease treatment, wound healing, and drug delivery (Gwon et al., 2021).

With the combination of UiO-66-NH$_2$ with PVA and serving the resulting MOF composite as a drug carrier of levofloxacin (LV@UiO-66-NH$_2$@PVA), bactericidal properties was achieved with 99% efficiency against *E. coli* and *S. aureus* and it showed promising results as a wound healing agent (Zhu et al., 2021).

Gold NPs enhance ROS release through bactericidal efficacy improvement. Conjugating Bi NPs with antibodies targeting specific microbes can amplify the bactericidal effect by reducing the distance between Bi NPs and bacterial cells (Luo et al., 2013). Silver ions dissolved in an aqueous solution support antimicrobial activity, but when Ag NPs are modified with other metals, metal oxides, or polymers can further enhance bactericidal activity. TiO$_2$ NPs generate ROS that disrupt bacterial cell walls, leading to bactericidal effects. CuO NPs demonstrate efficient bactericidal activity by releasing ions that damage bacterial cell walls (Pachaiappan et al., 2021).

Iron oxide NPs possess biocompatibility, large surface areas, crystalline structure, and magnetic properties, making them effective antibacterial agents. Combining silver or fibrillated cellulose with iron oxide enhances their bactericidal effects (Xiong et al., 2013).

Studies have demonstrated the synthesis of poly(lactic-*co*-glycolic) acid/silver nanocomposites, which ended in enhanced antibacterial and bactericidal properties. These nanocomposites exhibit high antioxidant activity and effectively combat bacterial pathogens (Parmar et al., 2019).

In another study, as shown in Fig. 11.3, researchers evaluated the biocompatibility and bactericidal activity of PLA/TiO$_2$ nanocomposites synthesized using green methods. The photocatalytic properties of TiO$_2$ could enhance the bactericidal activity, particularly when exposed to the light. Evidently, the porous NPs incorporated film demonstrates the ability to kill approaching bacteria effectively (Shebi & Lisa, 2019).

Co-based MOF have been developed for bactericidal activity, with Co interacting with bacterial cell walls to achieve rapid antibacterial effects (Zhuang et al., 2012).

Alonso and coworkers showed the bactericidal activity of the fibrous polymer/silver/cobalt nanocomposite. This material showed ideal bactericide features for being applied to bacterial disinfection of water compared with other nanocomposites only containing Ag or Co NPs, the capacity to kill a wide range of bacterial types (from coliforms to Gram-positive bacteria), and a long performance-time with an efficiency of 100% with 0% viability (Alonso et al., 2012).

Yuan et al. designed and synthesized a CeO_2/Nb_2C nanocomposite for bacterial infection with photothermal nanozyme catalytic activity. Due to the limitation of Nb_2C MXene for antibacterial therapy with only a single photothermal, it should be to endow the Nb_2C MXene with additional properties to achieve a lasting bactericidal effect. Also, it has been elucidated that the antioxidant activity of CeO_2 facilitates cell proliferation inside the wounds and accelerates wound closure through protection of cells from excessive ROS at the late stage of the wound-healing process. It has been found that the CeO_2/Nb_2C nanocomposite, with near-infrared radiation, can perform multiple functions for infection control and wound healing. The results show that the CeO_2/Nb_2C nanocomposite exhibits good biosafety in vitro and in vivo and effective bactericidal activities without being affected by bacterial resistance (Yuan et al., 2023) (Fig. 11.4).

These findings highlight diverse applications and mechanisms of composite structures as bactericidal agents, including nanocomposites, metal-based materials, NPs and MOFs composites, in combating bacterial infections. The development

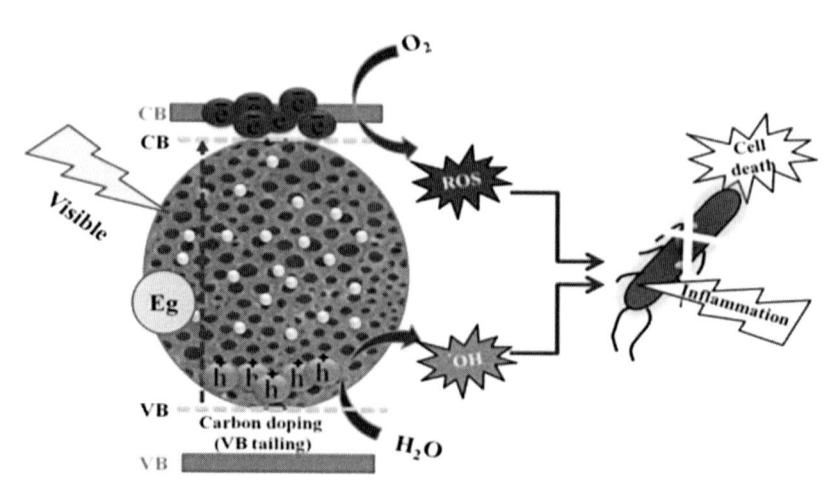

FIGURE 11.4 Schematic representation of generation of ROS in n-GST (Shebi & Lisa, 2019). *ROS*, reactive oxygen species.

of novel bactericidal agents holds promise for addressing the challenges posed by antibiotic-resistant bacteria and advancing public health initiatives (Fig. 11.5).

Metal-organic framework composites in antibacterial applications

Fundamentals

The antibacterial activity of MOFs has justified its use in research investigations for healing and antibacterial treatments. MOFs exhibited flexible interactions with either host of the guest molecules, with controlled release potential due to strong interaction with bacterial membranes, and high loading capacities for agents like ions or antibiotics. The organic linkers in the microstructure of MOFs act as antibiotic agents or photosensitizers, generating ROS upon light exposure to effectively combat bacteria. Overall, investigations have shown that Gram-negative bacteria display higher sensitivity to the antibacterial effects of MOFs compared to Gram-positive bacteria.

A variety of metals, including but not limited to Ag, Zr, Fe, and Co have been utilized in synthesis of MOFs, which revealed acceptable antibacterial activities when compared with metal ions. The enhanced antibacterial effects have been reported upon engineering of microstructure of MOFs attributed to

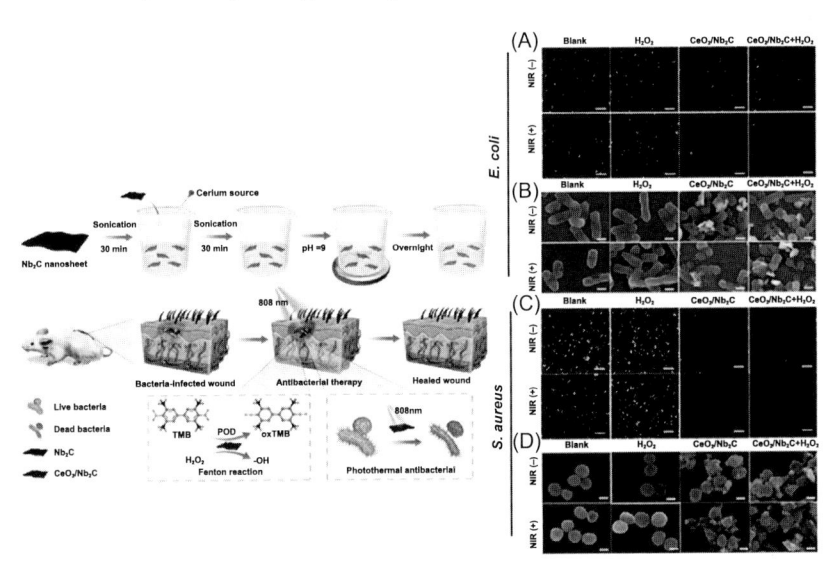

FIGURE 11.5 Schematic representation of the CeO$_2$/Nb$_2$C nanocomposite used to kill pathogens (left), and (A) live/dead bacterial viability assessment of *E. coli* and (C) *S. aureus* under various conditions (scale bar = 10 μm); (B) Scanning electron microscope images of *E. coli* and (D) *S. aureus* treated with various conditions (scale bar = 1 μm) (right) (Yuan et al., 2023).

the release of metal ions, crystalline structure of MOFs, their components, and coordination chemistry of frameworks. Moreover, post-synthesis modifications of MOFs have led to significant advancements in antibacterial applications (Sheta et al., 2018). Table 11.5 displays various studies highlighting antibacterial activities of MOF composites.

The investigations revealed that MOFs in the form of composite store metal ions in their frameworks to release them gradually, resulting in sustained antibacterial effects, which is mechanistically similar to the behavior of metal or metal oxide antibacterial agents. Furthermore, MOF composites have found extensive applications in antibacterial therapy by carrying antibacterial drugs. The variety of microstructure in MOF composites is responsible for higher specific surface area and porosity for loading bactericidal agents, controlled release triggered by stimuli, selection of nontoxic precursors for biocompatible MOFs, and abundant surface activity groups compared to traditional antibacterial materials, which allows for precise targeting at infection sites.

Surprisingly, MOF composites can be engineered structurally for antibacterial applications due to their designable structure, adjustable size, and tunable properties. Based on their different metals, MOF composites can primarily be categorized into six groups, such as (Ag)-based MOFs, (Zn)-based MOFs, (Cu)-based MOFs, (Co)-based MOFs, (Fe)-based MOFs, and (Zr)-based MOFs, each offering unique functionalities and properties for effective antibacterial therapy. In the following sub-sections of chapter, these classes of MOF composites are overviewed in terms of their antibacterial activity and mechanism.

Silver-based metal-organic frameworks

Ag-based MOFs containing Ag ions have been at the forefront of antibacterial applications. These crystalline materials, formed through the coordination of Ag ions with organic ligands, exhibit efficient antibacterial activities through various mechanisms. The interaction of Ag ions with functional groups in DNA and cell membranes of bacteria leads to bacterial inactivation (Liu et al., 2010). Studies have highlighted antibacterial effects of Ag-MOFs superior to the commercial Ag NPs. Furthermore, the combination of Ag-MOFs with other agents makes MOF composites demonstrating higher antibacterial properties. For instance, the use of Ag-MOF and GO-Ag-MOF composites resulted in a significant eradication of live bacterial cells (Firouzjaei et al., 2018). Comparisons between Ag-MOFs and Ag-MOFs@CS composites have also demonstrated that the latter exhibited higher antibacterial properties and longer shelf life against various bacterial strains, attributed to the slow release of Ag^+. The stability and reduced silver toxicity of Ag-MOFs@CS were additionally reflected upon 168 h of release process, demonstrating higher aqueous stability alongside lower silver toxicity compared to those of Ag-MOFs. The interaction between Ag ions and active groups of CS was found to contribute to inhibiting mold growth and achieving prolonged preservation effects (Zhang, Lin, et al., 2022).

TABLE 11.5 Various investigations made into MOF composites with different antibacterial properties.

MOF	Application	Function	Refs.
Ag-MOF	Water treatment, Antibacterial and antifouling	Good antibacterial activity, Strong resistance to biofouling	Tan et al. (2022), Pejman et al. (2020)
Ag-MOFs@CS	Fruit fresh-keeping	Good antibacterial properties, Releasing Ag^+	Zhang, Lin, et al. (2022)
GO-Ag-MOF	Biomedical	The extirpation of 95 % of live bacteria cells	Firouzjaei et al. (2018)
Ag@MOF-PVDF[1]	Anti-bacterial and Antifouling	Antimicrobial activity, Enhance surface hydrophilicity to prevent biofouling.	Xu, Zhuang, et al. (2020)
polyCu-MOF@AgNPs	Wound healing	Improved the healing efficiency, Promoting skin regeneration and dense collagen deposition.	Guo, Cheng, Liang, Zhang, Jia, et al. (2022)
Cu-MOF@PDMS	Antibacterial and tissue engineering	High antibacterial potential against *E. coli* and *S. aureus* cells.	Ansari-Asl et al. (2022)
Cu-MOF	Anticancer therapy, Antibacterial activities	Excellent antibacterial efficacy against both Gram-positive and Gram-negative microorganism	Chakraborty et al. (2022), Jo et al. (2019)
Cu/H_3BTC[2] MOF	Antibacterial therapeutic	Hinder the growth of pathogenic microbes, Releasing Cu^{2+}	Shams et al. (2020)
NO^3@Cu-MOFs@PC[4]	Anticoagulation and antibacterial	Excellent mechanical stability, Anticoagulation, Antibacterial, Good	Zhang, Ke, et al. (2022)

(Continued)

TABLE 11.5 (Continued)

MOF	Application	Function	Refs.
Copper/dipicolinic acid–metal-organic framework cross-linked oxidized pectin and chitosan (Cu/DPA-MOF/OP/CS)	Antimicrobial	biocompatibility, but releasing was not reported	Jasim Al-Khafaji et al. (2023)
Fe$_3$O$_4$@Cu-MOF/core-shell	Antibacterial activities	Higher antimicrobial properties than drugs	Azizabadi et al. (2021)
PS[5]@Cu-MOF	Antibacterial activity and Biomedical	Good antibacterial activities against both Gram-positive and Gram-negative bacteria.	Gwon et al. (2021)
M[6]Cu-MOF	Antimicrobial Activity	Low cytotoxicity, High bactericidal, Treatment for skin disease, Wound healing, Drug delivery	Abdelmoaty, El-Beih and Hanna (2022)
St[7]/Fe$_3$O$_4$/MIL-88(Fe)	Antibacterial carrier	Antimicrobial effects on gram-positive and gram-negative bacteria and fungi	Abbasian and Khayyatalimohammadi (2023)
Fe-MOF-Ag	Antibacterial	Highly sustained release of antibiotic agents, Releasing tetracycline	Hu et al. (2021)
Fe-MIL-88NH$_2$	Bactericidal and Antibacterial	Excellent antibacterial properties against both Gram-negative and Gram-positive bacteria	Haseena et al. (2022)
MOF-53(Fe)	Biomedical and Antibacterial	Effective for reversing bacterial resistance, Releasing Vancomycin and Methicillin	
		Exhibited lasting antibacterial effect, High antibacterial efficiency of 99.3% without cytotoxicity, Releasing vancomycin	

Material	Property	Description	Reference
Ag@MIL-53(Fe)	Antibacterial property	Excellent antibacterial activity, Releasing Ag^{2+}	Huang et al. (2021)
ZIF-8	Antibacterial properties, Antibacterial treatment	Superior antibacterial properties, Bactericidal efficiency, Releasing Zn^{2+}	Taheri et al. (2021), (Wang et al. (2022)
GOD[8]/Ag@ZIF-HA[9]	Anticancer and Antibacterial Therapy	High anticancer and antibacterial performance	Li, Gao, et al. (2021)
ZIF-8@SA[10]	Antibacterial textiles	High performance antibacterial textiles for medical dressings, Surgical sutures and masks	Zheng et al. (2020)
ZIF-8@ZIF-67	Antibiotic decomposition, Antibacterial activities	Significant activities for E.coli reduction	Sadiq et al. (2023)
ZnO@ZIF-8	Antibacterial activity	Antibacterial activity against the S. aureus bacteria	Kalati and Akhbari (2021)
Fe_3O_4@PAA[11]@ZIF-8	Drug delivery, Antibacterial activity	Improve the antibacterial properties of the drug, Releasing ciprofloxacin	Esfahanian, Ghasemzadeh and Razavian (2019)
AgNPs@ZIF-8	Antibacterial performance	Significant efficacy in eradicating pathogenic bacterial biofilms	He et al. (2023)
ZIF-8/GO-NH			Ahmad et al. (2020)

(Continued)

TABLE 11.5 (Continued)

MOF	Application	Function	Refs.
γ-Fe$_2$O$_3$@SiO$_2$@ZIF-8-Ag	Ultrafiltration membrane	Increased the antibacterial properties via the deformation of bacterial cells by the sharp edges of GO nanosheets and the attachment of Zn^{+2} ions on the negatively charged bacterial cell wall	Rahmati et al. (2020)
TiZ5/ZIF-8 and MC[12]/ZIF-8	Antibacterial activity	The antibacterial activity improved by doping silver ions, Releasing Ag$^+$	Ravinayagam and Rehman (2020)
SPEEK@ZIF-8@Ag	Antibacterial characterization	Antibacterial component as a natural antioxidants and anticancer drug	Yang et al. (2021)
Ag NPs/Zn-MOFs	Antibacterial activity	Excellent antimicrobial activity, Releasing Zn^{2+} and Ag$^+$	Sacourbaravi et al. (2020)
Au/Ir@Cu/Zn-MOF	Antibacterial Agents	A potential scaffold for the development of antibacterial agents.	Zhong et al. (2022)
Cur[13]@Zn-MOF	Antibacterial treatment	Precise photothermal chemodynamic synergistic antibacterial efficacy in vitro	Yan et al. (2023)
zinc imidazolate frameworks (ZIF-8-CTS[14])	Antibacterial activity for wound healing	In vivo wound healing effect	Xu, Fang, et al. (2020)
	Antibacterial infection	Bactericidal effect with antibacterial ratio of up to 100% for Gram-negative E. coli and Gram-positive S. aureus.	

Ag/Zn-MOFs	Antibacterial activity for non-woven fabrics	Improve antibacterial mechanisms	Zhu et al. (2022)
Zn-MOF	Antibacterial activity	High antibacterial performance is due to high amounts of Zn^{2+} ions, and their greater release in comparison to inactive MOFs	Nakhaei et al. (2021)
Zr-MOF	Antibacterial wound dressing	Antibacterial activity against E. coli and S. aureus because of diverse antibacterial components	Rezaee et al. (2022)
UiO-66-NH$_2$	Antibacterial activity and biofilm inhibition	Strong antibiofilm, Releasing Cefazolin	Dastneshan et al. (2023)
CA[15]@UiO-66	Antibacterial mechanism	Antimicrobial activity and low toxicity, Antibacterial coating for food packaging products, Releasing Caffeic acid	Zhou et al. (2023)
Zr-MOF (MOF-801)	Antibacterial, Anti-inflammation, Osteogenesis	Improved antibacterial activity with an increase in the content and release of fluorine	Yan, Tan, et al. (2022)
Cu$_2$O/Zr-Fc[16]-MOF	Antibacterial therapy	Excellent antibacterial effect on Gram-positive bacteria and Gram-negative bacteria, Promoting wound healing in vivo	Zhao et al. (2022)
CIP[17]-UiO-66	Antibacterial activities	A reduction occurred in the frequency of drug use per day for the patient, which is considered as a promising result	Nasrabadi, Ghasemzadeh and Monfared (2019)

(Continued)

TABLE 11.5 (Continued)

MOF	Application	Function	Refs.
UiO-66-NH$_2$	Photodynamic therapy and chemotherapy	Predictable phototoxicity and limited dark toxicity and achieves synergistic PDT and CT, making it an ideal antibacterial agent	Lv et al. (2020)
CMC[18]/TC[19]@UiO-66	Antibacterial wound dressing	Good cytocompatibility toward Human skin fibroblast (HFF-1) cells and a significant activity against both E. coli and S. aureus, Releasing tetracycline	Javanbakht et al. (2021)
LV@UiO-66-NH$_2$@PVANFMs[20]	Antibacterial therapy for wound healing	Superior wound healing properties, Releasing levofloxacin	Zhu et al. (2021)
Ag-MOFs/CMFP	Fruit fresh-keeping	Outstanding antibacterial activity which prolongs the shelf-life of tomatoes and peaches, Releasing Ag^{2+} due to the good antibacterial ability	Chen et al. (2024)

PVDF, Poly (vinylidene fluoride); BTC, 1, 3, 5-benzenetricarboxylic acid; NO, Nitric oxide; PC, polyurethane prepolymer coating; PS, Polystyrene; M, melamine; St, Starch; GOD, glucose oxidase; HA, hyaluronic acid; SA, sodium alginate; PAA, polyacrylic acid; MC, mesoporous carbon; Cur, Curcumin-regulated; CTS, chitin sponge; CA, caffeic acid; Zr-Fc, ferrocene; CIP, ciprofloxacin; CMC, carboxymethyl cellulose; TC, tetracycline; PVANFMs, nanofibrous membrane.

In another investigation, Ag-MOFs/carboxymethyl filter paper (CMFP) composites exhibited good antibacterial activity beneficial for fruit preservation. The cross-linking of active groups in CMFP and Ag$^+$ ions in Ag-MOFs resulted in enhanced aqueous stability, low toxicity, and effective antibacterial properties (Chen et al., 2024).

In another survey, modified membranes incorporating Ag-MOFs have demonstrated long-lasting antibacterial activities, biocompatibility, low environmental toxicity, and effectiveness in aqueous purification. These membranes provided a practical solution for maintaining antibacterial properties for extended periods, contributing to water purification and environmental safety (Tan et al., 2022; Xu, Zhuang, et al., 2020).

Copper-based metal-organic frameworks

Copper-based MOFs, such as Cu-MOF, have also gained attention for their antibacterial properties, which boosted the antibacterial and antifungal activities contributed from biological properties of copper. The high specific surface area and nano-sized structures of Cu-MOFs enabled efficient antibacterial action, primarily identified to undertake the ROS mechanism. The release of Cu ions from Cu-MOFs composite structure disrupts bacterial cell membranes, interferes with DNA synthesis, and induces cell death (Chakraborty et al., 2022; Jasim Al-Khafaji et al., 2023; Jo et al., 2019).

Further studies have demonstrated the antibacterial efficacy of Cu-MOFs, such as Cu/H$_3$BTC MOF, leading to damage to cell membranes, intracellular component leakage, and bacterial cell death. Additionally, the incorporation of a polymeric coating in NO@CuMOFs@PC enhances stability, reduces the release rate of Cu in aqueous environments, and significantly improves antibacterial activity against *E. coli* and *S. aureus* (Shams et al., 2020).

The addition of Fe$_3$O$_4$ to Cu-MOFs as core–shell structures was also found to enhance the stability and sustains Cu ion release, thereby increasing antibacterial activity (Azizabadi et al., 2021).

Some reports on application of polymers evidenced that the effectiveness of Cu-MOFs as antibacterial agents could be enhanced, showcasing biocompatibility, low toxicity, and bactericidal effects against *E. Coli* and *S. aureus*, making them promising candidates for biomedical treatments and tissue engineering applications (Ansari-Asl et al., 2022; Guo, Cheng, Liang, Zhang, Jia, et al., 2022; Gwon et al., 2021). Therefore, it can be concluded that Cu-based MOFs exhibit potent antibacterial activities through mechanisms that disrupt cell membranes, interfere with DNA synthesis, and induce bacterial cell death. Their versatility, biocompatibility, and targeted antibacterial effects have simultaneously made them able to combat bacterial infections, also well-known for addressing antibiotic resistance challenges in various biomedical applications.

Iron-based metal-organic frameworks

Fe-based MOFs and MOF composites have been in charge of bio-related applications over the last decade. These structures offer high safety levels owing to their low cytotoxicity and excellent biocompatibility, making them extensively applied in various biomedicine including drug delivery, bioimaging, and disease treatments, with special potential towards antibacterial properties. In fact, Fe ions present in Fe-MOFs brings about an inherent antibacterial activity. Utilizing Fe-MOFs as carriers for antibacterial drugs further enhances their antibacterial activity, resulting in a synergistic effect. The pivotal role of Fe ions in antibacterial activities within Fe-MOFs is well established in literature. Studies indicate that increasing the concentration of Fe ions can improve the antibacterial properties and cellular viability of Fe-MOFs. Evidently, an enhanced cellular activity of osteoblasts was reported upon higher amount of Fe ion incorporation. In addition, incorporating Fe-MOFs into the composite structures can yield desirable properties. For instance, synthesis of a nanocomposite based on triplet of starch, Fe_3O_4, and MIL-88(Fe) was responsible for the detection of an interaction between the resulting hybrid MOF composite and cargoes leading to enhanced biodegradability and hydrophilicity. The resulting composite, namely St/Fe_3O_4/MIL-88(Fe), is well-suited for high content loading and sustained drug release. The synergistic effect of these three materials enhanced the range of possible interactions between tetracycline (TC) and St/Fe_3O_4/MIL-88(Fe), demonstrating potential for antibacterial activity against *E. Coli* and *S. aureus* (Abbasian & Khayyatalimohammadi, 2023).

Zeolitic imidazolate framework-based metal-organic frameworks

ZIF-based MOFs exhibit a number of advantages compared to other types of MOFs. For example, ZIF-8 enjoy from low cost, preserving a large surface area that arises from its porous structure, high thermal and chemical stability, crystallinity, active sites for reactions, easy synthesis, customizable dimensions, stability, simple functionalization, and environmental friendliness.

ZIF-8, which is composed of transition metal Zn and imidazolate linkers, can gradually degrade in an acidic environment, a kind of characteristic of cancerous cells, releasing abundant Zn^{2+} ions that induce chemical harm to bacteria. Importantly, the pH-dependent release of Zn^{2+} from ZIF-8 correlates positively with its antibacterial efficacy. Zn ions can lead to cell deformation, cell wall rupture, and the development of an alkaline microenvironment that hampers bacterial growth, making them potent antibacterial agents. Furthermore, ZIF-8 NPs can generate ROS through which bacterial cell membrane proteins can be affected, by which they can disrupt bacterial structures, and exhibit robust antibacterial performance. The porous nature of ZIF-8 and its derivatives endow them with significant potential in antibacterial applications. Research has shown that combinations of ZIF-8 with NPs like ZIF-8/Au/GOx, ZnO@ZIF-8, and Ag@ZIF-8 display exceptional bactericidal properties. These MOFs/NPs show the ability to produce ROS and release Zn ions hindering the proliferation, which damages bacterial membranes, induces bacterial cell death, and showcases high bactericidal and antibacterial activity (He et al., 2023; Kalati & Akhbari, 2021; Wang et al., 2022).

In a separate study, polyethersulfone membranes modified with ZIF-8/GO/NH were employed in separation processes to combat severe biofouling issues. This modified membrane effectively inhibited the growth of *E. coli*, improved membrane hydrophilicity, and enhanced water flux performance. The sharp edges of GO nanosheets and the attachment of Zn ions augmented the membrane's antibacterial properties, leading to bacterial death upon contact with the membrane surface (Ahmad et al., 2020).

Additionally, the release of other agents from ZIF-8 contributes to its antibacterial properties. For example, studies have indicated that the release of Ag ions from ZIF-8 enhances antibacterial activity. Moreover, when Ag and Zn ions cooperate, they interact with cytoplasmic membrane phospholipid molecules, disrupting the bacterial membrane and causing leakage of bacterial contents (Rahmati et al., 2020; Yang et al., 2021).

Other studies have demonstrated that ZIF-8 nanocomposites exhibit outstanding antibacterial performance (Sadiq et al., 2023). Some studies suggest that ZIF-8 composites reduce Zn size, enhancing antibacterial activity by promoting the penetration of nanosized Zn into cell walls, ultimately leading to cell death (Ravinayagam & Rehman, 2020).

Zirconium-based metal-organic frameworks

Zr-based MOFs represent a highly promising class of MOF materials due to their thermal, chemical, and mechanical stability, as well as their large pore size and surface area. These characteristics make Zr-MOFs versatile for various applications, including adsorption, photocatalysis, catalytic oxidation, and drug delivery https://doi.org/10.1177/11786221221080183.

Zr, being biocompatible and low-toxic, has enabled the utilization of Zr-MOFs in biomedicine. Zr-MOFs have demonstrated significant antibacterial activity, leading to their extensive use in applications, such as targeted drug delivery, photothermal therapy (PTT), photodynamic therapy (PDT), and bioimaging.

UiO-66, a notable Zr-based MOF, comprises $Zr_6O_4(OH)_4$ metal clusters and 1,4-benzene dicarboxylate organic linkers. On the other hand, UiO-66-NH$_2$ consists of Zr^{4+} as the metal ion and 2-amino-terephthalic acid as the coordination system. These MOF materials are well-regarded for their potential in antibiotic delivery applications. They have the ability to establish electrostatic and hydrophobic interactions with antimicrobial compounds, enabling them to adsorb these compounds into their pores and onto their surfaces during antibacterial treatment. This property enhances their efficacy in combating bacterial infections and underscores their potential in biomedical applications.

The simultaneous release of both Zr and drugs from Zr-based MOFs has been reported to induce physical damage to bacterial cell walls and lead to bacterial death. This mechanism can be highly effective in combating bacterial infections as it targets the structural integrity of the bacteria, ultimately causing their demise.

Moreover, studies have indicated that Zr-based MOFs, such as UiO-66 exhibit low cytotoxicity and excellent biocompatibility. These properties make them ideal candidates as nanocarriers for drug delivery to various cells and tissues. The biocompatibility of Zr-based MOFs enhances their safety profile and reduces the risk of adverse effects when used in biomedical applications (Filippousi et al., 2016).

Furthermore, Zr-based MOFs like UiO-66 have shown potential in inhibiting and controlling pathogens, such as *E. coli* and *S. aureus*. Their ability to deliver antibiotics effectively to target bacteria, coupled with their low cytotoxicity and biocompatibility, makes them promising vehicles for antibiotic delivery. It is intended to optimize the properties of Zr-MOFs to combat bacterial infections and enhance therapeutic outcomes in medicine (Dastneshan et al., 2023; Nasrabadi et al., 2019).

Another study has demonstrated that combining UiO-66 with polymers like PVA confers antibacterial properties to the MOF system. It has been shown that MOF/polymer composites can reduce the size of wound areas. Additionally, this composite exhibits hydrophilicity and excellent bactericidal effectiveness against *E. coli* and *S. aureus* (Zhu et al., 2021).

Connecting carboxylic zinc phthalocyanine as a photosensitizer to UiO-66-NH_2 and loading linezolid antibiotic into pores, along with a lysozyme coating on the surface, demonstrates the antibacterial effectiveness of PDT and chemotherapy. The oxygen produced by these agents leads to strong antibacterial activity against *E. coli*. Furthermore, this nanomaterial exhibits predictable and low toxicity levels (Lv et al., 2020).

Challenges with antibacterial metal-organic framework composites

MOFs are a modern class of porous materials constructed from metal ions or clusters coordinated to organic ligands. Their unique structural properties which includes high surface area, tunable pore sizes, and versatile functionality have made them highly attractive for various applications. These applications range in between antibacterial activities and other biomedical features. However, several challenges must be addressed to fully harness the potential of MOF composites, especially in biological contexts.

A significant difficulty with MOF composites in biomedical applications is the potential toxicity of their components. Many MOFs utilize organic ligands and metal ions that can be harmful to living organisms. For instance, metal ions like Cu^{2+} and Zn^{2+}, often used in MOF synthesis for their antimicrobial properties, can be toxic at or beyond certain concentrations. The organic ligands themselves may also exhibit low biocompatibility, limiting the safe use of MOF composites in medical applications. To overcome this challenge, the incorporation of biologically active molecules or biocompatible cations in MOF composites construction can be considered. Biological ligands like amino acids and nucleobases have shown promise in enhancing biocompatibility and reducing cytotoxicity, offering a pathway to safer antibacterial MOF composites.

Various biological MOF composites examples, such as cyclodextrin-based MOF or BiO-MIL, have shown promise as safe and effective systems for drug delivery and antimicrobial effects. However, these MOF composites have been reported to face limitations, such as structural degradation, low stability, and a gradual decline in antimicrobial efficacy over time.

The limited stability of MOFs and MOF-based composites, particularly in the presence of humidity where bacteria thrive, results in a reduced active lifespan of MOFs. Surface modification has been suggested as a strategy to enhance the biomedical interaction or compatibility of MOFs and MOF-based composite systems. For instance, coating MOF surfaces with mesoporous silica or organic polymers can improve their stability, aqueous dispersion, and biocompatibility, thereby their antimicrobial properties.

Currently, most MOF-based composites are synthesized using the hydrothermal/solvothermal method, which comes with drawbacks, such as toxicity, high costs on an industrial scale, and negative environmental impacts. To address these issues, the adoption of "green synthesis methods" for MOF composites construction is recommended. This approach involves using nontoxic metal sources and biocompatible organic ligands, leading to reduced energy consumption during synthesis and utilizing water instead of toxic and costly solvents.

It is accepted that a precise control over the physical and chemical properties of MOF-based composites seems a crucial prerequisite in designing MOF composites with high reproducibility and reliability, which itself impacts other applications. For instance, the design of ultra-small nanoscale MOF composites is desirable for many antibacterial applications. However, this approach introduces additional challenges in the synthesis process, requiring meticulous precision and control.

Conclusion and future direction

Bacterial infections and diseases have emerged as significant challenges worldwide, impacting various sectors, such as the environment, food, medicine, and water. Among the available treatments, MOFs and MOF-based composites have gained attention for their potential in antibacterial activities. MOFs and MOF-based composites offer numerous advantages, including high surface area, design flexibility, tunable physical and chemical properties, high loading capacity, porous structures, biocompatibility, and biodegradability, making them ideal candidates for antibacterial agents.

With incorporating NPs or polymer materials, such CS, PVA, PAA, etc., into MOFs can significantly boost their antibacterial effectiveness, this MOF composites can physically damage bacterial cell membranes, causing leakage of intracellular contents and cell death. In addition, with the stabilizing of materials can enhance the structural integrity and durability of MOF composites, extending their antimicrobial activity in various environments.

To further enhance the antibacterial capabilities of MOF composites, additional modifications and optimizations are necessary to overcome biological barriers for treating various diseases. Before utilizing the antibacterial properties of MOF composites in applications, the potential toxicity of the composites must be carefully evaluated. Given the toxicity of NPs on mammalian systems, it is vital for MOF-based composite antibacterial nanosystems to selectively target and eliminate pathogenic bacteria without harming beneficial bacteria or normal human cells.

Moreover, the development of green chemistry synthesis methods for MOF composites is crucial for large-scale manufacturing and clinical applications. Green synthesis approaches aim to reduce energy consumption, eliminate harsh manufacturing conditions, and mitigate hazards associated with conventional MOF composites synthesis. This shift toward sustainable and environmentally friendly practices will be important in advancing the utilization of MOFs and MOF-based composites in antibacterial applications.

Since MOF composites are diverse in view of the variety of structural elements, it appears a cumbersome process to develop novel composites with precise microstructure based on previous datasets. In this regard, AI, and particularly ML approaches, can be considered in exploring and generalizing the current knowledge about linkers, surface, and bulk modifiers, thereby defining non-experienced scenarios for controlling the porosity and functionality of MOF composite structures towards a target structure. Nevertheless, antibacterial activity of such structures could not necessarily be predicted in view of group contribution theories or antibacterial characteristics of MOF partners in a MOF composite structure. Providing such a database and AI-based tools would bring additional difficulties, but precise characterization and optimization of composite structures should be considered as a main part of future developments.

References

Abbasian, M., & Khayyatalimohammadi, M. (2023). Ultrasound-assisted synthesis of MIL-88(Fe) conjugated starch-Fe_3O_4 nanocomposite: A safe antibacterial carrier for controlled release of tetracycline. *International Journal of Biological Macromolecules, 234*, 123665. https://doi.org/10.1016/j.ijbiomac.2023.123665.

Abdelmoaty, A. S., El-Beih, A. A., & Hanna, A. A. (2022). Synthesis, characterization and antimicrobial activity of copper-metal organic framework (Cu-MOF) and its modification by melamine. *Journal of Inorganic and Organometallic Polymers and Materials, 32*(5), 1778–1785. https://doi.org/10.1007/s10904-021-02187-8.

Ahmad, N., Samavati, A., Nordin, N. A. H. M., Jaafar, J., Ismail, A. F., & Malek, N. A. N. N. (2020). Enhanced performance and antibacterial properties of amine-functionalized ZIF-8-decorated GO for ultrafiltration membrane. *Separation and Purification Technology, 239*. https://doi.org/10.1016/j.seppur.2020.116554.

Alonso, A., Muñoz-Berbel, X., Vigués, N., MacAnás, J., Muñoz, M., Mas, J., & Muraviev, D. N. (2012). Characterization of fibrous polymer silver/cobalt nanocomposite with enhanced bactericide activity. *Langmuir: The ACS Journal of Surfaces and Colloids, 28*(1), 783–790. https://doi.org/10.1021/la203239d.

Ansari-Asl, Z., Shahvali, Z., Sacourbaravi, R., Hoveizi, E., & Darabpour, E. (2022). Cu(II) metal-organic framework@polydimethylsiloxane nanocomposite sponges coated by chitosan for antibacterial and tissue engineering applications. *Microporous and Mesoporous Materials, 336*, 111866. https://doi.org/10.1016/j.micromeso.2022.111866.

Azizabadi, O., Akbarzadeh, F., Danshina, S., Chauhan, N. P. S., & Sargazi, G. (2021). An efficient ultrasonic assisted reverse micelle synthesis route for Fe_3O_4@Cu-MOF/core-shell nanostructures and its antibacterial activities. *Journal of Solid State Chemistry, 294*, 121897. https://doi.org/10.1016/j.jssc.2020.121897.

Bieniek, A., Terzyk, A. P., Wiśniewski, M., Roszek, K., Kowalczyk, P., Sarkisov, L., Keskin, S., & Kaneko, K. (2021). MOF materials as therapeutic agents, drug carriers, imaging agents and biosensors in cancer biomedicine: Recent advances and perspectives. *Progress in Materials Science, 117*, 100743. https://doi.org/10.1016/j.pmatsci.2020.100743.

Cai, J., & Liu, R. (2020). Introduction to antibacterial biomaterials. *Biomaterials Science, 8*(24), 6812–6813. https://doi.org/10.1039/d0bm90100h.

Cai, W., Chu, C. C., Liu, G., & Wáng, Y. X. J. (2015). Metal-organic framework-based nanomedicine platforms for drug delivery and molecular imaging. *Small, 11*(37), 4806–4822. https://doi.org/10.1002/smll.201500802.

Chakraborty, D., Musib, D., Saha, R., Das, A., Raza, M. K., Ramu, V., Chongdar, S., Sarkar, K., & Bhaumik, A. (2022). Highly stable tetradentate phosphonate-based green fluorescent Cu-MOF for anticancer therapy and antibacterial activity. *Materials Today Chemistry, 24*, 100882. https://doi.org/10.1016/j.mtchem.2022.100882.

Chen, N., Wang, C., Kong, F., & Wang, S. (2024). In situ facile synthesis and antibacterial activity of Ag-MOFs/cellulose filter paper composites for fruit fresh-keeping. *International Journal of Biological Macromolecules, 256*, 128424. https://doi.org/10.1016/j.ijbiomac.2023.128424.

Chen, T. T., Yi, J. T., Zhao, Y. Y., & Chu, X. (2018). Biomineralized metal-organic framework nanoparticles enable intracellular delivery and endo-lysosomal release of native active proteins. *Journal of the American Chemical Society, 140*(31), 9912–9920. https://doi.org/10.1021/jacs.8b04457.

Chen, X., Zhuang, Y., Rampal, N., Hewitt, R., Divitini, G., O'Keefe, C. A., Liu, X., Whitaker, D. J., Wills, J. W., Jugdaohsingh, R., Powell, J. J., Yu, H., Grey, C. P., Scherman, O. A., & Fairen-Jimenez, D. (2021). Formulation of metal–organic framework-based drug carriers by controlled coordination of methoxy PEG phosphate: Boosting colloidal stability and redispersibility. *Journal of the American Chemical Society, 143*(34), 13557–13572. https://doi.org/10.1021/jacs.1c03943.

Chernousova, S., & Epple, M. (2013). Silver as antibacterial agent: Ion, nanoparticle, and metal. *Angewandte Chemie International Edition, 52*(6), 1636–1653. https://doi.org/10.1002/anie.201205923.

Dastneshan, A., Rahiminezhad, S., Naderi Mezajin, M., Nouri Jevinani, H., Akbarzadeh, I., Abdihaji, M., Qahremani, R., Jahanbakhshi, M., Asghari Lalami, Z., Heydari, H., Noorbazargan, H., & Mostafavi, E. (2023). Cefazolin encapsulated UIO-66-NH_2 nanoparticles enhance the antibacterial activity and biofilm inhibition against drug-resistant *S. aureus*: In vitro and in vivo studies. *Chemical Engineering Journal, 455*, 140544. https://doi.org/10.1016/j.cej.2022.140544.

Deria, P., Mondloch, J. E., Karagiaridi, O., Bury, W., Hupp, J. T., & Farha, O. K. (2014). Beyond post-synthesis modification: evolution of metal–organic frameworks via building block replacement. *Chemical Society Reviews, 43*(16), 5896–5912. https://doi.org/10.1039/c4cs00067f.

Esfahanian, M., Ghasemzadeh, M. A., & Razavian, S. M. H. (2019). Synthesis, identification and application of the novel metal-organic framework Fe$_3$O$_4$@PAA@ZIF-8 for the drug delivery of ciprofloxacin and investigation of antibacterial activity. *Artificial Cells, Nanomedicine and Biotechnology, 47*(1), 2024–2030. https://doi.org/10.1080/21691401.2019.1617729.

Filippousi, M., Turner, S., Leus, K., Siafaka, P. I., Tseligka, E. D., Vandichel, M., Nanaki, S. G., Vizirianakis, I. S., Bikiaris, D. N., Van Der Voort, P., & Van Tendeloo, G. (2016). Biocompatible Zr-based nanoscale MOFs coated with modified poly (ε-caprolactone) as anticancer drug carriers. *International Journal of Pharmaceutics, 509*(1-2), 208–218. https://doi.org/10.1016/j.ijpharm.2016.05.048.

Firouzjaei, M. D., Shamsabadi, A. A., Sharifian, M., Gh, Rahimpour, A., & Soroush, M. (2018). A novel nanocomposite with superior antibacterial activity: A silver-based metal organic framework embellished with graphene oxide. *Advanced Materials Interfaces, 5*(11). https://doi.org/10.1002/admi.201701365, http://onlinelibrary.wiley.com/journal/10.1002/(ISSN)2196-7350.

Gao, F., Zhou, H., Shen, Z., Zhu, G., Hao, L., Chen, H., Xu, H., & Zhou, X. (2020). Long-lasting anti-bacterial activity and bacteriostatic mechanism of tea tree oil adsorbed on the amino-functionalized mesoporous silica-coated by PAA. *Colloids and Surfaces B: Biointerfaces, 188*, 110784. https://doi.org/10.1016/j.colsurfb.2020.110784.

Ghaffar, I., Imran, M., Perveen, S., Kanwal, T., Saifullah, S., Bertino, M. F., Ehrhardt, C. J., Yadavalli, V. K., & Shah, M. R. (2019). Synthesis of chitosan coated metal organic frameworks (MOFs) for increasing vancomycin bactericidal potentials against resistant *S. aureus* strain. *Materials Science and Engineering C, 105*. https://doi.org/10.1016/j.msec.2019.110111, https://www.journals.elsevier.com/materials-science-and-engineering-c.

Guo, C., Cheng, F., Liang, G., Zhang, S., Duan, S., Fu, Y., Marchetti, F., Zhang, Z., & Du, M. (2022). Multimodal antibacterial platform constructed by the schottky junction of curcumin-based bio metal–organic frameworks and Ti$_3$C$_2$T$_x$ MXene nanosheets for efficient wound healing. *Advanced NanoBiomed Research, 2*(10). https://doi.org/10.1002/anbr.202200064.

Guo, C., Cheng, F., Liang, G., Zhang, S., Jia, Q., He, L., Duan, S., Fu, Y., Zhang, Z., & Du, M. (2022). Copper-based polymer-metal–organic framework embedded with Ag nanoparticles: Long-acting and intelligent antibacterial activity and accelerated wound healing. *Chemical Engineering Journal, 435*, 134915. https://doi.org/10.1016/j.cej.2022.134915.

Gwon, K., Kim, Y., Cho, H., Lee, S., Yang, S. H., Kim, S. J., & Lee, D. N. (2021). Robust copper metal–organic framework-embedded polysiloxanes for biomedical applications: Its antibacterial effects on MRSA and in vitro cytotoxicity. *Nanomaterials, 11*(3), 1–13. https://doi.org/10.3390/nano11030719, https://www.mdpi.com/2079-4991/11/3/719/pdf.

Haddad, S., Abánades Lázaro, I., Fantham, M., Mishra, A., Silvestre-Albero, J., Osterrieth, J. W. M., Kaminski Schierle, G. S., Kaminski, C. F., Forgan, R. S., & Fairen-Jimenez, D. (2020). Design of a functionalized metal–organic framework system for enhanced targeted delivery to mitochondria. *Journal of the American Chemical Society, 142*(14), 6661–6674. https://doi.org/10.1021/jacs.0c00188.

Haseena, Shah, M., Rehman, K., Khan, A., Farid, A., Marini, C., Di Cerbo, A., & Shah, M. R. (2022). Characterization and antibacterial evaluation of biodegradable mannose-conjugated Fe-MIL-

88NH$_2$ composites containing vancomycin against methicillin-resistant *Staphylococcus aureus* strains. *Polymers, 14*(13), 2712. https://doi.org/10.3390/polym14132712.

Hatamie, S., Ahadian, M. M., Soufi Zomorod, M., Torabi, S., Babaie, A., Hosseinzadeh, S., Soleimani, M., Hatami, N., & Wei, Z. H. (2019). Antibacterial properties of nanoporous graphene oxide/cobalt metal organic framework. *Materials Science and Engineering C, 104*. https://doi.org/10.1016/j.msec.2019.109862.

He, C., Lu, K., Liu, D., & Lin, W. (2014). Nanoscale Metal–Organic Frameworks for the Co-delivery of cisplatin and pooled siRNAs to enhance therapeutic efficacy in drug-resistant ovarian cancer cells. *Journal of the American Chemical Society, 136*(14), 5181–5184. https://doi.org/10.1021/ja4098862.

He, Z., Yang, H., Gu, Y., Xie, Y., Wu, J., Wu, C., Song, J., Zhao, M., Zong, D., Du, W., Qiao, J., Pang, Y., & Liu, Y. (2023). Green synthesis of MOF-mediated pH-sensitive nanomaterial AgNPs@ZIF-8 and its application in improving the antibacterial performance of AgNPs. *International Journal of Nanomedicine, 18*, 4857–4870. https://doi.org/10.2147/IJN.S418308, https://www.dovepress.com/getfile.php?fileID=92297.

Hidalgo, T., Giménez-Marqués, M., Bellido, E., Avila, J., Asensio, M. C., Salles, F., Lozano, M. V., Guillevic, M., Simón-Vázquez, R., González-Fernández, A., Serre, C., Alonso, M. J., & Horcajada, P. (2017). Chitosan-coated mesoporous MIL-100(Fe) nanoparticles as improved bio-compatible oral nanocarriers. *Scientific Reports, 7*. https://doi.org/10.1038/srep43099, www.nature.com/srep/index.html.

Horcajada, P., Chalati, T., Serre, C., Gillet, B., Sebrie, C., Baati, T., Eubank, J. F., Heurtaux, D., Clayette, P., Kreuz, C., Chang, J.-S., Hwang, Y. K., Marsaud, V., Bories, P.-N., Cynober, L., Gil, S., Férey, G., Couvreur, P., & Gref, R. (2010). Porous metal–organic-framework nanoscale carriers as a potential platform for drug delivery and imaging. *Nature Materials, 9*(2), 172–178. https://doi.org/10.1038/nmat2608.

Hu, W. C., Younis, M. R., Zhou, Y., Wang, C., & Xia, X. H. (2020). In situ fabrication of ultrasmall gold nanoparticles/2D MOFs hybrid as nanozyme for antibacterial therapy. *Small, 16*(23). https://doi.org/10.1002/smll.202000553, http://onlinelibrary.wiley.com/journal/10.1002/(ISSN)1613-6829.

Hu, Z., & Zhao, D. (2015). De facto methodologies toward the synthesis and scale-up production of UiO-66-type metal–organic frameworks and membrane materials. *Dalton Transactions, 44*(44), 19018–19040. https://doi.org/10.1039/c5dt03359d.

Hu, Z., Liu, X., Jiao, L., Wei, X., Wang, Z., Huang, N., & Li, J. (2021). Ag-doped Fe-metal–organic framework nanozymes for efficient antibacterial application. *New Journal of Chemistry, 45*(38), 17772–17776. https://doi.org/10.1039/d1nj02088a.

Huang, X., Yu, S., Lin, W., Yao, X., Zhang, M., He, Q., Fu, F., Zhu, H., & Chen, J. (2021). A metal-organic framework MIL-53(Fe) containing sliver ions with antibacterial property. *Journal of Solid State Chemistry, 302*, 122442. https://doi.org/10.1016/j.jssc.2021.122442.

Hurdle, J. G., O'neill, A. J., Chopra, I., & Lee, R. E. (2011). Targeting bacterial membrane function: an underexploited mechanism for treating persistent infections. *Nature Reviews Microbiology, 9*(1), 62–75. https://doi.org/10.1038/nrmicro2474.

Jasim Al-Khafaji, H. H., Alsalamy, A., Abed Jawad, M., Ali Nasser, H., Dawood, A. H., Hasan, S. Y., Ahmad, I., Gatea, M. A., & Younis Albahadly, W. K. (2023). Synthesis of a novel Cu/DPA-MOF/OP/CS hydrogel with high capability in antimicrobial studies. *Frontiers in Chemistry, 11*. https://doi.org/10.3389/fchem.2023.1236580, http://journal.frontiersin.org/journal/chemistry.

Javanbakht, S., Nabi, M., Shadi, M., Amini, M. M., & Shaabani, A. (2021). Carboxymethylcellulose/tetracycline@UiO-66 nanocomposite hydrogel films as a potential antibacterial wound dressing. *International Journal of Biological Macromolecules, 188*, 811–819. https://doi.org/10.1016/j.ijbiomac.2021.08.061.

Ji, H., Sun, H., & Qu, X. (2016). Antibacterial applications of graphene-based nanomaterials: Recent achievements and challenges. *Advanced Drug Delivery Reviews, 105*, 176–189. https://doi.org/10.1016/j.addr.2016.04.009.

Jo, J. H., Kim, H. C., Huh, S., Kim, Y., & Lee, D. N. (2019). Antibacterial activities of Cu-MOFs containing glutarates and bipyridyl ligands. *Dalton Transactions, 48*(23), 8084–8093. https://doi.org/10.1039/c9dt00791a.

Kalati, M., & Akhbari, K. (2021). Optimizing the metal ion release and antibacterial activity of ZnO@ZIF-8 by modulating its synthesis method. *New Journal of Chemistry, 45*(48), 22924–22931. https://doi.org/10.1039/d1nj04534b.

Karakeçili, A., Topuz, B., Ersoy, F.Ş., Şahin, T., Günyakti, A., & Demirtaş, T. T. (2022). UiO-66 metal-organic framework as a double actor in chitosan scaffolds: Antibiotic carrier and osteogenesis promoter. *Biomaterials Advances, 136*. https://doi.org/10.1016/j.bioadv.2022.212757.

Krystyjan, M., Khachatryan, G., Grabacka, M., Krzan, M., Witczak, M., Grzyb, J., & Woszczak, L. (2021). Physicochemical, bacteriostatic, and biological properties of starch/chitosan polymer composites modified by graphene oxide, designed as new bionanomaterials. *Polymers, 13*(14), 2327. https://doi.org/10.3390/polym13142327.

Kumar, R., Anandan, S., Hembram, K., & Narasinga Rao, T. (2014). Efficient ZnO-based visible-light-driven photocatalyst for antibacterial applications. *ACS Applied Materials & Interfaces, 6*(15), 13138–13148. https://doi.org/10.1021/am502915v.

Lai, W. F., Deng, R., He, & Wong, W. T. (2021). Bioinspired, sustained-release material in response to internal signals for biphasic chemical sensing in wound therapy. *Advanced Healthcare Materials, 10*(2), 2001267. https://doi.org/10.1002/adhm.202001267.

Lee, Y. R., Kim, J., & Ahn, W. S. (2013). Synthesis of metal-organic frameworks: A mini review. *Korean Journal of Chemical Engineering, 30*(9), 1667–1680. https://doi.org/10.1007/s11814-013-0140-6.

Li, S., & Huo, F. (2015). Metal–organic framework composites: From fundamentals to applications. *Nanoscale, 7*(17), 7482–7501. https://doi.org/10.1039/C5NR00518C.

Li, S., Shi, X., Wang, H., & Xiao, L. (2021). A multifunctional dual-shell magnetic nanocomposite with near-infrared light response for synergistic chemo-thermal tumor therapy. *Journal of Biomedical Materials Research, Part B: Applied Biomaterials, 109*(6), 841–852. https://doi.org/10.1002/jbm.b.34749.

Li, Y., Gao, Z., Zhang, Y., Chen, F., An, P., Wu, H., You, C., & Sun, B. (2021). MOF-shielded and glucose-responsive ultrasmall silver nano-factory for highly-efficient anticancer and antibacterial therapy. *Chemical Engineering Journal, 416*, 127610. https://doi.org/10.1016/j.cej.2020.127610.

Lian, X., Fang, Y., Joseph, E., Wang, Q., Li, J., Banerjee, S., Lollar, C., Wang, X., & Zhou, H.-C. (2017). Enzyme–MOF (metal–organic framework) composites. *Chemical Society Reviews, 46*(11), 3386–3401. https://doi.org/10.1039/C7CS00058H.

Liu, B., Zhu, X, Feng, C., Huang, J., Yan, D, & Wang (2023). Bacteriostatic effect of Ag@TiO$_2$-poly(p-dioxanone)-coated gauzes in vitro and in vivo on otitis media pathogens. *Heliyon, 9*(9), e19375. https://doi.org/10.1016/j.heliyon.2023.e19375.

Liu, J., Chamakura, K., Perez-Ballestero, R., & Bashir, S. (2012). Historical overview of the first two waves of bactericidal agents and development of the third wave of potent disinfectants. *ACS Symposium Series, 1119*, 129–154. https://doi.org/10.1021/bk-2012-1119.ch006.

Liu, J., Liu, T., Du, P., Zhang, L., & Lei, J. (2019). Metal–organic framework (MOF) hybrid as a tandem catalyst for enhanced therapy against hypoxic tumor cells. *Angewandte Chemie International Edition, 131*(23), 7890–7894. https://doi.org/10.1002/ange.201903475.

Liu, X., Hu, M., Wang, M., Song, Y., Zhou, N., He, L., & Zhang, Z. (2019). Novel nanoarchitecture of Co-MOF-on-TPN-COF hybrid: Ultralowly sensitive bioplatform of electrochemical aptasensor toward ampicillin. *Biosensors and Bioelectronics, 123*, 59–68. https://doi.org/10.1016/j.bios.2018.09.089.

Liu, Y., Liu, Z., Huang, D., Cheng, M., Zeng, G., Lai, C., Zhang, C., Zhou, C., Wang, W., Jiang, D., Wang, H., & Shao, B. (2019). Metal or metal-containing nanoparticle@MOF nanocomposites as a promising type of photocatalyst. *Coordination Chemistry Reviews, 388*, 63–78. https://doi.org/10.1016/j.ccr.2019.02.031.

Liu, Y., Xu, X., Xia, Q., Yuan, G., He, Q., & Cui, Y. (2010). Multiple topological isomerism of three-connected networks in silver-based metal-organoboron frameworks. *Chemical Communications, 46*(15), 2608–2610. https://doi.org/10.1039/B923365B.

Luo, Y., Hossain, M., Wang, C., Qiao, Y., An, J., Ma, L., & Su, M. (2013). Targeted nanoparticles for enhanced X-ray radiation killing of multidrug-resistant bacteria. *Nanoscale, 5*(2), 687–694. https://doi.org/10.1039/c2nr33154c.

Lv, H., Zhang, Y., Chen, P., Xue, J., Jia, X., & Chen, J. (2020). Enhanced synergistic antibacterial activity through a smart platform based on UiO-66 combined with photodynamic therapy and chemotherapy. *Langmuir: The ACS Journal of Surfaces and Colloids, 36*(15), 4025–4032. https://doi.org/10.1021/acs.langmuir.0c00292.

Mao, Z., Hao, W., Wang, W., Ma, F., Ma, C., & Chen, S. (2023). BiOI@CeO$_2$@Ti$_3$C$_2$ MXene composite S-scheme photocatalyst with excellent bacteriostatic properties. *Journal of Colloid and Interface Science, 633*, 836–850. https://doi.org/10.1016/j.jcis.2022.11.140.

Meng, J., Chen, X., Tian, Y., Li, Z., & Zheng, Q. (2017). Nanoscale metal–organic frameworks decorated with graphene oxide for magnetic resonance imaging guided photothermal therapy. *Chemistry – A European Journal, 23*(69), 17521–17530. https://doi.org/10.1002/chem.201702573.

Miller, K. P., Wang, L., Benicewicz, B. C., & Decho, A. W. (2015). Inorganic nanoparticles engineered to attack bacteria. *Chemical Society Reviews, 44*(21), 7787–7807. https://doi.org/10.1039/C5CS00041F.

Mori, W., & Takamizawa, S. (2000). Microporous materials of metal carboxylates. *Journal of Solid State Chemistry, 152*(1), 120–129. https://doi.org/10.1006/jssc.2000.8675.

Moritz, M., & Geszke-Moritz, M. (2013). The newest achievements in synthesis, immobilization and practical applications of antibacterial nanoparticles. *Chemical Engineering Journal, 228*, 596–613. https://doi.org/10.1016/j.cej.2013.05.046.

Morris, W., Briley, W. E., Auyeung, E., Cabezas, M. D., & Mirkin, C. A. (2014). Nucleic acid-metal organic framework (MOF) nanoparticle conjugates. *Journal of the American Chemical Society, 136*(20), 7261–7264. https://doi.org/10.1021/ja503215w.

Nakhaei, M., Akhbari, K., Kalati, M., & Phuruangrat, A. (2021). Antibacterial activity of three zinc-terephthalate MOFs and its relation to their structural features. *Inorganica Chimica Acta, 522*, 120353. https://doi.org/10.1016/j.ica.2021.120353.

Nasrabadi, M., Ghasemzadeh, M. A., & Monfared, M. R. Z. (2019). The preparation and characterization of UiO-66 metal-organic frameworks for the delivery of the drug ciprofloxacin and an evaluation of their antibacterial activities. *New Journal of Chemistry, 43*(40), 16033–16040. https://doi.org/10.1039/C9NJ03216A.

Pachaiappan, R., Rajendran, S., Show, P. L., Manavalan, K., & Naushad, M. (2021). Metal/metal oxide nanocomposites for bactericidal effect: A review. *Chemosphere, 272*, 128607. https://doi.org/10.1016/j.chemosphere.2020.128607.

Pang, L., Dai, C., Bi, L., Guo, Z., & Fan, J. (2017). Biosafety and antibacterial ability of graphene and graphene oxide in vitro and in vivo. *Nanoscale Research Letters, 12*(1). https://doi.org/10.1186/s11671-017-2317-0.

Park, J., Jiang, Q., Feng, D., Mao, L., & Zhou, H. C. (2016). Size-controlled synthesis of porphyrinic metal-organic framework and functionalization for targeted photodynamic therapy. *Journal of the American Chemical Society, 138*(10), 3518–3525. https://doi.org/10.1021/jacs.6b00087.

Parmar, A., Kaur, G., Kapil, S., Sharma, V., Sachar, S., Sandhir, R., & Sharma, S. (2019). Green chemistry mediated synthesis of PLGA-Silver nanocomposites for antibacterial synergy: Introspection of formulation parameters on structural and bactericidal aspects. *Reactive and Functional Polymers, 141*, 68–81. https://doi.org/10.1016/j.reactfunctpolym.2019.04.018.

Pejman, M., Dadashi Firouzjaei, M., Aghapour Aktij, S., Das, P., Zolghadr, E., Jafarian, H., Arabi Shamsabadi, A., Elliott, M., Sadrzadeh, M., Sangermano, M., Rahimpour, A., & Tiraferri, A. (2020). In situ Ag-MOF growth on pre-grafted zwitterions imparts outstanding antifouling properties to forward osmosis membranes. *ACS Applied Materials & Interfaces, 12*(32), 36287–36300. https://doi.org/10.1021/acsami.0c12141.

Pelgrift, R. Y., & Friedman, A. J. (2013). Nanotechnology as a therapeutic tool to combat microbial resistance. *Advanced Drug Delivery Reviews, 65*(13-14), 1803–1815. https://doi.org/10.1016/j.addr.2013.07.011.

Pooresmaeil, M., Asl, E. A., & Namazi, H. (2021). A new pH-sensitive CS/Zn-MOF@GO ternary hybrid compound as a biofriendly and implantable platform for prolonged 5-Fluorouracil delivery to human breast cancer cells. *Journal of Alloys and Compounds, 885*, 160992. https://doi.org/10.1016/j.jallcom.2021.160992.

Raghunath, A., & Perumal, E. (2017). Metal oxide nanoparticles as antimicrobial agents: A promise for the future. *International Journal of Antimicrobial Agents, 49*(2), 137–152. https://doi.org/10.1016/j.ijantimicag.2016.11.011.

Rahimpour, A., Seyedpour, S. F., Aghapour Aktij, S., Dadashi Firouzjaei, M., Zirehpour, A., Arabi Shamsabadi, A., Khoshhal Salestan, S., Jabbari, M., & Soroush, M. (2018). Simultaneous improvement of antimicrobial, antifouling, and transport properties of forward osmosis membranes with immobilized highly-compatible polyrhodanine nanoparticles. *Environmental Science and Technology, 52*(9), 5246–5258. https://doi.org/10.1021/acs.est.8b00804.

Rahmati, Z., Abdi, J., Vossoughi, M., & Alemzadeh, I. (2020). Ag-doped magnetic metal organic framework as a novel nanostructured material for highly efficient antibacterial activity. *Environmental Research, 188*, 109555. https://doi.org/10.1016/j.envres.2020.109555.

Ravinayagam, V., & Rehman, S. (2020). Zeolitic imidazolate framework-8 (ZIF-8) doped TiZSM-5 and mesoporous carbon for antibacterial characterization. *Saudi Journal of Biological Sciences, 27*(7), 1726–1736. https://doi.org/10.1016/j.sjbs.2020.05.016.

Rezaee, R., Montazer, M., Mianehro, A., & Mahmoudirad, M. (2022). Single-step synthesis and characterization of Zr-MOF onto wool fabric: Preparation of antibacterial wound dressing with high absorption capacity. *Fibers and Polymers, 23*(2), 404–412. https://doi.org/10.1007/s12221-021-0211-y.

Sacourbaravi, R., Ansari-Asl, Z., Kooti, M., Nobakht, V., & Darabpour, E. (2020). Fabrication of Ag NPs/Zn-MOF nanocomposites and their application as antibacterial agents. *Journal of Inorganic and Organometallic Polymers and Materials, 30*(11), 4615–4621. https://doi.org/10.1007/s10904-020-01601-x.

Sadiq, S., Khan, I., Humayun, M., Wu, P., Khan, A., Khan, S., Khan, A., Khan, S., Alanazi, A. F., & Bououdina, M. (2023). Synthesis of metal–organic framework-based ZIF-8@ZIF-67 nanocomposites for antibiotic decomposition and antibacterial activities. *ACS Omega, 8*(51), 49244–49258. https://doi.org/10.1021/acsomega.3c07606.

Safaei, M., Foroughi, M. M., Ebrahimpoor, N., Jahani, S., Omidi, A., & Khatami, M. (2019). A review on metal-organic frameworks: Synthesis and applications. *TrAC Trends in Analytical Chemistry, 118*, 401–425. https://doi.org/10.1016/j.trac.2019.06.007.

Seidi, F., Arabi Shamsabadi, A., Dadashi Firouzjaei, M., Elliott, M., Saeb, M. R., Huang, Y., Li, C., Xiao, H., & Anasori, B. (2023). MXenes antibacterial properties and applications: A review and perspective. *Small, 19*(14). https://doi.org/10.1002/smll.202206716, http://onlinelibrary.wiley.com/journal/10.1002/(ISSN)1613-6829.

Seoane, B., Dikhtiarenko, A., Mayoral, A., Tellez, C., Coronas, J., Kapteijn, F., & Gascon, J. (2015). Metal organic framework synthesis in the presence of surfactants: Towards hierarchical MOFs? *CrystEngComm, 17*(7), 1693–1700. https://doi.org/10.1039/c4ce02324b.

Seyedpour, S. F., Arabi Shamsabadi, A., Khoshhal Salestan, S., Dadashi Firouzjaei, M., Sharifian Gh, M., Rahimpour, A., Akbari Afkhami, F., Shirzad Kebria, M. R., Elliott, M. A., Tiraferri, A., Sangermano, M., Esfahani, M. R., & Soroush, M. (2020). Tailoring the biocidal activity of novel silver-based metal azolate frameworks. *ACS Sustainable Chemistry & Engineering, 8*(20), 7588–7599. https://doi.org/10.1021/acssuschemeng.0c00201.

Shams, S., Ahmad, W., Memon, A. H., Shams, S., Wei, Y., Yuan, Q., & Liang, H. (2020). Cu/H$_3$BTC MOF as a potential antibacterial therapeutic agent against: *Staphylococcus aureus* and *Escherichia coli*. *New Journal of Chemistry, 44*(41), 17671–17678. https://doi.org/10.1039/D0NJ04120C.

Shebi, A., & Lisa, S. (2019). Evaluation of biocompatibility and bactericidal activity of hierarchically porous PLA-TiO$_2$ nanocomposite films fabricated by breath-figure method. *Materials Chemistry and Physics, 230*, 308–318. https://doi.org/10.1016/j.matchemphys.2019.03.045.

Sheta, M., Sheta, S. M., El-Sheikh, M. M., & Abd-Elzaher (2018). Simple synthesis of novel copper metal–organic framework nanoparticles: Biosensing and biological applications. *Dalton Transactions, 47*(14), 4847–4855. https://doi.org/10.1039/C8DT00371H.

Silva, J. Y. R., Proenza, Y. G., da Luz, L. L., de Sousa Araújo, S., Filho, M. A. G., Junior, S. A., Soares, T. A., & Longo, R. L. (2019). A thermo-responsive adsorbent-heater-thermometer nanomaterial for controlled drug release: (ZIF-8, EuxTby)@AuNP core-shell. *Materials Science and Engineering C, 102*, 578–588. https://doi.org/10.1016/j.msec.2019.04.078.

Sun, L., Du, T., Hu, C., Chen, J., Lu, J., Lu, Z., & Han, H. (2017). Antibacterial activity of graphene oxide/g-C$_3$N$_4$ composite through photocatalytic disinfection under visible light. *ACS Sustainable Chemistry & Engineering, 5*(10), 8693–8701. https://doi.org/10.1021/acssuschemeng.7b01431.

Taheri, M., Ashok, D., Sen, T., Enge, T. G., Verma, N. K., Tricoli, A., Lowe, A., Nisbet, D. R., & Tsuzuki, T. (2021). Stability of ZIF-8 nanopowders in bacterial culture media and its implication for antibacterial properties. *Chemical Engineering Journal, 413*, 127511. https://doi.org/10.1016/j.cej.2020.127511.

Tan, Z. K., Gong, J. L., Fang, S. Y., Li, J., Cao, W. C., & Chen, Z. P. (2022). Outstanding antibacterial thin-film composite membrane prepared by incorporating silver-based metal–organic framework (Ag-MOF) for water treatment. *Applied Surface Science, 590*. https://doi.org/10.1016/j.apsusc.2022.153059.

Teixeira-Santos, R., Gomes, M., Gomes, L. C., & Mergulhão, F. J. (2021). Antimicrobial and anti-adhesive properties of carbon nanotube-based surfaces for medical applications: A systematic review. *iScience, 24*(1). https://doi.org/10.1016/j.isci.2020.102001.

Velásquez-Hernández, M. D. J., Linares-Moreau, M., Astria, E., Carraro, F., Alyami, M. Z., Khashab, N. M., Sumby, C. J., Doonan, C. J., & Falcaro, P. (2021). Towards applications of bioentities@MOFs in biomedicine. *Coordination Chemistry Reviews, 429*, 213651. https://doi.org/10.1016/j.ccr.2020.213651.

Wang, M., Zhou, X., Li, Y., Dong, Y., Meng, J., Zhang, S., Xia, L., He, Z., Ren, L., Chen, Z., & Zhang, X. (2022). Triple-synergistic MOF-nanozyme for efficient antibacterial treatment. *Bioactive Materials, 17*, 289–299. https://doi.org/10.1016/j.bioactmat.2022.01.036.

Wang, S., Mcguirk, C. M., Aquino, A., Mason, J. A., & Mirkin, C. A. (2018). Metal-organic framework nanoparticles. *Advanced Materials, 30*(37), 1–14. https://doi.org/10.1002/adma.201800202.

Wang, W., Feng, H., Liu, J., Zhang, M., Liu, S., Feng, C., & Chen, S. (2020). A photo catalyst of cuprous oxide anchored MXene nanosheet for dramatic enhancement of synergistic antibacterial ability. *Chemical Engineering Journal, 386*, 124116. https://doi.org/10.1016/j.cej.2020.124116.

Wang, X., Xu, J., Yang, D., Sun, C., Sun, Q., He, F., Gai, S., Zhong, C., Li, C., & Yang, P. (2018). Fe_3O_4@MIL-100(Fe)-UCNPs heterojunction photosensitizer: Rational design and application in near infrared light mediated hypoxic tumor therapy. *Chemical Engineering Journal, 354*, 1141–1152. https://doi.org/10.1016/j.cej.2018.08.070.

Wang, X., Zhou, X., Yang, K., Li, Q., Wan, R., Hu, G., Ye, J., Zhang, Y., He, J., Gu, H., Yang, Y., & Zhu, L. (2021). Peroxidase- and UV-triggered oxidase mimetic activities of the UiO-66-NH_2/chitosan composite membrane for antibacterial properties. *Biomaterials Science, 9*(7), 2647–2657. https://doi.org/10.1039/d0bm01960g.

Wang, Y., Qi, W., Mao, Z., Wang, J., Zhao, R. C., & Chen, H. (2023). rPDAs doped antibacterial MOF-hydrogel: Bio-inspired synergistic whole-process wound healing. *Materials Today Nano, 23*, 100363. https://doi.org/10.1016/j.mtnano.2023.100363.

Wang, Y., Sun, W., Li, Y., Zhuang, X., Tian, C., Luan, F., & Fu, X. (2021). Imidazole metal-organic frameworks embedded in layered $Ti_3C_2T_x$ Mxene as a high-performance electrochemiluminescence biosensor for sensitive detection of HIV-1 protein. *Microchemical Journal, 167*, 106332. https://doi.org/10.1016/j.microc.2021.106332.

Wuttke, S., Braig, S., Preiß, T., Zimpel, A., Sicklinger, J., Bellomo, C., Rädler, J. O., Vollmar, A. M., & Bein, T. (2015). MOF nanoparticles coated by lipid bilayers and their uptake by cancer cells. *Chemical Communications, 51*(87), 15752–15755. https://doi.org/10.1039/c5cc06767g.

Wyszogrodzka, G., Marszałek, B., Gil, B., & Dorozyński, P. (2016). Metal-organic frameworks: Mechanisms of antibacterial action and potential applications. *Drug Discovery Today, 21*(6), 1009–1018. https://doi.org/10.1016/j.drudis.2016.04.009, www.elsevier.com/locate/drugdiscov.

Xiong, R., Lu, C., Wang, Y., Zhou, Z., & Zhang, X. (2013). Nanofibrillated cellulose as the support and reductant for the facile synthesis of Fe_3O_4/Ag nanocomposites with catalytic and antibacterial activity. *Journal of Materials Chemistry A, 1*(47), 14910. https://doi.org/10.1039/c3ta13314a.

Xu, W., Zhuang, H., Xu, Z., Huang, M., Gao, S., Li, Q., & Zhang, G. (2020). Design and construction of Ag@MOFs immobilized PVDF ultrafiltration membranes with anti-bacterial and antifouling properties. *Advances in Polymer Technology, 2020*, 1–11. https://doi.org/10.1155/2020/5456707.

Xu, Y., Fang, Y., Ou, Y., Yan, L., Chen, Q., Sun, C., Chen, J., & Liu, H. (2020). Zinc metal–organic framework@chitin composite sponge for rapid hemostasis and antibacterial infection. *ACS Sustainable Chemistry & Engineering, 8*(51), 18915–18925. https://doi.org/10.1021/acssuschemeng.0c06044.

Yan, B., Tan, J., Zhang, H., Liu, L., Chen, L., Qiao, Y., & Liu, X. (2022). Constructing fluorine-doped Zr-MOF films on titanium for antibacteria, anti-inflammation, and osteogenesis. *Biomaterials Advances, 134*, 112699. https://doi.org/10.1016/j.msec.2022.112699.

Yan, F., Cheng, F., Guo, C., Liang, G., Zhang, S., Fang, S., & Zhang, Z. (2023). Curcumin-regulated constructing of defective zinc-based polymer-metal-organic framework as long-acting antibacterial platform for efficient wound healing. *Journal of Colloid and Interface Science, 641*, 59–69. https://doi.org/10.1016/j.jcis.2023.03.050.

Yan, L., Gopal, A., Kashif, S., Hazelton, P., Lan, M., Zhang, W., & Chen, X. (2022). Metal organic frameworks for antibacterial applications. *Chemical Engineering Journal, 435*, 134975. https://doi.org/10.1016/j.cej.2022.134975.

Yang, G., Yin, H., Liu, W., Yang, Y., Zou, Q., Luo, L., Li, H., Huo, Y., & Li, H. (2018). Synergistic Ag/TiO$_2$-N photocatalytic system and its enhanced antibacterial activity towards *Acinetobacter baumannii*. *Applied Catalysis B: Environmental, 224*, 175–182. https://doi.org/10.1016/j.apcatb.2017.10.052.

Yang, X., Chai, H., Guo, L., Jiang, Y., Xu, L., Huang, W., Shen, Y., Yu, L., Liu, Y., & Liu, J. (2021). In situ preparation of porous metal-organic frameworks ZIF-8@Ag on poly-ether-ether-ketone with synergistic antibacterial activity. *Colloids and Surfaces B: Biointerfaces, 205*, 111920. https://doi.org/10.1016/j.colsurfb.2021.111920.

Yin, J., Meng, Q., Cheng, D., Fu, J., Luo, Q., Liu, Y., & Yu, Z. (2020). Mechanisms of bactericidal action and resistance of polymyxins for Gram-positive bacteria. *Applied Microbiology and Biotechnology, 104*(9), 3771–3780. https://doi.org/10.1007/s00253-020-10525-y.

Yuan, H., Hong, X., Ma, H., Fu, C., Guan, Y., Huang, W., Ma, J., Xia, P., Cao, M., Zheng, L., Xu, X., Xu, C., Liu, D., Li, Z., Geng, Q., & Wang, J. (2023). MXene-based dual functional nanocomposite with photothermal nanozyme catalytic activity to fight bacterial infections. *ACS Materials Letters, 5*(3), 762–774. https://doi.org/10.1021/acsmaterialslett.2c00771.

Zhang, F., & Cheng, W. (2022). The mechanism of bacterial resistance and potential bacteriostatic strategies. *Antibiotics*. 11, 1215 https://doi.org/10.3390/antibiotics11091215.

Zhang, J., Ke, X., Huang, M., Pei, X., Gao, S., Wu, D., Chen, J., & Weng, Y. (2022). NO released via both a Cu-MOF-based donor and surface-catalyzed generation enhances anticoagulation and antibacterial surface effects. *Biomaterials Science, 11*(1), 322–338. https://doi.org/10.1039/d2bm01515c.

Zhang, L., Liu, Z., Deng, Q., Sang, Y., Dong, K., Ren, J., & Qu, X. (2021). Nature-Inspired Construction of MOF@COF nanozyme with active sites in tailored microenvironment and pseudopodia-like surface for enhanced bacterial inhibition. *Angewandte Chemie International Edition, 60*(7), 3469–3474. https://doi.org/10.1002/anie.202012487.

Zhang, N., Zhang, X., Zhu, Y., Wang, D., Li, R., Li, S., et al. (2023). Bimetal–organic framework-loaded PVA/chitosan composite hydrogel with interfacial antibacterial and adhesive hemostatic features for wound dressings. *Polymers, 15*(22), 4362. https://doi.org/10.3390/polym15224362.

Zhang, Q., Yan, S., Yan, X., & Lv, Y. (2023). Recent advances in metal-organic frameworks: Synthesis, application and toxicity. *Science of The Total Environment, 902*, 165944. https://doi.org/10.1016/j.scitotenv.2023.165944.

Zhang, X., Peng, F., & Wang, D. (2022). MOFs and MOF-derived materials for antibacterial application. *Journal of Functional Biomaterials, 13*(4), 215. https://doi.org/10.3390/jfb13040215.

Zhang, Y., Lin, Z., He, Q., Deng, Y., Wei, F., Xu, C., Fu, L., & Lin, B. (2022). Enhanced aqueous stability and long-acting antibacterial of silver-based MOFs via chitosan-crosslinked for fruit fresh-keeping. *Applied Surface Science, 571*, 151351. https://doi.org/10.1016/j.apsusc.2021.151351.

Zhang, N., Zhang, X., Zhu, Y., Wang, D., Liu, W., Chen, D., Li, R., & Li, S. (2024). MOF/MXene-loaded PVA/chitosan hydrogel with antimicrobial effect and wound healing promotion under electrical stimulation and improved mechanical properties. *International journal of biological macromolecules, 264*, 130625. https://doi.org/10.1016/j.ijbiomac.2024.130625.

Zhao, H. X., Zou, Q., Sun, S. K., Yu, C., Zhang, X., Li, R. J., & Fu, Y. Y. (2016). Theranostic metal-organic framework core-shell composites for magnetic resonance imaging and drug delivery. *Chemical Science, 7*(8), 5294–5301. https://doi.org/10.1039/c6sc01359g, http://pubs.rsc.org/en/Journals/JournalIssues/SC.

Zhao, X., He, X., Hou, A., Cheng, C., Wang, X., Yue, Y., Wu, Z., Wu, H., Liu, B., Li, H., Shen, J., Tan, C., Zhou, Z., & Ma, L. (2022). Growth of Cu_2O nanoparticles on two-dimensional Zr-ferrocene-metal-organic framework nanosheets for photothermally enhanced chemodynamic antibacterial therapy. *Inorganic Chemistry, 61*(24), 9328–9338. https://doi.org/10.1021/acs.inorgchem.2c01091.

Zheng, X., Zhang, Y., Zou, L., Wang, Y., Zhou, X., & Yao, L. (2020). Robust ZIF-8/alginate fibers for the durable and highly effective antibacterial textiles. *Colloids Surfaces B Biointerfaces, 193*, 111127. https://doi.org/10.1016/j.colsurfb.2020.111127.

Zheng, X., Wang, L., Pei, Q., He, S., Liu, S., & Xie, Z. (2017). Metal-organic framework@porous organic polymer nanocomposite for photodynamic therapy. *Chemistry of Materials, 29*(5), 2374–2381. https://doi.org/10.1021/acs.chemmater.7b00228.

Zhong, Y., Zheng, X. T., Li, Q. L., Loh, X. J., Su, X., & Zhao, S. (2022). Antibody conjugated Au/Ir@Cu/Zn-MoF probe for bacterial lateral flow immunoassay and precise synergistic antibacterial treatment. *SSRN, China SSRN*. https://doi.org/10.1016/j.bios.2022.115033.

Zhou, J., Guo, M., Wu, D., Shen, M., Liu, D., & Ding, T. (2023). Synthesis of UiO-66 loaded-caffeic acid and study of its antibacterial mechanism. *Food Chemistry, 402*, 134248. https://doi.org/10.1016/j.foodchem.2022.134248.

Zhou, X., Zhao, W., Wang, M., Zhang, S., Li, Y., Hu, W., Ren, L., Luo, S., & Chen, Z. (2020). Dual-modal therapeutic role of the lactate oxidase-embedded hierarchical porous zeolitic imidazolate framework as a nanocatalyst for effective tumor suppression. *ACS Applied Materials & Interfaces, 12*(29), 32278–32288. https://doi.org/10.1021/acsami.0c05783.

Zhu, J., Qiu, W., Yao, C., Wang, C., Wu, D., Pradeep, S., Yu, J., & Dai, Z. (2021). Water-stable zirconium-based metal-organic frameworks armed polyvinyl alcohol nanofibrous membrane with enhanced antibacterial therapy for wound healing. *Journal of Colloid and Interface Science, 603*, 243–251. https://doi.org/10.1016/j.jcis.2021.06.084.

Zhu, Q.-L., & Xu, Q. (2014). Metal–organic framework composites. *Chemical Society Reviews, 43*(16), 5468–5512. https://doi.org/10.1039/c3cs60472a.

Zhu, Y., Mao, K., Rong, J., Zheng, Y., Zhang, T., Yang, D., & Qiu, F. (2022). In situ deposition of Ag/Zn-MOFs on the surface of non-woven fabrics for effective antibacterial activity. *Journal of Industrial Textiles, 52*. https://doi.org/10.1177/15280837221109643. 152808372211096.

Zhuang, W., Yuan, D., Li, J.-R., Luo, Z., Zhou, H.-C., Bashir, S., & Liu, J. (2012). Highly potent bactericidal activity of porous metal-organic frameworks. *Advanced Healthcare Materials, 1*(2), 225–238. https://doi.org/10.1002/adhm.201100043.

Zirehpour, A., Rahimpour, A., Arabi Shamsabadi, A., Sharifian, M. G., & Soroush, M. (2017). Mitigation of thin-film composite membrane biofouling via immobilizing nano-sized biocidal reservoirs in the membrane active layer. *Environmental Science and Technology, 51*(10), 5511–5522. https://doi.org/10.1021/acs.est.7b00782.

Zou, X., Zhang, L., Wang, Z., & Luo, Y. (2016). Mechanisms of the antimicrobial activities of graphene materials. *Journal of the American Chemical Society, 138*(7), 2064–2077. https://doi.org/10.1021/jacs.5b11411.

Chapter 12

Metal-organic framework composites for biomedical applications

Maryam Poostchi[1], Justyna Kucińska-Lipka[1], Mohsen Khodadadi Yazdi[2,3], and Mohammad Reza Saeb[4]
[1]*Department of Polymer Technology, Faculty of Chemistry, Gdańsk University of Technology, Gdańsk, Poland,* [2]*Division of Electrochemistry and Surface Physical Chemistry, Faculty of Applied Physics and Mathematics, Gdańsk University of Technology, Gdańsk, Poland,* [3]*Advanced Materials Center, Gdańsk University of Technology, Gdańsk, Poland,* [4]*Department of Pharmaceutical Chemistry, Medical University of Gdańsk, Gdańsk, Poland*

Introduction

Metal-organic frameworks (MOFs) are polymers obtained by coordination chemistry, which are crystalline networks composed of metal clusters coordinated to organic ligands. MOFs benefit from large surface areas and substantial porosity supporting a wide range of applications. MOFs are usually known by a generic name, which specifically addresses their structural features or laboratory or institute in which they are synthesized (Yusuf et al., 2024). They can be classified based on their unique characteristics, in terms of structure (pore size, shape, and linker), surface chemistry, functionality, or synthesis methods. Typically, MOFs exhibit miscellaneous structures and tailorable properties because of wide range of pore size from microporous (< 2 nm) to mesoporous (2–50 nm) and even macroporous (> 50 nm) (Guan et al., 2018). It is possible to tailor-make MOF porosity by adjusting parent metal centers and organic ligands, which determines the crystalline structure of MOFs in the form of cubic, hexagonal, or layered architectures. The combination of metal ions and organic linkers also affects the stability and functionality of MOFs (Ma et al., 2022). For instance, larger pores are the result of using longer organic linkers, which end in microstructures hosting macromolecules such as drugs and proteins (Kai et al., 2023). Moreover, it is possible to conduct surface engineering of MOFs as postsynthesis modification to anchor functional groups to organic linkers (Gadzikwa & Matseketsa, 2024). In this way, the degree of interaction between

MOFs and guest molecules or biological environment can be supported, thereby and enhanced biocompatibility, for example, with biological tissues, can be achieved (Singh et al., 2021). It also supports the release of encapsulated moieties in biomedicine, and enhances the biosensing and diagnostic performances (Ahmadi et al., 2021).

For biomedical applications, attention should be paid to adjust the structure of MOF by choosing proper inorganic metal ions and organic linker. Moreover, microstructural features such as the size, shape, and chemical affinity of pores can assist in mitigating the toxicity of MOFs. Therefore exploring the correlation between MOF structure and MOF toxicity mechanism in living organism seems essential prior to biomedical engineering (Wang et al., 2023). It is also possible to modify the metal clusters or organic linkers in order to induce fluorescence, magnetism, or other specific biological activities to MOFs (Zhong et al., 2021). Synthesis of MOF itself determines and can be performed in a manner to widen its biomedical applications, such that it affects the morphology, crystalline structure, and functionality of MOFs. For instance, although solvothermal synthesis is more popular because of high-quality crystals obtained, it is not recommended for biomedical applications. On the other hand, hydrothermal, microwave-assisted, and electrochemical methods are safter in view of higher purity and lower toxicity (Wiśniewska et al., 2023). MOFs are able to host various functional groups as well as inorganic and organic moieties. In this regard, modification of MOF has been in the core of attention to make MOFs prioritized and personalized for imaging, drug delivery, and biosensing applications. For instance, conductivity of MOF can be adjusted through surface and/or bulk modification to adapt the size and function of drug molecules with the pore size and shape of MOF whether drug loading or release are concurrently supported. Thus, either externally (surface) or internally (bulk) modified MOFs, hereafter referred to as "MOF composites" are progressed to widen their biomedical engineering window (Li, Ashrafizadeh, et al., 2024; Wang et al., 2024).

More than thousand types of MOFs are recognized worldwide, among which are several MOF composites synthesized, characterized, and used in different fields. In this chapter, we merely investigated, classified, and tabulated the ones already applied in biomedical engineering. A serious difficulty in classification of MOF composite arises from diversity in names being chosen by the researchers hyperlinked to MOF composites. Therefore, there was a need for systematic yet simple classification of MOF composites used in biomedical engineering. Accordingly, three main groups, namely MOF composites resulting from surface modification (I), bulk modification (II), and hybrid (concurrent surface and bulk) modification (III) are defined here and considered in literature classification. A closer observation revealed that dopants used in bulk modification were mainly metals and metal oxides, while surface modification or occasionally called functionalization of MOFs was demonstrative of cases where nanoparticles (NPs) or polymers like polysaccharides, RNA molecules, and dopamine were used as the most frequent modifiers. Among the reports collected and assessed, some were also available in which terms such as coating or encapsulation were ascribed to a polymer or NPs (either organic, or inorganic and organic–inorganic hybrid) anchored to the surface of MOFs, which we considered as the aforementioned group (I) of MOF composites. Hybrid cases, group (III) also comprise MOF structures sandwiched in between two precursors or any combination of MOF with other organic or mineral precursors whatever the size and source. This chapter brings about insight into the correlation between the structure and biomedical application of MOF composites.

Application of metal-organic frameworks in biomedical engineering

Fig. 12.1 shows trend in the number of publications on application of MOFs and also MOF composites in biomedical applications over the last decade. The diagram suggests that research on MOF composites overtook that of unmodified or neat MOFs over the past three years, which outlines the importance of modification of MOFs for biomedical engineering applications. As previously mentioned, MOF composites provide the user with a wider possibility window in order to tailoring the properties of MOF for biomedical engineering. MOF composites of the aforementioned types of (I), (II), and (III) can systematically be considered for different sectors of biomedical engineering. For instance, surface modification of MOFs through attachment of functional groups to the surface, thermal or solvent treatments, or other molecules leading for creation of type (I) MOF can induce chemical tendency and physical affinity of the external surface of MOFs. This process amends the interaction of MOFs with the surrounding environment, but the core structure and properties often remain unchanged. By contrast, modification of the bulk of MOFs through doping introduces dopants or other elements within the internal structure of MOFs, leading to formation of type (II) MOF. This structural modification changes more significantly the properties of MOFs by integrating the characteristics of the guest dopant(s) directly into that of MOF framework. Furthermore, MOF composites are also obtained through the combination of MOFs with other materials including other types of MOFs, minerals and organic moieties. The resulting versatile structures with tailored functionality and porosity, so-called type (III) MOF, are hybrid composites relying on rigorous applications. Thus, versatility and wide windows of innovation brought about by MOF composites can be understood—the reason why MOF composites overtook MOFs in research participation.

Overall, applications of MOF composites in biomedicine and the related fields can be classified into six subcategories. Drug delivery systems (DDS) come the first, where MOF composites appear promising and effective candidates for controlled drug release, due to their large surface area, tunable

414 Applications of Metal-Organic Framework Composites

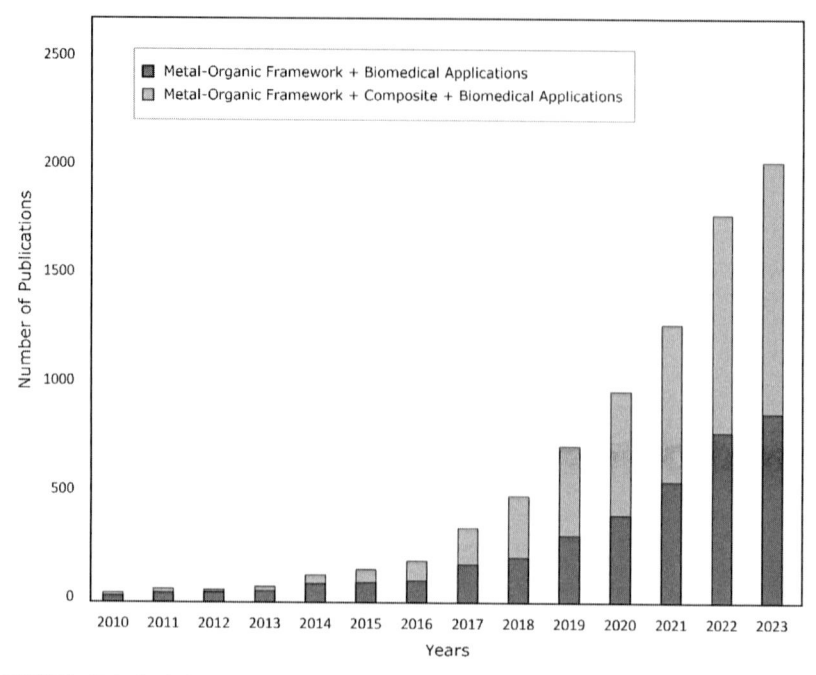

FIGURE 12.1 Statistics on metal-organic frameworks (MOFs) and metal-organic framework (MOF) composites used in biomedical applications. The similar growth interests for both MOFs and MOF composite are observed with gradual increase in publications on MOF composites in biomedical application. Data based on Scopus database; accessed September 2024. Search within title, abstract, and keywords for metal-organic framework+biomedical applications as: ("metal-organic framework" OR "MOF" OR "MIL" OR "ZIF" OR "PCN" OR "UiO" OR "HKUST") AND ("biomedical application" OR "biomedical" OR "drug delivery" OR "biosensing" OR "bioimaging" OR "theragnostic" OR "tissue engineering" OR "antibacterial" OR "antimicrobial" OR "cancer therapy" OR "biocompatibility" OR "biomedical devices" OR "diagnostics" OR "therapeutics") AND NOT ("composite" OR "blend" OR "surface functionalization" OR "surface modification" OR "grafting" OR "@" OR "particle" OR "nanoparticle" OR "nanocomposite" OR "hybrid material" OR "surface engineering" OR "dop") and for metal-organic framework+composite+biomedical applications as: ("metal-organic framework" OR "MOF" OR "MIL" OR "ZIF" OR "PCN" OR "UiO" OR "HKUST") AND ("biomedical application" OR "biomedical" OR "drug delivery" OR "biosensing" OR "bioimaging" OR "theragnostic" OR "tissue engineering" OR "antibacterial" OR "antimicrobial" OR "cancer therapy" OR "biocompatibility" OR "biomedical devices" OR "diagnostics" OR "therapeutics") AND ("composite" OR "blend" OR "surface functionalization" OR "surface modification" OR "grafting" OR "@" OR "particle" OR "nanoparticle" OR "nanocomposite" OR "hybrid material" OR "surface engineering" OR "dop").

pore sizes, and ability to carry large drug payloads. Surface modifications can help target specific tissues or cells (Priya et al., 2022). Second class of MOF composites used in biomedicine are those directed toward biosensing application, by which one could detect biomolecules, based on size exclusion and functionality of their porosity and functional groups (Mohanty et al., 2024). In this class, MOF composites used were resulting from hybridization of MOF with NPs, which has progressively employed as highly sensitive biosensors for diagnostic purposes (Wang et al., 2021). Third class of MOF composites used in biomedicine contains those used in imaging in order to enhance contrast bioimaging. Among imaging techniques applied in this regard are magnetic resonance imaging (MRI) and fluorescence imaging frequently used for the purpose of early diagnosis or monitoring therapeutic processes (Zhang, Sang, et al., 2019). Tissue engineering application comes as the fourth class of MOF composites applied as scaffolds in tissue regeneration. The highly porous, large surface area, and tunable functionality of MOF composites allow for tailored biocompatibility and encapsulation of growth factors or other bioactive molecules required for tissue repair purposes (Yao et al., 2024). Antibacterial and antiviral agents based on MOF composites are the fifth class, which are mainly based on silver and copper widely employed in wound healing applications (Liu, Zhou, et al., 2021). The sixth class of MOF composites applied in biomedical engineering refers to targeted cancer therapies, as structures encapsulating chemotherapeutic agents to directly access tumor sites, minimizing side effects while maximizing therapeutic efficacy (Yang et al., 2023). Fig. 12.2 shows ageneral view of application of MOF composites in the field of biomedical engineering.

Table 12.1 presents an overview of the structural elements and key properties of MOFs relevant to their use in biomedical applications. Although most of the cases presented are applied in biomedical contexts, some were also rationalized as being useful in view of their biocompatibility. Brunauer–Emmett–Teller (BET) surface area and pore volume of MOF structures are two main characteristics determining the potential of MOFs for modification with polymers and biomolecules applicable in biomedical applications (Du et al., 2021). In other words, the surface area quantified using BET measurement is a measure for the amount of surface available for interactions, which can be varied in between 1000 m^2/g and over 7000 m^2/g depending on the structure of MOFs (Chen, Kirlikovali, et al., 2022). A higher surface area results in more active sites, making MOFs more suitable for functionalization, which is often equivalent of higher biocompatibility. The pore volume, another critical parameter, measures the amount of void space within the porous structure, which is directly related to the capacity of MOF structures to adsorb other molecules, crucial for applications such as encapsulation and sensing (Ahmadi et al., 2021). These parameters along with the inorganic metal cluster and organic linker reveal the potential of MOFs in terms of biocompatibility, biodegradation rate, mechanical and chemical stability in various solution, magnetic properties, and capacity to encapsulate large or small molecules which are essential for biomedical applications (Hefayathullah et al., 2024). All in all, MOFs have been recently used in different fields of biomedicine, including tissue engineering, drug delivery, imaging, and cancer treatment. Nevertheless, low compatibility and sensible toxicity of MOFs together with

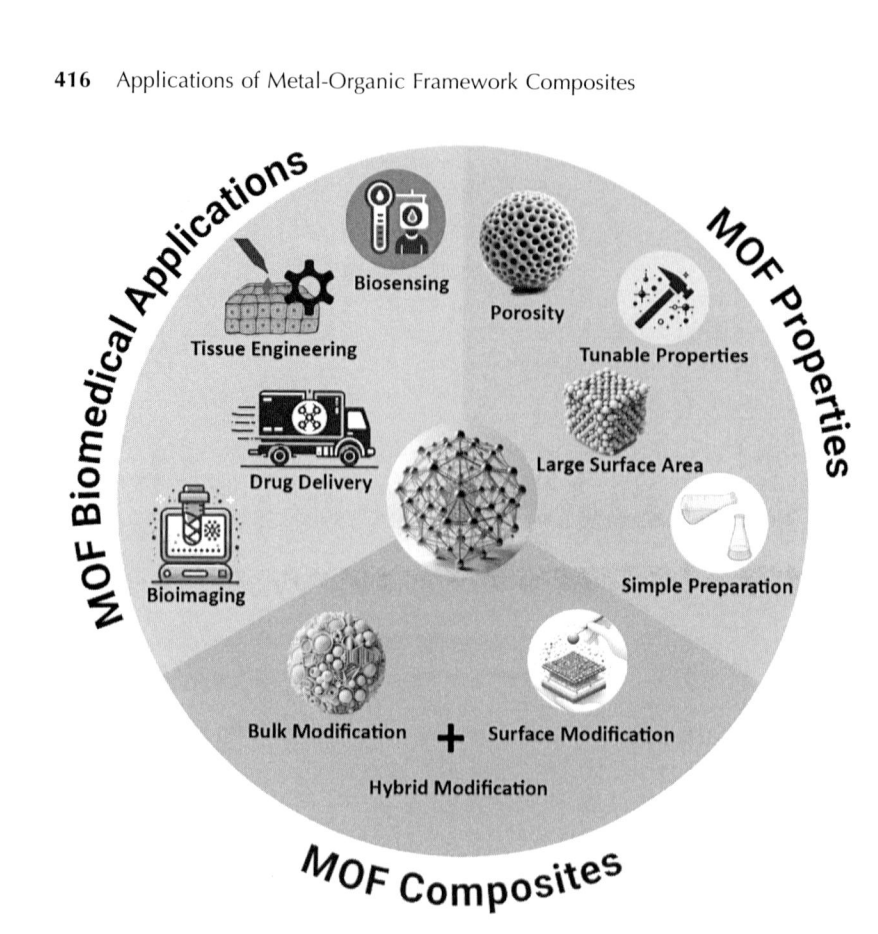

FIGURE 12.2 Metal-organic framework (MOF) properties, overview of MOF composites, and biomedical applications of MOFs.

limited interaction with the biological media outline the need for modification of the surface and bulk properties of MOFs, which rationalizes the use of MOF composites.

Application of metal-organic framework composites in biomedical engineering

As highlighted earlier, MOF composites modified through surface and/or bulk treatments, i.e., types (I), (II), and (III) MOFs, can compensate for poor compatibility arising from limited functionality and stability, and also poor antibacterial properties of MOFs. Based on the literature, the main application of MOF composites in biomedical engineering can be generally grouped into

TABLE 12.1 Characteristics of metal-organic frameworks (MOFs) applied or to be intended (potentially suitable) for biomedical applications.

MOF group	MOF	Metal ion	Linker in chemical structure	BET area (m^2/g)	Pore volume (cm^3/g)	Pros in biomedical applications	Cons in biomedical applications	Refs.
MOF	MOF-5	Zn^{2+}	Benzene-1,4-dicarboxylate (BDC)	2500–4000	1.55	High structural stability High surface area	Toxicity for cells in the size of nano Slow degradation rate Limited adaption to encapsulated drug	Chen et al. (2010), and Ruyra et al. (2015)
	MOF-177	Zn^{2+}	1,3,5-Benzenetribenzoate (BTB)	4500–5400	1.89	High surface area Structural tuning	Low flexible Functionalization difficulty	Santos et al. (2019)
	MOF-200	Zn^{2+}	1,3,5-Tris(4-carboxy [1,2-biphenyl]-4-yl)-benzene) (H$_3$BBC)	4200–4800	Up to 3.6	High surface area Large pore volume	Sensitivity to moisture Functionalization difficulty	Abu Ghalia (2016), and Ullah et al. (2020)
	MOF-199 (HKUST-1)	Cu^{2+}	Benzene-1,3,5-tricarboxylic acid	970–1700	0.69	Antibacterial activity Adsorption capacity of cationic molecules	Low biocompatibility	Liu, Yu, et al. (2021); Nguyen Thi et al. (2013); and Yang et al. (2024)
	MOF-210	Zn^{2+}	1,3,5-Benzenetricarboxylic acid	6000–6500	3.6	Ultrahigh surface area Large pore volume	Sensitivity to moisture Low biocompatibility	Furukawa et al. (2010)

(Continued)

TABLE 12.1 (Continued)

MOF group	MOF	Metal ion	Linker in chemical structure	BET area (m²/g)	Pore volume (cm³/g)	Pros in biomedical applications	Cons in biomedical applications	Refs.
	Bio-MOF-100	Zn^{2+}	Biphenyldicarboxylate (BPDC)	4300	4.3	Accessible mesopore structure Controlled release pH-sensitive drug release Biocompatibility	Low mechanical properties	Alves et al. (2021)
	Al-soc-MOF-1	Al^{3+}	4,4',4''-s-Triazine-2,4,6-triyl-tribenzoic acid (H₃TATB)	5300–5800	2.2	High surface area	Sensitivity to moisture	Ahmed et al. (2022)
MIL	Fe-MIL-100	Fe^{3+}	Trimesic acid (H₃BTC)	800–2200	0.9	Capacity of large amount of drug encapsulation Stability on physiologic solution	Low biocompatibility	Mahmoodi et al. (2018), and Quijia et al. (2021)
	Cr-MIL-101	Cr^{3+}	Terephthalic acid (H₂BDC)	2500–4100	1.22	High storage capacity Controlled drug release Easy functionalization	Toxicity due to Cr	Zou et al. (2022)
	NH₂-MIL-88(Fe)	Fe^{3+}	2-Aminoterephthalic acid (NH₂-BDC)	200–600	0.27	Sustained drug release	Low pore size	Kim et al. (2018)

PCN	PCN-223	Zr^{4+}	Tetrakis (4-carboxyphenyl) porphyrin (TCPP)	1000–2000	0.87	Capacity to capsulate drug through physical interactions (without need to change in drug composition) Biocompatibility	Low surface area	Chun et al. (2020)
	PCN-222	Zr^{4+}	Tetrakis(4-carboxyphenyl) porphyrin (TCPP)	2000–3000	1.6	Capacity to encapsulate large biomolecule Responsivity to the acidic pH Biocompatibility High water stability	Slow degradation rate	Leng et al. (2018), and Zhang et al. (2016)
	PCN-224	Zr^{4+}	Tetrakis(4-carboxyphenyl) porphyrin (TCPP)	2500–3500	0.9–1.2	Capacity to encapsulate drug Large surface area Easy functionalization Chemical stability	Slow degradation rate	Park et al. (2016)
UiO	Zr-UiO-66-NH$_2$	Zr^{4+}	2-Amino-1,4-benzenedicarboxylic acid	830	0.8–1.2	Low cytotoxicity Biodegradability Excellent optical properties Easy functionalization	Low surface area Reduced chemical stability	Canivet et al. (2014), Pourmadadi et al. (2022), and Wang et al. (2020)
	UiO-66	Zr^{4+}	1,4-Benzenedicarboxylic Acid	1200–1600	0.4–0.6	Biocompatibility Chemical stability	Sensitivity to moisture Limited pore size	Treger et al. (2023)

(Continued)

TABLE 12.1 (Continued)

MOF group	MOF	Metal ion	Linker in chemical structure	BET area (m²/g)	Pore volume (cm³/g)	Pros in biomedical applications	Cons in biomedical applications	Refs.
ZIF	ZIF-8	Zn^{2+}	2-Methylimidazole	1200–2100	0.1–0.8	Biocompatibility Stability under different conditions Responsivity to the acidic pH Antibacterial activity	Sensitivity to moisture Small pore size	Paul et al. (2022), and Zhang et al. (2021)
DUT	DUT-23-Co	Co^{2+}	Tricarboxylates or tetracarboxylates	4800–5200	0.5–1	Chemical stability Magnetic properties	Sensitivity to moisture	Ahmed et al. (2022)

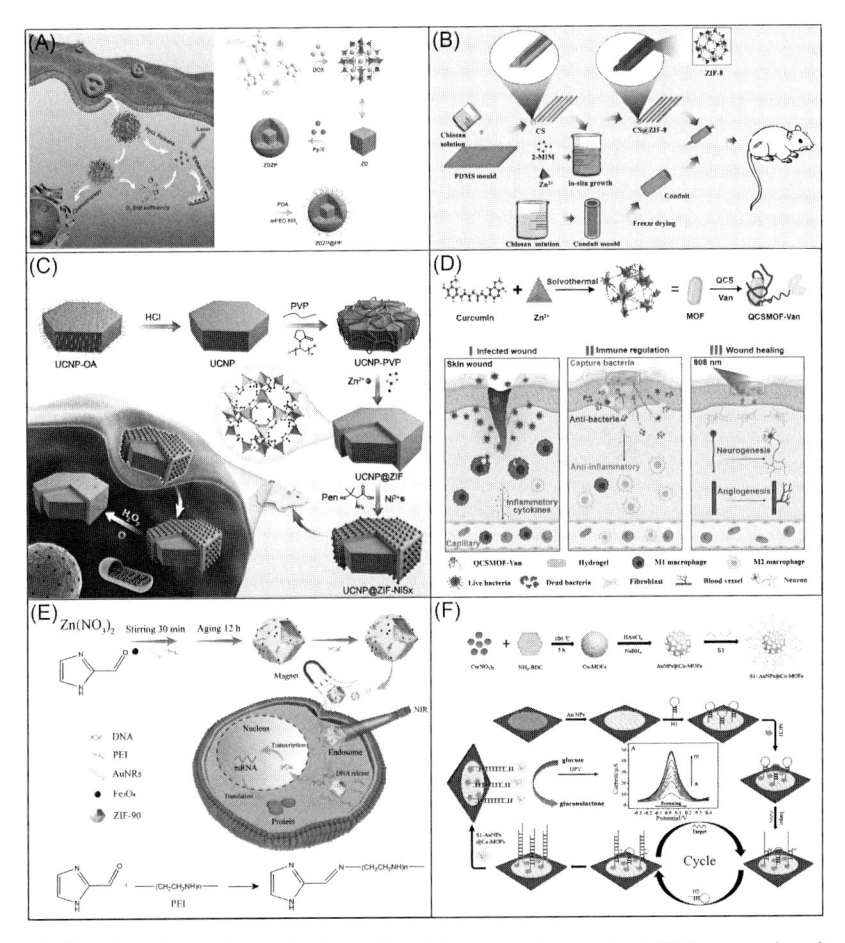

FIGURE 12.3 Schematic applications of metal-organic framework (MOF) composites in medicine (biomedical engineering). (A) Stimuli responsive drug delivery: Surface modification of ZIF-67 by ZIF-8 (ZIF-62@ZIF-8), preparation pH stimuli responsive DDS, and mechanism of photodynamic therapy of (ZIF-62@ZIF-8) (Ren et al., 2020). (B) Tissue engineering: Preparation of chitosan-ZIF-8 composite-based nerve conduit for neural tissue engineering, which ZIF-8 NPs were grafted onto the chitosan surface (Zhang et al., 2024). (C) Bioimaging: Bulk modification of ZIF-NiSx by upconversion nanoparticle (UCNP) and ROS detection mechanism through UCNP@ZIF-NiSx (Hao et al., 2019). (D) Antimicrobial activity: Surface modification of Bio-MOF by quaternary ammonium salt chitosan (QCS) and mechanism of antimicrobial activity of MOF in wound healing (Huang et al., 2022). (E) Gene delivery: Fabrication of AuNRs@Fe$_3$O$_4$@ZIF-90-PEI25k (AFZP25k)/DNA NPs, transporting gene onto the targeted cells (Liu et al., 2024). (F) Biosensing: Surface modification of Cu-MOF by single strand of DNA (S1) and Au nanoparticles (AUNPs) for detection of glucose biosensor (Wang et al., 2018).

several classes, which are mainly including stimuli-responsive drug delivery, tissue engineering, bioimaging, antimicrobial activity, gene delivery, and biosensing (Fig. 12.3).

The major biomedical applications of MOF composites have been already featured by stimuli-responsive DDS, which are fueled by versatile and highly tunable structures of MOF composites compared to unmodified MOFs. These composites have been surface- or bulk-engineered to respond to specific stimuli, mainly including pH, temperature, light, or redox conditions (Musarurwa & Tavengwa, 2022). MOF composites obtained through surface modification of MOFs can enlarge drug loading capacity, enable controlling the rate of drug release, and improve its biocompatibility. For instance, surface functionalization ensures ligand targeting that supports selective delivery of drugs and therapeutic agents to the specific cells or tissues, along with minimizing side effects (Demir Duman et al., 2022; Li, Yuan, et al., 2024). MOF composites can also respond to the acidic environments, such as tumor tissues or infection sites, more sensitively compared to the unmodified MOFs. An engineered MOF composite can degrade easier supporting the release of loaded drugs precisely where needed (Toh et al., 2024). Additionally, the ability of MOF composites to carry a large amount of drugs, combined with their improved biocompatibility provides support for their high effectiveness in treating diseases while reducing side effects. This unique feature of smart MOF composite-based DDS makes them more efficient and targeted for cancer treatment (Cai et al., 2019). On the other hand, modifying the bulk structure of MOF allows for delivery of either chemotherapeutics or diagnostic agents, individually and simultaneously (so-called combination therapy), which is highly needed in cancer therapy. In conclusion, MOF composites enable personalized medicine by targeted delivery of drugs, also genes, individually or combined with drugs. MOF composites could also play a key role in gene delivery, for example, DNA or RNA applied for gene therapy, which is based on larger capacity and precise functionality that provide support for efficient encapsulation and controlled release, respectively (He et al., 2023).

Due to their highly tunable properties, versatility of their microstructure, and higher biocompatibility compared to unmodified MOFs, MOF composites have been used as versatile biomaterials for tissue engineering applications. These nano-scale biomaterials as scaffolds support mimicking the extracellular matrix (ESM) for cell growth and tissue regeneration. MOF composites-based scaffolds take the privilege of higher mechanical properties and biological interactions thanks to surface or bulk modification of MOF, which supports tailoring the target tissue, frequently surveyed for bone and cartilage or muscle regeneration. Once its surface chemistry is elaborated, MOFs become composite structures with the ability of binding to bioactive molecules or growth factors. These structures support cell adhesion, growth, and differentiation, which is known as triplet of tissue repair (Zhuang et al., 2017). Once bulk modified, MOFs can promote osteoinductive performance needed for bone tissue engineering (Asadniaye Fardjahromi et al., 2022). Moreover, highly porous structure of MOF composites enabling enlarged capacity of loading makes it possible to control the rate of delivery of therapeutic agents like growth factors for precisely targeting at the injury site. Therefore bulk-modified MOF composites promote and accelerate wound healing while minimizing infectious risks (Chen et al., 2021). Furthermore, MOF composites with the ability of incorporating signaling molecules into the microstructure support stimulation of specific cellular responses, which promotes the repair of tissue. Thus MOF composites are more versatile thanks to their functionality to be served as platforms in tissue engineering to treat damaged tissues as well as drug delivery performance.

Biomedical engineering supported by MOF composites takes the benefits of higher porosity, tunable frameworks, and surface functionality of MOFs, which support bioimaging processes/tools (Gupta et al., 2024). It is well-accepted that the higher porosity and loading capacity, more surface functionality and/or higher mechanical strength of MOF composites compared to unmodified MOFs promotes the signal detection efficiency through encapsulation of imaging agents, among which are radionuclides, fluorescent dyes, biomarkers, and contrast agents (Sohrabi et al., 2021). For instance, higher contrasts of images are the result of modification of MOFs, where both organic linkers and metal clusters can be tailored through a bewildering array of possibilities. To improve the clarity of image, some MOF composites are designed and structurally engineered to enhance the contrast in different imaging methods like computed tomography (CT) scans, MRI, and fluorescence imaging (Shahzaib et al., 2023). Moreover, MOF composites support a wider biocompatibility as well as multifunctionality spectrum with respect to unmodified MOFs, which positively boost imaging and theragnostic, particularly when targeting a tissue through noninvasive diagnostics. For instance, MOF composites can specifically enhance the selectivity, sensitivity, and specificity of imaging modalities. It is also possible to benefit from targeted bioimaging through surface functionalization of MOF, considering that specific ligands capable of recognizing and binding to particular biomarkers anchored to target tissues or cells are achieved by surface engineering of MOFs (Nazari et al., 2022).

MOF composites also take greater credit for real-time monitoring of biological processes (Gong et al., 2024). MOF composites are superior to unmodified MOFs in view of antibacterial activity as well. This is due to their customizable structures, which are created through surface or bulk modifications. This strategy is supported by incorporating inherently antibacterial metal ions such as silver, copper, and zinc into framework of MOFs effectively inhibiting the growth of bacteria (Guo et al., 2024). The release of such antibacterial metal ions interrupts growth of bacterial cell membranes, thereby causing cell death. Furthermore, improved porosity of MOF composites in comparison with unmodified MOFs allows for gradual release of antibacterial agents over a prolonged period (Livesey et al., 2023). MOF composites supporting selective toxicity are also flourished over the last years, such that supported targeting bacteria without harming human cells.

When used as biosensing materials, MOF composites have obvious edge over pure or unmodified MOFs. For instance, MOF composites enjoy from higher stability, selectivity, and conductivity compared to pure MOFs for biosensing applications (Mohanty et al., 2024). Such beneficial features are contributed from polymers, NPs, or carbonaceous counterparts amalgamated by MOFs, which facilitate conductivity, enzymatic function, biomolecular detectability, mechanical strength, durability, or conductivity of biosensors (Rabiee et al., 2022). Thus it can be concluded that the biomedical applications of MOF composites are obviously broadened through modification of surface or bulk of MOFs compared to the pure MOFs.

In view of the message of this chapter, a well-classified table including biomedical engineering of MOF composites would be beneficial. Typically, examples are to some extent personalized rather than realized, such that even exaggerated reports could also be found. Therefore, the authors of this chapter has skipped fuzzy or less-trustable reports, instead they collected data from reliable investigations. Moreover, there were some other biomedical applications detected for MOF composites, like regenerative medicine, which were beyond tissue engineering and drug delivery. This approach could potentially enable organ repair or replacement, but the perspective deemed somewhat blurry. Nevertheless, growth of complex tissues based on the use of MOF composites may be a hot topic for the future research. Therefore, Table 12.2 summarizes and aligns biomedical applications of MOF composites, which are categorized in terms of the type of MOF composite (surface modification, bulk modification, or hybrid modification). As outlined in this table, there have been continued efforts in order to synthesize diverse MOF composites by engineering of surface and/or bulk characteristics of MOFs. Surface modification of MOFs with biomolecules such as dopamine and hyaluronic acid has been frequently considered as a utilitarian approach to enhance properties of MOFs in terms of stability in aqueous environment, protection from encapsulated drug, targeting release, and sometimes to impart magnetic property to the MOFs. For instance, Fe_3O_4 is a biocompatible NPs, which has been widely utilized to enhance magnetic properties of MOFs, making them suitable for bioimaging and stimuli-responsive targeting systems (Zhang et al., 2019). The high surface area of MOFs makes it possible to develop MOF composites as a carrier for encapsulation of therapeutic molecules for DDS, offering controlled and targeted release, high encapsulation capacity, and cellular uptake (Valizadeh Harzand et al., 2023). Additionally, antibacterial activity of some metal clusters such as zinc and copper imparts significant potential to MOF composites needed for wound healing. Doping MOFs with polymers, inorganic materials, and other MOFs is also a common well-practised method to improve their biocompatibility and functionality. A deep analysis of reports may be useful for a review paper in the future, but what we summarized here may provide a basis for the future analysis.

TABLE 12.2 Applications of metal-organic framework (MOF) composites[a] in biomedicine.

MOF	Composite materials[a]	Type of composite/material	Biomedical applications	Key findings	Refs.
MOF-199 (HKUST-1)	Nimesulide@Fe$_3$O$_4$@MOF-199	Bulk modification/Fe$_3$O$_4$	Drug delivery	Showed controlled release of nimesulide Showed magnetic behavior	Guo et al. (2024)
HKUST-1	Fe$_3$O$_4$@SiO$_2$@HKUST-1@letrozole	Surface modification/Fe$_3$O$_4$-SiO$_2$	Drug delivery	Controlled release of letrozole in acidic environment	Hashemipour & Ahmad Panahi (2017)
Bio-MOF	Folic acid-chitosan@Fe$_3$O$_4$@Bio-MOF@curcumin	Hybrid modification/folic acid-chitosan/Fe$_3$O$_4$	Tumor targeted delivery	Showed excellent biocompatibility Selective drug release High cellular uptake	Nejadshafiee et al. (2019)
MIL-100 (Fe)	Hydroxyapatite@Fe-MOF@Fe$_3$O$_4$	Hybrid modification/hydroxyapatite/Fe$_3$O$_4$	Drug delivery	Controlled drug released at acidic environment Revealed enhanced biocompatibility Showed magnetic behavior	Yang et al. (2017)
MIL-100	Fe$_2$O$_3$@MIL-100- USPIO	Hybrid modification/Fe$_2$O$_3$/USPIO	Image-guided therapy	Exhibited excellent colloidal stability High encapsulation capacity Controlled release Showed magnetic behavior	Sene et al. (2017)

(Continued)

TABLE 12.2 (Continued)

MOF	Composite materials[a]	Type of composite/material	Biomedical applications	Key findings	Refs.
MIL-100	Dopamine- hyaluronic acid@MIL-100@curcumin	Surface modification/dopamine- hyaluronic acid	Image-guided therapy	Exhibited excellent colloidal stability; Showed targeted drug delivery	Zhang et al. (2018)
MIL-101	MIL-101(Cr)@M (M: Cu, Co, Cr or Fe)	Bulk modification/M (M: Cu, Co, Cr or Fe)	Antibacterial activity	Enhanced antibacterial properties	Hachemaoui et al. (2021)
MIL-100	Fe$_3$O$_4$@MIL-100@DOX	Surface modification/Fe$_3$O$_4$	Drug delivery	Controlled release of DOX	Bellusci et al. (2018)
MOF-74	Dopamine@MOF-74	Surface modification/dopamine	Drug delivery	Lower homolytic activity of DOX; Controlled drug delivery	Kazemi et al. (2024)
MIL-53 NH$_2$-MIL-53	Al@MIL-53 Al@NH$_2$-MIL-53	Surface modification/Al	Antibiofilm	Higher antibacterial activity	Gecgel et al. (2022)
MIL-88A	Fe$_3$O$_4$@MIL-88A@dopamine	Hybrid modification/Fe$_3$O$_4$/dopamine	Drug delivery	Enhanced biocompatibility; Controlled release of dopamine	Pinna et al. (2018)
PCN-224	DNA@PCN-224@NaYF$_4$	Surface modification/DNA	Photodynamic therapy	Enhanced targeted delivery; Showed excellent cellular uptake	He et al. (2017)

UiO-66-NH$_2$	Iodine@UiO-66-NH$_2$@PCL	Hybrid modification/iodine/PCL	Antibacterial activity	Enhanced immobilization of iodine to the MOF	Chen, Zhu, et al. (2022)
UiO-66	Carboxymethyl cellulose@tetracycline@UiO-66	Surface modification/carboxymethyl cellulose	Antibacterial activity Drug delivery	Enhanced biocompatibility Improved mechanical strength Showed antibacterial activity Controlled release of tetracycline	Javanbakht et al. (2021)
UiO-66 ZIF-8 MIL-101	Aptamer-dopamine@UiO-66@DOX Aptamer-dopamine@ZIF-8@DOX Aptamer-dopamine@MIL-101@DOX	Surface modification/dopamine	Chemo-photothermal therapy	Demonstrated superior biocompatibility Showed targeted drug delivery	Feng et al. (2019)
ZIF-8	ZIF-8@chitosan	Bulk modification/chitosan	Nerve tissue engineering	Showed biocompatibility Induced nerve regeneration	Zhang et al. (2024)
ZIF-8	PCL@collagen@ZIF-8	Bulk modification/PCL-collagen	Bone tissue engineering	Enhanced angiogenesis and osteogenesis	Xue et al. (2021)
ZIF-8	Sodium alginate@curcumin@ZIF-8	Surface modification/sodium alginate	Drug delivery Antibacterial activity	Enhanced stability of hydrogel Strengthened the encapsulation of CCM inside ZIF-8	Li et al. (2023)

(Continued)

TABLE 12.2 (Continued)

MOF	Composite materials[a]	Type of composite/material	Biomedical applications	Key findings	Refs.
ZIF-8	Zn-ZIF-8@methacrylated hyaluronic acid	Surface modification/methacrylated hyaluronic acid	Wound healing Drug delivery	Enhanced biocompatibility Sustainable release Accelerated epithelial regeneration	Yao et al. (2021)
ZIF-8	Lactobionic acid—AuNR@ZIF-8@DOX	Surface modification/lactobionic acid-AuNR	Image-guided therapy	Exhibited high contrast imaging	Zhang, Zhang, et al. (2019)
ZIF-67	ZIF-67@sodium alginate	Bulk modification/sodium alginate	Antibacterial activity	Showed excellent antibacterial activity Enhanced mechanical strength	Zheng et al. (2020)

USPIO, ultra-small superparamagnetic iron oxide; PCN, porous coordination network; PCL, polycaprolactone; DOX, doxorubicin.
[a] Composite structure defines composite components from outer layer to inner layers.

Concluding remarks and future direction

This chapter summarizes research outcomes on biomedical applications of MOF composites, highlighting their advantages over neat or unmodified MOFs, alongside showcasing engineering of their microstructure. In conclusion, there are a number of beneficial features when one uses MOF composites rather than neat MOFs in biomedical applications. Enhanced biocompatibility was one of the main reasons for modification of either the surface or the bulk of MOF structures. The main purpose was to obtain MOF composites with enhanced biocompatibility towards biological environments through tailoring the functionality and extending the surface area of MOF structures. The resulting composite structures were shown to give rise to reduced toxicity, which minimizes harmful interactions with biological environments, making them significantly safer for use in DDS and implantation. Moreover, neat MOFs have been shown to be potentially degradable in physiological environments, demonstrating need for incorporating stabilizing agents like polymers and NPs into MOF structures. This makes MOFs highly resistant against external stmuli, e.g., moisture, enzymes, and pH fluctuations prolonging their stability for biomedical applications. Precise porosity and functionality of MOFs can guarantee a well-controlled kinetics of drug release provided by MOF composites. Under such circumstances, drug delivery to specific sites and cells in the human body would be possible under precise and regulated drug release rate supporting higher therapeutic efficacy. Depending on functionality, MOF composites are potent to integrate several capabilities such as drug delivery, imaging, and biosensing into a single platform, which condiserably expands their biomedical application window. It is also possible to enhance the mechanical strength of MOFs through surface modification in addition to affecting their biological and therapeutic features. In this regard, incorporating dopants as secondary materials into the structure of MOF composites expands their versatility for tissue engineering or implant applications where both biocompatibility and mechanical stability are required.

Although the aforementioned beneficial features look promising for developing MOF composites targeted at biomedicine, there are some challenges that need to be considered for further developments in this field. For instance, metal ions appended to MOF composites are not essentially non-toxic to human cells, which adversely affect the biological functions of MOFs. The stability of MOF composites in biological environments over a prolonged service may also hinder delivery of drugs, especially for highly porous structures which degrade rapidly in biological environments. Moreover, synthesizing MOF composites experiences at early stage of development with complications associated with their synthesis, characterization, and drug release kinetics for clinical purposes. The toxicity issues are considerable, such that even copper or zinc may trigger cytotoxic responses in living organisms. Therefore, balancing the functionality, stability, porosity, and the antibacterial neutrality of MOF composites is essential. Since MOF composites are surface and/or bulk-engineered structures with tailored properties and structural characteristics, they can target specific tissues or cell types more precisely rather than pure MOFs. This appears as a specific advantage in cancer therapy or site-specific drug delivery, where precision is the key for minimizing side effects. In the case of hybrid MOFs where MOFs are combined with NPs or functional groups, the resulting tailored structures can target several functions simultaneously, such as drug loading, controlled release, imaging, and biosensing; enhancing their utility in biomedical applications. Such features look promising, but individually or simultaneously hinder their transplantability, particularly issues with stabilizing and delivering vaccines, which necessitates in-depth analyses. From a sustainability standpoint, MOF composites should focus on environmental safety by using fewer renewable resources, having a shorter lifecycle, consuming less energy, and promoting a circular biomedical economy. The use of green solvents in the production of MOF composites and their adaptability with decarbonization policy should seriously be taken into account. Recycling of MOF composites should also be more complicated rather than neat or unmodified MOFs, also their approval for clinical test necessitates consideration of diverse possibilities. These all highlight future challenging features of developing MOF composites for biomedical application.

References

Abu Ghalia, M. (2016). Development and evaluation of zeolites and metal–organic frameworks for carbon dioxide separation and capture. *Energy Technology, 5*, 18. https://doi.org/10.1002/ente.201600359.

Ahmadi, M., Ayyoubzadeh, S. M., Ghorbani-Bidkorbeh, F., Shahhosseini, S., Dadashzadeh, S., Asadian, E., Mosayebnia, M., & Siavashy, S. (2021). An investigation of affecting factors on MOF characteristics for biomedical applications: A systematic review. *Heliyon, 7*(4), e06914. https://doi.org/10.1016/j.heliyon.2021.e06914.

Ahmed, A., McHugh, D., & Papatriantafyllopoulou, C. (2022). Synthesis and biomedical applications of highly porous metal–organic frameworks. *Molecules, 27*(19), 6585. https://www.mdpi.com/1420-3049/27/19/6585.

Alves, R. C., Schulte, Z. M., Luiz, M. T., Bento da Silva, P., Frem, R. C. G., Rosi, N. L., & Chorilli, M. (2021). Breast cancer targeting of a drug delivery system through postsynthetic modification of curcumin@N3-bio-MOF-100 via click chemistry. *Inorganic Chemistry, 60*(16), 11739–11744. https://doi.org/10.1021/acs.inorgchem.1c00538.

Asadniaye Fardjahromi, M., Nazari, H., Ahmadi Tafti, S. M., Razmjou, A., Mukhopadhyay, S., & Warkiani, M. E. (2022). Metal-organic framework-based nanomaterials for bone tissue engineering and wound healing. *Materials Today Chemistry, 23*, 100670. https://doi.org/10.1016/j.mtchem.2021.100670.

Bellusci, M., Guglielmi, P., Masi, A., Padella, F., Singh, G., Yaacoub, N., Peddis, D., & Secci, D. (2018). Magnetic metal–organic framework composite by fast and facile mechanochemical process. *Inorganic Chemistry, 57*(4), 1806–1814. https://doi.org/10.1021/acs.inorgchem.7b02697.

Cai, W., Wang, J., Chu, C., Chen, W., Wu, C., & Liu, G. (2019). Metal–organic framework-based stimuli-responsive systems for drug delivery. *Advanced Science, 6*(1), 1801526. https://doi.org/10.1002/advs.201801526.

Canivet, J., Bonnefoy, J., Daniel, C., Legrand, A., Coasne, B., & Farrusseng, D. (2014). Structure–property relationships of water adsorption in metal–organic frameworks. *New Journal of Chemistry, 38*(7), 3102–3111. https://doi.org/10.1039/C4NJ00076E.

Chen, B., Wang, X., Zhang, Q., Xi, X., Cai, J., Qi, H., Shi, S., Wang, J., Yuan, D., & Fang, M. (2010). Synthesis and characterization of the interpenetrated MOF-5. *Journal of Materials Chemistry, 20*(18), 3758–3767. https://doi.org/10.1039/B922528E.

Chen, S., Lu, J., You, T., & Sun, D. (2021). Metal-organic frameworks for improving wound healing. *Coordination Chemistry Reviews, 439*, 213929. https://doi.org/10.1016/j.ccr.2021.213929.

Chen, W., Zhu, P., Chen, Y., Liu, Y., Du, L., & Wu, C. (2022). Iodine immobilized UiO-66-NH2 metal-organic framework as an effective antibacterial additive for poly(ε-caprolactone). *Polymers, 14*(2), 283. https://www.mdpi.com/2073-4360/14/2/283.

Chen, Z., Kirlikovali, K. O., Li, P., & Farha, O. K. (2022). Reticular chemistry for highly porous metal–organic frameworks: The chemistry and applications. *Accounts of Chemical Research, 55*(4), 579–591. https://doi.org/10.1021/acs.accounts.1c00707.

Chun, N. Y., Kim, S.-N., Choi, Y. S., & Choy, Y. B. (2020). PCN-223 as a drug carrier for potential treatment of colorectal cancer. *Journal of Industrial and Engineering Chemistry, 84*, 290–296. https://doi.org/10.1016/j.jiec.2020.01.010.

Demir Duman, F., Monaco, A., Foulkes, R., Becer, C. R., & Forgan, R. S. (2022). Glycopolymer-functionalized MOF-808 nanoparticles as a cancer-targeted dual drug delivery system for carboplatin and floxuridine. *ACS Applied Nano Materials, 5*(10), 13862–13873. https://doi.org/10.1021/acsanm.2c01632.

Du, R., Wu, Y., Yang, Y., Zhai, T., Zhou, T., Shang, Q., Zhu, L., Shang, C., & Guo, Z. (2021). Porosity engineering of MOF-based materials for electrochemical energy storage. *Advanced Energy Materials, 11*(20), 2100154. https://doi.org/10.1002/aenm.202100154.

Feng, J., Xu, Z., Dong, P., Yu, W., Liu, F., Jiang, Q., Wang, F., & Liu, X. (2019). Stimuli-responsive multifunctional metal–organic framework nanoparticles for enhanced chemo-photothermal therapy. *Journal of Materials Chemistry B, 7*(6), 994–1004. https://doi.org/10.1039/C8TB02815J.

Furukawa, H., Ko, N., Go, Y. B., Aratani, N., Choi, S. B., Choi, E., Yazaydin, A.Ö., Snurr, R. Q., O'Keeffe, M., Kim, J., & Yaghi, O. M. (2010). Ultrahigh porosity in metal-organic frameworks. *Science, 329*(5990), 424–428. https://doi.org/10.1126/science.1192160.

Gadzikwa, T., & Matseketsa, P. (2024). The post-synthesis modification (PSM) of MOFs for catalysis. *Dalton Transactions, 53*(18), 7659–7668. https://doi.org/10.1039/D4DT00514G.

Gecgel, C., Gonca, S., Turabik, M., & Özdemir, S. (2022). An aluminum-based MOF and its amine form as novel biological active materials for antioxidant, DNA cleavage, antimicrobial, and biofilm inhibition activities. *Materials Today Sustainability, 19*, 100204. https://doi.org/10.1016/j.mtsust.2022.100204.

Gong, L., Chen, L., Lin, Q., Wang, L., Zhang, Z., Ye, Y., & Chen, B. (2024). Nanoscale metal–organic frameworks as a photoluminescent platform for bioimaging and biosensing applications. *Small, 20*, 2402641. https://doi.org/10.1002/smll.202402641.

Guan, H.-Y., LeBlanc, R. J., Xie, S.-Y., & Yue, Y. (2018). Recent progress in the syntheses of mesoporous metal–organic framework materials. *Coordination Chemistry Reviews, 369*, 76–90. https://doi.org/10.1016/j.ccr.2018.05.001.

Guo, L., Kong, W., Che, Y., Liu, C., Zhang, S., Liu, H., Tang, Y., Yang, X., Zhang, J., & Xu, C. (2024). Research progress on antibacterial applications of metal-organic frameworks and their biomacromolecule composites. *International Journal of Biological Macromolecules, 261*, 129799. https://doi.org/10.1016/j.ijbiomac.2024.129799.

Gupta, D. K., Kumar, S., & Wani, M. Y. (2024). MOF magic: Zirconium-based frameworks in theranostic and bio-imaging applications. *Journal of Materials Chemistry B, 12*(11), 2691–2710. https://doi.org/10.1039/D3TB02562D.

Hachemaoui, M., Mokhtar, A., Ismail, I., Mohamedi, M. W., Iqbal, J., Taha, I., Bennabi, F., Zaoui, F., Bengueddach, A., Hamacha, R., & Boukoussa, B. (2021). M (M: Cu, Co, Cr or Fe) nanoparticles-loaded metal-organic framework MIL-101(Cr) material by sonication process: Catalytic activity and antibacterial properties. *Microporous and Mesoporous Materials, 323*, 111244. https://doi.org/10.1016/j.micromeso.2021.111244.

Hao, C., Wu, X., Sun, M., Zhang, H., Yuan, A., Xu, L., Xu, C., & Kuang, H. (2019). Chiral core–shell upconversion nanoparticle@MOF nanoassemblies for quantification and bioimaging of reactive oxygen species in vivo. *Journal of the American Chemical Society, 141*(49), 19373–19378. https://doi.org/10.1021/jacs.9b09360.

Hashemipour, S., & Ahmad Panahi, H. (2017). Fabrication of magnetite nanoparticles modified with copper based metal organic framework for drug delivery system of letrozole. *Journal of Molecular Liquids, 243*, 102–107. https://doi.org/10.1016/j.molliq.2017.07.127.

He, L., Brasino, M., Mao, C., Cho, S., Park, W., Goodwin, A. P., & Cha, J. N. (2017). DNA-assembled core-satellite upconverting-metal–organic framework nanoparticle superstructures for efficient photodynamic therapy. *Small, 13*(24), 1700504. https://doi.org/10.1002/smll.201700504.

He, Y., Li, D., Wu, L., Yin, X., Zhang, X., Patterson, L. H., & Zhang, J. (2023). Metal-organic frameworks for gene therapy and detection. *Advanced Functional Materials, 33*(12), 2212277. https://doi.org/10.1002/adfm.202212277.

Hefayathullah, M., Singh, S., Ganesan, V., & Maduraiveeran, G. (2024). Metal-organic frameworks for biomedical applications: A review. *Advances in Colloid and Interface Science*, 103210. https://doi.org/10.1016/j.cis.2024.103210.

Huang, K., Liu, W., Wei, W., Zhao, Y., Zhuang, P., Wang, X., Wang, Y., Hu, Y., & Dai, H. (2022). Photothermal hydrogel encapsulating intelligently bacteria-capturing bio-MOF for infectious wound healing. *ACS Nano, 16*(11), 19491–19508. https://doi.org/10.1021/acsnano.2c09593.

Javanbakht, S., Nabi, M., Shadi, M., Amini, M. M., & Shaabani, A. (2021). Carboxymethyl cellulose/tetracycline@UiO-66 nanocomposite hydrogel films as a potential antibacterial wound dressing. *International Journal of Biological Macromolecules, 188*, 811–819. https://doi.org/10.1016/j.ijbiomac.2021.08.061.

Kai, M., Wang, S., Gao, W., & Zhang, L. (2023). Designs of metal-organic framework nanoparticles for protein delivery. *Journal of Controlled Release, 361*, 178–190. https://doi.org/10.1016/j.jconrel.2023.07.056.

Kazemi, A., Afshari, M. H., Baesmat, H., Bozorgnia, B., Manteghi, F., Nabipour, H., Rohani, S., Aliabadi, H. A. M., Adibzadeh, S., & Saeb, M. R. (2024). Polydopamine-coated Zn-MOF-74 nanocarriers: Versatile drug delivery systems with enhanced biocompatibility and cancer therapeutic efficacy. *Journal of Inorganic and Organometallic Polymers and Materials*. https://doi.org/10.1007/s10904-024-03173-6.

Kim, S.-N., Park, C. G., Huh, B. K., Lee, S. H., Min, C. H., Lee, Y. Y., Kim, Y. K., Park, K. H., & Choy, Y. B. (2018). Metal-organic frameworks, NH2-MIL-88(Fe), as carriers for ophthalmic delivery of brimonidine. *Acta Biomaterialia, 79,* 344–353. https://doi.org/10.1016/j.actbio. 2018.08.023.

Leng, X., Huang, H., Wang, W., Sai, N., You, L., Yin, X., & Ni, J. (2018). Zirconium-Porphyrin PCN-222: pH-responsive controlled anticancer drug oridonin. *Evidence-Based Complementary and Alternative Medicine, 2018*(1), 3249023. https://doi.org/10.1155/2018/3249023.

Li, B., Ashrafizadeh, M., & Jiao, T. (2024). Biomedical application of metal-organic frameworks (MOFs) in cancer therapy: Stimuli-responsive and biomimetic nanocomposites in targeted delivery, phototherapy and diagnosis. *International Journal of Biological Macromolecules, 260,* 129391. https://doi.org/10.1016/j.ijbiomac.2024.129391.

Li, B., Yuan, D., Chen, H., Wang, X., Liang, Y., Wong, C. T. T., & Xia, J. (2024). Site-selective antibody-lipid conjugates for surface functionalization of red blood cells and targeted drug delivery. *Journal of Controlled Release, 370,* 302–309. https://doi.org/10.1016/j.jconrel.2024.04.038.

Li, J., Yan, Y., Chen, Y., Fang, Q., Hussain, M. I., & Wang, L.-N. (2023). Flexible curcumin-loaded Zn-MOF hydrogel for long-term drug release and antibacterial activities. *International Journal of Molecular Sciences, 24*(14), 11439. https://www.mdpi.com/1422-0067/24/14/11439.

Liu, L., Qi, G., Wang, M., He, J., Zheng, Y., Guan, J., Lv, P., & Zeng, D. (2024). Construction of intelligent response gene vector based on MOF/Fe$_3$O$_4$/AuNRs for tumor-targeted gene delivery. *International Journal of Biological Macromolecules, 277,* 134313. https://doi.org/10.1016/j.ijbiomac.2024.134313.

Liu, Q., Yu, H., Zeng, F., Li, X., Sun, J., Li, C., Lin, H., & Su, Z. (2021). HKUST-1 modified ultrastability cellulose/chitosan composite aerogel for highly efficient removal of methylene blue. *Carbohydrate Polymers, 255,* 117402. https://doi.org/10.1016/j.carbpol.2020.117402.

Liu, Y., Zhou, L., Dong, Y., Wang, R., Pan, Y., Zhuang, S., Liu, D., & Liu, J. (2021). Recent developments on MOF-based platforms for antibacterial therapy. *RSC Medicinal Chemistry, 12*(6), 915–928. https://doi.org/10.1039/D0MD00416B.

Livesey, T. C., Mahmoud, L. A. M., Katsikogianni, M. G., & Nayak, S. (2023). Metal–organic frameworks and their biodegradable composites for controlled delivery of antimicrobial drugs. *Pharmaceutics, 15*(1), 274. https://www.mdpi.com/1999-4923/15/1/274.

Ma, M., Lu, X., Guo, Y., Wang, L., & Liang, X. (2022). Combination of metal-organic frameworks (MOFs) and covalent organic frameworks (COFs): Recent advances in synthesis and analytical applications of MOF/COF composites. *Trends in Analytical Chemistry, 157,* 116741. https://doi.org/10.1016/j.trac.2022.116741.

Mahmoodi, N. M., Abdi, J., Oveisi, M., Alinia Asli, M., & Vossoughi, M. (2018). Metal-organic framework (MIL-100 (Fe)): Synthesis, detailed photocatalytic dye degradation ability in colored textile wastewater and recycling. *Materials Research Bulletin, 100,* 357–366. https://doi.org/10.1016/j.materresbull.2017.12.033.

Mohanty, B., Kumari, S., Yadav, P., Kanoo, P., & Chakraborty, A. (2024). Metal-organic frameworks (MOFs) and MOF composites based biosensors. *Coordination Chemistry Reviews, 519,* 216102. https://doi.org/10.1016/j.ccr.2024.216102.

Musarurwa, H., & Tavengwa, N. T. (2022). Smart metal-organic framework (MOF) composites and their applications in environmental remediation. *Materials Today Communications, 33,* 104823. https://doi.org/10.1016/j.mtcomm.2022.104823.

Nazari, M., Saljooghi, A. S., Ramezani, M., Alibolandi, M., & Mirzaei, M. (2022). Current status and future prospects of nanoscale metal–organic frameworks in bioimaging. *Journal of Materials Chemistry B, 10*(43), 8824–8851. https://doi.org/10.1039/D2TB01787C.

Nejadshafiee, V., Naeimi, H., Goliaei, B., Bigdeli, B., Sadighi, A., Dehghani, S., Lotfabadi, A., Hosseini, M., Nezamtaheri, M. S., Amanlou, M., Sharifzadeh, M., & Khoobi, M. (2019). Magnetic bio-metal-organic framework nanocomposites decorated with folic acid conjugated chitosan as a promising biocompatible targeted theranostic system for cancer treatment. *Materials Science and Engineering: C, 99*, 805–815. https://doi.org/10.1016/j.msec.2019.02.017.

Nguyen Thi, T. V., Luu, C. L., Hoang, T. C., Nguyen, T., Bui, T. H., Duy Nguyen, P. H., & Pham Thi, T. P. (2013). Synthesis of MOF-199 and application to CO_2 adsorption. *Advances in Natural Sciences: Nanoscience and Nanotechnology, 4*(3), 035016. https://doi.org/10.1088/2043-6262/4/3/035016.

Park, J., Jiang, Q., Feng, D., Mao, L., & Zhou, H.-C. (2016). Size-controlled synthesis of porphyrinic metal–organic framework and functionalization for targeted photodynamic therapy. *Journal of the American Chemical Society, 138*(10), 3518–3525. https://doi.org/10.1021/jacs.6b00007.

Paul, A., Banga, I. K., Muthukumar, S., & Prasad, S. (2022). Engineering the ZIF-8 pore for electrochemical sensor applications—a mini review. *ACS Omega, 7*(31), 26993–27003. https://doi.org/10.1021/acsomega.2c00737.

Pinna, A., Ricco, R., Migheli, R., Rocchitta, G., Serra, P. A., Falcaro, P., Malfatti, L., & Innocenzi, P. (2018). A MOF-based carrier for in situ dopamine delivery. *RSC Advances, 8*(45), 25664–25672. https://doi.org/10.1039/C8RA04969F.

Pourmadadi, M., Eshaghi, M. M., Ostovar, S., Shamsabadipour, A., Safakhah, S., Mousavi, M. S., Rahdar, A., & Pandey, S. (2022). UiO-66 metal-organic framework nanoparticles as gifted MOFs to the biomedical application: A comprehensive review. *Journal of Drug Delivery Science and Technology, 76*, 103758. https://doi.org/10.1016/j.jddst.2022.103758.

Priya, S., Desai, V. M., & Singhvi, G. (2022). Surface modification of lipid-based nanocarriers: A potential approach to enhance targeted drug delivery. *ACS Omega, 8*(1), 74–86. https://doi.org/10.1021/acsomega.2c05976.

Quijia, C. R., Lima, C., Silva, C., Alves, R. C., Frem, R., & Chorilli, M. (2021). Application of MIL-100(Fe) in drug delivery and biomedicine. *Journal of Drug Delivery Science and Technology, 61*, 102217. https://doi.org/10.1016/j.jddst.2020.102217.

Rabiee, N., Atarod, M., Tavakolizadeh, M., Asgari, S., Rezaei, M., Akhavan, O., Pourjavadi, A., Jouyandeh, M., Lima, E. C., Hamed Mashhadzadeh, A., Ehsani, A., Ahmadi, S., & Saeb, M. R. (2022). Green metal-organic frameworks (MOFs) for biomedical applications. *Microporous and Mesoporous Materials, 335*, 111670. https://doi.org/10.1016/j.micromeso.2021.111670.

Ren, S.-Z., Wang, B., Zhu, X.-H., Zhu, D., Liu, M., Li, S.-K., Yang, Y.-S., Wang, Z.-C., & Zhu, H.-L. (2020). Oxygen self-sufficient core–shell metal–organic framework-based smart nanoplatform for enhanced synergistic chemotherapy and photodynamic therapy. *ACS Applied Materials and Interfaces, 12*(22), 24662–24674. https://doi.org/10.1021/acsami.0c08534.

Ruyra, À., Yazdi, A., Espín, J., Carné-Sánchez, A., Roher, N., Lorenzo, J., Imaz, I., & Maspoch, D. (2015). Synthesis, culture medium stability, and in vitro and in vivo zebrafish embryo toxicity of metal-organic framework nanoparticles. *Chemistry, 21*(6), 2508–2518. https://doi.org/10.1002/chem.201405380.

Santos, K. M. C., Santos, R. J. O., De Araújo Alves, M. M., De Conto, J. F., Borges, G. R., Dariva, C., Egues, S. M., Santana, C. C., & Franceschi, E. (2019). Effect of high pressure CO_2 sorption on the stability of metalorganic framework MOF-177 at different temperatures. *Journal of Solid State Chemistry, 269*, 320–327. https://doi.org/10.1016/j.jssc.2018.09.046.

Sene, S., Marcos-Almaraz, M. T., Menguy, N., Scola, J., Volatron, J., Rouland, R., Grenèche, J.-M., Miraux, S., Menet, C., Guillou, N., Gazeau, F., Serre, C., Horcajada, P., & Steunou, N.

(2017). Maghemite-nanoMIL-100(Fe) bimodal nanovector as a platform for image-guided therapy. *Chem, 3*(2), 303–322. https://doi.org/10.1016/j.chempr.2017.06.007.

Shahzaib, A., Shaily, Kamran, L. A., & Nishat, N. (2023). The biomolecule-MOF nexus: Recent advancements in biometal-organic frameworks (Bio-MOFs) and their multifaceted applications. *Materials Today Chemistry, 34*, 101781. https://doi.org/10.1016/j.mtchem.2023.101781.

Singh, N., Qutub, S., & Khashab, N. M. (2021). Biocompatibility and biodegradability of metal organic frameworks for biomedical applications. *Journal of Materials Chemistry B, 9*(30), 5925–5934. https://doi.org/10.1039/D1TB01044A.

Sohrabi, H., Javanbakht, S., Oroojalian, F., Rouhani, F., Shaabani, A., Majidi, M. R., Hashemzaei, M., Hanifehpour, Y., Mokhtarzadeh, A., & Morsali, A. (2021). Nanoscale metal-organic frameworks: Recent developments in synthesis, modifications and bioimaging applications. *Chemosphere, 281*, 130717. https://doi.org/10.1016/j.chemosphere.2021.130717.

Toh, J. E., Lee, C. S., Lim, W. H., Pichika, M. R., & Chua, B. W. (2024). Stimulus-responsive MOF–hydrogel composites: Classification, preparation, characterization, and their advancement in medical treatments. *Open Chemistry, 22*(1). https://doi.org/10.1515/chem-2024-0061.

Treger, M., Hannebauer, A., Schaate, A., Budde, J. L., Behrens, P., & Schneider, A. M. (2023). Tuning the optical properties of the metal–organic framework UiO-66 via ligand functionalization. *Physical Chemistry Chemical Physics, 25*(8), 6333–6341. https://doi.org/10.1039/D2CP03746G.

Ullah, S., Bustam, M. A., Al-Sehemi, A. G., Assiri, M. A., Abdul Kareem, F. A., Mukhtar, A., Ayoub, M., & Gonfa, G. (2020). Influence of post-synthetic graphene oxide (GO) functionalization on the selective CO_2/CH_4 adsorption behavior of MOF-200 at different temperatures; an experimental and adsorption isotherms study. *Microporous and Mesoporous Materials, 296*, 110002. https://doi.org/10.1016/j.micromeso.2020.110002.

Valizadeh Harzand, F., Mousavi Nejad, S. N., Babapoor, A., Mousavi, S. M., Hashemi, S. A., Gholami, A., Chiang, W.-H., Buonomenna, M. G., & Lai, C. W. (2023). Recent advances in metal-organic framework (MOF) asymmetric membranes/composites for biomedical applications. *Symmetry, 15*(2), 403. https://www.mdpi.com/2073-8994/15/2/403.

Wang, H., Jian, Y., Kong, Q., Liu, H., Lan, F., Liang, L., Ge, S., & Yu, J. (2018). Ultrasensitive electrochemical paper-based biosensor for microRNA via strand displacement reaction and metal-organic frameworks. *Sensors and Actuators B: Chemical, 257*, 561–569. https://doi.org/10.1016/j.snb.2017.10.188.

Wang, N., Xie, M., Wang, M., Li, Z., & Su, X. (2020). UiO-66-NH2 MOF-based ratiometric fluorescent probe for the detection of dopamine and reduced glutathione. *Talanta, 220*, 121352. https://doi.org/10.1016/j.talanta.2020.121352.

Wang, A., Walden, M., Ettlinger, R., Kiessling, F., Gassensmith, J. J., Lammers, T., Wuttke, S., & Peña, Q. (2023). Biomedical metal–organic framework materials: Perspectives and challenges. *Advanced Functional Materials, 2308589*. https://doi.org/10.1002/adfm.202308589.

Wang, X., Wang, Y., & Ying, Y. (2021). Recent advances in sensing applications of metal nanoparticle/metal–organic framework composites. *Trends in Analytical Chemistry, 143*, 116395. https://doi.org/10.1016/j.trac.2021.116395.

Wang, D., Yao, H., Ye, J., Gao, Y., Cong, H., & Yu, B. (2024). Metal-organic frameworks (MOFs): Classification, synthesis, modification, and biomedical applications. *Small, 20*, 2404350. https://doi.org/10.1002/smll.202404350.

Wiśniewska, P., Haponiuk, J., Saeb, M. R., Rabiee, N., & Bencherif, S. A. (2023). Mitigating metal-organic framework (MOF) toxicity for biomedical applications. *Chemical Engineering Journal,* 144400. https://doi.org/10.1016/j.cej.2023.144400.

Xue, Y., Zhu, Z., Zhang, X., Chen, J., Yang, X., Gao, X., Zhang, S., Luo, F., Wang, J., Zhao, W., Huang, C., Pei, X., & Wan, Q. (2021). Accelerated bone regeneration by MOF modified multifunctional membranes through enhancement of osteogenic and angiogenic performance. *Advanced Healthcare Materials, 10*(6), 2001369. https://doi.org/10.1002/adhm.202001369.

Yang, J., Dai, D., Zhang, X., Teng, L., Ma, L., & Yang, Y.-W. (2023). Multifunctional metal-organic framework (MOF)-based nanoplatforms for cancer therapy: From single to combination therapy. *Theranostics, 13*(1), 295. https://doi.org/10.7150/thno.80687.

Yang, L., Wang, K., Guo, L., Hu, X., & Zhou, M. (2024). Unveiling the potential of HKUST-1: Synthesis, activation, advantages and biomedical applications. *Journal of Materials Chemistry B, 12*(11), 2670–2690. https://doi.org/10.1039/D3TB02929H.

Yang, Y., Xia, F., Yang, Y., Gong, B., Xie, A., Shen, Y., & Zhu, M. (2017). Litchi-like Fe_3O_4@Fe-MOF capped with HAp gatekeepers for pH-triggered drug release and anticancer effect. *Journal of Materials Chemistry B, 5*(43), 8600–8606. https://doi.org/10.1039/C7TB01680H.

Yao, S., Chi, J., Wang, Y., Zhao, Y., Luo, Y., & Wang, Y. (2021). Zn-MOF encapsulated antibacterial and degradable microneedles array for promoting wound healing. *Advanced Healthcare Materials, 10*(12), 2100056. https://doi.org/10.1002/adhm.202100056.

Yao, X., Chen, X., Sun, Y., Yang, P., Gu, X., & Dai, X. (2024). Application of metal-organic frameworks-based functional composite scaffolds in tissue engineering. *Regenerative Biomaterials, 11,* rbae009. https://doi.org/10.1093/rb/rbae009.

Yusuf, V. F., Malek, N. I., & Kailasa, S. K. (2024). Correction to "review on metal–organic framework classification, synthetic approaches, and influencing factors: Applications in energy, drug delivery, and wastewater treatment". *ACS Omega, 9*(27), 29947–29950. https://doi.org/10.1021/acsomega.4c03457.

Zhang, G. Y., Zhuang, Y. H., Shan, D., Su, G. F., Cosnier, S., & Zhang, X. J. (2016). Zirconium-based porphyrinic metal-organic framework (PCN-222): Enhanced photoelectrochemical response and its application for label-free phosphoprotein detection. *Analytical Chemistry, 88*(22), 11207–11212. https://doi.org/10.1021/acs.analchem.6b03484.

Zhang, H., Zhang, Q., Liu, C., & Han, B. (2019). Preparation of a one-dimensional nanorod/metal organic framework Janus nanoplatform via side-specific growth for synergistic cancer therapy. *Biomaterials Science, 7*(4), 1696–1704. https://doi.org/10.1039/C8BM01591K.

Zhang, X., Qi, T., Sun, Y., Chen, X., Yang, P., Wei, S., Cheng, X., & Dai, X. (2024). Chitosan nerve conduit filled with ZIF-8-functionalized guide microfibres enhances nerve regeneration and sensory function recovery in sciatic nerve defects. *Chemical Engineering Journal, 480,* 147933. https://doi.org/10.1016/j.cej.2023.147933.

Zhang, Y., Li, T.-T., Shiu, B.-C., Lin, J.-H., & Lou, C.-W. (2021). Two methods for constructing ZIF-8 nanomaterials with good bio compatibility and robust antibacterial applied to biomedical. *Journal of Biomaterials Applications, 36*(6), 1042–1054. https://doi.org/10.1177/08853282211033682.

Zhang, Y., Wang, L., Liu, L., Lin, L., Liu, F., Xie, Z., Tian, H., & Chen, X. (2018). Engineering metal-organic frameworks for photoacoustic imaging-guided chemo-/photothermal combinational tumor therapy. *ACS Applied Materials and Interfaces, 10*(48), 41035–41045. https://doi.org/10.1021/acsami.8b13492.

Zhang, Z., Sang, W., Xie, L., & Dai, Y. (2019). Metal-organic frameworks for multimodal bioimaging and synergistic cancer chemotherapy. *Coordination Chemistry Reviews, 399,* 213022. https://doi.org/10.1016/j.ccr.2019.213022.

Zheng, X., Zhang, Y., Wang, Z., Wang, Y., Zou, L., Zhou, X., Hong, S., Yao, L., & Li, C. (2020). Highly effective antibacterial zeolitic imidazolate framework-67/alginate fibers. *Nanotechnology, 31*(37), 375707. https://doi.org/10.1088/1361-6528/ab978a.

Zhong, Y., Liu, W., Rao, C., Li, B., Wang, X., Liu, D., Pan, Y., & Liu, J. (2021). Recent advances in Fe-MOF compositions for biomedical applications. *Current Medicinal Chemistry, 28*(30), 6179–6198. https://doi.org/10.2174/0929867328666210511014129.

Zhuang, J., Young, A. P., & Tsung, C.-K. (2017). Integration of biomolecules with metal–organic frameworks. *Small, 13*(32), 1700880. https://doi.org/10.1002/smll.201700880.

Zou, M., Dong, M., & Zhao, T. (2022). Advances in metal-organic frameworks MIL-101(Cr). *International Journal of Molecular Sciences, 23*(16), 9396. https://www.mdpi.com/1422-0067/23/16/9396.

Chapter 13

Metal-organic framework composites for catalysis

Ramin Ebrahimi, and Kamran Akhbari
School of Chemistry, College of Science, University of Tehran, Tehran, Iran

Introduction

Metal-organic frameworks (MOFs) are a class of crystalline materials characterized by their high porosity and surface area. The MOFs consist of paired units (metal ions or metal-oxo clusters) coordinated by organic ligands. The term "MOF" was first introduced in 1990. Due to their structures and properties, MOFs can be used in various applications, including drug delivery (Alavijeh & Akhbari, 2020, 2024; Cao et al., 2020; Parsaei & Akhbari, 2022a, 2022b, 2022c; Parsaei et al., 2024), antibacterial (Amidi & Akhbari, 2024; Davoodi et al., 2023; Karimi Alavijeh et al., 2018, 2022; Soltani & Akhbari, 2022; Wu et al., 2024), absorption, and separation of gases (Akhbari & Morsali, 2013; Alavijeh et al., 2019; Mahdipoor et al., 2023; Parsaei et al., 2022, 2023; Salimi et al., 2022, 2024; Song et al., 2024), carbon dioxide conversion (Salimi et al., 2023; Sullivan et al., 2021), the preparation of nanomaterials (Mirzadeh & Akhbari, 2016; Moeinian & Akhbari, 2015; Shahangi Shirazi & Akhbari, 2015), and catalytic processes (Kong et al., 2021; Konnerth et al., 2020; Noori & Akhbari, 2017; Shan et al., 2023; Yang et al., 2020).

MOF composites merge the properties of MOFs and functional materials, resulting in new chemical and physical properties. They address the limitations of pure MOFs, such as low mechanical strength and poor stability, while enhancing performance. The synergistic interaction between the components promotes catalytic activity by modulating electronic structures and reducing energy barriers. Tailoring the composition and structure of the composites allows for efficient charge transfer and enables a broader range of reactions. MOF composites offer advantages such as ordered porosity, tunable pore size, and diverse functionalities. They have applications in various fields and expand the catalytic repertoire, leading to more efficient and sustainable systems (Abazari et al., 2020; Song et al., 2016). Various types of functional

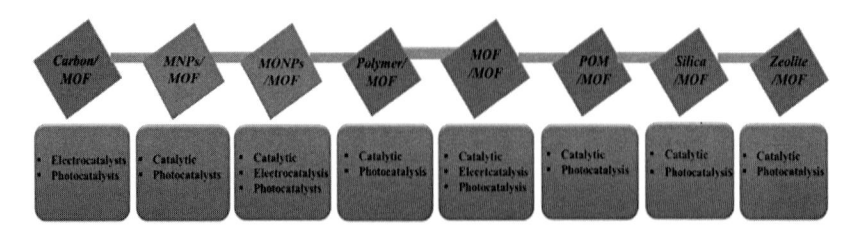

FIGURE 13.1 Schematic summary of metal-organic framework composites.

materials, such as metal nanoparticles (MNPs), metal oxide nanoparticles (MONPs), polyoxometalates (POMs), silica, polymers, carbon, zeolite, or MOFs themselves, have been integrated with the MOF to create MOF composites. Fig. 13.1 depicts the most important MOF composites and their applications as catalysts.

MOF composites combine the host MOF and guest components to create materials with enhanced or tailored characteristics. The incorporation of an additional component into MOF structures endows them with novel properties and expands their application possibilities. The strategic selection of the secondary unit profoundly influences the characteristics of the resulting composite, especially in catalytic applications. The primary motivation for synthesizing MOF composites is to improve the catalytic performance of the parent MOF. This requires a synergistic effect between the MOF and the introduced species, where both components work together to facilitate the desired reaction. The cooperativity can manifest through interactions between the MOF's metal nodes, organic linkers, porous framework, and the auxiliary moieties introduced by the secondary component (Guo et al., 2016). The strength of the connection between a MOF and a secondary unit, such as nanoparticles (NPs), is crucial for the stability of composite materials. Leaching tests are commonly used to evaluate the robustness of the MOF-secondary unit connection (Chen et al., 2014). Sinha et al. developed MOF-polymer composites by integrating MOF particles into polymer matrices. These composites exhibited improved water stability compared to pristine MOFs, effectively addressing the water instability issues of unmodified MOFs. The polymer environment provided protective properties, allowing the composites to withstand exposure to aqueous conditions (Liu et al., 2005). In another study, Kuang et al. created a multifunctional composite catalyst by using a self-sacrificial templating approach. The resulting PtNi/MOF-74-POM material functioned as a bifunctional metal/acid catalyst. The MOF-74 component played dual roles by encapsulating PtNi alloy NPs and facilitating the embedding of phosphotungstic acid ($H_3O_{40}PW_{12}$) clusters within its cavities. This composite catalyst exhibited superior conversion and selectivity in the synthesis of benzocaine compared to control catalysts lacking either the POM or PtNi components. The synergistic combination of the MOF scaffold,

MNPs, and POM clusters contributed to the enhanced catalytic performance (Chen, Zhang, et al., 2019). Below is a chart showing the significant advances in the catalytic applications of MOF composites (Fig. 13.2).

The design and synthesis of MOF composites play an important role in their catalytic application. MOF composites can be classified based on the dimensions of the secondary component included in the MOF matrix (Table 13.1).

In chemistry, catalysts are used to speed up chemical reactions or lower the activation energy. Catalysts are recyclable, selective, and lead to high product

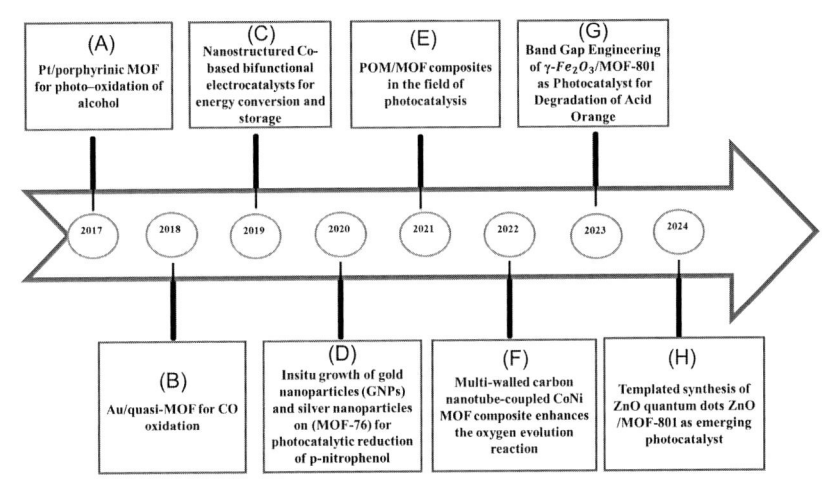

FIGURE 13.2 Timeline of important breakthroughs in the catalytic applications of metal-organic framework composites. (A) Chen, Wang et al. (2017), (B) Tsumori et al. (2018), (C) Li et al. (2019), (D) Singh et al. (2020), (E) Liu et al. (2021), (F) Yu et al. (2022), (G) Ghasemzadeh and Akhbari (2023c), and (H) Ghasemzadeh and Akhbari (2024).

TABLE 13.1 Metal-organic frameworks composition in different spatial dimensions.

Dimension	Example
Zero-dimension (0D)	NPs and quantum dots (QDs)
One-dimension (1D)	Nanowires or nanotubes.
Two-dimension (2D)	Combination of nano sheet and MOF
Three-dimension (3D)	Yolk-shell composites

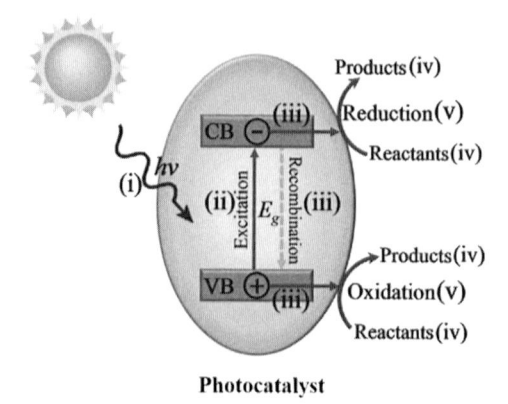

FIGURE 13.3 Schematic illustration of a conventional photocatalyst (Low et al., 2017).

yields. Photocatalysts and electrocatalysts are a subset of catalysts. Photocatalysis involves receiving light energy for chemical reactions. It includes the absorption of light by photosensitive units, the separation of electron-hole pairs, the transfer and recombination of these pairs, and redox reactions on the catalyst surface that enable the transformation of light energy to chemical energy (Fig. 13.3) (Low et al., 2017). Photocatalysis benefits from a narrow band gap for enhanced light absorption and efficient generation of electrons and holes (Ghasemzadeh & Akhbari, 2024). However, a narrow band gap can also promote recombination, reducing overall efficiency. To design efficient photocatalysts, it is important to tailor their composition and structure, adjust band gaps, improve electron-hole segregation, incorporate accessible catalytic sites, and enhance substrate affinity. Optimization of these factors enables the development of high-performance photocatalysts (Wu et al., 2023).

Electrocatalysts participate in electrochemical reactions. Electrocatalysts are a type of catalyst that work on the electrode surface. The electrocatalyst can be heterogeneous, such as a platinum surface or NPs, or homogeneous, such as an enzyme or complex compound. Electrocatalysts accelerate reactions in two general ways: by adding a chemical conversion intermediate defined by a half-reaction or by participating in electron transfer between the electrode and the reactants (Lipkowski & Ross, 1998).

Metal-organic framework composites for catalysis

Carbon/metal-organic framework composites

Carbon/MOF composites are highly attractive due to the unique properties of both MOFs and carbon-based materials. MOFs have high surface area and tunable pore size, while carbon materials have properties such as electrical

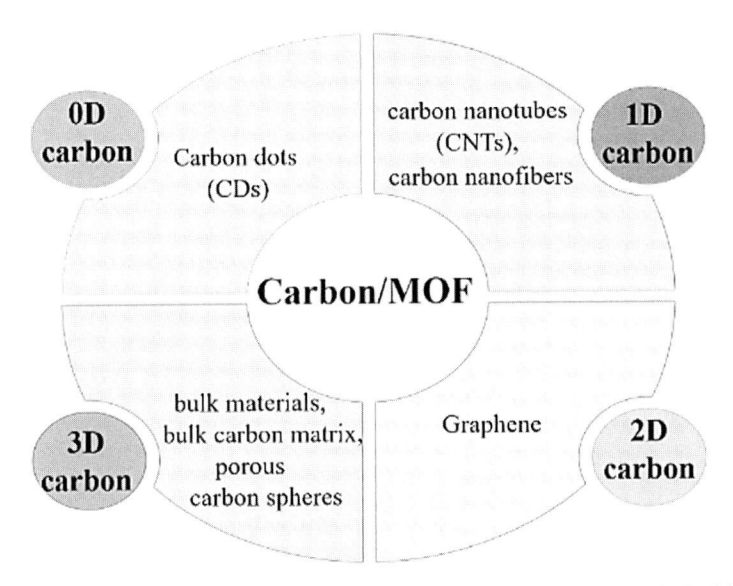

FIGURE 13.4 The combination of carbon materials with metal-organic frameworks in different spatial dimensions.

conductivity, mechanical strength, and high chemical and thermal stability. These properties make carbon materials well-suited for applications such as catalysts, energy storage systems, waste disposal, and drug delivery systems. By combining MOFs with carbon matrices, composites can leverage the high surface area and adjustable porosity of MOFs, leading to enhanced catalytic performance. Overall, carbon/MOF composites offer a promising approach to overcoming the limitations of MOFs and enabling a wide range of applications. In Fig. 13.4, the types of carbons are divided in terms of dimensions, which have different catalytic applications (Su et al., 2023; Wang, Kim et al., 2020).

Preparation methods of carbon/metal-organic framework composites

The synthesis methods of carbon/MOF composites are briefly mentioned based on the spatial dimensions of carbon in Table 13.2.

Catalytic performances of carbon/metal-organic framework composites

Electrocatalysis

Carbon/MOF hybrids offer distinct advantages and have potential applications in various fields. The integration of carbon and MOF components in these

TABLE 13.2 Types of synthesis method carbon/metal-organic framework composites and examples of each method.

Carbon/ MOF composite	Method name	Example	References
0D	Ship-in-a-bottle	CDs/NH$_2$-MIL-125(Ti) MIL-53(Fe)/CQDs	Wang et al. (2020) and Lin et al. (2018)
	Bottle-around-the-ship	MOF/CCQDs/NiF Co-ZIF/CDs/CC	Hou, Liu, et al. (2020) and Hong et al. (2023)
1D	One-step	MOFs-MWCNTs	Yu et al. (2022)
	Stepwise	ZIF-67/CNT UiO-66-NO$_2$/CoCNT	Jung et al. (2023) and Zeng et al. (2019)
		MCCF/NiMnMOFs	Cheng et al. (2020)
	Ex situ synthesis strategies	Sn-MOF/CNT	Rani et al. (2020)
2D	Solvothermal (In situ synthesis method)	Pristine-graphene-templated MOF (GMOF)	Hu et al. (2019)
	Solvothermal (ex situ synthesis)	2D Ni(C$_4$O$_4$) (H$_2$O)$_2$/graphene NSs	Zheng et al. (2019)
	Sonochemical	Cu-BDC-NH$_2$/GO	Dastbaz et al. (2019)
	Electrochemical	α-CD-rGO/Ni-MOF/TM	Xu, Zhang, et al. (2022)
	Mechanical	N-G/MOF	Zhuang et al. (2018)
3D	In situ synthesis	MOF/MCCS	Shen et al. (2021)
		ZIF-67/HMCS	Xiong et al. (2020)

hybrids results in synergistic effects, leading to improved properties and new functionalities compared to pristine MOFs alone. Key factors influencing the performance of carbon/MOF composites as electrocatalysts include conductivity, stability, surface characteristics, textural properties, and morphology. Stronger binding and increased contact area between carbon and MOF components enhance conductivity and stability. The surface characteristics, textural properties, and morphology of the composites determine the number of active sites and facilitate mass transfer. A larger surface area and hierarchical porous structures are beneficial for catalytic processes. The synergistic effects between MOFs and carbon-based materials enhance the number and intrinsic activity of active sites in carbon/MOF composites. The spatial dimension of carbon materials plays a significant role in enhancing the electrocatalytic performance of carbon/MOF composites. The use of conductive 0D carbon dots (CDs) is primarily aimed at overcoming the inherent electrical insulation of MOFs. This enables improved electron conduction and electrocatalytic efficiency of the composites (Wang et al., 2023). However, 1D carbon materials, with their distinctive structure and large specific surface area, offer even better solutions to the limitations of MOFs compared to 0D CDs (Chronopoulos et al., 2022). The network structure created by 1D carbon materials like carbon fibers (CF) and carbon nanotubes (CNTs) facilitates the formation of an array structure when combined with MOFs. This arrangement is advantageous as it exposes numerous active sites and enhances mass transfer within the composite. On the other hand, graphene-based materials, which exhibit carboxyl groups, pyridine groups, and aromatic sp^2 structural domains, can not only engage in bonding interactions with MOFs but also act as structural nodes and even electrocatalytic active components. Consequently, graphene-based materials can promote coordination bonds and induce the growth of MOFs into a more beneficial structure (Jayaramulu et al., 2022). Although three-dimensional (3D) carbon/MOF composites are relatively uncommon in electrocatalysis studies thus far, they deserve further exploration due to their unique 3D structure. The subsequent sections of the study focus on the electrochemical performances of carbon/MOF composites in various catalytic processes such as the oxygen evolution reaction (OER), oxygen reduction reaction (ORR), hydrogen evolution reaction (HER), and bifunctional catalysis (overall water splitting, ORR, and OER). Some of the relevant results are compiled in Table 13.3.

HER electrocatalysis: The electrocatalytic HER is a very important step in the process of water splitting. Previous studies have highlighted that the efficiency of the HER is determined by two key factors: the chemisorption (adsorption) and desorption of intermediates on the catalyst's surface. Catalysts that exhibit a strong binding affinity with hydrogen (H) atoms but a weak binding with the H_2 product are considered favorable for efficient HER. The specific mechanisms of the HER can vary depending on the electrolyte conditions. Under acidic conditions, the formation of H atom intermediates involves the coupling of a proton from the solution with an electron from the electrode. Subsequently, two neighboring H atoms can combine, or one H atom can combine with another proton, resulting in the formation of H_2, which then desorbs from the catalyst's surface. In contrast, under neutral or alkaline electrolytes, the absence of protons necessitates the formation of H atoms through the dissociation of absorbed water (H_2O) molecules. This distinction

TABLE 13.3 Summary of carbon/metal-organic framework composites for electrocatalysis.

Nos.	MOF	Material	Carbon	Durability (hours)	η_{10}/E1/2 (V)	Tafel slope (mV/dec)	References
HER	Hf$_{12}$-CoDBP	Hf$_{12}$-CoDBP/CNT	CNT	7	η_{10} 0.65	178	Micheroni et al. (2018)
	CoFe-MOF-74	CoFe-MOF-74/Co/CC	CC	35	0.094	81.9	Zha et al. (2020)
	Bi-TCPP MOF	RuO$_2$/CNT/MOF	CNT	24	0.050	61.665	Xu et al. (2023)
OER	NiFe-BTC	NiFe-BTC//G	G	150	106	55	Lyu et al. (2022), Li et al. (2023)
	ZIF-67	ZIF-67/HMCS	HMCS	–	0.47	99.3	Lyu et al. (2022)
	CoNi-MOF	CoNi-MOF/rGO CDs/MOF	rGO CDs	48 40	0.318 0.32	48 62	Zheng et al. (2019), Rehman et al. (2021)
	Ni-BTC						
	Ni-MOFs	Ni-MOF/CNT	CNT	–	0.37	138.2	Sreekanth et al. (2021)
	CoNi MOF	CoNi MOFs-CNTs	CNT	15	0.306	42	Yu et al. (2022)
	Ni-HMOF	Ni-HMOF/GE-PBA	GE	120	0.143@η_{50}	42	Hai et al. (2021)
	cMOF	cMOF/LDH/CC	CC	24	0.216@η_{50}	34.1	Wang et al. (2021)
ORR	ZIF-67	ZIF-67/HMCS	HMCS	7.8	$E_{1/2}$ 0.823	49.8	Xiong et al. (2020), Zheng et al. (2019), Kim, Kang, et al. (2019), Zeng et al. (2019)
	CoNi-MOF	CoNi-MOF/rGO	rGO	–	0.72	67	
	Cu-MOF/mC	HKUST-1	mC	–	0.664	–	
	UiO-66-NO$_2$	UiO-66-NO$_2$/CoCNT	CoCNT	–	0.865	8.3	

Note: η_{10}, η_{20}, and η_{50} indicate the overpotential at the current densities of 10, 20, and 50 mA/cm^2, respectively. $E_{1/2}$ refers to the half-wave potential. *DBP*, 5,15-Di(p-benzoato) porphyrin; *TCPP*, tetrakis (4-carboxyphenyl) porphyrin; *HMCS*, hollow mesoporous carbon spheres; *G*, graphene; *CC*, carbon cloth; *CNT*, carbon nanotube; *rGO*, reduced graphene oxide; *CDs*, carbon dots; *mC*, microporous carbons; *MOF*, metal-organic framework.

in the mechanism can lead to degradation of the catalytic properties under alkaline or neutral conditions (Li, Hao et al., 2019; Liao et al., 2018). Currently, platinum (Pt) is widely regarded as the most efficient electrocatalyst for the HER across a broad pH range. However, significant attempts have been made by researchers to enhance the catalytic performance of alternative cheap metals or metal-free electrocatalysts, such as carbon-based materials and MOFs. MOFs, in particular, possess unique properties that can be leveraged to reduce the free energy of hydrogen adsorption when modifying electrodes with MOF-based catalysts. This presents an opportunity for these catalysts to potentially surpass the benchmark performance of 20% Pt/C. In carbon/MOF composites, the catalytically active sites for HER primarily originate from the metal nodes within the MOFs and other active guest materials. The role of the carbon component is mainly to control the growth of MOFs, thereby exposing more active sites, as well as to enhance the conductivity and stability of the electrocatalysts. While the number of MOF-carbon electrocatalysts for HER is currently limited, their significant potential justifies further investigation and study (Hao et al., 2024).

OER electrocatalysis: The OER is a very important process in energy conversion devices such as water electrolysis cells and metal–air batteries. However, the mechanism of OER in different electrolytes is still under debate. An important factor in OER is the binding energy between the electrocatalyst and the oxygen-containing intermediates. Additionally, during the OER process, metal active centers often transform into oxides or hydroxides. Further research is necessary to fully understand the OER mechanism in different electrolytes and improve its efficiency (Jiao et al., 2015; Li, Shao, et al., 2018; Liao et al., 2018). OER is a slow and energy-intensive process that requires highly active electrocatalysts to overcome its sluggish kinetics (McCrory et al., 2013). To enhance the conductivity and activity of MOFs, combining them with carbon materials has been proven effective. In these carbon/MOF composites, the metal nodes in the MOFs serve as active sites, while the carbon component acts as a stable substrate, improving conductivity and stability while facilitating desirable MOF morphology. The synergistic effect between MOFs and carbon enables the development of electrocatalysts with excellent OER properties (Zhao et al., 2016). 0D carbons, which possess different functional groups on their surfaces, have been found to facilitate hybridization and doping with various materials. When CDs are doped into MOFs, they significantly enhance electron transfer efficiency and improve the electrocatalytic activity of the composite for the OER. The activity of the electrocatalyst was improved and showed excellent stability with an initial potential of 1.5 V and an additional potential of 320 mV at a current density of 10 mA/cm^2. The enhanced OER catalytic activity can be attributed to the synergistic effect of CDs/MOF. The incorporated CDs not only increased the active surface area but also decreased resistance and improved conductivity in the composite. Furthermore, the hybridization of CDs led to stronger and more negative oxidation peaks of Ni^{2+} to Ni^{3+} species, indicating the presence of more highly active nickel species in the composite. This further contributed to the improved OER activity (Mou et al., 2022). 1D and 2D carbon materials have been found to be more advantageous for achieving superior electrocatalysts for the OER compared to 0D carbon materials. These carbon materials have features such as structural flexibility, good electrical conductivity, high mechanical strength, and wide surface area. They also facilitate the formation of desirable morphologies, such as 2D MOF nanosheet arrays, which further enhance the OER catalytic activity. For example, Sreekanth et al. integrated marigold flower-like Ni-MOFs onto multiwalled carbon nanotubes (MWCNTs) to form a composite. This composite exhibited improved OER activity with a reduced extrapotential of 370 mV to reach a flow density of 10 mA/cm^2. The enhanced performance was attributed to the hierarchical micro-flower morphology of the composite. The high conductivity of MWCNTs improved the electronic conductivity of the composite and facilitated ion transport through the linked network-like structure during redox reactions. Additionally, the flower-like MOF structure provided a high surface area, offering more active sites for the OER (Sreekanth et al., 2021).

ORR electrocatalysis: The ORR is an important process in fuel cells and metal–air batteries. It involves the reduction of oxygen molecules to hydroxide or water through multiple electron transfers. The ORR mechanism is complex, with various intermediate products such as O*, OH*, and OOH*. There are two main pathways for ORR: a 4-electron pathway and a 2-electron pathway. The 4-electron pathway directly reduces O_2 to H_2O, while the 2-electron pathway forms intermediates like H_2O_2 or HO- before producing H_2O. The 4-electron pathway is faster but requires higher energy to break the O–O bond. Understanding these pathways is crucial for developing efficient catalysts and electrode materials in metal–air batteries and fuel cells (Li et al., 2019; Wu et al., 2012). The use of carbon-MOF composites for ORR is still relatively limited; there is significant potential for designing and achieving carbon/MOF composites with high catalytic performance for ORR. These composites can leverage the synergistic effects of MOFs and carbon materials, offering improved conductivity, stability, and enhanced catalytic properties for ORR. Zhong et al. successfully designed a highly efficient ORR electrocatalyst by combining a 2D conjugated MOF (PcCu-(OH)$_8$) with CNTs. The composite exhibited excellent catalytic performance, primarily attributed to the active sites of Co-O$_4$ centers and the synergistic effects between CuN$_4$ units and Co-O$_4$ centers. The presence of CNTs improved the overall conductivity of the composite, enabling efficient electron conduction (Zhong et al., 2019).

Photocatalysis

Carbon-based materials, including nitrogen-doped carbon graphite (g-C$_3$N$_4$), graphene-based materials (graphene, graphene oxide [GO], reduced graphene oxide [rGO]), CNTs, and fullerenes, are widely utilized as substrates for the development of materials with desirable electrical properties. These materials possess stable structures, efficient electron transfer capabilities, and exceptional light absorption and electron buffering capacities, enabling them to facilitate the transfer of photogenerated electrons in composite systems. Given the inherent poor conductivity of most MOFs, carbon materials have been increasingly employed as cocatalysts in the design of photocatalytic MOF composites to enhance charge separation. For instance, researchers introduced g-C_3N_4 during the growth process of Co-Mn-MOF-74, resulting in the synthesis of a Z-scheme heterojunction photocatalyst (Wen et al., 2023). This modification significantly improved the efficiency of photocatalytic degradation, particularly for doxycycline hydrochloride, surpassing the performance of single-component catalysts. Additionally, C60 (C60 is a fullerene), known for its strong electron-accepting properties, was incorporated into the pores of NU-901, inducing a robust built-in electric field through host-guest interactions. This interaction greatly enhanced charge separation efficiency and achieved highly efficient photocatalytic hydrogen production under 420 nm light irradiation (Liu, Meng, et al., 2023). In addition, the addition of CNTs during the synthesis of MOF-808, in concentrations of 0.1% or 0.3% by weight, was done by the solvothermal method. It has been used for the photocatalytic degradation of carbamazepine (CBZ) and diazinon. Structure MOF-808 alone can transfer a limited number of electrons from the valence bond (VB) to the conduction band (CB) with light irradiation. Still, when CNTs are placed in the structure of MOF-808, CNTs themselves have electrons and can produce peroxide radicals that contribute to the photocatalytic decomposition of CBZ and diazinon helps accelerate the process (Samy et al., 2021). Another study involved anchoring rGO on the surface of MOF via a simple one-step hydrothermal method, yielding MOF-5/rGO composites with high photocatalytic efficiency (Thi et al., 2020). Generally, carbon-based materials serve as valuable components in composite materials, either as photosensitive components (e.g., g-C_3N_4) or co-catalysts. As cocatalysts, they demonstrate excellent charge transfer ability, acting as transmission channels for photogenerated charges produced by MOFs while also offering multiple active sites for enhanced performance.

Metal nanoparticle/metal-organic framework composites

MNPs, particularly small-sized MNPs, are receiving growing attention due to their crucial role in catalysis. The MNPs have desirable properties for various applications but are prone to aggregation due to their high surface free energy (Liu, Iocozzia et al., 2017; Prieto et al., 2013; Seh et al., 2017). To address this challenge, fixing MNPs on porous supports will be able to prevent aggregation. A confined space is created by using MOFs as a host, which limits the growth of particles and prevents their aggregation (Chen & Xu, 2019; Falcaro et al., 2016; Yang et al., 2017). By loading MNPs into the porous structure of MOFs, unique properties can be achieved that are not present in individual materials (Dhakshinamoorthy & Garcia, 2012). Additionally, fine adjustment of the size and shape of the MNPs within the MOF pores can further enhance their catalytic activity. Numerous MNP/MOF composites have been created and utilized in catalytic reactions. In these composites, MNPs serve as the active centers for catalytic conversions, while MOFs play a crucial role in stabilizing the MNPs within their pores and modulating the electronic environment to enhance catalytic activity. As a result, MNP/MOF composites demonstrate a synergistic effect, leading to raised activity compared to individual MNPs or physical mixtures of components (Bavykina et al., 2020).

There has been a lot of interest in the use of metal in recent years, as green heterogeneous catalysts encapsulated within MOFs (Han et al., 2019; Xia et al., 2019). These composite materials have been employed in various chemical transformations, such as reductions (Singh et al., 2020) and oxidation (Bassen et al., 2023), which have been reported in the literature. Due to the high surface energies and S_{BET}-to-volume ratio of MNPs, they have a tendency to aggregate. The specific choice of MOF and MNPs depends on the desirable properties and intended use of the composite material.

Noble MNPs (such as gold or Au NPs) encapsulated in MOF crystals, particularly in a core-shell configuration known as Au NP/HKUST-1. Efforts have been made to construct well-defined structures of MNP/MOF, aiming to prevent the agglomeration of MNPs on the outer surface of the MOF crystals. Additionally, the incorporation of multiple types of NPs within MOFs has also been a topic of interest. These endeavors seek to achieve controlled and confined environments for the NPs within the MOF framework, enabling enhanced stability, catalytic activity, and potential synergy between different types of NPs.

Preparation methods for metal nanoparticle/metal-organic framework composites

There are various methods to confine MNPs in the pores of MOFs. The bottle-around-ship, one-step synthesis, and ship-in-bottle are three common procedures used to construct these composite materials. These methods allow for the controlled incorporation of MNPs into MOFs, resulting in composites with diverse applications. For example, palladium (Yang, Yao, et al., 2019), gold (Dai et al., 2022), ruthenium (Wang et al., 2022), copper, platinum (Zhang, Shi et al., 2020), nickel (Xu, Liu, et al., 2022), silver (Shen et al., 2020) NP/MOFs, etc. (Fig. 13.5A–C).

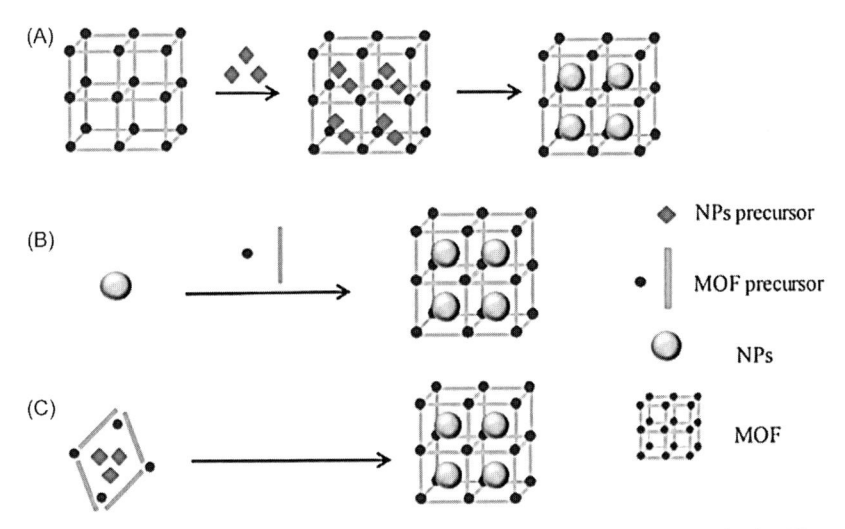

FIGURE 13.5 (A) Ship-in-bottle, (B) Bottle-around-ship, and (C) One-step synthesis (Xiang et al., 2017).

Catalytic performances of metal nanoparticle/metal-organic framework composites

Common catalytic performances

In the initial findings, the combination of MNP/MOF composites has demonstrated effectiveness in catalyzing liquid-phase reduction reactions and gas-phase oxidation reactions. Specifically, a zeolite-type MOF was utilized as a support for synthesizing nanoparticulate Au catalysts through a straightforward solid grinding technique. This approach, which is the first of its kind, resulted in the production of catalysts that exhibited significant activity in gas-phase carbon monoxide (CO) oxidation. These findings open up new possibilities for developing high-performance gold catalysts by utilizing rapidly advancing MOFs as supports. Ongoing research is focused on investigating the detailed catalytic mechanism and extending the application of MOFs, particularly ZIF-8, to other host MOFs for the incorporation of Au NPs and exploring their catalytic properties (Xiang et al., 2017). Typically, in catalysis reactions involving MNP/MOF composites, the MNPs serve as the active centers responsible for catalytic transformations, while MOFs play a dual role as stabilizers and regulators. MOFs enhance the stability of MNPs, protecting them from aggregation or deactivation (Chen et al., 2018; Li, Zhao, et al., 2018).

MOFs possess highly flexible and adjustable pore structures, which enable them to exhibit a molecular sieve (MS)-like efficacy for selective transformations. By synthesizing MOFs with appropriate pore sizes, they can selectively permit the diffusion of substrates smaller than the MOF pores into the pores, where they can reach the MNPs (active sites) and undergo selective catalysis. Gao et al. (2021) conducted a study in which they synthesized core-shell structures consisting of MNPs encapsulated within ZIF-8. By combining the catalytic properties of the MNPs (specifically, Pt NPs) with the molecular sieving ability of ZIF-8, they investigated the hydrogenation function of these nanocomposites in the context of *cis*-cyclooctene and *n*-hexene. The experimental results indicated that the conversion of *n*-hexene reached 13.3%, which was higher compared to the conversion of *cis*-cyclooctene at 1.7%. This difference in reactivity can be attributed to the molecular sieving effect of ZIF-8. Since *cis*-cyclooctene has a larger size than the pore size of ZIF-8, it was unable to reach the active sites located within the MOF's pores. Consequently, catalytic activity for *cis*-cyclooctene was lower compared to n-hexene, which could distribute into the pores and interact with the active sites more effectively. Another example involves the use of MNPs/MOF composites as catalysts for the selective hydrogenation of cinnamaldehyde. In particular, Pt/UiO-66-NH$_2$ composites were investigated, where Pt NPs were encapsulated within the pores of UiO-66-NH$_2$. These composites exhibited both high selectivity and conversion in transforming cinnamaldehyde into cinnamyl alcohol. On the other hand, Pt/UiO-66-NH$_2$ composites, where Pt NPs were loaded onto the external surface of UiO-66-NH$_2$, showed much lower selectivity toward cinnamyl alcohol. This difference in selectivity can be attributed to the steric hindrance imposed by UiO-66-NH$_2$. The C=C bond in the middle of the cinnamaldehyde molecule makes it difficult to reach the platinum surface compared to the C=O bond at the end of the molecule (Guo et al., 2024).

Electrocatalysis

HER electrocatalysis: To date, there have been fewer reported catalysts for the HER in alkaline environments compared to acidic ones. However, in alkaline media, HER catalysts can be broadly classified into three main groups. The first group comprises noble metals and their alloy-based catalysts, such as Pt, Ir, Ru, Pd, and Ag. The second group consists of low-cost transition metals and their diverse nanostructures, including Fe, Cu, Mn, Co, Mo, W, and Ni. Lastly, the third group comprises nonmetal-based catalysts, which encompass elements such as B, P, C, N, S, and their alloys. Monama et al. prepared CuPc/MOF composites using a simple impregnation method followed by palladium chemical plating to synthesize Pd/CuPc/MOF. The resulting Pd/CuPc/MOF composite exhibited excellent catalytic properties for the HER. It displayed a low Tafel slope of Mv/dec and remained stable even in 0.300 mol/L H$_2$SO$_4$, with a high exchange current density of 8.9 A/m^2. These results indicate that the catalyst exhibited well HER activity and stability. The presence of CuPc in the composite played a crucial role in enhancing electron transfer and increasing the electrochemically accessible surface area. This facilitated efficient charge transfer at the catalyst/electrolyte interface, ultimately resulting in improved HER catalytic activity. The incorporation of CuPc in the MOF composite brought about enhanced functionalities and contributed to the overall performance enhancement of the catalyst (Koçyiğit et al., 2017; Monama et al., 2018; Wen & Guan, 2019). A summary of the HER activity of MOF composites in acidic/neutral/alkaline conditions is listed in Table 13.4.

The use of transition metals for HER activity still needs to be investigated. Transition metals have demonstrated good catalytic activity for HER (Wen & Guan, 2019). Therefore, the synthesis of MNP/MOF composites based on transition metals holds great potential for HER applications.

OER electrocatalysis: The electrical conductivity and catalytic activity of fresh MOFs have posed significant limitations on their electrocatalytic applications, despite the considerable interest in utilizing MOF-based materials as OER catalysts. The incorporation of MNPs to form MNP/MOF composites is an approach to increase the electrocatalytic efficiency of MOFs (Liang et al., 2020). The catalytic performance of OER is improved by placing Pt NPs in two dimensions (NiNSMOFs) and synergistic interaction between metal and platinum (Xia et al., 2019). The Pt/NiNSMOFs composite, with a Pt loading of 3.2 wt.%, exhibits outstanding OER performance. It achieves an extrapotential of only 55 mV at a current density of 10 mA/cm^2, which is significantly lower compared to the extrapotential of commercial RuO$_2$, a commonly used OER catalyst. This demonstrates the superior catalytic activity of Pt/NiNSMOFs for the OER. In the field of OER, researchers have also developed catalysts based on MOFs. Ma et al. conducted a study in which they synthesized two metallic MOF-encapsulated Cu-Ni NPs (Ni-Cu/Cu-Ni-MOF) through in situ etching using Cu-Ni circular nanostructures as precursors. By comparing the performance of Ni-Cu/Cu-Ni-MOF with Cu-Ni seeds, Ni-MOF-74, and Cu-MOF-74, they found significant improvements. At an overpotential of 624 mV, the flow density of Ni-Cu/Cu-Ni-MOF was 1.48, 1.60, and 71.18 times higher than that of Cu-Ni seeds, Ni-MOF-74, and Cu-MOF-74, respectively. In addition, the stability test (after 30,000 seconds) showed that Ni-Cu/Cu-Ni-MOF retained 87% of its current density (6.6 mA/cm^2), approximately two times that of Cu-Ni grains (3.4 mA/cm^2). The encapsulation of two metallic NPs within the MOFs enhanced both the electrical conductivity and stability of the MOFs. Furthermore, the two metallic system provided additional active sites, leading to enhanced OER activity (Ma, Qi, et al., 2019). The development of MNP/MOF electrocatalysts for the OER is an ongoing research area. To make these electrocatalysts practically applicable, there is a need for reducing the extrapotential required for the OER and improving their stability. Considering the high cost and limited availability of precious metals, there is a growing expectation for the development of effective MNP/MOF catalysts based on mediator metals for OER in the future. Researchers are actively exploring alternative materials and formulations to achieve these goals and pave the way

TABLE 13.4 Examples of metal-organic framework composite in different conditions for activity hydrogen evolution reaction.

Sample	Electrolyte	Tafel slope (mV/dec)	Extrapotential (mV)	Reference
Cu/MOF composite		84	209 at 30 mA/cm^2	Jahan et al. (2013)
CTGU-5	0.5 M H$_2$SO$_4$	−125	−388	Zaman et al. (2021)
CTGU-6	0.5 M H$_2$SO$_4$	−176	−425	Zaman et al. (2021)
AB&CTGU-5	0.5 M H$_2$SO$_4$	−45	−44	Zaman et al. (2021)
UiO-66-NH$_2$/Mo-5	0.5 M H$_2$SO$_4$	−59	−200	Dai et al. (2016)
NENU-500	0.5 M H$_2$SO$_4$	−96	−237	Qin et al. (2015)
CoP/BCN-1	1.0 M KOH	−52	−215	Tabassum et al. (2017)

for sustainable and economically viable OER electrocatalysis. Progress in the development of MNP/MOF electrocatalysis for the OER is still ongoing. To achieve practical application of these catalysts, further efforts must be made to reduce the overpotential and enhance their stability. Due to the high price and rarity of precious metals, it is expected that efficient MNP/MOF electrocatalysis based on transition metals for the oxygen evolution reaction will be introduced in the future.

CO_2 reduction reaction: The development of efficient electrocatalysts is crucial for improving the CO_2RR process. MNPs supported by MOFs offer a promising approach to address the challenges in CO_2RR, providing stability and increased catalytic performance. The choice of catalyst and active center determines the products obtained in CO_2RR. MNPs offer a high surface area-to-volume ratio, providing a large number of active sites (Wang, Liu et al., 2021). Nonetheless, small MNPs with high surface energy tend to agglomerate during catalytic processes. MOFs emerge as promising carriers for stabilizing MNPs due to their high porosity and tunable chemical composition and structure (Wang et al., 2021; Xin et al., 2021). Depending on the active center of the catalyst, different products (such as formic acid, methanol, CO, formaldehyde, methane, and ethylene) can be obtained.

CO_2 electroreduction to CO: The electrochemical reduction of CO_2 to CO can achieve high current density with minimal energy input and a high efficiency in terms of Faraday yield. CO, as a chemical building block, holds potential as a fuel source, and the resulting CO_2 from its combustion can be recycled, creating a closed carbon cycle. Consequently, the selective conversion of CO_2 to CO represents a promising approach (Chen, Khosrowabadi Kotyk, et al., 2018; Hou, Liang, et al., 2020; Hu et al., 2022). Guntern et al. (2019) synthesized composites consisting of silver nanoparticles (Ag NCs) supported on an aluminum-based porous metal-organic framework (Al-PMOF) and investigated their catalytic performance for CO_2 reduction. The close interfacial contact between Ag NCs and Al-PMOF, achieved by removing the native ligand using acetone, enables efficient electron transfer from Al-PMOF to Ag NCs. Electron-rich silver NCs enhance the activation of CO_2 molecules by facilitating electron movement from active sites to CO_2 antibonding orbitals, resulting in an improved selectivity for CO_2 reduction over the competing HER. Across the entire potential range studied, the Ag/Al-PMOF composite demonstrated enhanced selectivity for CO production. Specifically, at a potential of −1.1 V versus the reversible hydrogen electrode (RHE), the highest recorded Faraday efficiency for CO production with Ag/Al-PMOF reached 55.8%, representing a 2.2-fold increase compared to bare Ag NCs.

Nitrogen reduction reaction: The conventional Haber-Bosch process for ammonia production requires high pressure and temperature, resulting in significant carbon emissions and energy consumption (Lin et al., 2022; Qiang et al., 2023; Suryanto et al., 2021). As an alternative, the nitrogen reduction reaction (NRR) offers a more environmentally friendly approach. NRR replaces the steam reforming step, reducing energy consumption and gas emissions. It allows for the conversion of renewable energy sources into portable chemicals, eliminates CO_2 emissions, enables on-site ammonia production, and offers a modular technology for electrochemical applications. The NRR holds promise as a sustainable and efficient method for ammonia production, and ongoing research aims to optimize this process under ambient conditions (Liu, He, et al., 2023; Xie et al., 2023; Yao et al., 2023). The NRR has garnered attention for its potential to occur under mild conditions, reducing energy consumption and emissions (Ren et al., 2021). However, the progress of electrochemical NRR has been hindered by competition from the HER. MOFs and their composites have emerged as promising electrocatalysts for NRR. They offer advantages such as effective regulation and the ability to concentrate nitrogen gas (N_2) within the catalytic system (Chanda et al., 2021). Introducing hydrophobicity to prevent water from entering the catalytic system has proven successful in suppressing the HER. These advancements in MOFs and their composites contribute to more efficient and selective electrochemical NRR systems, with potential applications in fertilizer production and energy storage (Cai et al., 2020; He, Wen, et al., 2022). He et al. synthesized a hydrophobic composite by encapsulating ultrafine Au NPs within a MOF crystal with disulfide trimer units. The composite was further surface-modified using silicone to enhance its hydrophobicity. This hydrophobic Au/MOF composite demonstrated efficient loading of small-sized Au NPs and suppressed the HER. The composite exhibited a high ammonia yield of 49.5 μg/(h·mgcat) at −0.3 V versus RHE, with a Faradaic efficiency (FE) of 60.9%. These values were significantly higher than those of the unmodified Au/MOF. The study highlights the potential of hydrophobic MOF-based composites for improving the performance of electrocatalytic NRR and advancing sustainable ammonia production (He, Zhu, et al., 2022). MNPs embedded in MOF catalysts have shown promising activity for the NRR while effectively mitigating the competing HER. However, for wide application of these MNP/MOF catalysts, it is crucial to develop cost-effective and simple synthesis methods.

Photocatalysis

CO_2 reduction: Photocatalytic CO_2 reduction using MOF catalysts is a promising approach to address the energy crisis and reduce CO_2 diffusion (Inoue et al., 1979; Wang & Wang, 2016). However, the selectivity and efficiency of pure MOFs are low. To improve performance, MNPs can be incorporated into MOFs, creating MNP/MOF photocatalysts. These hybrid materials enable the reduction of CO_2 to produce valuable C1 products such as CO, CH_3OH, and $HCOO^-$. By combining the CO_2 capture capability and electronic properties of MOFs with the catalytic properties of MNPs, MNP/MOF photocatalysts offer enhanced efficiency and selectivity in CO_2 reduction. This approach has the potential to advance sustainable energy production and CO_2 mitigation through photocatalysis (Liang, Qu et al., 2018; Zhang et al., 2020).

Hydrogen production: Photocatalytic hydrogen production, particularly in the visible light range, offers a sustainable solution to environmental and energy challenges (Bie et al., 2022; Li et al., 2015). Modification of organic binders in MOFs enables them to respond to UV and visible light, while MNPs stabilized by MOFs act as efficient electron acceptors, facilitating the separation of photogenerated charges. The high photocatalytic activity for hydrogen production depends on the synthesis methods of MNP/MOF composites. Designing photocatalysts with superior performance under visible light irradiation is a desirable goal (Rahman et al., 2020; Shi et al., 2016). MOFs are considered semiconductors capable of transitioning into a charge-separated state upon photon absorption; it has a hole in the valence band (positive) and an electron in the conduction band (negative) (Shi et al., 2019; Silva, Corma, et al., 2010). While pristine MOFs can be used for photocatalytic hydrogen production, their hydrogen production rates, such as UiO-66 and UiO-66-NH_2, are relatively low (Silva, Luz, et al., 2010). The incorporation of Pt NPs into MOFs greatly enhances water-splitting activity (Wang, Zhen, et al., 2018; Yue et al., 2018). For instance, Pt/UiO-66-NH_2, where ultrafine Pt NPs are encapsulated within the MOF, exhibits significantly improved catalytic performance compared to UiO-66-NH_2 and Pt/UiO-66-NH_2. The Pt/UiO-66-NH_2 catalyst demonstrates a production rate about 150 times higher than UiO-66-NH_2 and five times higher than Pt/UiO-66-NH_2. The presence of Pt NPs allows for easy access to protons and efficient electron transfer, resulting in enhanced activity and stability (Xiao et al., 2016). Similarly, Pt clusters with ultra-small sizes encapsulated within MIL-125-NH-CH_2OH MOFs show a synergistic effect that promotes charge transfer and separation. The Pt/MIL-125-NH-CH_2OH catalyst achieves a significantly higher H_2 production yield compared to pristine MIL-125-NH-CH_2OH (Huang et al., 2021).

Ni/MOF-5, consisting of small nickel particles (9 nm) dispersed within MOF-5, shows promise as a hydrogen evolution catalyst. Under visible light irradiation, it achieves a hydrogen evolution rate of 30.22 mmol/(h·g), similar to Pt/MOF-5. Ni/MOF-5 exhibits excellent stability, retaining 99% activity in the second cycle, indicating its potential for efficient and durable photocatalytic hydrogen production (Zhen et al., 2016). Efficient charge transfer and separation are crucial for achieving high photocatalytic activity and improved performance in photocatalytic reactions (Xiao et al., 2018). Introducing a new guest to form a mixed or heterogeneous structure to strengthen generation H_2, such as PNPMOF (POMs and metal NPs incorporated into MOFs). In PNPMOF, POMs facilitate catalytic H_2 evolution by promoting electron transfer between NH_2-MIL-53 and catalytic Pt NPs. This strategy accelerates the photocatalytic process and enhances overall hydrogen production (Guo et al., 2016).

The factor that slows down the production of hydrogen is the direct decoration of the electron intermediate ferrocene carboxylic acid (Fc) as a guest on the outer surface of Pt/UiO-66-NH$_2$ (Pt/UiO-66-NH$_2$/Fc). In this case, the transfer of electrons primarily occurred from the MOF to Fc rather than Pt. However, when Fc was introduced into the pores of UiO-66-NH$_2$, forming Pt Fc/UiO-66-NH$_2$, the Fe portion in Fc was oxidized, leading to a higher photocatalytic H$_2$ production rate of 514.8 mmol/(g.h) compared to Pt/UiO-66-NH$_2$/Fc (102.6 mmol/(g·h), Xu et al., 2021). The presence of heterojunctions in photocatalysts plays a crucial role in enhancing photocatalytic activity. The formation of p–n heterojunctions between porous Co$_3$O$_4$ and BiVO$_4$ in ZIF-67 MOFs-derived materials improves charge transfer and reduces electron-hole recombination, leading to enhanced performance (Hou et al., 2015). Additionally, a highly reactive Au/CdS/MIL-101 heterostructure demonstrates a high H$_2$ yield of 25,000 μmol/(h.g) under visible light irradiation, attributed to enhanced electron separation and transfer (Wang, Zhang et al., 2016). Zhang et al. (2020) designed a Pt/NH$_2$-UiO-66/MnO$_x$ heterostructured photocatalyst with cocatalysts Pt and MnO$_x$, achieving a hydrogen production rate of 1340.6 μmol/(g.h). The spatial separation of Pt and MnO$_x$ in NH$_2$-UiO-66 inhibits electron-hole recombination, while the MnO$_x$ promotes the transfer of photogenerated holes and inhibits recombination. These heterostructures significantly enhance photocatalytic performance and hold promise for efficient hydrogen production.

CO$_2$ photoreduction to CO: The incorporation of MNPs, such as Pt, Ag, Au, and Cu, has been found to enhance the catalytic performance of MOFs by providing active sites and promoting light absorption (Guo et al., 2021; Najafi et al., 2021). For example, Deng et al. (2019) inserted Ag NPs with hollow structures into Co-MOF-74 for photocatalytic reduction of CO$_2$. AgNPs/MOF-74 showed 3.8 times more activity than MOF-74, which was attributed to the localized surface plasmon resonance (LSPR) of Ag NPs. Another effective approach to improve catalytic performance is by tuning the structure of MOFs to facilitate effective charge transfer in hybrid catalysis (Li, Cui et al., 2022). Guo, Yang, et al. (2019) synthesized nanoscale MIL-101(Cr)-Ag with tunable size, resulting in good photocatalytic performance. The introduction of Ag-induced defect states in MIL-101(Cr)-Ag, which inhibited charge recombination, promoted charge separation, and ultimately improved the photocatalytic activity for CO$_2$ reduction. These findings highlight the potential of incorporating MNPs and optimizing MOF structures to enhance the catalytic performance of MOFs for various applications. By utilizing semiconductor-like NH$_2$-UiO-68 as a carrier with amino-functionalized linkers and permanent porosity, it becomes possible to embed Pt NPs inside MOFs while maintaining good charge-transfer ability (Guo, Wei, et al., 2019). This approach, denoted as Pt/NH$_2$-UiO-68, demonstrates enhanced photocatalytic activity

(400.2 mmol/g) compared to externally deposited Pt NPs (Pt/NH$_2$-UiO-68) (121.2 mmol/g). The effective charge transfer from NH$_2$-UiO-68 to the excited Pt NPs effectively suppresses the recombination of photogenerated charge carriers, resulting in an overall enhancement of photocatalytic performance.

CO$_2$ photoreduction to CH$_3$OH: The production of CO$_2$ in an environmentally friendly manner involves either a one-step or two-step reaction to obtain methanol. However, this process faces challenges related to enhancing charge separation and achieving efficient visible light absorption (Hu et al., 2021; Onishi & Himeda, 2022; Yang et al., 2022). To address these challenges, a study conducted by Becerra et al. (2020) focused on depositing Au NPs on stable ZIF-67 to improve the semiconductor properties through the LSPR effect of Au NPs. This resulted in improved photocurrent and raised charge separation. The hybrid material, Au20/ZIF-67, exhibited a maximum methanol production rate of 2.5 mmol/(g.h) and an ethanol generation rate of 0.5 mmol/(g.h). In contrast, pristine ZIF-67 did not produce any products under the same conditions. These findings highlight the effectiveness of incorporating Au NPs onto ZIF-67 to enhance the photocatalytic activity and promote the conversion of CO$_2$ into valuable chemicals such as methanol. In a study conducted by Ostad et al. (2022), Cu- and Au-modified ZIF-8 nanocomposites, namely Cu/ZIF-8 and Au/ZIF-8, were synthesized. These nanocomposites exhibited a narrower band gap energy compared to pristine ZIF-8, enabling them to absorb a wider range of visible light for photocatalytic reactions. Both Au/ZIF-8 and Cu/ZIF-8 photocatalysts demonstrated exceptional yields of 2650 and 2240 μmol/g, respectively, when a 0.1 M Na$_2$SO$_3$ solution was used. These yields were significantly higher than that of pristine ZIF-8 (1300 μmol/g) under the same conditions. The results indicate that the introduction of Cu and Au into ZIF-8 enhances the photocatalytic activity, leading to improved yields in the conversion of reactants. This highlights the potential of Cu/ZIF-8 and Au/ZIF-8 nanocomposites as effective photocatalysts for various applications, including the production of valuable chemicals from CO$_2$.

Hydrogenation reactions: The selective hydrogenation of nitroaromatics to produce aniline derivatives is a significant reaction in modern industry (Li et al., 2018; Murugesan et al., 2020). However, achieving high selectivity for nitro hydrogenation is challenging due to the simultaneous presence of nitro and other unsaturated groups in most cases, which conventional catalysts struggle to address (Wang, Guan, et al., 2018; Zhang et al., 2017). In yolk-shell PdAg/ZIF-8 complexes, both Pd and Ag play crucial roles. Pd exhibits excellent hydrogen and nitroaromatic activation ability, while Ag possesses strong visible light absorption properties (Li, Li et al., 2022). The LSPR properties of Ag enable effective photothermal conversion. Under visible light irradiation, the composite exhibits a remarkable selectivity of 97.5% for the hydrogenation of nitro styrene to amino styrene. DFT (density functional theory) calculations further show that the adsorption of the -NO$_2$ functional group is facilitated by highly electronegative PdAg nanocage structures and causes the selectivity of the catalyst.

These findings demonstrate the potential of yolk-shell PdAg/ZIF-8 complexes as efficient catalysts for achieving highly selective hydrogenation of nitroaromatics, especially under visible light conditions.

Oxidation reactions: MOFs have found applications in photocatalytic organic oxidation reactions as well. In a previous study by Shen et al. (2015), three M/MIL-125(Ti) composites (M=Au, Pd, Pt) were synthesized, incorporating highly dispersed MNPs. These M/MIL-125(Ti) composites exhibited enhanced activity compared to the original MIL-125(Ti) in the photooxidation of benzyl alcohol to benzaldehyde, with selectivity exceeding 99%. The improved photoactivity can be attributed to the formation of a Schottky junction between the MNPs and MIL-125(Ti), which extends the spatial separation and lifetime of the photogenerated carriers. In recent years, Cu has been found to display surface plasmon resonance upon visible light irradiation, enabling light absorption in the visible region (Lou et al., 2018; Wang, Song et al., 2019). UiO-66, a similar semiconductor with a wide band gap, was chosen as a carrier to load Cu NPs, resulting in the formation of Cu/UiO-66. This hybrid material extended light absorption to the visible region while preserving the redox capability of the MOF. The formation of a Schottky junction in Cu/UiO-66 facilitated efficient photoexcitation electron transfer while suppressing the complexation of photoinduced electrons and holes (Fig. 13.6) (Xiao et al., 2019). The conversion of benzyl alcohol over UiO-66 and 0.1% Cu/UiO-66 was 30.8% and 53.3%, respectively. The catalytic performance of 0.1% Cu/UiO-66 was improved compared to the UiO-66 precursor. These findings demonstrate the potential of utilizing MOFs, such as M/MIL-125(Ti) composites and Cu/UiO-66, for photocatalytic organic oxidation reactions, achieving enhanced catalytic performance and selectivity compared to their pristine counterparts.

Photocatalytic coupling reaction: In the Suzuki-coupling reaction, a C=C bond is formed between aryl halides and boronic acids. Pd catalysts have demonstrated high activity in Suzuki-coupling reactions. By utilizing NH_2-UiO-66 as a carrier, well-dispersed Pd NPs with a high number of active sites can be

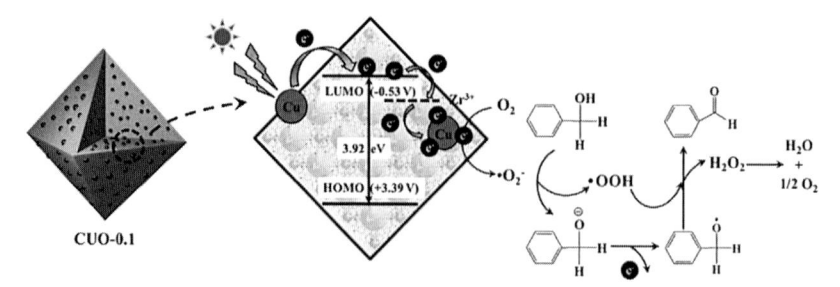

FIGURE 13.6 Migration of electrons in CuO-0.1 according to energy levels and proposed mechanism for the selective oxidation of benzyl alcohol (Xiao et al., 2019).

obtained (Sun & Li, 2016). The resulting Pd/NH$_2$-UiO-66(Zr) composite exhibited remarkable catalytic activity under visible light irradiation, achieving 90.4% conversion of iodobenzene within 30 minutes. Even after expanding the reaction eight times, Pd/NH$_2$-UiO-66(Zr) maintained its excellent catalytic performance, converting more than 99% of iodobenzene to biphenyl with a high turnover number of 2514. This indicates that the scalability of the reaction did not compromise its catalytic efficiency. To further enhance the transfer of photoexcited electrons from the excited state of NH$_2$-UiO-66(Zr) to the Pd nanocatalysts. Sun et al. introduced Cu into Pd/NH$_2$-UiO-66(Zr) (Sun et al., 2018). With an extended reaction time of 4 hours and an increased amount of triethylamine (TEA) (0.45 mmol), the CuPd/NH$_2$-UiO-66(Zr) composite achieved nearly complete conversion (99%) with a selectivity exceeding 99%. The presence of Cu as an electron mediator facilitated the transfer of photoexcited electrons from NH$_2$-UiO-66(Zr) to Pd, resulting in electron-rich Pd species and higher catalytic activity. This suggests that the interaction between metals and MOFs can modulate the electronic properties of the metal, leading to improved catalytic performance. In the photocatalytic coupling of benzylamines, which is significant in the pharmaceutical and fine chemical industries, Schottky junctions can be formed between Pt and ZIF-8 to enhance photogenerated carrier separation and improve photocatalytic performance (Sun et al., 2022). The aniline conversion using Pt/ZIF-8 was significantly improved to 99.7%, surpassing that of Pt/UiO-66 and Pt/MIL-125. The high energy gap of ZIF-8 prevents backward injection of excited electrons from the Pt bands into the MOF, resulting in increased electron density on Pt and improved catalytic performance (Fig. 13.7).

Another important reaction is the direct alkylation of amines using alcohols as alkylating agents, which is an efficient method for synthesizing N-alkyl amines (Fertig et al., 2018; Guillena et al., 2010). Wang and Li (2016) encapsulated Pd NPs in MIL-100(Fe) and applied them to the photoinduced alkylation of alcohols and amines. After 24 hours of visible light irradiation,

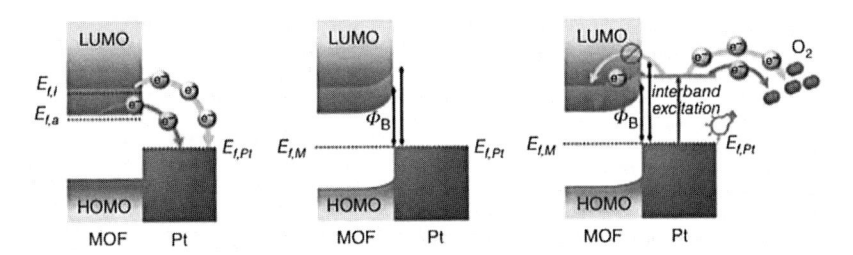

FIGURE 13.7 Schematic illustration of band diagrams for the Pt/metal-organic framework and the migration of electrons originated from interband excitation of Pt nanoparticles under light irradiation (Sun et al., 2022).

Pd/MIL-100(Fe) achieved 88% conversion of aniline, with a selectivity of 76% toward N-benzylaniline. MIL-100, acting as a semiconductor, could generate electrons to transfer into the Pd NPs, leading to the formation of electron-rich Pd species and improved catalytic performance.

MNP/MOF photocatalytic organic reactions can follow two main pathways: enhancing the photothermal effect to drive organic reactions and promoting the generation of photogenerated carriers to drive organic reactions. The tunable properties of MOF carriers and MNPs contribute to the increasing use of MNP/MOF composites in photocatalytic organic reactions, offering high activity and selectivity. Exploring different combinations of MOFs and MNPs holds great potential for the development of new composites for photocatalytic organic reactions.

Ghasemzadeh et al. (2024) reported a new MOF (MUT-16) based on cobalt with the chemical formula of $Co_2(DCITPA)_2(DABCO)\cdot(DMF)_4$, (DCITPA= 2,5-dichloroterephthalic acid and DABCO=1,4-diazabicyclo[2.2.2]octane), which has been designed and prepared by the solvothermal method. Using the via photoreduction route (PR) method, Ag NPs were loaded into porous MUT-16. The photocatalytic activity of the Ag/MUT-16 nanocomposite, after 30 minutes under visible light irradiation, was 87.5% in the degradation of quinoline yellow (QY). The distinctive characteristics of the Ag/MUT-16 nanocomposite that play significant roles in the photocatalytic degradation of QY, such as the Fenton-like effect of Co^{2+} ions, surface plasmon resonance (SPR) of Ag NPs, Schottky junction at interfaces between Ag NPs and MUT-16, and reduction of electron-hole recombination through electron trapping by Ag NPs as co-catalyst.

Ghasemzadeh and Akhbari (2023b) also synthesized Ag/MOF-801/MIL-88A(Fe) heterogeneous structure by a simple and cost-effective method. MIL-88A(Fe) was internally grown on MOF-801. Then, Ag NPs were integrated using a dual solvent method along with a single photoreduction pathway. It showed the highest photocatalytic activity of 91.72% for the degradation of Reactive Black 5 under visible light irradiation after 30 minutes. This improvement in catalytic activity is related to the increase in the intensity of light absorption through SPR of silver nanoparticles and the reduction of electron-hole recombination by silver nanoparticles that act as electron acceptors.

Metal oxide nanoparticle/metal-organic framework composites

Metal oxides in MOFs increase the number of active sites, which increases their absorption and catalytic ability (Behera et al., 2022; Kalati & Akhbari, 2021, 2022; Panda et al., 2023; Subudhi et al., 2020). Researchers used copper-based metal-organic frameworks (Cu-based MOFs) as templates to synthesize CuO/Cu_2O NPs while preserving aspects of the original MOF framework. Specifically, Bagheri et al. discovered that by transforming the Cu-based MOFs into CuO, they could effectively promote the catalytic conversion of 4-nitrophenol to

4-aminophenol. Interestingly, this CuO catalyst retained some of the original crystal structure of the HKUST-1 MOF, which is considered a quasi-MOF (Bagheri et al., 2022). The construction of heterogeneous bonds can accelerate the electron transfer of materials and improve the catalytic performance (Li, Huang et al., 2023). MONPs exhibit unique physical, chemical, optical, and electronic properties with respect to those of their bulk counterparts due to the quantum confinement and more availability of the surface atoms than interior atoms for participating in any reaction. Applications of MONPs depend upon their properties such as surface area, shape, size, stability, crystallinity, anticorrosiveness, conductivity, and photocatalytic activity, etc. Metal oxides play a very important role in the field of materials science and many other fields (Arshadi Edlo & Akhbari, 2023; Edlo & Akhbari, 2024; Falcaro et al., 2016). MONPs have extraordinary physical and chemical properties. Some of the known metal oxides include TiO_2, ZnO, CuO, SnO_2, VO_x, and MoO_x.

Preparation methods of metal oxide nanoparticle/metal-organic framework composites

For the synthesis of MONP/MOF, various methods have been presented in the articles, which add metal oxide before synthesis, during synthesis, or after synthesis (Table 13.5). Of course, metal oxide can also be synthesized alone, the methods of which are mentioned in Fig. 13.8 (Rao et al., 2017).

TABLE 13.5 Types of synthesis method metal oxide/metal-organic framework composites and examples of each method.

MONP/MOF composites	MOF synthesis method	MONPs synthesis method	Reference
Cu_2O/HKUST-1	Solvothermal	Oxidation Cu to CuO and Cu_2O	Li et al. (2023)
CuO/HKUST-1–20/12	Solvothermal	CuO was prepared	Jin et al. (2019)
Cu/ZnO_x/UiO-bpy	Solvothermal	Cu and ZnO was prepared	An et al. (2017)
ZnOQDs/MOF-801	Reflux	ZnO was prepared	Ghasemzadeh and Akhbari (2024)
Cu_2OQDs/MOF-801	Solvothermal	Cu_2O was prepared	Ghasemzadeh and Akhbari (2023a)
γ-Fe_2O_3QDs/MOF-801	Solvothermal	Fe_2O_3 was prepared	Ghasemzadeh and Akhbari (2023c)

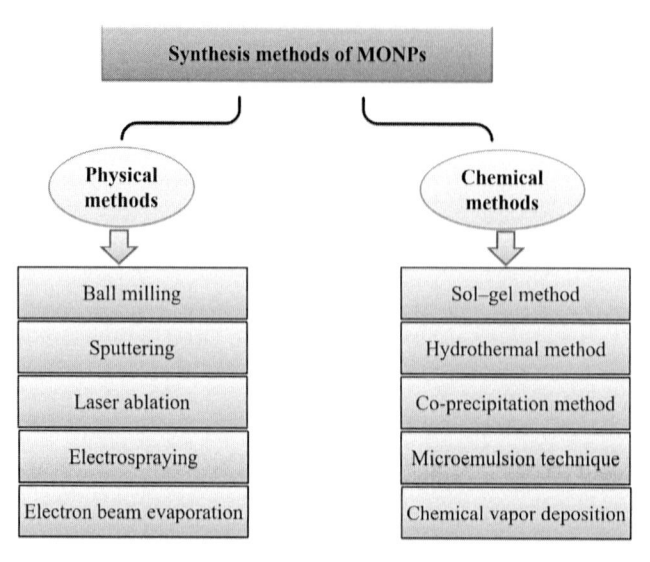

```
                    Synthesis methods of MONPs
                              │
            ┌─────────────────┴─────────────────┐
            ▼                                   ▼
      Physical                            Chemical
      methods                             methods
            │                                   │
            ▼                                   ▼
    ┌──────────────────┐              ┌──────────────────────┐
    │   Ball milling   │              │   Sol–gel method     │
    ├──────────────────┤              ├──────────────────────┤
    │   Sputtering     │              │ Hydrothermal method  │
    ├──────────────────┤              ├──────────────────────┤
    │  Laser ablation  │              │ Co-precipitation method │
    ├──────────────────┤              ├──────────────────────┤
    │  Electrospraying │              │ Microemulsion technique │
    ├──────────────────┤              ├──────────────────────┤
    │ Electron beam evaporation │     │ Chemical vapor deposition │
    └──────────────────┘              └──────────────────────┘
```

FIGURE 13.8 Different physical and chemical methods used for the synthesis of metal oxide nanoparticles (Rao et al., 2017).

Catalytic performance of metal oxide nanoparticle/metal-organic framework composites

Common catalytic performances

Metal/metal oxide heterogeneous catalysts are a crucial class of materials for various chemical transformations. The interfaces between the metal and metal oxide components are believed to play a critical role in their catalytic activity and selectivity. MOFs can serve as excellent functionalized supports to confine ultrafine MNPs and MONPs within their porous frameworks. This approach helps stabilize the metal-metal oxide interface and prevent phase separation during the catalytic process. An et al. (2017) used a UiO-type MOF constructed from 2,2'-bipyridine-5,5'-dicarboxylate (bpydc) and Zr_6 nodes as the support for Cu/ZnO_x catalysts. This Cu/ZnO_x/UiO-bpy composite exhibited exceptional performance for the hydrogenation of CO_2 to methanol. The Cu/ZnO_x/UiO-bpy catalyst showed a space-time yield to methanol (STY_{MeOH}) of 2.59 $g_{MeOH}/(kg_{Cu} \cdot h)$ at a gas hourly space velocity of 18,000 1/h. This remarkable performance significantly outperformed the commercial $Cu/ZnO/Al_2O_3$ catalyst, which had a STY_{MeOH} of only 0.83 $g_{MeOH}/(kg_{Cu} \cdot h)$ under similar conditions. The superior catalytic activity of the Cu/ZnO_x/UiO-bpy catalyst was attributed to the rich interfacial sites created between the well-dispersed Cu, ZnO_x, and Zr_6 nodes within the UiO-bpy MOF framework. These interfacial sites are believed to facilitate the adsorption and activation of

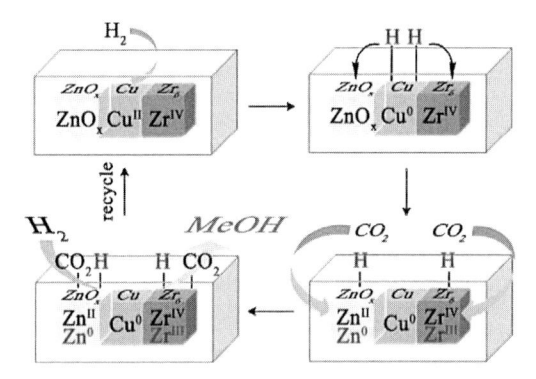

FIGURE 13.9 Schematic of the performance of different surface sites in the catalytic hydrogenation of CO_2 and active sites enclosed in metal-organic frameworks (An et al., 2017).

the reactants, H_2 and CO_2, thereby enhancing the overall efficiency of the CO_2 hydrogenation reaction (Fig. 13.9).

The integration of MONPs into MOFs has garnered significant interest due to the broad range of potential applications for these composite materials. The integration of metal/MONPs into MOFs has garnered significant interest due to the broad range of potential applications for these composite materials. In a breakthrough, a research team led by Bo Li reported a novel template protection-sacrifice (TPS) method that successfully encapsulated metastable Cu_2O NPs into MOFs in 2018. In this approach, SiO_2 was used as both a protective shell for the Cu_2O nanocubes and a sacrificial template to facilitate the formation of a yolk-shell structure. The resulting Cu_2O/ZIF-8 composite material exhibited excellent cyclic stability and high catalytic activity in the hydrogenation of 4-nitrophenol. This represents the first report of a Cu_2O/MOF-type composite, demonstrating the potential of the TPS method to encapsulate other unstable, yet highly active, MONPs into MOF frameworks (Li et al., 2018). The TPS method provides an efficient and versatile strategy for incorporating metastable active NPs into MOFs or other porous materials, overcoming a significant barrier that has hindered the development of these advanced composite catalysts. In a separate study, a magnetic bifunctional SrO-ZnO/MOF catalyst was synthesized and supported on a MOF for one-step biodiesel production from high acid value oils. The catalyst possessed both basicity (2.84 mmol/g) and acidity (0.02 mmol/g) due to the active components of $Sr_3Fe_2O_6$ and ZnO. The MOF support increased the catalyst's specific surface area by 3.7 times and provided a Fe_3O_4 component for magnetic separation, which facilitated the catalyst's recovery and reuse. The catalyst demonstrated excellent activity, achieving a remarkable soybean biodiesel yield of 99.5% at 80°C with three reaction cycles. Importantly, the catalyst also exhibited high resistance to free fatty acids, achieving a biodiesel yield of

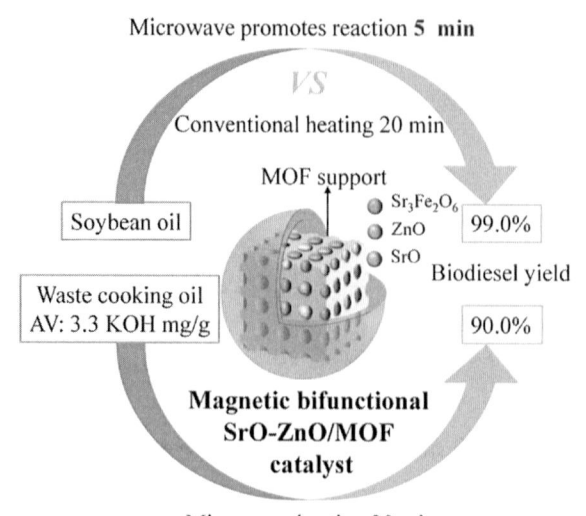

FIGURE 13.10 Schematic of the photocatalytic process SrO-ZnO/metal-organic framework catalyst (Yang et al., 2023).

90.0% from waste cooking oil (with an acid value of 3.3 mg KOH/g) under slightly longer reaction conditions (Fig. 13.10) (Yang et al., 2023).

Photocatalysis

Although MOFs are promising photocatalysts due to their large surface area and tunable pore structure, the rapid recombination of photogenerated charge carriers often results in low photocatalytic activity. To address this challenge, various strategies have been explored to incorporate MOFs with light-harvesting semiconductor materials to improve their photocatalytic performance. One effective photocatalyst design was developed by incorporating a Cu-MOF with ZnO for the photocatalytic degradation of Rose Bengal. This composite photocatalyst exhibited an excellent degradation efficiency of 97.4% within 45 minutes under natural sunlight, using a catalyst dosage of 320 mg/L. Further studies revealed that the degradation process followed first-order kinetics with a rate constant of 0.077869 1/min. The degradation mechanism was investigated using a combination of techniques, including photoluminescence (PL) spectroscopy, X-ray photoelectron spectroscopy (XPS), zeta potential measurements, and scavenger experiments. These analyses provided insights into the charge transfer dynamics and the roles of various reactive species in the photocatalytic process. Importantly, the fabricated Cu-MOF/ZnO composite displayed good recovery and reusability,

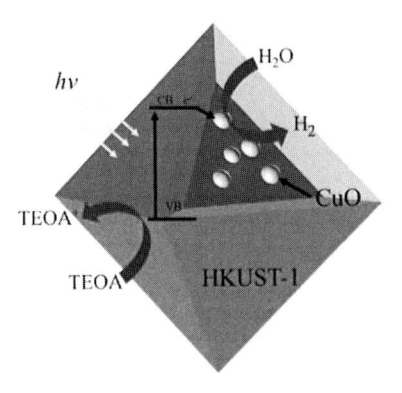

FIGURE 13.11 Schematic diagram of the photocatalytic H_2 generation over the CuO/HKUST-1 heterojunctions (Jin et al., 2019).

maintaining its performance for up to five reaction cycles, as confirmed by X-ray diffraction (XRD) analysis. This demonstrates the potential of this MOF-based photocatalyst for environmental remediation applications (Roy et al., 2023).

In a separate study, CuO/HKUST-1 nanocomposites were successfully fabricated by a facile solvent-free, one-pot reaction method. The surface morphology, porosity, and particle size of the trapped CuO NPs could be easily tuned by adjusting the ratio of reactants. Compared to the pristine HKUST-1 and commercial CuO NPs, the CuO/HKUST-1 nanocomposites exhibited improved photocatalytic activity for hydrogen evolution under visible light irradiation. The enhanced photocatalytic performance of the CuO/HKUST-1 nanocomposites was attributed to the efficient charge transfer from the CB of HKUST-1 to the CB of CuO, facilitating the water splitting reaction and hydrogen production under visible light illumination (Fig. 13.11) (Jin et al., 2019).

HKUST-1 has been explored as a potential catalyst for various applications; it was found to be ineffective for the oxidation of CO. However, the transformation of HKUST-1 at temperatures above 230°C proved to be the key to its improved catalytic activity. The transformation process involves a gradual segregation of CuO NPs from the HKUST-1 structure, which occurs simultaneously with a loss of MOF crystallinity. This structural change appears to be the critical factor in enhancing the catalytic performance of the material toward CO oxidation. To further improve the catalytic properties, researchers incorporated a cerium (Ce) precursor into the HKUST-1 network, followed by activation. This approach yielded a highly dispersed mixture of CuO and CeO_2 NPs with a high degree of interaction between the two metal oxide components. This Ce/HKUST-1 composite material exhibited enhanced catalytic activity for CO oxidation compared to the original HKUST-1. The researchers also demonstrated an effective route for synthesizing immobilized

CeO$_2$-CuO NPs in a monolithic reactor by anchoring and activating the HKUST-1 and Ce/HKUST-1 coatings. This approach allowed the preservation of the high catalytic activity of the NPs while also providing a robust, immobilized catalyst system (Zamaro et al., 2012).

In another work, researchers designed CeO$_2$/MIL-101(Fe) catalysts by introducing different amounts of CeO$_2$ into the MIL-101(Fe) framework. The materials were extensively characterized using various analytical techniques. An optimized photocatalysis-extraction oxidation desulfurization system (PEODS) was selected, comprising CeO$_2$/MIL-101(Fe) catalyst, acetonitrile as the extractant, and H$_2$O$_2$ as the oxidant. This system was employed to remove dibenzothiophene (DBT) from model oil under visible-light irradiation. The evaluation results showed that 90% of DBT in the oil phase can be converted within 2 hours under these conditions: (1) CeO$_2$ addition amount: 60 mg, (2) Model oil to acetonitrile volume ratio: 2:1, and (3) H$_2$O$_2$ to DBT molar ratio: 3:1. Radical scavenger experiments confirmed that hydroxyl radicals (•OH) and superoxide radicals (•O$_2^-$) were the main reactive species driving the process. The high photocatalytic activity was attributed to the synergistic effect between the active sites of CeO$_2$ and the active surface of MIL-101(Fe), forming a heterostructure in the catalyst. This heterostructure broadened the optical response range and facilitated photoinduced charge separation and transfer, leading to the enhanced performance (Fig. 13.12) (Huo et al., 2021).

Defect-induced CeO$_2$/MIL-53 heterojunction was performed for advanced photocatalysis. The researchers fabricated a heterojunction between defect-induced CeO$_2$ and an iron-based MOF (MIL-53). They adopted a simple

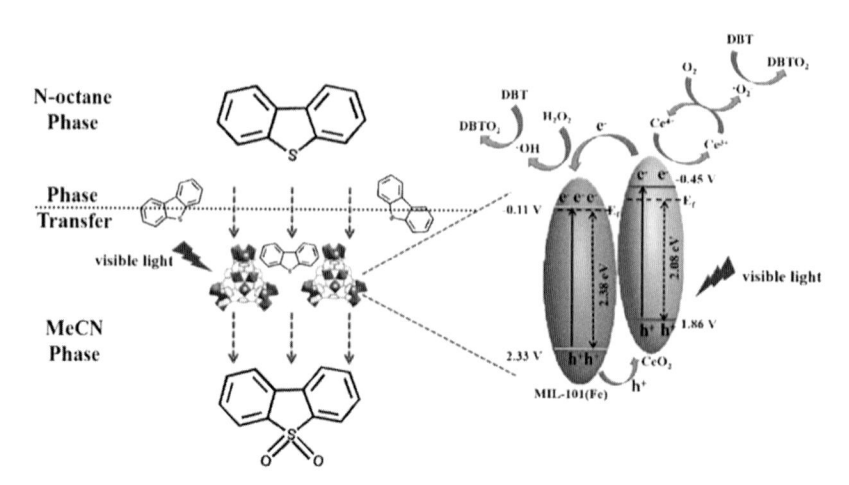

FIGURE 13.12 Schematic of the mechanism of MIL-101(Fe) modified with CeO$_2$ for desulfurization oxidation extraction by photocatalytic extraction of model oil under visible light irradiation (Huo et al., 2021).

chemical redox etching methodology to introduce oxygen vacancies in pristine CeO_2, thereby narrowing its band gap. The photocatalytic performance of the resulting defect-induced CeO_2/MIL-53 (MCO-X) heterojunction was investigated for bisphenol A (BPA) degradation and photocatalytic hydrogen generation from water splitting. The significantly improved photocatalytic activity of the MCO-X heterojunction was attributed to the switching of the charge dynamics mechanism from Type-I to Type-II due to the introduction of defects in the pristine CeO_2. This change in the charge transfer mechanism enhanced the visible light harnessing capacity and overall photocatalytic activity of the system. The optimal photocatalyst, MCO-30, displayed the highest photocatalytic BPA degradation rate constant (0.045 1/min) and the highest hydrogen evolution rate (3286.2 μmol/(h.g)) among the tested samples (Fig. 13.13). This study provides a comprehensive analysis of how defect engineering in pristine CeO_2 within the MCO-X heterojunction can lead to a switch in the charge transfer mechanism, ultimately resulting in remarkable photocatalytic performance under visible light irradiation (Sahoo et al., 2024).

By using the loading of metal oxide QDs inside MOFs, the band gap can be engineered, and this can affect the response to visible light in photocatalytic processes. Ghasemzadeh and Akhbari (2023c) reported that after loading of γ-Fe_2O_3 QDs inside MOF-801, the band gap energy was reduced to 3.1 eV for γ-Fe_2O_3QDs/MOF-801 compared to 4.4 eV for MOF-801. The efficiency of photocatalytic degradation of acid orange 7 for γ-Fe_2O_3QDs/MOF-801 and MOF-801 was 84.15% and 39.97%, respectively, thus a significant increase in

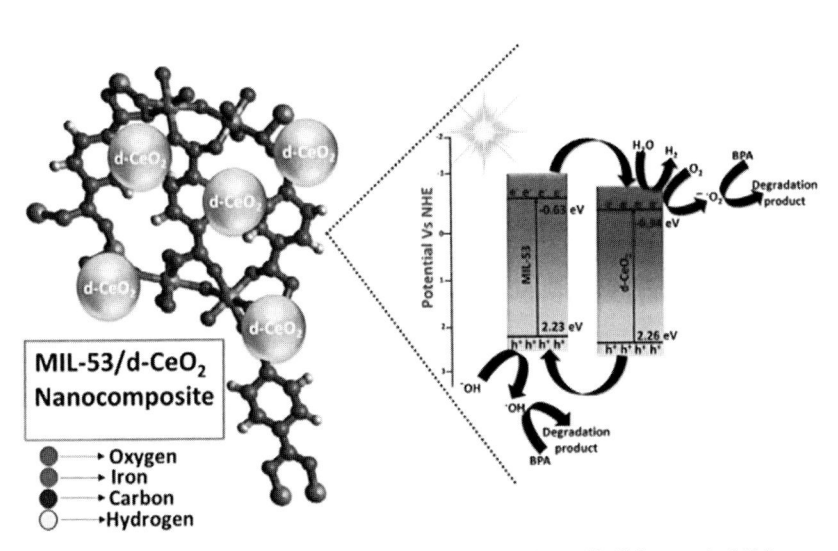

FIGURE 13.13 Schematic of photocatalytic activity of CeO_2/MIL-53 (Sahoo et al., 2024).

activity is observed for γ-Fe$_2$O$_3$QDs/MOF-801. This increase in activity can be attributed to the band gap, the shortening of the charge transfer path, and the reduction of the rate of electron-hole recombination.

In another study, Ghasemzadeh and Akhbari (2024) focused on developing a cost-effective and environmentally friendly technique for synthesizing QDs without the need for harsh synthetic conditions. In this study, MOF-801 was used as a template for the templated synthesis of ZnO QDs. The synthesis was performed using a dual solvent method (DSM) combined with an in situ reduction route (ISRR), which allowed control over the shape and size of ZnO QDs. The results showed that the crystal structure of MOF-801 was retained even in the presence of a strong reducing agent (NaBH$_4$). Characterization techniques, including Fourier-transform infrared spectroscopy (FT-IR), energy-dispersive X-ray spectroscopy (EDS), inductively coupled plasma optical emission spectroscopy (ICP-OES), and XPS, confirmed the successful synthesis of the ZnO QDs/MOF-801 nanocomposite. The Brunauer–Emmett–Teller (BET) surface area and pore volume of the nanocomposite decreased, and transmission electron microscopy (TEM) analysis revealed that ZnO QDs, approximately 3 nm in size, were successfully encapsulated inside the MOF-801 template. The absorption edge of the ZnO QDs/MOF-801 nanocomposite was blue-shifted compared to bulk ZnO, with a band edge at 375 nm, due to the quantum confinement effect. The researchers then evaluated the ZnO QDs/MOF-801 nanocomposite as an emerging photocatalyst for the photocatalytic degradation of Acid Blue 25 under UV light. The results showed that the ZnO QDs embedded in the MOF-801 matrix exhibited improved photocatalytic activity compared to bare ZnO QDs. This enhancement was attributed to the quantum confinement effect, the large surface-to-volume ratio, the reduction in electron-hole pair recombination, the multiple exciton generation, and the shortened electron and hole transfer distances.

In another work, Ghasemzadeh and Akhbari (2023a) successfully synthesized a Cu$_2$O QDs/MOF-801 nanocomposite using a simple and low-cost route. In this approach, Cu$_2$O QDs were immobilized inside the MOF-801 matrix via a DSM coupled with an in situ reduction route (ISRR), which was reported for the first time. The results showed that the formation of the Cu$_2$O QDs/MOF-801 nanocomposite led to improvements in optical, electronic, and redox properties, as well as the lifetime of charge carriers and photocatalytic activity, compared to the MOF-801 material alone. In particular, the band gap energy of Cu$_2$O QDs/MOF-801 nanocomposites decreased from 4.4 to 2.8 eV compared to MOF-801, making the nanocomposite active in the visible light range. The photocatalytic degradation performance of Cu$_2$OQDs/MOF-801 nanocomposite was evaluated for the degradation of yellow acid 23. The results showed that the degradation percentage was 83.57%, and the optical degradation rate was 0.00648 1/min, which was 1.62 times higher than MOF-81.

Polymer/metal-organic framework composites

Polymers have various unique properties, such as softness and thermal and chemical stability. The combination of MOFs and polymers can produce new and versatile materials that exhibit collective properties to stabilize the framework and enhance activity (Kitao et al., 2017). Polymers on the nanometer scale show amazing properties. These polymers confined in the MOF pores/channels can have abundant interfaces with the MOF backbone to exhibit synergistic catalysis (Aguila et al., 2018).

Preparation methods of polymer/metal-organic framework composites

Typically, polymer/MOF composites are created using a postsynthesis method. Depending on the starting point of polymerization, this method can be classified as either "grafting to" or "grafting from" Additionally, the interactions between the MOF and polymer can be covalent and/or noncovalent, depending on the functional groups present on both the MOF and polymer (Li & Huo, 2015; Parsapour et al., 2024). In Fig. 13.14, you can see the possible methods for preparing polymer/MOF composites.

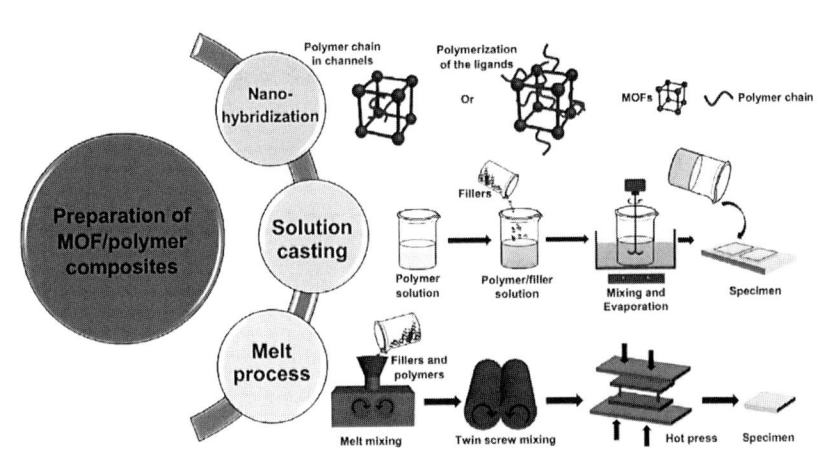

FIGURE 13.14 Schematic of possible methods for preparing polymer/metal-organic framework composites (Li, Zhang et al., 2020).

Catalytic performances of polymer/metal-organic framework composites

Common catalytic performances

MOFs and polymers are both well-defined heterogeneous catalysts. MOFs have adjustable and defined structural units, while polymers exhibit high performance in chemical transformations (Yang, Peng, et al., 2019). However, polymers also have undesirable properties that can lead to side reactions. In recent years, the combination of MOFs and polymers has gained significant attention in research. Synergistic effects between MOFs and polymers have been observed, resulting in a significant increase in catalytic activity compared to either component alone. This research area holds promise for the development of new and improved catalytic systems (Sherrington, 1980). Table 13.6 shows some of the catalytic applications of polymer/MOF composites.

In the next part, the application of polymer/MOF composites as catalysts for organic transformations is explained. Polymers as catalysts can impart various types of interesting structural and physical properties to MOFs. For example, polymers can be used as supports for MOFs or inserted into the pores of MOFs to catalyze multistep reactions, chemical synthesis, and chiral transformations. Even polymers can be used to increase the porosity and increase the stability of NPs in MOFs.

Imparting basicity to the metal-organic framework catalyst: Chemical warfare agents (CWA), such as organophosphates, pose a lethal threat even in small amounts. To counter this, scalable and functional solutions are needed to adsorb and convert CWA into harmless substances. Zirconium-based MOFs are commonly used due to their stability and Zr-OH sites. Incorporating basic buffer solutions into fabrics and protective gear is challenging, but introducing basic polymers to MOFs can eliminate the need for liquid buffers (Moon et al., 2016). Recent advancements in polymer/MOF composites have shown promising catalytic activity in decomposing CWAs, offering potential solutions for their detoxification. Extensive research has been devoted to polymer/MOF composites, leading to rapid degradation of CWA as summarized in Fig. 13.15. Zirconium-based MOFs containing Zr(IV) Lewis acidic sites have demonstrated rapid hydrolysis of nerve agents under basic conditions. N-ethylmorpholine (NEM) is commonly used as a buffer in aqueous solutions to create the necessary basic environment. Polymer/MOF composites can be processed using techniques such as electrospinning, thin-film formation, and encapsulation within microcapsules. Researchers have explored various polymers for the decontamination of CWA using polymer/MOF composites. In a specific study, UiO-66-NH$_2$ MOFs were grown on polypropylene (PP) fibers with metal oxide nucleation sites. The fibers were coated with metal oxide films using atomic layer deposition (ALD) and subsequently incorporated with MOF crystals through solvothermal methods. The adhesion of MOF

TABLE 13.6 Literature review indicating the prevalence of polymer/metal-organic framework composites for various catalytic applications.

MOF	Polymer	Catalytic reaction	Enhanced performance	Reference
MIL-101(Cr)	Polystyrenesulfonate and polyvinyl methylamino pyridine	Tandem deacetalization-Knoevenagel condensation	Enhanced catalytic performance	Zhao et al. (2018)
MIL-101 (Cr)	Poly(proline) methacrylate	Aldol condensation	Enantio-selectivity	Dong et al. (2018)
MIL-101(Cr)	Poly(proline) methacrylate	Aldol condensation	Enantio-selective conversion of the reagents	Wu et al. (2019)
ZIF-8	Polystyrene	[3+3] cycloaddition reactions	Enhanced catalytic performance	Zhang et al. (2014)
UiO-66	Polypyrrole	Reduction of 4-nitrophenol	Model reaction	Zhao et al. (2019)
HKUST-1	Polyacrylic acid sodium salt	Oxidation of dibenzylamine	Enhanced catalytic performance	Jiang et al. (2011)
NU-901	Linear, branched PEI(Poly(Ethyleneimine) and PEI dendrimers	Nerve agent to organics	Improved speed of degradation of toxic species	Chen, Islamoglu, et al. (2019)
M_2(NDISA)	Polydopamine	Suzuki coupling	Improved catalytic performance	Peng et al. (2019)
UiO-66-NH_2	Nylon	Nerve agent to organics	Improved speed of degradation of toxic species	Kalaj et al. (2019)
UiO-66-NH_2	Non-woven polypropylene	Nerve agent to organics	Improved speed of degradation of toxic species	Lee et al. (2017)
UiO-66-NH_2	Poly(ethylene glycol) (PEG) methyl ether methacrylate	Reduction of 4-nitrophenol	Model reaction	Xie et al. (2015)
UiO-66	(Polymethyl) methacrylate	Hydrolysis of methyl paraoxon	Improved speed of degradation of toxic species	McCarthy et al. (2017)

NDISA, Naphthalene diimide salicylic acid.

474 Applications of Metal-Organic Framework Composites

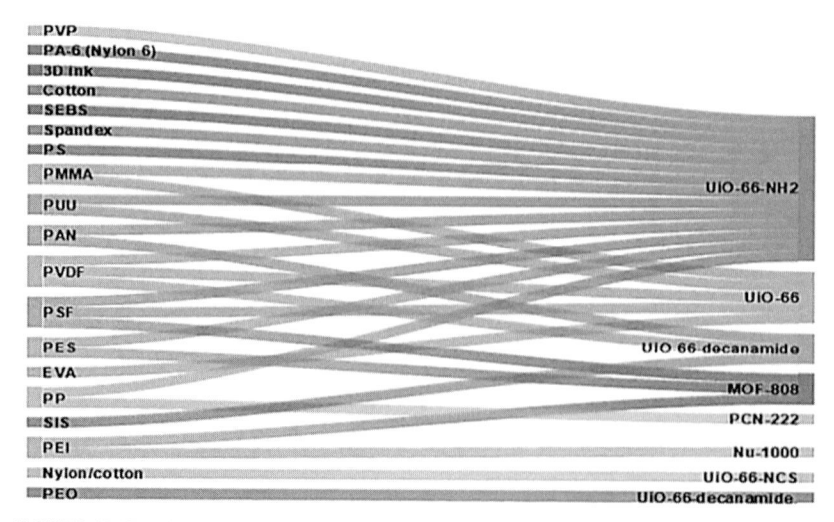

FIGURE 13.15 Diagram illustrating the wide range of polymers used so far with MOF catalysts for CWA decontamination. Polymers are listed on the left, and MOF catalysts on the right. The various ribbons indicate with which MOF the polymer was combined. List of terms: *PVP*, Polyvinyl-pyrrolidone; *PA-6,* polyamide 6; *SEBS*, styrene-ethylene-butylene styrene; *PS*, polystyrene; *PMMA*, polymethyl methacrylate; *PUU*, polyurethane urea; *PAN*, polyacrylonitrile; *PVDF*, polyvinylidene fluoride; *PSF*, polysulfone; *PES*, polyether sulfone; *EVA*, ethyl-vinyl acetate; *PP*, polypropylene; *SIS*, styreneisoprene-styrene; *PEI*, polyethyleneimine; *PE*, polyethylene glycol (Snider & Hill, 2023).

to the fibers and the rate of CWA removal were influenced by the metal oxide composition, while the turnover frequency (TOF) remained unaffected. For more details, refer to Table 13.7.

Imparting multiple functionalities to the metal-organic framework catalyst for multistep reactions: Polymer/MOF composites have been proposed as catalysts for the Knoevenagel condensation reaction, which involves the formation of C=C bonds (Corma & Iborra, 2006). These composites can replace large amounts of basic catalysts traditionally used in the reaction, reducing environmental concerns. By functionalizing MOFs with polymers, multistep cascade reactions, including the Knoevenagel condensation, can be catalyzed. In a study by Zhao et al., a bifunctional catalyst was developed by incorporating two different polymers with distinct functionalities into separate compartments of a MOF structure. The combination of acidic and basic polymers exhibited remarkable efficiency in deacetalization and Knoevenagel condensation reactions, achieving faster reactions and milder conditions while achieving complete conversions. This approach allows for the immobilization of diverse polymers within a MOF framework, creating accessible acidic and basic sites for the target substrates and enabling the generation of two distinct chemical environments within a single MOF material (Zhao et al., 2018). In

TABLE 13.7 Examples of composite polymer sample with metal-organic framework and its catalytic performance.

Material	Catalyst	Substrate(s)	Conditions	TON[a] or catalytic mol%	Results	Reference
PMMA/Ti(OH)₄/TiO₂/MOF	UiO-66-NH₂	DMNP	Aq. Nethylmorpholine buffer (pH 10, 0.45 M)	18[b]	$t_{1/2}$=26 minutes	Dwyer et al. (2018)
PMMA/Ti(OH)₄/UiO-66 composite	UiO-66	DMNP	Aq. Nethylmorpholine buffer at 25°C	3	Hydrolysis to 4-nitrophenolate anion and dimethyl phosphate, $t_{1/2}$=29 minutes	McCarthy et al. (2017)
Polypropylene/ZnO/UiO-66-NH₂	UiO-66-NH₂	DMNP	Aq. Nethylmorpholine buffer at 25°C	11	Hydrolysis to 4-nitrophenolate anion and phosphate within 90 minutes, $t_{1/2}$=10 minutes TOF=0.019 1/s	Lee et al. (2017)
Polypropylene/TiO₂/UiO-66-NH₂	UiO-66-NH₂	DMNP	Aq. Nethylmorpholine buffer at 25°C	16	Hydrolysis to 4-nitrophenolate anion and phosphate within 90 minutes, $t_{1/2}$=15 minutes TOF=0.018 1/s	Lee et al. (2017)
Polypropylene/Al₂O₃/UiO-66-NH₂	UiO-66-NH₂	DMNP	Aq. Nethylmorpholine buffer at 25°C	70	Hydrolysis to 4-nitrophenolate anion and phosphate within 90 minutes, $t_{1/2}$=78 minutes TOF=0.015 1/s	Lee et al. (2017)
Cotton/UiO-66-NH₂	UiO-66-NH₂	DMNP	Aq. Nethylmorpholine buffer (0.45 M)	108	$t_{1/2}$=17 minutes	Bunge et al. (2018)
Cotton/UiO-66-NH₂	UiO-66-NH₂	DMNP	Buffered solution (details not provided)	Not reported	$t_{1/2}$=4 minutes	Bunge et al. (2020)

(Continued)

TABLE 13.7 (Continued)

Material	Catalyst	Substrate(s)	Conditions	TON[a] or catalytic mol%	Results	Reference
Cotton/UiO-66-NH$_2$	UiO-66-NH$_2$	DMNP	Aq. Nethylmorpholine buffer (0.45 M)	Not reported	$t_{1/2}$ ~ 30 minutes	Couzon et al. (2022)
Zr-MOF Filter (UiO-66-NH$_2$/PVP)	UiO-66-NH$_2$	DMNP	Aq. Nethylmorpholine buffer (pH 10, 0.45 M)	10	$t_{1/2}$=2.4 minutes TOF: 0.068 1/s	Liang et al. (2018)
PA-6/Ti(OH)$_4$/TiO$_2$/MOF	UiO-66-NH$_2$	DMNP	Aq. Nethylmorpholine buffer (pH 10, 0.45 M)	42	$t_{1/2}$=45 minutes	Dwyer et al. (2018)
PA/UiO-66-NH$_2$	UiO-66-NH$_2$	DMNP	Aq. Nethylmorpholine buffer (0.45 M)	Not reported	$t_{1/2}$ ~ 30 minutes	Couzon et al. (2022)
PVDF/UiO-66-NH$_2$ (MOFabric-33%)	UiO-66-NH$_2$	DMNP	GD GD: 50% RH DMNP: Aq. Nethylmorpholine buffer (0.45 M)	Not reported	GD $t_{1/2}$=131 minutes DMNP t=12 minutes	Lu et al. (2017)
UiO-66/crosslinked polymer (3D printed)	UiO-66	DMNP	Rehydrated composites (16 h) Aq. Nethylmorpholine (0.45 M) buffer at 25°C for 200 min	Not reported	90% hydrolysis after 114 minutes $t_{1/2}$=16 TOF=0.79 1/s	Young et al. (2019)

Nu-1000/PEI	Zr6-MOF (Nu-1000)	DMNP GD VX	DMNP: Aq. Nethylmorpholine buffer at 25°C GD/VX: Aqueous solution at 25°C	DMNP: 16 GD: 10 VX: 10	DMNP: 100% hydrolysis, $t_{1/2}$=2 minutes Recycled, had lower activity GD: 90% conversion, $t_{1/2}$=5 minutes VX: 80% conversion, $t_{1/2}$=13 minutes	Moon et al. (2016)
MOF-808/PEI/fiber	MOF-808	GD VX DMNP	50% RH at RT	Not reported	GD: $t_{1/2}$=12 minutes, degradation to PMPA after 1 hour DMNP: $t_{1/2}$=0.4 hours VX: full degradation in 1 hour	Chen, Ma, et al. (2019)
Spandex/UiO-66-NH$_2$	UiO-66-NH$_2$	DMNP GD	DMNP: Aq. Nethylmorpholine buffer (0.45 M) GD: 50% RH over 24 h	Not reported	DMNP: $t_{1/2}$=~55 minutes GD: ~50% removed in 24 hours	Morgan et al. (2021)
PA-6/UiO-66-NH$_2$ (Nylon/UiO-66-NH$_2$)	UiO-66-NH$_2$	DMNP GD	DMNP: Aq. Nethylmorpholine buffer (0.45 M) GD: 50% RH over 24 hours	Not reported	DMNP: $t_{1/2}$=7.4 minutes GD: ~65% removed in 24 hours	Morgan et al. (2021)
UiO-66-NH$_2$/nylon	UiO-66-NH$_2$	DMNP	Aq. Nethylmorpholine buffer (pH 8, 20 mM M)	Not reported	Report hydrolysis rate of 34 mM/s×10^{-6}	Kalaj et al. (2019)
UiO-6/UiO-66-NH$_2$	UiO-66-NH$_2$ and UiO-66	DMNP	Aq. Nethylmorpholine buffer (pH 8, 22mM)	Not reported	Reported hydrolysis rate of 5×10^{-6} mM/s.mg	Palomba et al. (2021)

(Continued)

TABLE 13.7 (Continued)

Material	Catalyst	Substrate(s)	Conditions	TON[a] or catalytic mol%	Results	Reference
Zr-MOF polysulfone MMM	MOF-808, UiO-66, or UiO-66-NH$_2$	DMNP	DMNP passed through MMM at 0.1 mL/min flow rate. Aq. Nethylmorpholine buffer (pH 11.8, 0.4 M)	Not reported	MOF-808: 97% conversion UiO-66-NH$_2$: 68% conversion UiO-66: 18% conversion	Lee, Kim, et al. (2020)
MOFwich4 [SEBS G1642/ UiO-66-NH$_2$(50%)+ SEBS G1642/ Zr(OH)$_4$ (50%)]	UiO-66-NH$_2$ and Zr(OH)$_4$	GD VX	Neat liquid agents dropped onto the composite. Degradation rates measured via ^{31}P MAS NMR	Not reported	GD: hydrolysis generates pinacolyl methyl phosphonic acid (PMPA), finish in 54 minutes VX: hydrolysis generates ethyl methyl phosphonic acid (EMPA), $t_{1/2}$=71 and 158 minutes	Peterson et al. (2018)
MOF-808/ polyester fibers	MOF-808	DMNP GD	aq. Nethylmorpholine buffer (pH 10, 0.45 M) at 25°C	Not reported	DMNP: $t_{1/2}$=1.5 minutes GD: $t_{1/2}$=2 minutes	Ma, Islamoglu, et al. (2019)

[a] TON: Turnover number=moles of substrate/moles of catalyst.
[b] Calculated from TOF at $t_{1/2}$.

another study, Dong et al. utilized a chiral polymer to introduce enantioselectivity to catalytic reactions. They immobilized a chiral proline-based polymer within the pores of MIL-101(Cr) through in situ polymerization of chiral monomers. The proximity of the active chiral polymer to Lewis acidic Cr sites facilitated the aldol condensation of aromatic aldehydes with cyclohexanone. Both symmetric and asymmetric aldol condensations were achieved with good recyclability and high enantiomeric excess values, indicating the preferential formation of one enantiomer. This work represents a significant advancement in the development of enantioselective MOF catalysts (Dong et al., 2018).

Imparting meso- and macro-porosity to metal-organic framework catalysts: In catalytic processes involving MOFs, restricted diffusion of guest molecules can be a challenge due to the microporous structure of MOFs. To address this issue, polymers can be used as templates to create meso- or macroporous architectures within MOF-based catalysts. This approach increases the pore dimensions, allowing for improved transport of guest molecules to the active sites and enhancing the overall catalytic efficiency. For example, Zhang et al. (2014) used PS as a template to produce hollow MOF shells for catalytic applications. They functionalized PS beads with -COOH groups and coated them with ZIF-8. By dissolving the polymer, hollow ZIF-8 particles were obtained, which exhibited high activity in [3+3] cycloaddition reactions. Similarly, Song et al. reported a Pd/ZIF-8 hollow nanosphere catalyst using PS nanospheres as a soluble template. The uniform pore size of the ZIF-8 shell provided excellent size-selective catalytic properties in the liquid-phase hydrogenation of olefins (Jongert et al., 2024; Wang & Li, Cao, et al., 2016; Xin et al., 2024). In 2018, Li et al. introduced SOM-ZIF-8 (SOM: Single-crystal Ordered Macropore), a structured porous material with a well-organized hierarchy of macro- and micropores. They grew ZIF-8 onto carboxylate-terminated PS nanospheres and then removed the PS template, resulting in SOM-ZIF-8. This material exhibited exceptional catalytic activity and recyclability, particularly in reactions involving bulky substrates. The improved performance was attributed to the superior mass diffusion properties of the hierarchical framework and its single-crystalline nature (Shen et al., 2018). Polymers can be utilized as effective tools for structuring MOF particles, enabling the creation of pathways that facilitate rapid diffusion of guest molecules. Chen et al. employed a macroporous polyacrylamide impregnated with various MOFs to produce hierarchically porous composites with both macro- and micro-porosity. The MOF particles grew on the surfaces of the macro-porous polymer, enhancing diffusion efficiency toward the catalytically active MOF particles. The resulting composite of HKUST-1/polyacrylamide exhibited remarkable activity in the conversion of α-pinene oxide to campholenic aldehyde. This approach demonstrates the potential of polymer-structured MOF composites for achieving enhanced catalytic performance through improved diffusion pathways (Chen, Ding et al., 2019). Polymers can be used as struts to mechanically reinforce and stabilize mesoporous MOF catalysts. By introducing polydopamine (PDA) through in situ polymerization in an isostructural series of MOFs (M$_2$[NDISA]), researchers successfully prevented the collapse of these mechanically unstable architectures. The addition of PDA improved the stability of the porous structure, resulting in highly porous composites suitable for host–guest applications. The researchers demonstrated the potential of these composites by immobilizing small Pd NPs using the PDA polymer. The resulting Ni$_2$(NDISA)-PDA-Pd composite exhibited excellent catalytic activity, particularly in the Suzuki–Miyaura cross-coupling reaction. This approach offers a promising strategy for enhancing the stability and catalytic performance of mesoporous MOF catalysts for various chemical transformations (Peng et al., 2019).

Photocatalysis

The combination of flexible, disordered, multichain polymers with crystalline, ordered, porous MOFs has attracted significant attention from researchers. This has led to the construction of composite materials where polymer chains are immobilized within the tunable nanopores of MOFs, opening up new possibilities for various applications (Kitao et al., 2017). In the field of photocatalysis, the introduction of polymers into MOF-based photocatalysts has shown potential in enhancing stability and catalytic activity. One example of such a composite is the work by Brahmi et al., who combined two MOFs (MIL-53 and HKUST-1) with an acrylate polymer network through photopolymerization. The resulting catalyst exhibited high stability comparable to classical crosslinked polymers, along with efficient photocatalytic activity inherited from the MOFs. These polymer-MOF composites demonstrated excellent performance in the repeated and stable photodegradation of organic dyes (Brahmi et al., 2021). Hou et al. proposed an innovative approach using de-doped polyaniline (DPANI) as a mediating layer during the deposition of four MOFs on cellulose fibers. This approach promoted the in situ growth and nanosizing process of the MOFs. Among the catalysts developed, the MIL-100(Fe)/DPANI/CelF composite showed remarkable photodegradation efficiency in the degradation of ciprofloxacin (CIP), achieving up to 105.96 mg/g of CIP degradation (Hou et al., 2021). Sadjadi and colleagues utilized a functionalized natural polymer called chitin (Ch), which can be easily extracted from shrimp, along with Fe-based MOFs to synthesize a composite called MOF-Ch through an amino substitution reaction. The resulting MOF-Ch catalyst exhibited photocatalytic degradation ability for methylene blue (MB) under light irradiation. This approach demonstrated the synergy between renewable biopolymers and MOFs, holding promise for enhancing photocatalysis and potentially reducing the cost of synthesizing high-performance photocatalysts (Seh et al., 2017). Overall, the integration of polymers with MOFs in composite materials offers opportunities to leverage the stability and catalytic properties of polymers while harnessing the unique features of MOFs for efficient photocatalysis. These advancements hold significant potential for various applications in the field of photocatalysis.

Metal-organic framework/metal-organic framework composites

MOF/MOF hybridization is a method of creating hybrid MOFs by integrating different types of MOFs. This can be done by growing guest MOFs onto host MOFs or using growth kinetic-controlled routes. The approach allows for diverse compositions and structural variations within the MOFs, leading to enhanced functionality and tailored properties. It has gained attention due to its potential applications in gas storage, separation, drug delivery, and sensing. The guest MOFs often possess different compositions and/or properties compared to the host MOFs. This integration not only expands the range of available ligands and metal centers but also enhances the structural variations and functionalities of MOFs (Ji et al., 2018; Luo et al., 2019; Yao et al., 2019).

MOF/MOF hybrids have demonstrated superior performance compared to individual MOF units in various applications such as gas adsorption/separation (Gu et al., 2017), detection (Yao et al., 2019), and catalysis (Zhao et al., 2016; Zhao, Yuan et al., 2016). They have shown remarkable potential as precursor materials for synthesizing other functional composite materials, similar to single MOFs. Specifically, they have been used to synthesize metal oxides (Zhang, Cao et al., 2018) and carbon-based materials (Liu, Wang et al., 2017; Pan et al., 2018), resulting in improved performance in energy storage/conversion, separation processes, and heterogeneous catalysis. These hybrid materials possess enhanced properties and functionalities, making them valuable for advancing technologies in these areas. Fig. 13.16 provides an overview of the synthetic strategies, structures, and applications of MOF/MOF hybrids.

Preparation methods of metal-organic framework/metal-organic framework composites

Various fabrication methods are employed to create these types of composites, which can exhibit different structures such as core-shell, yolk-shell, epitaxial growth, and more. Zhou et al. presented a schematic (Fig. 13.17) that showed a comparison between a core-shell structure of MOF on MOF and the hierarchical organization of a natural peach's exocarp–mesocarp–endocarp system (Feng et al., 2020). Since the pioneering work by Furukawa et al. (2009) on core-shell MOF/MOF hybrids, significant progress has been made in this field. Researchers like Yu and Zhang have achieved the synthesis of well-defined and adjustable core-satellite MOF/MOF hybrids through site-selective epitaxial growth (Li, Zhao et al., 2023; Zhao et al., 2020). Yu et al. have also expanded the scope to ternary heterostructures using a multiple selective assembly strategy. Various architectures, including core-shell, yolk-shell,

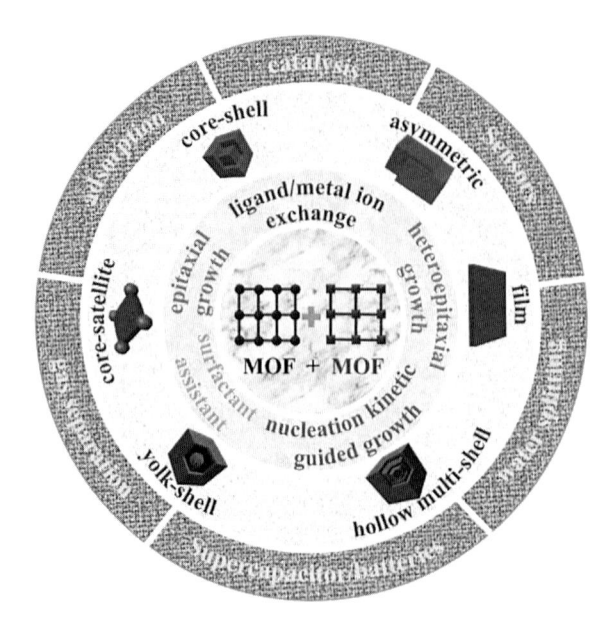

FIGURE 13.16 An overview of the synthetic strategies, structures, and applications of metal-organic frameworks (Liu, Wang, et al., 2021).

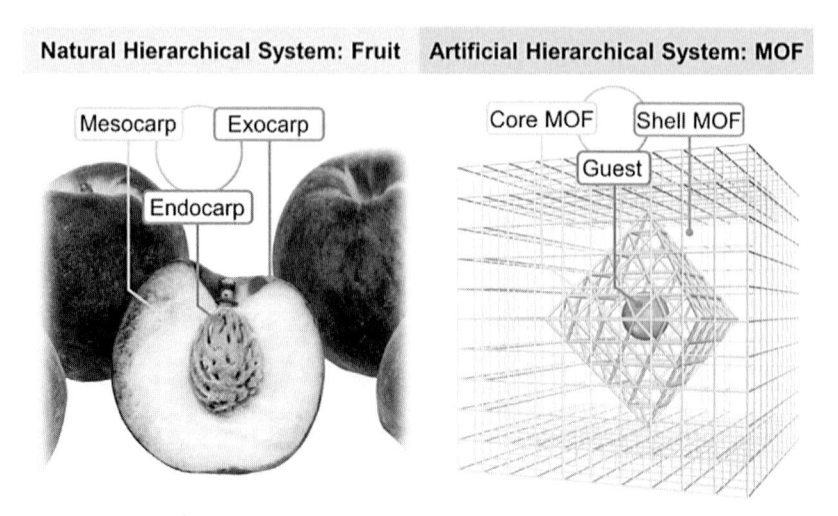

FIGURE 13.17 The figure depicts the hierarchical arrangement of the exocarp–mesocarp–endocarp system in a natural peach (Feng et al., 2020).

core-satellite, hollow multishell, asymmetric structures, and film-on-film, have been successfully developed. Synthesis strategies like epitaxial growth, surfactant-assisted growth, heteroepitaxial growth, ligand exchange, and nucleation kinetic-guided growth have been employed. These advancements have increased the versatility and complexity of MOF/MOF hybrids (Lee, Lee, et al., 2020; Wang, Xu et al., 2020; Yao et al., 2019).

The typical formation of a MOF/MOF composite involves the construction of a MOF shell using a low-valent metal and a carboxylic-based linker under mild conditions. In contrast, the MOF core contains a high-valent metal and is presynthesized under harsher conditions. The development of these structures is influenced by the functional groups present in both MOFs and their corresponding interactions. The lattice parameters of the MOFs also play a crucial role in the epitaxial growth process, ensuring that the mismatch between the two lattice parameters does not exceed 15% for successful composite formation (Liu et al., 2020).

Catalytic performances of metal-organic framework/metal-organic framework composites

Common catalytic performance

MOF structures have shown significant potential as catalyst supports for heterogeneous catalysis reactions. Among these structures, yolk-shell NPs have gained attention as nanoreactors, exhibiting enhanced activity and selectivity compared to solid materials (Li, Zhu, Li et al., 2019). For example, Yang et al. (2015) used yolk-shell Co-ZIF/ZIF-8 as a substrate to load Pd NPs as nanoreactors for acetylene conversion, demonstrating higher activity and selectivity compared to core-shell catalysts and pure Pd. The advantages of the yolk-shell MOF-on-MOF structure can be attributed to the molecular sieving effect of the ZIF-8 shell, the high dispersity of Pd on the Co-ZIF core, and enhanced mass transfer through the hollow cavity.

MOF/MOF-derived metal/carbon composites have shown remarkable catalytic activity in heterogeneous catalysis. In a study by Chen et al. (2019), the catalytic performance of Co/N-doped carbon materials derived from MOF/MOF structures was investigated for the selective hydrogenation of furfural to cyclopentanol (CPL) (Fig. 13.18). The researchers also examined the impact of the composite's structure and shell number on its catalytic performance. The quadruple-shell hollow structure exhibited superior activity compared to solid, single, double, and triple-shell materials. The enhanced performance was attributed to the multishell hollow structure, high surface area, and effective dispersal of the active species within the composite.

In recent studies, several intriguing MOF/MOF architectures have been investigated, featuring both a sieving region and a catalytically active region within a single system. This unique combination leads to synergistic effects that enhance both selectivity and activity in various catalytic processes.

484 Applications of Metal-Organic Framework Composites

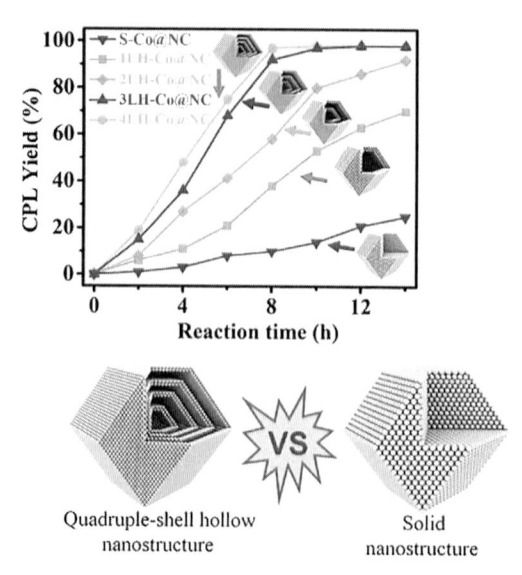

FIGURE 13.18 Plots of the cyclopentanol yields and selectivity versus the reaction time catalyzed by metal/microporous carbon (Chen et al., 2019).

Yang et al. (2018) developed size-selective catalysts for olefin epoxidation using a core-shell MOF/MOF architecture. By controlling nucleation kinetics, they synthesized PCN-222(Fe)/UiO-67, consisting of red needle-shaped PCN-222(Fe) crystals embedded in colorless UiO-67 crystals. This design enabled the integration of a catalytically active component (PCN-222(Fe)) and a size-selective framework (UiO-67), resulting in improved catalytic activity and selectivity for olefin epoxidation reactions. Zhou et al. prepared additional MOF/MOF architectures, PCN-222(Fe)/Zr-NDC and PCN-222(Fe)/Zr-AZDC, by altering the linker composition. These structures combined catalytic activity and size selectivity, enabling effective size-selective catalysis in olefin epoxidation reactions.

In 2018, the same group of researchers investigated size-selective catalysis using a different synthetic approach to create MOF/MOF composites (Feng et al., 2018). They synthesized various MOF-on-MOF structures, such as PCN-222(Fe)/ZIF-8, PCN-222/MOF-5, and HKUST-1/MOF-5, using surface functionalization and retrosynthetic stability principles. The focus was on size-selective oxidation of organic molecules using PCN-222(Fe)/ZIF-8, where the PCN-222(Fe) core provided catalytic activity, and the ZIF-8 shell acted as a molecular sieve. To confirm size selectivity, they compared the catalytic activity of PCN-222(Fe) and PCN-222(Fe)/ZIF-8 in the oxidation of two different substrates: o-phenylenediamine (o-PDA) and 2, 2'-azino-bis (3-ethylbenzothiazoline-6-sulfonic acid) (ABTS). Although the overall reactivity of PCN-222(Fe)/ZIF-8 was lower than that of the original PCN-222(Fe) crystal, it exhibited better size selectivity. The smaller molecule, o-PDA, easily diffused through the ZIF-8 pores and reached the inner catalytic site, while the larger substrate, ABTS, was sterically hindered and showed almost no catalytic activity when the shell was present. In a recent study, Kim and their group developed a core-shell MOF called UiO-67-TEMPO-CS (TEMPO: (2,2,6,6 tetramethylpiperidin-1-yl)oxyl; CS: core-shell) for aerobic alcohol oxidation (Kim et al., 2020). Unlike previous MOF/MOF composites, the aim was to enhance size selectivity by deactivating the surface active sites and confining reactions within the pores. This was achieved through postsynthetic ligand exchange, resulting in UiO-67-TEMPO-CS with a TEMPO-bearing UiO-67 core and a pure UiO-67 shell. The catalytic activity of TEMPO-functionalized MOFs, in the presence of $Eu(NO_3)_3$, showed improved efficiency and a wider substrate range. While UiO-67-TEMPO exhibited moderate size selectivity toward larger alcohols, UiO-67-TEMPO-CS demonstrated enhanced size selectivity. When the molecular size increased, the reactivity dramatically decreased, and certain substrates did not convert even with extended reaction times, confirming the superior size selectivity of UiO-67-TEMPO-CS. This effective size selectivity was achieved through simple surface deactivation, highlighting the potential of hierarchical MOF/MOF structures for size-selective catalysis (Hoover et al., 2013; Kim, Kim, et al., 2019; Lagerblom et al., 2018).

In a study by Gu et al. (2017), a modular 3D core-satellite hybrid MOF architecture called MIL-101(Cr)/NH_2-MIL-125(Ti) was prepared using the internal extended growth method (IEGM). The hybrid structure exhibited enhanced adsorption and photocatalytic activity for Cr^+ reduction to Cr^{3+}. The researchers used polyvinylpyrrolidone as a nucleation-promoting source to create a tablet-shaped MIL-101(Cr) core surrounded by cake-like NH_2-MIL-125(Ti). MIL-101(Cr)/NH_2-MIL-125(Ti) showed improved Cr^V adsorption capacity compared to NH_2-MIL-125(Ti) and MIL-101(Cr) alone. This was attributed to the additional pathways created by the introduction of MIL-101(Cr). Furthermore, the channel through MIL-101(Cr) served as a pathway for Cr^{VI}, the catalytic substrate, to reach the photocatalytic active site within NH_2-MIL-125(Ti), as well as an exit for Cr, the catalytic product. As a result, MIL-101(Cr)/NH_2-MIL-125(Ti) exhibited a 72% reduction of Cr, surpassing the reduction efficiencies of NH_2-MIL-125(Ti) (47%) and MIL-101(Cr) (13%).

In a study by Guo et al. (2018), the photocatalytic ability of ZIF-67/Co-MOF-74 for visible-light-driven H_2O oxidation was investigated. ZIF-67 was transformed into ZIF-67/Co-MOF-74 through surface ligand exchange using 2,5-dihydroxyterephthalate (DHTP), which forms hard coordination bonding. By adjusting the amount of DHTP, the thickness of the shell was controlled. The researchers compared the catalytic performances of pure ZIF-67 and Co-MOF-74 with a series of ZIF-67/Co-MOF-74 samples of varying shell thicknesses. The core-shell MOFs exhibited enhanced oxygen evolution as the shell thickness increased. The core-shell structure with a 50 nm thickness generated 15 μmol of oxygen. By optimizing the reaction conditions, the oxygen evolution of ZIF-67/Co-MOF-74 (shell: 50 nm) reached 122 µmol with a quantum yield (QY) of 11.3%, while the single MOFs, ZIF-67 and Co-MOF-74, had lower QYs of 10.1% and 10.3%, respectively. The researchers suggested that this improvement in performance could be attributed to the efficient contact between the reaction system and the exposed active metal center during the ligand exchange process, smoother collection of H_2O molecules facilitated by the hydroxyl and carboxyl groups of DHTP on the core-shell surface, and the conductive interface of the core-shell structure enabling more efficient separation and transfer of photocatalytic holes and electrons.

Electrocatalysis

Electrocatalytic reactions like OER, HER, and ORR are crucial for applications such as water splitting devices, fuel cells, and zinc–air batteries. The challenge is to find cost-effective catalysts that exhibit both high activity and stability (Shao et al., 2016). Researchers are exploring MOF/MOF derived metal or metal phosphide/carbon materials as multifunctional electrocatalysts with high reactivity and stability for clean energy conversion technologies. Qiu et al. synthesized hollow double-shelled hybrid nanocages (DSNC) by directly carbonizing a core-shell ZIF-8/ZIF-67 precursor. The outer shell of the DSNC consisted of Co, N-doped graphitic carbon (Co-NGC), while the inner shell was composed of N-doped microporous carbon (NC). This structure demonstrated promising electrocatalytic properties (Liu et al., 2017). DSNC exhibited impressive electrocatalytic performance, comparable to commercial Pt/C and RuO_2 catalysts. It showed a half-wave potential ($E_{1/2}$) of 0.82 V for ORR and high activity for OER. The overall oxygen electrode activity (ΔE) of DSNC was 0.82 V, demonstrating its potential as a cost-effective electrocatalyst for clean energy conversion.

Photocatalysis

Liu et al., 2020 recently prepared a MIL-125/ZIF-8 heterostructure for the photocatalytic degradation of the dye pollutant orange II under visible light irradiation (Fig. 13.19). The MIL-125/ZIF-8 heterostructure exhibited remarkable performance, with a fast degradation rate of orange II. In just 120 minutes, it achieved a removal rate of 97.3%, surpassing the performance of individual MIL-125 (54.6%) and ZIF-8 (8.4%). This superior performance can be attributed to the synergistic effect of MIL-125 and ZIF-8 in the photocatalytic reaction. These findings highlight the potential of MOF/MOF heterostructures as efficient photocatalysts for the degradation of organic pollutants, such as dyes.

Ghasemzadeh and Akhbari (2023b) synthesized MIL-88A(Fe)-on-MOF-801 photocatalyst, which was environmentally friendly and cost-effective, using the in-house extended growth method (IEGM). Then, Ag NPs were

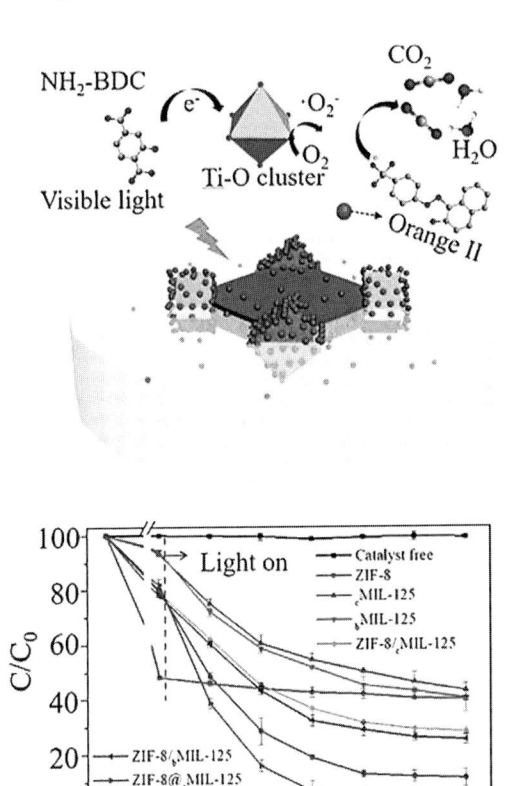

FIGURE 13.19 Illustration of the enhanced photocatalytic degradation process toward orange II using ZIF-8/MIL-125 for dye degradation (Liu et al., 2020).

incorporated into the MOF-801/MIL-88A(Fe) structure using the DSM followed by a PR process. Succeeded in synthesizing an Ag/MOF-801/MIL-88A (Fe) nanocomposite. Because MIL-88A(Fe) was outside in MOF-801 and the pores were blocked, and in addition to the presence of Ag NPs inside their pores, the surface area and pore volume of Ag/MOF-801/MIL-88A (Fe) pores are less of MOF-801 and MOF-801/MIL-88A (Table 13.8).

It showed the lowest band gap energy of the MOF-801/MIL-88A nanocomposite (2.6 eV) and the highest photocatalytic activity of 91.72% for the degradation of reactive black 5 after 30 minutes under visible light irradiation. Increasing light absorption intensity through SPR of Ag NPs and reducing electron-hole recombination improved the photocatalytic activity of the Ag/MOF-801/MIL-88A(Fe) nanocomposite (Fig. 13.20).

488 Applications of Metal-Organic Framework Composites

TABLE 13.8 Summary of the physicochemical parameters of the photocatalysts.

Sample	BET surface area (m²/g)	Total pore volume (cm³/g)	Average pore diameter (nm)
MOF-801	610.1	0.38	2.50
MIL-88A(Fe)	5.6	0.02	16.37
MOF-801/MIL-88A(Fe)	384.4	0.27	2.90
Ag/ MOF-801/MIL-88A(Fe)	145.2	0.16	4.56

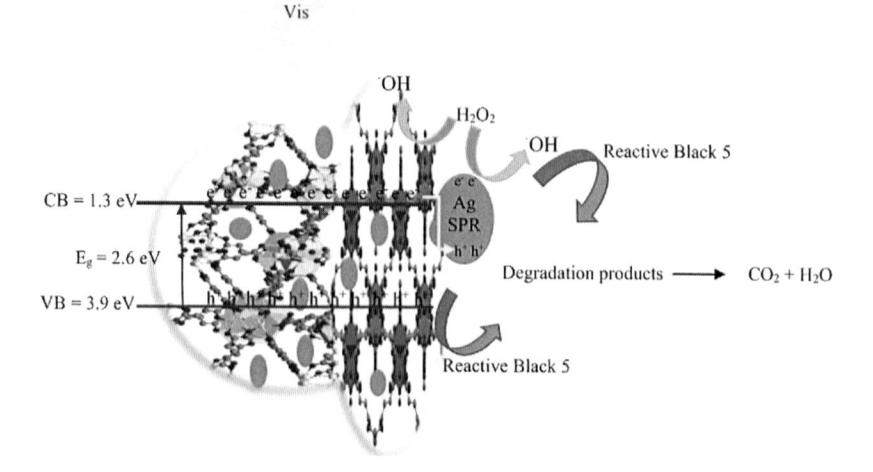

FIGURE 13.20 Proposed mechanism for the photocatalytic degradation of reactive black 5 over Ag/MOF-801/MIL 88 A(Fe) under visible light irradiation (Ghasemzadeh & Akhbari, 2023b).

Polyoxometalate/metal-organic framework composites

POMs constitute a remarkable family of discrete anionic metal–oxygen clusters, renowned for their rich chemical diversity, tunable shapes and sizes, variable solubilities, redox potentials, and strong acidities. These unique properties endow POMs with immense potential for catalyzing a wide range of transformations, particularly acid and oxidation reactions. However, their practical applications have been hindered by their inherently low specific surface area and limited stability. A promising strategy to overcome these limitations involves the immobilization of POMs within MOFs, thereby stabilizing and optimizing their catalytic properties (Du et al., 2014; Ye & Wu, 2016). Owing to their compositional diversity and structural versatility, POMs can serve as versatile building blocks for the construction of POM-based MOFs, acting as nodes, pillars, or templates within the framework's cages. Alternatively, POMs can be encapsulated within the pores of preformed MOFs through host–guest interactions, resulting in the formation of POM/MOF composite materials.

Preparation methods of polyoxometalate/metal-organic framework composites

POM/MOF hybrid materials with catalytic activity are commonly synthesized through encapsulation methods such as solvothermal, mechanochemical, coprecipitation, or impregnation, where the POMs become part of the MOF framework (Buru & Farha, 2020). The resulting hybrid materials' crystallinity, structure, defects, size, acidity, and POM location are influenced by the synthesis conditions and postsynthetic handling (Chen et al., 2017). Consequently, POM/MOF materials prepared using different procedures often exhibit variable catalytic properties and activity (Ye & Wu, 2016). There are two main synthetic strategies for POM/MOFs described in the literature. The first strategy involves encapsulating POMs within the MOFs, where the POMs act as guest molecules. The second strategy involves direct connections between POMs and organic linkers or attachment to metal centers and organic ligands as integral parts of the MOF framework. Both types of materials are commonly referred to as POMOFs in the reviewed literature, regardless of whether the POMs are guest molecules or building units in the framework (Fig. 13.21) (Liu, Tang et al., 2021).

Catalytic performances of polyoxometalate/metal-organic framework composites

Common catalytic performances

POMs are highly researched for their acidity, redox activity, and thermal stability (Misono, 2013), finding applications in industrial processes. MOFs also possess diverse catalytic characteristics. The first POMOF composite was successfully prepared in 1998, combining polyoxomolybdate clusters and copper–organoamine, exhibiting unique properties. Lacunary heteropolytungstate was later incorporated into MOF MIL-101 (Férey et al., 2005). In recent years, numerous POMOF composite-based catalysts have been reported for various catalytic reactions. These composites show excellent performance in organic transformations photo- and electro-catalytic activities (Choi et al.,

490 Applications of Metal-Organic Framework Composites

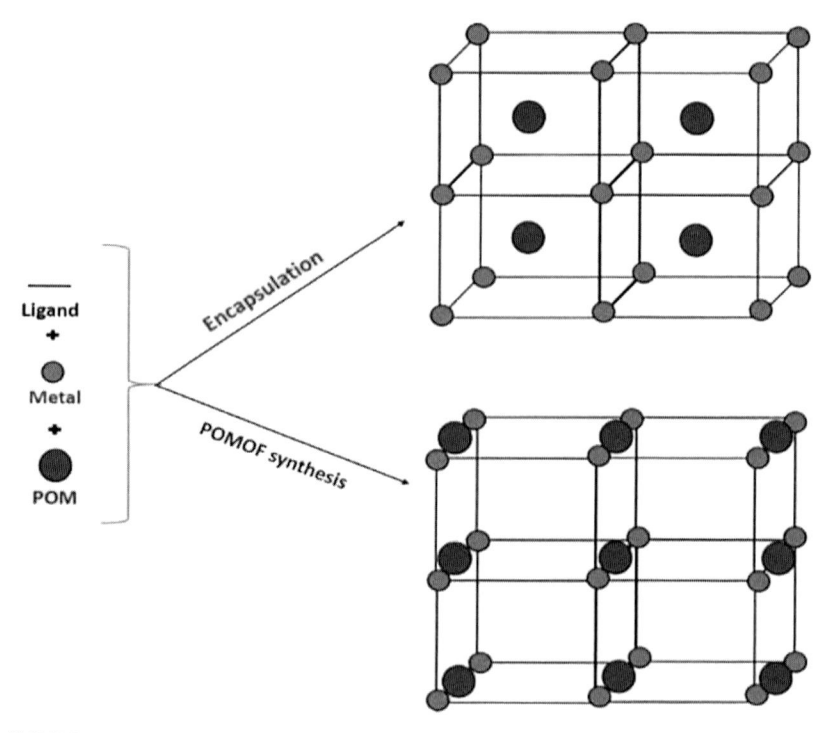

FIGURE 13.21 Synthesis strategy for polyoxometalate/metal-organic frameworks (Maru et al., 2022).

2021; Liu, Xu, et al., 2021). The synergistic combination of POMs and MOFs offers new opportunities in catalysis, leveraging their distinct features to achieve desirable catalytic outcomes.

Oxidation reactions are widely studied, and there is a significant body of research on various catalytic systems. POMOF hybrid materials are particularly attractive as oxidative catalysts due to the presence of acidic sites within MOFs and the strong acidity and redox capabilities of POMs (Jin et al., 2020). Several well-known MOFs, including MIL (Cr, Fe, or Al), NH_2-MIL-101 (Cr or Al), NH_2-MILL-53 (Al), Zr-MOF UiO-66, the ZIF series, Cu-BTC, and NU-1000, have been utilized for encapsulating POMs in oxidation transformations. The selective oxidation of alcohols and olefins, as well as oxidative desulfurization (ODS), has been extensively investigated as oxidation reactions using POMOF catalysts (Granadeiro et al., 2016; Song et al., 2019). Epoxides play a crucial role as intermediates in industrial organic synthesis (Palomeque et al., 2002). The selective epoxidation of alkenes, a common process in the fine chemicals industry, can be achieved using POM/MOF composites with stoichiometric amounts. These composites have gained recognition as active and selective catalysts, utilizing H₂O₂ as the primary oxidant (Jin et al., 2020). In 2008, Kholdeeva et al. developed a composite consisting of Ti and Co Keggin heteropolyanions ($[PW_{11}CoO_{39}]^{5-}$ and $[PW_{11}TiO_{40}]^{5-}$) and MOF MIL-101 (Ti-POM·MIL-101 and Co-POM·MIL-101) for the oxidation of various alkenes such as α-pinene, caryophyllene, and cyclohexene. The Ti and Co heteropolyanions were electrostatically linked to the MOF surface while retaining their structures. Ti-POM/MIL-101 demonstrated catalytic performance in allylic oxidation of α-pinene and selective epoxidation of caryophyllene using aqueous H₂O₂ under mild conditions in CH₃CN, while Co-POM·MIL-101 showed allylic oxidation of α-pinene in the presence of molecular oxygen (Fig. 13.22). Later, polyoxotungstates (PW₄O₂₄

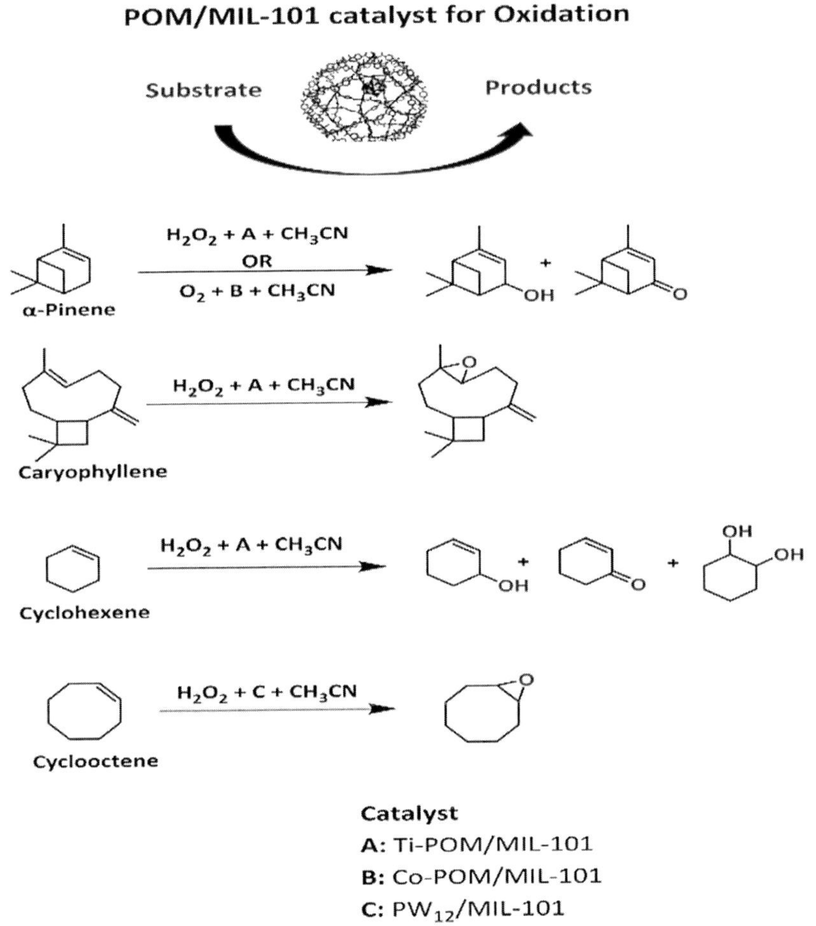

FIGURE 13.22 Selectivity alkene oxidation over polyoxometalate/MIL-101 (Maru et al., 2022).

and $PW_{12}O_{40}$) were incorporated into MIL-101 to create POMOF composites (PW_4/MIL-101 and PW_{12}/MIL-101) for selective alkene oxidation. These composites exhibited comparable or superior catalytic activity to homogeneous PW_X catalysts and outperformed traditional supported PW_4 catalysts, highlighting their potential as efficient catalysts for alkene epoxidation (Maksimchuk et al., 2008, 2010).

In a study by Granadeiro et al. (2014), a MIL-101-based POMOF composite called PW_9/MIL-101 was utilized for the selective oxidation of monoterpenes (geraniol and R-(+)-limonene) and oxidative desulfurization (ODS) of a model oil. The oxidation of geraniol demonstrated high chemoselectivity, yielding 2, 3-epoxygeraniol with a selectivity of 99%. However, the oxidation of limonene was less reactive, initially producing 1,2-epoxide within the first 10 minutes of the reaction, followed by the formation of other products such as diepoxide and 1,2-diol. Further details can be found in Table 13.9 (Granadeiro et al., 2014).

Duan et al. developed chiral POMOF materials (ZnWPYIs) for the asymmetric transformation of olefins into cyclic carbonates. The combination of a chiral organocatalyst and a polyoxometalate oxidation catalyst enabled efficient conversion of CO_2 into enantiomerically pure cyclic carbonates, offering a one-step, stereoselective process (Han et al., 2015). Han et al. (2018) developed chiral CuWPYIs, which are utilized for the simultaneous transformation of aldehyde and ketone into chiral epoxy ketones and the epoxidation of alkenes via a one-pot synthesis technique. CuWPYIs combine a copper (II) Lewis acid catalyst, a chiral organocatalyst pyrrolidine, and an oxidation catalyst polyoxometalate into a single porous coordination network.

TABLE 13.9 Time-dependent chemoselectivity of PW_9/MIL-101 for the oxidation of limonene.

R-(+)-limonene → (A) 1,2-Epoxide + (B) Diepoxide + (C) 1,2-diol (via PW_9@MIL-101, H_2O_2)

Time	Selectivity (A)%	Selectivity (B)%	Selectivity (C)%
10 minutes	99	0.5	0.5
1 hours	70	10	20
4 hours	55	25	20

The two-center catalysts activate carbonyl groups of ketones, leading to nucleophilic intermediates that react with electrophilic benzaldehyde molecules. The catalytic cycle involves hydroxyl and unsaturated intermediates, followed by peroxotungstate intermediate formation and direct spiro attack by the olefin double bond, resulting in cinnamaldehyde epoxidation. Hydrolysis of the intermediates yields chiral products, completing the catalytic cycle. EPR spectroscopic investigations confirmed the preservation of the Cu(II) valence state, supporting the proposed "two-center catalysis" mechanism (Fig. 13.23) (Wang, Liu et al., 2019).

In addition to the epoxidation of alkenes, metal alloys have been used in the oxidation of styrene to aldehydes, dihydroxylation, and oxidation of alcohols (Maru et al., 2022). A summary of the catalytic reaction and types of MOF and POM used is summarized in Table 13.10.

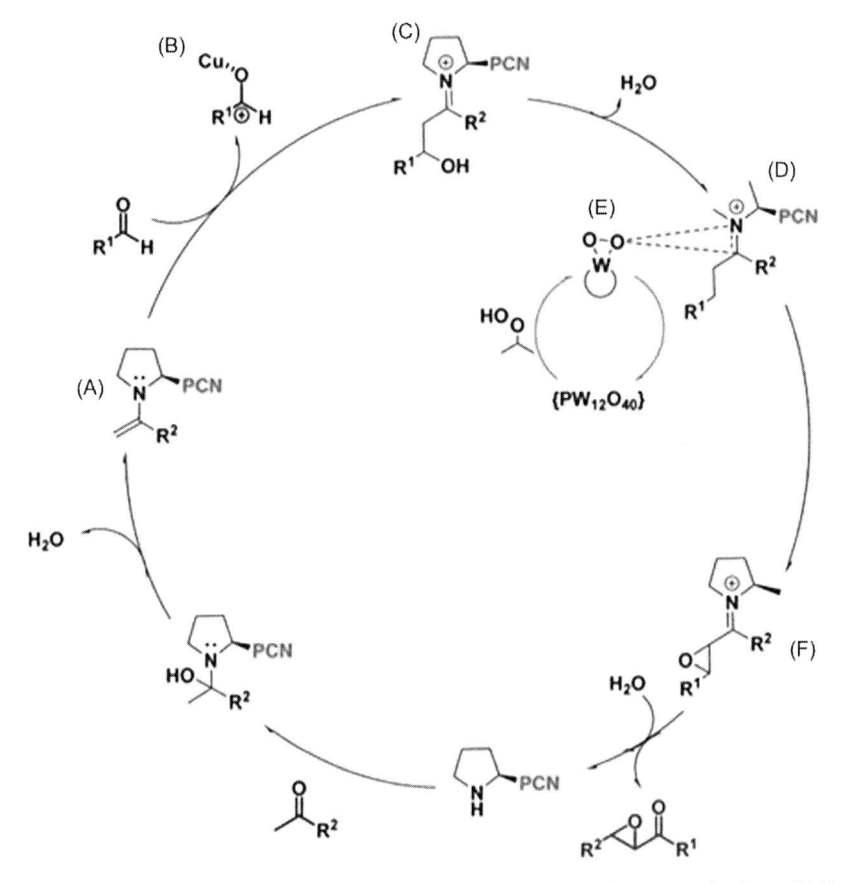

FIGURE 13.23 Potential mechanism of two-center catalysis in cascade asymmetric direct aldol/epoxidation reaction (Maru et al., 2022).

TABLE 13.10 Application of polyoxometalate/metal-organic framework materials in heterogeneous catalysis.

Entry	MOF	POM	Catalytic reaction	Reference
1	MIL-100(Fe)	$H_{3+x}PMo_{12-x}V_xO_{40}$ (x=0, 1, 2)	Oxidation of cyclohexene	Tong et al. (2017)
2	MIL-101(Cr)	$[PW_{11}CoO_{39}]^{5-}$ $[PW_{11}TiO_{40}]^{5-}$	Oxidation of alkenes	Maksimchuk et al. (2008)
3	MIL-101(Cr)	$H_5PV_2Mo_{10}O_{40}$	Oxidation of 2-chloroethyl ethyl sulfide	Li et al. (2018)
4	Cu-BTC	$[PTi_2W_{10}O_4]^{7-}$	CO_2 reduction	Liu et al. (2018)
5	MIL-101(Cr)	α-$PW_{15}V_3$, α-$P2W_{17}Ni$, α-$P_2W_{17}Co$	H_2 evolution	Tian et al. (2016)
6	Ag-based metal-organic nanotubes[a]	$H_3S\ PW_{12}O_{40}$ $H_4SiW_{12}O_{40}$	H_2 evolution	Li et al. (2018)
7	Cu-based metal-organic nanotubes[b]	$K_6P2W_{18}O_{62}$ $H_6As_2W_{18}O_{62}$	H_2 evolution	Zhang et al. (2018)
8	Cu-based MOF[c]	$H_4SiMo_{12}O_{40}$	Degradation of Rhodamine B	Chen, Liu, et al. (2018)
9	ZIF-67	$H_3PW_{12}O_{40}$	H_2O oxidation	Mukhopadhyay et al. (2020)
10	MOF-808	$H_3PW_{12}O_{40}$	Friedel-Crafts acylation of anisole with benzoyl chloride	Ullah et al. (2018)
11	MIL-100(Fe)	$[PMo_{11}Mn(H_2O)O_{39}]^{5-}$	Reduction of p-nitrophenol	Shah et al. (2018)

12	UiO-66	$H_4SiW_{12}O_{40}$	Hydrogenation of methyl levulinate/ transesterification of methyl-3-hydroxyvalerate	Cai et al. (2019)
13	NU-1000	$H_3PW_{12}O_{40}$	Isomerization/disproportionation of o-xylene	Ahn et al. (2018)
14	MOF-74	$H_3PW_{12}O_{40}$	Hydrogenation–esterification tandem reaction	Chen, and Zhang, Zhou, et al. (2019)
15	UiO-66-2COOH	$H_3PW_{12}O_{40}$	Ransesterification-esterification of acidic vegetable oils	Xie and Wan (2019)

[a] $Ag_{10}(\mu_4\text{-tta})_4(H_2O)_4$, $Ag_{10}(\mu_4\text{-tta})_4(H_2O)_4$.
[b] $C_{60}H_{46}N_{42}Cu_6$.
[c] $[(Cu_4Cl)(CPT)_4]\cdot3NO_3\cdot5NMP\cdot3.5H_2O$} (HCPT = 4-(4-carboxyphenyl)-1,2,4-triazole and NMP = N-methyl pyrrolidone), and {[$(Cu_4Cl)(CPT)_4$]·$(HSiW_{12}O_{40})\cdot31H_2O$}.

Photocatalysis

CO₂ reduction, organic synthesis, and water pollutant degradation are some of the examples of POMOF photocatalysis reactions (Ghahramaninezhad et al., 2018; Huo et al., 2016; Yang et al., 2021).

CO₂ reduction: A significant and sustainable method for natural carbon recycling is photoreduction of CO_2 into highly valuable compounds (such as CH_3OH and $HCOOH$) and hydrocarbon fuels (such as CH_4, C_2H_4, and C_2H_2). In 2020, Mellot-Draznieks et al. coimmobilized the Keggin POM ($PW_{12}O_{40}{}^{3-}$) and the catalytic complex Cp*Rh(bpydc)Cl₂ (bpyde 2,2'-bipyridine-5,5'-dicarboxylic acid) in the Zr(IV)-based MOF (UiO-67). In situ synthesis encapsulates POM within the MOF cavities, and postsynthesis linker exchange introduces the Rh catalytic complex into the POMOF composite ((PW_{12},Cp*Rh)/UiO-67). They used this catalyst, (PW_{12},Cp*Rh)/UiO-67, in the photocatalytic reduction of CO_2, and the product $HCOO^-$ was formed with the evolution of H_2 gas. The photoreduction of CO_2 is doubled when compared to the POM-free Cp*Rh/UiO-67 catalyst. Overall, the (PW_{12},Cp*Rh)/UiO-67 catalyst shows high activity toward the photoreduction of CO_2 under visible light (Benseghir et al., 2020). In the last decade, several POMOF-based catalysts have been synthesized and utilized in the photocatalytic reduction of CO_2 (Fig. 13.24) (Ghahramaninezhad et al., 2018; Hu, Zhou et al., 2019; Huang & Liu, 2020; Liu et al., 2018).

POM/MOF-based photocatalysts in organic synthesis: In addition, POM/MOF composites are effective photocatalysts for the selective oxidation of organic compounds to fine chemicals, such as the conversion of alcohols

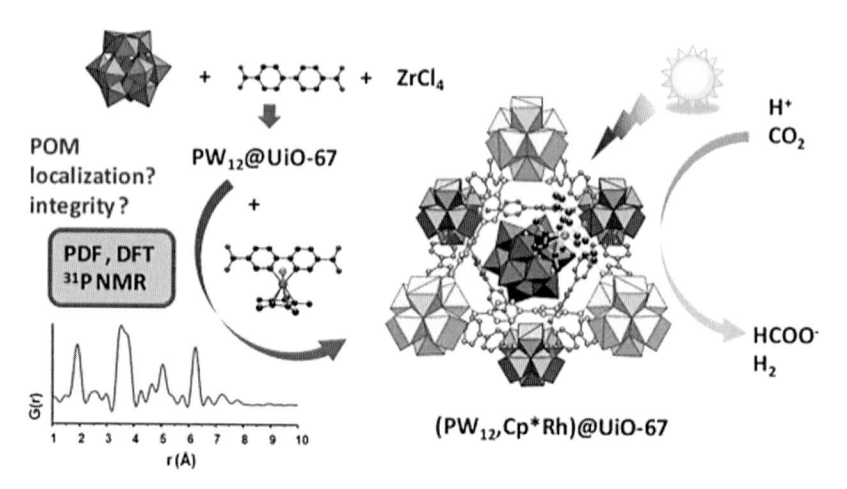

FIGURE 13.24 Photocatalysis CO_2 reduction over (PW_{12}, Cp*Rh)/UiO-67) (Benseghir et al., 2020).

to aldehydes and sulfides to sulfoxides (Fu et al., 2022; Zhao et al., 2017). Shi et al. in 2018 prepared Ag-based POMOF composite materials via the hydrothermal method, namely $(Ag_8(mttz)_4(H_2O)[PW^V W_{11}^{VI} O_{40}]H_2O$ (Parsaei et al., 2024) and $Ag_8(mttz)_4(H_2O) [PMO^V MO^{VII} O_{40}]$ (Parsaei & Akhbari, 2022a) (mttz: 5-methylthio-2Htetrazole). They used these hybrid materials for the selective oxidation of *cis*-cyclooctene to epoxy cyclooctane. Compound 1 shows higher catalytic performance than compound 2. The high activity of compound 1 is attributed to the abundant exposed active edge sites of Ag and the synergy of the components (Lin, Zhang et al., 2018). Later, Niu et al. prepared $\{[Zn(HPYI)_3]_2(DPNDI)\}[BW_{12}O_{40}]_2(ZnW\text{-}DPNDI\text{-}PYI)$ and used it in photocatalytic coupling of primary amines and olefins epoxidation with air under visible-light conditions (Fig. 13.25) (He et al., 2019).

Xu et al. have prepared three vanadium-substituted (Keggin-type $PMo_{12-n}V_nO_{40}$ ($n=1-3$)) POMOF composites using rho-ZIF by the mechanochemical method (Fig. 13.26). All PMoV/rho-ZIF systems showed higher selectivity as well as conversion for the photocatalytic conversion of thioanisole to sulfoxides. The $PMo_{11}VO_{40}$/rho-ZIF hybrid composite shows the highest catalytic selectivity for sulfoxides (93%), which is higher than those of $PMo_{11}VO_{40}$ (51%) and rho-ZIF (22%). For the oxidation of a variety of thioanisoles, this photocatalyst showed high activity and selectivity, and it could be recycled for several cycles without substantial loss of activity. However, when the size of the substrate increases, the conversion and

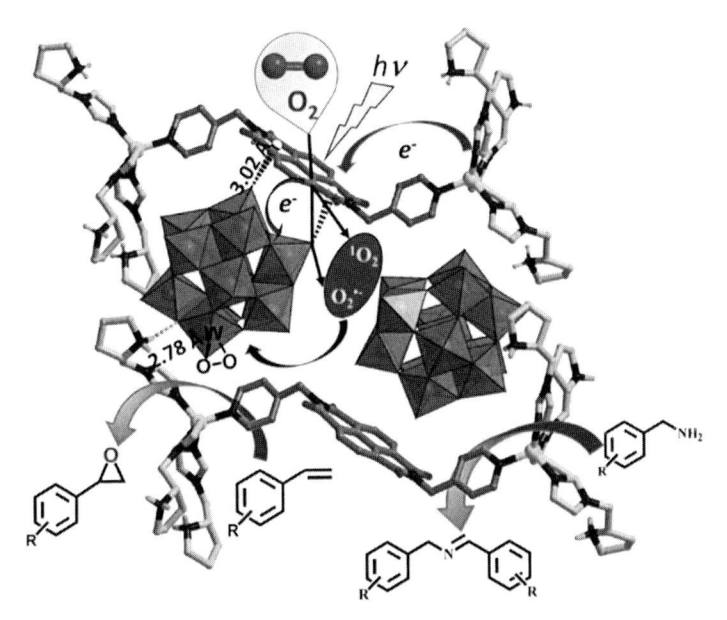

FIGURE 13.25 Photocatalytic organic transformation over ZnW-DPNDI-PYI (He et al., 2019).

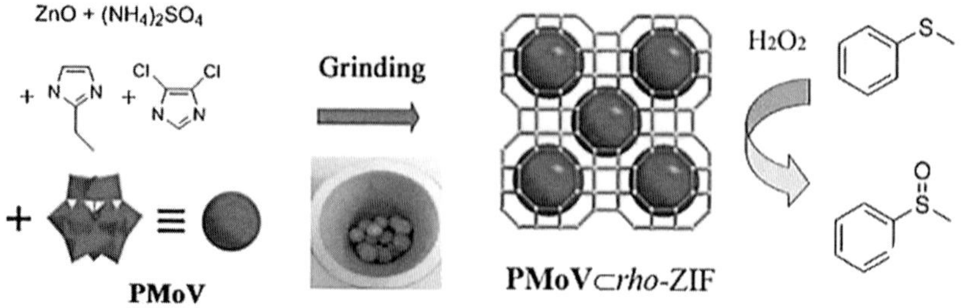

FIGURE 13.26 Conversion of sulfoxides from thioanisoles over $PMo_{11}VO_{40}$/rho-ZIF (Zhao et al., 2017).

selectivity may decrease significantly due to the increased steric hindrance (Zhao et al., 2017). In the last decade, a few different POMOF catalysts have been used in the field of photocatalytic organic transformations (Liao et al., 2015).

Metal-organic framework/silica composites

Silica NPs and nanostructures have garnered significant interest in catalytic applications due to their unique properties such as porosity, stability, and hydrophilicity. The integration of silica with MOFs offers the opportunity to combine the advantages of both materials and explore new applications. Currently, there are two main types of MOF-silica composites: SiO_2/MOFs and MOFs/SiO_2. SiO_2/MOFs involve incorporating dispersed silica NPs within the pores or channels of MOFs or growing an MOF shell on preformed silica spheres (Uemura et al., 2008). On the other hand, MOFs/SiO_2 employ silica as a coating shell grown on the surface of MOFs or as a support to facilitate the growth of MOF particles (Jo et al., 2011; Kou & Sun, 2018). These composite materials have the potential to enable innovative catalytic processes and provide enhanced performance and functionality.

Preparation methods of metal-organic framework/silica composites

In recent times, great efforts have been made to create MOF–silica composites. The MOF–silica composite materials with different morphologies can be prepared by adjusting the synthetic conditions. Different morphologies of the prepared MOF–silica composites lead to the distinct properties of the materials, so they can be applied in varying fields. Classify the synthetic strategies of the reported MOF–silica composites into three categories according to the construction, including MOFs/SiO_2, SiO_2/MOFs, and other MOFs/SiO_2 hybrids (Yuan et al., 2020).

Catalytic performances of metal-organic framework/silica composites

Common catalytic performances

Several articles were published that used MOF-silica composites as a catalyst (Table 13.11). Like other composites, silica is used to increase stability and increase porosity. Meanwhile, these composites can effectively increase the catalytic performance of MOFs to a great extent. Therefore, increasing attention has been focused on the catalytic performance of MOF-silica composites (Liang et al., 2015; Ying et al., 2018).

Karimi and Morsali (2013) prepared MOF-5/SBA-15 composites with different morphologies by adjusting the silica concentration. They concluded that as the concentration of SAB-15 increases (from 1% to 5%), the morphology of MOF-5 changes from the usual cubic structure to nanorods. By adjusting the concentration of silica, the crystallization and growth of MOFs can be well controlled. By adding SBA-15 to MOF-5, the BET surface area gradually increased, and the catalytic performance of MOF-5/SBA-15 for Friedel-Crafts alkylation increased by 50% compared to pure MOF-5. Research on MOF-silica composite materials for catalytic oxidation reactions has also been carried out. Pascanu et al. (2015) reported a double-supported Pd/MIL-88B-NH$_2$/nano-SiO$_2$ nano-palladium catalyst (8 wt.% Pd). In this study, a catalyst was developed by incorporating palladium metallic NPs into MIL-88B-NH$_2$, resulting in the formation of Pd/MOF. To enhance its stability, a protective layer of mesoporous SiO$_2$ was applied to the Pd/MOF material. The resulting composite, Pd/MIL-88B-NH$_2$/nano-SiO$_2$, exhibited a high BET surface area of 603 ± 4 m^2/g. When used as a catalyst in the aerobic oxidation of benzylic alcohols, it demonstrated improved efficiency compared to previous catalysts. Scientists have shown interest in using MOF-silica composites for catalytic desulfurization. Li et al. (2018) reported the preparation of a mesoporous composite called SRL-POM/MOF-199/MCM-41 using a surfactant-type heteropolyacid (SRL-POM), MCM-41, and MOF-199 through a one-pot method. The study revealed that the BET specific surface area of the composite decreased when the modified heteropolyacid was incorporated into the larger pores of MCM-41. The resulting catalyst was tested for its ability to oxidize DBT to sulfones. Remarkably, the SRL-POM/MOF-199/MCM-41 catalyst exhibited 100% selectivity in the oxidation of DBT to sulfones. In addition, MOF-silica composites have shown promising catalytic effects in coupling reactions. Yang et al. (2016) successfully synthesized porous OMS/Pd-ZnMOF (OMS refers to ordered mesoporous silicas) through cooperative template-directed self-assembly. The Pd-ZnMOF materials were prepared by adjusting the Pd/(Pd + Zn) molar ratio and then grown on the surface of silica spheres. The resulting OMS/Pd-ZnMOF-5% composite exhibited a BET surface area of 239 m^2/g and a total pore volume of 0.308 cm^3/g. The catalytic

TABLE 13.11 Several reports that use metal-organic framework-silica composites as catalysts.

Type of composite	Reactant	Reaction	S_{BET} (m^2/g)	Pore volume (cm^3/g)	Catalytic efficiency	Reference
MOF-5/SBA-15	Benzyl bromide	Friedel–Crafts alkylation	816	0.419	89%	Karimi and Morsali (2013)
Ni/SiO$_2$/a$_m$Ni-MOF-74	Nitrobenzene benzaldehyde	Tandem reaction	–	–	100%	Li and Zeng (2019)
OMS/Pd-ZnMOF-5%	Iodobenzene	Suzuki–Miyaura coupling reactions	239	0.308	79.1%	Yang et al. (2016)
SRL-POM/MOF-199/MCM-41	Dibenzothiophene	Oxidesulfurization	281	0.12	100%	Li et al. (2018)
Pd/MIL-88B-NH$_2$/nano-SiO$_2$ (8 wt.% Pd)	1-phenylethanol	Catalytic oxidation	603	1.83	7%	Pascanu et al. (2015)

a$_m$Ni-MOF-74, Amorphous Ni-MOF-74; OMS, ordered mesoporous silicas.

performance of the OMS/Pd-ZnMOF composites was evaluated in Suzuki–Miyaura coupling reactions, demonstrating highly active catalytic properties, particularly in these coupling reactions. Subsequently, researchers explored the application of this composite in tandem reaction systems. Sorribas et al. (2012) developed Ni/SiO$_2$/a$_m$Ni-MOF-74, where low-cost transition-metal nickel NPs were used as a solid precursor to synthesize Ni/SiO$_2$ hollow spheres. These spheres were then coated with an amorphous Ni-MOF-74 (a$_m$Ni-MOF-74) shell, creating a core-shell structured hollow catalyst. The catalytic efficiency of the tandem imination of nitrobenzene with benzaldehyde was evaluated, and the core-shell-structured hollow catalyst exhibited enhanced catalytic ability by suppressing the occurrence of side reactions. The hydrophilic nature of silica is influenced by the presence of silanol groups on its surface. Reducing the size of silica particles can increase the number of surface silanol groups, thereby enhancing the hydrophilicity of the material and improving its ability to adsorb hydrophilic molecules. In a study conducted by Kitagawa et al., they utilized porous CPL-5([Cu$_2$(pzdc)$_2$(L)]$_n$) (a MOF composed of copper and pzdc=pyrazine-2,3-dicarboxylate, L=pillar organic ligands) as a template to produce subnanosized silica NPs within the one-dimensional channels. This was achieved through a sol-gel reaction involving tetramethoxysilane. By employing this method, they were able to create hydrophilic silica NPs with enhanced surface functionality for the adsorption of hydrophilic molecules (Uemura et al., 2009). In their research, Kitagawa and his team provided evidence that the incorporation of silica NPs into the pores of porous coordination polymers (PCPs) led to significant alterations in the properties of the PCPs. This encapsulation process resulted in remarkable modifications to the characteristics and behavior of the PCPs. The adsorption measurements conducted on CPL-5 pores that were modified with hydrophilic silica revealed that the presence of silica enhanced the adsorption capacity for water and 1,4-dioxane. However, the diffusion of cyclohexane, which is a hydrophobic molecule, was significantly limited. This observation is noteworthy because many chemical reactions involve hydrophilic molecules. Therefore, this research has the potential to contribute to the advancement of MOFs in catalytic applications by facilitating the diffusion rate of hydrophilic molecules into the pores of MOFs. This can ultimately enhance their catalytic performance by increasing the accessibility of hydrophilic reactants within the MOF structure (Uemura et al., 2011). The integration of MOF NPs within mesoporous silica provides several advantages. Confining MOFs within silica maximizes the exposure of active sites on the outer surface of MOFs, enhances their chemical and mechanical stability, and offers protection. De Vos and colleagues demonstrated the effectiveness of this approach by confining a Zr-based MOF, UiO-66(NH$_2$), within SBA-15 channels. The resulting MOF-silica composites showed enhanced catalytic activity and reusability as heterogeneous catalysts for the synthesis of steroid derivatives. The smaller size of the MOF within SBA-15 promoted higher catalytic activity, particularly for bulky substrates that could not diffuse into the MOF's pores. Moreover, the composite exhibited excellent stability with minimal leaching of Zr, making it a promising catalyst for practical applications (Cirujano et al., 2017). Galarneau and colleagues successfully addressed the challenges of using fragile MOF crystals as robust catalysts by immobilizing them within structured silica monoliths. They achieved in situ synthesis of CuBTC NPs in a silica monolith and used them as catalysts for the Friedländer reaction under continuous flow conditions. The resulting CuBTC-MonoSil catalysts showed excellent performance, with a steady-state conversion rate of 85% and a productivity of 2.2 mmol/(min.g_{CuBTC}).The hierarchical structure of the silica monolith allowed for efficient and homogeneous flow, minimizing pressure drop. In comparison, using commercial CuBTC powder in a batch reactor yielded lower productivity. The immobilization of CuBTC within the silica monolith improved catalytic performance and flow characteristics, offering a promising solution for enhancing the practicality of MOF catalysts in continuous flow reactions (Sachse et al., 2012).

Photocatalysis

Li et al. presented a facial route for the one-step synthesis of $Cu_3(BTC)_2/SiO_2$ core-shell nanoparticles, which was used as a new way for photodegradation of organic materials in recent years. They presented a route for one-step synthesis of $Cu_3(BTC)_2/SiO_2$ core-shell NPs, which has been used as a new method for photodegradation of organic materials in recent years. The surface area of the composite was 1092.14 m^2/g. They investigated the photocatalytic activity of $Cu_3(BTC)_2/SiO_2$ thin films through the photodegradation of phenol in an aqueous solution and concluded that the composites showed greater degradation of phenol under visible light irradiation at pH=4, and the degradation rate was up to 93.1% (Li et al., 2013). Liu et al. (2005) used SOS (spheres-on-sphere) silica modified with different functional groups (–SH, –COOH, and –NH_2) on which ZIF-8 could grow, leading to the production of SOS/ZIF-8 composites. They obtained more convincing results that the size and shape of MOFs change in the presence of SOS microspheres. They derived SOS/ZIF-8 (SOS/ZnO), which enhanced the photocatalytic degradation of methyl orange in water.

Zeolite/metal-organic framework composites

Combining zeolite with MOF materials is a crucial approach to overcoming their individual limitations. Zeolite provides thermal, mechanical, and structural stability to the zeolite/MOF composites, while the MOF imparts functionality and flexibility. This integration results in superior performance compared to their respective parent materials (Xue et al., 2020). To address the limitations of MOFs, various strategies have been employed. These include

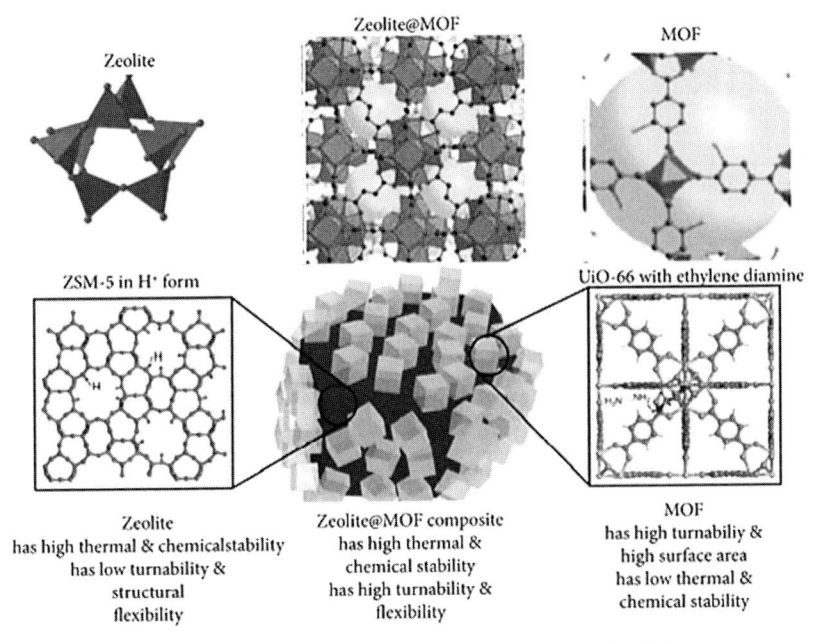

FIGURE 13.27 Zeolite/metal-organic framework composite (Derbe et al., 2023).

grafting active sites, modifying with functional groups, using chelating ligands, changing organic linkers, and impregnating suitable active materials. Additionally, researchers have focused on compositing MOFs with suitable materials like zeolite to enhance stability and electrical conductivity (Fig. 13.27). For example, Zhu et al. synthesized a ZSM-5/UiO-66 MOF composite through a solvothermal method. The UiO-66 MOF was grown over the surface of presynthesized zeolite to create a bifunctional acid-base catalyst for cascade reactions. The researchers observed enhanced catalytic performance of the ZSM-5/UiO-66 MOF composite compared to its individual components, ZSM-5 and UiO-66 MOF (Zhu et al., 2014).

Preparation methods of zeolite/metal-organic framework composites

Zeolite and MOF materials can be synthesized using a variety of methods, such as hydrothermal, sol-gel, coprecipitation, microwave-assisted, and biosynthesis. Zeolites can be functionalized with organic groups to enhance compatibility with MOFs. Similarly, MOFs can be synthesized through methods like solvothermal, spray-drying, microwave-assisted, hydrothermal, ionothermal, mechanochemical, and sonochemical methods. The synthesis of zeolite/MOF composites poses challenges due to competitive nucleation and crystal growth processes. A postsynthesis method, where zeolites are first synthesized and then MOFs are grown on their surface, is often employed to mitigate these challenges. Various synthesis methods exist for zeolite/MOF composites, allowing for the optimization of their properties and functionalities (Fallah & Sohrabnezhad, 2019). Some synthesis methods for the synthesis of zeolite/MOF composites are mentioned in Table 13.12.

Catalytic performances of zeolite/metal-organic framework composites

Common catalytic performances

The integration of zeolites and MOFs results in the formation of zeolite/MOF composites, which offer stability, recyclability, and high selectivity as bifunctional catalysts for various reactions. The catalytic activity is attributed to functional sites derived from organic bonds, heterocyclic skeletons, or metal nodes within the MOFs. This integration enables efficient catalytic processes and holds promise for diverse applications (Liao et al., 2020; Pascanu et al., 2019; Singh et al., 2022). The combination of zeolites and MOFs leads to the formation of bifunctional catalysts, as illustrated in Fig. 13.28. These catalysts, known as zeolite/MOF composites, find applications in various reactions such as cyclization, isomerization, condensation, Knoevenagel condensation, and dehydration of oxygen-containing molecules like alcohols, aldehydes, and acids. The synergistic effects of zeolites and MOFs in these composites enable enhanced catalytic performance and broaden the scope of catalytic transformations for these types of molecules (Fallah & Sohrabnezhad, 2019).

Singh et al. (2022) conducted a study where they synthesized a composite material called ZSM-5/Zn-MOF-1 for the production of renewable aromatics from isopropyl alcohol (IPA). By combining H-ZSM-5 zeolites with Zn-MOF-1, they created conditions that were highly selective for the production of lower aromatics from IPA. Interestingly, the reaction between IPA and ZSM-5/Zn-MOF-1 occurred predominantly within the zeolite framework, with minimal involvement of the MOF component. The composite material, with its appropriate density of Brønsted and Lewis acidic sites and hierarchical pore structures, exhibited high aromatic selectivity, yield, and stability, making it a promising catalyst for the production of renewable aromatics. In a study by Rani and Srivastava (2019), the catalytic conversion of sucrose into 2-((5-(hydroxymethyl) furan-2-yl) methylene) malononitrile was investigated using beta zeolite, Zr-based MOFs (UiO-66 and UiO-66-NH_2), and their composite zeolite/MOF. The authors compared the catalytic activity of these materials and found that the Beta zeolite, when combined with UiO-66-NH_2 in a composite structure (Beta [10%]/UiO66-NH_2), exhibited Broensted acid sites that facilitated the hydrolysis and isomerization of sucrose into glucose and fructose. On the other hand, the base group

TABLE 13.12 Synthesis methods of zeolite/metal-organic framework composites.

Method	Prepared zeolite	Prepared MOF	Composite	Reference
Solvothermal	Hydrothermal method	Solvothermal method. ([Zn₃(BTC)₂])	Zeolite/[Zn₃(BTC)₂]	Ayub and Rafique (2017)
	Solvothermal	Solvothermal (Zr-BDC)	Beta zeolite/Zr-BDC	Rani and Srivastava (2018)
	Solvothermal (ZSM-5)	Solvothermal (UiO-66)	ZSM-5/UiO-66	Zhu et al. (2014)
Hydrothermal	Hydrothermal (mordenite)	Hydrothermal (MIL-101(Cr))	Modernie zeolite/MIL-101 (Cr)	Fallah and Sohrabnezhad (2019)
	Hydrothermal (zeolite-5A)	Hydrothermal (MOF-74)	Zeolite/MOF-74	Al-Naddaf et al. (2018)
Reflux	Hydrothermal (mordenite)	Hydrothermal (MIL-101(Cr))	Mordenite/MIL-101R	Fallah and Sohrabnezhad (2019)
Microwave-assisted synthesis	—	—	MCM-41/Cu(BDC)	Tari et al. (2016)

MCM-41 refers to a mesoporous material consisting of silicate and aluminosilicate, mobil composition of matter no. 41.

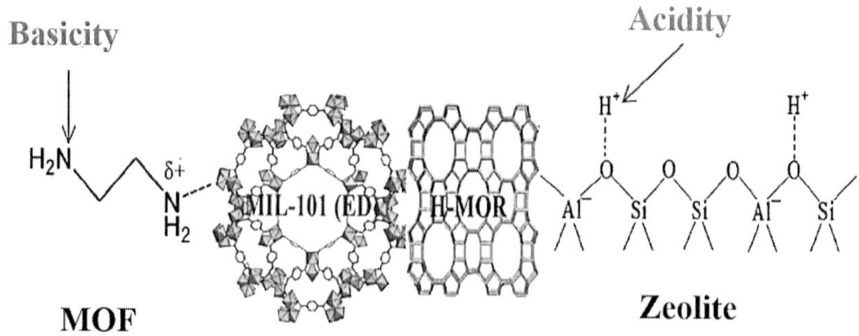

FIGURE 13.28 Schematic of zeolite/metal-organic framework composite and bifunctionality (acidity part of zeolite and basicity metal-organic framework) (Fallah & Sohrabnezhad, 2019).

present in Beta (10%)-UiO-66-NH$_2$ was responsible for the conversion of 5-(hydroxymethyl) furfural (HMF) to 2-((5-(hydroxymethyl) furan-2-yl) methylene) malononitrile through Knoevenagel condensation. The Beta zeolite demonstrated the highest activity in the conversion of sucrose to HMF in the first step, while UiO-66-NH$_2$ exhibited the best activity in the second step of Knoevenagel condensation. The proton (H$^+$) sites in Beta zeolite and the Zr^{4+}/Al^{3+} sites in Beta (10%)/UiO-66-NH$_2$ provided the necessary acidity for the hydrolysis and isomerization processes leading to the formation of HMF. Additionally, the amine group in the framework of UiO-66-NH$_2$ also aided in the isomerization of glucose to fructose. In a study by Zhu et al. (2014), the catalytic activity of a ZSM-5/UiO-66-NH$_2$ composite was investigated. This composite demonstrated promising bifunctional acid–base catalytic properties. The basic sites in the composite were introduced through postsynthesis modification of UiO-66 with amine groups, while the acid sites were derived from the tetrahedrally coordinated framework aluminum in the ZSM-5 zeolite. The catalytic process studied involved the synthesis of benzylidene malononitrile from malononitrile and benzaldehyde dimethyl acetal. The reaction proceeded through the hydrolysis of the acetal, catalyzed by the Brønsted acid sites in the ZSM-5 core, followed by the Knoevenagel condensation reaction catalyzed by the basic sites in the UiO-66-NH$_2$. This clearly demonstrated the bifunctional activity of the ZSM-5/UiO-66-NH$_2$ composite, where both acid and base sites synergistically contributed to the catalytic process. Li et al. (2020) conducted a study on the conversion of xylose to γ-valerolactone (GVL) using Zr-Al-SCM-1/zeolite as a bifunctional catalyst. The authors observed five cyclic steps involving acid-catalyzed reactions and Meerwein–Ponndorf–Verley (MPV) reduction reactions. The five steps in the conversion process are as follows: (I) Dehydration of xylose to furfural; (II) Hydrogenation of furfural to furfural alcohol and ether (FOL/FE); (III) Hydrolysis of FOL/FE to levulinic acid and esters (LA/

IPL); (IV) Hydrogenation of LA/IPL to 4-hydroxypentanoates; (V) Lactonization of 4-hydroxypentanoates to γ-valerolactone (GVL). In the production of GVL from xylose, the Zr-Al-SCM-1/zeolite catalyst plays a crucial role as an acid-base catalyst, serving as both Lewis and Brønsted acids. This bifunctional catalyst facilitates the various reactions involved in the conversion process. Pan et al. (2020) synthesized a composite material called HZSM-5/UiO-66-NH$_2$/Pd for the hydrogenation of CO_2. The synthesis involved coating UiO-66-NH$_2$ onto HZSM-5 to form HZSM-5/UiO-66-NH$_2$, followed by the fixation of Pd NPs onto the composite structure (Fig. 13.29). In this composite, UiO-66-NH$_2$ serves as an anchor for the metal NPs, while HZSM-5 provides structural support to prevent collapse. This arrangement helps to prevent the agglomeration of Pd NPs and enhances the stability of the catalyst. The presence of HZSM-5 zeolite contributes to the improved stability of the catalyst. The HZSM-5/UiO-66-NH$_2$ composite exhibited enhanced CO selectivity (92.2%) and improved catalyst stability for the hydrogenation of CO_2. The mechanism of the catalytic process involves the adsorption of CO_2 on HZSM-5/UiO-66-NH$_2$, followed by the dissociation of CO_2 into CO and O_2 on the Pd NP sites. The activated

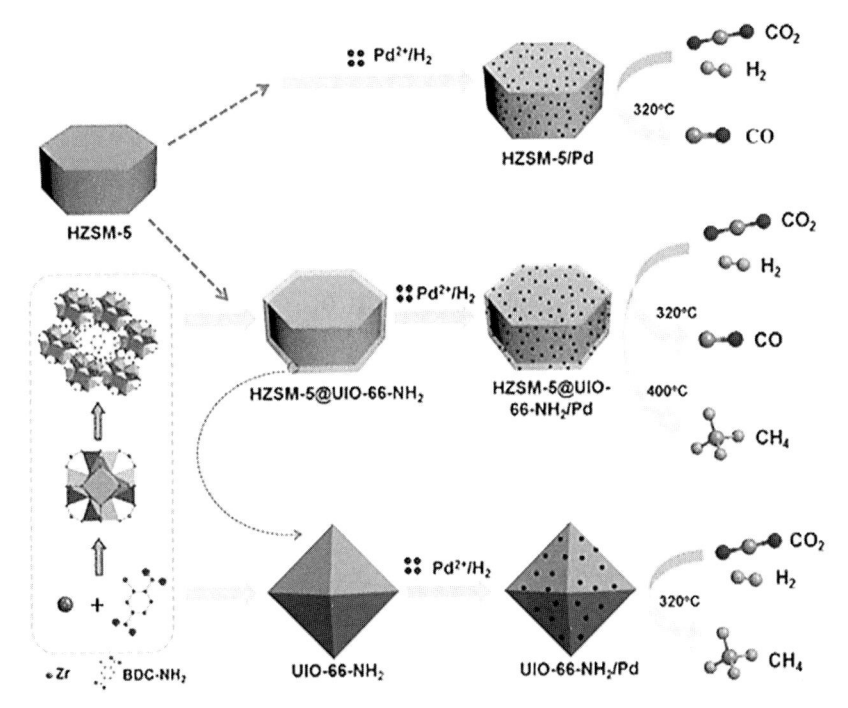

FIGURE 13.29 Assembling strategy for synthesis of HZSM-5/Pd, HZSM-5/UiO-66-NH$_2$/Pd, UiO-66-NH$_2$/Pd and their effects of CO_2 hydrogenation (Pan et al., 2020).

hydrogen atoms then react with the adsorbed oxygen to form water (H_2O). Finally, CO desorbs from the carrier and enters the gas phase, resulting in the formation of the final products.

$$CO_2 + Pd \rightarrow PdO_xCO$$

$$PdO_x + H_2 \rightarrow Pd + H_2O$$

$$CO_2 + 4H_2 \xrightarrow{HZM-5/UiO-66-NH_2/Pd} CH_4 + 2H_2O$$

Photocatalysis

Zeolite/MOF is used in the degradation of organic dyes (methyl blue and methyl orange) for many applications such as dyes, paintings, and cosmetics (Albouyeh et al., 2021). Improper disposal of these organic dyes has led to environmental threats. Among the existing methods, photodegradation has attracted attention due to its potential to convert pollutants into less toxic molecules, recovery, and rapid efficiency. In the photodegradation mechanism, when a photocatalyst such as zeolite/MOF is exposed to light, electrons in the material are excited to the CB, while holes are generated in the VB (Sadjadi et al., 2021). These photo-excited electrons and holes have two possible pathways: they can either recombine, releasing thermal energy, or they can migrate to the surface of the photocatalyst and interact with adsorbed dye molecules (Huang et al., 2019; Sadjadi et al., 2021). As a result of these interactions, reactive radical species are generated (Rani & Srivastava, 2018). The reactive free radical species, such as hydroxyl radicals (•OH), possess strong oxidative properties and play a key role in the degradation process. They can effectively decompose organic pollutants into CO_2 and H_2O molecules. The photocatalytic efficiency of a photocatalyst relies on factors such as charge separation, charge recombination, stability, reusability, band gap, and interfacial electron transfer (Cheng et al., 2021; Singh et al., 2021). Photodegradation can occur directly or through sensitization, depending on the band gaps between the CB and VB (Sadjadi et al., 2021; Singh et al., 2021). If the energy of the dye molecule's LUMO (lowest unoccupied molecular orbital) exceeds the CB of the semiconductor, photo-excited electrons transfer to the CB, reducing molecular oxygen to superoxide radicals (•O_2^-). Hydroxyl radicals (•OH) and superoxide radicals effectively degrade surface-bound dyes into carbon dioxide, water, and other small molecules (Fig. 13.30) (Singh et al., 2021).

The photodegradation process of dyes using zeolite/MOF composites begins with the conversion of photon energy into chemical energy. In a reaction, when exposed to light, the valence band electrons (e^-) absorb photon energy and are excited to the conduction band, resulting in the formation of counterbalance holes (h^+) in the valence band (Albouyeh et al., 2021; Sadjadi

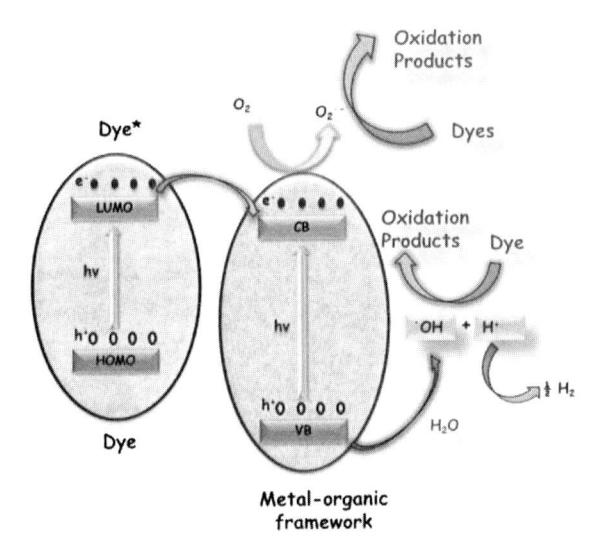

FIGURE 13.30 Schematic representation of sensitization-mediated superoxide radicals ($\cdot O_2^-$) generation and dye oxidation process, modified from Singh et al. (2021).

et al., 2021; Zhao & Cai, 2021). The photo-generated electrons and holes are further excited to the LUMO and HOMO (highest occupied molecular orbital), respectively (Cheng et al., 2021; Singh et al., 2021; Zhao & Cai, 2021). Within the conduction band, the photo-excited electrons react with O_2, leading to the formation of free radicals such as $\cdot O_2$ and $\cdot OH$, which are responsible for the degradation of dyes (reaction b) (Albouyeh et al., 2021; Han et al., 2018; Zhao & Cai, 2021). These free radicals play a crucial role in the degradation process. Zeolite/MOF composite materials help alleviate the rapid recombination of electron-hole pairs in MOFs, thereby enhancing the efficiency of the photocatalytic process. Albouyeh et al. (2021) conducted a study on the photocatalytic degradation of methylene blue in wastewater using a ZSM-12/MOF-74 core-shell composite under visible light. In this composite, the semiconductor Zn-based MOF-74 is incorporated within the ZSM-12 framework. Upon illumination, the VB electrons of MOF-74 are excited to the CB, generating active free radicals. These excited electrons and holes can then react with the methylene blue dye, water molecules, or hydroxyl ions (OH^-) to form hydroxyl free radicals. These hydroxyl radicals effectively decompose the dye molecules adsorbed on the composite's surface (Fig. 13.31). The presence of Zn^{2+} ions helps prevent the rapid recombination of charges, thereby enhancing the overall efficiency of the photocatalytic process. In their study, the authors achieved the highest removal efficiencies for the synthesized core/shell (80.37%), MOF-74 (Zn) (47.50%), and ZSM-12 (18.37%) at pH 9. The BET surface area was determined to be 750 m^2/g for MOF-74 and 980 m^2/g

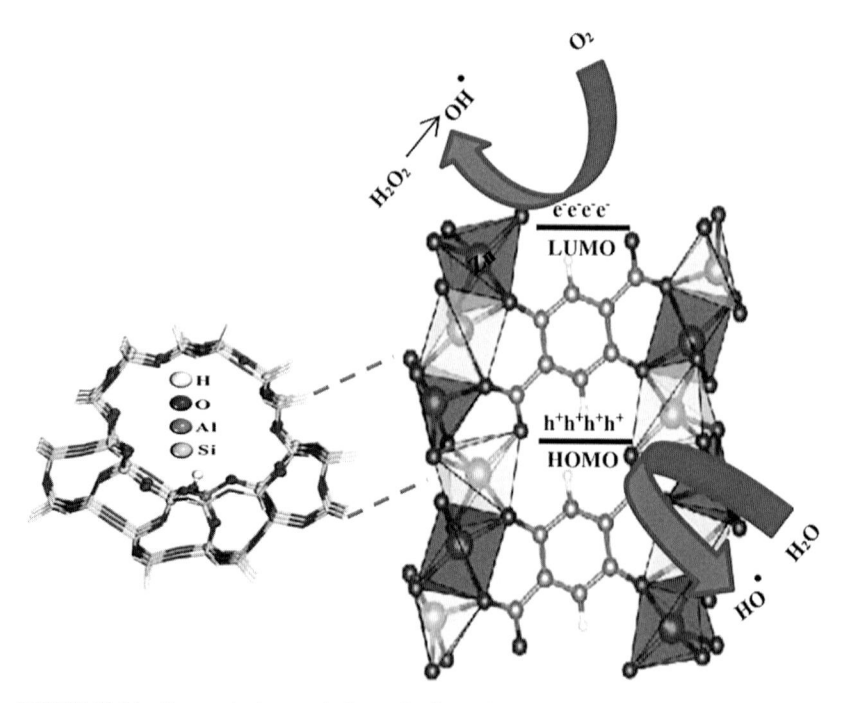

FIGURE 13.31. Proposed photocatalytic mechanism of ZSM-12/MOF-74 core/shell (Albouyeh et al., 2021).

for ZSM-12/MOF-74 core/shell. The authors concluded that ZSM-12/MOF-74 exhibited a higher removal percentage (67.87%) compared to its parent materials, zeolite ZSM-12 (18.37%) and MOF-74 (47.50%), under the same conditions (Table 13.13). The synthesized catalysts demonstrated high adsorption capacity and degradation at pH > 5, attributed to the electrostatic interaction between the negatively charged catalyst surface and the cationic methylene blue dye. The removal efficiency of the cationic dye followed the order: ZSM-12/MOF-74 > MOF-74 > ZSM-12. Additionally, the ZSM-12/MOF-74 composite exhibited the highest catalytic activity during five cycle tests, indicating its potential for reuse in the degradation of methylene blue through a simple washing process (Fallah, Sohrabnezhad & Abedini, 2019; Imyen et al., 2020; Wu, Ge et al., 2019).

a) Zeolite/MOF + hv → $h_{VB}^+ + e_{CB}^-$

b) $h_{VB}^+ + OH^- \rightarrow \cdot OH$

$h_{VB}^+ + H_2O \rightarrow \cdot OH + H^+$

TABLE 13.13 The photocatalytic performances of the MOF-74(Zn) and ZSM-12/MOF-74, at methylene blue=10 ppm, dosage: 0.04 g/L, at 30 minutes dark and 30 minutes under visible light irradiation (Albouyeh et al., 2021).

Sample	Adsorbed dye[a] (%)	Degraded dye[b] (%)	Removed dye[c] (%)
MOF-74 (pH=3)	14.40	23.23	37.63
MOF-74 (pH=7)	21.23	35.75	56.98
MOF-74 (pH=9)	28.98	36.15	65.13
ZSM-12 (pH=3)	9.32	7.38	16.70
ZSM-12 (pH=7)	17.54	15.27	32.81
SZM-12 (pH=9)	20.51	18.35	38.86
ZSM-12/MOF-74 (pH=3)	15.64	35.47	51.11
ZSM-12/MOF-74 (pH=7)	28.72	46.74	75.46
ZSM-12/MOF-74 (pH=9)	32.87	47.50	80.37

[a] Determined with UV–vis after 30 min of stirring in dark.
[b] Determined with UV–vis after 30 min of irradiation with visible-light.
[c] The total amount of dye removed after 60 min of reaction.

Conclusion

Combining MOFs and various functional materials, such as metal NPs, polymers, carbons, metal oxides, zeolites, silicas, and POMs, has been realized through various methods developed to prepare MOF composites. MOF composites have wide catalytic applications with activity (MNPs, MONPs), selectivity (zeolites, POMs), and stability (polymers, zeolites), which shows synergistic effects between components. MOF composites can act as multifunctional catalysts for tandem reactions.

MNP/MOF catalysts have shown superior performance in electrocatalysis, photocatalysis, and catalytic activities compared to their single counterparts. However, the rarity of precious metals limits their use. MONPs in MOFs increase the number of active sites, which increases their adsorption and catalytic ability. Carbon/MOF can increase composite conductivity stability and activity for electrocatalysis. MOFs have attracted considerable interest in various fields to improve performance. The strategy of MOF/MOF synthesis is very important in improving their performance; for example, the core-shell puzzle improves the catalytic performance due to the synergistic effect between the intrinsic properties of both MOFs. In the zeolite/MOF composites, zeolite provides thermal, mechanical, and structural stability to the zeolite/MOF composites, while the MOF imparts functionality and flexibility. The combination of zeolites and MOFs leads to the formation of bifunctional catalysts. Polymers have various unique properties, such as softness and thermal and chemical stability. Polymers at the nanometer scale show amazing properties. These polymers confined in the pores/channels of the MOF can have abundant interfaces with the MOF backbone to exhibit synergistic catalysis. One recent application of polymer/MOF composites is their promising catalytic activity in the degradation of CWA, such as organophosphates. The integration of silica with MOFs combines the unique properties of both materials and leads to new applications; for example, they provide a powerful substrate for many nanoscale functions (such as porosity, stability, and hydrophilicity) that provide significant performance and have attracted attention in catalytic applications. POMs have tunable shapes and sizes, solubility, redox potential, and strong acidity. This is important for catalytic transformations, especially in acidic and oxidation reactions. Of course, their application is limited due to their low specific surface and low stability.

AI disclosure

During the preparation of this work, the author(s) used Grammarly. This artificial intelligence has been used only to correct grammatical errors and make the text more readable. After using this tool/service, the author(s) reviewed and edited the content as needed and take(s) full responsibility for the content of the publication.

References

Abazari, R., Morsali, A., & Dubal, D. P. (2020). An advanced composite with ultrafast photocatalytic performance for the degradation of antibiotics by natural sunlight without oxidizing the source over TMU-5@Ni-Ti LDH: Mechanistic insight and toxicity assessment. *Inorganic Chemistry Frontiers, 7*(12), 2287–2304. https://doi.org/10.1039/d0qi00050g.

Aguila, B., Sun, Q., Wang, X., O'Rourke, E., Al-Enizi, A. M., Nafady, A., & Ma, S. (2018). Lower activation energy for catalytic reactions through host–guest cooperation within metal-organic frameworks. *Angewandte Chemie, 130*(32), 10264–10268. https://doi.org/10.1002/ange.201803081.

Ahn, S., Nauert, S. L., Buru, C. T., Rimoldi, M., Choi, H., Schweitzer, N. M., Hupp, J. T., Farha, O. K., & Notestein, J. M. (2018). Pushing the limits on metal-organic frameworks as a catalyst support: NU-1000 supported tungsten catalysts for o-xylene isomerization and disproportionation. *Journal of the American Chemical Society, 140*(27), 8535–8543. https://doi.org/10.1021/jacs.8b04059.

Akhbari, K., & Morsali, A. (2013). Modulating methane storage in anionic nano-porous MOF materials via post-synthetic cation exchange process. *Dalton Transactions, 42*(14), 4786–4789. https://doi.org/10.1039/c3dt32846e.

Alavijeh, R. K., & Akhbari, K. (2020). Biocompatible MIL-101(Fe) as a smart carrier with high loading potential and sustained release of curcumin. *Inorganic Chemistry, 59*(6), 3570–3578. https://doi.org/10.1021/acs.inorgchem.9b02756.

Alavijeh, R. K., & Akhbari, K. (2024). Cancer therapy by nano MIL-n series of metal-organic frameworks. *Coordination Chemistry Reviews, 503*, 215643.

Alavijeh, R. K., Akhbari, K., & White, J. (2019). Solid–liquid conversion and carbon dioxide storage in a calcium-based metal-organic framework with micro- and nanoporous channels. *Crystal Growth & Design, 19*(12), 7290–7297. https://doi.org/10.1021/acs.cgd.9b01174.

Albouyeh, A., Pourahmad, A., & Kefayati, H. (2021). Synthesis of MTW@MOF nanocomposite for removal of methylene blue. *Journal of Coordination Chemistry, 74*(13), 2174–2184. https://doi.org/10.1080/00958972.2021.1954173.

Al-Naddaf, Q., Thakkar, H., & Rezaei, F. (2018). Novel zeolite-5A@MOF-74 composite adsorbents with core-shell structure for H_2 purification. *ACS Applied Materials and Interfaces, 10*(35), 29656–29666. https://doi.org/10.1021/acsami.8b10494.

Amidi, D. M., & Akhbari, K. (2024). Iodine-loaded ZIF-7-coated cotton substrates show sustained iodine release as effective antibacterial textiles. *New Journal of Chemistry, 48*(5), 2016–2027. https://doi.org/10.1039/d3nj05198f.

An, B., Zhang, J., Cheng, K., Ji, P., Wang, C., & Lin, W. (2017). Confinement of ultrasmall Cu/ZnO_x nanoparticles in metal-organic frameworks for selective methanol synthesis from catalytic hydrogenation of CO_2. *Journal of the American Chemical Society, 139*(10), 3834–3840. https://doi.org/10.1021/jacs.7b00058.

Arshadi Edlo, A., & Akhbari, K. (2023). Modulating the antibacterial activity of a CuO@HKUST-1 nanocomposite by optimizing its synthesis procedure. *New Journal of Chemistry, 47*(45), 20770–20776. https://doi.org/10.1039/d3nj03914e.

Ayub, N., & Rafique, U. (2017). Synthesis and characterization of aluminosilicates [Zn_3 (BTC)$_2$] hybrid composite materials. *Journal of Silicate Based & Composite Materials, 69*(3), 98.

Bagheri, M., Melillo, A., Ferrer, B., Masoomi, M. Y., & Garcia, H. (2022). Quasi-HKUST prepared via postsynthetic defect engineering for highly improved catalytic conversion of 4-nitrophenol. *ACS Applied Materials and Interfaces, 14*(1), 978–989. https://doi.org/10.1021/acsami.1c19862.

Bassen, M., Nchimi, K., & Terraschke, H. (2023). *Nanostructured materials: Applications, synthesis and in-situ characterization*.

Bavykina, A., Kolobov, N., Khan, I. S., Bau, J. A., Ramirez, A., & Gascon, J. (2020). Metal-organic frameworks in heterogeneous catalysis: Recent progress, new trends, and future perspectives. *Chemical Reviews, 120*(16), 8468–8535. https://doi.org/10.1021/acs.chemrev.9b00685.

Becerra, J., Nguyen, D. T., Gopalakrishnan, V. N., & Do, T. O. (2020). Plasmonic Au nanoparticles incorporated in the zeolitic imidazolate framework (ZIF-67) for the efficient sunlight-driven photoreduction of CO_2. *ACS Applied Energy Materials, 3*(8), 7659–7665. https://doi.org/10.1021/acsaem.0c01083.

Behera, P., Subudhi, S., Tripathy, S. P., & Parida, K. (2022). MOF derived nano-materials: A recent progress in strategic fabrication, characterization and mechanistic insight towards divergent photocatalytic applications. *Coordination Chemistry Reviews, 456*, 214392.

Benseghir, Y., Lemarchand, A., Duguet, M., Mialane, P., Gomez-Mingot, M., Roch-Marchal, C., Pino, T., Ha-Thi, M. H., Haouas, M., Fontecave, M., Dolbecq, A., Sassoye, C., & Mellot-Draznieks, C. (2020). Co-immobilization of a Rh catalyst and a keggin polyoxometalate in the UiO-67 Zr-based metal-organic framework: In depth structural characterization and photocatalytic properties for CO2 reduction. *Journal of the American Chemical Society, 142*(20), 9428–9438. https://doi.org/10.1021/jacs.0c02425.

Bie, C., Wang, L., & Yu, J. (2022). Challenges for photocatalytic overall water splitting. *Chem, 8*(6), 1567–1574. https://doi.org/10.1016/j.chempr.2022.04.013.

Brahmi, C., Benltifa, M., Vaulot, C., Michelin, L., Dumur, F., Millange, F., Frigoli, M., Airoudj, A., Morlet-Savary, F., & Bousselmi, L. (2021). New hybrid MOF/polymer composites for the photodegradation of organic dyes. *European Polymer Journal, 154*, 110560.

Bunge, M. A., Davis, A. B., West, K. N., West, C. W., & Glover, T. G. (2018). Synthesis and characterization of UiO-66-NH$_2$ metal-organic framework cotton composite textiles. *Industrial and Engineering Chemistry Research, 57*(28), 9151–9161. https://doi.org/10.1021/acs.iecr.8b01010.

Bunge, M. A., Pasciak, E., Choi, J., Haverhals, L., Reichert, W. M., & Glover, T. G. (2020). Ionic liquid welding of the UIO-66-NH$_2$ MOF to cotton textiles. *Industrial and Engineering Chemistry Research, 59*(43), 19285–19298. https://doi.org/10.1021/acs.iecr.0c03763.

Buru, C. T., & Farha, O. K. (2020). Strategies for incorporating catalytically active polyoxometalates in metal-organic frameworks for organic transformations. *ACS Applied Materials and Interfaces, 12*(5), 5345–5360. https://doi.org/10.1021/acsami.9b19785.

Cai, W., Han, Y., Li, H., Qi, W., Xu, J., Wu, X., Zhao, H., Zhang, X., Lai, J., & Wang, L. (2020). Significantly enhanced electrocatalytic N$_2$ reduction to NH$_3$ by surface selenization with multiple functions. *Journal of Materials Chemistry A, 8*(39), 20331–20336. https://doi.org/10.1039/d0ta06991d.

Cai, X., Xu, Q., Tu, G., Fu, Y., Zhang, F., & Zhu, W. (2019). Synergistic catalysis of ruthenium nanoparticles and polyoxometalate integrated within single UiO–66 microcrystals for boosting the efficiency of methyl levulinate to γ-valerolactone. *Frontiers in Chemistry, 7*. https://doi.org/10.3389/fchem.2019.00042.

Cao, J., Li, X., & Tian, H. (2020). Metal-organic framework (Mof)-based drug delivery. *Current Medicinal Chemistry, 27*(35), 5949–5969. https://doi.org/10.2174/0929867326666190618152518.

Chanda, D., Xing, R., Xu, T., Liu, Q., Luo, Y., Liu, S., Tufa, R. A., Dolla, T. H., Montini, T., & Sun, X. (2021). Electrochemical nitrogen reduction: Recent progress and prospects. *Chemical Communications, 57*(60), 7335–7349. https://doi.org/10.1039/d1cc01451j.

Chen, C., Khosrowabadi Kotyk, J. F., & Sheehan, S. W. (2018). Progress toward commercial application of electrochemical carbon dioxide reduction. *Chem, 4*(11), 2571–2586. https://doi.org/10.1016/j.chempr.2018.08.019.

Chen, D. M., Liu, X. H., Zhang, N. N., Liu, C. S., & Du, M. (2018). Immobilization of polyoxometalate in a cage-based metal–organic framework towards enhanced stability and highly effective dye degradation. *Polyhedron, 152*, 108–113. https://doi.org/10.1016/j.poly.2018.05.059.

Chen, H., Shen, K., Tan, Y., & Li, Y. (2019). Multishell hollow metal/nitrogen/carbon dodecahedrons with precisely controlled architectures and synergistically enhanced catalytic properties. *ACS Nano, 13*(7), 7800–7810. https://doi.org/10.1021/acsnano.9b01953.

Chen, J., Shen, K., & Li, Y. (2017). Greening the processes of metal–organic framework synthesis and their use in sustainable catalysis. *ChemSusChem, 10*(16), 3165–3187. https://doi.org/10.1002/cssc.201700748.

Chen, L., Ding, X., Huo, J., El Hankari, S., & Bradshaw, D. (2019). Facile synthesis of magnetic macroporous polymer/MOF composites as separable catalysts. *Journal of Materials Science, 54*(1), 370–382. https://doi.org/10.1007/s10853-018-2835-x.

Chen, L., Luque, R., & Li, Y. (2018). Encapsulation of metal nanostructures into metal-organic frameworks. *Dalton Transactions, 47*(11), 3663–3668. https://doi.org/10.1039/c8dt00092a.

Chen, L., & Xu, Q. (2019). Metal-organic framework composites for catalysis. *Matter, 1*(1), 57–89. https://doi.org/10.1016/j.matt.2019.05.018.

Chen, L., Zhang, X., Zhou, J., Xie, Z., Kuang, Q., & Zheng, L. (2019). A nano-reactor based on PtNi@metal–organic framework composites loaded with polyoxometalates for hydrogenation–esterification tandem reactions. *Nanoscale, 11*(7), 3292–3299. https://doi.org/10.1039/c8nr08734b.

Chen, Y., Han, S., Li, X., Zhang, Z., & Ma, S. (2014). Why does enzyme not leach from metal-organic frameworks (MOFs)? Unveiling the interactions between an enzyme molecule and a MOF. *Inorganic Chemistry, 53*(19), 10006–10008. https://doi.org/10.1021/ic501062r.

Chen, Y. Z., Wang, Z. U., Wang, H., Lu, J., Yu, S. H., & Jiang, H. L. (2017). Singlet oxygen-engaged selective photo-oxidation over pt nanocrystals/porphyrinic MOF: The roles of photothermal effect and pt electronic state. *Journal of the American Chemical Society, 139*(5), 2035–2044. https://doi.org/10.1021/jacs.6b12074.

Chen, Z., Islamoglu, T., & Farha, O. K. (2019). Toward base heterogenization: A zirconium metal-organic framework/dendrimer or polymer mixture for rapid hydrolysis of a nerve-agent simulant. *ACS Applied Nano Materials, 2*(2), 1005–1008. https://doi.org/10.1021/acsanm.8b02292.

Chen, Z., Ma, K., Mahle, J. J., Wang, H., Syed, Z. H., Atilgan, A., Chen, Y., Xin, J. H., Islamoglu, T., & Peterson, G. W. (2019). Integration of metal-organic frameworks on protective layers for destruction of nerve agents under relevant conditions. *Journal of the American Chemical Society, 141*(51), 20016–20021.

Cheng, W., Lu, X. F., Luan, D., & Lou, X. W. (2020). NiMn-based bimetal–organic framework nanosheets supported on multi-channel carbon fibers for efficient oxygen electrocatalysis. *Angewandte Chemie - International Edition, 59*(41), 18234–18239. https://doi.org/10.1002/anie.202008129.

Cheng, Y., Wang, X., Mei, Y., Wang, D., & Ji, C. (2021). ZnCDs/ZnO@ZIF-8 zeolite composites for the photocatalytic degradation of tetracycline. *Catalysts, 11*(8), 934.

Choi, S., Jung, W. J., Park, K., Kim, S. Y., Baeg, J. O., Kim, C. H., Son, H. J., Pac, C., & Kang, S. O. (2021). Rapid exciton migration and amplified funneling effects of multi-porphyrin arrays in a Re(I)/porphyrinic MOF hybrid for photocatalytic CO_2 reduction. *ACS Applied Materials and Interfaces, 13*(2), 2710–2722. https://doi.org/10.1021/acsami.0c19856.

Chronopoulos, D. D., Saini, H., Tantis, I., Zbořil, R., Jayaramulu, K., & Otyepka, M. (2022). Carbon nanotube based metal–organic framework hybrids from fundamentals toward applications. *Small, 18*(4), 2104628.

Cirujano, F. G., Luz, I., Soukri, M., Goethem, C. V., Vankelecom, I. F. J., Lail, M., & De Vos, D. E. (2017). Boosting the catalytic performance of metal–organic frameworks for steroid transformations by confinement within a mesoporous scaffold. *Angewandte Chemie, 129*(43), 13487–13491. https://doi.org/10.1002/ange.201706721.

Corma, A., & Iborra, S. (2006). Optimization of alkaline earth metal oxide and hydroxide catalysts for base-catalyzed reactions. *Advances in Catalysis, 49*, 239–302. https://doi.org/10.1016/S0360-0564(05)49004-5.

Couzon, N., Ferreira, M., Duval, S., El-Achari, A., Campagne, C., Loiseau, T., & Volkringer, C. (2022). Microwave-assisted synthesis of porous composites MOF-textile for the protection against chemical and nuclear hazards. *ACS Applied Materials and Interfaces, 14*(18), 21497–21508. https://doi.org/10.1021/acsami.2c03247.

Dai, S., Ngoc, K. P., Grimaud, L., Zhang, S., Tissot, A., & Serre, C. (2022). Impact of capping agent removal from Au NPs@MOF core-shell nanoparticle heterogeneous catalysts. *Journal of Materials Chemistry A, 10*(6), 3201–3205. https://doi.org/10.1039/d1ta09108e.

Dai, X., Liu, M., Li, Z., Jin, A., Ma, Y., Huang, X., Sun, H., Wang, H., & Zhang, X. (2016). Molybdenum polysulfide anchored on porous Zr-metal organic framework to enhance the performance of hydrogen evolution reaction. *The Journal of Physical Chemistry C, 120*(23), 12539–12548. https://doi.org/10.1021/acs.jpcc.6b02818.

Dastbaz, A., Karimi-Sabet, J., & Moosavian, M. A. (2019). Sonochemical synthesis of novel decorated graphene nanosheets with amine functional Cu-terephthalate MOF for hydrogen adsorption: Effect of ultrasound and graphene content. *International Journal of Hydrogen Energy, 44*(48), 26444–26458. https://doi.org/10.1016/j.ijhydene.2019.08.116.

Davoodi, A., Akhbari, K., & Alirezvani, M. (2023). Prolonged release of silver and iodine from ZIF-7 carrier with great antibacterial activity. *CrystEngComm, 25*(27), 3931–3942. https://doi.org/10.1039/d3ce00529a.

Deng, X., Yang, L., Huang, H., Yang, Y., Feng, S., Zeng, M., Li, Q., & Xu, D. (2019). Shape-defined hollow structural Co-MOF-74 and metal nanoparticles@Co-MOF-74 composite through a transformation strategy for enhanced photocatalysis performance. *Small, 15*(35), 1902287.

Derbe, T., Sani, T., Amare, E., & Girma, T. (2023). Mini review on synthesis, characterization, and application of zeolite@MOF composite. *Advances in Materials Science and Engineering*. https://doi.org/10.1155/2023/8760967.

Dhakshinamoorthy, A., & Garcia, H. (2012). Catalysis by metal nanoparticles embedded on metal–organic frameworks. *Chemical Society Reviews, 41*(15), 5262–5284. https://doi.org/10.1039/c2cs35047e.

Dong, X. W., Yang, Y., Che, J. X., Zuo, J., Li, X. H., Gao, L., Hu, Y. Z., & Liu, X. Y. (2018). Heterogenization of homogeneous chiral polymers in metal-organic frameworks with enhanced catalytic performance for asymmetric catalysis. *Green Chemistry, 20*(17), 4085–4093. https://doi.org/10.1039/c8gc01323c.

Du, D. Y., Qin, J. S., Li, S. L., Su, Z. M., & Lan, Y. Q. (2014). Recent advances in porous polyoxometalate-based metal-organic framework materials. *Chemical Society Reviews, 43*(13), 4615–4632. https://doi.org/10.1039/c3cs60404g.

Dwyer, D. B., Lee, D. T., Boyer, S., Bernier, W. E., Parsons, G. N., & Jones, W. E. (2018). Toxic organophosphate hydrolysis using nanofiber-templated UiO-66-NH$_2$ metal-organic framework polycrystalline cylinders. *ACS Applied Materials and Interfaces, 10*(30), 25794–25803. https://doi.org/10.1021/acsami.8b08167.

Edlo, A., & Akhbari, K. (2024). Modulated antibacterial activity in ZnO@ MIL-53 (Fe) and CuO@ MIL-53 (Fe) nanocomposites prepared by simple thermal treatment process. *Applied Organometallic Chemistry, 38*(2), e7326.

Falcaro, P., Ricco, R., Yazdi, A., Imaz, I., Furukawa, S., Maspoch, D., Ameloot, R., Evans, J. D., & Doonan, C. J. (2016). Application of metal and metal oxide nanoparticles at MOFs. *Coordination Chemistry Reviews, 307*, 237–254. https://doi.org/10.1016/j.ccr.2015.08.002.

Fallah, M., & Sohrabnezhad, S. (2019). Study of synthesis of mordenite zeolite/MIL-101 (Cr) metal–organic framework compounds with various methods as bi-functional adsorbent. *Advanced Powder Technology, 30*(2), 336–346. https://doi.org/10.1016/j.apt.2018.11.011.

Fallah, M., Sohrabnezhad, S., & Abedini, M. (2019). Synthesis of chromene derivatives in the presence of mordenite zeolite/MIL-101 (Cr) metal–organic framework composite as catalyst. *Applied Organometallic Chemistry, 33*(4), e4801.

Feng, L., Wang, K. Y., Willman, J., & Zhou, H. C. (2020). Hierarchy in metal-organic frameworks. *ACS Central Science, 6*(3), 359–367. https://doi.org/10.1021/acscentsci.0c00158.

Feng, L., Yuan, S., Li, J. L., Wang, K. Y., Day, G. S., Zhang, P., Wang, Y., & Zhou, H. C. (2018). Uncovering two principles of multivariate hierarchical metal-organic framework synthesis via retrosynthetic design. *ACS Central Science, 4*(12), 1719–1726. https://doi.org/10.1021/acscentsci.8b00722.

Fertig, R., Irrgang, T., Freitag, F., Zander, J., & Kempe, R. (2018). Manganese-catalyzed and base-switchable synthesis of amines or imines via borrowing hydrogen or dehydrogenative condensation. *ACS Catalysis, 8*(9), 8525–8530. https://doi.org/10.1021/acscatal.8b02530.

Fu, W., Yi, J., Cheng, M., Liu, Y., Zhang, G., Li, L., Du, L., Li, B., Wang, G., & Yang, X. (2022). When bimetallic oxides and their complexes meet Fenton-like process. *Journal of Hazardous Materials, 424*(Part B), 127419.

Furukawa, S., Hirai, K., Nakagawa, K., Takashima, Y., Matsuda, R., Tsuruoka, T., Kondo, M., Haruki, R., Tanaka, D., Sakamoto, H., Shimomura, S., Sakata, O., & Kitagawa, S. (2009). Heterogeneously hybridized porous coordination polymer crystals: Fabrication of heterometallic core–shell single crystals with an in-plane rotational epitaxial relationship. *Angewandte Chemie, 121*(10), 1798–1802. https://doi.org/10.1002/ange.200804836.

Férey, G., Mellot-Draznieks, C., Serre, C., Millange, F., Dutour, J., Surblé, S., & Margiolaki, I. (2005). A chromium terephthalate-based solid with unusually large pore volumes and surface area. *Science, 309*(5743), 2040–2042. https://doi.org/10.1126/science.1116275.

Gao, C., Lyu, F., & Yin, Y. (2021). Encapsulated metal nanoparticles for catalysis. *Chemical Reviews, 121*(2), 834–881. https://doi.org/10.1021/acs.chemrev.0c00237.

Ghahramaninezhad, M., Soleimani, B., & Niknam Shahrak, M. (2018). A simple and novel protocol for Li-trapping with a POM/MOF nano-composite as a new adsorbent for CO_2 uptake. *New Journal of Chemistry, 42*(6), 4639–4645. https://doi.org/10.1039/c8nj00274f.

Ghasemzadeh, R., & Akhbari, K. (2023a). Heterostructured Ag@MOF-801/MIL-88A(Fe) nanocomposite as a biocompatible photocatalyst for degradation of reactive black 5 under visible light. *Inorganic Chemistry, 62*(43), 17818–17829. https://doi.org/10.1021/acs.inorgchem.3c02616.

Ghasemzadeh, R., & Akhbari, K. (2023b). Band gap engineering of MOF-801 via loading of γ-Fe_2O_3 quantum dots inside it as a visible light-responsive photocatalyst for degradation of acid orange 7. *Crystal Growth & Design, 23*(9), 6359–6368. https://doi.org/10.1021/acs.cgd.3c00272.

Ghasemzadeh, R., & Akhbari, K. (2023). Embedding of copper(i) oxide quantum dots in MOF-801 for the photocatalytic degradation of acid yellow 23 under visible light. *New Journal of Chemistry, 47*(33), 15760–15770. https://doi.org/10.1039/d3nj01395b.

Ghasemzadeh, R., & Akhbari, K. (2024). Templated synthesis of ZnO quantum dots via double solvents method inside MOF-801 as emerging photocatalyst for photodegradation of Acid Blue 25 under UV light. *Journal of Photochemistry and Photobiology A: Chemistry, 448*, 115306.

Ghasemzadeh, R., Akhbari, K., & Kawata, S. (2024). Ag@MUT-16 nanocomposite as a Fenton-like and plasmonic photocatalyst for degradation of Quinoline Yellow under visible light. *Dalton Transactions, 53*(26), 11094–11111. https://doi.org/10.1039/d4dt00322e.

Granadeiro, C. M., Barbosa, A. D. S., Ribeiro, S., Santos, I. C. M. S., De Castro, B., Cunha-Silva, L., & Balula, S. S. (2014). Oxidative catalytic versatility of a trivacant polyoxotungstate incorporated into MIL-101(Cr). *Catalysis Science and Technology, 4*(5), 1416–1425. https://doi.org/10.1039/c3cy00853c.

Granadeiro, C. M., Nogueira, L. S., Julião, D., Mirante, F., Ananias, D., Balula, S. S., & Cunha-Silva, L. (2016). Influence of a porous MOF support on the catalytic performance of Eu-polyoxometalate based materials: Desulfurization of a model diesel. *Catalysis Science and Technology, 6*(5), 1515–1522. https://doi.org/10.1039/c5cy01110h.

Gu, Y., Wu, Y.-nan, Li, L., Chen, W., Li, F., & Kitagawa, S. (2017). Controllable modular growth of hierarchical MOF-on-MOF architectures. *Angewandte Chemie, 129*(49), 15864–15868. https://doi.org/10.1002/ange.201709738.

Guillena, G., Ramón, D. J., & Yus, M. (2010). Hydrogen autotransfer in the N-alkylation of amines and related compounds using alcohols and amines as electrophiles. *Chemical Reviews, 110*(3), 1611–1641. https://doi.org/10.1021/cr9002159.

Guntern, Y. T., Pankhurst, J. R., Vávra, J., Mensi, M., Mantella, V., Schouwink, P., & Buonsanti, R. (2019). Nanocrystal/metal–organic framework hybrids as electrocatalytic platforms for CO_2 conversion. *Angewandte Chemie, 131*(36), 12762–12769. https://doi.org/10.1002/ange.201905172.

Guo, C., Guo, J., Zhang, Y., Wang, D., Zhang, L., Guo, Y., Ma, W., & Wang, J. (2018). Synthesis of core-shell ZIF-67@Co-MOF-74 catalyst with controllable shell thickness and enhanced photocatalytic activity for visible light-driven water oxidation. *CrystEngComm, 20*(47), 7659–7665. https://doi.org/10.1039/c8ce01266k.

Guo, D., Chen, L., & Li, Y. (2024). Catalysis in confined frameworks: Synthesis. *Characterization, and Applications*, 219–271.

Guo, F., Wei, Y. P., Wang, S. Q., Zhang, X. Y., Wang, F. M., & Sun, W. Y. (2019). Pt nanoparticles embedded in flowerlike NH_2-UiO-68 for enhanced photocatalytic carbon dioxide reduction. *Journal of Materials Chemistry A, 7*(46), 26490–26495. https://doi.org/10.1039/c9ta10575a.

Guo, F., Yang, S., Liu, Y., Wang, P., Huang, J., & Sun, W. Y. (2019). Size engineering of metal-organic framework MIL-101(Cr)-Ag hybrids for photocatalytic CO_2 reduction. *ACS Catalysis, 9*(9), 8464–8470. https://doi.org/10.1021/acscatal.9b02126.

Guo, J., Wan, Y., Zhu, Y., Zhao, M., & Tang, Z. (2021). Advanced photocatalysts based on metal nanoparticle/metal-organic framework composites. *Nano Research, 14*(7), 2037–2052. https://doi.org/10.1007/s12274-020-3182-1.

Guo, W., Lv, H., Chen, Z., Sullivan, K. P., Lauinger, S. M., Chi, Y., Sumliner, J. M., Lian, T., & Hill, C. L. (2016). Self-assembly of polyoxometalates, Pt nanoparticles and metal-organic frameworks into a hybrid material for synergistic hydrogen evolution. *Journal of Materials Chemistry A, 4*(16), 5952–5957. https://doi.org/10.1039/c6ta00011h.

Hai, G., Tao, Z., Gao, H., Zhao, J., Jia, D., Huang, X., Chen, X., Xue, X., Feng, S., & Wang, G. (2021). *Nano Energy, 79*.

Han, Q., Qi, B., Ren, W., He, C., Niu, J., & Duan, C. (2015). Polyoxometalate-based homochiral metal-organic frameworks for tandem asymmetric transformation of cyclic carbonates from olefins. *Nature Communications, 6*, 10007.

Han, Y., Shi, H., Bai, C., Zhang, L., Wu, J., Meng, H., Xu, Y., & Zhang, X. (2018). Ag_3PO_4-MIL-53(Fe) composites with visible-light-enhanced photocatalytic activities for rhodamine B degradation. *ChemistrySelect, 3*(28), 8045–8050. https://doi.org/10.1002/slct.201800404.

Han, Y., Xu, H., Su, Y., Xu, Zl, Wang, K., & Wang, W. (2019). Noble metal (Pt, Au@Pd) nanoparticles supported on metal organic framework (MOF-74) nanoshuttles as high-selectivity CO_2 conversion catalysts. *Journal of Catalysis, 370*, 70–78. https://doi.org/10.1016/j.jcat.2018.12.005.

Hao, T. T., Guan, S.-J., Zhang, D., Zhang, P., Cao, Y., Hou, J., & Suen, N.-T. (2024). Correlation between d electrons and the sweet spot for the hydrogen evolution reaction: Is platinum always the best electrocatalyst? *Inorganic Chemistry, 63*(11), 5076–5082.

He, H., Wen, H.-M., Li, H.-K., & Zhang, H.-W. (2022). *Coordination Chemistry Reviews, 471*.

He, H., Zhu, Q.-Q., Yan, Y., Zhang, H.-W., Han, Z.-Y., Sun, H., Chen, J., Li, C.-P., Zhang, Z., & Du, M. (2022). Metal–organic framework supported Au nanoparticles with organosilicone coating for high-efficiency electrocatalytic N_2 reduction to NH_3. *Applied Catalysis B: Environmental, 302*, 120840.

He, J., Han, Q., Li, J., Shi, Z., Shi, X., & Niu, J. (2019). Ternary supramolecular system for photocatalytic oxidation with air by consecutive photo-induced electron transfer processes. *Journal of Catalysis, 376*, 161–167. https://doi.org/10.1016/j.jcat.2019.06.040.

Hong, Q., Wang, Y., Wang, R., Chen, Z., Yang, H., Yu, K., Liu, Y., Huang, H., Kang, Z., & Menezes, P. W. (2023). *Small, 19*.

Hoover, J. M., Ryland, B. L., & Stahl, S. S. (2013). Mechanism of copper(I)/TEMPO-catalyzed aerobic alcohol oxidation. *Journal of the American Chemical Society, 135*(6), 2357–2367. https://doi.org/10.1021/ja3117203.

Hou, C. C., Li, T. T., Chen, Y., & Fu, W. F. (2015). Improved photocurrents for water oxidation by using metal-organic framework derived hybrid porous Co_3O_4@Carbon/$BiVO_4$ as a Photoanode. *ChemPlusChem, 80*(9), 1465–1471. https://doi.org/10.1002/cplu.201500058.

Hou, X., Sun, L., Hu, Y., An, X., & Qian, X. (2021). De-doped polyaniline as a mediating layer promoting in-situ growth of metal–organic frameworks on cellulose fiber and enhancing adsorptive-photocatalytic removal of ciprofloxacin. *Polymers, 13*(19), 3298.

Hou, Y., Liang, Y.-L., Shi, P.-C., Huang, Y.-B., & Cao, R. (2020). Atomically dispersed Ni species on N-doped carbon nanotubes for electroreduction of CO_2 with nearly 100% CO selectivity. *Applied Catalysis B: Environmental, 271*, 118929.

Hou, Y., Liu, Z., Tong, L., Zhao, L., Kuang, X., Kuang, R., & Ju, H. (2020). One-step electrodeposition of the MOF@CCQDs/NiF electrode for chiral recognition of tyrosine isomers. *Dalton Transactions, 49*(1), 31–34. https://doi.org/10.1039/c9dt04354c.

Hu, A., Pang, Q., Tang, C., Bao, J., Liu, H., Ba, K., Xie, S., Chen, J., Chen, J., Yue, Y., Tang, Y., Li, Q., & Sun, Z. (2019). Epitaxial growth and integration of insulating metal-organic frameworks in electrochemistry. *Journal of the American Chemical Society, 141*(28), 11322–11327. https://doi.org/10.1021/jacs.9b05869.

Hu, C., Wang, Y., Chen, J., Wang, H. F., Shen, K., Tang, K., Chen, L., & Li, Y. (2022). Main-group metal single-atomic regulators in dual-metal catalysts for enhanced electrochemical CO_2 reduction. *Small, 18*(22), 2201391.

Hu, C. Y., Zhou, J., Sun, C. Y., Chen, Mm, Wang, X. L., & Su, Z. M. (2019). HKUST-1 derived hollow C-$Cu_{2-x}S$ nanotube/g-C_3N_4 composites for visible-light CO_2 photoreduction with H_2O vapor. *Chemistry - A European Journal, 25*(1), 379–385. https://doi.org/10.1002/chem.201804925.

Hu, J., Yu, L., Deng, J., Wang, Y., Cheng, K., Ma, C., Zhang, Q., Wen, W., Yu, S., Pan, Y., Yang, J., Ma, H., Qi, F., Wang, Y., Zheng, Y., Chen, M., Huang, R., Zhang, S., Zhao, Z., ... Deng, D. (2021). Sulfur vacancy-rich MoS_2 as a catalyst for the hydrogenation of CO_2 to methanol. *Nature Catalysis, 4*(3), 242–250. https://doi.org/10.1038/s41929-021-00584-3.

Huang, X., Li, X., Luan, Q., Zhang, K., Wu, Z., Li, B., Xi, Z., Dong, W., & Wang, G. (2021). Highly dispersed Pt clusters encapsulated in MIL-125-NH2 via in situ auto-reduction method for photocatalytic H_2 production under visible light. *Nano Research, 14*(11), 4250–4257. https://doi.org/10.1007/s12274-021-3597-3.

Huang, Y., & Liu (2020). *International Journal of Biological Macromolecules*, 53–63.

Huang, Y., Su, W., Wang, R., & Zhao, T. (2019). Removal of typical industrial gaseous pollutants: From carbon, zeolite, and metal-organic frameworks to molecularly imprinted adsorbents. *Aerosol and Air Quality Research, 19*(9), 2130–2150. https://doi.org/10.4209/aaqr.2019.04.0215.

Huo, M., Yang, W., Zhang, H., Zhang, L., Liao, J., Lin, L., & Lu, C. (2016). A new POM-MOF hybrid microporous material with ultrahigh thermal stability and selective adsorption of organic dyes. *RSC Advances, 6*(112), 111549–111555. https://doi.org/10.1039/C6RA10422C.

Huo, Q., Liu, G., Sun, H., Fu, Y., Ning, Y., Zhang, B., Zhang, X., Gao, J., Miao, J., & Zhang, X. (2021). CeO$_2$-modified MIL-101(Fe) for photocatalysis extraction oxidation desulfurization of model oil under visible light irradiation. *Chemical Engineering Journal, 422*, 130036.

Imyen, T., Znoutine, E., Suttipat, D., Iadrat, P., Kidkhunthod, P., Bureekaew, S., & Wattanakit, C. (2020). Methane utilization to methanol by a hybrid zeolite@metal-organic framework. *ACS Applied Materials and Interfaces, 12*(21), 23812–23821. https://doi.org/10.1021/acsami.0c02273.

Inoue, T., Fujishima, A., Konishi, S., & Honda, K. (1979). Photoelectrocatalytic reduction of carbon dioxide in aqueous suspensions of semiconductor powders [3]. *Nature, 277*(5698), 637–638. https://doi.org/10.1038/277637a0.

Jahan, M., Liu, Z., & Loh, K. P. (2013). A graphene oxide and copper-centered metal organic framework composite as a tri-functional catalyst for HER, OER, and ORR. *Advanced Functional Materials, 23*(43), 5363–5372. https://doi.org/10.1002/adfm.201300510.

Jayaramulu, K., Mukherjee, S., Morales, D. M., Dubal, D. P., Nanjundan, A. K., Schneemann, A., Masa, J., Kment, S., Schuhmann, W., Otyepka, M., Zbořil, R., & Fischer, R. A. (2022). Graphene-based metal-organic framework hybrids for applications in catalysis, environmental, and energy technologies. *Chemical Reviews, 122*(24), 17241–17338. https://doi.org/10.1021/acs.chemrev.2c00270.

Ji, H., Lee, S., Park, J., Kim, T., Choi, S., & Oh, M. (2018). Improvement in crystallinity and porosity of poorly crystalline metal-organic frameworks (MOFs) through their induced growth on a well-crystalline MOF template. *Inorganic Chemistry, 57*(15), 9048–9054. https://doi.org/10.1021/acs.inorgchem.8b01055.

Jiang, D., Mallat, T., Krumeich, F., & Baiker, A. (2011). Polymer-assisted synthesis of nanocrystalline copper-based metal-organic framework for amine oxidation. *Catalysis Communications, 12*(7), 602–605. https://doi.org/10.1016/j.catcom.2010.12.010.

Jiao, Y., Zheng, Y., Jaroniec, M., & Qiao, S. Z. (2015). Design of electrocatalysts for oxygen- and hydrogen-involving energy conversion reactions. *Chemical Society Reviews, 44*(8), 2060–2086. https://doi.org/10.1039/c4cs00470a.

Jin, M., Niu, Q., Liu, G., Lv, Z., Si, C., & Guo, H. (2020). Encapsulation of ionic liquids into POMs-based metal–organic frameworks: Screening of POMs-ILs@MOF catalysts for efficient cycloolefins epoxidation. *Journal of Materials Science, 55*(19), 8199–8210. https://doi.org/10.1007/s10853-020-04611-9.

Jin, M., Qian, X., Gao, J., Chen, J., Hensley, D. K., Ho, H. C., Percoco, R. J., Ritzi, C. M., & Yue, Y. (2019). Solvent-free synthesis of CuO/HKUST-1 composite and its photocatalytic application. *Inorganic Chemistry, 58*(13), 8332–8338. https://doi.org/10.1021/acs.inorgchem.9b00362.

Jo, C., Lee, H. J., & Oh, M. (2011). One-pot synthesis of silica@coordination polymer core–shell microspheres with controlled shell thickness. *Advanced Materials, 23*(15), 1716–1719.

Jongert, T. K., Slowinski, I. A., Dao, B., Cortez, V. H., Gredig, T., Plascencia, N. D., & Tian, F. (2024). Zeta potential and size analysis of zeolitic imidazolate framework-8 nanocrystals prepared by surfactant-assisted synthesis. *Langmuir, 40*, 6138–6148.

Jung, H. B., Kim, Y., Lim, J., Cho, S., Seo, M., Kim, I.-S., Kim, M., Lee, C., Lee, Y.-W., & Yoo, C.-Y. (2023). ZIF-67 metal-organic frameworks synthesized onto CNT supports for oxygen evolution reaction in alkaline water electrolysis. *Electrochimica Acta, 439*, 141593.

Kalaj, M., Denny, M. S., Bentz, K. C., Palomba, J. M., & Cohen, S. M. (2019). Nylon–MOF composites through postsynthetic polymerization. *Angewandte Chemie, 131*(8), 2358–2362. https://doi.org/10.1002/ange.201812655.

Kalati, M., & Akhbari, K. (2021). Optimizing the metal ion release and antibacterial activity of ZnO@ZIF-8 by modulating its synthesis method. *New Journal of Chemistry, 45*(48), 22924–22931. https://doi.org/10.1039/d1nj04534b.

Kalati, M., & Akhbari, K. (2022). Copper(II) nitrate and copper(II) oxide loading on ZIF-8; synthesis, characterization and antibacterial activity. *Journal of Porous Materials, 29*(6), 1909–1917. https://doi.org/10.1007/s10934-022-01302-5.

Karimi, Z., & Morsali, A. (2013). Modulated formation of metal-organic frameworks by oriented growth over mesoporous silica. *Journal of Materials Chemistry A, 1*(9), 3047–3054. https://doi.org/10.1039/c2ta01565j.

Karimi Alavijeh, R., Akhbari, K., & White, J. M. (2022). A Ca-based nano bio-coordination polymer providing reversible structural conversion with ability to enhance cytotoxicity of curcumin and induce apoptosis in human gastric cancer AGS cells. *CrystEngComm, 24*(40), 7125–7136. https://doi.org/10.1039/d2ce01258h.

Karimi Alavijeh, R., Beheshti, S., Akhbari, K., & Morsali, A. (2018). Investigation of reasons for metal–organic framework's antibacterial activities. *Polyhedron, 156*, 257–278. https://doi.org/10.1016/j.poly.2018.09.028.

Kim, H. S., Kang, M. S., & Yoo, W. C. (2019). Boost-up electrochemical performance of MOFs: Via confined synthesis within nanoporous carbon matrices for supercapacitor and oxygen reduction reaction applications. *Journal of Materials Chemistry A, 7*(10), 5561–5574. https://doi.org/10.1039/c8ta12200h.

Kim, S., Kim, Y., Jin, H., Park, M. H., Kim, Y., Lee, K. M., & Kim, M. (2019). Europium-catalyzed aerobic oxidation of alcohols to aldehydes/ketones and photoluminescence tracking. *Advanced Synthesis & Catalysis, 361*(6), 1259–1264. https://doi.org/10.1002/adsc.201801499.

Kim, S., Lee, J., Jeoung, S., Moon, H. R., & Kim, M. (2020). Surface-deactivated core–shell metal–organic framework by simple ligand exchange for enhanced size discrimination in aerobic oxidation of alcohols. *Chemistry - A European Journal, 26*(34), 7568–7572. https://doi.org/10.1002/chem.202000933.

Kitao, T., Zhang, Y., Kitagawa, S., Wang, B., & Uemura, T. (2017). Hybridization of MOFs and polymers. *Chemical Society Reviews, 46*(11), 3108–3133. https://doi.org/10.1039/c7cs00041c.

Kong, X. J., He, T., Zhou, J., Zhao, C., Li, T. C., Wu, X. Q., Wang, K., & Li, J. R. (2021). In situ porphyrin substitution in a Zr(IV)-MOF for stability enhancement and photocatalytic CO_2 reduction. *Small, 17*(22), 2005357.

Konnerth, H., Matsagar, B. M., Chen, S. S., Prechtl, M. H., Shieh, F.-K., & Wu, K. C.-W. (2020). Metal-organic framework (MOF)-derived catalysts for fine chemical production. *Coordination Chemistry Reviews, 416*, 213319.

Kou, J., & Sun, L. B. (2018). Fabrication of metal-organic frameworks inside silica nanopores with significantly enhanced hydrostability and catalytic activity. *ACS Applied Materials and Interfaces, 10*(14), 12051–12059. https://doi.org/10.1021/acsami.8b01652.

Koçyiğit, N., Özen, Ü. E., Özer, M., Salih, B., Özkaya, A. R., & Bekaroğlu, Ö. (2017). Electrocatalytic activity of novel ball-type metallophthalocyanines with trifluoro methyl linkages in oxygen reduction reaction and application as Zn-air battery cathode catalyst. *Electrochimica Acta, 233*, 237–248. https://doi.org/10.1016/j.electacta.2017.03.035.

Lagerblom, K., Keskiväli, J., Parviainen, A., Mannisto, J., & Repo, T. (2018). Selective aerobic oxidation of alcohols with NO_3-activated nitroxyl radical/manganese catalyst system. *ChemCatChem, 10*(13), 2908–2914. https://doi.org/10.1002/cctc.201800438.

Lee, D. T., Zhao, J., Oldham, C. J., Peterson, G. W., & Parsons, G. N. (2017). UiO-66-NH_2 metal-organic framework (MOF) nucleation on TiO_2, ZnO, and Al_2O_3 atomic layer deposition-treated polymer fibers: Role of metal oxide on MOF growth and catalytic hydrolysis of chemical warfare agent simulants. *ACS Applied Materials and Interfaces, 9*(51), 44847–44855. https://doi.org/10.1021/acsami.7b15397.

Lee, G., Lee, S., Oh, S., Kim, D., & Oh, M. (2020). Tip-to-middle anisotropic MOF-on-MOF growth with a structural adjustment. *Journal of the American Chemical Society, 142*(6), 3042–3049. https://doi.org/10.1021/jacs.9b12193.

Lee, J., Kim, E.-Y., Chang, B.-J., Han, M., Lee, P.-S., & Moon, S.-Y. (2020). Mixed-matrix membrane reactors for the destruction of toxic chemicals. *Journal of Membrane Science, 605*, 118112.

Li, A., Zhu, W., Li, C., Wang, T., & Gong, J. (2019). Rational design of yolk-shell nanostructures for photocatalysis. *Chemical Society Reviews, 48*(7), 1874–1907. https://doi.org/10.1039/c8cs00711j.

Li, B., Ma, J.-G., & Cheng, P. (2018). Silica-protection-assisted encapsulation of Cu_2O nanocubes into a metal–organic framework (ZIF-8) to provide a composite catalyst. *Angewandte Chemie, 130*(23), 6950–6953. https://doi.org/10.1002/ange.201801588.

Li, B., & Zeng, H. C. (2019). Synthetic chemistry and multifunctionality of an amorphous Ni-MOF-74 shell on a Ni/SiO_2 hollow catalyst for efficient tandem reactions. *Chemistry of Materials, 31*(14), 5320–5330. https://doi.org/10.1021/acs.chemmater.9b02070.

Li, F.-L., Shao, Q., Huang, X., & Lang, J.-P. (2018). Nanoscale trimetallic metal–organic frameworks enable efficient oxygen evolution electrocatalysis. *Angewandte Chemie, 130*(7), 1906–1910. https://doi.org/10.1002/ange.201711376.

Li, G., Zhao, S., Zhang, Y., & Tang, Z. (2018). Metal–organic frameworks encapsulating active nanoparticles as emerging composites for catalysis: Recent progress and perspectives. *Advanced Materials, 30*(51), 1800702.

Li, H., Zhang, X., Sun, J., & Dong, X. (2020). Dynamic resource levelling in projects under uncertainty. *International Journal of Production Research*, 1–21. https://doi.org/10.1080/00207543.2020.1788737.

Li, J., Cui, Z., Zheng, Y., Liu, X., Li, Z., Jiang, H., Zhu, S., Zhang, Y., Chu, P. K., & Wu, S. (2022). Atomic-layer Fe_2O_3-modified 2D porphyrinic metal-organic framework for enhanced photocatalytic disinfection through electron-withdrawing effect. *Applied Catalysis B: Environmental, 317*, 121701.

Li, J., Huang, R., Chen, L., Xia, Y., Yan, G., & Liang, R. (2023). Mixed valence copper oxide composites derived from metal-organic frameworks for efficient visible light fuel denitrification. *RSC Advances, 13*(51), 36477–36483. https://doi.org/10.1039/d3ra07532j.

Li, L., Li, Y., Jiao, L., Liu, X., Ma, Z., Zeng, Y. J., Zheng, X., & Jiang, H. L. (2022). Light-induced selective hydrogenation over PdAg nanocages in hollow MOF microenvironment. *Journal of the American Chemical Society, 144*(37), 17075–17085. https://doi.org/10.1021/jacs.2c06720.

Li, S., Hao, X., Abudula, A., & Guan, G. (2019). Nanostructured Co-based bifunctional electrocatalysts for energy conversion and storage: Current status and perspectives. *Journal of Materials Chemistry A, 7*(32), 18674–18707. https://doi.org/10.1039/c9ta04949e.

Li, S., & Huo, F. (2015). Metal-organic framework composites: From fundamentals to applications. *Nanoscale, 7*(17), 7482–7501. https://doi.org/10.1039/c5nr00518c.

Li, S., Zhang, L., Lan, Y., O'Halloran, K. P., Ma, H., & Pang, H. (2018). Polyoxometalate-encapsulated twenty-nuclear silver-tetrazole nanocage frameworks as highly active electrocatalysts for the hydrogen evolution reaction. *Chemical Communications, 54*(16), 1964–1967. https://doi.org/10.1039/c7cc09223g.

Li, S., Zhao, L., Yao, Y., Gu, Z., Liu, C., Hu, W., Zhang, Y., Zhao, Q., & Yu, C. (2023). MOF-on-MOF heterostructures with core–shell and core–satellite structures via controllable nucleation of guest MOFs. *CrystEngComm, 25*(2), 284–289. https://doi.org/10.1039/d2ce01272c.

Li, S. W., Yang, Z., Gao, R. M., Zhang, G., & Zhao, Js (2018). Direct synthesis of mesoporous SRL-POM@MOF-199@MCM-41 and its highly catalytic performance for the oxidesulfurization of DBT. *Applied Catalysis B: Environmental, 221*, 574–583. https://doi.org/10.1016/j.apcatb.2017.09.044.

Li, X., Yuan, X., Xia, G., Liang, J., Liu, C., Wang, Z., & Yang, W. (2020). Catalytic production of γ-valerolactone from xylose over delaminated Zr-Al-SCM-1 zeolite via a cascade process. *Journal of Catalysis, 392*, 175–185. https://doi.org/10.1016/j.jcat.2020.10.004.

Li, Y., Gao, Q., Zhang, L., Zhou, Y., Zhong, Y., Ying, Y., Zhang, M., Huang, C., & Wang, Y. (2018). H5PV2Mo10O40 encapsulated in MIL-101(Cr): Facile synthesis and characterization of rationally designed composite materials for efficient decontamination of sulfur mustard. *Dalton Transactions, 47*(18), 6394–6403. https://doi.org/10.1039/c8dt00572a.

Li, Z., Wang, Q., Kong, C., Wu, Y., Li, Y., & Lu, G. (2015). Interface charge transfer versus surface proton reduction: Which is more pronounced on photoinduced hydrogen generation over sensitized Pt cocatalyst on RGO? *The Journal of Physical Chemistry C, 119*(24), 13561–13568. https://doi.org/10.1021/acs.jpcc.5b00746.

Li, Z. Q., Wang, A., Guo, C. Y., Tai, Y. F., & Qiu, L. G. (2013). One-pot synthesis of metal-organic framework@SiO_2 core-shell nanoparticles with enhanced visible-light photoactivity. *Dalton Transactions, 42*(38), 13948–13954. https://doi.org/10.1039/c3dt50845e.

Liang, H., Yao, A., Jiao, X., Li, C., & Chen, D. (2018). Fast and sustained degradation of chemical warfare agent simulants using flexible self-supported metal-organic framework filters. *ACS Applied Materials and Interfaces, 10*(24), 20396–20403. https://doi.org/10.1021/acsami.8b02886.

Liang, Q., Chen, J., Wang, F., & Li, Y. (2020). Transition metal-based metal-organic frameworks for oxygen evolution reaction. *Coordination Chemistry Reviews, 424*, 213488.

Liang, R., Shen, L., Jing, F., Qin, N., & Wu, L. (2015). Preparation of MIL-53(Fe)-reduced graphene oxide nanocomposites by a simple self-assembly strategy for increasing interfacial contact: Efficient visible-light photocatalysts. *ACS Applied Materials and Interfaces, 7*(18), 9507–9515. https://doi.org/10.1021/acsami.5b00682.

Liang, Z., Qu, C., Guo, W., Zou, R., & Xu, Q. (2018). Pristine metal–organic frameworks and their composites for energy storage and conversion. *Advanced Materials, 30*(37), 1702891.

Liao, J. Z., Zhang, H. L., Wang, S. S., Yong, J. P., Wu, X. Y., Yu, R., & Lu, C. Z. (2015). Multifunctional radical-doped polyoxometalate-based host-guest material: Photochromism and photocatalytic activity. *Inorganic Chemistry, 54*(9), 4345–4350. https://doi.org/10.1021/acs.inorgchem.5b00041.

Liao, P. Q., Shen, J. Q., & Zhang, J. P. (2018). Metal–organic frameworks for electrocatalysis. *Coordination Chemistry Reviews, 373*, 22–48. https://doi.org/10.1016/j.ccr.2017.09.001.

Liao, S., Liu, J., Xiao, X., Fu, D., Wang, G., & Jin, L. (2020). Modified gradient neural networks for solving the time-varying Sylvester equation with adaptive coefficients and elimination of matrix inversion. *Neurocomputing, 379*, 1–11. https://doi.org/10.1016/j.neucom.2019.10.080.

Lin, R., Li, S., Wang, J., Xu, J., Xu, C., Wang, J., Li, C., & Li, Z. (2018). Facile generation of carbon quantum dots in MIL-53(Fe) particles as localized electron acceptors for enhancing their photocatalytic Cr(vi) reduction. *Inorganic Chemistry Frontiers, 5*(12), 3170–3177. https://doi.org/10.1039/c8qi01164h.

Lin, W., Chen, H., Lin, G., Yao, S., Zhang, Z., Qi, J., Jing, M., Song, W., Li, J., & Liu, X. (2022). Creating frustrated Lewis pairs in defective boron carbon nitride for electrocatalytic nitrogen reduction to ammonia. *Angewandte Chemie, 134*(36), e202207807.

Lin, X., Zhang, M., Zhu, M., Huang, H., Shi, C., Liu, Y., & Kang, Z. (2018). Engineering a polyoxometalate-based metal organic framework with more exposed active edge sites of Ag for visible light-driven selective oxidation of: Cis-cyclooctene. *Inorganic Chemistry Frontiers, 5*(10), 2493–2500. https://doi.org/10.1039/c8qi00648b.

Lipkowski, J, & Ross, P. N. (1998).

Liu, C., Lin, L., Sun, Q., Wang, J., Huang, R., Chen, W., Li, S., Wan, J., Zou, J., & Yu, C. (2020). Site-specific growth of MOF-on-MOF heterostructures with controllable nano-architectures: Beyond the combination of MOF analogues. *Chemical Science, 11*(14), 3680–3686. https://doi.org/10.1039/d0sc00417k.

Liu, C., Wang, J., Wan, J., & Yu, C. (2021). MOF-on-MOF hybrids: Synthesis and applications. *Coordination Chemistry Reviews, 432*, 213743.

Liu, D., Xu, H., Wang, C., Shang, H., Yu, R., Wang, Y., Li, J., Li, X., & Du, Y. (2021). 3D porous Ru-doped NiCo-MOF hollow nanospheres for boosting oxygen evolution reaction electrocatalysis. *Inorganic Chemistry, 60*(8), 5882–5889. https://doi.org/10.1021/acs.inorgchem.1c00295.

Liu, J., He, L., Zhao, S., Li, S., Hu, L., Tian, J. Y., Ding, J., Zhang, Z., & Du, M. (2023). *Advanced Science, 10*.

Liu, L., Meng, H., Chai, Y., Chen, X., Xu, J., Liu, X., Liu, W., Guldi, D. M., & Zhu, Y. (2023). Enhancing built-in electric fields for efficient photocatalytic hydrogen evolution by encapsulating C_{60} fullerene into zirconium-based metal-organic frameworks. *Angewandte Chemie, 135*(11), e202217897.

Liu, S., Wang, Z., Zhou, S., Yu, F., Yu, M., Chiang, C. Y., Zhou, W., Zhao, J., & Qiu, J. (2017). Metal-organic-framework-derived hybrid carbon nanocages as bifunctional electrocatalyst for oxygen reduction and evolution. *Advanced Materials, 29*(31), 1–10.

Liu, S. M., Zhang, Z., Li, X., Jia, H., Ren, M., & Liu, S. (2018). Ti-substituted Keggin-type polyoxotungstate as proton and electron reservoir encaged into metal–organic framework for carbon dioxide photoreduction. *Advanced Materials Interfaces, 5*(5), 1801062.

Liu, X., Iocozzia, J., Wang, Y., Cui, X., Chen, Y., Zhao, S., Li, Z., & Lin, Z. (2017). Noble metal-metal oxide nanohybrids with tailored nanostructures for efficient solar energy conversion, photocatalysis and environmental remediation. *Energy and Environmental Science, 10*(2), 402–434. https://doi.org/10.1039/c6ee02265k.

Liu, Y., Sinha, S., McDonald, O. G., Shang, Y., Hoofnagle, M. H., & Owens, G. K. (2005). Kruppel-like factor 4 abrogates myocardin-induced activation of smooth muscle gene expression. *Journal of Biological Chemistry, 280*(10), 9719–9727. https://doi.org/10.1074/jbc.M412862200.

Liu, Y., Tang, C., Cheng, M., Chen, M., Chen, S., Lei, L., Chen, Y., Yi, H., Fu, Y., & Li, L. (2021). Polyoxometalate@metal-organic framework composites as effective photocatalysts. *ACS Catalysis, 11*(21), 13374–13396. https://doi.org/10.1021/acscatal.1c03866.

Lou, Y., Zhang, Y., Cheng, L., Chen, J., & Zhao, Y. (2018). A stable plasmonic Cu@Cu2O/ZnO heterojunction for enhanced photocatalytic hydrogen generation. *ChemSusChem, 11*(9), 1505–1511. https://doi.org/10.1002/cssc.201800249.

Low, J., Yu, J., Jaroniec, M., Wageh, S., & Al-Ghamdi, A. A. (2017). Heterojunction photocatalysts. *Advanced Materials, 29*(20), 1601694.

Lu, A. X., McEntee, M., Browe, M. A., Hall, M. G., Decoste, J. B., & Peterson, G. W. (2017). MOFabric: Electrospun nanofiber mats from PVDF/UiO-66-NH$_2$ for chemical protection and decontamination. *ACS Applied Materials and Interfaces, 9*(15), 13632–13636. https://doi.org/10.1021/acsami.7b01621.

Luo, T. Y., Liu, C., Gan, X. Y., Muldoon, P. F., Diemler, N. A., Millstone, J. E., & Rosi, N. L. (2019). Multivariate stratified metal-organic frameworks: Diversification using domain building blocks. *Journal of the American Chemical Society, 141*(5), 2161–2168. https://doi.org/10.1021/jacs.8b13502.

Lyu, S., Guo, C., Wang, J., Li, Z., Yang, B., Lei, L., Wang, L., Xiao, J., Zhang, T., & Hou, Y. (2022). Exceptional catalytic activity of oxygen evolution reaction via two-dimensional graphene multilayer confined metal-organic frameworks. *Nature Communications, 13*, 6171.

Ma, K., Islamoglu, T., Chen, Z., Li, P., Wasson, M. C., Chen, Y., Wang, Y., Peterson, G. W., Xin, J. H., & Farha, O. K. (2019). Scalable and template-free aqueous synthesis of zirconium-based metal-organic framework coating on textile fiber. *Journal of the American Chemical Society, 141*(39), 15626–15633. https://doi.org/10.1021/jacs.9b07301.

Ma, X., Qi, K., Wei, S., Zhang, L., & Cui, X. (2019). In situ encapsulated nickel-copper nanoparticles in metal-organic frameworks for oxygen evolution reaction. *Journal of Alloys and Compounds, 770*, 236–242. https://doi.org/10.1016/j.jallcom.2018.08.096.

Mahdipoor, H. R., Ebrahimi, R., Babakhani, E. G., Halladj, R., Safari, N., & Ganji, H. (2023). Investigating the selective adsorption of CO_2 by MIL-101 (Cr)-NH_2 and modeling the equilibrium data using a new three-parameter isotherm. *Colloids and Surfaces A: Physicochemical and Engineering Aspects, 675*, 131971.

Maksimchuk, N., Timofeeva, M., Melgunov, M., Shmakov, A., Chesalov, Y., Dybtsev, D., Fedin, V., & Kholdeeva, O. (2008). Heterogeneous selective oxidation catalysts based on coordination polymer MIL-101 and transition metal-substituted polyoxometalates. *Journal of Catalysis, 257*(2), 315–323. https://doi.org/10.1016/j.jcat.2008.05.014.

Maksimchuk, N. V., Kovalenko, K. A., Arzumanov, S. S., Chesalov, Y. A., Melgunov, M. S., Stepanov, A. G., Fedin, V. P., & Kholdeeva, O. A. (2010). Hybrid polyoxotungstate/MIL-101 materials: Synthesis, characterization, and catalysis of H_2O_2-based alkene epoxidation. *Inorganic Chemistry, 49*(6), 2920–2930. https://doi.org/10.1021/ic902459f.

Maru, K., Kalla, S., & Jangir, R. (2022). MOF/POM hybrids as catalysts for organic transformations. *Dalton Transactions, 51*(32), 11952–11986. https://doi.org/10.1039/d2dt01895k.

McCarthy, D. L., Liu, J., Dwyer, D. B., Troiano, J. L., Boyer, S. M., Decoste, J. B., Bernier, W. E., & Jones, W. E. (2017). Electrospun metal-organic framework polymer composites for the catalytic degradation of methyl paraoxon. *New Journal of Chemistry, 41*(17), 8748–8753. https://doi.org/10.1039/c7nj00525c.

McCrory, C. C. L., Jung, S., Peters, J. C., & Jaramillo, T. F. (2013). Benchmarking heterogeneous electrocatalysts for the oxygen evolution reaction. *Journal of the American Chemical Society, 135*(45), 16977–16987. https://doi.org/10.1021/ja407115p.

Micheroni, D., Lan, G., & Lin, W. (2018). Efficient electrocatalytic proton reduction with carbon nanotube-supported metal-organic frameworks. *Journal of the American Chemical Society, 140*(46), 15591–15595. https://doi.org/10.1021/jacs.8b09521.

Mirzadeh, E., & Akhbari, K. (2016). Synthesis of nanomaterials with desirable morphologies from metal-organic frameworks for various applications. *CrystEngComm, 18*(39), 7410–7424. https://doi.org/10.1039/c6ce01076h.

Misono, M. (2013). Catalysis of heteropoly compounds (polyoxometalates). *Studies in Surface Science and Catalysis, 176*, 97–155. https://doi.org/10.1016/B978-0-444-53833-8.00004-1.

Moeinian, M., & Akhbari, K. (2015). How the guest molecules in nanoporous Zn(II) metal-organic framework can prevent agglomeration of ZnO nanoparticles. *Journal of Solid State Chemistry, 225*, 459–463. https://doi.org/10.1016/j.jssc.2015.02.017.

Monama, G. R., Mdluli, S. B., Mashao, G., Makhafola, M. D., Ramohlola, K. E., Molapo, K. M., Hato, M. J., Makgopa, K., Iwuoha, E. I., & Modibane, K. D. (2018). Palladium deposition on copper(II) phthalocyanine/metal organic framework composite and electrocatalytic activity of the modified electrode towards the hydrogen evolution reaction. *Renewable Energy, 119*, 62–72. https://doi.org/10.1016/j.renene.2017.11.084.

Moon, S. Y., Proussaloglou, E., Peterson, G. W., DeCoste, J. B., Hall, M. G., Howarth, A. J., Hupp, J. T., & Farha, O. K. (2016). Detoxification of chemical warfare agents using a Zr6-based metal–organic framework/polymer mixture. *Chemistry - A European Journal, 22*(42), 14864–14868. https://doi.org/10.1002/chem.201603976.

Morgan, S. E., O'Connell, A. M., Jansson, A., Peterson, G. W., Mahle, J. J., Eldred, T. B., Gao, W., & Parsons, G. N. (2021). Stretchable and multi-metal-organic framework fabrics via high-yield rapid sorption-vapor synthesis and their application in chemical warfare agent hydrolysis. *ACS Applied Materials and Interfaces, 13*(26), 31279–31284. https://doi.org/10.1021/acsami.1c07366.

Mou, Q., Wang, X., Xu, Z., Zul, P., Li, E., Zhao, P., Liu, X., Li, H., & Cheng, G. (2022). A synergy establishment by metal-organic framework and carbon quantum dots to enhance electrochemical water oxidation. *Chinese Chemical Letters, 33*(1), 562–566. https://doi.org/10.1016/j.cclet.2021.08.028.

Mukhopadhyay, S., Basu, O., Kar, A., & Das, S. K. (2020). Efficient electrocatalytic water oxidation by Fe(salen)-MOF composite: Effect of modified microenvironment. *Inorganic Chemistry, 59*(1), 472–483. https://doi.org/10.1021/acs.inorgchem.9b02745.

Murugesan, K., Senthamarai, T., Chandrashekhar, V. G., Natte, K., Kamer, P. C. J., Beller, M., & Jagadeesh, R. V. (2020). Catalytic reductive aminations using molecular hydrogen for synthesis of different kinds of amines. *Chemical Society Reviews, 49*(17), 6273–6328. https://doi.org/10.1039/c9cs00286c.

Najafi, M., Abednatanzi, S., Yousefi, A., & Ghaedi, M. (2021). Photocatalytic activity of supported metal nanoparticles and single atoms. *Chemistry - A European Journal, 27*(72), 17999–18014. https://doi.org/10.1002/chem.202102877.

Noori, Y., & Akhbari, K. (2017). Post-synthetic ion-exchange process in nanoporous metal-organic frameworks; an effective way for modulating their structures and properties. *RSC Advances, 7*(4), 1782–1808. https://doi.org/10.1039/c6ra24958b.

Onishi, N., & Himeda, Y. (2022). Homogeneous catalysts for CO_2 hydrogenation to methanol and methanol dehydrogenation to hydrogen generation. *Coordination Chemistry Reviews, 472*, 214767.

Ostad, M. I., Shahrak, M. N., & Galli, F. (2022). The effect of different reaction media on photocatalytic activity of Au-and Cu-decorated zeolitic imidazolate Framework-8 toward CO_2 photoreduction to methanol. *Journal of Solid State Chemistry, 315*, 123514.

Palomba, J. M., Wirth, D. M., Kim, J. Y., Kalaj, M., Clarke, E. M., Peterson, G. W., Pokorski, J. K., & Cohen, S. M. (2021). Strong, ductile MOF-poly(urethane urea) composites. *Chemistry of Materials, 33*(9), 3164–3171. https://doi.org/10.1021/acs.chemmater.0c04874.

Palomeque, J., Lopez, J., & Figueras, F. (2002). Epoxydation of activated olefins by solid bases. *Journal of Catalysis, 211*(1), 150–156. https://doi.org/10.1006/jcat.2002.3706.

Pan, X., Xu, H., Zhao, X., & Zhang, H. (2020). Metal-organic framework-membranized bicomponent core-shell catalyst HZSM-5@UIO-66-NH_2/Pd for CO_2 selective conversion.

ACS Sustainable Chemistry and Engineering, 8(2), 1087–1094. https://doi.org/10.1021/acssuschemeng.9b05912.

Pan, Y., Sun, K., Liu, S., Cao, X., Wu, K., Cheong, W. C., Chen, Z., Wang, Y., Li, Y., Liu, Y., Wang, D., Peng, Q., Chen, C., & Li, Y. (2018). Core-shell ZIF-8@ZIF-67-derived CoP nanoparticle-embedded N-doped carbon nanotube hollow polyhedron for efficient overall water splitting. *Journal of the American Chemical Society, 140*(7), 2610–2618. https://doi.org/10.1021/jacs.7b12420.

Panda, J., Tripathy, S. P., Dash, S., Ray, A., Behera, P., Subudhi, S., & Parida, K. (2023). Inner transition metal-modulated metal organic frameworks (IT-MOFs) and their derived nanomaterials: A strategic approach towards stupendous photocatalysis. *Nanoscale, 15*(17), 7640–7675. https://doi.org/10.1039/d3nr00274h.

Parsaei, M., & Akhbari, K. (2022a). MOF-801 as a nanoporous water-based carrier system for in situ encapsulation and sustained release of 5-FU for effective cancer therapy. *Inorganic Chemistry, 61*(15), 5912–5925. https://doi.org/10.1021/acs.inorgchem.2c00380.

Parsaei, M., & Akhbari, K. (2022b). Smart multifunctional UiO-66 metal-organic framework nanoparticles with outstanding drug-loading/release potential for the targeted delivery of quercetin. *Inorganic Chemistry, 61*(37), 14528–14543. https://doi.org/10.1021/acs.inorgchem.2c00743.

Parsaei, M., & Akhbari, K. (2022c). Synthesis and application of MOF-808 decorated with folic acid-conjugated chitosan as a strong nanocarrier for the targeted drug delivery of quercetin. *Inorganic Chemistry, 61*(48), 19354–19368. https://doi.org/10.1021/acs.inorgchem.2c03138.

Parsaei, M., Akhbari, K., Tylianakis, E., & Froudakis, G. E. (2023). Computational simulation of a three-dimensional Mg-based metal–organic framework as nanoporous anticancer drug carrier. *Crystal Growth & Design, 23*(11), 8396–8406. https://doi.org/10.1021/acs.cgd.3c01058.

Parsaei, M., Akhbari, K., Tylianakis, E., & Froudakis, G. E. (2024). Effects of fluorinated functionalization of linker on quercetin encapsulation, release and hela cell cytotoxicity of Cu-based MOFs as smart pH-stimuli nanocarriers. *Chemistry–A European Journal, 30*(1), e202301630.

Parsaei, M., Akhbari, K., & White, J. (2022). Modulating carbon dioxide storage by facile synthesis of nanoporous pillared-layered metal-organic framework with different synthetic routes. *Inorganic Chemistry, 61*(9), 3893–3902. https://doi.org/10.1021/acs.inorgchem.1c03414.

Parsapour, F., Moradi, M., Safarifard, V., & Sojdeh, S. (2024). Polymer/MOF composites for metal-ion batteries: A mini review. *Journal of Energy Storage, 82*, 110487.

Pascanu, V., Bermejo Gómez, A., Ayats, C., Platero-Prats, A. E., Carson, F., Su, J., Yao, Q., Pericàs, M. A., Zou, X., & Martín-Matute, B. (2015). Double-supported silica-metal-organic framework palladium nanocatalyst for the aerobic oxidation of alcohols under batch and continuous flow regimes. *ACS Catalysis, 5*(2), 472–479. https://doi.org/10.1021/cs501573c.

Pascanu, V., González Miera, G., Inge, A. K., & Martín-Matute, B. (2019). Metal-organic frameworks as catalysts for organic synthesis: A critical perspective. *Journal of the American Chemical Society, 141*(18), 7223–7234. https://doi.org/10.1021/jacs.9b00733.

Peng, L., Yang, S., Jawahery, S., Moosavi, S. M., Huckaba, A. J., Asgari, M., Oveisi, E., Nazeeruddin, M. K., Smit, B., & Queen, W. L. (2019). Preserving porosity of mesoporous metal-organic frameworks through the introduction of polymer guests. *Journal of the American Chemical Society, 141*(31), 12397–12405. https://doi.org/10.1021/jacs.9b05967.

Peterson, G. W., Lu, A. X., Hall, M. G., Browe, M. A., Tovar, T., & Epps, T. H. (2018). MOFwich: Sandwiched metal-organic framework-containing mixed matrix composites for chemical warfare agent removal. *ACS Applied Materials and Interfaces, 10*(8), 6820–6824. https://doi.org/10.1021/acsami.7b19365.

Prieto, G., Zečević, J., Friedrich, H., De Jong, K. P., & De Jongh, P. E. (2013). Towards stable catalysts by controlling collective properties of supported metal nanoparticles. *Nature Materials, 12*(1), 34–39. https://doi.org/10.1038/nmat3471.

Qiang, S., Wu, F., Yu, J., Liu, Y. T., & Ding, B. (2023). Complementary design in multicomponent electrocatalysts for electrochemical nitrogen reduction: Beyond the leverage in activity and selectivity. *Angewandte Chemie International Edition, 62*(15), e202217265.

Qin, J. S., Du, D. Y., Guan, W., Bo, X. J., Li, Y. F., Guo, L. P., Su, Z. M., Wang, Y. Y., Lan, Y. Q., & Zhou, H. C. (2015). Ultrastable polymolybdate-based metal-organic frameworks as highly active electrocatalysts for hydrogen generation from water. *Journal of the American Chemical Society, 137*(22), 7169–7177. https://doi.org/10.1021/jacs.5b02688.

Rahman, M. Z., Kibria, M. G., & Mullins, C. B. (2020). Metal-free photocatalysts for hydrogen evolution. *Chemical Society Reviews, 49*(6), 1887–1931. https://doi.org/10.1039/c9cs00313d.

Rani, P., & Srivastava, R. (2018). Integration of a metal–organic framework with zeolite: a highly sustainable composite catalyst for the synthesis of γ-valerolactone and coumarins. *Sustainable Energy & Fuels, 2*(6), 1287–1298. https://doi.org/10.1039/C8SE00098K.

Rani, P., & Srivastava, R. (2019). Multi-functional metal-organic framework and metal-organic framework-zeolite nanocomposite for the synthesis of carbohydrate derived chemicals via one-pot cascade reaction. *Journal of Colloid and Interface Science, 557*, 144–155. https://doi.org/10.1016/j.jcis.2019.09.008.

Rani, S., Sharma, B., Malhotra, R., Kumar, S., Varma, R. S., & Dilbaghi, N. (2020). Sn-MOF@CNT nanocomposite: An efficient electrochemical sensor for detection of hydrogen peroxide. *Environmental Research, 191*, 110005.

Rao, B. G., Mukherjee, D., & Reddy, B. M. (2017). *Novel approaches for preparation of nanoparticles. Nanostructures for novel therapy: Synthesis, characterization and applications*. India: Elsevier Inc, 1–36.

Rehman, M. Yur, Manzoor, S., Nazar, N., Abid, A. G., Qureshi, A. M., Chughtai, A. H., Joya, K. S., Shah, A., & Ashiq, M. N. (2021). *Journal of Alloys and Compounds, 856*.

Ren, Y., Yu, C., Tan, X., Huang, H., Wei, Q., & Qiu, J. (2021). Strategies to suppress hydrogen evolution for highly selective electrocatalytic nitrogen reduction: Challenges and perspectives. *Energy and Environmental Science, 14*(3), 1176–1193. https://doi.org/10.1039/d0ee03596c.

Roy, S., Darabdhara, J., & Ahmaruzzaman, M. (2023). ZnO-based Cu metal–organic framework (MOF) nanocomposite for boosting and tuning the photocatalytic degradation performance. *Environmental Science and Pollution Research, 30*(42), 95673–95691. https://doi.org/10.1007/s11356-023-29105-4.

Sachse, A., Ameloot, R., Coq, B., Fajula, F., Coasne, B., De Vos, D., & Galarneau, A. (2012). In situ synthesis of Cu–BTC (HKUST-1) in macro-/mesoporous silica monoliths for continuous flow catalysis. *Chemical Communications, 48*(39), 4749–4751. https://doi.org/10.1039/c2cc17190b.

Sadjadi, S., Koohestani, F., Mahmoodi, N. M., & Rabeie, B. (2021). Composite of MOF and chitin as an efficient catalyst for photodegradation of organic dyes. *International Journal of Biological Macromolecules, 182*, 524–533. https://doi.org/10.1016/j.ijbiomac.2021.04.034.

Sahoo, U., Pattnayak, S., Choudhury, S., Aparajita, P., Pradhan, D. K., & Hota, G. (2024). Facile synthesis of defect induced CeO_2/MIL-53(Fe) nanocatalyst: Strategically switching the charge transfer dynamics for remarkable enhancement of photocatalytic Bisphenol A degradation and H_2 evolution. *Applied Catalysis B: Environmental, 343*, 123624.

Salimi, S., Akhbari, K., Farnia, S. M. F., Tylianakis, E., E. Froudakis, G., & M. White, J. (2024). Nanoporous metal–organic framework based on furan-2,5-dicarboxylic acid with high potential in selective adsorption and separation of gas mixtures. *Crystal Growth & Design, 24*(10), 4220–4231. https://doi.org/10.1021/acs.cgd.4c00349.

Salimi, S., Akhbari, K., Farnia, S. M. F., & White, J. M. (2022). Multiple construction of a hierarchical nanoporous manganese(II)-based metal–organic framework with active sites for regulating N_2 and CO_2 trapping. *Crystal Growth & Design, 22*(3), 1654–1664. https://doi.org/10.1021/acs.cgd.1c01183.

Salimi, S., Farnia, S. M. F., Akhbari, K., & Tavasoli, A. (2023). Engineered catalyst based on MIL-68(Al) with high stability for hydrogenation of carbon dioxide and carbon monoxide at low temperature. *Inorganic Chemistry, 62*(43), 17588–17601. https://doi.org/10.1021/acs.inorgchem.3c01094.

Samy, M., Ibrahim, M. G., Fujii, M., Diab, K. E., ElKady, M., & Alalm, M. G. (2021). CNTs/MOF-808 painted plates for extended treatment of pharmaceutical and agrochemical wastewaters in a novel photocatalytic reactor. *Chemical Engineering Journal, 406*, 127152.

Seh, Z. W., Kibsgaard, J., Dickens, C. F., Chorkendorff, I., Nørskov, J. K., & Jaramillo, T. F. (2017). Combining theory and experiment in electrocatalysis: Insights into materials design. *Science, 355*(6321), eaad4998.

Shah, W. A., Noureen, L., Nadeem, M. A., & Kögerler, P. (2018). Encapsulation of Keggin-type manganese-polyoxomolybdates in MIL-100 (Fe) for efficient reduction of p-nitrophenol. *Journal of Solid State Chemistry, 268*, 75–82. https://doi.org/10.1016/j.jssc.2018.08.024.

Shahangi Shirazi, F., & Akhbari, K. (2015). Preparation of zinc oxide nanoparticles from nanoporous metal-organic framework with one-dimensional channels occupied with guest water molecules. *Inorganica Chimica Acta, 436*, 1–6. https://doi.org/10.1016/j.ica.2015.07.025.

Shan, Y., Zhang, G., Shi, Y., & Pang, H. (2023). *Cell Reports Physical Science, 4*.

Shao, M., Chang, Q., Dodelet, J. P., & Chenitz, R. (2016). Recent advances in electrocatalysts for oxygen reduction reaction. *Chemical Reviews, 116*(5), 3594–3657. https://doi.org/10.1021/acs.chemrev.5b00462.

Shen, K., Zhang, L., Chen, X., Liu, L., Zhang, D., Han, Y., Chen, J., Long, J., Luque, R., Li, Y., & Chen, B. (2018). Ordered macro-microporous metal-organic framework single crystals. *Science, 359*(6372), 206–210. https://doi.org/10.1126/science.aao3403.

Shen, L., Luo, M., Huang, L., Feng, P., & Wu, L. (2015). A clean and general strategy to decorate a titanium metal-organic framework with noble-metal nanoparticles for versatile photocatalytic applications. *Inorganic Chemistry, 54*(4), 1191–1193. https://doi.org/10.1021/ic502609a.

Shen, M., Forghani, F., Kong, X., Liu, D., Ye, X., Chen, S., & Ding, T. (2020). Antibacterial applications of metal–organic frameworks and their composites. *Comprehensive Reviews in Food Science and Food Safety, 19*(4), 1397–1419. https://doi.org/10.1111/1541-4337.12515.

Shen, Y., Li, Z. F., Guo, S. Y., Shao, Y. R., & Hu, T. L. (2021). Encapsulation of ultrafine metal-organic framework nanoparticles within multichamber carbon spheres by a two-step double-solvent strategy for high-performance catalysts. *ACS Applied Materials and Interfaces, 13*(10), 12169–12180. https://doi.org/10.1021/acsami.1c01451.

Sherrington, D. C. (1980). Polymers as catalysts. *British Polymer Journal, 12*(2), 70–74. https://doi.org/10.1002/pi.4980120206.

Shi, L., Li, P., Zhou, W., Wang, T., Chang, K., Zhang, H., Kako, T., Liu, G., & Ye, J. (2016). n-type boron phosphide as a highly stable, metal-free, visible-light-active photocatalyst for hydrogen evolution. *Nano Energy, 28,* 158–163. https://doi.org/10.1016/j.nanoen.2016.08.041.

Shi, Y., Yang, A. F., Cao, C. S., & Zhao, B. (2019). Applications of MOFs: Recent advances in photocatalytic hydrogen production from water. *Coordination Chemistry Reviews, 390,* 50–75. https://doi.org/10.1016/j.ccr.2019.03.012.

Silva, C. G., Corma, A., & García, H. (2010). Metal-organic frameworks as semiconductors. *Journal of Materials Chemistry, 20*(16), 3141–3156. https://doi.org/10.1039/b924937k.

Silva, C. G., Luz, I., Llabrés, F. X., Xamena, I., Corma, A., & García, H. (2010). Water stable Zr-benzenedicarboxylate metal-organic frameworks as photocatalysts for hydrogen generation. *Chemistry - A European Journal, 16*(36), 11133–11138. https://doi.org/10.1002/chem.200903526Spain.

Singh, A., Singh, A. K., Liu, J., & Kumar, A. (2021). Syntheses, design strategies, and photocatalytic charge dynamics of metal-organic frameworks (MOFs): A catalyzed photo-degradation approach towards organic dyes. *Catalysis Science and Technology, 11*(12), 3946–3989. https://doi.org/10.1039/d0cy02275f.

Singh, K., Kukkar, D., Singh, R., Kukkar, P., Bajaj, N., Singh, J., Rawat, M., Kumar, A., & Kim, K. H. (2020). In situ green synthesis of Au/Ag nanostructures on a metal-organic framework surface for photocatalytic reduction of p-nitrophenol. *Journal of Industrial and Engineering Chemistry, 81,* 196–205. https://doi.org/10.1016/j.jiec.2019.09.008.

Singh, O., Agrawal, A., Abraham, B. M., Goyal, R., Pendem, C., & Sarkar, B. (2022). Integration of zeolite@metal–organic framework: A composite catalyst for isopropyl alcohol conversion to aromatics. *Materials Today Chemistry, 24,* 100796.

Snider, V. G., & Hill, C. L. (2023). Functionalized reactive polymers for the removal of chemical warfare agents: A review. *Journal of Hazardous Materials, 442,* 130015.

Soltani, S., & Akhbari, K. (2022). Embedding an extraordinary amount of gemifloxacin antibiotic in ZIF-8 framework with one-step synthesis and measurement of its H_2O_2-sensitive release and potency against infectious bacteria. *New Journal of Chemistry, 46*(40), 19432–19441. https://doi.org/10.1039/d2nj02981b.

Song, J. Y., Ahmed, I., Seo, P. W., & Jhung, S. H. (2016). UiO-66-type metal-organic framework with free carboxylic acid: Versatile adsorbents via H-bond for both aqueous and nonaqueous phases. *ACS Applied Materials and Interfaces, 8*(40), 27394–27402. https://doi.org/10.1021/acsami.6b10098.

Song, R., Wu, X., Liu, L., & Han, Z. (2024). Amino-functionalized rare earth hexanuclear cluster based MOFs for CO_2/CH_4 separation. *Inorganic Chemistry Communications, 163,* 112373.

Song, X., Hu, D., Yang, X., Zhang, H., Zhang, W., Li, J., Jia, M., & Yu, J. (2019). Polyoxomolybdic cobalt encapsulated within Zr-based metal-organic frameworks as efficient heterogeneous catalysts for olefins epoxidation. *ACS Sustainable Chemistry and Engineering, 7*(3), 3624–3631. https://doi.org/10.1021/acssuschemeng.8b06736.

Sorribas, S., Zornoza, B., Téllez, C., & Coronas, J. (2012). Ordered mesoporous silica-(ZIF-8) core-shell spheres. *Chemical Communications, 48*(75), 9388–9390. https://doi.org/10.1039/c2cc34893d.

Sreekanth, T., Dillip, G., Nagajyothi, P., Yoo, K., & Kim, J. (2021). Integration of marigold 3D flower-like Ni-MOF self-assembled on MWCNTs via microwave irradiation for high-performance electrocatalytic alcohol oxidation and oxygen evolution reactions. *Applied Catalysis B: Environmental, 285,* 119793.

Su, Z., Huang, Q., Guo, Q., Hoseini, S. J., Zheng, F., & Chen, W. (2023). Metal–organic framework and carbon hybrid nanostructures: Fabrication strategies and electrocatalytic application for the water splitting and oxygen reduction reaction. *Nano Research Energy, 2*(4), e9120078.

Subudhi, S., Swain, G., Tripathy, S. P., & Parida, K. (2020). UiO-66-NH$_2$ metal-organic frameworks with embedded MoS$_2$ nanoflakes for visible-light-mediated H$_2$ and O$_2$ evolution. *Inorganic Chemistry, 59*(14), 9824–9837. https://doi.org/10.1021/acs.inorgchem.0c01030.

Sullivan, I., Goryachev, A., Digdaya, I. A., Li, X., Atwater, H. A., Vermaas, D. A., & Xiang, C. (2021). Coupling electrochemical CO$_2$ conversion with CO$_2$ capture. *Nature Catalysis, 4*(11), 952–958. https://doi.org/10.1038/s41929-021-00699-7.

Sun, D., & Li, Z. (2016). Double-solvent method to Pd nanoclusters encapsulated inside the cavity of NH$_2$–Uio-66(Zr) for efficient visible-light-promoted Suzuki coupling reaction. *The Journal of Physical Chemistry C, 120*(35), 19744–19750. https://doi.org/10.1021/acs.jpcc.6b06710.

Sun, D., Xu, M., Jiang, Y., Long, J., & Li, Z. (2018). Small-sized bimetallic cupd nanoclusters encapsulated inside cavity of NH$_2$-UiO-66(Zr) with superior performance for light-induced Suzuki coupling reaction. *Small Methods, 2*(12), 1800164.

Sun, Z. X., Sun, K., Gao, M. L., Metin, Ö., & Jiang, H. L. (2022). *Angewandte Chemie, 134*.

Suryanto, B. H. R., Matuszek, K., Choi, J., Hodgetts, R. Y., Du, H. L., Bakker, J. M., Kang, C. S. M., Cherepanov, P. V., Simonov, A. N., & MacFarlane, D. R. (2021). Nitrogen reduction to ammonia at high efficiency and rates based on a phosphonium proton shuttle. *Science, 372*(6547), 1187–1191. https://doi.org/10.1126/science.abg2371.

Tabassum, H., Guo, W., Meng, W., Mahmood, A., Zhao, R., Wang, Q., & Zou, R. (2017). Metal-organic frameworks derived cobalt phosphide architecture encapsulated into B/N Co-doped graphene nanotubes for all pH value electrochemical hydrogen evolution. *Advanced Energy Materials, 7*(9).

Tari, N. E., Tadjarodi, A., Tamnanloo, J., & Fatemi, S. (2016). One pot microwave synthesis of MCM-41/Cu based MOF composite with improved CO$_2$ adsorption and selectivity. *Microporous and Mesoporous Materials, 231*, 154–162. https://doi.org/10.1016/j.micromeso.2016.05.027.

Thi, Q. V., Tamboli, M. S., Ta, Q. T. H., Kolekar, G. B., & Sohn, D. (2020). A nanostructured MOF/reduced graphene oxide hybrid for enhanced photocatalytic efficiency under solar light. *Materials Science and Engineering: B, 261*, 114678.

Tian, J., Xu, Z.-Y., Zhang, D.-W., Wang, H., Xie, S.-H., Xu, D.-W., Ren, Y.-H., Wang, H., Liu, Y., & Li, Z.-T. (2016). Supramolecular metal-organic frameworks that display high homogeneous and heterogeneous photocatalytic activity for H$_2$ production. *Nature Communications, 7*, 11580.

Tong, J., Wang, W., Su, L., Li, Q., Liu, F., Ma, W., Lei, Z., & Bo, L. (2017). Highly selective oxidation of cyclohexene to 2-cyclohexene-1-one over polyoxometalate/metal-organic framework hybrids with greatly improved performances. *Catalysis Science and Technology, 7*(1), 222–230. https://doi.org/10.1039/c6cy01554a.

Tsumori, N., Chen, L., Wang, Q., Zhu, Q. L., Kitta, M., & Xu, Q. (2018). Quasi-MOF: Exposing inorganic nodes to guest metal nanoparticles for drastically enhanced catalytic activity. *Chem, 4*(4), 845–856. https://doi.org/10.1016/j.chempr.2018.03.009.

Uemura, T., Hiramatsu, D., Yoshida, K., Isoda, S., & Kitagawa, S. (2008). Sol-gel synthesis of low-dimensional silica within coordination nanochannels. *Journal of the American Chemical Society, 130*(29), 9216–9217. https://doi.org/10.1021/ja8030906.

Uemura, T., Kadowaki, Y., Kim, C. R., Fukushima, T., Hiramatsu, D., & Kitagawa, S. (2011). Incarceration of nanosized silica into porous coordination polymers: Preparation, characterization, and adsorption property. *Chemistry of Materials, 23*(7), 1736–1741. https://doi.org/10.1021/cm102610r.

Uemura, T., Yanai, N., & Kitagawa, S. (2009). Polymerization reactions in porous coordination polymers. *Chemical Society Reviews, 38*(5), 1228–1236. https://doi.org/10.1039/b802583p.

Ullah, L., Zhao, G., Xu, Z., He, H., Usman, M., & Zhang, S. (2018). 12-Tungstophosphoric acid niched in Zr-based metal-organic framework: A stable and efficient catalyst for Friedel-Crafts acylation. *Science China Chemistry, 61*(4), 402–411. https://doi.org/10.1007/s11426-017-9182-0.

Wang, C., Kim, J., Tang, J., Kim, M., Lim, H., Malgras, V., You, J., Xu, Q., Li, J., & Yamauchi, Y. (2020). New strategies for novel MOF-derived carbon materials based on nanoarchitectures. *Chem, 6*(1), 19–40. https://doi.org/10.1016/j.chempr.2019.09.005.

Wang, D., & Li, Z. (2016). Coupling MOF-based photocatalysis with Pd catalysis over Pd@MIL-100(Fe) for efficient N-alkylation of amines with alcohols under visible light. *Journal of Catalysis, 342*(1), 151–157. https://doi.org/10.1016/j.jcat.2016.07.021.

Wang, H., Liu, G., Chen, C., Tu, W., Lu, Y., Wu, S., O'Hare, D., & Xu, R. (2021). Single-Ni sites embedded in multilayer nitrogen-doped graphene derived from amino-functionalized MOF for highly selective CO_2 electroreduction. *ACS Sustainable Chemistry and Engineering, 9*(10), 3792–3801. https://doi.org/10.1021/acssuschemeng.0c08749.

Wang, L., Guan, E., Zhang, J., Yang, J., Zhu, Y., Han, Y., Yang, M., Cen, C., Fu, G., & Gates, B. C. (2018). Single-site catalyst promoters accelerate metal-catalyzed nitroarene hydrogenation. *Nature Communications, 9*, 1362.

Wang, M., Zhen, W., Tian, B., Ma, J., & Lu, G. (2018). The inhibition of hydrogen and oxygen recombination reaction by halogen atoms on over-all water splitting over Pt-TiO_2 photocatalyst. *Applied Catalysis B: Environmental, 236*, 240–252. https://doi.org/10.1016/j.apcatb.2018.05.031.

Wang, Q., Yang, G., Fu, Y., Li, N., Hao, D., & Ma, S. (2022). *ChemNanoMat, 8*.

Wang, S., Liu, Y., Zhang, Z., Li, X., Tian, H., Yan, T., Zhang, X., Liu, S., Sun, X., Xu, L., Luo, F., & Liu, S. (2019). One-step template-free fabrication of ultrathin mixed-valence polyoxovanadate-incorporated metal-organic framework nanosheets for highly efficient selective oxidation catalysis in air. *ACS Applied Materials and Interfaces, 11*(13), 12786–12796. https://doi.org/10.1021/acsami.9b00908.

Wang, S., & Wang, X. (2016). Imidazolium ionic liquids, imidazolylidene heterocyclic carbenes, and zeolitic imidazolate frameworks for CO_2 capture and photochemical reduction. *Angewandte Chemie - International Edition, 55*(7), 2308–2320. https://doi.org/10.1002/anie.201507145.

Wang, X., Li, M., Cao, C., Liu, C., Liu, J., Zhu, Y., Zhang, S., & Song, W. (2016). Surfactant-free palladium nanoparticles encapsulated in ZIF-8 hollow nanospheres for size-selective catalysis in liquid-phase solution. *ChemCatChem, 8*(20), 3224–3228. https://doi.org/10.1002/cctc.201600846.

Wang, X. G., Xu, L., Li, M. J., & Zhang, X. Z. (2020). Construction of flexible-on-rigid hybrid-phase metal–organic frameworks for controllable multi-drug delivery. *Angewandte Chemie - International Edition, 59*(41), 18078–18086. https://doi.org/10.1002/anie.202008858.

Wang, Y., Yan, L., Dastafkan, K., Zhao, C., Zhao, X., Xue, Y., Huo, J., Li, S., & Zhai, Q. (2021). *Advanced Materials, 33*.

Wang, Y., Zhang, Y., Jiang, Z., Jiang, G., Zhao, Z., Wu, Q., Liu, Y., Xu, Q., Duan, A., & Xu, C. (2016). Controlled fabrication and enhanced visible-light photocatalytic hydrogen production of Au at CdS/MIL-101 heterostructure. *Applied Catalysis B: Environmental, 185*, 307–314. https://doi.org/10.1016/j.apcatb.2015.12.020.

Wang, Z., Jin, X., Yan, L., Yang, Y., & Liu, X. (2023). Recent research progress in CDs@MOFs composites: Fabrication, property modulation, and application. *Microchimica Acta, 190*, 28.

Wang, Zj, Song, H., Pang, H., Ning, Y., Dao, T. D., Wang, Z., Chen, H., Weng, Y., Fu, Q., Nagao, T., Fang, Y., & Ye, J. (2019). Photo-assisted methanol synthesis via CO$_2$ reduction under ambient pressure over plasmonic Cu/ZnO catalysts. *Applied Catalysis B: Environmental, 250*, 10–16. https://doi.org/10.1016/j.apcatb.2019.03.003.

Wen, Q., Li, D., Li, H., Long, M., Gao, C., Wu, L., Song, F., & Zhou, J. (2023). Synergetic effect of photocatalysis and peroxymonosulfate activated by Co/Mn-MOF-74@g-C$_3$N$_4$ Z-scheme photocatalyst for removal of tetracycline hydrochloride. *Separation and Purification Technology, 313*, 123518.

Wen, X., & Guan, J. (2019). Recent progress on MOF-derived electrocatalysts for hydrogen evolution reaction. *Applied Materials Today, 16*, 146–168. https://doi.org/10.1016/j.apmt.2019.05.013.

Wu, B., Ge, J., Zhang, Z., Huang, C., Li, X., Tan, Z., Fang, X., & Sun, J. (2019). Combination of sodium selenite and doxorubicin prodrug Ac-Phe-Lys-PABC-ADM affects gastric cancer cell apoptosis in xenografted mice. *BioMed Research International, 2019*, 1–8. https://doi.org/10.1155/2019/2486783.

Wu, J., Wang, Q., Deng, Z., Zhang, S., & Jiao, Z. (2024). A novel design of superparamagnetic iron oxide-based metal–organic framework for SSI-assisted drug delivery system†. *New Journal of Chemistry, 48*(9), 4118–4125. https://doi.org/10.1039/d3nj05247h.

Wu, K., Liu, X. Y., Cheng, P. W., Xie, M., Lu, W., & Li, D. (2023). Metal-organic frameworks as photocatalysts for aerobic oxidation reactions. *Science China Chemistry, 66*(6), 1634–1653. https://doi.org/10.1007/s11426-022-1519-x.

Wu, X., Han, X., Zhang, J., Jiang, H., Hou, B., Liu, Y., & Cui, Y. (2019). Metal- and covalent organic frameworks threaded with chiral polymers for heterogeneous asymmetric catalysis. *Organometallics, 38*(18), 3474–3479. https://doi.org/10.1021/acs.organomet.9b00277.

Wu, Z. S., Yang, S., Sun, Y., Parvez, K., Feng, X., & Müllen, K. (2012). 3D nitrogen-doped graphene aerogel-supported Fe$_3$O$_4$ nanoparticles as efficient electrocatalysts for the oxygen reduction reaction. *Journal of the American Chemical Society, 134*(22), 9082–9085. https://doi.org/10.1021/ja3030565.

Xia, Z., Fang, J., Zhang, X., Fan, L., Barlow, A. J., Lin, T., Wang, S., Wallace, G. G., Sun, G., & Wang, X. (2019). Pt nanoparticles embedded metal-organic framework nanosheets: A synergistic strategy towards bifunctional oxygen electrocatalysis. *Applied Catalysis B: Environmental, 245*, 389–398. https://doi.org/10.1016/j.apcatb.2018.12.073.

Xiang, W., Zhang, Y., Lin, H., & Liu, C. (2017). Nanoparticle/metal–organic framework composites for catalytic applications: Current status and perspective. *Molecules, 22*(12), 2103.

Xiao, J. D., Han, L., Luo, J., Yu, S. H., & Jiang, H. L. (2018). Integration of plasmonic effects and schottky junctions into metal–organic framework composites: Steering charge flow for enhanced visible-light photocatalysis. *Angewandte Chemie - International Edition, 57*(4), 1103–1107. https://doi.org/10.1002/anie.201711725.

Xiao, J. D., Shang, Q., Xiong, Y., Zhang, Q., Luo, Y., Yu, S. H., & Jiang, H. L. (2016). Boosting photocatalytic hydrogen production of a metal–organic framework decorated with platinum nanoparticles: The platinum location matters. *Angewandte Chemie - International Edition, 55*(32), 9389–9393. https://doi.org/10.1002/anie.201603990.

Xiao, L., Zhang, Q., Chen, P., Chen, L., Ding, F., Tang, J., Li, Y. J., Au, C. T., & Yin, S. F. (2019). Copper-mediated metal-organic framework as efficient photocatalyst for the partial oxidation of aromatic alcohols under visible-light irradiation: Synergism of plasmonic effect and schottky junction. *Applied Catalysis B: Environmental, 248*, 380–387. https://doi.org/10.1016/j.apcatb.2019.02.012.

Xie, K., Fu, Q., He, Y., Kim, J., Goh, S. J., Nam, E., Qiao, G. G., & Webley, P. A. (2015). Synthesis of well dispersed polymer grafted metal–organic framework nanoparticles. *Chemical Communications, 51*(85), 15566–15569. https://doi.org/10.1039/C5CC06694H.

Xie, K., Liu, X., Li, H., Fang, L., Xia, K., Yang, D., Zou, Y., & Zhang, X. (2023). Heteroatom tuning in agarose derived carbon aerogel for enhanced potassium ion multiple energy storage. *Carbon Energy, 6*(3), e427.

Xie, W., & Wan, F. (2019). Immobilization of polyoxometalate-based sulfonated ionic liquids on UiO-66-2COOH metal-organic frameworks for biodiesel production via one-pot transesterification-esterification of acidic vegetable oils. *Chemical Engineering Journal, 365*, 40–50. https://doi.org/10.1016/j.cej.2019.02.016.

Xin, Y., Cao, Y., Yang, J., Guo, X., Shen, K., & Yao, W. (2024). Mesopore and macropore engineering in metal-organic frameworks for energy environment-related applications. *Journal of Materials Chemistry A, 12*(9), 4931–4970. https://doi.org/10.1039/d3ta07697k.

Xin, Z., Liu, J., Wang, X., Shen, K., Yuan, Z., Chen, Y., & Lan, Y. Q. (2021). Implanting polypyrrole in metal-porphyrin MOFs: Enhanced electrocatalytic performance for CO2RR. *ACS Applied Materials and Interfaces, 13*(46), 54959–54966. https://doi.org/10.1021/acsami.1c15187.

Xiong, W., Li, H., You, H., Cao, M., & Cao, R. (2020). Encapsulating metal organic framework into hollow mesoporous carbon sphere as efficient oxygen bifunctional electrocatalyst. *National Science Review, 7*(3), 609–619. https://doi.org/10.1093/nsr/nwz166.

Xu, M., Li, D., Sun, K., Jiao, L., Xie, C., Ding, C., & Jiang, H.-L. (2021). Interfacial microenvironment modulation boosting electron transfer between metal nanoparticles and MOFs for enhanced photocatalysis. *Angewandte Chemie, 133*(30), 16508–16512. https://doi.org/10.1002/ange.202104219.

Xu, T., Zhang, Y., Liu, M., Wang, H., Ren, J., Tian, Y., Liu, X., Zhou, Y., Wang, J., & Zhu, W. (2022). In-situ two-step electrodeposition of α-CD-rGO/Ni-MOF composite film for superior glucose sensing. *Journal of Alloys and Compounds, 923*, 166418.

Xu, W., Liu, M., Wang, S., Peng, Z., Shen, R., & Li, B. (2022). Interfacial ensemble effect of copper nanoparticles and nickel metal-organic framework on promoting hydrogen generation. *International Journal of Hydrogen Energy, 47*(55), 23213–23220. https://doi.org/10.1016/j.ijhydene.2022.05.108.

Xu, Y., Yang, C., Deng, Q., Zhou, Y., Mao, C., Song, Y., Zhu, M., & Zhang, Y. (2023). Synthesis and electrochemical properties of multi-layered SnO/rGO composite as anode materials for sodium ion batteries. *Applied Surface Science, 612*, 155859.

Xue, C., Wei, X., Zhang, Z., Bai, Y., Li, M., & Chen, Y. (2020). Synthesis and characterization of LSX zeolite/AC composite from elutrilithe. *Materials, 13*(16), 34–69.

Yang, j, Cong, W., Zhu, Z., Miao, Z., Wang, Y.-T., Nelles, M., & Fang, Z. (2023). Microwave-assisted one-step production of biodiesel from waste cooking oil by magnetic bifunctional SrO–ZnO/MOF catalyst. *Journal of Cleaner Production, 395*. https://doi.org/10.1016/j.jclepro.2023.136182.

Yang, J., Zhang, F., Lu, H., Hong, X., Jiang, H., Wu, Y., & Li, Y. (2015). Hollow Zn/Co ZIF particles derived from core–shell ZIF-67@ZIF-8 as selective catalyst for the semi-hydrogenation of acetylene. *Angewandte Chemie, 127*(37), 11039–11043. https://doi.org/10.1002/ange.201504242.

Yang, Q., Xu, Q., & Jiang, H. L. (2017). Metal-organic frameworks meet metal nanoparticles: Synergistic effect for enhanced catalysis. *Chemical Society Reviews, 46*(15), 4774–4808. https://doi.org/10.1039/c6cs00724d.

Yang, Q., Yao, F., Zhong, Y., Chen, F., Shu, X., Sun, J., He, L., Wu, B., Hou, K., & Wang, D. (2019). *Particle & Particle Systems Characterization, 36*.

Yang, S., Peng, L., Bulut, S., & Queen, W. L. (2019). Recent advances of MOFs and MOF-derived materials in thermally driven organic transformations. *Chemistry – A European Journal, 25*(9), 2161–2178. https://doi.org/10.1002/chem.201803157.

Yang, X., Xie, X., Li, S., Zhang, W., Zhang, X., Chai, H., & Huang, Y. (2021). The POM@MOF hybrid derived hierarchical hollow Mo/Co bimetal oxides nanocages for efficiently activating peroxymonosulfate to degrade levofloxacin. *Journal of Hazardous Materials, 419*, 126360.

Yang, X., Yuan, S., Zou, L., Drake, H., Zhang, Y., Qin, J., Alsalme, A., & Zhou, H.-C. (2018). One-step synthesis of hybrid core–shell metal–organic frameworks. *Angewandte Chemie, 130*(15), 3991–3996. https://doi.org/10.1002/ange.201710019.

Yang, Y., Cong, D., & Hao, S. (2016). Template-directed ordered mesoporous silica@palladium-containing zinc metal-organic framework composites as highly efficient Suzuki coupling catalysts. *ChemCatChem, 8*(5), 900–905. https://doi.org/10.1002/cctc.201501314.

Yang, Y., Pan, Y.-X., Tu, X., & Liu, C. (2022). Nitrogen doping of indium oxide for enhanced photocatalytic reduction of CO_2 to methanol. *Nano Energy, 101*, 107613.

Yang, Y., Zhang, X., Kanchanakungwankul, S., Lu, Z., Noh, H., Syed, Z. H., Farha, O. K., Truhlar, D. G., & Hupp, J. T. (2020). Unexpected "spontaneous" evolution of catalytic, MOF-supported single Cu(II) cations to catalytic, MOF-supported Cu(0) nanoparticles. *Journal of the American Chemical Society, 142*(50), 21169–21177. https://doi.org/10.1021/jacs.0c10367.

Yao, M.-S., Xiu, J.-W., Huang, Q.-Q., Li, W.-H., Wu, W.-W., Wu, A.-Q., Cao, L.-A., Deng, W.-H., Wang, G.-E., & Xu, G. (2019). Van der Waals heterostructured MOF-on-MOF thin films: Cascading functionality to realize advanced chemiresistive sensing. *Angewandte Chemie, 131*(42), 15057–15061. https://doi.org/10.1002/ange.201907772.

Yao, Z., Liu, S., Liu, H., Ruan, Y., Hong, S., Wu, T. S., Hao, L., Soo, Y. L., Xiong, P., & Li, M. M. J. (2023). Pre-adsorbed H-assisted N_2 activation on single-atom cadmium-O_5 decorated In_2O_3 for efficient NH_3 electrosynthesis. *Advanced Functional Materials, 33*(5), 2209843.

Ye, J. J., & Wu, C. D. (2016). Immobilization of polyoxometalates in crystalline solids for highly efficient heterogeneous catalysis. *Dalton Transactions, 45*(25), 10101–10112. https://doi.org/10.1039/c6dt01378c.

Ying, J., Herbst, A., Xiao, Y. X., Wei, H., Tian, G., Li, Z., Yang, X. Y., Su, B. L., & Janiak, C. (2018). Nanocoating of hydrophobic mesoporous silica around MIL-101Cr for enhanced catalytic activity and stability. *Inorganic Chemistry, 57*(3), 899–902. https://doi.org/10.1021/acs.inorgchem.7b01992.

Young, A. J., Guillet-Nicolas, R., Marshall, E. S., Kleitz, F., Goodhand, A. J., Glanville, L. B. L., Reithofer, M. R., & Chin, J. M. (2019). Direct ink writing of catalytically active UiO-66 polymer composites. *Chemical Communications, 55*(15), 2190–2193. https://doi.org/10.1039/c8cc10018g.

Yu, S., Wu, Y., Xue, Q., Zhu, J.-J., & Zhou, Y. (2022). A novel multi-walled carbon nanotube-coupled CoNi MOF composite enhances the oxygen evolution reaction through synergistic effects. *Journal of Materials Chemistry A, 10*(9), 4936–4943. https://doi.org/10.1039/d1ta10681c.

Yuan, N., Zhang, X., & Wang, L. (2020). The marriage of metal–organic frameworks and silica materials for advanced applications. *Coordination Chemistry Reviews, 421*, 213442.

Yue, X., Yi, S., Wang, R., Zhang, Z., & Qiu, S. (2018). Synergistic effect based NixCo1-x architected Zn0.75Cd0.25S nanocrystals: An ultrahigh and stable photocatalysts for hydrogen evolution from water splitting. *Applied Catalysis B: Environmental, 224,* 17–26. https://doi.org/10.1016/j.apcatb.2017.10.010.

Zaman, N., Noor, T., & Iqbal, N. (2021). Recent advances in the metal-organic framework-based electrocatalysts for the hydrogen evolution reaction in water splitting: A review. *RSC Advances, 11*(36), 21904–21925. https://doi.org/10.1039/d1ra02240g.

Zamaro, J. M., Pérez, N. C., Miró, E. E., Casado, C., Seoane, B., Téllez, C., & Coronas, J. (2012). HKUST-1 MOF: A matrix to synthesize CuO and CuO-CeO$_2$ nanoparticle catalysts for CO oxidation. *Chemical Engineering Journal, 195-196,* 180–187. https://doi.org/10.1016/j.cej.2012.04.091.

Zeng, S., Lyu, F., Sun, L., Zhan, Y., Ma, F. X., Lu, J., & Li, Y. Y. (2019). UiO-66-NO$_2$ as an oxygen "pump" for enhancing oxygen reduction reaction performance. *Chemistry of Materials, 31*(5), 1646–1654. https://doi.org/10.1021/acs.chemmater.8b04934.

Zha, Q., Li, M., Liu, Z., & Ni, Y. (2020). Hierarchical Co,Fe-MOF-74/Co/carbon cloth hybrid electrode: Simple construction and enhanced catalytic performance in full water splitting. *ACS Sustainable Chemistry and Engineering, 8*(32), 12025–12035. https://doi.org/10.1021/acssuschemeng.0c02993.

Zhang, F., Wei, Y., Wu, X., Jiang, H., Wang, W., & Li, H. (2014). Hollow zeolitic imidazolate framework nanospheres as highly efficient cooperative catalysts for [3+3] cycloaddition reactions. *Journal of the American Chemical Society, 136*(40), 13963–13966. https://doi.org/10.1021/ja506372z.

Zhang, J., Bai, T., Huang, H., Yu, M. H., Fan, X., Chang, Z., & Bu, X. H. (2020). Metal–organic-framework-based photocatalysts optimized by spatially separated cocatalysts for overall water splitting. *Advanced Materials, 32*(49), 2004747.

Zhang, J., Wang, L., Shao, Y., Wang, Y., Gates, B. C., & Xiao, F.-S. (2017). A Pd@zeolite catalyst for nitroarene hydrogenation with high product selectivity by sterically controlled adsorption in the zeolite micropores. *Angewandte Chemie, 129*(33), 9879–9883. https://doi.org/10.1002/ange.201703938.

Zhang, L., Li, S., Gómez-García, C. J., Ma, H., Zhang, C., Pang, H., & Li, B. (2018). Two novel polyoxometalate-encapsulated metal-organic nanotube frameworks as stable and highly efficient electrocatalysts for hydrogen evolution reaction. *ACS Applied Materials and Interfaces, 10*(37), 31498–31504. https://doi.org/10.1021/acsami.8b10447.

Zhang, W., Shi, W., Ji, W., Wu, H., Gu, Z., Wang, P., Li, X., Qin, P., Zhang, J., Fan, Y., Wu, T., Fu, Y., Zhang, W., & Huo, F. (2020). Microenvironment of MOF channel coordination with Pt NPs for selective hydrogenation of unsaturated aldehydes. *ACS Catalysis, 10*(10), 5805–5813. https://doi.org/10.1021/acscatal.0c00682.

Zhang, Z., Cao, S., Luo, X., Shi, W., & Zhu, Y. (2018). Research on the thrust of a high-pressure water jet propulsion system. *Ships and Offshore Structures, 13*(1), 1–9. https://doi.org/10.1080/17445302.2017.1321713.

Zhao, D., & Cai, C. (2021). Cerium-based UiO-66 metal-organic framework for synergistic dye adsorption and photodegradation: A discussion of the mechanism. *Dyes and Pigments, 185,* 108957.

Zhao, D., Wang, Y., & Han, D. (2016). Periosteal distraction osteogenesis: An effective method for bone regeneration. *BioMed Research International. 2016,* 1–10. https://doi.org/10.1155/2016/2075317.

Zhao, J. H., Yang, Y., Che, J. X., Zuo, J., Li, X. H., Hu, Y. Z., Dong, X. W., Gao, L., & Liu, X. Y. (2018). Compartmentalization of incompatible polymers within metal–organic frameworks towards homogenization of heterogeneous hybrid catalysts for tandem reactions. *Chemistry - A European Journal, 24*(39), 9903–9909. https://doi.org/10.1002/chem.201801416.

Zhao, M., Chen, J., Chen, B., Zhang, X., Shi, Z., Liu, Z., Ma, Q., Peng, Y., Tan, C., Wu, X. J., & Zhang, H. (2020). Selective epitaxial growth of oriented hierarchical metal-organic framework heterostructures. *Journal of the American Chemical Society, 142*(19), 8953–8961. https://doi.org/10.1021/jacs.0c02489.

Zhao, M., Yuan, K., Wang, Y., Li, G., Guo, J., Gu, L., Hu, W., Zhao, H., & Tang, Z. (2016). Metal-organic frameworks as selectivity regulators for hydrogenation reactions. *Nature, 539*(7627), 76–80. https://doi.org/10.1038/nature19763.

Zhao, X., Duan, Y., Yang, F., Wei, W., Xu, Y., & Hu, C. (2017). Efficient mechanochemical synthesis of polyoxometalate⊂ZIF complexes as reusable catalysts for highly selective oxidation. *Inorganic Chemistry, 56*(23), 14506–14512. https://doi.org/10.1021/acs.inorgchem.7b02163.

Zhao, Y., Li, Y., Pang, H., Yang, C., & Ngai, T. (2019). Controlled synthesis of metal-organic frameworks coated with noble metal nanoparticles and conducting polymer for enhanced catalysis. *Journal of Colloid and Interface Science, 537*, 262–268. https://doi.org/10.1016/j.jcis.2018.11.031.

Zhen, W., Ma, J., & Lu, G. (2016). Small-sized Ni(1 1 1) particles in metal-organic frameworks with low over-potential for visible photocatalytic hydrogen generation. *Applied Catalysis B: Environmental, 190*, 12–25. https://doi.org/10.1016/j.apcatb.2016.02.061.

Zheng, X., Cao, Y., Liu, D., Cai, M., Ding, J., Liu, X., Wang, J., Hu, W., & Zhong, C. (2019). Bimetallic metal-organic-framework/reduced graphene oxide composites as bifunctional electrocatalysts for rechargeable Zn-air batteries. *ACS Applied Materials and Interfaces, 11*(17), 15662–15669. https://doi.org/10.1021/acsami.9b02859.

Zhong, H., Ly, K. H., Wang, M., Krupskaya, Y., Han, X., Zhang, J., Zhang, J., Kataev, V., Büchner, B., Weidinger, I. M., Kaskel, S., Liu, P., Chen, M., Dong, R., & Feng, X. (2019). A phthalocyanine-based layered two-dimensional conjugated metal–organic framework as a highly efficient electrocatalyst for the oxygen reduction reaction. *Angewandte Chemie - International Edition, 58*(31), 10677–10682. https://doi.org/10.1002/anie.201907002.

Zhu, G., Graver, R., Emdadi, L., Liu, B., Choi, K. Y., & Liu, D. (2014). Synthesis of zeolite@metal-organic framework core-shell particles as bifunctional catalysts. *RSC Advances, 4*(58), 30673–30676. https://doi.org/10.1039/c4ra03129f.

Zhuang, S., Singh, H., Nunna, B. B., Mandal, D., Boscoboinik, J. A., & Lee, E. S. (2018). Nitrogen-doped graphene-based catalyst with metal-reduced organic framework: Chemical analysis and structure control. *Carbon, 139*, 933–944. https://doi.org/10.1016/j.carbon.2018.07.068.

Chapter 14

Metal-organic framework composites as electrocatalyst for carbon dioxide conversion

Khatereh Roohi, Jasper Coppen, Arjan Mol, and Peyman Taheri
Department of Materials Science and Engineering, Faculty of Mechanical Engineering, Delft University of Technology, Mekelweg, Delft, The Netherlands

Introduction

The escalating concentration of CO_2 in the atmosphere, rising from 280 ppm in the early 1800s to 405 ppm in recent years, poses a severe threat to global climate stability (Li, Wang, et al., 2020). This issue is mostly due to combustion of fossil fuels, causing global warming, ocean acidification, and the risk of catastrophic climate events. To mitigate the CO_2 impacts and prevent further global warming, it is crucial to reduce net anthropogenic CO_2 emissions to zero (Wang et al., 2021; Zhang et al., 2021). While fossil fuels are expected to remain one of the major energy sources in the foreseeable future, the development of technologies for carbon capture, utilization, conversion, and storage is essential. Among promising approaches, electrochemical CO_2 reduction reaction (CO_2RR) offers a pathway to convert CO_2 into fuels and value-added chemicals. Employing renewable energy resources, like wind and solar power, CO_2RR provides a simple way to close the carbon-neutral cycle, as shown in Fig. 14.1 (Kibria et al., 2019; Zhou et al., 2024).

Importance of the electrochemical carbon dioxide reduction reaction

Carbon dioxide or CO_2 can be converted into valuable products through various methods including chemical conversion, biological transformation, and photocatalytic and electrocatalytic reduction (Garg et al., 2020). Among these, the electrochemical CO_2RR stands out due to its:

- Compatibility with renewable energy sources like solar and wind power (Zheng et al., 2018),

FIGURE 14.1 Schematic of carbon-neutral cycle.

- Ability to produce valuable fuels and chemicals, such as carbon monoxide, formic acid, alcohols, and hydrocarbons (Zou & Wang, 2021),
- Simple, scalable process that can operate under mild conditions (Kumar et al., 2016),
- Highly manageable reaction steps, enabling the adjustment of electrochemical parameters with a relatively high level of efficiency in conversion (Long et al., 2019),
- Products are heavily utilized in petrochemical-based processes—for example, syngas as a building block—or as fuels, thus no new infrastructure is needed (Hernández et al., 2017).

The direct electroreduction of CO_2 is particularly promising as it allows for selective conversion into products like formate, ethanol, ethylene, and syngas, which can be integrated into existing carbon recycling systems and offer opportunities for renewable energy storage (Zhang et al., 2020). Despite the stability of the CO_2 molecule due to its linear and centrosymmetric structure, which typically requires harsh conditions for conversion (Zhang et al., 2017), advances in electrocatalysis have demonstrated the potential of CO_2RR as an environmentally compatible and economically feasible solution for reducing CO_2 emissions and utilizing CO_2 as a resource.

The development of CO_2RR is although, hindered due to a number of reasons, one, as mentioned, being the thermodynamic stability of the CO_2 molecule. The stable linear structure of CO_2, with its strong C=O bonds and zero dipole moment, requires significant overpotentials to break, making the reaction kinetically sluggish (Fan et al., 2018; Gong et al., 2023). Moreover, the competitive hydrogen evolution reaction (HER) leads to challenges in achieving high CO$_2$RR activity and selectivity (Goyal et al., 2020). Catalysts are crucial in activating CO$_2$ and determining the reaction pathways. Metal-based electrocatalysts have been extensively studied and classified based on the products they could produce, such as formate, CO, hydrocarbons, or HER. However, the interaction between CO$_2$ and catalyst surfaces influences the efficiency of the reaction, overcoming these kinetic barriers while achieving high selectivity at low overpotentials is essential (Choi et al., 2020). Despite advances, further research is needed to design catalysts with improved structure–property relationships, especially those that can operate efficiently through multiple conversion cycles.

Metal-organic framework composites for carbon dioxide reduction reaction

Metal-organic frameworks (MOFs) are crystalline, porous structures composed of inorganic metal nodes and organic linkers (Zhan et al., 2022). MOFs, owing to their high tunability, have attracted great attention in research in recent years, including heterogeneous catalysis (see Fig. 14.2) (Liang et al., 2018). Moreover, their distinct and beneficial characteristics—such as a high surface area, customizable pore structures, and controllable, facile synthesis—make them excellent candidates for catalysts in various fields, including CO$_2$RR (Furukawa et al., 2013; Rowsell & Yaghi, 2005).

However, the catalytic centers in pristine MOFs are typically confined to coordinatively unsaturated metal sites, which function as Lewis acid centers, and the active groups on the organic linkers, usually consisting of acid or base sites. This limitation restricts the range of catalytic reactions that MOFs can effectively perform (Jiao et al., 2018). The structure of MOFs can be modified in several ways to improve their performances as electrocatalysts. For instance, the inherent surface of MOFs can serve as an active support for loading functional guest components, such as metal/metal oxide nanoparticles

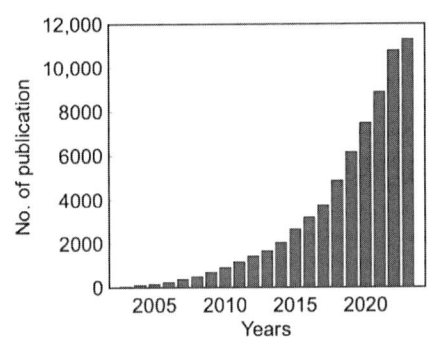

FIGURE 14.2 Number metal-organic framework-related research articles in the last 25 years using Scopus database (Adegoke et al., 2024).

(Singh et al., 2020). Additionally, given the presence of organic molecules in their structures, postsynthesis calcination of MOFs could generate high surface area functional materials, including doped and pure carbon-based nanostructures and metal/carbon (M/C) hybrids. Beyond creating functional materials directly from MOFs, various synthesis approaches enable the customization of MOFs on highly conductive substrates, such as carbon nanotubes (CNTs), hexagonally-ordered mesoporous carbon, and graphene (Vinoth et al., 2020). By effectively combining the benefits of both MOFs and functional materials, MOF composites exhibit abundant functions and significantly enhanced performance compared with the pristine MOFs, in many catalytic processes (Chen et al., 2024).

Scopes and objectives

This chapter aims to review the fundamentals of CO_2RR and explore the catalytic mechanism behind the reduction of CO_2. With a focus on MOF composites as electrocatalysts, recent advances on catalytic design and performance improvement will be discussed. The current state-of-the-art MOF composites including metal and metal oxide nanoparticle@MOFs, metal/metal oxide@pyrolyzed MOF, MOF@conductive substrates, MOF@carbon substrates, MOF@MOF, MOF@polymers, and polyoxometalates and MOF–MXene composites are discussed and their potential performance in CO_2RR are summarized. This chapter aims to highlight the benefits of MOF composites, such as flexible composition and structure, porosity, surface area, active single-site structure, and synthesis through design.

Electrocatalytic mechanism of carbon dioxide reduction reaction

Electrochemical CO_2RR is one of the most promising carbon capture and utilization strategies (Kou et al., 2022). This technology enables the production of various chemical feedstocks or materials for recycling within industrial plants, including carbon monoxide (CO), methane (CH_4), ethylene (C_2H_4), methanol (CH_3OH), ethanol (CH_3CH_2OH), and formic acid (HCOOH) (Sajna et al., 2023). Products, such as alcohols and formic acid can be sold on commodity markets, while CO or syngas can be recycled as carbon feedstock to lower production costs and CO_2 emissions (Xu & Carter, 2019). A notable application is in iron–steel mills, where CO_2 produced from blast furnaces could be electrochemically converted to CO using low-carbon energy sources (e.g., solar or wind) to further reduce iron ores in the blast furnace, helping achieve carbon neutrality (Li, Garg, et al., 2020). CO_2RR is a multielectron process that results in variety of products depending on the numbers of electrons and protons present in the reaction which are listed in Table 14.1 (Kortlever et al., 2015). The particular product or range of products varies based on the nature of the catalytic

TABLE 14.1 Standard redox potentials for carbon dioxide reduction related half reactions (Kortlever et al., 2015; Zhang, Hu, et al., 2018).

Half reactions	E° (V vs RHE)
$CO_2 + 2H^+ + 2e^- \rightarrow HCOOH$	−0.2 (for pH < 4)
$CO_2 + 2H^+ + 2e^- \rightarrow CO + H_2O$	−0.1
$CO_2 + 4H^+ + 4e^- \rightarrow HCHO + H_2O$	−0.07
$CO_2 + 6H^+ + 6e^- \rightarrow CH_3OH + H_2O$	0.02
$CO_2 + 8H^+ + 8e^- \rightarrow CH_4 + 2H_2O$	0.17
$CO_2 + 12H^+ + 12e^- \rightarrow CH_2CH_2 + 4H_2O$	0.08
$CO_2 + 12H^+ + 12e^- \rightarrow CH_3CH_2OH + 3H_2O$	0.09

material. For instance, 16 different products have been identified in the case of metallic copper as electrocatalyst (Gu et al., 2018).

Additionally, CO₂RR involves a complex mechanism with sluggish reaction kinetics. Given that CO₂ molecule has an extraordinarily stable linear structure, substantial negative potential is needed to turn adsorbed CO₂ molecules into *CO₂⁻, which may further be transformed into a variety of chemicals or intermediates (Jin et al., 2021). This step is usually considered to be the rate determining step (RDS) in CO₂ conversion reaction. Another key challenge of CO₂RR is the competitive HER, which as schematically shown in Fig. 14.3, occurs at the interface of the cathode/electrolyte in aqueous media together with CO₂RR. HER occupies catalytic active sites that could otherwise interact with CO₂, significantly reducing the efficiency of CO₂ reduction (Kou et al., 2022). Moreover, as a multielectron (2–12 electrons) transfer process, with multi intermediates, CO₂ thermodynamically could be reduced to a mixture of C₁, C₂, and C₃ products, being selective toward one specific product is challenging (Jin et al., 2021). To decrease the postreaction separation costs, selectivity is the key parameter.

Given these challenges, the main focus of CO₂RR research has shifted toward finding suitable electrocatalyst materials that can efficiently and selectively convert CO₂ into valuable chemicals. The ideal catalyst would be highly active and selective, optimizing the conversion rate and favoring a specific reaction pathway. Although noble metals like Au, Ag, and Pd exhibit strong selectivity for CO₂RR over HER, their limited availability and high cost make them impractical for industrial use (Lu et al., 2018). Therefore the consensus is that effective CO₂RR electrocatalysts should be affordable, offer abundant active sites, possess high electrical conductivity, preferentially drive CO₂RR over HER, and have a large surface area, while ideally producing a single CO₂RR product.

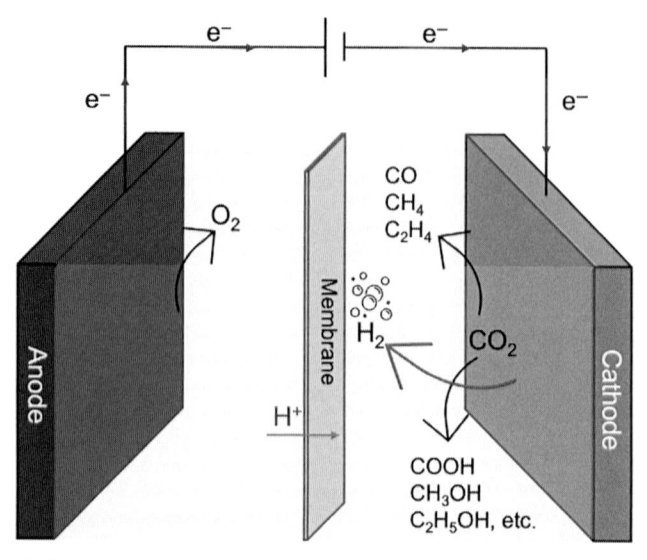

FIGURE 14.3 Schematic of a carbon dioxide reduction reaction electrolyzer (Kou et al., 2022).

Electrochatalyst materials for carbon dioxide reduction reaction

In analyzing the performance of electrocatalysts for electrochemical CO_2 reduction, key parameters include Faraday efficiency (FE), overpotential, and current density. FE quantifies the selectivity of the catalyst for producing specific products using the following equation:

$$FE = n_i\ NF/jt$$

where n_i represents the number of moles of the detected product, N is the number of the transferred electrons in a reaction, F is the Faraday constant, j is the current density, and t is the time of the measurement. Overpotential reflects the difference between the applied and required thermodynamic potential to initiate the reaction. Additionally, current density, representing the production rate normalized by the electrochemical active surface area. An ideal catalyst should exhibit high current density and low overpotential to reduce the cost of scaling up the process.

Based on Hori's research (Hori et al., 1994), in which the catalytic CO_2RR for the first time was investigated, transition metal (TM) catalysts are classified into four groups:

- TMs producing CO,
- Metals producing formate,
- Metals with minimal activity, and
- Copper, which generates a range of products.

The performance of these catalysts is influenced by their electronic structural properties and surface-active sites. Further, progress in electrochemical CO_2 reduction (ECR) research has shifted from material properties to structural effects on catalyst performance. The geometry of the metal catalyst in terms of the size and the number of active centers, seems to be crucial to improving the efficiency of the process and blocking HER to promote CO_2RR (Cai et al., 2018). Nanostructured electrocatalysts demonstrate enhanced activity, stability, and selectivity compared to bulk materials due to their increased surface-active sites. Reducing the particle size, particularly for metals like copper, gold, and silver, improves current density and selectivity (Tekalgne et al., 2020; Xie et al., 2018). Moreover, while TMs have shown high FE for CO_2RR, their high electron mobility under reaction conditions along with low long-term stability and high price, hamper their industrial utilization. Thus TM oxide-derived catalysts are explored by increasing surface roughness and/or providing subsurface oxygen species to improve long-term stability and increase charge transfer capabilities, respectively (Tayyebi et al., 2018). For instance, it has been shown that the oxygen vacancy in Cu_2O crystals, is the active site for C_2H_4 formation, indicating that oxide catalysts could be useful for selective production of C_2-plus (C_{2+}) products (Handoko et al., 2016). Furthermore, nonmetallic catalysts have also been reported in literature, such as MoS_2 showing comparable efficiency in both HER and CO_2RR. The molybdenum-terminated edge of MoS_2 exhibits high activity in reducing CO_2 to CO, especially in an ionic liquid-modified electrolyte (Lu & Jiao, 2016). As an example, Asadi et al. (2014) demonstrated that bulk Mo_2S can achieve equal or superior CO_2 reduction performance compared to state-of-the-art metallic catalysts. Density functional theory (DFT) calculations revealed that Mo edge atoms, with their d-orbital electron states near the Fermi level, are responsible for the high catalytic activity, though further experimental studies are needed to validate these findings. Recent advancements in nanostructure synthesis and the manipulation of compositional and structural properties have expanded the capabilities of metallic and metal oxide nanomaterials, paving the way for more efficient and selective CO_2RR catalysts.

Mechanistic study of carbon dioxide reduction reaction

Over the years, both experimental and theoretical methods have been employed to propose mechanisms for the reduction of CO_2 in various products. These methods include theoretical DFT as well as several in situ techniques, such as Raman spectroscopy, Fourier-transform infrared spectroscopy (FT-IR), and X-ray photoelectron spectroscopy (XPS), all of which help identify the intermediate species formed during the reduction process (Lu & Jiao, 2016). CO_2 is an amphoteric molecule with a Lewis acidic carbon center and basic oxygen atoms, making it more effective as an electron acceptor than a donor.

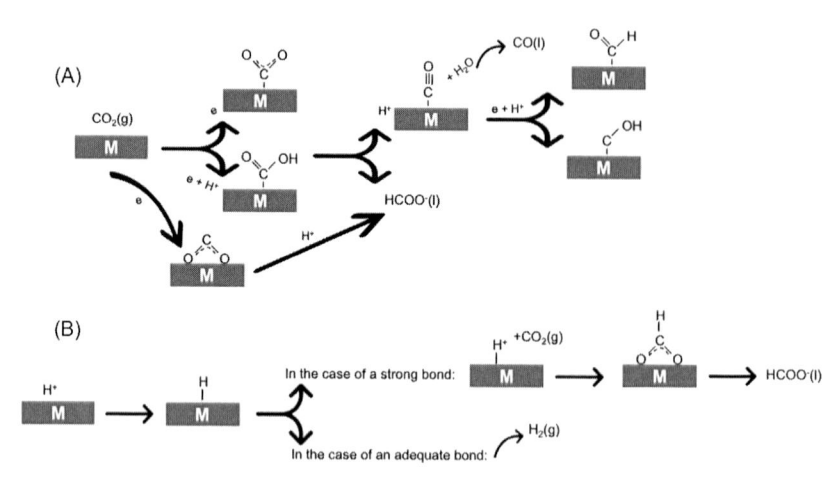

FIGURE 14.4 Plausible pathways of carbon dioxide reduction reaction (M = metal). (A) Pathways to carbon monoxide and formic acid. (B) Pathway of carbon dioxide reduction to formate and hydrogen evolution reaction in the case of H⁺ adsorption on the metal surface (Saha et al., 2022). http://doi.org/10.1021/acs.accounts.1c00678, http://pubs.acs.org/journal/achre4

This electrophilic nature of carbon dominates in CO_2RR, where most reduction pathways involve either the binding of CO_2 to the metal center (see Fig. 14.4A) of the catalyst or insertion into a metal–hydride bond (see Fig. 14.4B) (Saha et al., 2022). C_2 production pathway, the critical step for forming the C–C bond involves CO dimerization, resulting in the formation of a *OCCO– intermediate. Proton transfer occurs after the negatively charged CO dimer has been adsorbed, which dimerization considered to be the RDS in the C_2 production process (Kortlever et al., 2015).

Metal-organic framework composites as electrocatalysts

A composite is a solid material made up of two or more distinct substances, each maintaining its own characteristics while enhancing the overall properties of the combined material. MOFs are a class of highly porous materials consisting of inorganic ions or clusters linked to organic ligands. These structures form a framework with an extraordinarily large surface area, reaching thousands of square meters per gram of material (Furukawa et al., 2013). Although an important research stream on MOFs focuses on synthesizing novel two/three dimensional structure using innovative approaches, a new research trend investigates the combination of MOFs with a variety of nano- and microparticles known as MOF composites (Yu et al., 2017). MOF composites are combination of functional materials, such as carbon substrates, nanoparticles, polymers, and enzymes with MOF which have further improved physico-chemical properties like mechanical stability, conductivity, and catalytic performance. The reported MOF composites through the years are:

- Metal and metal oxide nanoparticles (Falcaro et al., 2016),
- Polyoxometalates (POMs) and polymers (Kalaj et al., 2020),
- Graphene and CNTs (Yan et al., 2020),
- Enzymes and bio molecules (Lian et al., 2017),
- Quantum dots (Wu, Liu, et al., 2020).

In the context of CO_2RR electrocatalysis, to compensate for the limited electrical conductivity of pure MOFs, MOF composites are utilized. In most cases, pristine MOFs are employed as supporting materials for a variety of species, including metal/metal oxide/metal complexes nano/microparticles. This is mostly due to high porosity of MOFs which leads to maintaining higher active surface area for the reaction. Although many MOFs have been directly used as a precursor to derive catalysts for CO_2RR (Deng et al., 2010; Liu et al., 2023).

Recent advances on metal-organic framework composites for electrocatalytic carbon dioxide reduction reaction

In this section, different classes of MOF composites will be discussed, with a focus on their applications in CO_2RR and their performance as electrocatalysts.

Pristine metal-organic frameworks

It is well known that downsizing metal nanostructures into atomically distributed centers, known as single atom catalysis, is an effective approach for metal catalysts (Li, Wang, et al., 2020). MOFs, due to their metal atomic centers, are categorized as single-atom catalysts for CO_2RR. MOFs tend to produce lighter products like CO and formic acid because their complex structure and single-atom dispersion hinder C–C coupling in most cases, limiting the formation of heavier C_{2+} products. Moreover, the coordination environment of the metal active sites, influenced by MOF linkers, plays a critical role in catalytic selectivity (Kornienko et al., 2015). By altering the coordination environment using different linkers, MOFs can break the linear scaling relationship existing between key intermediates, making it possible to tune selectivity toward specific products. Thus tunable coordination environment of MOFs offers a unique advantage for controlling product formation in CO_2RR (Ko et al., 2018). However, despite their advantages in improving the selectivity, the use of MOFs in electrocatalyst applications has been hindered by poor conductivity, therefore low activity (Downes & Marinescu, 2017). This issue arises mainly from the presence of carboxylate linkers in the multidimensional frameworks. The high electronegativity of the oxygen atoms in these carboxylates increases the potential barrier for electron transfer through the organic linkers (Sun et al., 2016). As a result, the overlap between oxygen atoms and metal d-orbitals is reduced, causing the low conductivity typically seen in most MOFs (Xie et al., 2020). This limitation leads to the production of different MOF composites to tune the conductivity and electrochemical activity.

Metal-organic frameworks as supports

This group of MOF composites has been extensively researched for CO_2RR, utilizing materials in which the MOF acts as a porous support with active catalytic sites, such as metal and metal oxide nanoparticles (MNPs) deposited on its surface (Zhao et al., 2018). The incorporation of MOFs, improves the electrocatalytic performance of MNPS due to their porous structures and large specific surface area. Cu_2O particles, known for their tunable crystal facets and relatively stable bulk-phase structure, are considered promising candidates for studying the mechanism of CO_2 reduction. However, significant hydrogen production during the electrocatalytic process often results in a reduced FE for hydrocarbons. To improve the overall hydrocarbon FE, it is crucial to enhance the CO_2 adsorption capacity of Cu_2O. The Cu_2O@Cu-MOF catalyst, when used as an electrocatalyst for the electrochemical reduction of CO_2, demonstrates exceptional performance, achieving a total FE for hydrocarbons (CH_4 and C_2H_4) of 79.4%, with a notable FE of 63.2% for CH_4 (Tan et al., 2019). The multifunctionality of the synthesized Cu_2O@Cu-MOF, encompassing adsorption, activation, and catalysis, stems from the synergistic interaction between Cu-MOF and Cu_2O.

In addition, metallic nanoparticles are recognized as suitable catalysts for CO_2RRs; however, their tendency to agglomerate diminishes their catalytic efficiency (Liu et al., 2023). MOFs, with their high surface area and notable porosity, can prevent this agglomeration and enhance the catalytic performance of NPs encapsulated within them. Silver is one of the most extensively studied metals for ECR, producing CO as the main product due to weak CO adsorption on its surface (Zhang, Jin, et al., 2018). Both CO_2 electroreduction to CO and the competing HER take place concurrently on Ag NPs. In this regard, Jiang et al. (2017) investigated the Ag/ZIF-7 MOF composite as a CO_2 electroreduction catalyst and higher CO FE and current density was observed (80.5% and 26.2 mA/cm^2 at -1.2 V) compared to its pristine components. This improved performance is attributed to the large accessible surface area and the synergistic interaction between Ag nanoparticles and ZIF-7 MOF.

As another example of MOF supports, in a study by Yi et al. (2020) uniform $Cu_2O(111)$ NPs were fabricated on a conductive CuHHTP (HHTP= 2,3,6,7,10,11-hexahydroxytriphenylene) framework via electrochemical treatment, resulting in conversion of some Cu^{2+} centers within MOF into 3.5 nm Cu_2O NPs with exposed (111) crystal planes. The Cu_2O@CuHHTP composite demonstrated high conductivity and excellent performance in CO_2 electroreduction to CH_4, achieving 73% FE and 10.8 mA/cm^2 partial current density. The small size of the MNPs, exposed crystal planes, and released hydroxyl groups (OH$^-$) from the HHTP ligand contributed to the high CH_4 selectivity, while the CuHHTP support enhanced electron transfer for better current density. Furthermore, in a more recent study zirconium-based PCN-222 MOFs with metalloporphyrin Cu centers and incorporated gold nanoneedles were successfully synthesized using ligand carboxylates as reducing agents. Compared to similar MOF components, AuNN@PCN-222(Cu) exhibited a significantly enhanced ethylene production with a FE of 52.5% (Xie et al., 2022) (Fig. 14.5A–D). Some of the other metal/metal oxide–MOF composites and their CO_2RR performance are listed in Table 14.2.

Metal/metal oxide MOF composites could be synthesized with various methods, such as liquid impregnation. In this method, desolvated MOFs are soaked in metal precursor solutions and then reduced to form MNPs within the pores. Incipient wetness impregnation involves using a solution with a volume equal to the MOF's pore volume for precise control over metal loading. Gas-phase infiltration (MOCVD) is another method which exposes desolvated MOFs to volatile metal precursors under vacuum, followed by reduction or decomposition to produce MNPs. The double solvent method employs both hydrophilic and hydrophobic solvents to minimize MNP aggregation on the MOF surface. Redox-active MOFs use the inherent redox properties of MOF to reduce metal ions into nanoparticles without additional reducing agents. Photochemical reduction leverages light to facilitate MNP formation within MOFs, while colloidal deposition and urea precipitation methods, though effective, often result in surface aggregation of MNPs (Zhu & Xu, 2014). One method of preparing metal-based MOF-derived composites is pyrolysis or carbonization of the pristine MOF.

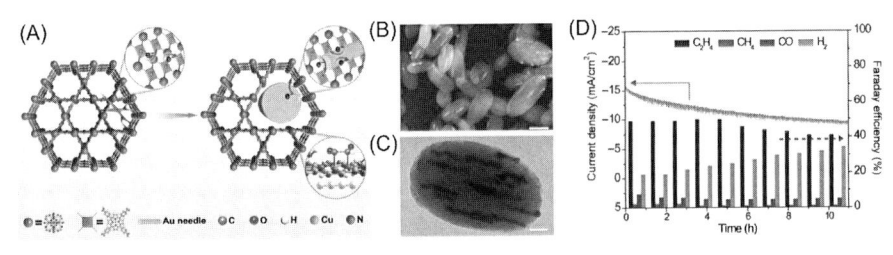

FIGURE 14.5 (A) Incorporation of Au nanoneedles into PCN-222(Cu) to alter the charge transfer and pathway of the carbon dioxide reduction reaction; (B) Scanning electron microscope (SEM) image of the composite (scale bar: 250 nm); (C) Transmission electron microscopy (TEM) image of the composite (scale bar: 50 nm); (D) Chronoamperometry stability tests for AuNN@PCN-222(Cu) at −1.2 V vs. RHE.

TABLE 14.2 List of studied metal/metal oxide–metal-organic framework composites and their performance for carbon dioxide reduction reaction (Adegoke et al. (2024), Liu et al. (2023), Huang et al. (2023).

MOF composite	Electrolyte	Potential	Product(s)	FE (%)	Reference
Ag/Co-MOF	0.1M KHCO$_3$	−1.8 V versus SCE	CO	55.6	Zhang, Jin, et al. (2018)
CuO/Cu-MOF	0.1M KHCO$_3$	−1.1 V	C$_2$H$_4$	50	Skoulidas (2004)
Ag@Al-PMOF	0.1M KHCO$_3$	−1.1 V	CO	55.8 ± 2.8	Guntern et al. (2019)
Cu$_2$O/Cu-MOF	0.1M KHCO$_3$	−1.71 V versus RHE	CH$_4$	63.2	Tan et al. (2019)
Cu$_2$O/Cu-MOF	0.1M KHCO$_3$	−1.71 V versus RHE	CH$_4$	63.2	Tan et al. (2019)
Bi$_2$O$_3$/2D-Zr-TATB (triaminotrinitrobenzene)-MOF	0.5M KHCO$_3$	−0.97 V versus RHE	HCOOH	≈85	Liu et al. (2022)
SnO$_2$/2D-Zr-TATB-MOF	0.5M KHCO$_3$	−0.97 V versus RHE	HCOOH	≈35	Liu et al. (2022)
In$_2$O$_3$/2D-Zr-TATB-MOF	0.5M KHCO$_3$	−0.97 V versus RHE	HCOOH	≈45	Liu et al. (2022)

FE, Faraday efficiency; *MOF*, metal-organic framework; *2D*, two-dimensional.

Carbonization is a process frequently used to produce carbon rods and other carbon nano-sized materials. In this process, by increasing the temperature, bonds between carbon and other elements (i.e., O, N, S, and TMs) in the organic building unit within MOF structure, break. These elements leave the material, allowing carbon atoms to bond with each other, resulting in a framework composed of pure carbon (Wang et al., 2011). Metallic centers, on the other hand, do not evaporate, resulting in a TM-decorated porous carbon framework. Since pure carbon structures have much higher electrical conductivity than organic linkers, the metal-doped nanoporous carbon exhibits higher conductivity than pristine MOF. Additionally, nanoporous carbons are known for their chemical stability (Bhadra et al., 2019). To ensure a desired structure, proper carbonization settings are crucial. Factors, such as ramp-up temperature speed, carbonization temperature, process duration, and inert gas flow influence the final product, among which, the holding temperature is the most significant (Abdelkader-Fernández et al., 2020).

Fig. 14.6 represents the scanning electron microscopic images of carbonized HKUST-1 MOF known as Cu-BTC, at the holding temperature of 650°C. As can be seen, the copper clusters form on the surface of the carbon framework with similar morphology as the pristine MOF crystals. These clusters act now as catalytic sites whereas the carbon framework in the substrate helps with better charge transfer resulting in higher current density than the pristine MOF. In this regard, Zhao et al. (2017) carbonized Cu-BTC at 900°C, 1000°C, and 1100°C. The results showed that the 1000°C sample has an FE of about 45% for CH_3OH and 27 % for C_2H_5OH at a potential of −0.3 V versus standard hydrogen electrode. The electrode was made of an ink containing active material, high purity water, and Nafion deposited on commercial carbon paper using a spin coating technique. A two-compartment electrolysis cell was filled with 0.1 M $KHCO_3$ electrolyte and equipped with a Nafion membrane for ion transport. They also found that with increasing

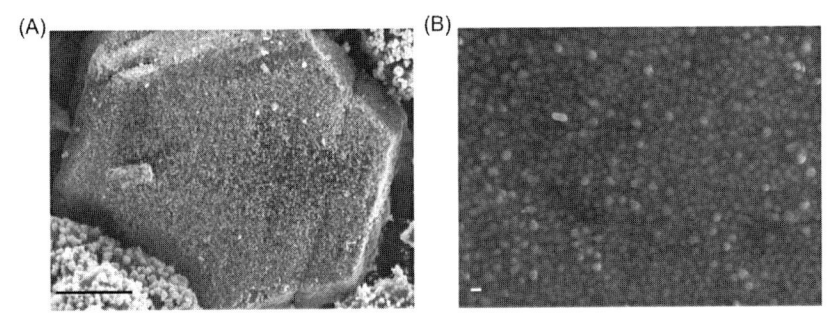

FIGURE 14.6 SEM figures of carbonized Cu-BTC at 650°C in different magnifications. (A) 10 μm and (B) 100 nm. Formation of copper clusters smaller than 100 nm on the surface of metal-organic framework is visible (results from our group) (Zhao et al., 2017).

temperature, the pore size and thus the specific surface area decreases. Nevertheless, the 1000°C sample showed better performance for CO_2RR than the 900°C one. This was attributed to the lower charge transfer resistance of the 1000°C sample, showing that the specific surface area is not the only important factor to consider for a CO_2RR catalyst when considering the charge transfer parameter.

Among the electrocatalytic materials produced by MOF template carbonization for CO_2RR, ZIF-derived carbons are predominant, with those derived from ZIF-8 being particularly noteworthy. In studies on carbonized MOFs, since zinc is not an electrocatalytically active metal for CO_2RR, it is removed during high-temperature carbonization. As a result, various strategies have been developed to incorporate active TM into the structure. Consequently, several studies have combined Fe with ZIF-8 to produce Fe-containing N-doped carbon electrocatalysts for CO_2-to-CO conversion (Abdelkader-Fernández et al., 2020). This material achieved 97% FE for CO at −0.56 V versus RHE, whereas its Fe-free counterpart reached 100% FE(C)O but required a higher potential of versus 0.86 V versus RHE.

Metal-organic frameworks on conductive substrates

One of the limitations of MOFs is their low electrical conductivity, which restricts their electrochemical applications. Graphene which is a one atom thick carbon lattice with a cellular structure, offers exceptional electronic conductivity and a large surface area (Peng et al., 2022). These properties enhance interactions with MOFs, making them an ideal anchor material with significant potential for improving the performance of MOF-based electrocatalysts (Li & Huo, 2015). Moreover, the two-dimensional (2D) nonmetallic semiconductor g-C_3N_4 has garnered significant attention in recent scientific research. g-C_3N_4 possesses remarkable characteristics, such as excellent physicochemical properties, an ideal band gap, a distinctive optoelectronic structure, high stability in air (up to 600°C), low cost, and abundant availability in nature. Although this specific composite has mostly been utilized as a photo(electro)catalyst in CO_2RR due to its small bandgap (Usman et al., 2022). In addition to these materials, CNTs are emerging as another highly promising candidate for functional applications. CNTs are highly structured carbon allotropes with a large aspect ratio. They exist in two primary forms: single-walled CNTs, with diameters ranging from 0.4 to 2 nm, and multiwalled CNTs, with diameters spanning 2–100 nm (Kempahanumakkagari et al., 2018). Both types are characterized by exceptional tensile strength, ultra-lightweight nature, and remarkable chemical and thermal stability. One example of this composite is [PCN-222(Fe)/CNTs], synthesized via a solvothermal method by incorporating iron porphyrin-centered PCN-222(Fe) onto CNTs. This composite exhibited excellent performance in CO_2 electroreduction, achieving a FE of 90% for CO production at an overpotential of –0.6 V, at the optimal ratio of m(Fe-TCPP):m (CNTs)=1:30. The enhanced performance was attributed to the conductive support provided by CNTs, which increased electrical conductivity and improved the overall structure of the MOF composite (Xu et al., 2022).

In another example, a ReL(CO)$_3$Cl (L=2,2-bipyridine-5,5-dicarboxylic acid) thin film was grown on the surface of a fluorine-doped tin oxide (FTO) conductive glass electrode, exhibiting 93% ± 5% Faradaic efficiency for CO production, which is the highest reported so far and exceeds those reported for covalent organic frameworks thin films (Ye et al., 2016). In the context of MOF–FTO composites, in a study by Kornienko et al. (2015), examination of a cobalt–porphyrin MOF, Al$_2$(OH)$_2$TCPP-Co (TCPP-H$_2$—4,4′,4″,4‴-(porphyrin-5,10,15,20-tetrayl)tetrabenzoate) revealed a selectivity for CO production in excess of 76% and stability over 7 hours (Fig. 14.7A and B).

Metal-organic frameworks on porous materials

MOFs could be also incorporated inside other porous structures, such as porous carbon, Ni foam, and in the form of core shell with other MOFs (Liu et al., 2021). Considering the inherent diversity of MOFs, combining different MOFs offers significant potential for advancements in synthesis, structural design, and performance applications (Xue et al., 2019). Building MOFs@MOFs is not only of interest for structural research but also an efficient approach to boost application performance by harnessing the synergistic properties of two distinct MOFs. In a 2020 review study by Wu, Liu et al., it is well discussed that for selective catalysis, many MOFs@MOFs catalysts are typically designed with active metal nanoparticle-loaded MOFs as core materials and MOFs with tailored pore sizes serving as the filtering shell (Wu, Liu, et al., 2020). While limited research has focused on creating selective MOFs@MOFs catalysts electrochemical CO$_2$RR, this approach holds significant potential for further development. An example of MOF@MOF composites is CuBi-based MOFs (HKUST-1 and CAU-17) supported in gas diffusion electrodes showed the FE of 28.3% for production of ethanol. The findings initially revealed that incorporating bismuth as a cocatalyst reduced the required overpotential, indicating a synergistic interaction between copper and bismuth in the CO$_2$ reduction pathway (Albo et al., 2019). A similar case applies to the incorporation of MOFs into porous carbon materials, such as carbon black. In a study by Raut et al., it was demonstrated that MOF-5 on carbon black supports exhibits significantly higher conductivity compared to either pristine MOF-5 or carbon black alone in alkaline solutions, which could be advantageous for CO$_2$RR (Raut et al., 2021). Another example of a MOF–porous carbon composite is a recent study by Sathiyan et al. (2023), which demonstrates the high efficiency of a composite of HKUST-1 and activated carbon as an electrocatalyst for CO$_2$ reduction. In this study, HKUST-1 NPs were embedded into a conductive porous carbon, creating strong π–π interactions between the aromatic linkers of

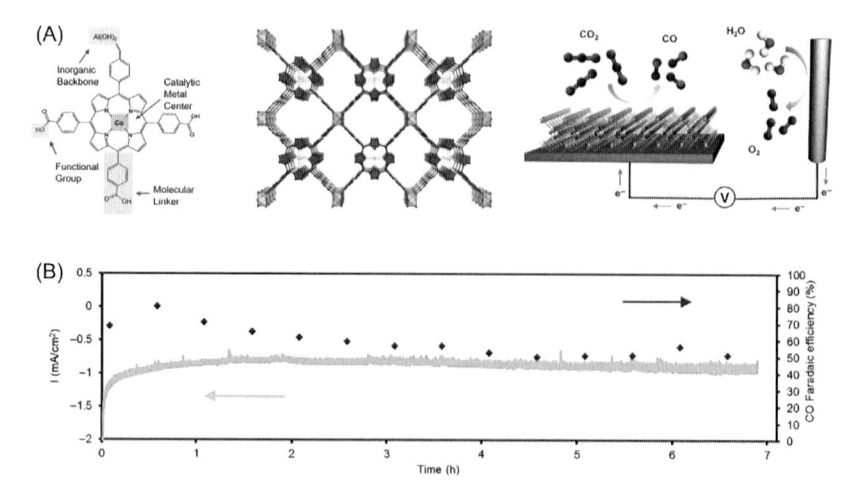

FIGURE 14.7 (A) The organic building units, in the form of cobalt-metalated TCPP, are assembled into a 3D metal-organic framework, Al$_2$(OH)$_2$TCPP-Co, with variable inorganic building blocks. In the structure (middle), each carboxylate is bound to the aluminum inorganic backbone. The metal-organic framework is integrated with a conductive substrate to form a functional carbon dioxide electrochemical reduction system (right). (B) Stability of the metal-organic framework catalyst evaluated by chronoamperometric measurements and faradaic efficiency (Kornienko et al., 2015).

MOF and the graphitic carbon, leading to a lateral conductivity of 17.2 S/m. The composite demonstrated a high electroactive surface coverage of 155 nmol/cm² and exhibited excellent catalytic performance under CO$_2$-saturated conditions. The onset potential for CO$_2$ reduction was –0.31 V versus RHE, with a high reduction current density of –18 mA/cm² at –1.0 V. Additionally, the composite remained stable over 12 hours of operation, with formic acid identified as the main product.

Metal-organic framework-polymer/polyoxometalate composites

The commercialization of MOFs has been hindered by their crystalline or microcrystalline forms (e.g., powders), which limit their integration into various technologies (Chen et al., 2021). To address this, efforts have been made to develop processable materials that combine MOFs with polymers, enhancing their practical utility. These materials are being developed using two approaches: a top-down method where MOFs are synthesized first and then integrated into polymers, and a bottom-up method where hybrid materials are formed concurrently with MOF synthesis. Common polymers used in the synthesis of MOF/polymer composites are shown in Fig. 14.8. The use of MOF/polymer composites in CO$_2$RR requires more scientific attention. In a

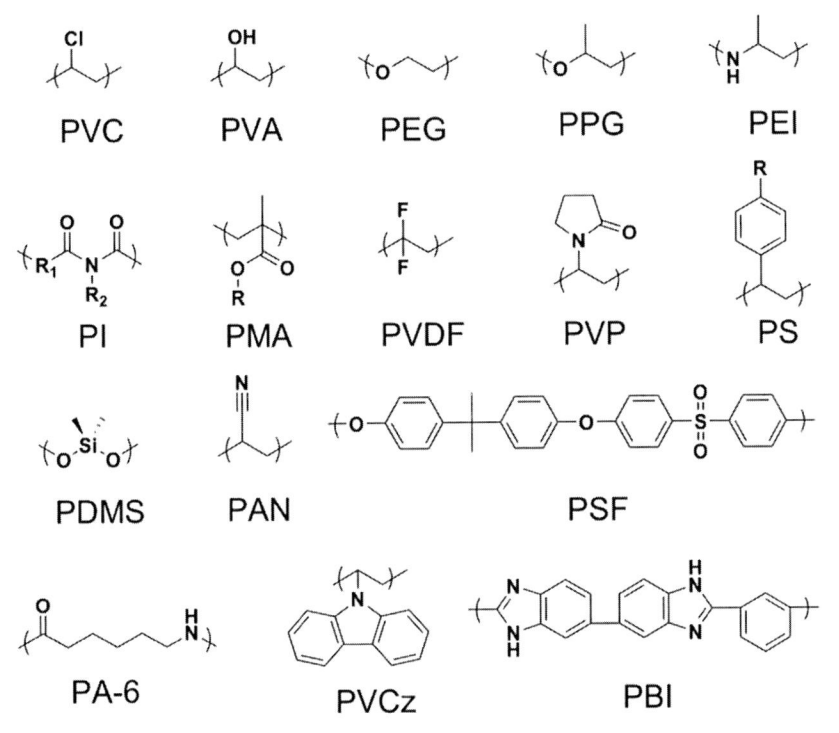

FIGURE 14.8 Common polymers used in the synthesis of metal-organic framework-polymer hybrid materials (Kalaj et al., 2020).

2020 study, Yang et al. used a Ni–MOF–polymer composite as a precursor for pyrolysis to prepare a catalyst consisting of highly dispersed Ni–N$_x$ species. The study demonstrated that incorporating a polymer within the MOF pores helps stabilize the collapsible MOF structure and prevents nickel aggregation during pyrolysis. This process results in the formation of single-atom nickel species within nitrogen-doped carbon, significantly enhancing activity, CO selectivity, and stability (Deeraj et al., 2023; Yang et al., 2020).

POMs are nanoscale, discrete metal-oxygen clusters composed of earth-abundant metals and exhibit unique structures. Often referred to as electronic sponges (Liu et al., 2021). POMs can hold a large number of electrons and engage in reversible, stepwise multielectron transfer processes while maintaining their structural integrity. Incorporating POMs into MOF channels can shorten the electron transport distance and improve the efficiency of electron transfer from the electrode to the active sites of the electrocatalyst, thereby enhancing the effectiveness of CO$_2$RR (Du et al., 2022; Li et al., 2019). Research into POMs and MOFs is currently gaining momentum due to their unique structural features, catalytic potential, and other inherent properties. For instance, due to their high electron mobility, PCN-222(Co) has increasingly been selected as a supportive framework for Co-metalloporphyrins, which consist of active single cobalt metal sites and conjugated π-electron porphyrin planes (Liu et al., 2023).

As an example, a directional electron-transfer channel was successfully constructed at a molecular level by synthesizing mixed-valence POM@MOFs composites through postmodification. The combination of POMs and catalytic single-metal site Co in the porphyrin-based MOF resulted in a composite with enhanced electron-transfer efficiency. Catalytic CO_2RR using H-POM@PCN-222(Co) exhibited a high FE for CO (96.2%) and satisfying stability for over 10 hours. DFT calculations confirmed that POM introduction accelerated multielectron transfer, enriching the electron density at the Co center and reducing RDS energy in the CO_2RR process (Sun et al., 2021).

Furthermore, assembling reductive POMs and metaloporphyrins to construct MOFs is shown to be promising candidates to enhance the efficiency and selectivity of CO_2RR. POM-metalloporphyrin organic frameworks (PMOFs) facilitate efficient electron transfer in electrocatalytic CO_2 reduction due to the direct communication between the POM unit and metalloporphyrin. As reported in a novel study by Wang et al. (2018), Co-PMOF stands out, selectively converting CO_2 to CO with a remarkable 99% FE, at –0.8 V (Fig. 14.9), and excellent stability for over 36 hours. Moreover, DFT calculations highlight the superior performance of Co-PMOF and the synergistic effect between reductive POM and Co-porphyrin.

Metal-organic framework/MXene composites

MOFs offer the combined benefits of both heterogeneous and homogeneous catalysts. Integrating MOFs with conductive 2D MXene nanosheets can further enhance their electrocatalytic efficiency. The MOF/MXene composites exhibit superior performance due to their increased surface area, greater number of accessible active sites, enhanced catalytic activity, and improved stability (Xu et al., 2023). An example of this is Fe–Nx/N/Ti_3C_2, which was synthesized by pyrolyzing MOFs supported on Ti_3C_2 MXene. The negative charge of Ti_3C_2 MXene allows MOFs to anchor firmly on its surface, preventing aggregation, while its metallic conductivity enhances electron transfer. The strong interaction between Fe–N_x–C and Ti_3C_2 MXene also helps regulate the catalytic activity (Gu et al., 2022). However, there are limited studies on the use of MOF/MXene composites for CO_2RR; nevertheless, this example highlights the high potential of these materials.

Conclusion and outlook

MOF composites, due to their unique structural features, are increasingly being developed to replace conventional electrocatalysts for converting CO_2 into

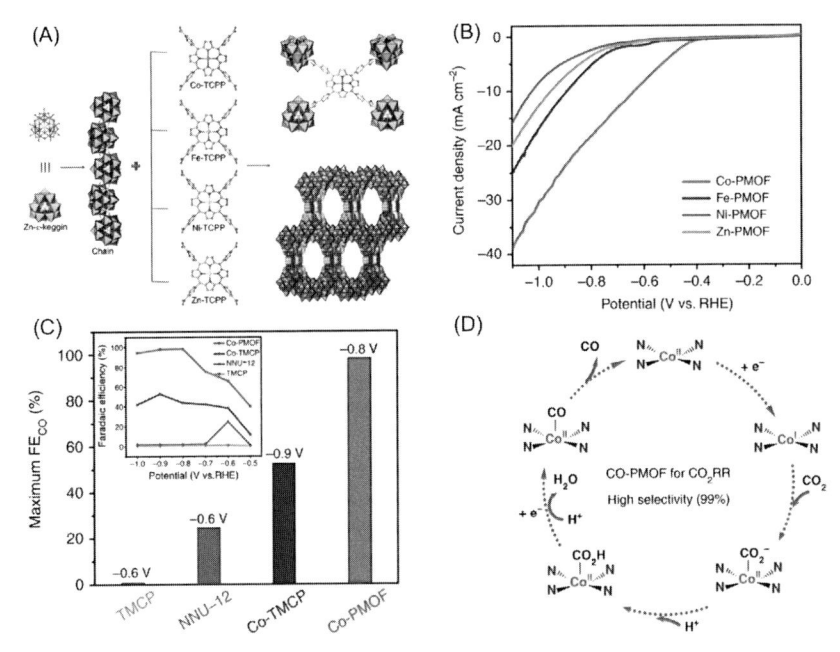

FIGURE 14.9 (A) Schematic illustration of the structures of M-polyoxometalate-metalloporphyrin organic frameworks (PMOFs) (M=Co, Fe, Ni, Zn). Linker=TCPP and zigzag polyoxometalate chains, (B) Linear sweep voltammetric curves showing electrocatalytic performances of different composites, (C) Maximum FECO of different composites, and (D) Proposed mechanistic scheme for the carbon dioxide reduction reaction on Co-PMOF (Wang et al., 2018).

valuable products. While the application of MOF composites for CO_2RR has shown promise, there are not many studies on this topic. In this regard, a better understanding of the performance of MOF composites as electrocatalysts will help in designing more efficient compounds. Moreover, most CO_2RR studies using MOF composites have only been conducted in laboratories, and scaling this process to an industrial level is crucial for addressing atmospheric CO_2. Notably, MOF composite electrocatalysts are considered cost-effective catalyst materials. Thus further studies are essential to refine these materials for ECR. Current research indicates that MOF composites have significantly improved electrocatalytic performance for CO_2RR compared to conventional catalyst materials. Despite advancements, their use in this area is still in the early stages. Practical applications are hindered by challenges, such as limited recycling and stability under certain reaction conditions. Therefore, the development of MOF composites with more stable structures is necessary for broader applications.

Additionally, while many methods for synthesizing MOF composites have been reported, each comes with limitations. It is essential to explore new, versatile approaches to constructing functional MOF materials, particularly for controlling guest species within MOF matrices. Significant attention has also been given to MOFs that are responsive to environmental stimuli, opening up the possibility of creating smart, responsive electrocatalysts. Further research will clarify the relationship between the structure and function of MOF composites and aid in the design of innovative materials for CO_2RR and other energy conversion applications.

In summary, addressing the current challenges through more in-depth research will pave the way for the development of low-cost, high-performance MOF composites. This progress will ultimately enable benchmark performance for ECR in practical renewable energy applications.

References

Abdelkader-Fernández, V. K., Fernandes, D. M., & Freire, C. (2020). Carbon-based electrocatalysts for CO_2 electroreduction produced via MOF, biomass, and other precursors carbonization: A review. *Journal of CO_2 Utilization, 42*, 101350. https://doi.org/10.1016/j.jcou.2020.101350, http://www.journals.elsevier.com/journal-of-co2-utilization/.

Adegoke Kayode, A., et al. (2024). Metal-organic framework composites for electrochemical CO_2 reduction reaction. Separation and Purification Technology 341, 126532.

Albo, J., Perfecto-Irigaray, M., Beobide, G., & Irabien, A. (2019). Cu/Bi metal-organic framework-based systems for an enhanced electrochemical transformation of CO_2 to alcohols. *Journal of CO_2 Utilization, 33*, 157–165. https://doi.org/10.1016/j.jcou.2019.05.025.

Asadi, M., Kumar, B., Behranginia, A., Rosen, B. A., Baskin, A., Repnin, N., Pisasale, D., Phillips, P., Zhu, W., Haasch, R., Klie, R. F., Král, P., Abiade, J., & Salehi-Khojin, A. (2014). Robust carbon dioxide reduction on molybdenum disulphide edges. *Nature Communications, 5*. https://doi.org/10.1038/ncomms5470, http://www.nature.com/ncomms/index.html.

Bhadra, B. N., Vinu, A., Serre, C., & Jhung, S. H. (2019). MOF-derived carbonaceous materials enriched with nitrogen: Preparation and applications in adsorption and catalysis. *Materials Today, 25*, 88–111. https://doi.org/10.1016/j.mattod.2018.10.016, http://www.journals.elsevier.com/materials-today/.

Cai, Z., Wu, Y., Wu, Z., Yin, L., Weng, Z., Zhong, Y., Xu, W., Sun, X., & Wang, H. (2018). Unlocking bifunctional electrocatalytic activity for CO_2 reduction reaction by win-win metal-oxide cooperation. *ACS Energy Letters, 3*(11), 2816–2822. https://doi.org/10.1021/acsenergylett.8b01767, http://pubs.acs.org/journal/aelccp.

Chen, D., Zheng, Y.-T., Huang, N.-Y., & Xu, Q. (2024). Metal-organic framework composites for photocatalysis. *EnergyChem, 6*(1), 100115. https://doi.org/10.1016/j.enchem.2023.100115.

Chen, Z., Wasson, M. C., Drout, R. J., Robison, L., Idrees, K. B., Knapp, J. G., Son, F. A., Zhang, X., Hierse, W., Kühn, C., Marx, S., Hernandez, B., & Farha, O. K. (2021). The state of the field: From inception to commercialization of metal-organic frameworks. *Faraday Discussions, 225*, 9–69. https://doi.org/10.1039/d0fd00103a, http://pubs.rsc.org/en/journals/journal/fd.

Choi, W., Won, D. H., & Hwang, Y. J. (2020). Catalyst design strategies for stable electrochemical CO_2 reduction reaction. *Journal of Materials Chemistry A, 8*(31), 15341–15357. https://doi.org/10.1039/d0ta02633f, http://pubs.rsc.org/en/journals/journal/ta.

Deeraj, B. D. S., Jayan, J. S., Raman, A., Asok, A., Paul, R., Saritha, A., & Joseph, K. (2023). A comprehensive review of recent developments in metal-organic framework/polymer composites and their applications. *Surfaces and Interfaces, 43*, 103574. https://doi.org/10.1016/j.surfin.2023.103574.

Deng, H., Doonan, C. J., Furukawa, H., Ferreira, R. B., Towne, J., Knobler, C. B., Wang, B., & Yaghi, O. M. (2010). Multiple functional groups of varying ratios in metal-organic frameworks. *Science, 327*(5967), 846–850. https://doi.org/10.1126/science.1181761.

Downes, C. A., & Marinescu, S. C. (2017). Electrocatalytic metal–organic frameworks for energy applications. *ChemSusChem, 10*(22), 4374–4392. https://doi.org/10.1002/cssc.201701420, http://www.interscience.wiley.com/jpages/1864-5631.

Du, J., Ma, Y. Y., Cui, W. J., Zhang, S. M., Han, Z. G., Li, R. H., Han, X. Q., Guan, W., Wang, Y. H., Li, Y. Q., Liu, Y., Yu, F. Y., Wei, K. Q., Tan, H. Q., Kang, Z. H., & Li, Y. G. (2022). Unraveling photocatalytic electron transfer mechanism in polyoxometalate-encapsulated metal-organic frameworks for high-efficient CO_2 reduction reaction. *Applied Catalysis B: Environmental, 318*. https://doi.org/10.1016/j.apcatb.2022.121812, http://www.elsevier.com/inca/publications/store/5/2/3/0/6/6/index.htt.

Falcaro, P., Ricco, R., Yazdi, A., Imaz, I., Furukawa, S., Maspoch, D., Ameloot, R., Evans, J. D., & Doonan, C. J. (2016). Application of metal and metal oxide nanoparticles at MOFs. *Coordination chemistry reviews, 307*, 237–254. https://doi.org/10.1016/j.ccr.2015.08.002, http://www.journals.elsevier.com/coordination-chemistry-reviews/.

Fan, Q., Zhang, M., Jia, M., Liu, S., Qiu, J., & Sun, Z. (2018). Electrochemical CO_2 reduction to C^{2+} species: Heterogeneous electrocatalysts, reaction pathways, and optimization strategies. *Materials Today Energy, 10*, 280–301. https://doi.org/10.1016/j.mtener.2018.10.003, http://www.journals.elsevier.com/materials-today-energy.

Furukawa, H., Cordova, K. E., O'Keeffe, M., & Yaghi, O. M. (2013). The chemistry and applications of metal-organic frameworks. *Science, 341*(6149). https://doi.org/10.1126/science.1230444, http://www.sciencemag.org/content/341/6149/1230444.full.pdf.

Garg, S., Li, M., Weber, A. Z., Ge, L., Li, L., Rudolph, V., Wang, G., & Rufford, T. E. (2020). Advances and challenges in electrochemical CO_2 reduction processes: An engineering and design perspective looking beyond new catalyst materials. *Journal of Materials Chemistry A, 8*(4), 1511–1544. https://doi.org/10.1039/c9ta13298h, http://pubs.rsc.org/en/journals/journal/ta.

Gong, S., Yang, S., Wang, W., Lu, R., Wang, H., Han, X., Wang, G., Xie, J., Rao, D., Wu, C., Liu, J., Shao, S., & Lv, X. (2023). Promoting CO_2 dynamic activation via micro-engineering technology for enhancing electrochemical CO_2 reduction. *Small, 19*(26). https://doi.org/10.1002/smll.202207808, http://onlinelibrary.wiley.com/journal/10.1002/.

Goyal, A., Marcandalli, G., Mints, V. A., & Koper, M. T. M. (2020). Competition between CO_2 reduction and hydrogen evolution on a gold electrode under well-defined mass transport conditions. *Journal of the American Chemical Society, 142*(9), 4154–4161. https://doi.org/10.1021/jacs.9b10061, http://pubs.acs.org/journal/jacsat.

Gu, W., Wu, M., Xu, J., & Zhao, T. (2022). MXene boosted metal-organic framework-derived Fe–N–C as an efficient electrocatalyst for oxygen reduction reactions. *International Journal of Hydrogen Energy, 47*(39), 17224–17232. https://doi.org/10.1016/j.ijhydene.2022.03.229, http://www.journals.elsevier.com/international-journal-of-hydrogen-energy/.

Gu, Z., Shen, H., Shang, L., Lv, X., Qian, L., & Zheng, G. (2018). Nanostructured copper-based electrocatalysts for CO_2 reduction. *Small Methods, 2*(11). https://doi.org/10.1002/smtd.201800121, http://onlinelibrary.wiley.com/journal/23669608.

Guntern, Y. T., Pankhurst, J. R., Vávra, J., Mensi, M., Mantella, V., Schouwink, P., & Buonsanti, R. (2019). Nanocrystal/metal–organic framework hybrids as electrocatalytic platforms for CO_2 conversion. *Angewandte Chemie - International Edition, 131*(36), 12762–12769. https://doi.org/10.1002/ange.201905172.

Handoko, A. D., Ong, C. W., Huang, Y., Lee, Z. G., Lin, L., Panetti, G. B., & Yeo, B. S. (2016). Mechanistic insights into the selective electroreduction of carbon dioxide to ethylene on Cu_2O-derived copper catalysts. *Journal of Physical Chemistry C, 120*(36), 20058–20067. https://doi.org/10.1021/acs.jpcc.6b07128, http://pubs.acs.org/journal/jpccck.

Hernández, S., Farkhondehfal, M. A., Sastre, F., Makkee, M., Saracco, G., & Russo, N. (2017). Syngas production from electrochemical reduction of CO_2: Current status and prospective implementation. *Green Chemistry, 19*(10), 2326–2346. https://doi.org/10.1039/c7gc00398f, http://pubs.rsc.org/en/journals/journal/gc.

Hori, Y., Wakebe, H., Tsukamoto, T., & Koga, O. (1994). Electrocatalytic process of CO selectivity in electrochemical reduction of CO_2 at metal electrodes in aqueous media. *Electrochimica Acta, 39*(11–12), 1833–1839. https://doi.org/10.1016/0013-4686(94)85172-7.

Huang, Jian-Mei, et al. (2023). MOF-based materials for electrochemical reduction of carbon dioxide. Coordination Chemistry Reviews 494, 215333.

Jiang, X., Wu, H., Chang, S., Si, R., Miao, S., Huang, W., Li, Y., Wang, G., & Bao, X. (2017). Boosting CO_2 electroreduction over layered zeolitic imidazolate frameworks decorated with Ag_2O nanoparticles. *Journal of Materials Chemistry A, 5*(36), 19371–19377. https://doi.org/10.1039/c7ta06114e, http://pubs.rsc.org/en/journals/journalissues/ta.

Jiao, L., Wang, Y., Jiang, H.-L., & Xu, Q. (2018). Metal–organic frameworks as platforms for catalytic applications. *Advanced Materials, 30*(37), e1703663.

Jin, S., Hao, Z., Zhang, K., Yan, Z., & Chen, J. (2021). Advances and challenges for the electrochemical reduction of CO_2 to CO: From fundamentals to industrialization. *Angewandte Chemie - International Edition, 133*(38), 20795–20816. https://doi.org/10.1002/ange.202101818.

Kalaj, M., Bentz, K. C., Ayala, S., Palomba, J. M., Barcus, K. S., Katayama, Y., & Cohen, S. M. (2020). MOF-polymer hybrid materials: From simple composites to tailored architectures. *Chemical Reviews, 120*(16), 8267–8302. https://doi.org/10.1021/acs.chemrev.9b00575, http://pubs.acs.org/journal/chreay.

Kempahanumakkagari, S., Vellingiri, K., Deep, A., Kwon, E. E., Bolan, N., & Kim, K. H. (2018). Metal–organic framework composites as electrocatalysts for electrochemical sensing applications. *Coordination Chemistry Reviews, 357*, 105–129. https://doi.org/10.1016/j.ccr.2017.11.028, http://www.journals.elsevier.com/coordination-chemistry-reviews/.

Kibria, M. G., Edwards, J. P., Gabardo, C. M., Dinh, C. T., Seifitokaldani, A., Sinton, D., & Sargent, E. H. (2019). Electrochemical CO_2 reduction into chemical feedstocks: From mechanistic electrocatalysis models to system design. *Advanced Materials, 31*(31). https://doi.org/10.1002/adma.201807166, http://onlinelibrary.wiley.com/journal/10.1002/.

Ko, M., Mendecki, L., & Mirica, K. A. (2018). Conductive two-dimensional metal-organic frameworks as multifunctional materials. *Chemical Communications, 54*(57), 7873–7891. https://doi.org/10.1039/c8cc02871k, http://pubs.rsc.org/en/journals/journal/cc.

Kornienko, N., Zhao, Y., Kley, C. S., Zhu, C., Kim, D., Lin, S., Chang, C. J., Yaghi, O. M., & Yang, P. (2015). Metal-organic frameworks for electrocatalytic reduction of carbon dioxide. *Journal of the American Chemical Society, 137*(44), 14129–14135. https://doi.org/10.1021/jacs.5b08212, http://pubs.acs.org/journal/jacsat.

Kortlever, R., Shen, J., Schouten, K. J. P., Calle-Vallejo, F., & Koper, M. T. M. (2015). Catalysts and reaction pathways for the electrochemical reduction of carbon dioxide. *Journal of Physical Chemistry Letters, 6*(20), 4073–4082. https://doi.org/10.1021/acs.jpclett.5b01559, http://pubs.acs.org/journal/jpclcd.

Kou, Z., Li, X., Wang, T., Ma, Y., Zang, W., Nie, G., & Wang, J. (2022). Fundamentals, on-going advances and challenges of electrochemical carbon dioxide reduction. *Electrochemical*

Energy Reviews, 5(1), 82–111. https://doi.org/10.1007/s41918-021-00096-5, http://springer.com/journal/41918.

Kumar, B., Brian, J. P., Atla, V., Kumari, S., Bertram, K. A., White, R. T., & Spurgeon, J. M. (2016). New trends in the development of heterogeneous catalysts for electrochemical CO_2 reduction. *Catalysis Today, 270*, 19–30. https://doi.org/10.1016/j.cattod.2016.02.006, http://www.sciencedirect.com/science/journal/09205861.

Li, M., Garg, S., Chang, X., Ge, L., Li, L., Konarova, M., Rufford, T. E., Rudolph, V., & Wang, G. (2020). Toward excellence of transition metal-based catalysts for CO_2 electrochemical reduction: An overview of strategies and rationales. *Small Methods, 4*(7). https://doi.org/10.1002/smtd.202000033, http://onlinelibrary.wiley.com/journal/23669608.

Li, M., Wang, H., Luo, W., Sherrell, P. C., Chen, J., & Yang, J. (2020). Heterogeneous single-atom catalysts for electrochemical CO_2 reduction reaction. *Advanced Materials, 32*(34). https://doi.org/10.1002/adma.202001848, http://onlinelibrary.wiley.com/journal/10.1002/.

Li, S., & Huo, F. (2015). Metal-organic framework composites: From fundamentals to applications. *Nanoscale. 7*(17), 7482–7501. https://doi.org/10.1039/c5nr00518c, http://pubs.rsc.org/en/journals/journal/nr.

Li, X. X., Liu, J., Zhang, L., Dong, L. Z., Xin, Z. F., Li, S. L., Huang-Fu, X. Q., Huang, K., & Lan, Y. Q. (2019). Hydrophobic polyoxometalate-based metal-organic framework for efficient CO_2 photoconversion. *ACS Applied Materials and Interfaces, 11*(29), 25790–25795. https://doi.org/10.1021/acsami.9b03861, http://pubs.acs.org/journal/aamick.

Lian, X., Fang, Y., Joseph, E., Wang, Q., Li, J., Banerjee, S., Lollar, C., Wang, X., & Zhou, H. C. (2017). Enzyme-MOF (metal-organic framework) composites. *Chemical Society reviews, 46*(11), 3386–3401. https://doi.org/10.1039/c7cs00058h, http://pubs.rsc.org/en/journals/journal/cs.

Liang, Z., Qu, C., Guo, W., Zou, R., & Xu, Q. (2018). Pristine metal–organic frameworks and their composites for energy storage and conversion. *Advanced Materials, 30*(37), e1702891.

Liu, H., Wang, H., Song, Q., Küster, K., Starke, U., van Aken, P. A., & Klemm, E. (2022). Assembling metal organic layer composites for high-performance electrocatalytic CO_2 reduction to formate. *Angewandte Chemie - International Edition, 61*(9). https://doi.org/10.1002/anie.202117058, http://onlinelibrary.wiley.com/journal/10.1002/.

Liu, J., Chen, C., Zhang, K., & Zhang, L. (2021). Applications of metal–organic framework composites in CO_2 capture and conversion. *Chinese Chemical Letters, 32*(2), 649–659. https://doi.org/10.1016/j.cclet.2020.07.040, http://www.elsevier.com/wps/find/journaldescription.cws_home/997/description#description.

Liu, K.-G., Bigdeli, F., Panjehpour, A., Larimi, A., Morsali, A., Dhakshinamoorthy, A., & Garcia, H. (2023). Metal organic framework composites for reduction of CO_2. *Coordination Chemistry Reviews, 493*, 215257. https://doi.org/10.1016/j.ccr.2023.215257.

Liu, Y., Tang, C., Cheng, M., Chen, M., Chen, S., Lei, L., Chen, Y., Yi, H., Fu, Y., & Li, L. (2021). Polyoxometalate@metal-organic framework composites as effective photocatalysts. *ACS Catalysis, 11*(21), 13374–13396. https://doi.org/10.1021/acscatal.1c03866, http://pubs.acs.org/page/accacs/about.html.

Long, C., Li, X., Guo, J., Shi, Y., Liu, S., & Tang, Z. (2019). Electrochemical reduction of CO_2 over heterogeneous catalysts in aqueous solution: Recent progress and perspectives. *Small Methods, 3*(3). https://doi.org/10.1002/smtd.201800369, http://onlinelibrary.wiley.com/journal/23669608.

Lu, Q., & Jiao, F. (2016). Electrochemical CO_2 reduction: Electrocatalyst, reaction mechanism, and process engineering. *Nano Energy, 29*, 439–456. https://doi.org/10.1016/j.nanoen.2016.04.009, http://www.journals.elsevier.com/nano-energy/.

Lu, X., Wu, Y., Yuan, X., Huang, L., Wu, Z., Xuan, J., Wang, Y., & Wang, H. (2018). High-performance electrochemical CO_2 reduction cells based on non-noble metal catalysts. *ACS Energy Letters, 3*(10), 2527–2532. https://doi.org/10.1021/acsenergylett.8b01681, http://pubs.acs.org/journal/aelccp.

Peng, Y., Xu, J., Xu, J., Ma, J., Bai, Y., Cao, S., Zhang, S., & Pang, H. (2022). Metal-organic framework (MOF) composites as promising materials for energy storage applications. *Advances in Colloid and Interface Science, 307*, 102732. https://doi.org/10.1016/j.cis.2022.102732.

Raut, V., Bera, B., Neergat, M., & Das, D. (2021). Metal-organic framework and carbon black supported MOFs as dynamic electrocatalyst for oxygen reduction reaction in an alkaline electrolyte. *Journal of Chemical Sciences, 133*(2). https://doi.org/10.1007/s12039-021-01900-x, http://www.springer.com/chemistry/journal/12039.

Rowsell, J. L. C., & Yaghi, O. M. (2005). Strategies for hydrogen storage in metal-organic frameworks. *Angewandte Chemie - International Edition, 44*(30), 4670–4679. https://doi.org/10.1002/anie.200462786.

Saha, P. P., Amanullah, S., & Dey, A. (2022). Selectivity in electrochemical CO_2 reduction. *Accounts of Chemical Research, 55*(2), 134–144. https://doi.org/10.1021/acs.accounts.1c00678, http://pubs.acs.org/journal/achre4.

Sajna, M. S., Zavahir, S., Popelka, A., Kasak, P., Al-Sharshani, A., Onwusogh, U., Wang, M., Park, H., & Han, D. S. (2023). Electrochemical system design for CO_2 conversion: A comprehensive review. *Journal of Environmental Chemical Engineering, 11*(5), 110467. https://doi.org/10.1016/j.jece.2023.110467.

Sathiyan, K., Dutta, A., Marks, V., Fleker, O., Zidki, T., Webster, R. D., & Borenstein, A. (2023). Nano-encapsulation: Overcoming conductivity limitations by growing MOF nanoparticles in meso-porous carbon enables high electrocatalytic performance. *NPG Asia Materials, 15*(1). https://doi.org/10.1038/s41427-022-00459-4, http://www.nature.com/am/.

Singh, B. K., Lee, S., & Na, K. (2020). An overview on metal-related catalysts: Metal oxides, nanoporous metals and supported metal nanoparticles on metal organic frameworks and zeolites. *Rare Metals, 39*(7), 751–766. https://doi.org/10.1007/s12598-019-01205-6, http://www.springer.com/materials/special+types/journal/12598.

Skoulidas, A. I. (2004). Molecular dynamics simulations of gas diffusion in metal-organic frameworks: Argon in CuBTC. *Journal of the American Chemical Society, 126*(5), 1356–1357. https://doi.org/10.1021/ja039215.

Sun, L., Campbell, M. G., & Dincǎ, M. (2016). Electrically conductive porous metal-organic frameworks. *Angewandte Chemie - International Edition, 55*(11), 3566–3579. https://doi.org/10.1002/anie.201506219, http://onlinelibrary.wiley.com/journal/10.1002/.

Sun, M. L., Wang, Y. R., He, W. W., Zhong, R. L., Liu, Q. Z., Xu, S., Xu, J. M., Han, X. L., Ge, X., Li, S. L., Lan, Y. Q., Al-Enizi, A. M., Nafady, A., & Ma, S. (2021). Efficient electron transfer from electron-sponge polyoxometalate to single-metal site metal–organic frameworks for highly selective electroreduction of carbon dioxide. *Small, 17*(20). https://doi.org/10.1002/smll.202100762, http://onlinelibrary.wiley.com/journal/10.1002/.

Tan, X., Yu, C., Zhao, C., Huang, H., Yao, X., Han, X., Guo, W., Cui, S., Huang, H., & Qiu, J. (2019). Restructuring of Cu_2O to Cu_2O@Cu-metal-organic frameworks for selective electrochemical reduction of CO_2. *ACS Applied Materials and Interfaces, 11*(10), 9904–9910. https://doi.org/10.1021/acsami.8b19511, http://pubs.acs.org/journal/aamick.

Tayyebi, E., Hussain, J., Abghoui, Y., & Skúlason, E. (2018). Trends of electrochemical CO_2 reduction reaction on transition metal oxide catalysts. *The Journal of Physical Chemistry C, 122*(18), 10078–10087. https://doi.org/10.1021/acs.jpcc.8b02224.

Tekalgne, M. A., Do, H. H., Hasani, A., Van Le, Q., Jang, H. W., Ahn, S. H., & Kim, S. Y. (2020). Two-dimensional materials and metal-organic frameworks for the CO_2 reduction reaction. *Materials Today Advances, 5,* 100038. https://doi.org/10.1016/j.mtadv.2019.100038.

Usman, M., Zeb, Z., Ullah, H., Suliman, M. H., Humayun, M., Ullah, L., Shah, S. N. A., Ahmed, U., & Saeed, M. (2022). A review of metal-organic frameworks/graphitic carbon nitride composites for solar-driven green H_2 production, CO_2 reduction, and water purification. *Journal of Environmental Chemical Engineering, 10*(3). https://doi.org/10.1016/j.jece.2022.107548, http://www.journals.elsevier.com/journal-of-environmental-chemical-engineering/.

Vinoth, S., Ramaraj, R., & Pandikumar, A. (2020). Facile synthesis of calcium stannate incorporated graphitic carbon nitride nanohybrid materials: A sensitive electrochemical sensor for determining dopamine. *Materials Chemistry and Physics, 245,* 122743. https://doi.org/10.1016/j.matchemphys.2020.122743.

Wang, Q., Cai, C., Dai, M., Fu, J., Zhang, X., Li, H., Zhang, H., Chen, K., Lin, Y., Li, H., Hu, J., Miyauchi, M., & Liu, M. (2021). Recent advances in strategies for improving the performance of CO_2 reduction reaction on single atom catalysts. *Small Science, 1*(2). https://doi.org/10.1002/smsc.202000028, http://onlinelibrary.wiley.com/journal/26884046.

Wang, R.-M., Zheng, S.-R., & Zheng, Y.-P. (2011). *Reinforced materials.* Elsevier BV, 29–548. https://doi.org/10.1533/9780857092229.1.29.

Wang, Y. R., Huang, Q., He, C. T., Chen, Y., Liu, J., Shen, F. C., & Lan, Y. Q. (2018). Oriented electron transmission in polyoxometalate-metalloporphyrin organic framework for highly selective electroreduction of CO_2. *Nature Communications, 9*(1). https://doi.org/10.1038/s41467-018-06938-z, http://www.nature.com/ncomms/index.html.

Wu, M. X., Wang, Y., Zhou, G., & Liu, X. (2020). Core-shell MOFs@MOFs: Diverse designability and enhanced selectivity. *ACS Applied Materials and Interfaces, 12*(49), 54285–54305. https://doi.org/10.1021/acsami.0c16428, http://pubs.acs.org/journal/aamick.

Wu, T., Liu, X., Liu, Y., Cheng, M., Liu, Z., Zeng, G., Shao, B., Liang, Q., Zhang, W., He, Q., & Zhang, W. (2020). Application of QD-MOF composites for photocatalysis: Energy production and environmental remediation. *Coordination Chemistry Reviews, 403,* 213097. https://doi.org/10.1016/j.ccr.2019.213097.

Xie, H., Wang, T., Liang, J., Li, Q., & Sun, S. (2018). Cu-based nanocatalysts for electrochemical reduction of CO_2. *Nano Today, 21,* 41–54. https://doi.org/10.1016/j.nantod.2018.05.001, http://www.elsevier.com/wps/find/journaldescription.cws_home/706735/description#description.

Xie, L. S., Skorupskii, G., & Dincă, M. (2020). Electrically conductive metal-organic frameworks. *Chemical Reviews, 120*(16), 8536–8580. https://doi.org/10.1021/acs.chemrev.9b00766, http://pubs.acs.org/journal/chreay.

Xie, X., Zhang, X., Xie, M., Xiong, L., Sun, H., Lu, Y., Mu, Q., Rummeli, M. H., Xu, J., Li, S., Zhong, J., Deng, Z., Ma, B., Cheng, T., Goddard, W. A., & Peng, Y. (2022). Au-activated N motifs in non-coherent cupric porphyrin metal organic frameworks for promoting and stabilizing ethylene production. *Nature Communications, 13*(1). https://doi.org/10.1038/s41467-021-27768-6, http://www.nature.com/ncomms/index.html.

Xu, L.-W., Qian, S.-L., Dong, B.-X., Feng, L.-G., & Li, Z.-W. (2022). The boosting of electrocatalytic CO_2-to-CO transformation by using the carbon nanotubes-supported PCN-222 (Fe) nanoparticles composite. Journal of Materials Science, 57, 526–537. .

Xu, S., & Carter, E. A. (2019). Theoretical insights into heterogeneous (photo)electrochemical CO_2 reduction. *Chemical Reviews, 119*(11), 6631–6669. https://doi.org/10.1021/acs.chemrev.8b00481, http://pubs.acs.org/journal/chreay.

Xu, T., Wang, Y., Xue, Y., Li, J., & Wang, Y. (2023). Mxenes@metal-organic framework hybrids for energy storage and electrocatalytic application: Insights into recent advances. *Chemical Engineering Journal, 470*, 144247.

Xue, Y., Zheng, S., Xue, H., & Pang, H. (2019). Metal-organic framework composites and their electrochemical applications. *Journal of Materials Chemistry A, 7*(13), 7301–7327. https://doi.org/10.1039/C8TA12178H, http://pubs.rsc.org/en/journals/journal/ta.

Yan, Y., Bo, X., & Guo, L. (2020). MOF-818 metal-organic framework-reduced graphene oxide/multiwalled carbon nanotubes composite for electrochemical sensitive detection of phenolic acids. *Talanta, 218*, 121123. https://doi.org/10.1016/j.talanta.2020.121123.

Yang, S., Zhang, J., Peng, L., Asgari, M., Stoian, D., Kochetygov, I., Luo, W., Oveisi, E., Trukhina, O., Clark, A. H., Sun, D. T., & Queen, W. L. (2020). A metal-organic framework/polymer derived catalyst containing single-atom nickel species for electrocatalysis. *Chemical Science, 11*(40), 10991–10997. https://doi.org/10.1039/d0sc04512h, http://pubs.rsc.org/en/journals/journal/sc.

Ye, L., Liu, J., Gao, Y., Gong, C., Addicoat, M., Heine, T., Wöll, C., & Sun, L. (2016). Highly oriented MOF thin film-based electrocatalytic device for the reduction of CO_2 to CO exhibiting high faradaic efficiency. *Journal of Materials Chemistry A, 4*(40), 15320–15326. https://doi.org/10.1039/c6ta04801c, http://pubs.rsc.org/en/journals/journalissues/ta.

Yi, J.-D., Xie, R., Xie, Z.-L., Chai, G.-L., Liu, T.-F., Chen, R.-P., Huang, Y.-B., & Cao, R. (2020). Highly selective CO_2 electroreduction to CH_4 by in situ generated Cu_2O single-type sites on a conductive MOF: Stabilizing key intermediates with hydrogen bonding. *Angewandte Chemie - International Edition, 132*(52), 23849–23856. https://doi.org/10.1002/ange.202010601.

Yu, J., Mu, C., Yan, B., Qin, X., Shen, C., Xue, H., & Pang, H. (2017). Nanoparticle/MOF composites: Preparations and applications. *Materials Horizons, 4*(4), 557–569. https://doi.org/10.1039/c6mh00586a, http://rsc.li/materials-horizons.

Zhan, T., Zou, Y., Yang, Y., Ma, X., Zhang, Z., & Xiang, S. (2022). Two-dimensional metal-organic frameworks for electrochemical CO_2 reduction reaction. *ChemCatChem, 14*(3). https://doi.org/10.1002/cctc.202101453, http://onlinelibrary.wiley.com/journal/10.1002/.

Zhang, B., Jiang, Y., Gao, M., Ma, T., Sun, W., & Pan, H. (2021). Recent progress on hybrid electrocatalysts for efficient electrochemical CO_2 reduction. *Nano Energy, 80*, 105504. https://doi.org/10.1016/j.nanoen.2020.105504.

Zhang, L., Zhao, Z. J., & Gong, J. (2017). Nanostructured materials for heterogeneous electrocatalytic CO_2 reduction and their related reaction mechanisms. *Angewandte Chemie - International Edition, 56*(38), 11326–11353. https://doi.org/10.1002/anie.201612214, http://onlinelibrary.wiley.com/journal/10.1002/.

Zhang, S., Fan, Q., Xia, R., & Meyer, T. J. (2020). CO_2 reduction: From homogeneous to heterogeneous electrocatalysis. *Accounts of Chemical Research, 53*(1), 255–264. https://doi.org/10.1021/acs.accounts.9b00496, http://pubs.acs.org/journal/achre4.

Zhang, S. Y., Yang, Y. Y., Zheng, Y. Q., & Zhu, H. L. (2018). Ag-doped Co_3O_4 catalyst derived from heterometallic MOF for syngas production by electrocatalytic reduction of CO_2 in water. *Journal of Solid State Chemistry, 263*, 44–51. https://doi.org/10.1016/j.jssc.2018.04.007, http://www.elsevier.com/inca/publications/store/6/2/2/8/9/8/index.htt.

Zhang, W., Hu, Y., Ma, L., Zhu, G., Wang, Y., Xue, X., Chen, R., Yang, S., & Jin, Z. (2018). Progress and perspective of electrocatalytic CO_2 reduction for renewable carbonaceous fuels and chemicals. *Advanced Science, 5*(1). https://doi.org/10.1002/advs.201700275, http://onlinelibrary.wiley.com/journal/10.1002/.

Zhang, X. G., Jin, X., Wu, D. Y., & Tian, Z. Q. (2018). Selective electrocatalytic mechanism of CO_2 reduction reaction to CO on silver electrodes: A unique reaction intermediate. *Journal of*

Physical Chemistry C, 122(44), 25447–25455. https://doi.org/10.1021/acs.jpcc.8b08170, http://pubs.acs.org/journal/jpccck.

Zhao, K., Liu, Y., Quan, X., Chen, S., & Yu, H. (2017). CO_2 electroreduction at low overpotential on oxide-derived Cu/carbons fabricated from metal organic framework. *ACS Applied Materials and Interfaces, 9*(6), 5302–5311. https://doi.org/10.1021/acsami.6b15402, http://pubs.acs.org/journal/aamick.

Zhao, X., Xu, H., Wang, X., Zheng, Z., Xu, Z., & Ge, J. (2018). Monodisperse metal-organic framework nanospheres with encapsulated core-shell nanoparticles Pt/Au@Pd@{$Co_2(oba)_4(3\text{-bpdh})_2$}$4H_2O$ for the highly selective conversion of CO_2 to CO. *ACS Applied Materials and Interfaces, 10*(17), 15096–15103. https://doi.org/10.1021/acsami.8b03561, http://pubs.acs.org/journal/aamick.

Zheng, T., Jiang, K., & Wang, H. (2018). Recent advances in electrochemical CO_2-to-CO conversion on heterogeneous catalysts. *Advanced Materials, 30*(48). https://doi.org/10.1002/adma.201802066, http://onlinelibrary.wiley.com/journal/10.1002/.

Zhou, Y., Wang, K., Zheng, S., Cheng, X., He, Y., Qin, W., Zhang, X., Chang, H., Zhong, N., & He, X. (2024). Advancements in electrochemical CO_2 reduction reaction: A review on CO_2 mass transport enhancement strategies. *Chemical Engineering Journal, 486*, 150169. https://doi.org/10.1016/j.cej.2024.150169.

Zhu, Q. L., & Xu, Q. (2014). Metal-organic framework composites. *Chemical Society Reviews, 43*(16), 5468–5512. https://doi.org/10.1039/c3cs60472a, http://pubs.rsc.org/en/journals/journal/cs.

Zou, Y., & Wang, S. (2021). An investigation of active sites for electrochemical CO_2 reduction reactions: From in situ characterization to rational design. *Advanced Science, 8*(9). https://doi.org/10.1002/advs.202003579, http://onlinelibrary.wiley.com/journal/10.1002/.

Chapter 15

Metal-organic framework composites for catalytic desulfurization and denitrogenation of fuels

Peiwen Wu
School of Chemistry and Chemical Engineering, Jiangsu University, Zhenjiang, Jiangsu Province, P.R. China

Background

With the ongoing advancement of modern industry, the demand for energy continues to increase. Despite the increasing popularity of emerging energy sources like wind, solar, hydrogen, and nuclear power in recent years, their industrial-scale production capabilities do not yet meet the increasing applications. Therefore, traditional energy sources such as petroleum, natural gas, and coal remain the primary energy sources for modern industry. Fossil fuels still constitute over 70% of global energy consumption, with petroleum alone accounting for half of this share. Petroleum, often referred to as the "lifeblood of industry," originates from the decomposition of animal remains over millennia. It serves as a crucial raw material for producing various fuels for internal combustion engines (including gasoline, diesel, kerosene, aviation fuels, boiler fuels, and so on) and a myriad of other petroleum products such as coke, asphalt, lubricants, paraffin, and organic solvents. Whether in daily national production or military applications, petroleum stands as an indispensable energy source. Consequently, it has long been a strategic reserve asset pursued by nations worldwide. By the end of 2015, it was confirmed that global petroleum reserves had amassed to 1691.5 billion barrels, with new oil fields continually being discovered and offshore petroleum resources developed. As such, gasoline and diesel are expected to remain the principal energy sources for internal combustion engines for the foreseeable future.

As gasoline and diesel consumption expands, the environmental issues stemming from their combustion have become increasingly severe. Notably, the combustion of sulfur and nitrogen compounds in fuel generates emissions of SO_x and NO_x, which pose significant threats to human respiratory health, skin integrity, and overall well-being (Zhao et al., 2011). Moreover, under the influence of suspended particles, SO_x and NO_x contribute to acid rain, causing damage to vegetation, buildings, and environmental pollution in water bodies and soils, thus resulting in considerable economic losses. The worsening smog conditions are closely linked to SO_x and NO_x emissions, as studies have established a direct correlation between particulate matter in vehicle exhaust and the sulfur and nitrogen content in the fuel oils.

Additionally, the production of SO_x and NO_x can deactivate catalytic converters employed for vehicle emissions control, thereby deactivating their effectiveness in treating exhaust gases and indirectly exacerbating environmental issues (Cai et al., 2018). To address these concerns, nations worldwide have implemented increasingly stringent fuel quality standards, including restrictions on sulfur content in gasoline and diesel. For instance, many regions and countries, such as China, European Union, the United States, currently regulate sulfur contents in fuel oils to be below 10 parts per million (ppm), with California in the United States setting an even higher standard at 5 ppm for gasoline. With global environmental challenges growing more urgent, it is anticipated that future standards for sulfur content in gasoline and diesel will become even more stringent, potentially requiring sulfur content below 5 ppm or even achieving sulfur-free gasoline and diesel. Such advancements impose stricter demands on the refining processes of petroleum industries and require higher standards for fuel desulfurization technologies.

Unlike strict regulations on sulfur content in fuel oils, most regions and countries presently lack specific requirements for nitrogen content in fuel oils. However, the escalating environmental regulations may lead to stringent standards for nitrogen content in the future. Furthermore, nitrogen compounds in fuel production can poison hydrofining catalysts, severely impacting their efficiency and increasing energy consumption, hydrogen consumption, and carbon emissions. Consequently, efficient removal of nitrogen compounds from fuels is also in urgent demand.

Based on their characteristics, desulfurization and denitrogenation processes are broadly classified into two main types: (1) hydrodesulfurization (HDS) and hydrodenitrogenation (HDN), which require the use of H_2 and are widely employed in industry, and (2) nonhydrogenation pathways, which are also sought to complement the limitations of HDS and HDN.

Presently, the drawbacks of commonly used HDS and HDN processes include: (1) high pressure and temperature requirements, which increase both reaction and fuel production costs; (2) a lack of selective catalysts, making HDS and HDN processes more complex, particularly for the hydrogenation of aromatic sulfur/nitrogen compounds; (3) the emission of gases like H_2S and

NH$_3$ during HDS and HDN processes, contributing to secondary environmental pollution and nonreusability; and (4) the need for thick-walled reactors due to the high temperature and pressure requirements of HDS and HDN, leading to significantly increased costs.

Therefore, developing a suitable non-HDS/HDN method to overcome the limitations of HDS and HDN can serve as a posttreatment solution for fuel desulfurization and denitrogenation following hydrogenation. Oxidative desulfurization (ODS) and oxidative denitrogenation (ODN) methods are considered among the most promising approaches due to their mild reaction conditions, high removal efficiency, remarkable activity toward aromatic sulfur/nitrogen compounds, low cost, and lack of requirement for H$_2$. The principle of ODS and ODN involves the oxidation of aromatic sulfur/nitrogen compounds using suitable oxidants (such as hydrogen peroxide, peroxides, oxygen, and air) to generate corresponding highly polar products, which can then be separated using adsorbents or extractants (Fig. 15.1). Hence, the ODS/ODN process primarily involves two aspects: (1) oxidation of sulfur/nitrogen compounds; (2) liquid-phase extraction or adsorption. Current research in ODS and ODN primarily focuses on two areas: the design of novel catalysts and the selection of lower-cost, higher safety-factor oxidants.

Metal-organic frameworks (MOFs) represent a class of porous materials whose structures comprise inorganic nodes and multidentate organic compounds interconnected through covalent or coordination interactions (Annamalai et al., 2022). MOFs consist of metal nodes (metal ions or metal oxide clusters), organic ligands, and nanoscale internal void spaces (formed through self-assembly of metal and ligand components). Among these elements, metal nodes are crucial because they possess open coordination sites to which Lewis base molecules tend to bind. The construction of MOFs involves processes such as reticular chemistry, topology-based design and synthesis, ligand exchange or insertion, and postsynthesis modification.

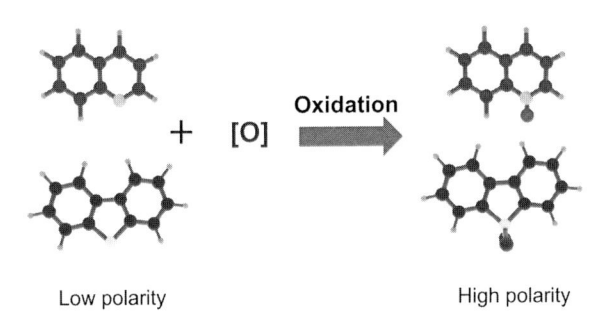

FIGURE 15.1 Schematic diagram of the oxidative desulfurization and oxidative denitrogenation processes

Reticular design focuses on connecting organic and inorganic molecular building units through strong directional chemical bonds to produce stable crystalline extended structures. This methodology allows control over the pore size of MOFs by adjusting the length of the organic linking unit without altering the fundamental framework topology (Freund et al., 2021). MOFs offer advantages such as high crystallinity, large surface area, high porosity, controllable surface pore structure, customizable framework design, and diverse structural characteristics (Xu et al., 2023). Exploiting the distinctive advantages of MOFs, these highly ordered nanoporous materials have emerged as exceptionally promising support for catalytic applications. MOFs, with their extraordinarily large surface areas, offer a platform for the dispersion of catalytic active sites, allowing these sites to be extensively distributed and fully exposed (Liu et al., 2022). This facilitates an enhanced contact area with substrate molecules, thereby significantly increasing the accessibility and efficiency of catalytic active sites while reducing the quantity of required active centers. Consequently, this optimization lowers both the economic cost and environmental impact of catalytic processes. Especially, the interaction between MOFs and catalytic active sites extends beyond mere physical adsorption. Through meticulous design and synthesis, metal ions or organic ligands within MOFs can form strong interactions with catalytic species, thereby facilitating an efficient charge transfer. This charge transfer enhances the adsorption and activation of reactant molecules at catalytic sites and accelerates the transformation of reaction intermediates, markedly improving overall catalytic performance (Granadeiro et al., 2016). In practical applications, researchers have successfully integrated MOFs into diverse catalytic systems, particularly by incorporating catalytically active species such as noble metals (Fortea-Pérez et al., 2017; Zhang et al., 2016), metal oxides (Abney et al., 2017; Falcaro et al., 2016), ionic liquids (Kinik et al., 2017; Sun, Huang, et al., 2018), and polyoxometalates (Zhu et al., 2015), resulting in a series of functionalized MOF composites. These composites not only retain the inherent porosity and tunable structure of MOFs but also integrate the distinctive properties of various catalytic active sites, leading to exceptional catalytic efficiency. For example, in environmental applications such as fuel desulfurization or denitrogenation, MOF composites serve as highly selective and efficient catalysts for the removal of sulfur- or nitrogen-containing compounds (Ahmed et al., 2013; Li, Yang, et al., 2020; Li, Wang, et al., 2020; Mondol et al., 2023). The high dispersion and synergistic effects between MOFs and catalytic active sites significantly enhance desulfurization and denitrogenation activities compared to the use of catalytic species alone. This advancement not only reduces sulfur and nitrogen content in fuels to comply with stringent environmental regulations but also presents more economical and efficient solutions for industry, thereby advancing green chemistry and sustainable development.

Metal-organic framework composites for oxidative desulfurization of fuel oils

Metal-organic framework composites for oxidative desulfurization with H_2O_2 as the oxidant

With the continuous advancement of ODS technology, a variety of catalysts, including metal oxides, heteropoly acids, and zeolites, have been utilized in ODS reactions. Despite their widespread applications, these catalysts exhibit several limitations in ODS systems. For example, the bulk structure of metal oxides can restrict the interaction between active sites and reactants; heteropoly acids are prone to leaching from their supports, resulting in poor recyclability; and silicon-based zeolites suffer from high synthesis costs and less efficient oxidant activation. Among the various oxidants employed in ODS systems, H_2O_2 stands out as a promising candidate due to its high active oxygen content per unit mass (47%), with water as the sole by-product, making it green, nontoxic, environmentally friendly, and cost-effective. MOFs, characterized by their large surface area, high porosity, and structural diversity, have attracted considerable interest in catalysis. Up to date, numerous studies have demonstrated that MOF materials exhibit high catalytic efficiency and excellent recyclability in ODS systems utilizing H_2O_2 as the oxidant. Among the various MOFs, zirconium-based MOFs (Zr-MOFs) are frequently employed as catalysts in ODS reactions due to their microporous structures, Lewis acid centers, and structural diversity (Li et al., 2023). Viana et al. (2021) prepared defect-containing monometallic UiO-66(Zr) frameworks, and the UiO-66(Zr) underwent postsynthetic modifications using three different salts (TiCl$_4$, ZrCl$_4$, and TiBr$_4$) to incorporate chloride or bromide ions into the MOFs' inorganic clusters. Notably, the pure UiO-66(Zr) exhibited limited sulfur removal performance in ODS system. Meanwhile, the UiO-66(Zr)-TiCl$_4$ catalyst, with incorporation of chloride ions, achieved a sulfur removal activity of over 87% for DBT (initial sulfur content is 1000 ppm) within 1 hour at 50°C using H_2O_2 as the oxidant, demonstrating excellent desulfurization performance. Furthermore, the catalyst maintained its activity after three cycles. Additionally, the incorporation of chlorides into the UiO-66(Zr) framework did not retain cationic Ti(IV), and the possible interactions between the chlorides and the UiO-66(Zr) framework enhanced the overall catalytic activity. The catalyst did not undergo structural degradation even after several reaction cycles.

To further improve the ODS performance, researchers have developed bimetallic MOFs for ODS. With the formation of bimetallic MOFs, the electron transfer between different metals or defects can be expected in the bimetallic MOFs, further improving the catalytic performance. Ye et al. (2021) introduced a rapid and environmentally friendly method for synthesizing a bimetallic organic framework, UiO-66(Hf-Zr). The UiO-66(Hf-Zr) was characterized to be a catalyst with highly dispersed Hf/Zr-OH defect sites.

This synthesis involved combining $ZrOCl_2·8H_2O$, $HfOCl_2·8H_2O$, and 1,4-benzenedicarboxylic acid (BDC) in a mortar, followed by crystallization under high temperature and pressure. Computational studies revealed that the exposed Hf-OH sites are more reactive with H_2O_2, forming Hf-OOH intermediates that enhance ODS performance. Using H_2O_2 as the oxidant, UiO-66(Hf-Zr) showed excellent ODS performance for DBT and 4,6-DMDBT (initial sulfur concentrations are 1000 ppm) at 30°C. The catalytic activity of UiO-66(Hf-Zr) increased with the increment of O/S molar ratio, achieving a desulfurization rate of 98.2% within 30 minutes at an O/S ratio of 2. This rate increased to 99.3% within 20 minutes at an O/S ratio of 3, and to 99.8% within 15 minutes at an O/S ratio of 4, demonstrating the high efficiency of UiO-66(Hf-Zr) even at low H_2O_2 dosages. Furthermore, the catalyst maintained excellent catalytic performance and structural stability after 4 reaction cycles.

Compared to other Zr-MOFs, the novel Zr-MOF known as MOF-808 exhibits an outstanding turnover frequency (TOF) of 42.7 1/h and a low Arrhenius activation energy of 22.0 kJ/mol in ODS reactions (Zheng et al., 2019). However, the unsaturated Zr_6 clusters in MOF-808 are prone to interact with H_2O molecules generated from H_2O_2 during the ODS process, further leads to the coordination with ·OH groups. Such a coordination would lead to the collapse of the framework, limiting its further application in ODS processes. Under this circumstance, introducing different transition metals into the framework can enhance structural stability and chemical diversity, significantly improving the catalytic performance and stability of monometallic MOFs. Li et al. (2023) employed a traditional sol-gel method to anchor Cu^+ onto MOF-808, resulting in the formation of Cu(I)-MOF-808. The Cu(I)-MOF-808 was further employed as a catalyst in ODS reactions with H_2O_2 as the oxidant, and a deep desulfurization of 99.0% can be gained on DBT with an initial concentration of 1000 ppm, under the reaction conditions of 60°C, 120 minutes. This exceptional performance is attributed to the adsorption effect of Zr^{4+} to DBT and the activation effect of Cu^+ on H_2O_2. Additionally, Cu(I)-MOF-808 retained 99.2% of its ODS activity after 12 cycles without experiencing framework collapse or metal leaching. Thus it can be concluded that the incorporation of Cu^+ prevents the framework collapse of MOF-808, thereby enhancing the catalytic stability of Cu(I)-MOF-808 in ODS reactions.

However, by only using MOFs as catalysts, limited ODS performance can be expected. Under these circumstances, to further improve the catalytic ODS performances of MOFs, a strategy of introducing catalytically active species into MOFs to form MOF composites has been widely employed. Such a strategy can take full advantage of MOFs such as tunable structures and high specific surface areas, which can induce high dispersion as well as excellent exposure of catalytically active species, favoring the catalytic ODS performance. For instance, a type of polyoxometalate (POM) with high acidity, flexible shape, and a unique cage structure has been encapsulated within the cavities of MOFs via a host–guest interaction to form POM@MOF composites. These composites exhibit superior desulfurization performance and higher load capacity compared to conventional materials. Pioneer work by Haruna et al. reported the preparation of POM-based MOF composites, specifically PW_{11}@MOF-808 and PW_{12}@MOF-808, using this encapsulation method, which can significantly enhance the stability of POMs (Fig. 15.2) (Haruna, Merican, Musa, & Rahman, 2023). They applied these composites to the catalytic ODS of liquid fuels with H_2O_2 as the oxidant. Experimental results demonstrated that pure POM or MOF-808 alone did not exhibit ideal desulfurization activity in ODS systems. However, an appropriate amount of

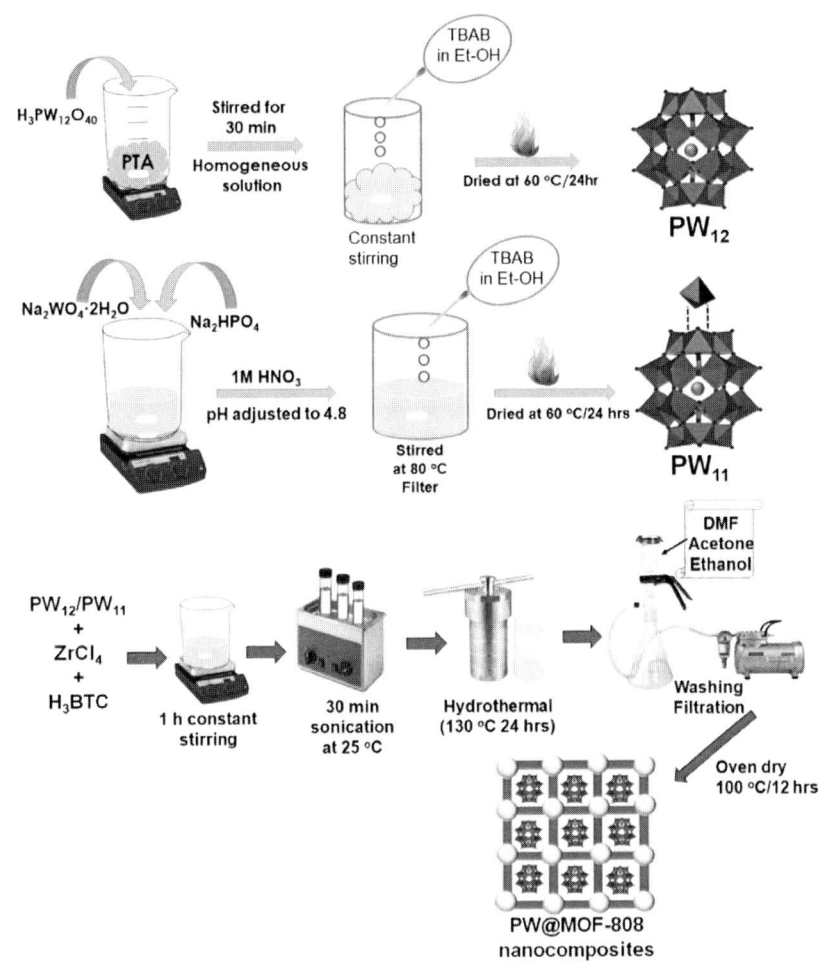

FIGURE 15.2 Schematic preparation methods of the PW_{11}@MOF-808 and PW_{12}@MOF-808 (Haruna, Merican, Musa, & Rahman, 2023).

POM, serving as a single active site, endowed MOF-808 with abundant catalytic active sites, thereby promoting the ODS reaction and enhancing the catalytic performance of the composite. Specifically, PW_{11}@MOF-808 achieved a desulfurization rate of 99.32% for DBT within 30 minutes at 60°C using H_2O_2 as the oxidant and maintained catalytic activity even after 7 cycles.

Zeolitic imidazolate frameworks (ZIFs), a subclass of MOFs, possess excellent physical and chemical properties such as high porosity, crystallinity, tunable structures, and exceptional chemical and thermal stability. Jafari et al. (2024) developed a novel framework material integrating zinc/cobalt-based ZIFs with TiO_2 for catalytic ODS of DBT-containing fuel oils with H_2O_2 as the oxidant. They explored the coordination effect between metal centers (Zn^{2+} or Co^{2+}) within ZIF-8 and ZIF-67 frameworks and TiO_2 as well as their effects on catalytic performance. The findings revealed that the size of ZIF crystals and the type of metal ions significantly influence their coordination with TiO_2, particularly affecting the strength of N-Ti-O bonds. The optimized ZIF-8-W@TiO_2 composite catalyst exhibited outstanding catalytic performance, achieving a maximum ODS efficiency of 98.6% with H_2O_2 as the oxidant. This superior performance was attributed to its high specific surface area, uniform TiO_2 distribution, and stable N-Ti-O bonds. Additionally, the catalyst showed remarkable recyclability, indicating its potential for practical applications in fuel desulfurization.

Table 15.1 summarizes the ODS efficiencies and recyclability of various MOF composites using H_2O_2 as the oxidant. The mechanisms of H_2O_2 activation by MOF composites for catalytic ODS generally involve several key aspects: (1) *Interaction with active sites:* The active sites in MOF composites can interact with H_2O_2 to form highly reactive intermediates, such as ·OH radicals. These radicals possess strong oxidizing properties, enabling the efficient oxidation of sulfur-containing compounds (Peng et al., 2018). (2) *Enhanced active sites in MOF composites:* Compared to traditional MOFs, MOF composites enhance the number of active sites, promote internal electron transfer, and alter activation pathways through the synergistic effect between MOFs and catalytically active sites. (3) *Introduction of structural defects:* Introducing structural defects increases the number of unsaturated metal sites, thereby improving the activation efficiency of the catalyst. Overall, these mechanisms collectively contribute to the high efficiency of MOFs in activating H_2O_2 for catalytic ODS (Haruna et al., 2022; Piscopo et al., 2020).

Metal-organic frameworks composites for oxidative desulfurization with O_2 as the oxidant

Compared to H_2O_2, oxygen (O_2) as an oxidant in the catalytic ODS process offers several notable advantages. Firstly, O_2 is environmentally friendly, producing minimal harmful by-products during use, which greatly reduces its environmental effects (Dong et al., 2019). Secondly, O_2 is stable and does not

TABLE 15.1 Oxidative desulfurization performances of DBT using different catalysts with H_2O_2 as the oxidant.

Catalysts	DBT concentration (ppm S)	O/S	Temperature (°C)	Time (min)	DBT conversion (%)	Ref.
VO-MoO$_3$@NPC	200	3	60	20	99.1	An et al. (2024)
PMA/HP-UiO-66-NH$_2$-9	1000	8	50	70	99.9	Xu et al. (2024)
0.045PW$_{11}$Mn@MOF-808	500	5	50	30	100	Haruna, Merican, and Musa (2023)
PMA/H-UiO-66-1	500	3	80	30	100	Zong et al. (2023)
W/UiO-66(Zr)	1000	4	30	30	100	Ye et al. (2021)
[mim(CH$_2$)$_3$COO]$_3$PW@UiO-66	1000	5	70	60	100	Qi et al. (2020)
PW/UiO-66(Zr)-green	1000	6	25	25	99.7	Ye et al. (2020)
PTA@MIL-101(Cr)	500	5	60	120	98.6	Sun, Chen, et al. (2018)
0.2-PMA@MOF-808-H	1000	7	50	90	100	Wang et al. (2022)
42%PTA@MOF-808A	1000	5	60	30	100	Lin et al. (2018)

NPC, Nitrogen-doped porous carbon; *PMA*, phosphomolybdic acid; *PTA*, phosphotungstic acid.

readily decompose, ensuring safer storage and handling. Additionally, O_2 is abundant and readily available from the air, ensuring a nearly limitless supply for reactions (Rao et al., 2007). However, despite these benefits, the practical application of O_2 encounters inevitable challenges, primarily related to its activation. O_2 molecules in their ground state exist in a triplet state, which complicates the formation of highly reactive intermediates during activation (Zhang et al., 2014). Overcoming this hurdle requires the development of more efficient catalysts. Thus efficient catalyst design is crucial for enhancing O_2 activation efficiency and facilitating the formation of reactive intermediates in ODS processes. This approach improves overall reaction efficiency and effectiveness. Continued advancements in catalyst technology are expected to lead to significant breakthroughs in applications of ODS processes utilizing O_2 as the oxidant, thereby advancing green chemistry and environmental protection (Eseva et al., 2020).

Nano-metal oxides typically exhibit high catalytic activity in catalytic ODS processes, with molybdenum trioxide (MoO_3) and tungsten trioxide (WO_3) demonstrating particularly efficient catalytic performance and highly tunable characteristics under O_2 activation. However, single metal oxide catalysts often suffer from instability during the ODS reactions. To address this issue, Li et al. used MOF-199 as a support to encapsulate MoO_3 and WO_3 within the MOF cavities and introduced magnetic Fe_3O_4 into the MOF-199, synthesizing a highly efficient desulfurization catalyst, which was denoted as Fe@W-Mo@MOF (Li, Wang, & Zhao, 2020). The introduction of Fe_3O_4 facilitates easier separation of the catalyst from the reaction system after the ODS process, enhancing product recovery and catalyst recycling. Meanwhile, in this catalyst, W-modified MoO_3 serves as the active component, while the MOF functions as both an adsorbent and a support material. The Fe@W-Mo@MOF catalyst exhibits exceptionally high activity in the ODS system using O_2 as the oxidant. Under conditions of 40°C, the catalyst achieved 100% of desulfurization activity to DBT (the initial sulfur concentration was 400 ppm) within 60 minutes. Remarkably, the catalyst maintains over 95% recovery after 18 cycles and can be reused for over 20 cycles. This high performance is primarily attributed to the enhanced catalytic efficiency induced by the W atoms and the magnetic properties conferred by Fe_3O_4. The design of this catalyst not only improves the activation efficiency of O_2 but also significantly simplifies the recovery and reuse process by introducing the magnetic Fe_3O_4 material. This innovative approach highlights the great potential of MOFs supported catalysts in ODS processes, offering a new pathway for achieving efficient and sustainable industrial desulfurization.

Li et al. (2016) developed a catalyst named POM@MOF-199@MCM-41 (PMM) using a one-pot synthesis method to combine MCM-41 molecular sieve and MOF-199 as composite supports for loading $PMo_{12-x}W_xO_{40} \cdot nH_2O$. This catalyst efficiently catalyzes the ODS of sulfides in model oil using molecular oxygen as the oxidant. Under optimized experimental conditions, the PMM catalyst achieved a catalytic oxidative desulfurization rate of 98.5% for DBT, maintaining its catalytic activity over 10 reuse cycles. This study highlights the importance of loading polyoxometalates (POMs) in advancing catalytic ODS systems that use molecular O_2 as the oxidant. The PMM catalyst demonstrated excellent desulfurization performance and stable recyclability, indicating its significant potential for industrial applications. By leveraging the benefits of MCM-41 and MOF-199, the PMM catalyst not only possesses a high specific surface area but also offers enhanced stability. This research underscores the promising approach of incorporating POMs to improve catalytic ODS of fuel oils utilizing molecular O_2 as the oxidant. Their further research prepared a series of metal-POM@MOF-199@MCM-41 (metal-PMM) for ODS of fuel oils with O_2 as the oxidant, and the reaction rate was significantly enhanced to 99.1% using Co-PMM as the catalyst (Li et al., 2017). Further investigation revealed that the oxidative desulfurization by Co-PMM with O_2 as the oxidant is primarily due to the following mechanism: Initially, $M(O)_n^-$ (where M represents Mo or W) in Co-POM reacts with O_2 molecules to generate active peroxo species $M(O_2)_n^-$. Subsequently, DBT molecules interact with $M(O_2)_n^-$ to form sulfoxide. Finally, the sulfoxide can be rapidly further oxidized by $M(O_2)_n^-$ to yield sulfone, while $M(O_2)_n^-$ is reduced to $M(O)_n^-$, thus completing the catalytic cycle (Fig. 15.3).

Li et al. (2021) also developed a novel spininess-like MOF supported heteropolyacid catalyst, Fe_3O_4@MOF-PMoW (abbreviated as FeMP, where

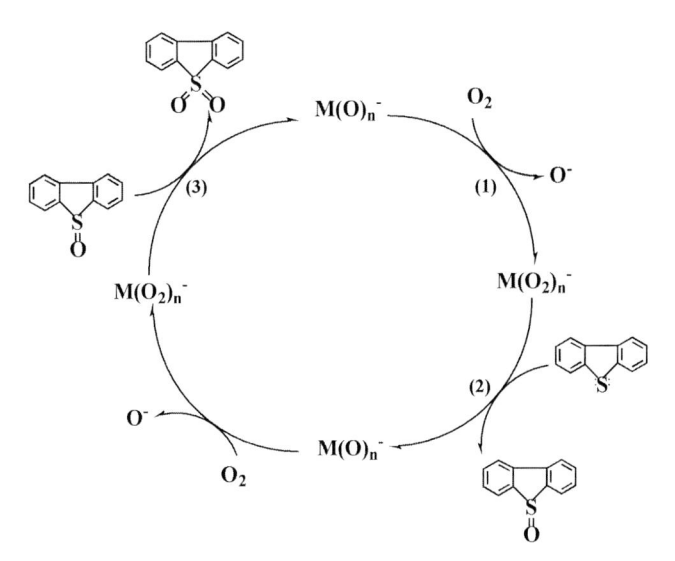

FIGURE 15.3 Possible oxidative desulfurization mechanism with Co-PMM as the catalyst and O_2 as the oxidant (Li et al., 2017).

M=Cr, Cu, Zr). The catalyst features a spininess-like surface morphology that increases the contact area with sulfur compounds, leading to excellent desulfurization performance of fuel oils. The study found that the selected MOF materials, including MIL-101(Cr), MOF-199(Cu), and UiO-66(Zr), did not inherently possess high-efficiency ultra-deep desulfurization performance. However, upon introducing the $H_3PMo_6W_6O_{40}$ heteropolyacid (denoted as PMoW) as the catalytically active species, the ODS activities of FeCrP, FeCuP, and FeZrP to DBT-containing fuel oils (the initial sulfur concentrations were 1000 ppm) reached 100%, 97.0%, and 91.6%, respectively. Detailed reaction mechanism studies revealed that the ODS activities of the MOF-supported heteropolyacid catalysts primarily depends on the high catalytic activity of the coordinated Mo and W atoms. The Mo and W atoms in the PMoW heteropolyacid are converted into peroxo ions, which act as key intermediates in the MOF-supported heteropolyacid-catalyzed ODS system, enhancing O_2 activation. Additionally, the incorporation of magnetic Fe_3O_4 improves the recovery efficiency as well as recyclability of the catalyst.

Similarly, Zhou et al. (2022) encapsulated phosphomolybdic acid (HPMo) into the MOF-199 (HPMo-x@MOF-199) to obtain a stable catalyst for catalytic ODS with O_2 as the oxidant. Experimental results indicated that HPMo could stably exist within the pores of MOF-199, not only increasing the loading amount of HPMo but also enhancing the stability of the catalytic active centers during the ODS reaction. The primary aromatic sulfur compounds in fuel oils are DBT and its derivatives, with 4,6-DMDBT (4,6-dimethyldibenzothiophene) being particularly challenging to be removed deeply in desulfurization process due to its large molecular size and significant steric hindrance. Remarkably, the HPMo-x@MOF-199 exhibited excellent performance and stability in catalytic oxidation of 4,6-DMDBT for ODS of fuel oils with O_2 as the oxidant. Under an oxygen atmosphere at 120°C, the catalyst achieved complete removal of 4,6-DMDBT (the initial sulfur concentration was 200 ppm) within 75 minutes and could be easily reused for 9 cycles. Additionally, the ODS product was characterized to be only 4,6-dimethyldibenzothiophene sulfone (4,6-$DMDBTO_2$), which is a high-value product. The 4,6-$DMDBTO_2$ was easily separable in this ODS system. The catalyst also demonstrated satisfactory performance in real diesel fuel oils, achieving a desulfurization efficiency of 90.2%, showcasing its potential application prospects. The research on this novel MOFs-supported heteropolyacid catalysts not only provides new methods for efficient desulfurization but also demonstrates the significant potential of material design in enhancing catalyst performance. As this field continues to develop, more efficient and stable ODS catalysts, especially the ones can readily activate O_2, are expected to be developed.

To further improve the catalytic ODS performance, modification of MOFs has been developed, which not only can promote the dispersion of catalytically active sites, but also enhance the adsorptive performance, which favors the ODS process. Zhao et al. developed a hybrid catalyst, which was named as CNTs@

MOF-199-POM, using a one-pot synthesis method (Gao et al., 2018). In this system, the employed POM was a vanadium-substituted Dawson-type heteropolyacid ($H_{6+n}P_2Mo_{18-n}VnO_{62} \cdot mH_2O$ (n=1–5)). Meanwhile, the CNTs@MOF-199 consists of MOF-199 doped with acid-treated carbon nanotubes (CNTs). Experimental results demonstrated that the integration of CNTs increased the specific surface area and adsorption capacity of MOF-199, while loading POM onto CNTs@MOF-199 introduced active Mo and V atoms, significantly enhancing the catalytic activity and stability of the MOF material. Under optimal conditions, the catalyst CNTs@MOF-199-Mo$_{16}$V$_2$ achieved a remarkable 98.30% of desulfurization efficiency to DBT-containing fuel oils (the initial sulfur concentrations were 2000 ppm). Moreover, the catalyst showed remarkable reusability, with only a slight decrease in catalytic efficiency after 7 cycles.

Furthermore, other studies have shown that incorporating inorganic porous materials into MOF-supported POM catalysts yields bifunctional catalysts with outstanding performance. This enhancement stems from the higher loading of active POM species within the porous materials and the protective benefits provided by the bifunctional porous material supports. For instance, Zhao et al. developed a highly active Keggin-type POM-based inorganic–organic compound catalyst, which was named as POM@MOF-199@LZSM-5 (PMZ) (Li et al., 2018). The PMZ catalyst was further applied to the catalytic ODS process with O_2 as the oxidant. The results indicated that this bifunctional porous catalyst exhibited 1.63 times higher ODS efficiency to DBT-containing fuel oils (the initial sulfur concentration was 2000 ppm) compared to MOF-supported POM catalysts under identical optimal conditions. These studies highlight that integrating CNTs or other inorganic porous materials into MOF structures can significantly enhance specific surface area, adsorption capacities, and catalytic activities. Additionally, incorporating POM containing Mo and V as active species further enhances catalyst stability and efficiency. Designing bifunctional catalysts not only boosts catalytic performance but also ensures catalyst recyclability, showcasing significant potential for industrial ODS applications, especially the ones with O_2 as the oxidant.

In addition to strategies like morphological control and the introduction of porous materials into MOFs, incorporating multimetal active sites within MOFs has been found to effectively enhance catalytic performances of MOFs. Introducing multiple metal species diversifies and increases catalytic active centers in MOFs, thereby improving their catalytic efficiency. Bimetallic MOFs, which contain two or more metals, have been successfully synthesized and applied in energy storage and conversion systems, catalysis, and adsorption due to their mesoporous structure and synergistic interactions between different metal ions, demonstrating superior performance compared to monometallic counterparts (Shi et al., 2024). For instance, Zhu et al. achieved significant results in the efficient removal of tetracycline (TC) using Zn/Co-MOF@rGO-600 as a catalyst (Qi et al., 2023). By employing Zn/Co-MOF@rGO-600 as a catalyst for peroxymonosulfate (PMS), the reaction rate constant

(k) increased by 14.88 times, achieving a TC removal efficiency of 91.66% and mineralization rates of 45.04%. Lang et al. introduced a third metal ion (Co^{2+} or Mn^{2+}) or utilized nickel foam (NF) to prepare ternary metal MOFs (Fe/Ni/Co(Mn)-MIL-53 and Fe/Ni/Co(Mn)-MIL-53/NF), which exhibited enhanced performance in the oxygen evolution reaction (OER) compared to bimetallic MOFs (Li et al., 2018). Particularly noteworthy is that by introducing five or more metals as central metal, "high-entropy MOFs" can be formed, which combine advantages of MOFs with those of high-entropy materials, including high temperature stability, increased configurational entropy with more elements, high disorder, and a stable crystal structure. With the formation of high-entropy MOFs, the high entropy structure would significantly enhance catalytic ODS performance by activating O_2 molecules as the oxidant, although literature on this topic remains limited (Liu et al., 2023; Xu et al., 2022). Our research group synthesized a high surface area, high porosity high-entropy zeolitic imidazolate framework (HE-ZIF) via a mechanical ball milling method (Liu et al., 2024). The HE-ZIF incorporated five metal elements, including Co, Zn, Ni, Cu, and Mo into ZIF-8, with Mo identified as the primary active element for activating O_2 molecules during the ODS process. Preparation of the HE-ZIF catalyst provides numerous catalytic sites and increased reaction surface area on the ZIF supports, thereby enhancing reaction rates and efficiency in catalytic ODS process with O_2 as the oxidant. Under optimal conditions, the HE-ZIF catalyst demonstrated remarkable catalytic ODS performance, which can almost totally convert DBT to dibenzothiophene dioxide ($DBTO_2$) within just 1.5 hours (the initial sulfur concentration was 200 ppm), showcasing its potential for ultra-deep desulfurization of fuel oils. Compared to low-entropy catalysts, HE-ZIF catalysts offer advantages such as shorter reaction times, higher conversion efficiency, and excellent selectivity.

The utilization of O_2 as an oxidant in ODS of fuel oils presents advantages such as cost-effectiveness, abundant oxidant sources, and environmental friendliness. However, the widespread adoption of these systems is often impeded by the requirement for high-performance catalysts. Enhancing the efficiency of catalytic active intermediates' generation through external field enhancement methods represents a promising approach to improve ODS performance significantly. Techniques such as applying magnetic fields, electric fields, or ultrasound can effectively enhance active site formation and strengthen the interaction between the catalyst and reactants. These methods accelerate reaction rates and overall catalytic activity, thereby enhancing the effectiveness and feasibility of ODS systems for industrial applications.

Metal-organic framework composites for photo-catalytic oxidative desulfurization of fuel oils

In more recent years, photocatalytic oxidative desulfurization (PODS) technology has emerged as a highly promising method for desulfurization, leveraging unique advantages in efficiency and environmental friendliness. Such a PODS approach involves utilizing photocatalysts that absorb photon energy from UV or visible light irradiation. When the photon energy matches or exceeds the band gap of the photocatalyst, electrons are excited from the valence band (VB) to the conduction band (CB), generating electron–hole pairs. Electrons on the photocatalyst surface exhibit reducing properties, reacting with adsorbed O_2 or H_2O_2 to form superoxide radicals ($•O_2^-$) or hydroxyl radicals ($•OH$) (Wang et al., 2024). Meanwhile, holes on the surface possess oxidizing properties, capable of oxidizing sulfur-containing compounds into sulfones, sulfoxides, and other easily separable products. A significant advantage of PODS technology is its operation at ambient temperature and pressure, which eliminates the need for additional energy-intensive equipment (Zhou et al., 2021). Such a characteristic not only reduces energy consumption and costs but also aligns with green chemistry principles by using natural light sources, minimizing environmental impact. While MOFs typically exhibit less promising photocatalytic performance alone, their efficiency can be significantly enhanced through structural adjustments or synergies with other semiconductor materials. Strategies include modifying MOFs' light absorption capability and improving electron–hole separation efficiency by introducing diverse metal centers or functionalizing organic ligands. Additionally, combining MOFs with traditional semiconductors such as TiO_2 (Liu et al., 2018) or zinc oxide (ZnO) (Nadzim et al., 2022) further boosts photocatalytic efficiency, enabling effective ODS of fuel oils. In addition, future advancements in PODS technology will focus on optimizing photocatalyst structures, leveraging advanced materials science and nanotechnology. This includes enhancing MOFs' sulfur compound adsorption capacity through precise control of pore size and surface properties and developing photocatalysts with broader spectral response ranges to utilize more sunlight efficiently.

Beshtar et al. (2024) synthesized titanium-modified UiO-66(Zr) nanophotocatalysts using an improved one-step solvothermal method and evaluated their performances in PODS of fuel oils with H_2O_2 as the oxidant. The introduction of titanium species altered the structure and properties of the catalysts. Their experiments demonstrated that the loading of titanium played a critical role in the formation of titanium species and the structure of the synthesized photocatalyst. Multiple parameters, including temperature, catalyst dosage, oxidant-to-sulfur ratio (O/S), and solvent-to-fuel ratio, were optimized in the PODS reaction. Under optimal conditions (temperature of 50°C, catalyst dosage of 2 g/L, oxidant-to-sulfur molar ratio (O/S) of 8, and solvent-to-fuel ratio of 1), the conversion rate of DBT reached 100%, achieving deep desulfurization of fuel oils. Additionally, the catalyst exhibited remarkable stability, with a total conversion rate decrease of only 18% after 6 cycles of reuse. Their further kinetic studies indicated that the PODS reaction of DBT followed pseudo-first-order kinetics, and the activation energy of the PODS process was determined to be 55.1 kJ/mol. This suggests that the reaction proceeds at a relatively stable rate at a given temperature, and the titanium-modified UiO-66(Zr) nanophotocatalyst exhibited excellent catalytic performance and stability in the photocatalytic oxidative desulfurization process. Their study demonstrates that modifying UiO-66(Zr) with titanium species significantly enhances its activity and stability in PODS of fuel oils. They also investigated the mechanism of PODS: Hydroxyl groups in the reaction solution lead to the capture of h$^+$ radicals in the valence band, resulting in the generation of ·OH radicals. Electrons produced by photocatalysis can react with oxygen molecules to form ·O$_2^-$. Ultimately, through the interactions among organosulfur compounds, H$^+$, ·OH, and ·O$_2^-$, the oxidation of aromatic organosulfur compounds occurs, converting them into more polar sulfones, which are then removed (Fig. 15.4). This proposed mechanism of photocatalytic oxidative desulfurization provides new insights and methods for developing efficient and stable photocatalytic desulfurization catalysts.

Furthermore, during the ODS of fuel oils, compared with H$_2$O$_2$, O$_2$ molecule is always a more ideal oxidant because of low cost, environmental friendliness, green characteristics, abundant source, etc. Thus in PODS systems, researchers are dedicated to designing and synthesizing MOF-based photocatalysts capable of activating O$_2$ molecules under light irradiation. Upon absorbing photon energy, these MOF-based photocatalysts can excite electrons from the valence band to the conduction band, leaving behind holes in the valence band. The resulting electron–hole pairs exhibit strong reducing and oxidizing properties, respectively. These pairs can react with O$_2$ molecules

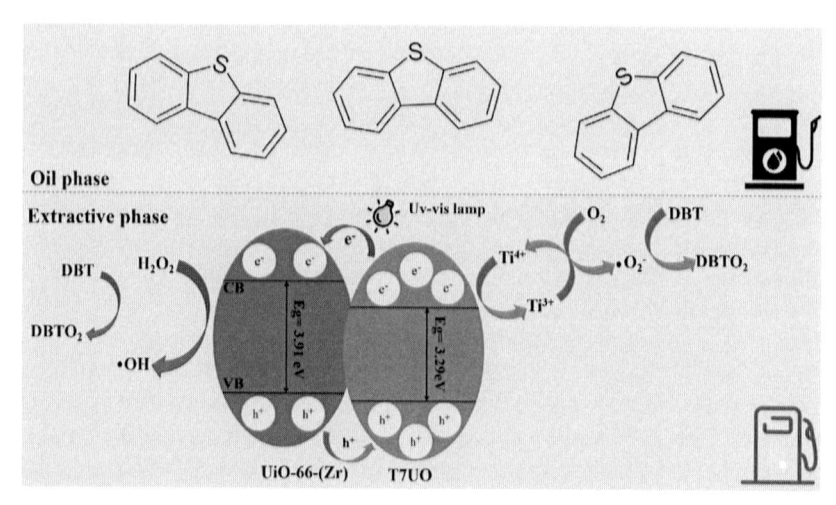

FIGURE 15.4 Proposed mechanism of photocatalytic oxidative desulfurization by titanium-modified UiO-66(Zr) (Beshtar et al., 2024).

adsorbed on the catalyst surface, generating superoxide radicals (•O$_2^-$) or hydroxyl radicals (•OH), which play a crucial role in the photocatalytic reaction. These intermediates can effectively oxidize sulfur compounds in fuel, such as DBT, converting them into easily separable sulfoxides or sulfones, thereby achieving desulfurization (Zhou et al., 2021; Zhou et al., 2024).

Most importantly, traditional ODS systems of fuel oils typically convert sulfur compounds into sulfoxides or sulfones, followed by separation by adsorption or extraction. However, this method fails to completely resolve sulfur removal issues and still suffers from high processing costs and low efficiency. If oxygen can be used as an oxidant in a PODS system to completely mineralize sulfur compounds into sulfur dioxide (SO$_2$) or SO$_4^{2-}$ and carbon dioxide (CO$_2$), then complete desulfurization can be achieved (Chandra Srivastava, 2012). The PODS system utilizes photocatalysts that activate O$_2$ molecules under light irradiation, generating highly oxidative active intermediates. These intermediates can directly oxidize sulfur compounds in fuel into SO$_2$ and CO$_2$. This approach not only ensures thorough desulfurization but also significantly reduces energy consumption and operational costs as the reaction occurs under ambient temperature and pressure.

Ali Morsali et al. were the first to apply MOFs for PODS of sulfur compounds in fuel oils (Bagheri et al., 2017). They incorporated an appropriate amount of MoO$_3$ into the [Zn(oba)(4-bpdh)0.5]$_n$·1.5DMF (TMU-5) to design a composite photocatalyst (MT-5) and investigated its PODS performance. The study indicated that the addition of 3 wt.% MoO$_3$ to TMU-5 acted as a crystal growth inhibitor. Under mild and green reaction conditions, a 5 wt.% MoO$_3$-TMU-5 composite material (MT-5) exhibited remarkable photocatalytic activity in the PODS reaction of model fuel oils, with only 3% of the MoO$_3$ content leaching out during the reaction. The MT-5 photocatalyst induce a PODS efficiency of 95.6% for DBT within approximately 1 hour. To determine the catalytically active species during the PODS process, free radical scavenging experiments and terephthalic acid fluorescence characterizations were carried out. Their results demonstrate that during the PODS process, •O$_2^-$ and •OH are the main active species, which are responsible for the photocatalytic degradation of DBT. The active surface of TMU-5, with the organic linker acting as a media, and the active sites of MoO$_3$ likely enhanced the PODS activity of the MT-5 photocatalyst through synergistic effects. This study achieved the first conversion of DBT into mineral compounds (CO$_2$ and SO$_2$) via PODS under UV and/or visible light irradiation with the photoactive MoO$_3$-TMU-5 composite material. However, its recyclability as a long-life photocatalyst remains relatively low.

Ahmad et al. (2023) reported a novel heterostructured photocatalyst by coating heterometallic ZIF-8 (h-ZIF-8) onto Cu/Ni/ZnO and further supported onto CNTs, which was named as h-ZIF-8@Cu/Ni/ZnO@CNTs, designed for selective PODS of coal and fuel oils. The preparation process involved stabilizing the Cu/Ni/ZnO ternary mixed metal oxide on the CNT sponge, followed by the in situ reaction of metal particles with 2-methylimidazole to form the h-ZIF-8 shell on Cu/Ni/ZnO. The resulting h-ZIF-8@Cu/Ni/ZnO@CNTs exhibited a high surface area (1283.47 m²/g), a mesoporous/microporous framework, a narrow band gap (2.52 eV), and numerous active sites. The growth of h-ZIF-8@Cu/Ni/ZnO on the conductive CNT sponge enhanced charge transfer due to the Schottky junction effect, thereby improving the photocatalytic activity. Additionally, the incorporation of metals altered the band structure, increasing the absorption capability across the visible spectrum. Compared to the control group (ZIF-8@ZnO@CNTs), h-ZIF-8@Cu/Ni/ZnO@CNTs demonstrated superior PODS performance. The h-ZIF-8@Cu/Ni/ZnO@CNTs photocatalyst showed excellent photocatalytic response in removing organosulfur compounds, with a removal rate of 0.76 kg(m·day), significantly higher than the control group's 0.4 kg/(m.day). For DBT, the removal rate further increased to 3.3 kg/(m.day), confirming the catalyst's higher PODS potential for recalcitrant organosulfur compounds. Their research demonstrated that the PODS process is mainly controlled by sulfur adsorption on h-ZIF-8@Cu/Ni/ZnO@CNTs. The high efficiency and reusability of the h-ZIF-8@Cu/Ni/ZnO@CNTs photocatalyst make it a promising material for coal and fuel desulfurization. Their work demonstrates that the formation of bimetallic MOFs, addition of metals for reduced band gap energy, addition of CNTs for enhanced charge transfer, as well as synergistic effects of higher surface area, are favorable for the promotion of PODS performance. Most importantly, the h-ZIF-8@Cu/Ni/ZnO@CNTs photocatalyst can be reused for 10 times with desulfurization efficiency decreased to ~80%.

Although the design of MOF-based photocatalysts and the construction of PODS systems have been extensively researched, several challenges remain in practical applications. Firstly, enhancing photocatalytic efficiency, particularly under real sunlight, is a critical issue. Additionally, reducing or avoiding secondary pollution of the oil during the desulfurization process is essential, as such pollution can severely impact the economic viability and environmental friendliness of the process. Another key challenge is achieving selective oxidation of sulfur compounds to sulfones. In the PODS process, the lack of selectivity often results in complete oxidation of sulfur compounds or the formation of other harmful by-products (SO_2 and CO_2), diminishing the economic and safety benefits of desulfurization. Therefore, it is crucial to develop catalysts with high selectivity that can efficiently convert sulfur compounds to sulfoxides under mild conditions. These issues limit the widespread industrial application of PODS technology. Researchers are thus tasked with improving photocatalytic efficiency while ensuring high selectivity for the oxidation of sulfur compounds, minimizing secondary pollution, and achieving efficient desulfurization under real sunlight irradiation. Addressing these challenges is critical for advancing PODS technology and achieving sustainable industrial desulfurization solutions.

Metal-organic framework composites-derived catalysts for oxidative desulfurization of fuel oils

MOFs are crystalline materials composed of metal ions and organic ligands, known for their ordered porous structures and customizable chemical compositions. When subjected to thermal or other treatments, MOFs and MOF composites can undergo transformations where the original organic components are removed or altered, yielding derivatives such as metal oxides, metal sulfides, or various metal-carbon compounds (Li et al., 2020; Qian et al., 2023). This process modifies surface properties, pore structure, thermal stability, and chemical reactivity of the material, thereby introducing new functionalities and applications for MOF-derived materials. This versatility significantly broadens the diversity and potential applications of MOF composite-based materials (Hayat et al., 2024).

After calcination, MOF-derived carbon materials retain the structural integrity and morphology of the parent MOFs, expanding the range of structural and performance characteristics available in carbon materials. Moreover, direct calcination of MOFs can avoid the need for complex synthesis steps typically associated with producing porous carbon materials, making MOFs highly suitable precursors for such applications. A common synthesis approach for MOF-derived materials involves carbonization at specific temperatures under an appropriate atmosphere (often N_2 or Ar), followed by simple acid treatment or high-temperature formation of metal atoms with MOF-derived catalysts. In addition to pyrolysis in inert atmospheres, MOFs can also be pyrolyzed in O_2 or air, where non-metal atoms (C, N, and O) from organic ligands are oxidized into gases, resulting in MOF-derived metal oxides (Wang & Astruc, 2020; Yang et al., 2019).

Compared to conventional methods for producing metal oxides, the framework structure of MOFs prevents metal particle agglomeration, thereby enhancing the catalytic activity of resulting catalysts. MOF derivatives maintain the porous characteristics of their parent MOFs while exhibiting enhanced stability, electrical conductivity, and catalytic performance, making them valuable materials across various industrial applications (Li et al., 2019; Mukoyoshi & Kitagawa, 2022).

Gao et al. (2023) utilized a Mo-based (Mo-MOF) containing highly dispersed metal sites as a source of designing molybdenum oxide catalysts. They integrated CNTs, known for their excellent electron transfer properties, acceptance capabilities, and high adsorption capacity for organic sulfur compounds, as supports. In their study, urea was employed as a modifier, and subsequent calcination under an Ar atmosphere was conducted to prepare nanostructured MoO_2 supported on CNTs (MoO_2@CNT-Ux). By adjusting the amount of urea and the calcination temperature, the nitrogen content and particle size of MoO_2 in the catalyst could be effectively controlled. Detailed experiments showed that increased N content and reduced MoO_2 particle size were found to significantly enhance the catalyst's catalytic activity. Experimental findings revealed that compared to catalysts without urea modification, MoO$_2$@CNT-Ux modified with urea exhibited superior performance in ODS. The removal efficiencies for typical sulfur-containing compounds using H$_2$O$_2$ as the oxidant, such as DBT, 4-MDBT, and 4,6-DMDBT, all exceeded 99% within 40 minutes of reaction time. Additionally, the catalyst demonstrated excellent cycling stability, showing only slight decreases in sulfur removal rate after 8 cycles. This high stability suggests promising potential for industrial applications. This study not only presents an effective approach for fuel oil desulfurization but also underscores the possibility of recycling oxidation products as valuable resources, highlighting the sustainability and economic feasibility of the proposed method.

Bhadra et al. (2017) synthesized a novel MOF composite material, (ZIF-8(x)@H$_2$N-MIL-125 (ZIF(x)@MOF, where x stands for the composition of ZIF-8), using a strategy in which one MOF is embedded within another MOF, tailored for specific compositions. They employed H$_2$N-MIL-125 (MOF) and ZIF(30)@MOF to prepare mesoporous carbon and TiO$_2$ nanoparticle composite materials via pyrolysis to obtain final products named MDC-P and MDC-C, respectively. Characterization results revealed that MDC-C consisted of crystalline TiO$_2$ supported on a mesoporous carbon material, exhibiting larger pore sizes and smaller, more uniformly dispersed TiO$_2$ nanoparticles compared to MDC-P. The MOF-derived carbon catalyst, MDC-C, demonstrated exceptional catalytic performance in the deep desulfurization of DBT in diesel oils under optimized conditions using H$_2$O$_2$ as the oxidant. It achieved an impressive 99.5% removal of DBT (the initial sulfur concentration was 1000 ppm) within 120 minutes. Notably, even at very low catalyst dosages, MDC-C exhibited a threefold higher kinetic rate for the ODS reaction compared to MDC-P, accompanied by a lower activation energy (19 kJ/mol). The large mesopores and well-dispersed TiO$_2$ nanoparticles in MDC-C were identified as key factors contributing to its outstanding catalytic conversion efficiency. Furthermore, the MDC-C catalyst displayed excellent recyclability, as it could be reused multiple times with simple acetone washing after each reaction cycle. This underscores MDC-C as a promising catalyst for ODS reactions due to its low activation energy, ease of recovery, high kinetics, and efficient throughput. Ultimately, the combination of MOF composition and subsequent pyrolysis represents a competitive method for fabricating functional catalysts, particularly those featuring large pore sizes and high mesoporosity for effective dispersion of active sites.

Beside the formation of metal oxide by pyrolysis of MOFs, by the addition of N precursors and changing the protection gas during the pyrolysis process, different types of metal catalysts can be gained. For example, Sung Hwa Jhung et al. reported the synthesis of carbon-supported vanadium nitride (VN@C) catalysts via pyrolysis of urea-loaded MIL-100(V) MOFs, which can avoid the requirement for external nitrogen sources such as ammonia (Ahmed et al.,

2023). Two variants were prepared: VN@C and carbon-supported vanadium oxide (VO@C), depending on the presence or absence of urea as a nitrogen source. Optimized VN@C catalysts achieved nearly complete conversion of DBT (the initial sulfur content was 1000 ppm) within 30 minutes under mild reaction conditions at 60°C. Notably, VN@C exhibited the highest reported performance and turnover frequency among vanadium-based catalysts. Additionally, the activation energy for DBT oxidation on VN@C was determined to be only 29.9 kJ/mol, lower than other reported V-based catalysts. Their further study revealed that •OH radicals generated by H_2O_2 on VN@C catalysts played a pivotal role in the oxidation reaction. Crucially, nitrogen atoms in VN catalysts were found to enhance H_2O_2 activation for •OH radical production compared to conventional oxygen atoms in oxides, significantly enhancing the efficiency of DBT oxidation and desulfurization. Furthermore, the catalyst could be regenerated via straightforward acetone washing and maintained stable performance over at least 5 ODS cycles with minimal decrease in desulfurization efficiency. They also utilized density functional theory (DFT) calculations to elucidate why the catalytic activity of VN@C is significantly higher than that of VO@C. Frontier molecular orbital theory indicates that during the oxidation reaction, electrons from the highest occupied molecular orbital (HOMO) of the catalyst (such as VN and V_2O_3) transfer to the lowest unoccupied molecular orbital (LUMO) of H_2O_2, thereby activating H_2O_2 and generating reactive oxygen species. The calculations show that the HOMO of VN (-4.31 eV) is closer to the LUMO of H_2O_2 (-0.97 eV) compared to the HOMO of V_2O_3 (-11.96 eV) (Fig. 15.5A). As a result, VN@C exhibits superior catalytic ODS performance in activating H_2O_2 compared to VO@C. Afterward, the VN@C can efficiently activate H_2O_2 to oxidize DBT to $DBTO_2$ for desulfurization (Fig. 15.5B).

While employing MOF composites as precursors for the pyrolytic synthesis of uniformly small-sized supported nanoparticle catalysts often offers high catalytic activity for ODS, the complex and costly synthesis of MOF composites limits the practical application of this strategy. Thus developing low-cost and straightforward synthesis methods for MOFs that can be utilized in pyrolysis to produce high-performance catalysts is a crucial direction for future research. Additionally, the pyrolysis conditions, such as temperature, atmosphere, and duration, significantly influence the structures and catalytic ODS performance of the final catalysts. Therefore another key research focus will be to develop a universal pyrolysis strategy for efficiently synthesizing uniformly small-sized supported nanoparticles. This involves systematically studying and optimizing various parameters during pyrolysis to identify the optimal conditions, ensuring that the nanoparticles exhibit uniform size, morphology, and good dispersibility. In conclusion, although using MOF composites as precursors to prepare high-performance nanoparticle catalysts holds great promise, achieving widespread application of this strategy requires breakthroughs in reducing MOFs synthesis costs, simplifying the synthesis

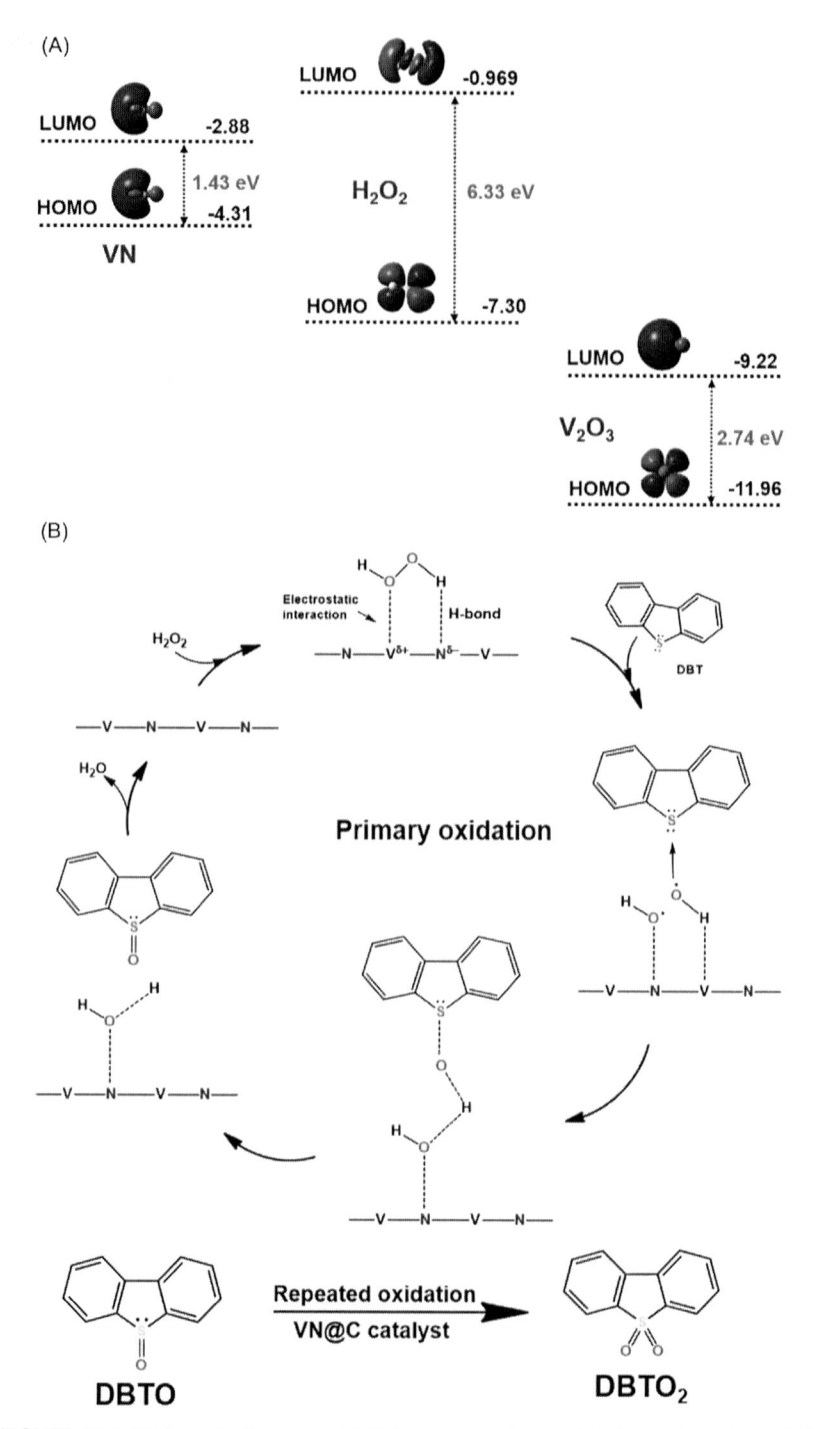

FIGURE 15.5 ODS mechanism using VN@C as the catalyst and H$_2$O$_2$ as the oxidant. (A) Calculated frontier orbital HOMO and LUMO energy levels of VN, H$_2$O$_2$, and V$_2$O$_3$. (B) A possible ODS mechanism for the oxidation of DBT over the VN@C catalysts including the effective activation of H$_2$O$_2$ over the VN (Ahmed et al., 2023).

process, and optimizing pyrolysis conditions. Advancements in these areas will not only propel the development of this field but also provide new solutions for applications in industrial catalysis and energy conversion.

Metal-organic framework composites-based porous ionic liquids for oxidative desulfurization of fuel oils

In recent years, ionic liquids (ILs) have garnered significant attention as catalysts and solvents in catalytic ODS systems for desulfurization of fuel oils. These salts, composed entirely of cations and anions, remain liquid at room temperature and are also known as room temperature ionic liquids, nonaqueous ionic liquids, and liquid organic salts. ILs have become a research hotspot in catalysis due to their tunable structures, particularly in ODS. As excellent extractants, ILs can integrate the catalytic and extraction steps in the ODS process, simplifying the process and reducing costs. This approach overcomes the limitations of traditional extraction desulfurization methods, such as low extraction efficiency and difficulties in separation and recovery. By combining extraction and catalytic oxidation, an extraction-coupled catalytic oxidative desulfurization (ECODS) system can be designed. This system first extracts sulfur compounds into the IL phase, where the higher concentration of sulfur compounds facilitates the catalytic oxidation reaction. Studies have demonstrated that ILs are superior extractants compared to traditional solvents like acetonitrile, DMF, and ethanol (Bhutto et al., 2016; Liu et al., 2021).

In 2007, James et al. introduced the concept of porous liquids (PLs) (O'Reilly et al., 2007), which combines the advantages of solid porous materials and ILs. In recent years, porous ionic liquids (PILs), a subclass of PLs composed of cations and anions, have shown significant advantages in gas capture, highlighting their broad research and application prospects. Especially, the type III PILs is mainly prepared by dispersing porous materials (such as MOFs) uniformly in ILs (Costa Gomes et al., 2018). Due to the electrostatic interactions, hydrogen bonding, and other intermolecular interactions between the organic functional groups of ILs and MOFs, MOFs can be stably dispersed in ILs, forming stable PILs. These interactions ensure the uniform distribution of MOFs within the ILs while preserving their porous structure. Moreover, because the molecular size of ILs is larger than the pore sizes of MOFs, ILs do not penetrate the internal pores of MOFs. This characteristic allows PILs to retain their permanent cavity structure, thereby enhancing their performance in various applications. Through this method, PILs combine the high surface area and structural tunability of MOFs with the unique physicochemical properties of ILs, such as low volatility and high thermal stability. Due to their dual advantages, PILs have garnered extensive attention and have promising applications in catalysis. Compared to traditional homogeneous catalysts, the pores in PILs significantly enhance the dispersion of catalytic active sites. Unlike traditional porous heterogeneous catalysts, the pores in PILs remain open, enhancing mass transfer (Wu et al., 2024; Yang et al., 2022).

Yin et al. (2023) dispersed UiO-66 in a [P$_{6,6,6,14}$][NTF$_2$] IL (trihexyl(tetradecyl)phosphonium bis(trifluoromethylsulfonyl)imide), forming the PILs UiO-66-T3PIL, which remained stable for several months. UiO-66-T3PIL exhibits the catalytic activity of homogeneous catalysts and the ease of separation characteristic of heterogeneous catalysts. In the extraction combined oxidative desulfurization (ECODS) of fuel oils process, UiO-66-T3PIL showed excellent desulfurization performance, achieving a desulfurization rate of 95.3% under optimal conditions. Similarly, we dispersed UiO-66 into a Brønsted acidic ionic liquid ([BSPy][CF$_3$SO$_3$]) to design bifunctional porous ionic liquids (UiO-66-BAPILs) for ODS (Yin et al., 2024). Experiments demonstrated that UiO-66-BAPILs retained some permanent cavities of UiO-66, achieving a desulfurization rate of 99.5% under optimal conditions using H$_2$O$_2$ as the oxidant. Furthermore, UiO-66-BAPILs acted as both the extractant and the catalyst in the ECODS process. These findings pave the way for developing bifunctional Brønsted acidic PILs to enhance catalytic oxidation performance.

We further analyzed the limitations of PILs in the ODS system and found that the application of PILs in liquid–liquid biphasic reactions is constrained in extraction-coupled desulfurization systems using H$_2$O$_2$ as the oxidant. The mass transfer and diffusion of substrate molecules and oxidants between the oil phase, PIL phase, and H$_2$O$_2$ phase require precise adjustment of the hydrophilicity of PILs to improve the contact interface between H$_2$O$_2$ and the oil phase with PILs. Additionally, many PILs lack sufficient catalytic active sites, limiting their catalytic activity. Therefore we propose constructing a class of MOF-confined heteropoly acids as porous frameworks, which not only can disperse heteropoly acids to provide abundant catalytic active sites but also modify the internal surface of MOFs from lipophilic to hydrophilic, enhancing contact with H$_2$O$_2$ for efficient activation. Further, these can be dispersed in lipophilic ILs to form a class of PILs. Using the organic guests of ILs to efficiently extract and enrich sulfur compounds in fuel oils, and utilizing the hydrophilic internal surface of the porous framework and abundant catalytic active sites to efficiently activate H$_2$O$_2$ to generate reactive oxygen species for sulfur compound oxidation, breaks the extraction equilibrium and achieves deep ECODS. To validate this approach, we used ZIF-8-confined HPMo (HPMo@ZIF-8) as the porous framework and dispersed it in [Bpy][NTf$_2$] IL to construct the type III PILs (PILS-M). Studies showed that dispersing HPMo in ZIF-8 pores not only enhanced the dispersion of HPMo and improved the performance of ECODS but also prevented the loss of catalytic active sites during reactions. Additionally, the hydrophilicity of HPMo allowed for the hydrophilicity adjustment of ZIF-8 internal surface, facilitating selective entry and activation of H$_2$O$_2$ within the pores. Using PILS-M for extraction-coupled desulfurization achieved a 100%

desulfurization rate. Compared to nonconfined PILs, the recycling performance was significantly improved, maintaining stable desulfurization performance after 8 cycles, whereas nonconfined PILs showed a significant decrease in desulfurization rate after 4 cycles (Wu et al., 2024).

Metal-organic framework composites for oxidative denitrogenation of fuel oils

Unlike the explicit standards set by many countries and regions for sulfur content in fuel oils, most countries and regions do not have direct regulations for nitrogen compounds in fuel oils. However, nitrogen compounds in fuel oils reduce the activity of HDS catalysts, thus impeding the desulfurization process. Thus it is in urgent need to remove the nitrogen compounds before desulfurization. Nitrogen compounds in fuel oils are generally classified into basic and nonbasic nitrogen compounds based on whether the nitrogen atom's electrons form a delocalized system with the aromatic ring. Currently, the mainstream industrial technology for nitrogen removal is HDN, which involves breaking the C-N bonds in nitrogen-containing compounds (NCCs) and converting nitrogen into NH_3. However, HDN is not suitable for aromatic nitrogen compounds and operates under stringent conditions, prompting researchers to explore cost-effective ODN approaches.

MOFs and MOF composites have been proposed as catalysts for ODS and ODN due to their high porosity, tunable pore structure, and ease of modification and functionalization. However, compared to ODS, progress in ODN research is much slower and requires further development. In the MOF composite-based ODS and ODN studies, the oxidation process of NCCs is more complex than that of sulfur compounds (SCCs) because of the different chemical properties of N and S atoms and the significant influence of oxidant selectivity on the oxidation process (Bhadra & Jhung, 2019). Consequently, ODN requires different catalytic strategies.

Researchers have primarily explored MOF composites-based ODN methods using many common oxidants such as H_2O_2 and O_2. Particularly, MOFs or MOF composites-derived nanomaterials, such as carbon-supported metal oxide nanoparticles, which are constructed from MOFs or MOF composites, have proven to be effective heterogeneous catalysts (Bhadra et al., 2019). Therefore, this character also focuses on the current research on MOF-derived and MOF composites-derived catalysts in ODN. H_2O_2-based ODN research mainly focuses on utilizing its strong oxidizing properties to promote the ODN of NCCs through MOF catalysts. H_2O_2 can generate highly active oxidative intermediates at the active sites of MOFs, which can effectively break C-N bonds and remove nitrogen compounds. However, although O_2 is a green, cost-effective, and source abundant oxidant, ODN using O_2 as the oxidant is still rarely reported. ODN research with O_2 as the oxidant employs MOFs or MOF composites to activate O_2 molecules under light irradiation, generating superoxide radicals ($•O_2^-$) and hydroxyl radicals ($•OH$) (Fakhri et al., 2021). These radicals can oxidize NCCs into easily removable products. Compared to H_2O_2, O_2 as an oxidant has advantages such as wide availability, low cost, and environmental friendliness, but its lower reactivity requires more efficient catalysts to improve reaction efficiency. MOF-derived and MOF composites-derived catalysts, such as supported metal oxide nanoparticles, leverage the framework structure of MOFs and the superior performance of derived materials to further enhance ODN reaction efficiency (Liang et al., 2020). These derived materials show promise in improving catalyst stability, enhancing oxidation performance, and optimizing reaction conditions.

This chapter provides a systematic review of research on technologies involving MOFs, MOF composite, and their derived catalysts for proposing ODN methods, analyzing the applications and advantages of mainstream oxidants H_2O_2 and O_2 in the ODN process. Future research directions include optimizing the structure and performance of MOF composites, exploring new oxidants and reaction conditions, and improving ODN efficiency to provide more feasible solutions for industrial applications.

Metal-organic framework composites for oxidative denitrogenation of fuel oils using H_2O_2 as the oxidant

H_2O_2, with its relatively high active oxygen content of approximately 47% and environmentally friendly by-product (water) during the oxidation process, is considered a green oxidant and is widely used in oxidation research. In current studies on MOF-based ODN, H_2O_2 is the predominant choice of oxidant. However, the inherently low activity of MOFs in activating H_2O_2 for ODN necessitates structural tuning of MOFs to enhance their catalytic performance in ODN. In recent years, MOF composites with active sites such as encapsulated nanoparticles and supported polyoxometalates have shown catalytic effects in ODN similar to those observed in ODS. These MOF-based catalysts achieve efficient catalytic oxidative removal of nitrogen compounds through structural optimization and functional modifications. Defective MOFs introduce structural defects to increase the number of active sites, significantly enhancing catalytic performance. Encapsulated nanoparticles or other catalytically active sites in MOFs utilize the high surface area and adsorption properties of MOFs coupled with the high activity of nanoparticles, thereby improving ODN efficiency. MOFs supported with polyoxometalates enhance adsorption and oxidation activity through the synergistic effects of multiple metals. These improvements enable MOFs to efficiently activate H_2O_2 in the ODN process, generating highly reactive oxidative intermediates that can oxidize nitrogen compounds, thus achieving nitrogen compound removal.

The catalytic active sites of Zr-based MOFs, specifically the Zr-OH groups, demonstrate high stability, substantial voids, and high porosity. Within the UiO-66 family of MOFs, research has shown that substituting Zr with Hf can significantly enhance the ODN activity for nitrogen compounds such as indole and quinoline. This increase in ODN activity is primarily due to alterations in the Brønsted acid sites in MOFs with Hf clusters. Additionally, the oxophilicity of Hf(IV) generates defect sites within the MOFs, leading to the formation of linker vacancies. These vacancies are compensated by Hf-OH and Hf-H$_2$O groups, which facilitate interactions among NCCs, H$_2$O$_2$, and the catalyst, thereby improving catalytic efficiency. It has been reported that at 70°C, using 0.11 mol of 30% H$_2$O$_2$, 5 mg of UiO-66(Hf) can achieve a denitrogenation efficiency of 97% after 5 hours of reaction (both indole and quinoline concentrations in the study were 200 ppm, respectively), compared to 80% of the ODN activity for UiO-66(Zr) (Faria et al., 2021).

Zr-based MOFs are particularly suitable for loading modification due to their high stability. For example, manganese oxide nanoparticles (MnO$_2$-NPs) encapsulated in UiO-66(Zr) have been employed for ODN with sodium hypochlorite (NaClO) as the oxidant (Subhan et al., 2021). In the ODN system with H$_2$O$_2$ as the oxidant, a novel nano UiO-808(Zr) encapsulating polyoxometalates (PMo$_{12}$) has been synthesized via a simple in situ solvothermal one-pot method based on the "bottle-around-ship" approach. This catalyst exhibited excellent ODN activity for indole and quinoline in the ODN system with H$_2$O$_2$ as the oxidant. Under optimized conditions of 70°C, with the addition of 130 mL of 30% H$_2$O$_2$, and equal amounts of 780 mL of model oil and extractant ([BMIM]PF$_6$), UiO-808(Zr) with 1.5 μmol of PMo$_{12}$ active centers achieves complete desulfurization and denitrogenation within 1 hour (the total sulfur content in the model oil was 1500 ppm, and the nitrogen content was 800 ppm). Additionally, the catalyst demonstrates high cyclic stability, attributed to the match between the size of PMo$_{12}$ and the pore size of UiO-808(Zr) (Fernandes et al., 2021).

Studies have reported the employment of MIL-100(Fe) encapsulating PMo$_{12}$ as an active center via a "bottle-around-ship" method (Fig. 15.6) to achieve effective co-catalytic oxidation removal of nitrogen and sulfur compounds. In a multicomponent model fuel containing various sulfur and nitrogen compounds, complete desulfurization and denitrogenation were achieved within 30 minutes. This study was conducted in an ECOD system, specifically examining the effects of two different extractants, ethanol and [BMIM]PF$_6$, on the system. It was found that both extractants achieved similar removal efficiencies. However, ethanol enabled complete desulfurization and denitrogenation within 30 minutes, whereas the [BMIM]PF$_6$ system required 1 hour to achieve the same performance. Under the conditions including equal volumes of extractant and model oil, a temperature of 70°C, 0.7 mmol of 30% H$_2$O$_2$, and a catalyst amount of 3 μmol, total removals of sulfur and nitrogen compounds were gained (the total sulfur content in the model oil was 2000 ppm, with indole and quinoline concentrations both at 300 ppm, respectively) (Silva et al., 2023).

It is noteworthy that the inclusion of hydrophilic [BMIM]PF$_6$/ethanol as extractants in the H$_2$O$_2$ system not only accelerates the contact between H$_2$O$_2$, nitrogen compounds, and the catalyst, thereby reducing reaction time, but also enhances the solubility of oxidized nitrogen compounds in the extractant due to

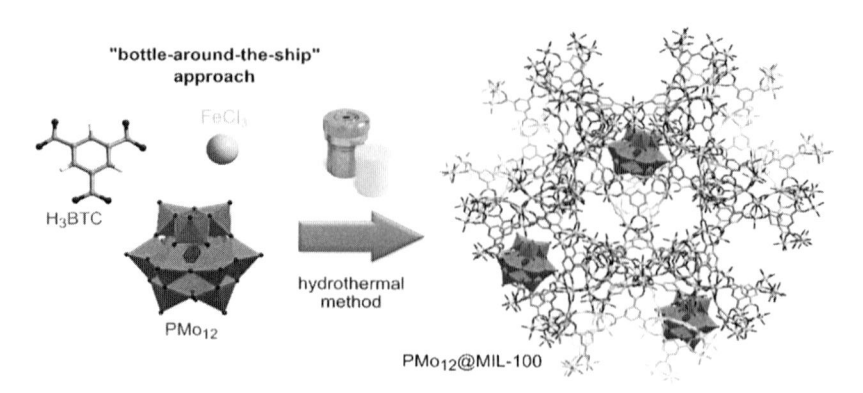

FIGURE 15.6 The preparation of the PMo$_{12}$@MIL-100(Fe) nanocomposite by the "bottle-around-the ship" approach (Silva et al., 2023).

their increased polarity, facilitating the recovery process. In the extractive oxidative desulfurization and denitrogenation system, literature reports the use of an ionic-layered coordination polymer based on a flexible triphosphonic acid linker to form layered MOF ([Gd(H$_4$nmp)(H$_2$O)$_2$]Cl$_2$·H$_2$O) for the simultaneous removal of sulfur and nitrogen compounds in a model oil/H$_2$O$_2$/[BMIM]PF$_6$ system. Under optimized conditions of 70°C, with 0.43 mmol of 30 wt.% H$_2$O$_2$ and a 2:1 volume ratio of model oil to extractant ([BMIM]PF$_6$), 20 mg of the catalyst enabled the complete oxidative removal of indole and quinoline within 1 hour, with both compounds initially at 200 ppm. Various synthesis methods for this catalyst were also investigated, showing no significant difference in catalytic activity, only catalyst characterization revealed that crystalline transformation did not affect the catalyst's initial layered structure, thereby maintaining its catalytic activity (Mirante et al., 2020).

Moreover, encapsulating ionic liquids (imidazole-based polyoxometalates) into ZIF-8 to form [BMIM]PMo$_{12}$@ZIF-8 demonstrated significant oxidative activity for both SCCs and NCCs in an H$_2$O$_2$/[BMIM]BF$_6$ extraction-oxidation system. Under conditions of 70°C and equal molar volumes of model oil and extractant, 3 mmol of the catalyst achieved 100% removal of SCCs and NCCs within 1 hour. The catalyst could be reused for 10 cycles without performance decrease (Silva et al., 2022).

The layered structure and pore size of MOFs significantly influence their catalytic activity. Research has demonstrated that synthesizing titanium-based MOFs (Ti-MOFs), specifically MIL-125 nanocrystals with structural defects using a coordination modulation method, results in enhanced catalytic properties compared to the original microporous MIL-125. These defect-rich MIL-125 nanocrystals exhibit a hierarchical micro-, meso-, and macroporous structure. The defects in the linker units provide an abundance of coordinatively unsaturated sites (CUSs), which greatly enhance intrinsic catalytic activity. The findings indicate that the defect-rich, hierarchical porous MIL-125 nanocrystals exhibit high catalytic performance for ODN. This superior catalytic performance is attributed to the hierarchical pore structure that overcomes the mass transfer limitations of large molecules, significantly increasing the accessibility of active sites. In an ODN system with H_2O_2 as the oxidant, the unmodified MIL-125 achieves denitrogenation rates of only 77.8% and 77.2% for indole and 2-methylquinoline, respectively, after 30 minutes of reaction. In contrast, the modified defect-rich MIL-125 achieves complete 100% removal of these nitrogen compounds within just 15 minutes under the same conditions (Lv et al., 2023).

In the context of ODN systems of fuel oils using H_2O_2 as the oxidant, various strategies have been employed to enhance the catalytic performance of MOFs. These strategies include altering the central metal of the MOFs, creating and regulating defect sites, tuning the pore system to encapsulate catalytic active substances, modifying the layered structure, and synthesizing new layered compounds. These approaches have not only yielded excellent research outcomes but also laid a solid foundation for future ODN research involving MOFs, providing valuable insights for researchers in the field.

Metal-organic framework composites for oxidative denitrogenation of fuel oils using O_2 as the oxidant

Compared to other oxidants, molecular O_2 is a green, nontoxic, inexpensive, and highly safe alternative. Currently, research on MOFs activating O_2 for ODN is limited to the field of photocatalysis. Studies suggest that chemical protonation and impregnation of graphitic carbon nitride (g-C_3N_4) can serve as feasible methods to synthesize novel composites of g-C_3N_4 derivatives with MOF materials such as MIL-100(Fe) for photocatalytic ODN. It has been found that suitably protonated g-C_3N_4 can uniformly coat the MIL-100(Fe) framework, exhibiting strong interactions. This coating effect provides a platform for superior electron transfer and the separation of photogenerated electron (e^-) and hole (h^+) pairs. The combination of h^+ and the main active species $\cdot O_2^-$ facilitates the catalytic conversion of pyridine to pyridine-N-oxide. As a result, this composite material acts as an effective photocatalyst for the oxidation and removal of pyridine under visible light (Huang et al., 2018).

However, aside from the aforementioned study, the use of O_2 as an oxidant in ODN remains largely unexplored and urgently requires further investigation by researchers in the field.

Metal-organic framework composites-derived catalysts for oxidative denitrogenation of fuel oils

MOFs-derived carbon materials present promising applications in catalysis, addressing potential limitations of MOFs such as limited thermal or chemical stability. First developed from MOF-5 by Liu et al. (2008), MOFs-derived nitrogen-rich/doped carbons have a unique set of nitrogen atoms. The carbonization temperature affects the carbon composition, and the properties and applications of doped porous carbons depend highly on the MOF components and carbonization conditions.

Studies have shown that TiO_2 catalysts supported on porous carbon materials (TiO_2@C), being derived from [ZIF-8(30)@H_2N-MIL-125], can be used for the ODN of fuel oils containing various NCCs. The ODN performances for NCCs were found to be: indole > 2-methylindole > 3-methylindole > 1-methylindole > pyrrole > carbazole. This order is due to the different electron densities on the nitrogen atoms of NCCs, similar to the oxidation order of SCCs. Despite carbazole's high electron density on the nitrogen atom, the stability of carbazole due to its aromatic rings (benzene and pyrrole) is determined by its band gap. TiO_2@C showed turnover frequencies (TOFs) 8–17 times higher than pure TiO_2 nanoparticles. Particularly, under conditions of 50°C, with H_2O_2 as the oxidant, and 50 mg of the catalyst, a 90% of denitrogenation efficiency of indole was achieved in 2 hours. In addition, they clarified the ODN processes of different nitrides, and the results demonstrated that different nitrides underwent different reaction processes to generate different products, and detailed reaction processes are summarized in Fig. 15.7 (Bhadra et al., 2019).

ODN by the TiO_2@C catalyst, obtained by pyrolysis of ZIF-8(x)@H_2N-MIL-125, under ultrasonic conditions has also been reported (Bhadra et al., 2021). Under ultrasonic irradiation, 5 mg of TiO_2@C achieved a 90% denitrogenation efficiency to indole within 30 minutes with H_2O_2 as the oxidant. The oxidation reaction occurs at the nitrogen atom, with the oxygen atom transferring to the adjacent carbon to form a more stable oxidation product. Similarly, a tungsten nitride porous carbon (W_2N@C) based on phosphotungstic acid and MAF-6 was used for ultrasonic-assisted ODN with H_2O_2 as the oxidant (Bhadra et al., 2021). With 5 mg of the catalyst, a 60% denitrogenation efficiency of stubborn carbazole was achieved in 120 minutes. The TOF for carbazole was reported to be 80–147 times higher than previous reports, highlighting the dominant role of nitrogen atom density in the ODN process.

It is evident that titanium-based and nitride porous carbon materials exhibit excellent performance in ODN. Research has also reported the carbonization of a mixture of titanium-based MOF (MIL-125(Ti)) and melamine to prepare nitrogen-rich carbon-supported titanium nitride catalysts (TiN@CN-2) (Mondol et al., 2022). The optimal catalyst in this series showed significant performance in ODN, with a low activation energy for indole (31.0 kJ/mol), a high TOF (58.1 1/h), and complete removal of 2000 ppm indole using only 10 mg of catalyst at 60°C.

In conclusion, MOF-derived metal oxides and porous carbon materials show great potential in the ODN. Especially, titanium-based and nitride porous

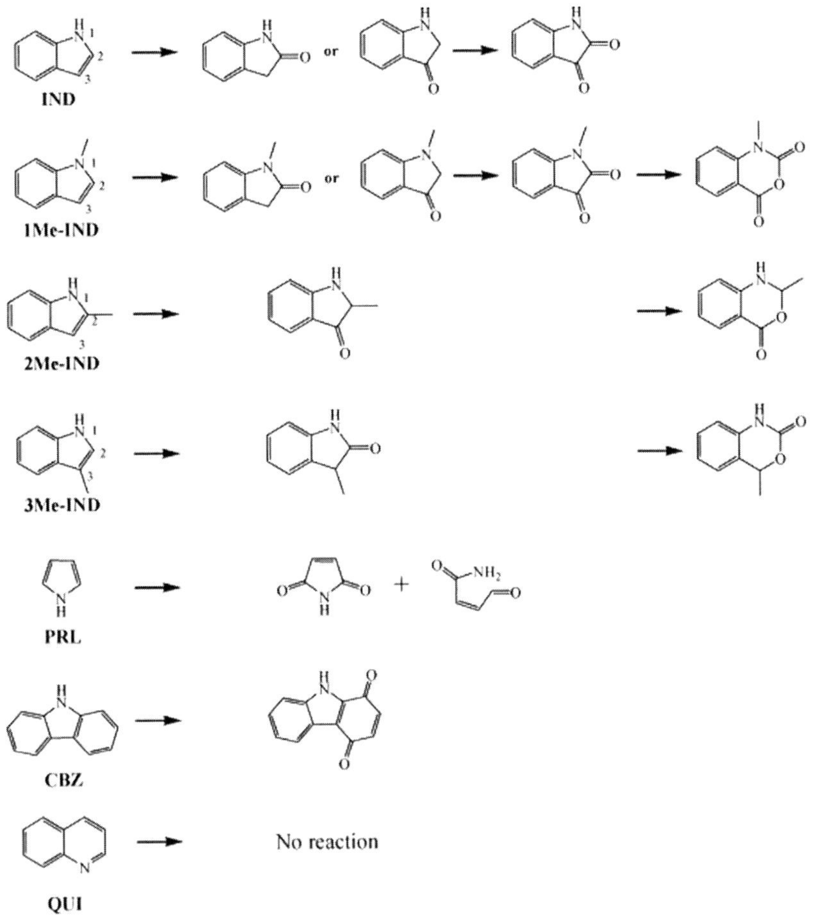

FIGURE 15.7 Chemical structures of the nitrides and the obtained products or intermediates after oxidation reactions (Bhadra et al., 2019).

carbon materials have proven effective in catalytic ODN of fuel oils. Furthermore, the structural modifications of MOF-derived materials significantly impact ODN performance. Future research can expand on NCCs and oxidants, focusing on the electronic density of nitrogen atoms and achieving higher denitrogenation efficiency, thereby building on the foundation laid by current studies and inspiring subsequent researchers to explore related work.

Conclusions and outlook

In conclusion, MOF composites have emerged as a pivotal class of catalysts in the pursuit of more efficient ODS and ODN technologies for producing clean fuel oils. These alternatives to traditional HDS and HDN methods offer significant advantages, such as improved selectivity to aromatic sulfides and nitrides, milder reaction conditions, and lower energy consumption, making them well-suited for achieving ultra-deep sulfur and nitrogen removal. The versatility of MOF composites lies in their ability to host a wide array of active catalytic species while providing high surface areas, tunable pore structures, and tailored chemical properties. This enables them to address the limitations of conventional processes, particularly in the removal of stubborn SCCs and NCCs from fuel oils, which are resistant to HDS and HDN treatments.

The use of MOF composites in ODS has seen significant progress, particularly in their ability to facilitate the oxidation of aromatic sulfur compounds, which are typically more challenging to remove. Through structural modification and the incorporation of catalytic active centers, the catalysts can achieve higher catalytic activity and stability, even in harsher industrial environments. This adaptability has enabled MOFs to outcompete traditional catalysts in terms of both efficiency and environmental impact. However, challenges remain in optimizing these systems for a wider variety of sulfur-containing compounds, particularly nonaromatic sulfur species, where the efficiency of MOF catalysts tends to decrease. Additionally, the exploration of catalyst recovery and reusability continues to be an area of active research, as the ability to regenerate and recycle MOF composites without significant loss of activity is crucial for their industrial viability.

ODN research, while not as developed as ODS, is now gaining momentum, driven by the growing need to address NCCs in fuel oils, such as indoles and quinolines, which are particularly resistant to removal by traditional methods. MOFs and MOF composites offer significant potential for ODN catalysis due to their inherent structural flexibility, allowing for precise control over the active sites and catalytic environments necessary for nitrogen compound oxidation. Although the oxidation mechanisms of ODS and ODN share similarities, the higher electron density on nitrogen atoms in NCCs requires distinct catalytic strategies. MOF composites have begun to demonstrate their capacity to meet these challenges by offering tunable acidity, basicity, and redox properties. However, much work remains to be done in fully understanding the mechanistic aspects of ODN, and future research must focus on unlocking these molecular-level insights to achieve catalytic efficiency on par with ODS.

Moreover, MOF composites provide an almost limitless platform for innovation in catalytic material design. The modular nature of MOFs allows for the integration of various active species, such as ionic liquids, nanoparticles, and heteropoly acids, which can be incorporated into the framework or hosted within its pores. These hybrid catalysts have demonstrated unique properties, such as enhanced stability, increased surface area, and the ability to operate under mild conditions, making them particularly attractive for both ODS and ODN. This versatility extends to MOF derivatives as well, with materials such as MOF-derived metal oxides, porous carbon structures, and PILs showing great potential for improving catalytic performance. The development of these materials will play a key role in addressing the evolving demands of the energy and chemical industries, particularly as stricter environmental regulations necessitate the development of cleaner, more sustainable processes.

As MOF composite research progresses, several areas are ripe for further exploration. The scalability of these materials for industrial applications remains a critical concern, particularly in terms of cost-effectiveness, long-term stability, and ease of recovery and regeneration. While laboratory studies have demonstrated the feasibility of MOF-based catalysts and MOF composites for ODS and ODN, translating these successes into commercial processes will require overcoming challenges related to material synthesis, process optimization, and system integration. Additionally, future research should focus on enhancing the recyclability of MOF catalysts, with particular emphasis on green synthesis techniques that minimize waste and energy consumption. Such advances are vital for aligning MOF technologies with global sustainability goals, especially in light of the increasing emphasis on carbon neutrality and reducing the environmental footprint of industrial processes.

Furthermore, understanding the deeper mechanistic insights into how MOFs facilitate the oxidation of sulfur and nitrogen compounds will be key to driving future innovation. Advanced characterization techniques and computational modeling can shed light on the molecular-level interactions that govern catalytic activity, enabling the rational design of new MOF structures tailored for specific applications. By focusing on these fundamental aspects, researchers can develop next-generation MOF composites that not only deliver superior performance but also offer enhanced selectivity, stability, and durability.

In summary, the field of MOF composites for ODS and ODN stands at the forefront of catalytic innovation, offering a versatile and sustainable solution for the purification of fuel oils. The ability of MOF composites to outperform conventional catalysts in terms of efficiency, selectivity, and environmental impact positions them as a key technology in the transition to cleaner energy systems. However, there remains a pressing need for continued research to address the challenges of scalability, cost, and recyclability. As the world moves toward stricter environmental regulations and greater demand for sustainable technologies, MOF-based catalysts are poised to play a transformative role in the production of ultra-low sulfur and nitrogen fuels. Future developments in this field will likely focus on refining the molecular understanding of catalytic processes, developing multifunctional catalysts, and integrating green chemistry principles into material synthesis, ultimately paving the way for more sustainable and energy-efficient fuel purification processes.

AI disclosure

During the preparation of this work, the author(s) used ChatGPT in order to improve language and grammar, and did not involve any scientific content or generate any text content. Furthermore, author(s) carefully reviewed the modified content to avoid any scientific errors. After using this tool/service, the author(s) reviewed and edited the content as needed and take(s) full responsibility for the content of the publication.

References

Abney, C. W., Patterson, J. T., Gilhula, J. C., Wang, L., Hensley, D. K., Chen, J., Shiou Foo, G., Wu, Z., & Dai, S. (2017). Controlling interfacial properties in supported metal oxide catalysts through metal-organic framework templating. *Journal of Materials Chemistry A, 5*(26), 13565–13572. https://doi.org/10.1039/C7TA03894A.

Ahmad, M., Yousaf, M., Cai, W., & Zhao, Z.-P. (2023). Formulation of heterometallic ZIF-8@Cu/Ni/ZnO@CNTs heterostructure photocatalyst for ultra-deep desulphurization of coal and fuels. *Chemical Engineering Journal, 453*. https://doi.org/10.1016/j.cej.2022.139846.

Ahmed, I., Abedin Khan, N., Hasan, Z., & Jhung, S. H. (2013). Adsorptive denitrogenation of model fuels with porous metal-organic framework (MOF) MIL-101 impregnated with phosphotungstic acid: Effect of acid site inclusion. *Journal of Hazardous Materials, 250-251*, 37–44. https://doi.org/10.1016/j.jhazmat.2013.01.024.

Ahmed, I., Kim, C.-U., & Jhung, S. H. (2023). Carbon-supported vanadium nitride catalyst, prepared from urea-loaded MIL-100(V) in the absence of external ammonia flow, having good performance in oxidative desulfurization. *Journal of Cleaner Production, 384*, 135509. https://doi.org/10.1016/j.jclepro.2022.135509.

An, X., Gao, X., Yin, J., Xu, L., Zhang, B., He, J., Li, H., Li, H., & Jiang, W. (2024). MOF-derived amorphous molybdenum trioxide anchored on nitrogen-doped porous carbon by facile in situ annealing for efficient oxidative desulfurization. *Chemical Engineering Journal, 480*, 147879. https://doi.org/10.1016/j.cej.2023.147879.

Annamalai, J., Murugan, P., Ganapathy, D., Nallaswamy, D., Atchudan, R., Arya, S., Khosla, A., Barathi, S., & Ashok, K. (2022). Sundramoorthy, Synthesis of various dimensional metal organic frameworks (MOFs) and their hybrid composites for emerging applications – a review. *Chemosphere, 298*. https://doi.org/10.1016/j.chemosphere.2022.134184.

Bagheri, M., Masoomi, M. Y., & Morsali, A. (2017). A MoO3-metal-organic framework composite as a simultaneous photocatalyst and catalyst in the PODS process of light oil. *ACS Catalysis, 7*(10), 6949–6956. https://doi.org/10.1021/acscatal.7b02581.

Beshtar, M., Larimi, A., Akbar Asgharinezhad, A., & Khorasheh, F. (2024). Ultra-deep photocatalytic oxidative desulfurization of model fuel using Ti-UiO-66(Zr) metal–organic framework. *Catalysis Letters, 154*(6), 2633–2647. https://doi.org/10.1007/s10562-023-04506-9.

Bhadra, B. N., Baek, Y. S., Choi, C. H., & Jhung, S. H. (2021). How neutral nitrogen-containing compounds are oxidized in oxidative-denitrogenation of liquid fuel with TiO2@carbon. *Physical Chemistry Chemical Physics, 23*(14), 8368–8374. https://doi.org/10.1039/d1cp00633a.

Bhadra, B. N., & Jhung, S. H. (2019). Oxidative desulfurization and denitrogenation of fuels using metal-organic framework-based/-derived catalysts. *Applied Catalysis B: Environmental, 259*. https://doi.org/10.1016/j.apcatb.2019.118021.

Bhadra, B. N., Song, J. Y., Khan, N. A., & Jhung, S. H. (2017). TiO2-containing carbon derived from a metal-organic framework composite: A highly active catalyst for oxidative desulfurization. *ACS Applied Materials and Interfaces, 9*(36), 31192–31202. https://doi.org/10.1021/acsami.7b10336.

Bhadra, B. N., Baek, Y. S., Kim, S., Choi, C. H., & Jhung, S. H. (2021). Oxidative denitrogenation of liquid fuel over W2N@carbon catalyst derived from a phosphotungstinic acid encapsulated metal–azolate framework. *Applied Catalysis B: Environmental, 285*. https://doi.org/10.1016/j.apcatb.2020.119842.

Bhadra, B. N., Song, J. Y., Uddin, N., Khan, N. A., Kim, S., Choi, C. H., & Jhung, S. H. (2019). Oxidative denitrogenation with TiO2@porous carbon catalyst for purification of fuel: Chemical aspects. *Applied Catalysis B: Environmental, 240*, 215–224. https://doi.org/10.1016/j.apcatb.2018.09.004.

Bhutto, A. W., Abro, R., Gao, S., Abbas, T., Chen, X., & Yu, G. (2016). Oxidative desulfurization of fuel oils using ionic liquids: A review. *Journal of the Taiwan Institute of Chemical Engineers, 62*, 84–97. https://doi.org/10.1016/j.jtice.2016.01.014.

Cai, L.-X., Li, S.-C., Yan, D.-N., Zhou, L.-P., Guo, F., & Sun, Q.-F. (2018). Water-soluble redox-active cage hosting polyoxometalates for selective desulfurization catalysis. *Journal of the American Chemical Society, 140*(14), 4869–4876. https://doi.org/10.1021/jacs.8b00394.

Chandra Srivastava, V. (2012). An evaluation of desulfurization technologies for sulfur removal from liquid fuels. *RSC Advances, 2*(3), 759–783. https://doi.org/10.1039/C1RA00309G.

Costa Gomes, M., Pison, L., Červinka, C., & Padua, A. (2018). Porous ionic liquids or liquid metal–organic frameworks? *Angewandte Chemie International Edition, 57*(37), 11909–11912. https://doi.org/10.1002/anie.201805495.

Dong, Y., Zhang, J., Ma, Z., Xu, H., Yang, H., Yang, L., Bai, L., Wei, D., Wang, W., & Chen, H. (2019). Preparation of Co-Mo-O ultrathin nanosheets with outstanding catalytic performance in aerobic oxidative desulfurization. *Chemical Communications, 55*(93), 13995–13998. https://doi.org/10.1039/c9cc07452j.

Eseva, Akopyan, Anisimov, & Maksimov, A. (2020). Oxidative desulfurization of hydrocarbon feedstock using oxygen as oxidizing agent (a review). *Petroleum Chemistry, 60*(9), 979–990. https://doi.org/10.1134/S0965544120090091.

Fakhri, H., Esrafili, A., Farzadkia, M., Boukherroub, R., Srivastava, V., & Sillanpää, M. (2021). Preparation of tungstophosphoric acid/cerium-doped NH2-UiO-66 Z-scheme photocatalyst: A new candidate for green photo-oxidation of dibenzothiophene and quinoline using molecular oxygen as the oxidant. *New Journal of Chemistry, 45*(24), 10897–10906. https://doi.org/10.1039/D1NJ00328C.

Falcaro, P., Ricco, R., Yazdi, A., Imaz, I., Furukawa, S., Maspoch, D., Ameloot, R., Evans, J. D., & Doonan, C. J. (2016). Application of metal and metal oxide nanoparticles at MOFs. *Coordination Chemistry Reviews, 307*, 237–254. https://doi.org/10.1016/j.ccr.2015.08.002.

Faria, R. G., Julião, D., Balula, S. S., & Cunha-Silva, L. (2021). Hf-Based UiO-66 as adsorptive compound and oxidative catalyst for denitrogenation processes. *Compounds, 1*(1), 3–14. https://doi.org/10.3390/compounds1010002.

Fernandes, S. C., Viana, A. M., de Castro, B., Cunha-Silva, L., & Balula, S. S. (2021). Synergistic combination of the nanoporous system of MOF-808 with a polyoxomolybdate to design an effective catalyst: Simultaneous oxidative desulfurization and denitrogenation processes. *Sustainable Energy and Fuels, 5*(16), 4032–4040. https://doi.org/10.1039/d1se00522g.

Fortea-Pérez, F. R., Mon, M., Ferrando-Soria, J., Boronat, M., Leyva-Pérez, A., Corma, A., Herrera, J. M., Osadchii, D., Gascon, J., Armentano, D., & Pardo, E. (2017). The MOF-driven synthesis of supported palladium clusters with catalytic activity for carbene-mediated chemistry. *Nature Materials, 16*(7), 760–766. https://doi.org/10.1038/nmat4910.

Freund, R., Zaremba, O., Arnauts, G., Ameloot, R., Skorupskii, G., Dincă, M., Bavykina, A., Gascon, J., Ejsmont, A., Goscianska, J., Kalmutzki, M., Lächelt, U., Ploetz, E., Diercks, C. S., & Wuttke, S. (2021). The current status of MOF and COF applications. *Angewandte Chemie - International Edition, 60*(45), 23975–24001. https://doi.org/10.1002/anie.202106259.

Gao, X., Jiang, W., An, X., Xu, L., He, J., Li, H., Zhang, M., Zhu, W., & Li, H. (2023). Construction of Mo-MOF-derived molybdenum dioxide on carbon nanotubes with tunable nitrogen content and particle size for oxidative desulfurization. In: *Fuel Processing Technology, 239.* https://doi.org/10.1016/j.fuproc.2022.107526.

Gao, Y., Lv, Z., Gao, R., Zhang, G., Zheng, Y., & Zhao, J. (2018). Oxidative desulfurization process of model fuel under molecular oxygen by polyoxometalate loaded in hybrid material CNTs@MOF-199 as catalyst. *Journal of Hazardous Materials, 359*, 258–265. https://doi.org/10.1016/j.jhazmat.2018.07.008.

Granadeiro, C. M., Nogueira, L. S., Julião, D., Mirante, F., Ananias, D., Balula, S. S., & Cunha-Silva, L. (2016). Influence of a porous MOF support on the catalytic performance of Eu-polyoxometalate based materials: Desulfurization of a model diesel. *Catalysis Science and Technology, 6*(5), 1515–1522. https://doi.org/10.1039/C5CY01110H.

Haruna, A., Merican, Z. M. A., & Musa, S. G. (2022). Recent advances in catalytic oxidative desulfurization of fuel oil – a review. *Journal of Industrial and Engineering Chemistry, 112*, 20–36. https://doi.org/10.1016/j.jiec.2022.05.023.

Haruna, A., Merican, Z. M. A., Musa, S. G., & Rahman, M. B. A. (2023). MOF-808(Zr)-supported with Keggin polyoxometalates as an efficient oxidative desulfurization catalyst. *Journal of the Taiwan Institute of Chemical Engineers, 147*, 104919. https://doi.org/10.1016/j.jtice.2023.104919.

Haruna, A., Merican, Z. M. A., & Musa, S. G. (2023). Remarkable stability and catalytic performance of PW11M@MOF-808 (M=Mn and Cu) nanocomposites for oxidative desulfurization of fuel oil. *Molecular Catalysis, 541.* https://doi.org/10.1016/j.mcat.2023.113079.

Hayat, A., Rauf, S., Al Alwan, B., Jery, A. E., Almuqati, N., Melhi, S., Amin, M. A., Al-Hadeethi, Y., Sohail, M., Orooji, Y., & Lv, W. (2024). Recent advance in MOFs and MOF-based composites: Synthesis, properties, and applications. *Materials Today Energy, 41.* https://doi.org/10.1016/j.mtener.2024.101542.

Huang, J., Zhang, X., Song, H., Chen, C., Han, F., & Wen, C. (2018). Protonated graphitic carbon nitride coated metal-organic frameworks with enhanced visible-light photocatalytic activity for contaminants degradation. *Applied Surface Science, 441*, 85–98. https://doi.org/10.1016/j.apsusc.2018.02.027.

Jafari, Z., Golshan, M., & Akbari, A. (2024). (Zn/Co)-ZIFs@TiO2 composite catalysts for oxidative desulfurization: Impacts of Zn2+/Co2+ on TiO2 interactions. *Journal of Environmental Chemical Engineering, 12*(1). https://doi.org/10.1016/j.jece.2024.111874.

Kinik, F. P., Uzun, A., & Keskin, S. (2017). Ionic liquid/metal–organic framework composites: From synthesis to applications. *ChemSusChem, 10*(14), 2842–2863. https://doi.org/10.1002/cssc.201700716.

Li, B., Ma, J.-G., & Cheng, P. (2019). Integration of metal nanoparticles into metal–organic frameworks for composite catalysts: Design and synthetic strategy. *Small, 15*(32). https://doi.org/10.1002/smll.201804849.

Li, F.-L., Shao, Q., Huang, X., & Lang, J.-P. (2018). Nanoscale trimetallic metal–organic frameworks enable efficient oxygen evolution electrocatalysis. *Angewandte Chemie International Edition, 57*(7), 1888–1892. https://doi.org/10.1002/anie.201711376.

Li, J., Yang, Z., Hu, G., & Zhao, J. (2020). Heteropolyacid supported MOF fibers for oxidative desulfurization of fuel. *Chemical Engineering Journal, 388*. https://doi.org/10.1016/j.cej.2020.124325.

Li, J., Xiaolei, J., Mingyuan, Z., & Dai (2023). Cu(I) anchoring in MOF-808 as a stable catalyst in ultra-deep oxidation desulfurization. *Fuel, 341*. https://doi.org/10.1016/j.fuel.2023.127674.

Li, S.-W., Gao, R.-M., Zhang, R.-L., & Zhao, J.-she (2016). Template method for a hybrid catalyst material POM@MOF-199 anchored on MCM-41: Highly oxidative desulfurization of DBT under molecular oxygen. *Fuel, 184*, 18–27. https://doi.org/10.1016/j.fuel.2016.06.132.

Li, S.-W., Gao, R.-M., Zhang, W., Zhang, Y., & Zhao, J.-she (2018). Heteropolyacids supported on macroporous materials POM@MOF-199@LZSM-5: Highly catalytic performance in oxidative desulfurization of fuel oil with oxygen. *Fuel, 221*, 1–11. https://doi.org/10.1016/j.fuel.2017.12.093.

Li, S.-W., Li, J.-R., Gao, Y., Liang, L.-L., Zhang, R.-L., & Zhao, J.-she (2017). Metal modified heteropolyacid incorporated into porous materials for a highly oxidative desulfurization of DBT under molecular oxygen. *Fuel, 197*, 551–561. https://doi.org/10.1016/j.fuel.2017.02.064.

Li, S.-W., Wang, W., & Zhao, J.-S. (2020). Highly effective oxidative desulfurization with magnetic MOF supported W-MoO3 catalyst under oxygen as oxidant. *Applied Catalysis B: Environmental, 277*, 119224. https://doi.org/10.1016/j.apcatb.2020.119224.

Li, S.-W., Zhang, H.-Y., Han, T.-H., Wu, W.-Q., Wang, W., & Zhao, J.-S. (2021). A spinous Fe3O4@MOF-PMoW catalyst for the highly effective oxidative desulfurization under oxygen as oxidant. *Separation and Purification Technology, 264*, 118460. https://doi.org/10.1016/j.seppur.2021.118460.

Li, X., Zhang, L., & Sun, Y. (2020). Titanium-modified mil-101(Cr) derived titanium-chromium-oxide as highly efficient oxidative desulfurization catalyst. *Catalysts, 10*(9), 1–10. https://doi.org/10.3390/catal10091091.

Liang, R., Liang, Z., Chen, F., Xie, D., Wu, Y., Wang, X., Yan, G., & Wu, L. (2020). Sodium dodecyl sulfate-decorated MOF-derived porous Fe2O3 nanoparticles: High performance, recyclable photocatalysts for fuel denitrification. *Chinese Journal of Catalysis, 41*(1), 188–199. https://doi.org/10.1016/S1872-2067(19)63402-9.

Lin, Z.-J., Zheng, H.-Q., Chen, J., Zhuang, W.-E., Lin, Y.-X., Su, J.-W., Huang, Y.-B., & Cao, R. (2018). Encapsulation of phosphotungstic acid into metal-organic frameworks with tunable window sizes: Screening of PTA@MOF catalysts for efficient oxidative desulfurization. *Inorganic Chemistry, 57*(20), 13009–13019. https://doi.org/10.1021/acs.inorgchem.8b02272.

Liu, B., Shioyama, H., Akita, T., & Xu, Q. (2008). Metal-organic framework as a template for porous carbon synthesis. *Journal of the American Chemical Society, 130*(16), 5390–5391. https://doi.org/10.1021/ja7106146.

Liu, F., Yu, J., Basit Qazi, A., Zhang, L., & Liu, X. (2021). Metal-based ionic liquids in oxidative desulfurization: A critical review. *Environmental Science and Technology, 55*(3), 1419–1435. https://doi.org/10.1021/acs.est.0c05855.

Liu, J., Goetjen, T. A., Wang, Q., Knapp, J. G., Wasson, M. C., Yang, Y., Syed, Z. H., Delferro, M., Notestein, J. M., Farha, O. K., & Hupp, J. T. (2022). MOF-enabled confinement and related effects for chemical catalyst presentation and utilization. *Chemical Society Reviews, 51*(3), 1045–1097. https://doi.org/10.1039/D1CS00968K.

Liu, J., Li, X.-M., He, J., Wang, L.-Y., & Lei, J.-D. (2018). Combining the photocatalysis and absorption properties of core-shell Cu-BTC@TiO2 microspheres: Highly efficient desulfurization of thiophenic compounds from fuel. *Materials, 11*(11). https://doi.org/10.3390/ma11112209.

Liu, Y., Deng, C., Wu, P., Liu, H., Liu, F., Liu, R., Zhu, W., & Xu, C. (2024). High-entropy zeolitic imidazolate framework for efficient oxidative desulfurization of diesel fuel: Towards complete sulfur removal and valuable sulfone production. *Fuel, 359*, 130375. https://doi.org/10.1016/j.fuel.2023.130375.

Liu, Z., Xu, J., Zhang, F., Ji, L., & Shi, Z. (2023). Defect-rich high-entropy oxide nanospheres anchored on high-entropy MOF nanosheets for oxygen evolution reaction. *International Journal of Hydrogen Energy, 48*(39), 14622–14632. https://doi.org/10.1016/j.ijhydene.2022.12.333.

Lv, H.-T., Yang, P., Li, N., & Fan, Y. (2023). Defective MIL-125 nanocrystals with enhanced catalytic performance for oxidative denitrogenation. *Journal of Cluster Science, 34*(3), 1445–1451. https://doi.org/10.1007/s10876-022-02316-4.

Mirante, F., Mendes, R. F., Paz, F. A. A., & Balula, S. S. (2020). High catalytic efficiency of a layered coordination polymer to remove simultaneous sulfur and nitrogen compounds from fuels. *Catalysts, 10*(7), 1–15. https://doi.org/10.3390/catal10070731.

Mondol, M. M. H., Ahmed, I., Lee, H. J., Morsali, A., & Jhung, S. H. (2023). Metal–organic frameworks and metal–organic framework-derived materials for denitrogenation of liquid fuel via adsorption and catalysis. *Coordination Chemistry Reviews, 495*. https://doi.org/10.1016/j.ccr.2023.215382.

Mondol, M. M. H., Kim, C. U., & Jhung, S. H. (2022). Titanium nitride@nitrogen-enriched porous carbon derived from metal–organic frameworks and melamine: A remarkable oxidative catalyst to remove indoles from fuel. *Chemical Engineering Journal, 450*. https://doi.org/10.1016/j.cej.2022.138411.

Mukoyoshi, M., & Kitagawa, H. (2022). Nanoparticle/metal-organic framework hybrid catalysts: Elucidating the role of the MOF. *Chemical Communications, 58*(77), 10757–10767. https://doi.org/10.1039/D2CC03233C.

Nadzim, U. K. H. M., Hairom, N. H. H., Hamdan, M. A. H., Khairul Ahmad, M., Abdul Jalil, A., Jusoh, N. W. C., & Hamzah, S. (2022). Effects of different zinc oxide morphologies on photocatalytic desulfurization of thiophene. *Journal of Alloys and Compounds, 913*. https://doi.org/10.1016/j.jallcom.2022.165145.

O'Reilly, N., Giri, N., & James, S. L. (2007). Porous liquids. *Chemistry - A European Journal, 13*(11), 3020–3025. https://doi.org/10.1002/chem.200700090.

Peng, Y.-L., Liu, J., Zhang, H.-F., Luo, D., & Li, D. (2018). A size-matched POM@MOF composite catalyst for highly efficient and recyclable ultra-deep oxidative fuel desulfurization. *Inorganic Chemistry Frontiers, 5*(7), 1563–1569. https://doi.org/10.1039/C8QI00295A.

Piscopo, Granadeiro, Balula, & Bošković (2020). Metal-organic framework-based catalysts for oxidative desulfurization. *ChemCatChem, 12*(19), 4721–4731. https://doi.org/10.1002/cctc.202000688.

Qi, M., Lin, P., Shi, Q., Bai, H., Zhang, H., & Zhu, W. (2023). A metal-organic framework (MOF) and graphene oxide (GO) based peroxymonosulfate (PMS) activator applied in pollutant removal. *Institution of Chemical Engineers, China Process Safety and Environmental Protection, 171*, 847–858. https://doi.org/10.1016/j.psep.2023.01.069 Available from: http://www.elsevier.com/wps/find/journaldescription.cws_home/713889/description#description.

Qi, Z., Huang, Z., Wang, H., Li, L., Ye, C., & Qiu, T. (2020). In situ bridging encapsulation of a carboxyl-functionalized phosphotungstic acid ionic liquid in UiO-66: A remarkable catalyst for oxidative desulfurization. *Chemical Engineering Science, 225*. Available from: https://doi.org/10.1016/j.ces.2020.115818, http://www.journals.elsevier.com/chemical-engineering-science/.

Qian, Y., Zhang, F., Zhao, S., Bian, C., Mao, H., Kang, D. J., & Pang, H. (2023). Recent progress of metal-organic framework-derived composites: Synthesis and their energy conversion applications. *Nano Energy, 111.* https://doi.org/10.1016/j.nanoen.2023.108415.

Rao, T. V., Sain, B., Kafola, S., Nautiyal, B. R., Sharma, Y. K., Nanoti, S. M., & Garg, M. O. (2007). Oxidative desulfurization of HDS diesel using the aldehyde/molecular oxygen oxidation system. *Energy and Fuels, 21*(6), 3420–3424. https://doi.org/10.1021/ef700245g.

Shi, G., Liang, Y., Nie, L., Huang, R., Yang, Z., Liu, X., Zhou, J., Zhang, Q., & Ye, G. (2024). One-pot preparation of nitro-functionalized bimetallic UiO-66(Zr-Hf) with hierarchical porosity for oxidative desulfurization performance. *Inorganic Chemistry, 63*(35), 16554–16564. https://doi.org/10.1021/acs.inorgchem.4c02959.

Silva, D. F., Faria, R. G., Santos-Vieira, I., Cunha-Silva, L., Granadeiro, C. M., & Balula, S. S. (2023). Simultaneous sulfur and nitrogen removal from fuel combining activated porous MIL-100(Fe) catalyst and sustainable solvents. *Catalysis Today, 423.* https://doi.org/10.1016/j.cattod.2023.114250 Available from: http://www.sciencedirect.com/science/journal/09205861.

Silva, D. F., Viana, A. M., Santos-Vieira, I., Balula, S. S., & Cunha-Silva, L. (2022). Ionic liquid-based polyoxometalate incorporated at ZIF-8: A sustainable catalyst to combine desulfurization and denitrogenation processes. *Molecules, 27*(5). https://doi.org/10.3390/molecules27051711.

Subhan, S., Yaseen, M., Ahmad, B., Tong, Z. F., Subhan, F., Ahmad, W., & Sahibzada, M. (2021). Fabrication of MnO2NPs incorporated UiO-66 for the green and efficient oxidative desulfurization and denitrogenation of fuel oils. *Journal of Environmental Chemical Engineering, 9*(3). https://doi.org/10.1016/j.jece.2021.105179.

Sun, M., Chen, W.-C., Zhao, L., Wang, X.-L., & Su, Z.-M. (2018). A PTA@MIL-101(Cr)-diatomite composite as catalyst for efficient oxidative desulfurization. *Inorganic Chemistry Communications, 87,* 30–35. https://doi.org/10.1016/j.inoche.2017.11.008.

Sun, Y., Huang, H., Vardhan, H., Aguila, B., Zhong, C., Perman, J. A., Al-Enizi, A. M., Nafady, A., & Ma, S. (2018). Facile approach to graft ionic liquid into MOF for improving the efficiency of CO2 chemical fixation. *ACS Applied Materials and Interfaces, 10*(32), 27124–27130. https://doi.org/10.1021/acsami.8b08914.

Viana, A. M., Julião, D., Mirante, F., Faria, R. G., de Castro, B., Balula, S. S., & Cunha-Silva, L. (2021). Straightforward activation of metal-organic framework UiO-66 for oxidative desulfurization processes. *Catalysis Today, 362,* 28–34. https://doi.org/10.1016/j.cattod.2020.05.026 Available from: http://www.sciencedirect.com/science/journal/09205861.

Wang, C., Li, A., Ma, Y., & Qing, S. (2022). Preparation of formate-free PMA@MOF-808 catalysts for deep oxidative desulfurization of model fuels. *Environmental Science and Pollution Research, 29*(26), 39427–39440. https://doi.org/10.1007/s11356-022-18685-2.

Wang, L., Man, Z., Dai, X., Wang, K., Wang, W., Feng, X., Liu, D., & Xiao, H. (2024). Graphene-decorated novel Z-scheme Bi@MOF composites promote high performance photocatalytic desulfurization. *Chemical Engineering Journal, 485.* https://doi.org/10.1016/j.cej.2024.149977.

Wang, Q., & Astruc, D. (2020). State of the art and prospects in metal-organic framework (MOF)-based and MOF-derived nanocatalysis. *Chemical Reviews, 120*(2), 1438–1511. https://doi.org/10.1021/acs.chemrev.9b00223.

Wu, P., Wang, B., Chen, L., Zhu, J., Yang, N., Zhu, L., Deng, C., Hua, M., Zhu, W., & Xu, C. (2024). Tailoring type III porous ionic liquids for enhanced liquid-liquid two-phase catalysis. *Advanced Science, 11*(18). https://doi.org/10.1002/advs.202401996.

Xu, H., Wu, L., Zhao, X., Yang, S., Yao, Y., Liu, C., Chang, G., & Yang, X. (2024). Hierarchically porous amino-functionalized nanoMOF network anchored phosphomolybdic acid for oxidative desulfurization and shaping application. *Journal of Colloid and Interface Science, 658*, 313–323. https://doi.org/10.1016/j.jcis.2023.12.081.

Xu, L., An, X., She, J., Li, H., Zhu, L., Zhu, W., Li, H., & Jiang, W. (2023). Molybdenum-based metal–organic frameworks as highly efficient and stable catalysts for fast oxidative desulfurization of fuel oil. *Separation and Purification Technology, 326*. https://doi.org/10.1016/j.seppur.2023.124699.

Xu, S., Li, M., Wang, H., Sun, Y., Liu, W., Duan, J., & Chen, S. (2022). High-entropy metal-organic framework arrays boost oxygen evolution electrocatalysis. *Journal of Physical Chemistry C, 126*(33), 14094–14102. https://doi.org/10.1021/acs.jpcc.2c05083.

Yang, N., Lu, L., Zhu, L., Wu, P., Tao, D., Li, X., Gong, J., Chen, L., Chao, Y., & Zhu, W. (2022). Phosphomolybdic acid encapsulated in ZIF-8-based porous ionic liquids for reactive extraction desulfurization of fuels. *Inorganic Chemistry Frontiers, 9*(1), 165–178. https://doi.org/10.1039/D1QI01255J.

Yang, S., Peng, L., Bulut, S., & Queen, W. L. (2019). Recent advances of MOFs and MOF-derived materials in thermally driven organic transformations. *Chemistry, 25*(9), 2161–2178. https://doi.org/10.1002/chem.201803157.

Ye, G., Wang, H., Zeng, X., Wang, L., & Wang, J. (2021). Defect-rich bimetallic UiO-66(Hf-Zr): Solvent-free rapid synthesis and robust ambient-temperature oxidative desulfurization performance. *Applied Catalysis B: Environmental, 299*. Available from https://doi.org/10.1016/j.apcatb.2021.120659, www.elsevier.com/inca/publications/store/5/2/3/0/6/6/index.htt.

Ye, G., Hu, L., Gu, Y., Lancelot, C., Rives, A., Lamonier, C., Nuns, N., Marinova, M., Xu, W., & Sun, Y. (2020). Synthesis of polyoxometalate encapsulated in UiO-66(Zr) with hierarchical porosity and double active sites for oxidation desulfurization of fuel oil at room temperature. *Journal of Materials Chemistry A, 8*(37), 19396–19404. https://doi.org/10.1039/d0ta04337k.

Ye, G., Wang, H., Chen, W., Chu, H., Wei, J., Wang, D., Wang, J., & Li, Y. (2021). In situ implanting of single tungsten sites into defective UiO-66(Zr) by solvent-free route for efficient oxidative desulfurization at room temperature. *Angewandte Chemie - International Edition, 60*(37), 20318–20324. https://doi.org/10.1002/anie.202107018.

Yin, J., Fu, W., Zhang, J., Liu, X., Zhang, X., Wang, C., He, J., Jiang, W., Li, H., & Li, H. (2023). UiO-66(Zr)-based porous ionic liquids for highly efficient extraction coupled catalytic oxidative desulfurization. *Chemical Engineering Journal, 470*. https://doi.org/10.1016/j.cej.2023.144290.

Yin, J., Fu, W., Zhang, J., Zhang, X., Qiu, W., Jiang, W., Zhu, L., Li, H., & Li, H. (2024). Bifunctional pyridinium-based Brønsted acidic porous ionic liquid for deep oxidative desulfurization. *Chemical Engineering Journal, 492*, 152349. https://doi.org/10.1016/j.cej.2024.152349.

Zhang, W., Zhang, H., Xiao, J., Zhao, Z., Yu, M., & Li, Z. (2014). Carbon nanotube catalysts for oxidative desulfurization of a model diesel fuel using molecular oxygen. *Green Chemistry, 16*(1), 211–220. https://doi.org/10.1039/C3GC41106K.

Zhang, Y., Zhou, Y., Zhao, Y., & Liu, C.-J. (2016). Recent progresses in the size and structure control of MOF supported noble metal catalysts. *Catalysis Today, 263*, 61–68. https://doi.org/10.1016/j.cattod.2015.10.022.

Zhao, Y., Han, Y., Ma, T., & Guo, T. (2011). Simultaneous desulfurization and denitrification from flue gas by Ferrate(VI). *Environmental Science and Technology, 45*(9). https://doi.org/10.1021/es103857g 4060–5.

Zheng, H. Q., Zeng, Y. N., Chen, J., Lin, R. G., Zhuang, W. E., Cao, R., & Lin, Z. J. (2019). Zr-based metal-organic frameworks with intrinsic peroxidase-like activity for ultradeep oxidative desulfurization: Mechanism of H2O2 decomposition. *American Chemical Society, China*

Inorganic Chemistry, 58(10), 6983–6992. https://doi.org/10.1021/acs.inorgchem.9b00604 Available from: http://pubs.acs.org/journal/inocaj.

Zhou, S., He, J., Wu, P., He, L., Tao, D., Lu, L., Yu, Z., Zhu, L., Chao, Y., & Zhu, W. (2022). Metal-organic framework encapsulated high-loaded phosphomolybdic acid: A highly stable catalyst for oxidative desulfurization of 4,6-dimethyldibenzothiophene. *Fuel, 309*, 122143. https://doi.org/10.1016/j.fuel.2021.122143.

Zhou, X., Wang, T., He, D., Chen, P., Liu, H., Lv, H., Wu, H., Su, D., Pang, H., & Wang, C. (2024). Efficient photocatalytic desulfurization in air through improved photogenerated carriers separation in MOF MIL101/carbon dots-g-C3N4 nanocomposites. *Angewandte Chemie - International Edition, 63*(35). https://doi.org/10.1002/anie.202408989.

Zhou, X., Wang, T., Liu, H., Gao, X., Wang, C., & Wang, G. (2021). Desulfurization through photocatalytic oxidation: A critical review. *ChemSusChem, 14*(2), 492–511. https://doi.org/10.1002/cssc.202002144.

Zhu, J., Wang, P.-C., & Lu, M. (2015). Study on the one-pot oxidative esterification of glycerol with MOF supported polyoxometalates as catalyst. *Catalysis Science and Technology, 5*(6), 3383–3393. https://doi.org/10.1039/C5CY00102A.

Zong, M.-Y., Zhao, Z., Fan, C.-Z., Xu, J., & Wang, D.-H. (2023). Design of hierarchically porous Zr-MOFs with reo topology and confined PMA for ultra-efficient oxidation desulfurization. *Molecular Catalysis, 538*. https://doi.org/10.1016/j.mcat.2023.113007.

Chapter 16

Metal-organic framework composites for environmental remediation

Muhammad Altaf Nazir, and Sami Ullah
Institute of Chemistry, The Islamia University of Bahawalpur, Bahawalpur, Pakistan

Introduction to wastewater treatment

Industrial waste and water pollution

Water is necessary for life on Earth, and all animals, including humans, need clean drinking water to survive (Nazir, 2024). With the rapid increase of humanity, society, science, and technology, the environmental disorder with a major pollution problem has become one of the most critical challenge in the last half-century (Huang et al., 2011). Numerous businesses, including those in food processing, textiles, paper, plastics, printing, leather, cosmetics, and other fields, produce large amounts of colored effluents and wastewater. Because of their detrimental effects on aquatic vegetation's ability to photosynthesize and their negative effects on human health, including liver, renal, respiratory, brain, reproductive, and central nervous system issues, these effluents have raised serious concerns about the environment. A certain quantity of wastewater containing different contaminants is produced by the majority of companies. An estimated 7 million tons of different dyes are manufactured, and a significant portion of those dyes are dumped into aquatic environment without being properly treated. Most of these dyes and their compounds are hazardous, non-biodegradable, carcinogenic, and mutagenic (Mahmoodi, 2015). Heavy metals including lead (Pb), cadmium (Cd), and copper (Cu) must be removed from industrial wastewaters before being released into surface water because they have detrimental effects on both human and animal health (Zhang et al., 2020). The harmful effects of pharmaceuticals and personal care products (PCPPs) on human health and the aquatic environment have led to their identification as common emerging contaminants. Some of these effects include interference with the endocrine system and disruption of hormonal activity (Bhadra & Jhung, 2018). Numerous additional contaminants may be found in wastewater streams, and in order to comply with sustainability standards, these pollutants must be effectively cleaned before being released into the environment.

Current state of wastewater treatment

Wastewater treatment is becoming increasingly difficult as the demand for clean water rises globally. Pollution, the overuse of water resources, and global warming are placing significant pressure on our limited freshwater supplies. Wastewater is a vital component of the water management cycle, but in many parts of the globe, it is released into the environment without sufficient or any cleaning. This fact directly affects aquatic ecosystems, influencing both their healthy growth and the profitable and productive activities that result from the use of water resources (Dulio et al., 2018; Ryder, 2017). Effective methods for eliminating very hazardous chemical molecules from water have attracted a lot of attention. Numerous methods, such as coagulation, filtration in conjunction with coagulation, precipitation, ozonation, adsorption, ion exchange, reverse osmosis, and advanced oxidation processes (AOPs), have been used to remove organic pollutants from polluted water and wastewater. Due to the high initial and ongoing costs associated with these systems, certain restrictions have been identified. However, ion exchange and reverse osmosis are more appealing techniques, as they enable the recovery of pollutant values in addition to their removal from wastewater. From a commercial perspective, reverse osmosis, ion exchange, and advanced oxidation technologies seem unfeasible due to their high startup costs and continuous operating expenses. Of the various methods for treating water, the adsorption process utilizing solid adsorbents appears to be one of the most promising approaches for treating and eliminating organic contaminants in wastewater. Adsorption is superior to other techniques because of its straightforward construction and its ability to use less starting material and land. Scholars are increasingly interested in the adsorption process, which is widely used to eliminate both organic and inorganic contaminants from industrial effluent. Recently, there has been greater competition in the search for inexpensive pollutant-binding adsorbents (Crini, 2005). This method is very effective when using adsorbents with the right properties. Because of this, it is critical to investigate the textural, physicochemical, and morphological characteristics of adsorbents in order to get relevant data for choosing the best adsorbents for upcoming uses. Adsorption is a technologically viable approach that may be used to remove harmful and dangerous metals from wastewater (Allen & Brown, 1995). The advantages of the adsorption process are its straightforward operation and simply adjustable architecture. Furthermore, adsorption is frequently reversible, and the appropriate desorption process may allow the adsorbents to be recycled (Fu & Wang, 2011). Furthermore, photocatalysis is among the most attractive methods for removing organic pollutants from wastewater.

The photocatalyst material and impurities react with light support during the photocatalytic process. First, the photocatalyst material absorbs light energy and produces excited electron and hole pairs that further initiate additional chemical processes and breakdown organic contaminants (Thuan, 2023).

This chapter emphasizes on metal-organic framework (MOF) composites, a specific class of materials that have been the subject of much recent study for the adsorptive treatment of wastewater in a sustainable manner. MOFs were chosen mainly because of the strong interest in this subject for study, particularly with regard to the materials' structural and chemical tailorability. According to the scientists, these substances show promise as highly specialized water treatment materials when a highly tailored adsorbent is needed. MOF could be particularly effective for this purpose because their pore size and openings can be adjusted to fit the size of the substances they need to adsorb. Additionally, the chemical properties of the pores can be made compatible with the target substances. Considering the breadth of this book and its emphasis on sustainability, the use of MOF composites significantly improves the sustainability of the water treatment process.

Metal-organic frameworks

MOFs are crystalline materials that are distinguished by the way that metal ions and organic ligands combine to form a unique porous structure (Nazir et al., 2020). MOFs consist of interconnected (1D, 2D, and 3D) dimensional frameworks that are created by the interaction of organic ligands with metallic clusters/nodes. Various metal ions or clusters are combined with organic linkers in the right solvent to produce MOFs. Between these precursors, coordination polymerization takes place, producing a cross-linked network with possible vacancies. Because of this coordination, the structure is very porous and has a large surface area (Nazir et al., 2021). Fig. 16.1 shows the general schematic representations for the MOF preparation. More than 20,000

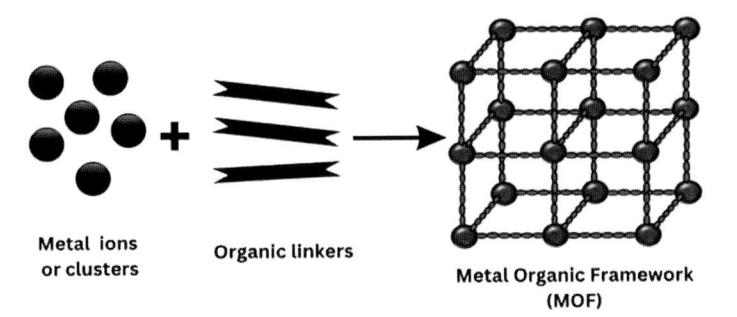

FIGURE 16.1 General schematic diagram for MOF preparation (Ullah, Rehman, *et al.*, 2024).

different kinds of MOF materials are available, and a wide variety of organic ligands and solvents may be used to create MOFs (Zhang, Liu, et al., 2018). Unlike traditional porous materials such as carbon, zeolite, and porous silica, MOFs have unique morphological features such as well-defined crystalline structures, large surface areas, and flexible pore environments. These characteristics offer flexibility for creating customized MOF structures (Corma et al., 2004). Customized MOF design is made possible by carefully choosing organic ligands and metal clusters and using the right synthetic techniques or postsynthetic alterations. These frameworks can display specified topological sizes and topologies, pore shapes, and chemical affinities that are tailored for specific uses. Consequently, MOFs have emerged as a new class of adaptable crystalline porous materials that exhibit exceptional performance in a variety of applications, especially in well-established domains like water treatment (Sarker et al., 2019). On the other hand, a lot of MOF compounds become unstable in humid air or water. For instance, when IRMOF-1 (IRMOF: iso-reticular metal-organic framework) comes into contact with water molecules, it loses its structure. IRMOF-1 transitions from a high surface area to a low surface area when water molecules take the place of carboxylic groups to interact with zinc centers, MOF-69c occurs (Wu et al., 2010). For water treatment applications, water-stable MOFs as MIL-100 or UiO-66 should be used. On the other hand, functionalization of MOFs has shown to be a successful tactic for enhancing water stability and controlling wettability (Rubin & Reynolds, 2017).

Zeolitic imidazolate frameworks

A subclass of MOFs with persistent porosity, zeolitic imidazolate frameworks (ZIFs) provide development and modification possibilities due to their relatively excellent thermal and chemical stability (James & Lin, 2017; Yao & Wang, 2014). ZIFs are composed of zinc, copper, and other metal atoms. Imidazolate linkers' nitrogen atoms serve as a bridge between these metal atoms. M-Im-M (Im: imidazolates) bridges with M being Zn, Cu, or Co with a matching bond angle of 145θ replace T-O-T bridges with T being Si, P, or Al in the analogous frameworks of ZIFs and zeolites (Hayashi et al., 2007). By altering the metal ions in addition to the imidazolate chemical molecules, various ZIF structures may be produced (Banerjee et al., 2008). Two of the ZIF structures, ZIF-8 and ZIF-11, have shown exceptional thermal and chemical stability even in aqueous alkaline solutions and water (Jiang, Yan, et al., 2016; Zhang et al., 2015). Testing revealed that ZIF-8 and ZIF-11 both maintained their structures in water at 50°C for 7 days. After 7 days, ZIF-8 was the only material to stay stable in boiling water and maintain its original structure; after 3 days, ZIF-11 underwent a structural change. Furthermore, ZIF-8 maintained its stability in 0.1 and 8 M aqueous sodium hydroxide for up to 24 hours at 100°C (Park et al., 2006). This illustrates ZIF-8 remarkable resistance among

MOF solids. The reason for this is that imidazolate linkers form stronger connections between metal atoms and ligands due to their higher basicity compared to carboxylate linkers. ZIF-8 superior stability over other ZIFs may also be attributed to its hydrophobicity (Ge & Lee, 2011). ZIF-8 (Zn(2-methylimidazolate)$_2$) is a zeolite with a structure similar to sodalites, but with a pore size of 1.14 nm, which is much bigger, and a pore window with nine members that is 0.34 nm. Because of ZIF-8 unique properties namely, its large surface area of 1300–1600 m^2/g and high porosity, it is essential for adsorption applications (Jiang et al., 2013).

Carboxylate frameworks

Metal centers and organic linkers may be changed to form a wide variety of MOFs by altering the surface area, pore size, and surface functionality. Between 70 and 75 thousand MOF structures are thought to have been deposited at the Cambridge Crystallographic Data Center (Moghadam et al., 2017; Øien-Ødegaard et al., 2017). The majority of these structures fall under the category of carboxylate frameworks since they include linkers with two or more carboxylate functions in them. The framework's nods are formed by the carboxylate group's strong coordination connections with the metal ions. Carboxylate framework synthesis allows additional functionalities such as alkyl, amino, nitro, halo-genic, or sulfonic groups on the carboxylate linker. Thus during the synthesis, functionalized MOFs may be readily produced by utilizing the appropriate carboxylic acids (Huang et al., 2013). Two unusual metal carboxylate pyrazolate clusters were used by Wang et al. to create a novel porous MOF (Wang et al., 2015). Pyrazolate and carboxylate readily form a cluster with condensing metal density when they coordinate with metal centers, which improves structural stability and facilitates interactions with CO_2. Moreover, a lengthy biphenyl spacer of 2,2′-bipyridine-5,5′-dicarboxylic acid (H$_2$bpydc) connected the cluster to provide a permeable framework. Moreover, a heterogeneous copper catalyst has also been produced by the use of an aromatic carboxylate ligand. In a catalytic process, carboxylate functions as a ligand to increase catalytic activity (Qi et al., 2015).

Nanoporous carbons derived from metal-organic frameworks

Nanostructured porous carbon materials are intriguing because of their unique physical and chemical properties that make them perfect for application as sorbents in gas or liquid adsorption, catalyst support, electrodes, and fuel cells (Jiang et al., 2011; Jiang, Cao, et al., 2016). Because of their affinity for organic contaminations, these materials may find use in environmental applications, such as the adsorptive removal of dangerous toxins (Angamuthu et al., 2017; Torad et al., 2014). These porous materials have a substantial surface area, a restricted range of pore sizes, and good chemical and thermal stability. The carbons that are nanoporous are produced by carbonization or chemical vapor deposition of carbon sources using the hard templates. They may also be made by carbonizing polymeric carbon gels using an organic–organic soft templating approach (Chaikittisilp et al., 2013). Because they are resistant to changes like hydrothermal carbonization, certain materials combine to produce a low surface area, nonporous carbonaceous network. Specifically, these materials do not have the micro porosity needed to have significant adsorption power. Conventional activation techniques are required to enhance the pore capacity and specific surface area of these materials. These materials include crystalline cellulose and lignin, which are both produced from lignocellulosic biomass and resistant to the conditions of hydrothermal carbonization (Unur, 2013). Recently, there has been a lot of interest in porous MOFs, which are becoming a new class of crystalline porous materials with many functions (Li et al., 1999). Because of their large pore volumes, large surface areas, and diverse topologies, MOFs have been considered as alternative resources for the synthesis of nanoporous carbons. Since Liu et al.'s first publication, MOFs have been shown to function as precursors of nitrogen-doped porous carbon (NPC) materials (Liu et al., 2008). Newly developed porous materials known as MOF-derived NPCs have a well-defined form and a large surface area, making them excellent options for a variety of uses. Salunkhe et al. created a composite material using nanoporous carbons produced from MOF. The results demonstrated the benefits of ZIF-8-derived NPC, which satisfies the needs of having a long cycle life and strong rate capacity (Salunkhe et al., 2014). Nanoporous carbon produced from a MOF was described by Liu et al. as a novel type of adsorbent for the dispersive solid-phase extraction of pesticides, including benzoylureas. The resultant nanoporous carbon, or MOF-C, has a large pore volume and a high specific surface area. The findings show that the composite preconcentrated trace amounts of benzoylurea pesticides from water and tangerine samples effectively (Liu et al., 2015). Several MOFs, including MOF-5, Al-PCP (PCP: porous coordination polymer), and ZIF-8, have been utilized to produce nanoporous carbons with the perfect properties for adsorption, sensing, catalysis, and electrochemical capacitance applications (Chaikittisilp et al., 2013; Li et al., 1999; Park et al., 2006). Because of their high carbon content, MOFs may be directly carbonized into nanoporous carbons without the need for an external carbon supply. Al-PCP has been converted into nanoporous carbon at 80°C using the direct carbonization technique, a simple one-step process. The study shows that the resulting nanoporous carbon has a large pore volume of 4.3 cm^3/g and surface area of more than 5000 m^2/g. It was discovered that in order to achieve the required large pore volume and high surface area, the carbonization temperature was critical (Hu et al., 2012).

Recently, Lim et al. provided evidence of the direct carbonization of MOFs using zinc atoms. They emphasized that the rearrangement of carbon atoms might produce very nanoporous carbons even from nonporous MOF materials.

This is an intriguing discovery that highlights the better properties and functionality of MOF materials when compared to other chemicals (Lim et al., 2012).

Adsorption mechanism

Adsorption has been deemed better than other decontamination methods due to its relatively low cost, broad use, ease of design, ease of operation, minimal production of hazardous byproducts, and ease of adsorbent renewal. A porous adsorbent's capacity to selectively adsorb certain molecules from the environment or refinery streams is the foundation of adsorptive removal. Given their easy access to the solid sorbents' pores, the compounds with the appropriate size and shape can be eliminated via adsorption. Physical or chemical adsorption can be distinguished based on the kinds of interactions that occur between an adsorbate and a porous sorbent (Babich & Moulijn, 2003). Adsorbent adsorption is the common term for physical adsorption, while reactive adsorption is the term used for chemical adsorption. Adsorbates are often confined inside the pores of solid adsorbents by weak (van der Waals) pressures in the case of adsorptive removal. As a result, simple solvent exchange or other physical processes like sonication and calcination can readily replenish the adsorbent. Reactive adsorption, on the other hand, happens when the adsorbates and the adsorbent really establish chemical connections. Chemical treatments are often used to regenerate the wasted adsorbent. The adsorption capacity, selectivity for certain chemicals, durability, and regenerability of the adsorbents are the primary factors that define the efficacy of adsorptive removal (Khan et al., 2013).

Adsorption efficiency is usually explained by both kinetic models and adsorption equilibrium isotherms. The adsorption isotherm, which is a function of the equilibrium concentration of adsorbate, displays the amount of material adsorbed per mass of adsorbent. The equation states that for a given concentration of an adsorbate, a given volume of solution at a given mass of adsorbent reaches equilibrium. The mass balance for adsorption experiments is introduced here (Silva et al., 2016).

$$Qe = 1 + \frac{C_o - C_e \times V}{m}$$

where C_o and C_e are the initial and final concentrations of the adsorbate (mg/L), respectively; V is the solution volume (L), and m is the mass of the adsorbent (g).

Because of MOFs' large surface area, adjustable porosity, and variety of functional group types, they are being extensively researched as hazardous chemical adsorbents (Lin et al., 2019; Zhao, Azhar, et al., 2019; Zhao, Lun, et al., 2019). A variety of interactions, including π-complexation, hydrogen bonding, acid–base interaction, and π–π contact, can lead to adsorption on

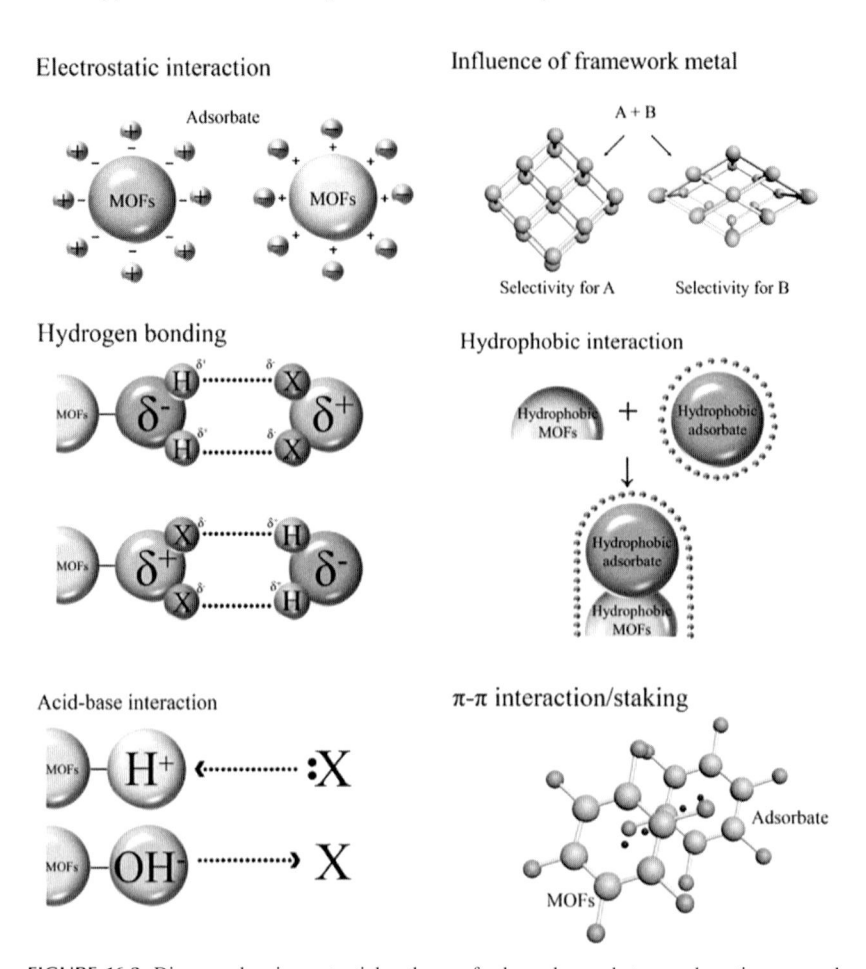

FIGURE 16.2 Diagram showing potential pathways for hazardous substance adsorptive removal over MOFs (Dhaka et al., 2019).

MOFs. The potential interactions between the MOF and the adsorbates are summarized in Fig. 16.2. In order to achieve an increased adsorption between the MOF and certain adsorbates, MOFs can be developed and modified to contain a variety of adsorption sites, organic linker functions, open metal sites, or loaded active species. The literature has examined MOFs extensively in relation to the adsorption of various contaminants from water.

Influential factors on adsorption

Numerous variables, including the adsorbent dosage, the starting metal ion concentration, the pH of the solution, temperature, and the duration of the interaction, can influence the adsorption of contaminants. By maximizing these variables, the amount of pollutants that are adsorbed may be significantly increased.

A substance known as an adsorbent is one that, while staying unaltered, may adsorb molecules of solids, liquids, and gases. In order to achieve the maximum adsorption capacity under operating circumstances, it is crucial to monitor this parameter for the adsorbent concentration (Zvinowanda et al., 2009). Since there are more sites available for adsorption when the adsorbent dosage is increased, the adsorption percentage usually rises as well. However, the adsorption capacity falls as a result of adsorption sites aggregating, which reduces the surface area available for the adsorption of pollutants (Ananpattarachai & Kajitvichyanukul, 2016). It is feasible to choose an acceptable adsorbent concentration for the greatest pollutant removal with the lowest necessary adsorbent dosage by optimizing the adsorbent dose.

An adsorptive molecule's degree of ionization and surface properties can be determined by optimizing pH. If all other experimental conditions are constant, the initial pH of the solution may lead to varying removal percentages and adsorption capabilities. Additionally, it might control the ionization of adsorbent functional groups as well as the adsorbent-metal ion attraction and repulsion to either promote or reduce adsorption. pH is a crucial parameter because it determines how much electrostatic charge metal ions have, which in turn influences the rate of adsorption (Abdullah et al., 2019). Because positive binding sites and positive metal ions interact repulsively, lower pH causes the adsorbent's binding sites to protonate, which positively charges the adsorbent's surface and decreases adsorption. A higher pH causes more positive metal ions to cling to the negative surface of the adsorbent, increasing adsorption (Krishnani et al., 2013; Lin et al., 2021).

According to Radovic et al. (1997), the adsorption of aromatic solutes on carbon compounds appears to be dominated by $\pi-\pi$ dispersion interactions. They came to the conclusion that weak aromatic systems' equilibrium uptakes can be influenced and controlled by both dispersive and electrostatic interactions. Another work by Ghaffar and Younis (2015) shows that van der Waals forces, hydrophobic interaction, hydrogen bonding, $\pi-\pi$ electron donor–acceptor, electrostatic and covalent interactions, and hydrophobic interaction can all have an impact on the adsorption process for methylene blue (MB) on carbon nanotubes (CNTs) either simultaneously or independently.

The degree of hydration of the molecules and the adsorption of water are both influenced by the temperature of a solution (Dąbrowski et al., 2005). Gas-phase adsorption on open sites is always exothermic, according to the van Hoff's adsorption equation. Water treatment and other liquid phase procedures involving wetted surfaces never have totally empty adsorption sites. Adsorption of a solute from the liquid phase is always connected to the desorption of solvent molecules occupying the site. This increases the complexity of these processes' thermodynamic effects. Endothermic activity has often been reported in the adsorption of dyes in solutions (Hong et al., 2009). Viscosity and diffusion rate are two other kinetic parameters that are dependent on the solution temperature. Consequently, at higher temperature, adsorption equilibria are reached more quickly. Furthermore, some adsorbents are desirable for temperature swing adsorption because of their unique thermoresponsive properties (Yamagiwa et al., 1997).

ZIF-8/Fly ash composite was produced by Wang et al. for the purpose of adsorbing Cu^{2+}, Zn^{2+}, and Ni^{2+} from aqueous solutions. Moreover, they examine the impact of several variables on adsorption efficiency. It has been noted that as the pH of the aqueous solution rises, so does the rate at which heavy metal ions are removed. In a similar vein, when ZIF-8/FA (FA: fly ash) is increased, the clearance rate rises and the quantity of adsorption falls as adsorbent concentration rises. Up to three consecutive cycles, the composite material demonstrated outstanding reusability (Wang et al., 2020). CS-ZIF-8 (CS: chitosan) composite beads were created by Wang et al. to effectively remove Pb^{2+} and Cu^{2+} from water. An ideal pH of 5 was reached by condition adjustment, and the adsorption of Cu^{2+} took 5 hours to equilibrate, whereas Pb^{2+} took 2 hours. Cu^{2+} and Pb^{2+} adsorption capacities grew progressively as pH climbed from 1 to 5. As the adsorbent dosage increases, it is evident that the removal efficiency of Pb^{2+} increases from 21.2% to 95% and Cu^{2+} increases from 29.2% to 98.9% (Wang et al., 2022).

Photocatalytic degradation mechanism

In the degradation of organic pollutants, photocatalysis is a possible AOP that is often used due to its exceptional efficiency, simplicity, ease of handling, and outstanding reproducibility (Ali et al., 2017). The photocatalysis technique, which is a cheap and efficient method of decomposing organic pollutants, was able to completely mineralize organic pollutants under ambient settings and produce safe products by using just atmospheric oxygen as the chemical ingredient (Sudhaik et al., 2018). It is therefore a low-energy, non-toxic, and environmentally friendly method of eliminating hazardous pollutants. Strong oxidizing and reducing agents (h^+ and e^-) created by UV or visible light on photocatalyst surfaces interact with organic pollutants to form the basis of photocatalysis technology (Hasija et al., 2019).

The semiconductor (photocatalysts) absorbs energy (hυ) equivalent to or more than its energy bandgap (E_g) during the photocatalytic activity, which results in the valence band (VB) to conduction band (CB) excitation of electrons in the semiconductor. At zero temperature, the electron-filled VB has a lower energy than the empty CB, which has a greater energy. Its optical properties and color are determined by the size of the energy gap that separates the VB and CB bands. For instance, semiconductors with a bandgap of 1.5–3.0 eV may absorb energy in the visible light spectrum and have a color spectrum ranging from red to violet. Charge carrier separation is ensured by

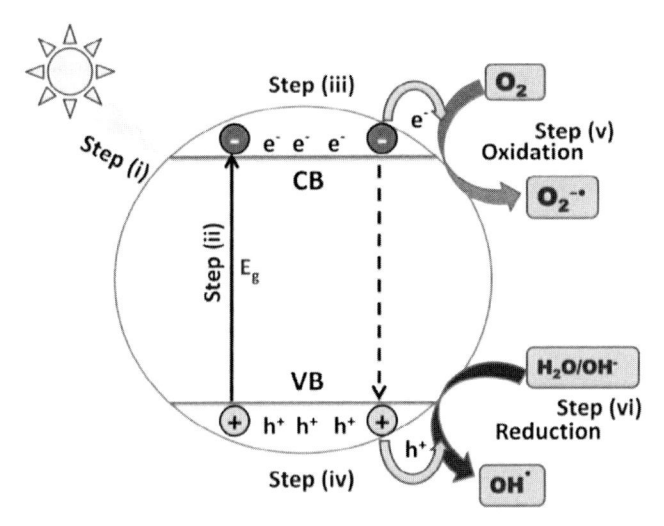

FIGURE 16.3 Diagrammatic representation of the whole photodegradation process (Pattanayak et al., 2023).

this, enabling their involvement in surface oxidation-reduction processes. Ideality e⁻ and h⁺ take part in oxidation-reduction reactions by adsorbing electron donors and acceptors at the semiconductor's active surface. Radical species are produced when the organic molecule that is adsorbed at the semiconductor's surface by interfacial charges is reduced and oxidized by the e⁻ and h⁺. Fig. 16.3 shows a straightforward schematic diagram of the whole photodegradation process, which consists of many phases (Pattanayak et al., 2023).

Since they can break down organic dyes significantly, MOFs, are a novel class of photocatalytic materials that have garnered a lot of interest in the purification of wastewater. This is due to MOFs' exceptional porosity, robust mechanical stability, huge specific surface area, and designable architecture (Liu et al., 2014; Wu & Lin, 2007). The electronic properties of ground-level molecules and their excited states may be predicted using density functional theory (DFT) calculations, which have shown to be a very useful theoretical technique. The data from the calculations agree well with experimental findings (Pramanik et al., 2011). Based on band structure predictions, Wang et al. have devised a procedure that indicates MOF may be triggered to create electron–hole pairs when exposed to visible light. This process can oxidize organic contaminants by generating superoxide ions (O_2^-) and other reactive oxygen species (Li et al., 2016). By enhancing the organic ligands, MOFs may enhance their photocatalytic activity even more by encouraging electron localization, the transition from metal-ligand cluster charge transfer to the ligand → metal cluster charge transition in the conjugated ring, the $\pi \rightarrow \pi^*$

transition, and the ligand → ligand* transition. Metal clusters are activated by mobile charge carriers, allowing a variety of photo-redox processes (Zhang & Lin, 2014).

Factors influencing the photocatalytic degradation

The efficacy of the photocatalytic system and the rates of oxidation are greatly influenced by a variety of operational parameters that regulate the photodegradation of organic molecules (Gnanaprakasam et al., 2015; Kumar & Pandey, 2017). Operational parameters have been shown to be important in several research investigations.

Because of the direct relationship between organic molecules and the surface coverage of the photocatalyst, particle and agglomerate sizes, or surface shape, are important elements to consider in the photocatalytic degradation process (Kormann et al., 1988). The amount of photons that strike the photocatalyst determines the pace of reaction, indicating that the reaction only occurs during the photocatalyst's absorbed phase (Zhu et al., 2006). There is a relationship between degradation and photocatalyst quantity. Heterogeneous photocatalysis has the characteristic that photodegradation rises with increasing catalyst concentration. As the amount of catalyst grows, more active sites are created on the photocatalyst surface, which leads to a rise in the production of •OH radicals. The solution turns turbid above a specific catalyst concentration, blocking UV light from the process and causing the percentage of degradation to start falling (Coleman et al., 2007).

Dyes must adhere to the photocatalyst's surface in order for photocatalysis to occur. The amount of dye present in the majority of the solution does not affect the photocatalysis process; only the amount adsorbed on the surface of the photocatalyst does. The dye's starting concentration affects how well it adsorbs. In each particular photocatalytic process, the starting dye concentration is a critical element that must be considered. While maintaining a constant quantity of catalyst, the percentage degradation generally decreases as dye concentration rises (Reza et al., 2017). When the concentration of dyes with the photocatalyst decreases, the rate of dye photodegradation also decreases (Azad & Gajanan, 2017). Reaction temperature over 80°C encourage the recombination of charge carriers and hinder the adsorption of organic molecules, but they also typically lead to greater photocatalytic activity (Hashimoto et al., 2005). Reaction temperatures below 80°C are favorable for adsorption, whereas further cooling the reaction to 0°C raises the apparent activation energy (Peral et al., 1997). Thus the ideal temperature range for the efficient photo mineralization of organic material has been identified as 20°C–80°C (Mamba et al., 2014).

A unique composite photocatalyst called BiOBr/ZIF-67 (BiOBr: bismuth oxybromide) was created, according to Zhong et al., for improved visible-light photocatalytic degradation of RhB (RhB: rhodamine B). The quick electron transport at the BiOBr/ZIF-67 interface is responsible for the boost in photocatalytic efficiency. The intensity of RhB's absorption peaks clearly reduced in the presence of 30% BiOBr/ZIF-67 as the reaction time increased, suggesting that BiOBr/ZIF-67 possesses strong photocatalytic activity. It is discovered that the composite catalyst may continue to degrade at a rate of 92.1% in the second reaction and is reusable (Zhong et al., 2022). Similar to this, Chen et al. created heterogeneous Fenton-like catalysts by fabricating magnetic $Fe_3O_4@ZIF-67$ composites. Several reactive species, including radicals, were involved in the degradation process. The breakdown of TBBPA (tetrabromobisphenol A) was caused by -OH, SO_4^- radicals, and $^1O_2$, with $^1O_2$ being the main key factor. When x was raised from 0.1 to 1, it was discovered that the degradation rate constant k of TBBPA rose from 61.0 to 110.3 L/g.min, as more active catalytic center was given by the increased ZIF-67 concentration (Chen et al., 2021).

Adsorption of pollutants

Adsorptive removal of organic dye by metal-organic framework composites

The adsorption process is only influenced by the amount of dye adsorbed on the surface of the adsorbent, not by the amount in the majority of the solution (Ullah, Shah, et al., 2024). Reusing adsorbents, which may process and absorb water contaminants and potentially recover precious resources, is made possible by the regeneration process, which is especially remarkable (Wu et al., 2021). Making effective sorbents is one of the most crucial phases in the adsorption process. However, structural and morphological changes can be an effective tactic for boosting reaction sites and, consequently, adsorption capacity for a particular adsorbent (Sriram et al., 2022; Verma et al., 2010; Wei et al., 2019). Catalytic dyes produced in wastewater decay slowly due to their high toxicity, aromatic character, and high water solubility. They are hence enduring contaminants in the environment (Zhao, Chen, et al., 2017). In general, cationic dyes including those with strong color compounds are exceedingly hazardous to the environment (Sriram et al., 2022). A well-known example of a cationic dye is MB, which has an absorption maximum wavelength (λ_{max}) of 663 nm and a molecular weight of 319.8 g/mol. Long-term exposure to elevated environmental MB concentrations can cause major health concerns, such as jaundice, perspiration, heart palpitations, irritated eyes, vomiting, quadriplegia, nausea, and cyanosis (Dong et al., 2021). Methyl orange (MO), a member of the sulphonated Azo family, is a very stable anion-based Azo pigment (Carolin et al., 2021; Liu et al., 2020). MO possesses strong color, exceptional chemical durability, harmful effects, and poor biodegradable qualities due to its Azo group, sulfur group, and fragrant qualities (Gómez-Obando et al., 2019). Malachite green causes eye burning, fast breathing, excessive sweating, and cancer by harming the central nervous system. For many aquatic and terrestrial species, it is exceedingly destructive. Development and reproduction rates are decelerated. The liver, spleen, kidney, skin, eyes, lungs, bones, and heart are among the organs impacted by it. It is both toxic and carcinogenic to mammalian cells. Its presence in the water hinders sunlight from penetrating, decreases photosynthesis, and negatively impacts aquatic life (Raval et al., 2017). Wastewater contains a wide range of other colors that can be harmful, including tartrazine, azo, basic, acid, MO, brilliant blue, sulfur, Allura red, and brilliant blue dyes. It is essential to remove all of these dyes from water while keeping all of these factors in mind. The general consensus is that adsorption is better than alternative decontamination techniques due to its low cost, wide range of uses, environmental friendliness, and simplicity in regenerating the adsorbents (Veisi et al., 2015). Dyes are extracted from wastewater using MOFs made of coordinated ligands and metallic ions by a combination of catalysis and adsorption (Huang et al., 2017).

Because of MOF's poor chemical and thermal stabilities, its instability in water poses a serious problem for water filtration. Researchers have been looking for ways to improve the characteristics of MOFs in order to overcome this problem. Rasidi Sule et al. looked at the impacts of adding CNTs, graphite, and functionalized graphene nanosheets to HKUST-1 in order to enhance the device's capacity to remove MB from wastewater in a paper published in 2019. According to the study, when exposed to water, the MOF $Cu_3(BTC)_2$ was vulnerable to surface decrease, surface degradation, and crystal alterations. To solve these issues, the researchers developed a HKUST-1 composite that was impregnated with multiwall carbon nanotubes (FMWCNTs), or acid-treated FMWCNTs. This resulted in a significant increase in the composite's surface area, which was 1131.2 m^2/g. In comparison to the HKUST-1 sample, the composite showed greater pore volume and size. The original MOF's pore volume rose from 0.76 to 1.93 cm^3/g upon the addition of FMWCNTs. Barrett, Joyner, and Halenda (BJH) desorption yielded an overall pore size measurement of 6.97 nm for the composite. The composites' maximal adsorption capacity was determined to be more than 100 mg/g at 298 K. The high adsorption may have resulted from an electrostatic interaction between the positively charged MB and the negatively charged surface of FMWCNT/HKUST-1. Furthermore, the inclusion of a carboxylate group and a hydroxyl site in the adsorbent may have contributed to the exceptional removal efficiency that was achieved. The results suggest the potential use of the FMWCNT/HKUST-1 nanocomposite as an adsorbent for the extraction of MB from synthetic dye solutions (Sule & Mishra, n.d.). The manganese-based Mn@ZIF-8 nanocomposite was produced by Nazir et al. using solvothermal techniques. The ability of the produced Mn@ZIF-8 nanostructure to absorb the MO dye from water was evaluated. The microporous structure of the Mn@ZIF-8 and virgin ZIF-8 nanocomposite is formed like a rhombic dodecahedral.

Mn@ZIF-8 has a greater MO removal capacity at pH 3, measuring 87.44 mg/g (Fig. 16.4A). After a 6-hours contact period, Mn@ZIF-8 absorbed 91.71% of MO (Fig. 16.4B). The pseudo first-order (PFO) and second-order (PSO) kinetics are shown in Fig. 16.4C and D. During the adsorption process, the π electrons in the imidazole ring interact with the MO dyes through π–π interactions. Additionally, there is a hydrogen bond between the dye molecule and the imidazolate ring (Fig. 16.4E). The remarkable adsorption ability of Mn@ZIF-8 (q_{max}=406.50 mg/g) allows it to effectively remove MO from wastewater. Additionally, the synthesized Mn@ZIF-8 material exhibits remarkable reusability of up to 92% when compared to the initial cycle (Nazir, 2024).

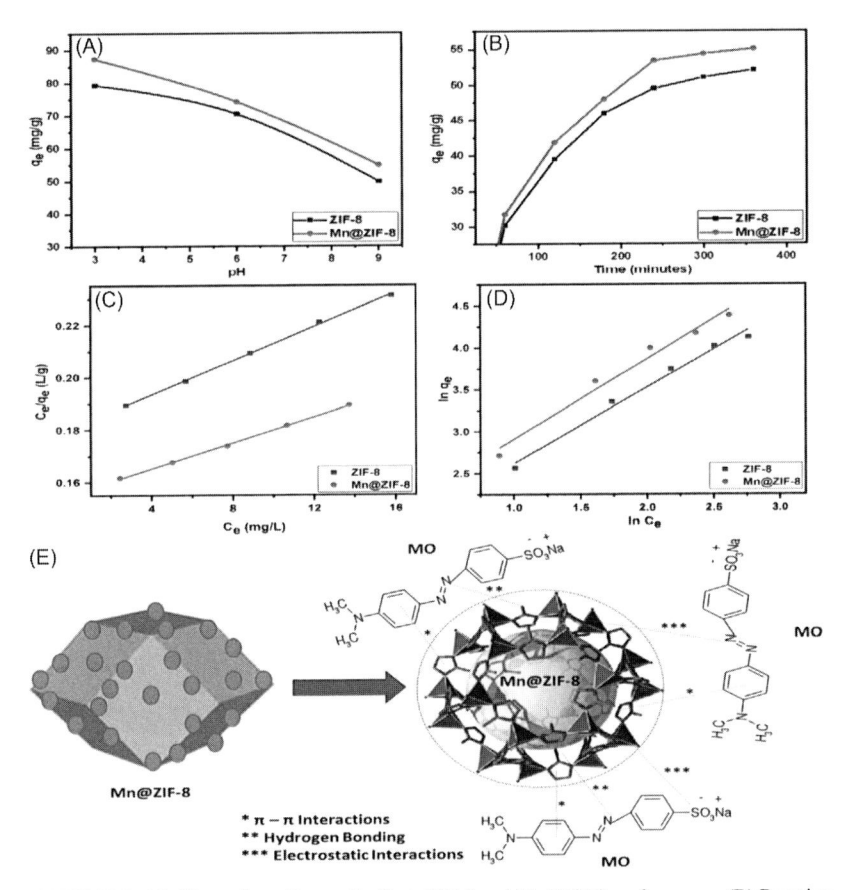

FIGURE 16.4 (A) Illustration of how pH affects ZIF-8 and Mn@ZF-8 performance; (B) Reaction time effect; (C) PFO kinetics; (D) PSO kinetics; (E) Expected mechanism for MO adsorption by Mn@ZIF-8 (Nazir, 2024).

Although a lot of study has been done on the use of MOF composites to remove anionic and cationic dyes, there is not much information on how these composites are used to remove neutral dyes. In contrast to anionic and/or cationic dyes, it has been demonstrated that MOF-based composites are less capable of adsorbing neutral dyes. The ability of MOF-based composites to adsorb neutral dyes is often ascribed to the π–π stacking interactions between the MOF framework and the dye molecules. However, because neutral dyes do not include charged groups, electrostatic interactions between the MOF network and the dye molecules are not feasible. Studies on MOF-based composite adsorption have used dimethyl yellow (DY), methyl red (MR), and neutral red (NR) as neutral dyes. One such composite that showed a high MR adsorption capacity (1.25 mg/g) in these trials was MIL-101(Fe)@PDopa@Fe$_3$O$_4$. Additionally, this composite effectively removed MR from textile effluent samples from a single textile mill with an 89.9% removal rate. The PFO, PSO, and intraparticle diffusion models were used to assess the mechanism of MR and MG adsorption onto MIL-101(Fe)@PDopa@Fe$_3$O$_4$. The composite is a practical and environmentally friendly choice for wastewater treatment because of its magnetic component (Fe$_3$O$_4$), which makes it simple to recover from contaminated water using a magnet. Moreover, the adsorbent's persistent adhesion to the magnetic material is ensured by the use of poly(3,4-dihydroxy-L-phenylalanine) (PDopa) as a bonding agent, which allows for up to five cycles of adsorption and desorption (Rojas & Horcajada, 2020).

Adsorptive removal of pharmaceutical compounds by metal-organic framework composites

Worldwide, a new class of developing pollutants known as pharmaceutical contaminants has been regularly found in a variety of aquatic ecosystems and sediment (Kairigo et al., 2020). Pharmaceutical manufacturing and use have grown significantly on a worldwide scale in recent years. When these medications leak or are released into the environment without being treated, aquatic life is at danger of toxicity and human health may suffer as a result (Oladipo et al., 2021). Especially, they tend to accumulate in aquatic creatures, which present a concern to the environment. In the secondary wastewater discharge, for instance, the antipyretic medication paracetamol and ibuprofen are developing pollutants with the greatest amounts that may disturb the endocrine system. Antibiotics, on the other hand, such as the most often given amoxicillin, raise the risk of becoming highly resistant microbial strains since their presence in soil and wastewater has been linked to a mutation in superbugs (Sivaselvam et al., 2020). As a result, eliminating pharmaceutical pollutants from the environment is a crucial step in protecting both environmental and public health. Because of their distinct physicochemical characteristics and low concentrations, removing various medicines from the aquatic system effectively has proven to be an extremely difficult task (Cristóvão et al., 2020). Because of its excellent target absorption, high removal rate, and ease of use, adsorption technology has been found to be particularly successful and cost-effective among the several methods investigated for eliminating pharmaceutical pollutants from water. MOFs are among several adsorptive materials that have been researched to date for the elimination of medicines. The majority of MOFs have poor resistance to heat, water, and strong electron beams in addition to their lack of conductivity. Low mechanical strength and a tendency to be found in powder form make their configurations unsuitable for further processing. Many strategies may be used to address the aforementioned issues, and one of them is the creation of composites using MOFs. The exceptional qualities of these composites, which successfully address the drawbacks of MOFs, include great mechanical strength, durability, reduced toxicity, low weight, and in certain cases economic efficiency (Ullah et al., 2024). MOF composites are very interesting because they have favorable kinetics and/or high adsorption capacity in addition to their enhanced water stability.

Zhang et al. successfully synthesized the magnetic UiO-66-NH$_2$ composite, which they then used as an adsorbent to efficiently extract salicylic acid (SA) and acetylsalicylic acid (ASA) from an aqueous solution. The magnetic UiO-66-NH$_2$ demonstrated excellent adsorption capacities for both SA and ASA, and the Langmuir isotherm model and PSO kinetics were used to properly characterize the adsorption process. It was previously believed that Zr–O cluster affinities for carboxyl groups, hydrogen bonds, and electrostatic interactions were the main adsorption mechanisms. The magnetic UiO-66-NH$_2$ adsorbents show tremendous promise as effective SA and ASA removers due to their exceptional adsorption capacity, simple magnetic separation approach, high removal rate, and outstanding regenerability (Zhang et al., 2019). Similar to this, a solvent-free process was used to synthesize the novel class of porous materials MIL-100(Cr) in order to assess the removal of pharmaceutical pollutants such as metformin, ibuprofen, carbamazepine, or paracetamol from aqueous sources. The characterization results showed that the solvent-free process utilized during the synthesis of the material resulted in an amorphous solid structure type. Metformin had a lower value of 12 mg/L on MIL-100(Cr) according to adsorption tests, whereas the values of carbamazepine, paracetamol, and ibuprofen were higher at 21, 20, and 17 mg/L, respectively. π–π, electrostatic, and hydrogen bond interactions between the medicinal compounds and MIL-100(Cr) were among the interactions that explained the variations in adsorption values. Because synthesis solvent-free procedures are used, MIL-100(Cr) has minimum crystallinity but yet has the necessary structural and physicochemical properties to be used as an adsorbent of pharmaceutical pollutants from aqueous solutions (Matus et al., 2023).

By using the sono crystallization technique, a unique zeolite/ZIF-8 composite was created to remove azithromycin, a pharmaceutical contaminant, from polluted water. The drug adsorption followed PSO kinetics, according to the

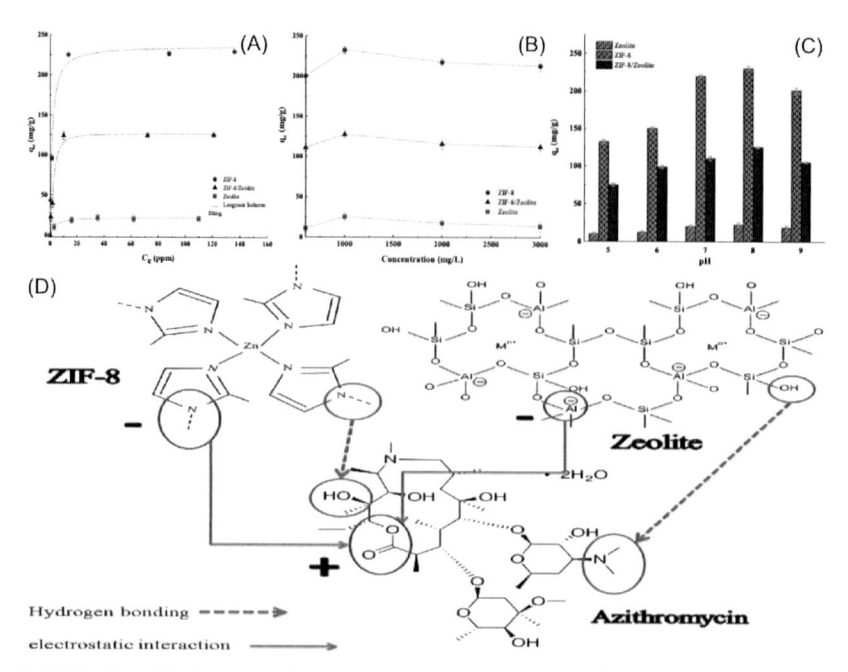

FIGURE 16.5 (A) Azithromycin's adsorption capacity in terms of adsorbent dose; (B) The influence of pH on azithromycin adsorption; (C) pH's impact on azithromycin; (D) Azithromycin adsorption mechanism on ZIF-8/zeolite.

kinetic studies, and the adsorbent achieved equilibrium at pH=8 in 60 minutes. Because of the electrostatic interactions, pH had a significant impact on the adsorption outcome. Increased entropy was linked to the spontaneous, endothermic adsorption process. Fig. 16.5A shows the Langmuir isotherm fitting diagram. The composite ZIF-8/zeolite has an adsorption capability of 131 mg/g (Fig. 16.5B). Adsorbent recovery showed that the composite may be employed 10 times with 85% removal efficiency. The number of cycles used for adsorption and reuse in comparison to ZIF-8 demonstrated the stability of the composite in an aqueous solution (Liu, Bahadoran, et al., 2023). Fig. 16.5C shows the pH impact on the performance of the composite material. Fig. 16.5D shows the azithromycin adsorption process on ZIF-8/zeolite, where hydrogen bonds and electrostatic attractions serve as the primary driving factors.

Adsorptive removal of heavy metals by metal-organic framework composites

Naturally occurring elements with large atomic weights and densities are known as heavy metals. Certain heavy metals, including iron, zinc, and copper, are essential for life in minimal amounts, but when they are present in excess, they may be extremely hazardous. Heavy metals can have a number of negative consequences when they go into wastewater (Nassef et al., 2024). Because heavy metals are indigestible, they can readily build up in the body and cause serious illnesses including kidney and liver failure (Mahmoodi et al., 2019). Some of them even at extremely low doses cause significant harm. For this reason, guidelines for water quality standards have been established by the World Health Organization. Prevalent catalysts, such Pd, Au, and Fe, are frequently used to address issues with precious metal leakage or ferrous ion leakage, which can result in secondary pollution (Shao et al., 2017). Aquatic environments are susceptible to heavy metal accumulation, which can harm aquatic life and disrupt food webs. This can lead to high concentrations in predators when aquatic species absorb heavy metals from the water and transfer them up the food chain. Fish, plants, and other aquatic organisms can be adversely affected by heavy metal toxicity, potentially resulting in a decline in biodiversity. When polluted sediments are deposited on land, heavy metals can persist in soils and sediments, impacting terrestrial ecosystems. Conventional wastewater treatment methods may be hindered by heavy metal contamination, as it can be challenging to effectively remove these metals because they can adhere to treatment equipment or form insoluble precipitates (Santoso et al., 2023). It is crucial to remove heavy metals in order to reduce the hazards they provide.

Recently, a variety of applications such as adsorption have made use of MOF-based materials with regulated design structures, strong metal skeletons, ultra-high adsorption capacity, specific surface area, excellent selectivity, and permanently functionalizable pore structures. Several researches have shown the effectiveness of MOFs materials in the adsorption removal of contaminants, including heavy metals (Mo et al., 2022). Guo significantly improved the material's adsorption efficiency for metal ions by using green cellulose aerogels and magnesium-based MOF composites to address the issue of MOF nanoparticle aggregation and recycling challenges. The primary removal mechanism is the significant complexation of Pb^{2+}, Cd^{2+}, and Cu^{2+} with the surface sites of Mg-MOF-74@CA (CA: cellulose aerogel). Moreover, Pb^{2+}, Cd^{2+}, and Cu^{2+} could have greater replacement capabilities than Na^+ ions, even if ion exchange also contributes to the elimination (Guo, 2023). Wang and colleagues utilized a composite material consisting of magnetic elements and MOF to enhance the material's adsorption capacity for cadmium ions, exhibiting remarkable stability (Wang et al., 2021). Pb^{2+} adsorption on $Ni_{0.6}Fe_{2.4}O_4$-UiO-66-PEI (PEI: polyethyleneimine) may be explained by Pb^{2+} chelating with imine/amine functional groups in the matrix of the material, while Cr^{6+} adsorption may result via a mix of redox reaction, chelation, and electrostatic attraction. After five regeneration cycles, this material maintained a rather excellent binding capacity and demonstrated good selectivity for competing ions. These composite materials have the potential to overcome some of the limitations of MOFs, enhance their performance in a range of applications, and spur innovative scientific and technological advancements.

Ji et al. created a magnetic nanocomposite generated from graphene oxide and Fe-MOFs for improved As^{5+} adsorption in aqueous solution. The novel nanocomposites adsorbed As^{5+} through chemisorption, controlled by liquid film and intraparticle diffusion. As shown in Fig. 16.6A and B, coexisting ions and pH levels control the adsorption process. Fig. 16.6C and D shows the PSO, PFO, and Temkin model adsorption kinetic models that were used. The adsorption of As^{5+} by the new nanocomposites was primarily facilitated by electrostatic interactions, hydrogen bonding, and chemical complexation (Fig. 16.6E). The composite demonstrated its maximum capacity for adsorption up to 26.11 mg/g. To effectively overcome the limitations of MOFs and broaden their range of applications, the study promotes the rational

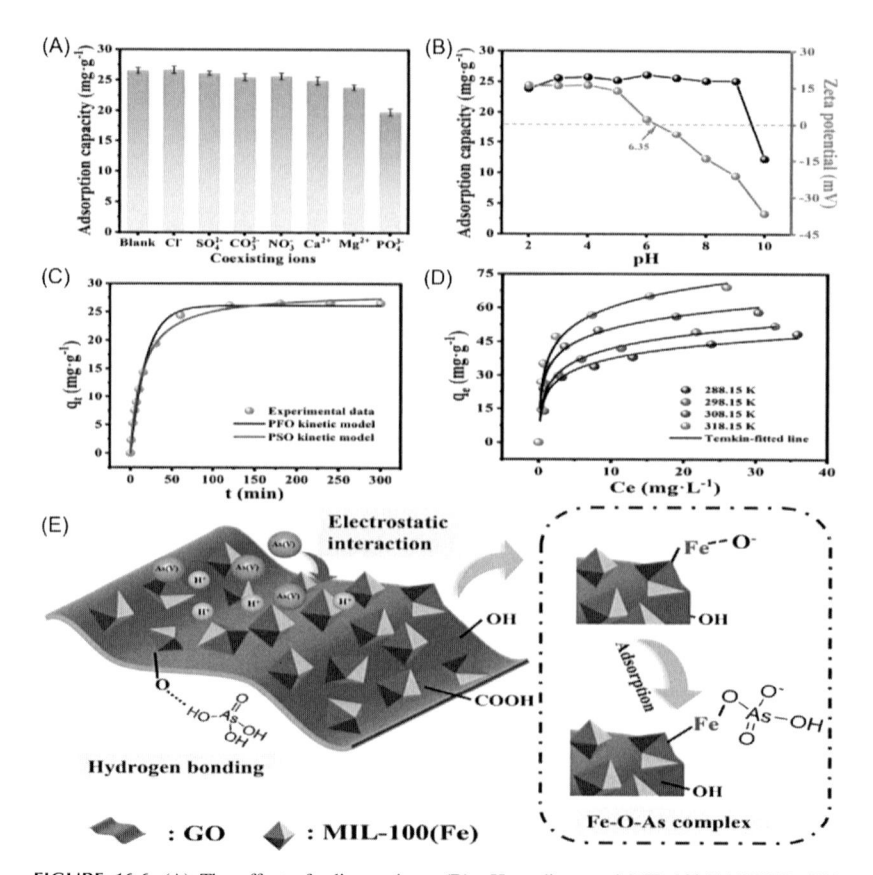

FIGURE 16.6 (A) The effect of adjacent ions; (B) pH readings and MIL-100(Fe)/1%GO-400's associated zeta potential; (C) Fitting curves to PFO and PSO models using adsorption kinetic data; (D) Temkin model, circumstances of experimentation: pH = 6.0, T = 288.15–318.15, C_0 = 5–50 mg/L, 0.4 mg/L for the $C_{adsorbent}$, and t = 5 hours; (E) Diagrammatic representation of the suggested process of adsorption (Ji et al., 2024).

development of multifunctional MOFs-derived composites (Ji et al., 2024). In a separate study, Fe-MOFs were synthesized by solvent thermal method and immobilized in the nanofibrous membrane by coelectrospinning in situ. Using the acquired PAN/MOFs ENFMs (PAN: polyacrylonitrile; ENFNs: electrospun nanofibrous membranes), water was treated to eliminate Cr^{6+}. At pH 4.0, the composite's maximum capacity to extract Cr^{6+} from water was 127.70 mg/g. Cr^{6+} was adsorbing endothermically and spontaneously. PAN/MOFs ENFMs were less sensitive to solution pH than Fe-MOFs particles and could adsorb Cr^{6+} in water across a larger pH range. The adsorption mechanisms for Cr^{6+} in water include coprecipitation, redox reaction, coordination effect, and electrostatic contact. The composite material was the most effective choice for heavy metal removal from wastewater using adsorption (Miao et al., 2022).

Oil water separation by metal-organic framework composites

Resources are wasted and marine habitats are often destroyed by oil exploration and transportation disasters. In addition, large amounts of oily wastewater are produced everyday by the steel, petrochemical, mining, and food industries. Many highly hydrophobic materials have been produced in the last few decades. Depending on the kind of separation, materials for oil–water separation may be divided into three groups (Xue et al., 2014). The components used for oil removal include materials that are both hydrophobic and oleophilic, showing a strong affinity for oil. These characteristics make them exceptionally effective at absorbing and filtering oil from water. This makes them suitable for the "oil removal" category. Using the materials' various water affinities is the concept underlying "oil removal" type separation materials. Filtration or absorption may be used to effectively separate oil from water by using the typically hydrophobic/oleophilic surface (Liu, Lin, et al., 2023).

Wettability is a critical property of solid surfaces; it is the wetting phenomena that happen when a liquid comes into contact with a solid surface. The properties of the special wetting material include superhydrophobic, superoleophilic, superhydrophilic, and superoleophobic properties. Superoleophilic and superhydrophobic materials reject water, yet allow the oil phase to freely spread, penetrate, and split from the water phase (Yang et al., 2022). But these materials, together with absorbed oil and lipophilic substance waste, are vulnerable to recycling restrictions during reprocessing. Oil–water separation is only one of the several applications for the separation technique based on MOFs and their composites, among many other materials. Several researchers have generated superoleophilic and superhydrophobic MOF materials with success. The potential of MOFs as adsorbents or separation materials seems to be enhanced by their special porous material qualities, such as their rich structure and huge specific surface area (Hu et al., 2022). Given that a large number of MOFs are moisture-sensitive, the development of hydrophobic MOFs boosts their water resistance. If the applications are more reliable, more people may use them (Xie, 2020). Because these adsorbent materials are so stable, extrusion may be employed to collect the adsorbed oil as it is being separated (Deng et al., 2021). Furthermore, due to their simple synthesis, superhydrophobic-superoleophilic MOFs and their hybrids are great choices for the creation of state-of-the-art oil/water separation systems. Studies on oil–water separations have turned to materials based on MOFs, which are hydrophobic-superoleophilic MOFs used to separate oil from water, but also possess the added properties of being superhydrophilic and submerged superoleophobic (Gu et al., 2019). These MOF materials' submerged superoleophobic properties increase separation efficiency and filter life because oil does not clog the pores of MOF-based filtering devices.

Due to ZIF-8's chemical and thermal stability, it has received a lot of interest in the super wettability domains (Li et al., 2021). Ye et al. (2019) created self-healing superhydrophobic coatings utilizing copolymer dimethylsiloxane and ZIF-8 modified with CNTs. In a similar vein, cotton textiles based on MOFs with switchable wettability were created and their efficacy in oil–water separation and self-cleaning was thoroughly investigated. The fabric has a great deal of potential for use in the purification of oily wastewater since it could effectively separate oil–water combinations with varying densities under high flux conditions after being wetted with ethanol. The fabric's efficiency and flux were above 15,300 L/m^2.h and 98.6%, respectively. Additionally, the superhydrophobic fabric offered a significant benefit for real-world applications due to its superior chemical stability and mechanical abrasion resistance (Zhang et al., 2022). Raj and colleagues synthesized a composite known as strontium metal-organic framework (Sr-MOF) for use in oil and water separation. Because Sr-MOF is highly hydrophilic and underwater oleophobic, it effectively separates the oil–water mixture. When it comes to separating the oil–water mixture, Sr-MOF adheres to the Cassie–Baxter state. Under extreme conditions, Sr-MOF remains stable across a range of pH solutions and salinity concentrations. Depending on the substrate on which Sr-MOF is deposited, the rejection of oil exceeds 99% and demonstrates a high flux. The flow increases and remains steady as the number of reuse cycles increases, maintaining consistent oil rejection throughout (Raj et al., 2023).

To achieve extremely effective oil and water separation, two essential components of the membrane are its building block structures and surface chemical composition. In order to separate oil from water, He et al. developed Cu-CAT-1 MOFs with special two-dimensional (2D) hierarchical structures that were grown directly on copper mesh (Cu-CAT-1@CM) (where Cu-CAT-1=Cu triphenylene catecholate and CM=copper mesh). The membrane shows strong separation capabilities via testing with several oils, including xylene and actual crude oil, which are recognized to be the most notorious pollutants in the water. Less than 24.6 mg/L of residual oil is present in six samples. By identifying the proper hierarchical structures of the membrane surface as a code for the excessive wettability, the mechanism analysis was able to significantly increase the separation efficiency. Additionally, a high-performing membrane may be readily synthesized using an electrochemical technique in about 20 minutes at ambient temperature, highlighting the advantages of a straightforward and inexpensive production process. Overall, the energy-efficient gravity-driven membrane and ultra-high permeability enable the quick separation of water and oil, suggesting potential uses for addressing the issue of oil pollution (He et al., 2022).

Piao et al. developed a flame-retardant flexible polyurethane sponge (FPUF) modified with layered double hydroxide (LDH) derived from MOF for high-performance oil–water separation. After grafting MOF-LDH with hexadecyltrimethoxysilane, FPUF@MOF-LDH@HDTMS (HDTM: hexadecyltrimethoxysilane) exhibited exceptional superhydrophobic/superoleophilic performance (water contact angle=153° and oil contact angle=0°). Because of its distinctive bionic structure, the synthesized composite may swiftly and readily absorb oily substances that have settled or suspended in water. Additionally, it has a great capacity for absorbing oil and organic solvents; even after 20 cycles of usage, it still has this ability (Fig. 16.7A and B). Significantly, Fig. 16.7C explains the adsorption process in detail and illustrates how continuous oil–water separation using composite material may achieve up to 99.1% separation efficacy. Additionally, because of the bionic superhydrophobic sponge's remarkable flame retardancy, there will be fewer polyurethane sponge-related fires in the future. Therefore a novel approach to designing the more efficient oil–water separation sponges is provided by the biomimetic micro-nano composite structure (Piao et al., 2023).

Degradation of pollutants

Degradation of organic dye by metal-organic framework composites

Currently, one of the most important issues facing the globe is environmental pollution. Both human health and society suffer greatly from environmental contamination, which also causes significant financial losses (Manisalidis et al., 2020). Herbicides and dyes are only two examples of the numerous organic, inorganic, and biological pollutants found in soil and water. The growth of the dye business in recent years has had a disastrous effect on both human health and the environment. It is crucial to develop a plan to treat and eliminate this color from polluted water (Manzoor & Sharma, 2020). The photodegradation process is the most advantageous of all the useful techniques because of its straightforward procedures, low energy consumption, strong chemical stability, affordability, and absence of secondary pollutants. During photodegradation, light and active radicals created by photocatalysts cause the

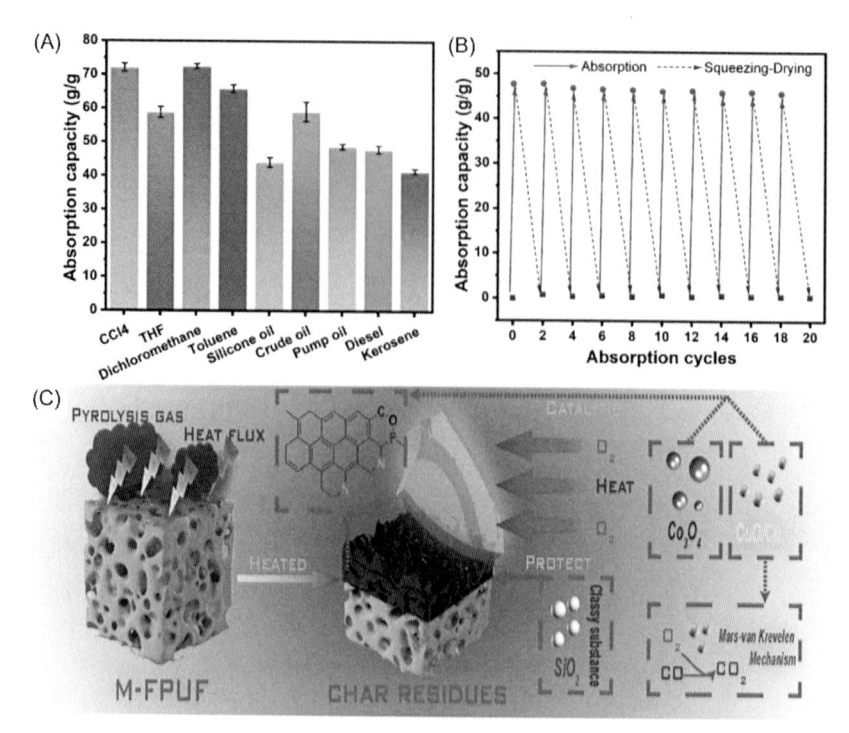

FIGURE 16.7 (A) The FPUF@MOF-LDH@HDTMS's capability to absorb different types of oil. (B) The FPUF@MOF-LDH@HDTMS's ability to be recycled during 20 cycles of absorption, squeezing, and drying; and (C) A schematic representation of the flame-retardant mechanism of FPUF@MOF-LDH@HDTMS (M-FPUF).

pollutant's molecular structure to be photochemically degraded (Mahmoodi et al., 2011). The photocatalyst bandgap should be considered while selecting the appropriate light. The electron in the VB is excited by the photocatalyst's light, travels to the CB, and subsequently degrades the pollutants' chemical structure to produce radicals (Najafi et al., 2022).

The efficiency of MOFs as adsorbents and/or photocatalysts in the elimination or degradation of organic pollutants has attracted a lot of attention lately (Nazir et al., 2022; Shahid et al., 2024). One of the challenges in the remediation process is that MOFs release metal ions and ligands into the solution, which might contaminate new water. As a result, more and more studies have looked at doping composite materials with elements including magnetic particles, Ag/AgCl, graphene oxide, activated carbon, and others. Enhancing MOFs' photocatalytic and adsorptive properties, structural stability, and separability is the aim of this strategy (Oladoye et al., 2021).

The synergistic integration of porous MOF onto MXene nanosheets (MXOF composite) may preserve or even enhance the inherent advantages of both components. Leveraging the superior specific surface area and low bandgap energy of MOFs, the MXOF composite also offers an adjustable bandgap and suitable photocatalytic activity. Using MXene nanosheets to construct MOF crystals in situ, a recent study developed the MXOF composite and examined its photocatalytic activity in degrading organic dyes. The MXOF composite has a sheet-like structure and high stability. The addition of MOF reduced MXene's bandgap energy and increased its electron recombination deficiency. The MXOF composite was optically analyzed using photoluminescence (PL) spectra and electrochemical impedance spectroscopy (EIS) spectra, which revealed that, under exposure to mercury vapor light (250 W), the composite effectively forms electron–hole pairs, exhibits low electronic current resistance, and enables rapid interfacial charge transfer. Under optimal conditions, the photocatalytic dye degradation process eliminated 35% of DR31 and 62% of MB dyes. The radicals h+, OH, and O_2^- were found to participate in the photodegradation of MB; however, the O_2^- radical was ineffective for DR31 dye. Additionally, the recyclability of the MXOF composite suggests it may be used for at least four photodegradation cycles (Far, 2023).

Najafidoust et al. effectively synthesized Ag_2CrO_4/MIL-53(Fe) by use of an ultrasonic-assisted precipitation technique. The AO7 (acid orange 7), MB (methylene blue), and CR (congo red) dyes were effectively removed from aquatic solutions using the produced composites. Although Ag_2CrO_4's specific surface area makes photocatalysis challenging, immobilizing MIL-53(Fe) enhanced the activity. ACM-50 (ACM: (Ag_2CrO_4: MIL-53(Fe)) demonstrated superior photocatalytic activity, achieving a 91.5% removal of AO7 dye compared to ACM-25 and ACM-75 composites. This was observed under conditions of pH=6, with a concentration of 30 mg/L of AO7, and under simulated sunlight irradiation, according to the results of the photo-degradation experiment. Additionally, after four cycles, the ACM-50 exhibited great stability, according to the reusability data. Therefore a unique Ag_2CrO_4 heterostructure based on MIL-53(Fe) may be the best option for eliminating organic dyes released into the environment. The near surface area of Ag_2CrO_4 was improved and the electron-hole recombination rate was lowered, according to DRS (diffuse reflectance spectroscopy) study utilizing MIL-53(Fe) (Najafidoust et al., 2022). The synthesis of CDs@ZIF-8@AgI (CD: carbon dots), a visible light-sensitive photocatalyst, was successfully accomplished using an in situ ion deposition technique and a solvent-free method. The photocatalyst may be able to degrade antibiotics or other organic dyes in addition to having a 96% catalytic degradation efficiency for AO dye. The heterojunction material's photocatalytic activity was greatly enhanced by the addition of CDs, which also increased the catalyst's ability to absorb visible light and the efficiency of electron transport. The indirect Z-scheme heterojunction type has been postulated as the mechanism of photocatalytic degradation. The paper presents a workable method for creating photocatalysts to treat environmental pollutants using solar energy, in addition to a heterojunction material for the elimination of AO dye in environmental wastewater under visible light radiation (Xia et al., 2023).

BiOBr@Zn/Fe-MOF composites were produced by Shi et al. to degrade RhB when exposed to visible light. When it came to removing RhB, the combination of BiOBr and ZnFe-MOF functioned better than either substance alone; in only 90 days, the degradation efficiency went from 84.9% to 99%. The enhanced photo-degradation performance is ascribed to the heterojunction structure including BiOBr and ZnFe-MOF, in addition to the cooperative influence of distinct metal and functional groups. Throughout the trial cycle, the BiOBr@ZnFe-MOF composite demonstrated remarkable stability and reusability (Fig. 16.8A). Fig. 16.8B displays the UV–Vis absorption spectra for RhB

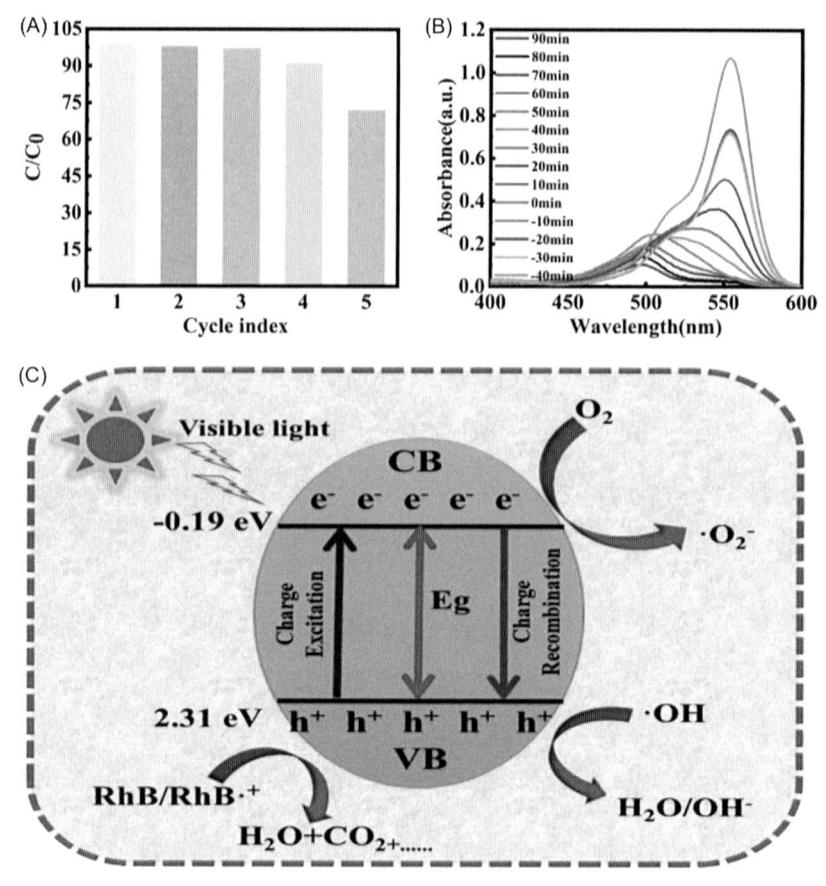

FIGURE 16.8 (A) BiOBr@ZnFe-MOF composites' reusability for photocatalytic RhB degradation; (B) UV–Vis absorption spectra for RhB degradation; (C) A potential mechanism for RhB's photodegradation when exposed to visible light.

degradation. The RhB degradation process, reaction mechanism, and potential photocatalytic mechanism were all significantly impacted by O_2^- and H^+, as shown by the free radical capture experiment (Fig. 16.8C) (Shi et al., 2022).

Degradation of pharmaceutical compounds by metal-organic framework composites

The abuse of antidepressant medications and the ineffectiveness of wastewater treatment plant processing have drawn significant attention to pharmaceutical products as one class of emerging pollutants in recent years. As a result, their concentration in the environment has been steadily rising and could seriously endanger both human health and the ecological balance (Pino-Otín et al., 2017). Numerous treatment techniques have been used, but solar-photocatalysis which involves the interaction of solar light with the heterogeneous catalyst interface is the most economical of all the techniques (Ibrahim et al., 2020). It is worthwhile to focus more on creating efficient photocatalysts in order to remove harmful pollutants and increase solar energy use. These catalysts also need to produce reactive oxygen species, or ROS. Numerous innovative photocatalysts have been created, but MOF-based materials have garnered a lot of interest because of their broad functional groups, high specific surface area, increased pore volume, and possible catalytic capabilities (Cai et al., 2021). However, certain water-stable MOFs have poor catalytic efficiency for the breakdown of organic contaminants. One such framework is the most widely used UiO-66. Improving photocatalytic efficiency and lowering electron-hole recombination are crucial. With 12 linked organic linkers, this architecture is stable in water and provides a large number of homogeneous pores, including two different kinds of octahedral and tetrahedral cages (Singh et al., 2018). Because of the presence of Ce^{3+} and Ce^{4+} ions in the framework, cerium-MOFs demonstrated exceptional catalytic activity. Nevertheless, few research has been done to assess their efficacy in purifying water, most likely because of the lack of assurance regarding the water's stability and catalytic performance.

Using a solvothermal method, the g-C_3N_4/NH_2-MIL-101(Fe) composite was created and then used as a powerful photocatalyst to degrade AAP (acetaminophen) in the presence of solar light. The composite was characterized using field emission scanning electron microscopy (FESEM), UV-Vis spectroscopy, and X-ray diffraction (XRD) analysis. When comparing the composite to the bare g-C_3N_4 and pure Fe-MOF, the photoluminescence (PL) analysis verified that the composite's charge recombination rate was lower. The rate at which AAP degrades was raised to 4.5 times by the addition of H_2O_2 during the photocatalytic process. The ROS and electron paramagnetic resonance (EPR) tests confirmed that O_2^- was essential to the degradation of AAP. The results of the cyclic test analysis show that even after 10 cycles, the photocatalyst remained extremely stable and maintained its catalytic effectiveness (Pattappan et al., 2022). A large Hal′@MIL-PMA photocatalyst for the photooxidation of tetracycline (TC) was produced using a hydrothermal process. The photocatalytic activity of this photocatalyst for decontaminating antibiotics was 97.11%, 90.87%, and 95.31% under optimal operating conditions, respectively, due to the photo-decontamination rates of TC, RIF (rifampicin), and SSZ (sulfasalazine). The photocatalyst's effectiveness was verified by a number of techniques, including light absorption, bandgap reduction, and effective e^-/h^+ pair separation. Additionally, the thermodynamic experiments demonstrated that the photocatalyst was very effective in the presence of visible light, with the decontamination process being endothermic and having a low activation energy (E_a=21.74 kJ/mol). The findings of the mineralization evaluation, along with high-performance liquid chromatography-mass spectrometry (HPLC-MS) analysis, unveiled a unique approach to TC photo decontamination. An evaluation of the purification of pharmaceutical wastewater revealed an impressive total organic carbon (TOC) removal rate of 63.23%. Additionally, after five consecutive cycles, the reusability research showed a mild reduction of around 17%, indicating a very good stability of the photocatalyst. These results show how well the photocatalyst Hal′@MIL-PMA-1x performs in breaking down difficult-to-remove contaminants from aquatic environments (Mohammadi et al., 2024).

Ensuring the availability and sustainable management of water that is threatened by persistent pollutants has become a social need that has gained importance in recent years. Many physical and chemical approaches can be used to treat wastewater, but electrochemical advanced oxidation techniques (electro- and photoelectro-Fenton) have emerged as practical choices. It is commonly recognized that the resistant organic pollutants in wastewater may be effectively degraded using electro-Fenton technology. Fe-MOF was used as a catalyst and electro catalyst by Fdez-Sanromán et al. to create a heterogeneous electro-Fenton system for the degradation of pharmaceuticals. These approaches yielded fibers and composites with antipyrine removal rates of 82.5% and 75.4%, respectively, which demonstrated promising outcomes. The extremely hydrophilic substance provided by the porous fibers increases the accessibility of Fe-MOF for the reactions (Fdez-Sanromán et al., 2023).

Yousefi et al. present a new carbohydrate-based nanocomposite (Fe_3O_4/MOF/AmCs-Alg) as an effective photocatalytic agent for the degradation of ciprofloxacin (CIP) in aqueous solutions. Under optimized conditions, the nanocomposite demonstrated an impressive 95.85% degradation efficiency of CIP. Fig. 16.9A illustrates the effects of light intensity on the photocatalytic degradation of CIP (at pH 5, 10 mg/L of CIP, and 0.4 g/L of nanocomposite dosage). The nanocomposite maintained consistently high photocatalytic efficiency for CIP degradation even after five cycles, with only a slight 16% reduction in performance (Fig. 16.9B), highlighting its advantageous reusability. Additionally, the study confirmed that the $O_2^{.-}$ radical was the primary active species in CIP degradation, as supported by trapping tests. Treatment of tap water and treated wastewater samples provided significant evidence of the

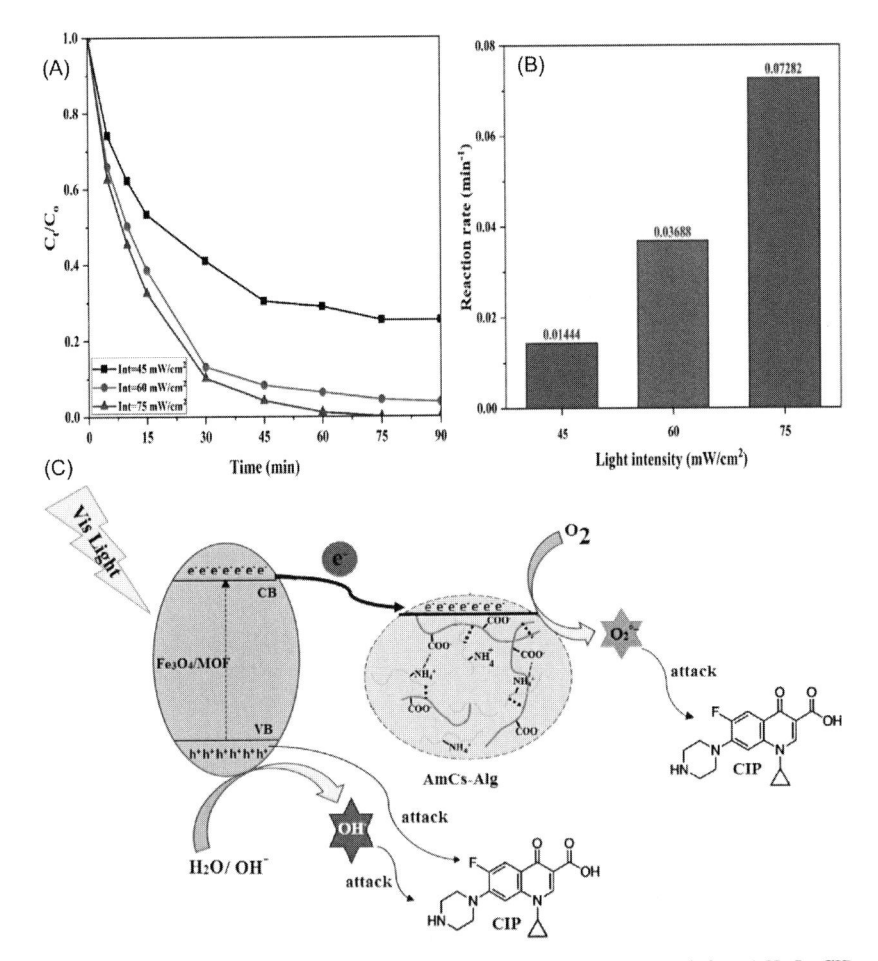

FIGURE 16.9 (A) Light intensity's effect on CIP photocatalytic degradation (pH=5, CIP concentration=10 mg/L, and nanocomposite dosage=0.4 g/L). (B) Reusability of the nanocomposite for CIP degradation over five cycles. (C) Diagram showing the electron transfer processes needed to clean CIP (Yousefi et al., 2024).

Fe_3O_4/MOF/AmCs-Alg/Vis system's viability and effectiveness. The mineralization rate of CIP was found to be 51.21% of the TOC. Fig. 16.9C presents a schematic of the electron transfer processes involved in the CIP degradation process (Yousefi et al., 2024).

Degradation of heavy metals by metal-organic framework composites

Water contaminated by heavy metals is a global issue that has received a lot of attention. Cations (such as Hg^{2+}, Pb^{2+}, Cd^{2+}, and Cu^{2+}) and oxygen anions (such as $CrO_4^{2-}/Cr_2O_7^{2-}$, SeO_3^{2-}/SeO_4^{2-}, and $HAsO_4^{2-}/AsO_4^{3-}$) are the most prevalent heavy metal pollutants detected in water (Upadhyay et al., 2021). Compared to organic contaminants, heavy metal ions in water are more challenging to biodegrade into environmentally safe chemicals. Biological organisms can easily absorb heavy metal-contaminated water, allowing these compounds to enter the food chain and accumulate within organisms. This accumulation poses a significant threat to the health of humans and other living creatures (Zheng et al., 2020). Because hazardous metal ions are diverse and changeable, eliminating them from actual water bodies remains a challenge undertaking. Investigating practical techniques to extract heavy metal ions from actual water bodies is thus essential. Effectively removing heavy metal ions is necessary due to their varying levels of toxicity, which depend substantially on their valence. Fortunately, by applying photocatalytic technology, light energy can be transformed into chemical energy in a sustainable and eco-friendly manner. This process enables the breakdown and reduction of contaminants in water (Qin et al., 2021). Photocatalytic reduction is a useful technique for changing extremely toxic, high-valence heavy metal ions into low-valence, low-toxicity, or nontoxic heavy metal ions. This method uses light energy to accelerate the transformation of dangerous heavy metal ions into less dangerous ones.

MOFs may absorb light through metal centers or organic linkers and are photoresponsive. The benefits of MOFs as a photocatalyst include their high specific surface area and excellent structure, which can encourage target molecule translocation and good adjustment (Zhang, 2020). There is significant potential for the use of photocatalytic reduction in addressing heavy metal ion contamination. A critical factor in the effectiveness of MOFs for this purpose is their water stability. In environments where materials are susceptible to water, it can negatively impact their structure and function. However, water-stable metal-organic frameworks (WMOFs) are able to effectively extract heavy metal ions from water because they preserve both structural and functional stability in aquatic conditions (Yao et al., 2016). Composite materials can be a useful tool for increasing the stability of MOF.

Numerous photo-induced processes allow MOFs to function very well in the context of heavy metal removal by photocatalysis. A number of intricate processes, such as light absorption, charge carrier production, migration and separation, metal ion adsorption, and photocatalytic redox reactions, take place in response to light irradiation (Ke et al., 2023). Some of the characteristics that can have a big influence on the photocatalytic degradation efficiency of organic pollutants and heavy metals include pH, temperature, catalyst dose, and starting contaminant concentration (Hasanpour & Hatami, 2020). By affecting the surface charge of MOFs and the speciation of target pollutants, which in turn changes the adsorption capabilities, pH influences photocatalytic reaction rates (Wang et al., 2022). According to the Arrhenius equation, higher temperatures generally improve reaction kinetics; yet, too much heating might reduce absorptivity. More active sites are produced by increasing the dosage of the photocatalyst; however, excessive doses may decrease degradation efficiency because of things like light scattering. Inhibiting photocatalytic activity can occur when high initial pollutant concentrations saturate the catalytic active sites (Abdollahzadeh et al., 2023).

Zinc-based metal-organic framework (NNU-36), a novel visible light photocatalyst, has been produced by the pillared-layer synthetic approach. According to research, NNU-36 works very well as a heterogeneous photocatalyst to reduce Cr^{6+} in aqueous solutions. Experiments show that the photocatalytic reduction of Cr^{6+} to Cr^{3+} is strongly pH dependent. Additionally, the photocatalytic activity is greatly increased when methanol is used as a hole scavenger. Additionally, the statistics show that NNU-36 is stable, which makes it a feasible wastewater treatment alternative (Zhao, Xia, et al., 2017). MIL-101(Fe) was grown in situ on TiO_2 to create the MIL-101(Fe)/TiO_2 composite. To accomplish Cr(VI) photoreduction, it is used with MC-LR (microcystin-leucine arginine) photooxidation. The synergistic photocatalytic degradation process boosts Cr^{6+} reduction activity and marginally increases the MC-LR oxidation rate in comparison to bare TiO_2 and MIL-101(Fe). Fig. 16.10A and B shows the photocatalytic reduction of Cr^{6+} as well as the kinetics of Cr^{6+} reduction by different photocatalysts. In this case, photo-induced separation of charge carriers is improved by MC-LR acting as "hole scavengers". Furthermore, by absorbing photoexcited electrons, TiO_2 effectively inhibits charge carrier recombination when added to MIL-101(Fe), according to the electrochemical investigation (Fig. 16.10C). This study contributes to a better knowledge of how MOFs and semiconductor heterojunctions are constructed for the removal of pollutants in multicomponent contaminated systems (Wang, 2021).

Degradation of oil spills by metal-organic framework composites

Since human activity discharges liquid petroleum hydrocarbons into the ocean or other coastal regions, it is primarily to blame for most marine oil spills. These include the discharge of crude oil from drilling rigs, tankers, wells, and offshore platforms. Crude oil spills have the potential to inflict massive financial losses as well as harm to communities, public health, and natural systems (Zhang, Matchinski, et al., 2018). In addition to being used as fuel and gasoline, crude oil may be utilized as a raw material for manufacturing goods like plastics, insecticides, and fertilizers (Luo et al., 2022). Aquatic oil spills are broken down and spread by a variety of processes, including emulsification, dissolution, absorption, mixing, evaporation, photodegradation, biodegradation, and chemical reactions (King et al., 2014). Oil is converted into carbon dioxide, water, and salts by a process called photocatalytic oxidation, which uses light as its main energy source. Hydroxyl radicals are the end result

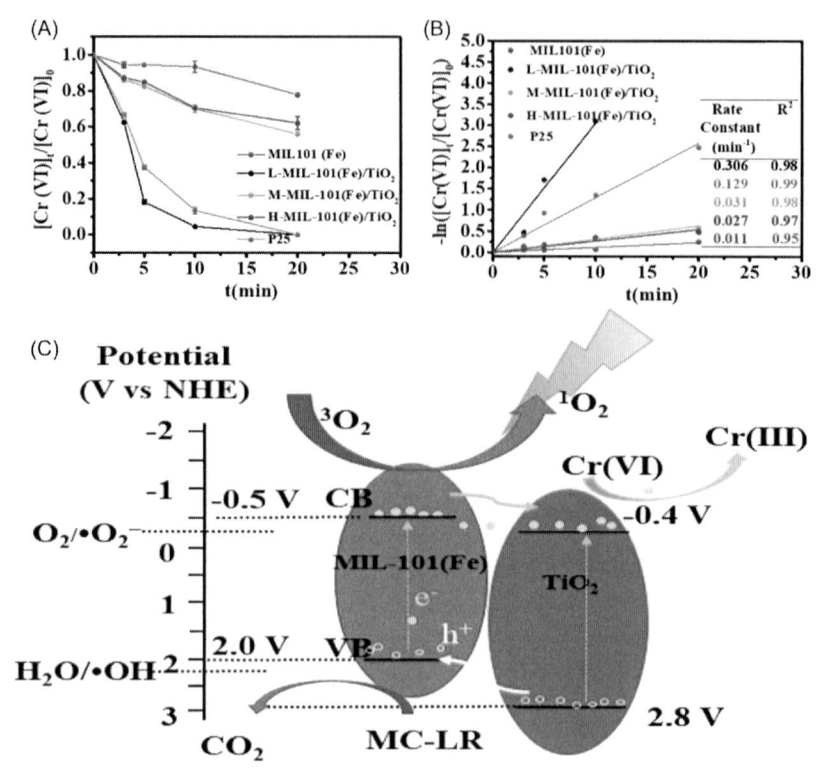

FIGURE 16.10 (A) Cr^{6+} reduction by photocatalysis using various photocatalysts; (B) The kinetics of the reduction of Cr^{6+} by different photocatalysts; (C) The schematic procedure for Cr^{6+} reduction and MC-LR oxidation occurring simultaneously over 0.1M/P under UV light irradiation.

of AOPs carried out in aqueous conditions. These radicals may react with a range of oil pollutants due to their strong reactivity with nonselective chemicals and high oxidation potential (E_o=2.8 V) (Shivaraju et al., 2016).

MOFs and their composites have demonstrated significant advantages in degrading oil spills. Titanium-based metal-organic framework (MIL-125) photocatalysts were developed and successfully used for the photocatalytic degradation of petroleum oil spills in seawater using a simple solvothermal approach. The findings show that the sample formed at 150°C and 48 hours reaction time (MT48) has an optical bandgap of 3.2 eV and a surface area of 301 m²/g. It has a 7 nm low crystal size. When exposed to UV–Vis light, the as-prepared MOFs showed improved photocatalytic activity in clearing crude oil spills in saltwater. It has been shown that applying 250 ppm of MT48 photocatalyst under UV-Vis irradiation can remove 99% of oil spills in water after 2 hours of exposure (Showman et al., 2022).

To produce a superhydrophobic aerogel that could separate oil and water and degrade when ZIF-8 and BiOBr were present, the functional composite material just needed to be dipped into the melamine sponge's support surface. Its superhydrophobicity and superoleophilicity can be used to alleviate the issue of surface oil contamination. Under light irradiation, the contaminants dissolved in water can break down thanks to the presence of BiOBr in aerogels. DRS and the Mott–Schottky curve were used to calculate the bandgap width of BiOBr (E_g=2.6 eV) and the density of the photogenerated carriers (5.21×10^{23} 1/cm^3), respectively. The material's great mechanical stability was demonstrated by the confirmation of the functional coating's resistance against the substrate during 100 compression tests and 4000 friction (1.06×10^5 Pa) tests (Ge et al., 2020). Using a solvothermal technique, a stable MOF, MIL-101(Fe), was created and is being used as a cutting-edge photocatalyst to break down crude oil in wastewater from oil fields. By fine-tuning the reaction conditions, the optimal parameters were determined to be a pH of 5.5, 150 mg/L of catalyst, a reaction temperature of 303.15 K, 30 minutes for the dark reaction, and 30 minutes for the light reaction (Fig. 16.11A and B). The clearance rate under these reaction circumstances was an astounding 94.73%. Notably, MIL-101(Fe) is efficiently recyclable and demonstrated exceptional endurance in mildly acidic settings. These findings offer crucial information on the possible industrial application of MIL-101(Fe) as a material for the removal of crude oil from water contaminated with oil using photocatalytic degradation. The suggested mechanism for the breakdown of spilled oil under visible light irradiation is depicted in Fig. 16.11C (Liang et al., 2024).

Conclusion and perspective

Various treatment technologies have been employed to solve the complicated task of removing organic contaminants from wastewater. Porous hybrid materials based on MOFs have demonstrated potential in the removal of dangerous pollutants. The challenges of creating MOF composites to remove organic pollutants, including pigments, pharmaceuticals, heavy metals, and spilled oils, are discussed in this research. There are two distinct treatment approaches that may be used to remove organic contaminants: adsorption and degradation. Although MOF composites have the potential to effectively breakdown and eliminate organic pollutants in a cost-effective and user-friendly manner, adsorption requires the identification of suitable sorbents. In order to enhance the effective and focused removal of organic analytes, MOF composites may be developed and produced to have better regeneration, more surface area, higher stability, and more suitable supramolecular interactions than bare MOFs. H-bonding, ion exchange, π–π interaction, physical adsorption, electrostatic interactions, and van der Waals interactions are significant participants in the adsorption process. Strong oxidizing and reducing agents (h^+ and e^-) created by UV or visible light on photocatalyst

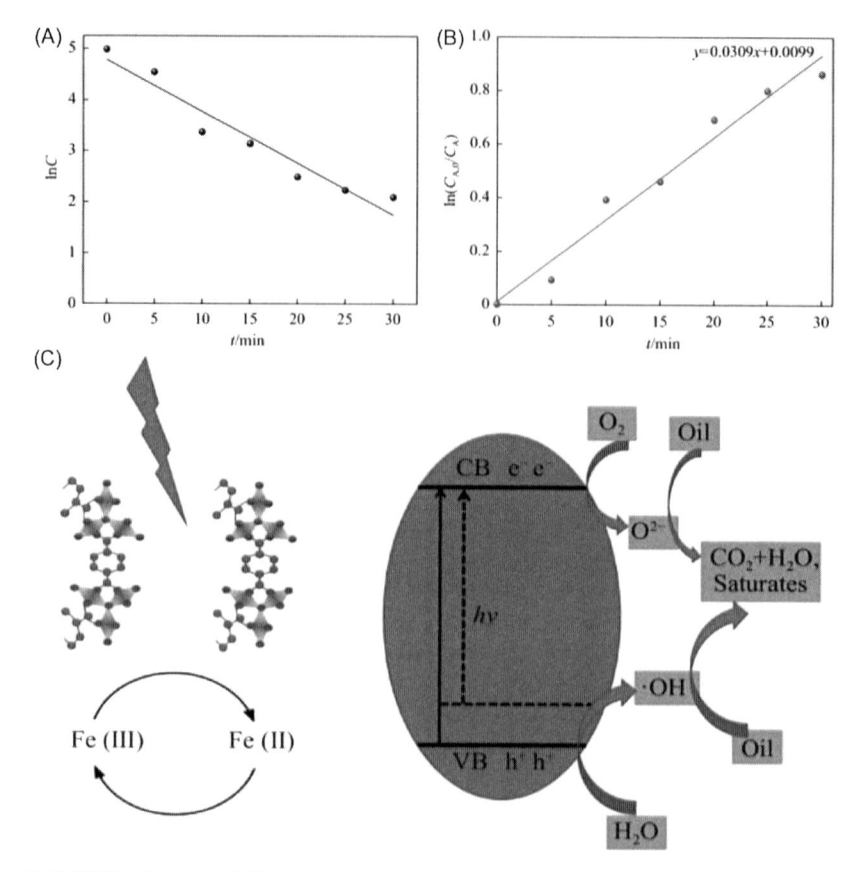

FIGURE 16.11 (A and B) Under ideal circumstances, an experimental kinetic investigation of the photocatalytic breakdown of crude oil utilizing MIL-101(Fe), and (C) suggested mechanical route when exposed to visible light (Liang et al., 2024).

surfaces interact with organic pollutants to form the basis of photocatalysis technology. Wastewater pollutant removal should be considered throughout the design and manufacturing of MOF composites, taking into account the different types of impurities and their possible interactions. It may be advantageous to choose the second composite component logically and to construct the MOF with certain functional groups. The creation of appropriate supramolecular interactions is essential to the creation of a successful MOF composite. MOF composites can be used in a number of straightforward ways, such as surface modification, functionalized material immobilization, grafting or substituting relevant functional groups to organic ligands or metal sites, combining or modifying metal sites, and composite construction using appropriate materials, to effectively remove organic contaminants from surfaces (Sharifzadeh & Morsali, 2022). Even if composite MOFs perform better, they frequently have drawbacks.

Leaching of active components, which lowers the composite's reusability, is one significant problem. In industrial applications where cost is a primary consideration, this might be a serious problem. The synthesis conditions and component cost are important factors to take into account when choosing a MOF composite for actual water decontamination. The price and environmental friendliness of the material can be increased by manufacturing MOFs at room temperature or at a modest pressure with inexpensive linkers and nontoxic, widely available metal. Additionally, MOFs may be composed to become more stable and better suited for repeated applications. When compared to employing solely MOFs, this method can minimize side costs and maximize the usage of low-synthesis-efficiency MOFs. The effect that the materials being utilized will have on the environment is another crucial factor. In order to design MOF composites that completely eradicate environmental pollution, especially water pollution, green materials with low microorganism risk must be used. Even while OF (organic-functionalized) composites often outperform bare MOFs in terms of adsorption capacity and degradation efficiency, this improved performance necessitates more care and attention in the future. In addition to finding new applications where individual MOFs cannot be utilized, MOF composites may be employed in every sector where MOFs can be used. The only restrictions on composites are one's creativity and the ability to prepare them for new applications when functional materials are appropriate. Further demands on the unique properties of these materials will arise in the future, and MOF composites are able to meet these needs. The functionality, porosity, synthesis ease, and thermal, magnetic, and electrical characteristics of MOFs may all be enhanced to satisfy particular needs by mixing them with appropriate elements. Thus it can be said that although while research into these composite materials is still in its early stages, the opportunities are enormous and will only grow.

References

Abdollahzadeh, S., Sayadi, M. H., & Shekari, H. (2023). Synthesis of biodegradable antibacterial nanocomposite (metal–organic frameworks supported by chitosan and graphene oxide) with high stability and photocatalytic activities. *Inorganic Chemistry Communications, 156*. https://doi.org/10.1016/j.inoche.2023.111302, http://www.journals.elsevier.com/inorganic-chemistry-communications/.

Abdullah, N. H., Shameli, K., Abdullah, E. C., & Abdullah, L. C. (2019). Solid matrices for fabrication of magnetic iron oxide nanocomposites: Synthesis, properties, and application for the adsorption of heavy metal ions and dyes. *Composites Part B: Engineering, 162*, 538–568. https://doi.org/10.1016/j.compositesb.2018.12.075.

Ali, T., Tripathi, P., Azam, A., Raza, W., Ahmed, A. S., Ahmed, A., & Muneer, M. (2017). Photocatalytic performance of Fe-doped TiO_2 nanoparticles under visible-light irradiation. *Materials Research Express, 4*(1), 015022. https://doi.org/10.1088/2053-1591/aa576d.

Allen, S. J., & Brown, P. A. (1995). Isotherm analyses for single component and multi-component metal sorption onto lignite. *Journal of Chemical Technology and Biotechnology, 62*(1), 17–24. https://doi.org/10.1002/jctb.280620103.

Ananpattarachai, J., & Kajitvichyanukul, P. (2016). Enhancement of chromium removal efficiency on adsorption and photocatalytic reduction using a bio-catalyst, titania-impregnated chitosan/xylan hybrid film. *Journal of Cleaner Production, 130*, 126–136. https://doi.org/10.1016/j.jclepro.2015.10.098.

Angamuthu, M., Satishkumar, G., & Landau, M. V. (2017). Precisely controlled encapsulation of Fe3O4 nanoparticles in mesoporous carbon nanodisk using iron based MOF precursor for effective dye removal. *Microporous and Mesoporous Materials, 251*, 58–68. https://doi.org/10.1016/j.micromeso.2017.05.045.

Azad, K., & Gajanan (2017). Photodegradation of methyl orange in aqueous solution by the visible light active Co: La: TiO2 nanocomposite. *Journal of Chemical Sciences, 8*(3), 1000164–1000174 2017.

Babich, I. V., & Moulijn, J. A. (2003). Science and technology of novel processes for deep desulfurization of oil refinery streams: A review. *Fuel, 82*(6), 607–631. https://doi.org/10.1016/S0016-2361(02)00324-1.

Banerjee, R., Phan, A., Wang, B., Knobler, C., Furukawa, H., O'Keeffe, M., & Yaghi, O. M. (2008). High-throughput synthesis of zeolitic imidazolate frameworks and application to CO2 capture. *Science, 319*(5865), 939–943. https://doi.org/10.1126/science.1152516.

Bhadra, B. N., & Jhung, S. H. (2018). Adsorptive removal of wide range of pharmaceuticals and personal care products from water using bio-MOF-1 derived porous carbon. *Microporous and Mesoporous Materials, 270*, 102–108. https://doi.org/10.1016/j.micromeso.2018.05.005, www.elsevier.com/inca/publications/store/6/0/0/7/6/0.

Cai, G., Yan, P., Zhang, L., Zhou, H. C., & Jiang, H. L. (2021). Metal-organic framework-based hierarchically porous materials: Synthesis and applications. *Chemical Reviews, 121*(20), 12278–12326. https://doi.org/10.1021/acs.chemrev.1c00243, http://pubs.acs.org/journal/chreay.

Carolin, C. F., Kumar, P. S., & Joshiba, G. J. (2021). Sustainable approach to decolourize methyl orange dye from aqueous solution using novel bacterial strain and its metabolites characterization. *Clean Technologies and Environmental Policy, 23*(1), 173–181. https://doi.org/10.1007/s10098-020-01934-8, https://link.springer.com/journal/10098.

Chaikittisilp, W., Ariga, K., & Yamauchi, Y. (2013). A new family of carbon materials: Synthesis of MOF-derived nanoporous carbons and their promising applications. *Journal of Materials Chemistry A, 1*(1), 14–19. https://doi.org/10.1039/c2ta00278g.

Chen, M., Wang, N., Wang, X., Zhou, Y., & Zhu, L. (2021). Enhanced degradation of tetrabromobisphenol A by magnetic Fe3O4@ZIF-67 composites as a heterogeneous Fenton-like catalyst. *Chemical Engineering Journal, 413*, 127539. https://doi.org/10.1016/j.cej.2020.127539.

Coleman, H. M., Vimonses, V., Leslie, G., & Amal, R. (2007). Degradation of 1,4-dioxane in water using TiO2 based photocatalytic and H2O2/UV processes. *Journal of Hazardous Materials, 146*(3), 496–501. https://doi.org/10.1016/j.jhazmat.2007.04.049.

Corma, A., Rey, F., Rius, J., Sabater, M. J., & Valencia, S. (2004). Supramolecular self-assembled molecules as organic directing agent for synthesis of zeolites. *Nature, 431*(7006), 287–290. https://doi.org/10.1038/nature02909.

Crini, G. (2005). Recent developments in polysaccharide-based materials used as adsorbents in wastewater treatment. *Progress in Polymer Science, 30*(1), 38–70. https://doi.org/10.1016/j.progpolymsci.2004.11.002.

Cristóvão, M. B., Janssens, R., Yadav, A., Pandey, S., Luis, P., Van der Bruggen, B., Dubey, K. K., Mandal, M. K., Crespo, J. G., & Pereira, V. J. (2020). Predicted concentrations of anticancer drugs in the aquatic environment: What should we monitor and where should we treat? *Journal of Hazardous Materials, 392*, 122330. https://doi.org/10.1016/j.jhazmat.2020.122330.

Dąbrowski, A., Podkościelny, P., Hubicki, Z., & Barczak, M. (2005). Adsorption of phenolic compounds by activated carbon—a critical review. *Chemosphere, 58*(8), 1049–1070. https://doi.org/10.1016/j.chemosphere.2004.09.067.

Deng, Y., Wu, Y., Chen, G., Zheng, X., Dai, M., & Peng, C. (2021). Metal-organic framework membranes: Recent development in the synthesis strategies and their application in oil-water separation. *Chemical Engineering Journal, 405*, 127004. https://doi.org/10.1016/j.cej.2020.127004.

Dhaka, S., Kumar, R., Deep, A., Kurade, M. B., Ji, S.-W., & Jeon, B.-H. (2019). Metal–organic frameworks (MOFs) for the removal of emerging contaminants from aquatic environments. *Coordination Chemistry Reviews, 380*, 330–352. https://doi.org/10.1016/j.ccr.2018.10.003.

Dong, X., Lin, Y., Ren, G., Ma, Y., & Zhao, L. (2021). Catalytic degradation of methylene blue by Fenton-like oxidation of Ce-doped MOF. *Colloids and Surfaces A: Physicochemical and Engineering Aspects, 608*, 125578. https://doi.org/10.1016/j.colsurfa.2020.125578.

Dulio, V., van Bavel, B., Brorström-Lundén, E., Harmsen, J., Hollender, J., Schlabach, M., Slobodnik, J., Thomas, K., & Koschorreck, J. (2018). Emerging pollutants in the EU: 10 years of NORMAN in support of environmental policies and regulations. *Environmental Sciences Europe, 30*(1). https://doi.org/10.1186/s12302-018-0135-3.

Far, S. (2023). Synthesis of MXene/metal-organic framework (MXOF) composite as an efficient photocatalyst for dye contaminant degradation. *Inorganic Chemistry Communications, 152* 2023.

Fdez-Sanromán, A., Pazos, M., Angeles Sanromán, M., & Rosales, E. (2023). Heterogeneous electro-Fenton system using Fe-MOF as catalyst and electrocatalyst for degradation of pharmaceuticals. *Chemosphere, 340*, 139942. https://doi.org/10.1016/j.chemosphere.2023.139942.

Fu, F., & Wang, Q. (2011). Removal of heavy metal ions from wastewaters: A review. *Journal of Environmental Management, 92*(3), 407–418. https://doi.org/10.1016/j.jenvman.2010.11.011, https://www.sciencedirect.com/journal/journal-of-environmental-management.

Ge, B., Yang, H., Xu, X., Ren, G., Zhao, X., Pu, X., & Li, W. (2020). Facile synthesis of superhydrophobic ZIF-8/bismuth oxybromide photocatalyst aerogel for oil/water separation and hazardous pollutant degradation. *Applied Nanoscience, 10*(5), 1409–1419. https://doi.org/10.1007/s13204-020-01296-z.

Ge, D., & Lee, H. K. (2011). Water stability of zeolite imidazolate framework 8 and application to porous membrane-protected micro-solid-phase extraction of polycyclic aromatic hydrocarbons from environmental water samples. *Journal of Chromatography A, 1218*(47), 8490–8495. https://doi.org/10.1016/j.chroma.2011.09.077.

Ghaffar, A., & Younis, M. N. (2015). Interaction and thermodynamics of methylene blue adsorption on oxidized multi-walled carbon nanotubes. *Green Processing and Synthesis, 4*(3), 209–217. https://doi.org/10.1515/gps-2015-0009, http://www.degruyter.com/view/j/gps.

Gnanaprakasam, A., Sivakumar, V. M., Sivayogavalli, P. L., & Thirumarimurugan, M. (2015). Characterization of TiO2 and ZnO nanoparticles and their applications in photocatalytic degradation of azodyes. *Ecotoxicology and Environmental Safety, 121*, 121–125. https://doi.org/10.1016/j.ecoenv.2015.04.043.

Gómez-Obando, V. A., García-Mora, A. M., Basante, J. S., Hidalgo, A., & Galeano, L. A. (2019). CWPO degradation of methyl orange at circumneutral pH: Multi-response statistical optimization, main intermediates and by-products. *Frontiers in Chemistry, 7.* https://doi.org/10.3389/fchem.2019.00772, http://journal.frontiersin.org/journal/chemistry.

Gu, J., Fan, H., Li, C., Caro, J., & Meng, H. (2019). Robust superhydrophobic/superoleophilic wrinkled microspherical MOF@rGO composites for efficient oil–water separation. *Angewandte Chemie, 131*(16), 5351–5355. https://doi.org/10.1002/ange.201814487.

Guo, Z. (2023). Recyclable Mg-MOF-74@ cellulose aerogel composites for efficient removal of heavy metals from wastewater. *Journal of Solid State Chemistry, 323,* 124059.

Hasanpour, M., & Hatami, M. (2020). Photocatalytic performance of aerogels for organic dyes removal from wastewaters: Review study. *Journal of Molecular Liquids, 309,* 113094. https://doi.org/10.1016/j.molliq.2020.113094.

Hashimoto, K., Irie, H., & Fujishima, A. (2005). TiO2 photocatalysis: A historical overview and future prospects. *Japanese Journal of Applied Physics, 44*(12R), 2005.

Hasija, V., Raizada, P., Sudhaik, A., Sharma, K., Kumar, A., Singh, P., Jonnalagadda, S. B., & Kumar Thakur, V. (2019). Recent advances in noble metal free doped graphitic carbon nitride based nanohybrids for photocatalysis of organic contaminants in water: A review. *Applied Materials Today, 15,* 494–524. https://doi.org/10.1016/j.apmt.2019.04.003.

Hayashi, H., Côté, A. P., Furukawa, H., O'Keeffe, M., & Yaghi, O. M. (2007). Zeolite A imidazolate frameworks. *Nature Materials, 6*(7), 501–506. https://doi.org/10.1038/nmat1927.

He, X.-T., Li, B.-Y., Liu, J.-X., Tao, W.-Q., & Li, Z. (2022). Facile fabrication of 2D MOF-based membrane with hierarchical structures for ultrafast oil-water separation. *Separation and Purification Technology, 297,* 121488. https://doi.org/10.1016/j.seppur.2022.121488.

Hong, S., Wen, C., He, J., Gan, F., & Ho, Y.-S. (2009). Adsorption thermodynamics of Methylene Blue onto bentonite. *Journal of Hazardous Materials, 167*(1-3), 630–633. https://doi.org/10.1016/j.jhazmat.2009.01.014.

Hu, M., Reboul, J., Furukawa, S., Torad, N. L., Ji, Q., Srinivasu, P., Ariga, K., Kitagawa, S., & Yamauchi, Y. (2012). Direct carbonization of Al-based porous coordination polymer for synthesis of nanoporous carbon. *Journal of the American Chemical Society, 134*(6), 2864–2867. https://doi.org/10.1021/ja208940u.

Hu, T., Tang, L., Feng, H., Zhang, J., Li, X., Zuo, Y., Lu, Z., & Tang, W. (2022). Metal-organic frameworks (MOFs) and their derivatives as emerging catalysts for electro-Fenton process in water purification. *Coordination Chemistry Reviews, 451,* 214277. https://doi.org/10.1016/j.ccr.2021.214277.

Huang, C. H., Chang, K. P., Ou, H. D., Chiang, Y. C., & Wang, C. F. (2011). Adsorption of cationic dyes onto mesoporous silica. *Microporous and Mesoporous Materials, 141*(1-3), 102–109. https://doi.org/10.1016/j.micromeso.2010.11.002.

Huang, J., Song, H., Chen, C., Yang, Y., Xu, N., Ji, X., Li, C., & You, J.-A. (2017). Facile synthesis of N-doped TiO2 nanoparticles caged in MIL-100(Fe) for photocatalytic degradation of organic dyes under visible light irradiation. *Journal of Environmental Chemical Engineering, 5*(3), 2579–2585. https://doi.org/10.1016/j.jece.2017.05.012.

Huang, K., Liu, S., Li, Q., & Jin, W. (2013). Preparation of novel metal-carboxylate system MOF membrane for gas separation. *Separation and Purification Technology, 119,* 94–101. https://doi.org/10.1016/j.seppur.2013.09.008.

Ibrahim, I., Kaltzoglou, A., Athanasekou, C., Katsaros, F., Devlin, E., Kontos, A. G., Ioannidis, N., Perraki, M., Tsakiridis, P., Sygellou, L., Antoniadou, M., & Falaras, P. (2020). Magnetically separable TiO2/CoFe2O4/Ag nanocomposites for the photocatalytic reduction of hexavalent chromium pollutant under UV and artificial solar light. *Chemical Engineering Journal, 381*, 122730. https://doi.org/10.1016/j.cej.2019.122730.

James, J. B., & Lin, Y. S. (2017). Thermal stability of ZIF-8 membranes for gas separations. *Journal of Membrane Science, 532*, 9–19. https://doi.org/10.1016/j.memsci.2017.02.017, www.elsevier.com/locate/memsci.

Ji, W., Li, W., Wang, Y., Zhang, T. C., & Yuan, S. (2024). Fe-MOFs/graphene oxide-derived magnetic nanocomposite for enhanced adsorption of As(V) in aqueous solution. *Separation and Purification Technology, 334*, 126003. https://doi.org/10.1016/j.seppur.2023.126003.

Jiang, H., Yan, Q., Chen, R., & Xing, W. (2016). Synthesis of Pd@ZIF-8 via an assembly method: Influence of the molar ratios of Pd/Zn2+ and 2-methylimidazole/Zn2+. *Microporous and Mesoporous Materials, 225*, 33–40. https://doi.org/10.1016/j.micromeso.2015.12.010, www.elsevier.com/inca/publications/store/6/0/0/7/6/0.

Jiang, H. L., Liu, B., Lan, Y. Q., Kuratani, K., Akita, T., Shioyama, H., Zong, F., & Xu, Q. (2011). From metal-organic framework to nanoporous carbon: Toward a very high surface area and hydrogen uptake. *Journal of the American Chemical Society, 133*(31), 11854–11857. https://doi.org/10.1021/ja203184k.

Jiang, J. Q., Yang, C. X., & Yan, X. P. (2013). Zeolitic imidazolate framework-8 for fast adsorption and removal of benzotriazoles from aqueous solution. *ACS Applied Materials and Interfaces, 5*(19), 9837–9842. https://doi.org/10.1021/am403079n, http://pubs.acs.org/journal/aamick.

Jiang, M., Cao, X., Zhu, D., Duan, Y., & Zhang, J. (2016). Hierarchically porous N-doped carbon derived from ZIF-8 nanocomposites for electrochemical applications. *Electrochimica Acta, 196*, 699–707. https://doi.org/10.1016/j.electacta.2016.02.094.

Kairigo, P., Ngumba, E., Sundberg, L.-R., Gachanja, A., & Tuhkanen, T. (2020). Occurrence of antibiotics and risk of antibiotic resistance evolution in selected Kenyan wastewaters, surface waters and sediments. *Science of the Total Environment, 720*, 137580. https://doi.org/10.1016/j.scitotenv.2020.137580.

Ke, F., Pan, A., Liu, J., Liu, X., Yuan, T., Zhang, C., Fu, G., Peng, C., Zhu, J., & Wan, X. (2023). Hierarchical camellia-like metal–organic frameworks via a bimetal competitive coordination combined with alkaline-assisted strategy for boosting selective fluoride removal from brick tea. *Journal of Colloid and Interface Science, 642*, 61–68. https://doi.org/10.1016/j.jcis.2023.03.137.

Khan, N. A., Hasan, Z., & Jhung, S. H. (2013). Adsorptive removal of hazardous materials using metal-organic frameworks (MOFs): A review. *Journal of Hazardous Materials, 244-245*, 444–456. https://doi.org/10.1016/j.jhazmat.2012.11.011.

King, S. M., Leaf, P. A., Olson, A. C., Ray, P. Z., & Tarr, M. A. (2014). Photolytic and photocatalytic degradation of surface oil from the Deepwater Horizon spill. *Chemosphere, 95*, 415–422. https://doi.org/10.1016/j.chemosphere.2013.09.060, http://www.elsevier.com/locate/chemosphere.

Kormann, C., Bahnemann, D. W., & Hoffmann, M. R. (1988). Photocatalytic production of H2O2 and organic peroxides in aqueous suspensions of TiO2, ZnO, and desert sand. *Environmental Science and Technology, 22*(7), 798–806. https://doi.org/10.1021/es00172a009.

Krishnani, K. K., Srinives, S., Mohapatra, B. C., Boddu, V. M., Hao, J., Meng, X., & Mulchandani, A. (2013). Hexavalent chromium removal mechanism using conducting polymers. *Journal of Hazardous Materials, 252-253*, 99–106. https://doi.org/10.1016/j.jhazmat.2013.01.079.

Kumar, A., & Pandey, G. (2017). Synthesis of La: Co: TiO2 nanocomposite and photocatalytic degradation of tartaric acid in water at various parameters. *American Journal of Nano Research and Applications, 5*(4), 40–48 2017.

Li, H., Eddaoudi, M., O'Keeffe, M., & Yaghi, O. M. (1999). Design and synthesis of an exceptionally stable and highly porous metal-organic framework. *Nature, 402*(6759), 276–279. https://doi.org/10.1038/46248.

Li, Y., Lin, Z., Wang, X., Duan, Z., Lu, P., Li, S., Ji, D., Wang, Z., Li, G., Yu, D., & Liu, W. (2021). High-hydrophobic ZIF-8@PLA composite aerogel and application for oil-water separation. *Separation and Purification Technology, 270*, 118794.

Li, X., Wang, K., Shen, J., Kumari, S., Wu, F., & Hu, Y. (2016). An enhanced biometrics-based user authentication scheme for multi-server environments in critical systems. *Journal of Ambient Intelligence and Humanized Computing, 7*(3), 427–443. https://doi.org/10.1007/s12652-015-0338-z.

Liang, Y., Wang, B., Li, S., Chi, W., Bi, M., Liu, Y., Wang, Y., Yao, M., Zhang, T., & Chen, Y. (2024). Enhanced photocatalysis using metal-organic framework MIL-101(Fe) for crude oil degradation in oil-polluted water. *Journal of Fuel Chemistry and Technology, 52*(4), 607–618. https://doi.org/10.1016/s1872-5813(23)60396-2.

Lim, S., Suh, K., Kim, Y., Yoon, M., Park, H., Dybtsev, D. N., & Kim, K. (2012). Porous carbon materials with a controllable surface area synthesized from metal-organic frameworks. *Chemical Communications, 48*(60), 7447–7449. https://doi.org/10.1039/c2cc33439a.

Lin, S., Zhao, Y., Bediako, J. K., Cho, C.-W., Sarkar, A. K., Lim, C.-R., & Yun, Y.-S. (2019). Structure-controlled recovery of palladium(II) from acidic aqueous solution using metal-organic frameworks of MOF-802, UiO-66 and MOF-808. *Chemical Engineering Journal, 362*, 280–286. https://doi.org/10.1016/j.cej.2019.01.044.

Lin, W. H., Kuo, J., & Lo, S. L. (2021). Effect of light irradiation on heavy metal adsorption onto microplastics. *Chemosphere, 285*. https://doi.org/10.1016/j.chemosphere.2021.131457, http://www.elsevier.com/locate/chemosphere 131457.

Liu, X. L., Wang, C., Wang, Z. C., Wu, Q. H., & Wang, Z. (2015). Nanoporous carbon derived from a metal organic framework as a new kind of adsorbent for dispersive solid phase extraction of benzoylurea insecticides. *Microchimica Acta, 182*(11), 1903–1910. https://doi.org/10.1007/s00604-015-1530-8.

Liu, G., Yang, H. G., Pan, J., Yang, Y. Q., Lu, G. Q. M., & Cheng, H.-M. (2014). Titanium dioxide crystals with tailored facets. *Chemical Reviews, 114*(19), 9559–9612. https://doi.org/10.1021/cr400621z.

Liu, Y., Lin, Z., Luo, Y., Wu, R., Fang, R., Umar, A., Zhang, Z., Zhao, Z., Yao, J., & Zhao, S. (2023). Superhydrophobic MOF based materials and their applications for oil-water separation. *Journal of Cleaner Production, 420*, 138347. https://doi.org/10.1016/j.jclepro.2023.138347.

Liu, B., Shioyama, H., Akita, T., & Xu, Q. (2008). Metal-organic framework as a template for porous carbon synthesis. *Journal of the American Chemical Society, 130*(16), 5390–5391. https://doi.org/10.1021/ja7106146.

Liu, Y., Song, L., Du, L., Gao, P., Liang, N., Wu, S., Minami, T., Zang, L., Yu, C., & Xu, X. (2020). Preparation of polyaniline/emulsion microsphere composite for efficient adsorption of organic dyes. *Polymers, 12*(1), 167. https://doi.org/10.3390/polym12010167.

Liu, Z., Bahadoran, A., Alizadeh, A.'ad, Emami, N., Al-Musaw, T. J., Alawadi, A. H. R., Aljeboree, A. M., Shamsborhan, M., Najafipour, I., Mousavi, S. E., Mosallanezhad, M., & Toghraie, D. (2023). Sonocrystallization of a novel ZIF/zeolite composite adsorbent with high chemical stability for removal of the pharmaceutical pollutant azithromycin from contaminated water. *Ultrasonics Sonochemistry, 97*, 106463. https://doi.org/10.1016/j.ultsonch.2023.106463.

Luo, C., Hu, X., Bao, M., Sun, X., Li, F., Li, Y., Liu, W., & Yang, Y. (2022). Efficient biodegradation of phenanthrene using Pseudomonas stutzeri LSH-PAH1 with the addition of sophorolipids: Alleviation of biotoxicity and cometabolism studies. *Environmental Pollution, 301*, 119011. https://doi.org/10.1016/j.envpol.2022.119011.

Mahmoodi, N. M. (2015). Surface modification of magnetic nanoparticle and dye removal from ternary systems. *Journal of Industrial and Engineering Chemistry, 27*, 251–259. https://doi.org/10.1016/j.jiec.2014.12.042, http://www.sciencedirect.com/science/journal/1226086X.

Mahmoodi, N. M., Arami, M., & Zhang, J. (2011). Preparation and photocatalytic activity of immobilized composite photocatalyst (titania nanoparticle/activated carbon). *Journal of Alloys and Compounds, 509*(14), 4754–4764. https://doi.org/10.1016/j.jallcom.2011.01.146.

Mahmoodi, N. M., Taghizadeh, M., Taghizadeh, A., Abdi, J., Hayati, B., & Shekarchi, A. A. (2019). Bio-based magnetic metal-organic framework nanocomposite: Ultrasound-assisted synthesis and pollutant (heavy metal and dye) removal from aqueous media. *Applied Surface Science, 480*, 288–299. https://doi.org/10.1016/j.apsusc.2019.02.211, http://www.journals.elsevier.com/applied-surface-science/.

Mamba, G., Mamo, M. A., Mbianda, X. Y., & Mishra, A. K. (2014). Nd,N,S-TiO2 decorated on reduced graphene oxide for a visible light active photocatalyst for dye degradation: Comparison to its MWCNT/Nd,N,S-TiO2 analogue. *Industrial and Engineering Chemistry Research, 53*(37), 14329–14338. https://doi.org/10.1021/ie502610y.

Manisalidis, I., Stavropoulou, E., Stavropoulos, A., & Bezirtzoglou, E. (2020). Environmental and health impacts of air pollution: a review. *Frontiers in public health, 8*, 505570. https://doi.org/10.3389/fpubh.2020.00014.

Manzoor, J., & Sharma, M. (2020). Impact of textile dyes on human health and environment. In Impact of textile dyes on public health and the environment. *IGI Global*, 162–169. https://doi.org/10.4018/978-1-7998-0311-9.ch008.

Matus, C., Baeza, P., Serrano-Lotina, A., Pastén, B., Fernanda Ramírez, M., Ojeda, J., & Camú, E. (2023). Removal of pharmaceuticals from an aqueous matrix by adsorption on metal–organic framework MIL-100(Cr). *Water, Air, and Soil Pollution, 234*(11). https://doi.org/10.1007/s11270-023-06736-4.

Miao, S., Guo, J., Deng, Z., Yu, J., & Dai, Y. (2022). Adsorption and reduction of Cr(VI) in water by iron-based metal-organic frameworks (Fe-MOFs) composite electrospun nanofibrous membranes. *Journal of Cleaner Production, 370*, 133566. https://doi.org/10.1016/j.jclepro.2022.133566.

Mo, Z., Tai, D., Zhang, H., & Shahab, A. (2022). A comprehensive review on the adsorption of heavy metals by zeolite imidazole framework (ZIF-8) based nanocomposite in water. *Chemical Engineering Journal, 443*, 136320. https://doi.org/10.1016/j.cej.2022.136320.

Moghadam, P. Z., Li, A., Wiggin, S. B., Tao, A., Maloney, A. G. P., Wood, P. A., Ward, S. C., & Fairen-Jimenez, D. (2017). Development of a Cambridge structural database subset: A collection of metal-organic frameworks for past, present, and future. *Chemistry of Materials, 29*(7), 2618–2625. https://doi.org/10.1021/acs.chemmater.7b00441, http://pubs.acs.org/journal/cmatex.

Mohammadi, A., Kazemeini, M., & Sadjadi, S. (2024). A developed novel Hal′@MIL-PMA triple composite utilized as a robust photocatalyst for activated removal of antibiotic contaminants: A new chemical mechanism for Tetracycline degradation. *Journal of Environmental Chemical Engineering, 12*(3), 112941. https://doi.org/10.1016/j.jece.2024.112941.

Najafi, M., Akbarzadeh, A. R., Rahimi, R., & Keshavarz, M. H. (2022). QSPR model for estimation of photodegradation average rate of the porphyrin-TiO2 complexes and prediction of their biodegradation activity and toxicity: Engineering of two annihilators for water/waste contaminants. *Journal of Molecular Structure, 1249*, 131463. https://doi.org/10.1016/j.molstruc.2021.131463.

Najafidoust, A., Abdollahi, B., Sarani, M., Darroudi, M., & Vala, A. M. (2022). MIL-(53)Fe metal-organic framework (MOF)-based Ag2CrO4 hetrostructure with enhanced solar-light degradation of organic dyes. *Optical Materials, 125*, 112108. https://doi.org/10.1016/j.optmat.2022.112108.

Nassef, H. M., Al-Hazmi, G. A. A. M., Alayyafi, A. A. A., El-Desouky, M. G., & El-Bindary, A. A. (2024). Synthesis and characterization of new composite sponge combining of metal-organic framework and chitosan for the elimination of Pb(II), Cu(II) and Cd(II) ions from aqueous solutions: Batch adsorption and optimization using Box-Behnken design. *Journal of Molecular Liquids, 394*, 123741. https://doi.org/10.1016/j.molliq.2023.123741.

Nazir, M. A. (2024). Synthesis of bimetallic Mn@ ZIF–8 nanostructure for the adsorption removal of methyl orange dye from water. *Inorganic Chemistry Communications, 165*, 2024, 112294.

Nazir, M. A., Bashir, M. A., Najam, T., Javed, M. S., Suleman, S., Hussain, S., Kumar, O. P., Shah, S. S. A., & Rehman, A. U. (2021). Combining structurally ordered intermetallic nodes: Kinetic and isothermal studies for removal of malachite green and methyl orange with mechanistic aspects. *Microchemical Journal, 164*, 105973. https://doi.org/10.1016/j.microc.2021.105973.

Nazir, M. A., Khan, N. A., Cheng, C., Shah, S. S. A., Najam, T., Arshad, M., Sharif, A., Akhtar, S., & Rehman, A. U. (2020). Surface induced growth of ZIF-67 at Co-layered double hydroxide: Removal of methylene blue and methyl orange from water. *Applied Clay Science, 190*, 105564. https://doi.org/10.1016/j.clay.2020.105564.

Nazir, M. A., Najam, T., Shahzad, K., Wattoo, M. A., Hussain, T., Tufail, M. K., Shah, S. S. A., & Rehman, A. U. (2022). Heterointerface engineering of water stable ZIF-8@ZIF-67: Adsorption of rhodamine B from water. *Surfaces and Interfaces, 34*, 102324. https://doi.org/10.1016/j.surfin.2022.102324.

Øien-Ødegaard, S., Shearer, G. C., Wragg, D. S., & Lillerud, K. P. (2017). Pitfalls in metal–organic framework crystallography: Towards more accurate crystal structures. *Chemical Society Reviews, 46*(16), 4867–4876. https://doi.org/10.1039/C6CS00533K.

Oladipo, A. C., Abodunrin, T. O., Bankole, D. T., Oladeji, O. S., Egharevba, G. O., & Bello, O. S. (2021). Environmental applications of metal-organic frameworks and derivatives: Recent advances and challenges. *ACS Symposium Series, 1394*, 257–298. https://doi.org/10.1021/bk-2021-1394.ch011, http://global.oup.com/academic/content/series/a/acs-symposium-series-acsss/?cc=in&lang=en.

Oladoye, P. O., Adegboyega, S. A., & Giwa, A. R. A. (2021). Remediation potentials of composite metal-organic frameworks (MOFs) for dyes as water contaminants: A comprehensive review of recent literatures. *Environmental Nanotechnology, Monitoring and Management, 16*. https://doi.org/10.1016/j.enmm.2021.100568, http://www.journals.elsevier.com/environmental-nanotechnology-monitoring-and-management/, 100568.

Park, K. S., Ni, Z., Côté, A. P., Choi, J. Y., Huang, R., Uribe-Romo, F. J., Chae, H. K., O'Keeffe, M., & Yaghi, O. M. (2006). Exceptional chemical and thermal stability of zeolitic imidazolate frameworks. *Proceedings of the National Academy of Sciences, 103*(27), 10186–10191. https://doi.org/10.1073/pnas.0602439103.

Pattanayak, D. S., Pal, D., Mishra, J., & Thakur, C. (2023). Noble metal–free doped graphitic carbon nitride (g-C3N4) for efficient photodegradation of antibiotics: Progress, limitations, and future directions. *Environmental Science and Pollution Research, 30*(10), 25546–25558. https://doi.org/10.1007/s11356-022-20170-9, https://www.springer.com/journal/11356.

Pattappan, D., Kavya, K. V., Vargheese, S., Kumar, R. T. R., & Haldorai, Y. (2022). Graphitic carbon nitride/NH2-MIL-101(Fe) composite for environmental remediation: Visible-light-assisted photocatalytic degradation of acetaminophen and reduction of hexavalent chromium. *Chemosphere, 286*, 131875. https://doi.org/10.1016/j.chemosphere.2021.131875.

Peral, J., Domènech, X., & Ollis, D. F. (1997). Heterogeneous photocatalysis for purification, decontamination and deodorization of air. *Journal of Chemical Technology and Biotechnology, 70*(2), 117–140. https://doi.org/10.1002/(SICI)1097-4660(199710)70:2 %3C 117::AID-JCTB746 %3E 3.0.CO;2-F.

Piao, J., Lu, M., Ren, J., Wang, Y., Feng, T., Wang, Y., Jiao, C., Chen, X., & Kuang, S. (2023). MOF-derived LDH modified flame-retardant polyurethane sponge for high-performance oil-water separation: Interface engineering design based on bioinspiration. *Journal of Hazardous Materials, 444*, 130398. https://doi.org/10.1016/j.jhazmat.2022.130398.

Pino-Otín, M. R., Muñiz, S., Val, J., & Navarro, E. (2017). Effects of 18 pharmaceuticals on the physiological diversity of edaphic microorganisms. *Science of the Total Environment, 595*, 441–450. https://doi.org/10.1016/j.scitotenv.2017.04.002, www.elsevier.com/locate/scitotenv.

Pramanik, S., Zheng, C., Zhang, X., Emge, T. J., & Li, J. (2011). New microporous metal−organic framework demonstrating unique selectivity for detection of high explosives and aromatic compounds. *Journal of the American Chemical Society, 133*(12), 4153–4155. https://doi.org/10.1021/ja106851d.

Qi, Y., Luan, Y., Yu, J., Peng, X., & Wang, G. (2015). Nanoscaled copper metal–organic framework (MOF) based on carboxylate ligands as an efficient heterogeneous catalyst for aerobic epoxidation of olefins and oxidation of benzylic and allylic alcohols. *Chemistry, 21*(4), 1589–1597. https://doi.org/10.1002/chem.201405685.

Qin, L., Wang, Z., Fu, Y., Lai, C., Liu, X., Li, B., Liu, S., Yi, H., Li, L., Zhang, M., Li, Z., Cao, W., & Niu, Q. (2021). Gold nanoparticles-modified MnFe2O4 with synergistic catalysis for photo-Fenton degradation of tetracycline under neutral pH. *Journal of Hazardous Materials, 414*, 125448. https://doi.org/10.1016/j.jhazmat.2021.125448.

Radovic, L. R., Silva, I. F., Ume, J. I., Menéndez, J. A., Leon, C. A., Leon, Y., & Scaroni, A. W. (1997). An experimental and theoretical study of the adsorption of aromatics possessing electron-withdrawing and electron-donating functional groups by chemically modified activated carbons. *Carbon, 35*(9), 1339–1348. https://doi.org/10.1016/s0008-6223(97)00072-9.

Raj, A., Rego, R. M., Ajeya, K. V., Jung, H.-Y., Altalhi, T., Neelgund, G. M., Kigga, M., & Kurkuri, M. D. (2023). Underwater oleophobic-super hydrophilic strontium-MOF for efficient oil/water separation. *Chemical Engineering Journal, 453*, 139757. https://doi.org/10.1016/j.cej.2022.139757.

Raval, N. P., Shah, P. U., & Shah, N. K. (2017). Malachite green "a cationic dye" and its removal from aqueous solution by adsorption. *Applied Water Science, 7*(7), 3407–3445. https://doi.org/10.1007/s13201-016-0512-2, https://www.springer.com/journal/13201.

Reza, K. M., Kurny, A., & Gulshan, F. (2017). Parameters affecting the photocatalytic degradation of dyes using TiO2: A review. *Applied Water Science, 7*(4), 1569–1578. https://doi.org/10.1007/s13201-015-0367-y, https://www.springer.com/journal/13201.

Rojas, S., & Horcajada, P. (2020). Metal-organic frameworks for the removal of emerging organic contaminants in water. *Chemical Reviews, 120*(16), 8378–8415. https://doi.org/10.1021/acs.chemrev.9b00797, http://pubs.acs.org/journal/chreay.

Rubin, H. N., & Reynolds, M. M. (2017). Functionalization of metal-organic frameworks to achieve controllable wettability. *Inorganic Chemistry, 56*(9), 5266–5274. https://doi.org/10.1021/acs.inorgchem.7b00373, http://pubs.acs.org/journal/inocaj.

Ryder, G. (2017). Wastewater: Tthe untapped resource. The United Nations world water development report.

Salunkhe, R. R., Kamachi, Y., Torad, N. L., Hwang, S. M., Sun, Z., Dou, S. X., Kim, J. H., & Yamauchi, Y. (2014). Fabrication of symmetric supercapacitors based on MOF-derived nanoporous carbons. *Journal of Materials Chemistry A, 2*(46), 19848–19854. https://doi.org/10.1039/C4TA04277H.

Santoso, S. P., Kurniawan, A., Angkawijaya, A. E., Shuwanto, H., Warmadewanthi, I. D. A. A., Hsieh, C.-W., Hsu, H.-Y., Soetaredjo, F. E., Ismadji, S., & Cheng, K.-C. (2023). Removal of heavy metals from water by macro-mesoporous calcium alginate–exfoliated clay composite sponges. *Chemical Engineering Journal, 452*, 139261. https://doi.org/10.1016/j.cej.2022.139261.

Sarker, M., Shin, S., Jeong, J. H., & Jhung, S. H. (2019). Mesoporous metal-organic framework PCN-222(Fe): Promising adsorbent for removal of big anionic and cationic dyes from water. *Chemical Engineering Journal, 371*, 252–259. https://doi.org/10.1016/j.cej.2019.04.039.

Shahid, M. U., Najam, T., Islam, M., Hassan, A. M., Assiri, M. A., Rauf, A., Rehman, A. U., Shah, S. S. A., & Nazir, M. A. (2024). Engineering of metal organic framework (MOF) membrane for waste water treatment: Synthesis, applications and future challenges. *Journal of Water Process Engineering, 57*, 104676. https://doi.org/10.1016/j.jwpe.2023.104676.

Shao, F. Q., Feng, J. J., Lin, X. X., Jiang, L. Y., & Wang, A. J. (2017). Simple fabrication of AuPd@Pd core-shell nanocrystals for effective catalytic reduction of hexavalent chromium. *Applied Catalysis B: Environmental, 208*, 128–134. https://doi.org/10.1016/j.apcatb.2017.02.051, www.elsevier.com/inca/publications/store/5/2/3/0/6/6/index.htt.

Sharifzadeh, Z., & Morsali, A. (2022). Amine-functionalized metal-organic frameworks: From synthetic design to scrutiny in application. *Coordination Chemistry Reviews, 459*, 214445. https://doi.org/10.1016/j.ccr.2022.214445.

Shi, Y., Wang, L., Dong, S., Miao, X., Zhang, M., Sun, K., Zhang, Y., Cao, Z., & Sun, J. (2022). Wool-ball-like BiOBr@ZnFe-MOF composites for degradation organic pollutant under visible-light: Synthesis, performance, characterization and mechanism. *Optical Materials, 131*, 112580. https://doi.org/10.1016/j.optmat.2022.112580.

Shivaraju, H. P., Muzakkira, N., & Shahmoradi, B. (2016). Photocatalytic treatment of oil and grease spills in wastewater using coated N-doped TiO2 polyscales under sunlight as an alternative driving energy. *International Journal of Environmental Science and Technology, 13*(9), 2293–2302. https://doi.org/10.1007/s13762-016-1038-8.

Showman, M. S., El-Aziz, A. M. A., & Yahya, R. (2022). Efficient photocatalytic degradation of petroleum oil spills in seawater using a metal-organic framework (MOF). *Scientific Reports, 12*(1). https://doi.org/10.1038/s41598-022-26295-8, https://www.nature.com/srep/.

Silva, T. L., Ronix, A., Pezoti, O., Souza, L. S., Leandro, P. K. T., Bedin, K. C., Beltrame, K. K., Cazetta, A. L., & Almeida, V. C. (2016). Mesoporous activated carbon from industrial laundry sewage sludge: Adsorption studies of reactive dye Remazol Brilliant Blue R. *Chemical Engineering Journal, 303*, 467–476. https://doi.org/10.1016/j.cej.2016.06.009, www.elsevier.com/inca/publications/store/6/0/1/2/7/3/index.htt.

Singh, S., Sharma, S., Umar, A., Jha, M., Mehta, S. K., & Kansal, S. K. (2018). Nanocuboidal-shaped zirconium based metal organic framework for the enhanced adsorptive removal of nonsteroidal anti-inflammatory drug, ketorolac tromethamine, from aqueous phase. *New Journal of Chemistry, 42*(3), 1921–1930. https://doi.org/10.1039/C7NJ03851H.

Sivaselvam, S., Premasudha, P., Viswanathan, C., & Ponpandian, N. (2020). Enhanced removal of emerging pharmaceutical contaminant ciprofloxacin and pathogen inactivation using morphologically tuned MgO nanostructures. *Journal of Environmental Chemical Engineering, 8*(5), 104256. https://doi.org/10.1016/j.jece.2020.104256.

Sriram, G., Bendre, A., Mariappan, E., Altalhi, T., Kigga, M., Ching, Y. C., Jung, H.-Y., Bhaduri, B., & Kurkuri, M. (2022). Recent trends in the application of metal-organic frameworks (MOFs) for the removal of toxic dyes and their removal mechanism-a review. *Sustainable Materials and Technologies, 31*, e00378. https://doi.org/10.1016/j.susmat.2021.e00378.

Sudhaik, A., Raizada, P., Shandilya, P., Jeong, D.-Y., Lim, J.-H., & Singh, P. (2018). Review on fabrication of graphitic carbon nitride based efficient nanocomposites for photodegradation of aqueous phase organic pollutants. *Journal of Industrial and Engineering Chemistry, 67*, 28–51. https://doi.org/10.1016/j.jiec.2018.07.007.

Sule, R., & Mishra, A. K. (n.d.). Synthesis of mesoporous MWCNT/HKUST-1 composite for wastewater treatment. *Applied Sciences, 9*.

Thuan (2023). Photodegradation of hazardous organic pollutants using titanium oxides -based photocatalytic: A review. *Environmental Research, 229*, 2023.

Torad, N. L., Hu, M., Ishihara, S., Sukegawa, H., Belik, A. A., Imura, M., Ariga, K., Sakka, Y., & Yamauchi, Y. (2014). Direct synthesis of MOF-derived nanoporous carbon with magnetic Co nanoparticles toward efficient water treatment. *Small (Weinheim an der Bergstrasse, Germany), 10*(10), 2096–2107. https://doi.org/10.1002/smll.201302910.

Ullah, S., Rehman, A. U., Najam, T., Hossain, I., Anjum, S., Ali, R., Shahid, M. U., Shah, S. S. A., & Nazir, M. A. (2024). Advances in metal-organic framework@activated carbon (MOF@AC) composite materials: Synthesis, characteristics and applications. *Journal of Industrial and Engineering Chemistry*. https://doi.org/10.1016/j.jiec.2024.03.041.

Ullah, S., Shah, S. S. A., Altaf, M., Hossain, I., El Sayed, M. E., Kallel, M., El-Bahy, Z. M., Rehman, A. U., Najam, T., & Altaf Nazir, M. (2024). Activated carbon derived from biomass for wastewater treatment: Synthesis, application and future challenges. *Journal of Analytical and Applied Pyrolysis, 179*, 106480. https://doi.org/10.1016/j.jaap.2024.106480.

Unur, E. (2013). Functional nanoporous carbons from hydrothermally treated biomass for environmental purification. *Microporous and Mesoporous Materials, 168*, 92–101. https://doi.org/10.1016/j.micromeso.2012.09.027.

Upadhyay, U., Sreedhar, I., Singh, S. A., Patel, C. M., & Anitha, K. L. (2021). Recent advances in heavy metal removal by chitosan based adsorbents. *Carbohydrate Polymers, 251*, 117000. https://doi.org/10.1016/j.carbpol.2020.117000.

Veisi, H., Sedrpoushan, A., & Hemmati, S. (2015). Palladium supported on diaminoglyoxime-functionalized Fe3O4 nanoparticles as a magnetically separable nanocatalyst in Heck coupling reaction. *Applied Organometallic Chemistry, 29*(12), 825–828. https://doi.org/10.1002/aoc.3375, http://www3.interscience.wiley.com/journal/2676/home.

Verma, S., Mishra, A. K., & Kumar, J. (2010). The many facets of adenine: Coordination, crystal patterns, and catalysis. *Accounts of Chemical Research, 43*(1), 79–91. https://doi.org/10.1021/ar9001334India, http://pubs.acs.org/doi/pdfplus/10.1021/ar9001334.

Wang, C., Xiong, C., He, Y., Yang, C., Li, X., Zheng, J., & Wang, S. (2021). Facile preparation of magnetic Zr-MOF for adsorption of Pb(II) and Cr(VI) from water: Adsorption characteristics and mechanisms. *Chemical Engineering Journal, 415*, 128923. https://doi.org/10.1016/j.cej.2021.128923.

Wang, D., Li, Y., Jiang, Y., Cai, X., & Yao, X. (2022). Perspectives on surface chemistry of nanostructured catalysts for heterogeneous advanced oxidation processes. *Environmental Functional Materials, 1*(2), 182–186. https://doi.org/10.1016/j.efmat.2022.08.003.

Wang, H. H., Jia, L. N., Hou, L., Shi, W. J., Zhu, Z., & Wang, Y. Y. (2015). A new porous MOF with two uncommon metal-carboxylate-pyrazolate clusters and high CO2/N2 selectivity. *Inorganic Chemistry, 54*(4), 1841–1846. https://doi.org/10.1021/ic502733v, http://pubs.acs.org/journal/inocaj.

Wang, Y. (2021). TiO2@ MOF photocatalyst for the synergetic oxidation of microcystin-LR and reduction of Cr (VI) in aqueous media. *Catalysts, 11* 2021.

Wang, C., Yang, R., & Wang, H. (2020). Synthesis of ZIF-8/Fly Ash Composite for Adsorption of Cu2+, Zn2+ and Ni2+ from Aqueous Solutions. *Materials, 13*(1), 214. https://doi.org/10.3390/ma13010214.

Wei, F. H., Ren, Q. H., Liang, Z., & Chen, D. (2019). Synthesis of graphene oxide/metal-organic frameworks composite materials for removal of congo red from wastewater. *ChemistrySelect, 4*(19), 5755–5762. https://doi.org/10.1002/slct.201900363, http://onlinelibrary.wiley.com/journal/10.1002/(ISSN)2365-6549.

Wu, C. D., & Lin, W. (2007). Heterogeneous asymmetric catalysis with homochiral metal-organic frameworks: Network-structure-dependent catalytic activity. *Angewandte Chemie - International Edition, 46*(7), 1075–1078. https://doi.org/10.1002/anie.200602099.

Wu, T., Shen, L., Luebbers, M., Hu, C., Chen, Q., Ni, Z., & Masel, R. I. (2010). Enhancing the stability of metal-organic frameworks in humid air by incorporating water repellent functional groups. *Chemical Communications, 46*(33), 6120–6122. https://doi.org/10.1039/c0cc01170c, http://pubs.rsc.org/en/journals/journal/cc.

Wu, Y., Liu, Z., Bakhtari, M. F., & Luo, J. (2021). Preparation of GO/MIL-101(Fe, Cu) composite and its adsorption mechanisms for phosphate in aqueous solution. *Environmental Science and Pollution Research, 28*(37), 51391–51403. https://doi.org/10.1007/s11356-021-14206-9.

Xia, Q., Hao, Y., Deng, S., Yang, L., Wang, R., Wang, X., Liu, Y., Liu, H., & Xie, M. (2023). Visible light assisted heterojunction composite of AgI and CDs doped ZIF-8 metal-organic framework for photocatalytic degradation of organic dye. *Journal of Photochemistry and Photobiology A: Chemistry, 434*, 114223. https://doi.org/10.1016/j.jphotochem.2022.114223.

Xie, L. H. (2020). Hydrophobic metal–organic frameworks: assessment, construction, and diverse applications. *Advanced Science, 7*(4) 1901758.

Xue, Z., Cao, Y., Liu, N., Feng, L., & Jiang, L. (2014). Special wettable materials for oil/water separation. *Journal of Materials Chemistry A, 2*(8), 2445–2460. https://doi.org/10.1039/c3ta13397d.

Yamagiwa, K., Katoh, M., Yoshida, M., Ohkawa, A., & Ichijo, H. (1997). Potentiality of temperature-swing adsorption for removal of hydrophobic contaminants in water. *Water Science and Technology, 35*(7), 213–218. https://doi.org/10.2166/wst.1997.0279.

Yang, Y., Guo, Z., & Liu, W. (2022). Special superwetting materials from bioinspired to intelligent surface for on-demand oil/water separation: A comprehensive review. *Small (Weinheim an der Bergstrasse, Germany), 18*(48). https://doi.org/10.1002/smll.202204624, http://onlinelibrary.wiley.com/journal/10.1002/(ISSN)1613-6829.

Yao, J., & Wang, H. (2014). Zeolitic imidazolate framework composite membranes and thin films: Synthesis and applications. *Chemical Society Reviews, 43*(13), 4470–4493. https://doi.org/10.1039/c3cs60480b, http://pubs.rsc.org/en/journals/journal/cs.

Yao, L., Zhang, L., Wang, R., Chou, S., & Dong, Z. L. (2016). A new integrated approach for dye removal from wastewater by polyoxometalates functionalized membranes. *Journal of Hazardous Materials, 301*, 462–470. https://doi.org/10.1016/j.jhazmat.2015.09.027.

Ye, H., Chen, D., Li, N., Xu, Q., Li, H., He, J., & Lu, J. (2019). Durable and robust self-healing superhydrophobic Co-PDMS@ZIF-8-coated MWCNT films for extremely efficient emulsion separation. *ACS Applied Materials and Interfaces, 11*(41), 38313–38320. https://doi.org/10.1021/acsami.9b13539.

Yousefi, M., Farzadkia, M., Mahvi, A. H., Kermani, M., Gholami, M., & Esrafili, A. (2024). Photocatalytic degradation of ciprofloxacin using a novel carbohydrate-based nanocomposite from aqueous solutions. *Chemosphere, 349*, 140972. https://doi.org/10.1016/j.chemosphere.2023.140972.

Zhang, B., Matchinski, E. J., Chen, B., Ye, X., Jing, L., & Lee, K. (2018). *Marine oil spills-oil pollution, sources and effects. World seas: An environmental evaluation volume III: Ecological issues and environmental impacts*. Canada: Elsevier, 391–406. http://www.sciencedirect.com/science/book/9780128050521, https://doi.org/10.1016/B978-0-12-805052-1.00024-3.

Zhang, G., Liu, Y., Chen, C., Long, L., He, J., Tian, D., Luo, L., Yang, G., Zhang, X., & Zhang, Y. (2022). MOF-based cotton fabrics with switchable superwettability for oil–water separation. *Chemical Engineering Science, 256*, 117695. https://doi.org/10.1016/j.ces.2022.117695.

Zhang, H., Liu, X., Wu, Y., Guan, C., Cheetham, A. K., & Wang, J. (2018). MOF-derived nanohybrids for electrocatalysis and energy storage: Current status and perspectives. *Chemical Communications, 54*(42), 5268–5288. https://doi.org/10.1039/c8cc00789f.

Zhang, H., Liu, D., Yao, Y., Zhang, B., & Lin, Y. S. (2015). Stability of ZIF-8 membranes and crystalline powders in water at room temperature. *Journal of Membrane Science, 485*, 103–111. https://doi.org/10.1016/j.memsci.2015.03.023.

Zhang, R., Wang, Z., Zhou, Z., Li, D., Wang, T., Su, P., & Yang, Y. (2019). Highly effective removal of pharmaceutical compounds from aqueous solution by magnetic Zr-based MOFs composites. *Industrial and Engineering Chemistry Research, 58*(9), 3876–3884. https://doi.org/10.1021/acs.iecr.8b05244.

Zhang, T., & Lin, W. (2014). Metal-organic frameworks for artificial photosynthesis and photocatalysis. *Chemical Society Reviews, 43*(16), 5982–5993. https://doi.org/10.1039/c4cs00103f, http://pubs.rsc.org/en/journals/journal/cs.

Zhang, X., Yan, L., Li, J., & Yu, H. (2020). Adsorption of heavy metals by l-cysteine intercalated layered double hydroxide: Kinetic, isothermal and mechanistic studies. *Journal of Colloid and Interface Science, 562*, 149–158. https://doi.org/10.1016/j.jcis.2019.12.028.

Zhang, Y. (2020). Recent advances in photocatalysis over metal–organic frameworks-based materials. *Solar RRL, 2020*(5).

Zhao, C., Zhao, L., Meng, L., Liu, X., & Liu, C. (2019). A Zn-MOF with 8-fold interpenetrating structure constructed with N,N′-bis(4-carbozylbenzyl)-4-aminotoluene ligands, sensors and selective adsorption of dyes. *Journal of Solid State Chemistry, 274*, 86–91. https://doi.org/10.1016/j.jssc.2019.03.014.

Zhao, H., Xia, Q., Xing, H., Chen, D., & Wang, H. (2017). Construction of pillared-layer MOF as efficient visible-light photocatalysts for aqueous Cr(VI) reduction and dye degradation. *ACS Sustainable Chemistry and Engineering, 5*(5), 4449–4456. https://doi.org/10.1021/acssuschemeng.7b00641.

Zhao, L., Azhar, M. R., Li, X., Duan, X., Sun, H., Wang, S., & Fang, X. (2019). Adsorption of cerium (III) by HKUST-1 metal-organic framework from aqueous solution. *Journal of Colloid and Interface Science, 542*, 421–428. https://doi.org/10.1016/j.jcis.2019.01.117.

Zhao, S., Chen, D., Wei, F., Chen, N., Liang, Z., & Luo, Y. (2017). Removal of Congo red dye from aqueous solution with nickel-based metal-organic framework/graphene oxide composites prepared by ultrasonic wave-assisted ball milling. *Ultrasonics Sonochemistry, 39*, 845–852. https://doi.org/10.1016/j.ultsonch.2017.06.013.

Zheng, S., Wang, Q., Yuan, Y., & Sun, W. (2020). Human health risk assessment of heavy metals in soil and food crops in the Pearl River Delta urban agglomeration of China. *Food Chemistry, 316*, 126213. https://doi.org/10.1016/j.foodchem.2020.126213.

Zhong, R., Liao, H., Deng, Q., Zou, X., & Wu, L. (2022). Preparation of a novel composite photocatalyst BiOBr/ZIF-67 for enhanced visible-light photocatalytic degradation of RhB. *Journal of molecular structure, 1259*, 132768. https://doi.org/10.1016/j.molstruc.2022.132768.

Zhu, J., Deng, Z., Chen, F., Zhang, J., Chen, H., Anpo, M., Huang, J., & Zhang, L. (2006). Hydrothermal doping method for preparation of Cr^{3+}-TiO_2 photocatalysts with concentration gradient distribution of Cr^{3+}. *Applied Catalysis B: Environmental, 62*(3-4), 329–335. https://doi.org/10.1016/j.apcatb.2005.08.013.

Zvinowanda, C. M., Okonkwo, J. O., Shabalala, P. N., & Agyei, N. M. (2009). A novel adsorbent for heavy metal remediation in aqueous environments. *International Journal of Environmental Science and Technology, 6*(3), 425–434. https://doi.org/10.1007/bf03326081.

Chapter 17

Metal-organic framework composites for sensing applications

Ali Aghakhani
Department of Food Science, Engineering and Technology, Faculty of Agriculture, College of Agriculture & Natural Resources University of Tehran, Alborz, Karaj, Iran

Introduction to sensors

Sensors are continually being researched and developed, with advances in physics, chemistry, biochemistry, and new equipment driving their progress. In chemical analysis, techniques like chromatography and spectroscopy, despite their high precision and accuracy, face issues such as high costs, low portability, skilled labor requirements, and time-consuming preparations (Hajivand et al., 2024). These factors make them unsuitable for online monitoring of processes, especially for volatile organic compounds (VOCs). There's a shift toward affordable, user-friendly devices, exemplified by blood glucose sensors and pregnancy tests. An effective sensor should have a suitable sampling method, replaceable and renewable interfaces, good selectivity and sensitivity, and reliable responses (Swager & Mirica, 2019). Sensors provide cost-effective and efficient alternatives for chemical detection using various signal transduction techniques, classified into optical, dielectric, mechanical, and electrical. However, these sensors often face challenges like poor selectivity, sensitivity, moisture interference, and short lifespan, which need to be addressed.

Metal-organic frameworks (MOFs) are complex crystalline materials consisting of metal nodes linked by organic ligands, notable since 1999. They feature adjustable structures, high surface area, porosity, strong adsorption capabilities, chemical and size selectivity, as well as remarkable thermal and chemical stability. MOFs find extensive use in diverse applications, including gas storage, separation, catalysis, drug delivery, and sensors (Safaei et al., 2019; Zhou et al., 2019). Tailoring MOF properties involves selecting suitable metal nodes and organic linkers to achieve varied geometries, dimensions, and functional groups. Interactions such as electrostatic forces, π–π stacking, and hydrogen bonding play roles in how analyte molecules interact with MOFs (Hajivand et al., 2024). MOFs possess unique properties that make them excellent candidates as sensors. Unlike many metal oxides used in sensors that require high temperatures or lack selectivity, MOFs offer advantages such as high surface area, high porosity, adjustable size and shape, fast response times, operation at room temperature, and reversible responses. Key factors like high surface area and tailored pore sizes greatly enhance MOF-based sensor sensitivity and selectivity. MOFs can perform preconcentration due to their inherent porosity, unlike less porous materials like metal oxides or polymers (Zhu et al., 2021).

Combining MOFs with other materials can improve their weaknesses, like poor chemical stability and low conductivity, and introduce new properties. Reported composites include MOFs with nanoparticles (NPs), quantum dots (QDs), oxides, polymers, graphene, carbon nanotubes (CNTs), and biomolecules (Zhu & Xu, 2014). For example, 0D materials like QDs and NPs enhance dispersability, stability, and chemical activity. 1D materials such as CNTs improve dispersability, porosity, and electrical conductivity. 2D materials like graphene support MOF decoration, while 3D materials increase the contact surface area and improve interactions between reactants and the composite surface (Xue et al., 2019). Interactions between target molecules and MOF composites can modify material properties, affecting optical, dielectric, electrical, or mechanical characteristics, which are exploited for sensor signal generation. In MOF-based optical sensors, guest analyte molecules enter the host MOF structure and alter various optical properties of the host–guest system, such as absorption and color (vapochromic sensors) or emission (luminescence sensors), and dielectric properties like refractive index interferometry, surface-enhanced Raman scattering (SERS), and localized surface-plasmon resonance (LSPR spectroscopy) (Zhu et al., 2019).

In this chapter, we will examine the role of MOF composites in sensors across the four aforementioned categories and demonstrate how advancements in MOF science and technology can address the existing challenges in each type of sensor.

Metal-organic frameworks-based optical sensors

Vapochromism sensors based on metal-organic frameworks

In vapochromism sensors, analytes entering the MOFs host structure cause changes in its absorbance or emission properties, resulting in a color change of MOFs. This color change can be visually observed for measurement purposes (Zhu et al., 2021). The presence of a chromophore group in the MOF's subunits is necessary for this effect. Host–guest interactions, such as van der Waals forces, π–π interactions, hydrogen bonding, or coordination with metal centers, can induce these color changes. Open metal sites are particularly effective for interacting with gas molecules and causing chromophoric property alterations. While the vapochromism method is simple and attractive, it may lose sensitivity over time with prolonged vapor exposure. In the following, some interesting examples of this method will be discussed. MOF-based composites are employed in vapochromic optical sensors to generate color change signals. These composites interact with analytes in various ways to induce color changes. For instance, MOFs can catalyze color-forming reactions like artificial enzymes (Yin et al., 2016) or inhibit these reactions when combined with analytes specifically captured by aptamers (Wang et al., 2016). Composites can also be doped with color indicators that respond to analytes (Hashemian et al., 2023) or contain metals that catalyze polymerization, leading to color changes (Hu et al., 2023). Strategies like dye@MOF are used for creating sensor arrays (Lin et al., 2023), and MOF@textile composites enable colorimetric sensors that react upon contact with analytes (Abdelhamid, 2023). Luminescent metal-organic frameworks (LMOFs) are also used, enhancing or quenching fluorescence to produce detectable color changes (Huo et al., 2023). In the following, examples of each of these strategies are examined.

Lee et al. (2024) have developed a dual-purpose MOF compound named Co_2(m-DOBDC) (m-DOBDC4$^-$=4,6-dioxo-1,3-benzenedicarboxylate) for chemisorption and monitoring of cyanogen chloride, a chemical warfare agent. Unlike traditional TEDA (triethylenediamine)-impregnated carbon materials, which produce harmful by-products, Co_2(m-DOBDC) adsorbs cyanogen chloride without generating dangerous substances. In Co_2(m-DOBDC), the electron-rich C5 site on the m-DOBDC^{4-} ligand is substituted with cyanogen, forming a stable C–C bond. Additionally, chloride ions (Cl$^-$) coordinate with the Co^{2+} site as a σ donor. So, it can absorb up to five times more cyanogen chloride than TEDA-impregnated materials. Upon adsorption, a color change from dark purple to green occurs as a result of coordination of Cl$^-$ with the Co^{2+} site, making Co_2(m-DOBDC) an effective colorimetric sensor for detecting cyanogen chloride (Fig. 17.1).

Yin et al. (2016) developed hemeprotein-MOF composites (H-MOFs) that act as artificial enzymes. They combined bovine hemoglobin (BHb) with ZIF-8 (ZIF-8@BHb), resulting in 423% higher peroxidase-like catalytic activity compared to free BHb. ZIF-8@BHb can catalytically react with chromogenic compounds such as 3,3',5,5'-tetramethylbenzidine (TMB) or 4-aminoantipyrine in the presence of H_2O_2 to produce a colored product. Using the composite's unique pore structure, they created a colorimetric biosensor for detecting H_2O_2 and phenol with a detection limit of 1 μM and a wide linear range (0–800 μM for H_2O_2 and 0–200 μM for phenol). This sensor was used for measuring H_2O_2 in cells and phenol in wastewater, showing high repeatability and stability. Absorption and calibration curves, as well as color changes for varying H_2O_2 concentrations, were demonstrated in Fig. 17.2.

Wang et al. (2016) improved the enzyme-mimicking activity of MOFs for the catalytic color change of TMB in the presence of H_2O_2 or glucose by

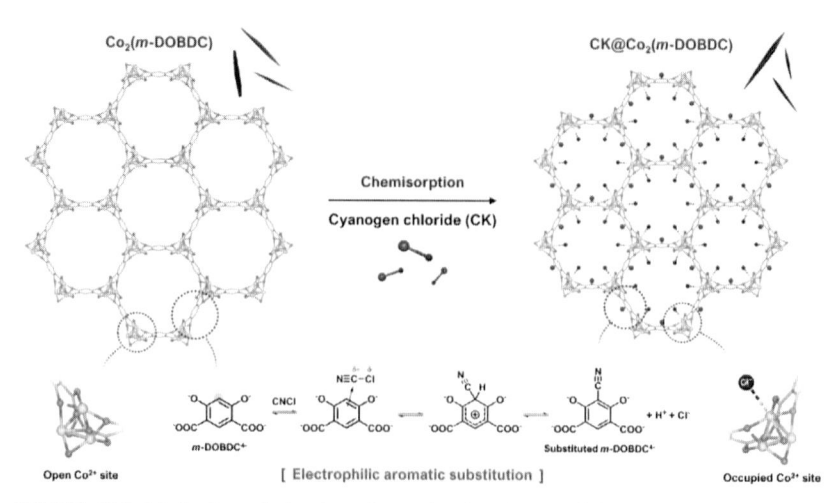

FIGURE 17.1 Mechanism of chemisorption and color change of Co$_2$(m-DOBDC) from dark purple to green upon reaction with cyanogen chloride (Lee et al., 2024).

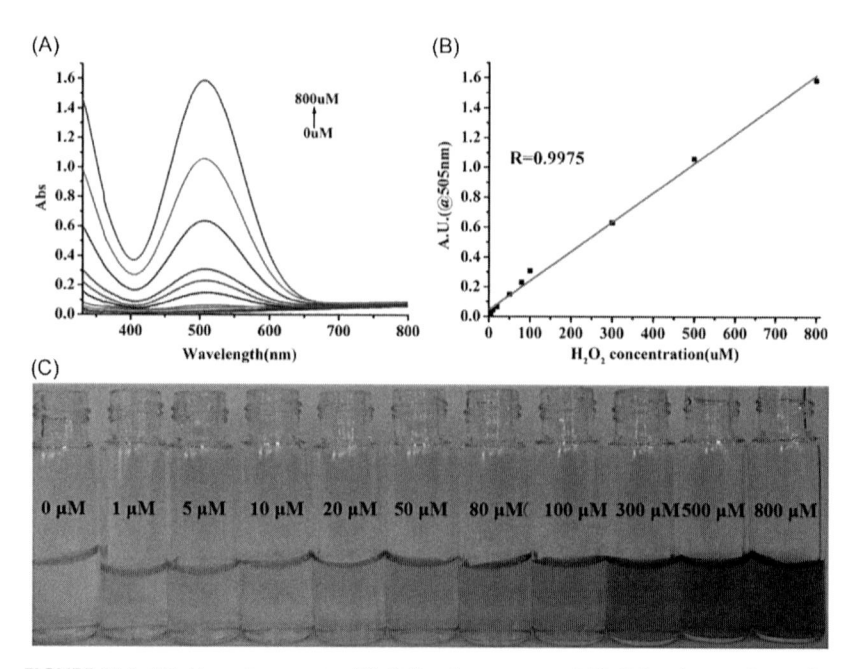

FIGURE 17.2 (A) Absorption spectra, (B) Calibration curves, and (C) Color changes for varying H$_2$O$_2$ concentrations catalyzed by ZIF-8@BHb are observed from colorless to dark red.

hybridizing MOFs with aptamers. Fe-MIL-88A can convert TMB to blue, but this activity is inhibited in the presence of a target molecule like thrombin and its aptamer, allowing colorimetric detection. They achieved a detection limit of 10 nM for thrombin, visible to the naked eye, and used this method to measure thrombin in human serum samples. In fact, this method extends the strategy explained by the work of Yin et al. (2016) to other analytes. For this purpose, an analyte that is specifically captured by the aptamer inhibits the MOF-catalyzed color-forming reaction, and the detection is recorded through the change in color.

Hashemian et al. (2023) developed a portable, on-site solid-state colorimetric sensor using a composite of cellulose acetate film and HKUST-1, doped with methyl red (MR) and brilliant cresyl blue (BCB) (MR/BCB-HKUST-1/CA). This sensor visually detects ammonia levels, indicating food spoilage. Upon ammonia exposure, the film changes from red to green, assessable by eye or mobile phone. The composite is stable, water-permeable, and selective for ammonia over other VOCs. It detects ammonia with a range of 1–100 ppb and 0.1–1340 ppm and a detection limit of 0.02 ppm. Fig. 17.3 illustrates the colorimetric detection of fish spoilage over five days at 25°C and 4°C. Spoilage occurs at room temperature by the second day and slightly by the fifth day at 4°C.

Hu et al. (2023) developed a rapid and sensitive colorimetric sensor for detecting acetylene (C_2H_2) using a Pt@MOF composite, UiO-68-SMe. They synthesized UiO-68-SMe with the ligand 2′,5′-dimethylthio-[1,1′:4′,1″-terphenyl]-4,4″-dicarboxylic acid (H_2L) and $ZrCl_2$. This MOF contains –SCH_3 groups that can attach Pt to the linkers through postsynthetic modification (Fig. 17.4A). The sensor changes color within 15 seconds of C_2H_2 exposure, maintaining stability in various environments, including water, acetone,

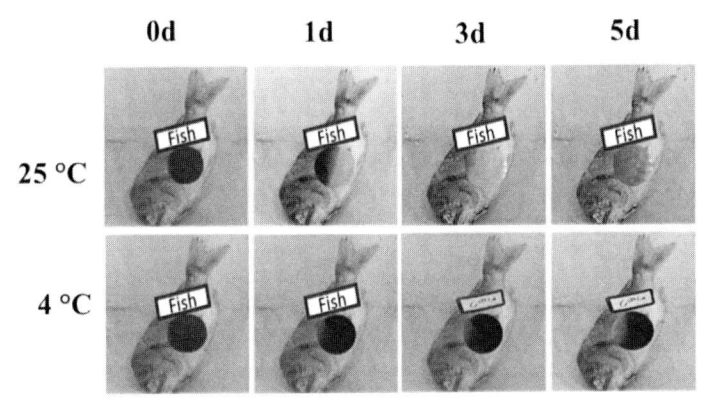

FIGURE 17.3 Monitoring fish freshness using the solid-state composite sensor MR/BCB-HKUST-1/CA at two different temperatures (Hashemian et al., 2023).

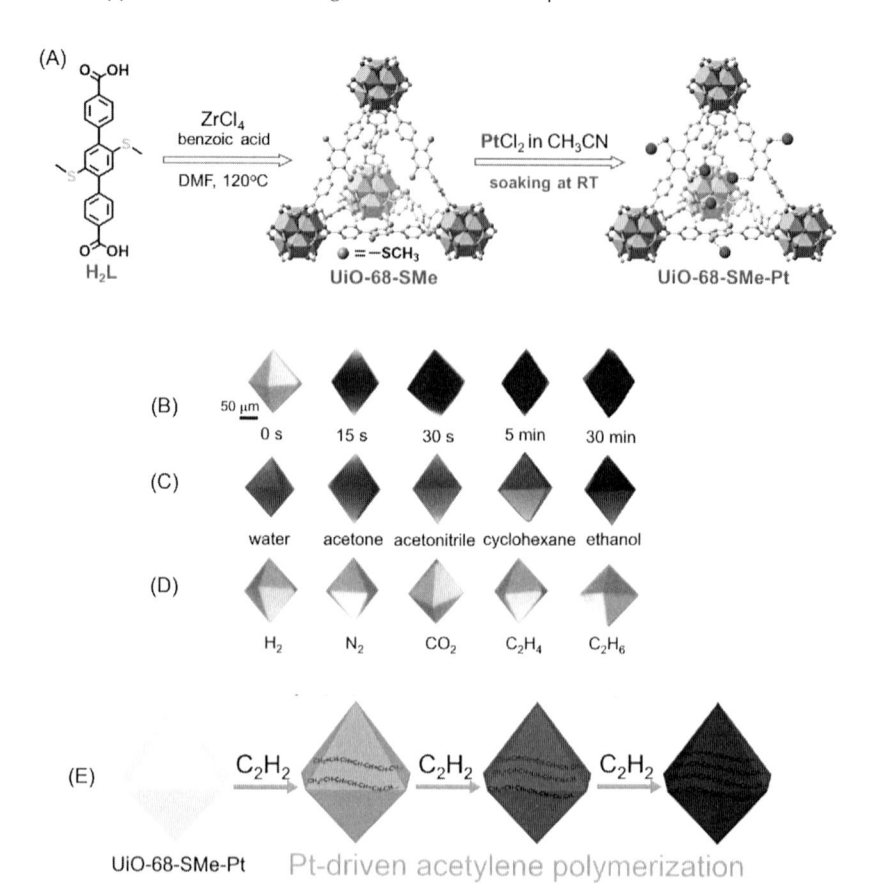

662 Applications of Metal-Organic Framework Composites

FIGURE 17.4 (A) Preparation method of UiO-68-SMe-Pt composite, (B) Color change of the sensor from yellow to dark red at different contact times, (C) Color change of the sensor in different solvents, (D) Selectivity of the sensor, and (E) Mechanism of the sensor's color change from yellow to dark red.

acetonitrile, ethanol, and cyclohexane (Fig. 17.4B and C). The selective color change, due to a polymerization reaction catalyzed with Pt, enables the sensor to detect selectively C_2H_2 with a detection limit of 4.1 ppm, making it effective for monitoring acetylene production in transformer oil (Fig. 17.4D and E).

Lin et al. (2023) created a rapid colorimetric sensor array to detect VOCs from wheat contaminated with *Aspergillus flavus*. The array used three nanocomposites, including dye@MOFs, dye@porous silica nanoparticle (PSN), and dye@polystyrene-co-acrylic acid (PSA), each combined with boron dipyrromethene (BODIPY) as dyes. The MOF@BODIPY composite was prepared by synthesizing the MOF structure $[Cu(L)(I)]_{2n} \cdot 2nDMF \cdot nMeCN$ (L=4-(4-methoxyphenyl)-4,2′:6′,4′-terpyridine), dispersing it with BODIPY in

DMF, and heating at 90°C for 1 hour. Each nanocomposite changed color differently upon exposure to VOCs like 1-octen-3-ol. These changes were analyzed using image analysis, linear discriminant analysis (LDA), and k-nearest neighbor (KNN) algorithms, effectively distinguishing contaminated samples from healthy ones. Fig. 17.5 shows the preparation and interaction mechanisms of the nanocomposites.

Huo et al. (2023) developed a turn-on fluorescent sensor using H_4BINDI-based Ln-MOFs (Ln=Eu, Sm, H_4BINDI=N,N'-bis(5-isophthalic acid)-1,4,5,8-naphthalenediimide) and created a paper strip sensor with polyvinyl alcohol (PVA). This sensor detects volatile amines through visible color changes under UV light (Fig. 17.6A). Its effectiveness is due to the unique fluorescence properties of lanthanides with their 4f electrons, such as long lifetimes, large Stokes shifts, and high stability, and the use of electron-deficient linkers like H_4BINDI. These ligands enhance interaction with electron-rich guest molecules, such as aliphatic amines, leading to turn-on ratiometric sensing.

Abdelhamid (2023) developed textile-based MOF sensors for selective vapochromic detection of pyridine. Recognizing the challenges of using MOFs in powder form across various applications, they directly grew CuBDC (BDC: benzene-1,4-dicarboxylic acid) onto cotton fibers (CuBDC@Textile) to create a solid-state sensor and adsorbent. This material showed a high adsorption capacity of 139 mg/g (Fig. 17.6B), with a detection limit for pyridine of 0.0098 ppm using a colorimetric method.

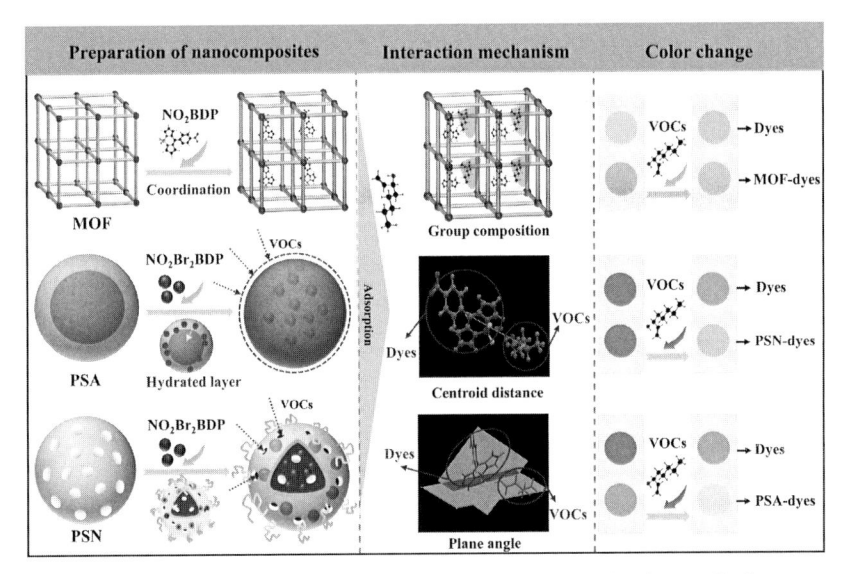

FIGURE 17.5 The method for fabricating nanocomposites and their color changes (Amber orange for MOF-dye, peach pink for PSN-dye and PSA-dye) upon exposure to volatile organic compounds for constructing the colorimetric sensor array (Lin et al., 2023).

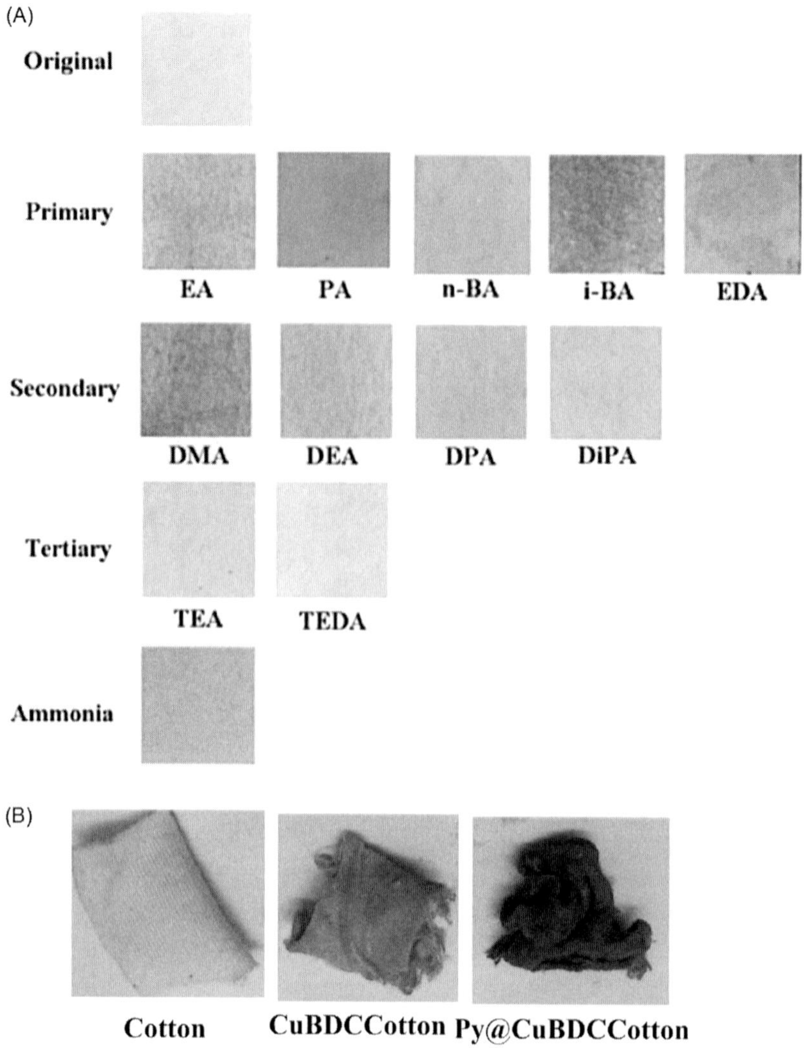

FIGURE 17.6 Vapochromic sensors based on various metal-organic frameworks: (A) Color change of paper strip sensor based on H₄BINDI-based Ln-MOFs before and after exposure to amine vapors from linen to various darker shades for each amine (Dong et al., 2019) and (B) Color changes of CuBDC@Cotton due to interaction with pyridine from sky blue to indigo color (Abdelhamid, 2023).

Möslein et al. (2023) developed the ZnQ@OX-1 composite, consisting of a guest molecule (ZnQ=zinc(II) bis(8-hydroxyquinoline)) encapsulated within the cavities of the OX-1 scaffold (a zinc-based MOF with BDC as the organic linker) that changes color in the presence of acetone vapors. The encapsulation of ZnQ during synthesis leads to a distorted framework, which alters its emission properties (green color) from that of free ZnQ (yellow color). When an organic molecule like acetone comes into contact with the composite, it disrupts the guest–host interaction, causing the ZnQ to revert to its normal state (green color), thereby inducing fluorescence quenching, which was studied to achieve a detection limit of 50 ppm.

Luminescence sensors-based on metal-organic framework composites

Luminescent MOFs (LMOFs) are ideal for optical applications like sensing and bioimaging due to their unique luminescent properties, which can originate from the metal center, organic ligand, or guest molecule within the structure (Allendorf et al., 2009; Haldar et al., 2020; Liu et al., 2019). Their crystalline nature enhances luminescence by reducing nonradiative pathways and increasing quantum yield. Their optical properties are sensitive to geometry and molecular environment, allowing for the creation of various sensors through analyte interactions. LMOFs can interact with guest analytes in ways that either quench (turn off), enhance (turn on), or proportionally (ratiometrically) alter their luminescence (Fig. 17.7A–D). Among these, most LMOF-based optical sensors are of the turn-off type. However, turn-on sensors are more sensitive to background interferences, offering potentially greater

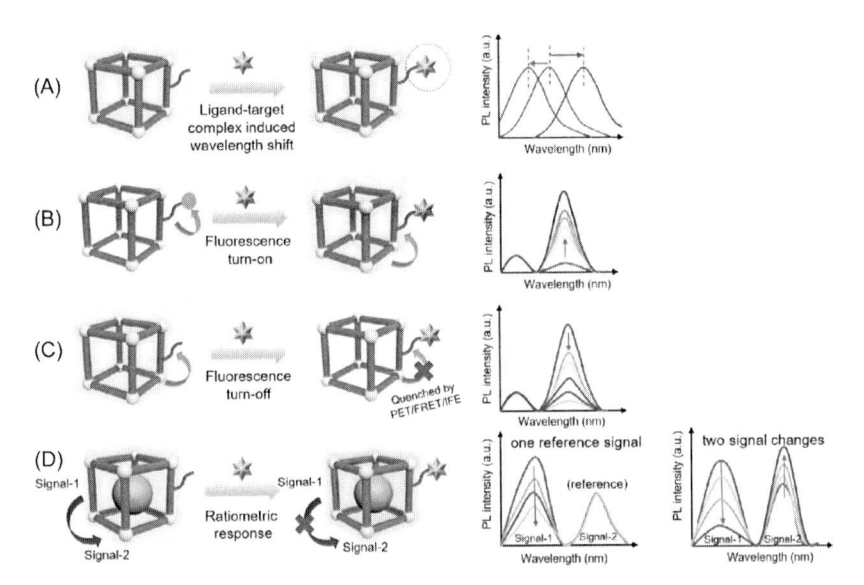

FIGURE 17.7 Different sensing mechanisms in the measurement of substances by luminescent metal-organic frameworks. (A) Wavelength shift in ligand-MOF complex, (B) Fluorescence turn-on, (C) Fluorescence turn-off, and (D) Ratiometric fluorescence (Du et al., 2021).

precision in some applications. Ratiometric sensors, which utilize two emission bands with one acting as an internal standard, provide highly accurate measurements due to their self-referencing property. These sensors can exhibit three distinct outcomes: both emission bands can either increase or decrease, one can increase while the other decreases, or one may increase while the other remains unchanged. They can be classified based on the source of luminescence: linker, metal center, or host molecule, and are used in various sensing technologies due to their versatile and efficient luminescent behavior.

LMOFs in composite form exhibit new capabilities that are less commonly observed in pure LMOFs. For example, their composites with polymers lead to increased chemical and mechanical stability while also potentially enhancing the ligand-to-metal center antenna effect, resulting in improved sensitivity (Rozenberga et al., 2024). Incorporating LMOFs into foams like melamine foam not only improves mechanical stability but also, due to the foam's higher absorption properties, enhances sensor sensitivity. MOFs in the form of colloidal crystals, when composited with luminescent color-emitting compounds, can increase the sensitivity of fluorescence sensors by confining photons within their optical cavities and coherently scattering the emitted light. Many of these new composites are used as ratiometric fluorescence sensors, in which different components within the composite can emit two to three emission lines. By using the ratio of these emission lines and their changes due to interactions with the analyte(s), more selective and sensitive sensors can be obtained due to their self-calibrating properties. In most of these composites, along with LMOFs, other luminescent compounds such as Tb^{3+} (Li et al., 2023), carbon dots (CDs) (Sun et al., 2022), and Rhodamine B (RhB) (Yang et al., 2020) have been used. One of the features of these composites is that they prevent the quenching effect of aqueous solvents on the luminescent material, thereby increasing quantum yield (Sun et al., 2022). In these structures, limited access to water, due to specific structures like spinning networks (Sun et al., 2022) or particular pores and channels (e.g., in MOF-to-MOFs) (Li et al., 2023), obstructs water molecules, resulting in higher fluorescence quantum yield. In the following, we will examine some examples demonstrating the performance of sensors based on composites containing LMOFs.

Sun et al. (2022) developed a composite of CDs-decorated hydroxyapatite nanowires and lanthanide MOF (HAPNWs-CDs-Tb/MOF) for ratiometric fluorescence detection of dopamine. The composite emits green fluorescence from Tb^{3+} at 543 nm and blue fluorescence from CDs at 426 nm (Fig. 17.8A). The composite is easy to prepare. They first synthesized hydroxyapatite nanowires using the solvothermal method. Subsequently, Tb-MOF nanoparticles were grown in situ on the nanowires. Finally, the CDs were mixed with the composite and decorated on their surfaces. While dopamine causes fluorescence quenching of Tb^{3+} at 543 nm in the Tb-MOF structure, in the HAPNWs-CDs-Tb/MOF composite, fluorescence increases upon interaction with dopamine.

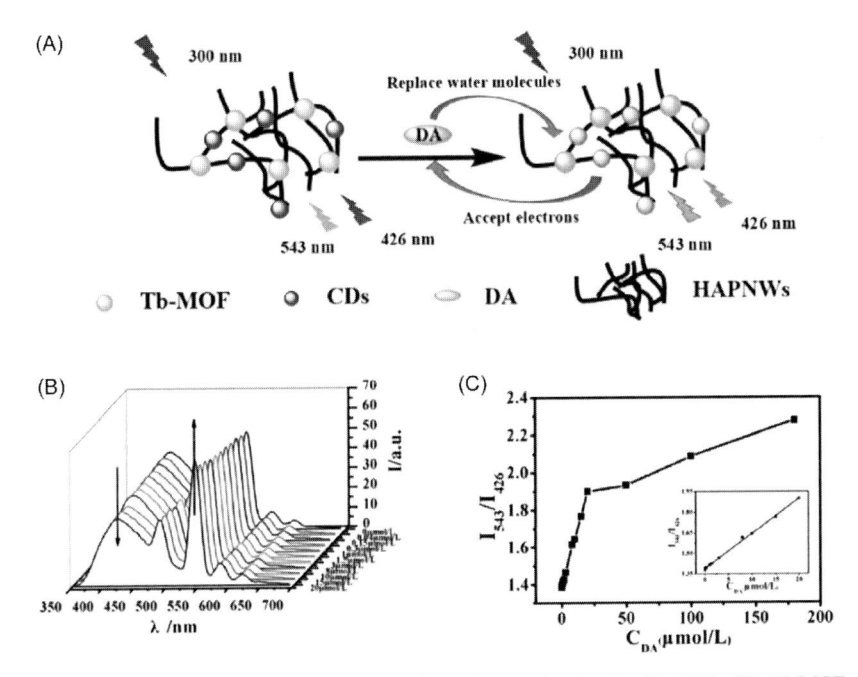

FIGURE 17.8 (A) Schematic illustration of dopamine sensing by the HAPNWs-CDs-Tb/MOF composite fluorescence probe, (B) Emission spectra of the composite with increasing dopamine concentration, and (C) The sensor response as a function of dopamine concentration.

This is because, in the composite, Tb-MOF forms a spinning network structure on hydroxyapatite, providing more action sites and steric space. Upon the addition of dopamine, the composite obstructs water molecules and reduces nonradiative energy loss, resulting in increased fluorescence. In contrast, the Tb-MOF alone, with its rod-like structure, offers fewer action sites and steric space. When dopamine obstructs the water molecules, it also affects charge transfer between the organic ligand and Tb^{3+}, further enhancing intramolecular energy transfer in the composite and boosting fluorescence. On the other hand, when CDs are used as a standalone probe, their fluorescence behavior is similar to that in the composite. The fluorescence at 436 nm gradually decreases with increasing dopamine concentration, as dopamine is oxidized to dopaquinone in an alkaline solution, leading to CDs quenching. The results indicate that in the composite probe, the I_{543}/I_{426} ratio increases with increasing dopamine concentration (Fig. 17.8B and C). Compared to single-wavelength sensors, this result is more accurate and more resistant to interference. The sensor offers advantages like high stability, selectivity, and fast dopamine detection in human serum with a linear range of 0.04–20 μM and a detection limit of 12.26 nM, showing a relative recovery of 100.8%–103.3% and a relative standard deviation of 3.82%.

Rozenberga et al. (2024) developed a composite of europium MOF and polymer for detecting Fe^{3+} ions in mining leaching solutions. They used 2-methoxyterephthalic acid (MTA) as the ligand and europium (Eu) ions to synthesize (EuMTA) MOF. The MOF powder was then incorporated into a solution with monomer, initiator, and cross-linker and polymerized under UV light to form a thin composite film. The presence of Fe^{3+} leads to fluorescence quenching in the composite, with a linear relationship to Fe^{3+} concentration (Fig. 17.9A and B). Embedding EuMTA in crosslinked poly(2-hydroxyethyl methacrylate) (PHEMA) significantly enhances the ligand-to-Eu^{3+} antenna effect, increasing fluorescence intensity 30-fold and extending the Fe^{3+} sensing range from 2 to 120 ppm. The polymer matrix also protects the MOF, improving sensor stability under acidic conditions. The MOF-polymer composite shows excellent luminescence, long-term stability, and reusability. While EuMTA suspensions show up to 15% quenching with Cu^{2+} and Fe^{2+}, they experience 56% quenching with Fe^{3+}. The PHEMA-EuMTA composite is highly selective for Fe^{3+}, with minimal interference from Cu^{2+} and Fe^{2+}. The PHEMA-EuMTA composite exhibits less than 5% quenching in the presence of Cu^{2+} and Fe^{2+} but shows up to 90% quenching with 120 ppm Fe^{3+}, without being affected by other common alkali and transition metal ions.

Yang et al. (2020) developed a dual-emission dye@MOF composite, comprising a Zr-based MOF and encapsulated RhB. The composite functions as a self-calibrating fluorescence sensor by measuring the intensity difference between two emission peaks. It is used to detect cations, nitroaromatics, and pesticides. Energy transfer occurs from the Zr-MOF to RhB within the composite, but this energy transfer is interrupted when the composite interacts with an electron-deficient analyte, leading to fluorescence quenching. The composite

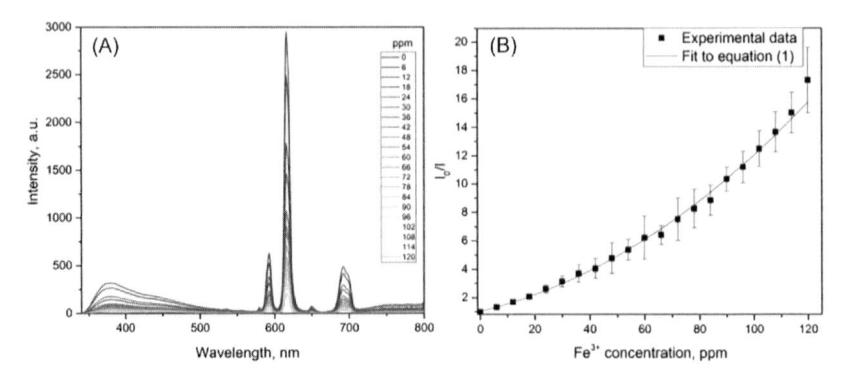

FIGURE 17.9 The sensing response of the PHEMA-EuMTA composite to Fe^{3+} ions: (A) Luminescence emission spectra of the composite at an excitation wavelength of 320 nm at different Fe^{3+} concentrations in water and (B) Calibration curve of the emission band at 616 nm (Rozenberga et al., 2024).

shows high sensitivity, selectivity, and recyclability, with detection limits of 1.6 µM for Fe^{3+}, 0.9 µM for 4-nitrophenol, and 0.2 µM for nitenpyram.

Li et al. (2023) developed a MOF-on-MOF composite encapsulating Tb^{3+} ions for ratiometric fluorescence detection of fluoride ions (F^-). The composite, composed of Tb^{3+}@UiO66/MOF801, serves as a built-in fluorescent probe with two emission peaks at 375 nm (UiO66/MOF801) and 544 nm (Tb^{3+}). Under 300 nm excitation, the 544 nm emission is sensitive to F^-, while the 375 nm emission is not (Fig. 17.10A). This self-calibrating detection is based on unequal energy transfer to the emission centers. Since Tb^{3+} only emits when encapsulated in UiO66/MOF801 and shows much weaker emission in UiO66 or none in solution (Fig. 17.10B), its effectiveness is due to the protective role of UiO66/MOF801, which prevents Tb^{3+} quenching by water. The positively charged Tb^{3+}@UiO66/MOF801 composite attracts F^-, forming a photosensitive complex that enhances light absorption. The composite demonstrates high sensitivity, strong antiinterference capability, fast response (under 30 seconds), and a low detection limit, making it a

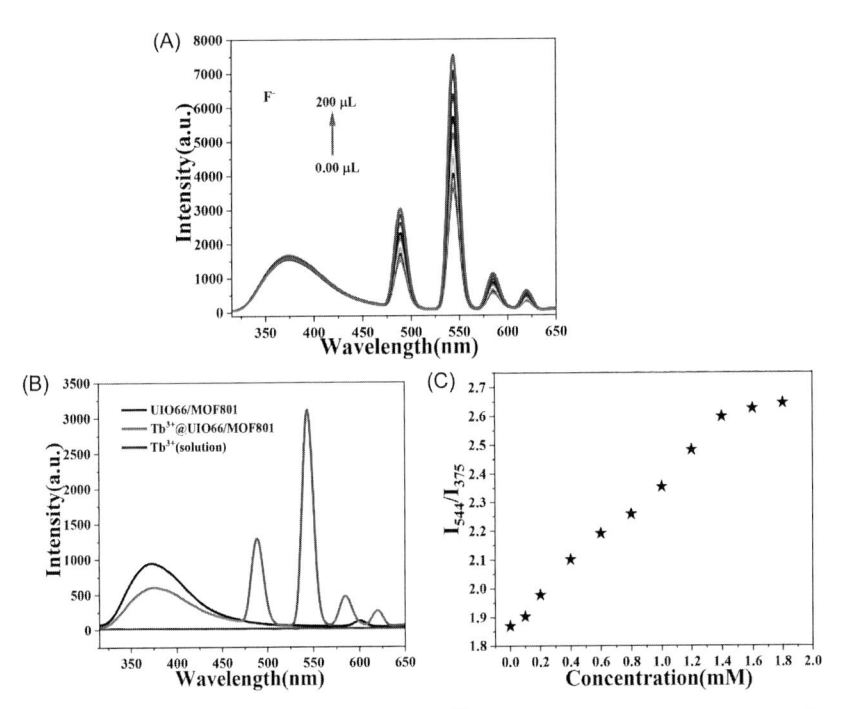

FIGURE 17.10 (A) Fluorescence spectra of Tb^{3+}@UiO66/MOF801 at different dropwise additions of F^-, (B) Fluorescence spectra of UiO66/MOF801, Tb^{3+}@UiO66/MOF801 and Tb^{3+} (solution) (λ_{ex} = 300 nm), and (C) The I_{544}/I_{375} value of Tb^{3+}@UiO66/MOF801 and the concentration of F^- at room temperature (Li et al., 2023).

promising fluorescence probe (Fig. 17.10C). The detection limit is 76.6 ppb, significantly lower than the WHO guideline for drinking water (1 ppm).

Jiang et al. (2024) developed a tri-channel luminescent composite array using Gd-MOF for the first-time detection and discrimination of epilepsy drugs (pregabalin) and their biomarkers (epilepsy-DNA-1 and human serum albumin (HSA)). This sensor, utilizing principal component analysis (PCA) and hierarchical cluster analysis (HCA), can help clinicians rapidly and accurately identify these compounds in human serum. The Gd-MOF structure, known as $[Gd_2L_3(DMF)(H_2O)_2]_n$ and (H_2L=5-(4-(imidazol-1-yl)phenyl)isophthalic acid), was functionalized to create three composites: Gd-MOF-1 with 3-bromo-L-phenylalanine (L-3-Br-PHE-OH) for pregabalin detection, Gd-MOF-2 with thioflavin T (ThT) for epilepsy-DNA-1 detection, and Gd-MOF-3 with anti-HSA/fluorescein isothiocyanate (FITC) for HSA detection (Fig. 17.11). To prepare Gd-MOF-1, the chiral compound L-3-Br-PHE-OH is incorporated to effectively recognize pregabalin through interactions with its amino and carboxyl groups. In Gd-MOF-2, ThT, a fluorescent dye with high selectivity and sensitivity to G-rich DNA sequences, is used to detect epilepsy-DNA-1 by forming G-quadruplex compounds, which increases fluorescence intensity. Each of the Gd-MOF-1, Gd-MOF-2, and Gd-MOF-3 composites specifically interacts with the analytes pregabalin, epilepsy-DNA-1, and HSA, respectively, leading to the separate quenching of all three emission lines. This dual ratiometric sensing array can selectively and accurately measure all three analytes. Each composite exhibited high selectivity and sensitivity, with detection limits of 45.41 nM for pregabalin, 5.56 nM for epilepsy-DNA-A, and 5.11 nM for HSA.

Li et al. (2023) developed a tri-channel fluorescent sensor array using a composite material based on the tri-emissive MOF (PCN-222), incorporating a porphyrin primary ligand, two auxiliary naphthyl ligands (2,6-naphthalenedicarboxylic acid (H_2ndc) and 4,8-disulfo-2,6-naphthalenedicarboxylic acid (H_4dsndc)), and the guest molecule RhB. The all-in-one strategy used in this study integrates multiple sensing elements within a single MOF, focusing on distinct luminescence emission sites to assess different properties of the target analyte and simplifying the process of analytical operations. The composite sensor, with three emission lines at 368, 566, and 676 nm, exhibits fluorescence quenching upon exposure to tetracycline (TC) solutions (Fig. 17.12A). The quenching, particularly at 368 nm, is attributed to Forster resonance energy transfer (FRET) and follows a static pathway, indicating preassociation between the analyte and sensor. The smaller reduction in RhB emission is due to partial displacement by TC molecules, with its higher emission intensity compensating for the quenching effect. This sensor detects analytes by responding to their redox properties, adsorption affinity, and absorption/exchanging capabilities. The sensor array is effective in detecting and discriminating six TC homologs and has been successfully applied to real food samples, such as milk (Fig. 17.12B).

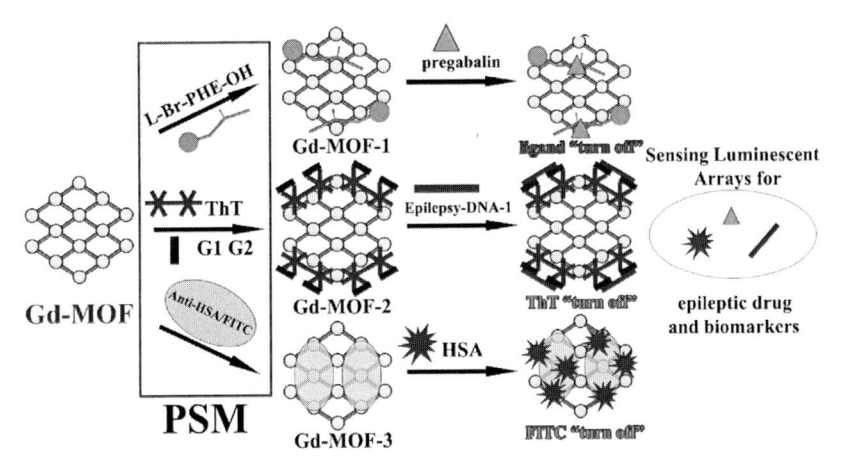

MOF composites for sensing applications Chapter | 17 671

FIGURE 17.11 The synthesis pathway of Gd-MOF composites and their luminescence sensing process (Jiang et al., 2024).

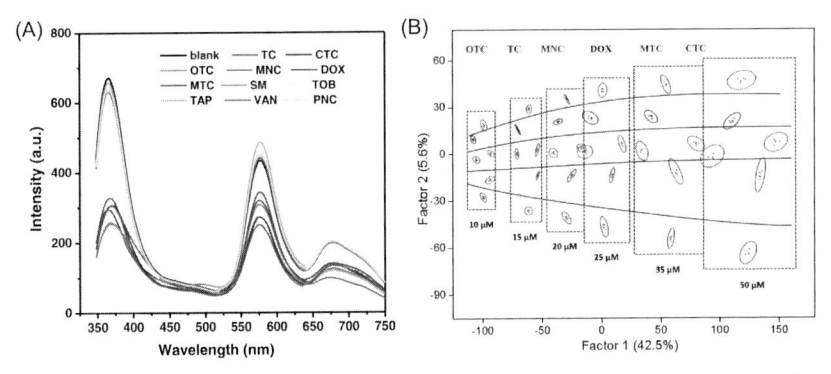

FIGURE 17.12 (A) Emission spectrum of the RhB@PCN-22-ndc/dsndc composite exposed to various antibiotics and interfering compounds and (B) Linear discriminant analysis (LDA) plot for the detection of TCs at different concentrations (Li et al., 2023).

Zhao et al. (2018) introduced a MOF composite with three white-light-emitting (WLE) colors as the first multidimensional ratiometric luminescent probe. This composite was synthesized through simple ion exchange by simultaneously incorporating two luminescent cation complexes, $[Ir(CF_3\text{-}ppyF_2)_2(bpy)]^+$ (red emission) and $[Ru(bpy)_3]^{2+}$ (green emission), as encapsulated luminescent modules (ELMs) into a porous anionic luminescent MOF ((Me_2NH_2) $[Zn_2(L)(H_2O)]\cdot 4DMA$ (H_5L=2,5-(6-(3-carboxyphenylamino)-1,3,5-triazine-2,4-diyl-diimino)diterephthalic acid); DMA=N,N-dimethylacetamide) with blue emission (Fig. 17.13A). When the composite is exposed to

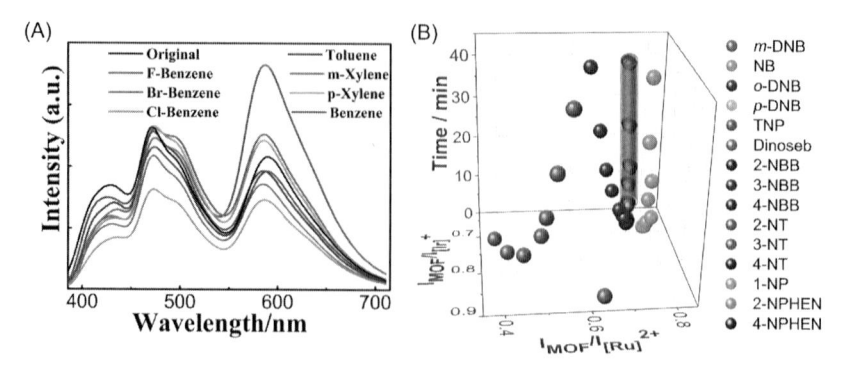

FIGURE 17.13 (A) Emission spectrum of the composite in contact with vapors of aromatic organic molecules and (B) Three-dimensional sensor response of the composite to explosive nitroaromatic compounds (Zhao et al., 2018).

selective volatile organic solvents (VOSs) or explosive nitroaromatics (NACs), the ratio of the emission peak heights between MOF-to-ELM1 and MOF-to-ELM2 changes depending on the analyte. This results in a significant color change in the luminescence of the composite, making it a highly sensitive and selective probe in a two-dimensional detection method. If these color changes are monitored and recorded over time, the method can be expanded into a three-dimensional detection scheme (Fig. 17.13B). Such a two-dimensional ratiometric luminescence approach offers advantages over one-dimensional methods because, upon interaction with the analyte, both I_{MOF}/I_{ELM1} and I_{MOF}/I_{ELM2} ratios change simultaneously, producing a unique color in the visible spectrum that is distinctly different from the initial white color and can be seen with the naked eye. This sensing mechanism eliminates the need for strong guest-analyte interactions, enabling the development of more sensitive luminescent probes. Clearly, this method can double or triple the output data, significantly enhancing the sensitivity and selectivity of the measurements. Compared to three physically mixed probes, the energy transfer between the MOF framework and ELM is more effectively modulated by the analyte in this approach, which amplifies the changes in the two ratios and leads to greater sensitivity.

Self-assembled structures like colloidal crystals (CCs) offer a method to enhance fluorescence signals. These structures, composed of submicrometer-sized particle arrays, can scatter light within a photonic stop band (PSB). Fluorescence enhancement in dye molecules doped within colloidal crystals occurs through two mechanisms. Firstly, optical cavities of colloidal crystals can confine photons, intensifying excitation experienced by emitting molecules and accelerating emission rates. Secondly, emitted light can be coherently scattered by colloidal crystals, directing it out of the crystal into surrounding space (leaky mode of colloidal crystal), further boosting fluorescence. Greater enhancement is achieved when emission wavelengths align closely with PSB wavelengths. Olorunyomi et al. (2020) presented the first colloidal crystal layers of pure ZIF-8 MOFs, achieving up to 200-fold fluorescence enhancement for Nile red (NR). These layers effectively interacted with volatile analytes, altering fluorescence intensity at low concentrations. NR~ccZIF-8 layers were utilized to detect acetone, methanol, toluene, and xylenes at room temperature, demonstrating strong selectivity, especially with a distinct fluorescence response to toluene and p-xylene due to subnanometer ZIF-8 nanopores that enable molecular sieving effects.

Guo et al. (2020) developed a Zr-MOF composite named MPDB-PCN for detecting the toxic mycotoxin 3-nitropropionic acid (3-NPA) from moldy sugarcane, using a pH-modulated ratiometric luminescent switch (MPDB=(2-(6-(4-methylpiperazin-1-yl)-1,3-dioxo-1H-benzo[de]isoquinolin-2(3H)-yl) acetic acid)). MPDB-PCN also responds well to pH changes across the acidic and basic range. The MOFs were synthesized using two luminescent donor ligands: First, the Zr-MOF structure PCN-224 was constructed using a porphyrinic linker ligand, which emits at a wavelength of 650 nm. In the second step, the MPDB ligand (with a size of 15.4 Å), derived from naphthalimide, is incorporated into PCN-224 cavities with a size of 19 Å. The carboxyl group of MPDB coordinates with the unsaturated sites of the Zr_6 cluster. The incorporation of MPDB changes the color of PCN-224 from dark purple to light purple (Fig. 17.14A–E). The fabricated composite can detect pH changes in the range of 2.5–8.6. It can also detect the toxic mycotoxin 3-NPA in the complex sugarcane matrix with a detection limit of 15 μM.

Gai et al. (2020) employed three-dimensional Zn-based MOFs named ROD-Zn1 and ROD-Zn2 with rod second building units (SBUs) for the removal and detection of antibiotics from aqueous samples. These MOFs exhibit stability in aqueous environments (both acidic and basic) and high kinetic absorption rates. Additionally, they prepared hybrid MOF-melamine foam (MF) composites (Fig. 17.15A and B). Compositing with MF increases mechanical stability and enhances absorption efficiency, resulting in improved sensitivity and stability of the sensor. The synthesized MOFs demonstrate excellent fluorescent properties and undergo linear fluorescence quenching in the presence of low concentrations of antibiotics in water in the range of 0–45 μM (Fig. 17.15C). The quenching percentage of the composite varies with different TCs as shown in Fig. 17.15D. The highest suppression is for TC, nitrofurazone (NZF), and nitrofurantoin (NFT), which can achieve up to 90% suppression.

Dielectric sensor based on metal-organic framework composites

Optical fibers originally developed for low-loss, long-distance telecommunications (Elsherif et al., 2022) have transitioned into sensor applications (Elsherif et al., 2022). Among these, sensors utilizing refractive index (RI) properties

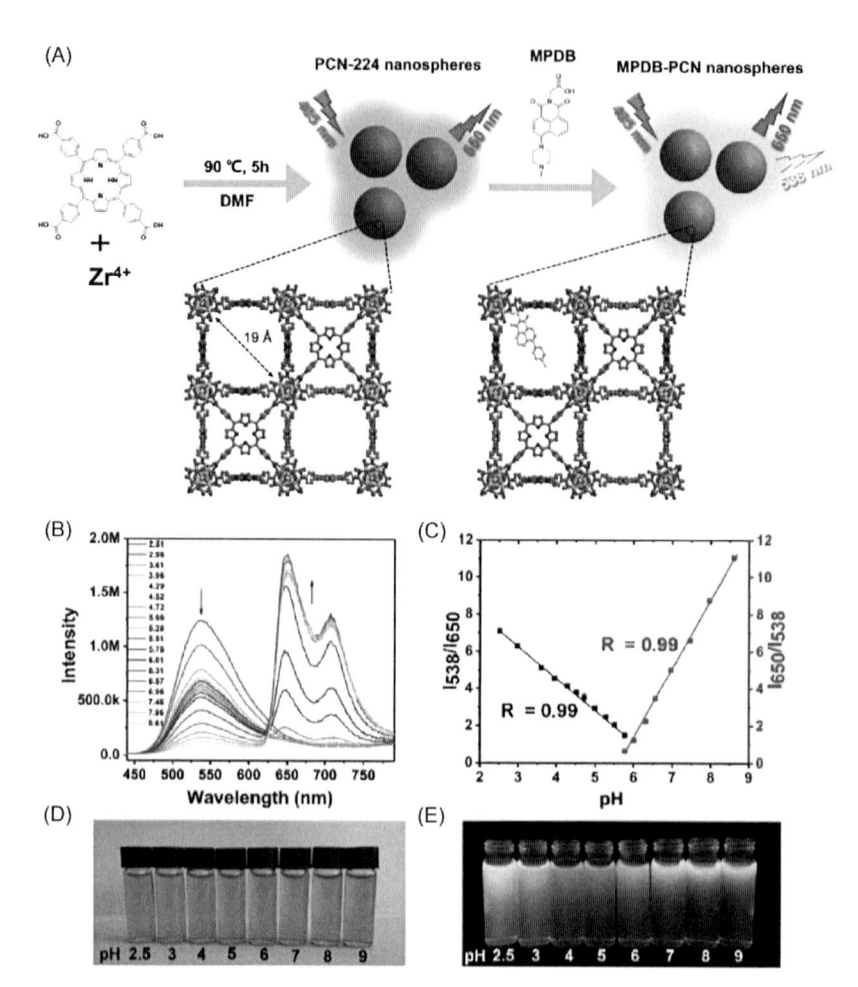

FIGURE 17.14 (A) The solvothermal synthesis of nano-pores in PCN-224 followed by its postsynthetic modification to produce MPDB-PCN and (B–E) Their fluorescence response to pH changes (Guo et al., 2020).

are prominent due to their small size, high sensitivity, electromagnetic interference resilience, and remote operation capabilities. By altering the surrounding medium's RI, properties of guided light in optical fibers, such as intensity, wavelength, phase, and polarization, can be modified. Challenges such as selectivity and sensitivity limitations have hindered this sensing approach. However, MOFs with high porosity and large surface area offer a promising solution by selectively absorbing target molecules and significantly altering the RI. This capability enhances both selectivity and sensitivity in RI-based optical fiber sensors. Techniques involving MOF coatings on optical

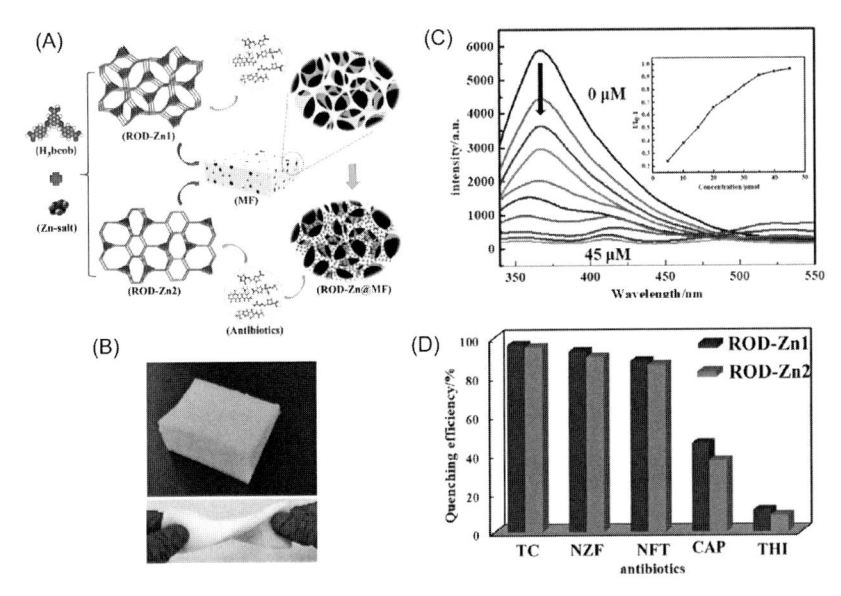

FIGURE 17.15 (A) Synthetic route of ROD-Zn1 and ROD-Zn2 and MOF@MF composites, (B) Melamine foam, (C) Emission spectra of ROD-Zn1 in TC, and (D) Luminescence quenching efficiencies of ROD-Zn1 and ROD-Zn2 with certain antibiotics (TC, Tetracycline; NFZ, Nitrofurazone; NFT, Nitrofurantoin; CAP, Chloramphenicol; THI, Thiamphenicol) (Gai et al., 2020).

fibers allow precise control over size, thickness, and morphology. Various sensing mechanisms like interferometry, evanescent field sensing (long-period gratings), transmission spectroscopy, and (localized) surface plasmon resonance utilize these principles. This section focuses on the application of MOF composites in optical fiber-MOF (OF-MOF) sensors, which utilize these changes in dielectric properties for sensing purposes. Various examples of relevant research in this field will be examined.

Fiber-optic interferometric sensors based on metal-organic framework composites

Fiber-optic interferometric sensors utilize various types of interferometers, such as the Michelson interferometer (MI), Mach-Zehnder interferometer (MZI), Sagnac interferometer (SI), and Fabry-Perot interferometer (FPI). Among these, the FPI stands out for its simplicity, small size, and ease of fabrication. These interferometers consist of two partial reflectors spaced apart. A MOFs material is attached to the end of an optical fiber (Fig. 17.16A), where light partially reflects (I_1) at the fiber-MOFs interface and again (I_2) at the MOFs-environment interface. The interference of these reflected beams produces an output signal (I) intensity dependent on the light wavelength

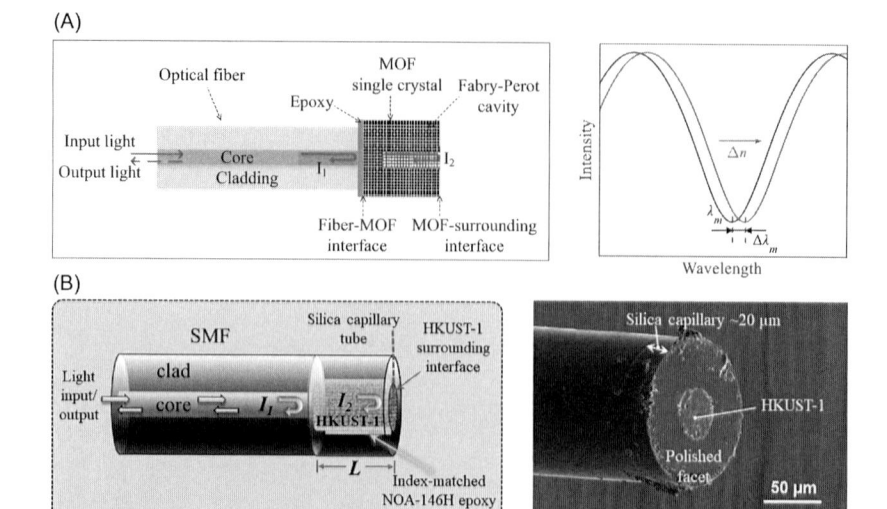

FIGURE 17.16 (A) Schematic of the optical fiber sensor modified with metal-organic frameworks and representation of the original (blue, left side) and shifted (red, right side) interferograms (Zhu et al., 2019) and (B) Concept demonstration of a miniature FPI containing a single crystal of HKUST-1, and scanning electron microscopy image of end facet of the sensor (Mumtaz et al., 2024).

(λ), initial phase difference (φ), and medium RI (n). Absorption of guest molecules in the MOFs alters its RI, causing a wavelength shift correlating with analyte concentration (Zhu et al., 2019). The MOFs layer in FBI-based OF-MOFs sensors consists of three main parts: a sensing cavity, an analyte receptor, and an analyte concentrator. By selectively capturing and concentrating the analyte within the sensing cavity, MOFs cause substantial changes in the RI. In fiber optic sensors, MOFs are used at the fiber's end to enhance sensitivity to RI changes. MOF crystals can be attached to the fiber by bonding or direct growth. Composites, like MOF/polydimethylsiloxane (PDMS), are commonly used to adjust analyte solubility and film permeability, with the hydrophobic PDMS layer also reducing humidity sensitivity and aiding in the selective detection of hydrophobic molecules like methane (Cao et al., 2020). MOF/PDMS composites are also effective in sensors for gas pressure by using MOFs with suitable functional groups and larger pore sizes (Jing et al., 2024). Enzyme@MOF composites have also been used in sensor construction. In this type of composite, the enzyme exhibits a high degree of encapsulation within the MOF structure, resulting in long-term stability and activity, allowing it to function in harsh environments. Examples of such sensors include glucose oxidase (Zhu et al., 2019) and urease (Zhu et al., 2019) composite sensors for the detection of glucose and urea, respectively. To address the moisture sensitivity issues of MOF sensors, an alkylamine@MOF composite has been developed (Kim et al., 2019), where the thin alkylamine layer makes the MOF nonpolar, increasing the sensor's stability and reducing its sensitivity to humidity. Some related examples are discussed below.

The first MOF-based FBI sensor was reported by Lu and Hupp (2010). They coated a 1 μm-thick layer of ZIF-8 onto the end of an optical fiber using the layer-by-layer (LBL) method. They reported a detection limit of 0.3 vol.% ethanol in water (equivalent to 100 ppm ethanol vapor).

Zhu et al. (2019) developed a real-time optical sensor using a single-crystal MOF, such as HKUST-1, attached to an optical fiber with epoxy. The MOF absorbs the target compound, and the analyte concentration is measured by analyzing shifts in interference spectra from the FPI (Fig. 17.16B). This sensor effectively detected nitrobenzene, an explosive simulator, at very low concentrations. The HKUST-1 crystal, with 9×9 Å windows, can accommodate nitrobenzene molecules (approximately 4.0 Å) through coordination of its nitro group with Cu(II) or interaction of its aromatic ring with BTC^{3-} ligands in the MOFs structure.

Cao et al. (2020) developed a gas-sensitive composite coating for optical fiber sensor using a PDMS/nanocrystalline MOF (ZIF-8) composite, designed to function as a methane sensor for monitoring in natural gas facilities. The PDMS serves as a host material for ZIF-8, enhancing methane solubility and film permeability. It demonstrates selective response to methane over nitrogen and oxygen due to methane's higher adsorption potential and solubility in the PDMS/ZIF-8 composite. Additionally, the hydrophobic nature of PDMS and ZIF-8 minimizes humidity interference. However, the sensor's response time is relatively slow due to the gradual gas adsorption process. The sensor exhibits a linear range from 1% to 50% methane (Fig. 17.17A and B), with a detection limit of 1%, making it suitable for detecting leaks in natural gas pipelines.

Pressure sensitivity and gas desorption time are two critical parameters determining the performance of a pressure sensor. PDMS films have shown great potential for highly sensitive pressure sensors; however, balancing high sensitivity with a low desorption time remains a challenge. Jing et al. (2024) developed a highly sensitive gas pressure sensor with rapid desorption, using a PDMS/MOF composite coated on a tapered single-mode fiber (SMF). NH_2-MIL-101 (Cr) was chosen for its highly porous structure, large surface area, and the presence of amino functional groups, which can interact with gas molecules through van der Waals forces or hydrogen bonding, thereby enhancing the gas adsorption capability of the composite film. Moreover, NH_2-MIL-101 (Cr) contains large pores that facilitate fast gas molecule transport channels, leading to quicker gas desorption times. As a result, the PDMS/MOF composite film exhibits significant changes in volume and RI in response to external gas pressure variations. With increased MOF content doped into PDMS, the sensor's pressure sensitivity improves by up to 5%, and the desorption time decreases. The sensor's pressure sensitivity reaches 228.32 nm/MPa, with a gas desorption time as short as 16 seconds, capable of operating within a pressure range of 0–100 KPa.

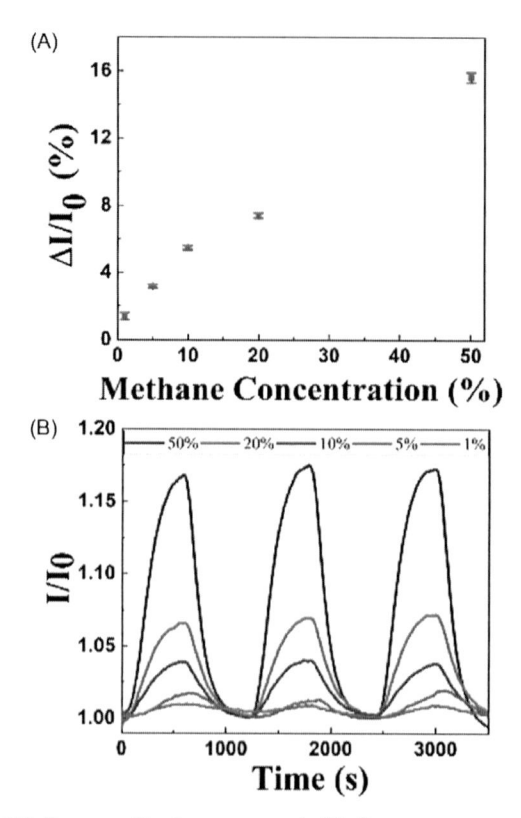

FIGURE 17.17 (A) Sensor calibration curve and (B) Sensor response graph at different concentrations and repetitions (Cao et al., 2020).

Since MOFs can be used as carriers for enzyme encapsulation, Zhu et al. (2019) prepared a composite of ZIF-8/Glucose Oxidase (GO$_x$), integrated it into a long-period grating (LPG), and presented a label-free fiber optic biosensor for selective glucose measurement. The ZIF-8/GO$_x$ composite is directly grown on the LPG fibers. For this purpose, a mixture of Zn^{2+}, 2-methylimidazole (2-MIM), and GO$_x$ is prepared, and the LPG fiber is immersed in it for half an hour. The composite exhibits a high degree of protein encapsulation, along with high activity and stability under harsh conditions. In the prepared optical sensor, ZIF-8 not only acts as a protective layer for GO$_x$ but also serves as an analyte collector on the fiber. As shown in Fig. 17.18A and B, wavelength shifts are linearly correlated to glucose concentration in the range of 1–8 mM, with a sensitivity of 0.5 nm/mM. Additionally, as shown in Fig. 17.18C, the composite demonstrates high selectivity in an environment containing ten times the glucose concentration of other interfering sugars.

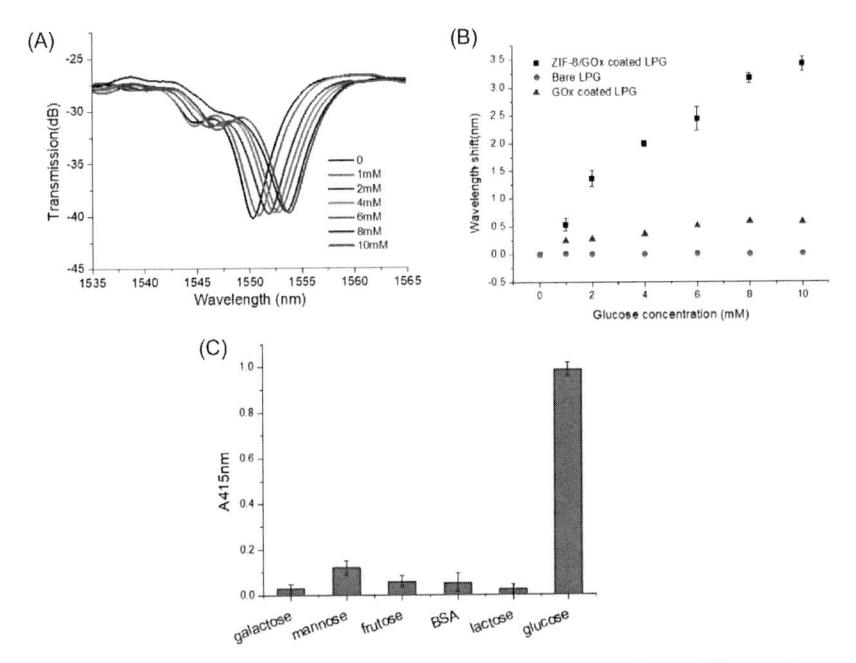

FIGURE 17.18 (A) Transmission spectra of the ZIF-8/GOx composite at different glucose concentrations, (B) Calibration curve of the sensor for the ZIF-8/GOx composite, GOx, and bare long period grating, and (C) Selectivity of the ZIF-8/GOx composite (Zhu et al., 2019).

Zhu et al. (2019) developed a highly sensitive tapered single-mode coreless single-mode (SCS) structure for RI sensing. They created a selective label-free optical biosensor for urea detection by encapsulating urease enzyme in a ZIF-8 structure grown in situ on coreless fiber. The online monitoring of urea was achieved by its binding to urease within the MOF composite, resulting in a wavelength shift in the fiber optic biosensor due to changes in RI (Fig. 17.19A and B). ZIF-8 provided a high surface area for urease immobilization, resulting in a sensor with rapid response, label-free detection, and excellent sensitivity and selectivity in real samples. The sensor showed a linear response to urea concentrations between 1 and 10 mM with a sensitivity of 0.8 nm/RIU (RI unit). As shown in Fig. 17.19C, the response of the composite is significantly more sensitive than that of each individual component.

Sun et al. (2018) developed a graphene oxide-nickel (GO-Ni) MOF composite with micrometer thickness using an electrochemical method on an MZI fiber. In the first step, they coated the MZI fiber with GC and then grew Ni-MOF on it electrochemically. This composite was used to create an optical fiber MZI sensor. The GO-Ni MOF demonstrated stability at temperatures up to 125°C and exhibited good hydrogen adsorption capabilities, retaining hydrogen molecules for up to 11 hours after absorbing 1% hydrogen

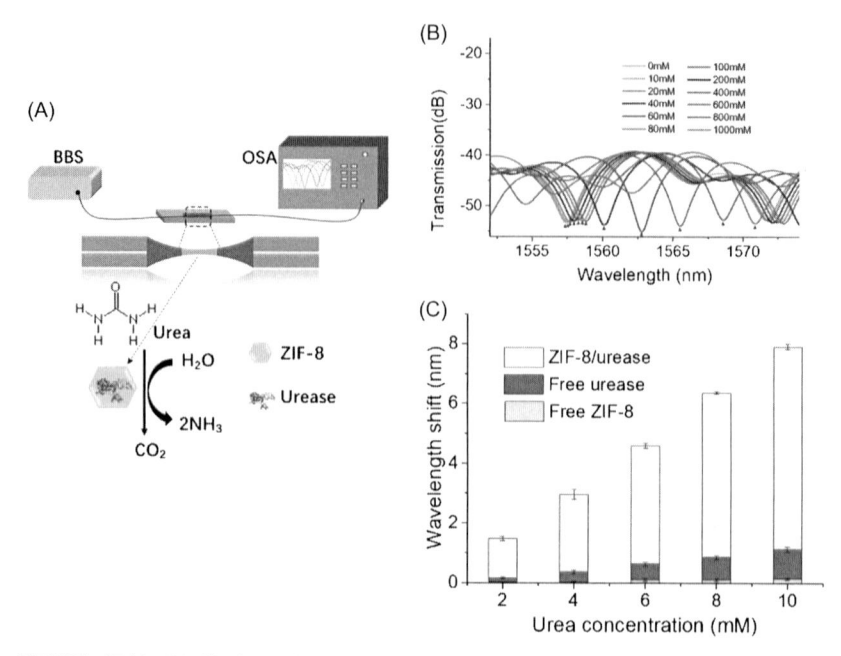

FIGURE 17.19 (A) Single-mode coreless single-mode sensor setup and urea measurement mechanism, (B) Transmission spectra of the single-mode coreless single-mode at different urea concentrations, and (C) Calibration curve of the sensor for the composite and its individual components (Zhu et al., 2019).

at room temperature. Hydrogen adsorption was linear within the concentration range of 0.1%–0.6%.

Kim et al. (2019) utilized a thin layer of cobalt-doped ZIF-8 (Co/ZIF-8) modified with alkylamine to develop an in situ optical transmission spectroscopy fiber optic sensor for CO_2 detection under wet conditions. Oleylamine (OLA), a primary alkylamine, strongly complexes with the metal sites of the MOF through its NH_2 functional group, forming a thick aliphatic layer on the MOF surface and imparting hydrophobic properties. The study results showed that the composite retained its pore structure and porosity without affecting the sensor's sensitivity to CO_2 while effectively preventing the negative effects of moisture, such as MOF instability, cross-sensitivity, and baseline shift.

Surface-enhanced Raman scattering based in metal-organic framework composites

SERS enhances the faint molecular signals of Raman spectroscopy using noble metal nanostructures with surface plasmon resonance (SPR) or oxygen vacancies in semiconductors, boosting signal intensities by factors ranging from 10^{14} to 10^{15}. This enables ultra-sensitive detection down to single-molecule levels (Nie &

Emory, 1997). SERS offers advantages such as minimal sample consumption, rapid response, nondestructive and noninvasive analysis, and high specificity while addressing traditional Raman spectroscopy limitations. Its effectiveness relies on the surface properties of active materials, typically plasmonic nanoparticles (PNPs) made of gold, silver, or copper, which enhance signals through electromagnetic (EM) or chemical (CM) mechanisms. Effective SERS materials must effectively handle interfacial effects in real samples, concentrate target molecules onto PNPs rapidly, and exhibit selective absorption capabilities (Lai et al., 2020).

MOFs have distinct properties that make them appealing for SERS applications. Their high surface area and uniform nanoporous structures enable them to function effectively as enhancer surfaces, selectively concentrating molecules based on pore size. Moreover, MOFs exhibit desirable characteristics such as chemical and thermal stability, diverse pore shapes and sizes, and a variety of functional groups, all of which enhance their suitability for SERS applications. MOF-based composites are used in SERS sensors to enhance sensitivity. In MOF@Au (Qiao et al., 2018) and MOF@Ag (Koh et al., 2018) composites, the MOF traps and concentrates analyte molecules, improving their interaction with gold or silver nanoparticles. Advanced core/shell composites like Au@MOF on gold films utilize the MOF's adsorption properties and thin MOF shells, which act as plasmon-coupled nanogaps to further amplify the SERS signal (Liu et al., 2024). These examples will be examined in greater detail in the following.

The SERS technique, capable of detecting single molecules, holds promise for early cancer detection via exhaled breath analysis. Aldehyde compounds released from tumor tissues serve as potential lung cancer biomarkers. However, challenges include weak Raman scattering of many VOC biomarkers and low adsorption on solid substrates due to high gas molecule mobility. Qiao et al. (2018) addressed these challenges by coating self-assembled gold superparticles (GSPs) with a ZIF-8 layer. This layer slows down the flow rate of gas biomarkers and reduces electromagnetic field decay around the GSP surface. ZIF-8 facilitates the capture of aldehyde biomarkers onto GSPs, enabled by a chemical reaction with 4-aminothiophenol previously adsorbed onto the gold GSPs, achieving a detection limit of up to 10 ppb (Fig. 17.20).

Koh et al. (2018) introduced a plasmonic nose for measuring VOC compounds with a detection limit in the ppm range. They used Ag@ZIF composite particles as the SERS substrate. To achieve higher sensitivity, they optimized the density of Ag nanocubes, finding that as the density increased and the spacing between them decreased, plasmonic coupling occurred more intensely, resulting in stronger electromagnetic fields and hot spots, which improved SERS by up to five times (Fig. 17.21). Additionally, the presence of a ZIF layer on the Ag nanocubes concentrated analyte molecules within the electromagnetic field range, doubling the signal strength. The developed sensor was employed to measure polycyclic aromatic hydrocarbons (PAHs) and

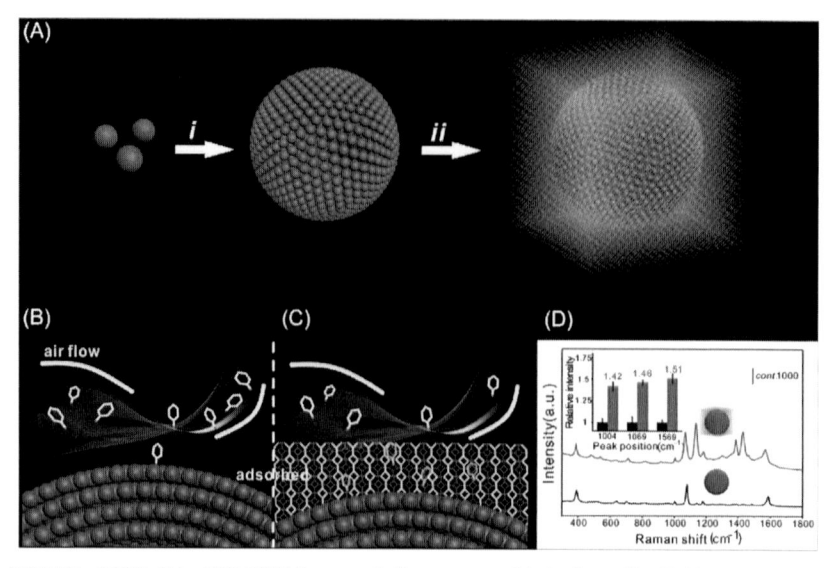

FIGURE 17.20 (A) GSP@ZIF-8 core–shell structure fabrication: (i) Gold nanoparticles assembled into GSPs, (ii) ZIF-8 shell coated on gold superparticle surface; Schematic illustration of (B) Gold superparticles and (C) GSPs@ZIF-8 with gas collisions; (D) Surface-enhanced Raman scattering spectra of p-aminothiophenol on gold superparticles (black, bottom curve) and GSPs@ZIF-8 (red, top curve) (Qiao et al., 2018).

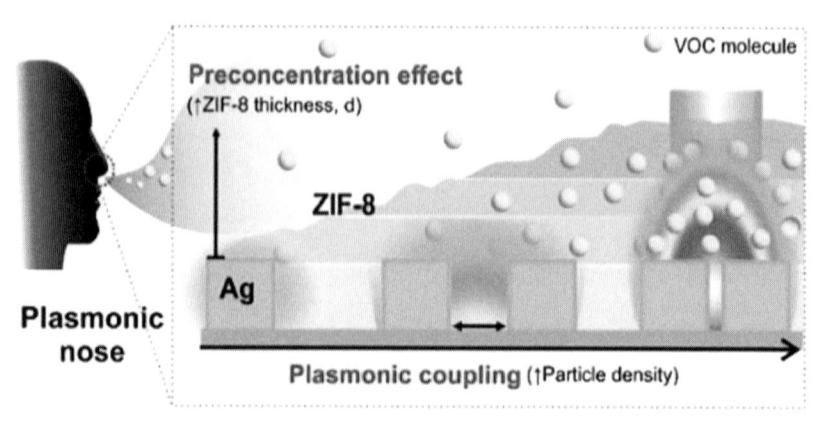

FIGURE 17.21 Conceptual illustration of a plasmonic nose based on tuning plasmonic hotspots and metal-organic framework thickness (Koh et al., 2018).

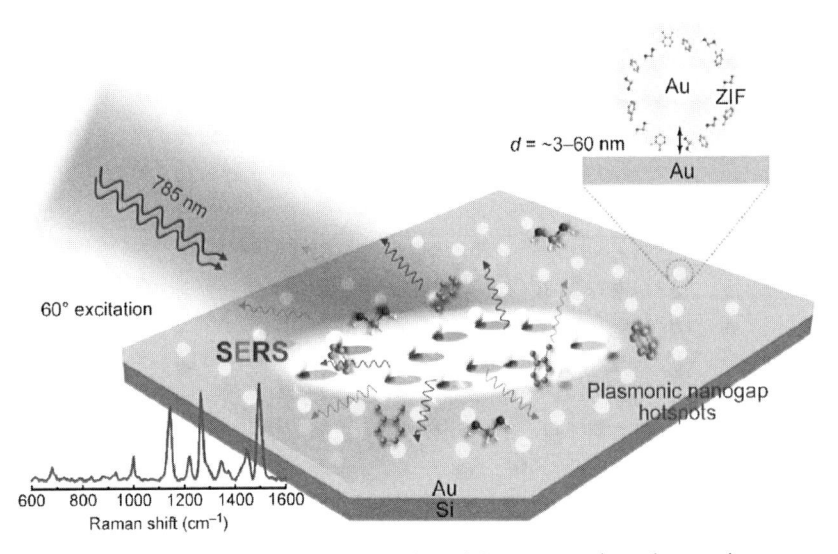

FIGURE 17.22 A plasmon-coupled surface-enhanced Raman scattering substrate, the nanogap hotspots are filled with ZIF-8 to trap analytes (Liu et al., 2024).

certain VOCs such as chloroform and 2-naphthalenethiol. For 2-naphthalenethiol, a detection limit of 50 ppb was achieved, which is 15 times lower than the exposure limit of 0.7 ppm.

Liu et al. (2024) developed an ultrasensitive SERS method by combining enhanced plasmonic electric fields with MOF's concentrating properties. They encapsulated gold nanoparticles within ZIF-8 shells, coupling them onto a gold film to create plasmon-coupled nanogaps. Adjusting the ZIF-8 layer thickness (10 nm) enhanced the electric field and altered the plasmon resonance wavelength (Fig. 17.22). This setup captured active analyte molecules within the ZIF-8 layer, resulting in robust SERS signals with a detection limit of around 10 ng/L for VOCs.

Localized surface-plasmon resonance sensor based metal-organic framework composites

SPR spectroscopy is an advanced optical sensing technique that measures small changes in the RI. Using plasmonic nanoparticles and thin films, it detects coherent oscillations of conduction band electrons induced by light. Light absorption or scattering significantly increases at frequencies that excite LSPR. This interaction results in a characteristic extinction spectrum, which varies based on the nanoparticles' size, shape, composition, and surrounding RI. Changes in the dielectric environment are observed as shifts in the extinction spectrum.

Localized surface plasmon (LSP) occurs when surface plasmons are confined within a nanoparticle comparable in size to the incident light.

684 Applications of Metal-Organic Framework Composites

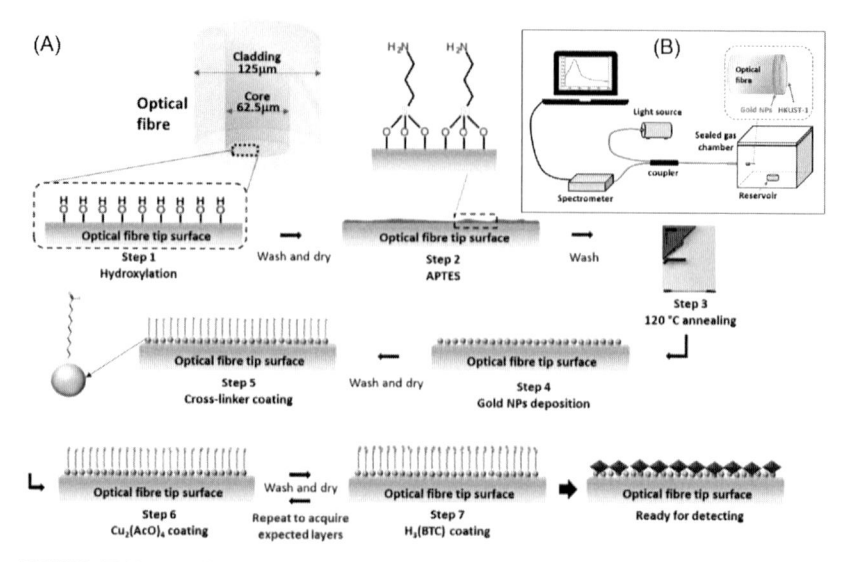

FIGURE 17.23 (A) Tip-based fiber optic coating fabrication scheme and (B) Localized surface-plasmon resonance sensor set-up (He et al., 2021).

While LSPR effectively detects large biological analytes due to significant RI changes, sensing small molecules is challenging due to smaller RI changes. Kreno et al. (2010) enhanced LSPR gas sensing by coating Ag nanoparticles with HKUST-1, achieving a 14-fold increase in sensitivity for CO_2 detection. The sensing signals come from the nanoparticle extinction spectrum, allowing various MOFs to be used for selective gas adsorption.

He et al. (2021) developed an LSPR optical fiber sensor for VOC detection. The schematic of the synthesis method and the measurement device are shown in Fig. 17.23(A and B), respectively. They activated a silica surface, added amino-propyl silane groups, and coated it with gold nanoparticles. Carboxyl groups were then attached, followed by the LBL growth of HKUST-1 by alternately immersing in copper acetate and $H_3(BTC)$ solutions. The sensor showed a red spectral shift when exposed to VOCs like acetone, ethanol, and methanol, with reversible responses to concentrations between 0% and 6%. For acetone, the sensitivity was 13.7 nm/% for 120 MOF layers, with a detection limit of 50 ppm. While suitable for industrial applications, the sensor needs improved sensitivity for breath analysis.

Mechanical sensors based on metal-organic framework composites

Most MOFs are not luminescent and do not change color, making optical signal changes impractical for all MOF-based sensing applications (Zhu et al.,

2021). Instead, mechanical sensors like surface acoustic wave (SAW) sensors, quartz crystal microbalance (QCM), and microcantilever (MCL) can be used universally with all MOFs. In these sensors, the MOF is deposited on the surface of an electromechanical device. These sensors measure weight increases in MOFs (in the nanogram range) when they adsorb target analytes, causing shifts in oscillator frequency that correspond to analyte concentration. The sensor's selectivity is determined by the pore size, geometry, and functional groups of the coated MOFs, while the sensor's sensitivity is determined by its high surface area (Yan et al., 2020).

QCM is a widely used mass sensor that can measure particles from ng to μg and is suitable for low-concentration gas detection (Na Songkhla & Nakamoto, 2021). It consists of an AT-cut piezoelectric quartz crystal between two electrodes, operating on the inverse piezoelectric effect. When an alternating voltage is applied between the electrodes, it generates an electric field, which induces mechanical oscillations in the quartz crystal due to the resulting strain (Balasingam et al., 2023). In a QCM, the resonant acoustic waves generated in the quartz crystal propagate perpendicular to the crystal surface (Fig. 17.24A). Changes in oscillation frequency due to analyte adsorption on the MOF-coated surface serve as the sensing signal. QCM is inexpensive, stable, and can directly measure analyte mass in the ng range (Son et al., 2021). SAW sensors propagate acoustic waves parallel to the surface of a piezoelectric substrate (Fig. 17.24B). The binding of analytes alters the wave's velocity, phase, and amplitude, with changes corresponding to the analyte's mass. SAW sensors are

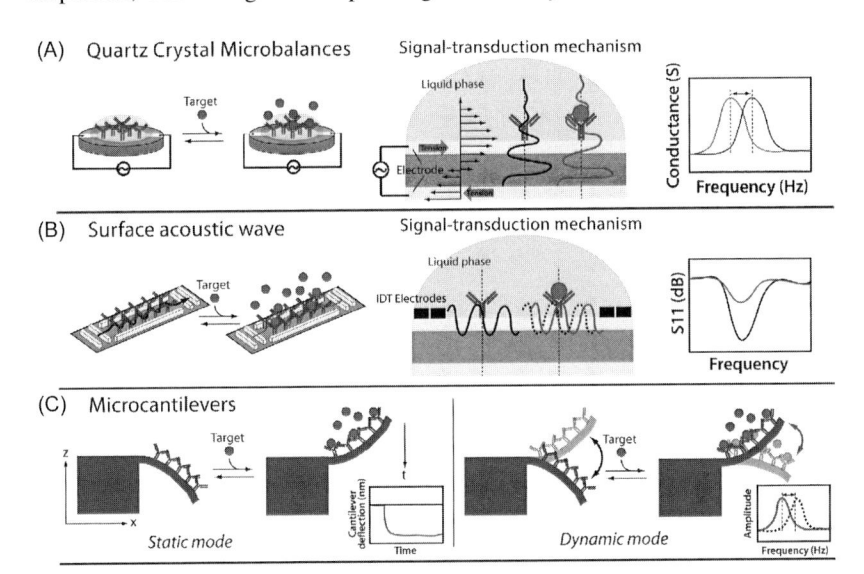

FIGURE 17.24 Types of mechanical sensors: (A) Quartz crystal microbalance, (B) Surface acoustic wave, and (C) Microcantilever (Idili et al., 2022).

more sensitive than QCM due to higher operating frequencies. Cantilever sensors use mechanical transducers in microscale or nanoscale cantilevers that bend when a substance binds to them. The bending is measured by the angle of a laser beam reflected from the cantilever, proportional to the mass on the surface (Fig. 17.24C).

MOFs can be used as sensor coatings in QCM by direct deposition, creating composites, or as templates for porous materials. Coating methods include drop coating, spin coating, and spraying (Aghakhani et al., 2019). Various types of composites have been used in QCM sensors, where the required selectivity is often achieved by combining MOF with factors such as aptamers (Yang et al., 2022) and molecularly imprinted polymers (MIPs) (Qian et al., 2016). Additionally, pore engineering, such as MOF-on-MOF structures (Wannapaiboon et al., 2015), can also contribute to the desired selectivity. Sensitivity control is achieved through the high surface area of MOFs or the use of various nanoparticles such as CNTs (Chappanda et al., 2018), polyaniline (PANI) (Abuzalat et al., 2019), and ionic liquids (Noorani et al., 2023). The following section introduces some related examples.

Yang et al. (2022) developed a highly sensitive QCM sensor based on an aptamer@MOFs composite for detecting TC. They coated a QCM crystal with HKUST-1 suspension, followed by deposition of gold nanoparticles on the MOFs using cyclic voltammetry in $HAuCl_4$ solution. Finally, they applied aptamer onto the electrode surface to attach it via Au–S bonds (Fig. 17.25). The developed sensor has a linear response over a wide range from 0.1 ng/mL to 10 µg/mL, with a detection limit as low as 8 pg/mL. The sensor exhibits excellent selectivity and sensitivity.

Wannapaiboon et al. (2015) developed a novel method known as continuous stepwise liquid-phase epitaxial (LPE) growth to synthesize two-dimensional MOF-on-MOF composites directly on the surface of QCM crystals. They synthesized heterostructured MOFs using two different types with distinct structural and chemical properties by carefully controlling the

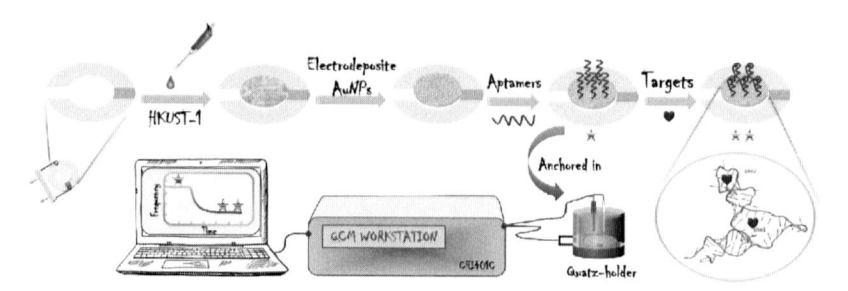

FIGURE 17.25 Schematic of the preparation and measurement of the aptamer@metal-organic framework composite quartz crystal microbalance sensor (Yang et al., 2022).

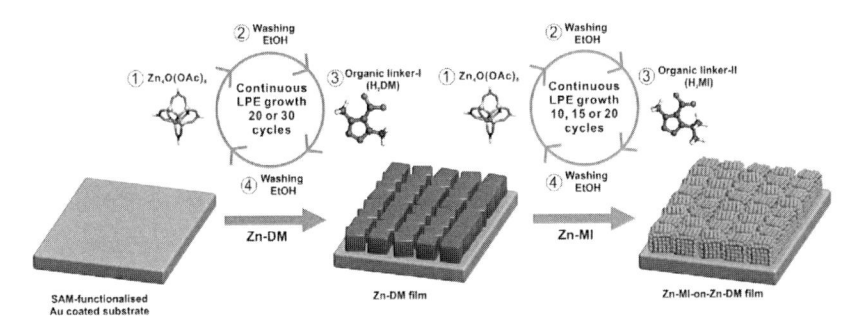

FIGURE 17.26 Schematic of preparing a size-selective layer by stepwise epitaxial growth (Wannapaiboon et al., 2015).

growth stages. Initially, they grew layers of large-pore MOFs (Zn-DM and Zn-ME) directly on the gold-coated QCM crystal. Subsequently, smaller-pore MOFs (Zn-MI and Zn-DE) were synthesized on top of these layers (Fig. 17.26). The resulting sensor exhibited selective detection capabilities based on pore size, demonstrating differentiation between alcohols and also between water and methanol.

Noorani et al. (2023) prepared an ionic liquid (IL)@MOF composite and demonstrated that the QCM sensor based on this composite has a higher sensitivity capacity than the sensor based on MOF alone. They grew $Cu_2(BDC)_2$ and $Cu_3(H_2N\text{-}BDC)_2$ MOFs on QCM gold electrodes and used the vacuum impregnation method to introduce the cholinium-amino acid IL. Their findings indicated that this modification could increase CO_2 adsorption by up to three times.

Qian et al. (2016) innovatively combined MOFs with MIPs to develop a sensitive and selective MOF@MIP composite coating for the QCM-based detection of metolcarb in pear juice solutions. They synthesized MIL-101 nanoparticles and polymerized MIP in their presence, then applied MIL@MIP nanoparticles as a sensor layer on QCM crystals using drop-coating. The composite nanoparticles have a specific surface area of 1579.43 m^2/g. The sensor exhibits a linear response in the range of 0.1–0.9 mg/L, with a detection limit of 0.0689 mg/L. By combining the high surface area of MOF with the selectivity of MIP, this sensor achieves an exceptionally low detection limit and high selectivity.

Similarly, Yao et al. (2020) developed UiO-66@MIP composites with a high surface area of 994 m^2/g. Fig. 17.27 illustrates the schematic process of preparing MIP@MOF coatings, emphasizing customized binding sites for specific molecule shapes. The method has a detection limit of 61.65 µg/L with a linear range between 80 and 500 µg/L. The sensor's response to tyramine is six times greater than to similar amines. The sensor exhibits excellent selectivity, absorptivity, and chemical stability, making it suitable for the rapid and accurate measurement of biogenic amines in food samples.

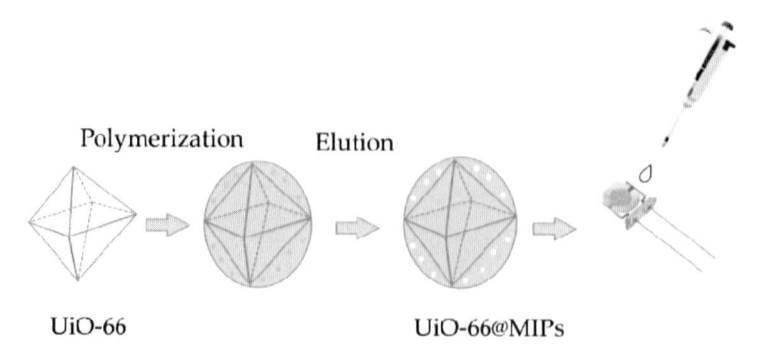

FIGURE 17.27 Schematic of preparing molecularly imprinted polymer@metal-organic framework coating (Yao et al., 2020).

Chappanda et al. (2018) employed CNT/MOF composites for a humidity sensor. The CNT/HKUST-1 composite showed about 230% greater sensitivity than a sensor using HKUST-1 film alone, while CNT alone had a weak humidity response. The authors attribute this sensitivity improvement to differences in the crystal size of the deposited films, which significantly affect adsorption kinetics. The presence of CNT reduces the crystal size of HKUST-1 in the composite, leading to enhanced adsorption kinetics and sensor sensitivity.

Abuzalat et al. (2019) developed a hydrogen sensor under ambient conditions by growing a Cu-BTC/PANI nanocomposite film on a QCM surface. The researchers first coated the QCM surface with a layer of copper, oxidized it, and then exposed it to a solution containing PANI, BTC ligand, and copper ions. The nanocomposite was formed under intense pulsed light irradiation. The composite showed high selectivity for hydrogen gas over methane and carbon dioxide. The advantages of this composite sensor include a rapid response time of 2–5 seconds and operation at room temperature. The composite effectively eliminates the humidity-related poisoning effect on the PANI or Cu-BTC layer. The sensor has a detection limit of 15 ppm.

Electrical sensor based on metal-organic framework composites

Various materials, such as semiconductor metal oxides, polymers, nanostructured carbon materials, and zeolites, are used in electronic sensors, each with its own advantages and disadvantages. Semiconductor metal oxides need high temperatures to operate, leading to low selectivity and high costs. Polymers work at low temperatures and are cost-effective but have low selectivity and reduced stability over time. Carbon materials like nanotubes and graphene offer high sensitivity at room temperature but are expensive and require modifications for better selectivity. Zeolites face issues with low porosity and structural limitations.

MOFs, however, are well-suited for sensors due to their structural diversity, tunable pore size, reversible adsorption at room temperature, thermal and chemical stability, high surface area, and permanent porosity. These properties allow MOFs to serve as both adsorbents and sensors with adjustable selectivity. Open metal sites in MOFs that function as Lewis acid centers can enhance selectivity. Furthermore, the ability to trap guest molecules or nanoparticles within the pores can be utilized in sensor mechanisms (Chidambaram & Stylianou, 2018).

MOF-based electronic sensors detect analyte interactions by measuring changes in electrical properties like impedance, resistance, capacitance, or work function, providing the desired signal. A significant challenge in using MOFs in electronic and electrochemical sensors is their lack of sufficient electrical conductivity (Zhu et al., 2021). Efforts have been made to address this deficiency by developing composites. Various types of composites have been created where the second component helps to improve electrical conductivity. Examples of such composites include MOFs combined with materials like nickel foam (NF), covalent organic frameworks (COFs), MXenes, CNTs, Ag NPs, Au NPs, ZnO NPs, Cu NPs, SnS_2 NPs, GO, partially reduced GO, organic semiconductors like tetrathiafulvalene (TTF), conductive polymers, and ILs. Additionally, to enhance selectivity, materials such as MIPs, aptamers, antibodies, and beta-cyclodextrins have been incorporated into these composites. In the following, examples of each of the above-mentioned composites will be reviewed.

Vijayaraghavan et al. (2023) created an immunosensor based on the FeCo-MOF@NF composite for oral cancer screening using electrochemical impedance spectroscopy (EIS). They functionalized the FeCo-MOF/NF substrate with IL-1RA antibodies via surface carboxyl and amine groups, enabling precise detection of IL-1RA, a biomarker found in human patients at concentrations ranging from 100 to 400 pg/mL (Fig. 17.28A and B). The immunosensor operates effectively across a wide concentration range, from 10 fg/mL to 10 ng/mL. The combination of MOF and NF significantly enhances the electrochemical response due to NF's high surface area and excellent conductivity, promoting uniform MOF growth and improving sensor performance for rapid and highly sensitive detection.

MOF-based sensors can achieve selectivity by coating their surfaces with aptamers, which are versatile oligonucleotides used for capturing and detecting specific molecules. Unlike antibodies, aptamers are easily synthesized, modified, and stable over time. When immobilized on MOF surfaces, they interact via strong electrostatic forces, π–π stacking, and hydrogen bonding. Zhou et al. (2019) developed an impedance sensor by directly growing Ce-MOF on COFs and then immobilizing selective aptamers for oxytetracycline (OTC) detection. The COF support enhanced electrochemical currents, boosting sensitivity. The Ce-MOF@COF composite structure ensured aptamers remain predominantly on the surface, optimizing contact with analytes and thereby improving both the sensitivity and selectivity of the sensor (Fig. 17.29).

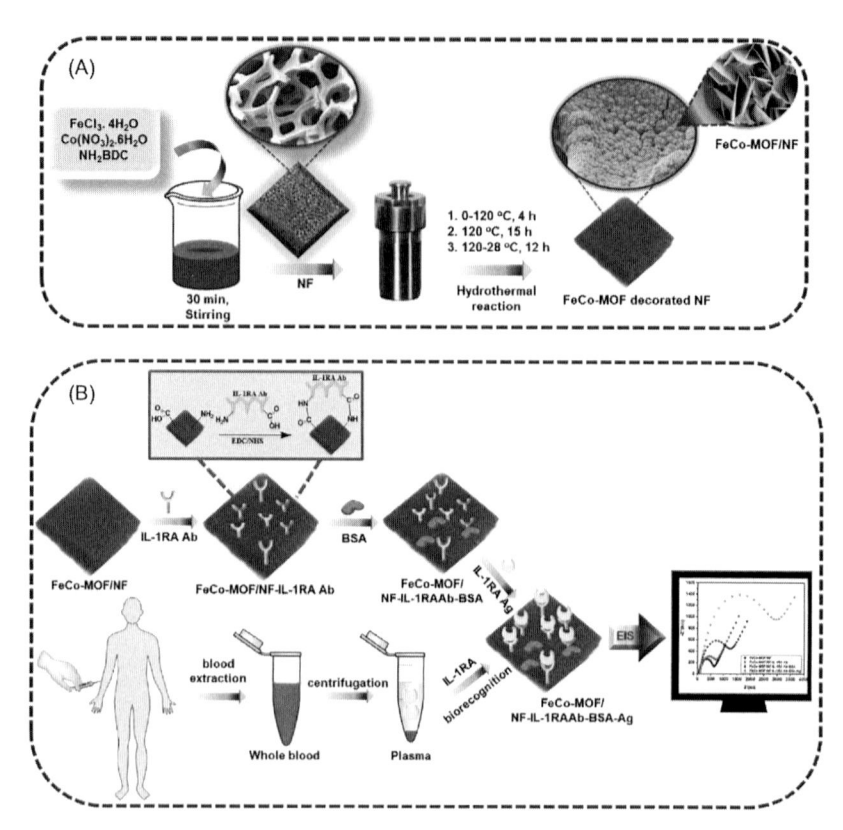

690 Applications of Metal-Organic Framework Composites

FIGURE 17.28 (A) The direct growth method of FeCo-MOF on nickel foam substrates and (B) Its impedance immunosensing technique (Vijayaraghavan et al., 2023).

Lin et al. (2024) developed a multifunctional biosensor for real-time monitoring of metabolic diseases, focusing on simultaneous measurement of uric acid (UA) and glucose (Glu) levels in sweat, alongside capturing electrophysiological signals and delivering electrostimulation therapy. The biosensor integrates highly conductive MXene and porous c-MOF (c-MOF, $Ni_3(HITP)_2$, (HITP=triphenylene-2,3,6,7,10,11-hexaamine)) within a nonwoven nanocellulose substrate, tailored for wearable applications in patients with lower limb muscle dysfunction due to hyperuricemia or hyperglycemia (Fig. 17.30). Compared to commercial Ag/AgCl electrodes, MOF/MXene composite electrodes offer lower impedance, high stability, and a superior signal-to-noise ratio (S/N). The combination exploits MXene's high surface area and conductivity with c-MOF's tunable porous structure and catalytic sites, enabling effective electrochemical detection and muscle therapy via electrostimulation. This integrated biosensor holds promise for noninvasive

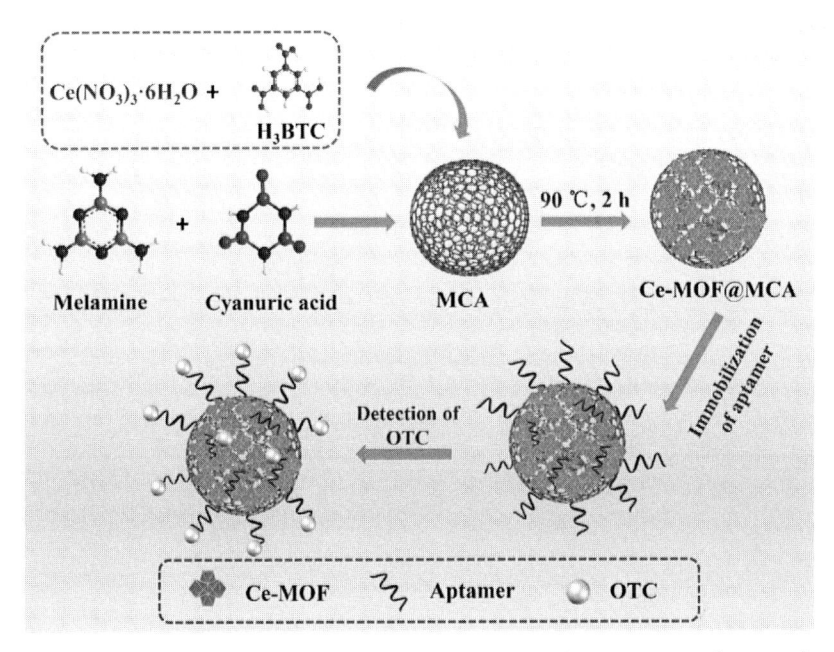

FIGURE 17.29 Schematic representation of the method for preparing an aptasensor for measuring oxytetracycline (Zhou et al., 2019).

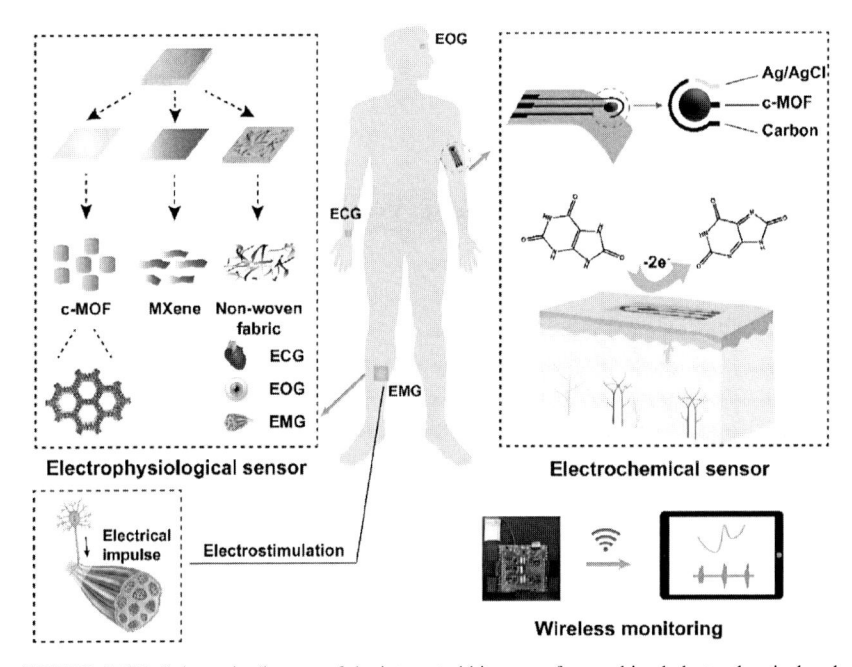

FIGURE 17.30 Schematic diagram of the integrated biosensor for combined electrochemical and electrophysiological sensing (Lin et al., 2024).

daily muscle theranostics and monitoring, with wireless data transmission capabilities to mobile terminals.

Ghanbarian et al. (2018) developed a MIL-53 (Cr-Fe)/Ag/CNTs nanocomposite to create a resistive sensor for detecting VOCs like methanol, ethanol, and isopropanol. MIL-53(Cr-Fe) was synthesized via a sonochemical method, resulting in uniform nanometric dimensions. The low electrical conductivity of the MOF composite is improved due to the presence of Ag NPs and CNTs. Additionally, the presence of Ag helps in the spread of electrons through the MOF structure. Due to the nanoscale nature of the composite, its thin film exhibits uniformity, adhesion, and stability properties. The sensor has a fast, reversible response with a low detection limit (30.5 ppm for methanol) and shows a linear response in the range of 10–500 ppm.

More et al. (2024) prepared a composite of Ce-MOF/GO and coated it onto an indium tin oxide electrode. The composite coating was irradiated with swift heavy ions (SHIs), 100 MeV Au$^+$ ions. SHI irradiation affects the electrical properties of GO by creating or annealing defects due to high electronic energy transfer, leading to localized reduction of GO. As a result, the conductivity of the composite increases and creates sites for H_2S adsorption, enhancing the sensor's efficiency for H_2S detection. The SHI-irradiated Ce-BTC/GO composite exhibits excellent response time (42 seconds) and recovery time (69 seconds), outstanding reproducibility, good repeatability, and stability at room temperature. The resistive sensor has a linear response range of 10–100 ppm with a detection limit of 10 ppm.

Strauss et al. (2019) developed a resistive sensor based on a guest@MOF composite, specifically using TTF within Co-MOF-74. The introduction of the TTF semiconductor molecule into the composite significantly increases the electrical conductivity of the MOF, which can affect its gas adsorption properties. CO_2 molecules can be adsorbed by the composite through interactions with Co centers. Although the CO_2 adsorption capacity decreases due to the partial occupation of the MOF's internal space by TTF, the sensor's sensitivity greatly increases due to the higher conductivity resulting from TTF.

Garg et al. (2023) prepared a MOF@polymer composite consisting of Zr-MOF (UiO-66-NH_2) and PEDOT:PSS (poly(3,4-ethylenedioxythiophene):poly (styrene sulfonate)) for toluene detection using a resistive sensor method. PEDOT:PSS is a polymer composed of short, positively charged PEDOT chains electrostatically bound to long, negatively charged PSS chains, providing good electrical conductivity. UiO-66-NH_2@PEDOT:PSS composite is a porous, hydrophobic, and semiconducting material with a high surface area and numerous active sites for toluene adsorption. The resulting sensor selectively responds to toluene in the presence of other VOCs such as benzene, toluene, xylene, ammonia, ethanol, and formaldehyde. The sensor has relatively short response and recovery times of 362 and 106 seconds, respectively, and exhibits a linear response in the concentration range of 1–20 ppm with a detection limit of 172 ppb. Additionally, the sensor demonstrates repeatable and stable responses.

MOF composites for sensing applications **Chapter | 17** 693

Capacitive sensors detect changes in the dielectric constant or thickness of the dielectric material due to gas analyte adsorption. MOFs enhance selectivity and sensitivity by allowing structural tuning through the choice of metal nodes, linkers, and pore sizes. However, their dielectric responses are often insufficient, requiring sensitivity enhancement strategies. Gonçalves et al. (2024) improved MOF-based capacitive sensors by incorporating ILs into the MOF structure, creating various MOF:IL composites. Using ZIF-8 for its hydrophobic nature and affinity for imidazolium-based ILs, they developed hybrids with different porosities and selectivities for VOCs. Fig. 17.31 illustrates the evolution of the local structure of MOF:IL composites as the molar ratio of IL and MOF is modified. When the composition is 100% MOF, the structure is a porous solid insulator. As the percentage of IL increases, it first forms a porous solid ionic conductor, then a porous ink ionic conductor, and finally, at 100% IL, a liquid ionic conductor is obtained. Fig. 17.31 also shows that with 100% IL, the highest sensor response is achieved, while the lowest response corresponds to ZIF-8. As the IL percentage increases, both electrical conductivity and sensor response improve. Also, as the IL percentage increases, it provideds recoverable and fast sensor responses, which were suitable for printed sensor technology.

Mukundan et al. (2023) developed a nonenzymatic electrochemical sensor using a ZnO nanoparticle-Cu MOF@NF composite for detecting serotonin (5-HT) in blood serum. The electrochemical activity of Cu-MOF@NF is enhanced

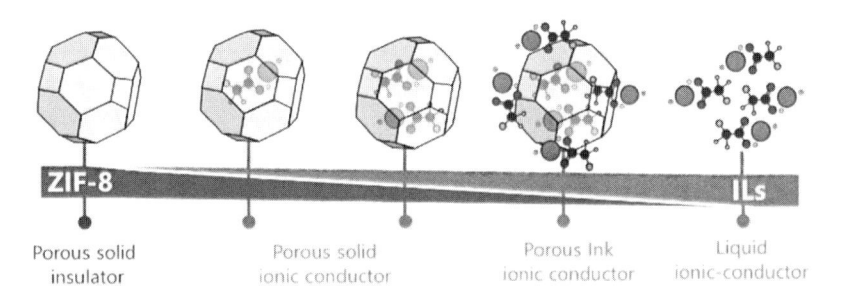

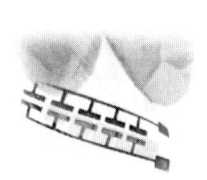

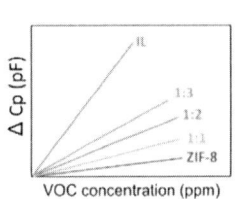

FIGURE 17.31 Schematic of the various possible states for metal-organic framework:ionic liquid hybrids in the construction of capacitive sensors (Gonçalves et al., 2024).

by the addition of ZnO nanoparticles. Differential pulse voltammetry analysis under optimal conditions resulted in a method with a linear range between 1 ng/mL and 1 mg/mL and a detection limit of 0.49 ng/mL, which is below the minimum physiological concentration and can be used for studying various neurological and psychological disorders. It demonstrated excellent sensitivity to serotonin amidst various biological interferents (glucose, dopamine, epinephrine, uric acid, and ascorbic acid), with high recovery rates (99.25%–102.5%) in blood serum. The sensor also exhibited strong longevity and reproducibility, proving its potential for broader electrochemical sensing applications.

Wang et al. (2023) developed a 2D CNT@Ce-MOF composite for constructing an electrochemical sensor. In this work, CNT was used as a backbone to create layered MOF nanosheets, employing a low ligand-metal ratio strategy controlled by the ligand loaded onto the CNT surface to successfully produce 2D layered MOF (Fig. 17.32A). The composite, featuring a unique layered structure and hierarchical channels, offers 49 times greater surface area than Ce-MOF and improved conductivity, lower mass transfer resistance, and more active sites. As shown in Fig. 17.32B, the electrochemical activity of the composite is greater than that of CNT and Ce-MOF. Ce serves as the electrocatalytic material for nitrite sensing. As Fig. 17.32C shows, the current increases with nitrite concentration and exhibits two wide linear detection ranges (0.65–3.25 µM and 3.25–7000 µM), along with a detection limit of 0.12 µM, as well as high repeatability, stability, and selectivity.

Sun et al. (2024) developed a SnS$_2$/ZnCo-MOF nanocomposite attached to a glassy carbon electrode (GCE) using gold nanoparticles (AuNPs) and created specific cavities for chlortetracycline (CTC) using the MIP method.

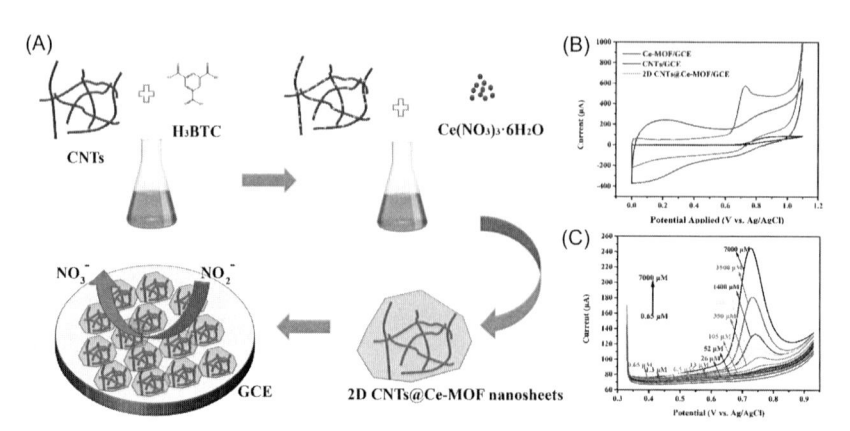

FIGURE 17.32 (A) The method for preparing the CNTs@Ce-MOF composite, (B) CV curves of the CNTs@Ce-MOF composite compared to carbon nanotubes and Ce-MOF in the presence of nitrite, and (C) The electrochemical performance of the composite sensor at different nitrite concentrations (Wang et al., 2023).

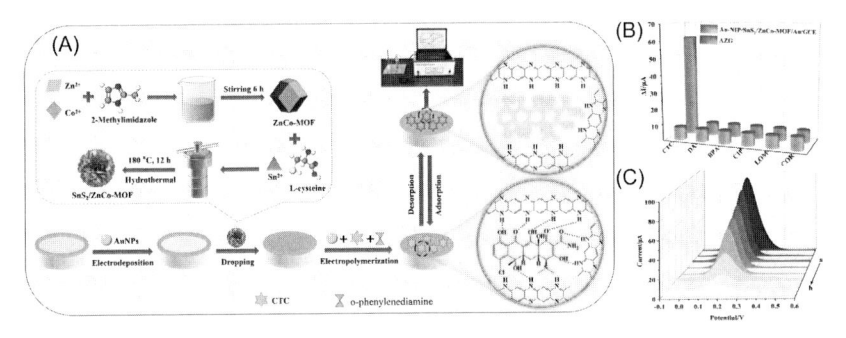

FIGURE 17.33 (A) Steps for fabricating the Au-MIP/SnS$_2$/ZnCo-MOF/Au/GCE composite sensor, (B) Sensor selectivity, and (C) Sensor response to different concentrations of chlortetracycline (Sun et al., 2024).

Fig. 17.33A shows the method for fabricating the composite. The Au-MIP/SnS$_2$/ZnCo-MOF/Au/GCE (AZG) system was used for selective CTC detection, with SnS$_2$ increasing surface area and ZnCo-MOF enhancing sensor stability (Fig. 17.33B). The conductivity of AuNPs improves electron transfer between the probe and the electrode through the insulating MIP layer. The stabilization of AuNPs and MIP via electropolymerization contributes to the sensor's selective detection and strengthens its output signal. The AZG sensor demonstrated a wide linear range (0.1–100 μM) (Fig. 17.33C), a low detection limit (0.072 nM), and high repeatability, reversibility, and accuracy (recovery 96.08%–104.6%) in food samples such as milk, eggs, and meat.

Ma et al. (2022) developed a nonenzymatic electrochemical sensor for glucose detection using a Cu@Co-MOF composite. Initially, they synthesized Co-MOF using the ligand terephthalic acid (H$_2$BDC) and 1,3-dimethylurea (DMU) with cobalt ions in a deep eutectic solvent. The resulting Co-MOF was then immersed in a copper ion solution, allowing Cu^{2+} ions to deposit on the structure (Cu^{2+}@Co-MOF). Subsequently, by adding NaBH$_4$, the adsorbed copper ions in Co-MOF were reduced, yielding the Cu@Co-MOF composite. Finally, a suspension of the composite was coated onto a GCE using Nafion. The presence of copper nanoparticles in the composite significantly increased its conductivity. While Co-MOF itself exhibited no sensitivity to glucose, the Cu@Co-MOF composite showed strong catalytic oxidation in the presence of glucose, producing a pronounced redox peak. The Cu@Co-MOF sensor demonstrated a linear chronoamperometric response to glucose in the ranges of 0.005–0.4 and 0.4–1.8 mM, with a detection limit of 1.6 μM. The composite exhibited a rapid current response (7 seconds), high antiinterference properties (against mannose, fructose, dopamine, urea, uric acid, and aspartic acid), and good stability and reproducibility in glucose measurement. The sensor was also used to measure glucose in human serum and orange juice, showing excellent relative recovery (93%–100%) with high accuracy.

Chen et al. (2023) developed a Ni-MOF@β-cyclodextrin composite sensor for sensitive and rapid dopamine detection. The sensor combines the biocompatibility and guest-recognition ability of β-cyclodextrin with the catalytic efficiency of Ni-MOF, resulting in enhanced electrochemical response. It demonstrates high selectivity, detecting dopamine even in the presence of various interfering substances (KCl, NaCl, fructose, glucose, sodium citrate, L-leucine, alanine, uric acid, ascorbic acid, and oxalic acid), and achieved recovery rates of 94.3%–102.3% in human serum. The sensor also offers a wide linear range (0.7–310 μM) and a detection limit of 0.277 μM.

Mouafo-Tchinda et al. (2023) synthesized an organoclay/Cu-MOF composite. A thin film of the composite was formed on a GCE without the need for any polymer. The composite electrode exhibited electrocatalytic behavior toward deoxyepinephrine (DXEP) and could simultaneously detect three species—DXEP, acetaminophen, and tyrosine—by displaying well-defined oxidation peaks at voltages of 0.2, 0.42, and 0.72 V, respectively. Under optimal conditions, the sensor demonstrated large calibration curves with ranges of 5–138 μM for DXEP, 4–153 μM for acetaminophen, and 1–29.4 μM for tyrosine, with detection limits of 0.4, 0.7, and 0.2 μM, respectively.

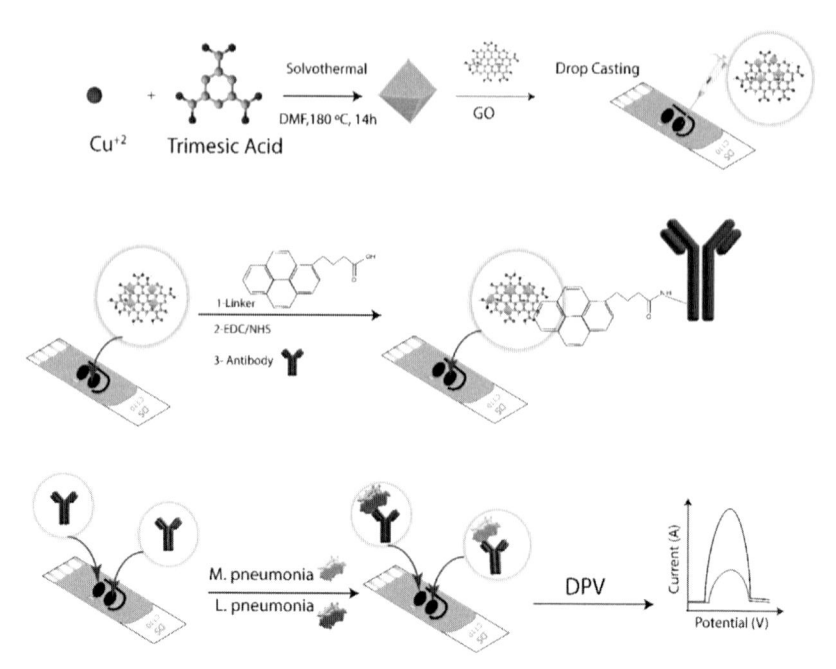

FIGURE 17.34 Preparation method of the GO/Cu-MOF composite and the fabrication of an immunosensor for the detection of target antigens using differential pulse voltammetry (Pandiyaraj et al., 2024).

Pandiyaraj et al. (2024) developed an antibody@GO/Cu-MOF nanocomposite dual immunosensor for the simultaneous detection of biomarkers related to lower respiratory infections. The sensor was created by drop-casting the nanocomposite onto a screen-printed electrode and functionalizing it with a pyrene linker. Monoclonal antibodies for Mycoplasma pneumoniae and Legionella pneumophila were covalently immobilized using EDC-NHS chemistry. Fig. 17.34 showed the method of synthesis and fabrication of the immunosensor. The sensor detected these biomarkers via differential pulse voltammetry (DPV), showing a strong correlation across a wide concentration range (1 pg/mL–100 ng/mL) and high selectivity for antigens of different respiratory pathogens. It demonstrated good recovery in spiked water samples (95%–105%). The high sensitivity of the immunosensor is attributed to the enhanced electrocatalytic properties, stability, and conductivity of the GO-MOF composite, as well as the synergistic interaction between GO and MOF, which improved redox behavior, stability, and conductivity. Proper surface functionalization of the composite allowed for efficient antibody immobilization, facilitating the strong formation of immune complexes with target antigens. This immunosensor offers rapid analytical response, simplicity in fabrication, and portability, making it a viable platform for on-field monitoring of pathogens in environmental samples.

Conclusions

MOF-containing composites possess unique properties derived from the combination of two or more components with different characteristics, utilizing their synergistic effects to overcome the limitations of MOFs in sensing applications. In vapochromic optical sensors, the composite components participate in the generation or suppression of color, similar to an artificial enzyme, to create a color change signal. These reactions can be customized by incorporating specific aptamers. Dye@MOF composites have also been widely used in the construction of colorimetric sensor arrays. The stabilization of MOF composites on textiles and paper strips is one of the strategies for using these types of sensors. Composites containing LMOFs often exhibit better chemical and mechanical stability, with higher quantum yields in luminescent compounds. Most luminescent MOF composite sensors are of the ratiometric fluorescence type, where various components are added besides the LMOF to produce two or more emission lines within the composite. The interaction with analytes causes changes in the intensity of these lines, providing a self-calibration feature and enhancing the sensitivity and accuracy of measurements. In this context, luminescent compounds such as dyes, CDs, and ions like Eu^{3+}, have been employed. MOF composites have been used at the ends of optical fibers to enhance RI changes due to interactions with analytes, contributing to the creation of highly sensitive sensors. Examples include MOF/PDMS and enzyme@MOF composites.

MOF@Au or MOF@Ag composites have been utilized in SERS sensors, aiding in analyte trapping at hotspots to produce highly sensitive sensors. Mechanical sensors like QCM have been developed using MOF composites, where various components such as CNTs, conducting polymers, ILs, aptamers, and MIPs have been added to enhance mechanical and chemical stability, longevity, and selectivity. In electrical and electrochemical sensors, to address the stability issues and lack of electrical conductivity in MOFs, their composites with various materials such as NF, COFs, CNTs, Mxenes, GO, nanoparticles like Au, Cu, ZnO, SnS_2, or semiconducting compounds like TTF, conducting polymers, and ILs have greatly contributed to the development of more sensitive sensors. The selectivity of electrochemical sensors has been improved with MOF composites containing MIPs, aptamers, antibodies, and cyclodextrins. Therefore, it can be said that MOF-containing composites, while leveraging the advantages and strengths of MOFs, also overcome their limitations, resulting in more sensitive, stable, selective, and versatile sensors. The synergistic interactions between MOFs and other composite components are key to achieving the desired performance in advanced sensing applications.

AI disclosure

During the preparation of this work, the author(s) used ChatGPT in order to In the preparation of this chapter, ChatGPT has been used to improve the language. After using this tool/service, the author(s) reviewed and edited the content as needed and take(s) full responsibility for the content of the publication.

References

Abdelhamid, H. N. (2023). MOFTextile: Metal-organic frameworks nanosheets incorporated cotton textile for selective vapochromic sensing and capture of pyridine. *Applied Organometallic Chemistry, 37*(5), e7078. https://doi.org/10.1002/aoc.7078, http://onlinelibrary.wiley.com/journal/10.1002/(ISSN)1099-0739.

Abuzalat, O., Wong, D., Park, S. S., & Kim, S. (2019). High-performance, room temperature hydrogen sensing with a Cu-BTC/polyaniline nanocomposite film on a quartz crystal microbalance. *IEEE Sensors Journal, 19*(13), 4789–4795. https://doi.org/10.1109/JSEN.2019.2904870.

Aghakhani, A., Mohamadi, F., & Ghadimi, J. (2019). Novel alcohol vapour sensor based on the mixed-ligand modified MOF-199 coated quartz crystal microbalance. *International Journal of Environmental Analytical Chemistry,* 1–18.

Allendorf, M. D., Bauer, C. A., Bhakta, R. K., & Houk, R. J. T. (2009). Luminescent metal-organic frameworks. *Chemical Society Reviews, 38*(5), 1330–1352. https://doi.org/10.1039/b802352m.

Balasingam, J. A., Swaminathan, S., Nazemi, H., Love, C., Birjis, Y., & Emadi, A. (2023). *Chemical sensors: Acoustic gas sensors.* Elsevier BV, 209–225. https://doi.org/10.1016/b978-0-12-822548-6.00001-7.

Cao, R., Ding, H., Kim, K.-J., Peng, Z., Wu, J., Culp, J. T., Ohodnicki, P. R., Beckman, E., & Chen, K. P. (2020). Metal-organic framework functionalized polymer coating for fiber optical methane sensors. *Sensors and Actuators B: Chemical, 324*, 128627. https://doi.org/10.1016/j. snb.2020.128627, https://www.sciencedirect.com/science/article/pii/S0925400520309734.

Chappanda, K. N., Shekhah, O., Yassine, O., Patole, S. P., Eddaoudi, M., & Salama, K. N. (2018). The quest for highly sensitive QCM humidity sensors: The coating of CNT/MOF composite sensing films as case study. *Sensors and Actuators, B: Chemical, 257*, 609–619. https://doi.org/10.1016/j.snb.2017.10.189.

Chen, C., Ren, J., Zhao, P., Zhang, J., Hu, Y., & Fei, J. (2023). A novel dopamine electrochemical sensor based on a β-cyclodextrin/Ni-MOF/glassy carbon electrode. *Microchemical Journal, 194*, 109328. https://doi.org/10.1016/j.microc.2023.109328, https://www.sciencedirect.com/science/article/pii/S0026265X23009475.

Chidambaram, A., & Stylianou, K. C. (2018). Electronic metal-organic framework sensors. *Inorganic Chemistry Frontiers, 5*(5), 979–998. https://doi.org/10.1039/c7qi00815e, http://pubs.rsc.org/en/journals/journal/qi.

Dong, Z. P., Zhao, J. J., Liu, P. Y., Liu, Z. L., & Wang, Y. Q. (2019). A metal-organic framework constructed by a viologen-derived ligand: Photochromism and discernible detection of volatile amine vapors. *New Journal of Chemistry, 43*(23), 9032–9038. https://doi.org/10.1039/c9nj01380f, http://pubs.rsc.org/en/journals/journal/nj.

Du, T., Huang, L., Wang, J., Sun, J., Zhang, W., & Wang, J. (2021). Luminescent metal-organic frameworks (LMOFs): An emerging sensing platform for food quality and safety control. *Trends in Food Science and Technology, 111*, 716–730. https://doi.org/10.1016/j.tifs.2021.03.013, http://www.elsevier.com/wps/find/journaldescription.cws_home/601278/description#description.

Elsherif, M., Salih, A. E., Muñoz, M. G., Alam, F., AlQattan, B., Antonysamy, D. S., Zaki, M. F., Yetisen, A. K., Park, S., Wilkinson, T. D., & Butt, H. (2022). Optical fiber sensors: Working principle, applications, and limitations. *Advanced Photonics Research, 3*(11), 2100371. https://doi.org/10.1002/adpr.202100371, https://onlinelibrary.wiley.com/journal/26999293.

Gai, S., Zhang, J., Fan, R., Xing, K., Chen, W., Zhu, K., Zheng, X., Wang, P., Fang, X., & Yang, Y. (2020). Highly stable zinc-based metal-organic frameworks and corresponding flexible composites for removal and detection of antibiotics in water. *ACS Applied Materials and Interfaces, 12*(7), 8650–8662. https://doi.org/10.1021/acsami.9b19583, http://pubs.acs.org/journal/aamick.

Garg, G., Garg, N., Deep, A., & Soni, D. (2023). Zr-MOF and PEDOT:PSS composite sensor for chemoresistive sensing of toluene at room temperature. *Journal of Alloys and Compounds, 956*, 170309. https://doi.org/10.1016/j.jallcom.2023.170309, https://www.sciencedirect.com/science/article/pii/S0925838823016122.

Ghanbarian, M., Zeinali, S., & Mostafavi, A. (2018). A novel MIL-53(Cr-Fe)/Ag/CNT nanocomposite based resistive sensor for sensing of volatile organic compounds. *Sensors and Actuators B: Chemical, 267*, 381–391. https://doi.org/10.1016/j.snb.2018.02.138, https://www.sciencedirect.com/science/article/pii/S0925400518304179.

Gonçalves, B. F., Fernández, E., Valverde, A., Gaboardi, M., Salazar, H., Petrenko, V., Porro, J. M., Cavalcanti, L. P., Urtiaga, K., Esperança, J. M. S. S., Correia, D. M., Fernandez-Alonso, F., Lanceros-Mendez, S., & Fernández de Luis, R. (2024). Exploring the compositional space of a metal-organic framework with ionic liquids to develop porous ionic conductors for enhanced signal and selectivity in VOC capacitive sensors. *Journal of Materials Chemistry A, 12*(24), 14595–14607. https://doi.org/10.1039/d4ta00959b, http://pubs.rsc.org/en/journals/journal/ta.

Guo, X., Zhu, N., Lou, Y., Ren, S., Pang, S., He, Y., Chen, X. B., Shi, Z., & Feng, S. (2020). A stable nanoscaled Zr-MOF for the detection of toxic mycotoxin through a pH-modulated ratiometric luminescent switch. *Chemical Communications, 56*(40), 5389–5392. https://doi.org/10.1039/d0cc01006e, http://pubs.rsc.org/en/journals/journal/cc.

Hajivand, P., Carolus Jansen, J., Pardo, E., Armentano, D., Mastropietro, T. F., & Azadmehr, A. (2024). Application of metal-organic frameworks for sensing of VOCs and other volatile biomarkers. *Coordination Chemistry Reviews, 501*. https://doi.org/10.1016/j.ccr.2023.215558, http://www.journals.elsevier.com/coordination-chemistry-reviews/.

Haldar, R., Bhattacharyya, S., & Maji, T. K. (2020). Luminescent metal–organic frameworks and their potential applications. *Journal of Chemical Sciences, 132*, 99.

Hashemian, H., Ghaedi, M., Dashtian, K., Mosleh, S., Hajati, S., Razmjoue, D., & Khan, S. (2023). Cellulose acetate/MOF film-based colorimetric ammonia sensor for non-destructive remote monitoring of meat product spoilage. *International Journal of Biological Macromolecules, 249*, 126065. https://doi.org/10.1016/j.ijbiomac.2023.126065, https://www.sciencedirect.com/science/article/pii/S0141813023029604.

He, C., Liu, L., Korposh, S., Correia, R., & Morgan, S. P. (2021). Volatile organic compound vapour measurements using a localised surface plasmon resonance optical fibre sensor decorated with a metal-organic framework. *Sensors, 21*(4), 1420. https://doi.org/10.3390/s21041420.

Hu, J., Chen, S., Liu, Z., Li, J.-R., Huang, J.-H., Jiang, Z., Ou, W., Liao, W.-M., Lu, J., & He, J. (2023). MOF-based colorimetric sensor for rapid and visual readout of trace acetylene. *Journal of Materials Chemistry A, 11*(20), 10577–10583. https://doi.org/10.1039/D3TA00582H, http://doi.org/10.1039/D3TA00582H.

Huo, R., Wang, C., Wang, M. Y., Sun, M. Y., Jiang, S., Xing, Y. H., & Bai, F. Y. (2023). Preparation of naphthalenediimide-decorated electron-deficient photochromic lanthanide (III)-MOF and paper strip as multifunctional recognition and ratiometric luminescent turn-on sensors for amines and pesticides. *Inorganic Chemistry, 62*(17), 6661–6673. https://doi.org/10.1021/acs.inorgchem.3c00144, http://pubs.acs.org/journal/inocaj.

Idili, A., Montón, H., Medina-Sánchez, M., Ibarlucea, B., Cuniberti, G., Schmidt, O. G., Plaxco, K. W., & Parolo, C. (2022). Continuous monitoring of molecular biomarkers in microfluidic devices. *Progress in Molecular Biology and Translational Science, 187*(1), 295–333. https://doi.org/10.1016/bs.pmbts.2021.07.027, http://www.elsevier.com/books/book-series/progress-in-molecular-biology-and-translational-science#.

Jiang, Y., Fang, X., Ni, Y., Huo, J., Wang, Q., Liu, Y., Wang, X., & Ding, B. (2024). Gd-MOF composites luminescent arrays for highly sensitive detection of epileptic drug and biomarkers. *Chemical Engineering Journal, 479*, 147232. https://doi.org/10.1016/j.cej.2023.147232, https://www.sciencedirect.com/science/article/pii/S1385894723059636.

Jing, C., Xing, M., Zhou, Q., Yao, W., Huang, H., Wu, H., Wen, H., Zhou, A., & Zhao, Y. (2024). Enhancing gas adsorption/desorption performance of optical fiber pressure sensor using MOF composite. *Journal of Lightwave Technology, 42*(2), 914–920. https://doi.org/10.1109/JLT.2023.3320169.

Kim, K.-J., Culp, J. T., Ohodnicki, P. R., Cvetic, P. C., Sanguinito, S., Goodman, A. L., & Kwon, H. T. (2019). Alkylamine-integrated metal–organic framework-based waveguide sensors for efficient detection of carbon dioxide from humid gas streams. *ACS Applied Materials & Interfaces Journal, 11*(36), 33489–33496. https://doi.org/10.1021/acsami.9b12052, https://doi.org/10.1021/acsami.9b12052.

Koh, C. S. L., Lee, H. K., Han, X., Sim, H. Y. F., & Ling, X. Y. (2018). Plasmonic nose: Integrating the MOF-enabled molecular preconcentration effect with a plasmonic array for recognition of molecular-level volatile organic compounds. *Chemical Communications,*

54(20), 2546–2549. https://doi.org/10.1039/c8cc00564h, http://pubs.rsc.org/en/journals/journal/cc.

Kreno, L. E., Hupp, J. T., & Van Duyne, R. P. (2010). Metal-organic framework thin film for enhanced localized surface plasmon resonance gas sensing. *Analytical Chemistry, 82*(19), 8042–8046. https://doi.org/10.1021/ac102127p.

Lai, H., Li, G., Xu, F., & Zhang, Z. (2020). Metal-organic frameworks: Opportunities and challenges for surface-enhanced Raman scattering-a review. *Journal of Materials Chemistry C, 8*(9), 2952–2963. https://doi.org/10.1039/d0tc00040j, http://pubs.rsc.org/en/journals/journal/tc.

Lee, B., Bae, J., Go, B., Kim, M. K., & Park, J. (2024). Dual-functional metal-organic framework for chemisorption and colorimetric monitoring of cyanogen chloride. *Chemosphere, 362*. https://doi.org/10.1016/j.chemosphere.2024.142633, https://www.sciencedirect.com/science/journal/00456535.

Li, J., Liu, M., Li, J., & Liu, X. (2023). A MOF-on-MOF composite encapsulating sensitized Tb (III) as a built-in self-calibrating fluorescent platform for selective sensing of F ions. *Talanta, 259*, 124521. https://doi.org/10.1016/j.talanta.2023.124521, https://www.sciencedirect.com/science/article/pii/S0039914023002722.

Li, W.-T., Wang, J.-S., Pang, M., Li, Y., & Ruan, W.-J. (2023). Fluorescent sensor array for tetracyclines discrimination using a single Dye@MOF composite sensor. *Sensors and Actuators B: Chemical, 381*, 133375. https://doi.org/10.1016/j.snb.2023.133375, https://www.sciencedirect.com/science/article/pii/S0925400523000904.

Lin, H., Wang, F., Lin, J., Yang, W., Kang, W., Jiang, H., Adade, S. Y.-S. S., Cai, J., Xue, Z., & Chen, Q. (2023). Detection of wheat toxigenic *Aspergillus flavus* based on nano-composite colorimetric sensing technology. *Food Chemistry, 405*, 134803. https://doi.org/10.1016/j.foodchem.2022.134803, https://www.sciencedirect.com/science/article/pii/S0308814622027650.

Lin, X., Song, D., Shao, T., Xue, T., Hu, W., Jiang, W., Zou, X., & Liu, N. (2024). A multifunctional biosensor via MXene assisted by conductive metal–organic framework for healthcare monitoring. *Advanced Functional Materials, 34*(11), 2311637. https://doi.org/10.1002/adfm.202311637, http://onlinelibrary.wiley.com/journal/10.1002/(ISSN)1616-3028.

Liu, Y., Xie, X. Y., Cheng, C., Shao, Z. S., & Wang, H. S. (2019). Strategies to fabricate metal-organic framework (MOF)-based luminescent sensing platforms. *Journal of Materials Chemistry C, 7*(35), 10743–10763. https://doi.org/10.1039/c9tc03208h, http://pubs.rsc.org/en/journals/journal/tc.

Liu, Y., Chui, K. K., Fang, Y., Wen, S., Zhuo, X., & Wang, J. (2024). Metal-organic framework-enabled trapping of volatile organic compounds into plasmonic nanogaps for surface-enhanced Raman scattering detection. *ACS Nano, 18*(17), 11234–11244. https://doi.org/10.1021/acsnano.4c00208, http://pubs.acs.org/journal/ancac3.

Lu, G., & Hupp, J. T. (2010). Metal-organic frameworks as sensors: A ZIF-8 based fabry-pérot device as a selective sensor for chemical vapors and gases. *Journal of the American Chemical Society, 132*(23), 7832–7833. https://doi.org/10.1021/ja101415b.

Ma, Z.-Z., Wang, Y.-S., Liu, B., Jiao, H., & Xu, L. (2022). A non–enzymatic electrochemical sensor of Cu@Co–MOF composite for glucose detection with high sensitivity and selectivity. *Chemosensors*, 10(10), 416. https://doi.org/10.3390/chemosensors10100416.

More, M. S., Bodkhe, G. A., Singh, F., Dole, B. N., Tsai, M.-L., Hianik, T., & Shirsat, M. D. (2024). Hydrogen sulfide chemiresistive sensor based on swift heavy ion irradiated cerium-based metal–organic framework/graphene oxide composite. *Synthetic Metals, 306*, 117622.

https://doi.org/10.1016/j.synthmet.2024.117622, https://www.sciencedirect.com/science/article/pii/S0379677924000845.
Mouafo-Tchinda, E., Kemmegne-Mbouguen, J. C., Nanseu-Njiki, C. P., Langmi, H. W., Kowenje, C., Musyoka, N. M., & Mokaya, R. (2023). Solvothermal synthesis of organoclay/Cu-MOF composite and its application in film modified GCE for simultaneous electrochemical detection of deoxyepinephrine, acetaminophen and tyrosine. *RSC Advances, 13*(30), 20816–20829. https://doi.org/10.1039/D3RA03850E, https://doi.org/10.1039/D3RA03850E.
Mukundan, G., Ganapathy, N., & Badhulika, S. (2023). ZnO nanoparticles-copper metal-organic framework composite on 3D porous nickel foam: A novel electrochemical sensing platform to detect serotonin in blood serum. *Nanotechnology, 34*(40), 405501. https://doi.org/10.1088/1361-6528/ace368, https://doi.org/10.1088/1361-6528/ace368.
Mumtaz, F., Zhang, B., Subramaniyam, N., Roman, M., Holtmann, P., Hungund, A. P., O'Malley, R., Spudich, T. M., Davis, M., Gerald, R. E., & Huang, J. (2024). Miniature optical fiber fabry-perot interferometer based on a single-crystal metal-organic framework for the detection and quantification of benzene and ethanol at low concentrations in nitrogen gas. *ACS Applied Materials and Interfaces, 16*(10), 13071–13081. https://doi.org/10.1021/acsami.3c18702, http://pubs.acs.org/journal/aamick.
Möslein, A. F., Gutiérrez, M., Titov, K., Donà, L., Civalleri, B., Frogley, M. D., Cinque, G., Rudić, S., & Tan, J. C. (2023). A multimodal study on the unique sensing behavior of a guest@metal-organic framework material for the detection of volatile acetone. *Advanced Materials Interfaces, 10*(3), 2201401. https://doi.org/10.1002/admi.202201401, http://onlinelibrary.wiley.com/journal/10.1002/(ISSN)2196-7350.
Na Songkhla, S., & Nakamoto, T. (2021). Overview of quartz crystal microbalance behavior analysis and measurement. *Chemosensors, 9*(12), 350. https://doi.org/10.3390/chemosensors9120350.
Nie, S., & Emory, S. R. (1997). Probing single molecules and single nanoparticles by surface-enhanced Raman scattering. *Science (New York, N.Y.), 275*(5303), 1102–1106. https://doi.org/10.1126/science.275.5303.1102.
Noorani, N., Mehrdad, A., & Darbandi, M. (2023). CO_2 adsorption on ionic liquid–modified cupper terephthalic acid metal organic framework grown on quartz crystal microbalance electrodes. *Journal of the Taiwan Institute of Chemical Engineers, 145*, 104849. https://doi.org/10.1016/j.jtice.2023.104849.
Olorunyomi, J. F., Sadiq, M. M., Batten, M., Konstas, K., Chen, D., Doherty, C. M., & Caruso, R. A. (2020). Advancing metal-organic frameworks toward smart sensing: enhanced fluorescence by a photonic metal-organic framework for organic vapor sensing. *Advanced Optical Materials, 8*(19), 2000961. https://doi.org/10.1002/adom.202000961, http://onlinelibrary.wiley.com/journal/10.1002/(ISSN)2195-1071.
Pandiyaraj, K., Elkaffas, R. A., Mohideen, M. I. H., & Eissa, S. (2024). Graphene oxide/Cu–MOF-based electrochemical immunosensor for the simultaneous detection of *Mycoplasma pneumoniae* and *Legionella pneumophila* antigens in water. *Scientific Reports, 14*(1), 17172. https://doi.org/10.1038/s41598-024-68231-y.
Qian, K., Deng, Q., Fang, G., Wang, J., Pan, M., Wang, S., & Pu, Y. (2016). Metal-organic frameworks supported surface-imprinted nanoparticles for the sensitive detection of metolcarb. *Biosensors and Bioelectronics, 79*, 359–363. https://doi.org/10.1016/j.bios.2015.12.071, https://www.elsevier.com/locate/bios.
Qiao, X., Su, B., Liu, C., Song, Q., Luo, D., Mo, G., & Wang, T. (2018). Selective surface enhanced raman scattering for quantitative detection of lung cancer biomarkers in

superparticle@MOF structure. *Advanced Materials, 30*(5), 1702275. https://doi.org/10.1002/adma.201702275, http://www3.interscience.wiley.com/journal/119030556/issue.

Rozenberga, L., Bloch, W., Gillam, T. A., Lancaster, D. G., Skinner, W., Krasowska, M., Blencowe, A., & Beattie, D. A. (2024). Europium metal–organic framework polymer composites with enhanced luminescence and selective ferric ion sensing. *ACS Applied Polymer Materials Journal, 6*(6), 3544–3553. https://doi.org/10.1021/acsapm.4c00111.

Safaei, M., Foroughi, M. M., Ebrahimpoor, N., Jahani, S., Omidi, A., & Khatami, M. (2019). A review on metal-organic frameworks: Synthesis and applications. *TrAC – Trends in Analytical Chemistry, 118,* 401–425. https://doi.org/10.1016/j.trac.2019.06.007, https://www.elsevier.com/locate/trac.

Son, J., Ji, S., Kim, S., Kim, S., Kim, S. K., Song, W., Lee, S. S., Lim, J., An, K. S., & Myung, S. (2021). GC-like graphene-coated quartz crystal microbalance sensor with microcolumns. *ACS Applied Materials and Interfaces, 13*(3), 4703–4710. https://doi.org/10.1021/acsami.0c19010, http://pubs.acs.org/journal/aamick.

Strauss, I., Mundstock, A., Treger, M., Lange, K., Hwang, S., Chmelik, C., Rusch, P., Bigall, N. C., Pichler, T., Shiozawa, H., & Caro, J. (2019). Metal–organic framework Co-MOF-74-based host–guest composites for resistive gas sensing. *ACS Applied Materials & Interfaces Journal, 11*(15), 14175–14181. https://doi.org/10.1021/acsami.8b22002, https://doi.org/10.1021/acsami.8b22002.

Sun, M., Zhang, L., Xu, S., Yu, B., Wang, Y., Zhang, L., & Zhang, W. (2022). Carbon dots-decorated hydroxyapatite nanowires–lanthanide metal–organic framework composites as fluorescent sensors for the detection of dopamine. *Analyst, 147*(5), 947–955. https://doi.org/10.1039/D2AN00049K, http://doi.org/10.1039/D2AN00049K.

Sun, R., Han, S., Zong, W., Chu, H., Zhang, X., & Jiang, H. (2024). Ultrasensitive detection of chlortetracycline in animal-origin food using molecularly imprinted electrochemical sensor based on SnS_2/ZnCo-MOF and AuNPs. *Food Chemistry, 452,* 139537. https://doi.org/10.1016/j.foodchem.2024.139537, https://www.sciencedirect.com/science/article/pii/S0308814624011877.

Sun, Z., Liu, Z., Xiao, Y., Gong, J., Shuai, S., Lang, T., Zhao, C., & Shen, C. (2018). Thermal stability of optical fiber metal organic framework based on graphene oxide and nickel and its hydrogen adsorption application. *Optics Express, 26*(24), 31648–31656. https://doi.org/10.1364/OE.26.031648, https://opg.optica.org/oe/abstract.cfm?URI=oe-26-24-31648.

Swager, T. M., & Mirica, K. A. (2019). Introduction: Chemical sensors. *Chemical Reviews, 119*(1), 1–2. https://doi.org/10.1021/acs.chemrev.8b00764, http://pubs.acs.org/journal/chreay.

Vijayaraghavan, P., Wang, Y. Y., Palanisamy, S., Lee, L. Y., Chen, Y. K., Tzou, S. C., Yuan, S. S. F., & Wang, Y. M. (2023). Hierarchical ensembles of FeCo metal-organic frameworks reinforced nickel foam as an impedimetric sensor for detection of IL-1RA in human samples. *Chemical Engineering Journal, 458,* 141444. https://doi.org/10.1016/j.cej.2023.141444, www.elsevier.com/inca/publications/store/6/0/1/2/7/3/index.htt.

Wang, S., Xue, Y., Yu, Z., Huang, F., & Jin, Y. (2023). Layered 2D MOF nanosheets grown on CNTs substrates for efficient nitrite sensing. *Materials Today Chemistry, 30,* 101490. https://doi.org/10.1016/j.mtchem.2023.101490, https://www.sciencedirect.com/science/article/pii/S2468519423001179.

Wang, Y., Zhu, Y., Binyam, A., Liu, M., Wu, Y., & Li, F. (2016). Discovering the enzyme mimetic activity of metal-organic framework (MOF) for label-free and colorimetric sensing of biomolecules. *Biosensors and Bioelectronics, 86,* 432–438. https://doi.org/10.1016/j.bios.2016.06.036, https://www.sciencedirect.com/science/article/pii/S095656631630570X.

Wannapaiboon, S., Tu, M., Sumida, K., Khaletskaya, K., Furukawa, S., Kitagawa, S., & Fischer, R. A. (2015). Hierarchical structuring of metal-organic framework thin-films on quartz crystal microbalance (QCM) substrates for selective adsorption applications. *Journal of Materials Chemistry A, 3*(46), 23385–23394. https://doi.org/10.1039/c5ta05620a, http://pubs.rsc.org/en/journals/journalissues/ta.

Xue, Y., Zheng, S., Xue, H., & Pang, H. (2019). Metal–organic framework composites and their electrochemical applications. *Journal of Materials Chemistry A, 7*(13), 7301–7327. https://doi.org/10.1039/C8TA12178H, http://doi.org/10.1039/C8TA12178H.

Yan, X., Qu, H., Chang, Y., Pang, W., Wang, Y., & Duan, X. (2020). Surface engineering of metal-organic framework prepared on film bulk acoustic resonator for vapor detection. *ACS Applied Materials and Interfaces, 12*(8), 10009–10017. https://doi.org/10.1021/acsami.9b22407, http://pubs.acs.org/journal/aamick.

Yang, L., Liu, Y.-L., Liu, C.-G., Fu, Y., & Ye, F. (2020). A built-in self-calibrating luminescence sensor based on RhB@Zr-MOF for detection of cations, nitro explosives and pesticides. *RSC Advances, 10*(33), 19149–19156. https://doi.org/10.1039/D0RA02843F, http://doi.org/10.1039/D0RA02843F.

Yang, Y., Yang, L., Ma, Y., Wang, X., Zhang, J., Bai, B., Yu, L., Guo, C., Zhang, F., & Qin, S. (2022). A novel metal–organic frameworks composite-based label-free point-of-care quartz crystal microbalance aptasensing platform for tetracycline detection. *Food Chemistry, 392*, 133302.

Yao, C. X., Zhao, N., Liu, J. M., Fang, G. Z., & Wang, S. (2020). Ultra-stable UiO-66 involved molecularly imprinted polymers for specific and sensitive determination of tyramine based on quartz crystal microbalance technology. *Polymers, 12*(2), 281. https://doi.org/10.3390/polym12020281, https://res.mdpi.com/d_attachment/polymers/polymers-12-00281/article_deploy/polymers-12-00281.pdf.

Yin, Y., Gao, C., Xiao, Q., Lin, G., Lin, Z., Cai, Z., & Yang, H. (2016). Protein-metal organic framework hybrid composites with intrinsic peroxidase-like activity as a colorimetric biosensing platform. *ACS Applied Materials & Interfaces Journal, 8*(42), 29052–29061. https://doi.org/10.1021/acsami.6b09893, https://doi.org/10.1021/acsami.6b09893.

Zhao, H., Ni, J., Zhang, J.-J., Liu, S.-Q., Sun, Y.-J., Zhou, H., Li, Y.-Q., & Duan, C.-Y. (2018). A trichromatic MOF composite for multidimensional ratiometric luminescent sensing. *Chemical Science, 9*(11), 2918–2926. https://doi.org/10.1039/C8SC00021B, http://doi.org/10.1039/C8SC00021B.

Zhou, N., Ma, Y., Hu, B., He, L., Wang, S., Zhang, Z., & Lu, S. (2019). Construction of Ce-MOF@COF hybrid nanostructure: Label-free aptasensor for the ultrasensitive detection of oxytetracycline residues in aqueous solution environments. *Biosensors and Bioelectronics, 127*, 92–100. https://doi.org/10.1016/j.bios.2018.12.024, www.elsevier.com/locate/bios.

Zhu, C., Perman, J. A., Gerald, R. E., Ma, S., & Huang, J. (2019). Chemical detection using a metal-organic framework single crystal coupled to an optical fiber. *ACS Applied Materials and Interfaces, 11*(4), 4393–4398. https://doi.org/10.1021/acsami.8b19775, http://pubs.acs.org/journal/aamick.

Zhu, C., Gerald, R. E., & Huang, J. (2021). Metal-organic framework materials coupled to optical fibers for chemical sensing: A review. *IEEE Sensors Journal, 21*(18), 19647–19661. https://doi.org/10.1109/JSEN.2021.3094092, http://ieeexplore.ieee.org/xpl/RecentIssue.jsp?punumber=7361.

Zhu, G., Cheng, L., Qi, R., Zhang, M., Zhao, J., Zhu, L., & Dong, M. (2019). A metal-organic zeolitic framework with immobilized urease for use in a tapered optical fiber urea biosensor.

Microchimica Acta, 187(1), 72. https://doi.org/10.1007/s00604-019-4026-0, https://doi.org/10.1007/s00604-019-4026-0.

Zhu, G., Zhang, M., Lu, L., Lou, X., Dong, M., & Zhu, L. (2019). Metal-organic framework/enzyme coated optical fibers as waveguide-based biosensors. *Sensors and Actuators B: Chemical, 288*, 12–19. https://doi.org/10.1016/j.snb.2019.02.083, https://www.sciencedirect.com/science/article/pii/S0925400519302977.

Zhu, Q.-L., & Xu, Q. (2014). Metal–organic framework composites. *Chemical Society Reviews, 43*(16), 5468–5512. https://doi.org/10.1039/C3CS60472A, http://doi.org/10.1039/C3CS60472A.

Chapter 18

Metal-organic framework composites for enhancing fire safety of polymers

Hafezeh Nabipour, and Sohrab Rohani
Department of Chemical and Biochemical Engineering, University of Western Ontario, London, ON, Canada

Introduction

Polymers, which are fundamental to our modern lifestyle, play an integral role in numerous aspects of daily existence. These macromolecules, composed of repeating monomeric units, are categorized into two main groups: naturally occurring and synthetic polymers. Nature provides a rich variety of natural polymers such as silk, cotton, wool, and wood, valued for their longstanding contributions to clothing, textiles, and construction. Their inherent strengths, flexibility, and biodegradability make them invaluable resources. In addition to nature's offerings, synthetic polymers, products of human ingenuity primarily derived from petroleum oils, complement the spectrum. Materials like epoxy resin, polypropylene, polyurethanes, polyester, polyvinyl chloride, and polyethylene have revolutionized industries, becoming deeply ingrained in our daily routines through applications in cushions, clothing, construction, automotive sectors, and beyond.

Polymers are often considered highly combustible due to their molecular structure and composition. Many polymers contain carbon and hydrogen atoms bonded together in long chains, which can readily undergo combustion when exposed to heat or flame. The widespread use of highly flammable polymers unknowingly contributes to a higher proportion of fire hazards within households. This acceptance inadvertently raises the risk of fire incidents due to the combustible nature of these materials (Sharma et al., 2024; Wen et al., 2023). In a typical fire incident, several key factors must be present: heat, fuel, and sufficient oxygen levels, as illustrated in Fig. 18.1 (Laoutid et al., 2009). As the fire begins, the temperature of the polymers undergoing combustion steadily rises to a point where chemical bonds within the polymer chains start to break down. This breakdown initiates primarily on the surface of the polymer,

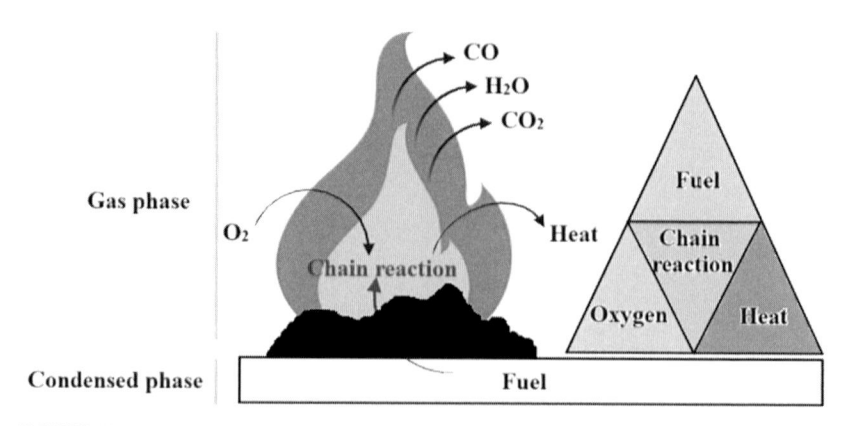

FIGURE 18.1 Schematic diagram of the combustion mechanism of polymer materials.

resulting in the formation of a carbonaceous layer, often referred to as char. The char layer can have varying effects depending on its composition: it may either catalyze further decomposition reactions, leading to the production of additional volatile compounds, or it may act as a barrier, slowing down the overall combustion process. Simultaneously, in the gas phase, the degraded polymer molecules react with oxygen in the surrounding atmosphere, yielding free radical species such as H· and OH·. These radicals subsequently combine with atmospheric oxygen to form more combustible mixtures, effectively serving as additional fuel sources that sustain the combustion process and generate additional heat. Over time, as the fire progresses, the heat released by the burning polymers continues to increase until it reaches a critical threshold. At this point, the polymers are capable of sustaining the burning process even in the absence of an external heat source. This self-sustaining combustion process is made possible by the continuous supply of fuel, heat, and oxidizing agents. Thus a complex interplay of chemical reactions in both the condensed and gas phases contributes to the progression and persistence of a fire incident involving polymers. Understanding these mechanisms is crucial for developing effective fire prevention and mitigation strategies in various settings (Laoutid et al., 2009; Lyon & Janssens, 2015).

It is crucial to understand that when polymers burn, they do not just produce flames and heat; they also release toxic gases into the air, which can cause significant harm to individuals, further complicating the situation. Fire gas toxicity, therefore, becomes a critical consideration when assessing the hazards associated with a fire incident. When polymers combust, they can emit a range of hazardous compounds, including isocyanates, nitrogen oxides (NO_x), dioxins and furans, carbon monoxide (CO), polycyclic aromatic hydrocarbons (PAHs), carbon dioxide, halogen halides, organic irritants, and soot. The specific compounds emitted and their concentrations depend on factors such as the polymer type, combustion conditions (temperature, oxygen availability), and the presence of other materials. If these toxic gases are inhaled, they can lead to immediate symptoms in victims, such as coughing, blurred vision, drowsiness, loss of muscle control, loss of consciousness, wheezing, coughing up blood, difficulty in breathing, and dizziness, exacerbating the dangers posed by the fire itself (Lazar et al., 2020; Stec et al., 2008; Wen et al., 2023; Wu et al., 2021).

The release of combustion gases during fires not only poses immediate risks to individuals but also has broader environmental implications, contributing to air pollution and irreversible damage to air quality. To mitigate these impacts, flame-retardant polymers can be employed in residential settings to suppress or minimize the spread of fires effectively. Flame retardants, when incorporated into polymeric materials, can reduce the likelihood of fire events or the rapid spread of flames through various mechanisms, which will be detailed further in subsequent sections. Therefore the development of flame-retardant materials goes beyond merely preventing fires; it also aims to address the secondary damages caused by toxic gases and smoke production during fire events. Specifically, flame-retardant materials with the ability to suppress smoke production or selectively adsorb toxic volatile compounds could significantly mitigate the environmental and health consequences of fires. By focusing on the development and implementation of these advanced materials, we can not only enhance fire safety but also reduce the long-term environmental impact of residential fires, promoting sustainable and resilient communities.

Flame retardants and mechanism of flame retardancy

One effective approach is the incorporation of flame retardants into polymeric materials, resulting in the creation of flame-retardant polymers. These specialized polymers are designed to reduce the likelihood of fire events or limit the spread of flames through various mechanisms, which will be discussed in detail later. Flame retardants can act in different ways to inhibit or slow down the combustion process. The major mechanisms include:

Gas phase mechanism

The gas phase mechanism is a vital strategy employed by certain flame retardants to impede or extinguish combustion in polymeric materials. It involves releasing inert gases or radicals that disrupt the combustion process, hindering flame propagation effectively. Inert gases like nitrogen or carbon dioxide dilute combustible gases and oxygen, starving the fire of essential elements for combustion. Melamine-based flame retardants, such as melamine polyphosphate, release a significant amount of nitrogen gas during thermal decomposition. This nitrogen gas dilutes the combustible gases and oxygen, effectively suppressing the flame propagation. Halogenated flame retardants, such as brominated or chlorinated compounds, release hydrogen halides (e.g., HBr, HCl) during combustion. These gases act as radical traps, interfering with the free-radical chain reactions involved in the combustion process, thereby inhibiting flame propagation. Factors like flame retardant type, concentration, polymer matrix, and combustion conditions influence effectiveness. While effective, gas phase mechanisms may produce harmful byproducts, requiring careful consideration for environmental and health impacts (Chattopadhyay & Raju, 2007; Qin et al., 2022).

Solid phase mechanism

The solid phase mechanism is a critical strategy utilized by certain flame retardants to hinder or suppress combustion in polymeric materials. This mechanism involves the creation of a protective char layer on the polymer surface, serving as a barrier to insulate the material from heat and oxygen, thus preventing further combustion. During combustion, specific flame retardants facilitate the formation of a carbonaceous, porous char layer through thermal decomposition and condensation reactions. This char layer acts as a physical barrier, effectively separating the polymer from heat and oxygen sources essential for combustion. For instance, intumescent flame retardants, such as ammonium polyphosphate (APP) and pentaerythritol (PER), undergo reactions leading to a thick, insulating char layer formation, shielding the material from further combustion. Essential characteristics of an effective char layer include thermal insulation, oxygen barrier properties, structural integrity, and adherence to the polymer surface. The effectiveness of the solid phase mechanism relies on various factors, such as flame retardant type and concentration, polymer matrix, and combustion conditions. Often, a combination of solid phase and other mechanisms is employed to achieve optimal flame retardancy (Bras et al., 1999; Velencoso et al., 2018).

The development of effective and environmentally friendly flame retardants for polymers has become a significant research focus due to the growing demand for fire safety and sustainability. While certain halogenated flame retardants were initially found to significantly enhance the flame retardancy of polymers, their use was subsequently discontinued due to their detrimental impact on the environment and human health. Concerns over the potential release of toxic substances, such as dioxins and furans, during the combustion or disposal of halogenated flame retardants, as well as their persistence and bioaccumulation in the environment, have led to increased scrutiny and eventual phase-out of these compounds. Additionally, studies have linked exposure to certain halogenated flame retardants to adverse health effects, including endocrine disruption, neurodevelopmental issues, and potential carcinogenic effects. Phosphorous-based flame retardants, for instance, have shown efficacy in reducing flammability, but their integration into polymer systems can lead to issues such as poor dispersion and compatibility. Similarly, nitrogen-based flame retardants, which often rely on the formation of char layers to inhibit combustion, may face challenges in achieving uniform distribution within polymer matrices, impacting their overall effectiveness. Metal oxide variants, including compounds like alumina and zinc oxide, have also been explored for their flame-retardant properties. However, their incorporation into polymers can pose challenges due to issues such as phase separation or adverse effects on mechanical properties (Bellayer et al., 2018; Yang et al., 2019; Zhang et al., 2020).

Metal-organic framework composites as flame retardants

In recent times, a new material known as metal-organic frameworks (MOFs) has demonstrated significant potential in overcoming the limitations of conventional flame retardants (Nabipour et al., 2020; Nabipour et al., 2022). MOFs are innovative composite materials characterized by their periodic network structures, which are created through coordination bonds between metal ions or clusters and organic ligands (Liu et al., 2018). As a novel type of porous material, MOFs have unique properties in supercapicitors (Gao et al., 2021), adsorption (Ghanbarian et al., 2020), sensing and detection (Hosseini & Zeinali, 2019), gas storage (Li et al., 2019), catalysis (Konnerth et al., 2020), fuel cells (Ponnada et al., 2021), batteries (Mehek et al., 2021), biological applications (Zhao et al., 2023), and drug delivery systems (Nabipour et al., 2023, 2024). In particular, MOFs are effective for fire safety (Nabipour et al., 2020, 2022) due to their freely adjustable pore size, superior specific surface area, and good porosity, customizable size, outstanding thermal and mechanical stability, and great chemical stability (Liu et al., 2023). MOFs combine with other materials to form composite materials. These MOFs-based composites exhibit unique nanostructures and integrated properties, playing a significant role in flame-retardant polymeric materials. The exceptional properties of MOF composites position them as highly attractive candidates for the development of advanced flame retardant systems. Unlike traditional inorganic flame retardants, the organic components within MOF composites enhance compatibility with polymer chains, reducing the need for additional organic modifications. Moreover, the organic structures of MOF composites lend themselves well to the incorporation of fire-resistant groups, such as phosphorus, nitrogen, and aromatic derivatives, through tailored modifications. Apart from the organic parts, transition metals inside MOF composites act as strong Lewis acids, showing impressive catalytic capabilities. These metals play a crucial role in reducing harmful flue gas emissions by facilitating the conversion of volatile organic compounds into less harmful substances. Moreover, during combustion events, they contribute to the formation of a protective char layer from the organic components, thereby impeding the spread of flames and reducing fire hazards. Additionally, metal species derived from MOF composites, such as metal hydroxides (copper oxide, cobalt oxide, cobalt monoxide), metal hydroxides (beta-iron hydroxide, aluminum hydroxide, magnesium hydroxide), and transition metal dichalcogenides, exhibit exceptional catalytic activity. These metal oxides and transition metal dichalcogenides demonstrate remarkable effectiveness in catalyzing reactions that suppress smog formation or reduce carbon monoxide levels, thereby further bolstering the fire safety properties of MOF-based flame retardant systems (Cao et al., 2024; Song et al., 2023; Wang et al., 2013). Therefore the catalytic capabilities of transition metals within MOF composites not only enhance fire safety but also contribute to environmental sustainability by reducing emissions of harmful pollutants. As such, the incorporation of these metal species into MOF composites-based flame retardant formulations represents a promising avenue for advancing fire safety technology while simultaneously mitigating environmental impacts.

This chapter systematically examines the effectiveness of MOF composites in enhancing the fire safety of various polymers, including epoxy resin (EP), thermoplastic polyurethane (TPU), rigid polyurethane foam (RPUF), flexible polyurethane foam (FPUF), polylactic acid (PLA), unsaturated polyester resin (UPR), polyurethane elastomer (PUE), polystyrene (PS), polyurea (PUA), ethylene vinyl acetate (EVA), polyamide 6 (PA6), polybutylene succinate (PBS), and polycarbonate (PC). It discusses the different flame-retardant mechanisms utilized by MOF composites and their impact on reducing parameters such as peak heat release rate (PHRR), total heat release (THR), peak CO production rate (PCO) and smoke factor (SF), smoke production rate (SPR), and total smoke release (TSR) during fires. Parameters such as UL-94 vertical burning test ratings, limiting oxygen index (LOI) test results, and char residue yields are considered when evaluating the performance of MOF composites.

Flame retardant mechanism of metal-organic framework composites

MOF composites demonstrate remarkable flame retardant and smoke suppression abilities due to their unique chemical compositions and thermal decomposition mechanisms. As shown in Fig. 18.2, when exposed to heat, these composites undergo several processes that contribute to their outstanding fire-resistant properties. Upon thermal decomposition, MOF composites form metal oxide particles that cover the polymer matrix's surface. This layer acts as a physical barrier, shielding the underlying polymer from further burning while efficiently adsorbing gases and smoke. The metal oxide layer serves as a heat shield, preventing heat and oxygen transfer to the polymer, thereby inhibiting combustion. Some MOFs release nonflammable gases, such as ammonia (NH_3), during decomposition. These gases dilute the concentration of flammable gases in the surrounding environment, effectively reducing the available fuel for combustion. By lowering the concentration of flammable

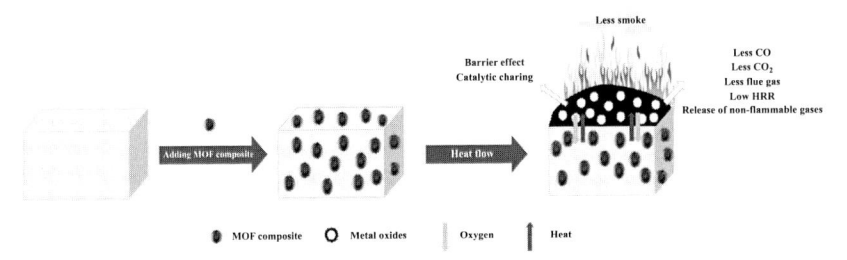

FIGURE 18.2 Mechanism of flame retardancy in MOF composites.

gases, the heat release rate is significantly reduced, enhancing the flame retardancy of the MOF composites-polymer composite. The thermal decomposition of MOF composites also promotes the formation of an expanded carbonaceous char or coating layer on the polymer surface. This char layer acts as an insulating barrier, limiting the exposure of the bulk polymer to air and preventing further oxidation and combustion. The char layer effectively shields the underlying polymer from heat and oxygen, decreasing the heat release rate and smoke formation. The formation of a protective metal oxide layer, the release of nonflammable gases, and the creation of an insulating char layer work in synergy to provide exceptional flame retardancy and smoke suppression properties to MOF composites and polymer composites. By reducing the available fuel, limiting heat and oxygen transfer, and creating physical barriers, MOF composites effectively inhibit combustion and minimize the risk of fire propagation. These unique characteristics make MOF composites highly attractive as flame retardant additives in polymer materials, offering enhanced fire safety and improved smoke suppression capabilities (Hou et al., 2017; Qi et al., 2019; Shi et al., 2017).

Metal-organic framework composites applied in fire retardancy of epoxy resin

EP, known for its superb adhesion, corrosion resistance, electrical insulation, and impressive mechanical properties, is a thermosetting polymer material extensively employed across diverse industrial sectors like adhesives, coatings, laminates, and semiconductor packaging materials. Despite its versatile applications, EP exhibits inadequate flame retardancy and susceptibility to combustion, presenting safety risks and constraining its usage potential (Xiong et al., 2013; Xu et al., 2022). Additionally, it has been noted that MOF composites with distinctive organic–inorganic hybrid structures can strengthen the cohesive forces between fillers and polymer chains, thereby improving the compatibility of nanofillers. Moreover, they have demonstrated flame-retardant properties for EP composites (Deng et al., 2023). In a research paper, Gong et al. (2022) prepared the leaf-shaped MXene-based flame retardant MXene@Bi-MOF and incorporated it into EP resin. The inclusion of the Bi-MOF component enhances the dispersion of MXene within the EP matrix by acting as a surface modifier. It also functions as a synergist to boost the flame retardancy and smoke suppression capabilities of MXene, thereby improving the thermal properties of the composites and reducing the maximum mass loss rate by 27.7% compared to pure EP. EP composites with just 2 wt.% MXene@Bi-MOF exhibited significantly enhanced fire safety, reducing the PHRR by 28.8%, PSPR by 45.3%, THR by 36.5%, PCO by 30.7%, and SF by 55.3% (Fig. 18.3). The synergistic effects between MXene and Bi-MOF, including the barrier effect of MXene layers, catalytic char formation, radical quenching by metal oxides, and gas dilution by Bi-MOF decomposition products, were responsible for the improved flame retardancy.

In another research, a novel flame retardant consisting of Fe-Co bimetallic MOF grown on hexagonal boron nitride (h-BN) sheets, creating Fe-Co MOF@BN hybrids, was used. These hybrids were then incorporated into EP composites along with APP. The release and patterns of gases produced during the burning of various composites, including pure EP, EP/APP, EP/APP/Fe-Co MOF, and EP/APP/Fe-Co MOF@BN, were examined

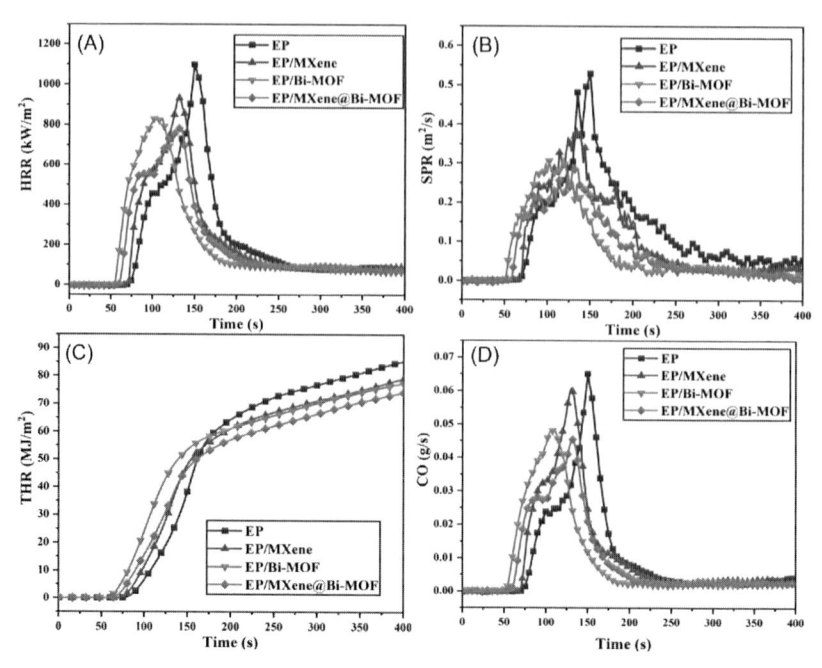

FIGURE 18.3 (A) PHRR, (B) THR, (C) SPR, (D) CO production rate curves from cone calorimeter tests of EP composites (Gong et al., 2022).

using thermogravimetric analysis-infrared spectrometry (TG-IR). According to the findings presented in Fig. 18.4A–D, the presence of these flame retardants significantly reduced the emission of gases. Cone calorimetry tests (CCT) showed a remarkable reduction in PHRR by 73.7%, COP by 70%, and PSPR by 71.4% in the EP/APP/Fe-Co MOF@BN composites compared to the standard EP. Additionally, an increase in the LOI to 30.2% was observed. Fig. 18.4E highlighted how the layered structure of Fe-Co MOF@BN, along with the catalytic effects of char formation, products from APP decomposition, improved cross-linking of the char layer, and synergistically enhanced the flame retardancy. Moreover, the Fe-Co MOF@BN enhanced the distribution of APP within the EP, expanding the potential uses of MOFs and h-BN. Overall, the EP/APP/Fe-Co MOF@BN composites exhibited enhanced fire safety, reduced smoke production, and provided excellent mechanical properties (Song et al., 2024).

Lv et al. (2020) investigated the use of ZIF-8@PZN, a composite material synthesized from zeolitic imidazolate framework (ZIF-8) and phosphazine (PZN), as an enhancer in an EP/APP system. Their findings revealed several improvements in fire retardancy. Substituting 3 wt.% of ZIF-8@PZN for APP in the EP system increased the LOI value of the EP by 4%, indicating better flame resistance. The composite also exhibited a 7.9% reduction in the CO emission rate compared to pure EP. The PSPR decreased significantly from 48.4% for pure EP to 72.6% for the composite, demonstrating enhanced smoke suppression. Additionally, the reduction in PHRR increased from 77.5% for pure EP to 80.8% for the composite, indicating better heat release control. These results highlight the strong synergistic effect between ZIF-8@PZN and APP in enhancing the fire retardancy of EP composites, making the material more fire-safe and environmentally friendly.

Hou and coworkers developed a novel hybrid flame retardant, termed MPOFs-P, which integrates polyhedral oligomeric silsesquioxane (POSS), phosphorus-containing compounds, and transition metals (Fig. 18.5A). This innovative approach combines multiple flame-retardant components into a single multifunctional material. The incorporation of just 2 wt.% of MPOFs-P into the composite material resulted in remarkable improvements in fire safety performance. Notably, the LOI value of the composite increased to 27.0%, achieving the stringent UL-94 V-0 rating (Fig. 18.5B). This rating is a widely recognized standard for assessing the flammability of polymeric materials, and the UL-94 V-0 classification indicates superior flame retardancy. In addition to the enhanced LOI value, the MPOFs-P hybrid flame retardant also demonstrated significant reductions in various combustion parameters. Compared to the unmodified composite, the PHRR decreased by 46.6%, the TSP reduced by 25.2%, and the emission of toxic CO decreased by 39.8% (Hou et al., 2022).

Huang et al. designed a novel MOF composite by combining NH_2-MIL-101(Al) MOF with a phosphorus-containing nitrogen ionic liquid ([DPP-NC(3)][PMO]). The ionic liquid, comprising an imidazole cation modified

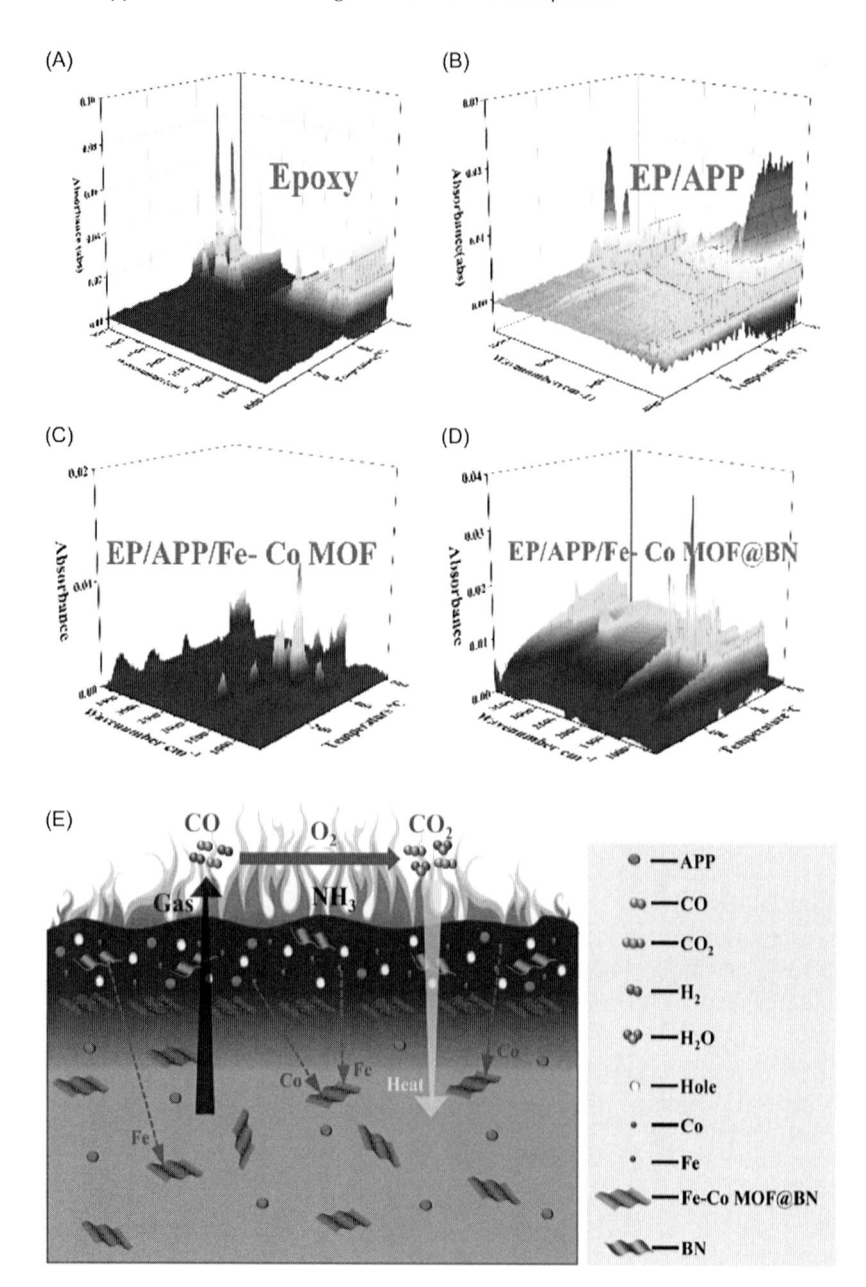

FIGURE 18.4 3D-TG-IR plots of (A) EP, (B) EP/APP, (C) EP/APP/Fe-Co MOF, (D) EP/APP/Fe-Co MOF@BN composites. (E) Depiction of the flame retardancy and smoke suppression mechanisms of Fe-Co MOF@BN and APP in EP (Song et al., 2024).

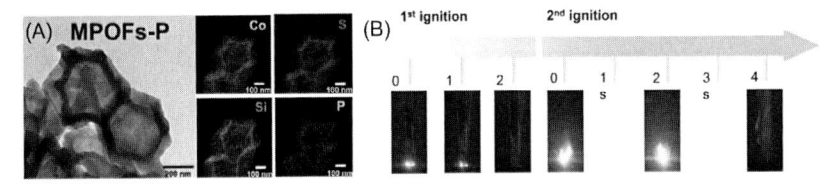

FIGURE 18.5 (A) TEM image and elemental mapping of MPOFs-P; (B) video screenshots of the EP/MPOFs-P sample bar in the UL-94 test (Hou et al., 2022).

with diphenylphosphinic group and phosphomolybdic acid anions, effectively traps degrading polymer radicals and reduces smoke emissions, while the MOF prevents ionic liquid agglomeration and enhances compatibility with EP. The composite demonstrated significant improvements in fire safety at a low addition of 3 wt.%, increasing the LOI value to 29.8% and decreasing HRR by 51.2%, SPR by 37.8%, and CO release rate by 44.8%. The enhanced performance is attributed to the combined effects of the ionic liquid in trapping radicals and promoting char formation, and the MOF framework acting as a heat and combustible material barrier (Huang et al., 2020).

Therefore, recent advancements in MOF composites have greatly improved the flame retardancy of EP materials. Incorporating hybrids like MXene@Bi-MOF, Fe-Co MOF@BN, and ZIF-8@PZN into EP enhances thermal properties, reduces flammability, and improves nanofiller dispersion. These enhancements lead to significant reductions in PHRR, THR, and smoke production, making EP composites safer and more effective for industrial applications.

Metal-organic framework composites applied in fire retardancy of polyurethanes

Polyurethanes are versatile polymers extensively used in foams, coatings, adhesives, and elastomers due to their flexibility, durability, and insulation properties. Despite these advantages, polyurethanes are highly flammable, igniting easily and releasing substantial heat and smoke, which poses a fire hazard. Their flammability is due to their chemical composition and thermal decomposition behavior. To enhance their fire resistance and comply with safety standards, MOF composites have been used as flame retardants (Nik Pauzi et al., 2014; Wang et al., 2019). For example, Shi et al. (2022) developed a novel flame-retardant hybrid material by combining Co-MOF with two-dimensional MXene nanosheets using a solvothermal method. These Co-MOF@MXene hybrids were then integrated into TPU as flame retardants. Experimental findings demonstrated that incorporating MXene into Co-MOF reduced its tendency to agglomerate, thereby significantly enhancing its compatibility with the TPU matrix and improving its dispersion. Thermal stability was assessed through TGA. When analyzing the composites in a nitrogen atmosphere, it was observed that the presence of both Co-MOFs and Co-MOF@MXene hybrids caused premature decomposition of TPU at lower temperatures, although the char residues of the composites were augmented at higher temperatures. Compared to pure TPU, the TPU/Co-MOF@MXene composites exhibited substantially higher residual content after combustion. The TPU composites with Co-MOF@MXene hybrids displayed significant improvements in fire safety, with reductions of 28.3% in PHRR, 14.5% in THR, 58.8% in SPR, and 47.5% in TSP compared to pure TPU. These enhancements are attributed to the barrier effects of MXene and the catalytic effect of Co, which improved the dispersion and compatibility of Co-MOF in the TPU matrix. The presence of Co-MOF@MXene hybrids also resulted in increased char residues, effectively forming a cohesive and dense char layer that mitigates fire hazards.

Wan et al. (2024) developed a novel multilayered nanomaterial coating, APP-PEI@MXene@ZIF-67, and incorporated it into TPU polymers using an eco-friendly assembly method. The findings revealed that APP-PEI@MXene@ZIF-67 was more effective in promoting catalytic char formation than unmodified APP, resulting in denser and more continuous residual char. Compared to pure TPU, the TPU/APP-PEI@MXene@ZIF-67-2BL composites showed significant reductions in PHRR by 72.75%, THR by 87.25%, PSPR by 59.58%, and TSP by 85.97%. These substantial decreases in heat and smoke emissions during combustion indicate enhanced fire safety of the modified TPU composites. The synergistic effects of gas-phase and condensed-phase flame-retardant mechanisms contributed to a potential approach for mitigating fire risk and reducing heat and smoke hazards in TPUs. This strategy enhances the flame retardancy of TPU polymers, promoting their widespread application in various fields through the use of multilayered nanomaterial coatings like APP-PEI@MXene@ZIF-67.

As illustrated in Fig. 18.6, researchers developed a novel synergistic flame retardant by combining ZIF-8 nanoparticles with sepiolite (SEP) clay, forming a ZIF-8@SEP hybrid. This hybrid was then incorporated into TPU with aluminum hypophosphite (AHP) to create the composite TPU/8.0AHP/1.0ZIF-8@SEP. The composite achieved LOI of 27.5% and met the UL-94 V-0 grade. Compared to control TPU, the PHRR and THR decreased by 78.9% and 39.1%, respectively. The increased smoke production caused by AHP was partially adsorbed by ZIF-8@SEP, resulting in the lowest smoke release in the CCT for the TPU/8.0AHP/1.0ZIF-8@SEP sample. AHP captured free radicals generated from the broken TPU chains via pyrolytic PO· and PO$_2$·, halting chain scission. Additionally, pyrophosphoric acid from AHP promoted TPU carbonization, and ZIF-8@SEP absorbed harmful gases like CO and HCN within its hollow structure. Importantly, this synergistic flame retardant caused less than a 2.0% loss in tensile strength and a 7.8% drop in elongation at break, maintaining the mechanical performance of TPU. These findings demonstrate

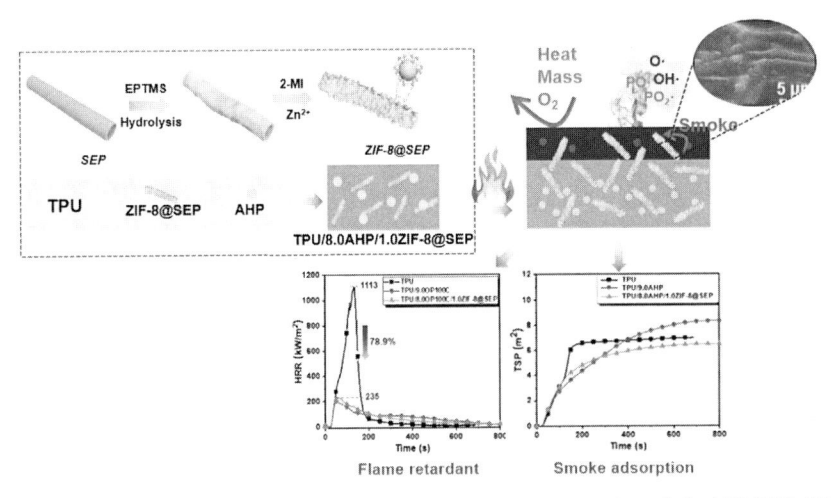

FIGURE 18.6 Creating a hybrid from a ZIF-8@SEP to mitigate fire hazards in AHP/THP (Li et al., 2022).

that ZIF-8@SEP is an excellent synergistic flame retardant for TPU, effectively enhancing flame retardancy while causing minimal mechanical damage (Li et al., 2022).

Huang and coworkers used reduced graphene oxide (RGO) and an iron-based MOF (MIL-101(Fe)) to enhance FPU (flexible polyurethane) foam through a layer-by-layer (LbL) assembly method, followed by a hydrazine hydrate vapor reduction process (Fig. 18.7A). Compared to the unmodified foam, the nano-coated foam with 10 layers showed notable reductions in key smoke and heat release parameters: a 26.8% decrease in PSPR, a 76.7% decrease in TSP, and a 12.4% decrease in THR (Fig. 18.7B). These findings suggest that the nano-coating significantly reduces smoke emission from coated foams and changes the combustion behavior of FPU foam. The main reason for this smoke suppression is that RGO sheets create a lamellar barrier on the FPU foam's surface. Graphene serves as an excellent barrier against the combustion of polymer composites. This physical barrier effectively prevents the release of thermal degradation products from the condensed phase into the gas phase, thus lowering smoke production during FPU foam combustion. Furthermore, the multilayer nano-coating amplifies this barrier effect. The smoke suppression properties of the MOF are due to its porous structure and catalytic charring effect. MIL-101(Fe) on the surface of the FPU foam adsorbs smoke particles and catalyzes their conversion into a char layer, thanks to its large specific surface area. Therefore the smoke suppression ability of the nano-coating is attributed to the combined effects of RGO and MIL-101(Fe) (Huang & Yuan, 2021).

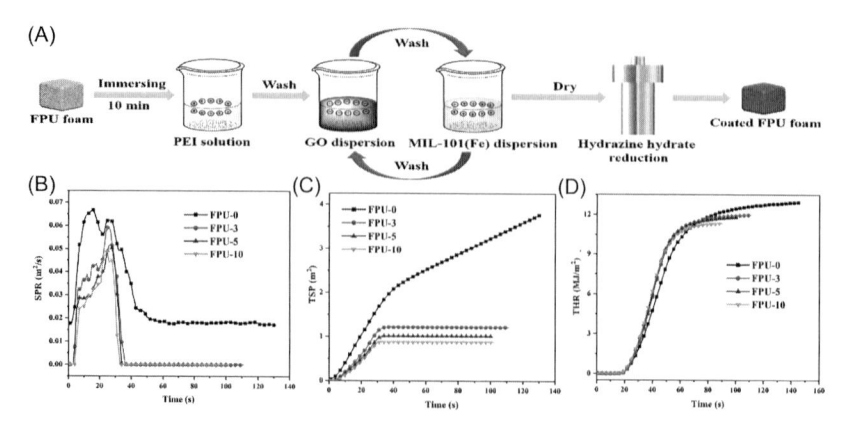

FIGURE 18.7 (A) Schematic illustration of the preparation of coated FPU foams using LbL self-assembly technology; (B) SPR curves, (C) TSP curves, and (D) THR curves for both uncoated FPU-0 and coated FPU foams (Huang & Yuan, 2021).

Researchers developed a new fire-resistant and low-smoke FPUF nanocomposite by combining MOF-LDH (LDH: layered double hydroxide) with $Ti_3C_2T_x$ MXene nanosheets. FPUFs with 6 wt.% $Ti_3C_2T_x$@MOF-LDH showed improved fire and smoke suppression properties: 16.1% less total smoke and 22.2% lower peak smoke production. Toxic gas emissions, including CO, CO_2, and aromatic compounds, were significantly reduced. Heat release rates decreased compared to pure FPUF. Alongside enhanced fire safety, the nanocomposite foams displayed excellent mechanical properties. This study offers a promising method for creating high-performance, fire-resistant, low-smoke polyurethane foam nanocomposites (Venkateshalu et al., 2023).

In other study, a ternary inorganic–organic hybrid flame retardant (A@P-Z) was created by combining APP, polydopamine (PDA), and ZIF-67. This hybrid flame retardant was incorporated into RPUF to enhance its fire safety (Fig. 18.8A). Results from cone calorimeter testing revealed significant improvements in fire retardancy for the RPUF/A@P-Z composite compared to pure RPUF, with notable reductions of 45.9% in THR, 53.4% in TSP, and 38.1% in pCOP. While unmodified RPUF exhibited a LOI of 19.7%, indicating high ignitability, the LOI of RPUF/A@P-Z increased to 25.6%. As shown in Fig. 18.8B–D, in UL-94 testing, neat RPUF received "no rating", whereas both RPUF/APP and RPUF/A-Z demonstrated a self-extinguishing behavior after the removal of the test flame. Ultimately, the composite achieved a UL-94 V-0 rating, highlighting its high flame-retardant efficiency. The excellent fire safety performance of RPUF/A@P-Z is due to the synergistic catalytic effects of the hybrid flame retardant, which formed a highly graphitized and robust char layer that protected the underlying material from further combustion. Additionally, the hybrid flame retardant demonstrated effective flame inhibition in the gas phase

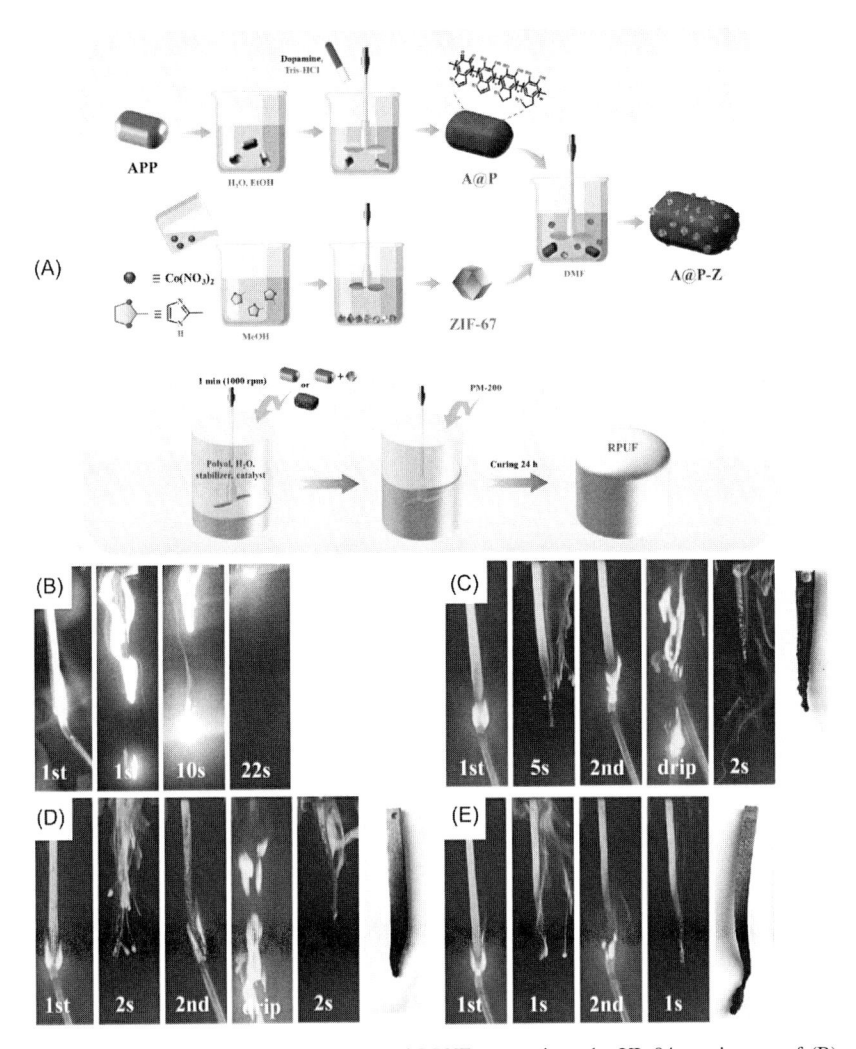

FIGURE 18.8 (A) Schematic of A@P-Z and RPUF composites; the UL-94 test images of (B) RPUF, (C) RPUF/APP, (D) RPUF/A-Z, and (E) RPUF/A@P-Z (Liu et al., 2023).

by releasing phosphorus-containing and nitrogen-containing compounds that quenched free radicals and diluted flammable gases. This study introduces a novel and effective method for creating high-performance RPUF composites with enhanced fire safety through the use of inorganic–organic hybrid flame retardants, making RPUF/A@P-Z suitable for applications requiring superior flame resistance (Liu et al., 2023).

In summary, studies have shown that incorporating MOF composites into polyurethane materials leads to significant reductions in key fire parameters like

PHRR, THR, PSPR, and TSP. These improvements are due to the formation of dense char layers and the synergistic effects of flame-retardant mechanisms. Overall, MOF composites hold promise for enhancing the fire safety of polyurethanes, making them suitable for diverse industrial applications.

Metal-organic framework composites applied in fire retardancy of polylactic acid

PLA, a biodegradable polymer, offers desirable mechanical properties, ease of fabrication, and excellent biocompatibility, making it suitable for applications in drug delivery, textiles, and packaging. However, its inherent flammability poses a significant limitation, restricting its use as an engineering material. To address this challenge, researchers have focused on developing effective flame retardants for PLA, with MOF composites gaining significant attention (Jia et al., 2022; Wu et al., 2022). For example, Zhang et al. synthesized nanosized zeolitic imidazolate framework-8/graphene oxide (ZIF-8/GO) hybrid materials and incorporated them into PLA to improve its flame-retardant properties. The addition of ZIF-8@GO hybrids to PLA resulted in significant enhancements in flame retardancy. Notably, the LOI value increased from 21% for neat PLA to 24% for the PLA/ZIF-8@GO nanocomposite. A higher LOI value indicates improved flame resistance. Furthermore, the PLA/ZIF-8@GO nanocomposite exhibited self-extinguishing behavior in the UL-94 vertical burning test, a widely recognized standard for assessing the flammability of polymeric materials. The nanocomposite achieved a V-2 rating, which is a desirable classification for materials with acceptable flame retardancy. The synergistic effects of the ZIF-8 and GO components contribute to the improved fire safety performance of the PLA nanocomposites (Zhang et al., 2018).

Hou et al. synthesized Co-MOF nanosheets using organic ligands with Schiff base structures and modified them with 9,10-dihydro-9-oxa-10-phos-phaphenanthrene-10-oxide (DOPO) to create DOPO@Co-MOF. These nanosheets were incorporated into a PLA matrix to prepare DOPO@Co-MOF blends, as shown in Fig. 18.9A. The DOPO@Co-MOF nanocomposites demonstrated significantly enhanced thermal stability, flame retardancy, and suppression of smoke and CO emissions compared to the individual components. This suggests a synergistic effect between DOPO and Co-MOF in improving the fire safety of PLA. The results, illustrated in Fig. 18.9B–D, show that the PLA composites exhibited a 27% reduction in PHRR, a 56% decrease in TSP, and a 20% reduction in total CO production. These findings underscore the potential of DOPO@Co-MOF nanocomposites as highly effective flame retardants for PLA, offering synergistic improvements in fire safety, and thermal stability (Hou et al., 2018).

Researchers developed an inorganic–organic flame retardant, UiO@TEPx, by encapsulating triethyl phosphate (TEP) in UiO66-NH$_2$ (Fig. 18.10A). The introduction of UiO@TEP$_{50}$ significantly improved the flame retardancy of

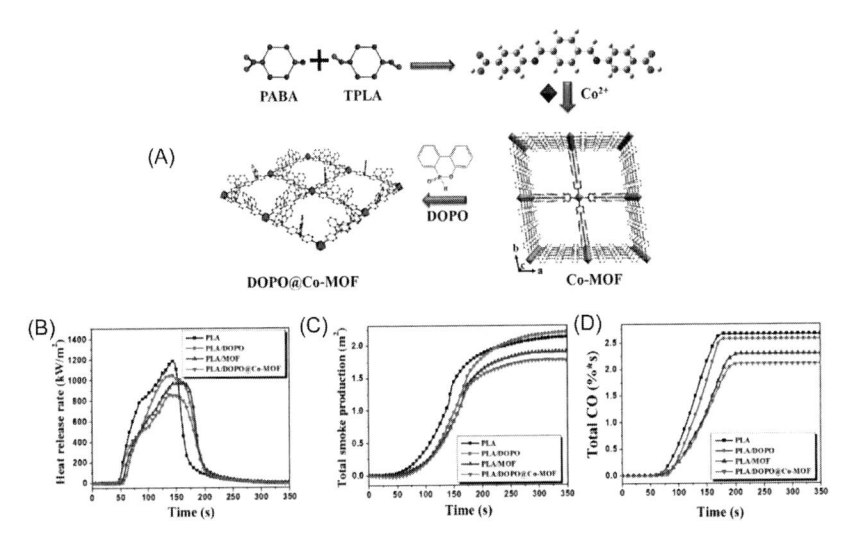

FIGURE 18.9 (A) Schematic diagram of the preparation process for DOPO@Co-MOF hybrids; (B) HRR, (C) TSP, and (D) total CO release curves for PLA and its composites (Hou et al., 2018).

PLA. Flammability tests showed that UiO@TEP$_{50}$ significantly enhanced flame retardancy in PLA compared to UiO66-NH$_2$. Adding 2.0% UiO@TEP$_{50}$ to PLA composites increased the UL-94 rating to V-0, raised the LOI to 24.2%, and reduced THR, PHRR, TSP, and TCOP by 9.8%, 14.2%, 29.5%, and 24.2%, respectively. UiO@TEP$_{50}$ exhibited a "thermosensitive release" flame-retardant mechanism, wherein the released TEP captured free radicals in the gas phase and promoted char formation in the condensed phase. Additionally, the reduction in smoke and CO production was attributed to the porous structure of UiO@TEP50 and the catalytic activity of ZrO$_2$ (Fig. 18.10B) (Wang et al., 2023).

In another study, researchers developed a highly efficient fire-resistant composite material made from poly(L-lactic acid) (PLLA) combined with piperazine pyrophosphate (PAPP) and a ZIF-67. The study reported a superior synergistic effect by incorporating both PAPP and the cobalt-based MOF. In the UL-94 fire test, pure PLA burned rapidly upon ignition, exhibiting severe melting and dripping, which can potentially ignite nearby materials, and received a V-2 rating, indicating poor fire resistance. However, the PLA/4.94PAPP/0.06MOF composite achieved a V-0 rating, showing minimal dripping and significantly improved fire resistance. The PLA/4.9PAPP/0.1MOF composite demonstrated excellent fire performance, with an LOI value of 31% (Fig. 18.11A). It exhibited a 28.3% reduction in PHRR and a 9.0% decrease in THR compared to pure PLA. After the CCT, the residue of the PLA/5PAPP sample showed some char formation but was insufficient to cover the entire sample holder. In contrast, the char residue of the PLA/

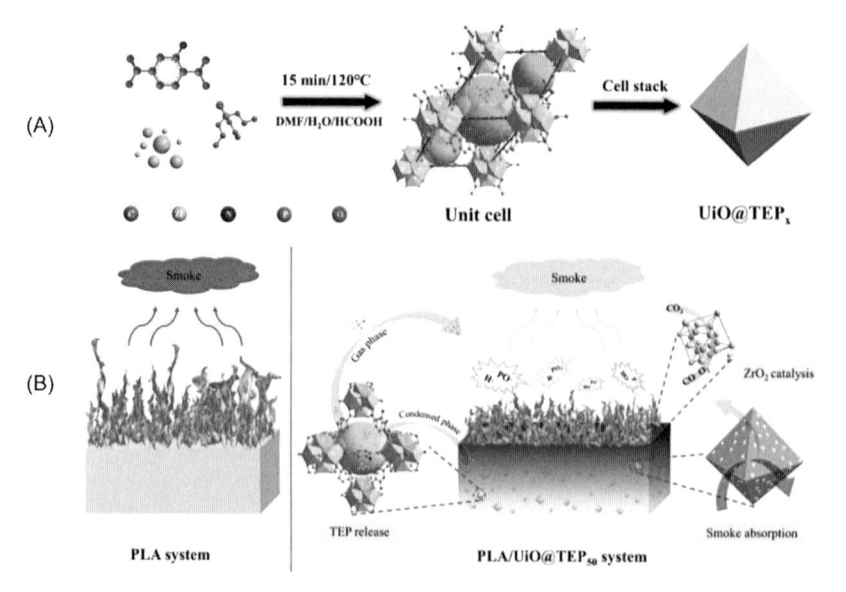

FIGURE 18.10 (A) The synthesis process of UiO@TEPX; (B) The schematic representation of the flame-retardant mechanism of UiO@TEP50 (Wang et al., 2023).

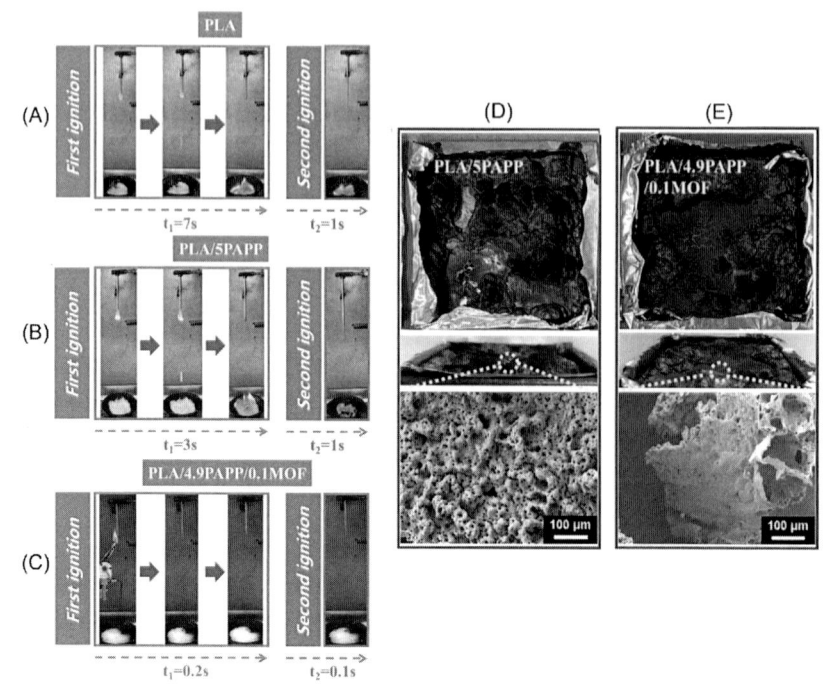

FIGURE 18.11 Burning process of (A) PLA, (B) PLA/5PAPP, and (C) PLA/4.9PAPP/0.1MOF in the UL-94; digital photos and SEM images of char residue for (D) PLA/5PAPP and (E) PLA/4.9PAPP/0.1MOF (Jin et al., 2023).

4.9PAPP/0.1MOF composite was integral, with significantly fewer pores and a denser structure, indicating better char formation and insulation properties (Fig. 18.11B) (Jin et al., 2023).

Therefore, MOF composites can significantly enhance the flame retardancy of PLA by reducing key metrics such as PHRR, THR, and SPR. These improvements are achieved through the synergistic effects of the MOF components, which promote char formation and capture free radicals, leading to better thermal stability and reduced flammability.

Metal-organic framework composites applied in fire retardancy of unsaturated polyester resin

UPR is a highly utilized thermosetting resin known for its excellent mechanical strength, corrosion resistance, and ease of processing, making it popular in industries like transportation, aerospace, and construction. However, UPR's high flammability poses a significant drawback, as its combustion releases substantial styrene monomers and oligomers, increasing fire hazards. To mitigate this issue, developing UPR with improved flame-retardant properties is essential (Chu et al., 2021; Reuter et al., 2019; Shanks et al., 2010). MOF composites have been utilized in UPR flame retardants due to their catalytic and adsorption properties. For instance, Du et al. (2024) have synthesized a new coating flame retardant, called APP@UiO, through the in situ synthesis of UiO-66-NH$_2$ on the surface of APP. This innovative flame retardant was then added to UPR. When 10 wt.% of APP@UiO was added, significant improvements in fire safety performance were observed in the UPR composites compared to pure UPR. Compared to APP and UiO-66-NH$_2$, APP@UiO exhibited notably enhanced thermal stability, with a 61.34% decrease in PHRR and a 37.5% reduction in maximum smoke yield (SPR$_{max}$). Additionally, the LOI value, an indicator of flame resistance, reached an impressive 31.2%. APP@UiO enhances UPR's flame retardancy by promoting carbon formation and forming a dense char layer that limits heat and oxygen transfer. During thermal decomposition, APP@UiO releases NH$_3$, which dilutes flammable gases, and PO· and PO$_2$·, which capture free radicals and hinder gas volatilization. Furthermore, it catalytically converts CO to CO$_2$, reducing toxic gas emissions and suppressing smoke. The organic ligands in APP@UiO improve compatibility with UPR, resulting in higher impact strength compared to using APP alone. As a result, APP@UiO not only boosts UPR's flame resistance but also enhances its mechanical properties.

In a research investigation, hybrids referred to as BMCN, consisting of Co-Cu bimetallic MOFs combined with graphitic carbon nitride (g-C$_3$N$_4$), were synthesized via a coprecipitation method. The study utilized TG-IR to investigate the pyrolysis process and resulting byproducts of UPR, aiming to uncover BMCN's flame-retardant mechanism (Fig. 18.12). Comparative analysis of 3D FT-IR (Fourier-transform infrared spectroscopy) and dynamic spectra revealed that both UPR and UPR/4BMCN exhibit similar pyrolysis

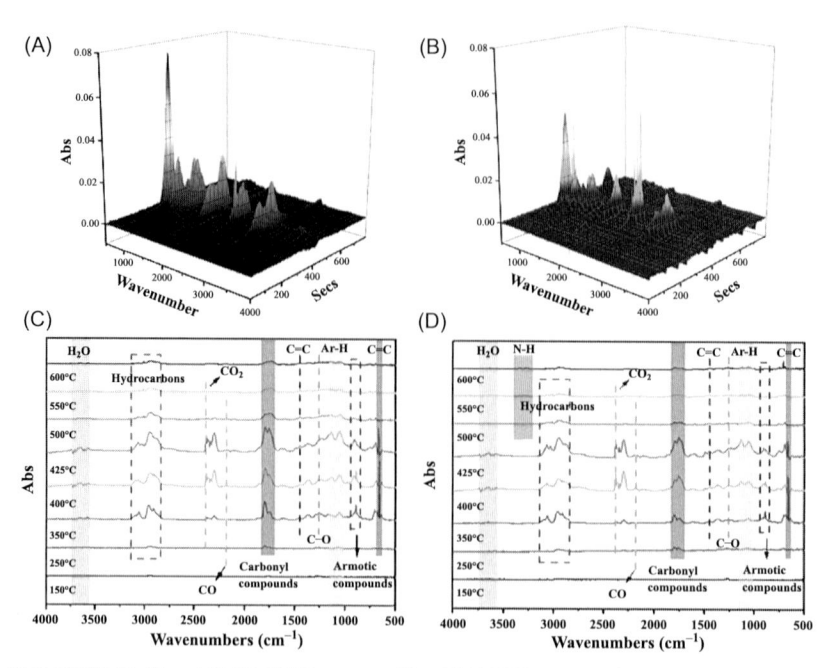

FIGURE 18.12 (A and B) 3D FT-IR spectra; (C and D) FT-IR spectra of UPR and UPR/4BMCN at different pyrolysis temperatures (Yang et al., 2023).

behaviors, with variations observed primarily in the N–H region. The presence of BMCN accentuated the –CH$_2$ signals, suggesting increased destabilization of thermal oxidation. Furthermore, BMCN accelerated the disappearance of various chemical peaks, indicating premature degradation of UPR. These findings align with TG results, highlighting BMCN's role in hastening UPR's degradation and facilitating the formation of a char layer, which effectively suppresses flame propagation. These hybrid materials enhance the compatibility between g-C$_3$N$_4$ and UPR, accelerate UPR's premature degradation and catalytic carbonization, and establish a dense carbonization layer that acts as a thermal barrier. Moreover, the catalytic properties of metal oxides contribute to mitigating the release of harmful gases (Yang et al., 2023).

A new flame retardant, MFeCN, comprising MIL-53 (Fe)@C/graphite carbon nitride hybrid, was synthesized via a hydrothermal process and then introduced into UPR. TG analysis revealed that the residues of UPR/MFeCN composites increased upon MFeCN addition, suggesting enhanced thermal stability through catalytic carbon formation. CCT results indicated reductions of 39.8%, 10.2%, 33.3%, and 14.5% in PHRR, THR, PSPR, and TSP, respectively, for UPR/MFeCN-4 compared to UPR alone. The proposed flame retardancy mechanism involves the generation of Fe$_2$O$_3$ during combustion, acting as a heat barrier that enhances char formation in the UPR/MFeCN composites. This char layer impedes the exchange of heat radiation between the UPR matrix and its surroundings, thereby improving flame retardancy relative to pure UPR and UPR/g-C$_3$N$_4$ composites (Chen et al., 2019).

Metal-organic framework composites applied in fire retardancy of polyurethane elastomer

PUE is primarily composed of hard segments (isocyanates and chain extenders) and soft segments (polyols). It is widely used in industries like mining, machinery, construction, and auto-making due to its excellent elasticity, radiation resistance, abrasion resistance, and low-temperature resistance. However, PUE is highly flammable and releases toxic gases such as HCN and CO during combustion, posing significant health and environmental risks. Consequently, researching the flame retardancy and smoke suppression properties of PUE is essential (Kultys & Rogulska, 2011; Shi et al., 2019).

ZIF-8 was effectively grown onto the surface of α-ZrP using a straightforward method. The synthesized ZIF-8/α-ZrP hybrid was incorporated into PUE to enhance its flame retardancy and smoke suppression performance. The TGA results indicated that there is no significant difference in $T_{50\%}$ and T_{max} between α-ZrP/PUE and ZIF-8/PUE. However, the $T_{50\%}$ and T_{max} of the ZIF-8/α-ZrP/PUE composites slightly increase, indicating that the surface modification of α-ZrP with ZIF-8 enhances the thermal stability of the PUE composites at high temperatures. CCT test results revealed that the PHRR, THR, SPR, and TSP of ZIF-8/α-ZrP/PUE decreased by 69.6%, 45.6%, 48.7%, and 16.6%, respectively. The enhanced flame retardancy and smoke suppression are primarily attributed to the physical barrier and catalytic carbonization effects of α-ZrP. Additionally, the metal oxide decomposed from ZIF-8 further promotes the formation of char residue and increases the compactness of the char layer (Xu et al., 2018).

In another study, researchers combined ZIF-8 with expandable graphite (EG) and incorporated them into PUE to investigate their flame retardant and smoke suppression properties. The best performance was achieved when the total flame retardant addition was 3.0 wt.% and the ratio of ZIF-8 to EG was 1:3. Compared to pure PUE, this composite showed significant improvements: an 83.4% reduction in PHRR, a 42.6% decrease in THR, and a 22.4% reduction in maximum smoke density. Additionally, the LOI value increased to 30.2%, indicating better flame retardancy, and the composite achieved a V-1 rating in the UL-94 vertical burning test. The CCT showed the char residue formed after burning the composites. Pure PUE had no char residue, while ZIF-8/PUE exhibited a slight increase in residue. EG/PUE generated an intumescent char residue, but it was not dense and tended to fall off the polymer surface. In contrast, the ZIF-8:EG(1:3)/PUE composite formed a more complete and compact char residue, contributing to its enhanced flame retardancy and smoke suppression performance (Wang et al., 2020).

In conclusion, MOF composites significantly improve the flame retardancy and smoke suppression properties of PUE. Incorporating MOF composites such as ZIF-8/α-ZrP and ZIF-8/EG into PUE composites resulted in substantial reductions in PHRR, THR, SPR, and TSP, along with enhanced thermal stability and increased char residue compactness, demonstrating their effectiveness in mitigating fire hazards.

Metal-organic framework composites applied in fire retardancy of polystyrene

PS, a synthetic polymer made from styrene, is widely used in packaging, insulation, electronics, and consumer products due to its low cost and good insulation properties. However, it is highly flammable, with a low ignition temperature and high heat release rate, releasing toxic gases when burned. Flame-retardant additives such as MOF composites can improve its fire resistance, but it remains combustible and should be handled cautiously in fire-prone environments. Fig. 18.13A shows that a cobalt-based MOF-71-NH$_2$ (MOF-NH$_2$) was synthesized and modified with phosphonitrilic chloride trimer (PCT) using a postsynthesis modification (PSM) strategy, creating PCT@MOF-NH$_2$ to improve the flame retardancy of PS (Xu et al., 2021). CCT results indicate that PS/PCT@MOF-NH$_2$-3.0% composites exhibit significantly better flame retardancy than neat PS, with reductions of 40.6% in PHRR and 31.7% in THR. TG-IR analysis shows a marked decrease in peak intensity compared to pure PS, demonstrating that the additive reduces decomposed volatiles. The addition of fillers also delays the release behavior of PS. The differences between PS composites and pure PS, especially between neat PS and PS/PCT@MOF-NH$_2$ samples, highlight the combined action of PCT and MOF-NH$_2$ in inhibiting aromatic compound release. The synergistic effect between PCT and MOF-NH$_2$ enhances PS fire resistance by reducing heat, smoke, and toxic gas emissions, with PCT forming inert substances and MOF-NH$_2$ accelerating char formation and acting as a barrier. These combined effects inhibit heat and mass transfer, effectively preventing fire spread (Fig. 18.13B).

Hou et al. synthesized a nickel-based MOF (Ni-MOF) with 2-methylimidazole ligands and combined it with GO to form a composite material called GOF. Incorporating just 1.0 wt.% of GOF filler into PS resulted in a reduction of over 33% in the PHRR during combustion. The PS/GOF composite also showed significant reductions of 21% in TSP and 52.3% in carbon monoxide generation compared to pure PS. The GO acted as an effective barrier, inhibiting heat and gaseous fuel transfer, thus enhancing thermal and flame resistance. Additionally, the distinct metal-nitrogen structures in Ni-MOF provided excellent stability against heat and chemicals. This PS/GOF

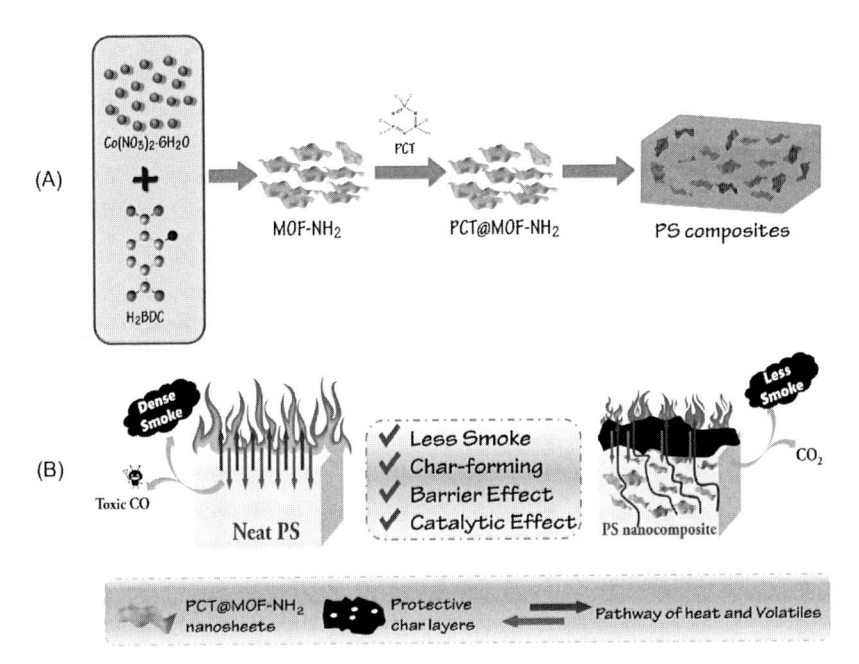

FIGURE 18.13 (A) Diagrammatic representation of PCT@MOF-NH₂ and PS composites. (B) Diagrammatic representation of the flame-retardant mechanism in PS/MOF composites (Xu et al., 2021).

composite significantly reduces smoke, toxic gases, and flammability, making it highly promising for fire safety applications (Hou et al., 2017).

Metal-organic framework composites applied in fire retardancy of polyurea

PUA, formed from isocyanate and amino compounds, is prized for its resilience and resistance to moisture, corrosion, and aging. Widely used in civilian, military, and marine sectors, its high flammability presents a barrier to further development and application. Therefore improving PUA's flame retardancy is crucial for enhancing safety and expanding its utility. Wang and colleagues synthesized a new flame-retardant material termed ZnO@MOF@PZS using a combination of hydrothermal synthesis and polycondensation. ZnO nanoflowers acted as a scaffold for the growth of the MOF@PZS structure, supplying the zinc component for the MOF. The SEM images exhibit the morphology of the resulting ZnO@MOF@PZS composite. Subsequently, the researchers integrated the produced ZnO@MOF@PZS into PUA to assess its flame retardant efficacy and ability to reduce toxic gas emissions. CCT revealed that the addition of just 3 wt.% of ZnO@MOF@

PZS to PUA markedly improved flame retardant characteristics, reducing PHRR by 28.30%, THR by 19.59%, and total CO yield by 36.65%. The enhanced flame retardancy of the PUA composites is predominantly ascribed to the development of a dense char layer facilitated by the presence of ZnO@MOF@PZS (Wang et al., 2021).

Biobased-functionalized MOFs (Bio-FUN-MOFs) offer promising flame retardancy through multiple mechanisms and green synthesis. However, functionalizing MOFs often compromises their inherent advantages, leading to resource wastage. This study presents an acid-base balance strategy to controllably prepare Bio-FUN-MOFs while maximizing flame retardant loading and preserving MOF structures. A buffer layer of biobased arginine is created on ZIF-67 to resist excessive etching by phytic acid during phosphorus loading, preserving internal crystal integrity. The resulting arginine and phytic acid-functionalized ZIF-67 (ZIF@Arg-Co-PA) with a yolk@shell structure acts as an efficient biobased flame retardant for PUA composites. At 5 wt.% loading, ZIF@Arg-Co-PA imparts a LOI of 23.2% and reduces PHRR, THR, and TSP significantly compared to neat polyurea. The composites also exhibit acceptable mechanical properties. The preserved ZIF-67 endows ZIF@Arg-Co-PA with a high surface area and hierarchical porous structure, enhancing smoke capture capability. The phytic acid-arginine-cobalt shell improves compatibility in the polymer matrix. A hollow structure (Arg-Co-PA) was also synthesized by increasing phytic acid dosage, showing slightly inferior flame retardancy due to the excellent shell-core matching in ZIF@Arg-Co-PA, avoiding template wastage from excessive etching (Song et al., 2024).

In summary, the quest for flame-retardant materials is vital to improve the safety and versatility of PUA and MOF composite exhibits efficient flame retardancy for polyurea composites, along with improved mechanical properties and smoke capture capabilities, advancing sustainable flame-retardant solutions for polymer materials.

Metal-organic framework composites applied in fire retardancy of other polymers

As mentioned in the introduction section, MOF composites have been used with various other polymers. However, due to the limited number of publications on this topic, we have cited all the available literature on the subject. Ethylene vinyl acetate (EVA) boasts versatile performance qualities, making it a staple in industries such as transportation, packaging, and construction. However, its high flammability poses significant fire hazards, necessitating the development of effective flame retardants like MOF composites to enhance safety and expand its utility (Huang et al., 2021). A novel combination of cerium with a metal organic framework (Ce-MOF) was employed alongside piperazine pyrophosphate (PAPP) to develop flame-retardant EVA copolymeric nanostructures. These EVA@PAPP@Ce-MOF nanostructures achieved a UL-94 V-0 rating with an LOI of 31.3% when 17 wt.% PAPP@Ce-MOF was incorporated at a mass fraction of 97:3. In comparison, EVA@PAPP with the same PAPP level achieved a UL-94 V-1 rating and an LOI of 26.9%. THR decreased by 25.6% from 132.8 MJ/m^2 for EVA@PAPP to 98.8 MJ/m^2 for EVA@PAPP@Ce-MOF. TSP fell by 36.5% from 11.5 m^2 for EVA@PAPP to 7.3 m^2 for EVA@PAPP@Ce-MOF. Moreover, the maximum smoke density witnessed a notable reduction from 488.8 for plain EVA and 375.9 for EVA@PAPP down to only 177.5 for the EVA@PAPP@Ce-MOF composite. This underscores that combining PAPP and Ce-MOF components yielded a synergistic effect, significantly enhancing the fire-retardant capabilities and smoke suppression of the EVA polymer matrix (Jiang et al., 2023).

PA6 is a widely used engineering plastic known for its mechanical strength, high abrasion resistance, chemical resistance, and good processability, making it suitable for automotive, electronics, and electrical industry applications. However, its flammability limits its use. To address this issue, flame retardant PA6 composites with MOF additives have been created (Weil & Levchik, 2009). In Fig. 18.14, Li et al. combined aluminum hypophosphite (AHP) with melamine cyanurate (MCA) and MIL-100 (Fe) to prepare PA6 composites (PA6/MAF) to inhibit PH$_3$ release during processing and improve fire safety. The addition of MCA and the MIL-100 (Fe) helped reduce the amount of phosphine gas released and delayed the initial decomposition temperature of the AHP blend. Under accelerated thermal testing conditions, the PA6/AHP composite released a significant quantity of phosphine gas measuring 137.4 ppm, while the concentration in the PA6/MAF composite was 0 ppm. During the processing stage, the PA6/AHP composite reached a maximum PH$_3$ concentration of 673 ppm at 510 seconds. In contrast, the PH$_3$

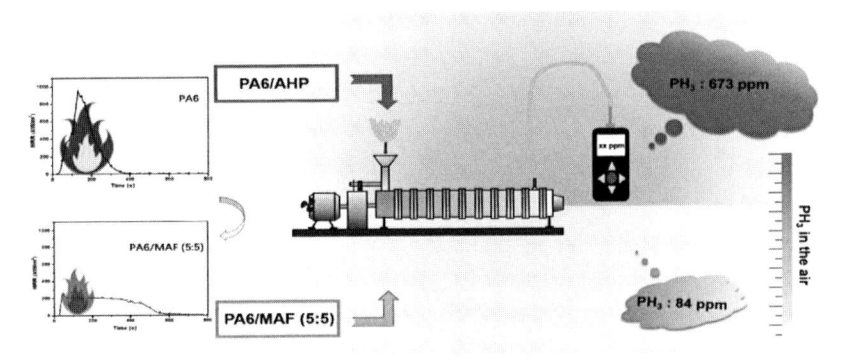

FIGURE 18.14 Reducing the release of toxic PH$_3$ and the fire hazard in PA6/AHP composites using MOFs (Li et al., 2020).

concentration from the PA6/MAF (5:5) composite was substantially lower, peaking at only 84 ppm at 1284 seconds. This represents an 87.5% reduction in PH$_3$ release compared to the PA6/MA composite without the addition of MIL-100 (Fe). Additionally, MCA and MIL-100 (Fe) reduced the PHRR from 962 to 260 kW/m² and decreased CO emissions by 11.78%. The amino groups in MCA reacted with PH$_3$, while MIL-100 (Fe) adsorbed PH$_3$ and promoted char formation, blocking heat and combustible gas release (Li et al., 2020).

PBS finds extensive application in various fields, including agriculture, biomedicine, packaging, and injection molding. However, its inherent flammability and tendency to drip when combusted restrict its broader usage, necessitating efforts to mitigate its fire hazard for practical applications (Georgousopoulou et al., 2016; Nobile et al., 2015). In other study (Li et al., 2024), a novel "three-in-one" flame retardant was created by grafting organophosphorus onto Fe-MOF functionalized Mg(OH)$_2$ (designated as MH@Fe-MOF-P), as shown in Fig. 18.15A. This compound was then incorporated into a PBS matrix to examine its effects on flame retardancy properties. The findings showed that PBS/20MH@Fe-MOF-P demonstrated outstanding flame retardancy, with a LOI of 29.6%, a UL-94 rating of V-0, and an 82.1% reduction in the PHRR and 53.2% reduction in the THR (Fig. 18.15B and C). The flame-retardant mechanism for PBS/MH@Fe-MOF-P involves a synergistic effect of its three components: MH, Fe-MOF, and phosphorus (P). These components work together to delay the decomposition of PBS and promote its carbonization, thereby enhancing the flame retardancy of the composite. The decomposition of MH releases steam, which decreases the surface temperature and dilutes flammable gases. The grafted diphenylphosphine pyrolyzes into P• and PO• radicals, helping to quench H• and HO• radicals in the gas phase, thereby inhibiting combustion. Fe-MOF catalyzes the carbonization of PBS-derived fragments, increasing the amount of char with a higher degree of graphitization. MgO particles from MH, along with the solid products from PBS chains and Fe-MOF derivatives, construct a dense char layer. This char layer provides a better barrier effect, isolating the permeation of heat and oxygen, delaying the diffusion of flammable gases, and inhibiting flame propagation with low fire intensity (Fig. 18.15D).

Vinyl resin, such as 430 LV, is widely used in marine architecture due to its exceptional durability, weather resistance, and cost-effectiveness. However, its inherent flammability poses significant safety concerns, especially in enclosed spaces. To address this issue, MOF composites have been developed to enhance its fire resistance. A MOF derived from zinc and p-hydroxybenzoic acid (PHBA) was synthesized using a solvothermal method. These Zn-PHBA nanorods were incorporated as flame retardants into vinyl resin (430 LV), commonly used in marine structures. Results from CCT demonstrated that the addition of Zn-PHBA nanorods significantly improved flame retardancy and reduced smoke production in the 430 LV material. With the inclusion of 10 wt. % Zn-PHBA, the LOI of 430 LV increased from 19.5% to 28.5%, achieving a

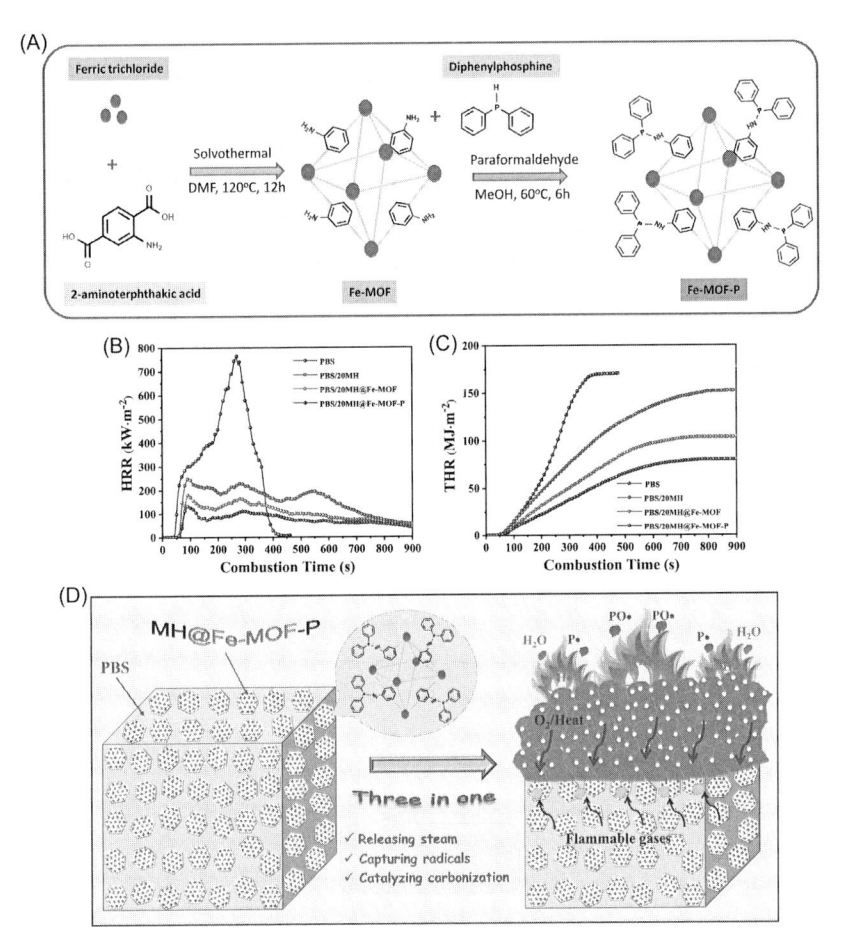

FIGURE 18.15 (A) Diagram illustrating the process of synthesizing Fe-MOF-P; (B) HRR, (C) THR curves of PBS specimens; (D) Potential flame-retardant mechanism of MH@Fe-MOF-P in PBS composites (Li et al., 2024).

V-0 rating in UL-94. The study observed notable decreases in PHRR, COP, CO$_2$P, and TSP values, indicating enhanced fire safety properties. These improvements are attributed to the nano-scale size of the MOF and the presence of transition metal elements, which promote the formation of a dense carbon layer during combustion, offering insulation and smoke suppression (Wang et al., 2023).

Textile fabrics and related materials are widely used across various applications due to their unique attributes, including affordability, diverse types, availability in both natural and synthetic forms, and flexibility. However, their significant flammability and low thermal stability present ongoing challenges for their safe use in various applications. To address these issues, MOF composites have been developed to reduce fire hazards (Ma et al., 2021). Attia et al. (2023) introduced a systematic approach to crafting multifunctional textile fabric coatings, leveraging UiO-66 as a porous crystalline scaffold, alongside spherical polypyrrole nanoparticles (PPY-NPs) and chitosan chains. Through varying mass loadings of UiO-66 and PPY-NPs, the study elucidated their impact on coated textile properties. The resulting coating displayed remarkable flame retardancy, reducing burning rates by 54% compared to untreated fabrics, while also enhancing thermal stability. Notably, during combustion, the PPY-NPs transformed into porous carbon nanoparticles, effectively absorbing and suppressing combustible gases, thereby impeding the burning process and heat transfer. This process led to the formation of a protective char layer on the fabric surface, contributing to further fire resistance (Fig. 18.16).

PC is a highly utilized thermoplastic engineering plastic, valued for its robustness and resistance to heat aging. Its applications span various industries, including high-speed railways and aviation, where it is used in wall panels and luggage racks. However, PC's tendency to burn rapidly and release toxic gases presents safety concerns. To address this, flame-retardant systems such as MOF composite has been developed to enhance PC's fire resistance and reduce the spread of flames and toxic emissions (Levchik & Weil, 2005). A novel halogen-free binary hybrid from Zr-BDC and cerium phenylphosphonate (CeHPP), Zr-BDC@CeHPP, with a lamellar and porous structure was synthesized via hydrothermal reaction and blended with PC to enhance its thermal resistance and fire safety (Fig. 18.17). Incorporating 2 wt.% Zr-BDC@CeHPP into PC increased the temperatures under peak decomposition rates (T_{max1} and T_{max2}) by 28°C and 34°C in an oxidizing atmosphere. The time to reach t-PHRR and TTI

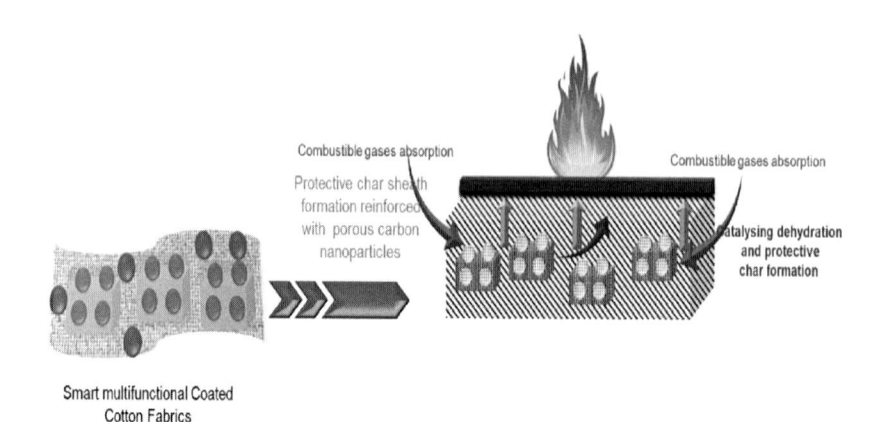

FIGURE 18.16 Flame retardant of metal-organic framework-coated fabrics (Attia et al., 2023).

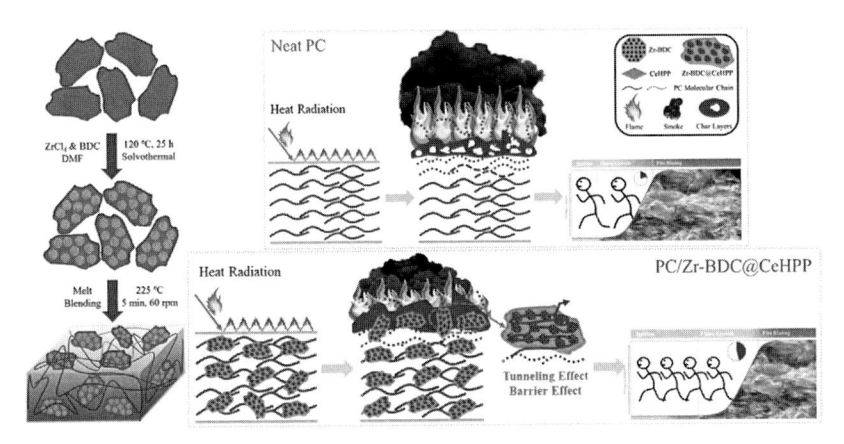

FIGURE 18.17 Deposition growth of Zr-BDC on CeHPP lamella for enhanced fire safety of PC (Sai et al., 2020).

were effectively extended to 145 and 18 seconds, respectively. This addition also reduced the PHRR, THR, peak specific extinction area (PSEA), and TSR by 45%, 20%, 74%, and 18%, respectively. The flame-retardant PC achieved a V-0 rating in the UL-94 vertical burning test, indicating its enhanced fire safety. The effectiveness of Zr-BDC@CeHPP is attributed to the π-π conjugation between PC's main chains and phenylphosphonate benzene rings, ensuring homogeneous dispersion. Zr-BDC@CeHPP delays and suppresses explosive combustion in the ignition and middle stages, and catalyzes char formation in severe combustion, creating a barrier effect. The porous structure of Zr-BDC prolongs TTI and t-PHRR, aiding in safe evacuation during fires (Fig. 18.17) (Sai et al., 2020).

Challenges and current limitations

Despite significant advancements and the potential of MOFs as flame retardants for polymers, several challenges and limitations need to be critically addressed to enhance their practical application, design efficiency, and future development. While MOFs exhibit potential flame-retardant features, their inherent efficiency as standalone flame retardants does not meet the high standards required for effective fire safety. This limitation necessitates the development of more effective MOF composites-based systems or their combination with other flame-retardant materials.

Although a vast number of MOF composites have been synthesized, only a small subset is suitable for use as flame retardants. This is due to specific requirements such as particular metal centers, organic linkers, pore dimensions, and the chemical and thermal stability of the MOF framework. These criteria limit the applicability of many MOF composites in flame retardancy.

Additionally, the effectiveness of MOF composites-based flame retardants varies significantly depending on the chemical structure and thermal degradation processes of different polymers. Consequently, the same MOF composite-based flame retardant can exhibit different mechanisms and performance levels across various polymer materials, complicating the general assessment and application of MOF composite flame retardants.

The scalability and cost-effectiveness of producing MOF composites are major barriers to their widespread commercial application. Current production methods are not sufficiently efficient or economical, hindering the practical use of MOF composites as flame retardants on a large scale. Moreover, inferior interfacial adhesion between MOF composites and polymeric matrices inhibits the design and fabrication of enhanced flame retardant performance in polymer composites. Many metallic derivatives are not compatible with a wide range of organic polymers, and the residual entities on the filler surface are often insufficient to induce strong adherence within the polymer matrix. During actual fire scenarios, high levels of smoke and heat release critically endanger lives and hinder rescue operations. While MOF composites can catalytically reduce flue gas release and promote char formation, their overall impact on smoke suppression and heat release needs further improvement. The formation of nanomaterial aggregates and agglomerates presents a significant challenge. The organic ligands of MOF composites can enhance compatibility with polymer matrices, but further strategies are needed to prevent aggregation and ensure uniform dispersion. To address these challenges, future research should focus on enhancing MOF/polymeric nanoarchitectures through hybridization and modification, promoting existing features and creating properties superior to traditional flame retardants. Developing functionalization strategies that incorporate both inorganic and organic flame retardant entities to fabricate diverse structures such as core-shell, 3-D multilayered, and sheet conveyor architectures is essential. Exploring synergies between MOF composites and other nanomaterials (e.g., graphene oxide, melamine cyanurate, ammonium polyphosphate) can improve flame retardant effectiveness. Additionally, gaining a deeper understanding of the structure–property relationships in MOF/polymer composites, including the roles of porous structures, transition metal species, and organic ligands, will be crucial. By addressing these limitations and exploring innovative solutions, MOF composites could significantly enhance the fire safety of polymers, opening new avenues for high-performance, fire-safe polymer composites across various industries such as energy storage, aerospace, maritime, automotive, and oil and gas.

Conclusion

The field of flame-retardant polymer synthesis has advanced significantly with the introduction of metal-containing complexes incorporating organic ligands and flame-retardant elements. MOFs, renowned for their inherent functionalities and architectural periodicity, offer versatile applications in catalysis, medical science, and gas storage. The catalytic effects of various metal centers in MOFs, including nickel, cobalt, iron, zinc, and copper, have demonstrated efficient char formation, enhancing flame-retardant capabilities. Recent studies have showcased the synergistic effects of MOFs with flame retardants, underscoring their potential in enhancing fire safety.

While MOF composites hold promise for flame retardant polymer composites, challenges persist in large-scale production, cost control, and functionalization complexity. Nevertheless, the incorporation of MOF composites as synergistic flame retardants presents a versatile solution for polymeric nanoarchitectures, offering enhanced flame retardant and multifunctional properties. Research efforts should focus on elucidating structure–property relationships and optimizing functionalization strategies to maximize flame retardant effectiveness.

Furthermore, the catalytic and adsorptive properties of MOF composites, along with their ability to generate char residues, contribute significantly to fire retardancy. The transformation of MOF composites into derivatives and their compatibility with polymer matrices offer additional avenues for exploration. Despite challenges such as high production costs and environmental concerns, the ongoing advancements in MOF-based flame retardants hold tremendous potential for revolutionizing polymer fire safety.

In conclusion, the development of metal-organic hybrids as flame retardants represents an evolving frontier, driven by innovations in surface modification and molecular design. By addressing challenges related to environmental impact, mass production, and efficiency, MOFs and their derivatives can overcome current limitations and establish themselves as indispensable components in enhancing polymer flame retardancy and mechanical properties.

References

Attia, N. F., Oh, H., & El Ashery, S. E. (2023). Design and fabrication of metal-organic-framework based coatings for high fire safety and UV protection, reinforcement and electrical conductivity properties of textile fabrics. *Progress in Organic Coatings, 179*, 107545.

Bellayer, S., Jimenez, M., Prieur, B., Dewailly, B., Ramgobin, A., Sarazin, J., Revel, B., Tricot, G., & Bourbigot, S. (2018). Fire retardant sol-gel coated polyurethane foam: Mechanism of action. *Polymer Degradation and Stability, 147*, 159–167.

Bras, M. L., Bourbigot, S., & Revel, B. (1999). Comprehensive study of the degradation of an intumescent EVA-based material during combustion. *Journal of Materials Science, 34*(23), 5777–5782.

Cao, J., Pan, Y.-T., Vahabi, H., Song, J.-I., Song, P., Wang, D.-Y., & Yang, R. (2024). Zeolitic imidazolate frameworks-based flame retardants for polymeric materials. *Materials Today Chemistry, 37*, 102015.

Chattopadhyay, D. K., & Raju, K. V. S. N. (2007). Structural engineering of polyurethane coatings for high performance applications. *Progress in Polymer Science, 32*(3), 352–418.

Chen, Z., Chen, T., Yu, Y., Zhang, Q., Chen, Z., & Jiang, J. (2019). Metal-organic framework MIL-53 (Fe)@C/graphite carbon nitride hybrids with enhanced thermal stability, flame retardancy, and smoke suppression for unsaturated polyester resin. *Polymers for Advanced Technologies, 30,* 2458–2467.

Chu, F., Xu, Z., Zhou, Y., Zhang, S., Mu, X., Wang, J., Hu, W., & Song, L. (2021). Hierarchical core–shell TiO2@LDH@Ni(OH)2 architecture with regularly-oriented nanocatalyst shells: Towards improving the mechanical performance, flame retardancy and toxic smoke suppression of unsaturated polyester resin. *Chemical Engineering Journal, 405,* 126650.

Deng, G. J., Sun, M. N., Shi, Y. Q., Feng, Y. Z., Lv, Y. C., Tang, L. C., Gao, J. F., & Song, P. A. (2023). Construction of Mxene/MOFs nano-coatings on PU sponge with enhanced interfacial interaction and fire resistance towards efficient removal of liquid hazardous chemicals. *Journal of Cleaner Production, 403,* 13.

Du, J. Y., Fan, X. L., Xin, F., Chen, Y., Feng, K. X., & Hu, J. Y. (2024). Novel coating flame retardant APP@UiO for preparation of flame retardant unsaturated polyester resin composites. *European Polymer Journal, 207,* 11.

Gao, H., Shen, H., Wu, H., Jing, H., Sun, Y., Liu, B., Chen, Z., Song, J., Lu, L., Wu, Z., & Hao, Q. (2021). Review of pristine metal–organic frameworks for supercapacitors: Recent progress and perspectives. *Energy and Fuels, 35,* 12884–12901.

Georgousopoulou, I. N., Vouyiouka, S., Dole, P., & Papaspyrides, C. D. (2016). Thermomechanical degradation and stabilization of poly(butylene succinate). *Polymer Degradation and Stability, 128,* 182–192.

Ghanbarian, M., Zeinali, S., Mostafavi, A., & Shamspur, T. (2020). Facile synthesis of MIL-53 (Fe) by microwave irradiation and its application for robust removal of heavy metals from aqueous solution by experimental design approach: Kinetic and equilibrium. *Analytical and Bioanalytical Chemistry Research, 7,* 263–280.

Gong, K. L., Cai, L., Shi, C. L., Gao, F. Y., Yin, L., Qian, X. D., & Zhou, K. Q. (2022). Organic-inorganic hybrid engineering MXene derivatives for fire resistant epoxy resins with superior smoke suppression. *Composites Part A: Applied Science and Manufacturing, 161,* 107109.

Hosseini, M., & Zeinali, S. (2019). Capacitive humidity sensing using a metal–organic framework nanoporous thin film fabricated through electrochemical in situ growth. *Journal of Materials Science: Materials in Electronics, 30,* 3701–3710.

Hou, B., Zhang, W., Lu, H., Song, K., Geng, Z., Ye, X., Pan, Y.-T., Zhang, W., & Yang, R. (2022). Multielement flame-retardant system constructed with metal POSS–organic frameworks for epoxy resin. *ACS Applied Materials and Interfaces, 14*(43), 49326–49337.

Hou, Y., Hu, W., Gui, Z., & Hu, Y. (2017). A novel Co(II)–based metal-organic framework with phosphorus-containing structure: Build for enhancing fire safety of epoxy. *Composites Science and Technology, 152,* 231–242.

Hou, Y., Hu, W., Zhou, X., Gui, Z., & Hu, Y. (2017). Vertically aligned nickel 2-methylimidazole metal–organic framework fabricated from graphene oxides for enhancing fire safety of polystyrene. *Industrial and Engineering Chemistry Research, 56,* 8778–8786.

Hou, Y. B., Liu, L. X., Qiu, S. L., Zhou, X., Gui, Z., & Hu, Y. (2018). DOPO-modified two-dimensional Co-based metal-organic framework: Preparation and application for enhancing fire safety of poly(lactic acid). *ACS Applied Materials and Interfaces, 10,* 8274–8286.

Huang, C., Zhao, Z. Y., Deng, C., Lu, P., Zhao, P. P., He, S., Chen, S. W., & Lin, W. (2021). Facile synthesis of phytic acid and aluminum hydroxide chelate-mediated hybrid complex toward fire safety of ethylene-vinyl acetate copolymer. *Polymer Degradation and Stability, 190,* 109659.

Huang, R., Guo, X. Y., Ma, S. Y., Xie, J. X., Xu, J. Z., & Ma, J. (2020). Novel phosphorus-nitrogen-containing ionic liquid modified metal-organic framework as an effective flame retardant for epoxy resin. *Polymers, 12*, 108.

Huang, Y. L., & Yuan, B. H. (2021). Reduced graphene oxide/iron-based metal-organic framework nano-coating created on flexible polyurethane foam by layer-by-layer assembly: Enhanced smoke suppression and oil adsorption property. *Materials Letters, 298*, 129974.

Jia, L., Huang, W., Zhao, Y., Wen, S., Yu, Z., & Zhang, Z. (2022). Ultra-light polylactic acid/combination composite foam: A fully biodegradable flame retardant material. *International Journal of Biological Macromolecules, 220*, 754–765.

Jiang, S., Li, S., Yang, X., Liu, L., Li, X., & Xu, M. (2023). A novel strategy for preparing ethylene-vinyl acetate composites with high effective flame retardant and smoke suppression performance by incorporating piperazine pyrophosphate and Ce-MOF. *Journal of Polymer Science, 61*(20), 2426–2439.

Jin, X., Zhang, J., Zhu, Y. L., Zhang, A. Y., Wang, R., Cui, M., Wang, D. Y., & Zhang, X. Q. (2023). Highly efficient metal-organic framework based intumescent poly(L-lactic acid) towards fire safety, ignition delay and UV resistance. *International Journal of Biological Macromolecules, 250*, 126127.

Konnerth, H., Matsagar, B. M., Chen, S. S., Prechtl, M. H. G., Shieh, F.-K., & Wu, K. C. W. (2020). Metal-organic framework (MOF)-derived catalysts for fine chemical production. *Coordination Chemistry Reviews, 416*, 213319.

Kultys, M., & Rogulska, H. (2011). The effect of soft-segement structure on the properties of novel thermoplastic polyurethane elastomers based on an unconventional chain extender. *Polymer International, 60*, 652–659.

Laoutid, F., Bonnaud, L., Alexandre, M., Lopez-Cuesta, J. M., & Dubois, P. (2009). New prospects in flame retardant polymer materials: From fundamentals to nanocomposites. *Materials Science and Engineering: R: Reports, 63*, 100–125.

Lazar, S. T., Kolibaba, T. J., & Grunlan, J. C. (2020). Flame-retardant surface treatments. *Nature Reviews Materials, 5*(4), 259–275.

Levchik, S. V., & Weil, E. D. (2005). Overview of recent developments in the flame retardancy of polycarbonates. *Polymer International, 54*(7), 981–998.

Li, C., Long, Y., Lou, Y., Huo, X., Ma, L., Tang, Y., Hao, C., & Wen, X. (2024). Three-in-one novel flame retardant through grafting organophosphorus onto Fe-MOF functionalized Mg(OH)$_2$ for improving fire safety and mechanical properties of PBS composites. *Sustainable Materials and Technologies, 40*, e00906.

Li, H., Li, L., Lin, R.-B., Zhou, W., Zhang, Z., Xiang, S., & Chen, B. (2019). Porous metal-organic frameworks for gas storage and separation: Status and challenges. *EnergyChem, 1*, 100006.

Li, H., Meng, D., Qi, P., Sun, J., Li, H., Gu, X., & Zhang, S. (2022). Fabrication of a hybrid from metal-organic framework and sepiolite (ZIF-8@SEP) for reducing the fire hazards in thermoplastic polyurethane. *Applied Clay Science, 216*, 106376.

Li, Y., Li, X., Pan, Y.-T., Xu, X., Song, Y., & Yang, R. (2020). Mitigation the release of toxic PH3 and the fire hazard of PA6/AHP composite by MOFs. *Journal of Hazardous Materials, 395*, 122604.

Liu, B., Liu, Z. C., Lu, X. J., Wu, P., Sun, Z. G., Chu, H. Q., & Peng, H. S. (2023). Controllable growth of drug-encapsulated metal-organic framework (MOF) on porphyrinic MOF for PDT/chemo-combined therapy. *Materials and Design, 228*, 111861.

Liu, X., Guo, P., Zhang, B., & Mu, J. (2023). A novel ternary inorganic–organic hybrid flame retardant containing biomass and MOFs for high-performance rigid polyurethane foam. *Colloids and Surfaces A: Physicochemical and Engineering Aspects, 671*, 131625.

Liu, Y., Xu, N., Chen, W., Wang, X., Sun, C., & Su, Z. (2018). Supercapacitor with high cycling stability through electrochemical deposition of metal-organic frameworks/polypyrrole positive electrode. *Dalton Transactions, 47*, 13472–13478.

Lv, X., Zeng, W., Yang, Z., Yang, Y., Wang, Y., Lei, Z., Liu, J., & Chen, D. (2020). Fabrication of ZIF-8@polyphosphazene core-shell structure and its efficient synergism with ammonium polyphosphate in flame-retarding epoxy resin. *Polymers for Advanced Technologies, 31*(5), 997–1006.

Lyon, R. E., & Janssens, M. L. (2015). Polymer flammability. In H. F. Mark (Ed.). *Encyclopedia of polymer science and technology* (pp. 1–70). (4th ed.). Hoboken: John Wiley and Sons.

Ma, Z., Zhang, Z., & Wang, Y. (2021). A novel efficient nonflammable coating containing gC3N4 and 8 intumescent flame retardant fabricated via layer-by-layer assembly on cotton fiber. *Journal of Materials Science, 56*, 9678–9691.

Mehek, R., Iqbal, N., & Noor, T. (2021). Metal-organic framework-based electrode materials for lithium-ion batteries: A review. *RSC Advances, 11*, 29247–29266.

Nabipour, H., Aliakbari, F., Volkening, K., Strong, M. J., & Rohani, S. (2023). New metal-organic framework coated sodium alginate for the delivery of curcumin as a sustainable drug delivery and cancer therapy system. *International Journal of Biological Macromolecules, 12*, 128875.

Nabipour, H., Wang, X., Song, L., & Hu, Y. (2020). Metal-organic frameworks for flame retardant polymers application: A critical review. *Composites Part A: Applied Science and Manufacturing, 139*, 106113.

Nabipour, H., Qiu, S., Wang, X., Song, L., & Hu, Y. (2022). Adenine as an efficient adsorbent for zinc ions removal from wastewater to in situ form bio-based metal–organic frameworks: A novel approach to preparing fire-safe polymers. *Composites Part A: Applied Science and Manufacturing, 161*, 107099.

Nabipour, H., Aliakbari, F., Volkening, K., Strong, M. J., & Rohani, S. (2024). The development of a bio-based metal-organic framework coated with carboxymethyl cellulose with the ability to deliver curcumin with anticancer properties. *Materials Today Chemistry, 37*, 101976.

Nik Pauzi, N. N. P., Majid, R. A., Dzulkifli, M. H., & Yahya, M. Y. (2014). Development of rigid bio-based polyurethane foam reinforced with nanoclay. *Composites Part B: Engineering, 67*, 521–526.

Nobile, M. R., Cerruti, P., Malinconico, M., & Pantani, R. (2015). Processing and properties of biodegradable compounds based on aliphatic polyesters. *Journal of Applied Polymer Science, 132*(48), 42481.

Ponnada, S., Kiai, M. S., & Gorle, D. B. (2021). Insight into the role and strategies of metal-organic frameworks in direct methanol fuel cells: A review. *Energy and Fuels, 35*, 15265–15284.

Qi, X. L., Zhou, D. D., Zhang, J., Hu, S., Haranczyk, M., & Wang, D. Y. (2019). Simultaneous improvement of mechanical and fire-safety properties of polymer composites with phosphonate-loaded MOF additives. *ACS Applied Materials and Interfaces, 11*(22), 20325–20332.

Qin, Y., Li, M., Huang, T., Shen, C., & Gao, S. (2022). A study on the modification of polypropylene by a star-shaped intumescent flame retardant containing phosphorus and nitrogen. *Polymer Degradation and Stability, 195*, 109801.

Reuter, J., Greiner, L., Schönberger, F., & Döring, M. (2019). Synergistic flame retardant interplay of phosphorus containing flame retardants with aluminum trihydrate depending on the specific surface area in unsaturated polyester resin. *Journal of Applied Polymer Science, 136*(13).

Sai, T., Ran, S., Guo, Z., Yan, H., Zhang, Y., Song, P., Zhang, T, Wang, H., & Fang, Z. (2020). Deposition growth of Zr-based MOFs on cerium phenylphosphonate lamella towards enhanced thermal stability and fire safety of polycarbonate. *Composites Part B: Engineering, 197*, 108064.

Shanks, R. A., Wong, S., & Preston, C. M. (2010). Ceramifying fire-retardant and fire-barrier unsaturated polyester composites. *Advanced Materials Research, 123*, 23–26.

Sharma, V., Agarwal, S., Mathur, A., Singhal, S., & Wadhwa, S. (2024). Advancements in nanomaterial based flame-retardants for polymers: A comprehensive overview. *Journal of Industrial and Engineering Chemistry, 133*, 38–52.

Shi, C. L., Wan, M., Hou, Z. B., Qian, X. D., Che, H. L., Qin, Y. P., Jing, J. Y., Li, J., Ren, F., Yu, B., & Hong, N. N. (2022). Co-MOF@MXene hybrids flame retardants for enhancing the fire safety of thermoplastic polyurethanes. *Polymer Degradation and Stability, 204*, 110119.

Shi, L., Zhang, R., Ying, W., Hu, H., Wang, Y., Guo, Y., Wang, W., Tang, Z., & Zhu, J. (2019). Polyether-polyester and HMDI based polyurethanes: Effect of PLLA content on structure and property. *Chinese Journal of Polymer Science, 37*, 1152–1161.

Shi, X., Dai, X., Cao, Y., Li, J., Huo, C., & Wang, X. (2017). Degradable poly(lactic acid)/metal–organic framework nanocomposites exhibiting good mechanical, flame retardant, and dielectric properties for the fabrication of disposable electronics. *Industrial and Engineering Chemistry Research, 56*(14), 3887–3894.

Song, J., Zhang, Y., Wang, J., Gu, C., Hu, J., Yin, P., Shi, X., & Feng, J. (2024). Self-assembly Fe-Co MOF@BN and epoxy resin nano composites with highly enhanced flame retardancy and smoke suppression properties. *Polymer Degradation and Stability, 225*, 110811.

Song, K., Zhang, H., Pan, Y.-T., Ur Rehman, Z., He, J., Wang, D.-Y., & Yang, R. (2023). Metal-organic framework-derived bird's nest-like capsules for phosphorous small molecules towards flame retardant polyurea composites. *Journal of Colloid and Interface Science, 643*, 489–501.

Song, K., Bi, X., Yu, C., Pan, Y. T., Xiao, P., Wang, J., Song, J.-I., He, J., & Yang, R. (2024). Structure of metal–organic frameworks eco-modulated by acid–base balance toward biobased flame retardant in polyurea composites. *ACS Applied Materials and Interfaces, 16*(12), 15227–15241.

Stec, A. A., Hull, T. R., Lebek, K., Purser, J. A., & Purser, D. A. (2008). The effect of temperature and ventilation condition on the toxic product yields from burning polymers. *Fire and Materials, 32*, 49–60.

Velencoso, M. M., Battig, A., Markwart, J. C., Schartel, B., & Wurm, F. R. (2018). Molecular firefighting—How modern phosphorus chemistry can help solve the challenge of flame retardancy. *Angewandte Chemie International Edition, 57*(33), 10450–10467.

Venkateshalu, S., Tomboc, G. R. M., Nagalingam, S. P., Kim, J., Sawaira, T., Sehar, K., Lee, K., Kim, J. Y., Nirmala Grace, A., & Lee, K. (2023). Synergistic MXene/LDH heterostructures with extensive interfacing as emerging energy conversion and storage materials. *Journal of Materials Chemistry A, 11*(27), 14469–14488.

Wan, M., Shi, C., Chen, L., Deng, L., Qin, Y., Che, H., Jing, J., Li, J., & Qian, X. (2024). Core-shell flame retardant/APP-PEI@MXene@ZIF-67: A nanomaterials self-assembly strategy towards reducing fire hazard of thermoplastic polyurethane. *Polymer Degradation and Stability,* 110821 12 May.

Wang, G., Xu, W., Chen, R., Li, W., Liu, Y., & Yang, K. (2020). Synergistic effect between zeolitic imidazolate framework-8 and expandable graphite to improve the flame retardancy and smoke suppression of polyurethane elastomer. *Journal of Applied Polymer Science, 137*, 48048.

Wang, G., Mei, Z., Li, Y., Chen, G., Xin, J., Sun, Z., Wang, Q., & Luo, X. (2023). Metal–organic frameworks derived from Zn and P -hydroxybenzoic acid for smoke suppression and flame retardation in vinyl resin. *Materials and Engineering, 308*(2), 2200461.

Wang, L., Song, L., Hu, Y., & Yuen, R. K. K. (2013). Influence of different metal oxides on the thermal, combustion properties and smoke suppression in ethylene-vinyl acetate. *Industrial and Engineering Chemistry Research, 52*(23), 8062–8069.

Wang, R. Z., Chen, Y., Liu, Y. Y., Ma, M. L., Tong, Z. Y., Chen, X. L., Bi, Y. X., Huang, W. B., Liao, Z. J., Chen, S. L., Zhang, X. Y., & Li, Q. Q. (2021). Metal-organic frameworks derived ZnO@MOF@PZS flame retardant for reducing fire hazards of polyurea nanocomposites. *Polymers for Advanced Technologies, 32*, 4700–4709.

Wang, X., Zhang, P., Huang, Z., Xing, W., Song, L., & Hu, Y. (2019). Effect of aluminum diethylphosphinate on the thermal stability and flame retardancy of flexible polyurethane foams. *Fire Safety Journal, 106*, 72–79.

Wang, X. G., Qi, P., Zhang, S. J., Jiang, S. L., Li, Y. C., Sun, J., Fei, B., Gu, X. Y., & Zhang, S. (2023). A novel flame-retardant modification strategy for UiO66-NH2 by encapsulating triethyl phosphate: Preparation, characterization, and multifunctional application in poly (lactic acid). *Materials Today Chemistry, 30*, 101550.

Weil, E. D., & Levchik, S. V. (2009). Current practice and recent commercial developments in flame retardancy of polyamides. In *Flame retardants*. Hanser, 85–104.

Wen, O. Y., Tohir, M. Z. M., Yeaw, T. C. S., Razak, M. A., Zainuddin, H. S., & Hamid, M. R. A. (2023). Fire-resistant and flame-retardant surface finishing of polymers and textiles: A state-of-the-art review. *Progress in Organic Coatings, 175*, 107330.

Wu, D., Li, Q., Shang, X., Liang, Y., Ding, X., Sun, H., Li, S., Wang, S., Chen, Y., & Chen, J. (2021). Commodity plastic burning as a source of inhaled toxic aerosols. *Journal of Hazardous Materials, 416*, 125820.

Wu, J., Yin, Z., Sun, X., Zhang, X., Zhu, Z., Xu, Z., Yang, J., Xie, Z., Li, Y., Yang, X., Huang, Q., Liu, J., & Wang, J. (2022). Enhanced fire-proofing performance and crystallizability of bio-based poly(L-lactic acid): Dual functions of a Schiff base-containing synergistic flame retardant. *International Journal of Biological Macromolecules, 222*, 305–324.

Xiong, Y., Jiang, Z., Xie, Y., Zhang, X., & Xu, W. (2013). Development of a DOPO-containing melamine epoxy hardeners and its thermal and flame-retardant properties of cured products. *Journal of Applied Polymer Science, 127*(6), 4352–4358.

Xu, B., Xu, W., Liu, Y., Chen, R., Li, W., Wu, Y., & Yang, Z. (2018). Surface modification of α-zirconium phosphate by zeolitic imidazolate frameworks-8 and its effect on improving the fire safety of polyurethane elastomer. *Polymers for Advanced Technologies, 29*(11), 2816–2826.

Xu, Y., Yang, W.-J., Zhou, Q.-K., Gao, T.-Y., Xu, G.-M., Tai, Q.-L., Zhu, S.-E., Lu, H.-D., Yuen, R. K. K., Yang, W., & Wei, C.-X. (2022). Highly thermo-stable resveratrol-based flame retardant for enhancing mechanical and fire safety properties of epoxy resins. *Chemical Engineering Journal, 450*, 138475.

Xu, Z., Xing, W., Hou, Y., Zou, B., Han, L., Hu, W., & Hu, Y. (2021). The combustion and pyrolysis process of flame-retardant polystyrene/cobalt-based metal organic frameworks (MOF) nanocomposite. *Combustion and Flame, 226*, 108–116.

Yang, B., Chen, Z., Yu, Y., Chen, T., Chu, Y., Song, N., Zhang, Q., Liu, Z., Liu, Z., & Jiang, J. (2023). Construction of bimetallic metal-organic frameworks/graphitic carbon nitride hybrids as flame retardant for unsaturated polyester resin. *Materials Today Chemistry, 30*, 101482.

Yang, H., Yu, B., Song, P., Maluk, C., & Wang, H. (2019). Surface-coating engineering for flame retardant flexible polyurethane foams: A critical review. *Composites Part B: Engineering, 176*, 107185.

Zhang, J., Li, Z., Qi, X.-L., & Wang, D.-Y. (2020). Recent progress on metal–organic framework and its derivatives as novel fire retardants to polymeric materials. *Nano-Micro Letters, 12*(1). https://doi.org/10.1007/s40820-020-00497-z.

Zhang, S., Yan, Y., Wang, W., Gu, X., Li, H., Li, J., & Sun, J. (2018). Intercalation of phosphotungstic acid into layered double hydroxides by reconstruction method and its application in intumescent flame retardant poly (lactic acid) composites. *Polymer Degradation and Stability, 147*, 142–150.

Zhao, L., Xu, G., Gao, C., & Song, P. (2023). A novel RhB@ MOF-808 fluorescent probe for the rapid detection of dopamine and Fe3+. *Analytical Biochemistry, 671*, 115154.

Chapter 19

Metal-organic framework composites as sensors for gases and volatile compounds

Alaa Bedair[1], Reda M. Abdelhameed[2], Marcello Locatelli[3], and Fotouh R. Mansour[4,5]

[1]Department of Analytical Chemistry, Faculty of Pharmacy, University of Sadat City, Sadat City, Monufia, Egypt, [2]Applied Organic Chemistry Department, Chemical Industries Research Institute, National Research Centre, Giza, Egypt, [3]Department of Science, University "G. d'Annunzio" of Chieti-Pescara, Chieti, Italy, [4]Medicinal Chemistry Department, Faculty of Pharmacy, King Salman International University (KSIU), Ras Sudr, South Sinai, Egypt, [5]Department of Pharmaceutical Analytical Chemistry, Faculty of Pharmacy, Tanta University, Tanta, Egypt

Introduction

Typically, ambient air contains O_2, N_2, noble gases, CO_2, and trace quantities of other gases. There is no need to indicate that air quality is critical for all living species on Earth. Human interference and geological dangers have affected the composition of air through the release of vapors and volatile organic compounds (VOCs). Many gaseous emissions including greenhouse gases, harmful gases (e.g., H_2S, NH_3, SO_2, NO_x, CO_x, CS_2), and VOCs, have significant negative impacts on the world's climate and environment (Barea et al., 2014; He et al., 2019; Woellner et al., 2018). So, there is an urgent need to develop successful gas and VOC sensors for monitoring the surrounding environment, protecting personnel from harmful chemicals, controlling the quality of food in agriculture-related industries, early detection of illnesses in medicine, and monitoring indoor air quality (IAQ) (Rasheed & Nabeel, 2019; Wang, Kaneti, et al. 2018; Wang, Lustig, et al., 2018; Wenger, 2013). While traditional gas sensors are extensively utilized, they suffer from certain limitations. Usually, for gas sensing applications, analytical methods including mass spectroscopy, ion mobility spectroscopy, gas chromatography (GC), and nuclear magnetic resonance (NMR) have long been employed. In addition to being expensive and sophisticated, the majority of these procedures need sample preparation, which makes real-time analysis challenging, and therefore are not appropriate for live monitoring. Sensitive solid-state instruments have so taken the role of traditional analytical methods. When it comes to real-time monitoring, "chemical gas sensors," also known as electronic noses (E-noses) or similar devices, are more effective and require less complex circuitry for analysis compared to previously available sensors (Beale et al., 2017; Garg et al., 2021; Wilson, 2018). These devices employ the transduction principle to alter the physical properties (such as resistance, capacitance, and current) of the sensor layer when it comes into contact with analytes. This alteration is subsequently converted into a measurable signal. Chemo-resistive gas sensors employ changes in the conductivity of electricity or resistance of the material being sensed to identify different gaseous or chemical analytes (Park et al., 2019). The performance of a gas sensor is determined by a number of sensing processes and phenomena that occur in chemo-resistive devices, including the transfer of charge, swelling effect, Schottky barrier modulation, and hydrogen bonding between the sensing layer and the analyte. An optimal sensor should possess the necessary sensitivity, long-term stability, selectivity, rapid response and recovery times, and cost-effective production. The sensor must be reversible and deliver consistent results under similar environmental circumstances. A vast range of materials, including carbon materials and their composites, metal oxides, conducting polymers, and organic and inorganic substances, have been utilized to fabricate gas sensors for the detection of dangerous VOCs and gases (Neri, 2015; Pandey, 2016; Park et al., 2017; Zamiri, 2020). Developing an efficient and selective chemo-resistive gas sensor requires careful consideration of sensing layer material, manufacturing process, nanostructured sensor pore size, and pore distribution. Developing smart sensors that work well for gases and VOCs is critical (Li, Zhao, et al., 2020). Advanced materials, including metal oxides, conductive polymers, carbon-based substances (e.g., carbon nanotubes, graphene), metallic nanoparticles, quantum dots, and metal-organic frameworks (MOFs), have been produced to fulfill current technological demands (Ghanbarian et al., 2018). Gas sensing applications benefit from composite materials due to their high sensitivity, long shelf life, responsiveness, and recovery time (DMello et al., 2018; Panes-Ruiz et al., 2018; Ren et al. 2018, 2019). Nevertheless, these materials are subject to one or more of the following limitations when utilized as a single component: instability, baseline drift, or elevated temperature requirements. Consequently, when two or more different nanomaterials combine to produce a composite, a new material with a unique structure and shape is created. This new material is expected to offer improved performance compared to the original components.

MOFs are porous crystalline structure substances that self-assemble from nodes of metal ions or clusters and organic ligands through coordination bonds (Abdelhameed et al., 2023; Hammad et al., 2024; Mansour, Abdelhameed, et al., 2024; Mansour, Hammad, et al., 2024). In this chapter, the benefits of using MOF-based hybrids for the sensing layer of chemo-resistive gas sensors are demonstrated. MOF composites provide improved functionality, more adsorption sites, higher porosity levels, improved stability, enhanced catalytic efficiency, and other features when compared to MOFs as a single component material. In this chapter, MOF-based composites as sensors for gases and volatile compounds will be discussed.

Characterization of metal-organic framework composites

MOFs are essentially organic linkers and metal nodes in three-dimensional coordination polymers that provide large surface area, adjustable pore structures, and flexibility toward post synthetic variations (Sahoo et al., 2019; Subudhi, Paramanik, et al., 2020). But practically, MOFs must demonstrate heat, mechanical, chemical, and water stability. Since aqueous stability is a crucial requirement, metal nodes and organic linkers should be chosen in order to preserve MOFs' aqueous stability in accordance with the HSAB principle, also known as the hard/soft acid/base principle (Subudhi, Swain, et al., 2020). Furthermore, it is imperative to employ functionalized linkers in conjunction with either an in situ approach or the post-synthetic modification (PSM) method to get visible active MOFs (Subudhi, Tripathy, & Parida, 2021). The physiological properties of any MOF composite must be analyzed, which is why this section presents a variety of reported characterization techniques, including structural analysis using electron microscopy, catalytic performance characterization, and mechanism analysis aspects of various MOF composites. The basic and crucial factors are topography and physical structure, which may be examined using experimental methods such as high-resolution transmission electron microscope (HRTEM) and field emission scanning electron microscopy (FESEM). Crystallographic characteristics can be identified using experimental techniques like as X-ray diffraction (XRD). X-ray analysis involves interpreting scattered X-ray intensity from crystal surfaces (Subudhi, Tripathy, & Parida, 2021). Surface chemistry is characterized by Fourier-transfer infrared spectroscopy (FT-IR). The Brunauer-Emmett-Teller (BET) analyzer utilizes nitrogen adsorption to determine the specific surface area and pore size distribution. Thermal stability is assessed by thermogravimetric analysis (TGA).

Zuo et al. (2020) developed MOF@rGO (rGO: reduced graphene oxide (Fig. 19.1A)) for water purification from chromate with the aid of an electrochemical strategy. SEM pictures of the MOF@rGO electrode surface revealed that a porous network is formed by the 20–30 μm-sized rGO sheets that were equally dispersed throughout the graphite sheet (Fig. 19.1B). The consistent distribution of N and Co over the rGO surface, as shown by the EDS data (Fig. 19.1C), suggesting that Co-MOF was successfully formed on the rGO surface. The FT-IR spectra of the GO, Co-MOF, and MOF@rGO powders were presented in Fig. 19.1D. The distinctive peaks of C=O stretching at 1733 cm^{-1}, O–H group at 1619 cm^{-1}, and C–O stretching at 1070 cm^{-1} show that the as-synthesized GO included an abundance of hydroxyl, epoxy, and carbonyl groups. These findings supported the existence of amine and

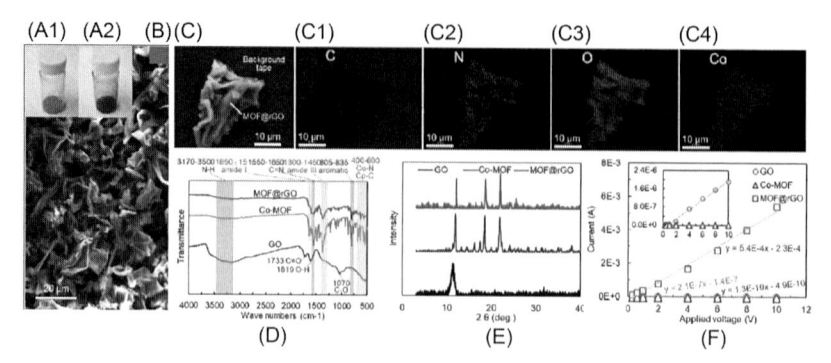

FIGURE 19.1 GO, Co-MOF, and MOF@rGO properties. (A) Photograph of Co-MOF and MOF@rGO powders; (B) SEM image of the electrode; (C) Zoomed-in SEM image of one rGO sheet with elemental mapping for C, N, O, and Co; (D) FT-IR; (E) Powder XRD; and (F) Electrical conductivity (Zuo et al., 2020).

carboxyl functional groups in the Co-MOF structure. The spectra of MOF@rGO is similar to that of Co-MOF, with weaker peaks. During the 3-day hydrothermal synthesis of MOF@rGO, the mass ratio of Co-MOF and GO was 1:1. However, the absence of prominent absorbance peaks in the GO spectrum suggested that GO underwent chemical transformation, most likely reduction. XRD measurements of GO, Co-MOF, and MOF@rGO powders indicate that GO decreased during MOF@rGO synthesis. Fig. 19.1E displayed the XRD spectrum of GO, which revealed a distinct peak at $2\theta = 11.7$ degrees. The Co-MOF powder had a similar XRD spectrum as previously described (Song et al., 2012), with a sharper peak at 2θ of 11.7 degrees, similar to the GO peak but with a smaller half peak breadth due to high crystallinity. The MOF@rGO spectrum has strong peaks at $2\theta = 11.7$, 18.4, and 21.8 degrees, similar to the Co-MOF spectrum. However, the new peak at $2\theta = 25.5$ degrees is a distinctive feature of rGO (Gurunathan et al., 2012). In addition to that the electrical conductivity of MOF@rGO (0.01 S m) is significantly greater than that of Co-MOF (2.48×10^{-9} S m) and GO (4.01×10^{-6} S m) (Fig 19.1F). All of these data point to the reduction of GO to rGO, which has a greater electrical conductivity (Jung et al., 2008). The MOF@rGO's strong electrical conductivity makes it suitable for use as a material for electrodes in electrochemical systems used in the purification of water.

Metal-organic framework-based graphene/graphene derivatives composites for gas sensing

Due to their high charge carrier movement, high surface-to-volume ratio, greater contact with gas molecules, and simple modification; graphene and its derivatives, such as GO and rGO, have been widely used in chemo-resistive sensor devices. These traits are necessary for high-performance chemical sensors. Additionally, epoxy, hydroxyl, and carboxylic groups on the sheet's surface and edges produce vacancies for target analytes (Bibi et al., 2013; Krishnan et al., 2019).

As indicated previously, the combination of MOFs with carbon-based substances offers benefits from both carbon-based substances and MOFs. Graphene and its derivatives have been recognized for their unusual structure, low level of toxicity, and exceptional electrical, thermal, electrochemical, and mechanical properties. Since the majority of the metal ions utilized in MOFs are electrochemically active, they offer considerable promise in the area of electrochemistry in addition to the uses already described (Sun et al., 2017). Crucially, MOFs may modify the locations of atomic metal centers, reducing mass consumption and increasing the interface of the electrode (Wang et al., 2014; Xing et al., 2018). Unfortunately, single MOFs still have several drawbacks that keep them from having the biggest influence on the development of chemical sensors. These include limited electrocatalytic activity, low mechanical stability, and poor electrical conductivity. Numerous studies have demonstrated that adding graphene-based compounds to other materials can result in surprisingly high gains in electrical conductivity and stability (Yang et al., 2016, 2017). So, adding graphene and graphene derivatives enhances MOFs' conductivity and selectivity (Yang et al., 2015).

Tung et al. (2020) developed three different pristine graphene MOFs for detection of various VOCs biomarkers including tetrahydrofuran, methanol, chloroform, ethanol, acetone and acetonitrile. In this study, three pristine graphene-MOF composites (pG-Cu BTC, pG-UiO 66, and pG-ZIF 8) were synthesized by shear mixing of graphene and MOF in a 2:1 molar ratio. These nanocomposites are capable of detecting VOC biomarkers with greater sensitivity and selectivity. This nanohybrid material improves sensing capability by combining graphene, an extremely conductive sensor element, with MOFs that have a large surface area and capability for adsorption, resulting in increased sensitivity and selectivity for the studied VOCs. The pG-Cu BTC sensor demonstrated the maximum sensitivity and selectivity for chloroform and methanol VOCs at 2.82–22.6 ppm levels.

More et al. (2023) developed Zn-BDC@rGO for discrimination between ammonia (NH_3), carbon monoxide (CO), and sulfur dioxide (SO_2). The synthesized Zn-BDC@rGO composite was tested for NH_3, CO, and SO_2 detection via chemiresistive modality after being deposited via drop casting method on electrodes made of copper on a glass substrate (100 μm gap) via the shadow mask method by the e-beam evaporator. The sensor that was created selectively discriminated among the CO, SO_2, and NH_3 gases and demonstrated a fast response/recovery time, that is, 60/120 s at 20 ppm. The linear voltage sweep approach is used to examine the current-voltage characteristics of the sensors (Fig. 19.2). GO has been found to have extremely low conductivity. Zn-BDC MOF exhibits almost insulating properties with resistances of 100 kΩ and 1 M, respectively; nevertheless, its resistance dropped to 1 kΩ during thermal reduction of GO. In contrast, the Zn-BDC@rGO composite's S-shape and 200 kΩ resistance

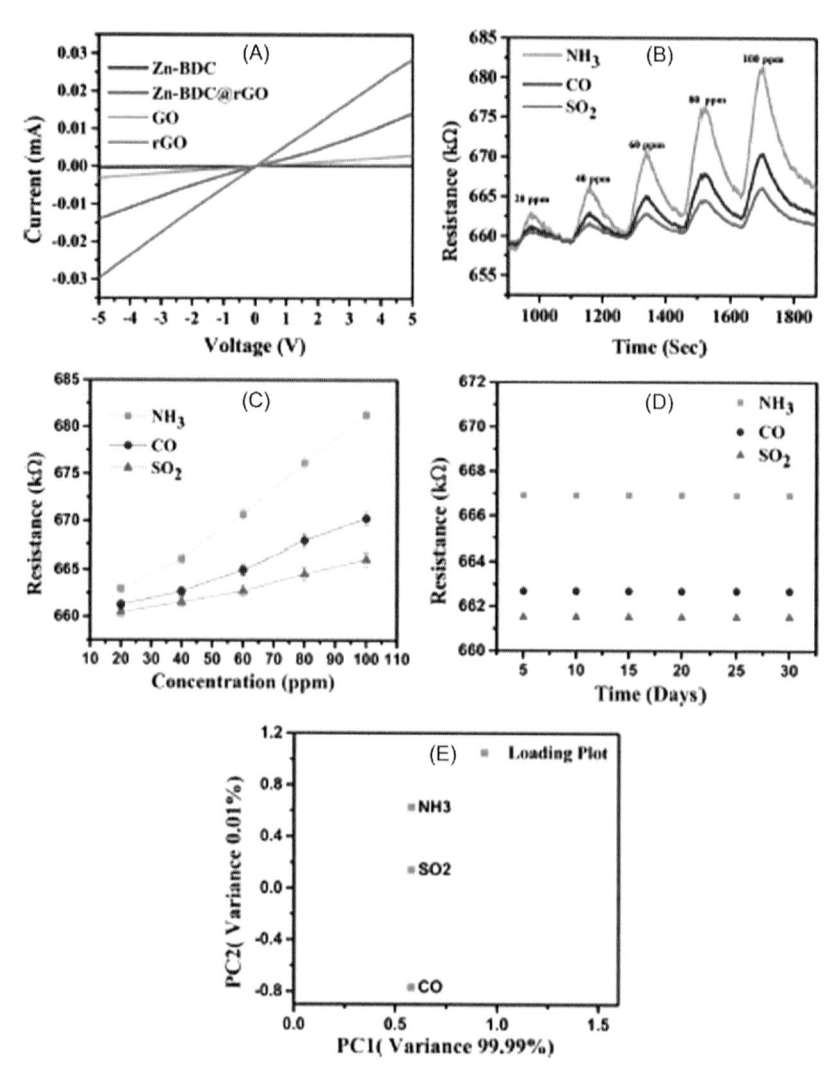

FIGURE 19.2 (A) Zn-BDC and Zn-BDC@rGO I–V characteristics, (B) Chemiresistive sensing response of the composite for NH_3, CO, and SO_2 gases, (C) Calibration plot, (D) Stability at 60 ppm, and (E) (PCA) of the sensor (More et al., 2023).

support the semiconducting behavior. A dynamic gas sensing system was used for the chemiresistive sensing study. Controlling the gas concentration and flow rate was achieved with Alicat Scientific mass flow controls. Measurements of the resistance variations with respect to shifts with time and concentrations of gases were made using the semiconductor parameter analyzer. The real-time detecting plot was shown in Fig. 19.2B. At 20 ppm, the manufactured sensor distinguishes between NH$_3$, CO, and SO$_2$ gases. For the NH$_3$, CO, and SO$_2$ gases, Fig. 19.2C displayed a linear dependence with extremely low divergence between resistance and analyte concentration. The relative correlation coefficient (R^2) values are 0.98447, 0.97875, and 0.98482, for NH$_3$, CO, and SO$_2$ gases, respectively. By monitoring the sensor's reaction to NH$_3$, CO, and SO$_2$ at 60 parts per million (ppm) for 30 days, the sensor's consistent behavior was examined (Fig. 19.2D). Every 5 days, measurements were made by stabilizing the baseline. The results indicated that the device remains stable for up to 30 days for all tested gases, including NH$_3$, CO, and SO$_2$. The sensor successfully differentiated between three gases (NH$_3$, CO, and SO$_2$), as demonstrated by the principal component analysis (PCA) plot (Fig. 19.2E). The sensing mechanism is most likely a result of the combined effect of zinc-based benzene dicarboxylate (Zn-BDC) and rGO. The porous structure, extensive surface area, and catalytic properties of Zn-BDC make it well-suited for gas absorption. Nevertheless, its low electrical conductivity restricts its usefulness in chemiresistive sensing. The epoxy group in partially reduced graphene oxide replaces the oxygen atom in ZnO$_4$ tetrahedra, resulting in the formation of a heterojunction between Zn-BDC and rGO. The addition of rGO to Zn-BDC improves the electrical conductivity and catalytic performance of the material. Exposure to an analyte enhances the charge transition capabilities and alters the electrical conductivity of this material (Petit & Bandosz, 2009).

Travlou et al. (2015) developed the Ca-BTC-GO hybrid for sensing of NH$_3$. The results indicated that combining MOFs with graphene improved their electrical transport capabilities, making them suitable for sensing applications. Despite the absence of the crystal-like porosity structure in reversible sensors, the amorphous phase (resulting from the collapse of MOF) can still adsorb NH$_3$ weakly in the vicinity of the graphene-based phase. This enables the detection and recording of changes in electrical signals. At this stage, adsorption occurs through the formation of NH$_3$ complexes at metal sites, acid and base reactions involving ligand carboxylic groups, and particular interactions with graphene phases. These processes lead to a lasting change in the signal. On the other hand, the process of physisorption occurring in unaltered pores, the interactions between graphene phases through dispersive forces, and the relatively weak connection between NH$_3$ and BTC acid at the GO scaffold result in a reversible sensing mechanism. In addition to that the developed composites had a synergistic influence on conductivity, allowing for carrier mobility. The electrical alterations caused by these interactions resulted in an increasing in electron-donating sites on the conducting phase.

Metal-organic framework-based carbon nanotube composites for gas sensing

Carbon nanotubes (CNTs) are extensively employed as carbon-based nanostructures. These structural forms can be categorized into three main types: single-walled, double-walled, and multi-walled. The three varieties of CNTs possess an sp^2-carbon structure that is hollow and cylindrical. They exhibit variations in terms of their length, diameter, and the number of walls they have. Due to their unique mechanical, optical, thermal, and electrical characteristics, they have attracted considerable attention and find application in various domains (Anik & Çevik, 2009; Timur et al., 2007). As indicated previously, MOF is characterized by crystalline organized structures, customizable pore diameters, high surface areas, chemical stability, and thermal stability (Wang et al., 2016). So, combining MOFs with CNTs in biosensors that are electrochemical enhances conductivity and electrocatalytic activity (Kempahanumakkagari et al., 2018; Wen et al., 2015; Zhou et al., 2014).

Ghanbarian et al. (2018) used MIL-53(Cr-Fe)/Ag/CNT hybrid to develop thin film for resistive sensors for sensing VOCs. In this work, a sonochemical approach was used to synthesize MIL-53(Cr-Fe) nanoparticles, a type of bimetallic MOF. Using these nanoparticles, a MIL-53(Cr-Fe)/Ag/CNT ternary nanocomposite was created to develop a resistive gas sensor device. This device can identify VOCs in the atmosphere, such as methanol, ethanol, and isopropanol, at 10% relative humidity and 25°C. This ternary nanocomposite showed significant reactivity to polar VOCs, especially methanol. For methanol, the lowest detection limit is 30.5 ppm. Gas sensing measurements were taken under ambient conditions using a homemade sensing chamber. The constructed resistive sensor was inserted into the sensing chamber, as shown in Fig. 19.3 Certain amounts of VOCs were delivered into the detecting chamber and placed over the microheater to achieve the correct vapor concentration.

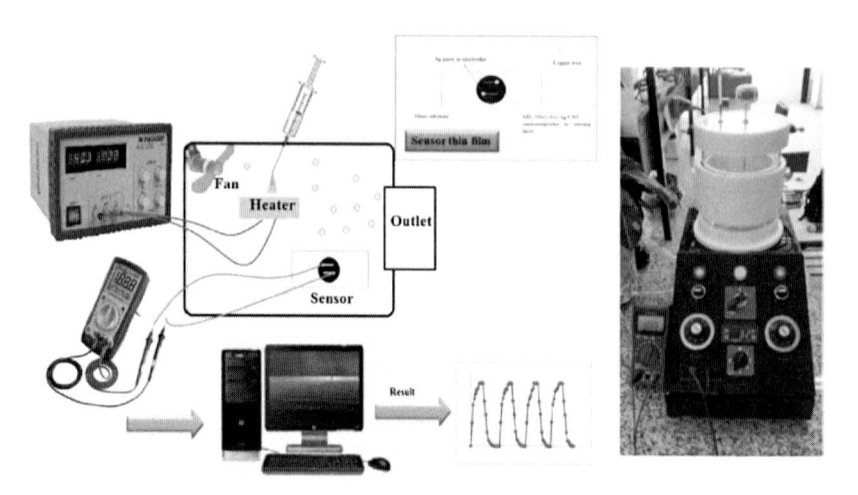

FIGURE 19.3 Diagrammatic illustration of the experimental arrangement for resistive gas sensing, along with the composition of the thin film sensor, made up of MIL-53(Cr-Fe)/Ag/CNT (Ghanbarian et al., 2018).

Homayoonnia and Kim (2023) designed a sensor by coating a quartz crystal microbalance (QCM) with ZIF-8/MWCNT. This sensor was able to detect VOCs at a concentration of 0.62 ppm under normal environmental conditions. The performance of this sensor was superior to using ZIF-8 alone. The selectivity for acetone, a crucial biomarker for diabetes, was diminished in the presence of other VOCs. This study investigated the impact of employing various sensing platforms and response mechanisms on achieving a high level of specificity for detecting acetone. The study utilized a ZIF-8/MWCNT-coated QCM and a chemiresistive sensor. The chemiresistive sensor exhibited excellent performance in detecting acetone, demonstrating high selectivity, minimum sensitivity to humidity, robust repeatability, perfect reversibility, and rapid response time.

The detection limit under ambient settings was 1.7 parts per billion (ppb), which is three orders of magnitude lower than the concentration observed in the exhaled breath of diabetes patients (1.8 ppm). The molecules were adsorbed into the pores and surfaces of the ZIF-8 material by van der Waals contact when exposed to different gases or vapors at room temperature (25°C). The process of adsorption caused an increase in the mass of the QCM crystal, leading to a reduction in its resonance frequency. The resistance changed due to the van der Waals interaction, which facilitates the flow of electrons between the ZIF-8/MWCNT surface and the gas/vapor molecules. The QCM sensor, covered with ZIF-8, exhibited negligible response ($\Delta F \cong 0$) to acetone, isopropyl alcohol (IPA), ethanol (EtOH), and methanol (MeOH) gases, as depicted in Fig. 19.4A. Fig. 19.4B showed that the QCM sensor coated with ZIF-8/MWCNT was very responsive to all VOCs at a concentration of 0.62 ppm. The composite's narrower energy gap facilitated electron sharing with gas molecules, leading to enhanced gas adsorption in comparison to pure ZIF-8. The chemiresistive sensor coated with ZIF-8/MWCNT demonstrated excellent specificity toward acetone and negligible response to other VOCs (Fig. 19.4C).

It should be noted that previous scientific studies have shown that alcohols and acetone have very low sorption levels in ZIFs at ambient pressure and room temperature (Cousinsaintremi et al., 2011; Zhang et al., 2013). As depicted in Fig. 19.4A, the QCM sensor covered with ZIF-8 exhibited negligible sensitivity to acetone, IPA, EtOH, and MeOH under ambient pressure and room temperature (25°C). The mass of the QCM sensor remains constant, suggesting that there is very little adsorption of the analyte in the pores of ZIF-8. Zhang and Snurr (2017) and Canivet et al. (2014) have found that the absence of hydrogen bond groups and uncoordinated centers in ZIF-8 material makes it unsuitable for adsorbing alcohols and acetone molecules at atmospheric pressure. ZIF-8 is distinct from other MOFs due to its absence of metal oxide clusters, which are commonly employed for the adsorption of alcohols and acetone by hydrogen bonding. Regrettably, the utilization of uncoated multi-walled carbon nanotubes (MWCNTs) in the ZIF-8/MWCNTs

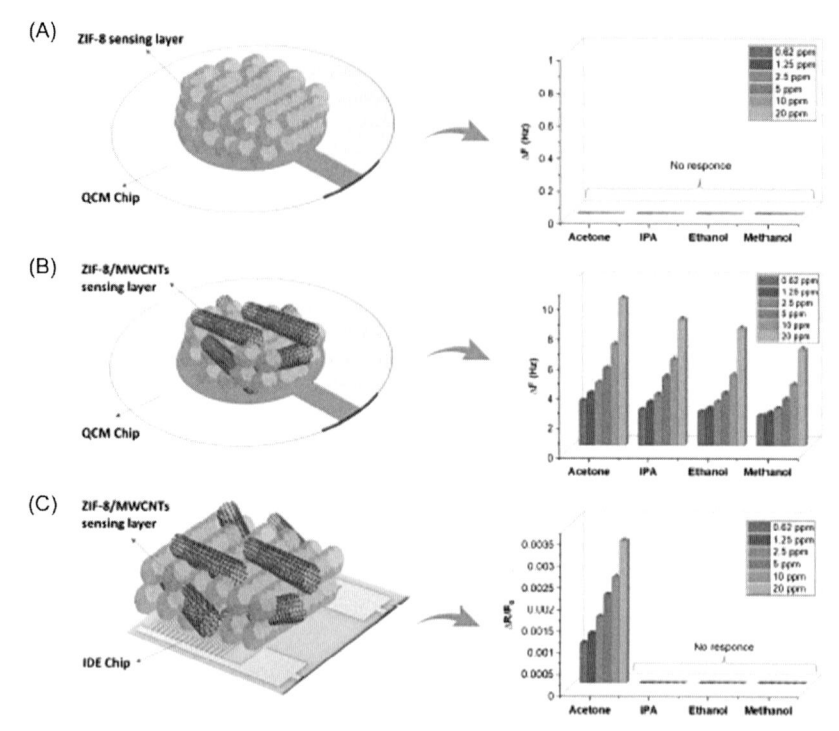

FIGURE 19.4 Responses of three different sensors: (A) ZIF-8-coated QCM, (B) ZIF-8/MWCNT-coated QCM, and (C) ZIF-8/MWCNT-coated IDE. The sensors were exposed to varied concentrations (ranging from 0.62 to 20 ppm) of acetone, IPA, EtOH, and MeOH at room temperature (Homayoonnia and Kim, 2023).

composite lacks hydrogen bond groups, which hinders the adsorption of trace amounts of alcohols and acetone under normal atmospheric pressure. Consequently, the disparities in surface area between ZIF-8 and CNT-embedded ZIF-8 are discernible solely when alcohols and acetone are present in high concentrations or when subjected to elevated pressure. At low levels of VOCs under atmospheric pressure, variations in BET surface area among materials have a minimal effect.

Ingle et al. (2021) developed chemiresistive gas sensor using Ni-MOF modified –OH-SWNTs and –OH-MWNTs for detection of SO_2. The results indicated that the addition of -OH-SWNTs and -OH-MWNTs significantly enhanced Ni-MOF's electrical and morphological characteristics and increased its capacity to sense SO_2 gas at ambient temperature. In addition to that the study examined the specific response of Ni-MOF/-OH-SWNTs and Ni-MOF/-OHMWNTs to different gases including SO_2, NO_2, NH_3, and CO analytes (0.5–15 ppm) by tracking alterations in the material's resistance to electricity at room temperature. According to this work, doping the MOF with -OH-SWNTs and -OH-MWNTs effectively increases the sensing properties.

Metal-organic framework-based metal and metal oxide composites for gas sensing

Metal oxide gas sensors, also known as solid-state resistive devices, have become prevalent in a variety of applications related to energy conservation and combustion process emission control, as well as health and safety (such as medical diagnostics, food processing, air quality monitoring, and the detection of toxic, flammable, and explosive gases) (Miller et al., 2014; Stetter & Li, 2008). Metal oxide-based gas sensors face the challenge of improving selectivity due to changes in resistance caused by various gas species, particularly VOCs. Numerous studies have been conducted to improve selectivity and other sensing parameters of resistive-type metal oxide gas sensors (Kim & Lee, 2014; Kolmakov & Moskovits, 2004; Na et al., 2011; Zhang et al., 2011). So, to reduce false-positive or interfering signals, new methods of improving the materials' selectivity are required. It is worth indicating that the pores inside MOFs can be functionalized by changing the pore diameters to make advantage of size-exclusive effects for the isolation (Chen et al., 2010; Jiang & Xu, 2011). As a result, MOF-metal oxide hybrid enhanced the selectivity in comparison with metal oxide.

Drobek et al. (2016) developed ZnO@ZIF-8 gas sensor, the hybrid exhibited a significantly more specific response to H_2 when compared to the unaltered zinc oxide nanowires (ZnO NWs) sensor. To enhance gas sensing selectivity, ZnO NWs were encapsulated with a MOF-based membrane in two steps: (1) growth on the sensor support and (2) surface conversion to create a thin ZIF-8-selective barrier (Fig. 19.5).

The technique was based on using ZnO NWs as a special supply of zinc, which was then utilized to build a thin ZIF-8 layer on the surface of the nanowires. Actually, using this technique allowed for uniform ZnO NWs coverage with a ZIF-8 selective barrier (membrane) that had strong molecular sieving capabilities and a high comparable surface area. The pure ZnO NWs surface was converted by placing it in contact with a methanolic solution of 2-methylimidazole (2-mim) at 100°C for 24 hours, resulting in the encased ZnO@ZIF-8 NWs. The objective was to preserve the linked ZnO network on the sensor support while developing a continuous ZIF-8 layer on the ZnO NWs. Indeed, a ZIF-8 layer grew to cover the whole surface of the NWs made of ZnO under ideal reaction circumstances at a 2-mim concentration of 10% wt %. ZnO is converted to ZIF-8 by a coordination process between imidazolium anions (mim$^-$) and Zn^{2+} cations (ZnO top-surface) in the solution. In the event that the dissolution and coordination rates of Zn^{2+} are properly balanced, leaching of metals is not anticipated to occur during the solvothermal synthesis. With just a limited (top-surface) conversion of ZnO to ZIF-8,

Growth of networked ZnO nanowires

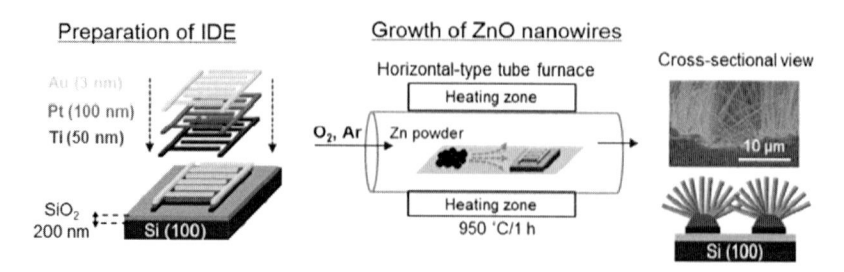

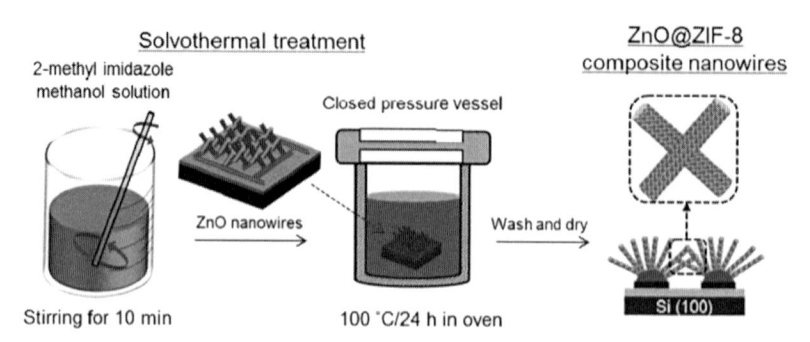

FIGURE 19.5 Preparation procedure of ZIF-membrane encapsulated ZnO NWs (Drobek et al., 2016).

relatively sluggish conversion rates were involved, given the low rate of dissociation of ZnO into Zn^{2+} at the mild solvothermal conditions used. ZnO@ZIF- 8-based nanocomposite sensor demonstrates dramatically selective response to H_2 in contrast with the pure ZnO NWs sensor, while demonstrating the insignificant sensing response to C_7H_8 and C_6H_6. Another work using ZnO@ZIF–8 was performed by Tian et al. (2016) for gas sensing. In this study, ZnO and ZIF-8 core–shell heterostructures were created utilizing a self-template technique. ZnO nanorods served as both the template and the source of Zn^{2+} ions for the ZIF-8 shell. Hydrothermal synthesis was used to install the ZIF-8 shell evenly, resulting in ZnO@ZIF-8 nanorods having core-shell heterostructures. The hybrid sensor demonstrated various gas responses for reducing gases with varying molecule sizes. The hybrid nanorods sensor enhanced its selectivity for detecting formaldehyde due to the ZIF-8 shell's aperture constraint. It is worth noting that the self-template approach is a viable option for creating MOF core-shell heterostructures. Metal oxide templates can be dissolved in solvents to give metal ions and then used to create MOFs with organic molecules on their surfaces. The template can easily regulate the size and shape of MOFs' core-shell heterostructures. ZnO nanorods are an ideal template for creating core-shell heterostructures in MOFs, as they are readily synthesized at the nanoscale. Fig. 19.6A illustrated the self-template technique for synthesizing ZnO@ZIF-8 nanorods with core-shell heterostructures. ZnO nanorods served as both the template and source of Zn^{2+} ions. ZIF-8, a shell coating on ZnO nanorods, was synthesized using Zn^{2+} and 2-mim in a DMF-H_2O solvent mixture. Fig. 19.6B depicted a TEM picture of ZnO@ZIF-8 nanorods with core-shell heterostructures. ZnO nanorods had a core diameter of 300 ± 50 nm and a ZIF-8 shell of 100 ± 50 nm on their rough surface. In the ZIF-8 structure, the Zn atom center is only controlled by the N atoms in the 1,3 positions of the five-membered ring. Fig. 19.6C showed the architecture of a substance that is porous with tiny apertures (3.4 Å) and huge cavities (11.6 Å). Fig. 19.6D showed that ZnO@ZIF-8 nanorods exhibited distinct gas responses to VOCs of varying molecular size. ZnO nanorods were covered with a porous ZIF–8 shell with two openings. At various temperatures, oxygen flowed through the aperture of the ZIF-8 shell and forms four types of oxygen species on the surface of ZnO nanorods: O_2^- (80°C), O_2^- (150°C), O_2^- (300°C–400°C), and O_2^- (550°C). When ZnO grains come into contact with reducing gases, they undergo a redox process that consumes

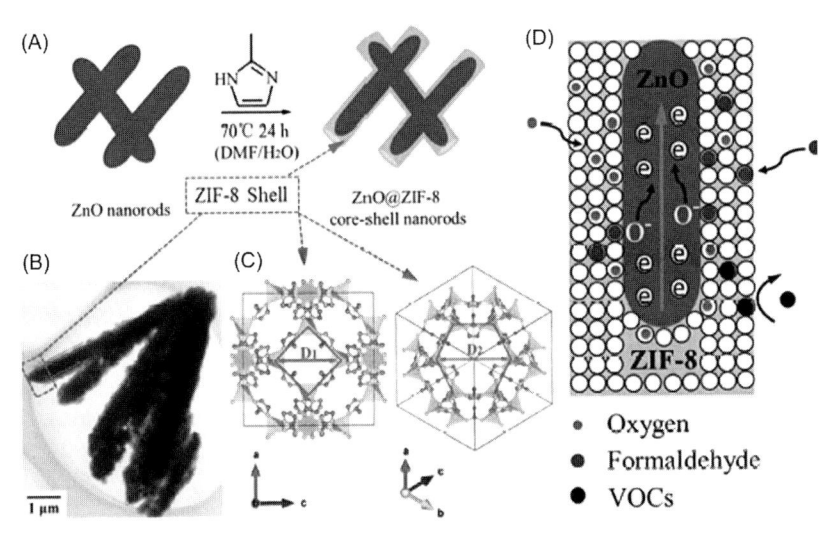

FIGURE 19.6 (A) ZnO@ZIF-8 nanorods, manufactured by using ZnO nanorods as a template in a mixed solvent at a temperature of 70°C for a duration of 24 hours. The schematic diagram illustrates this process. (B) Low-magnification TEM image showing the core-shell heterostructures of ZnO@ZIF-8 nanorods. (C) Two different aperture diameters of the ZIF-8 structure. (D) Diagram showing the ZnO@ZIF-8 nanorod sensor's ability to selectively detect various VOCs (Tian et al., 2016).

oxygen species and releases electrons back into the grains, resulting in decreased electrical resistance (Woo et al., 2012).

Zhou et al. (2018) studied the effect of the pore size of MOF on a gas sensor's selectivity feature by comparing the pore size of the ZIF-8 structure to that of other gases with comparable, larger, and smaller molecular diameters. Two zeolite-based MOFs with varying pore sizes ZIF-8 (~3.4 Å) and ZIF-71 (~4.8 Å) that were encased on ZnO nanorod surfaces were created. Target gases included in the investigation were acetone (4.60 Å), benzene (5.85 Å), H_2 (2.89 Å), NH_3 (2.90 Å), and EtOH (4.53 Å). The results indicated that although size of pores variation is effective for developing and producing selective gas sensors, when interfering gases have the same molecular diameter as the MOF, it is not effective for evaluating selectivity.

Yao et al. (2016) the ZnO@ZIF-CoZn gas sensor by depositing a layer of ZIF-CoZn thin film over a ZnO nanowire array using a simple solution approach (Fig. 19.7). ZnO@ZIF-CoZn was synthesized using a combination of Zn^{2+} dissolved from ZnO nanorods and a foreign Co^{2+} source. The competition between Co^{2+} and Zn^{2+} for coordination with organic ligands makes controlling the synthesis of ZnO@ZIF-CoZn more challenging.

Fig. 19.8A,E depicts the morphological characteristics of a ZnO NW array with lengths of 2–4 μm and diameters of 50–100 nm, respectively. All ZnO NWs (Fig. 19.8A,E) are single crystals.

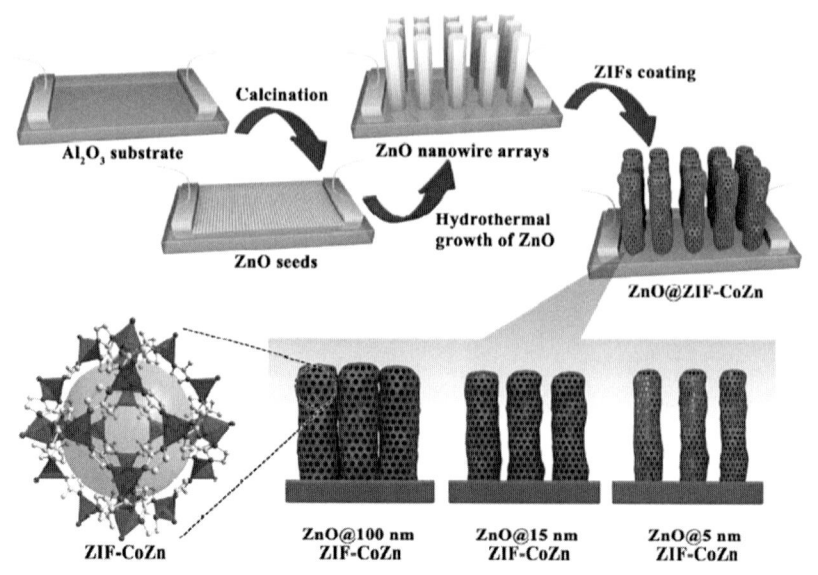

FIGURE 19.7 Diagrammatic demonstration of the synthesis of ZnO@ZIF-CoZn sensors (Yao et al., 2016)

After coating with ZIF-CoZn (Fig. 19.8B–D, F–H), the NWs remain upright on the substrate and have identical length. However, increasing the thickness of the ZIF-CoZn sheath leads to bigger diameters.

DMello et al. (2018) developed SnO_2@ZIF-67 for sensing of CO_2. The addition of ZIF-8 significantly enhances the sensitivity of the sensor to CO_2 gas concentration (50%) by 10 orders of magnitude, compared to a SnO_2 sensor without ZIF-8. The improvement in sensing capability can be attributed to the synergistic effects resulting from the formation of a heterojunction between SnO_2 and ZIF-67, as well as the increased surface area provided by the composite material (SnO_2@ZIF-67). The application of the MOF coating resulted in a notable decrease in the sensor's recovery time, reducing it from 208 to 22 seconds for 50% CO_2 concentration and from 96 to 25 seconds for 5000 ppm CO_2 concentration. Additionally, the MOF coating also enhanced the sensor's reaction time by around 10 seconds.

Wang et al. (2018) developed Au@ZnO@ZIF-8 hybrid for sensing of VOCs. In this work, anisotropic growth was used to develop and synthesize the Au@ZnO@ZIF-8 Janus nanostructure for the VOCs. The Janus nanostructure exhibited a great sensing performance with a selective detection toward formaldehyde at room temperature owing to the synergistic effects of the excellent conductivity of ZnO, the greater gas adsorption capability of ZIF-8, the uncontaminated interface between ZnO and ZIF-8, and the plasmonic resonance of gold nanorods. The results indicated that even in the presence of water and toluene molecules as interferences, it demonstrated good selective detection of formaldehyde (HCHO, as a typical VOC) at ambient temperature throughout a wide range of concentrations (from 0.25 to 100 ppm). In addition, HCHO was also observed to be partially oxidized into non-toxic formic acid simultaneously with detection. The working principle of this technology has

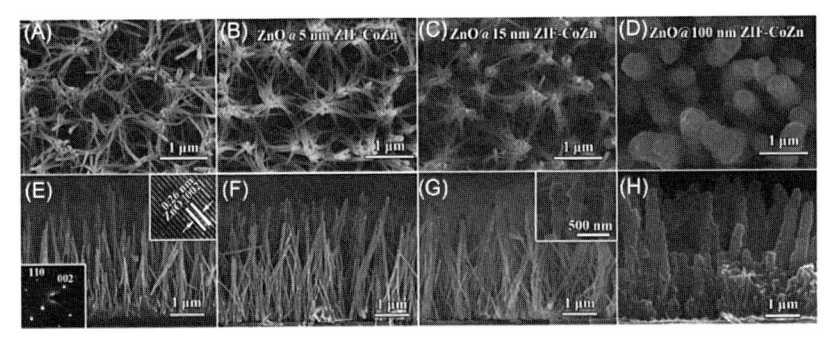

FIGURE 19.8 Diagram and side view of ZnO and ZnO@ZIF-CoZn NW arrays: (A,E) The image shows pure ZnO using HRTEM and SAED patterns of a single ZnO NW as an inset; (B,F) The image shows ZnO@5 nm ZIF-CoZn; (C,G) The image shows ZnO@15 nm ZIF-CoZn with a high magnification SEM image as an inset; (D,H) The image shows ZnO@100 nm ZIF-CoZn (Yao et al., 2016).

been determined by both theoretical computations and experimental measurements: ZnO preserved conductivity while ZIF-8 boosted selective gas adsorption; at room temperature, visible light-driven photocatalysis of ZnO is improved by Au nanorods' plasmonic effect.

Metal-organic framework-on-metal-organic framework composites for gas sensing

Effective design of MOFs with detailed structural characteristics is crucial for increased performance in many different fields. MOFs can be regulated structurally through four strategies (He et al., 2019): adjusting metal nodes and ligands (Barea et al., 2014), controlling morphology (Woellner et al., 2018), creating hybrids (Rasheed & Nabeel, 2019), and fabricating MOF-derived materials (Bernales et al., 2018; Pullen & Clever, 2018; Qiu et al., 2019; Yao et al., 2020). The first technique involves regulating MOFs structurally and compositionally using different metal nodes and ligands, including mixed linkers (Bernales et al., 2018; Pullen & Clever, 2018; Qiu et al., 2019; Yao et al., 2020) and nodes (Feng et al., 2019). The second technique involves controlling the morphology of MOFs, which has gained popularity. MOFs have many topologies and dimensionalities, including 0D particles (Min et al., 2019; Saliba et al., 2018; Wang et al., 2018), 1D wires (Pachfule et al., 2016; Xu et al., 2016; Zhang et al., 2014; Zou et al., 2018), 2D sheets (Duan et al., 2017; Zhao et al. 2015, 2016, 2018), and 3D hierarchical structures (Hong et al., 2019; Li et al., 2020). The main distinction between methods three and four is that while MOFs are changed into different combinations in materials created from MOFs, they are maintained within the framework of MOF-based hybrids. For the synthesis of carbon-based materials, metal oxides, metal hydroxides, metal sulfides, and metal phosphides, for instance, MOFs have been utilized as precursors (Hou et al., 2020; Reddy et al., 2020; Wen et al., 2020). As a new method of creating hybrid MOFs by integrating two or more distinct MOF types, a fascinating MOF-on-MOF hybridization design has drawn particular interest. The main methods used to achieve this integration are growth kinetic controlled approaches or growing guest MOFs on pre-synthesized host MOFs. The guest MOFs' properties and/or composition frequently differ from the hosts. Consequently, the MOF-on-MOF design can enhance the structural diversity (such as porosity, surface characteristics, and functionalities) of MOFs in addition to their composition (such as ligands and/or metal centers) (Gu et al., 2017; Qiu et al., 2016; Yao et al., 2019). High selectivity for the intended gas is necessary. Enhancing chemical affinity is also necessary for high sensitivity at low concentrations of the target gas, in addition to size selectivity. An intriguing gas sensing platform is generated by combining sensitive MOF-based materials with another MOF-based promoter. Matatagui et al. (2018) developed MOF-on-MOF composing of ZIF-8/ZIF-67 for gas sensing.

A nanocomposite sensor based on ZIF-8/ZIF-67 was developed to detect several VOCs, including toluene, EtOH, CO, H_2, and nitrogen dioxide. The constructed system utilized ZIF-67 as the core material, decorated using ZIF-8 throughout the whole surface. At 180°C, the MOF@MOF composite sensor demonstrated a baseline resistance of 500 KΩ, compared to ZIF-8 starting resistance of 120 MΩ. The sensors detected both reducing (toluene, EtOH, CO, H_2) and oxidizing (nitrogen dioxide) gases. The findings showed that, in comparison to ZIF alone, the ZIF combination sensor had 5.1 times the sensitivity and could detect target gas concentrations as low as 10 ppm. The reason for the enhanced sensing response upon encapsulation may be attributed to the availability of a greater number of active sites for the adsorption of target gas molecules. Unfortunately, it took a long time for either of the sensors to react (>2 to 6 min for each gas under test).

Perspectives

The unique features of MOF-based carbon nanofibers (CNFs) composites for gas sensing allow for numerous applications, including selective adsorption, polymer reinforcement, electrochemical catalysis, and hydrogen storage (Hammel et al., 2004; Wang et al., 2009; Zhang et al., 2016). The orientation of carbon layers in CNFs affects their mechanical characteristics. CNFs are sp^2-based discontinuous linear filaments with aspect ratios greater than 100:1 (Rauti et al., 2019). Most CNF have unaltered graphitic plane layers along their axis. CNF can vary in shape, as seen in Fig. 19.9, depending on the angle of the graphene layers that form the filament. In addition to platelet (Fig. 19.9A) and tubular or ribbon CNFs (Fig. 19.9B), there are also fishbone CNFs with graphene layers positioned at an angle between the primary and perpendicular axes (Cheng et al., 2012).

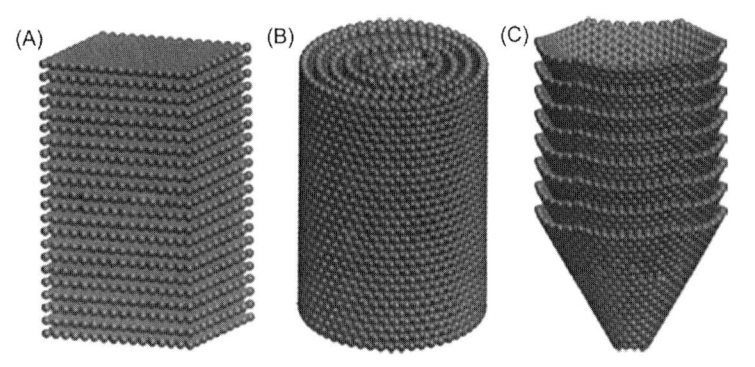

FIGURE 19.9 Diagrammatic demonstration of three CNFs with varying basal-to-edge surface area ratios: (A) Platelet-type CNF, (B) Tubular-type CNF, and (C) Fishbone-type CNF (Cheng et al., 2012).

The production of MOF-derived one-dimensional (1D) CNFs has been encouraged using the versatile electrospinning technique (Liu et al., 2016; Wang et al., 2018; Zhang et al., 2019). Kwon et al. (2022) developed MOF-CNF hybrid as a gas sensor for NO_2. As indicated previously, MOF materials have attracted significant interest for gas-sensing applications due to their elevated porosity. But there are still issues with MOFs' processability for application in dependable gas-sensing devices. This work demonstrated the coupling of organic thin-film transistor-type chemical sensors with the potent gas-adsorbing capabilities of MOF nanomaterials. The hybrid blend system including organic semiconductors and inorganic MOF nanoparticles displays thermodynamic instability due to the difficult-to-set self-aggregation of each phase in the absence of surface functionalization. A unique technique uses CNF as a framework to create an inorganic–organic hybrid sensor. The results indicated that CNF served two functions: attaching MOF nanoparticles to the fiber surface, stabilizing the embedded polymer layer, and preserving dependable conductivity for better gas sensing performance. The nanomorphology and nanocrystal structure revealed that MOF nanoparticles and CNF form a hybrid core-shell structure within the conjugated polymer matrix. The organic-inorganic hybrid system was integrated into a field-effect transistor device to detect hazardous NO_2 gas analytes in real-time with good sensitivity. The prototype chemical sensor has great potential for different chemical sensors.

Changes in the structure of the MOF have a considerable impact on the chemo-resistive response of the sensing layer. The study of Kwon et al. (2022) showcases the enhancement of porosity, chemical stability, and room-temperature gas detection capabilities of MOFs by their hybridization with other composite materials. Furthermore, when compared to metal oxides, hollow and porous nanostructures derived from MOF exhibit enhanced sensing responsiveness. Although the MOF-derived metal oxide nanostructure can function at temperatures below those required for traditional metal oxides, further research is necessary to achieve a working temperature at or below ambient temperature.

Conclusion

This chapter discusses the latest developments in chemo-resistive gas sensors that utilize MOF-based composite technology. For several years, MOFs have been offering the scientific community their unique and fascinating characteristics for their application in chemo-resistive gas sensing. There are various categories of MOFs, such as pure MOFs, 2-D conducting MOFs, and MOF composites which consist of graphene and its derivative, CNT, CNF, MOF-metal oxide, and MOF atop MOF. Due to their substantial surface area and porosity, undiluted MOFs can facilitate enhanced surface reactions, which are essential for the gas sensing performance of the sensor. This performance relies on the surface interaction between the sensing layer and the analyte.

However, because of its insulating properties at normal temperature, this particular surface reaction can only take place at elevated temperatures, hence hindering the pure MOFs from achieving their maximum capabilities. 2-D conducting MOFs were used to detect VOCs with high sensitivity and to differentiate between different interfering gases. 2D MOFs can be used directly without any modifications. Furthermore, they offer a wide range of new opportunities for improving sensor arrays, hence enhancing the performance characteristics of the sensors. Although these materials have shown a noticeable sensitivity to the analytes at room temperature, the level of sensitivity is not enough for industrial applications and commercialization. MOF membranes that form heterostructures with core-shell morphology, consisting of functional materials as the core and MOFs as the shell, are more attractive than pure MOFs for the development of selective chemoresistive gas sensors. The pore size of the MOF selectively allows only certain gases to interact with the sensing layer, and the combined effect of the core and the shell enhances the performance of the sensor. As a result, the presence of other gases in the system is minimized, leading to improved selectivity and a longer lifespan for the sensor.

References

Abdelhameed, R. M., Hammad, S. F., Abdallah, I. A., Bedair, A., Locatelli, M., & Mansour, F. R. (2023). A hybrid microcrystalline cellulose/metal–organic framework for dispersive solid phase microextraction of selected pharmaceuticals: A proof-of-concept. *Journal of Pharmaceutical and Biomedical Analysis, 235*, 115609. https://doi.org/10.1016/j.jpba.2023.115609.

Anik, Ü., & Çevik, S. (2009). Double-walled carbon nanotube based carbon paste electrode as xanthine biosensor. *Microchimica Acta, 166*(3–4), 209–213. https://doi.org/10.1007/s00604-009-0190-y.

Barea, E., Montoro, C., & Navarro, J. A. R. (2014). Toxic gas removal – metal-organic frameworks for the capture and degradation of toxic gases and vapours. *Chemical Society Reviews, 43*(16), 5419–5430. https://doi.org/10.1039/C3CS60475F.

Beale, D. J., Jones, O. A. H., Karpe, A. V., Dayalan, S., Oh, D. Y., Kouremenos, K. A., Ahmed, W., & Palombo, E. A. (2017). A review of analytical techniques and their application in disease diagnosis in breathomics and salivaomics research. *International Journal of Molecular Sciences, 18*(1). https://doi.org/10.3390/ijms18010024, http://www.mdpi.com/1422-0067/18/1/24/pdf.

Bernales, V., Ortuño, M. A., Truhlar, D. G., Cramer, C. J., & Gagliardi, L. (2018). Computational design of functionalized metal-organic framework nodes for catalysis. American Chemical Society, United States. *ACS Central Science, 4*(1), 5–19. http://pubs.acs.org/journal/acscii.

Bibi, S., Li, P., & Zhang, J. (2013). X-Shaped donor molecules based on benzo[2,1-b:3,4-b′] dithiophene as organic solar cell materials with PDIs as acceptors. *Journal of Materials Chemistry A, 1*(44), 13828. https://doi.org/10.1039/c3ta12421e.

Canivet, J., Bonnefoy, J., Daniel, C., Legrand, A., Coasne, B., & Farrusseng, D. (2014). Structure–property relationships of water adsorption in metal-organic frameworks. *New Journal of Chemistry, 38*(7), 3102–3111. https://doi.org/10.1039/C4NJ00076E.

Chen, B., Xiang, S., & Qian, G. (2010). Metal-organic frameworks with functional pores for recognition of small molecules. *Accounts of Chemical Research, 43*(8), 1115–1124. https://doi.org/10.1021/ar100023y.

Cheng, H.-Y., Zhu, Y.-A., Sui, Z.-J., Zhou, X.-G., & Chen, D. (2012). Modeling of fishbone-type carbon nanofibers with cone-helix structures. *Carbon, 50*(12), 4359–4372. https://doi.org/10.1016/j.carbon.2012.05.005.

Cousinsaintremi, J., Rémy, T., Vanhunskerken, V., Vandeperre, S., Duerinck, T., Maes, M., Devos, D., Gobechiya, E., Kirschhock, C. E. A., Baron, G. V., & Denayer, J. F. M. (2011). Biobutanol separation with the metal-organic framework ZIF-8. *ChemSusChem, 4*(8), 1074–1077. http://www.interscience.wiley.com/jpages/1864-5631.

DMello, M. E., Sundaram, N. G., & Kalidindi, S. B. (2018). Assembly of ZIF-67 metal–organic framework over tin oxide nanoparticles for synergistic chemiresistive CO_2 gas sensing. *Chemistry – A European Journal, 24*(37), 9220–9223. http://onlinelibrary.wiley.com/journal/10.1002/(ISSN)1521-3765.

Drobek, M., Kim, J. H., Bechelany, M., Vallicari, C., Julbe, A., & Kim, S. S. (2016). MOF-based membrane encapsulated ZnO nanowires for enhanced gas sensor selectivity. *ACS Applied Materials and Interfaces, 8*(13), 8323–8328. http://pubs.acs.org/journal/aamick.

Duan, J., Chen, S., & Zhao, C. (2017). Ultrathin metal-organic framework array for efficient electrocatalytic water splitting. *Nature Communications, 8*. http://www.nature.com/ncomms/index.html.

Feng, L., Wang, K. Y., Day, G. S., & Zhou, H. C. (2019). The chemistry of multi-component and hierarchical framework compounds. *Chemical Society Reviews, 48*(18), 4823–4853. https://doi.org/10.1039/c9cs00250b, http://pubs.rsc.org/en/journals/journal/cs.

Garg, N., Deep, A., & Sharma, A. L. (2021). Metal-organic frameworks based nanostructure platforms for chemo-resistive sensing of gases. *Coordination Chemistry Reviews, 445*, 214073. https://doi.org/10.1016/j.ccr.2021.214073.

Ghanbarian, M., Zeinali, S., & Mostafavi, A. (2018). A novel MIL-53(Cr-Fe)/Ag/CNT nanocomposite based resistive sensor for sensing of volatile organic compounds. *Sensors and Actuators B: Chemical, 267*, 381–391. https://doi.org/10.1016/j.snb.2018.02.138.

Gu, Y., Wu, Y.-N., Li, L., Chen, W., Li, F., & Kitagawa, S. (2017). Controllable modular growth of hierarchical MOF-on-MOF architectures. *Angewandte Chemie, 129*(49), 15864–15868. https://doi.org/10.1002/ange.201709738.

Gurunathan, S., Woong Han, J., Abdal Daye, A., Eppakayala, V., & Kim, J.-hoi (2012). Oxidative stress-mediated antibacterial activity of graphene oxide and reduced graphene oxide in *Pseudomonas aeruginosa*. *International Journal of Nanomedicine, 7*, 5901. https://doi.org/10.2147/ijn.s37397.

Hammad, S. F., Abdallah, I. A., Bedair, A., Abdelhameed, R. M., Locatelli, M., & Mansour, F. R. (2024). Metal organic framework-derived carbon nanomaterials and MOF hybrids for chemical sensing. *Egypt TrAC - Trends in Analytical Chemistry, 170*. https://doi.org/10.1016/j.trac.2023.117425., www.elsevier.com/locate/trac.

Hammel, E., Tang, X., Trampert, M., Schmitt, T., Mauthner, K., Eder, A., & Pötschke, P. (2004). Carbon nanofibers for composite applications. *Carbon, 42*(5–6), 1153–1158. https://doi.org/10.1016/j.carbon.2003.12.043.

He, C., Cheng, J., Zhang, X., Douthwaite, M., Pattisson, S., & Hao, Z. (2019). Recent advances in the catalytic oxidation of volatile organic compounds: A review based on pollutant sorts and sources. *Chemical Reviews, 119*(7), 4471–4568. https://doi.org/10.1021/acs.chemrev.8b00408, http://pubs.acs.org/journal/chreay.

Homayoonnia, S., & Kim, S. (2023). ZIF-8/MWCNT-nanocomposite based-resistive sensor for highly selective detection of acetone in parts-per-billion: Potential noninvasive diagnosis of diabetes. *Sensors and Actuators B: Chemical, 393*, 134197. https://doi.org/10.1016/j.snb. 2023.134197.

Hong, H., Liu, J., Huang, H., Atangana Etogo, C., Yang, X., Guan, B., & Zhang, L. (2019). Ordered macro-microporous metal-organic framework single crystals and their derivatives for rechargeable aluminum-ion batteries. *Journal of the American Chemical Society, 141*(37), 14764–14771. https://doi.org/10.1021/jacs.9b06957, http://pubs.acs.org/journal/jacsat.

Hou, C. C., Wang, H. F., Li, C., & Xu, Q. (2020). From metal-organic frameworks to single/dual-atom and cluster metal catalysts for energy applications. *Energy and Environmental Science, 13*(6), 1658–1693. https://doi.org/10.1039/c9ee04040d, http://pubs.rsc.org/en/journals/journal/ee.

Ingle, N., Sayyad, P., Deshmukh, M., Bodkhe, G., Mahadik, M., Al-Gahouari, T., Shirsat, S., & Shirsat, M. D. (2021). A chemiresistive gas sensor for sensitive detection of SO_2 employing Ni-MOF modified –OH-SWNTs and –OH-MWNTs. *Applied Physics A: Materials Science and Processing, 127*(2). https://doi.org/10.1007/s00339-021-04288-0, http://www.springer.com/materials/journal/339.

Jiang, H.-L., & Xu, Q. (2011). Porous metal–organic frameworks as platforms for functional applications. *Chemical Communications, 47*(12), 3351. https://doi.org/10.1039/c0cc05419d.

Jung, I., Dikin, D. A., Piner, R. D., & Ruoff, R. S. (2008). Tunable electrical conductivity of individual graphene oxide sheets reduced at "low" temperatures. *Nano Letters, 8*(12), 4283–4287. https://doi.org/10.1021/nl8019938.

Kempahanumakkagari, S., Vellingiri, K., Deep, A., Kwon, E. E., Bolan, N., & Kim, K. H. (2018). Metal–organic framework composites as electrocatalysts for electrochemical sensing applications. *Coordination Chemistry Reviews, 357*, 105–129. https://doi.org/10.1016/j.ccr. 2017.11.028, http://www.journals.elsevier.com/coordination-chemistry-reviews/.

Kim, H. J., & Lee, J. H. (2014). Highly sensitive and selective gas sensors using p-type oxide semiconductors: Overview. *Sensors and Actuators, B: Chemical, 192*, 607–627. https://doi.org/10.1016/j.snb.2013.11.005.

Kolmakov, A., & Moskovits, M. (2004). Chemical sensing and catalysis by one-dimensional metal-oxide nanostructures. *Annual Review of Materials Research, 34*, 151–180. https://doi.org/10.1146/annurev.matsci.34.040203.112141.

Krishnan, S. K., Singh, E., Singh, P., Meyyappan, M., & Singh Nalwa, H. (2019). A review on graphene-based nanocomposites for electrochemical and fluorescent biosensors. *RSC Advances, 9*(16), 8778–8881. https://doi.org/10.1039/c8ra09577a.

Kwon, E. H., Kim, M., Lee, C. Y., Kim, M., & Park, Y. D. (2022). Metal-organic-framework-decorated carbon nano fibers with enhanced gas sensitivity when incorporated into an organic semiconductor-based gas sensor. *ACS Applied Materials & Interfaces, 2022*.

Li, H. Y., Zhao, S. N., Zang, S. Q., & Li, J. (2020). Functional metal-organic frameworks as effective sensors of gases and volatile compounds. *Chemical Society Reviews, 49*(17), 6364–6401. https://doi.org/10.1039/c9cs00778d, http://pubs.rsc.org/en/journals/journal/cs.

Li, Q., Dai, Z., Wu, J., Liu, W., Di, T., Jiang, R., Zheng, X., Wang, W., Ji, X., Li, P., Xu, Z., Qu, X., Xu, Z., & Zhou, J. (2020). Fabrication of ordered macro-microporous single-crystalline MOF and its derivative carbon material for supercapacitor. *Advanced Energy Materials, 10*(33). https://doi.org/10.1002/aenm.201903750, http://onlinelibrary.wiley.com/journal/10.1002/(ISSN)1614-6840.

Liu, Y., Ma, J., Lu, T., & Pan, L. (2016). Electrospun carbon nanofibers reinforced 3D porous carbon polyhedra network derived from metal-organic frameworks for capacitive deionization. *Scientific Reports, 6*. https://doi.org/10.1038/srep32784, www.nature.com/srep/index.html.

Mansour, F. R., Abdelhameed, R. M., Hammad, S. F., Abdallah, I. A., Bedair, A., & Locatelli, M. (2024). A microcrystalline cellulose/metal-organic framework hybrid for enhanced ritonavir dispersive solid phase microextraction from human plasma. *Carbohydrate Polymer Technologies and Applications, 7*. https://doi.org/10.1016/j.carpta.2024.100453, https://www.sciencedirect.com/science/journal/26668939.

Mansour, F. R., Hammad, S. F., Abdallah, I. A., Bedair, A., Abdelhameed, R. M., & Locatelli, M. (2024). Applications of metal organic frameworks in point of care testing. *TrAC - Trends in Analytical Chemistry, 172*. https://doi.org/10.1016/j.trac.2024.117596, https://www.sciencedirect.com/science/journal/01659936.

Matatagui, D., Sainz-Vidal, A., Gràcia, I., Figueras, E., Cané, C., & Saniger, J. M. (2018). Chemoresistive gas sensor based on ZIF-8/ZIF-67 nanocrystals. *Sensors and Actuators B: Chemical, 274*, 601–608. https://doi.org/10.1016/j.snb.2018.07.137.

Miller, D. R., Akbar, S. A., & Morris, P. A. (2014). Nanoscale metal oxide-based heterojunctions for gas sensing: A review. *Sensors and Actuators, B: Chemical, 204*, 250–272. https://doi.org/10.1016/j.snb.2014.07.074.

Min, H., Wang, J., Qi, Y., Zhang, Y., Han, X., Xu, Y., Xu, J., Li, Y., Chen, L., Cheng, K., Liu, G., Yang, N., Li, Y., & Nie, G. (2019). Biomimetic metal-organic framework nanoparticles for cooperative combination of antiangiogenesis and photodynamic therapy for enhanced efficacy. *Advanced Materials, 31*, 1–11. https://doi.org/10.1002/adma.2018082002019.

More, M. S., Bodkhe, G. A., Singh, F., Kim, M., & Shirsat, M. D. (2023). Metal–organic framework-reduced graphene oxide (Zn-BDC@rGO) composite for selective discrimination among ammonia, carbon monoxide, and sulfur dioxide. *Applied Physics A: Materials Science and Processing, 129*(12). https://doi.org/10.1007/s00339-023-07103-0, https://www.springer.com/journal/339.

Na, C. W., Woo, H. S., Kim, I. D., & Lee, J. H. (2011). Selective detection of NO_2 and C_2H_5OH using a Co_3O_4-decorated ZnO nanowire network sensor. *Chemical Communications, 47*(18), 5148–5150. https://doi.org/10.1039/c0cc05256f.

Neri, G. (2015). First fifty years of chemoresistive gas sensors. *Chemosensors, 3*(1), 1–20. https://doi.org/10.3390/chemosensors3010001.

Pachfule, P., Shinde, D., Majumder, M., & Xu, Q. (2016). Fabrication of carbon nanorods and graphene nanoribbons from a metal-organic framework. *Nature Chemistry, 8*(7), 718–724. https://doi.org/10.1038/nchem.2515, http://www.nature.com/nchem/index.html.

Pandey, S. (2016). Highly sensitive and selective chemiresistor gas/vapor sensors based on polyaniline nanocomposite: A comprehensive review. *Journal of Science: Advanced Materials and Devices, 1*(4), 431–453. https://doi.org/10.1016/j.jsamd.2016.10.005.

Panes-Ruiz, L. A., Shaygan, M., Fu, Y., Liu, Y., Khavrus, V., Oswald, S., Gemming, T., Baraban, L., Bezugly, V., & Cuniberti, G. (2018). Toward highly sensitive and energy efficient ammonia gas detection with modified single-walled carbon nanotubes at room temperature. *ACS Sensors, 3*(1), 79–86. https://doi.org/10.1021/acssensors.7b00358, http://pubs.acs.org/journal/ascefj.

Park, S. Y., Kim, Y., Kim, T., Eom, T. H., Kim, S. Y., & Jang, H. W. (2019). Chemoresistive materials for electronic nose: Progress, perspectives, and challenges. *InfoMat, 1*(3), 289–316. https://doi.org/10.1002/inf2.12029 Available from: onlinelibrary.wiley.com/journal/25673165.

Park, S., Park, C., & Yoon, H. (2017). Chemo-electrical gas sensors based on conducting polymer hybrids. *Polymers, 9*(5), 155. https://doi.org/10.3390/polym9050155.

Petit, C., & Bandosz, T. J. (2009). MOF-graphite oxide nanocomposites: Surface characterization and evaluation as adsorbents of ammonia. *Journal of Materials Chemistry, 19*(36), 6521–6528. https://doi.org/10.1039/b908862h.

Pullen, S., & Clever, G. H. (2018). Mixed-ligand metal-organic frameworks and heteroleptic coordination cages as multifunctional scaffolds – A comparison. *Accounts of Chemical Research, 51*(12), 3052–3064. https://doi.org/10.1021/acs.accounts.8b00415, http://pubs.acs.org/journal/achre4.

Qiu, X., Zhong, W., Bai, C., & Li, Y. (2016). Encapsulation of a metal-organic polyhedral in the pores of a metal-organic framework. *Journal of the American Chemical Society, 138*(4), 1138–1141. https://doi.org/10.1021/jacs.5b12189, http://pubs.acs.org/journal/jacsat.

Qiu, Y. C., Yuan, S., Li, X. X., Du, D. Y., Wang, C., Qin, J. S., Drake, H. F., Lan, Y. Q., Jiang, L., & Zhou, H. C. (2019). Face-sharing archimedean solids stacking for the construction of mixed-ligand metal-organic frameworks. *Journal of the American Chemical Society, 141*(35), 13841–13848. https://doi.org/10.1021/jacs.9b05580, http://pubs.acs.org/journal/jacsat.

Rasheed, T., & Nabeel, F. (2019). Luminescent metal-organic frameworks as potential sensory materials for various environmental toxic agents. *Coordination Chemistry Reviews, 401*, 213065. https://doi.org/10.1016/j.ccr.2019.213065.

Rauti, R., Musto, M., Bosi, S., Prato, M., & Ballerini, L. (2019). Properties and behavior of carbon nanomaterials when interfacing neuronal cells: How far have we come? *Carbon, 143*, 430–446. https://doi.org/10.1016/j.carbon.2018.11.026, http://www.journals.elsevier.com/carbon/.

Reddy, R. C. K., Lin, J., Chen, Y., Zeng, C., Lin, X., Cai, Y., & Su, C.-Y. (2020). Progress of nanostructured metal oxides derived from metal–organic frameworks as anode materials for lithium–ion batteries. *Coordination Chemistry Reviews, 420*, 213434. https://doi.org/10.1016/j.ccr.2020.213434.

Ren, G., Li, Z., Yang, W., Faheem, M., Xing, J., Zou, X., Pan, Q., Zhu, G., & Du, Y. (2019). ZnO@ZIF-8 core-shell microspheres for improved ethanol gas sensing. *Sensors and Actuators, B: Chemical, 284*, 421–427. https://doi.org/10.1016/j.snb.2018.12.145, https://www.journals.elsevier.com/sensors-and-actuators-b-chemical.

Ren, H., Gu, C., Joo, S. W., Zhao, J., Sun, Y., & Huang, J. (2018). Effective hydrogen gas sensor based on NiO@rGO nanocomposite. *Sensors and Actuators, B: Chemical, 266*, 506–513. https://doi.org/10.1016/j.snb.2018.03.130, https://www.journals.elsevier.com/sensors-and-actuators-b-chemical.

Sahoo, M., Mansingh, S., Subudhi, S., Mohapatra, P., & Parida, K. (2019). A plasmonic AuPd bimetallic nanoalloy decorated over a GO/LDH hybrid nanocomposite: Via a green synthesis route for robust Suzuki coupling reactions: A paradigm shift towards a sustainable future. *Catalysis Science and Technology, 9*(17), 4678–4692. https://doi.org/10.1039/c9cy01085h, http://pubs.rsc.org/en/journals/journal/cy.

Saliba, D., Ammar, M., Rammal, M., Al-Ghoul, M., & Hmadeh, M. (2018). Crystal growth of zif-8, zif-67, and their mixed-metal derivatives. *Journal of the American Chemical Society, 140*(5), 1812–1823. https://doi.org/10.1021/jacs.7b11589, http://pubs.acs.org/journal/jacsat.

Song, P., Liu, B., Li, Y., Yang, J., Wang, Z., & Li, X. (2012). Two pillared-layer metal-organic frameworks constructed with Co(ii), 1,2,4,5-benzenetetracarboxylate, and 4,4′-bipyridine: Syntheses, crystal structures, and gas adsorption properties. *CrystEngComm, 14*(6), 2296–2301. https://doi.org/10.1039/c2ce05586d.

Stetter, J. R., & Li, J. (2008). Amperometric gas sensors – A review. *Chemical Reviews, 108*(2), 352–366. https://doi.org/10.1021/cr0681039.

Subudhi, S., Paramanik, L., Sultana, S., Mansingh, S., Mohapatra, P., & Parida, K. (2020). A type-II interband alignment heterojunction architecture of cobalt titanate integrated UiO-66-NH$_2$: A visible light mediated photocatalytic approach directed towards Norfloxacin degradation and green energy (Hydrogen) evolution. *Journal of Colloid and Interface Science, 568*, 89–105.

https://doi.org/10.1016/j.jcis.2020.02.043, http://www.elsevier.com/inca/publications/store/6/2/2/8/6/1/index.htt.

Subudhi, S., Swain, G., Tripathy, S. P., & Parida, K. (2020). UiO-66-NH$_2$ metal-organic frameworks with embedded MoS$_2$ nanoflakes for visible-light-mediated H$_2$ and O$_2$ evolution. *Inorganic Chemistry, 59*(14), 9824–9837. https://doi.org/10.1021/acs.inorgchem.0c01030, http://pubs.acs.org/journal/inocaj.

Subudhi, S., Tripathy, S. P., & Parida, K. (2021). Highlights of the characterization techniques on inorganic, organic (COF) and hybrid (MOF) photocatalytic semiconductors. *Catalysis Science and Technology, 11*(2), 392–415. https://doi.org/10.1039/d0cy02034f, http://pubs.rsc.org/en/journals/journal/cy.

Subudhi, S., Tripathy, S. P., & Parida, K. (2021). Metal oxide integrated metal organic frameworks (MO@MOF): Rational design, fabrication strategy, characterization and emerging photocatalytic applications. *Inorganic Chemistry Frontiers, 8*(6), 1619–1636. https://doi.org/10.1039/d0qi01117g, http://pubs.rsc.org/en/journals/journal/qi.

Sun, H., Tang, B., & Wu, P. (2017). Rational design of S-UiO-66@GO hybrid nanosheets for proton exchange membranes with significantly enhanced transport performance. *ACS Applied Materials and Interfaces, 9*(31), 26077–26087. https://doi.org/10.1021/acsami.7b07651, http://pubs.acs.org/journal/aamick.

Tian, H., Fan, H., Li, M., & Ma, L. (2016). Zeolitic imidazolate framework coated ZnO nanorods as molecular sieving to improve selectivity of formaldehyde gas sensor. *ACS Sensors, 1*(3), 243–250. https://doi.org/10.1021/acssensors.5b00236, http://pubs.acs.org/journal/ascefj.

Timur, S., Anik, U., Odaci, D., & Gorton, L. (2007). Development of a microbial biosensor based on carbon nanotube (CNT) modified electrodes. *Electrochemistry Communications, 9*(7), 1810–1815. https://doi.org/10.1016/j.elecom.2007.04.012.

Travlou, N. A., Singh, K., Rodríguez-Castellón, E., & Bandosz, T. J. (2015). Cu-BTC MOF-graphene-based hybrid materials as low concentration ammonia sensors. *Journal of Materials Chemistry A, 3*(21), 11417–11429. https://doi.org/10.1039/c5ta01738f, http://pubs.rsc.org/en/journals/journalissues/ta.

Tung, T. T., Tran, M. T., Feller, J. F., Castro, M., Van Ngo, T., Hassan, K., Nine, M. J., & Losic, D. (2020). Graphene and metal organic frameworks (MOFs) hybridization for tunable chemoresistive sensors for detection of volatile organic compounds (VOCs) biomarkers. *Carbon, 159*, 333–344. https://doi.org/10.1016/j.carbon.2019.12.010, http://www.journals.elsevier.com/carbon/.

Wang, C., Kaneti, Y. V., Bando, Y., Lin, J., Liu, C., Li, J., & Yamauchi, Y. (2018). Metal–organic framework-derived one-dimensional porous or hollow carbon-based nanofibers for energy storage and conversion. *Materials Horizons, 5*(3), 394–407. https://doi.org/10.1039/c8mh00133b.

Wang, D., Li, Z., Zhou, J., Fang, H., He, X., Jena, P., Zeng, J. B., & Wang, W. N. (2018). Simultaneous detection and removal of formaldehyde at room temperature: Janus au@zno@zif-8 nanoparticles. *Nano-Micro Letters, 10*(1), 1–11. https://doi.org/10.1007/s40820-017-0158-0, http://www.springer.com/engineering/journal/40820.

Wang, H., Lustig, W. P., & Li, J. (2018). Sensing and capture of toxic and hazardous gases and vapors by metal-organic frameworks. *Chemical Society Reviews, 47*(13), 4729–4756. https://doi.org/10.1039/c7cs00885f, http://pubs.rsc.org/en/journals/journal/cs.

Wang, K., Wang, Y., Wang, Y., Hosono, E., & Zhou, H. (2009). Mesoporous carbon nanofibers for supercapacitor application. *Journal of Physical Chemistry C, 113*(3), 1093–1097. https://doi.org/10.1021/jp807463uJapan, http://pubs.acs.org/doi/pdfplus/10.1021/jp807463u.

Wang, M. Q., Zhang, Y., Bao, S. J., Yu, Y. N., & Ye, C. (2016). Ni(II)-based metal-organic framework anchored on carbon nanotubes for highly sensitive non-enzymatic hydrogen peroxide sensing. *Electrochimica Acta, 190*, 365–370. https://doi.org/10.1016/j.electacta. 2015.12.199, http://www.journals.elsevier.com/electrochimica-acta/.

Wang, S., McGuirk, C. M., d'Aquino, A., Mason, J. A., & Mirkin, C. A. (2018). Metal–organic framework nanoparticles. *Advanced Materials, 30*(37). https://doi.org/10.1002/adma. 201800202.

Wang, X., Wang, Q., Wang, Q., Gao, F., Gao, F., Yang, Y., & Guo, H. (2014). Highly dispersible and stable copper terephthalate metal-organic framework-graphene oxide nanocomposite for an electrochemical sensing application. *ACS Applied Materials and Interfaces, 6*(14), 11573–11580. https://doi.org/10.1021/am5019918, http://pubs.acs.org/journal/aamick.

Wen, P., Gong, P., Sun, J., Wang, J., & Yang, S. (2015). Design and synthesis of Ni-MOF/CNT composites and rGO/carbon nitride composites for an asymmetric supercapacitor with high energy and power density. *Journal of Materials Chemistry A, 3*(26), 13874–13883. https://doi. org/10.1039/c5ta02461g, http://pubs.rsc.org/en/journals/journalissues/ta.

Wen, X., Zhang, Q., & Guan, J. (2020). Applications of metal–organic framework-derived materials in fuel cells and metal-air batteries. *Coordination Chemistry Reviews, 409*, 213214. https://doi.org/10.1016/j.ccr.2020.213214.

Wenger, O. S. (2013). Vapochromism in organometallic and coordination complexes: Chemical sensors for volatile organic compounds. *Chemical Reviews, 113*(5), 3686–3733. https://doi. org/10.1021/cr300396p.

Wilson, A. (2018). Application of electronic-nose technologies and VOC-biomarkers for the noninvasive early diagnosis of gastrointestinal diseases. *Sensors, 18*(8), 2613. https://doi.org/ 10.3390/s18082613.

Woellner, M., Hausdorf, S., Klein, N., Mueller, P., Smith, M. W., & Kaskel, S. (2018). Adsorption and detection of hazardous trace gases by metal–organic frameworks. *Advanced Materials, 30*(37). https://doi.org/10.1002/adma.201704679, http://onlinelibrary.wiley.com/journal/10. 1002/(ISSN)1521-4095.

Woo, H. S., Na, C. W., Kim, I. D., & Lee, J. H. (2012). Highly sensitive and selective trimethylamine sensor using one-dimensional ZnO-Cr_2O_3 hetero-nanostructures. *Nanotechnology, 23*(24). https://doi.org/10.1088/0957-4484/23/24/245501, http://iopscience. iop.org/0957-4484/23/24/245501/pdf/0957-4484_23_24_245501.pdf.

Xing, L. L., Huang, K. J., Cao, S. X., & Pang, H. (2018). Chestnut shell-like $Li_4Ti_5O_{12}$ hollow spheres for high-performance aqueous asymmetric supercapacitors. *Chemical Engineering Journal, 332*, 253–259. https://doi.org/10.1016/j.cej.2017.09.084, www.elsevier.com/inca/ publications/store/6/0/1/2/7/3/index.htt.

Xu, H. Q., Wang, K., Ding, M., Feng, D., Jiang, H. L., & Zhou, H. C. (2016). Seed-mediated synthesis of metal-organic frameworks. *Journal of the American Chemical Society, 138*(16), 5316–5320. https://doi.org/10.1021/jacs.6b01414, http://pubs.acs.org/journal/jacsat.

Yang, L., Tang, B., & Wu, P. (2015). Metal-organic framework-graphene oxide composites: A facile method to highly improve the proton conductivity of PEMs operated under low humidity. *Journal of Materials Chemistry A, 3*(31), 15838–15842. https://doi.org/10.1039/ c5ta03507d, http://pubs.rsc.org/en/journals/journalissues/ta.

Yang, Y., Lin, Z., Gao, S., Su, J., Lun, Z., Xia, G., Chen, J., Zhang, R., & Chen, Q. (2017). Tuning electronic structures of nonprecious ternary alloys encapsulated in graphene layers for optimizing overall water splitting activity. *ACS Catalysis, 7*(1), 469–479. https://doi.org/10. 1021/acscatal.6b02573, http://pubs.acs.org/page/accacs/about.html.

Yang, Z., Xu, X., Liang, X., Lei, C., Wei, Y., He, P., Lv, B., Ma, H., & Lei, Z. (2016). MIL-53(Fe)-graphene nanocomposites: Efficient visible-light photocatalysts for the selective oxidation of alcohols. *Applied Catalysis B: Environmental, 198*, 112–123. https://doi.org/10.1016/j.apcatb. 2016.05.041, www.elsevier.com/inca/publications/store/5/2/3/0/6/6/index.htt.

Yao, M. S., Zheng, J. J., Wu, A. Q., Xu, G., Nagarkar, S. S., Zhang, G., Tsujimoto, M., Sakaki, S., Horike, S., Otake, K., & Kitagawa, S. (2020). A dual-ligand porous coordination polymer chemiresistor with modulated conductivity and porosity. *Angewandte Chemie - International Edition, 59*(1), 172–176. https://doi.org/10.1002/anie.201909096, http://onlinelibrary.wiley.com/journal/10.1002/(ISSN)1521-3773.

Yao, M.-S., Xiu, J.-W., Huang, Q.-Q., Li, W.-H., Wu, W.-W., Wu, A.-Q., Cao, L.-A., Deng, W.-H., Wang, G.-E., & Xu, G. (2019). Van der Waals Heterostructured MOF-on-MOF thin films: Cascading functionality to realize advanced chemiresistive sensing. *Angewandte Chemie, 131*(42), 15057–15061. https://doi.org/10.1002/ange.201907772.

Yao, M.-S., Tang, W.-X., Wang, G.-E., Nath, B., & Xu, G. (2016). MOF thin film-coated metal oxide nanowire array: Significantly improved chemiresistor sensor performance. *Advanced Materials, 28*(26), 5229–5234. https://doi.org/10.1002/adma.201506457.

Zamiri, G. (2020). *Materials,* 2020.

Zhang, B., Kang, F., Tarascon, J.-M., & Kim, J.-K. (2016). Recent advances in electrospun carbon nanofibers and their application in electrochemical energy storage. *Progress in Materials Science, 76*, 319–380. https://doi.org/10.1016/j.pmatsci.2015.08.002.

Zhang, C. L., Lu, B. R., Cao, F. H., Wu, Z. Y., Zhang, W., Cong, H. P., & Yu, S. H. (2019). Electrospun metal-organic framework nanoparticle fibers and their derived electrocatalysts for oxygen reduction reaction. *Nano Energy, 55*, 226–233. https://doi.org/10.1016/j.nanoen.2018. 10.029, http://www.journals.elsevier.com/nano-energy/.

Zhang, H., & Snurr, R. Q. (2017). Computational study of water adsorption in the hydrophobic metal-organic framework ZIF-8: Adsorption mechanism and acceleration of the simulations. *Journal of Physical Chemistry C, 121*(43), 24000–24010. https://doi.org/10.1021/acs.jpcc. 7b06405, http://pubs.acs.org/journal/jpcccb.

Zhang, J., Liu, X., Wang, L., Yang, T., Guo, X., Wu, S., Wang, S., & Zhang, S. (2011). Synthesis and gas sensing properties of α-Fe$_2$O$_3$@ZnO core–shell nanospindles. *Nanotechnology, 22*(18), 185501. https://doi.org/10.1088/0957-4484/22/18/185501.

Zhang, K., Lively, R. P., Dose, M. E., Brown, A. J., Zhang, C., Chung, J., Nair, S., Koros, W. J., & Chance, R. R. (2013). Alcohol and water adsorption in zeolitic imidazolate frameworks. *Chemical Communications, 49*(31), 3245–3247. https://doi.org/10.1039/c3cc39116g.

Zhang, W., Wu, Z. Y., Jiang, H. L., & Yu, S. H. (2014). Nanowire-directed templating synthesis of metal-organic framework nanofibers and their derived porous doped carbon nanofibers for enhanced electrocatalysis. *Journal of the American Chemical Society, 136*(41), 14385–14388. https://doi.org/10.1021/ja5084128, http://pubs.acs.org/journal/jacsat.

Zhao, M., Huang, Y., Peng, Y., Huang, Z., Ma, Q., & Zhang, H. (2018). Two-dimensional metal-organic framework nanosheets: Synthesis and applications. *Chemical Society Reviews, 47*(16), 6267–6295. https://doi.org/10.1039/c8cs00268a, http://pubs.rsc.org/en/journals/journal/cs.

Zhao, M., Wang, Y., Ma, Q., Huang, Y., Zhang, X., Ping, J., Zhang, Z., Lu, Q., Yu, Y., Xu, H., Zhao, Y., & Zhang, H. (2015). Ultrathin 2D metal-organic framework nanosheets. *Advanced Materials, 27*(45), 7372–7378. https://doi.org/10.1002/adma.201503648, http://www3. interscience.wiley.com/journal/119030556/issue.

Zhao, S., Wang, Y., Dong, J., He, C. T., Yin, H., An, P., Zhao, K., Zhang, X., Gao, C., Zhang, L., Lv, J., Wang, J., Zhang, J., Khattak, A. M., Khan, N. A., Wei, Z., Zhang, J., Liu, S., Zhao, H., & Tang, Z. (2016). Ultrathin metal-organic framework nanosheets for electrocatalytic oxygen evolution. *Nature Energy, 1*(12). https://doi.org/10.1038/nenergy.2016.184, www.nature.com/nenergy/.

Zhou, E., Zhang, Y., Li, Y., & He, X. (2014). Cu(II)-based MOF immobilized on multiwalled carbon nanotubes: Synthesis and application for nonenzymatic detection of hydrogen peroxide with high sensitivity. *Electroanalysis, 26*(11), 2526–2533. https://doi.org/10.1002/elan.201400341, http://onlinelibrary.wiley.com/journal/10.1002/(ISSN)1521-4109.

Zhou, T., Sang, Y., Wang, X., Wu, C., Zeng, D., & Xie, C. (2018). Pore size dependent gas-sensing selectivity based on ZnO@ZIF nanorod arrays. *Sensors and Actuators, B: Chemical, 258*, 1099–1106. https://doi.org/10.1016/j.snb.2017.12.024.

Zou, L., Hou, C. C., Liu, Z., Pang, H., & Xu, Q. (2018). Superlong single-crystal metal-organic framework nanotubes. *Journal of the American Chemical Society, 140*(45), 15393–15401. https://doi.org/10.1021/jacs.8b09092, http://pubs.acs.org/journal/jacsat.

Zuo, K., Huang, X., Liu, X., Gil Garcia, E. M., Kim, J., Jain, A., Chen, L., Liang, P., Zepeda, A., Verduzco, R., Lou, J., & Li, Q. (2020). A hybrid metal-organic framework-reduced graphene oxide nanomaterial for selective removal of chromate from water in an electrochemical process. *Environmental Science and Technology, 54*(20), 13322–13332. https://doi.org/10.1021/acs.est.0c04703, http://pubs.acs.org/journal/esthag.

Chapter 20

Metal-organic framework composites for immobilized enzyme applications

Muhammad Rezki[1], Seiya Tsujimura[2], and K.S. Shalini Devi[1,2]
[1]*Graduate School of Pure and Applied Science, University of Tsukuba, Tsukuba, Ibaraki, Japan,*
[2]*Department of Material Sciences, Institute of Pure and Applied Sciences, University of Tsukuba, Tsukuba, Ibaraki, Japan*

Introduction

Metal-organic frameworks (MOFs), which are coordinated metal-ligand porous crystalline materials, have been intensively studied as promising versatile materials for various applications (Furukawa et al., 2013; Li et al., 2016). In early development, MOFs were applied in gas storage and separation owing to their ordered porous structure within an extremely large surface area that could accommodate large amounts of guest molecules (Bose et al., 2015; Li et al., 2019; Suh et al., 2012; Sumida et al., 2012). Furthermore, MOFs contain abundant active sites that make them applicable as catalysts in various types of sensors (Chen et al., 2022; Peng et al., 2023; Rezki et al., 2021), organic syntheses (Pascanu et al., 2019), and energy conversion (Qiu et al., 2020). However, due to their poor conductivity properties, the doping or modification of MOFs with highly conductive materials becomes necessary, particularly in electrochemistry-related applications. In contrast, several groups have utilized MOFs as templates in nanoarchitectural carbon (Xu et al., 2023), N-doped carbon (Shao et al., 2021), metal–carbon (Xue et al., 2021), metal phosphide (Tang et al., 2022), and metal sulfide syntheses (Liu et al., 2022). The structural tunability of MOFs, attributed to the diversity of metal and ligand options as well as various synthesis approaches, is a unique feature that has garnered considerable attention from researchers, driving ongoing investigation and innovative advancements in this material. To date, over a thousand MOF variations with distinct crystal structures and functionalities have been reported. Notably, even minor changes in MOF synthesis parameters, while using the same metal–ligand precursors, can result in different MOF properties. Postsynthetic modification further expands the diversity of the MOF family members.

In the 2010s, MOFs have been explored in biotechnological fields, particularly in biocatalytic applications (Sha et al., 2022; Saddique et al., 2024; Gröger et al., 2024). Although MOFs are still unable to replace natural enzymes as efficient biocatalysts, several types of MOFs with nanoenzyme properties have been successfully synthesized (Liu et al., 2021; Zhang et al., 2022; Niu et al., 2020). However, the requirement for special conditions to work optimally, such as basic or acidic pH, is a bottleneck in the development of MOF-based nanoenzymes. Thus, instead of replacing natural enzymes, using MOFs as a matrix for enzyme immobilization is a favorable approach (Wang et al., 2020; Devi et al., 2023; Li et al., 2020, 2020). Various strategies have been used for enzyme immobilization on MOFs, including surface adsorption, pore entrapment, covalent immobilization, and encapsulation (Devi et al., 2023). The immobilization of enzymes using MOFs successfully prevents enzyme unfolding and significantly improves enzyme stability (Liao et al., 2017). Moreover, the tight crystalline structure of the MOF surrounding the enzyme when the enzyme is encapsulated can act as armor for enzyme protection under extreme conditions, such as high temperatures (Singh et al., 2021). By utilizing MOFs for enzyme immobilization, the reusability, catalytic activity, and stability of enzymes can be significantly improved. In addition, this strategy overcomes the problems of leaching and inefficient loading during enzyme immobilization using other materials.

Choosing an appropriate MOF to immobilize an enzyme is an important factor to consider. The MOF functional groups, pore size, metal–ligand precursors, and synthesis procedures determine the catalytic activity of the immobilized enzyme. By integrating materials like polymers, metal nanoparticles, and carbon-based substances, MOF composites create a more stable and compatible environment for enzymes (Zheng et al., 2024; Li et al., 2020; Jabeen et al., 2024). This combination improves enzyme loading, retention, and activity, making the MOF composites-enzyme system highly suitable for applications in biosensing and biocatalysis.

In this chapter, we briefly summarize the preparation methods used to obtain enzyme–MOF biocomposites, as well as the effects of various parameters on the behavior of enzymes immobilized within MOFs, including the hydrophobicity of the MOF and the influence of ligands, metals, and buffers. Additionally, we discuss improvements in the biocatalytic activity and stability of immobilized enzymes, the orientation of the enzymes within the MOF, and the use of MOFs for the co-immobilization of multiple enzymes in cascade reactions. Finally, recent applications of enzyme-immobilized MOFs are also highlighted.

Preparation of enzyme-immobilized metal-organic framework composites

Enzyme–MOF composites are prepared using four different approaches: surface adsorption, pore entrapment, covalent immobilization, and in situ encapsulation (Fig. 20.1) (Devi et al., 2023; Ye et al., 2020). A combination of weak interactions, such as electrostatic, hydrophobic, or π–π interactions, is the main mechanism for the stable adsorption of enzymes on the MOF surface. However, the stability is limited to several hours or significantly decreases after one-time use (Nowroozi-Nejad et al., 2019). Therefore, this strategy is not recommended for long-term or repeated-use applications. In the case of immobilization by pore entrapment, the enzyme is immobilized inside the pores of the MOF. The advantage of pore immobilization in MOFs compared to other porous materials, such as mesoporous silica, is that the presence of additional interactions between the enzyme and the MOF pore wall improves enzyme stability (Chen et al., 2014). However, the drawback of this method is that most enzymes are larger than the microporous MOF apertures; thus, a pore adjustment strategy is commonly applied during MOF synthesis. A sacrificial templating method to leave mesopores or use a long ligand is a common approach (Zhang et al., 2017; Deng et al., 2012). The third method involves immobilizing the enzyme via covalent immobilization; an MOF that contains free functional ligands, such as amines (NH_2) or carboxyl groups (COOH), can be covalently crosslinked to an enzyme (Wang et al., 2022). An advantage of this approach is stable immobilization without restricting the substrate diffusion. However, the multistep preparation process and inability to fully

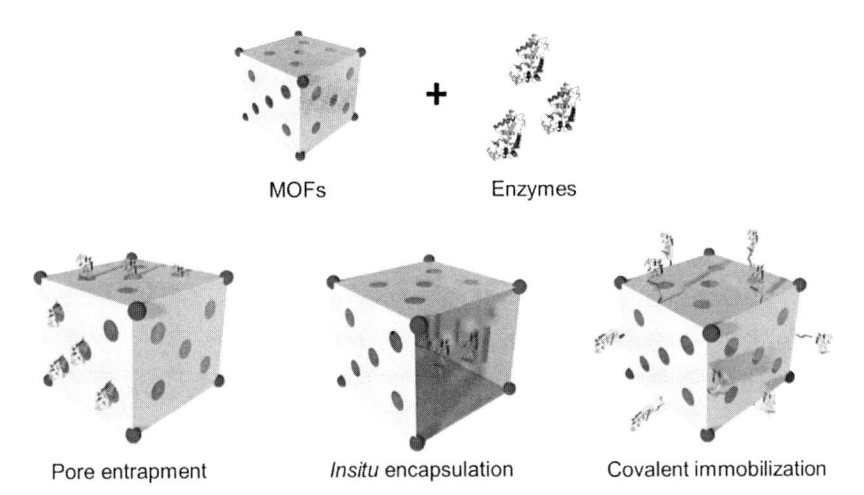

FIGURE 20.1 Diverse strategies for the preparation of enzyme-immobilized metal-organic framework composites (Devi et al., 2023)

protect the enzyme are the limitations of this method (Smith et al., 2020; Mohidem et al., 2023). The final strategy for enzyme immobilization using MOFs involves in situ encapsulation. This strategy can be performed via biomineralization or coprecipitation, with the main difference being the absence or presence of a precipitating agent (Liang et al., 2016; Nadar et al., 2020). Full enzyme protection that overcomes low stability and enzyme leaching problems is a valuable feature of this strategy. Moreover, the encapsulation of enzymes in porous crystalline structures is difficult to achieve using other materials, making this strategy a signature of enzyme immobilization using MOF-based materials. More recently, the in situ encapsulation of MOFs has been altered to produce partially buried enzymes, with some of the enzymes exposed on the MOF surface. This type of strategy can overcome the diffusion limitation problem of fully encapsulated enzymes and opens prospects for surface modification of enzymes related to the modulation of electron transfer (Pan, Li, Li, et al., 2021; Pan, Li, Lenertz, et al., 2021; Farmakes et al., 2020; Neupane et al., 2019).

Effect of metal-organic framework hydrophobicity on enzyme immobilization

Although the immobilization of enzymes using MOFs successfully protects the enzymes against extreme conditions, such as protein denaturation, organic solvents, and elevated temperatures, the interaction between the enzyme and the MOF carrier is rarely discussed. A crucial parameter that significantly determines the enzyme immobilization efficiency using this strategy is the hydrophobicity of the MOFs (Liang et al., 2019). Generally, proteins tend to be adsorbed on a hydrophobic surface through hydrophobic interactions; however, hydrophobic interactions are often responsible for conformational changes in the enzyme (Rabe et al., 2011). Enzyme immobilization on a more hydrophilic surface results in better enzyme activity and prevents conformational change (Ge et al., 2023). In the case of susceptible enzymes, such as laccase, immobilization using hydrophobic zeolitic imidazolate framework-8 (ZIF-8) deactivates the enzyme (Fig. 20.2A). Moreover, the distribution of enzymes in hydrophilic MOFs, such as ZIF-90 and metal azolate framework-7 (MAF-7) is reportedly more homogenous than that in ZIF-8 (Liang et al., 2019).

In addition, owing to its high hydrophilicity, MAF-7 showed better immobilized enzyme activity compared to ZIF-90, which was less hydrophilic (Liang et al., 2019). Ge et al. reported that the immobilization of glucose oxidase (GOx) using hydrophilic ZIF-90 retained 90% of its initial enzyme activity, compared to the encapsulation of GOx in ZIF-8, which only retained 20% of the initial enzyme activity (Ge et al., 2023). The enzyme-to-substrate affinity of GOx@ZIF-90 was twice that of GOx@ZIF-8 (Fig. 20.2B). Enzyme-to-substrate affinity is related to the Michaelis–Menten constant of the enzyme (K_m value),

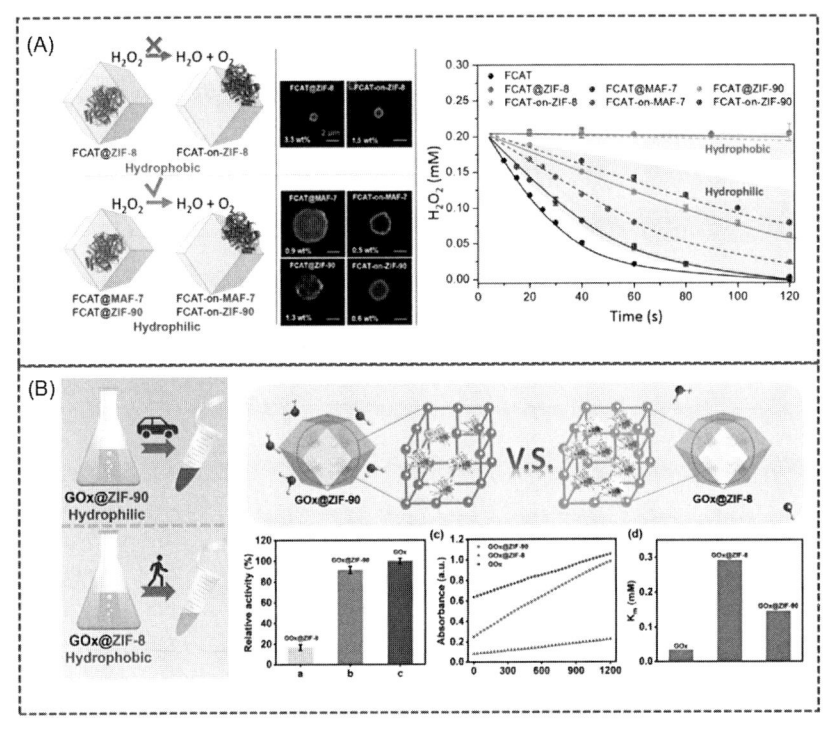

FIGURE 20.2 Comparison of enzyme immobilization using hydrophobic and hydrophilic MOFs. (A) Catalase immobilization (Liang et al., 2019) and (B) Glucose oxidase (GOx) immobilization (Ge et al., 2023).

which determines the saturation concentration value of the substrate; a higher enzyme-to-substrate affinity results in a lower K_m value. This is a crucial aspect for several applications, particularly as an enzymatic sensor platform. A lower K_m value indicates that the sensor has a narrow linear range of detection. Hence, high substrate affinity does not always indicate good performance in some specific applications. Recent studies have revealed that the traditional encapsulation of enzymes in the hydrophobic microenvironment of ZIF has several drawbacks, such as enzyme conformation, which leads to a decrease in enzyme activity (Liang et al., 2019; Zhu et al., 2022). Several robust enzymes, such as GOx, show good catalytic activity when immobilized on ZIF-8. For example, Huang et al. reported that ZIF-8 has an excellent ability to protect encapsulated GOx against organic solvents and enzyme digestion compounds because of its hydrophobic pore properties that can block the molecules from penetrating and reaching the enzyme inside (Zhu et al., 2022).

Alternative MOF materials and various modification strategies for obtaining hydrophilic MOF carriers are the focus of current research.

In addition to ZIF-90, MAF-7 is a favorable alternative for enzyme encapsulation, with a lower ligand price than that of the ZIF-90 ligand. Along with being hydrophilic, MAF-7 also has a structure similar to ZIF-8 with more available active sites (Zhao et al., 2024) and is more acid-tolerant (Zhu et al., 2022). These materials are a better choice than hydrophilic ZIF-8 that still suffers from low yield and efficiency (Wang et al., 2023). In another approach, He et al. used ZIF-8-L as a sacrificial template to produce GOx and horseradish peroxidase (HRP)-encapsulated ZnCo-layered double hydroxides (LDHs) (He et al., 2023). An LDH is a hydroxyl octahedral compound that coordinates with metal ions, with intercalated anions between the layers. The use of ZIF-8 as a template can produce three-dimensional (3D) LDH as an enzyme carrier with intriguing properties intended to replace conventional ZIF carriers. This strategy results in 5.56 times higher catalytic activity compared to the free enzyme or enzyme encapsulation using ZIF-8.

Effect of buffers

A buffered system is essential for maintaining the stability of enzymes during long-term storage or immobilization. Several questions have been raised regarding enzyme immobilization on MOFs.

Does the buffer affect the efficiency of the immobilization process? or Does the buffer affect the stability of the immobilized enzyme?

Ahmad et al. investigated the effect of a buffer during the surface adsorption of the bienzyme GOx and HRP on three different MOFs: UiO-66, UiO-67, and UiO-66–NH$_2$. Immobilization of the enzymes using the MOPSO (Zwitterionic aminosulfonate) buffer exhibited much higher activity compared to when immobilization was carried out using phosphate-buffered saline (PBS) for all MOFs. When PBS was used, the small phosphate ion molecules of the PBS interacted with the MOF surface and blocked the free sites for enzyme loading (Ahmad et al., 2023). Because phosphate ions have a negative charge, the electrostatic repulsion effect may hinder the enzyme adsorption process.

Shortall et al. investigated the stability of the MOF as an enzyme carrier toward several aqueous buffers, including acetate, citrate, phosphate, and tris-HCl buffers. A series of MOFs, including ZIF-Zni, Cu–trimesic acid (Cu-TMA), Ni–TMA, Co–TMA, and Fe–1,3,5-benzenetricarboxylate (Fe-BTC), were synthesized, and lipase was selected as the enzyme model. Metal release analysis was performed using inductively coupled plasma optical emission spectroscopy. The findings revealed that the citrate buffer had the worst effect on the MOF stability for all types, particularly Fe–BTC, which was fully dissolved in the citrate buffer. In contrast, other buffers were relatively friendly toward the MOFs, despite the percentage stability being dependent on the specific type of MOFs (Shortall et al., 2022). Thus, choosing an appropriate buffer is crucial, as the diverse metal–ligand coordination has different effects after the buffer treatment (Bůžek et al., 2021; Liu, Gong, et al., 2019; Gao et al., 2020). Notably, immobilized enzymes can easily leach if the MOF carrier loses its structural integrity.

Anionic buffers, such as phosphate, are regarded as ligands for the metal ions of the MOF, attracting the metal to form another coordination bond instead of the original ligand (Bunzen, 2021). In several cases, this mechanism compromises the framework structure of the MOF (Velásquez-Hernández et al., 2019); hence, the use of ionic buffers, such as PBS, for the long-term storage of MOF material is not recommended. However, pH also affects MOF stability. Although MOFs are stable within the physiological pH range, several types of MOFs, including those from the MAF family, degrade in acidic environments. A primary reason for this is the competitive coordination of H$^+$ ions with the ligand, which can replace the interacting metal (Bunzen, 2021). Interestingly, Gao et al. reported an improvement in the stability of ZIF in acidic media after the incorporation of biomolecules. They found that the coordination of carboxylate groups from the polypeptide or phosphate groups from DNA with zinc nodes acted as stabilizers for the ZIF. The Zn–N coordination bond strength increased by 27.82 kJ/mol in the presence of a carboxylate group coordinated with Zn. They also reported that the small-angle X-ray scattering pattern of ZIF-L disappeared after 20 minutes of treatment in acetate buffer, while the peak was still observable even after 3 hours with the incorporation of peptides (Gao et al., 2019).

Effect of ligands

Ligand selection plays a crucial role in enzyme immobilization using MOF-based materials, particularly in enzyme encapsulation strategies. Organic ligands determine the final properties of the synthesized MOF, whether hydrophilic or hydrophobic, and strongly affect the stability of the framework because of the varying metal–ligand coordination. Typically, various ligands have varying solubilities in different solvents, which means that not all ligands can be used for enzyme immobilization. In addition, the concentration of the ligand should be considered because a high amount of ligand initiates denaturation of the enzyme (Chen, Kou, et al., 2020). Huang et al. reported that ligand variations, including imidazole, 2-methyl imidazole, purine, imidazole 2-carbaldehyde, benzimidazole, 4,5-dichloroimidazole, and 2-nitroimidazole, significantly contributed to the biomineralization of the MOF; in this case, Zn was used as the metal ion for all the ligands (Huang et al., 2023). Among these ligands, only ZIF-Zni, 2-methylimidazole (ZIF-8), and imidazole carboxaldehyde (ZIF-90) maintained their enzyme activities after encapsulation. They reported enzyme immobilization efficiencies of 25.4%, 21.3%, and 42.05% for GOx@Zni, GOx@ZIF-8, and GOx@ZIF-90, respectively. The encapsulated enzyme activity of GOx@ZIF-8 was superior to those of GOx@ZIF-zni and GOx@ZIF-90, which was mainly due to the different ZIF skeleton structures resulting from ligand properties.

Gkaniatsou et al. investigated the effect of ligand functionalization in the immobilization of microperoxidase-8 (MP8) using MIL-101(Cr). The ligand containing the amino group (NH$_2$) demonstrated a higher enzyme immobilization efficiency than the unfunctionalized ligand. This was attributed to the electrostatic interaction between the COOH group of MP8 and the NH$_2$ group in the ligand. The sulfonate RSO$_3$H-functionalized ligand was used to provide negatively charged MIL-101(Cr), but MP8 immobilization failed due to the charge repulsion effect; lowering the pH to change the enzyme charge can solve this problem; however, the enzyme loses its activity at a more acidic pH (Gkaniatsou et al., 2020). The substitution of the original ligand with a specific compound or polymer can also improve the properties of the enzyme–MOF. Sun et al. reported that the incorporation of a polyphenol into ZIF-8 successfully enhanced the water-tolerant properties of the ZIF. In particular, a polyphenol—tannic acid—was introduced into an enzyme-encapsulated ZIF composite via the Zn–O interaction that partially substituted the metal–ligand coordination (Zn–N) of ZIF-8. This strategy significantly enhanced the reusability and storage stability of the encapsulated enzyme, with activity comparable to that without tannic acid modification (Sun et al., 2019).

Effect of metals

In addition to ligands, the effect of metal ions is a critical aspect to examine. Jordahl et al. studied the combination of five metals (Zn, Cu, Ni, Al, and Zr) and two ligands (benzene dicarboxylic acid [BDC] and biphenyl-4,4-dicarboxylate [BPDC]) for the immobilization of two enzymes—lipase and lysozyme (lys)—with different sizes and hydrophilicity. Both enzymes were successfully immobilized on 10 variations of the synthesized MOFs. In this study, the immobilization strategy was based on coprecipitation rather than biomineralization, resulting in partial exposure of the enzyme on the MOF surface. Both lipase and lys retained their activity after immobilization across all 10 MOFs, though with a slight decrease compared to the free enzyme. In a reusability study, lys immobilized on Zn–, Cu–, and Ni–BPDC exhibited greater stability than those immobilized on Al–BPDC and Zr–BPDC. Contrasting results were observed when the BDC ligand was used. Moreover, enzyme stability was found to be dependent on the combination of metal and ligand; for lys, Zn–BDC, Cu–BDC, Cu–BPDC, and Ni–BDC provided stability, while for lipase, Al–BDC, Al–BPDC, Zr–BDC, and Zr–BPDC were stable (Jordahl et al., 2022). Certain enzymes alter their properties after interacting with certain metal ions. For example, pyrolysin is destabilized after supplementation with Mg^{2+}, Ca^{2+}, or Na$^+$; however, its activity increases (Zeng et al., 2014). Lyu et al. reported that the interaction between Cyt C and Zn ion in ZIF-8 enhanced the enzyme substrate affinity toward H$_2$O$_2$, resulting in 10 times higher activity (Lyu et al., 2014). Yang et al. introduced metal ions, such as Ni^{2+}, to the immobilization process of pepsin in ZIF-8. Excess uninteracted nitrogen atoms in the methyl imidazole ligand were used to incorporate Ni^{2+}, resulting in a more positively charged surface of ZIF-8. Hence, negatively charged pepsin was more inclined to adsorb onto the positively charged ZIF-8–Ni surface via electrostatic interactions. In contrast, direct immobilization of pepsin to ZIF-8 in the absence of Ni led to a conformational change in pepsin due to massive hydrophobic interactions (Yang et al., 2023).

Metal-organic framework composites for improving enzyme immobilization

Enhancement of immobilized enzyme activity using metal-organic framework composites

Metal-organic framework polymer composites-enzyme

Most studies report a decrease in enzyme activity following immobilization in MOFs compared to free enzymes. This reduction is primarily due to two factors: limited substrate access caused by confinement within the MOF structure and conformational changes induced by unwanted interactions between the enzyme and the MOF carrier. To address these challenges, incorporating a polymer during immobilization can help protect the enzyme against denaturation and enhance enzyme activity by modifying its surface charge. Zhou et al. applied polyacrylic acid (PAA) to the enzyme surface to create a protective layer against the enzyme conformation. PAA was grafted onto the enzyme via electrostatic interactions, thus forming a negatively charged polymeric structure on the enzyme surface. Interestingly, the modification of the enzyme using PAA also created defects in the MOF carrier, resulting in better substrate diffusion to the confined enzyme. This synergetic effect enhances enzyme activity up to 23 times higher than direct enzyme immobilization without PAA (Zhou, Chao, et al., 2022). Chen et al. reported the use of a polypeptide, poly-L-glutamic acid (PLGA), to alter the structure of an enzyme–MOF biohybrid via an in situ synthesis method (Fig. 20.3A). They found that the obtained enzyme–MOF dimensionalities strongly determined the bioactivity. The rich carboxyl side group provided PLGA with an anionic charge. In the presence of biomolecules, PLGA forms electrostatic interactions with the surface of biomolecules. The interaction between PLGA and MOF metal ions resulted in the morphological evolution of the MOF biomolecule hybrid from a 3D to a two-dimensional (2D) structure by tuning the PLGA ratio. As explained by the obtained structure, the 2D mesoporous spindle shape provided the best activity compared with other synthesized enzyme–MOF composites (Chen et al., 2020).

Metal-organic framework/metal oxide or metal-organic framework/metal nanoparticle composites-enzyme

Nanoparticles are widely used to support MOFs, either as templates or as dopants, to enhance stability, conductivity, or catalytic activity for specific

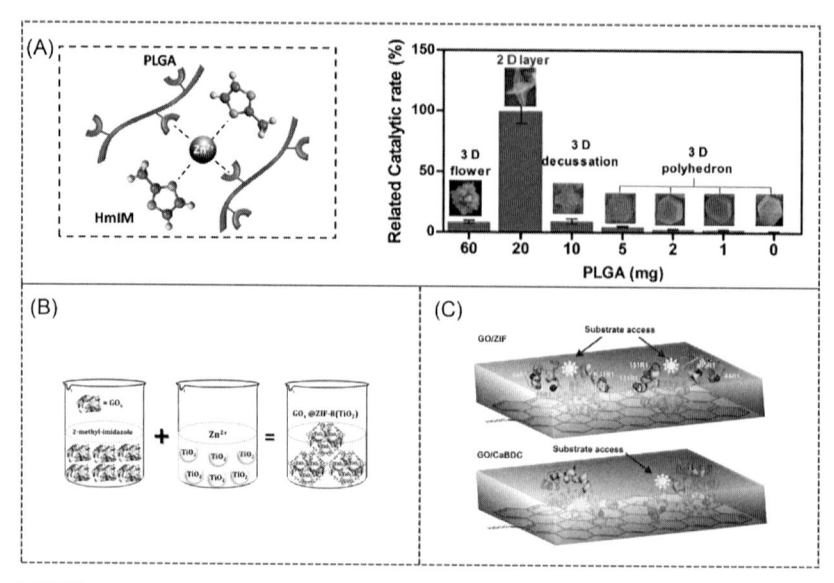

FIGURE 20.3 (A) Metal-organic framework polymer composite for improving the catalytic activity of the immobilized enzyme (Chen et al., 2020); (B) Synthesis process of GOx@ZIF-8(TiO$_2$) composite (Paul and Srivastava, 2018); (C) MOF-graphene composite for controlled immobilized enzyme orientation (Farmakes et al., 2020)

applications. Recent reports also indicate that MOF composites incorporating nanoparticles can improve enzyme immobilization and boost catalytic activity.

It has been reported that the activity of immobilized enzymes can be enhanced by increasing the flexibility of the encapsulated enzyme within the MOF structure (S.Y. Chen, Lo, et al., 2020). Applying this concept, Ma et al. immobilized enzymes onto silica (SiO$_2$) capsules using ZIF-8 as a template, demonstrating improved enzyme performance (Ma et al., 2021). The enzyme was first encapsulated in ZIF-8 via a one-pot synthesis method, followed by coating with a silica layer; ZIF-8 was then removed using an ethylenediamine acid solution. Using this strategy, the enzyme mimicked conditions similar to the eukaryotic cell microenvironment, and the enzyme immobilized in the hollow capsule demonstrated 15 times higher activity compared to when immobilization was carried out using ZIF-8. In addition, the cascade reaction efficiency improved when multiple enzymes were immobilized

In another study, TiO$_2$ nanoparticles were incorporated into the synthesis of GOx@ZIF-8 (Fig. 20.3B), enhancing the catalytic activity of the resulting biosensor. The synthesized TiO$_2$–MOF-enzyme nanocomposite demonstrates remarkable stability and can detect target molecules at low concentrations, reaching the nanomolar level (Paul & Srivastava, 2018).

Nanoparticles with enzyme-mimicking activity can also serve as tandem catalysts alongside natural enzymes (Metzger et al., 2021). As a result, MOF composites incorporating nanoparticles not only provide a stable immobilization matrix but also enable synergistic cascade reactions between the enzyme and nanoparticles. Coimmobilizing nanoparticles and enzymes within the MOF ensures their close proximity, enhancing reaction efficiency. For example, nickel-palladium nanoparticles (NiPd NPs) with peroxidase-like activity were combined with ZIF-8 and glucose oxidase (GOx) through an in situ synthesis process. This approach enables glucose detection via a colorimetric method, as the NiPd NPs can induce a color change in a chromogenic substrate upon enzymatic reaction.

Metal-organic framework carbon composites-enzyme

Carbon-based materials exhibit excellent conductivity, a large surface area, and the ability to be functionalized with various active groups, making them ideal candidates for enhancing MOF-based materials, especially in enzyme immobilization applications. Several studies have reported improved orientation and stability of immobilized enzymes in the presence of functionalized carbon-based materials, such as carbon nanotubes (CNTs) and graphene oxide (GO). These materials not only provide a stable support for the enzymes but also facilitate better electron transfer and enzyme activity.

Neupane et al. reported the oriented immobilization of enzymes on oxidized CNTs using MOF for large-substrate biocatalytic applications. This strategy avoided enzyme leaching and retained enzyme activity under both neutral and acidic conditions. Importantly, based on EPR characterization in the presence of oxidized CNTs, the active site of the enzyme was more exposed to the solution than to immobilization using only ZIF, making it more efficient for catalyzing large substrates. The main principle is the use of functionalized CNTs that provide a COOH group that can preimmobilize enzymes via electrostatic interactions. Finally, ZIF was used to trap and fix the enzyme and prevent its leaching from the CNT (Neupane et al., 2019).

Farmakes et al. compared ZIF with a calcium carboxylate-based MOF (CaBDC) as a matrix for the oriented immobilization of lys on a GO surface (Fig. 20.3C). They reported that the immobilization of lys on the GO surface using ZIF resulted in a more preferentially oriented enzyme compared to the randomly oriented immobilized enzyme when CaBDC was used. The use of GO provides advantages in large-substrate catalysis owing to the increase in the number of enzymes that are partially confined in the MOF structure (Farmakes et al., 2020).

Metal-organic framework carbon MXene composites-enzyme

MXenes are like sheets of metal reduced to just a few atomic layers, similar to how graphene is a single layer of carbon atoms. Their layered structure originates from "MAX phases," which are three-dimensional ceramics composed of transition metals, aluminum, and carbon or nitrogen. MXenes are known for their high electrical and thermal conductivity, large surface area, hydrophilicity, and tunable surface chemistry, making them ideal for a wide range of applications. One unique approach for improving confined enzyme activity of MOFs is the use of MXene nanosheets. Gu et al. used Ti_3C_2 to modulate the activity of Cyc C-embedded ZIF-8. The unique photothermal properties of Ti_3C_2 can increase the temperature in the solution up to 5°C, which depends on near-infrared (NIR) irradiation exposure, as a generally known temperature that mimics biological conditions can enhance enzyme activity. In addition, the hydrophilic properties of Ti_3C_2 confirmed the excellent microenvironment of the enzyme. Using this strategy, an increase in the enzymatic activity of up to 150% under NIR irradiation was successfully achieved (Gu et al., 2022).

Improvement of immobilized enzyme stability in metal-organic framework composites

The primary advantage of using MOFs or MOF composites over other materials for enzyme immobilization is the armored protection effect, particularly when the enzyme is encapsulated within the MOF or MOF composite structure. As reported by Liao et al., a catalase enzyme showed a significant improvement in stability after encapsulation in ZIF. The bioactivity of the embedded catalase was preserved after exposure to 6 M urea—an enzyme denaturing reagent—and exposure to a high temperature of 80°C (Liao et al., 2017). The immobilization of laccase in ZIF-8 also showed stability improvement toward organic solvents, including ethanol and dimethylformamide (DMF), at elevated temperatures (up to 70°C). After incubation in DMF, the immobilized enzyme retained 80% of its initial activity, whereas the free enzyme lost almost all of its activity. Incubation with ethanol decreased the immobilized enzyme activity to 50%, a result similar to that of the free enzyme. This result was due to size exclusion; large DMF molecules cannot enter the ZIF pores, while small ethanol molecules can easily penetrate and decrease the enzyme activity. A significant improvement in enzyme stability was observed with the variation in temperature. After 70°C treatment for 3 hours, free laccase lost its entire activity, while almost no loss of activity was observed for laccase immobilized in ZIF-8 (Knedel et al., 2019). The improvement in enzyme stability after encapsulation is primarily due to the size exclusion effect of the MOF or MOF composite carrier. This allows small molecules, such as substrates, to enter the pores and access the enzyme, while larger molecules, like DMF or proteinase K, which could denature the enzyme, are unable to penetrate the MOF or MOF composite carrier. In another study, the stability of urease was successfully improved after the encapsulation in ZIF-8; 40% initial activity was retained after incubation at 80°C in 30 minutes, while the free enzyme was deactivated (Liang et al., 2016). Liang et al. reported a high stability event in boiling water and DMF of HRP in ZIF-8, with

90% of the enzyme activity still preserved (Liang et al., 2015). Wu et al. reported that more than 80% of the remaining stability of coencapsulated graphite oxide (GO) and HRP inside ZIF-8; in comparison, free enzymes retained only 50% of their initial activity (Wu et al., 2015). The improved stability of the enzyme upon exposure to high temperatures has been reported to be due to the tightening of the MOF crystal structure around the enzyme and deagglomerated MOF crystals (Singh et al., 2021); hence, in some reported cases, the activity of the immobilized enzyme is also observed to increase. In real applications, such as in the industry, another factor that should be considered instead of biocatalyst stability is reusability. The recycling of biocatalysts will contribute significantly, particularly by reducing costs. Some enzymes, such as alcohol dehydrogenase from *Haloferax volcanii* (HvADH2), are already stable under harsh conditions, such as exposure to organic solvents, high salt concentrations, or high temperatures, owing to their natural sources. However, this enzyme has low reusability and stability in its soluble form. Carucci et al. overcame this problem through an enzyme entrapment strategy using a Fe–BTC MOF. The free enzyme exhibited the highest activity at 50°C, and after immobilization in the MOF, the optimum temperature changed to 60°C, and the observed enzyme activity was higher than that of the free enzyme. The enzyme activity decreased by approximately 75% when the temperature increased (60°C–80°C) in the case of the free enzyme and 71.5% in the case of the immobilized enzyme. Nevertheless, the activity of the immobilized enzyme was significantly higher than that of the free enzyme. Meanwhile, 75% of the initial enzyme activity was retained after four cycles for the immobilized enzyme (Carucci et al., 2018). In general, immobilized enzymes can be easily reused after centrifugal separation and washing. This is due to the insolubility of the MOF and MOF composites.

Recent fabrication and application of enzyme-immobilized metal-organic framework composites

The high reactivity of the metal to the ligand enables some types of MOF composites to be easily synthesized under mild conditions, particularly in the MAF groups. Additionally, crystallization is facilitated in the presence of biomolecules. This allows for the large-scale synthesis of enzyme-immobilized MOF composites with high precision using modern technologies. For example, injection printing technology has been employed to produce enzyme-encapsulated ZIF (Fig. 20.4A). In this case, the enzyme, ligand, and metal solutions are added to the ink cartridge, allowing the enzyme–MOF composite to be easily printed onto the substrate. As reported by Hou et al., printed Cyc-C@ZIF on paper showed a strong response to H_2O_2, indicated by a color change that depended on the H_2O_2 concentration in the colorimetric evaluation (Hou et al., 2017).

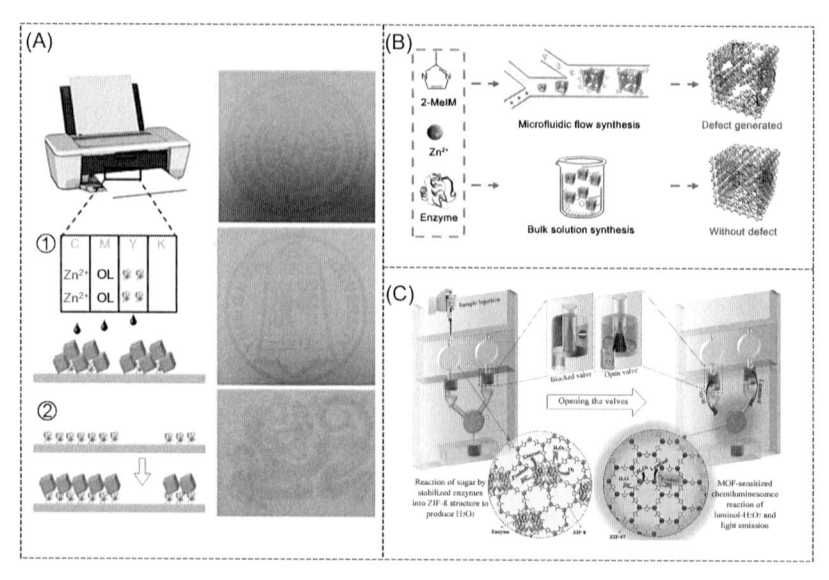

FIGURE 20.4 Fabrication and application of enzyme-encapsulated zeolitic imidazolate framework. (A) Enzyme-encapsulated zeolitic imidazolate framework prepared using the inkjet printing method (Hou et al., 2017); (B) Microfluidic laminar flow for the generation of defects in the enzyme-immobilized zeolitic imidazolate framework (Hu et al., 2020); (C) Application of enzyme-immobilized zeolitic imidazolate framework in lab-on-chip sensing design (Al Lawati et al., 2022)

Hu et al. reported a microfluidic flow synthesis method for producing an enzyme–MOF composite. Compared to bulk solution synthesis, this strategy could generate more defects on the enzyme–MOF composite surface due to the continuous change in MOF precursor concentration during microfluidic laminar flow synthesis (Fig. 20.4B). This defect improves substrate diffusion to the encapsulated enzyme, producing twice the catalytic activity compared to synthesis in a bulk solution (Hu et al., 2020). Lawati et al. developed a lab-on-chip system for sugar detection (Fig. 20.4C). The enzyme immobilized in ZIF-8 was used as the active material because of its higher stability over long storage times compared to the free enzyme. The chip contains two main parts: a sample reaction chamber and a luminol reaction chamber. A specific enzyme was immobilized on ZIF-8 to catalyze the oxidation of glucose, fructose, sucrose, maltose, and other sugar-containing compounds. The oxidation reaction of these substrates produces H_2O_2, which then flows and accesses the liminal reaction chamber containing luminol and 2D ZIF-67; hence, specific sugar detection can be achieved by observing the change in the luminol intensity of the chip (Al Lawati et al., 2022). Mohammad et al. fabricated an enzyme–MOF composite using surface patterning and biomineralization techniques for microfluidic biosensor applications. Polydopamine/polyethyleneimine was used to pattern an enzyme–ZIF-8 composite in a microfluidic channel. Using this strategy, the stability of the device was improved under acidic and high-temperature conditions compared to that without ZIF-8, which was mainly due to enzyme protection. Another advantage is that the obtained linear detection range increased with the use of encapsulated ZIF-8 due to substrate diffusion limitations, which is beneficial for biosensing applications (Mohammad et al., 2019).

Applications of enzyme-immobilized metal-organic framework composites

Biosensors

Different forms of MOFs and MOF composites can be employed as electrochemical biosensor matrices to covalently or physically connect to certain bioreceptor components or to enhance signals (optical, chemical, or electrical). MOF composite-based biosensors have been proven effective both in the environment and for firmly adsorbing heavy metal ions (Sohrabi et al., 2023). In the healthcare field, profound advancements have occurred in the detection of early biomarkers with good sensitivity and lower detection limits. The ZIF-8-NiPd-GOx composite was successfully applied for glucose detection using both low-voltage electrochemical methods and colorimetric analysis. This dual-mode detection capability allows for sensitive and versatile glucose sensing, highlighting the composite's potential for practical biosensing applications (Fig. 20.5A) (Wang et al., 2017). In another study, 3-mercaptophenylboronic acid was incorporated into polyvinylpyrrolidone (PVP) for the immobilization of GOx in ZIF-8 (Han et al., 2021). This strategy enhanced the affinity and catalytic efficiency for glucose sensing while also shortening

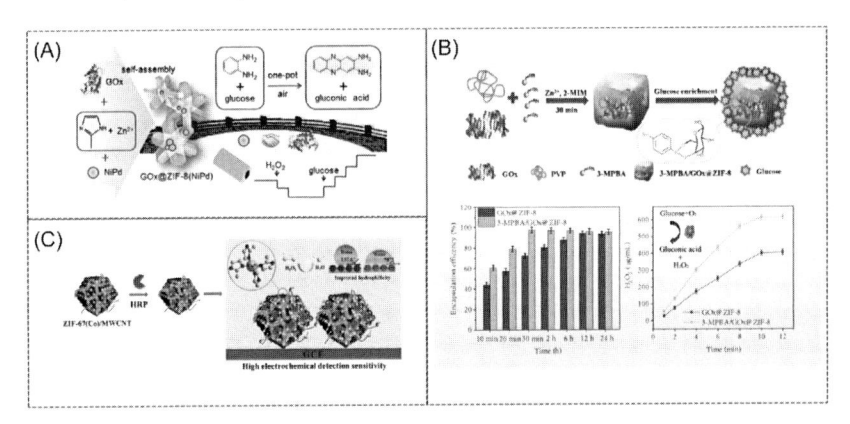

FIGURE 20.5 Applications of metal-organic framework composite-enzyme for glucose sensing: (A) Metal-organic framework nanoparticle-enzyme composite (Wang et al., 2017); (B) Metal-organic framework polymer-enzyme composite (Han et al., 2021); and (C) Metal-organic framework carbon-enzyme composite (Liu et al., 2019).

the encapsulation time (Fig. 20.5B). Notably, the method maintained high enzyme activity even with a reduced encapsulation duration (Chen, Kou, et al., 2020). The polymer-assisted immobilized enzyme also demonstrated excellent reusability, retaining approximately 80.8% of its activity after seven cycles, indicating robust stability and potential for repeated applications.

An enzyme-encapsulated MOF nanomesh biosensor exhibits exceptional sensitivity and stability in the detection of salivary glucose. The sensing interface of the biosensor was made of a custom-designed, necklace-like, conductive nanomesh (CNT/GOx@ZIF-8), which was generated by ZIF-8 integration of CNTs and GOx, with ZIF-8 grown by self-assembly along the CNTs with encapsulated GOx. Encapsulating GOx in ZIF-8 can increase its stability and allow the stable identification of glucose molecules (Wang et al., 2024). However, the low conductivity of MOF composites restricts their widespread use in biosensors. CNTs with high electron conductivity have been widely used in electrochemical biosensors; however, they suffer from poor electron transduction between the nanomaterials and enzymes. The serial network of ZIF-8 via CNTs increases electron transduction, compensating for the MOF-insulating properties. The fabricated biosensor has a high sensitivity of 86.86 μA/(mm.cm^2) and remains stable after 16 days.

Liu et al. utilized a ZIF-67-CNT composite to immobilize HRP enzyme, applying it as a highly sensitive electrochemical sensor for H_2O_2 (Fig. 20.5C) (Liu et al., 2019). By introducing hydrophilic carbon nanomaterials and forming MOF/carbon composites, both the conductivity of the matrix and the apparent affinity of the immobilized enzyme for hydrophilic substrates were significantly enhanced. This approach resulted in a 1.3-fold increase in sensitivity for H_2O_2 detection, highlighting the effectiveness of MOF/carbon composites in electrochemical biosensing applications.

Zhao et al. reported that the dual enzymes cholesterol oxidase (ChOx) and encapsulated HRP were immobilized on a water-soluble MOF with improved sensitivity and stability for the colorimetric detection of cholesterol (Zhao et al., 2019). PCN-333(Al) was used to coimmobilize ChOx and HRP on the MOF. Pore encapsulation of the bienzymes was achieved by considering the large surface area and concentration of the MOF mesoporous cages (PCN-333). The coimmobilized ChOx and HRP demonstrated strong catalytic activity in a cascade process. In addition, these bienzymes have long shelf lives and are highly resistant to protease digestion, organic solvents, pH, and heat fluctuations owing to their stiff MOF structures. A simple colorimetric approach for detecting cholesterol that is sensitive and selective was developed (Zhao et al., 2019).

Enzyme-immobilized metal-organic framework composites for the environment

Enzymes such as laccases, peroxidases, reductases, tyrosinases, and cytochromes are generally involved in dye degradation. MOFs, along with these enzymes, provide a unique surface area for wide adsorption, and enzyme immobilization is an efficient method for increasing recycling, reducing enzyme waste, and improving enzyme stability. Problems associated with single-enzyme-loaded MOFs include the rapid saturation of adsorbent failure and regeneration difficulties. Long et al. attempted a bimetallic approach using CoCu-MOFs via covalent binding to enhance enzyme loading and Congo Red degradation (Long et al., 2023). The enhanced hydrogen bonding and electrostatic interactions between the CoCu-MOF molecules and the enzyme after immobilization help maintain the conformational integrity of the enzyme molecule, thus providing long-term stability and reusability for dye degradation studies, which shows promise for industrial applications. Another study demonstrated a strategy for the coimmobilization of dual enzymes with MOF using HRP, GOx, and ZIF-8 nanostructures (NZIF-8s), which were designated as HRP/GOx@NZIF-8. This biocatalytic cascade system was fabricated to improve thermal and long-term stability. The fabricated cascade approach was also tested for the biodegradation of bisphenol A (BPA). The mesoporous network of the MOF provides a good surface area, and the conductivity improves the enzyme stability in its organic ligand formation (Babaei et al., 2023). The biodegradation observations showed that the bulk BPA content was reduced to 8.645 mg/L after 60 minutes of reaction. This remarkable removal efficiency may be partly attributed to the adsorption of BPA molecules adsorbing into the porous networks of NZIF-8 (adsorptive removal). In fact, some BPA molecules that entered the NZIF-8s remained intact and unoxidized in the porous networks. Thus, adsorptive clearance can be distinguished from enzymatic biodegradation.

Another remarkable degradation of organic dyes, such as methylene blue, neutral Sudan II, and methyl orange, was attempted by immobilizing enzyme MP8 in the mesoporous cages of polycarboxylate MOF (MIL-101(Cr)) nanoparticles. The entrapped enzyme was protected from acidic or oxidative environments because it was contained in the MOF matrix. This study also demonstrated for the first time that the MOF matrix may function in tandem with an enzyme to specifically enhance the oxidation of dyes. This increase was ascribed to reactant preconcentration via charge matching between the dye and the MOF (Gkaniatsou et al., 2018).

Metal-organic framework–enzyme composites for cancer therapy

MOF–enzyme composites are being explored in cancer therapy for their ability to target tumors, enhance drug delivery, and generate reactive oxygen species that induce cancer cell apoptosis, thereby improving treatment efficacy. Li et al. reported the encapsulation of GOx in a bimetallic Zn–Cu framework using 2-methylimidazole as a linker (GOx@Cu-ZIF-8) for application in chemodynamic therapy (CDT). This therapy relies on the induction of cellular apoptosis through the oxidative stress generated by hydroxyl radicals that are produced via the catalytic conversion of H_2O_2. However, the efficacy of CDT is often limited by the insufficient levels of H_2O_2 in tumor regions. ZIF-8 served as a carrier to deliver both GOx and Cu ions to the targeted tumor site. Owing to the pH-sensitive stability of ZIF-8, its structure disintegrates in the acidic environment of the tumor, releasing GOx and Cu ions. In the presence of oxygen, GOx then consumes glucose present in the tumor microenvironment, generating sufficient H_2O_2. Subsequently, the Cu ions catalyze a Fenton-like reaction, producing hydroxyl radicals. These radicals disrupt the oxidative balance within cancer cells, leading to efficient cellular apoptosis and enhanced CDT therapeutic efficacy (Li et al., 2023). The requirement for oxygen in the enzymatic production of H_2O_2 by GOx poses a significant challenge in the hypoxic microenvironment of tumor sites. This oxygen deficiency not only limits the efficacy of CDT but also significantly reduces the effectiveness of radiation therapy and some chemotherapies. These treatments often depend on oxygen to generate reactive species, such as hydroxyl radicals, through X-ray radiation to damage cancer cells. To address this issue, Zhao et al. developed an MOF carrier based on the high-atomic-number metal Hf (Hafnium). High-atomic-number metals can strongly absorb X-ray energy and produce secondary electrons, such as photoelectrons and Auger electrons, which enhance the production of hydroxyl radicals from the ionization of H_2O, where O_2 is essential in sustaining and amplifying this reaction. Hierarchically porous Hf-MOFs capable of loading hemoglobin were synthesized using a solvothermal process. Hemoglobin naturally acts as an oxygen transporter in blood cells, efficiently capturing and delivering oxygen. In this context, they oxygenated the immobilized hemoglobin on the porous Hf-MOF before treatment, thereby improving the efficiency of radiation therapy. Tetrakis (4-carboxyphenyl)porphyrin (TCPP) was used as an organic linker. TCPP is photosensitive; upon absorbing X-ray radiation energy, Hf metal emits energy as visible light, which activates TCPP and further enhances the therapeutic reaction (Zhao et al., 2023).

Metal-organic framework–enzyme composites for wound healing-related applications

Bacterial infections are a common challenge in wound healing and highlight the need for effective antibacterial agents. Although antibiotics are often used to treat or prevent such infections, prolonged use can lead to the development of antibiotic-resistant bacteria, which is a well-known and persistent issue in antibiotic therapy. To address this challenge, Liu et al. developed a 2D MOF, Cu–TCPP(Fe), which immobilizes GOx through physical adsorption. This composite continuously generates H_2O_2 and acts as a self-activated peroxidase catalyst. The Cu–TCPP(Fe)/GOx composite efficiently produces hydroxyl radicals (•OH), which can damage bacteria under physiological conditions—a feat that is difficult to achieve with other catalysts or composite materials that require an acidic environment (pH 3–4) to function optimally. The researchers explained that during the enzymatic reaction, immobilized GOx produces gluconic acid, which lowers the local pH of the material. This acidic microenvironment activates the TCPP(Fe) ligand, enabling it to mimic peroxidase activity and convert H_2O_2 into •OH, leading to bacterial death. This results in an efficient therapeutic effect with a negligible impact on normal tissue, as demonstrated in wound-healing tests conducted on mice. This innovative approach offers a promising solution for enhancing wound healing through targeted antibacterial activity (Liu, Yan, et al., 2019). Building on a similar concept but with a different MOF matrix, Zhou et al. utilized $MnFe_2O_4$@MIL to immobilize gold nanoparticles and GOx for wound-healing applications. Given that glutathione (GSH), which is commonly found in the microenvironment of bacterial infections, can neutralize •OH generated by peroxidase reactions, the depletion of GSH is crucial to enhance antibacterial efficacy. $MnFe_2O_4$ nanoparticles were incorporated to effectively deplete GSH, thereby increasing the production and efficiency of •OH radicals in damaging bacteria and ultimately improving the wound healing treatment (Zhou, Zhang, et al., 2022).

Metal-organic framework–enzyme composites in food and industrial applications

Lipases are enzymes commonly used in various food, chemical, and industrial applications because of their ability to catalyze diverse reactions such as esterification, hydrolysis, and alcoholysis. For instance, lipase-based reactions have been employed to produce food flavors through esterification. However, lipases are unstable enzymes sensitive to changes under environmental conditions. The performance and stability of lipase catalysts can be significantly improved by employing enzyme immobilization strategies. Suo et al. used magnetic ZIF-90 nanoparticles to immobilize lipases. ZIF-90 was grown on the surfaces of Fe_3O_4 particles modified with polydopamine. Immobilization of the enzyme on a magnetic support facilitates easy recovery of the product in a magnetic field and enhances the reusability of the biocatalyst. In their study, immobilized lipase was used to synthesize isoamyl acetate, a liquid flavor compound containing banana and pear notes. Remarkably, the immobilized lipase retained 90.5% of its original activity after seven reuse cycles. Additionally, the storage stability of the immobilized lipase was 88.7% after 35 days of storage. This approach improves productivity and reduces production costs, making it highly efficient for industrial applications (Suo et al., 2022). Zhu et al. employed multiple enzymes immobilized within ZIF-8 for methanol production via a CO_2 reduction reaction. The enzyme–MOF composite, which also contained NADH, was loaded onto a microporous membrane. This strategy significantly enhanced methanol production, achieving nearly twice the yield compared with the use of nonimmobilized enzymes on the membrane after

12 hours of use. Additionally, an immobilized multienzyme system retained 50% of its productivity, demonstrating improved stability and efficiency (Zhu et al., 2019). In response to the growing demand for environmentally friendly leather processing, Palanisamy et al. explored the use of an MOF–enzyme (Pro) system for the unhairing process as an alternative to traditional sodium sulfide and lime-based methods. Proenzymes assist in unhairing by breaking down keratin in hair, making it easier to remove it from the hide. In their study, a Zn-MOF was used to covalently immobilize Pro using a glutaraldehyde crosslinker. The advantage of using Zn-MOFs over other materials is their negligible impact on the color of the leather product. This approach also significantly improved the recyclability of the enzyme compared to that of its nonimmobilized counterpart. A study that utilized goatskin demonstrated that the MOF–enzyme composite was more effective in hair removal, maintaining a high efficiency even after four cycles of use (Palanisamy et al., 2024).

Metal-organic framework–enzyme composites for environmental application

Synthetic dyes are problematic and recognized as serious water contaminants. For example, Congo Red is a potential carcinogen that can cause cancer or tumors upon exposure. Certain enzymes, such as laccase, have been reported to be effective agents for dye degradation because of their ability to oxidize various aromatic compounds via Fenton-like reactions in the presence of oxygen and electron acceptors. However, the low stability of laccase limits its performance. To address this issue, Long et al. immobilized laccase within a bimetallic CoCu-MOF to remove Congo Red from wastewater. CoCu-MOF was functionalized with 3-aminopropyltriethoxysilane, and laccase was immobilized through covalent interactions facilitated by glutaraldehyde crosslinkers. In this system, the MOF plays a crucial role in efficiently adsorbing Congo Red, following which the immobilized laccase degrades the absorbed dye. This approach not only enhances the stability of laccase but also improves the efficiency of Congo Red removal for wastewater treatment applications (Long et al., 2023).

Conclusion and future perspectives

Although enzyme–MOF composites are commonly used in catalytic investigations, research on the synergistic catalysis of MOFs and enzymes has been insufficient. Encapsulated enzymes undergo minimal chemical changes, allowing optimum activity. This differs from standard enzyme cocatalysis investigations. Adding asymmetric groups to MOF surfaces may improve the enantiomeric excess during enzyme conversion. Functionalizing MOFs with hydrophilic groups can improve the interactions between hydrophilic substrates and hydrophobic enzymes, thereby improving conversion rates and yields. To improve the enzymatic activity of enzyme–MOF composites, balancing the anticipated loss of enzymatic activity following encapsulation within MOFs is necessary. Common tactics include changing the enzyme and MOF structures, fine-tuning enzyme–MOF interactions, and promoting substrate–product interchanges. The most prominent strategies for maximizing enzymatic activity involve adjusting the hydrophobic interface between the enzymes and MOFs via enzyme–MOF functionalization and designing enzyme active sites. Enzyme–MOF composites have a wide range of possible applications in biofuel cells, biocatalysts, biosensors, and biomedicine.

References

Ahmad, R., Rizaldo, S., Gohari, M., Shanahan, J., Shaner, S. E., Stone, K. L., & Kissel, D. S. (2023). Buffer effects in zirconium-based UiO metal-organic frameworks (MOFs) that influence enzyme immobilization and catalytic activity in enzyme/MOF biocatalysts. *ACS Omega, 8*(25), 22545–22555. https://doi.org/10.1021/acsomega.3c00703, https://pubs.acs.org/doi/10.1021/acsomega.3c00703.

Al Lawati, H. A. J., Hassanzadeh, J., & Bagheri, N. (2022). A handheld 3D-printed microchip for simple integration of the H_2O_2-producing enzymatic reactions with subsequent chemiluminescence detection: Application for sugars. *Food Chemistry, 383*, 132469. https://doi.org/10.1016/j.foodchem.2022.132469.

Babaei, H., Ghobadi Nejad, Z., Yaghmaei, S., & Farhadi, F. (2023). Co-immobilization of multi-enzyme cascade system into the metal–organic frameworks for the removal of bisphenol A. *Chemical Engineering Journal, 461*, 142050. https://doi.org/10.1016/j.cej.2023.142050.

Bose, P., Bai, L., Ganguly, R., Zou, R., & Zhao, Y. (2015). Rational design and synthesis of a highly porous copper-based interpenetrated metal–organic framework for high CO_2 and H_2 adsorption. *ChemPlusChem, 80*(8), 1259–1266. https://doi.org/10.1002/cplu.201500104.

Bunzen, H. (2021). Chemical stability of metal-organic frameworks for applications in drug delivery. *ChemNanoMat, 7*(9), 998–1007. https://doi.org/10.1002/cnma.202100226, http://onlinelibrary.wiley.com/journal/10.1002/(ISSN)2199-692X.

Bůžek, D., Adamec, S., Lang, K., & Demel, J. (2021). Metal-organic frameworks: Vs. buffers: Case study of UiO-66 stability. *Inorganic Chemistry Frontiers, 8*(3), 720–734. https://doi.org/10.1039/d0qi00973c, http://pubs.rsc.org/en/journals/journal/qi.

Carucci, C., Bruen, L., Gascón, V., Paradisi, F., & Magner, E. (2018). Significant enhancement of structural stability of the hyperhalophilic ADH from haloferax volcanii via entrapment on metal organic framework support. *Langmuir, 34*(28), 8274–8280. https://doi.org/10.1021/acs.langmuir.8b01037, http://pubs.acs.org/journal/langd5.

Chen, G., Huang, S., Kou, X., Zhu, F., & Ouyang, G. (2020). Embedding functional biomacromolecules within peptide-directed metal–organic framework (MOF) nanoarchitectures enables activity enhancement. *Angewandte Chemie-International Edition, 59*(33), 13947–13954. https://doi.org/10.1002/anie.202005529, http://onlinelibrary.wiley.com/journal/10.1002/.

Chen, G., Kou, X., Huang, S., Tong, L., Shen, Y., Zhu, W., Zhu, F., & Ouyang, G. (2020). Modulating the biofunctionality of metal–organic-framework-encapsulated enzymes through controllable embedding patterns. *Angewandte Chemie-International Edition, 59*(7), 2867–2874. https://doi.org/10.1002/anie.201913231, http://onlinelibrary.wiley.com/journal/10.1002/.

Chen, S. Y., Lo, W. S., Huang, Y. D., Si, X., Liao, F. S., Lin, S. W., Williams, B. P., Sun, T. Q., Lin, H. W., An, Y., Sun, T., Ma, Y., Yang, H. C., Chou, L. Y., Shieh, F. K., & Tsung, C. K. (2020). Probing interactions between metal-organic frameworks and freestanding enzymes in a hollow structure. *Nano Letters, 20*(9), 6630–6635. https://doi.org/10.1021/acs.nanolett.0c02265, http://pubs.acs.org/journal/nalefd.

Chen, Y., Han, S., Li, X., Zhang, Z., & Ma, S. (2014). Why does enzyme not leach from metal-organic frameworks (MOFs)? Unveiling the interactions between an enzyme molecule and a MOF. *Inorganic Chemistry, 53*(19), 10006–10008. https://doi.org/10.1021/ic501062r, http://pubs.acs.org/journal/inocaj.

Chen, Y., Yang, Z., Hu, H., Zhou, X., You, F., Yao, C., Liu, F. J., Yu, P., Wu, D., Yao, J., Hu, R., Jiang, X., & Yang, H. (2022). Advanced metal–organic frameworks-based catalysts in electrochemical sensors. *Frontiers in Chemistry, 10*.

Deng, H., Grunder, S., Cordova, K. E., Valente, C., Furukawa, H., Hmadeh, M., Gándara, F., Whalley, A. C., Liu, Z., Asahina, S., Kazumori, H., O'Keeffe, M., Terasaki, O., Stoddart, J. F., & Yaghi, O. M. (2012). Large-pore apertures in a series of metal-organic frameworks. *Science, 336*(6084), 1018–1023. https://doi.org/10.1126/science.1220131, http://www.sciencemag.org/content/336/6084/1018.full.pdf.

Devi, K. S. S., Rezki, M., & Tsujimura, S. (2023). Emerging electrochemical biosensing strategies using enzyme-incorporated metal-organic frameworks. *Talanta Open, 8*, 100263. https://doi.org/10.1016/j.talo.2023.100263.

Farmakes, J., Schuster, I., Overby, A., Alhalhooly, L., Lenertz, M., Li, Q., Ugrinov, A., Choi, Y., Pan, Y., & Yang, Z. (2020). Enzyme immobilization on graphite oxide (GO) surface via one-pot synthesis of GO/metal-organic framework composites for large-substrate biocatalysis. *ACS Applied Materials and Interfaces, 12*(20), 23119–23126. https://doi.org/10.1021/acsami.0c04101, http://pubs.acs.org/journal/aamick.

Furukawa, H., Cordova, K. E., O'Keeffe, M., & Yaghi, O. M. (2013). The chemistry and applications of metal-organic frameworks. *Science, 341*(6149). https://doi.org/10.1126/science.1230444, http://www.sciencemag.org/content/341/6149/1230444.full.pdf.

Gao, S., Hou, J., Deng, Z., Wang, T., Beyer, S., Buzanich, A. G., Richardson, J. J., Rawal, A., Seidel, R., Zulkifli, M. Y., Li, W., Bennett, T. D., Cheetham, A. K., Liang, K., & Chen, V. (2019). Improving the acidic stability of zeolitic imidazolate frameworks by biofunctional molecules. *Chem, 5*(6), 1597–1608. https://doi.org/10.1016/j.chempr.2019.03.025, http://www.cell.com/chem/home.

Gao, Y., Doherty, C. M., & Mulet, X. (2020). A systematic study of the stability of enzyme/zeolitic imidazolate framework-8 composites in various biologically relevant solutions. *ChemistrySelect, 5*(43), 13766–13774. https://doi.org/10.1002/slct.202003575, http://onlinelibrary.wiley.com/journal/10.1002/.

Ge, D., Li, M., Wei, D., Zhu, N., Wang, Y., Li, M., Zhang, Z., & Zhao, H. (2023). Enhanced activity of enzyme encapsulated in hydrophilic metal-organic framework for biosensing. *Chemical Engineering Journal, 469*, 144067. https://doi.org/10.1016/j.cej.2023.144067.

Gkaniatsou, E., Ricoux, R., Kariyawasam, K., Stenger, I., Fan, B., Ayoub, N., Salas, S., Patriarche, G., Serre, C., Mahy, J. P., Steunou, N., & Sicard, C. (2020). Encapsulation of microperoxidase-8 in MIL-101(Cr)-X nanoparticles: Influence of metal-organic framework functionalization on enzymatic immobilization and catalytic activity. *ACS Applied Nano Materials, 3*(4), 3233–3243. https://doi.org/10.1021/acsanm.9b02464, https://pubs.acs.org/journal/aanmf6.

Gkaniatsou, E., Sicard, C., Ricoux, R., Benahmed, L., Bourdreux, F., Zhang, Q., Serre, C., Mahy, J. P., & Steunou, N. (2018). Enzyme encapsulation in mesoporous metal–organic frameworks for selective biodegradation of harmful dye molecules. *Angewandte Chemie-International Edition, 57*(49), 16141–16146. https://doi.org/10.1002/anie.201811327, http://onlinelibrary.wiley.com/journal/10.1002/.

Gröger, H., Allahverdiyev, A., Yang, J., & Stiehm, J. (2024). Merging MOF chemistry & biocatalysis: A perspective for achieving efficient organic synthetic processes and applications in the chemical industry? Advanced Functional Materials, 34(43), 2304794.

Gu, C., She, Y., Chen, X. C., Zhou, B. Y., Zhu, Y. X., Ding, X. Q., Tan, P., Liu, X. Q., & Sun, L. B. (2022). Modulating the activity of enzyme in metal-organic frameworks using the photothermal effect of Ti_3C_2 nanosheets. *ACS Applied Materials and Interfaces, 14*(26), 30090–30098. https://doi.org/10.1021/acsami.2c06375, http://pubs.acs.org/journal/aamick.

Han, J., Huang, W., Zhao, M., Wu, J., Li, Y., Mao, Y., Wang, L., & Wang, Y. (2021). A novel enhanced enrichment glucose oxidase@ZIF-8 biomimetic strategy with 3-mercaptophenylboronic acid for highly efficient catalysis of glucose. *Colloids and Surfaces B: Biointerfaces, 208*, 112034. https://doi.org/10.1016/j.colsurfb.2021.112034.

He, W., Gan, Y., Qi, X., Wang, H., Song, H., Su, P., Song, J., & Yang, Y. (2023). Enhancing enzyme activity using hydrophilic hollow layered double hydroxides as encapsulation carriers. *ACS Applied Materials and Interfaces, 15*(29), 34513–34526. https://doi.org/10.1021/acsami.3c05237, http://pubs.acs.org/journal/aamick.

Hou, M., Zhao, H., Feng, Y., & Ge, J. (2017). Synthesis of patterned enzyme–metal–organic framework composites by ink-jet printing. *Bioresources and Bioprocessing, 4*(1), 40.

Hu, C., Bai, Y., Hou, M., Wang, Y., Wang, L., Cao, X., Chan, C. W., Sun, H., Li, W., Ge, J., & Ren, K. (2020). Defect-induced activity enhancement of enzyme-encapsulated metal-organic frameworks revealed in microfluidic gradient mixing synthesis. *Science Advances, 6*(5). https://doi.org/10.1126/sciadv.aax5785, https://advances.sciencemag.org/content/advances/6/5/eaax5785.full.pdf.

Huang, A., Tong, L., Kou, X., Gao, R., Li, Z. W., Huang, S., Zhu, F., Chen, G., & Ouyang, G. (2023). Structural and functional insights into the biomineralized zeolite imidazole frameworks. *ACS Nano, 17*(23), 24130–24140. https://doi.org/10.1021/acsnano.3c09118, http://pubs.acs.org/journal/ancac3.

Jabeen, R., Tajwar, M. A., Cao, C., Liu, Y., Zhang, S., Ali, N., & Qi, L. (2024). Confinement-induced biocatalytic activity enhancement of light- and thermoresponsive polymer@enzyme@MOF composites. *ACS Applied Materials and Interfaces, 16*(28), 36953–36961. https://doi.org/10.1021/acsami.4c05742, http://pubs.acs.org/journal/aamick.

Jordahl, D., Armstrong, Z., Li, Q., Gao, R., Liu, W., Johnson, K., Brown, W., Scheiwiller, A., Feng, L., Ugrinov, A., Mao, H., Chen, B., Quadir, M., Li, H., Pan, Y., & Yang, Z. (2022). Expanding the "library" of metal-organic frameworks for enzyme biomineralization. *ACS Applied Materials and Interfaces, 14*(46), 51619–51629. https://doi.org/10.1021/acsami.2c12998, http://pubs.acs.org/journal/aamick.

Knedel, T. O., Ricklefs, E., Schlüsener, C., Urlacher, V. B., & Janiak, C. (2019). Laccase encapsulation in ZIF-8 metal-organic framework shows stability enhancement and substrate selectivity. *ChemistryOpen, 8*(11), 1337–1344. https://doi.org/10.1002/open.201900146, http://onlinelibrary.wiley.com/journal/10.1002/.

Li, B., Wen, H. M., Cui, Y., Zhou, W., Qian, G., & Chen, B. (2016). Emerging multifunctional metal–organic framework materials. *Advanced Materials, 28*(40), 8819–8860. https://doi.org/10.1002/adma.201601133, http://www3.interscience.wiley.com/journal/119030556/issue.

Li, H., Li, L., Lin, R. B., Zhou, W., Zhang, Z., Xiang, S., & Chen, B. (2019). Porous metal-organic frameworks for gas storage and separation: Status and challenges. *EnergyChem, 1*(1), 100006.

https://doi.org/10.1016/j.enchem.2019.100006, http://www.journals.elsevier.com/energychem.

Li, Q., Pan, Y., Li, H., Alhalhooly, L., Li, Y., Chen, B., Choi, Y., & Yang, Z. (2020). Size-tunable metal-organic framework-coated magnetic nanoparticles for enzyme encapsulation and large-substrate biocatalysis. *ACS Applied Materials and Interfaces, 12*(37), 41794–41801. https://doi.org/10.1021/acsami.0c13148, http://pubs.acs.org/journal/aamick.

Li, Q., Yu, J., Lin, L., Zhu, Y., Wei, Z., Wan, F., Zhang, X., He, F., & Tian, L. (2023). One-pot rapid synthesis of Cu^{2+}-doped GOD@MOF to amplify the antitumor efficacy of chemodynamic therapy. *ACS Applied Materials and Interfaces, 15*(13), 16482–16491. https://doi.org/10.1021/acsami.3c00562, http://pubs.acs.org/journal/aamick.

Li, S. F., Zhai, X. J., Zhang, C., Mo, H. L., & Zang, S. Q. (2020). Enzyme immobilization in highly ordered macro-microporous metal-organic frameworks for rapid biodegradation of hazardous dyes. *Inorganic Chemistry Frontiers, 7*(17), 3146–3153. https://doi.org/10.1039/d0qi00489h, http://pubs.rsc.org/en/journals/journal/qi.

Liang, K., Coghlan, C. J., Bell, S. G., Doonan, C., & Falcaro, P. (2016). Enzyme encapsulation in zeolitic imidazolate frameworks: A comparison between controlled co-precipitation and biomimetic mineralisation. *Chemical Communications, 52*(3), 473–476. https://doi.org/10.1039/c5cc07577g, http://pubs.rsc.org/en/journals/journal/cc.

Liang, K., Ricco, R., Doherty, C. M., Styles, M. J., Bell, S., Kirby, N., Mudie, S., Haylock, D., Hill, A. J., Doonan, C. J., & Falcaro, P. (2015). Biomimetic mineralization of metal-organic frameworks as protective coatings for biomacromolecules. *Nature Communications, 6*, 7240. https://doi.org/10.1038/ncomms8240, http://www.nature.com/ncomms/index.html.

Liang, W., Xu, H., Carraro, F., Maddigan, N. K., Li, Q., Bell, S. G., Huang, D. M., Tarzia, A., Solomon, M. B., Amenitsch, H., Vaccari, L., Sumby, C. J., Falcaro, P., & Doonan, C. J. (2019). Enhanced activity of enzymes encapsulated in hydrophilic metal-organic frameworks. *Journal of the American Chemical Society, 141*(6), 2348–2355. https://doi.org/10.1021/jacs.8b10302, http://pubs.acs.org/journal/jacsat.

Liao, F. S., Lo, W. S., Hsu, Y. S., Wu, C. C., Wang, S. C., Shieh, F. K., Morabito, J. V., Chou, L. Y., Wu, K. C. W., & Tsung, C. K. (2017). Shielding against unfolding by embedding enzymes in metal-organic frameworks via a de novo approach. *Journal of the American Chemical Society, 139*(19), 6530–6533. https://doi.org/10.1021/jacs.7b01794, http://pubs.acs.org/journal/jacsat.

Liu, J., Ye, L. Y., Xiong, W. H., Liu, T., Yang, H., & Lei, J. (2021). A cerium oxide@metal-organic framework nanoenzyme as a tandem catalyst for enhanced photodynamic therapy. *Chemical Communications, 57*(22), 2820–2823. https://doi.org/10.1039/d1cc00001b, http://pubs.rsc.org/en/journals/journal/cc.

Liu, X., Chen, W., Lian, M., Chen, X., Lu, Y., & Yang, W. (2019). Enzyme immobilization on ZIF-67/MWCNT composite engenders high sensitivity electrochemical sensing. *Journal of Electroanalytical Chemistry, 833*, 505–511. https://doi.org/10.1016/j.jelechem.2018.12.027.

Liu, X., Li, Y., Cao, Z., Yin, Z., Ma, T., & Chen, S. (2022). Current progress of metal sulfides derived from metal-organic frameworks for advanced electrocatalysis: Potential electrocatalysts with diverse applications. *Journal of Materials Chemistry A, 10*(4), 1617–1641. https://doi.org/10.1039/d1ta09925f, http://pubs.rsc.org/en/journals/journal/ta.

Liu, X., Yan, Z., Zhang, Y., Liu, Z., Sun, Y., Ren, J., & Qu, X. (2019). Two-dimensional metal-organic framework/enzyme hybrid nanocatalyst as a benign and self-activated cascade reagent for in vivo wound healing. *ACS Nano, 13*(5), 5222–5230. https://doi.org/10.1021/acsnano.8b09501, http://pubs.acs.org/journal/ancac3.

Liu, Y., Gong, C. S., Dai, Y., Yang, Z., Yu, G., Liu, Y., Zhang, M., Lin, L., Tang, W., Zhou, Z., Zhu, G., Chen, J., Jacobson, O., Kiesewetter, D. O., Wang, Z., & Chen, X. (2019). In situ polymerization on nanoscale metal-organic frameworks for enhanced physiological stability and stimulus-responsive intracellular drug delivery. *Biomaterials, 218*, 119365. https://doi.org/10.1016/j.biomaterials.2019.119365, http://www.journals.elsevier.com/biomaterials/.

Long, H., Li, X., Liu, X., Wang, W., Yang, X., & Wu, Z. (2023). Immobilization of laccase by alkali-etched bimetallic CoCu-MOF to enhance enzyme loading and congo red degradation. *Langmuir, 39*(24), 8404–8413. https://doi.org/10.1021/acs.langmuir.3c00362, http://pubs.acs.org/journal/langd5.

Lyu, F., Zhang, Y., Zare, R. N., Ge, J., & Liu, Z. (2014). One-pot synthesis of protein-embedded metal-organic frameworks with enhanced biological activities. *Nano Letters, 14*(10), 5761–5765. https://doi.org/10.1021/nl5026419, http://pubs.acs.org/journal/nalefd.

Ma, X., Sui, H., Yu, Q., Cui, J., & Hao, J. (2021). Silica capsules templated from metal-organic frameworks for enzyme immobilization and catalysis. *Langmuir, 37*(10), 3166–3172. https://doi.org/10.1021/acs.langmuir.1c00065, http://pubs.acs.org/journal/langd5.

Metzger, K. E., Moyer, M. M., & Trewyn, B. G. (2021). Tandem catalytic systems integrating biocatalysts and inorganic catalysts using functionalized porous materials. *ACS Catalysis, 11*(1), 110–122. https://doi.org/10.1021/acscatal.0c04488, http://pubs.acs.org/page/accacs/about.html.

Mohammad, M., Razmjou, A., Liang, K., Asadnia, M., & Chen, V. (2019). Metal-organic-framework-based enzymatic microfluidic biosensor via surface patterning and biomineralization. *ACS Applied Materials and Interfaces, 11*(2), 1807–1820. https://doi.org/10.1021/acsami.8b16837, http://pubs.acs.org/journal/aamick.

Mohidem, N. A., Mohamad, M., Rashid, M. U., Norizan, M. N., Hamzah, F., & Mat, Hb (2023). Recent advances in enzyme immobilisation strategies: An overview of techniques and composite carriers. *Journal of Composites Science. 7*(12), 488. https://doi.org/10.3390/jcs7120488, http://www.mdpi.com/journal/jcs.

Nadar, S. S., Vaidya, L., & Rathod, V. K. (2020). Enzyme embedded metal organic framework (enzyme–MOF): De novo approaches for immobilization. *International Journal of Biological Macromolecules, 149*, 861–876. https://doi.org/10.1016/j.ijbiomac.2020.01.240, http://www.elsevier.com/locate/ijbiomac.

Neupane, S., Patnode, K., Li, H., Baryeh, K., Liu, G., Hu, J., Chen, B., Pan, Y., & Yang, Z. (2019). Enhancing enzyme immobilization on carbon nanotubes via metal-organic frameworks for large-substrate biocatalysis. *ACS Applied Materials and Interfaces, 11*(12), 12133–12141. https://doi.org/10.1021/acsami.9b01077, http://pubs.acs.org/journal/aamick.

Niu, X., Li, X., Lyu, Z., Pan, J., Ding, S., Ruan, X., Zhu, W., Du, D., & Lin, Y. (2020). Metal-organic framework based nanozymes: Promising materials for biochemical analysis. *Chemical Communications, 56*(77), 11338–11353. https://doi.org/10.1039/d0cc04890a, http://pubs.rsc.org/en/journals/journal/cc.

Nowroozi-Nejad, Z., Bahramian, B., & Hosseinkhani, S. (2019). A fast and efficient stabilization of firefly luciferase on MIL-53(Al) via surface adsorption mechanism. *Research on Chemical Intermediates, 45*(4), 2489–2501. https://doi.org/10.1007/s11164-019-03748-w, http://www.springer.com/chemistry/journal/11164.

Palanisamy, A., Palanisamy, T., & Vaidyanathan Ganesan, V. (2024). Metal organic framework enzyme-based unhairing of skins: A step toward sustainable leather processing. *ACS Sustainable Chemistry and Engineering, 12*(4), 1645–1654. https://doi.org/10.1021/acssuschemeng.3c07118, http://pubs.acs.org/journal/ascecg.

Pan, Y., Li, H., Lenertz, M., Han, Y., Ugrinov, A., Kilin, D., Chen, B., & Yang, Z. (2021). One-pot synthesis of enzyme@metal-organic material (MOM) biocomposites for enzyme biocatalysis. *Green Chemistry, 23*(12), 4466–4476. https://doi.org/10.1039/d1gc00775k, http://pubs.rsc.org/en/journals/journal/gc.

Pan, Y., Li, Q., Li, H., Farmakes, J., Ugrinov, A., Zhu, X., Lai, Z., Chen, B., & Yang, Z. (2021). A general Ca-MOM platform with enhanced acid-base stability for enzyme biocatalysis. *Chem Catalysis, 1*(1), 146–161. https://doi.org/10.1016/j.checat.2021.03.001, https://www.sciencedirect.com/journal/chem-catalysis.

Pascanu, V., González Miera, G., Inge, A. K., & Martín-Matute, B. (2019). Metal-organic frameworks as catalysts for organic synthesis: A critical perspective. *Journal of the American Chemical Society, 141*(18), 7223–7234. https://doi.org/10.1021/jacs.9b00733, http://pubs.acs.org/journal/jacsat.

Paul, A., & Srivastava, D. N. (2018). Amperometric glucose sensing at nanomolar level using MOF-encapsulated TiO_2 platform. *ACS Omega, 3*(11), 14634–14640. https://doi.org/10.1021/acsomega.8b01968, https://pubs.acs.org/doi/10.1021/acsomega.8b01968.

Peng, X., Wu, X., Zhang, M., & Yuan, H. (2023). Metal-organic framework coated devices for gas sensing. *ACS Sensors, 8*(7), 2471–2492. https://doi.org/10.1021/acssensors.3c00362, http://pubs.acs.org/journal/ascefj.

Qiu, T., Liang, Z., Guo, W., Tabassum, H., Gao, S., & Zou, R. (2020). Metal-organic framework-based materials for energy conversion and storage. *ACS Energy Letters, 5*(2), 520–532. https://doi.org/10.1021/acsenergylett.9b02625.

Rabe, M., Verdes, D., & Seeger, S. (2011). Understanding protein adsorption phenomena at solid surfaces. *Advances in Colloid and Interface Science, 162*(1-2), 87–106. https://doi.org/10.1016/j.cis.2010.12.007.

Rezki, M., Harimurti, S., Septiani, N. L. W., Amri, F., & Suyatman Nugraha Yuliarto, B. (2021). Facile synthesis of Cu-BDC MOF via solvothermal method for biosensor application,*AIP Conference Proceedings*, American Institute of Physics Inc., Indonesia, 2384. 10.1063/5.0077600 15517616,http://scitation.aip.org/content/aip/proceeding/aipcp.

Saddique, Z., Imran, M., Javaid, A., Rizvi, N. B., Akhtar, M. N., Iqbal, H. M. N., & Bilal, M. (2024). Enzyme-linked metal organic frameworks for biocatalytic degradation of antibiotics. *Catalysis Letters, 154*(1), 81–93. https://doi.org/10.1007/s10562-022-04261-3, https://www.springer.com/journal/10562.

Sha, M., Xu, W., Fang, Q., Wu, Y., Gu, W., Zhu, C., & Guo, S. (2022). Metal-organic-framework-involved nanobiocatalysis for biomedical applications. *Chem Catalysis, 2*(10), 2552–2589. https://doi.org/10.1016/j.checat.2022.09.005, https://www.sciencedirect.com/journal/chem-catalysis.

Shao, Y., Zhang, J., Jiang, H., & Chen, R. (2021). Well-defined MOF-derived hierarchically porous N-doped carbon materials for the selective hydrogenation of phenol to cyclohexanone. *Industrial and Engineering Chemistry Research, 60*(16), 5806–5815. https://doi.org/10.1021/acs.iecr.1c00422, http://pubs.acs.org/journal/iecred.

Shortall, K., Otero, F., Bendl, S., Soulimane, T., & Magner, E. (2022). Enzyme immobilization on metal organic frameworks: The effect of buffer on the stability of the support. *Langmuir, 38*(44), 13382–13391. https://doi.org/10.1021/acs.langmuir.2c01630, http://pubs.acs.org/journal/langd5.

Singh, R., Musameh, M., Gao, Y., Ozcelik, B., Mulet, X., & Doherty, C. M. (2021). Stable MOF@enzyme composites for electrochemical biosensing devices. *Journal of Materials Chemistry C, 9*(24), 7677–7688. https://doi.org/10.1039/d1tc00407g, http://pubs.rsc.org/en/journals/journal/tc.

Smith, S., Goodge, K., Delaney, M., Struzyk, A., Tansey, N., & Frey, M. (2020). A Comprehensive review of the covalent immobilization of biomolecules onto electrospun nanofibers. *Nanomaterials, 10*(11), 2142. https://doi.org/10.3390/nano10112142.

Sohrabi, H., Ghasemzadeh, S., Shakib, S., Majidi, M. R., Razmjou, A., Yoon, Y., & Khataee, A. (2023). Metal-organic framework-based biosensing platforms for the sensitive determination of trace elements and heavy metals: A comprehensive review. *Industrial and Engineering Chemistry Research, 62*(11), 4611–4627. https://doi.org/10.1021/acs.iecr.2c03011, http://pubs.acs.org/journal/iecred.

Suh, M. P., Park, H. J., Prasad, T. K., & Lim, D. W. (2012). Hydrogen storage in metal-organic frameworks. *Chemical Reviews, 112*(2), 782–835. https://doi.org/10.1021/cr200274s.

Sumida, K., Rogow, D. L., Mason, J. A., McDonald, T. M., Bloch, E. D., Herm, Z. R., Bae, T. H., & Long, J. R. (2012). Carbon dioxide capture in metal-organic frameworks. *Chemical Reviews, 112*(2), 724–781. https://doi.org/10.1021/cr2003272.

Sun, Y., Shi, J., Zhang, S., Wu, Y., Mei, S., Qian, W., & Jiang, Z. (2019). Hierarchically porous and water-tolerant metal-organic frameworks for enzyme encapsulation. *Industrial and Engineering Chemistry Research, 58*(28), 12835–12844. https://doi.org/10.1021/acs.iecr.9b02164, http://pubs.acs.org/journal/iecred.

Suo, H., Geng, X., Sun, Y., Zhang, L., Yang, J., Yang, F., Yan, H., Hu, Y., & Xu, L. (2022). Surface modification of magnetic ZIF-90 nanoparticles improves the microenvironment of immobilized lipase and its application in esterification. *Langmuir, 38*(49), 15384–15393. https://doi.org/10.1021/acs.langmuir.2c02672, http://pubs.acs.org/journal/langd5.

Tang, X., Li, N., & Pang, H. (2022). Metal–organic frameworks-derived metal phosphides for electrochemistry application. *Green Energy & Environment, 7*(4), 636–661. https://doi.org/10.1016/j.gee.2021.08.003.

Velásquez-Hernández, M. D. J., Ricco, R., Carraro, F., Limpoco, F. T., Linares-Moreau, M., Leitner, E., Wiltsche, H., Rattenberger, J., Schröttner, H., Frühwirt, P., Stadler, E. M., Gescheidt, G., Amenitsch, H., Doonan, C. J., & Falcaro, P. (2019). Degradation of ZIF-8 in phosphate buffered saline media. *CrystEngComm, 21*(31), 4538–4544. https://doi.org/10.1039/c9ce00757a, http://pubs.rsc.org/en/journals/journal/ce.

Wang, M., Wang, H., & Cheng, J. (2024). An enzyme-encapsulated metal-organic frameworks nanomesh biosensor for salivary glucose detection. *Advanced Materials Technologies, 9*(4), 2301678. https://doi.org/10.1002/admt.202301678, http://onlinelibrary.wiley.com/journal/10.1002/.

Wang, Q., Yu, Y., Chang, Y., Xu, X., Wu, M., Ediriweera, G. R., Peng, H., Zhen, X., Jiang, X., Searles, D. J., Fu, C., & Whittaker, A. K. (2023). Fluoropolymer-MOF hybrids with switchable hydrophilicity for 19F MRI-monitored cancer therapy. *ACS Nano, 17*(9), 8483–8498. https://doi.org/10.1021/acsnano.3c00694, http://pubs.acs.org/journal/ancac3.

Wang, Q., Zhang, X., Huang, L., Zhang, Z., & Dong, S. (2017). GOx@ZIF-8(NiPd) nanoflower: An artificial enzyme system for tandem catalysis. *Angewandte Chemie-International Edition, 56*(50), 16082–16085. https://doi.org/10.1002/anie.201710418, http://onlinelibrary.wiley.com/journal/10.1002/.

Wang, X., Lan, P. C., & Ma, S. (2020). Metal-organic frameworks for enzyme immobilization: Beyond host matrix materials. *ACS Central Science, 6*(9), 1497–1506. https://doi.org/10.1021/acscentsci.0c00687, http://pubs.acs.org/journal/acscii.

Wang, Z., Liu, Y., Li, J., Meng, G., Zhu, D., Cui, J., & Jia, S. (2022). Efficient immobilization of enzymes on amino functionalized MIL-125-NH$_2$ metal organic framework. *Biotechnology and Bioprocess Engineering, 27*(1), 135–144. https://doi.org/10.1007/s12257-020-0393-y, http://www.springerlink.com/content/1226-8372.

Wu, X., Ge, J., Yang, C., Hou, M., & Liu, Z. (2015). Facile synthesis of multiple enzyme-containing metal-organic frameworks in a biomolecule-friendly environment. *Chemical Communications, 51*(69), 13408–13411. https://doi.org/10.1039/c5cc05136c, http://pubs.rsc.org/en/journals/journal/cc.

Xu, S., Dong, A., Hu, Y., Yang, Z., Huang, S., & Qian, J. (2023). Multidimensional MOF-derived carbon nanomaterials for multifunctional applications. *Journal of Materials Chemistry A, 11*(18), 9721–9747. https://doi.org/10.1039/d3ta00239j, http://pubs.rsc.org/en/journals/journal/ta.

Xue, W., Zhou, Q., Cui, X., Jia, S., Zhang, J., & Lin, Z. (2021). Metal–organic frameworks-derived heteroatom-doped carbon electrocatalysts for oxygen reduction reaction. *Nano Energy, 86*, 106073.

Yang, X. G., Zhang, J. R., Tian, X. K., Qin, J. H., Zhang, X. Y., & Ma, L. F. (2023). Enhanced activity of enzyme immobilized on hydrophobic ZIF-8 modified by Ni^{2+} ions. *Angewandte Chemie-International Edition, 62*(7), e202216699. https://doi.org/10.1002/anie.202216699, http://onlinelibrary.wiley.com/journal/10.1002/.

Ye, N., Kou, X., Shen, J., Huang, S., Chen, G., & Ouyang, G. (2020). Metal-organic frameworks: A new platform for enzyme immobilization. *Chembiochem, 21*(18), 2585–2590. https://doi.org/10.1002/cbic.202000095, http://onlinelibrary.wiley.com/journal/10.1002/.

Zeng, J., Gao, X., Dai, Z., Tang, B., & Tang, X. F. (2014). Effects of metal ions on stability and activity of hyperthermophilic pyrolysin and further stabilization of this enzyme by modification of a Ca^{2+}-binding site. *Applied and Environmental Microbiology, 80*(9), 2763–2772. https://doi.org/10.1128/AEM.00006-14, http://aem.asm.org/content/80/9/2763.full.pdf.

Zhang, C., Wang, X., Hou, M., Li, X., Wu, X., & Ge, J. (2017). Immobilization on metal-organic framework engenders high sensitivity for enzymatic electrochemical detection. *ACS Applied Materials and Interfaces, 9*(16), 13831–13836. https://doi.org/10.1021/acsami.7b02803, http://pubs.acs.org/journal/aamick.

Zhang, K., Lu, L., Liu, Z., Cao, X., Lv, L., Xia, J., & Wang, Z. (2022). Metal-organic frameworks-derived bimetallic oxide composite nanozyme fiber membrane and the application to colorimetric detection of phenol. *Colloids and Surfaces A: Physicochemical and Engineering Aspects, 650*, 129662. https://doi.org/10.1016/j.colsurfa.2022.129662.

Zhao, M., Li, Y., Ma, X., Xia, M., & Zhang, Y. (2019). Adsorption of cholesterol oxidase and entrapment of horseradish peroxidase in metal-organic frameworks for the colorimetric biosensing of cholesterol. *Talanta, 200*, 293–299. https://doi.org/10.1016/j.talanta.2019.03.060, https://www.journals.elsevier.com/talanta.

Zhao, Q., Wu, D., Wang, Y., Meng, T., Sun, J., & Yang, X. (2024). Encapsulation of enzymes into hydrophilic and biocompatible metal azolate framework: Improved functions of biocatalyst in cascade reactions and its sensing applications. *Small, 20*(31), 2307192–.

Zhao, Y., Liang, C., Mei, Z., Yang, H., Wang, B., Xie, C., Xu, Y., & Tian, J. (2023). Oxygen-enriched MOF-hemoglobin X-ray nanosensitizer for enhanced cancer radio-radiodynamic therapy. *ACS Materials Letters, 5*(12), 3237–3247. https://doi.org/10.1021/acsmaterialslett.3c01158, https://pubs.acs.org/page/amlcef/about.html.

Zheng, X. T., Leoi, M. W. N., Yu, Y., Tan, S. C. L., Nadzri, N., Goh, W. P., Jiang, C., Ni, X. P., Wang, P., Zhao, M., & Yang, L. (2024). Co-encapsulating enzymes and carbon dots in metal–organic frameworks for highly stable and sensitive touch-based sweat sensors. *Advanced Functional Materials, 34*(10), 2310121. https://doi.org/10.1002/adfm.202310121, http://onlinelibrary.wiley.com/journal/10.1002/.

Zhou, X., Zhang, S., Liu, Y., Meng, J., Wang, M., Sun, Y., Xia, L., He, Z., Hu, W., Ren, L., Chen, Z., & Zhang, X. (2022). Antibacterial cascade catalytic glutathione-depleting MOF nanoreactors. *ACS Applied Materials and Interfaces, 14*(9), 11104–11115. https://doi.org/10.1021/acsami.1c24231, http://pubs.acs.org/journal/aamick.

Zhou, Z., Chao, H., He, W., Su, P., Song, J., & Yang, Y. (2022). Boosting the activity of enzymes in metal-organic frameworks by a one-stone-two-bird enzymatic surface functionalization strategy. *Applied Surface Science, 586*, 152815. https://doi.org/10.1016/j.apsusc.2022.152815.

Zhu, D., Ao, S., Deng, H., Wang, M., Qin, C., Zhang, J., Jia, Y., Ye, P., & Ni, H. (2019). Ordered coimmobilization of a multienzyme cascade system with a metal organic framework in a membrane: Reduction of CO_2 to methanol. *ACS Applied Materials and Interfaces, 11*(37), 33581–33588. https://doi.org/10.1021/acsami.9b09811, http://pubs.acs.org/journal/aamick.

Zhu, H., Li, X., He, Z., Chen, Y., & Zhu, J. J. (2022). Metal azolate coordination polymer-enabled high payload and non-destructive enzyme immobilization for biocatalysis and biosensing. *Analytical Chemistry, 94*(18), 6827–6832. https://doi.org/10.1021/acs.analchem.2c00637, http://pubs.acs.org/journal/ancham.

Chapter 21

Metal-organic framework composites: industrial applications

Fatemeh Shahrab, and Kamran Akhbari
School of Chemistry, College of Science, University of Tehran, Tehran, Iran

Introduction

Since the maturation of the metal-organic framework (MOF) field, characterized by the development of "designed" syntheses relying on "reticular chemistry" pioneered by Yaghi, compounds based on MOFs have demonstrated remarkable progress within their relatively short 20-year history (Feng et al., 2020; Yaghi et al., 2003). In recent years, researchers have produced a considerable number of compounds based on MOF and MOF composites using various techniques, illustrating extraordinary laboratory-scale (milligram) applications (Parsaei, Akhbari & Kawata, 2023; Parsaei, Akhbari & Tylianakis, Froudakis, et al., 2022; Salimi et al., 2024a). Some of these compounds have shown very promising results in various fields such as medicine and biomedical (Alavijeh & Akhbari, 2024; Arshadi Edlo & Akhbari, 2024; Bieniek et al., 2021; Davoodi et al., 2023; Ge et al., 2022; Karimi Alavijeh et al., 2018; Ma et al., 2021; Mohammadi Amidi et al., 2023; Parsaei & Akhbari, 2023; Parsaei, Akhbari & Tylianakis, & Froudakis, 2023), gas adsorption and storage of various gases (Ahmed & Jhung, 2014; Chuhadiya et al., 2021; Paitandi et al., 2023; Salimi et al., 2024b), particularly carbon capture (Ding et al., 2019; Gebremariam et al., 2023; Liu, Chen, et al., 2021), catalysts (Liu, Chen, et al., 2021; Murtaza et al., 2023; Shahrab & Tadjarodi, 2023; Wang et al., 2021), supercapacitors (Prajapati et al., 2023; Xu et al., 2022; Yue et al., 2021), sensors and biosensors (Amini et al., 2020; Liu et al., 2020; Wang et al., 2021), fuel cells (Annapragada et al., 2023; Nik Zaiman et al., 2022; Wen et al., 2020), solar cells (Annapragada et al., 2023; Lin et al., 2023; Sajid et al., 2024), water harvesting from the atmosphere (Gordeeva et al., 2021; Hu et al., 2023; Mohan et al., 2022), food applications (Magri et al., 2021; Sharanyakanth & Radhakrishnan, 2020; Shen, Liu, et al.,

2021), and more. Numerous research efforts have been conducted in the development of large-scale compounds based on MOF synthesis (He et al., 2022; Li et al., 2022; Pobłocki et al., 2022).

The domain of sales pertaining to MOFs is currently witnessing consistent expansion, reflecting escalating demand across diverse industrial sectors. It is widely acknowledged that market exigencies and product pricing are intricately entwined, thus rendering optimization of production methodologies imperative for future sustainability and cost-effectiveness. Nevertheless, it should be noted that cost-effectiveness alone does not exclusively dictate the dynamics of the MOF market. MOFs, owing to their enduring properties and considerable potential, possess the capacity to captivate investors and foster more sustainable and extensive production endeavors. Various parameters play a role in the analysis of the compounds based on the MOF market, such as secondary building unit (SBU) sales, linkers, product and service value in the domain of goods sold by manufacturers, and so on. All reports presented in the compounds based on the MOF sales domain unanimously indicate the increasing growth of the MOF market (He et al., 2022). As reported by Insight, the global market valuation of MOFs stood at approximately 7.5–7.9 billion dollars in 2022, with a projected surge to 19.5 billion dollars by 2030, boasting a compound annual growth rate (CAGR) of 12.7% (Insights, 2023).

When assessing the industrial viability of a technology, patent registration serves as a leading indicator for its future commercial activity. In general, it can be said that the growth of patents in the field of compounds based on MOF research began in 2011. As evident from the graphs, there has been a significant increase in patent registrations in recent years, showing nearly a 9-fold increase from about 78 patents in 2011 to 665 patents in 2019 (Fig. 21.1A). Approximately 50% of patent applicants are from the private sector, indicating that many companies have started commercializing compounds based on MOFs (Fig. 21.1B). Moreover, the majority of these innovations occur predominantly in China and the United States (Fig. 21.1C). The private sector's involvement in this field is diverse, encompassing various inventions from the battery and materials industries to chemical separations and packaging applications. These companies range from small commercial nuclei originating from university laboratory environments to startups and large commercial entities such as BASF. Large companies are typically multinational corporations (Chen, Wasson et al., 2021).

The major companies active in the production of compounds based on MOFs and MOF composites on a large scale (exceeding several kilograms) for commercial purposes include BASF, Framergy (USA), MOF Technologies (UK, now Nuada), MOFapps (Norway), Promethean Particles (UK), and novoMOF (Switzerland), Promethean, Particles, Immaterial, Porous Liquid Technologies, Tarsis Technology, Mosaic Materials, Coordination Pharmaceuticals, Matrix Sensors, Inmondo Tech, Water Harvesting, MPower, Nanoshel, Panacenano (US), ProfMOF (Norway), MOFwork (Australia),

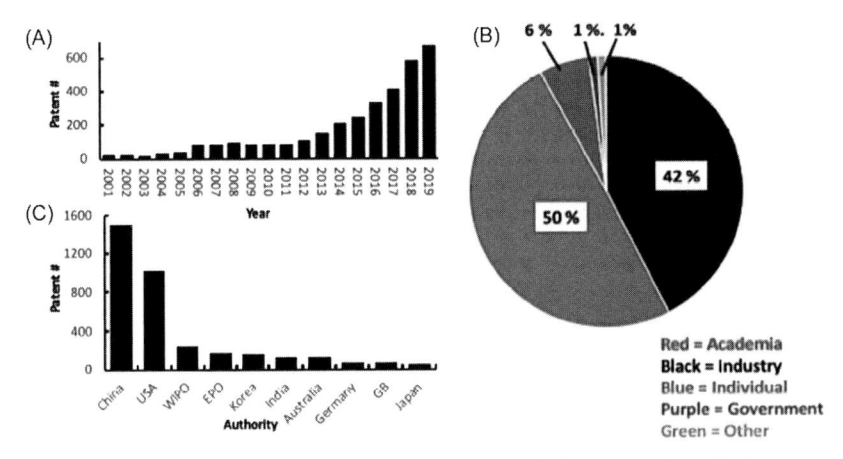

Metal-organic framework composites: industrial applications Chapter | 21 805

FIGURE 21.1 Global patent application trends in metal-organic frameworks (MOFs): A growing interest in advanced materials innovation (Chen et al., 2021). (A) Global patent application trends over time; (B) Standard patent assignee by type; (C) Top authorities for patents (2011–2019).

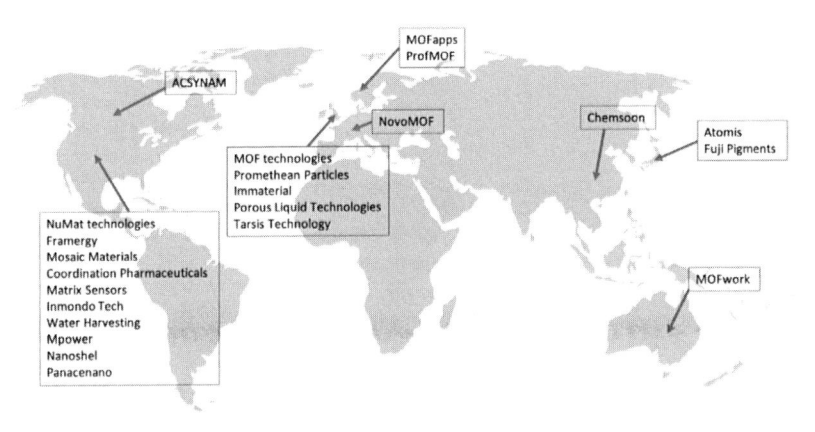

FIGURE 21.2 Global map for MOF companies (Ryu et al., 2021). Global distribution of companies involved in metal-organic framework research and development. Map lines delineate study areas and do not necessarily depict accepted national boundaries.

ACSYNAM (Canada), Atomis, Fuji Pigments (JP), Chemsoon (China) while companies like Sigma-Aldrich (Germany) or Strem Chemicals Inc. (USA) serve as distributors of MOFs for academic and small-scale commercial purposes (Fig. 21.2) (Chakraborty et al., 2023; Ryu et al., 2021).

In the synthesis of MOF composites, the industrial-scale production of MOFs as the primary substrate is essential. Table 21.1 enumerates the names of several MOFs synthesized industrially.

TABLE 21.1 Commercially sold metal-organic framework products produced via different companies and start-ups (Chakraborty et al., 2023).

Company	Production method	MOF products
BASF (Sigma Aldrich as retailer)	Electrochemical	Basolite Z1200 (ZIF-8)
		Basolite A100 (MIL-53(Al))
		Basolite C300
		Basolite F300
		Basolite Z377
Nuada (Previous name: MOF Technologies Ltd.)	Mechanochemistry	MOF-74 (Mg, Ni, Co, Cu)
		HKUST-1
		ZIF-8
		MIL-53(Al)
		Ni-TiF$_6$-Pyrazine
		Zn-SiF$_6$-Pyrazine
		Al(OH)fumarate
		CAU-10-H (CAU: Christian-Albrechts-University)
		Magnesium formate
Framergy	Atmospheric pressure	AYRSORB T125 (NH$_2$-MIL-125(Ti))
		AYRSORB F250 (Based on Fe-soc-MOF, MIL-127(Fe) or PCN250)
		AYRSORB F100 (MIL-100(Fe))
		AYRSORB P151 (PPN-151 and basal amine functionalized PPN-6)
MOFapps	Solvothermal, mechanochemistry, spray-drying, microwave-assisted process	UiO-66 and its derivatives
		MIL-53(Al)
		ZIF-8
		ZIF-67
		HKUST-1

(*Continued*)

TABLE 21.1 (Continued)

Company	Production method	MOF products
Promethean particles	Continuous flow synthesis (hydrothermal)	HKUST-1
		Fe-BTC
		MIL-53(Al)
		MIL-100(Fe)
		Aluminum fumarate
		ZIF-8
		ZIF-67
		MOF-74(Ni) or CPO-27(Ni)
		MOF-74(Zn) or CPO-27(Zn)
ProfMOF	N/A	UiO-66-ADC
		MOF-801/UiO-66-FA
		UiO-66-BDC
		UiO-66-BDC-NH$_2$
		UiO-66-BDC-COOH
		UiO-66-BDC-(COOH)$_2$
		UiO-67-BPDC
		UiO-67-BPY
		MOF-808
		CAU-10

Companies like BASF, Axel'One MOF, Framergy, Nuada, and MOFapps are now capable of producing various MOFs at scales exceeding kilograms, including MIL-100(Fe), UiO-66(Zr), Al-fum MOF, ZIF-8, ZIF-67, HKUST-1, and PCN-250(Fe) (PCN: Porous Coordination Network), or MIL-127(Fe). Recently, Shimizu and colleagues introduced CALF-20 (CALF: Calgary Frameworks), a microporous Zn oxalate triazolate, as a highly promising physical sorbent for CO_2 capture in the presence of humidity from postcombustion flue gas. BASF has scaled up the synthesis of CALF-20 to multiton levels, facilitating its use in pilot-scale studies for postcombustion CO_2 capture by Svante (Chakraborty et al., 2023; Ren et al., 2017; Silva et al., 2015). Fig. 21.3 illustrates one of NuMat's pilot plants utilized for process scale-up optimization.

FIGURE 21.3 NuMat's pilot plant units utilized for process scale-up optimization (Chen et al., 2021).

The focus on large- and pilot-scale synthesis and efficient, eco-friendly production methods for MOFs is a top priority for leading scientists in the field. Synthesis of Fe trimesate MIL-100(Fe) structure without the use of HF (Seo et al., 2012), electrochemical synthesis of microporous trimesate HKUST-1 (Mueller et al., 2011), production of microporous Alfumarate on a ton scale under ambient pressure conditions using water and aluminum sulfate (Gaab et al., 2012; Rubio-Martinez et al., 2016), continuous flow reactors for synthesize microporous Zr terephthalate UiO-66(Zr) for drug delivery applications (Tai et al., 2016), synthesis of MIL-53(Cr) with piezoelectric effect (Bayliss et al., 2014; Xue et al., 2021), synthesis of NOTT-400 and NOTT-401 (Ibarra et al., 2011), synthesis of scalable batch production zirconium MOFs (DUT-67, MOF-808, UiO-66-NH$_2$, UiO-66-(OH)$_2$) (Reinsch et al., 2016), synthesis of microfluidics MIL-88B type MOFs (Fe-MIL-88B-NH$_2$, Fe-MIL-88B, and Fe-MIL-88B-Br) (Paseta et al., 2013), low-cost and larger-scale production of MIL-160(Al) (Borges et al., 2017), MOF-74 (Das et al., 2016), MOF-74/174/184-M (Oh et al., 2017), ZIF-8 (Deacon et al., 2022; Polyzoidis et al., 2016; Quan et al., 2023), ZIF-4 (Hovestadt et al., 2017), ZIF-67 (Duan et al., 2022), CPO-27-Ni (Didriksen et al., 2017), PCN-250 (Fe) (Kirchon, 2020), MIL-101(Cr) (Zhao et al., 2015), self-assembly of MOF-801 for scale-up atmospheric water harvesting (He et al., 2023), scalable continuous solvothermal synthesis of MOF-5 (McKinstry et al., 2016), scalable synthesis of MIL-96(Al) (Gaab et al., 2012), upscaled green synthesis of MIL-91(Ti) (Benoit et al., 2016), scalable green synthesis and full-scale of CAU-10 (Lenzen et al., 2018), and others were reported.

When discussing composite synthesis, notable examples include pilot-scale synthesis of MOF composites and continuous flow reactor synthesis of mesoporous and ZIF-8 biocomposites (Carraro et al., 2020), synthesis of multicomponent MOFs include UMCM-1, MOF-5, MOF-177, and MUF-7 (He et al., 2021), synthesis of zirconium magnetic framework composite MgFe$_2$O$_4$@UiO-66-NH$_2$ for CO$_2$ capture (He et al., 2019), synthesis of Ag-nanocluster@MOF composite for visible-photocatalytic activity (Arenas-Vivo et al., 2021), synthesis of HKUST-1 crystals and composites (Sun et al., 2018), composites of ZIF-8 with magnetite nanoparticles (NPs), also Co$_3$BTC$_2$@Ni$_3$BTC$_2$, MOF-5@diCH$_3$-MOF-5, and Fe$_3$O$_4$@ZIF-8 core-shell MOFs were successfully synthesized using a novel two-step serial in-droplet microfluidic method (Faustini et al., 2013). Large-scale synthesis of ZIF-67/glycol-2-methylimidazole slurry for highly efficient carbon capture (Pan et al., 2015).

Pilot and large-scale production heavily relies on various parameters, including the type of reaction, reactor geometry, mixing speed, mass, and heat transfer. These factors significantly influence the feasibility of industrialization. Moreover, careful consideration must be given to the choice of solvent, reactants, and reaction method. This underscores the importance of techno-economic analysis, life-cycle assessment, and eco-design of MOFs (Ashworth et al., 2023; DeSantis et al., 2017; Grande et al., 2017; Hughes et al., 2021; Shi et al., 2021).

In MOF synthesis, four crucial components are solvents, metal sources, organic ligands, and acid/base catalysts. Material costs are divided into three groups: Standard ligand and metal (e.g., terephthalic acid and Zn/Cu/Zr), specialized ligand and standard metal (e.g., 2,5-dihydroxyterephthalic acid and Zn/Cu/Mg), and standard ligand and specialized metal (e.g., terephthalic acid and Hf/Pd) (Ryu et al., 2021). The space-time-yield (STY), representing the amount of MOF product generated per cubic meter of reaction mixture per day of synthesis, is a critical parameter in industrial MOF synthesis. Achieving a high STY is imperative for economic viability, as it directly impacts production costs. The STY is influenced by factors such as raw material expenses, reactor requirements, and the duration of synthesis. Challenges such as the need for pressure-sealed vessels, heating machinery, and handling chemical resistance issues contribute to intrinsic high costs and may require significant investments for effective large-scale implementation (Czaja et al., 2009; Silva et al., 2015).

In this chapter, a comprehensive exploration of MOF composites in industrial applications will be embarked upon. From elucidating their synthesis methodologies to assessing their performance across diverse sectors, we aim to provide insights into the transformative potential of these materials. By delving into real-world examples and market dynamics, we seek to underscore the pivotal role of MOF composites in driving innovation and addressing the evolving needs of industries worldwide.

Prominent industrial methods for metal-organic framework composites synthesis

The synthesis of MOFs and their compounds involves several conventional methods. These methods include solvothermal and hydrothermal techniques,

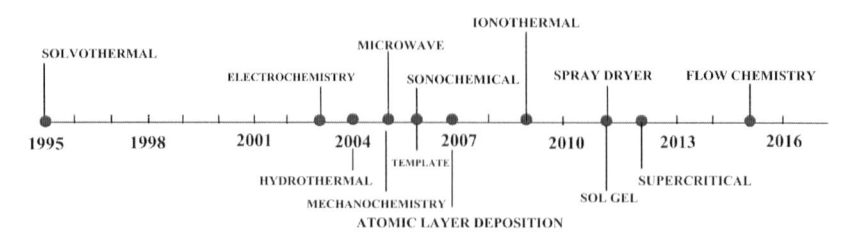

FIGURE 21.4 Timeline of the most common synthetic approaches for the synthesis of metal-organic framework (Olajire, 2018).

atomic layer deposition, sol-gel processes, sonochemistry, microwave synthesis, electrochemistry, mechanochemistry, spray drying, supercritical CO_2 dry heating, flow chemistry, and traditional heating. Each method has distinct advantages and considerations, which depend on the specific properties and requirements of the desired MOFs. The appropriate synthesis method is crucial for achieving the desired MOF characteristics and functionalities. In Fig. 21.4, the progression of MOF synthesis methods is symbolically illustrated (Chakraborty et al., 2023; He et al., 2022; Olajire, 2018; Pobłocki et al., 2022; Ren et al., 2017; Silva et al., 2015).

In pursuit of economically and environmentally viable MOF synthesis processes, several crucial factors must be considered, including the price, abundance, and environmental impact of raw materials, as well as the safety and toxicity of reactants. While metal nitrates and chlorides are commonly used due to their reactivity and solubility, their disadvantages, such as oxidative or corrosive properties, become apparent at larger scales, posing safety risks and increasing capital costs. Safer alternatives like acetates, carbonates, oxides, and sulfates are preferable, although they may suffer from poor solubility or hydrolysis issues. For instance, aluminum sulfate has been used successfully instead of aluminum chloride or nitrate, demonstrating the potential for safer large-scale production. Additionally, selecting metals with lower toxicity, such as Al, Ca, Fe, Mg, Zn, Ti, and Zr, is crucial for minimizing environmental impact. Limited studies on the life cycle assessment (LCA) of MOFs underscore the need for further research to evaluate their safe disposal, particularly concerning potential metal ion release over time. Overall, careful consideration of metal salt choice is essential for ensuring safety, handling, toxicity, and cost-effectiveness in MOF production, especially at larger scales and during industrialization. Large-scale production of MOFs also relies significantly on precursor costs, with ligands being the primary cost determinant. Commonly used ligands like terephthalic acid, trimesic acid, fumaric acid, and isophthalic acid offer cost-effectiveness and availability. Additionally, utilizing waste materials like polyethylene terephthalate (PET) bottles for ligand sources presents a sustainable alternative. Solvent choice also impacts environmental impact, safety, and cost, with nontoxic, nonflammable options being preferable for large-scale synthesis. Notably, replacing toxic solvents with greener alternatives enhances safety and environmental sustainability. Various synthesis routes have been explored for MOF production, with considerations for factors like solvent type, temperature, and reaction time. Simplifying synthesis parameters while maximizing yield and STY is crucial for scalability. Washing and activation steps are equally vital, ensuring proper purification and structural integrity of MOFs. Minimizing solvent usage, employing energy-efficient activation methods, and optimizing washing procedures contribute to sustainable large-scale production of MOFs. Ultimately, a holistic approach that integrates green solvents, sustainable chemicals, and low-energy processes is essential for cost-effective and scalable MOF production (Chakraborty et al., 2023; Pobłocki et al., 2022; Severino et al., 2021).

The most important challenge in the synthesis of scaling up MOF synthesis is that MOFs typically nucleate at the reaction surface, making the size of the reaction vessel a critical factor in synthesis conditions. This poses a limitation as reactions successful in small lab-scale setups may fail when scaled up to larger vessels. Consequently, scaled-up MOF chemistry is often restricted to a few robust MOFs, each necessitating specialized equipment. Therefore, a method allowing convenient scaling of reactions while maintaining vessel size is highly desirable for applied MOF chemistry, offering a versatile pathway to production. To overcome the challenges inherent in synthesizing MOF-based composites, researchers advocate for the utilization of both batch and continuous flow production methods for synthesizing MOFs. These methods offer adaptability to various synthetic approaches such as solvothermal, hydrothermal, mechanochemical, electrochemical, microwave (MW)-assisted synthesis, and sonochemical synthesis, thereby allowing for the customization of synthesis routes to suit specific application requirements and optimize material properties. This versatility empowers scientists to explore innovative strategies and enhance the performance of MOF-based composites across diverse fields, from catalysis to gas storage and beyond.

As shown in Fig. 21.5, batch process, foundational in chemical and bulk industries, orchestrates a sequence of operations on a fixed quantity of materials over time. It serves as a cornerstone for producing a myriad of products, spanning pharmaceuticals, agrochemicals, and fragrances. Typically, batch synthesis experiments are conducted under reflux conditions, leveraging the boiling point of the solvent. As industries seek efficiency and scalability, there's a discernible trend toward embracing flow methods in MOF production, promising refined control and productivity in manufacturing processes. Continuous flow reactors are highly favored for MOF production due to their versatility in accommodating solvent-based synthesis, facilitating the exploration of a broader range of MOF structures compared to batch methods. In contrast, batch synthesis techniques, particularly those involving milling, often limit control over crystal morphology and particle size distribution. HKUST-1,

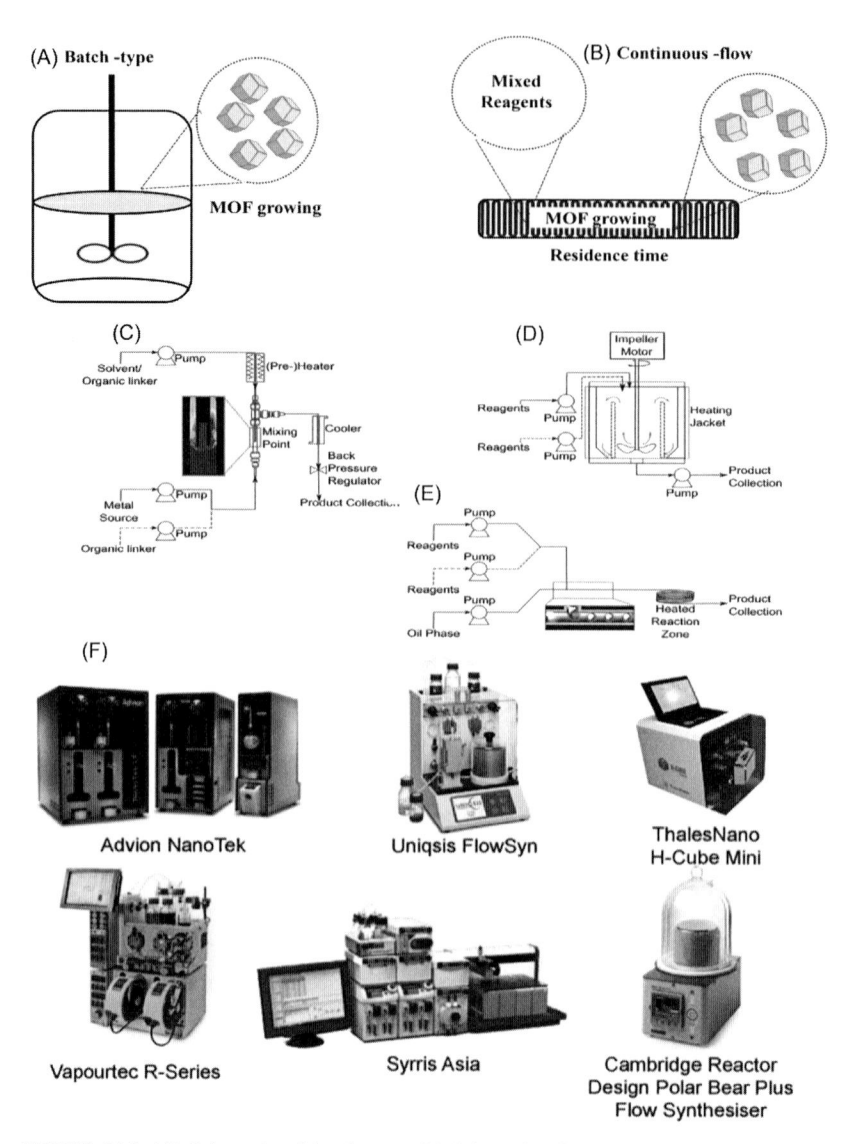

FIGURE 21.5 (A) Schematic of batch-type, (B) Schematic of continuous-flow production of metal-organic frameworks, (C) Continuous-flow (hydro) solvothermal synthesis (CFHS/ CFSS) (Chakraborty et al., 2023; Dunne et al., 2016), (D) Continuous-stirred tank reactors (CSTR) (Chakraborty et al., 2023; Dunne et al., 2016), (E) Micro- or millifluidic systems (Chakraborty et al., 2023; Dunne et al., 2016), and (F) Commercially available flow chemistry systems (Ryu et al., 2021).

CPO-27 (Ni) (also known as MOF-74), UiO-66, MIL-53 (Al), and ZIF-8 are among the common MOF structures that have been extensively investigated for their gas adsorption properties and various applications. These MOFs have been synthesized using continuous flow reactors on an industrial scale, showcasing the scalability and efficiency of this synthesis method. Continuous flow synthesis enables precise control over reaction conditions and facilitates the production of MOFs with consistent quality and properties, making them suitable for large-scale applications in various industries (Bayliss et al., 2014; Dunne et al., 2016).

Within the realm of continuous flow systems for MOF synthesis, three primary categories are recognized: continuous-flow (hydro)solvothermal synthesis (CFHS/CFSS), continuous-stirred tank reactors (CSTR), and microfluidic or millifluidic systems. A depiction of continuous flow is illustrated in Fig. 21.5A–F. Each of these systems brings unique advantages and applications to MOF synthesis, catering to diverse production requirements and scalability considerations. CFHS systems, originally designed for supercritical hydrothermal production, facilitate rapid mixing of solvent and precursor streams under elevated temperatures and pressures. The geometry of the mixing point is critical for controlling reaction kinetics and product quality, with postmixing isothermal zones often utilized to promote crystallization and growth. CSTRs, though less commonly employed for MOF synthesis, offer simplicity and scalability, making them suitable for specific continuous coprecipitation reactions. These traditional reactor designs feature heated tanks with mechanical stirring, providing straightforward operation and potential for scaled-up production. Microfluidic and millifluidic reactors, characterized by intricate nano/microchannels, provide precise control over mixing and reaction conditions. Leveraging advanced microfabrication techniques, these reactors enable well-defined flow paths, facilitating efficient heat and mass transfer. The incorporation of inert, immiscible phases enhances mixing and reaction control, making these systems invaluable for fine-tuning synthesis parameters and achieving reproducible results in MOF synthesis at various scales. Among the composites produced using this method, notable examples include the industrial synthesis of $Co_3BTC_2@Ni_3BTC_2$ and MOF-5@dimethyl-MOF-5 composites. Additionally, composites of ZIF-8 with magnetite NPs have been developed, enabling the immobilization or separation of active MOF catalysts through the utilization of multiple microfluidic reactors in series. Also, the utilization of continuous microfluidic technology has extended to the creation of MOF membranes (Cortés & Macéas, 2021; Dunne et al., 2016).

Fig. 21.6 presents a comprehensive visual comparison between pilot-scale and industrial-scale perspectives of numerous synthesized MOFs. This depiction offers valuable insights into the scalability and potential industrial applications of these MOFs across various production scales.

In Fig. 21.6A, ZIF-8 synthesis using supercritical CO_2 ($ScCO_2$) is illustrated. Solid reagents were ball-milled and mixed with $ScCO_2$, resulting

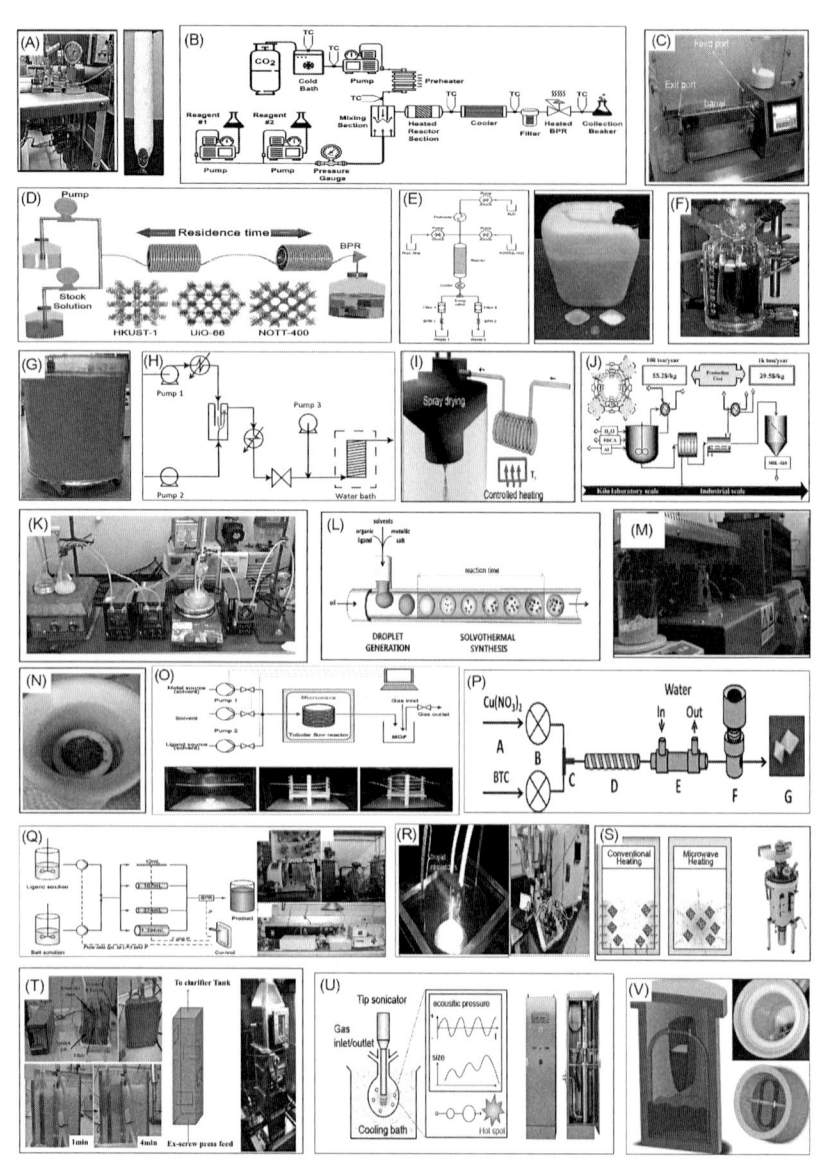

FIGURE 21.6 (A) ZIF-8 synthesis using supercritical of CO_2 (ScCO_2) (Marrett et al., 2018; Teo et al., 2021), (B) Synthesis of UiO-66 using supercritical CO_2 (Rasmussen et al., 2020; Teo et al., 2021), (C) Extruders for continuous synthesis of several metal-organic frameworks (MOFs) (Crawford et al., 2015; Ryu et al., 2021), (D) Continuous flow production for synthesizing HKUST-1, UiO-66, and NOTT-400 (Ren et al., 2022), (E) Producing MIL-53(Al) (Ren et al., 2022), (F) Electrochemical cell for synthesis HKUST-1 (Ryu et al., 2021), (G) Hydrothermal synthesis of MIL-100(Fe) (Ryu et al., 2021), (H) Large-scale hydrothermal production and activation of ZIF-8 (Ren et al., 2022), (I) Spray-drying synthesis of spherical hollow nanoMOFs (Ren et al., 2022), (J) Industrial-scale batch production of MIL-160(Al) (Chakraborty et al., 2023),

(*continued*)

in efficient MOF production within 5 minutes. Scaling up to a 100 g batch demonstrated the industrial viability of this method (Marrett et al., 2018). Fig. 21.6B illustrates a continuous-flow reactor for synthesizing the zirconium-based MOF UiO-66 using supercritical CO_2 (Rasmussen et al., 2020). Fig. 21.6C, showcases extruders for the continuous synthesis of various metal complexes and MOFs under solvent-free or low-solvent conditions. This approach achieves high production rates with properties similar to solvent-based methods, yielding up to 144×10^3 kg/(m³.day), significantly surpassing other MOF synthesis methods (Crawford et al., 2015). Fig. 21.6D provides a schematic depiction of the continuous flow production process employed for synthesizing MOFs such as HKUST-1, UiO-66, and NOTT-400. This method involves a continuous flow reactor system where the synthesis reactions occur in a continuous manner rather than in batch processes (Rubio-Martinez et al., 2014). Fig. 21.6E depicts a continuous flow method for producing MIL-53(Al) using only water as the reaction medium. This efficient process takes only 5–6 minutes to complete and achieves a high space-time yield of 1300 kg/(m³.day), demonstrating the feasibility of water-based synthesis for MOFs (Bayliss et al., 2014). Fig. 21.6F illustrates the experimental setup of an electrochemical cell operating within a laboratory glass reactor (Mueller et al., 2006). Fig. 21.6G depicts the hydrothermal synthesis of MIL-100(Fe); a 200 L reaction yielded MIL-100(Fe) powder. After purification with solvent extraction and NH_4F treatment, a highly porous F-free material was obtained, achieving a very high STY (> 1700 kg/(m³.day)) when contained in a 70 L poly (methyl methacrylate) box (Seo et al., 2012). In Fig. 21.6H, a new method for large-scale hydrothermal production and activation of ZIF-8 is introduced. Both lab-scale and pilot-scale production achieved activated ZIF-8 at rates of 27 and 810 g/h, respectively. The activated material exhibited a surface area of 1800 m²/g (Munn et al., 2015). Continuous ZIF-8 synthesis using microreaction technology, ensuring precise mass and heat control. Production reaches 640 g/day, with a space–time yield of 210,000 kg/(m³.day). The modular setup

◂ (K) Synthesis system of MOF-5 with CSTR (Ryu et al., 2021), (L) Microfluidic synthesizing MOFs crystals and hetero-structures (Ren et al., 2022), (M) Mechanochemical synthesis method for UiO-66, UiO-66-NH₂, MOF-801, and MOF-804 (Teo et al., 2021), (N) Dry-gel conversion synthesis process of Cr-MIL-101 (Ren et al., 2022), (O) Synthesis of UiO-66, MIL-53(Al), and HKUST-1 with a microwave method (Ryu et al., 2021), (P) Synthesis of Cu-BTC using a continuous-flow microreactor-assisted solvothermal system (Ren et al., 2022), (Q) Scale-up production flow chemistry four different flow reactors: 10, 107, 374, and 1394 mL (Ren et al., 2022), (R) NanoMOF synthesis droplet-based nanoreactors formed with PDMS and flow-focusing junction (Ren et al., 2022), (S) Large-scale synthesis of UiO-66 with a microwave system (Ryu et al., 2021), (T) Ultrasonic reactor set up commercially operating megasonic reactor for l-MOF (Ryu et al., 2021), (U) ZIF-8 which was synthesized with TEA and NaOH with a probe-type sonicator (Ryu et al., 2021), and (V) Vapor-assisted dry-gel conversion (DGC) method, Ni-MOF-74 containing FEP (Ren et al., 2022).

facilitates easy scale-up, while adjusting conditions allows tailored product morphology at both micro and macroscopic levels (Polyzoidis et al., 2016). The spray-drying method efficiently produces spherical hollow superstructures of nanoMOFs (Fig. 21.6I), including HKUST-1, Cu-bdc, NOTT-100, MIL-88A, MIL-88B, MOF-14, Zn-MOF-74, and UiO-66, all with diameters under 5 mm. This process eliminates the requirement for secondary solvents or surfactants. The resulting structures are ideal for various applications such as capsules, reactors, and composite materials (Carné-Sánchez et al., 2013). Fig. 21.6J illustrates industrial-scale batch production of MIL-160(Al) MOF, emphasizing cost calculations based on scale, raw materials, and process parameters. Costs range from $55/kg for 100 tons/year to $29.5/kg for 1 kton/year, with potential to drop below $10/kg using bio-derived ligands for bioplastic production (Severino et al., 2021). In a CSTR, MOF-5 was synthesized by Choline et al. (Fig. 21.6K). They confirmed that impurity-free MOF-5 could be obtained after 5 hours, and by adjusting the reaction time, they increased the yield. The resulting MOF-5 exhibited a surface area of 2302 m^2/g, surpassing the reference value of 2200 m^2/g (McKinstry et al., 2016). Fig. 21.6L introduces a novel microfluidic approach for swiftly synthesizing MOF crystals and hetero-structures. It enables rapid synthesis of various MOF types, including HKUST-1, MOF-5, IRMOF-3, and UiO-66, within minutes, showcasing faster kinetics than traditional methods. The method extends to producing complex structures like Ru_3BTC_2 under high-pressure conditions. Additionally, it synthesizes three core-shell MOF composites ($Co_3BTC_2@Ni_3BTC_2$, MOF-5@diCH_3-MOF-5, and Fe_3O_4@ZIF-8) in continuous flow mode. This microfluidic strategy offers a time-efficient and continuous route to high-quality MOF materials, serving as an alternative to conventional synthesis processes (Faustini et al., 2013). Fig. 21.6M presents a solvent-free, green, and rapid mechanochemical synthesis method for zirconium-based UiO-MOFs, including UiO-66, UiO-66-NH_2, MOF-801, and MOF-804. This approach is applicable on both laboratory and scalable industrial levels, achieved through planetary milling and twin-screw extrusion (TSE). It enables the production of more than 100 g of catalytically active UiO-66-NH_2 material per hour in a continuous process at a rate of 1.4 kg/h (Karadeniz et al., 2018). Fig. 21.6N, illustrating the dry-gel conversion synthesis process of Cr-MIL-101 facilitated by grinding. This method presents an innovative approach to MOF synthesis, converting a dry-gel precursor into the desired MOF structure, thereby offering an alternative to conventional synthesis techniques (Kim et al., 2013). Several MOFs, including UiO-66, MIL-53(Al), and HKUST-1, were synthesized using microwave tubular flow synthesis (Fig. 21.6O), achieving high STY values of 7204, 3618, and 64,800 kg/(m^3.day), respectively. Furthermore, each MOF exhibited a significantly higher surface area compared to the same MOF synthesized using conventional methods (Taddei et al., 2016). Fig. 21.6P illustrates the synthesis of Cu-BTC using a continuous-flow microreactor-assisted solvothermal system. This method enables high-rate production of Cu-BTC MOFs with exceptional properties, including a BET surface area exceeding 1600 m²/g and a production yield of 97%, all achieved within a total reaction time of 5 minutes (Kim et al., 2013). Fig. 21.6Q depicts the scale-up production of MOFs using flow chemistry. It includes schematics of four different flow reactors: 10, 107, 374, and 1394 mL, showcasing the versatility and scalability of flow chemistry in MOF synthesis (Ren et al., 2017). Fig. 21.6R shows a microfluidic chip for nanoMOF synthesis and modification, utilizing droplet-based nanoreactors formed with PDMS (polydimethylsiloxane). Automated setup drives the chip for synthesis and functionalization of nanosized MOFs. The chip design highlights key components like ports and flow-focusing junctions (Jambovane et al., 2016). Microwave irradiation enables precise control of MOF synthesis (Fig. 21.6S). Utilizing a CEM MARS 5 instrument, UiO-66 was synthesized with a significantly higher STY of 2241 kg/(m^3.day) in just 18 minutes, compared to 23 kg/(m^3.day) using conventional heating. The potential for industrial-scale synthesis using microwave reactors such as SAIREM is promising (Ryu et al., 2021). Ultrasonic separation is used to recover substances from suspensions (Fig. 21.6T). High-frequency ultrasound concentrates particles, facilitating their aggregation. For small or low-concentration MOFs, conventional methods may be inefficient, so ultrasonic separation reduces processing time and costs. It separates solids without chemical effects and can activate inner pores in porous structures like MOFs. For instance, Al-Fumarate and MIL-53 were separated within 10 minutes via 2 MHz sonication in a 1 L chamber (Ryu et al., 2021). Fig. 21.6U shows that ZIF-8 was synthesized using triethylamine (TEA) and sodium hydroxide (NaOH) as pH modulators in just 1 hour with a probe-type sonicator (at 60% power of 500 W, 20 kHz). The synthesis achieved an impressive yield of 85% and a high STY of 2140 kg/(m^3.day) (Ryu et al., 2021). Fig. 21.6V, provides an illustration of the vapor-assisted dry-gel conversion (DGC) method. It highlights essential components such as the solvent, FEP pouch, and Ni-MOF-74 reagents autoclave, offering insights into this scale up synthesis approach (Das et al., 2016).

In addition to the industrial and aforementioned methods, MOFs possess high versatility when combined and mixed with other materials, leading to the creation of composites. Various methods exist for preparing these composites. Functional species can be incorporated into MOF composites through various methods, such as formation by solution blending, electrospinning, in situ growth, surface coating with silica or polymers, and other techniques like incorporation into polyHIPEs (HIPEs: high internal phase emulsions), which offer user-defined geometry for MOF composites, enhancing their workability and applicability in various industries. MOFs can also serve as matrix materials, including methods such as impregnation of MOFs with functional species, encapsulation of presynthesized structures into MOFs (e.g., MOF shell formed by postgrowth), and chemical conversion of predeposited inorganic structures into MOFs. Additionally, techniques like optimizing pore size distributions of MOFs through the incorporation of other porous materials further expand the versatility and utility of MOF-based composites (Lin et al., 2023).

Method involves solution blending, where polymers and MOFs are mixed in a solution, followed by solvent vaporization to form a polymer-based composite. Examples include HKUST-1/PMMA composite (PMMA: poly (methyl methacrylate)) and ZIF-8/PI composite (PI: polyimide), but weak adhesion between polymers and MOF particles and the formation of defective membranes with voids are common issues. Surface modification of MOF particles can improve compatibility (Ben et al., 2012; Ordoñez et al., 2010). Another technique is electrospinning, which creates fibrous composites from MOF/polymer mixtures using high voltage. Examples include ZIF-8/PVP (PVP: polyvinyl pyrrolidone), MIL-101(Fe)/PS (PS: polystyrene) and HKUST-1/PS fibrous composites (Fan et al., 2012; Ostermann et al., 2011; Wu et al., 2012). This method offers tunability in fiber diameter and MOF loading, with potential applications in gas separation and personal protective systems. In situ growth of MOFs in matrices involves immersing matrices in MOF precursor solutions and heating to grow MOFs within them. Examples include HKUST-1 on silica microspheres and SIM-1 on alumina spheres. This method addresses low surface area issues seen in other methods, offering applications in catalysis and membranes. Additionally, MOFs can be coated with thin silica or polymer shells to improve stability and performance in bio-applications like bio-imaging and drug delivery. Examples include silica-coated MOF NPs and polymer-stabilized MOFs with improved targeting efficiency. Other methods include incorporating MOF particles into polyHIPEs through a three-step synthetic route, enhancing the workability of MOF composites by allowing for user-defined geometry. Overall, these methods offer various approaches to overcome the challenges associated with MOF processability and enhance their performance for specific applications (Ahmed et al., 2013; Ameloot et al., 2010; O'Neill et al., 2010).

MOFs as matrix materials in composites due to their unique properties: nanometer-scale cavities accessible to reactants, chemically tunable internal surfaces, and designed pore matrices for sieving. Impregnation of MOFs with functional species involves processes like direct adsorption, incipient wetness, and ion exchange, leading to composites such as Pd NP/MOF-5 and Pt NP/MIL-101 (Aijaz et al., 2012; Henschel et al., 2008; Pan et al., 2010). Encapsulation of pre-synthesized structures into MOFs utilizes methods like "build-bottle-around-ship," selective nucleation, and heteroepitaxial growth, resulting in composites like POM/MOF (POM: polyoxometalate) and Au NPs@HKUST-1 core-shell crystals (Juan-Alcañiz et al., 2010; Sun et al., 2009; Tsuruoka et al., 2011). Other methods, such as one-pot synthesis, solid grinding, and chemical conversion, yield composites like graphite oxide/MOF and GO/ZIF-8, with applications in gas storage, hydrogen uptake, and toxic gas removal (Kumar et al., 2013; Petit & Bandosz, 2012; Petit et al., 2010). In this field, the expression of several industrially synthesized compounds exhibits notable elegance. Fig. 21.7 showcases some of these synthesized compounds. By combining hydrophobic polymers with MOFs (HKUST-1, MIL-101 (Cr), ZIF-8, ZIF-67, MOF-74 (Ni), MOF$_{chemo}$, MOF$_{thermo}$) into a suspension, stable MOF inks can be created (Fig. 21.7A), resistant to air and water. The ink forms coatings upon solvent evaporation, exhibiting high contact angles and excellent hydrophobicity suitable for various surfaces. This method is versatile across different MOF types and enables coatings with chemical and thermal sensing properties. Nonetheless, this approach shows promise in producing functional MOF composite materials. Fig. 21.7B depicts 3D-printed models and nets of Cu-BTC@polymer. The study found that the adsorption capacity of the 3D-printed MOF remains intact, with notably enhanced hydrolytic stability compared to the MOF alone (Halevi et al., 2018). Fig. 21.7C illustrates the exploration of polymer solvent systems for electrospinning investigations with additional MOFs, including ZIF-8, UiO-66-NH$_2$, MOF-99, and MOF-74. Furthermore, it demonstrates the feasibility of using

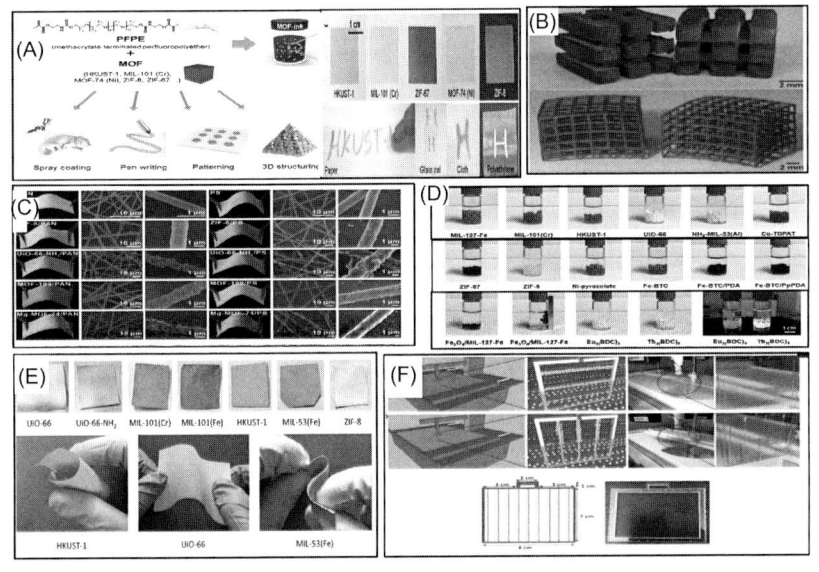

FIGURE 21.7 (A) Synthesis hydrophobic polymers@metal-organic frameworks (MOFs) of HKUST-1, MIL-101 (Cr), ZIF-8, ZIF-67, MOF-74 (Ni), MOF chemo, MOF thermo (Kim et al., 2019), (B) 3D-printed models and nets of Cu-BTC@polymer (Teo et al., 2021; Halevi et al., 2018), (C) Electrospinning synthesis of polymers@ZIF-8, UiO-66-NH2, MOF-99 and MOF-74 (Ryu et al., 2021; Zhang et al., 2016), (D) Biodegradable MOF-polymer composite poly acrylic acid (Teo et al., 2021; Yang et al., 2020), (E) Synthesis MOF-polymer mixed-matrix membranes ZIF-8, HKUST-1, MIL101, MIL53, UiO-66, and UiO-66-NH2 (Denny & Cohen, 2015; Teo et al., 2021), and (F) Single fiber-based devices made MOF (Rauf et al., 2021).

this method to coat a layer of MOFs onto various objects, such as gloves and face masks, highlighting the practical applicability of this technique (Zhang, Yuan et al., 2016). Fig. 21.7D illustrates the development of a general structuring method for preparing MOF-polymer composite beads using a straightforward polymerization strategy. This method utilizes biocompatible and biodegradable poly (acrylic acid) (PAA) and sodium alginate monomers, cross-linked with Ca^{2+} ions. The preparation procedure is water-based, rendering it nontoxic. Moreover, this universal method has been successfully applied to 12 different structurally diverse MOFs and three MOF-based composites. To validate its applicability, beads containing a MOF composite, specifically Fe-BTC/PDA, were fabricated (Yang et al., 2020). Fig. 21.7E presents processable films of various MOFs like ZIF-8, HKUST-1, MIL-101, MIL-53, UiO-66, and UiO-66-NH$_2$. The study describes MOF-polymer mixed-matrix membranes (MMMs) formed on different substrates, offering mechanically stable and pliable free-standing MMMs. The MOFs in these MMMs retain high crystallinity, porosity, and accessibility for further chemical modification. Overall, this versatile approach to preparing stable and functional MMMs is expected to significantly advance MOF applications (Denny & Cohen, 2015). Fig. 21.7F presents the Langmuir-Blodgett (LB) method for creating thin films of nanomaterials, vital for various applications. The study explores optimal coating configurations for porous MOF NPs on threads and nylon fiber. Customized holders are used for MOF deposition, revealing that film formation depends on deposition orientation and thread type. Cotton thread and nylon fiber show better coverage with vertical deposition, while conductive threads exhibit particle aggregation. These insights advance the development of single fiber-based devices (Rauf et al., 2021).

Industrial applications

The compounds-based MOF market is experiencing growth due to the expansion of major industries such as chemicals, healthcare, food and beverage, oil and gas, and packaging. Factors such as increasing healthcare expenses and the expansion of the food and beverage industry in developing economies are driving the adoption of innovative technologies like controlled drug delivery, medical imaging, and novel food packaging based on MOFs. Fig. 21.8 offers an overview of the compounds based on the MOF market, which we will now explore in more detail (Global Forecast to 2032, 2021).

MOF composites have emerged as a versatile class of materials with extensive industrial applications. These composites are created by integrating MOFs with a diverse range of materials, including metal NPs, organic and inorganic oxides, quantum dots, polymers, carbon-based materials, enzymes, silicon-based materials, and polyoxometalates. This integration enhances their properties and broadens their utility across various sectors. The industrial applications of MOF composites are vast and encompass several fields. One

Metal-organic framework composites: industrial applications Chapter | 21 821

- Gas Storage
- Sensing
- Others
- Drug Delivery
- Separation
- Catalysis
- Purification

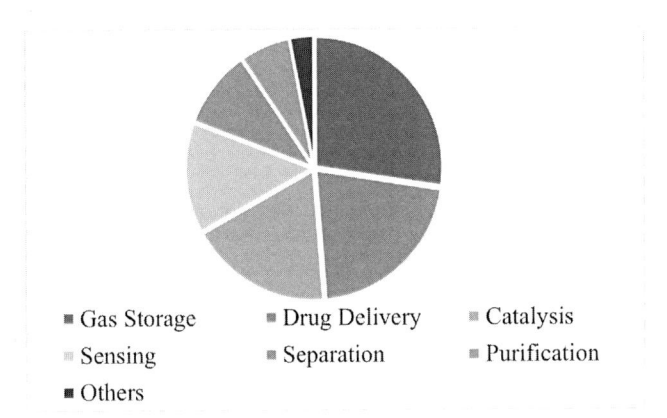

FIGURE 21.8 Industrial applications of metal-organic framework compounds across various market sectors.

significant area is gas storage and separation, where MOF composites play a crucial role in capturing and storing gases such as hydrogen and methane (Akhbari & Morsali, 2013, 2015). Additionally, they are instrumental in separating gas mixtures in industrial processes, contributing to efficient gas handling and storage solutions. In catalysis, MOF composites act as highly efficient catalysts for chemical reactions owing to their large surface area and customizable pore structures. Their unique properties make them suitable for accelerating reactions and improving reaction selectivity, thus driving advancements in industrial catalysis processes. MOF composites also find extensive use in drug delivery systems, facilitating targeted and controlled release of pharmaceuticals. By encapsulating drugs within MOF matrices, these composites enhance drug efficacy while minimizing side effects, offering significant benefits for pharmaceutical applications (Parsaei et al., 2024; Alavijeh & Akhbari, 2022; Amidi & Akhbari, 2024; Parsaei & Akhbari, 2022a, 2022b; Edlo & Akhbari, 2023). Furthermore, MOF composites serve as essential components in sensors designed for detecting gases, chemicals, and biomolecules with high sensitivity and selectivity. Their tailored properties enable precise detection, making them invaluable in various analytical and monitoring applications. In environmental remediation, MOF composites contribute to efforts aimed at removing pollutants from air and water. Their ability to selectively adsorb contaminants makes them effective tools for purifying environmental matrices, thereby supporting sustainability initiatives. Overall, the industrial applications of MOF composites continue to expand as researchers develop new synthesis methods and tailor their properties to meet the diverse demands of various industries. As advancements in materials science and engineering progress, MOF composites are expected to play an increasingly significant role in addressing complex industrial challenges. In the following discussion, we will examine each of these applications from an industrial perspective (Ahmed & Jhung, 2014; Gordeeva et al., 2021; Kim, Kim et al., 2019; Kirchon, 2020; Lin et al., 2023; Ma et al., 2021; Prajapati et al., 2023; Ryu et al., 2021; Yue et al., 2021; Zhang et al., 2019).

Chemical purification

Capture of CO_2

Porous MOF-based materials have emerged as a new type of functional CO_2 adsorbents, requiring specific characteristics for efficiency: (1) porosity, with good accessibility to the channels; (2) thermal stability; (3) the presence of organic ligands derived from nitrogen-containing heterocycles and/or (4) existence of functional groups (e.g., $-NH_2$ or $-OH$ groups) in the pores to interact with CO_2 and boost adsorption; (5) insertion of metal ions; and (6) the presence of open metal sites (Gebremariam et al., 2023; Janardan et al., 2021; Mirzaei & Amiri, 2023). Additionally, some MOFs exhibit selective CO_2 adsorption over other gases, including binary (e.g., CO_2/CO, CO_2/N_2, CO_2/CH_4) and ternary ($CO_2/N_2/CH_4$, $CO_2/N_2/H_2O$, $CO_2/N_2/O_2$) gas mixtures. For instance, ZIF-95 and ZIF-100 feature complex cages selectively adsorbing CO_2 from various gas mixtures. Porous Mg-MOF-74 material demonstrates significant CO_2 adsorption capabilities, particularly in separating CO_2 from CH_4, with subsequent easy release at lower temperatures. Numerous reports the remarkable CO_2 retention capacities of various MOFs and MOFs composite, such as MUT-1 (Alavijeh et al., 2019); (NTU-65) (Zhang, Song et al., 2022); MOF-74(CPO-27) (Choe et al., 2021); SIFSIX-2-Cu, SIFSIX-2-Cu-i, and SIFSIX-3-Zn (Burd et al., 2012); Mg-dobdc (Hu et al., 2019); UTSA-16 (Hu et al., 2019; Xiang et al., 2012); NOTT-122 (Alsmail et al., 2014; Deria et al., 2013; Yan et al., 2013); HKUST-1 (Chen, Wang et al., 2021; Gargiulo et al., 2023; Millward & Yaghi, 2005); MOF-5 (Kukulka et al., 2019; Ma et al., 2018; Millward & Yaghi, 2005); MOF-117 (Millward & Yaghi, 2005); MIL-100(Cr, Fe) (Lestari et al., 2021; Llewellyn et al., 2008; Teerachawanwong et al., 2023); MUT-5 (Salimi et al., 2022); MIL-101(Cr) (Chong et al., 2022; Habib et al., 2023; Park et al., 2022); NU-100 (Abazari et al., 2024; Farha et al., 2010); UMCM-1 (Olajire, 2018); MOF-210 and MOF-200 (Naveed et al., 2022); MOF-505 (Chen et al., 2017, 2019); NbOFFIVE-1-Ni (Bhatt et al., 2016); nickel face-cubic centered functional $[Ni_8(OH)_4(H_2O)_2(BDP^1)_6]$, $[Ni_8(OH)_4(H_2O)_2(BDP-OH)_6]$, and $[Ni_8(OH)_4(H_2O)_2(BDP-NH_2)_6]$ networks (Al-Rowaili et al., 2021; Colombo et al., 2012; López-Maya et al., 2014); and MUT-4 (Parsaei, Akhbari, & White, 2022).

---

1. *BDP*, 1,4-bis (pyrazol-4-yl) benzene-4.

Graphite oxide/metal-organic framework (GO/MOF) composites, in particular, have garnered considerable attention due to their synergistic effects in the reactive adsorption of CO_2. This is attributed to the active sites formed at the interface between graphene layers and MOFs. When compositing with GO or graphite oxide, MOFs like ZIF-8 and HKUST-1 demonstrate significantly enhanced CO_2 absorption capabilities. For instance, the CO_2 uptake increases from 27.2 wt.% (for pure ZIF-8) or 33 wt.% (for pure GO) to 72 wt.% for ZIF-8 composite with 20 wt.% GO at 195K in the case of GO/ZIF-8 composites (Kumar et al., 2013). Similarly, the graphite oxide/HKUST-1 composite exhibits a CO_2 uptake as high as 4.23 mmol/g at room temperature, which is twice that of the pristine HKUST-1 powder and 10 times higher than that of the graphite oxide sample (Zhao et al., 2013). An ultrafast room temperature synthesis method for producing GrO@HKUST-1 (GrO: graphite oxide) composites with high CO_2 adsorption capacity and CO_2/N_2 adsorption selectivity was reported (Xu et al., 2016). Also a thermally stable PM filter was designed by anchoring ZIF-8 onto a three-dimensional (3D) network of reduced graphene oxide aerogel (rGA) through natural drying. The filter exhibits high capture efficiencies for PM2.5 and PM10, exceeding 99.3% and 99.6%, respectively, at ambient conditions. Even under harsh conditions (200°C, flow velocity of 30 L/min), efficiencies remain high (>98.8% for PM2.5 and >99.1% for PM10). The filter can be regenerated through a simple washing process, making it effective for capturing particulate matter pollution at emission sources like car exhausts (Fig. 21.9) (Mao et al., 2019).

Also, the shaping of MIL-91(Ti)/graphene oxide (5 wt.%) composite has been reported through mixing the powder with 3 wt.% PVB (polyvinyl butyral) and vigorously mixing it in a spherical vessel (Muschi et al., 2021). Also, to enhance CO_2 separation from diverse gas mixtures, MMMs were fabricated.

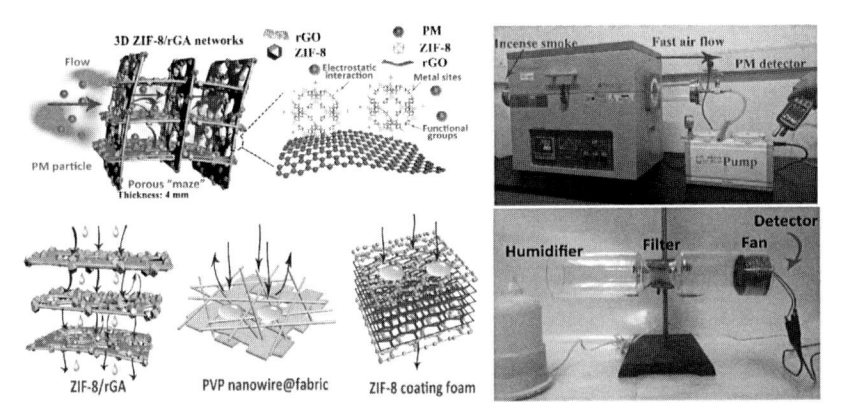

FIGURE 21.9 Schematic of ZIF-8 onto a three-dimensional network of reduced graphene oxide aerogel (Mao et al., 2019; Ryu et al., 2021).

These MMMs consisted of Zr-MOF, MIL-53(Al), ZIF-8, and HKUST-1 incorporated into a matrimid polymer matrix (Dai et al., 2023; Rangaraj et al., 2020; Zunita et al., 2022). Recently a novel coating machine featuring double blades and a surface cross-linking unit was designed for the industrial-scale production of multilayer composite membranes. By running the coating machine twice, the membranes were prepared effectively. The resulting multilayer composite membrane, such as MOF@MMM, demonstrated high and stable CO_2/N_2 performance, indicating promising potential for applications in flue gas CO_2 capture (Fig. 21.10) (Shen, Pan, et al., 2021).

The development of an MWCNT (multiwalled carbon nanotube)/MIL-101 composite has significantly enhanced CO_2 capture efficiency. This composite, exhibiting a morphology similar to pristine MIL-101, demonstrates an impressive 60% increase in CO_2 capture capacity, rising from 0.84 to 1.35 mmol/g. The enhancement is attributed to the enlarged micropore volume

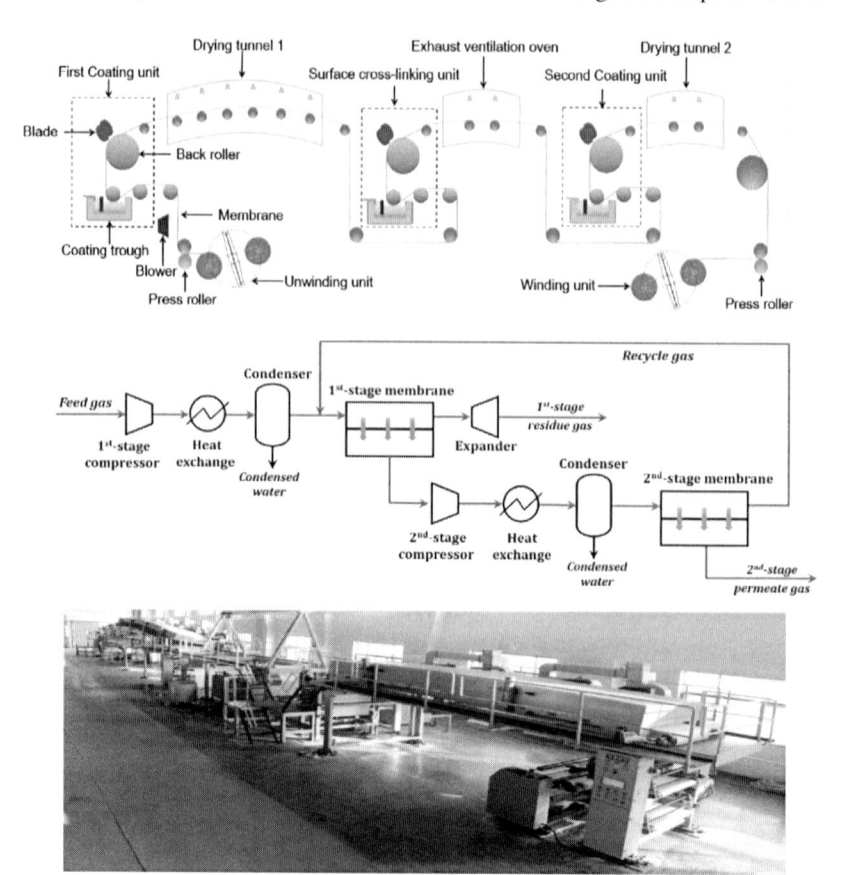

FIGURE 21.10 MMM demonstrated high and stable gas CO_2 capture (Sheng et al., 2021).

of MIL-101 facilitated by the integration of MWCNTs (Anbia & Hoseini, 2012). The cycloaddition of CO_2 and epoxides is a promising industrial process due to its mild temperature and pressure conditions. The resulting carbonate products are valuable chemical intermediates used in various fine chemical synthesis and functional material fabrication. Incorporating functional species into porous frameworks enhances the activity and selectivity of this reaction. For instance, the polyILs@MIL-101 composite (ILs=ionic liquids), which exhibited stronger CO_2 affinity and excellent catalytic activity for the cycloaddition reaction (Ding & Jiang, 2018; Liu, Sharifzadeh, et al., 2021). A scalable method for producing millimeter-scale, diamine-functionalized MOF-74/polymer composite beads was reported. Using the phase inversion process and loadings of 30, 40, and 50 wt.%, they combined diamine-free MOF with various concentrations of the hydrophobic polymer binder polyvinylidene fluoride (PVDF) and diamine-functionalized it using 1-ethylpropane 1,3-diamine (epn). This represents the first successful demonstration of diamine-functionalized MOF beads with diameters of 2–3 mm, suitable for practical applications in indoor CO_2 removal (Yuan et al., 2023). A new eco-friendly method was devised to rapidly grow a selective inorganic membrane on a polymeric hollow fiber support for postcombustion carbon capture (Fig. 21.11). Continuous ZIF-8 thin films were synthesized using water as a solvent, achieving a CO_2 permeance of 22 GPU and a record CO_2/N_2 selectivity of 52 (Marti et al., 2017).

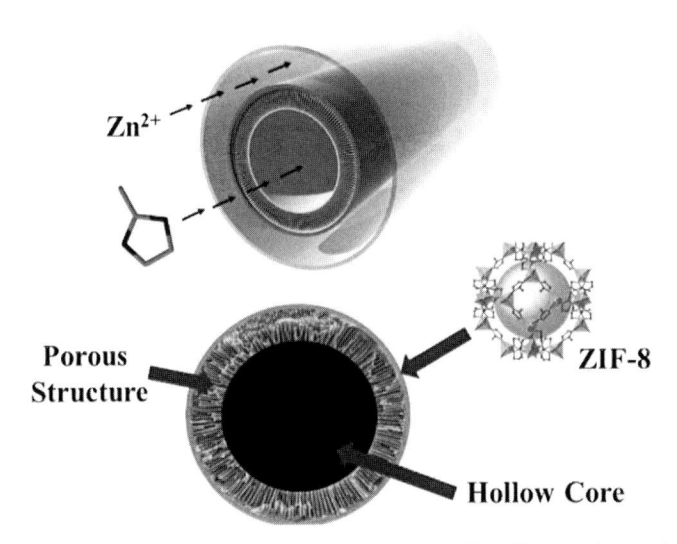

FIGURE 21.11 The diagram illustrates the supported ZIF-8 hollow fiber membrane, showcasing its hollow core, porous Torlon structure, and surface-coated ZIF-8 (Marti et al., 2017).

Company Novo MOF is actively engaged in carbon dioxide absorption and synthesized MIL-120 (Al) composite on a pilot scale. MIL-120(Al)-AP (ambient-pressure) composite shows significant promise in post-combustion CO_2 capture studies. Synthesizing at a scale exceeding 3 kg using a water-based and environmentally friendly protocol, its high CO_2 capture capacity at low pressure, comparable to benchmark amine-free CO_2 adsorbents, and cost-effective composition make it a compelling candidate for postcombustion CO_2 capture investigations (Chen et al., 2024). Also, MUF-17 with a polymeric binder, PVDF, was investigated. The resulting MUF-17/PVDF composite, in the form of pellets, showed comparable behavior in gas adsorption experiments to the pristine MOF (Qazvini & Telfer, 2020). In September 2021, Company MOFGEN launched a new composite based on zeolite imidazolate framework, ZIF-8, aimed at selectively absorbing CO_2 from industrial flue gas streams, thereby contributing to the reduction of greenhouse gas emissions. Similarly, the composite of aluminum-based MOF (MIL-160(Al)), which was synthesized from 2,5-furandicarboxylic acid obtained from biological sources, has shown promising results in CO_2 absorption, particularly in post-combustion scenarios. Pilot-scale synthesis of these MOFs was conducted using water-based protocols in 2 and 30 L glass-lined reactors at ambient pressure, yielding an absorption capacity of 185 kg/(m^3.day) at the larger scale (Severino et al., 2021). Recently, Mosaic Materials, Inc. has introduced proprietary filters utilizing MOF technology, capable of selectively capturing over 90% of CO_2 from air. Additionally, Shimizu and colleagues introduced CALF-20, a microporous Zn oxalate triazolate, known for its effectiveness in CO_2 absorption in the presence of moisture. These MOFs, produced by BASF at a scale exceeding 200 tons using environmentally friendly protocols, have been utilized by Svante for pilot-scale CO_2 absorption studies, achieving an annual productivity of 550 kg/(m^3.day). The estimated price of CALF-20 at a scale of 100 tons ranges from 20 to 30 dollars/kg (Chakraborty et al., 2023; Lin et al., 2021). Also, a successful pilot test on the separation of butane isomers by using ZIF-8 slurry is a critical step in approaching the industrial application of MOFs (Chen, 2022).

Capture of SO_2

Sulfur dioxide (SO_2) is a harmful gas emitted during combustion and industrial activities, presenting substantial health and environmental hazards. Recent studies have concentrated on crafting composite MOF-derived adsorbents for effective SO_2 removal from flue gases. Metal–carboxylate MOFs like MFM-300(In), MFM-600, and NOTT-202a, alongside metal–azolate MOFs such as Ba^{2+}-exchanged pyrazolate MOFs and fluorinated MOFs, notably SiF_6^{2-}-based SIFSIX materials, have demonstrated outstanding capacity and selectivity in absorbing SO_2 (Ryu et al., 2021). Recently, a stable porous Zr(IV)-MOF (HBU-20) has been synthesized with impressive characteristics. HBU-20 exhibits a high Brunauer–Emmett–Teller (BET) surface area of 1551.1 m^2/g and a pore volume of 0.896 cm^3/g. Notably, it demonstrates a significant SO$_2$ adsorption capacity of 6.69 mmol/g at 298 K and 100 kPa, with higher selectivities for SO$_2$/CO$_2$ (56.7) and SO$_2$/CH$_4$ (Kayal & Chakraborty, 2018) (v/v, 1:99). Theoretical simulations suggest efficient SO$_2$ capture is facilitated by host–guest interactions. Dynamic breakthrough experiments under ambient conditions confirm excellent separation performance, with experimental selectivities of SO$_2$/CO$_2$ and SO$_2$/CH$_4$ (v/v 1:99) reaching 81.1 and 117.6, respectively. HBU-20 shows promise as an absorbent for deep desulfurization of flue gas and natural gas (Ren et al., 2022).

To date, only a few fibrous MOF membranes have been studied for selectively capturing SO$_2$ under dynamic adsorption conditions, showing promising results. These membranes offer advantages over solid-state crystalline MOF materials, addressing issues like pipe clogging and recycling. Wang et al. focused on creating MOF filters by embedding MOF crystals into polymers via electrospinning. While several reports exist on this method, further investigations are needed to understand morphology tuning and surface functionality modification systematically. Analyzing polymer-MOF compatibility is crucial for developing efficient filters for SO$_2$ and particulate matter capture. Filters like UiO-66-NH$_2$/PAN and MOF-199/PAN exhibited dynamic SO$_2$ adsorption capacities of 0.019 and 0.014 g/g, respectively, with low gas resistance and pressure drop. The UiO-66-NH$_2$/PAN filter showed excellent durability, reusable for SO$_2$ removal. Another group reported MIL-53(Al)–NH$_2$/PAN (MGP), an electrospun nanofibrous material for selective SO$_2$ removal, achieving an uptake of 0.0028 g/g under dynamic conditions. The MGP filter reduced SO$_2$ concentration in a mixed gas from 7000 to 40 ppb. These studies on nanofibrous MOF-based filters inspire further exploration for practical and industrial applications (Fig. 21.12) (Wang, Xu, et al., 2019; Zhang et al., 2016).

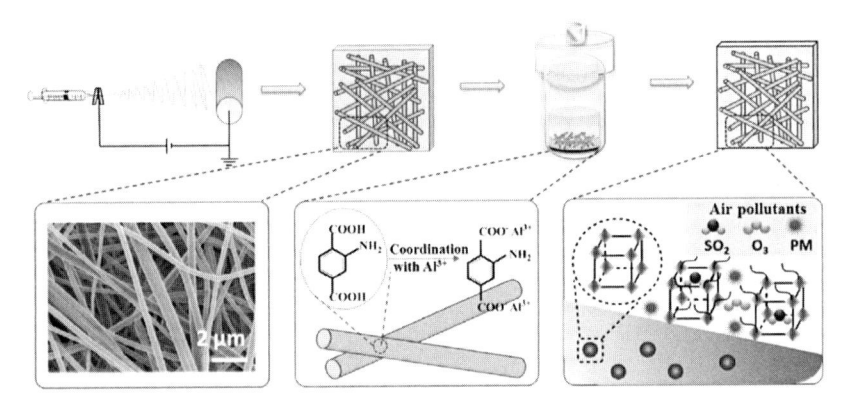

FIGURE 21.12 The synthesis protocol and adsorption processes of metal-organic framework-based fibrous membranes (Ryu et al., 2021; Wang et al., 2019).

Capture of NO$_x$

Capture of nitrogen oxides (NO$_x$) refers to the process of trapping NO$_x$, which are a group of highly reactive gases primarily emitted from combustion processes, such as vehicle engines and industrial activities. NO$_x$ emissions contribute to air pollution and are associated with adverse health effects and environmental damage. The adsorption of NO$_2$ gas onto UiO-66 (Zr)-COOH and UiO-67 (Zr)-COOH decreased from 40–73 to 3–10 g/kg and from 79–118 to 41–93 g/kg with melamine (melamine@UiO-66 (Zr)-COOH & melamine@ UiO-67 (Zr)-COOH) (Ebrahim & Bandosz, 2014). Additionally, a new oxalic acid-incorporated UiO-66 analog, UiO-66-ox, synthesized via a solvent-assisted ligand incorporation (SALI) approach, exhibited higher NO$_2$ uptake capacity (8.4 mmol/g) than UiO-66-vac (vacant sites in UiO-66) and UiO-66 (3.9 and 3.8 mmol/g, respectively) (DeCoste et al., 2015).

Another study investigated the Al-based MOF, MFM-300(Al), as an NO$_2$ adsorbent. MFM-300(Al) demonstrated a high reversible adsorption capacity of NO$_2$ (14 mmol/g) under ambient conditions (298K, 1.0 bar) and remarkable selectivity for NO$_2$ in NO$_2$/CO$_2$ and NO$_2$/SO$_2$ mixed gas conditions. The MOF effectively operates at very low NO$_2$ concentrations of 5000 ppm and even < 1 ppm NO$_2$, indicating strong interactions between the MOF and NO$_2$. Breakthrough experiments using kinetic, dynamic, and static methods revealed five soft supramolecular interactions responsible for the selective adsorption of NO$_2$. These advancements open avenues for further exploration into superior NO$_2$ absorption materials (Han et al., 2018).

Capture of volatile organic compounds

Capture of volatile organic compounds (VOCs) involves the removal or trapping of VOCs emitted from diverse sources like industrial processes, vehicles, and household products. VOCs contribute significantly to air pollution and pose health risks due to their potential to react with other pollutants and form harmful compounds like ozone and particulate matter. Various techniques, such as adsorption, absorption, and chemical reactions, are utilized to capture VOCs, aiming to mitigate their harmful effects on human health and the environment. A composite of chromium-based MIL-101 and graphite oxide (MIL-101@GO) was synthesized and tested for adsorption of VOCs (benzene, toluene, and ethylbenzene). Results from temperature-programmed desorption experiments showed that the composite had high uptake of aromatics, up to 1.8–6.0 times higher than conventional adsorbents. Desorption activation energies increased with the carbon number of aromatics. Ethylbenzene adsorption was highly reversible. Additionally, the MIL-101@GO composite exhibited enhanced thermal conductivity (up to 0.369 W/mK at 303 K). With its high adsorption capacity, improved thermal conductivity, and recyclability, it's a promising option for volatile organic compound adsorption (Sun et al., 2017). Also, a green and in situ MOF shaping strategy was devised to create cellulose foam HKUST-1 and modify it with polysaccharides for convenient adsorption applications. The resulting foam HKUST-1 maintains intrinsic thermal stability and a high surface area while exhibiting a unique crystal structure and interface. Furthermore, the surface and crystal size of foam HKUST-1s can be easily adjusted with natural polysaccharides, resulting in 5.2 times higher adsorption uptake (4.52 mmol/g) for vapor-toluene compared to MIL-101 (Cr) at low pressure P/P_o≈0.0026, with high selectivity (Jambovane et al., 2016) for acidic/basic dipeptides EE/RR. These foamed MOFs show promise for VOC adsorption under ambient conditions and improving acidic dipeptide enrichment in the liquid phase (Cui et al., 2019). MOFs incorporated into polyurethane foams (PUFs) can adsorb VOCs effectively, circumventing challenges associated with handling powdered MOFs. Using MIL-160(Al) and UiO-66(Zr)-(CF$_3$)$_2$, MOF/PUF composites were prepared in a single step, maintaining MOF characteristics and mechanical stability. These composites retained high VOC adsorption capacities for various compounds while allowing for considerable MOF particle incorporation (up to 200% w/w of foam) with a significant open-cell volume (32%). Demonstrating practicality, the composites effectively captured acetic acid in the presence of ambient moisture, showing promise for VOC removal in environmental applications (Freitas et al., 2024).

Petrochemicals purification

In recent years, there has been a growing focus on developing novel technologies for the separation and purification of essential gases and light hydrocarbon chemicals in various industrial applications. Traditional methods like cryogenic distillation, while effective, often consume high energy and are cost-intensive due to the similar physicochemical properties of the target compounds. Porous crystalline materials, such as MOFs, have emerged as promising candidates for selective adsorption and separation due to their unique properties. In particular, recent advances have shown promising applications in the selective separation of important chemical species like N_2/CH_4, C_2H_2/CO_2, and xylene isomer mixtures, offering potential solutions to existing challenges in industrial processes.

Composites of vanadium (II)-based MOFs, such as V$_2$Cl$_2$.8(btdd)(btdd=bis (1H-1,2,3-triazolo[4,5-b],[4′,5′-i])dibenzo[1,4]dioxin)) have demonstrated selective purification of N_2 over CH_4 in natural gas mixtures, as well as selective adsorption of olefins over paraffins. These materials exhibit exceptional reversible capacities over multiple cycles and high selectivity for ethylene over ethane-rich feeds, making them valuable for gas purification processes (Jaramillo et al., 2020). Similarly, Hofmann-based MOFs like Co(pyz)[Ni(CN)$_4$] (ZJU-74) have shown high selectivity for acetylene over CO_2, C_2H_4, and CH_4 at ambient conditions, with impressive C_2H_2 capturing capacities and selectivities (Pei et al., 2020). Furthermore, cobalt-based MOFs, such as Co$_2$(dobdc)(dobdc^{4-}=2,5-dioxido-1,4-benzenedicarboxylate), have demonstrated excellent separation of C_8 alkyl aromatics (xylenes isomers) via multiple adjacent unsaturated Co(II) metal sites (Gonzalez et al., 2018).

Gases storage

Industrial gases such as oxygen, hydrogen, nitrogen, and ammonia, which exist at room temperature and pressure, are used in various industries such as chemical production, power generation, medicine, electronics, aerospace, and food. The most established application of MOFs is gas storage for hydrogen, methane, acetylene, and SF_6. In recent years, significant research efforts have focused on developing MOF composites for various gas storage applications. MOF composites, which combine MOFs with other materials such as carbon nanotubes, polymers, or metal nanoparticles, further enhance gas storage capabilities by improving stability, selectivity, and overall performance. These composites leverage the unique properties of both MOFs and the additional components to achieve superior gas storage and separation efficiencies. Currently, North America dominates the market due to the presence of large MOF producers and research institutions exploring their potential applications. However, the Asia-Pacific region is expected to emerge as the fastest-growing regional market, as countries like China and India are heavily investing in MOF research and development (Ahmed & Jhung, 2014; Global Market Insights Inc., 2023; Nik Zaiman et al., 2022).

Hydrogen storage

Hydrogen storage represents a crucial application for MOF composites, especially in the context of reducing CO_2 emissions. As a clean energy source, hydrogen offers significant potential due to its high energy density and zero CO_2 emissions upon combustion. Fuel cell electric vehicles (FCEVs) employ a propulsion system similar to electric vehicles, where a fuel cell converts stored hydrogen (H_2) energy into electricity (Bakuru et al., 2019; Chen et al., 2020; Zhao et al., 2022). The Department of Energy (DOE) sets requirements for hydrogen storage materials, vital for fuel cell vehicles. By 2020, the DOE targets 4.5 wt.% (gravimetric working capacity) and 30 g/L (volumetric working capacity). Certain MOFs have, thus far, met the DOE's gravimetric target (DeSantis et al., 2017; Gómez-Gualdrón et al., 2017; Staffell et al., 2019). Since MOF-5 was first reported as a hydrogen adsorbent in 2003, a series of papers have been reported on hydrogen storage. MOFs demonstrate high hydrogen uptake capacities such as Ni_2(m-dobdc) (M_2(m-dobdc, M=Co, Ni; m-dobdc^{4-}=4,6-dioxido-1,3-benzenedicarboxylate and isomeric frameworks), with a usable volumetric capacity of 11.0 g/L at 25°C and 23.0 g/L with a temperature swing between −75°C and 25°C, over a pressure range of 100–5 bar (Kapelewski et al., 2018). NU-125 (49 g/L, 8.5 wt.%), NU-1000 (48 g/L, 8.3 wt.%), and UiO-68-Ant (47 g/L, 7.8 wt.%), NU-1103 (43 g/L, 12.6 wt.%), IRMOF-20 (51 g/L, 9.1 wt.%), NU-100 (47.6 g/L, 13.9 wt.%), and NU-1501-Al (46.2 g/L, 14 wt.%) (Ahmed et al., 2019; García-Holley et al., 2018).

MOFs are utilized for hydrogen backup power supply. This is succeeded by a charge phase where hydrogen is compressed and filled into the tanks. Subsequently, during the discharge phase, hydrogen is released to the fuel cell stacks, leaving 10% of hydrogen in the tank. After each discharge, the system remains at the storage temperature until the next cycle. This cyclic process continues until maintenance is required (Peng et al., 2022).

However, while MOFs demonstrate high hydrogen uptake capacities at low temperatures, they do not exhibit a significant advantage in hydrogen storage at room temperature compared to other porous materials like active carbons and zeolites. This limitation arises from weak interactions between H_2 molecules and MOF structures at room temperature. To address this challenge, new methods have been developed to enhance MOF interactions with H_2 by incorporating functional species, leading to improved hydrogen adsorption performances at room temperature (Ahmed et al., 2019; Molefe et al., 2021; Musyoka et al., 2017).

Moreover, the incorporation of other microporous materials, particularly carbon materials like carbon nanotubes (CNTs) and GO, into MOF composites has been explored to optimize the large void spaces in MOFs for hydrogen storage. These composite systems show enhanced gas storage capacities, with examples including the CNT/MOF-5 (Viditha et al., 2016; Yang et al., 2010) composite and the GO/HKUST-1 composite (Azim & Mohsin, 2019; Zhou et al., 2014). Additionally, the spillover effect, involving the dissociative chemisorption of hydrogen on metal NPs and its subsequent migration onto adjacent surfaces via diffusion, has been leveraged to enhance hydrogen capacities in metal NPs/MOF composites. This effect can be facilitated through the construction of carbon bridges between dissociative metals and MOFs or by directly doping dissociative metals into MOFs. Metal NPs/MOF composites, such as Pt@MOF-177 (Proch et al., 2008), Pd@HKUST-1 (Li et al., 2014), and Pd@MIL-100 (Zhu et al., 2023), have demonstrated higher hydrogen capacities compared to their counterparts (Kang et al., 2021; Malouche et al., 2019). A study reports the fabrication of core-shell nanocrystals by incorporating microporous UiO-66 into mesoporous MIL-101. The resulting core-shell nanocrystals showed hydrogen storage capacities 26% and 60% higher than pure-phase MIL-101 and UiO-66, respectively. Additionally, the core-shell MIL-101@UiO-66 structure demonstrated high moisture tolerance (Ren et al., 2014). PCN-610/NU-100 exhibits the highest usable pressure swing capacity overall, both in terms of gravimetric and volumetric measures. However, despite having the largest total gas capacity at 100 bar among the five MOFs studied, its usable capacity is the lowest. This suggests that total capacities may not accurately reflect the usable capacity under pressure swing conditions (Ahmed et al., 2019).

Fig. 21.13A–C shows a method of shaping Zr-MOF powder into spherical pellets using 10 wt.% sucrose as a binder and a granulator, producing kilogram batches in just 30 minutes. These pellets, suitable for small hydrogen storage tanks, underwent rigorous testing, showing minimal breakage. While there's a

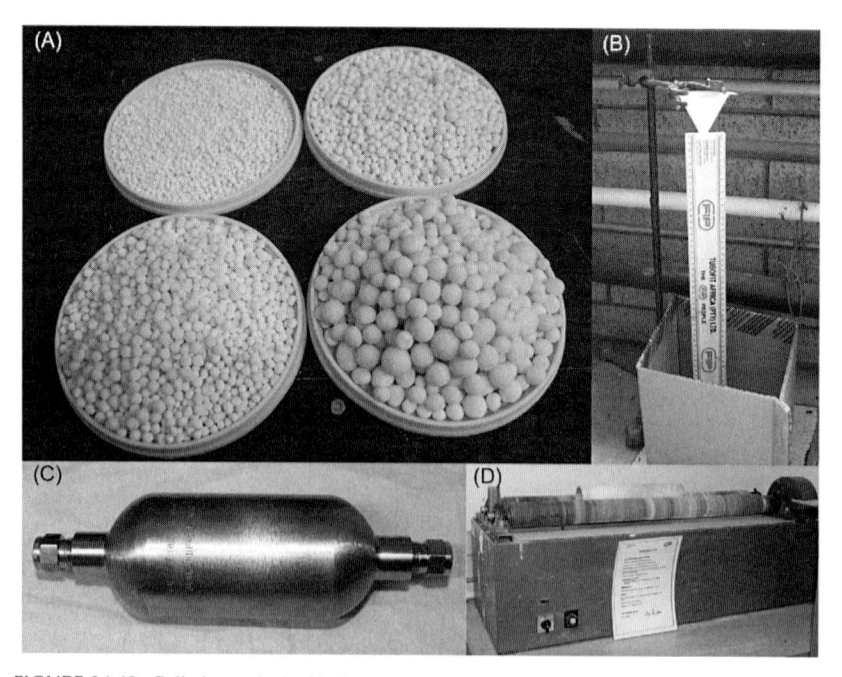

FIGURE 21.13 Cylinder packed with 5 mm Zr-MOF pellets and abrasion test set-up (Ren et al., 2015; Ryu et al., 2021). (A) Zr-MOF pellets featuring different diameters; (B) Experimental configuration for assessing mechanical strength via drop testing; (C) Cylinder containing packed Zr-MOF material; (D) Experimental setup for evaluating tumbling mechanical strength through abrasion testing.

slight decrease in surface area affecting hydrogen storage capacity, this method holds promise with the right binder choice (Ren et al., 2015; Ryu et al., 2021).

Since 2007, road tests on CNG and hydrogen vehicles, including passenger cars and large trucks, have been conducted. BASF has developed at least two registered prototypes for gas storage, namely BASOCUBE and BASOSTOR, which have also been studied for hydrogen storage (Jacoby, 2013). BASF showcased the hydrogen storage prowess of Basolite Z377 (MOF-177). In a 50-L prototype container, filling the tank took just 5 minutes at 50 bar and 77 K, achieving impressive volume-specific capacities. For gaseous hydrogen at 350 bar/298 K and 700 bar/298 K, the capacities were 23 and 37 g (H_2)/L, respectively. Moreover, a remarkable volume-specific capacity of 71 g(H_2)/L for liquid hydrogen at 1 bar/20K was attained. Collaborating with BASF, Ford Motor Company confirmed these results through validation tests on hydrogen uptake by various Basolite compounds. Notably, Basolite Z377 reached around 7 wt.% H_2 uptake, while Basolite Z100-H (MOF-5) achieved 8 wt.% H_2 uptake (Silva et al., 2015).

Ford Global Technologies filed patents for "Hydrogen Storage Materials" (US20110142752A1) (Yang & Hirano, 2011) and "Hydrogen Storage Systems and Methods using the same" (US20110142750A1) (Pulskamp et al., 2010). Additionally, the zero-emission Mercedes-Benz F125 research vehicle utilized hydrogen stored in MOFs. However, industrial implementation has been hindered by technical challenges, including system weight, volume, cost, efficiency, and durability, leading to delays in deployment.

ENI company collaborated in an academic–industry consortium focused on studying heterogeneous catalysts, including CPO-27-M (where M=Ni, Mg, Co, or Mn) and Zr-based UiO-66. The research demonstrated significant hydrogen storage for both families, with particularly high methane storage observed for CPO-27-Ni (MOFCAT Functional Metal Organic Frameworks as Heterogeneous Catalysts, 2022).

Long's lab has enhanced the adhesive properties of their MOFs by introducing low-coordinate metal cation sites. Although they successfully developed MOFs with suitable binding energies, they encountered a new challenge. The use of metal cations adds weight, limiting the amount of hydrogen these materials can hold relative to their weight. Therefore, their current focus is on creating MOFs with low-coordinate metal cations capable of interacting reversibly with two, three, or even four hydrogen molecules per cation, within the 15–20 kJ/mol range (Hydrogen storage gets real, 2022).

In January 2022, MOF Technologies, a company active in the field of MOF synthesis, introduced a new MOF adsorbent called MOF-210 for hydrogen storage applications. The volumetric hydrogen capacity of this framework is 70% more than today's compressed gas cylinders. For this reason, in March of this year, this company cooperated with SK Inc. It has partnered to advance the development and commercialization of MOFs for hydrogen storage. This strengthens its capability in clean energy technologies (Hydrogen production MOF Technologies Company, 2022). NU-1501 is also one of the other alloy metal frameworks of interest for hydrogen fuel cell storage and production in California. Toyota Mirai and Hyundai Nexo are two big car companies that have invested in this field.

Natural gas storage

Natural gas, primarily methane, plays a critical role in modern power supply. In typical adsorption equipment, a column packed with an adsorbent material separates components, with methane often retained strongly. Desorption then yields a methane-enriched stream. Additional separations may be needed for other components or particles. Swing adsorption methods, like temperature swing adsorption (TSA) or pressure swing adsorption (PSA), are common at large scale. TSA increases temperature for desorption, while PSA adjusts partial pressures or total pressure (Fig. 21.14). TSA is recommended for gas purification processes with low component concentrations ($<2\%$) (Ursueguía et al., 2021).

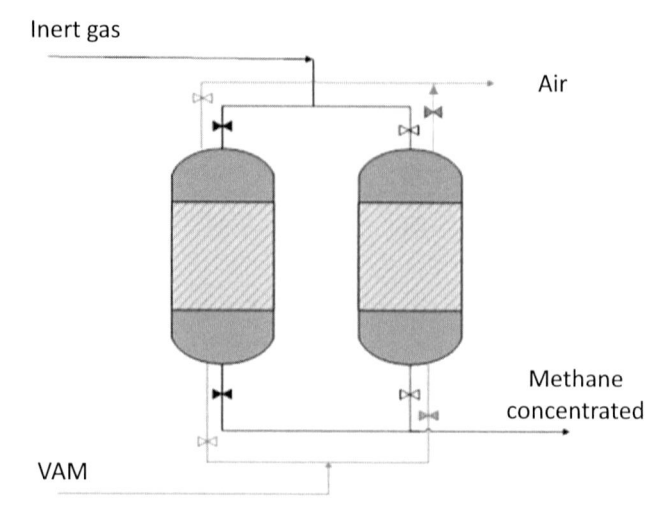

FIGURE 21.14 Schematic of an adsorption unit. Alternatively, one bed is in adsorption (gray line), whereas the other one is in regeneration mode (black line) (Ursueguía et al., 2021).

However, the storage and transportation of natural gas entail fire and explosion hazards, limiting its widespread use as a fuel. Utilizing porous materials such as MOFs offers a promising solution to meet U.S. DOE targets at moderate pressures (35–65 bar) and ambient temperatures. Some of the extensively studied MOFs include Al-soc-MOF-1, MOF-519, HKUST-1, MIL-101(Cr), UTSA-20, NU-111, NU-125, PCN-14, NOTT-122, NOTT-101, UTSA-76, MFM-112a, MFM-115a, UTSA-110, NJU-Bai, PCN-14 structure, and others (Ali & Mahmoud, 2021; Eyankware & Ateke, 2020; Huang, 2023; Poya & Fazl, 2023; Xian et al., 2023).

The utilization of MOFs in industrial applications, particularly for natural gas storage, necessitates meticulous consideration of various factors to ensure efficiency and reliability. One critical aspect is the impact of minor components present in natural gas, such as hydrocarbons, carbon dioxide, water, and sulfur compounds, on the adsorption process within MOFs. These components, adsorbed alongside methane, require additional energy for desorption, contributing to gradual poisoning of the adsorbent and reducing the overall adsorption capacity of the system (Poya & Fazl, 2023). The economic attributes of four MOFs in the natural gas storage domain have been explored in a recent techno-economic analysis. This study aimed to understand the cost implications of large-scale MOF synthesis and to find ways to achieve a production cost below $10 per kilogram of MOF. Four MOFs were examined: Ni-MOF-74, Mg-MOF-74, MOF-5, and HKUST-1, each with distinct metal centers and linkers. Baseline synthesis costs ranged from $35 to $71/kg, mainly due to organic solvent usage in solvothermal syntheses. Two

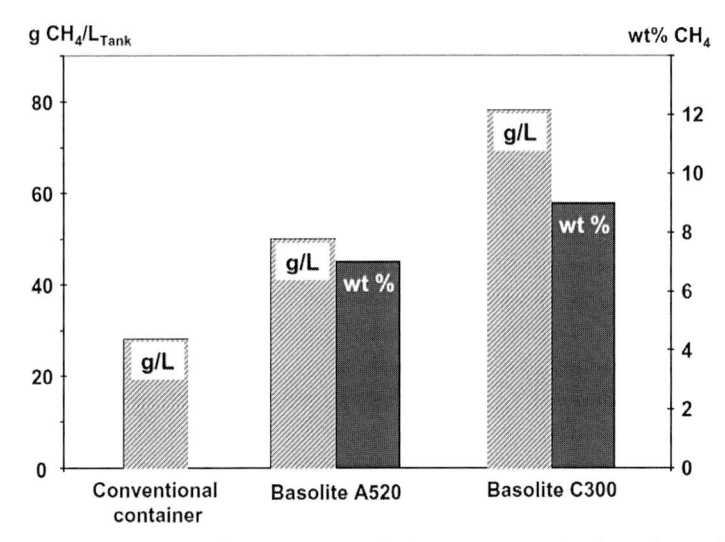

FIGURE 21.15 Comparison of CH$_4$-storage capacity between a conventional container and a tank filled with shaped bodies of metal-organic framework materials, under conditions of 35 bar and room temperature (Gaab et al., 2012).

alternative processes, liquid-assisted grinding (LAG) (Akhbari et al., 2018; Hasheminezhad et al., 2019; Kianimehr et al., 2020; Mirzadeh et al., 2018; Moeinian et al., 2016; Shahangi Shirazi et al., 2016) and aqueous synthesis, were evaluated to reduce solvent usage, resulting in cost projections between $13 and $36/kg (representing reductions of 34%–83%). Sensitivity studies rounded out the analysis, probing for additional pathways to meet the target production cost of less than $10/kg (DeSantis et al., 2017). Companies BASF, which produces MOFs such as Basolite2 A520 and Basolite2 C300, are at the forefront of developing these materials for industrial use (Fig. 21.15) (Gaab et al., 2012; Ursueguía et al., 2021).

Combining MOFs with other materials such as metal NPs (Yu et al., 2017), oxides (Bhoite et al., 2023), polymers (Tanvidkar et al., 2022), and CNTs (Liu et al., 2016) can result in composites with improved performance characteristics (Tanvidkar et al., 2022). In the realm of adsorbed natural gas (ANG) systems, the effectiveness of microporous carbon adsorbents depends on their physicochemical and mechanical properties, including mechanical strength, hydrophobicity, and high packing density. However, the poor operating characteristics of MOFs, such as low stability and hydrophilicity, limit their application in ANG systems (Al-Naddaf et al., 2018; Chong et al., 2023). To address this limitation, composites are fabricated by combining highly efficient yet unstable MOFs with carbon materials possessing desirable properties. One example is the MAX-MIL composite, which combines the high-performance carbon adsorbent Maxsorb III with MIL-101(Cr) (Kayal & Chakraborty,

2018). This composite exhibits enhanced methane adsorption characteristics compared to the individual components. The preparation involves mixing Cr $(NO_3)_3 \cdot 9H_2O$, terephthalic acid, and activated carbon, followed by a series of steps to remove unreacted components and activate the composite. The resulting material demonstrates improved hydrophobic properties, making MOFs more suitable for ANG systems. Carbon allotropes like graphene, fullerenes, and CNTs possess unique properties that make them ideal candidates for use in MOF composites (Al-Naddaf et al., 2018). For instance, the CNTs/MOF-5 composite, obtained by incorporating carbon nanotubes into the MOF-5 crystal lattice, exhibits enhanced stability and increased surface area compared to pristine MOF-5. Additionally, lithium cation-doped MCNTs@HKUST-1 composite shows improved methane uptake compared to pristine HKUST-1, attributed to the presence of ultramicropores at the CNTs/Li$^+$@HKUST-1 interface. Additionally, optimizing cooling techniques is crucial for efficient gas filling and maximizing storage capacity, as demonstrated by studies on MOF-based ANG systems. However, challenges persist in commercializing MOFs due to factors like weight, volume, limited service life, and high production costs. Policicchio et al. evaluated the performance of MOF-based vehicular adsorption storage systems using materials like Maxsorb-III (derived from oil coke) and Nuchar-SA (derived from wood), highlighting the potential of MOFs like HKUST-1 to enhance tank capacity and mileage. The EcoFuel Asia Tour project exemplifies the practical application of HKUST-1 adsorbent pellets, albeit facing hurdles in broad commercial adoption (Fig. 21.16A–D) (Tsivadze et al., 2019).

Framergy, in collaboration with Texas A&M University, has developed a technology to capture VOCs and methane emissions from oil and gas facilities.

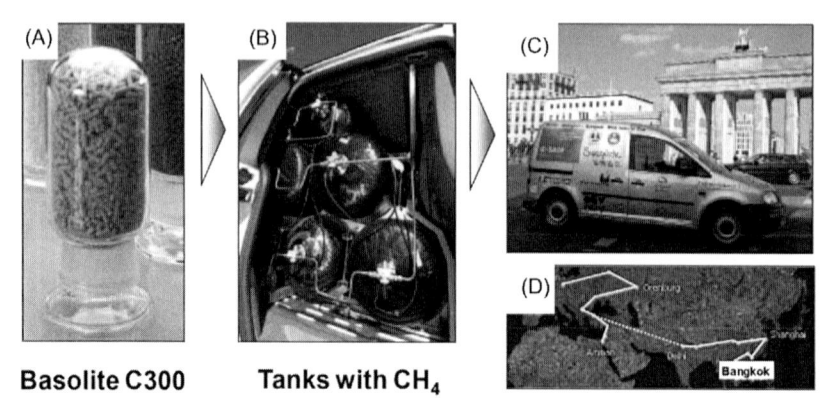

FIGURE 21.16 EcoFuel Asia Tour 2007 – Berlin to Bangkok: 32,000 km with Basolite C300 in tank: (A) Basolite C300 (HKUST-1), (B) MOF-enhanced fuel tanks with CH_4, (C) Volkswagen Caddy EcoFuel prototype car, and (D) Journey map (Yilmaz et al., 2012).

This solution combines VOC-philic solvents and novel MOFs to capture pollutants before they are released into the atmosphere. The technology is versatile and can be used in various facilities, such as storage tank vents and process equipment. Through optimization efforts, the cost-effectiveness of MOFs has been enhanced, making commercialization feasible. Bench-scale evaluations have demonstrated the efficacy of MOFs in capturing VOCs, and pilot-scale testing has confirmed their viability in real-world applications. Framergy is actively working on regulatory approvals and industry collaborations to deploy this innovative solution on a larger scale (Framergy Inc., 2022).

Atomis, Inc., Japan, is also an active company in this field, which investigates new possibilities for gases using porous polymer technology (PCP/MOF). The most important product of this company is an intelligent high-pressure gas chamber called CubiTan, in which MOF technology is employed (Matsuoka, 2022; Atomis Inc. Japan).

Catalysis

Chemical manufacturing heavily relies on catalytic processes, with homogeneous catalysts offering high activity but facing challenges at industrial scales. Heterogeneous catalysts address this with recyclability but often have ill-defined supports. MOFs enable the isolation of single-site heterogeneous catalysts (SSHCs) due to their long-range order. SSHCs feature one type of active site distributed uniformly, necessitating a well-defined support. In MOF-based catalysts, SSHCs can be either structural elements or guests within the pore structure. MOF activation exposes reactive metal sites as SSHCs. Initially, researchers focused on pristine MOFs, optimizing components, structures, and porosity for molecule diffusion and size-selective catalysis. Modifiers were introduced to control sizes, morphologies, and defects, expanding the range of active site species. MOF composites emerged, encapsulating various functional materials within pores, enriching active sites, and enabling diverse catalysis. Advancements in posttreatment methods for MOFs led to porous carbons, metal compounds, and single-atom catalysts (SACs) with dispersed active centers and excellent stability (Hou et al., 2020; Li, Xu, et al., 2019; Shen, Pan, et al., 2021; Wei et al., 2020).

Indeed, the industrial use of MOF composites as catalysts poses several challenges, particularly regarding the synthesis process and stability in various solvents. One significant challenge is the reliance on solvents during the synthesis of MOF compounds. Traditional methods often involve the use of organic solvents, which can be costly and environmentally harmful. Moreover, the presence of solvent molecules within the MOF structure can affect the properties and performance of the resulting catalyst. Table 21.2 summarizes some composites synthesized using MOF composites as green/sustainable catalysts (Liu, Sharifzadeh, et al., 2021).

TABLE 21.2 Summary of composite metal-organic frameworks as the green and sustainable catalyst (Liu, Sharifzadeh, et al., 2021).

Composite MOF	Green condition	Catalytic reaction
Pt/UiO-66	Green solvent and nanoparticle	Olefin hydrogenation
[TPA]$_2$[Zn$_2$(BDC)$_3$(DMA)$_2$]	Solvent-free	Knoevenagel condensation
[Co(BBPTZ)$_3$] [HPMo$_{12}$O$_{40}$]. 24H$_2$O	Green solvent	Oxidative desulfurization
DAIL-Fe$_3$O$_4$@NH$_2$-MIL-88B(Fe)	Ionic-liquid	Esterification
Au@ZIF-8	Green solvent and mild condition	Hydrogenation
MoO$_3$-TMU-5	Mild condition	Oxidative desulfurization
Cs-CA/Cu-MOF	Solvent-free	Knoevenagel condensation
ZrOTf-BTC	Mild condition	Friedel–Crafts acylation and Alkene hydroalkoxylation
MIL100/TNF	Nanoparticle	Photocatalytic degradation
PolyILs@MIL-101	Ionic-liquid	Cycloaddition of CO$_2$ and epoxides
Pt@NSMOFS	Nanoparticle	Oxygen electrocatalysis
PTA@MIL-101	Solvent-free	Biginelli condensation
PW$_{12}$@MFM-300(In)	Green solvent	Degradation of pharmaceutically active compound sulfamethazine (SMT)
PW$_{12}$@NU-1000	POM	Oxidation
Co-MOF-74@NDHPI	Solvent-free	Aerobic oxidation
BAIL/MIL-101	Ionic liquid	Acetalization
NNU-29	POM	CO$_2$ photoreduction
Pt/Au@Pd@1Co	Metal nanoparticle solvent-free	Conversion of CO$_2$ to CO
Pd@MIL-100(Fe)	Green solvent and mild conditions	Hydrogenation
Lipase-proline MOF	Enzyme	Biocatalysis

(*Continued*)

TABLE 21.2 (Continued)

Composite MOF	Green condition	Catalytic reaction
PdCl$_2$-ILs/CuBTC	Ionic-liquid	Oxidation of alkene
ZIF-8/717-resin	Mild condition	Knoevenagel condensation
Pt - Cu frame@HKUST-1	Green solvent	Hydrogenation of olefins
Pd NCS@ZIF-8	Solvent-free	Hydrogenation of alkene
TiO$_2$@UiO-68-CIL	Green-solvent and mild condition	Morita–Baylis–Hillman (MBH)
Pd C MIL-101(Fe)	Mild condition	Hydrogenation of CAL
Ru(bpy)$_3$@MIL-125-NH$_2$	Green solvent and mild condition	Oxidation benzyl alcohol
Au-1@NMOF-Ni	Green solvent and mild condition	Reduction of nitro-phenol
Pt@SALEM-2	Green solvent and mild condition	Hydrogenation
Pt@ZIF-71	Mild condition	Hydrogenation
CuO/HKUST 1	Green solvent	Photocatalytic hydrogen evolution
4-PW$_{12}$@MFM-300(In)	Green solvent	Degradation
Pd/UiO-66@PDMS-T	Green solvent	Hydrogenation
PSMIMHSO$_4$	Ionic-liquid	Desulfurization
Pd/IL/MOF	Mild condition and Ionic-liquid	Hydrogenation
IL/IRMOF-1	Ionic-liquid	CO$_2$ capture

TPA$^+$, Tetrapropylammonium; *H$_3$PW$_{12}$O$_{40}$, PTA*, Keggin-type phosphotungstic acid; *{Co$_2$(oba)$_4$(3-bpdh)$_2$}4H$_2$O (1Co)*, reverse water gas shift; *NSMOFs*, nanosheet metal-organic frameworks.

In the industry, Basolite MOFs, commercially supplied by BASF, offer diverse options for catalytic programs. Fe(BTC), marketed as Basolite F 300, contains 21.2% iron content and features Fe(III) as the nodal metal and tripodal BTC (benzene-1,2,4-tricarboxylic acid) as the ligand. Its crystal structure, likely a distorted form of MIL-100(Fe), remains unresolved. Despite this, Fe (BTC) demonstrates high structural stability and Lewis-acid sites, making it an effective solid acid catalyst. Comparative studies show its superior catalytic efficiency to MIL-100(Fe). Basolite C 300, or Cu$_3$(BTC)$_2$ (HKUST-1), features clusters of two copper ions coordinated with four carboxylate groups, forming paddle-wheel building units. Each copper ion is coordinated to one exchangeable water molecule, enhancing stability.

A Basolite 100, or Al(OH)(BDC) (MIL-53(Al)), comprises corner-sharing octahedral Al^{3+} ions linked by BDC (benzene-1,4-dicarboxylic acid) ligands. It may have limited catalytic activity unless structural defects are present, which can enhance activity, as observed in zeolite catalysis. Basolite Z 1200, or ZIF-8, features a linker without carboxylate groups, coordinating with metal nodes through basic nitrogen atoms. While it is easy to prepare, stability remains a concern, although nitrogen heterocycles can increase framework stability. The long-term productivity of ZIF-8 as a solid catalyst is yet to be determined. While these MOFs are available commercially and can be prepared for catalytic studies, using commercial samples mitigates variability in purity and crystallinity, ensuring reproducibility in heterogeneous catalysis. This spans from the classic Claisen–Schmidt reaction to the oxidation of amines. Studies have explored the intricacies of the ring opening of epoxides, acetalization of aldehydes, and acid-catalyzed selective hydrogenations, among others. These investigations aim to elucidate reaction pathways, understand stereochemistry, and optimize conditions for improved yields and selectivity. For instance, the isomerization of pinene oxide and N-methylation of aromatic amines have been subjects of exploration to uncover their underlying mechanisms and potential applications. Additionally, oxidation reactions, such as those involving benzylic compounds, alcohols, and thiols, as well as the aerobic oxidation of cycloalkanes and cycloalkenes, have been studied to develop efficient and selective oxidation methodologies (Dhakshinamoorthy et al., 2012).

In 2018, mesoporous MOFs were integrated with macroporous melamine foam (MF) using a one-pot process, generating a series of MOF/MF composite materials with preserved crystallinity, hierarchical porosity, and increased stability compared to melamine foam alone. The MOF nanocrystals were threaded through the melamine foam networks, resembling a ball-and-stick model overall. The resulting MOF/MF composite materials were employed as effective heterogeneous catalysts for the epoxidation of cholesteryl esters. Combining the advantages of interpenetrative mesoporous and macroporous structures, the MOF/MF composite provided higher dispersibility and greater accessibility of catalytic sites, exhibiting excellent catalytic performance (Huang et al., 2018).

The Green Science Alliance company has synthesized various types of MOFs and developed MOF-based heterogeneous catalysts with modified sulfone groups on the surface. These catalysts have demonstrated high activity comparable to commercial catalysts like Amberlyst, as confirmed by esterification reactions conducted at 80°C. The efficiency of the MOF-based solid acid catalyst remained consistent even after repeated use, with reaction efficiencies ranging from 73% to 92%. This breakthrough offers a sustainable and efficient solution for catalytic processes in the chemical industry (Green Science Alliance Co., Ltd., 2022).

Batteries and supercapacitors

The increasing demand for reliable, lightweight, intelligent, high-performance, sustainable, eco-friendly, and flexible energy storage devices arises from the rapid development of integrated electronic technologies and portable devices, along with factors such as rising oil prices, air pollution, and political tensions. Batteries and supercapacitors (SCs) are prominent energy storage devices, with significant efforts focused on enhancing their electrochemical performance to meet diverse application needs. While renewable power sources are still being developed, there is a growing need for more reliable, safe, flexible, and wearable energy storage solutions that align with lifestyle, economic growth, and energy supply requirements. SCs, also known as ultra-capacitors or electric double-layer capacitors (EDLCs), offer fast charging and discharging capabilities but face limitations in energy density compared to batteries. Efforts to enhance SC performance involve developing advanced electrode materials such as graphene, CNTs, and MOFs. Despite their low conductivity, MOFs hold promise when combined with conductive materials like graphene or metals, suggesting potential as electrode materials for SCs. In advanced energy applications, MOFs show promise as bifunctional catalysts for oxygen evolution and reduction reactions in metal–air batteries. In energy storage systems, MOFs offer potential applications in SCs, batteries, photocatalytic hydrogen evolution, sensors, and fuel cells. Their porous structure can serve as host matrices for encapsulation or as reaction sites, influencing product distribution and reaction rates. MOFs can also be used as support substrates for introducing sacrificial nanomaterials and functional metal oxides, enhancing their performance in energy-related applications. The synthesis and structural modifications of MOFs are areas of ongoing research, aiming to improve control over synthesis conditions, such as temperature and morphology, for reproducibility and commercial viability. Strategies to enhance electronic conductivity and charge transfer mechanisms are being explored to overcome MOFs' insulating nature and make them more suitable for energy storage applications. Additionally, there is growing interest in developing flexible and wearable energy storage systems using MOFs, with successful fabrication of flexible SCs and rechargeable batteries integrating MOF-derived materials with fiber-shaped substrates like carbon nanotubes. These flexible devices show promising performance characteristics and have potential applications in wearable electronics and portable energy storage devices (Prajapati et al., 2023; Shinde et al., 2021; Sundriyal et al., 2018; Yue et al., 2021).

The exploration of MOFs as electrodes in SCs represents a rich field of study. The application of MOFs in SCs encompasses various configurations, including pure MOFs engineered for enhanced conductivity and surface area. MOFs are combined with conductive polymers like POAP/chitosan/MOF-1 (POAP=poly ortho aminophenol) and ionic liquids (Ehsani et al., 2020), with carbon-based materials such as CNTs (Tan et al., 2017), carbon cloth (CC) (Li, Cai, et al., 2019), nitrogen-doped activated carbon (Liao et al., 2020; Zhang, Wu et al., 2022), graphene (Xin et al., 2020), and functionalized graphene oxide (Kumar et al., 2022). Moreover, MOFs are often combined with nanomaterials to enhance SC performance. Complex arrangements, like layered structures, demonstrate the versatility of MOFs (Shinde et al., 2020). Binary (Borhani et al., 2017) and mixed metal MOFs, including Co-Ni (Xu, Chen et al., 2020), Co-Fe (Zhang et al., 2023), Co-Mn (Kim, Sohail, et al., 2019), Ni-Mn (Li et al., 2023), Ni-Cu (Khokhar et al., 2022), Zn-Zr (Tian et al., 2021), Fe-Zr (Sanati et al., 2023), Zn-Co (Chen et al., 2022), and Fe-Co (Jin et al., 2022), are also utilized (Wang et al., 2016; Zhao, Wei et al., 2020).

As mentioned before, one of the parameters of industrialization is to examine the process of inventions. The analysis of patents related to MOF electrodes in SCs involves several key steps (Fig. 21.17A–C). Initially, about 135 patents were screened, resulting in 113 patents relevant to the subject. Time and cumulative patent filing processes were analyzed to understand the trend in patent filings related to MOF electrodes in SCs. The filing trend began in 2009, with an upward trend starting in 2013 and continuing until 2018. The cumulative filing process confirms this growth trend, with the fastest growth

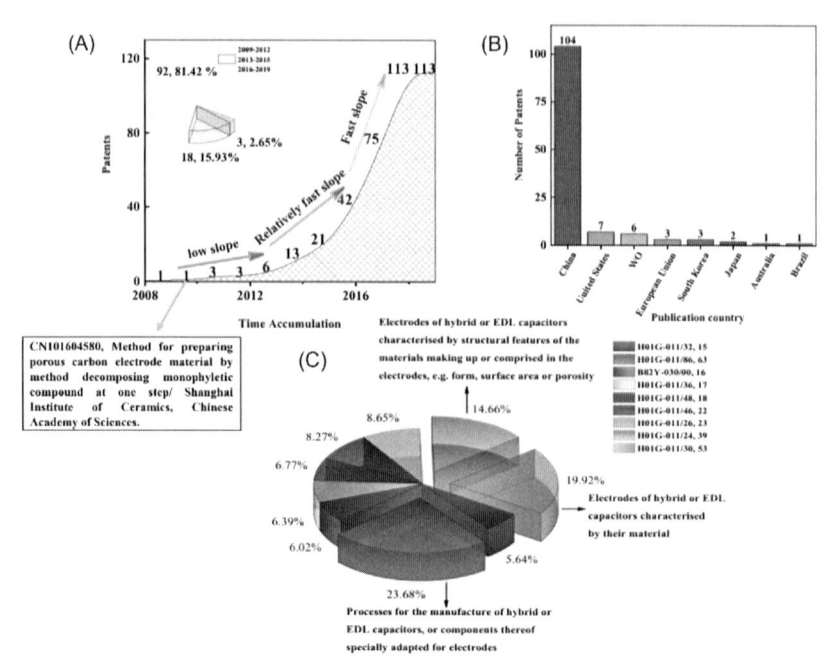

FIGURE 21.17 Patent analysis of metal-organic frameworks in supercapacitors, focusing on: (A) Time trends, (B) Geographic distribution by country, and (C) Application fields (Boorboor Ajdari et al., 2020).

observed from 2016 to 2019, during which about 92% of all patents in this field were filed. The legal status of the 113 filed patents indicates that approximately 27% of them (30 patents) were granted by at least one organization, reflecting their validity and high technical knowledge. Additionally, 92% of all filed patents (104 patents) are legally protected in at least one patent organization, signifying the novelty and recent focus on this technological field. Only 8% of filed patents are inactive due to lapsing by the applicants or being revoked. Among the surveyed patents, 86 legal owners from eight different countries were identified. China holds the majority of patents in this field, with 104 patents, followed by the United States with six patents. Other countries have fewer than two patents each, indicating China's significant role in patenting and competing in MOF applications for SCs. The top patent owners in terms of filing more than two patents are primarily knowledge-based institutes, with Fuzhou University leading with four patents. The top inventors with more than two patents are also predominantly from China, highlighting China's dominance in technical knowledge and patent filings in this field. Moreover, patents with high citation counts indicate their technical strength and importance. Several patents have received more than four citations, underscoring their significance in the field. For instance, patent CN101604580, which describes a method for producing porous carbon electrode materials for SCs, has 18 citations, demonstrating its importance and validity in the field. Overall, these findings underscore China's prominent role in patenting and advancing MOF applications in SCs while also highlighting the importance of certain patents in driving innovation and technological advancements in this domain.

The technical classification of patents filed in the realm of MOF electrodes for SCs reveals significant insights into the innovation landscape. Analyzing IPC codes shows a prevalence of manufacturing processes tailored for hybrid or EDL capacitors and their components, representing approximately 23% of all patents. Key metals like Co^{2+}, Ni^{2+}, Fe^{3+}, Cu^{2+}, and Zn^{2+}, along with commonly used ligands such as terephthalic acid and 1,4-phenylenediacetic acid, underscore the diverse approaches in MOF synthesis. Additionally, the wide use of inorganic electrolytes like KOH and H_2SO_4 highlights their accessibility and cost-effectiveness, while ionic liquids show promise for their high conductivity and other favorable attributes (Boorboor Ajdari et al., 2020).

The use of bilayer zwitterionic nanochannels (MOF-BZN) as solid-state electrolytes (SSEs) in solid-state batteries (SSBs) represents a groundbreaking innovation. MOF-BZN structures, featuring a rigid anionic MOF channel with chemically grafted soft multicationic oligomers (MCOs), enable selective superionic conduction. This design enhances Li^+ conductivity (8.76×10^{-4} S/cm) and achieves a high Li^+ transference number (0.75), with an electrochemical window of up to 4.9 V at 30°C. In practical use, SSBs employing flame-retarded MOF-BZN attain remarkable specific energy (419.6 Wh/kg$_{anode+cathode+electrolyte}$) under stringent conditions such as high cathode loading and limited lithium metal sources. This underscores MOF-BZN's potential as an SSE for highly efficient SSBs. Overall, integrating bilayer zwitterionic MOFs into SSBs marks a significant advancement, offering improved performance and safety for diverse applications (Ouyang et al., 2023).

A novel approach in developing SCs involved utilizing a bimetal oxide composite Zn–Co–O@CC and nanoporous carbon derived from bimetallic metal-organic frameworks (BMOFs) as positive and negative electrodes, respectively (Fig. 21.18A–C). This innovation enabled the creation of SCs with a wide operating voltage window of 0–2.0 V, a departure from conventional usage where BMOFs serve as positive electrodes. Remarkably, the resulting all-solid-state and flexible SCs, employing a polyvinyl alcohol/lithium-chloride gel electrolyte, exhibited robust performance even under bending conditions. Multiple flexible SCs connected in series demonstrated the capability to power LED lights and motor fans for extended durations (Javed et al., 2019). Notably, the highest reported performance involved Cu(NiCo)$_2$S$_4$/Ni$_3$S$_4$ connected in series, successfully illuminating a yellow LED for 30 min (Zhao, Yan et al., 2020). These findings underscore the

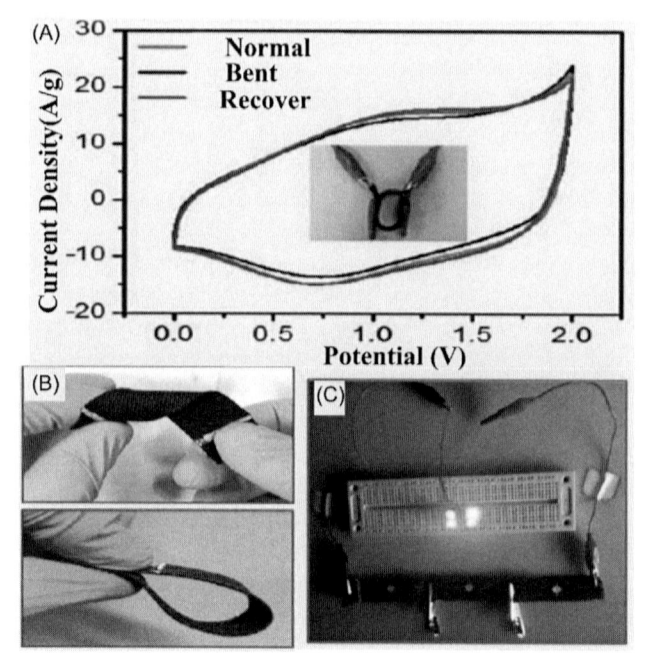

FIGURE 21.18 Summary of the merits of Zn–Co–O@CC//NPC@CC ASC: (A) CV curves of Zn–Co–O@CC//NPC@CC recorded under different bending conditions; the inset is the photograph taken during the flexibility test, (B) Digital photographs of the ASC bent at various positions demonstrating excellent flexibility without deformations in the structure of the device, and (C) Digital photograph of red LEDs illuminated by three ASCs connected in series (Javed et al., 2019).

potential of BMOF-based materials in advancing energy storage device technology.

The semiconductor industry has experienced significant advantages through the adoption of adsorbent-based storage and delivery systems for ultra-high-purity gases, which are integral to its operations. NuMat's ION-X product line represents a breakthrough in chemical delivery systems, leveraging MOF adsorbents to safely store, stabilize, and release hazardous electronic gases like arsine (AsH_3), phosphine (PH_3), and isotopically-enriched boron trifluoride ($^{11}BF_3$) on demand. Integrated into specialty gas cylinders, ION-X provides high-capacity, customized solutions for semiconductor fabs, reducing the health and environmental risks associated with gas releases. The higher surface area and design flexibility of MOFs offer advantages over traditional carbon adsorbents, allowing precise tuning of pore sizes to match gas molecules, enhance adsorption capacities, and stabilize self-reacting gases. NuMat's integrated workflow enabled the development of ION-X in approximately 18 months, from computational design to product qualification, showcasing the rapid advancement in material innovation. A global alliance with Versum Materials facilitated the distribution of ION-X products to semiconductor manufacturers worldwide (Fig. 21.19) (Arnó et al., 2018a, 2018b, 2022; Kerkel et al., 2018).

Sensor and biosensor

Accurate compound detection is crucial across industries such as medicine, environmental monitoring, food production, and forensic science. Whether diagnosing diseases, ensuring food safety, or investigating legal cases, precise detection methods are essential for maintaining standards and safeguarding health and quality. Compared to traditional porous materials like carbon, zeolites, and porous silicon, MOFs offer unique advantages, including customizable structures, well-ordered pores, and straightforward synthesis methods (Fang et al., 2018; Kajal et al., 2022; Zhang et al., 2021). These distinctive characteristics have positioned MOFs as promising candidates across various applications, notably in the development of chemical sensors for analytical purposes. The tunable nature of MOF structures allows for the incorporation of diverse chemical properties, facilitating the detection of changes induced by introduced targets, including luminescence and electrochemical catalytic activity. Additionally, MOFs serve as excellent matrices for immobilizing functional substances like noble metal nanoparticles and dye molecules, resulting in functionalized MOF composites with versatile chemical and physical properties. Furthermore, the crystalline structure of most MOFs enables preparation under mild synthesis conditions, broadening their utility in sensor applications. Despite the extensive exploration of MOF-based chemical sensors, their limited chemical selectivity has led to the integration of biomolecules with recognition functions into MOF-based sensors.

846 Applications of Metal-Organic Framework Composites

FIGURE 21.19 ON-X specification sheet from Versum Materials website (Chen et al., 2021).

Biomolecules such as enzymes, nucleic acids, antibodies, peptides, and phages offer specific target recognition, thereby enhancing sensor sensitivity and expanding the range of detectable targets. The integration of these biomolecules with MOFs to form MOF biocomposites represents a promising strategy for improving sensor performance and diversifying target species detection. These biomolecules offer high sensitivity and selectivity in recognizing diverse targets, while MOFs provide stability and reusability as an immobilization support. The resulting biocomposites enable the fabrication of various sensors, including colorimetric, fluorescent, and electrochemical sensors. Given their advantages, a significant number of MOF biocomposites have been

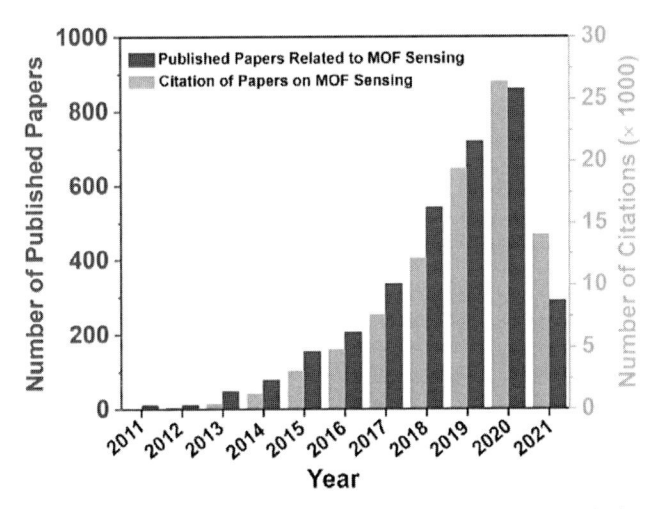

Metal-organic framework composites: industrial applications **Chapter | 21** **847**

FIGURE 21.20 The number of published papers with their corresponding citations on topics related to metal-organic framework sensing in the 2011 to 2021 period (Olorunyomi et al., 2021).

synthesized and categorized into enzyme-MOF, nucleic acid-MOF, antibody-MOF, and other biomolecule-MOF types. Considerable efforts have been directed toward developing simple and cost-effective synthesis methods for the production of biomimetic composites (Amini et al., 2020; Qiu et al., 2019; Wang et al., 2021).

Fig. 21.20 illustrates the publication trends and corresponding citations related to MOF sensing topics over the past decade (Olorunyomi et al., 2021). The number of articles published on the sensing applications of MOFs has increased steadily, with citations following a similar trajectory. The persistent rise in citations underscores the considerable potential of MOFs as sensitive materials for detecting diverse chemical and biological signals. Consequently, there is growing interest in developing MOF-based analytical devices capable of providing real-time information on specific substances in complex samples. Industries are shifting toward more affordable, user-friendly diagnostic systems that do not require complex equipment or specialized expertize. Luminescence sensors, particularly those employing "Turn-on" processes, smartphone sensors, point-of-care testing sensors, and wearable diagnostic sensors are gaining traction for their accessibility and ease of use. MOFs have emerged as key players in these areas, significantly contributing to advancements in diagnostic technology within the industry.

"Turn-on" processes

MOFs exhibit photoluminescence behavior, which arises from their metal ions, linker molecules, or entrapped guest molecules within their cavities. This

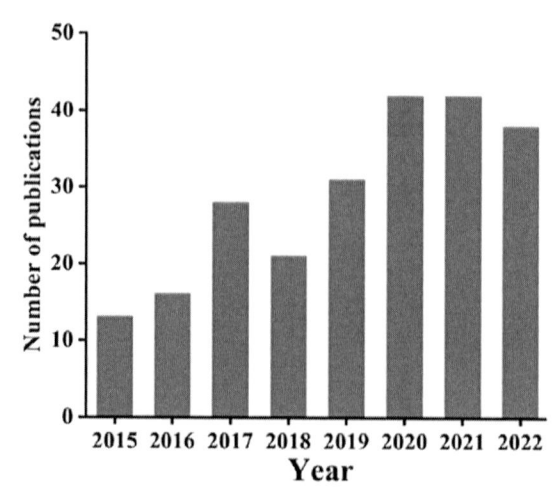

FIGURE 21.21 "Turn-on" sensor application by metal-organic frameworks for the recognition of diverse analytes (Pal, 2022).

tunable luminescent property enables MOFs to serve as sensors for the selective and sensitive detection of various analytes. Typically, the sensing of analytes by MOFs involves either a fluorescence "turn-on" process, where fluorescence intensity increases, or a "turn-on" process, where fluorescence intensity decreases. Among these, the "turn-on" process is more significant for real-time applications because it provides a clear signal against a dark background (Karmakar et al., 2019). Fig. 21.21 depicts a SciFinder survey illustrating the increasing trend in the number of "on" MOF-based sensors for detecting various analytes from 2015 to 2022. The graph indicates a gradual rise in the utilization of MOFs as "turn-on" sensors over the specified timeframe, highlighting the growing interest and development in this area of research (Pal, 2022).

In Table 21.3 certain MOF structures are outlined along with their respective detection limits and wavelengths.

Smartphone sensors

MOF-based chemical sensing can now be seamlessly integrated with smartphones, utilizing their built-in sensors and microprocessors. For instance, a ZIF-8/TiO$_2$ diffraction grating film, featuring a 200 nm structure replicated across 1 cm² of a substrate, was detected by a smartphone's charge-coupled device (CCD) camera. This setup monitored changes in diffraction efficiency resulting from variations in the refractive index of the ZIF-8/TiO$_2$ film caused by exposure to styrene and isopropanol vapors. The smartphone's CCD camera captured the changes in luminance of the film after vapor exposure, enabling

TABLE 21.3 Diverse metal-organic frameworks with their excitation, emission wavelength and respective LOD (limit of detection) values for the recognition of various analytes (neutral molecules, solvent molecules, biomarkers, biomolecules, hazardous compounds, and high-energy molecules) (Pal, 2022).

MOFs	metal in node/SBU	$\lambda_{excitation}$ (nm)	$\lambda_{emission}$ (nm)	Nature of analyte	LOD
IRMOF-3(-N$_3$)	Zn	395	430	H$_2$S	28.3 µM
UiO-66@NO$_2$	Zr	334	436	H$_2$S	188 µM
CAU-10-N$_3$	Al	330	405	H$_2$S	2.65 µM
Tb^{3+}@Cu-MOF	Tb	280	489, 544, 585 and 620	H$_2$S	1.20 µM
Fe^{III}-MIL-88-NH$_2$	Fe	333		H$_2$S	10 µM
DUT-52-(NO$_2$)$_2$	Zr	390	474	H$_2$S	20 µM
Zr(TBAPy)$_5$(TCPP)	Zr	365		H$_2$S gas and its derivatives S^{2-}	1 ppb
Al-MIL-53-NO$_2$ MMMs	Al	396	466	H$_2$S	92.31 µM
MOF-5-NH$_2$	Zn	365		SO$_2$ gas	0.168 ppm
Al-MIL-53-NO$_2$	Al	339		H$_2$S	69.3 µM
Sm-MOF@Fe^{3+}	Sm	300	562, 595, 643 and 706	TBHQ	5.6 ng/mL
UiO-66-MA	Zr			H$_2$S	3.3 nM

(Continued)

TABLE 21.3 (Continued)

MOFs	metal in node/SBU	λ excitation (nm)	λ emission (nm)	Nature of analyte	LOD
RhB/UiO-66-N$_3$	Zr	302	425 and 575	H$_2$S	29.9 µM
Eu^{3+}/Cu^{2+}@Znpda	Zn	408	614 and 386	H$_2$S	1.45 µM
$^3_\infty$[Ce (Im)$_3$ImH]·ImH	Ce	366	422	H$_2$O, O$_2$, and CH$_2$Cl$_2$	–
[Sm$_2$(abtc)$_{1.5}$(H$_2$O)$_3$(DMA)]·H$_2$O·DMA	Sm	400	612	Benzyl alcohol and benzaldehyde	–
[Me$_2$NH$_2$][EuL(H$_2$O)]·1.5H$_2$O	Eu	324	591, 613, 649 and 697	Methanol	–
SB1, SB2, SB3 and SB4	Cd	365	463 and 493	Water	–
Zn(hpi$_2$cf)(DMF)(H$_2$O)	Zn	365		Water	<0.05% v/v (ZnO-based film)
QC-loaded MOF	Zn	365	424 nm	Water in ethanol and DMF	0.015% and 0.030%
SXU-4	Cd	365	430	DMSO	–
{[Y$_{0.9}$Tb$_{0.1}$Mn$_{1.5}$(PDA)$_3$(H$_2$O)$_3$]·3.5H$_2$O}	Y and Mn	280	315,491, 545, 586 and 623	Trace water in EtOH, CH$_3$OH, CH$_3$CN, THF and n-heptane	1.12% (v/v), 0.47% (v/v), 0.04% (v/v), 0.13% (v/v), and 0.53% (v/v)
[(UO$_2$)(H$_2$DTATC)] (HNU-39)	U	373	430		–
Tb^{3+}@Ag-BTC	Ag	365	494, 547, 587 and 622	HCHO	1.9 mM
[In(BDC-NH$_2$)(OH)]$_n$	In	350	429	H$_2$O$_2$	0.42 µM
Zr-UiO-66-B(OH)$_2$	Zr	328	425	H$_2$O$_2$	–

PSM-1 and PSM-2	Cd	335	~435 and ~430	1,4-Dioxane	1.079 and 2.487 ppm
MIL-100(In)@Eu^{3+}/Cu^{2+}	In	285	594, 619, and 699	H$_2$S	0.535 ppm
Fe-MIL-88	Fe	326	445	GSH, Cys and Hcy	30, 40, and 40 nM
Fe-MIL-88-H$_2$O$_2$-o-phenylenedia-mine	Fe	433	576	Dopamine	46 nm
[Cu(mal)(bpy)]·2H$_2$O	Cu	480	518	Glycine and serine	0.81 and 1.51 µg/mL
[Cd(µ$_3$-abtz)·2I]$_n$	Cd	242	350	Dopamine	57 nM
Ag@Au nanoprism-MOF	Zn	532	550	Glucose	0.038 mM
CD-MONT-2	Pb	330	526	Uric acid	4.3 µM
Cu-BTC/Tb	Cu	238	488, 545, 583, and 621	Amyloid β-peptide (Aβ1-40)	0.3 nM
RhB@Cu-BTC	Cu	365	575	L-Cysteine	0.702 µmol/L
Zr-MOF; UiO-68-An/Ma	Zr	375	420	Biothiols (Cys, Hcy, and GSH)	Low concentration of 50 µmol/L
ZJU-108	Zn	323	419	Tryptophan	42.9 nM
NH$_2$-MOF-76(Eu)	Eu	385	400	Dipicolinic acid	3.8 mM
CrO$_4$$^{2-}$@Cd-MOFs	Cd	358	414	Ascorbic acid	7.27 ppm
ZIF-67	Co	540	585	Biothiols	31 nM
Eu^{3+}@Ga-MOF(MIL-61)	Ga	370	614	Ciprofloxacin	2.4 µg/mL

(Continued)

TABLE 21.3 (Continued)

MOFs	metal in node/SBU	$\lambda_{excitation}$ (nm)	$\lambda_{emission}$ (nm)	Nature of analyte	LOD
RSPh@EuBTC	Eu	306	546, 592, 615 and 692	DPA	0.52 µM
Eu^{3+}@[(Me)$_4$N]$_2$[Pb$_6$K$_6$(m-BDC)$_9$(OH)$_2$]·H$_2$O	Pb and K	377	612	Fleroxacin	43.91 ng/mL
UiO-67-sbdc	Zr	340	465	Glutathione (GSH)	107.2 µM
Eu^{3+}/Cu^{2+}@UiO-67-bpydc	Zr	372	592, 615, 651 and 700	GSH	54.3 nM
Cu(HBTC)-1 nanosheets	Cu		518	Oxytetracycline	0.40 µg/L
[(CH$_3$)$_2$NH$_2$]$_3$[(In$_3$Cl$_2$) (bpdc)$_5$]·(H$_2$O)$_5$(DMF)$_{2.5}$	In	340	410	Cysteine and 1-butanethiol	5.71×10^{-4} and 4.38×10^{-4} M
[Tb$_{0.43}$Eu$_{1.57}$(1,4-phda)$_3$(H$_2$O)] (H$_2$O)$_2$	Tb and Eu	260	544 and 614	Dipicolinic acid	5.9 nM
FCS-3	Zn	400	450	Fluoroquinolone	0.52 µM
UiO-HQCA-Al	Zr	339	438	Creatinine	4.7 nM
Eu-in-BTEC	Eu	365	526 and 617	Doxycycline	47 nM
Tb-MOF	Tb	394	506	Dicarboxylic acid	2.4 µM
Eu@ZIF-90-PA	Zn	463	579	Flumequine	0.24 ppm

open access=creativecommons CC-BY.

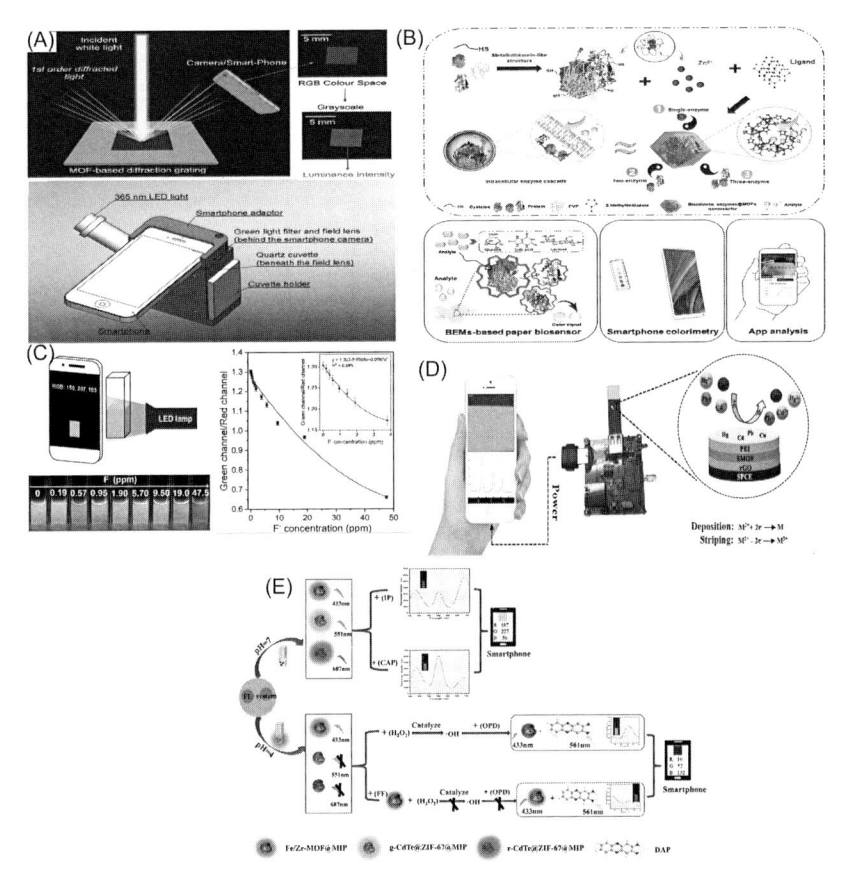

Metal-organic framework composites: industrial applications **Chapter | 21** 853

FIGURE 21.22 Illustration of the experimental set-ups for the detection using smartphone cameras: (A) Set-up for the detection by a ZIF-8/TiO$_2$ (Dalstein et al., 2016), (B) Schematic of smartphone-assisted biomimetic enzymes@MOFs (BEMs) based on colorimetry strategy (Kou et al., 2020), (C) Schematic diagram for the detection of F$^-$ using luminescent color changes Ln-MOFs probe with different concentrations of F$^-$ (Zeng et al., 2020), (D) Schematic representation of the quantitative POCT system and the simultaneous detection of multiple metal ions by the rGO/SMOF/PEI modified SPCEs (Xu et al., 2020), and (E) Schematic diagram tri-color fluorescence sensing platform for dual-mode visual intelligent detection of ibuprofen, chloramphenicol and florfenicol (Tang et al., 2023).

chemical sensing results comparable to those from benchtop ellipsometers (Fig. 21.22A) (Dalstein et al., 2016).

Additionally, smartphones can leverage their built-in light and image recognition capabilities for spectroscopic analysis. Various MOF-based sensing platforms have exploited these features, including colorimetric sensing, where smartphones capture color changes in MOF films upon interaction with analytes. Kou et al. combined a smartphone with their biomimetic MOF colorimetric paper for detecting biomolecules like glucose, uric acid, lactose, and urea, demonstrating its potential for point-of-care diagnostics (Fig. 21.22B) (Kou et al., 2020). Moreover, smartphone-based spectroscopic approaches, such as fluorescent detection, offer enhanced sensitivity. Zeng et al. achieved ppm-level detection of fluoride ions using a smartphone to capture emitted light from mixed lanthanide MOFs, enabling further detection in water samples. Other techniques, such as the excitation of fluorescent MOFs by UV LEDs and subsequent detection by smartphone light sensors, have also been demonstrated (Fig. 21.22C) (Zeng et al., 2020). In electrochemical sensing, smartphones serve as miniature analyzers when attached to specialized modules. Xu et al. developed a smartphone-controlled electrochemical sensor for detecting heavy metal ions using a composite of UiO-66-NH$_2$ and single-walled carbon nanotubes test strips (Fig. 21.22D) (Xu, Liu et al., 2020). Tang's team developed a novel sensing platform that integrates acid-sensitive fluorescence-imprinted polymers with a smartphone. The platform detects ibuprofen, chloramphenicol, and florfenicol using Fe/Zr-MOF@MIP, g-CdTe@ZIF-67@MIP, and r-CdTe@ZIF-67@MIP. The optimized sensor enables visual detection across various pH levels and can analyze RGB values (R: red channel, G: green channel, B: blue channel) for detection in water, meat, and urine, offering a promising approach for multitarget substance detection in real-world settings (Fig. 21.22E) (Tang et al., 2023).

Point-of-care testing sensors

These systems provide a straightforward means to analyze vital biological parameters, incorporating both smartphone-based and wearable technologies. In this section, we will explore additional methods, starting with paper-based systems such as lateral flow assays.

The nanocomposite Ab-Au/Ir@Cu/Zn-MOF serves as both a detection probe and an antibacterial agent. It provides high sensitivity in detecting *Staphylococcus aureus* and achieves complete sterilization at low concentrations through synergistic photothermal and chemodynamic effects (Fig. 21.23A) (Zhong et al., 2023). Similarly, MIL-101(Fe) demonstrates high sensitivity in detecting clinical transglutaminase 2 (TGM2) in urine, with a detection limit of 0.012 nM in Tris–HCl buffer. It outperforms Au NPs-based lateral flow assays by approximately 55-fold. This method offers rapid detection, excellent reproducibility, and stability, suggesting potential for TGM2 measurement in urine samples from both normal individuals and chronic kidney disease (CKD) patients (Supianto et al., 2024).

Additionally, a Cobalt-MOF-modified carbon cloth (CC)/paper hybrid electrochemical button sensor was developed for nonenzymatic glucose diagnostics. This compact and user-friendly electrochemical analytical chip (Co-MOF/CC/Paper) significantly enhances the specific surface area and catalytic sites compared to standard flat electrodes (Fig. 21.23B) (Wei et al., 2021). A transparent,

Metal-organic framework composites: industrial applications **Chapter | 21** **855**

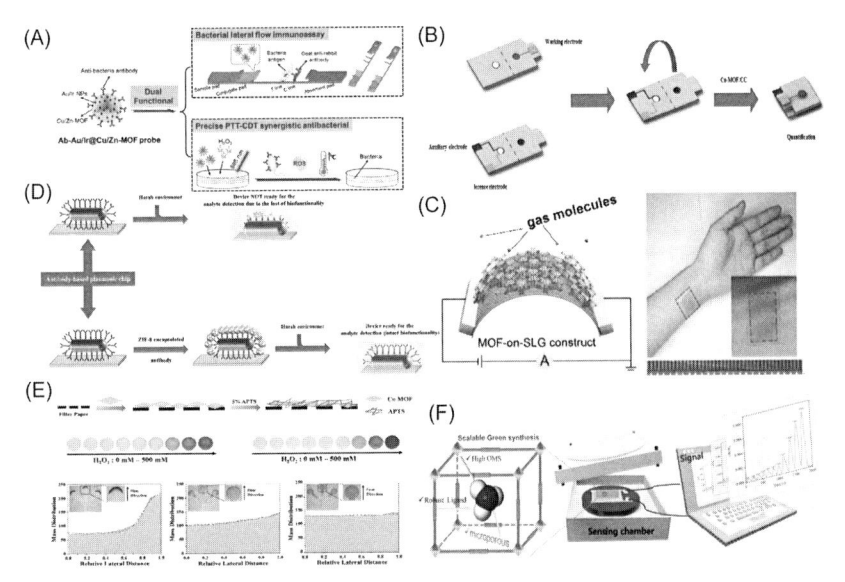

FIGURE 21.23 Several point-of-care testing sensors based on metal-organic frameworks (MOFs): (A) Schematic illustrations of Ab-Au/Ir@Cu/Zn-MOF probe-based bacterial lateral flow immunoassay (Zhong et al., 2023), (B) Schematic illustrations of Cobalt-MOF modified carbon cloth/paper hybrid electrochemical button-sensor for nonenzymatic glucose diagnostics (Mansour et al., 2024; Wei et al., 2021), (C) Schematic illustrations of Ni-CAT-1-on-SLG construct on transparent flexible substrates for personal electronics (Wu et al., 2020), (D) Schematic illustration of depicting that ZIF-8 encapsulated antibody (Mansour et al., 2024; Wang et al., 2018), (E) Diagrammatic sketch of the preparation of Ce-MOF anchored paper) (Luan et al., 2020), and, (F) Schematic illustration of stable microporous MOFs for NH_3 sensing (Wang et al., 2023).

electronic, and flexible Ni–MOF film was developed for detecting NH_3, CO, and O_2. The nickel catecholate (Ni-Cat-1) MOF was grown epitaxially on single-layer graphene (SLG) coated on PET and polydimethylsiloxane substrates, with controlled thickness to maintain transparency while ensuring flexibility (Fig. 21.23C) (Wu et al., 2020). Biochips with MOF coatings were developed for point-of-care testing (POCT), employing a plasmonic nanobiosensor with gold nanorods (AuNRs) as transducers. The MOF-coated biochips demonstrated resilience to adverse conditions and maintained antibody integrity and functionality (Fig. 21.23D) (Wang, Wang et al., 2018). The Ce-MOF-based OPSlipChip to enable selective detection of uric acid and glucose, thereby creating a point-of-care testing platform (Fig. 21.23E) (Luan et al., 2020). Fig. 21.23F shows that MOFs were used to develop a fast-response NH_3 sensor with low detection limit (25 ppb) and short response time (5 seconds) at room temperature. The sensor showed excellent selectivity, stability for over two months, and easy recovery by purging with nitrogen (Wang et al., 2023). Numerous research endeavors are underway in this field, listed in Table 21.4 (Olorunyomi et al., 2021).

TABLE 21.4 Applications of various field-deployable metal-organic framework composites devices to sense important biological, organic and inorganic analytes (Olorunyomi et al., 2021).

Application	Sensor	Analyte	Sensing platform
POC diagnostics	Co-MOF integrated carbon cloth/paper hybrid electrochemical button-sensor	Glucose	Electrochemical
POC diagnostics	Smartphone assisted rGO/SMOF/PEI	Heavy metals	Electrochemical
POC diagnostics	Acetylcholinesterase/Zn-MOF EμAD	Chlorpyrifos	Electrochemical
POC diagnostics	BC/e-MWCNTS/ZIF-8@LAC biofuel cell sensor	Bisphenol A	Electrochemical
POC diagnostics	CA/ZIF-8@LAC/MWCNTS/Au biofuel cell sensor	Glucose	Electrochemical
POC diagnostics	ACF-rGO/Cu(INA)$_2$ sensor	Glucose from sweat	Electrochemical
POC diagnostics	Trx-1 and ADAM17cyto-ZIF-8 IDE device	Protein-protein (ADAM17cyto) interactions	Electrochemical
POC diagnostics	Smartphone assisted HKUST-1-MoS$_2$ face mask sensor	Sleep apnea diagnosis	Electronic
POC diagnostics	ZIF-8@antibody plasmonic biochip	Goat anti-rabbit IgG	Optical
POC diagnostics	ZI-8@antibody plasmonic biochips	Neutrophil gelatinase associated lipocalin (NGAL)	Optical
POC diagnostics	Smartphone-assisted Ln^{3+}-MOF sensor	Phenylamine and 1-naphthol in human urine	Optical
POC diagnostics	Smartphone-assisted 1-OHP@CO/Tb-DPA sensor	pH	Optical
POC diagnostics	Smartphone-assisted 2-D Co-MOF sensor	Blood glucose	Optical
POC diagnostics	GOx-Eu^{3+}@UMOF Logic Detector	Glucose	Optical

(Continued)

TABLE 21.4 (Continued)

Application	Sensor	Analyte	Sensing platform
POC diagnostics	GOx@Zr-PCN-222(Fe) microfluidic μPADs	Glucose	Optical
POC diagnostics	Smartphone-assisted GOx/HRP@ZIF-8 paper biosensor	Glucose	Optical
POC diagnostics	Smartphone-assisted μPADS	Glucose	Optical
POC diagnostics	GOX/HRP@ZIF-8 microfluidic biosensor	Glucose	Optical
POC diagnostics	Smartphone-assisted NH_2-Cu-MOF sensors	Alkaline phosphatase activity	Optical
POC diagnostics	Apt/HRP@MAF-7 colorimetric device	Antibiotics	Optical
POC diagnostics	Smartphone-assisted Eu^{3+}/Sc-MOF	Phenylglyoxylic acid	Optical

SMOF, Nanocomposite consisting of single walled carbon nanotubes; PEI, polyethyleneimine; BC, bacterial cellulose; LAC, laccase; Cu(INA)₂, copper isonicotinate MOF; Apt, aptamer; HRP, horseradish peroxidase; ADAM17cyto, cytoplasmic domain of a disintegrn and metalloproteinase 17; μPADS, microfluidic paper-based analytical device.

Wearable diagnostic sensors

Wearable sensors are increasingly popular due to their exceptional performance and versatile applications, with market demand growing from $20 billion in 2016 to $80 billion in 2022. These sensors are used for tasks such as virus detection, cancer diagnosis, and pulse rate analysis. Integrating the sensing matrix into the working electrode of electrochemical sensors is crucial, with MOFs emerging as a promising matrix due to their sensitivity and versatility (Meskher et al., 2023). The most recent wearable sensor based on MOFs and MOF composites for healthcare applications is listed in Table 21.5.

In this field, a nonenzymatic electrochemical sensor for perspiration glucose analysis uses Pd NPs in a Co-based zeolitic imidazolate framework (ZIF-67). Operating without additional reagents under physiological pH, it enables long-term monitoring. Integrated into a sweatband with a flexible printed circuit board (FPCB), it offers real-time analysis and correlates well with commercial glucose meters. With stable sensitivity for up to 2 months

TABLE 21.5 Most recent wearable sensor based on metal-organic frameworks and metal-organic framework composites for health-care applications (Meskher et al., 2023).

Article title	MOF	Application and analytical performance	Outcomes
Design and fabrication of novel flexible sensor based on 2D Ni-MOF nanosheets as a preliminary step toward wearable sensor for onsite Ni (II) ions detection in biological and environmental samples	Ni-MOF Metal: Ni Ligand: Benzene-1,4-dicarboxylic acid (BDC)	Determination of nickel ions in biological and environmental samples. LR=1.0×10^{-5}–1.0×10^{-1} mol/L LOD=2.7×10^{-6} mol/L	• The prepared sensor exhibited excellent performance and fast response to nickel ions detection. • A lower LOD around 2.7 μM has been obtained. • The sensor exhibited great recoveries in real samples such water human saliva, sweat and tap water.
Highly stretchable wearable electrochemical sensor based on Ni–Co MOF nanosheet-decorated Ag/rGO/PU fiber for continuous sweat glucose detection	Ni–Co-MOF Metal: Ni–Co. Ligand: Benzene-1,4-dicarboxylic acid (BDC)	Detection of glucose in sweat Sensitivity=425.9 μA/(mM·cm^2) LR: 10 μM–0.66 mM.	• The fabricated electrode exhibited excellent electrochemical performance with high stretchability. • The presence of rGO in the sensing matrix has significantly improved the performance of the (Ni–Co MOF/Ag/rGO/PU) toward glucose detection with a sensitivity around 425.9 μA/(mM·cm^2). • The wearable sensor exhibited outstanding mechanical flexibility and stretching.
MOF-derived porous Ni/C material for high-performance hybrid nanogenerator and self-powered wearable sensor	Ni-MOF Metal: Ni–Co Ligand: Benzene-1,4-dicarboxylic acid (BDC)	Motion monitoring	• A hybrid nanogenerator integrated by a two-layer zigzag triboelectric nanogenerator and an EMG, has been fabricated. • The sensor harvested the mechanical energy generated from human walking for self-powered personnel positioning. • The effect of the TENG structure on the output is analyzed and showed potential for hand motion and gait monitoring.

Fabrication of a sensitive and fast response electrochemical glucose sensing platform based on Co-based metal-organic frameworks obtained from rapid in situ conversion of electrodeposited cobalt hydroxide intermediates	Co₃(BTC)₂ MOFs Metal: Co Ligand: Benzene-1,3,5-tricarboxylic acid (H₃BTC, C₉H₆O₆)	Glucose detection in human blood LR=1 μM–0.33 mM LOD=0.33 μM.	• A three steps cost-effective and eco-friendly synthesis procedure has been successfully proposed to synthesis crystalline Co₃(BTC)₂ MOFs. • The sensing matrix exhibited excellent performance toward glucose detection and provided outstanding analytical parameters.
Smartphone light-driven zinc porphyrinic MOF nanosheets-based enzyme-free wearable photoelectrochemical sensor for continuous sweat vitamin C detection	Zn-MOF Metal: Zn Ligand: Biphenyl-4,4'-dicarboxylic acid (BPDC)	Detection of vitamin C in human sweat. LR=10–1100 Mm. LOD=3.61 μM.	• A smartphone-connected enzymatic sensor to detect vitamin C in human sweat has been successfully developed. • The sensor exhibited great potentianl to ensure proper nutritional balance. • The non'enzymatic sensing platform is constructed by a two-dimensional zinc porphyrinic MOF nanosheets/multi-walled carbon nanotubes (2D-TCPP(Zn)//MCNTs).
A highly flexible Ni-Co MOF nanosheet coated Au/PDMS film based wearable electrochemical sensor for continuous human sweat glucose monitoring	Ni-Co-MOF Metal: Ni-Co Ligand: Benzene-1,4-dicarboxylic acid	Glucose Detection in sweat LR= 20 μM–790 μM. Sensitivity=205.1 μA/(mM·cm²).	• A stretchable sensor based on Ni–Co-MOF is prepared. • Operational parameters such as Ni:Co ratios were optimized. • The sensor exhibited excellent performance once it is attached to the skin to effectively detect glucose in sweat. • The sensing matrix exhibited excellent performance toward glucose detection and provided outstanding analytical parameters with a high sensitivity of 205.1 μA/(mM·cm²).

(Continued)

TABLE 21.5 (Continued)

Article title	MOF	Application and analytical performance	Outcomes
Fluorescent wearable platform for sweat Cl⁻ analysis and logic smart-device fabrication based on color adjustable lanthanide MOFs	DUT-101 Metal: Tb Ligand: Biphenyl-4,4'-dicarboxylic acid (BPDC)	Cl⁻ ions detection in sweat LOD=0.1 mM	• A fluorescence wearable sensor to detect Cl⁻ ions in human sweat is proposed. • The sensor exhibited lower limits of detection and quantification with high sensitivity and excellent selectivity to ward Cl- ions. • This system is a simple and effective solution for wearable sweat-based monitoring.
Chiral MOF derived wearable logic sensor for intuitive discrimination of physiologically active enantiomer	Chiral γ-cyclodextrin metal-organic framework (CDMOF)	Lactate enantiomers	• A dual responsive chiral sensor RT@CDMOF through in situ self-assembly of chiral γ-cyclodextrin MOF was prepared. • The embedded RGH and TCN inherit the chirality of host CDMOF, producing dual changes both in fluorescence and reflectance. • A flexible membrane sensor is successfully fabricated based on RT@CDMOF for wearable health monitoring. • Based on above, a chiral implication logic unit can be successfully achieved, demonstrating the promising potential of RT@CDMOF in design and assembly of novel smart devices.

Ultra-thin 2D bimetallic MOF nanosheets for highly sensitive and stable detection of glucose in sweat for dancer	NiMn-MOF Metal: Ni-Mn Ligand: Benzene-1,3,5-tricarboxylic acid (H_3BTC)	Glucose Detection in sweat Sensitivity=1576 µA/(mM·cm²) R=0–0.205 mM. LOD=0.28 µM.	• Accurate, fast response and sensitive wearable sensor during dancing is successfully fabricated. • The sensing matrix is based on a bimetallic NiMn-MOF and exhibited excellent catalytic activity. • The ultrathin nanosheet and heterogeneous metal ions in the structure optimize the electronic structure, which improves the electrical conductivity of MOFs. • The sensing matrix exhibited excellent performance toward glucose detection and provided outstanding analytical parameters with a high sensitivity of 1576 µA/(mM·cm²)
A wearable sweat electrochemical aptasensor based on the Ni–Co MOFnanosheet-decorated CNTs/PU film for monitoring of stress biomarker	Ni–Co-MOF Metal: Ni–Co Ligand: 2-Aminobenzene-1,4-dicarboxylic acid (NH_2-H_2BDC).	Cortisol detection LR=0.1–100 ng/mL. LOD=0.032 ng/mL.	• A three steps cost-effective and eco-friendly synthesis procedure has been successfully proposed to synthesis MOF/CCP. • The sensor exhibited excellent analytical performance toward cortisol with high sensitivity and with high repeatability and good selectivity. • The sensor exhibited promising potential for quantitative stress monitoring and management. • The fabricated sensor provided outstanding sensitivity due to the formation of the aptamer-cortisol complex.

open access=creative commons CC-BY.

under ambient conditions, it promises convenient wearable glucose monitoring for both clinical and sports applications (Fig. 21.24A) (Zhu et al., 2019). Recently, a novel wearable piezoelectric sensor utilizing PVDF nanofibers embedded with microporous zirconium-based MOFs has enhanced arterial pulse monitoring. This sensor achieves a peak-to-peak voltage of 600 mV under a 5 N force, outperforming traditional pressure sensors. Tested on the radial artery, the sensor exhibits superior output voltage (568 ± 76 mV) and sensitivity (0.118 V/N) compared to PVDF devices used for wrist pulse monitoring. This innovation introduces flexible piezoelectric nanofibrous sensors for sustainable energy generation and wearable healthcare monitoring (Fig. 21.24B) (Moghadam et al., 2020).

To collect local sweat effectively, the system incorporates iontophoresis capabilities, with the sensing area sealed to facilitate sweat collection using iontophoresis. The carefully constructed iontophoresis electrodes achieve

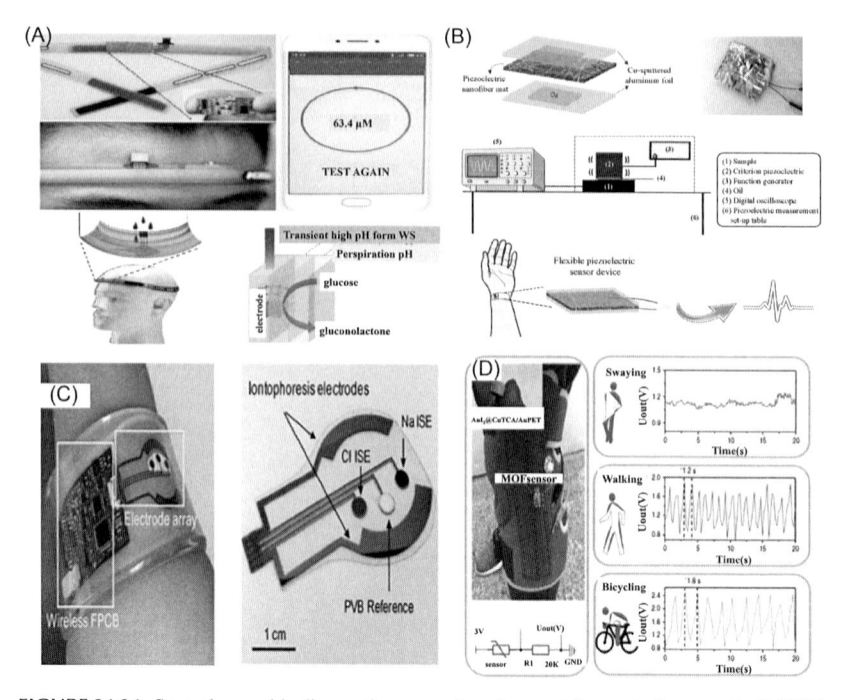

FIGURE 21.24 Several wearable diagnostic sensors based on metal-organic frameworks (MOFs): (A) Schematic illustrations of nonenzymatic electrochemical sensor ZIF-67 (Meskher et al., 2023; Zhu et al., 2019), (B) Schematic illustrations of wearable piezoelectric sensor, utilizing PVDF nanofibers embedded with microporous zirconium-based MOFs (Meskher et al., 2023; Moghadam et al., 2020), (C) Schematic illustrations of flexible wearable diagnostic sensors wireless FPCB (Meskher et al., 2023; Emaminejad et al., 2017), and (D) Schematic illustrations of flexible wearable AuI_2@CuTCA/AuPET (Pan et al., 2018).

constant secretion rates exceeding 100 nL/min/cm² , enabling sufficient sweat extraction for accurate in situ glucose analysis without compromising skin integrity or causing discomfort (Fig. 21.24C) (Emaminejad et al., 2017). Additionally, an innovative strain sensor designed for integration into wearable textiles aims to monitor body movements and detect potential injuries during physical activities. This sensor features a MOF as its core component, offering outstanding sensitivity, interference resistance, and durability. Its design enables precise signal detection while effectively filtering out noise, allowing accurate differentiation between subtle swaying and more vigorous activities, such as muscle hyperplasia. The sensor's construction and functionality are depicted in Fig. 21.24D (Pan et al., 2018).

Drug delivery

MOFs have emerged as promising materials in the realm of nano-sized therapeutics, particularly for cancer drug delivery applications (Kalati & Akhbari, 2021; Karimi Alavijeh & Akhbari, 2020; Parsaei & Akhbari, 2022c; Soltani & Akhbari, 2022a, 2022b). Their notable characteristics, including high drug uptake capacities and controlled-release mechanisms, make MOFs attractive vehicles for transporting sensitive drug molecules to target sites. MOFs incorporating biocompatible metals like Zn, Fe, or Zr (with LD50 values of 0.35, 4.1, and 0.45 g/kg, respectively) and organic linkers such as terephthalic acid, 1,3,5-benzenetricarboxylate, and 2-methylimidazole (with LD_{50} values of 1.13, 5.5, and 8.5 g/kg, respectively) are often unstable under acidic pH conditions or in the presence of phosphates. This instability facilitates effective clearance of the delivery vehicle from the patient's body. By incorporating other guest materials into MOFs to create composite materials, additional functionalities have been introduced, such as externally triggered drug release and improved pharmacokinetics and diagnostic capabilities. Incorporating guest materials into MOFs or encapsulating MOFs in host materials, known as MOF-composites, further enhances their performance. These composites, featuring NPs, polymers, hydrogel matrices, or biomolecules, offer improved drug release and imaging capabilities. Table 21.6 outlines several nanocomposites utilized within the realm of drug delivery (Osterrieth & Fairen-Jimenez, 2021).

In drug delivery, composites based on MOFs typically include NP-MOF composites, such as magnetic NPs@MOF, plasmonic NPs@MOF, upconverting NPs@MOF, quantum dots@MOF, and polymer-MOF composites like polypyrrole@MOF, polydopamine@MOF, and dyes@MOF. Additionally, biomolecule-MOF composites, such as lipid@MOF, protein@MOF, and nucleotide@MOF, are also utilized (Osterrieth & Fairen-Jimenez, 2021).

The preparation of UiO-66 NPs with controllable sizes is demonstrated in a scalable manner using a continuous flow microreactor. By carefully manipulating residence time, particle sizes ranging from hundreds to dozens of nanometers are achieved. Additionally, NH_2-functionalized nano-UiO-66

TABLE 21.6 Outlines several metal-organic framework nanocomposites utilized within the realm of drug delivery (Osterrieth & Fairen-Jimenez, 2021).

Composite material	MOF chemistry	Pore size (window size, if applicable) [nm]	Drug (size) [nm]	Toxicity (IC_{50} if stated in paper)
Fe_3O_4@HKUST-1(Cu)	Cu_3 btc	1.1–1.4	Nimesulide (1.1)	+++
Fe_3O_4 silica@HKUST-1(Cu)	Cu_3 btc	1.1–1.4	N/A	++++
Fe_3O_4@MIL-100(Fe)	$Fe_3O(OH)(H_2O)_2$ btc	2.4–2.9 (0.6–0.9)	DOX (1.5)	++
Fe_3O_4@MIL-100(Fe)	$Fe_3O(OH)(H_2O)_2$ btc	2.4–2.9 (0.6–0.9)	Ibuprofen* (1.3)	++
γ-Fe_2O_3 @SiO_2@HKUST-1(Cu)	Cu_3 btc	1.1–1.4	Letrozole (1.2)	++++
γ- Fe_2O_3@MIL-53(Al)	Al(OH) bdc	0.85	Ibuprofen (1.3)	+++
γ- Fe_2O_3@ZIF-8(Zn)	Zn 2-MIM	1.2 (0.34)	N/A	N/A
Fe_3O_4@MIL-88A(Fe)	Fe(III) fumarate	2.2–2.4	Dopamine (0.8)	N/A
Fe_3O_4@UiO-66-NH_2 (Zr)	Zr_6O_4 $(OH)_4$ bdc-NH_2	0.8–1.2	DOX (1.5)	+
Fe_3O_4@UiO-66-NH_2 (Zr)	Zr_6O_4 $(OH)_4$ bdc-NH_2	0.8–1.2	5-FU (0.5)	+
Fe_3O_4@IRMOF-3(Zn)	Zn_4O bdc-NH_2	1.9 (Yaghi et al., 2003)	Paclitaxel (1.5)	N/A
Fe_3O_4@BioMOF(Zn)	Zn_3 curcumin$_2$	1.1	Curcumin (1.9) 5-FU (0.5)	N/A
γ- Fe_2O_3 @MIL-100(Fe)	$Fe_3O(OH)(H_2O)_2$ btc	2.4–2.9 (0.6–0.9)	DOX (1.5)	++
AuNR@[Al(OH)(1,4-ndc)]	Al(OH) (1,4-ndc)	0.9–1	Anthracene* (1.1)	N/A
AuNP/polyoxometalate@MIL-101(Cr)	$Cr_3O(OH)(H_2O)_2$ bdc	2.9–3.4 (1.2–1.7)	N/A	N/A
AuNR@ZIF-8(Zn)	Zn 2-MIM	1.2 (0.34)	DOX (1.5)	+++

AuNR@ZIF-8(Zn)	Zn 2-MIM	1.2 (0.34)	N/A	4.45 µg/mL, MCF-7
AuNR/ZIF-8(Zn) Janus	Zn 2-MIM	1.2 (0.34)	DOX (1.5)	+++
CuS@ZIF-8(Zn)	Zn 2-MIM	1.2 (0.34)	Quercetin (1.3 nm)	+++
CuS@ZIF-8(Zn)	Zn 2-MIM	1.2 (0.34)	DOX (1.5)	+++
AuNR@MOF-545(Zr)	$Zr_6O_4(OH)_4$ tetracarboxyphenylporphyrin	3.7	Camptothecin (1.3)	N/A
AuNR@NU-901(Zr)	$Zr_6O_4(OH)_4$ tetra(p-benzoic acid) pyrene	3.1	N/A	N/A
NaYF$_4$@MIL-101-NH$_2$(Fe)	$Fe_3O(OH)(H_2O)_2$ bdc-NH$_2$	2.9–3.4 (1.2–1.7)	N/A	++
NaYF$_4$@PCN-224(Zr)	$Zr_6O_4(OH)_4$ tetracarboxyphenylporphyrin	2.2	N/A	N/A
NaYF: Yb^{3+}/Er^{3+}@MIL-100(Fe)	$Fe_3O(OH)(H_2O)_2$ btc	2.4–2.9 (0.6–0.9)	DOX (1.5)	++
NaYF:Yb^{3+}/Er^{3+}@ZIF-8(Zn)	Zn 2-MIM	1.2 (0.34)	5-FU (0.5)	+++
C-dot@ZIF-8(Zn)	Zn 2-MIM	1.2 (0.34)	5-FU (0.5)	+++
Fe$_3$O$_4$/C-dot@IRMOF-3(Zn)	Zn$_4$O bdc-NH$_2$	1.9 (Yaghi et al., 2003)	DOX (1.5)	10 µg mL, HeLa
AuNC@ZIF-8(Zn)	Zn 2-MIM	1.2 (0.34)	Camptothecin (1.3)	+++
PPy@MIL-100(Fe)	$Fe_3O(OH)(H_2O)_2$ btc	2.4–2.9 (0.6–0.9)	DOX (1.5)	++
PPy@MIL-53(Fe)	Fe(OH) bdc	0.85	DOX (1.5)	
PPy@MIL-100(Fe)	$Fe_3O(OH)(H_2O)_2$ btc	2.4–2.9 (0.6–0.9)	DOX (1.5)	++
Fe-soc-MOF@Ppy	Fe$_3$O 3,3',5,5'-azobenzenetetracarboxylate	0.67	N/A	N/A

(Continued)

TABLE 21.6 (Continued)

Composite material	MOF chemistry	Pore size (window size, if applicable) [nm]	Drug (size) [nm]	Toxicity (IC$_{50}$ if stated in paper)
MIL-100(Fe)@PDA	Fe$_3$O(OH)(H$_2$O)$_2$ btc	2.4–2.9 (0.6–0.9)	Curcumin (1.9)	++
ZIF-8(Zn)@PDA	Zn 2-MIM	1.2 (0.34)	DOX (1.5)	+++
PDA@MIL-100(Fe)	Fe$_3$O(OH)(H$_2$O)$_2$ btc	2.4–2.9 (0.6–0.9)	DOX (1.5)	++
ZIF-8(Zn)@PDA	Zn 2-MIM	1.2 (0.34)	DOX (1.5)	+++, 0.3 mg mL, HeLa
UiO-66(Zr)@PDA	Zr$_6$O$_4$(OH)$_4$btc	0.8–1.2	N/A	+
MIL-101-NH$_2$(Fe)@PDA	Fe$_3$O(OH)(H$_2$O)$_2$ bdc-NH$_2$	2.9–3.4 (1.2–1.7)	N/A	++
MnCo@PDA hydrogel	Mn$_3$[Co(CN)$_6$]$_2$	N/A	N/A	N/A
Fe-soc-MOF@ICG	Fe$_3$O 3,3',5,5'-azobenzenetetracarboxylate	0.67	N/A	N/A
Cypate@MIL-53(Fe)	Fe(OH) bdc	0.85	N/A	21.2 μm, A549l, 785
PAN@UiO-66(Zr)	Zr$_6$O$_4$(OH)$_4$ bdc	0.8–1.2	Fluorescein* (1.1)	+
PAA@ZIF-8(Zn)	Zn 2-MIM	1.2 (0.34)	DOX (1.5)	+++
MIL-100(Fe)@Heparin	Fe$_3$O(OH)(H$_2$O)$_2$ btc	1.2 (0.34)	Caffeine* (0.8)	++
UiO-66-NH$_2$(Zr)@PNIPAM	Zr$_6$O$_4$(OH)$_4$bdc-NH$_2$	0.8–1.2	Resorufin (1.0), Caffeine* (0.8), Procainamide (1.4)	+
PB@MIL-100(Fe)	Fe$_3$O(OH)(H$_2$O)$_2$ btc	2.4–2.9 (0.6–0.9)	Artemisinin (0.9)	++
PCN-224(Zr)@MnO$_2$	Zr$_6$O$_4$(OH)$_4$ tetracarboxyphenylporphyrin	2.2	N/A	N/A

Fe₃O₄@UiO-66-NH₂/graphdiyne	Zr₆O₄(OH)₄ bdc-NH₂	0.8–1.2	DOX (1.5)	+
ZIF-8(Zn)/Graphene oxide	Zn 2-MIM	1.2 (0.34)	Fluorescein* (1.1)	+++
Zr (IV) disuccinatocisplatin NCP@ DOPA:DOPC/DSPE-PEG	Zr disuccinatocisplatin	N/A	Cisplatin (0.5)	1.2 μm, A549 cell
Zn(II) biphosphonate NCP@ DOPA:DOPC/DSPE-PEG	Zn biphosphonate	N/A	Cisplatin (0.5) Oxaliplatin (0.5)	9.3 μm, CT26 10.5 μm, CT26
Zn(II) cisplatinate NCP@ DOPA:DOPC/pyrolipid/ DSPE-PEG	Zn- cis,cis,trans-[Pt(NH₃)₂ Cl₂(OCONHP(O)(OH)₂)₂]	N/A	Cisplatin (0.5)	1.3 μm, HN-SCC-135
MIL-100(Fe)@DOPC	Fe₃O(OH)(H₂O)₂ btc	2.4–2.9 (0.6–0.9)	N/A	++
MIL-101(Cr)@DOPC	Cr₃O(OH)(H₂O)₂ bdc	2.9–3.4 (1.2–1.7)	N/A	N/A
UiO-66(Zr)@DOPA	Zr₆O₄(OH)₄ bdc	0.8–1.2	N/A	+
UiO-67(Zr)@DOPA	Zr₆O₄(OH)₄ biphenyldicarboxilate	1.2–1.6	N/A	+
BUT-30(Zr)@DOPA	Zr₆O₄(OH)₄ 4,4'-(ethyne-1,2-diyl)dibenzoate	1.8	N/A	N/A
ZIF-4(Zn)@DOPC	Zn imidazole	0.5 (0.25)	Curcumin (1.9)	N/A
UiO-66(Zr)-DPGG	Zr₆O₄(OH)₄ bdc	0.8–1.2	N/A	+
ZIF-8(Zn)-DPGG	Zn 2-MIM	1.2 (0.34)	N/A	+++
ZIF-67(Co)-DPGG	Co 2-MIM	1.2 (0.34)	N/A	++++
HKUST-1(Cu)-DPGG	Cu₃ btc	1.1–1.4	N/A	N/A

(Continued)

TABLE 21.6 (Continued)

Composite material	MOF chemistry	Pore size (window size, if applicable) [nm]	Drug (size) [nm]	Toxicity (IC$_{50}$ if stated in paper)
MIL-101(Cr)-DPGG	Cr$_3$O(OH)(H$_2$O)$_2$ bdc	2.9–3.4 (1.2–1.7)	N/A	N/A
UiO-66(Zr)@TPP	Zr$_6$O$_4$(OH)$_4$ bdc	0.8–1.2	Dichloroacetate (0.4 nm)	+
Tyrosinase@PCN-333(Al)	Al3O 4,4′,4″-s-triazine-2,4,6-triyl- tribenzoate	3.4–5.5 (2.6–3)	Paracetamol* (0.9)	N/A
siRNA@NU-1000(Zr)	Zr$_6$O$_4$(OH)$_4$ 1,3,6,8-tetra(p-benzoic acid)pyrene	3.6	N/A	N/A
CRISPR/Cas9@ZIF-8(Zn)	Zn 2-MIM	1.2 (0.34)	N/A	+++
Zn(II) bisphosphonate NCP@ DOPA:DOPC/siRNA/ DSPE-PEG	Zn biphosphonate	N/A	Cisplatin (0.5)	45.8 μm, SKOV-3
ssDNA@IRMOF-74(Ni)	Zn 2,4-dihydroxyterephthalate	1.5	N/A	N/A
mRNA@UiO-66(Zr)	Zr$_6$O$_4$(OH)$_4$ bdc	0.8–1.2	N/A	+
siRNA@MIL-100(Fe)	Fe$_3$O(OH)(H$_2$O)$_2$ btc	2.4–2.9 (0.6–0.9)	N/A	++

Model drugs are highlighted with an*. Toxicity data based on MOF toxicity as analyzed by Ruyra et al. on zebrafish embryos, where "−" represents little toxicity and "+++" represents high toxicity.
btc, 1,3,5-Benzenetricarboxylate; *bdc*, 1,4-benzenedicarboxylate; *2-MIM*, 2-methylimidazole; *bdc-NH$_2$*, 2-aminobenzenedicarboxylate; *PPy*, polypyrrole; *ICG*, indocyanine green; *PDA*, polydopamine; *DPGG*, 1,2-dipalmitoyl-sn-glycero-3-galloyl; *PNIPAM*, poly(N-Isopropylacrylamide) polymer; *DOX*, doxorubicin; *5-FU*, 5-fluorouracil; *1·4 ndc*, 1,4-naphthalenedicarboxylate; *NCP*, nanoscale coordination polymer; *dopa*, 1,2-dioleoyl-sn-glycero-3-phosphate; *DOPC*, 1,2-dioleoyl-sn-glycero-3-phosphocholine; *DSPE–PEG*, 1,2-distearoyl-sn-glycero-3-phosphoethanolamine-N-(polyethylene glycol).
open access–creative commons CC-BY.

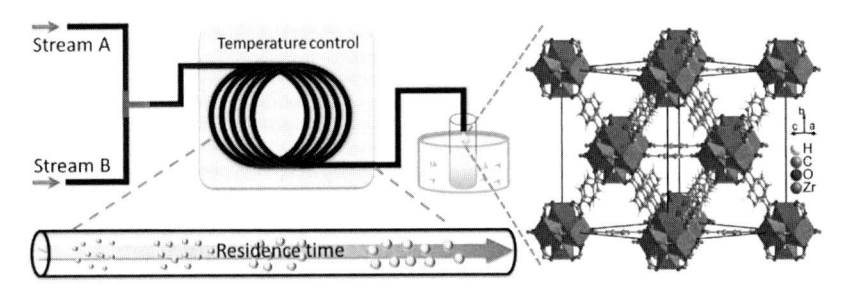

FIGURE 21.25 Nano-UiO-66 particles synthesized using a continuous flow microreactor (Ren et al., 2017; Tai et al., 2016).

particles are synthesized, and their efficacy in drug delivery, specifically with 5-Fu, is systematically investigated in phosphate-buffered saline (PBS) solution (Fig. 21.25) (Tai et al., 2016).

In recent years, industrial applications in the field of microrobots and MOFs in motion have gained significant attention. Microscale and nanoscale swimmers represent a groundbreaking development in biomedicine and on-the-fly chemistry applications. These small devices can move through various fluids by interacting with chemicals in their environment or by using external energy sources such as magnetic fields, electric fields, ultrasound, light, or a combination of these. Due to their mobility, they have been suggested for various roles, including therapeutic mobile carriers, biopsy agents, on-the-fly chemical sensors, microcleaners and nanocleaners for pollutants, and scavengers for metals (Terzopoulou, Nicholas, et al., 2020). Table 21.7 shows the list of MOF-based swimmers and actuators.

The first MOF-based swimmer was reported in 2012 by Matsui, Kitagawa, and their team. They designed MOF crystals containing self-assembling peptides like diphenylalanine (DPA) in their pores. The release of DPA induced motion in the MOF via the Marangoni effect, driven by a surface tension difference at the water/MOF interface. The study highlighted the importance of the porous crystalline network for generating motion (Fig. 21.26A) (Ikezoe et al., 2012). Troyano and coworkers introduced a preprogrammed shape design for MOFs that undergo shape deformation upon external stimulation, inspired by previous work on strain-engineered microtubes and self-folding microrobots. In their research, photothermal activation of various MOFs triggered motion by solvent removal from the pores upon exposure to UV-Vis light. Swelling behavior and reconfiguration of MIL-88A were observed. MIL-88A was incorporated into a passive PVDF matrix, creating passive and active domains in MIL-88A@PVDF films through selective etching of the MOF in certain areas (Fig. 21.26B) (Troyano et al., 2019). Li and their team introduced UiO-67-based swimmers with adjustable speed. By incorporating bipyridine ligands, they enabled the metalation of the MOF with active metal centers like Co and Mn, which propelled the MOFs via

TABLE 21.7 List of metal-organic framework composite-based swimmers and actuators, summarizing their underlying components, stimulus response, type of propulsion, and, where applicable, proof-of-principle (PoP) applications (Terzopoulou, Nicholas, et al., 2020).

MOF	Responsive component	Stimulus	Type of propulsion	PoP applications	Materials' biocompatibility/ toxicity	Environmental requirements
ZIF-8	Pt	H_2O_2 fuel	Catalytic	–	++	+
	Ag	H_2O_2 fuel	Catalytic	Water remediation	++	+
	MnO_2	H_2O_2 fuel	Catalytic	Water remediation	++	+
	Fe_3O_4 NPs	Magnetic field	Magnetic steering	N/A	N/A	N/A
	Pt	H_2O_2 fuel	Catalytic	Water remediation	++	+
	Fe_3O_4 NPs	Magnetic field	Magnetic steering	N/A	N/A	N/A
	Ni	Magnetic field	Magnetic steering	Drug delivery in vitro	+	+++
	Fe-based NPs	Magnetic field	Catalytic magnetic steering	Drug delivery in vitro	+++	+++
	Catalase	H_2O_2 fuel	Catalytic	Cancer treatment (ST and PDT) in vitro	+++	++
ZIF-67	Co-MOF Sites	H_2O_2 fuel	Catalytic	N/A	+	+
	Co-MOF	H_2O_2 fuel sites	Catalytic	Drug delivery in vitro	+	+
	Fe_3O_4 NPs	Magnetic field	Magnetic steering	N/A	N/A	N/A

				Drug delivery in vitro	
ZIF-L	Catalase	H₂O₂ fuel	Catalytic	N/A	++	+++
	β-lactoglobulin	pH	On/off chemical control	N/A	N/A	N/A
	PDPA polymer	pH	Vertical motion control	N/A	N/A	N/A
UiO-67	Co, Mn	H₂O₂ fuel	Catalytic	N/A	+	+
	EDTA, IDA	Chelation	Chemical speed control	N/A	N/A	N/A
UiO-66	Pt	H₂O₂ fuel	Catalytic	N/A	++	+
MIL-88A	MOF pores	UV-Light	Swelling/ Contraction of MOF pores	N/A	++	++
Cu₂L₂ted	MOF framework	EDTA – MOF degradation – Fuel release pH	Surface tension gradient Directional motion	N/A	+	++
PCN-222	MOF framework	Fuel release Shape Fuel type	Surface tension gradient Directional motion Speed control	N/A	+	++

open access=creative commons CC-BY.

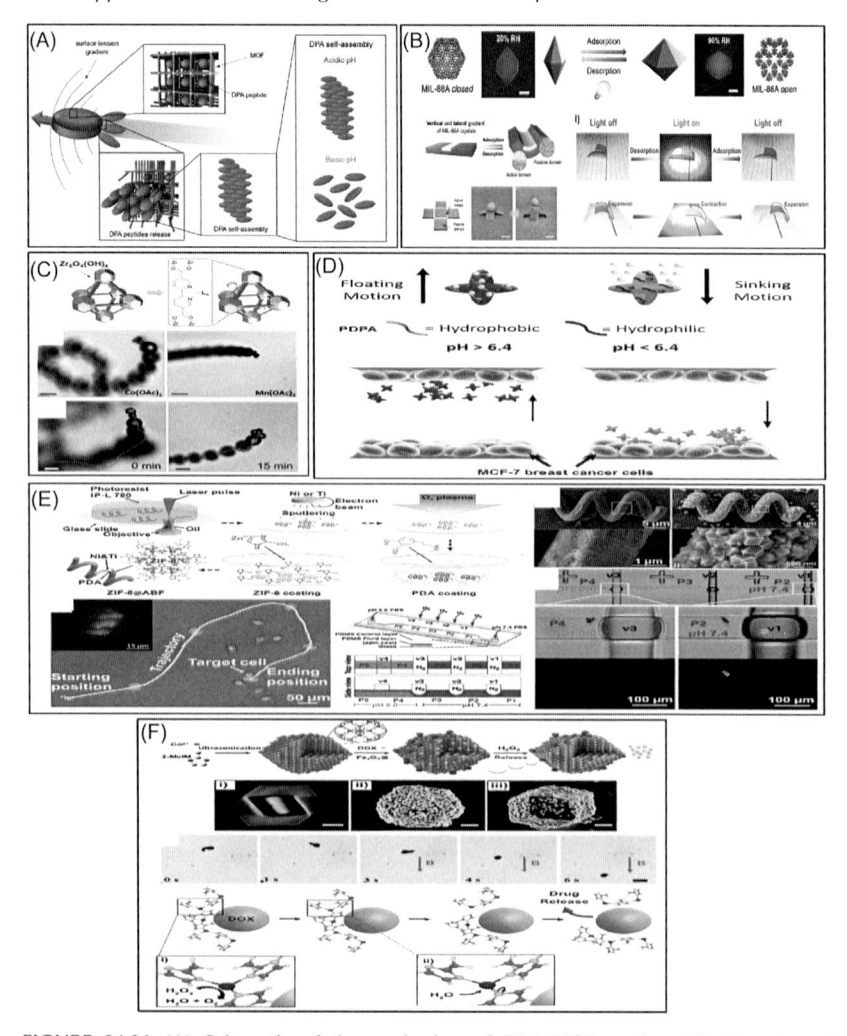

872 Applications of Metal-Organic Framework Composites

FIGURE 21.26 (A) Schematic of the mechanism of DPA-MOF motion, (B) Photothermal response of MIL-88A@PVDF films, (C) UiO-type MOF based microengines highlighting the metallization of the organic ligand, (D) Catalytically propelled, pH controlled ZIF-L based micromotor—CAT-PDPA@ZIF-L, (E) Magnetically driven MOF based swimmers for targeted drug delivery, (F) Catalytically propelled, magnetically controlled ZIF-67 based micromotor (Terzopoulou, Nicholas, et al., 2020).

H_2O_2 decomposition. Control over propulsion was achieved by adding chelating agents such as IDA (iminodiacetic acid) or EDTA (ethylenediamine-tetraacetic acid), which suppressed catalytic activity and propulsion. While this work pioneered chemical control over MOF propulsion, it did not address directional control (Fig. 21.26C) (Li et al., 2017). Guo and colleagues expanded on this concept with a pH-responsive MOF micromotor. By leveraging changes in pH, they controlled the vertical motion of the particles through buoyancy effects, allowing the micromotor to ascend or descend based on the pH environment. They functionalized ZIF-L with CAT for O_2 bubble generation and used poly(2-diisopropylamino) ethyl methacrylate (PDPA) to provide pH responsiveness. At pH > 6.4, deprotonation of PDPA's tertiary amine groups induced hydrophobicity, leading to O_2 bubble adsorption and upward motion. Conversely, at pH < 6.4, protonation occurred, causing hydrophilicity, O_2 desorption, and downward motion (Fig. 21.26D) (Guo et al., 2019).

MOFBOT, developed by Pane, Puigmarti, and their team, employed a layer-by-layer fabrication method on a helical polymeric chassis, coated with nanometer-thick layers of Ni and Ti for magnetic manipulation and biocompatible protection, respectively. The surface was treated with polydopamine to bind ZIF-8, which encapsulated a fluorescent model drug (Rhodamine-B) during synthesis. Utilizing rotating magnetic fields, MOFBOT exhibited corkscrew locomotion in biologically relevant fluids and targeted single cells via external magnetic navigation. Its pH-dependent drug release capabilities were demonstrated on a microfluidic platform (Fig. 21.26E) (Wang, Chen, et al., 2019). Wang et al. introduced a ZIF-based micromotor comprising ZIF-67 microcrystals decorated with Fe_3O_4 NPs, enabling magnetic steering. The cobalt(II) nodes of ZIF-67 facilitated catalytic propulsion, while the magnetite NPs allowed for magnetic control. These micromotors were loaded with doxorubicin (DOX), an anticancer drug. However, the catalytic decomposition of H_2O_2 and hydrolysis of ZIF-67 altered the MOF's structure, leading to premature drug release during motion and limiting its therapeutic effectiveness (Fig. 21.26F) (Wang, Zhu et al., 2018). Terzopoulou's team created the MOFBOT by merging Fe@ZIF-8 with a biodegradable gelatin methacryloyl (GelMA) hydrogel for guided drug delivery and eventual degradation in cell cultures. The MOFBOT features a GelMA core coated in Fe@ZIF-8, retaining the benefits of ZIF-8, such as high surface area and pH responsiveness. The soft magnetic properties of Fe@ZIF-8 enable magnetic motion control, ensuring effective drug delivery and reduced cell viability near DOX-carrying MOFBOTs. The system undergoes enzymatic degradation of the GelMA core, pH-triggered dissolution of Fe@ZIF-8, and oxidation of the magnetic components, ensuring nontoxic degradation and targeted drug release in acidic environments (Terzopoulou, Wang, et al., 2020).

Metal-organic framework-based nanocomposites for bioimaging

Magnetic nanoparticles, often enclosed within a MOF matrix via core-shell encapsulation, offer potential for magnetophoretic therapy and diagnostics, providing T2*-image contrast. Plasmonic nanoparticles like Au NRs, also encapsulated in core-shell structures, show near-infrared (nIR) absorbance and high photothermal energy conversion for controlled drug release. Luminescent

874 Applications of Metal-Organic Framework Composites

NP composites, such as quantum dots or upconversion nanoparticles (UCNPs), serve as diagnostic tools due to their strong emissive properties, although their toxicity requires careful consideration. Au NRs@MIL-88(Fe) core-shell nanostars are developed as triple-modality imaging probes (CT/MRI/PAI) for accurate and noninvasive glioma diagnosis (Wang, 2017). These nanoprobes exhibit low cytotoxicity, high contrast, deep penetration depth, and excellent spatial resolution (Fig. 21.27A) (Shang et al., 2016). Also, researchers synthesized an Fe_3O_4@UiO-66(Zr)-NH_2 core-shell composite, combining Fe_3O_4 NPs with UiO-66-NH_2 (Zr) shell, exhibiting enhanced mid-infrared (MIR) radiation absorption (Zhao et al., 2016). Similarly, a smart theragnostic nanoplatform combines nanovalve-operated MOF core-shell hybrids for targeted drug release, MRI guidance, sustained chemotherapy, and responsiveness to pH changes and temperature variations. Utilizing Fe_3O_4 particles for MRI and magnetic separation as the core and UiO-66 MOF with high loading capacity as the shell, the system achieves prolonged drug release over 7 days, enhancing anticancer efficacy at the lesion site (Wu et al., 2018). Nano metal-organic frameworks (NMOFs) with high-Z elements are explored

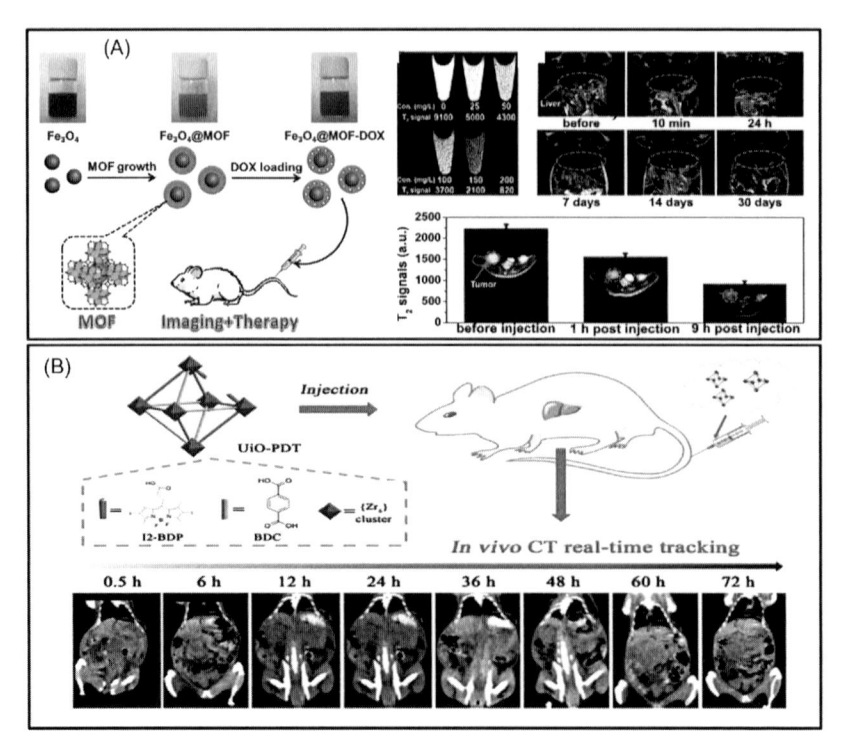

FIGURE 21.27 (A) Fe_3O_4@UiO-66(Zr)-NH_2 core-shell composite and (B) UiO-PDT nanocrystals for bioimaging applications (Shang et al., 2016; Zhang et al., 2017).

as potential CT contrast agents due to their strong X-ray attenuation. Zhang et al. developed iodine BODIPY (BODIPY=dipyrromethenoboron difluoride: 5,5-Difluoro-5H-4λ^5-dipyrrolo[1,2-c:2′,1′-*f*][1,3,2]diazaborinin-4-ylium-5-uide)-containing MOF nanocrystals for CT contrast. UiO-PDT nanocrystals, prepared via ligand exchange in UiO-type NMOFs, demonstrated good cytocompatibility and prolonged circulation. They preferentially accumulated in tumor sites, allowing clear tumor visualization without significant accumulation in surrounding tissues or organs in rats with hepatomas (Fig. 21.27B) (Zhang et al., 2017).

Atmospheric water-harvesting

Freshwater scarcity is a pressing global issue, affecting two-thirds of the population annually. Atmospheric water, largely untapped, offers promise for addressing this challenge. Sorption-based atmospheric water generators (AWGs) show potential in capturing moisture from the air, providing a universally accessible solution when integrated with renewable energy sources. However, commercial AWGs face challenges related to efficiency and scalability. Material densities are crucial for comparing adsorption capacities per unit volume. Advances in materials like MOFs offer improved water-harvesting systems. Selecting optimal sorbents, such as MOFs, is crucial for maximizing efficiency (Hanikel et al., 2020; Tu et al., 2018).

Sorbent-assisted water harvesting systems function on a shared fundamental principle: initially, water vapor is absorbed into the sorbent material. Subsequently, it is released from the sorbent by lowering the relative humidity of water in the surrounding environment, a process that can be induced by either temperature or pressure modulation. The resultant concentrated water vapor undergoes condensation to yield liquid water. This process may occur once daily (monocyclic) or multiple times throughout the day (multicyclic). Furthermore, external energy sources, such as electric heating of the sorbent material for vapor release and vapor compression or thermoelectric refrigeration for condensation, can facilitate water release and condensation in an active water harvesting system. Conversely, passive devices function solely under ambient sunlight and natural cooling conditions (Ahrestani et al., 2023; Song et al., 2023; Tu et al., 2018). Fig. 21.28 illustrates monocyclic and multicyclic systems (Hanikel et al., 2020).

Yaghi, Wang, and their team at Water Harvesting WaHa company have harnessed the unique properties of composite MOF-801 and MOF-303 to develop highly efficient devices for atmospheric water harvesting (Fig. 21.29A) (WaHab, 2024). Fig. 21.29B illustrates the progression from a proof-of-concept to the first and second generation devices. In point (A), the progression from proof-of-concept is depicted, while point (B) represents the first generation, point (C) the second generation, and point (D) the near-commercialization devices for water harvesting from air. Fig. 21.29 demonstrates advances in water

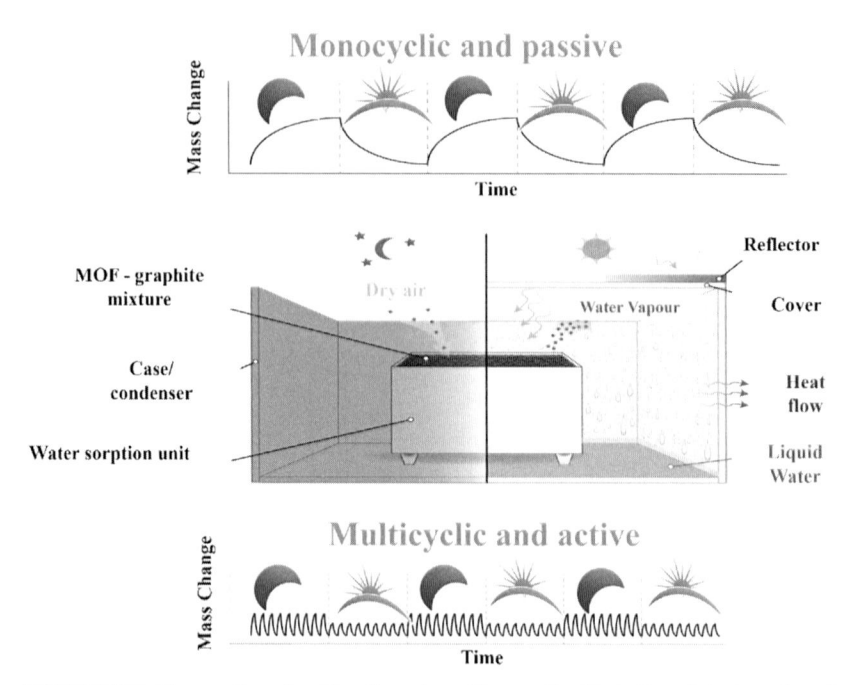

FIGURE 21.28 Monocyclic and multicyclic systems: Comparative illustration of single-cycle and multiple-cycle frameworks in chemical and material applications (Cattani et al., 2022; Hanikel et al., 2020).

production levels based on MOF-801 (A and B) and MOF-303 (C) (Xu & Yaghi, 2020).

MOF-801, known for its three distinct cavities capable of aggregating water molecules, served as the foundation for a MOF-based device designed to extract water from the air. This device, featuring a 1 mm MOF layer, demonstrated impressive performance, producing 2.8 L of water per kilogram per day under continuous harvesting conditions at 35°C and 1 kW per square meter of heat. Specifically, operating a MOF-801-based device under specific external conditions (30% RH and 22°C) corresponds to a dew point of 3.6°C. Heating applied during the desorption phase increases the dew point to 8.3°C and further to 11.2°C, significantly reducing condenser power consumption by 55% compared to direct cooling without MOF-801. Fig. 21.29C, depicts a MOF water harvester device operating in adaptive mode. It includes a schematic detailing the device's components: an air intake compartment with an air filter, fan, and electric heater; a sorption compartment housing the MOF-801 adsorbent on aluminum-lined trays; and a condensation compartment with a condenser and water collection funnel. Compared to silica gels, traditional absorbent materials, MOFs offer 1.5–2 times higher volumetric energy density, enabling device miniaturization, along with a 27% higher efficiency factor. These advancements underscore the

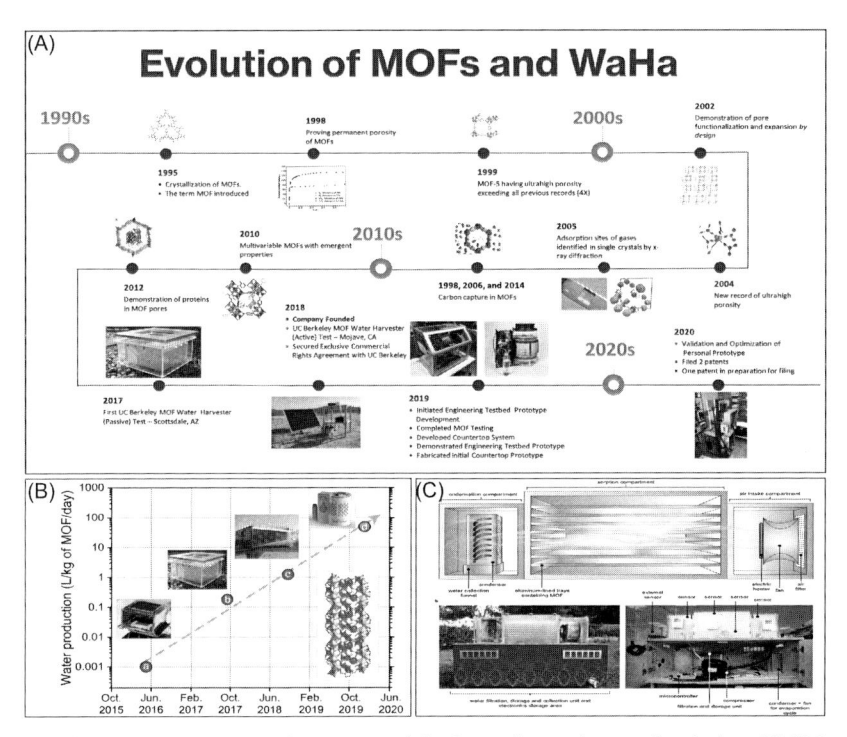

FIGURE 21.29 (A) Illustrates the progress of the first and second generation devices (WaHab, 2024), (B) Metal-organic framework water harvester device MOF 801 based (Xu & Yaghi, 2020), and (C) Schematic of the water harvester device – operating in adaptive mode (Almassad et al., 2022; Cheng et al., 2024).

potential of MOFs in revolutionizing various industrial applications (Almassad et al., 2022).

Similarly, MOF-303, a hydrophilic aluminum pyrazole dicarboxylate microporous structure, exhibited exceptional water scavenging capabilities, absorbing up to 40% of its weight in water within the relative humidity range of 20%–40% and completing rapid absorption and desorption cycles within minutes. Field tests in the Mojave Desert confirmed its effectiveness, yielding between 0.7 and 1 L of water per kilogram of MOF per day. Moreover, MOF-303 demonstrated efficient water extraction from barley, achieving a yield of 3.5 kg per pack with a remarkable efficiency rate of 94% (Xu & Yaghi, 2020). Environmentally friendly, water-based synthesis protocols for MOF-303 on small particles, along with comprehensive methods for large-scale synthesis, are detailed. Successful demonstration of scalability involved producing 3.5 kg batches of MOF-303 with a yield of 91%, maintaining similar crystallinity and water uptake capacity as those produced on a smaller scale. This underscores the efficiency and suitability of the green and water-based synthetic procedure for large-scale production. Overall, this approach offers a rapid and effective method for producing MOF-303, holding great promise for commercialization in water harvesting applications (Xu & Yaghi, 2020; Zheng et al., 2023).

Another hydrophilic aluminum MOF with promising commercial potential at an industrial scale is CAU-10-H, known for its exceptional cycle stability of 10,000 cycles under adsorption-based cooler (ADC) operating conditions. The pilot synthesis protocol for this structure involves using water with 5% ethanol and is based on isophthalic acid, eliminating the need for DMF synthesis. This method demonstrates an impressive efficiency of 95% (Chakraborty et al., 2023; Solovyeva et al., 2021). The University of California, Berkeley has developed an advanced water harvester that operates by blowing ambient air over a cartridge filled with a MOF, visible inside a plexiglass box. The MOF extracts water from arid air, which is then removed through mild heating. The concentrated water vapor is directed out through a tube to a condenser. This innovative process enables the production of drinkable water using only solar panels and a battery, even in extremely dry regions like the Mojave Desert (Berkeley, 2021). A portable, sun-powered, hand-held device was developed by the University of California, Berkeley, and Yaghi's team. This innovative device has the potential to assist in regions facing water scarcity (Song et al., 2023).

Table 21.8 compares the price of water produced by adaptive water harvesting (AWH) with that from commercial sources in representative lower-middle and upper-middle income countries. Considering both production costs and electricity expenses in these nations, the adaptive water harvesting device has the potential to generate substantial financial savings. Specifically, it could result in water cost reductions of up to 49%, 63%, 63%, and 46% in Morocco, Nigeria, Mexico, and Jordan, respectively (Almassad et al., 2022).

Food packaging

Ensuring food safety is critical, but current detection methods are often labor-intensive and costly. To address this, rapid and portable food sensing techniques, such as those based on MOFs, are gaining attention. MOF-based sensors can efficiently detect contaminants in food, offering advantages in portability, affordability, reliability, sensitivity, and stability. This topic has been extensively discussed in the sensor and biosensor fields. Additionally, addressing waste prevention and improving packaging in food processes is essential for overall food safety. Food spoilage is a major contributor to both food waste and food poisoning. Studies show that approximately one-third of the food produced—around 1.3 billion tons per year—is lost throughout the food supply chain. The primary factors contributing to food loss vary between developed and developing countries. In developed countries, food waste is often due to inadequate quality standards in food storage and packaging, whereas in developing countries, poor-quality food products, spoilage, and

TABLE 21.8 Price of water produced by adaptive water harvesting (Almassad et al., 2022).

Country	Population without SMDW (%)	Commercial drinking water price ($, USD)	Household electrical energy cost ($, USD 1/kWh)	Adaptive AWH drinking water cost ($, USD 1/L)[a,b]	Adaptive AWH drinking water cost ($, USD 1/L)[a,c]	Price reduction (%)[a,c]
Morocco	31–40	0.50	0.12	0.20–0.61	0.26–0.68	49
Nigeria	71–80	0.42	0.057	0.095–0.30	0.16–0.36	63
Mexico	51–60	0.56	0.087	0.145–0.46	0.21–0.52	63
Jordan	11–20	0.42	0.10	0.17–0.53	0.23–0.59	46

SMDW, Safely managed drinking water.
open access=creativecommons CC-BY.
[a] Dependent on environmental conditions.
[b] Excluding capital asset costs.
[c] Including capital asset costs (assuming 10-year lifespan).

contamination are the main issues. Supply chain management with traceability capabilities, which enables quality control of products, plays a significant role in the food industry. Innovative technologies, including MOFs, offer substantial assistance in addressing these challenges.

In developing countries, using antibacterial substances to prolong the shelf life of food and reduce waste is essential. Golmohamadpour et al. explored the incorporation of indocyanine green (ICG), an FDA-approved photosensitizer used in photodynamic therapy (PDT), into three types of MOFs (MIL-101(Fe), MIL-101(Al), and MIL-88(Fe)) to enhance antimicrobial activity against *Enterococcus faecalis* (*E. faecalis*). These composite materials exhibited improved aqueous stability and increased bacteriostatic efficiency under laser irradiation, outperforming pure MOFs or ICG alone. Similarly, Wu et al. integrated thymol, known for its antibacterial properties, into $Zn-BDC-NH_2$ (BDC: 2-amino-1,4-benzenedicarboxylic acid) MOFs, resulting in the inhibition of the growth of *Escherichia coli* (*E. coli*) O157 with nalidixic acid resistance. MOF delivery systems facilitated on-demand release, such as encapsulating iodine into ZIF-8 for pH-triggered antimicrobial action against acid-producing pathogens. Additionally, Huang et al. developed a recyclable antimicrobial carrier, HKUST-1@carboxymethyl chitosan (CMCS), loaded with dimethyl fumarate. This carrier demonstrated superior antimicrobial effects against *E. coli* and *Staphylococcus aureus* (*S. aureus*), particularly with phosphoric acid stimulation. The integration of MOFs with antimicrobial agents presents promising strategies for enhancing food preservation and combating bacterial proliferation (Li et al., 2021; Shen et al., 2020). Recent advancements include the development of CS-PEO (CS-PEO=chitosan polyethylene oxide) nanofiber mats with varying ZIF-8 loading using electrospinning techniques. These mats demonstrate satisfactory hydrophobicity and mechanical properties, with the CS-PEO-3% ZIF-8 mat exhibiting 100% antibacterial performance against *E. coli* and *S. aureus*. Consequently, the CS-PEO-3% ZIF-8 mat holds promise as a potential functional material for food coating and packaging. ZIF-8, produced using a stir bar synthesis method with specific reactants, has shown minimal degradation over at least 130 cycles, further enhancing its suitability for food packaging applications (Sharanyakanth & Radhakrishnan, 2020). MOFs can be made into films or integrated with existing ones. For instance, MOF-801 incorporated into a hydrophobic film demonstrated enhanced water barrier properties. Additionally, MOFs are used to develop antibacterial films, such as Cap-Fe^{3+}-HMOF-5/gel/chi (where Cap=capsaicin, Chi=chitosan, and GEL=gelatin), which improve water vapor permeability and exhibit antimicrobial activity against *E. coli*, offering a promising solution for food packaging (Zhao et al., 2020). MOFs help prevent food deterioration by scavenging moisture and oxygen from packaging. MIL-101(Cr)-encapsulated ferrocene effectively removes oxygen, ensuring food quality and stability (Zhang, Luo et al., 2016).

Metal-organic framework composites: industrial applications **Chapter | 21** **881**

In 2024, Rui Zhao and coworkers developed a cellulose nanocrystal/metal-organic framework (MC) composite to stabilize alkenyl succinic anhydride (ASA) Pickering emulsions, enhancing food packaging performance. As shown in Fig. 21.30, the MC-ASA composite provided superior emulsion stabilization, increased paper hydrophobicity, strong antimicrobial properties against *E. coli and S. aureus*, and extended strawberry freshness for seven days. This research underscores the potential of CNC/MOF composites for innovative food packaging, water-resistant materials, and environmental protection (Fig. 21.30) (Rui et al., 2024). Table 21.9 shows some of the MOF–polymer composites for antimicrobial food packaging applications.

In the field of food packaging technology, emphasis is placed on adsorptive properties, but its industrial adoption is hindered by concerns regarding the potential side effects of MOFs in direct contact with food. This apprehension arises from the scarcity of toxicity studies reported in the literature. However, two patented solutions aim to address these limitations. Mastertaste Inc., through inventors Herman Stephen and James Stuart, patented a MOF-based system for

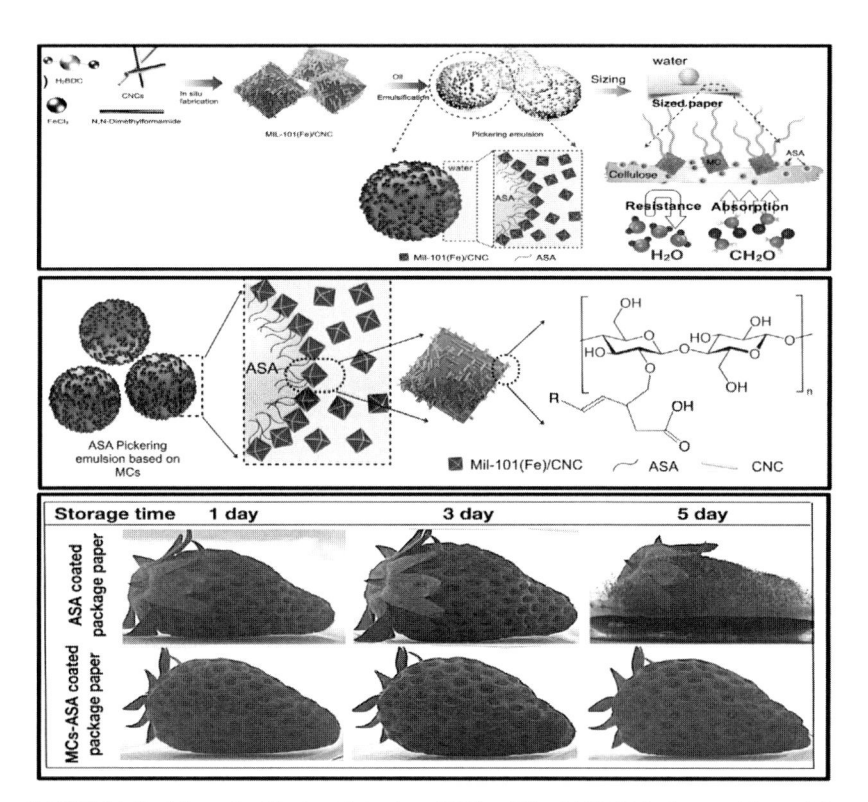

FIGURE 21.30 Schematic of in situ fabrication of MC-stabilized ASA Pickering emulsion and the sizing paper with water resistance and formaldehyde adsorption (Rui et al., 2024).

TABLE 21.9 Summary of metal-organic framework–polymer composites for antimicrobial food packaging applications.

Polymer/MOF	Composition/ synthesis method	Antimicrobial agent	Applications	Mechanism	Target organism	Antimicrobial efficiency	Ref
PCL/Cur@ ZIF-8	0%–35% MOF to PCL. Curcumin loaded during ZIF-8 synthesis, and solvent casting used to add PCL	Curcumin and ZIF-8 ROS Zn^{2+}	Antimicrobial food packaging	Curcumin release ~doubled when Poly-MOF exposed to pH 5 compared to a neutral pH following 72-h.	E. coli, S. aureus	99.9% decrease in the growth of E. coli and S. aureus when over 15% of Cur@ZIF-8 was loaded. Detachment of bacteria.	Cai et al. (2021)
MOF199@bamboo (carboxymethylated bamboo)	11.1 wt.% Cu^{2+} two stages synthesis to immobilize MOF-199	Cu^{2+} MOF composite	MOF-coated wood-based materials	Physical disinfection Surface active metal sites.	E. coli	Reduction in colony number by 38. 91.4% antibacterial ratio.	Su et al. (2021)
THY@PCN/PUL/PVA	Electrospinning	ROS/ photoirradiation thymol	Food packaging	Photodynamic therapy Cargo release.	E. coli, S. aureus	Inhibition of ~99% and ~98% for S. aureus and E. coli, upon irradiation, respectively.	Min et al. (2021)

Ag NPs@HKUST-1@CFs (carboxymethylated fibers)	Deposition ratio: 31.64% by weight Ag wt.%:4.79; Cu wt.%: 13.3 in situ preparation	Ag$^+$ and Cu^{2+}	Cellulose-based antibacterial materials (food and medical packaging)	Cargo release MOF disintegration.	S. aureus, E. coli	99.41% inhibition for S. aureus.	Duan et al. (2018)
TFC-Ag-MOF composites	In situ TFC functionalization	Ag$^+$	Antifouling membrane for FO applications	Ag$^+$ release.	P. aeruginosa	Bacterial mortality of 100% was nearly reached.	Seyedpour et al. (2019)

Cur, Curcumin; *PCL*, polycaprolacton; *ROS*, reactive oxygen species; *THY*, thymol; *PUL*, pullulan; *PVA*, polyvinyl alcohol; *FO*, forward osmosis; *TFC*, thin-film composite.

odor sequestration and fragrance delivery (Patent WO2007035596A2, filed in 2006). This innovation offers a reliable platform for addressing malodor situations, where MOFs containing various metals and linkers control or eliminate malodor molecules, even in the presence of customized molecules. Additionally, the same MOF architectures allow for the incorporation of fragrance compositions. Another example is a patent filed in 2010 by BASF (US20120016066A1), where Ulrich Müller and coworkers describe a biodegradable material composed of a polymer containing a porous MOF (ranging between 0.01% and 10% by weight of the polymer). This material, prepared as a foil or film, aims to adsorb ethene in food packaging to prolong the shelf life of fruits and vegetables. The versatility of this solution positions it as a competitor to traditional adsorbents like zeolites, silicas, or activated carbons (Silva et al., 2015).

Controlling the ripening process of fresh fruits and vegetables, particularly climacteric fruits, during storage and transportation poses a significant challenge in maintaining their quality. Tests comparing MOFs with conventional adsorbents like activated charcoal and silica gels have highlighted the potential of MOFs in selectively adsorbing bioactive compounds. Basolite C300 (MOF-199) and Basolite A520, both MOF-based materials, have shown promise in encapsulating ethylene for postharvest applications, thereby maintaining the quality of climacteric fruits (Chopra et al., 2017).

NovoMOF, for example, offers a solution for improving the shelf life of packaged food by absorbing ethylene. Therefore, the long-term benefits of high production and optimization of green synthesis protocols for MOFs extend beyond MOF chemistry (Chakraborty et al., 2023). Recently, Decco, a fruit packaging and preservation company, has employed MOFs to prevent premature fruit ripening and mitigate the effects of ethylene release, thereby reducing spoilage. The use of MOFs in engineered processes has significantly extended the shelf life of fruits. Fruits such as apples and pears produce ethylene during ripening. Methylcyclopropene, released from MOFs, binds to ethylene receptors in fruit and slows down the ripening process (Rubio-Martinez et al., 2017).

Conclusion

The exploration of MOF composites has revealed their substantial potential across a myriad of industrial applications. The introduction of MOF composites has ushered in new methodologies and technologies, significantly enhancing the capabilities of traditional materials. Prominent industrial methods for MOF composite synthesis have been meticulously developed, allowing for the fine-tuning of properties to meet specific application needs. These synthesis methods have paved the way for advancements in various industrial sectors, including chemical purification, gas storage, catalysis, energy storage, sensing, drug delivery, bioimaging, atmospheric water harvesting, and food packaging. In chemical purification, MOF composites have demonstrated exceptional efficacy in capturing harmful pollutants such as CO_2, SO_2, NO_x, and VOCs, as well as in petrochemical purification. For gas storage, particularly hydrogen and natural gas, the high surface area and tunable porosity of MOF composites have facilitated the development of efficient storage systems. Catalysis has benefited from MOF composites acting as highly selective and efficient catalysts, driving innovations in chemical manufacturing and environmental remediation. The energy storage sector has seen significant improvements with MOF composites in batteries and supercapacitors, contributing to enhanced energy density, stability, and charge--discharge cycles. In sensing and biosensing, MOF composites have enabled advanced sensor technologies, including "turn-on" processes, smartphone-integrated sensors, point-of-care testing sensors, and wearable diagnostic sensors, potentially revolutionizing healthcare diagnostics and environmental monitoring. Drug delivery systems have been transformed by MOF composites, offering controlled and targeted delivery of therapeutics, opening new avenues for personalized medicine and improved treatment efficacy. MOF-based nanocomposites have shown promise in bioimaging, providing high-resolution and noninvasive imaging techniques crucial for medical diagnostics and research. Atmospheric water harvesting has advanced with MOF composites efficiently capturing and releasing water from the air, addressing water scarcity in arid regions. In food packaging, MOF composites have demonstrated potential to enhance food preservation, extend shelf life, and reduce waste. In conclusion, the diverse industrial applications of MOF composites underscore their transformative impact on material science and engineering. Ongoing research and development efforts continue to address current limitations, driving the scalability, cost-effectiveness, and sustainability of MOF composite production. The future of MOF composites holds promise for further advancements, potentially unlocking new application domains and significantly contributing to various industrial processes.

References

Abazari, R., Sanati, S., Bajaber, M. A., Javed, M. S., Junk, P. C., Nanjundan, A. K., Qian, J., & Dubal, D. P. (2024). Design and advanced manufacturing of NU-1000 metal–organic frameworks with future perspectives for environmental and renewable energy applications. *Small (Weinheim an der Bergstrasse, Germany), 20*(15), 2306353. https://doi.org/10.1002/smll.202306353, http://onlinelibrary.wiley.com/journal/10.1002/(ISSN)1613–6829.

Ahmed, A., Forster, M., Clowes, R., Bradshaw, D., Myers, P., & Zhang, H. (2013). Silica SOS@HKUST-1 composite microspheres as easily packed stationary phases for fast separation. *Journal of Materials Chemistry A, 1*(10), 3276–3286. https://doi.org/10.1039/c2ta01125e.

Ahmed, A., Seth, S., Purewal, J., Wong-Foy, A. G., Veenstra, M., Matzger, A. J., & Siegel, D. J. (2019). Exceptional hydrogen storage achieved by screening nearly half a million metal-organic frameworks. *Nature Communications, 10*(1), 1568. https://doi.org/10.1038/s41467-019-09365-w, http://www.nature.com/ncomms/index.html.

Ahmed, I., & Jhung, S. H. (2014). Composites of metal-organic frameworks: Preparation and application in adsorption. *Materials Today, 17*(3), 136–146. https://doi.org/10.1016/j.mattod. 2014.03.002, http://www.journals.elsevier.com/materials-today/.

Ahrestani, Z., Sadeghzadeh, S., & Emrooz, H. B. M. (2023). An overview of atmospheric water harvesting methods, the inevitable path of the future in water supply. *RSC Advances, 13*(15), 10273–10307. https://doi.org/10.1039/d2ra07733g, http://pubs.rsc.org/en/journals/journal/ra.

Aijaz, A., Karkamkar, A., Choi, Y. J., Tsumori, N., Rönnebro, E., Autrey, T., Shioyama, H., & Xu, Q. (2012). Immobilizing highly catalytically active Pt nanoparticles inside the pores of metal-organic framework: A double solvents approach. *Journal of the American Chemical Society, 134*(34), 13926–13929. https://doi.org/10.1021/ja3043905.

Akhbari, K., Karami, S., Phuruangrat, A., & Saedi, Z. (2018). Irreversible replacement of sodium with thallium in sodium coordination polymer nanostructures by solid-state mechanochemical cation exchange process. *Journal of the Iranian Chemical Society, 15*(6), 1327–1335. https://doi.org/10.1007/s13738-018-1331-1, http://link.springer.com/journal/13738.

Akhbari, K., & Morsali, A. (2013). Modulating methane storage in anionic nano-porous MOF materials via post-synthetic cation exchange process. *Dalton Transactions, 42*(14), 4786–4789. https://doi.org/10.1039/c3dt32846e.

Akhbari, K., & Morsali, A. (2015). Needle-like hematite nano-structure prepared by directed thermolysis of MIL-53 nano-structure with enhanced methane storage capacity. *Materials Letters, 141*, 315–318. https://doi.org/10.1016/j.matlet.2014.11.110, http://www.journals.elsevier.com/materials-letters/.

Alavijeh, R. K., & Akhbari, K. (2022). Improvement of curcumin loading into a nanoporous functionalized poor hydrolytic stable metal-organic framework for high anticancer activity against human gastric cancer AGS cells. *Colloids and Surfaces B: Biointerfaces, 212*, 112340. https://doi.org/10.1016/j.colsurfb.2022.112340.

Alavijeh, R. K., & Akhbari, K. (2024). Cancer therapy by nano MIL-n series of metal-organic frameworks. *Coordination Chemistry Reviews, 503*, 215643. https://doi.org/10.1016/j.ccr. 2023.215643.

Alavijeh, R. K., Akhbari, K., & White, J. (2019). Solid–liquid conversion and carbon dioxide storage in a calcium-based metal–organic framework with micro- and nanoporous channels. *Crystal Growth & Design, 19*(12), 7290–7297. https://doi.org/10.1021/acs.cgd.9b01174.

Ali, L., & Mahmoud, E. (2021). Recent advances in the design of metal–organic frameworks for methane storage and delivery. *Journal of Porous Materials, 28*(1), 213–230. https://doi.org/10.1007/s10934-020-00984-z.

Almassad, H. A., Abaza, R. I., Siwwan, L., Al-Maythalony, B., & Cordova, K. E. (2022). Environmentally adaptive MOF-based device enables continuous self-optimizing atmospheric water harvesting. *Nature Communications, 13*(1), 4873. https://doi.org/10.1038/s41467-022-32642-0, http://www.nature.com/ncomms/index.html.

Al-Naddaf, Q., Al-Mansour, M., Thakkar, H., & Rezaei, F. (2018). MOF-GO hybrid nanocomposite adsorbents for methane storage. *Industrial and Engineering Chemistry Research, 57*(51), 17470–17479. https://doi.org/10.1021/acs.iecr.8b03638, http://pubs.acs.org/journal/iecred.

Al-Rowaili, F. N., Zahid, U., Onaizi, S., Khaled, M., Jamal, A., & AL-Mutairi, E. M. (2021). A review for metal-organic frameworks (MOFs) utilization in capture and conversion of carbon dioxide into valuable products. *Journal of CO_2 Utilization, 53*, 101715. https://doi.org/10.1016/j.jcou.2021.101715, http://www.journals.elsevier.com/journal-of-co2-utilization/.

Alsmail, N. H., Suyetin, M., Yan, Y., Cabot, R., Krap, C. P., Lü, J., Easun, T. L., Bichoutskaia, E., Lewis, W., Blake, A. J., & Schröder, M. (2014). Analysis of high and selective uptake of CO_2

in an oxamide-containing {Cu$_2$(OOCR)$_4$}-based metal-organic framework. *Chemistry - A European Journal, 20*(24), 7317–7324. https://doi.org/10.1002/chem.201304005, www.interscience.wiley.com.

Ameloot, R., Liekens, A., Alaerts, L., Maes, M., Galarneau, A., Coq, B., Desmet, G., Sels, B. F., Denayer, J. F. M., & Vos, D. E. D. (2010). Silica-MOF composites as a stationary phase in liquid chromatography. *European Journal of Inorganic Chemistry, 24,* 3735–3738. https://doi.org/10.1002/ejic.201000494, http://onlinelibrary.wiley.com/doi/10.1002/ejic.201000494/pdf.

Amidi, D. M., & Akhbari, K. (2024). Iodine-loaded ZIF-7-coated cotton substrates show sustained iodine release as effective antibacterial textiles. *New Journal of Chemistry, 48*(5), 2016–2027. https://doi.org/10.1039/d3nj05198f.

Amini, A., Kazemi, S., & Safarifard, V. (2020). Metal-organic framework-based nanocomposites for sensing applications – A review. *Polyhedron, 177,* 114260. https://doi.org/10.1016/j.poly.2019.114260.

Anbia, M., & Hoseini, V. (2012). Development of MWCNT@MIL-101 hybrid composite with enhanced adsorption capacity for carbon dioxide. *Chemical Engineering Journal, 191,* 326–330. https://doi.org/10.1016/j.cej.2012.03.025.

Annapragada, R., Vandavasi, K. R., & Kanuparthy, P. R. (2023). Metal-organic framework membranes for proton exchange membrane fuel cells: A mini-review. *Inorganica Chimica Acta, 546,* 121304. https://doi.org/10.1016/j.ica.2022.121304, https://www.sciencedirect.com/journal/inorganica-chimica-acta/issues.

Arenas-Vivo, A., Rojas, S., Ocaña, I., Torres, A., Liras, M., Salles, F., Arenas-Esteban, D., Bals, S., Ávila, D., & Horcajada, P. (2021). Ultrafast reproducible synthesis of a Ag-nanocluster@MOF composite and its superior visible-photocatalytic activity in batch and in continuous flow. *Journal of Materials Chemistry A, 9*(28), 15704–15713. https://doi.org/10.1039/d1ta02251b, http://pubs.rsc.org/en/journals/journal/ta.

Arnó, J., Farha, O., Morris, W., Siu, P., Tom, G., Weston, M., Fuller, P., McCabe, J., & Ameen, M. (2018a). Next generation dopant gas delivery system for ion implant applications. *Semiconductor Digest,* 27–30.

Arnó, J. Farha, O. K., Morris, W., Siu, P. W., Tom, G. M., Weston, M. H., & Fuller, P. E. (2018b). ION-X dopant gas delivery system performance characterization at Axcelis, *Proceedings of the international conference on ion implantation technology,* Institute of Electrical and Electronics Engineers Inc. United States. 227–230.. 10.1109/IIT.2018.8807983.

Arnó, J., Farha, O. K., Morris, W., Siu, P. W., Tom, G. M., Weston, M. H., & Fuller, P. E. (2022). Dopant gas purity and adsorbent stability. *MRS Advances, 7*(36), 1426–1430. https://doi.org/10.1557/s43580-022-00416-x, https://www.springer.com/journal/43580.

Arshadi Edlo, A., & Akhbari, K. (2024). Modulated antibacterial activity in ZnO@MIL-53(Fe) and CuO@MIL-53(Fe) nanocomposites prepared by simple thermal treatment process. *Applied Organometallic Chemistry, 38*(2), e7326. https://doi.org/10.1002/aoc.7326, http://onlinelibrary.wiley.com/journal/10.1002/(ISSN)1099-0739.

Ashworth, D. J., Driver, J., Sasitharan, K., Prasad, R. R. R., Nicks, J., Smith, B. J., Patwardhan, S. V., & Foster, J. A. (2023). Scalable and sustainable manufacturing of ultrathin metal–organic framework nanosheets (MONs) for solar cell applications. *Chemical Engineering Journal, 477,* 146871. https://doi.org/10.1016/j.cej.2023.146871, www.elsevier.com/inca/publications/store/6/0/1/2/7/3/index.htt.

Atomis Inc., Japan. https://www.atomis.co.jp/en/.

Azim, M. M., & Mohsin, U. (2019). Graphene oxide/transition metal oxide as a promising nanomaterial for hydrogen storage. In *Graphene-based nanotechnologies for energy and environmental applications*. South Africa: Elsevier, 121–144. https://www.sciencedirect.com/book/9780128158111, 10.1016/B978-0-12-815811-1.00007-7.

Bakuru, V. R., DMello, M. E., & Kalidindi, S. B. (2019). Metal-organic frameworks for hydrogen energy applications: Advances and challenges. *Chemphyschem: A European Journal of Chemical Physics and Physical Chemistry, 20*(10), 1177–1215. https://doi.org/10.1002/cphc.201801147, http://onlinelibrary.wiley.com/journal/10.1002/(ISSN)1439-7641.

Bayliss, P. A., Ibarra, I. A., Pérez, E., Yang, S., Tang, C. C., Poliakoff, M., & Schröder, M. (2014). Synthesis of metal-organic frameworks by continuous flow. *Green Chemistry, 16*(8), 3796–3802. https://doi.org/10.1039/c4gc00313f, http://pubs.rsc.org/en/journals/journal/gc.

Ben, T., Lu, C., Pei, C., Xu, S., & Qiu, S. (2012). Polymer-supported and free-standing metal-organic framework membrane. *Chemistry – A European Journal, 18*(33), 10250–10253. https://doi.org/10.1002/chem.201201574.

Benoit, V., Pillai, R. S., Orsi, A., Normand, P., Jobic, H., Nouar, F., Billemont, P., Bloch, E., Bourrelly, S., Devic, T., Wright, P. A., De Weireld, G., Serre, C., Maurin, G., & Llewellyn, P. L. (2016). MIL-91(Ti), a small pore metal-organic framework which fulfils several criteria: An upscaled green synthesis, excellent water stability, high CO_2 selectivity and fast CO_2 transport. *Journal of Materials Chemistry A, 4*(4), 1383–1389. https://doi.org/10.1039/c5ta09349j, http://pubs.rsc.org/en/journals/journalissues/ta.

Berkeley, M.P.U. (2021). *Aluminium makes water-harvesting MOF*. https://www.chemistryworld.com/news/aluminium-makes-water-harvesting-mof-10-times-thirstier/3010974.article.

Bhatt, P. M., Belmabkhout, Y., Cadiau, A., Adil, K., Shekhah, O., Shkurenko, A., Barbour, L. J., Eddaoudi, M., & Fine-Tuned, A. (2016). Fluorinated MOF addresses the needs for trace CO_2 removal and air capture using physisorption. *Journal of the American Chemical Society, 138*(29), 9301–9307. https://doi.org/10.1021/jacs.6b05345, http://pubs.acs.org/journal/jacsat.

Bhoite, A. A., Patil, K. V., Redekar, R. S., Jang, J. H., Sawant, V. A., & Tarwal, N. L. (2023). Recent advances in metal-organic framework (MOF) derived metal oxides and their composites with carbon for energy storage applications. *Journal of Energy Storage, 72*, 108557. https://doi.org/10.1016/j.est.2023.108557.

Bieniek, A., Terzyk, A. P., Wiśniewski, M., Roszek, K., Kowalczyk, P., Sarkisov, L., Keskin, S., & Kaneko, K. (2021). MOF materials as therapeutic agents, drug carriers, imaging agents and biosensors in cancer biomedicine: Recent advances and perspectives. *Progress in Materials Science, 117*, 100743. https://doi.org/10.1016/j.pmatsci.2020.100743, https://www.journals.elsevier.com/progress-in-materials-science.

Boorboor Ajdari, F., Kowsari, E., Niknam Shahrak, M., Ehsani, A., Kiaei, Z., Torkzaban, H., Ershadi, M., Kholghi Eshkalak, S., Haddadi-Asl, V., Chinnappan, A., & Ramakrishna, S. (2020). A review on the field patents and recent developments over the application of metal organic frameworks (MOFs) in supercapacitors. *Coordination Chemistry Reviews, 422*, 213441. https://doi.org/10.1016/j.ccr.2020.213441, http://www.journals.elsevier.com/coordination-chemistry-reviews/.

Borges, D. D., Normand, P., Permiakova, A., Babarao, R., Heymans, N., Galvao, D. S., Serre, C., De Weireld, G., & Maurin, G. (2017). Gas adsorption and separation by the Al-based metal-organic framework MIL-160. *Journal of Physical Chemistry C, 121*(48), 26822–26832. https://doi.org/10.1021/acs.jpcc.7b08856, http://pubs.acs.org/journal/jpccck.

Borhani, S., Moradi, M., Kiani, M. A., Hajati, S., & Toth, J. (2017). CoxZn1−x ZIF-derived binary Co_3O_4/ZnO wrapped by 3D reduced graphene oxide for asymmetric supercapacitor: Comparison of pure and heat-treated bimetallic MOF. *Ceramics International, 43*(16), 14413–14425. https://doi.org/10.1016/j.ceramint.2017.07.211.

Burd, S. D., Ma, S., Perman, J. A., Sikora, B. J., Snurr, R. Q., Thallapally, P. K., Tian, J., Wojtas, L., & Zaworotko, M. J. (2012). Highly selective carbon dioxide uptake by [Cu(bpy-n) 2(SiF 6)] (bpy-1=4,4′-bipyridine; Bpy-2=1,2-bis(4-pyridyl)ethene. *Journal of the American Chemical Society, 134*(8), 3663–3666. https://doi.org/10.1021/ja211340t.

Cai, Y., Guan, J., Wang, W., Wang, L., Su, J., & Fang, L. (2021). pH and light-responsive polycaprolactone/curcumin@zif-8 composite films with enhanced antibacterial activity. *Journal of Food Science, 86*(8), 3550–3562. https://doi.org/10.1111/1750-3841.15839, http://onlinelibrary.wiley.com/journal/10.1111/(ISSN)1750-3841.

Carné-Sánchez, A., Imaz, I., Cano-Sarabia, M., & Maspoch, D. (2013). A spray-drying strategy for synthesis of nanoscale metal-organic frameworks and their assembly into hollow superstructures. *Nature Chemistry, 5*(3), 203–211. https://doi.org/10.1038/nchem.1569.

Carraro, F., Williams, J. D., Linares-Moreau, M., Parise, C., Liang, W., Amenitsch, H., Doonan, C., Kappe, C. O., & Falcaro, P. (2020). Continuous-flow synthesis of ZIF-8 biocomposites with tunable particle size. *Angewandte Chemie, 132*(21), 8200–8204. https://doi.org/10.1002/ange.202000678.

Cattani, L., Magrini, A., & Leoni, V. (2022). Energy performance of water generators from gaseous mixtures by condensation: Climatic datasets choice. *Energies, 15*(20), 7581. https://doi.org/10.3390/en15207581.

Chakraborty, D., Yurdusen, A., Mouchaham, G., Nouar, F., & Serre, C. (2023). Large-scale production of metal–organic frameworks. *Advanced Functional Materials, 34*(43), 2309089.

Chen, B., Fan, D., Pinto, R. V., Dovgaliuk, I., Nandi, S., Chakraborty, D., García-Moncada, N., Vimont, A., McMonagle, C. J., Bordonhos, M., Al Mohtar, A., Cornu, I., Florian, P., Heymans, N., Daturi, M., Weireld, G. D., Pinto, M., Nouar, F., Maurin, G., ... Serre, C. (2024). A scalable robust microporous Al-MOF for post-combustion carbon capture. *Advanced Science, 11*(21), 2401070. https://doi.org/10.1002/advs.202401070.

Chen, C., Wang, H., Chen, Y., Wei, X., Zou, W., Wan, H., Dong, L., & Guan, G. (2021). Layer-by-layer self-assembly of hierarchical flower-like HKUST-1-based composite over amino-tethered SBA-15 with synergistic enhancement for CO_2 capture. *Chemical Engineering Journal, 413*, 127396. https://doi.org/10.1016/j.cej.2020.127396.

Chen, G.-J. (2022). Successful pilot test on the separation butane isomers by using ZIF-8 slurry-a critical step approaching the industrial application of metal-organic frameworks (MOFs). *Chemistry*. https://communities.springernature.com/posts/successful-pilot-test-on-the-separation-butane-isomers-by-using-zif-8-slurry-a-critical-step-approaching-the-industrial-application-of-metal-organic-frameworks-mofs.

Chen, H. C., Hou, L. Y., He, C., Laing, P. J., Huang, C. Y., & Kuo, W. S. (2022). Metal-organic framework-assisted synthesis of three-dimensional ZnCoS effloresced nanopillars@CNT paper for high-performance flexible all-solid-state battery-type supercapacitors with ultrahigh specific capacitance. *ACS Applied Energy Materials, 5*(7), 8262–8272. https://doi.org/10.1021/acsaem.2c00778, pubs.acs.org/journal/aaemcq.

Chen, Y., Lv, D., Wu, J., Xiao, J., Xi, H., Xia, Q., & Li, Z. (2017). A new MOF-505@GO composite with high selectivity for CO_2/CH_4 and CO_2/N_2 separation. *Chemical Engineering Journal, 308*, 1065–1072. https://doi.org/10.1016/j.cej.2016.09.138, www.elsevier.com/inca/publications/store/6/0/1/2/7/3/index.htt.

Chen, Y., Wu, H., Xiao, Q., Lv, D., Li, F., Li, Z., & Xia, Q. (2019). Rapid room temperature conversion of hydroxy double salt to MOF-505 for CO_2 capture. *CrystEngComm, 21*(1), 165–171. https://doi.org/10.1039/c8ce01489b, http://pubs.rsc.org/en/journals/journal/ce.

Chen, Z., Li, P., Anderson, R., Wang, X., Zhang, X., Robison, L., Redfern, L. R., Moribe, S., Islamoglu, T., Gómez-Gualdrón, D. A., Yildirim, T., Stoddart, J. F., & Farha, O. K. (2020).

Balancing volumetric and gravimetric uptake in highly porous materials for clean energy. *Science (New York, N.Y.), 368*(6488), 297–303. https://doi.org/10.1126/science.aaz8881, https://science.sciencemag.org/content/368/6488/297/tab-pdf.

Chen, Z., Wasson, M. C., Drout, R. J., Robison, L., Idrees, K. B., Knapp, J. G., Son, F. A., Zhang, X., Hierse, W., Kühn, C., Marx, S., Hernandez, B., & Farha, O. K. (2021). The state of the field: From inception to commercialization of metal-organic frameworks. *Faraday Discussions, 225*, 9–69. https://doi.org/10.1039/d0fd00103a, http://pubs.rsc.org/en/journals/journal/fd.

Cheng, L., Dang, Y., Wang, Y., & Chen, K. (2024). Recent advances in metal-organic frameworks for water absorption and their applications. *Materials Chemistry Frontiers, 8*, 1171–1194.

Choe, J. H., Kim, H., & Hong, C. S. (2021). MOF-74 type variants for CO_2 capture. *Materials Chemistry Frontiers, 5*(14), 5172–5185. https://doi.org/10.1039/d1qm00205h, rsc.li/frontiers-materials.

Chong, K. C., Ho, P. S., Lai, S. O., Lee, S. S., Lau, W. J., Lu, S.-Y., & Ooi, B. S. (2022). Solvent-free synthesis of MIL-101 (Cr) for CO_2 gas adsorption: the effect of metal precursor and molar ratio. *Sustainability, 14*(3), 1152.

Chong, K. C., Lai, S. O., Mah, S. K., Thiam, H. S., Chong, W. C., Shuit, S. H., Lee, S. S., & Chong, W. E. (2023). A review of HKUST-1 metal-organic frameworks in gas adsorption. *IOP conference series: Earth and environmental science*, Institute of Physics, Malaysia, 1135. https://doi.org/10.1088/1755-1315/1135/1/012030, https://iopscience.iop.org/journal/1755-1315.

Chopra, S., Dhumal, S., Abeli, P., Beaudry, R., & Almenar, E. (2017). Metal-organic frameworks have utility in adsorption and release of ethylene and 1-methylcyclopropene in fresh produce packaging. *Postharvest Biology and Technology, 130*, 48–55. https://doi.org/10.1016/j.postharvbio.2017.04.001, www.elsevier.com/inca/publications/store/5/0/3/1/3/index.htt.

Colombo, V., Montoro, C., Maspero, A., Palmisano, G., Masciocchi, N., Galli, S., Barea, E., & Navarro, J. A. R. (2012). Tuning the adsorption properties of isoreticular pyrazolate-based metal-organic frameworks through ligand modification. *Journal of the American Chemical Society, 134*(30), 12830–12843. https://doi.org/10.1021/ja305267m.

Cortés, P. H., & Macéas, S. R. (2021). Metal-organic frameworks in biomedical and environmental field. In *Metal-organic frameworks in biomedical and environmental field*. Spain: Springer International Publishing, 1–503. https://link.springer.com/book/10.1007/978-3-030-63380-6, 10.1007/978-3-030-63380-6.

Chuhadiya, S., Himanshu, Suthar, D., Patel, S. L., & Dhaka, M. S. (2021). Metal organic frameworks as hybrid porous materials for energy storage and conversion devices: A review. *Coordination Chemistry Reviews, 446*, 214115. https://doi.org/10.1016/j.ccr.2021.214115.

Crawford, D., Casaban, J., Haydon, R., Giri, N., McNally, T., & James, S. L. (2015). Synthesis by extrusion: Continuous, large-scale preparation of MOFs using little or no solvent. *Chemical Science, 6*(3), 1645–1649. https://doi.org/10.1039/c4sc03217a, http://pubs.rsc.org/en/Journals/JournalIssues/SC.

Cui, X., Sun, X., Liu, L., Huang, Q., Yang, H., Chen, C., Nie, S., Zhao, Z., & Zhao, Z. (2019). In-situ fabrication of cellulose foam HKUST-1 and surface modification with polysaccharides for enhanced selective adsorption of toluene and acidic dipeptides. *Chemical Engineering Journal, 369*, 898–907. https://doi.org/10.1016/j.cej.2019.03.129, www.elsevier.com/inca/publications/store/6/0/1/2/7/3/index.htt.

Czaja, A. U., Trukhan, N., & Müller, U. (2009). Industrial applications of metal–organic frameworks. *Chemical Society Reviews, 38*(5), 1284–1293. https://doi.org/10.1039/b804680h.

Dai, Y., Niu, Z., Luo, W., Wang, Y., Mu, P., & Li, J. (2023). A review on the recent advances in composite membranes for CO_2 capture processes. *Separation and Purification Technology, 307*, 122752. https://doi.org/10.1016/j.seppur.2022.122752.

Dalstein, O., Ceratti, D. R., Boissière, C., Grosso, D., Cattoni, A., & Faustini, M. (2016). Nanoimprinted, submicrometric, MOF-based 2D photonic structures: Toward easy selective vapors sensing by a smartphone camera. *Advanced Functional Materials, 26*(1), 81–90. https://doi.org/10.1002/adfm.201503016, http://onlinelibrary.wiley.com/journal/10.1002/(ISSN)1616-3028.

Das, A. K., Vemuri, R. S., Kutnyakov, I., McGrail, B. P., & Motkuri, R. K. (2016). An efficient synthesis strategy for metal-organic frameworks: Dry-gel synthesis of MOF-74 framework with high yield and improved performance. *Scientific Reports, 6*, 28050. https://doi.org/10.1038/srep28050, www.nature.com/srep/index.html.

Davoodi, A., Akhbari, K., & Alirezvani, M. (2023). Prolonged release of silver and iodine from ZIF-7 carrier with great antibacterial activity. *CrystEngComm, 25*(27), 3931–3942. https://doi.org/10.1039/d3ce00529a, http://pubs.rsc.org/en/journals/journal/ce.

Deacon, A., Briquet, L., Malankowska, M., Massingberd-Mundy, F., Rudić, S., Hyde, Tl, Cavaye, H., Coronas, J., Poulston, S., & Johnson, T. (2022). Understanding the ZIF-L to ZIF-8 transformation from fundamentals to fully costed kilogram-scale production. *Communications Chemistry, 5*(1), 18. https://doi.org/10.1038/s42004-021-00613-z, nature.com/commschem/.

DeCoste, J. B., Demasky, T. J., Katz, M. J., Farha, O. K., & Hupp, J. T. (2015). A UiO-66 analogue with uncoordinated carboxylic acids for the broad-spectrum removal of toxic chemicals. *New Journal of Chemistry, 39*(4), 2396–2399. https://doi.org/10.1039/c4nj02093f, http://pubs.rsc.org/en/journals/journal/nj.

Denny, M. S., & Cohen, S. M. (2015). In situ modification of metal-organic frameworks in mixed-matrix membranes. *Angewandte Chemie – International Edition, 54*(31), 9029–9032. https://doi.org/10.1002/anie.201504077, http://onlinelibrary.wiley.com/journal/10.1002/(ISSN)1521-3773.

Deria, P., Mondloch, J. E., Tylianakis, E., Ghosh, P., Bury, W., Snurr, R. Q., Hupp, J. T., & Farha, O. K. (2013). Perfluoroalkane functionalization of NU-1000 via solvent-assisted ligand incorporation: Synthesis and CO_2 adsorption studies. *Journal of the American Chemical Society, 135*(45), 16801–16804. https://doi.org/10.1021/ja408959g.

DeSantis, D., Mason, J. A., James, B. D., Houchins, C., Long, J. R., & Veenstra, M. (2017). Techno-economic analysis of metal-organic frameworks for hydrogen and natural gas storage. *Energy and Fuels, 31*(2), 2024–2032. https://doi.org/10.1021/acs.energyfuels.6b02510, http://pubs.acs.org/journal/enfuem.

Dhakshinamoorthy, A., Alvaro, M., & Garcia, H. (2012). Commercial metal–organic frameworks as heterogeneous catalysts. *Chemical Communications, 48*(92), 11275–11288. https://doi.org/10.1039/c2cc34329k.

Didriksen, T., Spjelkavik, A. I., & Blom, R. (2017). Continuous synthesis of the metal-organic framework cpo-27-ni from aqueous solutions. *Journal of Flow Chemistry, 7*(1), 13–17. https://doi.org/10.1556/1846.2016.00040, http://akademiai.com/doi/pdf/10.1556/1846.2016.00040.

Ding, M., Flaig, R. W., Jiang, H. L., & Yaghi, O. M. (2019). Carbon capture and conversion using metal-organic frameworks and MOF-based materials. *Chemical Society Reviews, 48*(10), 2783–2828. https://doi.org/10.1039/c8cs00829a, http://pubs.rsc.org/en/journals/journal/cs.

Ding, M., & Jiang, H. L. (2018). Incorporation of imidazolium-based poly(ionic liquid)s into a metal-organic framework for CO_2 capture and conversion. *ACS Catalysis, 8*(4), 3194–3201. https://doi.org/10.1021/acscatal.7b03404, http://pubs.acs.org/page/accacs/about.html.

Duan, C., Meng, J., Wang, X., Meng, X., Sun, X., Xu, Y., Zhao, W., & Ni, Y. (2018). Synthesis of novel cellulose- based antibacterial composites of Ag nanoparticles@ metal-organic frameworks@carboxymethylated fibers. *Carbohydrate Polymers, 193*, 82–88. https://doi.org/10.1016/j.carbpol.2018.03.089, http://www.elsevier.com/wps/find/journaldescription.cws_home/405871/description#description.

Duan, C., Yu, Y., & Hu, H. (2022). Recent progress on synthesis of ZIF-67-based materials and their application to heterogeneous catalysis. *Green Energy & Environment, 7*(1), 3–15. https://doi.org/10.1016/j.gee.2020.12.023.

Dunne, P. W., Lester, E., & Walton, R. I. (2016). Towards scalable and controlled synthesis of metal-organic framework materials using continuous flow reactors. *Reaction Chemistry and Engineering, 1*(4), 352–360. https://doi.org/10.1039/c6re00107f, www.rsc.org/journals-books-databases/about-journals/reaction-chemistry-engineering/.

Ebrahim, A. M., & Bandosz, T. J. (2014). Effect of amine modification on the properties of zirconium-carboxylic acid based materials and their applications as NO_2 adsorbents at ambient conditions. *Microporous and Mesoporous Materials, 188*, 149–162. https://doi.org/10.1016/j.micromeso.2014.01.009.

Edlo, A. A., & Akhbari, K. (2023). Modulating the antibacterial activity of a CuO@HKUST-1 nanocomposite by optimizing its synthesis procedure. *New Journal of Chemistry, 47*(45), 20770–20776. https://doi.org/10.1039/d3nj03914e, http://pubs.rsc.org/en/journals/journal/nj.

Ehsani, A., Bigdeloo, M., Assefi, F., Kiamehr, M., & Alizadeh, R. (2020). Ternary nanocomposite of conductive polymer/chitosan biopolymer/metal organic framework: Synthesis, characterization and electrochemical performance as effective electrode materials in pseudocapacitors. *Inorganic Chemistry Communications, 115*, 107885. https://doi.org/10.1016/j.inoche.2020.107885.

Emaminejad, S., Gao, W., Wu, E., Davies, Z. A., Nyein, H. Y. Y., Challa, S., Ryan, S. P., Fahad, H. M., Chen, K., Shahpar, Z., Talebi, S., Milla, C., Javey, A., & Davis, R. W. (2017). Autonomous sweat extraction and analysis applied to cystic fibrosis and glucose monitoring using a fully integrated wearable platform. *Proceedings of the National Academy of Sciences, 114*(18), 4625–4630. https://doi.org/10.1073/pnas.1701740114.

Eyankware, O. E., & Ateke, I. H. (2020). Methane and hydrogen storage in metal organic frameworks: A mini review. *Journal of Environmental and Earth Sciences, 2*(2), 56–68. https://doi.org/10.30564/jees.v2i2.2642, https://ojs.bilpublishing.com/index.php/jees/article/view/2642/2293.

Fan, L., Xue, M., Kang, Z., Li, H., & Qiu, S. (2012). Electrospinning technology applied in zeolitic imidazolate framework membrane synthesis. *Journal of Materials Chemistry, 22*(48), 25272–25276. https://doi.org/10.1039/c2jm35401b.

Fang, X., Zong, B., & Mao, S. (2018). Metal-organic framework-based sensors for environmental contaminant sensing. *Nano-micro Letters, 10*, 1–19.

Farha, O. K., Yazaydin, A. O., Eryazici, I., Malliakas, C. D., Hauser, B. G., Kanatzidis, M. G., Nguyen, S. T., Snurr, R. Q., & Hupp, J. T. (2010). De novo synthesis of a metal-organic framework material featuring ultrahigh surface area and gas storage capacities. *Nature Chemistry, 2*(11), 944–948. https://doi.org/10.1038/nchem.834.

Faustini, M., Kim, J., Jeong, G. Y., Kim, J. Y., Moon, H. R., Ahn, W. S., & Kim, D. P. (2013). Microfluidic approach toward continuous and ultrafast synthesis of metal-organic framework crystals and hetero structures in confined microdroplets. *Journal of the American Chemical Society, 135*(39), 14619–14626. https://doi.org/10.1021/ja4039642.

Feng, L., Wang, K. Y., Willman, J., & Zhou, H. C. (2020). Hierarchy in metal-organic frameworks. *ACS Central Science, 6*(3), 359–367. https://doi.org/10.1021/acscentsci.0c00158, http://pubs.acs.org/journal/acsii.

Framergy Inc. (2022). *A novel, nanostructured, metal-organic frameworks-based product loss prevention technology in the oil and natural gas sector.*

Freitas, C., Severino, M. I., Mohtar, A. A., Kolmykov, O., Pimenta, V., Nouar, F., Serre, C., & Pinto, M. (2024). Metal-organic frameworks polyurethane composite foams for the capture of volatile organic compounds. *ACS Materials Letters, 6*(1), 174–181. https://doi.org/10.1021/acsmaterialslett.3c01273, https://pubs.acs.org/page/amlcef/about.html.

Gaab, M., Trukhan, N., Maurer, S., Gummaraju, R., & Müller, U. (2012). The progression of Al-based metal-organic frameworks – From academic research to industrial production and applications. *Microporous and Mesoporous Materials, 157*, 131–136. https://doi.org/10.1016/j.micromeso.2011.08.016.

García-Holley, P., Schweitzer, B., Islamoglu, T., Liu, Y., Lin, L., Rodriguez, S., Weston, M. H., Hupp, J. T., Gómez-Gualdrón, D. A., Yildirim, T., & Farha, O. K. (2018). Benchmark study of hydrogen storage in metal-organic frameworks under temperature and pressure swing conditions. *Energy Letters, 3*(3), 748–754. https://doi.org/10.1021/acsenergylett.8b00154, http://pubs.acs.org/journal/aelccp.

Gargiulo, V., Policicchio, A., Lisi, L., & Alfe, M. (2023). CO_2 capture and gas storage capacities enhancement of HKUST-1 by hybridization with functionalized graphene-like materials. *Energy and Fuels, 37*(7), 5291–5302. https://doi.org/10.1021/acs.energyfuels.2c04289, http://pubs.acs.org/journal/enfuem.

Ge, X., Wong, R., Anisa, A., & Ma, S. (2022). Recent development of metal-organic framework nanocomposites for biomedical applications. *Biomaterials, 281*, 121322. https://doi.org/10.1016/j.biomaterials.2021.121322.

Gebremariam, S. K., Dumée, L. F., Llewellyn, P. L., Alwahedi, Y. F., & Karanikolos, G. N. (2023). Metal-organic framework hybrid adsorbents for carbon capture – A review. *Journal of Environmental Chemical Engineering, 11*(2), 109291. https://doi.org/10.1016/j.jece.2023.109291, http://www.journals.elsevier.com/journal-of-environmental-chemical-engineering/.

Global Forecast to 2032. (2021). Metal Organic Frameworks Market – By product (Aluminum, Copper, Iron, Zinc, Magnesium), By synthetic market (hydro thermal, microwave, ultrasonic, mechanochemical, electrochemical), By application (catalyst, carbon capture). https://www.gminsights.com/industry-analysis/metal-organic-frameworks-market.

Global Market Insights Inc. (2023). *By synthetic market (hydro thermal, microwave, ultrasonic, mechanochemical, electrochemical), by application (catalyst, carbon capture).* Global Forecast to. (2023).

Gonzalez, M. I., Kapelewski, M. T., Bloch, E. D., Milner, P. J., Reed, D. A., Hudson, M. R., Mason, J. A., Barin, G., Brown, C. M., & Long, J. R. (2018). Separation of Xylene Isomers through Multiple Metal Site Interactions in Metal-Organic Frameworks. *Journal of the American Chemical Society, 140*(9), 3412–3422. https://doi.org/10.1021/jacs.7b13825.

Gordeeva, L. G., Tu, Y. D., Pan, Q., Palash, M. L., Saha, B. B., Aristov, Y. I., & Wang, R. Z. (2021). Metal-organic frameworks for energy conversion and water harvesting: A bridge between thermal engineering and material science. *Energy. 84*, 105946. https://doi.org/10.1016/j.nanoen.2021.105946, http://www.journals.elsevier.com/nano-energy/.

Grande, C. A., Blom, R., Spjelkavik, A., Moreau, V., & Payet, J. (2017). Life-cycle assessment as a tool for eco-design of metal-organic frameworks (MOFs). *Sustainable Materials and Technologies, 14*, 11–18. https://doi.org/10.1016/j.susmat.2017.10.002.

Green Science Alliance Co., Ltd. (2022). https://www.gsalliance.co.jp/?lang=en.

Guo, Z., Wang, T., Rawal, A., Hou, J., Cao, Z., Zhang, H., Xu, J., Gu, C., Chen, V., & Liang, K. (2019). Biocatalytic self-propelled submarine-like metal-organic framework microparticles with pH-triggered buoyancy control for directional vertical motion. *Materials Today, 28*,

10–16. https://doi.org/10.1016/j.mattod.2019.04.022, http://www.journals.elsevier.com/materials-today/.

Gómez-Gualdrón, D. A., Wang, T. C., García-Holley, P., Sawelewa, R. M., Argueta, E., Snurr, R. Q., Hupp, J. T., Yildirim, T., & Farha, O. K. (2017). Understanding volumetric and gravimetric hydrogen adsorption trade-off in metal-organic frameworks. *Applied Materials and Interfaces, 9*(39), 33419–33428. https://doi.org/10.1021/acsami.7b01190, http://pubs.acs.org/journal/aamick.

Habib, N., Durak, O., Gulbalkan, H. C., Aydogdu, A. S., Keskin, S., & Uzun, A. (2023). Composite of MIL-101(Cr) with a pyrrolidinium-based ionic liquid providing high CO_2 selectivity. *ACS Applied Engineering Materials, 1*(6), 1473–1481. https://doi.org/10.1021/acsaenm.3c00010.

Halevi, O., Tan, J. M. R., Lee, P. S., & Magdassi, S. (2018). Hydrolytically stable MOF in 3D-printed structures. *Advanced Sustainable Systems, 2*(2), 1700150. https://doi.org/10.1002/adsu.201700150, www.advsustainsys.com.

Han, X., Godfrey, H. G. W., Briggs, L., Davies, A. J., Cheng, Y., Daemen, L. L., Sheveleva, A. M., Tuna, F., McInnes, E. J. L., Sun, J., Drathen, C., George, M. W., Ramirez-Cuesta, A. J., Thomas, K. M., Yang, S., & Schröder, M. (2018). Reversible adsorption of nitrogen dioxide within a robust porous metal–organic framework. *Nature Materials, 17*(8), 691–696. https://doi.org/10.1038/s41563-018-0104-7, http://www.nature.com/nmat/.

Hanikel, N., Prévot, M. S., & Yaghi, O. M. (2020). MOF water harvesters. *Nature Nanotechnology, 15*(5), 348–355. https://doi.org/10.1038/s41565-020-0673-x, http://www.nature.com/nnano/index.html.

Hasheminezhad, M., Akhbari, K., & Phuruangrat, A. (2019). Solid–solid and solid–liquid conversion of sodium and silver nano coordination polymers. *Polyhedron, 166*, 115–122. https://doi.org/10.1016/j.poly.2019.03.032, http://www.journals.elsevier.com/polyhedron/.

He, B., Macreadie, L. K., Gardiner, J., Telfer, S. G., & Hill, M. R. (2021). In situ investigation of multicomponent MOF crystallization during rapid continuous flow synthesis. *Applied Materials and Interfaces, 13*(45), 54284–54293. https://doi.org/10.1021/acsami.1c04920, http://pubs.acs.org/journal/aamick.

He, B., Sadiq, M. M., Batten, M. P., Suzuki, K., Rubio-Martinez, M., Gardiner, J., & Hill, M. R. (2019). Continuous flow synthesis of a Zr magnetic framework composite for post-combustion CO_2 capture. *Chemistry – A European Journal, 25*(57), 13184–13188. https://doi.org/10.1002/chem.201902560, http://onlinelibrary.wiley.com/journal/10.1002/(ISSN)1521-3765.

He, Q., Zhan, F., Wang, H., Xu, W., Wang, H., & Chen, L. (2022). Recent progress of industrial preparation of metal–organic frameworks: Synthesis strategies and outlook. *Materials Today Sustainability, 17*, 100104. https://doi.org/10.1016/j.mtsust.2021.100104.

He, Y., Fu, T., Wang, L., Liu, J., Liu, G., & Zhao, H. (2023). Self-assembly of MOF-801 into robust hierarchically porous monoliths for scale-up atmospheric water harvesting. *Chemical Engineering Journal, 472*, 144786. https://doi.org/10.1016/j.cej.2023.144786.

Henschel, A., Gedrich, K., Kraehnert, R., & Kaskel, S. (2008). Catalytic properties of MIL-101. *Chemical Communications, 35*, 4192–4194. https://doi.org/10.1039/b718371b, http://pubs.rsc.org/en/journals/journal/cc.

Hou, C. C., Wang, H. F., Li, C., & Xu, Q. (2020). From metal-organic frameworks to single/dual-atom and cluster metal catalysts for energy applications. *Energy and Environmental Science, 13*(6), 1658–1693. https://doi.org/10.1039/c9ee04040d, http://pubs.rsc.org/en/journals/journal/ee.

Hovestadt, M., Vargas Schmitz, J., Weissenberger, T., Reif, F., Kasperit, M., Schwieger, W., & Hartmann, M. (2017). Scale-up of the synthesis of zeolitic imidazolate framework ZIF-4.

Chemie-Ingenieur-Technik, *89*(10), 1374–1378. https://doi.org/10.1002/cite.201700105, http://onlinelibrary.wiley.com/journal/10.1002/(ISSN)1522-2640/.

Hu, Y., Ye, Z., & Peng, X. (2023). Metal-organic frameworks for solar-driven atmosphere water harvesting. *Chemical Engineering Journal, 452,* 139656. https://doi.org/10.1016/j.cej.2022.139656.

Hu, Z., Wang, Y., Shah, B. B., & Zhao, D. (2019). CO_2 capture in metal–organic framework adsorbents: An engineering perspective. *Advanced Sustainable Systems, 3*(1), 1800080.

Huang, J.-y (2023). Recent advancement in natural gas storage using metal-organic frameworks. *Highlights in Science, Engineering and Technology, 73,* 540–547. https://doi.org/10.54097/hset.v73i.14667.

Huang, N., Drake, H., Li, J., Pang, J., Wang, Y., Yuan, S., Wang, Q., Cai, P., Qin, J., & Zhou, H. C. (2018). Flexible and hierarchical metal–organic framework composites for high-performance catalysis. *Angewandte Chemie – International Edition, 57*(29), 8916–8920. https://doi.org/10.1002/anie.201803096, http://onlinelibrary.wiley.com/journal/10.1002/(ISSN)1521-3773.

Hughes, R., Kotamreddy, G., Ostace, A., Bhattacharyya, D., Siegelman, R. L., Parker, S. T., Didas, S. A., Long, J. R., Omell, B., & Matuszewski, M. (2021). Isotherm, kinetic, process modeling, and techno-economic analysis of a diamine-appended metal-organic framework for CO_2 capture using fixed bed contactors. *Energy and Fuels, 35*(7), 6040–6055. https://doi.org/10.1021/acs.energyfuels.0c04359, http://pubs.acs.org/journal/enfuem.

Hydrogen production MOF Technologies Company. (2022). https://www.coherentmarketinsights.com/industry-reports/metal-organic-framework-market.

Hydrogen storage gets real. (2022). https://www.chemistryworld.com/features/hydrogen-storage-gets-real/3010794.article.

Ibarra, I. A., Yang, S., Lin, X., Blake, A. J., Rizkallah, P. J., Nowell, H., Allan, D. R., Champness, N. R., Hubberstey, P., & Schröder, M. (2011). Highly porous and robust scandium-based metal-organic frameworks for hydrogen storage. *Chemical Communications, 47*(29), 8304–8306. https://doi.org/10.1039/c1cc11168j, http://pubs.rsc.org/en/journals/journal/cc.

Ikezoe, Y., Washino, G., Uemura, T., Kitagawa, S., & Matsui, H. (2012). Autonomous motors of a metal-organic framework powered by reorganization of self-assembled peptides at interfaces. *Nature Materials, 11*(12), 1081–1085. https://doi.org/10.1038/nmat3461, http://www.nature.com/nmat/.

Matsuoka, K. (2022). In Japan, start-up culture is starting to emerge. *Long given short shrift, young chemistry-related companies are winning government and business attention, 100,* 23–24.

Jacoby, M. (2013). Materials chemistry: Metal-organic frameworks go commercial. *Chemical and Engineering News,* 34–35.

Jambovane, S. R., Nune, S. K., Kelly, R. T., McGrail, B. P., Wang, Z., Nandasiri, M. I., Katipamula, S., Trader, C., & Schaef, H. T. (2016). Continuous, one-pot synthesis and post-synthetic modification of nanoMOFs using droplet nanoreactors. *Scientific Reports, 6,* 36657. https://doi.org/10.1038/srep36657, www.nature.com/srep/index.html.

Janardan, S., Eswara Rao, P. C. V. V, Manjunatha, H., Venkata Ratnam, K., Ratnamala, A., Chandra Babu Naidu, K., Sivarmakrishna, A., Khan, A., & Asiri, A. M. (2021). Permeable metal-organic frameworks for fuel (gas) storage applications. *Metal-Organic Frameworks for Chemical Reactions,* 111–126. 10.1016/b978-0-12-822099-3.00005-8.

Jaramillo, D. E., Reed, D. A., Jiang, H. Z. H., Oktawiec, J., Mara, M. W., Forse, A. C., Lussier, D. J., Murphy, R. A., Cunningham, M., Colombo, V., Shuh, D. K., Reimer, J. A., & Long, J. R. (2020). Selective nitrogen adsorption via backbonding in a metal-organic framework with exposed vanadium sites. *Nature materials, 19*(5), 517–521. https://doi.org/10.1038/s41563-019-0597-8.

Javed, M. S., Shaheen, N., Hussain, S., Li, J., Shah, S. S. A., Abbas, Y., Ahmad, M. A., Raza, R., & Mai, W. (2019). An ultra-high energy density flexible asymmetric supercapacitor based on hierarchical fabric decorated with 2D bimetallic oxide nanosheets and MOF-derived porous carbon polyhedra. *Journal of Materials Chemistry A, 7*(3), 946–957. https://doi.org/10.1039/c8ta08816k, http://pubs.rsc.org/en/journals/journal/ta.

Jin, X., Shan, Y., Sun, F., & Pang, H. (2022). Applications of transition metal (Fe, Co, Ni)-based metal–organic frameworks and their derivatives in batteries and supercapacitors. *Transactions of Tianjin University, 28*(6), 446–468. https://doi.org/10.1007/s12209-022-00340-z, https://www.springer.com/journal/12209.

Juan-Alcañiz, J., Ramos-Fernandez, E. V., Lafont, U., Gascon, J., & Kapteijn, F. (2010). Building MOF bottles around phosphotungstic acid ships: One-pot synthesis of bi-functional polyoxometalate-MIL-101 catalysts. *Journal of Catalysis, 269*(1), 229–241. https://doi.org/10.1016/j.jcat.2009.11.011.

Kajal, N., Singh, V., Gupta, R., & Gautam, S. (2022). Metal organic frameworks for electrochemical sensor applications: A review. *Environmental Research, 204*, 112320. https://doi.org/10.1016/j.envres.2021.112320.

Kalati, M., & Akhbari, K. (2021). Optimizing the metal ion release and antibacterial activity of ZnO@ZIF-8 by modulating its synthesis method. *New Journal of Chemistry, 45*(48), 22924–22931. https://doi.org/10.1039/d1nj04534b, http://pubs.rsc.org/en/journals/journal/nj.

Kang, P. C., Ou, Y. S., Li, G. L., Chang, J. K., & Wang, C. Y. (2021). Room-temperature hydrogen adsorption via spillover in Pt nanoparticle-decorated UiO-66 nanoparticles: Implications for hydrogen storage. *ACS Applied Nano Materials, 4*(10), 11269–11280. https://doi.org/10.1021/acsanm.1c02862, https://pubs.acs.org/journal/aanmf6.

Kapelewski, M. T., Runčevski, T., Tarver, J. D., Jiang, H. Z. H., Hurst, K. E., Parilla, P. A., Ayala, A., Gennett, T., Fitzgerald, S. A., Brown, C. M., & Long, J. R. (2018). Record high hydrogen storage capacity in the metal-organic framework Ni$_2$(m-dobdc) at near-ambient temperatures. *Chemistry of Materials, 30*(22), 8179–8189. https://doi.org/10.1021/acs.chemmater.8b03276, http://pubs.acs.org/journal/cmatex.

Karadeniz, B., Howarth, A. J., Stolar, T., Islamoglu, T., Dejanović, I., Tireli, M., Wasson, M. C., Moon, S. Y., Farha, O. K., Friščić, T., & Užarević, K. (2018). Benign by design: Green and scalable synthesis of zirconium UiO-metal-organic frameworks by water-assisted mechanochemistry. *ACS Sustainable Chemistry and Engineering, 6*(11), 15841–15849. https://doi.org/10.1021/acssuschemeng.8b04458, http://pubs.acs.org/journal/ascecg.

Karimi Alavijeh, R., & Akhbari, K. (2020). Biocompatible MIL-101(Fe) as a smart carrier with high loading potential and sustained release of curcumin. *Inorganic Chemistry, 59*(6), 3570–3578. https://doi.org/10.1021/acs.inorgchem.9b02756, http://pubs.acs.org/journal/inocaj.

Karimi Alavijeh, R., Beheshti, S., Akhbari, K., & Morsali, A. (2018). Investigation of reasons for metal–organic framework's antibacterial activities. *Polyhedron, 156*, 257–278. https://doi.org/10.1016/j.poly.2018.09.028, http://www.journals.elsevier.com/polyhedron/.

Karmakar, A., Samanta, P., Dutta, S., & Ghosh, S. K. (2019). Fluorescent "turn-on" sensing based on metal–organic frameworks (MOFs). *Chemistry – An Asian Journal, 14*(24), 4506–4519. https://doi.org/10.1002/asia.201901168, http://onlinelibrary.wiley.com/journal/10.1002/(ISSN)1861-471X.

Kayal, S., & Chakraborty, A. (2018). Activated carbon (type Maxsorb-III) and MIL-101(Cr) metal organic framework based composite adsorbent for higher CH$_4$ storage and CO$_2$ capture.

Chemical Engineering Journal, 334, 780–788. https://doi.org/10.1016/j.cej.2017.10.080, www.elsevier.com/inca/publications/store/6/0/1/2/7/3/index.htt.

Kerkel, K., Arno, J., Reichl, G., Feicht, J., Winzig, H., Farha, O. K., Morris, W., Siu, P. W., Tom, G. M., Weston, M. H., & Fuller, P. E. (2018).Evaluation of ION-X® hydride dopant gas sources on a Varian VIISion high current implanter. *Proceedings of the international conference on ion implantation technology*, Institute of Electrical and Electronics Engineers Inc., Germany. 223–226. https://doi.org/10.1109/IIT.2018.8807928.

Khokhar, S., Anand, H., & Chand, P. (2022). Current advances of nickel based metal organic framework and their nanocomposites for high performance supercapacitor applications: A critical review. *Journal of Energy Storage, 56,* 105897. https://doi.org/10.1016/j.est.2022.105897.

Kianimehr, A., Akhbari, K., White, J., & Phuruangrat, A. (2020). The mechanochemical conversion of potassium coordination polymer nanostructures to interpenetrated sodium coordination polymers with halogen bond, metal-carbon and metal-metal interactions. *CrystEngComm, 22*(5), 888–894. https://doi.org/10.1039/c9ce01861a, http://pubs.rsc.org/en/journals/journal/ce.

Kim, H., Sohail, M., Wang, C., Rosillo-Lopez, M., Baek, K., Koo, J., Seo, M. W., Kim, S., Foord, J. S., & Han, S. O. (2019). Facile one-pot synthesis of bimetallic Co/Mn-MOFs@rice husks, and its carbonization for supercapacitor electrodes. *Scientific Reports, 9*(1), 8984. https://doi.org/10.1038/s41598-019-45169-0, www.nature.com/srep/index.html.

Kim, J., Lee, Y. R., & Ahn, W. S. (2013). Dry-gel conversion synthesis of Cr-MIL-101 aided by grinding: High surface area and high yield synthesis with minimum purification. *Chemical Communications, 49*(69), 7647–7649. https://doi.org/10.1039/c3cc44559c.

Kim, J. O., Kim, J. Y., Lee, J. C., Park, S., Moon, H. R., & Kim, D. P. (2019). Versatile processing of metal-organic framework-fluoropolymer composite inks with chemical resistance and sensor applications. *ACS Applied Materials and Interfaces, 11*(4), 4385–4392. https://doi.org/10.1021/acsami.8b19630, http://pubs.acs.org/journal/aamick.

Kim, K. J., Li, Y. J., Kreider, P. B., Chang, C. H., Wannenmacher, N., Thallapally, P. K., & Ahn, H. G. (2013). High-rate synthesis of Cu-BTC metal-organic frameworks. *Chemical Communications, 49*(98), 11518–11520. https://doi.org/10.1039/c3cc46049e.

Kirchon, A. A. (2020). *Developing commercially scalable iron and titanium metal-organic frameworks for gas storage and water purification.* Electronic Theses, dissertations, and records of study.

Kou, X., Tong, L., Shen, Y., Zhu, W., Yin, L., Huang, S., Zhu, F., Chen, G., & Ouyang, G. (2020). Smartphone-assisted robust enzymes@MOFs-based paper biosensor for point-of-care detection. *Biosensors and Bioelectronics, 156,* 112095. https://doi.org/10.1016/j.bios.2020.112095.

Kukulka, W., Cendrowski, K., Michalkiewicz, B., & Mijowska, E. (2019). MOF-5 derived carbon as material for CO_2 absorption. *RSC Advances, 9*(32), 18527–18537. https://doi.org/10.1039/c9ra01786k, http://pubs.rsc.org/en/journals/journal/ra.

Kumar, N., Bansal, N., Yamauchi, Y., & Salunkhe, R. R. (2022). Two-dimensional layered heterostructures of nanoporous carbons using reduced graphene oxide and metal−organic frameworks. *Chemistry of Materials, 34*(11), 4946–4954. https://doi.org/10.1021/acs.chemmater.2c00160, http://pubs.acs.org/journal/cmatex.

Kumar, R., Jayaramulu, K., Maji, T. K., & Rao, C. N. R. (2013). Hybrid nanocomposites of ZIF-8 with graphene oxide exhibiting tunable morphology, significant CO_2 uptake and other novel properties. *Chemical Communications, 49*(43), 4947–4949. https://doi.org/10.1039/c3cc00136a.

Lenzen, D., Bendix, P., Reinsch, H., Fröhlich, D., Kummer, H., Möllers, M., Hügenell, P. P. C., Gläser, R., Henninger, S., & Stock, N. (2018). Scalable green synthesis and full-scale test of the metal–organic framework CAU-10-H for use in adsorption-driven chillers. *Advanced Materials, 30*(6), 1705869. https://doi.org/10.1002/adma.201705869, http://www3.interscience.wiley.com/journal/119030556/issue.

Lestari, W. W., Yunita, L., Saraswati, T. E., Heraldy, E., Khafidhin, M. A., Krisnandi, Y. K., Arrozi, U. S. F., & Kadja, G. T. M. (2021). Fabrication of composite materials MIL-100(Fe)/Indonesian activated natural zeolite as enhanced CO_2 capture material. *Chemical Papers, 75*(7), 3253–3263. https://doi.org/10.1007/s11696-021-01558-2, www.springer.com/11696.

Li, D., Xu, H. Q., Jiao, L., & Jiang, H. L. (2019). Metal-organic frameworks for catalysis: State of the art, challenges, and opportunities. *EnergyChem, 1*(1), 100005. https://doi.org/10.1016/j.enchem.2019.100005, www.journals.elsevier.com/energychem.

Li, G., Cai, H., Li, X., Zhang, J., Zhang, D., Yang, Y., & Xiong, J. (2019). Construction of hierarchical $NiCo_2O_4$@Ni-MOF hybrid arrays on carbon cloth as superior battery-type electrodes for flexible solid-state hybrid supercapacitors. *ACS Applied Materials and Interfaces. 11*(41), 37675–37684. https://doi.org/10.1021/acsami.9b11994, http://pubs.acs.org/journal/aamick.

Li, G., Kobayashi, H., Taylor, J. M., Ikeda, R., Kubota, Y., Kato, K., Takata, M., Yamamoto, T., Toh, S., Matsumura, S., & Kitagawa, H. (2014). Hydrogen storage in Pd nanocrystals covered with a metal-organic framework. *Nature Materials, 13*(8), 802–806. https://doi.org/10.1038/nmat4030, http://www.nature.com/nmat/.

Li, J., Yu, X., Xu, M., Liu, W., Sandraz, E., Lan, H., Wang, J., & Cohen, S. M. (2017). Metal-organic frameworks as micromotors with tunable engines and brakes. *Journal of the American Chemical Society, 139*(2), 611–614. https://doi.org/10.1021/jacs.6b11899, http://pubs.acs.org/journal/jacsat.

Li, R., Chen, T., & Pan, X. (2021). Metal-organic-framework-based materials for antimicrobial applications. *ACS Nano, 15*(3), 3808–3848. https://doi.org/10.1021/acsnano.0c09617, http://pubs.acs.org/journal/ancac3.

Li, R. Y., Wang, Z. S., Yuan, Z. Y., Van Horne, C., Freger, V., Lin, M., Cai, R. K., & Chen, J. P. (2022). A comprehensive review on water stable metal-organic frameworks for large-scale preparation and applications in water quality management based on surveys made since 2015. *Critical Reviews in Environmental Science and Technology, 52*(22), 4038–4071. https://doi.org/10.1080/10643389.2021.1975444, www.tandf.co.uk/journals/titles/10643389.asp.

Li, W., Zhang, W., Hao, S., & Wu, H. (2023). Bimetal metal-organic framework-derived Ni-Mn@carbon/reduced graphene oxide as a cathode for an asymmetric supercapacitor with high energy density. *Langmuir: The ACS Journal of Surfaces and Colloids, 39*(35), 12510–12519. https://doi.org/10.1021/acs.langmuir.3c01747, http://pubs.acs.org/journal/langd5.

Liao, M. D., Peng, C., Hou, S. P., Chen, J., Zeng, X. G., Wang, H. L., & Lin, J. H. (2020). Large-scale synthesis of nitrogen-doped activated carbon fibers with high specific surface area for high-performance supercapacitors. *Energy Technology, 8*(5), 1901477. https://doi.org/10.1002/ente.201901477, http://onlinelibrary.wiley.com/journal/10.1002/(ISSN)2194-4296.

Lin, J. B., Nguyen, T. T. T., Vaidhyanathan, R., Burner, J., Taylor, J. M., Durekova, H., Akhtar, F., Mah, R. K., Ghaffari-Nik, O., Marx, S., Fylstra, N., Iremonger, S. S., Dawson, K. W., Sarkar, P., Hovington, P., Rajendran, A., Woo, T. K., & Shimizu, G. K. H. (2021). A scalable metal-organic framework as a durable physisorbent for carbon dioxide capture. *Science (New York, N.Y.), 374*(6574), 1464–1469. https://doi.org/10.1126/science.abi7281, https://www.science.org/doi/10.1126/science.abi7281.

Lin, R., Chai, M., Zhou, Y., Chen, V., Bennett, T. D., & Hou, J. (2023). Metal-organic framework glass composites. *Chemical Society Reviews, 52*(13), 4149–4172. https://doi.org/10.1039/d2cs00315e, http://pubs.rsc.org/en/journals/journal/cs.

Liu, C.-S., Li, J., & Pang, H. (2020). Metal-organic framework-based materials as an emerging platform for advanced electrochemical sensing. *Coordination Chemistry Reviews, 410*, 213222. https://doi.org/10.1016/j.ccr.2020.213222.

Liu, J., Chen, C., Zhang, K., & Zhang, L. (2021). Applications of metal–organic framework composites in CO_2 capture and conversion. *Chinese Chemical Letters, 32*(2), 649–659. https://doi.org/10.1016/j.cclet.2020.07.040, http://www.elsevier.com/wps/find/journaldescription.cws_home/997/description#description.

Liu, K.-G., Sharifzadeh, Z., Rouhani, F., Ghorbanloo, M., & Morsali, A. (2021). Metal-organic framework composites as green/sustainable catalysts. *Coordination Chemistry Reviews, 436*, 213827. https://doi.org/10.1016/j.ccr.2021.213827.

Liu, X. W., Sun, T. J., Hu, J. L., & Wang, S. D. (2016). Composites of metal-organic frameworks and carbon-based materials: Preparations, functionalities and applications. *Journal of Materials Chemistry A, 4*(10), 3584–3616. https://doi.org/10.1039/c5ta09924b, http://pubs.rsc.org/en/journals/journal/ta.

Llewellyn, P. L., Bourrelly, S., Serre, C., Vimont, A., Daturi, M., Hamon, L., De Weireld, G., Chang, J. S., Hong, D. Y., Hwang, Y. K., Jhung, S. H., & Férey, G. (2008). High uptakes of CO_2 and CH_4 in mesoporous metal-organic frameworks MIL-100 and MIL-101. *Langmuir: The ACS Journal of Surfaces and Colloids, 24*(14), 7245–7250. https://doi.org/10.1021/la800227x, http://pubs.acs.org/journal/langd5.

Luan, X., Pan, Y., Zhou, D., He, B., Liu, X., Gao, Y., Yang, J., & Song, Y. (2020). Cerium metal organic framework mediated molecular threading for point-of-care colorimetric assays. *Biosensors and Bioelectronics, 165*, 112406. https://doi.org/10.1016/j.bios.2020.112406.

López-Maya, E., Montoro, C., Colombo, V., Barea, E., & Navarro, J. A. R. (2014). Improved CO_2 capture from flue gas by basic sites, charge gradients, and missing linker defects on nickel face cubic centered MOFs. *Advanced Functional Materials, 24*(39), 6130–6135. https://doi.org/10.1002/adfm.201400795, http://onlinelibrary.wiley.com/journal/10.1002/(ISSN)1616-3028.

Ma, X., Li, L., Chen, R., Wang, C., Li, H., & Wang, S. (2018). Heteroatom-doped nanoporous carbon derived from MOF-5 for CO_2 capture. *Applied Surface Science, 435*, 494–502. https://doi.org/10.1016/j.apsusc.2017.11.069, http://www.journals.elsevier.com/applied-surface-science/.

Ma, Y., Qu, X., Liu, C., Xu, Q., & Tu, K. (2021). Metal-organic frameworks and their composites towards biomedical applications. *Frontiers in Molecular Biosciences, 8*, 805228. https://doi.org/10.3389/fmolb.2021.805228, journal.frontiersin.org/journal/molecular-biosciences.

Magri, A., Petriccione, M., & Gutiérrez, T. J. (2021). Metal-organic frameworks for food applications: A review. *Food Chemistry, 354*, 129533. https://doi.org/10.1016/j.foodchem.2021.129533.

Malouche, A., Zlotea, C., & Szilágyi, P.Á. (2019). Interactions of hydrogen with Pd@MOF composites. *Chemphyschem: A European Journal of Chemical Physics and Physical Chemistry, 20*(10), 1282–1295. https://doi.org/10.1002/cphc.201801092, http://onlinelibrary.wiley.com/journal/10.1002/(ISSN)1439-7641.

Mansour, F. R., Hammad, S. F., Abdallah, I. A., Bedair, A., Abdelhameed, R. M., & Locatelli, M. (2024). Applications of metal organic frameworks in point of care testing. *TrAC – Trends in Analytical Chemistry, 172*, 117596. https://doi.org/10.1016/j.trac.2024.117596, https://www.sciencedirect.com/science/journal/01659936.

Mao, J., Tang, Y., Wang, Y., Huang, J., Dong, X., Chen, Z., & Lai, Y. (2019). Particulate matter capturing via naturally dried ZIF-8/graphene aerogels under harsh conditions. *iScience, 16*, 133–144. https://doi.org/10.1016/j.isci.2019.05.024, www.cell.com/iscience.

Marrett, J. M., Mottillo, C., Girard, S., Nickels, C. W., Do, J. L., Dayaker, G., Germann, L. S., Dinnebier, R. E., Howarth, A. J., Farha, O. K., Friščić, T., & Li, C. J. (2018). Supercritical carbon dioxide enables rapid, clean, and scalable conversion of a metal oxide into zeolitic metal-organic frameworks. *Crystal Growth and Design, 18*(5), 3222–3228. https://doi.org/10.1021/acs.cgd.8b00385, http://pubs.acs.org/journal/cgdefu.

Marti, A. M., Wickramanayake, W., Dahe, G., Sekizkardes, A., Bank, T. L., Hopkinson, D. P., & Venna, S. R. (2017). Continuous flow processing of ZIF-8 membranes on polymeric porous hollow fiber supports for CO$_2$ capture. *ACS Applied Materials and Interfaces, 9*(7), 5678–5682. https://doi.org/10.1021/acsami.6b16297, http://pubs.acs.org/journal/aamick.

McKinstry, C., Cathcart, R. J., Cussen, E. J., Fletcher, A. J., Patwardhan, S. V., & Sefcik, J. (2016). Scalable continuous solvothermal synthesis of metal organic framework (MOF-5) crystals. *Chemical Engineering Journal, 285*, 718–725. https://doi.org/10.1016/j.cej.2015.10.023, www.elsevier.com/inca/publications/store/6/0/1/2/7/3/index.htt.

Meskher, H., Belhaouari, S. B., & Sharifianjazi, F. (2023). Mini review about metal organic framework (MOF)-based wearable sensors: Challenges and prospects. *Heliyon, 9*(11), e21621. https://doi.org/10.1016/j.heliyon.2023.e21621, http://www.journals.elsevier.com/heliyon/.

Millward, A. R., & Yaghi, O. M. (2005). Metal-organic frameworks with exceptionally high capacity for storage of carbon dioxide at room temperature. *Journal of the American Chemical Society, 127*(51), 17998–17999. https://doi.org/10.1021/ja0570032.

Min, T., Sun, X., Zhou, L., Du, H., Zhu, Z., & Wen, Y. (2021). Electrospun pullulan/PVA nanofibers integrated with thymol-loaded porphyrin metal–organic framework for antibacterial food packaging. *Carbohydrate Polymers, 270*, 118391. https://doi.org/10.1016/j.carbpol.2021.118391, http://www.elsevier.com/wps/find/journaldescription.cws_home/405871/description#description.

Mirzadeh, E., Akhbari, K., & White, J. (2018). Mechanochemical conversion of nano potassium hydrogen terephthalate to thallium analogue nanoblocks with strong hydrogen bonding and straight chain metalophillic interactions. *Applied Organometallic Chemistry, 32*(5), e4313. https://doi.org/10.1002/aoc.4313, http://www3.interscience.wiley.com/journal/2676/home.

Mirzaei, M., & Amiri, A. (2023). Metal-organic frameworks in analytical chemistry. *Royal Society of Chemistry*, 470.

Moeinian, M., Akhbari, K., Boonmak, J., & Youngme, S. (2016). Similar to what occurs in biological systems; irreversible replacement of potassium with thallium in coordination polymer nanostructures. *Polyhedron, 118*, 6–11. https://doi.org/10.1016/j.poly.2016.07.039, http://www.journals.elsevier.com/polyhedron/.

MOFCAT (Functional Metal Organic Frameworks as Heterogeneous Catalysts) (2022). Final report summary. https://cordis.europa.eu/project/id/33335/reporting/it.

Moghadam, B. H., Hasanzadeh, M., & Simchi, A. (2020). Self-powered wearable piezoelectric sensors based on polymer nanofiber-metal-organic framework nanoparticle composites for arterial pulse monitoring. *ACS Applied Nano Materials, 3*(9), 8742–8752. https://doi.org/10.1021/acsanm.0c01551, https://pubs.acs.org/journal/aanmf6.

Mohammadi Amidi, D., Akhbari, K., & Soltani, S. (2023). Loading of ZIF-67 on silk with sustained release of iodine as biocompatible antibacterial fibers. *Applied Organometallic Chemistry, 37*(1), e6913. https://doi.org/10.1002/aoc.6913, http://onlinelibrary.wiley.com/journal/10.1002/(ISSN)1099-0739.

Mohan, B., Kumar, S., & Chen, Q. (2022). Obtaining water from air using porous metal–organic frameworks (MOFs). *Topics in Current Chemistry, 380*(6), 54. https://doi.org/10.1007/s41061-022-00410-9, https://www.springer.com/journal/41061.

Molefe, L. Y., Musyoka, N. M., Ren, J., Langmi, H. W., Mathe, M., & Ndungu, P. G. (2021). Effect of inclusion of MOF-polymer composite onto a carbon foam material for hydrogen storage application. *Journal of Inorganic and Organometallic Polymers and Materials, 31*(1), 80–88. https://doi.org/10.1007/s10904-020-01701-8, http://www.springerlink.com/content/r64737117kr4/, http://www.springer.com/east/home/generic/search/results?SGWID=5-40109-70-35505322-0.

Mueller, U., Puetter, H., Hesse, M., Schubert, M., Wessel, H., Huff, J., & Guzmann, M. (2011). Method for electrochemical production of a crystalline porous metal organic skeleton material. *Google Patents*.

Mueller, U., Schubert, M., Teich, F., Puetter, H., Schierle-Arndt, K., & Pastré, J. (2006). Metal-organic frameworks – Prospective industrial applications. *Journal of Materials Chemistry, 16*(7), 626–636. https://doi.org/10.1039/b511962f.

Munn, A. S., Dunne, P. W., Tang, S. V. Y., & Lester, E. H. (2015). Large-scale continuous hydrothermal production and activation of ZIF-8. *Chemical Communications, 51*(64), 12811–12814. https://doi.org/10.1039/c5cc04636j.

Murtaza, S. Z. M., Alqassem, H. T., Sabouni, R., & Ghommem, M. (2023). Degradation of micropollutants by metal organic framework composite-based catalysts: A review. *Environmental Technology and Innovation, 29*, 102998. https://doi.org/10.1016/j.eti.2022.102998, http://www.journals.elsevier.com/environmental-technology-and-innovation/.

Muschi, M., Devautour-Vinot, S., Aureau, D., Heymans, N., Sene, S., Emmerich, R., Ploumistos, A., Geneste, A., Steunou, N., Patriarche, G., De Weireld, G., & Serre, C. (2021). Metal-organic framework/graphene oxide composites for CO_2 capture by microwave swing adsorption. *Journal of Materials Chemistry A, 9*(22), 13135–13142. https://doi.org/10.1039/d0ta12215g, http://pubs.rsc.org/en/journals/journal/ta.

Musyoka, N. M., Ren, J., Langmi, H. W., North, B. C., Mathe, M., & Bessarabov, D. (2017). Synthesis of rGO/Zr-MOF composite for hydrogen storage application. *Journal of Alloys and Compounds, 724*, 450–455. https://doi.org/10.1016/j.jallcom.2017.07.040.

Naveed, H., Shaheen, H., Kumari, R., Lakra, R., Khan, A. L., & Basu, S. (2022). Sustainable metal-organic framework technologies for CO_2 Capture. In *Sustainable carbon capture: Technologies and applications*. Pakistan: CRC Press, 161–183. http://www.tandfebooks.com/doi/book/9781000537413, 10.1201/9781003162780-5.

Nik Zaiman, N. F. H., Shaari, N., & Harun, N. A. M. (2022). Developing metal-organic framework-based composite for innovative fuel cell application: An overview. *International Journal of Energy Research, 46*(2), 471–504. https://doi.org/10.1002/er.7198, http://onlinelibrary.wiley.com/journal/10.1002/(ISSN)1099-114X.

O'Neill, L. D., Zhang, H., & Bradshaw, D. (2010). Macro-/microporous MOF composite beads. *Journal of Materials Chemistry, 20*(27), 5720–5726. https://doi.org/10.1039/c0jm00515k.

Oh, H., Maurer, S., Balderas-Xicohtencatl, R., Arnold, L., Magdysyuk, O. V., Schütz, G., Müller, U., & Hirscher, M. (2017). Efficient synthesis for large-scale production and characterization for hydrogen storage of ligand exchanged MOF-74/174/184-M (M=Mg2+, Ni2+). *International Journal of Hydrogen Energy, 42*(2), 1027–1035. https://doi.org/10.1016/j.ijhydene.2016.08.153, http://www.journals.elsevier.com/international-journal-of-hydrogen-energy/.

Olajire, A. A. (2018). Synthesis chemistry of metal-organic frameworks for CO_2 capture and conversion for sustainable energy future. *Renewable and Sustainable Energy Reviews, 92*,

570–607. https://doi.org/10.1016/j.rser.2018.04.073, https://www.journals.elsevier.com/renewable-and-sustainable-energy-reviews.

Olorunyomi, J. F., Geh, S. T., Caruso, R. A., & Doherty, C. M. (2021). Metal-organic frameworks for chemical sensing devices. *Materials Horizons, 8*(9), 2387–2419. https://doi.org/10.1039/d1mh00609f, http://pubs.rsc.org/en/journals/journal/mh.

Ordoñez, M. J. C., Balkus, K. J., Ferraris, J. P., & Musselman, I. H. (2010). Molecular sieving realized with ZIF-8/Matrimid® mixed-matrix membranes. *Journal of Membrane Science, 361*(1-2), 28–37. https://doi.org/10.1016/j.memsci.2010.06.017, www.elsevier.com/locate/memsci.

Ostermann, R., Cravillon, J., Weidmann, C., Wiebcke, M., & Smarsly, B. M. (2011). Metal-organic framework nanofibers via electrospinning. *Chemical Communications, 47*(1), 442–444. https://doi.org/10.1039/c0cc02271c.

Osterrieth, J. W., & Fairen-Jimenez, D. (2021). Metal–organic framework composites for theragnostics and drug delivery applications. *Biotechnology Journal, 16*(2), 2000005.

Ouyang, Y., Gong, W., Zhang, Q., Wang, J., Guo, S., Xiao, Y., Li, D., Wang, C., Sun, X., Wang, C., & Huang, S. (2023). Bilayer zwitterionic metal-organic framework for selective all-solid-state superionic conduction in lithium metal batteries. *Advanced Materials, 35*(39), 2304685. https://doi.org/10.1002/adma.202304685, http://onlinelibrary.wiley.com/journal/10.1002/(ISSN)1521-4095.

Paitandi, R. P., Wan, Y., Aftab, W., Zhong, R., & Zou, R. (2023). Pristine metal–organic frameworks and their composites for renewable hydrogen energy applications. *Advanced Functional Materials, 33*(8), 2203224.

Pal, T. K. (2022). Metal-organic framework (MOF)-based fluorescence "turn-on" sensors. *Materials Chemistry Frontiers, 7*(3), 405–441. https://doi.org/10.1039/d2qm01070d.

Pan, L., Liu, G., Shi, W., Shang, J., Leow, W. R., Liu, Y., Jiang, Y., Li, S., Chen, X., & Li, R.-W. (2018). Mechano-regulated metal–organic framework nanofilm for ultrasensitive and anti-jamming strain sensing. *Nature Communications, 9*, 3813.

Pan, Y., Li, H., Zhang, X. X., Zhang, Z., Tong, X. S., Jia, C. Z., Liu, B., Sun, C. Y., Yang, L. Y., & Chen, G. J. (2015). Large-scale synthesis of ZIF-67 and highly efficient carbon capture using a ZIF-67/glycol-2-methylimidazole slurry. *Chemical Engineering Science, 137*, 504–514. https://doi.org/10.1016/j.ces.2015.06.069, http://www.journals.elsevier.com/chemical-engineering-science/.

Pan, Y., Yuan, B., Li, Y., & He, D. (2010). Multifunctional catalysis by Pd@MIL-101: One-step synthesis of methyl isobutyl ketone over palladium nanoparticles deposited on a metal-organic framework. *Chemical Communications, 46*(13), 2280–2282. https://doi.org/10.1039/b922061e, http://pubs.rsc.org/en/journals/journal/cc.

Park, J. M., Cha, G. Y., Jo, D., Cho, K. H., Yoon, J. W., Hwang, Y. K., Lee, S. K., & Lee, U. H. (2022). Amine and fluorine co-functionalized MIL-101(Cr) synthesized via a mixed-ligand strategy for CO_2 capture under humid conditions. *Chemical Engineering Journal, 444*, 136476. https://doi.org/10.1016/j.cej.2022.136476, www.elsevier.com/inca/publications/store/6/0/1/2/7/3/index.htt.

Parsaei, M., & Akhbari, K. (2022a). MOF-801 as a nanoporous water-based carrier system for in situ encapsulation and sustained release of 5-FU for effective cancer therapy. *Inorganic Chemistry, 61*(15), 5912–5925. https://doi.org/10.1021/acs.inorgchem.2c00380, http://pubs.acs.org/journal/inocaj.

Parsaei, M., & Akhbari, K. (2022b). Smart multifunctional UiO-66 metal-organic framework nanoparticles with outstanding drug-loading/release potential for the targeted delivery of

quercetin. *Inorganic Chemistry, 61*(37), 14528–14543. https://doi.org/10.1021/acs.inorgchem. 2c00743, http://pubs.acs.org/journal/inocaj.
Parsaei, M., & Akhbari, K. (2022c). Synthesis and application of MOF-808 decorated with folic acid-conjugated chitosan as a strong nanocarrier for the targeted drug delivery of quercetin. *Inorganic Chemistry, 61*(48), 19354–19368. https://doi.org/10.1021/acs.inorgchem.2c03138, http://pubs.acs.org/journal/inocaj.
Parsaei, M., & Akhbari, K. (2023). Magnetic UiO-66-NH$_2$ core-shell nanohybrid as a promising carrier for quercetin targeted delivery toward human breast cancer cells. *ACS Omega, 8*(44), 41321–41338. https://doi.org/10.1021/acsomega.3c04863, pubs.acs.org/journal/acsodf.
Parsaei, M., Akhbari, K., & Kawata, S. (2023). Computational simulation of CO$_2$/CH$_4$ separation on a three-dimensional Cd-based metal–organic framework. *Crystal Growth & Design, 23*(8), 5705–5718. https://doi.org/10.1021/acs.cgd.3c00366.
Parsaei, M., Akhbari, K., Tylianakis, E., & Froudakis, G. E. (2023). Computational simulation of a three-dimensional Mg-based metal–organic framework as nanoporous anticancer drug carrier. *Crystal Growth & Design, 23*(11), 8396–8406. https://doi.org/10.1021/acs.cgd.3c01058.
Parsaei, M., Akhbari, K., Tylianakis, E., & Froudakis, G. E. (2024). Effects of fluorinated functionalization of linker on quercetin encapsulation, release and hela cell cytotoxicity of Cu-based MOFs as smart pH-stimuli nanocarriers. *Chemistry – A European Journal, 30*(1), e202301630. https://doi.org/10.1002/chem.202301630, http://onlinelibrary.wiley.com/journal/10.1002/(ISSN)1521-3765.
Parsaei, M., Akhbari, K., Tylianakis, E., Froudakis, G. E., White, J. M., & Kawata, S. (2022). Computational study of two three-dimensional Co(II)-based metal–organic frameworks as quercetin anticancer drug carriers. *Crystal Growth & Design, 22*(12), 7221–7233. https://doi.org/10.1021/acs.cgd.2c00900.
Parsaei, M., Akhbari, K., & White, J. (2022). Modulating carbon dioxide storage by facile synthesis of nanoporous pillared-layered metal-organic framework with different synthetic routes. *Inorganic Chemistry, 61*(9), 3893–3902. https://doi.org/10.1021/acs.inorgchem.1c03414, http://pubs.acs.org/journal/inocaj.
Paseta, L., Seoane, B., Julve, D., Sebastián, V., Téllez, C., & Coronas, J. (2013). Accelerating the controlled synthesis of metal-organic frameworks by a microfluidic approach: A nanoliter continuous reactor. *ACS Applied Materials and Interfaces, 5*(19), 9405–9410. https://doi.org/10.1021/am4029872, http://pubs.acs.org/journal/aamick.
Pei, J., Shao, K., Wang, J. X., Wen, H. M., Yang, Y., Cui, Y., Krishna, R., Li, B., & Qian, G. (2020). A Chemically Stable Hofmann-Type Metal-Organic Framework with Sandwich-Like Binding Sites for Benchmark Acetylene Capture. *Advanced materials (Deerfield Beach. Fla.), 32*(24), e1908275. https://doi.org/10.1002/adma.201908275.
Peng, P., Anastasopoulou, A., Brooks, K., Furukawa, H., Bowden, M. E., Long, J. R., Autrey, T., & Breunig, H. (2022). Cost and potential of metal–organic frameworks for hydrogen back-up power supply. *Nature Energy, 7*(5), 448–458. https://doi.org/10.1038/s41560-022-01013-w, www.nature.com/nenergy/.
Petit, C., & Bandosz, T. J. (2012). Exploring the coordination chemistry of MOF-graphite oxide composites and their applications as adsorbents. *Dalton Transactions, 41*(14), 4027–4035. https://doi.org/10.1039/c2dt12017h.
Petit, C., Mendoza, B., & Bandosz, T. J. (2010). Hydrogen sulfide adsorption on MOFs and MOF/Graphite oxide composites. *Chemphyschem: A European Journal of Chemical Physics and Physical Chemistry, 11*(17), 3678–3684. https://doi.org/10.1002/cphc.201000689, http://onlinelibrary.wiley.com/journal/10.1002/(ISSN)1439-7641.

Pobłocki, K., Drzeżdżon, J., Gawdzik, B., & Jacewicz, D. (2022). Latest trends in the large-scale production of MoFs in accordance with the principles of green chemistry. *Green Chemistry, 24*(24), 9402–9427. https://doi.org/10.1039/d2gc03264c, http://pubs.rsc.org/en/journals/journal/gc.

Polyzoidis, A., Altenburg, T., Schwarzer, M., Loebbecke, S., & Kaskel, S. (2016). Continuous microreactor synthesis of ZIF-8 with high space–time-yield and tunable particle size. *Chemical Engineering Journal, 283*, 971–977. https://doi.org/10.1016/j.cej.2015.08.071.

Poya, M. T., & Fazl, F. (2023). A review of the application of metal-organic frameworks in the absorption, storage and release of methane. *Journal for Research in Applied Sciences and Biotechnology, 2*(6), 254–260. https://doi.org/10.55544/jrasb.2.6.35.

Prajapati, M., Singh, V., Jacob, M. V., & Ravi Kant, C. (2023). Recent advancement in metal-organic frameworks and composites for high-performance supercapatteries. *Renewable and Sustainable Energy Reviews, 183*, 113509. https://doi.org/10.1016/j.rser.2023.113509.

Proch, S., Herrmannsdörfer, J., Kempe, R., Kern, C., Jess, A., Seyfarth, L., & Senker, J. (2008). Pt@MOF-177: Synthesis, room-temperature hydrogen storage and oxidation catalysis. *Chemistry – A European Journal, 14*(27), 8204–8212. https://doi.org/10.1002/chem.200801043Germany, http://www3.interscience.wiley.com/cgi-bin/fulltext/121357743/PDFSTART.

Qazvini, O. T., & Telfer, S. G. (2020). A robust metal-organic framework for post-combustion carbon dioxide capture. *Journal of Materials Chemistry A, 8*(24), 12028–12034. https://doi.org/10.1039/d0ta04121a, http://pubs.rsc.org/en/journals/journal/ta.

Qiu, Q., Chen, H., Wang, Y., & Ying, Y. (2019). Recent advances in the rational synthesis and sensing applications of metal-organic framework biocomposites. *Coordination Chemistry Reviews, 387*, 60–78. https://doi.org/10.1016/j.ccr.2019.02.009, http://www.journals.elsevier.com/coordination-chemistry-reviews/.

Quan, Y., Parker, T. F., Hua, Y., Jeong, H. K., & Wang, Q. (2023). Process elucidation and hazard analysis of the metal-organic framework scale-up synthesis: A case study of ZIF-8. *Industrial and Engineering Chemistry Research, 62*(12), 5035–5041. https://doi.org/10.1021/acs.iecr.2c04570, http://pubs.acs.org/journal/iecred.

Rangaraj, V., Wahab, M. A., Reddy, K. S. K., Kakosimos, G., Abdalla, O., Favvas, E. P., Reinalda, D., Geuzebroek, F., Abdala, A., & Karanikolos, G. N. (2020). Metal organic framework—based mixed matrix membranes for carbon dioxide separation: Recent advances and future directions. *Frontiers in Chemistry, 8*, 534.

Rasmussen, E. G., Kramlich, J., & Novosselov, I. V. (2020). Scalable continuous flow metal-organic framework (MOF) synthesis using supercritical CO_2. *ACS Sustainable Chemistry and Engineering, 8*(26), 9680–9689. https://doi.org/10.1021/acssuschemeng.0c01429, http://pubs.acs.org/journal/ascecg.

Rauf, S., Andrés, M. A., Roubeau, O., Gascón, I., Serre, C., Eddaoudi, M., & Salama, K. N. (2021). Coating of conducting and insulating threads with porous MOF particles through langmuir-blodgett technique. *Nanomaterials, 11*(1), 160. https://doi.org/10.3390/nano11010160.

Reinsch, H., Waitschat, S., Chavan, S. M., Lillerud, K. P., & Stock, N. (2016). A facile "green" route for scalable batch production and continuous synthesis of zirconium MoFs. *European Journal of Inorganic Chemistry, 2016*(27), 4490–4498. https://doi.org/10.1002/ejic.201600295, http://onlinelibrary.wiley.com/journal/10.1002/(ISSN)1099-0682c.

Ren, J., Dyosiba, X., Musyoka, N. M., Langmi, H. W., Mathe, M., & Liao, S. (2017). Review on the current practices and efforts towards pilot-scale production of metal-organic frameworks

(MOFs). *Coordination Chemistry Reviews, 352*, 187–219. https://doi.org/10.1016/j.ccr.2017.09.005, http://www.journals.elsevier.com/coordination-chemistry-reviews/.

Ren, J., Musyoka, N. M., Langmi, H. W., North, B. C., Mathe, M., & Kang, X. (2014). Fabrication of core-shell MIL-101(Cr)@UiO-66(Zr) nanocrystals for hydrogen storage. *International Journal of Hydrogen Energy, 39*(27), 14912–14917. https://doi.org/10.1016/j.ijhydene.2014.07.056, http://www.journals.elsevier.com/international-journal-of-hydrogen-energy/.

Ren, J., Musyoka, N. M., Langmi, H. W., Swartbooi, A., North, B. C., & Mathe, M. (2015). A more efficient way to shape metal-organic framework (MOF) powder materials for hydrogen storage applications. *International Journal of Hydrogen Energy, 40*(13), 4617–4622. https://doi.org/10.1016/j.ijhydene.2015.02.011, http://www.journals.elsevier.com/international-journal-of-hydrogen-energy/.

Ren, Y.-B., Xu, H.-Y., Gang, S.-Q., Gao, Y.-J., Jing, X., & Du, J.-L. (2022). An ultra-stable Zr (IV)-MOF for highly efficient capture of SO_2 from SO_2/CO_2 and SO_2/CH_4 mixtures. *Chemical Engineering Journal, 431*(10), 134057.

Rubio-Martinez, M., Avci-Camur, C., Thornton, A. W., Imaz, I., Maspoch, D., & Hill, M. R. (2017). New synthetic routes towards MOF production at scale. *Chemical Society Reviews, 46*(11), 3453–3480. https://doi.org/10.1039/c7cs00109f, http://pubs.rsc.org/en/journals/journal/cs.

Rubio-Martinez, M., Batten, M. P., Polyzos, A., Carey, K. C., Mardel, J. I., Lim, K. S., & Hill, M. R. (2014). Versatile, high quality and scalable continuous flow production of metal-organic frameworks. *Scientific Reports, 4*, 5443. https://doi.org/10.1038/srep05443, www.nature.com/srep/index.html.

Rubio-Martinez, M., Hadley, T. D., Batten, M. P., Constanti-Carey, K., Barton, T., Marley, D., Mönch, A., Lim, K. S., & Hill, M. R. (2016). Scalability of continuous flow production of metal-organic frameworks. *ChemSusChem, 9*(9), 938–941. https://doi.org/10.1002/cssc.201501684, http://www.interscience.wiley.com/jpages/1864-5631.

Rui, Z., Yu, D., & Zhang, F. (2024). Novel cellulose nanocrystal/metal-organic framework composites: Transforming ASA-sized cellulose paper for innovative food packaging solutions. *Industrial Crops and Products, 207*, 117771. https://doi.org/10.1016/j.indcrop.2023.117771.

Ryu, U. J., Jee, S., Rao, P. C., Shin, J., Ko, C., Yoon, M., Park, K. S., & Choi, K. M. (2021). Recent advances in process engineering and upcoming applications of metal–organic frameworks. *Coordination Chemistry Reviews, 426*, 213544. https://doi.org/10.1016/j.ccr.2020.213544, http://www.journals.elsevier.com/coordination-chemistry-reviews/.

Sajid, M., Irum, G., Farhan, A., & Qamar, M. A. (2024). Role of metal-organic frameworks (MOF) based nanomaterials for the efficiency enhancement of solar cells: A mini-review. *Hybrid Advances, 5*, 100167. https://doi.org/10.1016/j.hybadv.2024.100167.

Salimi, S., Akhbari, K., Farnia, S. M. F., Tylianakis, E., Froudakis, G. E., & White, J. M. (2024a). Solvent-directed construction of a nanoporous metal-organic framework with potential in selective adsorption and separation of gas mixtures studied by Grand Canonical Monte Carlo simulations. *ChemPlusChem, 89*(1), e202300455. https://doi.org/10.1002/cplu.202300455, http://onlinelibrary.wiley.com/journal/10.1002/(ISSN)2192-6506.

Salimi, S., Akhbari, K., Farnia, S. M. F., Tylianakis, E., Froudakis, G. E., & White, J. M. (2024b). Nanoporous metal–organic framework based on furan-2,5-dicarboxylic acid with high potential in selective adsorption and separation of gas mixtures. *Crystal Growth & Design, 24*(10), 4220–4231. https://doi.org/10.1021/acs.cgd.4c00349.

Salimi, S., Akhbari, K., Morteza, S., Farnia, F., & White, J. M. (2022). Multiple construction of a hierarchical nanoporous manganese(II)-based metal–organic framework with active sites for regulating N$_2$ and CO$_2$ Trapping. *Crystal Growth & Design, 22*(3), 1654–1664. https://doi.org/10.1021/acs.cgd.1c01183.

Sanati, S., Morsali, A., & García, H. (2023). Metal-organic framework-based materials as key components in electrocatalytic oxidation and reduction reactions. *Journal of Energy Chemistry, 87*, 540–567. https://doi.org/10.1016/j.jechem.2023.08.042, elsevier.com/journals/journal-of-energy-chemistry/2095-4956.

Seo, Y. K., Yoon, J. W., Lee, J. S., Lee, U. H., Hwang, Y. K., Jun, C. H., Horcajada, P., Serre, C., & Chang, J. S. (2012). Large scale fluorine-free synthesis of hierarchically porous iron(III) trimesate MIL-100(Fe) with a zeolite MTN topology. *Microporous and Mesoporous Materials, 157*, 137–145. https://doi.org/10.1016/j.micromeso.2012.02.027.

Severino, M. I., Gkaniatsou, E., Nouar, F., Pinto, M. L., & Serre, C. (2021). MOFs industrialization: A complete assessment of production costs. *Faraday Discussions, 231*, 326–341. https://doi.org/10.1039/d1fd00018g, http://pubs.rsc.org/en/journals/journal/fd.

Seyedpour, S. F., Rahimpour, A., & Najafpour, G. (2019). Facile in-situ assembly of silver-based MOFs to surface functionalization of TFC membrane: A novel approach toward long-lasting biofouling mitigation. *Journal of Membrane Science, 573*, 257–269. https://doi.org/10.1016/j.memsci.2018.12.016, www.elsevier.com/locate/memsci.

Shahangi Shirazi, F., Akhbari, K., Kawata, S., & Kanazashi, K. (2016). Reversible liquid assisted mechanochemical conversion of sodium coordination polymer nanorods to organosilver coordination polymer nanosheets. *Inorganic Chemistry Communications, 74*, 31–34. https://doi.org/10.1016/j.inoche.2016.10.034, http://www.journals.elsevier.com/inorganic-chemistry-communications/.

Shahrab, F., & Tadjarodi, A. (2023). Novel magnetic nanocomposites BiFeO$_3$/Cu(BDC) for efficient dye removal. *Heliyon, 9*(11), e20689. https://doi.org/10.1016/j.heliyon.2023.e20689, http://www.journals.elsevier.com/heliyon/.

Shang, W., Zeng, C., Du, Y., Hui, H., Liang, X., Chi, C., Wang, K., Wang, Z., & Tian, J. (2016). Core-shell gold nanorod@metal-organic framework nanoprobes for multimodality diagnosis of glioma. *Advanced Materials, 29*(3), 27859713.

Sharanyakanth, P. S., & Radhakrishnan, M. (2020). Synthesis of metal-organic frameworks (MOFs) and its application in food packaging: A critical review. *Trends in Food Science & Technology, 104*, 102–116. https://doi.org/10.1016/j.tifs.2020.08.004.

Shen, M., Forghani, F., Kong, X., Liu, D., Ye, X., Chen, S., & Ding, T. (2020). Antibacterial applications of metal–organic frameworks and their composites. *Comprehensive Reviews in Food Science and Food Safety, 19*(4), 1397–1419. https://doi.org/10.1111/1541-4337.12515, http://onlinelibrary.wiley.com/journal/10.1111/(ISSN)1541-4337.

Shen, M., Liu, D., & Ding, T. (2021). Cyclodextrin-metal-organic frameworks (CD-MOFs): Main aspects and perspectives in food applications. *Current Opinion in Food Science, 41*, 8–15. https://doi.org/10.1016/j.cofs.2021.02.008, http://www.journals.elsevier.com/current-opinion-in-food-science/.

Shen, Y., Pan, T., Wang, L., Ren, Z., Zhang, W., & Huo, F. (2021). Programmable logic in metal–organic frameworks for catalysis. *Advanced Materials, 33*(46), 2007442.

Sheng, M., Dong, S., Qiao, Z., Li, Q., Yuan, Y., Xing, G., Zhao, S., Wang, J., & Wang, Z. (2021). Large-scale preparation of multilayer composite membranes for post-combustion CO$_2$ capture. *Journal of Membrane Science, 636*, 119595. https://doi.org/10.1016/j.memsci.2021.119595.

Shi, Z., Yuan, X., Yan, Y., Tang, Y., Li, J., Liang, H., Tong, L., & Qiao, Z. (2021). Techno-economic analysis of metal-organic frameworks for adsorption heat pumps/chillers: From directional computational screening, machine learning to experiment. *Journal of Materials*

Chemistry A, 9(12), 7656–7666. https://doi.org/10.1039/d0ta11747a, http://pubs.rsc.org/en/journals/journal/ta.

Shinde, P. A., Abdelkareem, M. A., Sayed, E. T., Elsaid, K., & Olabi, A. G. (2021). Metal organic frameworks (MOFs) in supercapacitors encyclopedia of smart materials. *Emirates, 414*–423. https://doi.org/10.1016/B978-0-12-815732-9.00152-2, https://www.sciencedirect.com/book/9780128157336.

Shinde, P. A., Seo, Y., Lee, S., Kim, H., Pham, Q. N., Won, Y., & Chan Jun, S. (2020). Layered manganese metal-organic framework with high specific and areal capacitance for hybrid supercapacitors. *Chemical Engineering Journal, 387*, 122982. https://doi.org/10.1016/j.cej.2019.122982, www.elsevier.com/inca/publications/store/6/0/1/2/7/3/index.htt.

Silva, P., Vilela, S. M. F., Tomé, J. P. C., & Almeida Paz, F. A. (2015). Multifunctional metal-organic frameworks: From academia to industrial applications. *Chemical Society Reviews, 44*(19), 6774–6803. https://doi.org/10.1039/c5cs00307e, http://pubs.rsc.org/en/journals/journal/cs.

Solovyeva, M. V., Shkatulov, A. I., Gordeeva, L. G., Fedorova, E. A., Krieger, T. A., & Aristov, Y. I. (2021). Water vapor adsorption on CAU-10- X: Effect of functional groups on adsorption equilibrium and mechanisms. *Langmuir: The ACS Journal of Surfaces and Colloids, 37*(2), 693–702. https://doi.org/10.1021/acs.langmuir.0c02729, http://pubs.acs.org/journal/langd5.

Soltani, S., & Akhbari, K. (2022a). Cu-BTC metal–organic framework as a biocompatible nanoporous carrier for chlorhexidine antibacterial agent. *JBIC Journal of Biological Inorganic Chemistry,* 1–7.

Soltani, S., & Akhbari, K. (2022b). Embedding an extraordinary amount of gemifloxacin antibiotic in ZIF-8 framework with one-step synthesis and measurement of its H_2O_2-sensitive release and potency against infectious bacteria. *New Journal of Chemistry, 46*(40), 19432–19441. https://doi.org/10.1039/d2nj02981b, http://pubs.rsc.org/en/journals/journal/nj.

Song, W., Zheng, Z., Alawadhi, A. H., & Yaghi, O. M. (2023). MOF water harvester produces water from Death Valley desert air in ambient sunlight. *Nature Water, 1*(7), 626–634. https://doi.org/10.1038/s44221-023-00103-7.

Staffell, I., Scamman, D., Velazquez Abad, A., Balcombe, P., Dodds, P. E., Ekins, P., Shah, N., & Ward, K. R. (2019). The role of hydrogen and fuel cells in the global energy system. *Energy and Environmental Science, 12*(2), 463–491. https://doi.org/10.1039/c8ee01157e, http://pubs.rsc.org/en/journals/journal/ee.

Su, M., Zhang, R., Li, J., Jin, X., Zhang, X., & Qin, D. (2021). Tailoring growth of MOF199 on hierarchical surface of bamboo and its antibacterial property. *Cellulose, 28*(18), 11713–11727. https://doi.org/10.1007/s10570-021-04265-z, www.springer.com/journal/10570.

Sun, C. Y., Liu, S. X., Liang, D. D., Shao, K. Z., Ren, Y. H., & Su, Z. M. (2009). Highly stable crystalline catalysts based on a microporous metal-organic framework and polyoxometalates. *Journal of the American Chemical Society, 131*(5), 1883–1888. https://doi.org/10.1021/ja807357rChina, http://pubs.acs.org/doi/pdfplus/10.1021/ja807357r.

Sun, J., Kwon, H. T., & Jeong, H. K. (2018). Continuous synthesis of high quality metal–organic framework HKUST-1 crystals and composites via aerosol-assisted synthesis. *Polyhedron, 153*, 226–233. https://doi.org/10.1016/j.poly.2018.07.022, http://www.journals.elsevier.com/polyhedron/.

Sun, X., Lv, D., Chen, Y., Wu, Y., Wu, Q., Xia, Q., & Li, Z. (2017). Enhanced adsorption performance of aromatics on a novel chromium-based MIL-101@graphite oxide composite. *Energy and Fuels, 31*(12), 13985–13990. https://doi.org/10.1021/acs.energyfuels.7b02665, http://pubs.acs.org/journal/enfuem.

Sundriyal, S., Kaur, H., Bhardwaj, S. K., Mishra, S., Kim, K. H., & Deep, A. (2018). Metal-organic frameworks and their composites as efficient electrodes for supercapacitor applications. *Coordination Chemistry Reviews, 369*, 15–38. https://doi.org/10.1016/j.ccr.2018.04.018, http://www.journals.elsevier.com/coordination-chemistry-reviews/.

Supianto, M., Yoo, D. K., Hwang, H., Oh, H. B., Jhung, S. H., & Lee, H. J. (2024). Linker-preserved iron metal-organic framework-based lateral flow assay for sensitive transglutaminase 2 detection in urine through machine learning-assisted colorimetric analysis. *ACS Sensors, 9*(3), 1321–1330. https://doi.org/10.1021/acssensors.3c02250, http://pubs.acs.org/journal/ascefj.

Taddei, M., Steitz, D. A., Van Bokhoven, J. A., & Ranocchiari, M. (2016). Continuous-flow microwave synthesis of metal-organic frameworks: A highly efficient method for large-scale production. *Chemistry – A European Journal, 22*(10), 3245–3249. https://doi.org/10.1002/chem.201505139, http://onlinelibrary.wiley.com/journal/10.1002/(ISSN)1521-3765.

Tai, S., Zhang, W., Zhang, J., Luo, G., Jia, Y., Deng, M., & Ling, Y. (2016). Facile preparation of UiO-66 nanoparticles with tunable sizes in a continuous flow microreactor and its application in drug delivery. *Microporous and Mesoporous Materials, 220*, 148–154. https://doi.org/10.1016/j.micromeso.2015.08.037, www.elsevier.com/inca/publications/store/6/0/0/7/6/0.

Tan, B., Wu, Z. F., & Xie, Z. L. (2017). Fine decoration of carbon nanotubes with metal organic frameworks for enhanced performance in supercapacitance and oxygen reduction reaction. *Science Bulletin, 62*(16), 1132–1141. https://doi.org/10.1016/j.scib.2017.08.011, http://link.springer.com/journal/11434.

Tang, K., Chen, Y., Wang, X., Zhou, Q., Lei, H., Yang, Z., & Zhang, Z. (2023). Smartphone-integrated tri-color fluorescence sensing platform based on acid-sensitive fluorescence imprinted polymers for dual-mode visual intelligent detection of ibuprofen, chloramphenicol and florfenicol. *Analytica Chimica Acta, 1260*, 341174. https://doi.org/10.1016/j.aca.2023.341174.

Tanvidkar, P., Appari, S., & Kuncharam, B. V. R. (2022). A review of techniques to improve performance of metal organic framework (MOF) based mixed matrix membranes for CO_2/CH_4 separation. *Reviews in Environmental Science and Biotechnology, 21*(2), 539–569. https://doi.org/10.1007/s11157-022-09612-5, http://www.kluweronline.com/issn/1569-1705.

Teerachawanwong, P., Dilokekunakul, W., Phadungbut, P., Klomkliang, N., Supasitmongkol, S., Chaemchuen, S., & Verpoort, F. (2023). Insights into the heat contributions and mechanism of CO_2 adsorption on metal–organic framework MIL-100 (Cr, Fe): Experiments and molecular simulations. *Fuel, 331*, 125863. https://doi.org/10.1016/j.fuel.2022.125863.

Teo, W. L., Zhou, W., Qian, C., & Zhao, Y. (2021). Industrializing metal–organic frameworks: Scalable synthetic means and their transformation into functional materials. *Materials Today, 47*, 170–186. https://doi.org/10.1016/j.mattod.2021.01.010, http://www.journals.elsevier.com/materials-today/.

Terzopoulou, A., Nicholas, J. D., Chen, X. Z., Nelson, B. J., Pane, S., & Puigmartí-Luis, J. (2020). Metal−organic frameworks in motion. *Chemical Reviews, 120*(20), 11175–11193. https://doi.org/10.1021/acs.chemrev.0c00535, http://pubs.acs.org/journal/chreay.

Terzopoulou, A., Wang, X., Chen, X. Z., Palacios-Corella, M., Pujante, C., Herrero-Martín, J., Qin, X. H., Sort, J., deMello, A. J., & Nelson, B. J. (2020). Biodegradable metal–organic framework-based microrobots (MOFBOTs). *Advanced Healthcare Materials, 9*(20), 2001031.

Tian, D., Wang, C., & Lu, X. (2021). Metal-organic frameworks and their derived functional materials for supercapacitor electrode application. *Advanced Energy and Sustainability Research. 2*(7), 2100024.

Troyano, J., Carné-Sánchez, A., & Maspoch, D. (2019). Programmable self-assembling 3D architectures generated by patterning of swellable MOF-based composite films. *Advanced Materials, 31*(21), 1808235. https://doi.org/10.1002/adma.201808235, http://onlinelibrary.wiley.com/journal/10.1002/(ISSN)1521-4095.

Tsivadze, A. Y., Aksyutin, O. E., Ishkov, A. G., Knyazeva, M. K., Solovtsova, O. V., Men'Shchikov, I. E., Fomkin, A. A., Shkolin, A. V., Khozina, E. V., & Grachev, V. A. (2019). Metal-organic framework structures: Adsorbents for natural gas storage. *Russian Chemical Reviews, 88*(9), 925–978. https://doi.org/10.1070/RCR4873, https://iopscience.iop.org/article/10.1070/RCR4873/pdf.

Tsuruoka, T., Kawasaki, H., Nawafune, H., & Akamatsu, K. (2011). Controlled self-assembly of metal-organic frameworks on metal nanoparticles for efficient synthesis of hybrid nanostructures. *ACS Applied Materials and Interfaces, 3*(10), 3788–3791. https://doi.org/10.1021/am200974t.

Tu, Y., Wang, R., Zhang, Y., & Wang, J. (2018). Progress and expectation of atmospheric water harvesting. *Joule, 2*(8), 1452–1475. https://doi.org/10.1016/j.joule.2018.07.015, https://www.journals.elsevier.com/joule.

Ursueguía, D., Díaz, E., & Ordóñez, S. (2021). Metal-organic frameworks (MOFs) as methane adsorbents: From storage to diluted coal mining streams concentration. *Science of The Total Environment, 790*, 148211. https://doi.org/10.1016/j.scitotenv.2021.148211.

Viditha, V., Srilatha, K., & Himabindu, V. (2016). Hydrogen storage studies on palladium-doped carbon materials (AC, CB, CNMs)@metal–organic framework-5. *Environmental Science and Pollution Research, 23*(10), 9355–9363. https://doi.org/10.1007/s11356-015-5194-z.

WaHab. (2024). *WaHa, a new solution selected by WEF-CAP to extract water from the atmosphere.*

Wang, C., Wang, L., Tadepalli, S., Morrissey, J. J., Kharasch, E. D., Naik, R. R., & Singamaneni, S. (2018). Ultrarobust biochips with metal-organic framework coating for point-of-care diagnosis. *ACS Sensors, 3*(2), 342–351. https://doi.org/10.1021/acssensors.7b00762, http://pubs.acs.org/journal/ascefj.

Wang, H., Zheng, F., Xue, G., Wang, Y., Li, G., & Tang, Z. (2021). Recent advances in hollow metal-organic frameworks and their composites for heterogeneous thermal catalysis. *Science China Chemistry, 64*(11), 1854–1874. https://doi.org/10.1007/s11426-021-1095-y.

Wang, H. S. (2017). Metal–organic frameworks for biosensing and bioimaging applications. *Coordination Chemistry Reviews, 349*, 139–155. https://doi.org/10.1016/j.ccr.2017.08.015, http://www.journals.elsevier.com/coordination-chemistry-reviews/.

Wang, L., Han, Y., Feng, X., Zhou, J., Qi, P., & Wang, B. (2016). Metal-organic frameworks for energy storage: Batteries and supercapacitors. *Coordination Chemistry Reviews, 307*, 361–381. https://doi.org/10.1016/j.ccr.2015.09.002, http://www.journals.elsevier.com/coordination-chemistry-reviews/.

Wang, L., Zhu, H., Shi, Y., Ge, Y., Feng, X., Liu, R., Li, Y., Ma, Y., & Wang, L. (2018). Novel catalytic micromotor of porous zeolitic imidazolate framework-67 for precise drug delivery. *Nanoscale, 10*(24), 11384–11391. https://doi.org/10.1039/c8nr02493f, http://pubs.rsc.org/en/journals/journal/nr.

Wang, S., Fu, Y., Wang, T., Liu, W., Wang, J., Zhao, P., Ma, H., Chen, Y., Cheng, P., & Zhang, Z. (2023). Fabrication of robust and cost-efficient Hoffmann-type MOF sensors for room temperature ammonia detection. *Nature Communications, 14*(1), 7261. https://doi.org/10.1038/s41467-023-42959-z, https://www.nature.com/ncomms/.

Wang, X., Chen, X. Z., Alcântara, C. C., Sevim, S., Hoop, M., Terzopoulou, A., Marco, C., Hu, C., de Mello, A. J., & Falcaro, P. (2019). MOFBOTS: Metal–organic-framework-based biomedical microrobots. *Advanced Materials, 31*(27), 1901592.

Wang, X., Wang, Y., & Ying, Y. (2021). Recent advances in sensing applications of metal nanoparticle/metal–organic framework composites. *TrAC Trends in Analytical Chemistry, 143*, 116395. https://doi.org/10.1016/j.trac.2021.116395.

Wang, X., Xu, W., Gu, J., Yan, X., Chen, Y., Guo, M., Zhou, G., Tong, S., Ge, M., Liu, Y., & Chen, C. (2019). MOF-based fibrous membranes adsorb PM efficiently and capture toxic gases selectively. *Nanoscale, 11*(38), 17782–17790. https://doi.org/10.1039/c9nr05795a, http://pubs.rsc.org/en/journals/journal/nr.

Wei, X., Guo, J., Lian, H., Sun, X., & Liu, B. (2021). Cobalt metal-organic framework modified carbon cloth/paper hybrid electrochemical button-sensor for nonenzymatic glucose diagnostics. *Sensors and Actuators B: Chemical, 329*, 129205. https://doi.org/10.1016/j.snb.2020.129205.

Wei, Y. S., Zhang, M., Zou, R., & Xu, Q. (2020). Metal-organic framework-based catalysts with single metal sites. *Chemical Reviews, 120*(21), 12089–12174. https://doi.org/10.1021/acs.chemrev.9b00757, http://pubs.acs.org/journal/chreay.

Wen, X., Zhang, Q., & Guan, J. (2020). Applications of metal–organic framework-derived materials in fuel cells and metal-air batteries. *Coordination Chemistry Reviews, 409*, 213214.

Wu, J., Chen, J., Wang, C., Zhou, Y., Ba, K., Xu, H., Bao, W., Xu, X., Carlsson, A., & Lazar, S. (2020). Metal–organic framework for transparent electronics. *Advanced Science, 7*, 1903003.

Wu, M.-X., Gao, J., Wang, F., Yang, J., Song, N., Jin, X., Mi, P., Tian, J., Luo, J., Liang, F., & Yang, Y.-W. (2018). Multistimuli responsive core–shell nanoplatform constructed from Fe$_3$O$_4$@MOF equipped with pillar[6]arene nanovalves. *Small (Weinheim an der Bergstrasse, Germany), 14*(17), e1704440. https://doi.org/10.1002/smll.201704440.

Wu, Y. N., Li, F., Liu, H., Zhu, W., Teng, M., Jiang, Y., Li, W., Xu, D., He, D., Hannam, P., & Li, G. (2012). Electrospun fibrous mats as skeletons to produce free-standing MOF membranes. *Journal of Materials Chemistry, 22*(33), 16971–16978. https://doi.org/10.1039/c2jm32570e.

Xian, S., Peng, J., Pandey, H., Thonhauser, T., Wang, H., & Li, J. (2023). Robust metal–organic frameworks with high industrial applicability in efficient recovery of C_3H_8 and C_2H_6 from natural gas upgrading. *Engineering, 23*, 56–63. https://doi.org/10.1016/j.eng.2022.07.017, http://www.journals.elsevier.com/engineering/.

Xiang, S., He, Y., Zhang, Z., Wu, H., Zhou, W., Krishna, R., & Chen, B. (2012). Microporous metal-organic framework with potential for carbon dioxide capture at ambient conditions. *Nature Communications, 3*, 954. https://doi.org/10.1038/ncomms1956.

Xin, N., Liu, Y., Niu, H., Bai, H., & Shi, W. (2020). In-situ construction of metal organic frameworks derived Co/Zn–S sandwiched graphene film as free-standing electrodes for ultra-high energy density supercapacitors. *Journal of Power Sources, 451*, 227772. https://doi.org/10.1016/j.jpowsour.2020.227772.

Xu, F., Chen, N., Fan, Z., & Du, G. (2020). Ni/Co-based metal organic frameworks rapidly synthesized in ambient environment for high energy and power hybrid supercapacitors. *Applied Surface Science, 528*, 146920. https://doi.org/10.1016/j.apsusc.2020.146920.

Xu, F., Yu, Y., Yan, J., Xia, Q., Wang, H., Li, J., & Li, Z. (2016). Ultrafast room temperature synthesis of GrO@HKUST-1 composites with high CO_2 adsorption capacity and CO_2/N_2 adsorption selectivity. *Chemical Engineering Journal, 303*, 231–237. https://doi.org/10.1016/j.cej.2016.05.143, www.elsevier.com/inca/publications/store/6/0/1/2/7/3/index.htt.

Xu, W., & Yaghi, O. M. (2020). Metal-organic frameworks for water harvesting from air, anywhere, anytime. *ACS Central Science, 6*(8), 1348–1354. https://doi.org/10.1021/acscentsci.0c00678, http://pubs.acs.org/journal/acscii.

Xu, Y., Li, Q., Guo, X., Zhang, S., Li, W., & Pang, H. (2022). Metal organic frameworks and their composites for supercapacitor application. *Journal of Energy Storage, 56*, 105819. https://doi.org/10.1016/j.est.2022.105819.

Xu, Z., Liu, Z., Xiao, M., Jiang, L., & Yi, C. (2020). A smartphone-based quantitative point-of-care testing (POCT) system for simultaneous detection of multiple heavy metal ions. *Chemical Engineering Journal, 394*, 124966. https://doi.org/10.1016/j.cej.2020.124966.

Xue, F., Cao, J., Li, X., Feng, J., Tao, M., & Xue, B. (2021). Continuous-flow synthesis of MIL-53(Cr) with a polar linker: Probing the nanoscale piezoelectric effect. *Journal of Materials Chemistry C, 9*(24), 7568–7574. https://doi.org/10.1039/d0tc06013e, http://pubs.rsc.org/en/journals/journal/tc.

Pulskamp, A., Yang, J.,Siegel, D. J., & Veenstra, M. J. (2010). *Hybrid hydrogen storage system and method using the same*. Application filed by Ford Global Technologies LLC.

Yaghi, O. M., O'Keeffe, M., Ockwig, N. W., Chae, H. K., Eddaoudi, M., & Kim, J. (2003). Reticular synthesis and the design of new materials. *Nature, 423*(6941), 705–714. https://doi.org/10.1038/nature01650.

Yan, Y., Suyetin, M., Bichoutskaia, E., Blake, A. J., Allan, D. R., Barnett, S. A., & Schröder, M. (2013). Modulating the packing of [Cu$_{24}$(isophthalate)$_{24}$] cuboctahedra in a triazole-containing metal–organic polyhedral framework. *Chemical Science, 4*(4), 1731–1736. https://doi.org/10.1039/c3sc21769h.

Yang, A. P. & Hirano, S. (2011). *Hydrogen storage materials*, in L. Assigned to FORD GLOBAL TECHNOLOGIES.

Yang, S., Peng, L., Syzgantseva, O. A., Trukhina, O., Kochetygov, I., Justin, A., Sun, D. T., Abedini, H., Syzgantseva, M. A., Oveisi, E., Lu, G., & Queen, W. L. (2020). Preparation of highly porous metal-organic framework beads for metal extraction from liquid streams. *Journal of the American Chemical Society, 142*(31), 13415–13425. https://doi.org/10.1021/jacs.0c02371, http://pubs.acs.org/journal/jacsat.

Yang, S. J., Cho, J. H., Nahm, K. S., & Park, C. R. (2010). Enhanced hydrogen storage capacity of Pt-loaded CNT@MOF-5 hybrid composites. *International Journal of Hydrogen Energy, 23*(35), 13062–13067. 10.1016/j.ijhydene.2010.04.066.

Yilmaz, B., Trukhan, N., & Müller, U. (2012). Industrial outlook on zeolites and metal organic frameworks. *Chinese Journal of Catalysis, 33*(1), 3–10. https://doi.org/10.1016/s1872-2067(10)60302-6.

Yu, J., Mu, C., Yan, B., Qin, X., Shen, C., Xue, H., & Pang, H. (2017). Nanoparticle/MOF composites: Preparations and applications. *Materials Horizons, 4*(4), 557–569. https://doi.org/10.1039/c6mh00586a, rsc.li/materials-horizons.

Yuan, J., Song, X., Yang, X., Yang, C., Wang, Y., Deng, G., Wang, Z., & Gao, J. (2023). Indoor carbon dioxide capture technologies: A review. *Environmental Chemistry Letters, 21*(5), 2559–2581. https://doi.org/10.1007/s10311-023-01620-3, https://www.springer.com/journal/10311.

Yue, T., Xia, C., Liu, X., Wang, Z., Qi, K., & Xia, B. Y. (2021). Design and synthesis of conductive metal-organic frameworks and their composites for supercapacitors. *ChemElectroChem, 8*(6), 1021–1034. https://doi.org/10.1002/celc.202001418, http://onlinelibrary.wiley.com/journal/10.1002/(ISSN)2196-0216.

Zeng, X., Hu, J., Zhang, M., Wang, F., Wu, L., & Hou, X. (2020). Visual detection of fluoride anions using mixed lanthanide metal-organic frameworks with a smartphone. *Analytical Chemistry, 92*(2), 2097–2102. https://doi.org/10.1021/acs.analchem.9b04598, http://pubs.acs.org/journal/ancham.

Zhang, B., Luo, Y., Kanyuck, K., Bauchan, G., Mowery, J., & Zavalij, P. (2016). Development of metal-organic framework for gaseous plant hormone encapsulation to manage ripening of climacteric produce. *Journal of Agricultural and Food Chemistry, 64*(25), 5164–5170. https://doi.org/10.1021/acs.jafc.6b02072, http://pubs.acs.org/journal/jafcau.

Zhang, J., Wu, D., Zhang, Q., Zhang, A., Sun, J., Hou, L., & Yuan, C. (2022). Green self-activation engineering of metal-organic framework derived hollow nitrogen-doped carbon spheres towards supercapacitors. *Journal of Materials Chemistry A, 10*(6), 2932–2944. https://doi.org/10.1039/d1ta10356c, http://pubs.rsc.org/en/journals/journal/ta.

Zhang, K., Huo, Q., Zhou, Y. Y., Wang, H. H., Li, G. P., Wang, Y. W., & Wang, Y. Y. (2019). Textiles/metal-organic frameworks composites as flexible air filters for efficient particulate matter removal. *ACS Applied Materials and Interfaces, 11*(19), 17368–17374. https://doi.org/10.1021/acsami.9b01734, http://pubs.acs.org/journal/aamick.

Zhang, L., Song, Y., Shi, J., Shen, Q., Hu, D., Gao, Q., Chen, W., Kow, K. W., Pang, C., Sun, N., & Wei, W. (2022). Frontiers of CO_2 capture and utilization (CCU) towards carbon neutrality. *Advances in Atmospheric Sciences, 39*(8), 1252–1270. https://doi.org/10.1007/s00376-022-1467-x, http://www.springerlink.com/content/0256-1530.

Zhang, L.-T., Zhou, Y., & Han, S.-T. (2021). The role of metal–organic frameworks in electronic sensors. *Angewandte Chemie, 133*(28), 15320–15340. https://doi.org/10.1002/ange.202006402.

Zhang, R., Zhang, Z., Jiang, J., & Pang, H. (2023). Recent electrochemical-energy-storage applications of metal–organic frameworks featuring iron-series elements (Fe, Co, and Ni). *Journal of Energy Storage, 65*, 107217. https://doi.org/10.1016/j.est.2023.107217.

Zhang, T., Wang, L., Ma, C., Wang, W., Ding, J., Liu, S., Zhang, X., & Xie, Z. (2017). BODIPY-containing nanoscale metal-organic frameworks as contrast agents for computed tomography. *Journal of Materials Chemistry B, 5*(12), 2330–2336. https://doi.org/10.1039/c7tb00392g, http://pubs.rsc.org/en/journals/journalissues/tb.

Zhang, Y., Yuan, S., Feng, X., Li, H., Zhou, J., & Wang, B. (2016). Preparation of nanofibrous metal-organic framework filters for efficient air pollution control. *Journal of the American Chemical Society, 138*(18), 5785–5788. https://doi.org/10.1021/jacs.6b02553, http://pubs.acs.org/journal/jacsat.

Zhao, D., Wang, X., Yue, L., He, Y., & Chen, B. (2022). Porous metal-organic frameworks for hydrogen storage. *Chemical Communications, 58*(79), 11059–11078. https://doi.org/10.1039/d2cc04036k, http://pubs.rsc.org/en/journals/journal/cc.

Zhao, H. X., Zou, Q., Sun, S. K., Yu, C., Zhang, X., Li, R. J., & Fu, Y. Y. (2016). Theranostic metal-organic framework core-shell composites for magnetic resonance imaging and drug delivery. *Chemical Science, 7*(8), 5294–5301. https://doi.org/10.1039/c6sc01359g, http://pubs.rsc.org/en/Journals/JournalIssues/SC.

Zhao, J., Wei, F., Xu, W., & Han, X. (2020). Enhanced antibacterial performance of gelatin/chitosan film containing capsaicin loaded MOFs for food packaging. *Applied Surface Science, 510*, 145418. https://doi.org/10.1016/j.apsusc.2020.145418.

Zhao, T., Jeremias, F., Boldog, I., Nguyen, B., Henninger, S. K., & Janiak, C. (2015). High-yield, fluoride-free and large-scale synthesis of MIL-101(Cr). *Dalton Transactions, 44*(38), 16791–16801. https://doi.org/10.1039/c5dt02625c, http://www.rsc.org/Publishing/Journals.

Zhao, W., Yan, G., Zheng, Y., Liu, B., Jia, D., Liu, T., Cui, L., Zheng, R., Wei, D., & Liu, J. (2020). Bimetal-organic framework derived Cu(NiCo)$_2$S$_4$/Ni$_3$S$_4$ electrode material with hierarchical hollow heterostructure for high performance energy storage. *Journal of Colloid and Interface Science, 565*, 295–304. https://doi.org/10.1016/j.jcis.2020.01.049, http://www.elsevier.com/inca/publications/store/6/2/2/8/6/1/index.htt.

Zhao, Y., Seredych, M., Zhong, Q., & Bandosz, T. J. (2013). Superior performance of copper based MOF and aminated graphite oxide composites as CO_2 adsorbents at room temperature. *ACS Applied Materials and Interfaces, 5*(11), 4951–4959. https://doi.org/10.1021/am4006989.

Zheng, Z., Nguyen, H. L., Hanikel, N., Li, K. K. Y., Zhou, Z., Ma, T., & Yaghi, O. M. (2023). High-yield, green and scalable methods for producing MOF-303 for water harvesting from desert air. *Nature Protocols, 18*(1), 136–156. https://doi.org/10.1038/s41596-022-00756-w, https://www.springer.com/journal/41596.

Zhong, Y., Zheng, X. T., Li, Q., Loh, X. J., Su, X., & Zhao, S. (2023). Antibody conjugated Au/Ir@ Cu/Zn-MOF probe for bacterial lateral flow immunoassay and precise synergistic antibacterial treatment. *Biosensors and Bioelectronics, 224*, 115033.

Zhou, H., Liu, X., Zhang, J., Yan, X., Liu, Y., & Yuan, A. (2014). Enhanced room-temperature hydrogen storage capacity in Pt-loaded graphene oxide/HKUST-1 composites. *International Journal of Hydrogen Energy, 39*(5), 2160–2167. https://doi.org/10.1016/j.ijhydene.2013.11.109.

Zhu, X., Yuan, S., Ju, Y., Yang, J., Zhao, C., & Liu, H. (2019). Water splitting-assisted electrocatalytic oxidation of glucose with a metal-organic framework for wearable nonenzymatic perspiration sensing. *Analytical Chemistry, 91*(16), 10764–10771. https://doi.org/10.1021/acs.analchem.9b02328, http://pubs.acs.org/journal/ancham.

Zhu, Z. Y., Wang, Y. D., Wang, X. W., Dai, G. L., Ma, S. J., Liu, X., Li, J. H., Jin, L., & Lin, Z. X. (2023). Pd/MIL-100(Fe) as hydrogen activator for FeIII/FeII cycle: Fenton removal of sulfamethazine. *Environmental Technology (United Kingdom), 44*(23), 3504–3517. https://doi.org/10.1080/09593330.2022.2064237, http://www.tandf.co.uk/journals/titles/09593330.asp.

Zunita, M., Natola O, W., David, M., & Lugito, G. (2022). Integrated metal organic framework/ionic liquid-based composite membrane for CO_2 separation. *Chemical Engineering Journal Advances, 11*, 100320. https://doi.org/10.1016/j.ceja.2022.100320.

Chapter 22

Perspectives and challenges of application of metal-organic framework composites

Hafezeh Nabipour, and Sohrab Rohani
Department of Chemical and Biochemical Engineering, University of Western Ontario, London, ON, Canada

Introduction

Metal-organic frameworks (MOFs) are highly porous materials formed by coordinating metal ions or clusters with organic ligands, creating highly organized structures (Nabipour, Rohani, et al., 2023). These frameworks possess key attributes such as high surface area, tunable pore sizes, permanent porosity, and structural flexibility, which contribute to their wide range of functionalities. MOFs are known for their remarkable chemical and thermal stability, making them suitable for various applications (Nabipour & Rohani, 2023; Nabipour, Aliakbari, et al., 2023). Their adaptability stems from a modular design, enabling customization for specific applications, including supercapacitors, carbon capture, fuel cells, solar cells, electromagnetic shielding, water splitting, food-related uses, adsorption, gas sorption and separation, drug delivery, catalysis, fire safety in polymers, and chemical sensing. Additionally, the crystalline structure of MOFs allows for precise control over both porosity and reactivity (Mohanty et al., 2024; Nabipour et al., 2024; Pramanik et al., 2023).

MOF composites, which combine MOFs with substrates such as carbon materials, metal oxides, metal particles, polyoxometalates, other MOFs, and polymers, show enhanced properties, expanding their applicability (Afzal et al., 2024; Shen et al., 2024; Yu et al., 2017). Despite their advantages, pure MOFs often face limitations such as weak mechanical strength, insufficient chemical and thermal stability, and low electrical conductivity. Research has focused on developing MOF composites to harness the synergistic interactions between components, which help overcome the shortcomings of individual materials (Shen et al., 2024). Since 2008, there has been a significant rise in publications on MOF composites (Fig. 22.1A). Ongoing efforts to improve these composites continue to yield advancements, highlighting their substantial potential across a wide array of applications (Fig. 22.1B).

By integrating MOFs with more stable materials such as polymers or metal oxides, these composites demonstrate improved stability, mechanical strength, and processability. This integration helps maintain structural integrity and functional properties under different environmental conditions and allows

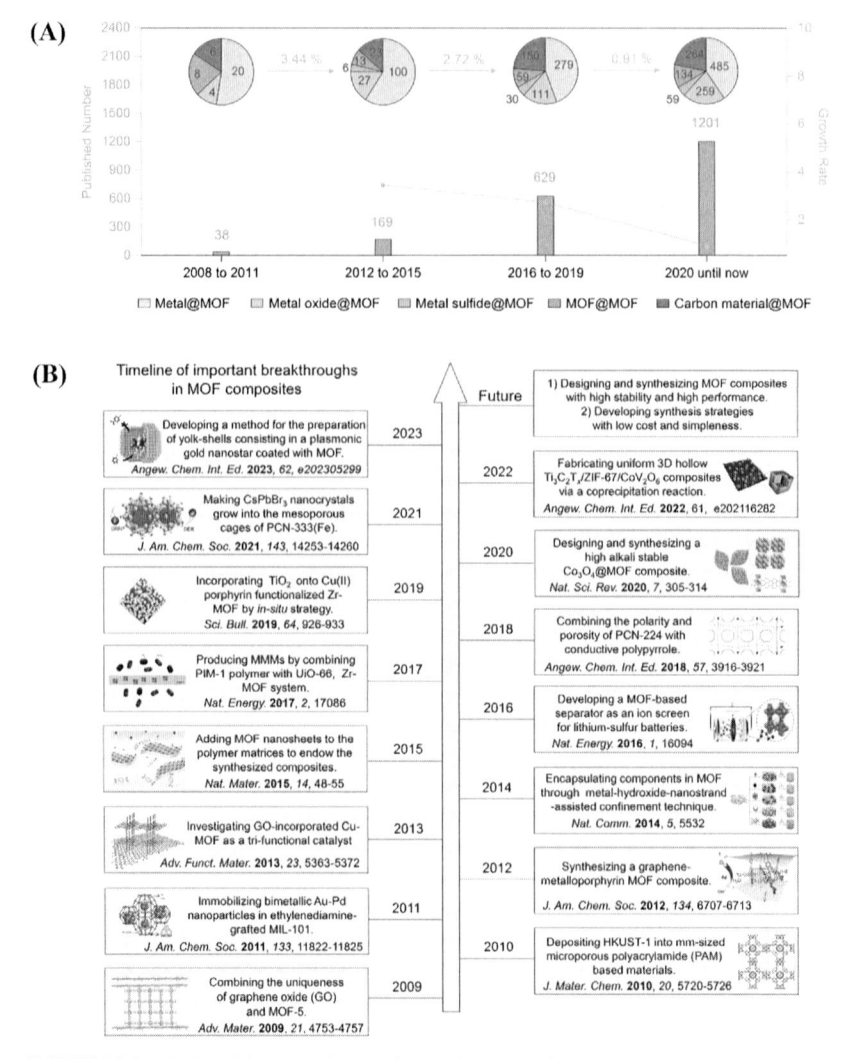

FIGURE 22.1 (A) Trends in research on various MOF composites over recent years. (B) Timeline highlighting significant breakthroughs in MOF composites (Shen et al., 2024).

MOFs to be processed into various forms, such as membranes, films, and bulk structures. The combination of MOFs and other materials in these composites frequently results in enhanced functionality. For example, in catalysis, MOF composites achieve greater catalytic activity and selectivity through the incorporation of metal nanoparticles or other active species. In energy storage, these composites improve the performance of batteries and supercapacitors by enhancing ion transport and electrical conductivity. Furthermore, MOF composites are promising in gas storage and separation, where their tunable porosity increases gas adsorption capacities and selectivity. Additionally, they offer multifunctional capabilities by combining attributes such as catalytic activity and magnetic properties, enhancing their utility and ease of recovery (Meskher, 2023; Wang et al., 2019; Yu et al., 2017).

In conclusion, MOF composites offer several key benefits, including enhanced stability, superior mechanical properties, improved processability, and synergistic functional capabilities. Their improved performance in energy storage, gas-related applications, and potential for multifunctional uses has established MOF composites as highly promising materials for tackling a wide range of technological challenges. These advantages continue to drive research and innovation in this evolving field.

Preparation of metal-organic framework composites

The development of MOF composites has attracted significant attention due to the superior properties and functionalities they provide compared to pure MOFs. Numerous synthesis methods have been established, each offering distinct advantages and addressing specific challenges. One widely used technique is the "ship in bottle" approach, where active components or nanoparticle precursors are encapsulated within pre-formed MOFs, followed by treatment to form the desired structure. This process uses methods such as solution infiltration, vapor deposition, and solid grinding to introduce the precursors, although controlling their spatial distribution and morphology can be difficult. To overcome these challenges, the double-solvent method (DSM) has been introduced. This method employs a combination of hydrophilic and hydrophobic solvents to prevent nanoparticle aggregation and enhance the integration of metallic and bimetallic nanoparticles into MOF frameworks. Another approach, known as the "bottle around ship" method, involves growing MOFs around pre-formed nanoparticles, offering greater control over nanoparticle size and distribution. However, this technique requires precise synthesis conditions to avoid damaging the nanoparticles. A more streamlined option is the one-step synthesis approach, which simultaneously forms the MOF structure while incorporating functional components, although this method requires careful optimization of reaction parameters. Additionally, MOF-derived glasses are being explored as matrices for composite materials, resulting in better interfacial bonding in MOF crystal-glass composites

(CGCs). Other synthesis techniques under investigation include electrochemical and sonochemical methods. Electrochemical methods provide milder synthesis conditions and shorter synthesis times, while sonochemical methods utilize ultrasonic radiation to control crystallization. The microemulsion method, though precise, relies on expensive and environmentally harmful surfactants. In summary, the synthesis of MOF composites involves a wide array of methods, each with its own advantages and drawbacks. Ongoing research is expected to generate new techniques that will further extend the applications of MOF composites across various fields (Chen et al., 2017; Li et al., 2015; Liu, Chang, et al., 2016; Liu, Sun, et al., 2016; Qiao et al., 2017; Zhan et al., 2013).

Metal-organic framework composites and their applications

As discussed in previous chapters, MOF composites represent cutting-edge hybrid materials created by incorporating MOFs with various functional components to either improve their properties or introduce new functionalities. These composites have garnered widespread interest due to their synergistic effects and vast potential applications. Each composite type offers specific benefits and characteristics, making them adaptable for a range of uses. The following sections highlight several MOF composites previously discussed and their diverse applications.

Carbon-based metal-organic framework composites

These composites combine carbon materials with MOFs to enhance their properties, making them versatile across various fields. In catalysis, carbon-based MOF composites increase catalytic activity by providing additional active sites and facilitating electron transfer. For example, they are used in the catalytic conversion of CO_2 into valuable chemicals, demonstrating higher efficiency than traditional catalysts. The high surface area and porosity of MOFs allow for better dispersion of catalytic species, improving reaction rates and selectivity. In energy storage, these composites serve as effective electrode materials for batteries and supercapacitors, where the combination of MOFs' large surface area with carbon materials' electrical conductivity results in enhanced charge storage capacity and cycling stability, often outperforming traditional electrodes. For gas storage and separation, carbon-based MOF composites take advantage of the tunable pore sizes and high surface areas of MOFs, while the addition of carbon materials enhances selectivity and capacity for specific gases, making them ideal for CO_2 capture from flue gases. In electromagnetic wave (EMW) absorption, these composites are highly conductive and offer tunable dielectric properties. The inclusion of carbon materials enhances interfacial polarization loss and optimizes impedance matching, making them suitable for electromagnetic interference (EMI) shielding and stealth technologies. Furthermore, these composites are valuable in environmental remediation, where their high surface area and porosity make them effective adsorbents for pollutants like heavy metals and organic contaminants. Their enhanced adsorption capacity and kinetics, provided by the carbon component, are advantageous for water treatment and air purification (Tshuma et al., 2020; Zhang et al., 2021).

Metal-based material@metal-organic framework composites

These composites incorporate metals, metal oxides, or metal sulfides into MOFs to create versatile materials with improved properties and functionalities. Metal-based@MOF composites often involve metal nanoparticles or clusters such as gold, iridium, titanium, and palladium embedded in or attached to the MOF structure. These composites exhibit superior catalytic activity, stability, and tunable electronic properties, making them highly effective in reactions like hydrogenation, oxidation, and coupling. The porous MOF structure offers a high surface area and confinement effects, while metal particles act as active catalytic sites. Metal-based MOF composites also hold promise for gas storage and separation due to their enhanced porosity and adjustable pore sizes (Duan et al., 2021; Guntern et al., 2019; Meskher, 2023; Yang et al., 2019).

Metal oxide/metal-organic framework composites

By integrating metal oxides like zinc oxide and titanium oxide with MOFs, these composites achieve exceptional performance in photocatalysis. They are applied in water splitting, CO_2 reduction, and the degradation of organic pollutants. Metal oxide/MOF composites also find applications in sensing and energy storage, where their improved electrical conductivity and stability make them suitable for supercapacitors and batteries (Xu et al., 2022; Zhai et al., 2022; Zheng et al., 2020).

Metal sulfide/metal-organic framework composites

Composites combining metal sulfides such as molybdenum disulfide, cadmium sulfide, and bismuth sulfide with MOFs show great potential. These materials demonstrate excellent performance in electrocatalysis, especially for hydrogen and oxygen evolution reactions, due to the synergy between metal sulfides and MOFs, enhancing both catalytic activity and stability. They also show promise in photocatalysis, such as for water splitting, pollutant degradation, and CO_2 reduction, where the combination of metal sulfides and MOFs improves light absorption and charge separation, leading to higher photocatalytic efficiency (Liu et al., 2020; Parsapour et al., 2024; Wang et al., 2021).

Polymer@metal-organic framework composites

These innovativce hybrid materials combine the unique properties of MOFs with the versatility of polymers. Polymer@MOF composites offer numerous advantages, such as enhanced mechanical stability, better processability of MOFs, and protection from degradation in harsh environments. These composites have improved gas separation and storage performance compared to pure polymer membranes, with higher gas selectivity and permeability. Their high adsorption capacity, coupled with polymer processability, also makes them ideal for water purification. In energy storage, incorporating MOFs into polymer electrolytes improves ionic conductivity and battery performance. In biomedical applications, polymer@MOF composites have been explored for drug delivery, using the high porosity of MOFs for drug loading and the polymer component for controlled release (Brahmi et al., 2021; Giliopoulos et al., 2020; Liu et al., 2021).

Polyoxometalate@metal-organic framework composites

These composites combine polyoxometalates (POMs) with MOFs, resulting in materials with enhanced stability, improved catalytic performance, and new functionalities. POM@MOF composites are particularly effective in catalysis, such as for oxidation reactions of alcohols and sulfides. The MOF structure provides high surface area and confinement effects, while POMs serve as active catalytic sites. These composites are also being studied for environmental remediation, where they combine MOFs' adsorption capacity with POMs' catalytic abilities to remove organic dyes and heavy metal ions from water. Additionally, POM@MOF composites show promise as electrocatalysts for hydrogen evolution reactions (HER) and oxygen evolution reactions (OER), where the synergistic effects between POMs and MOFs enhance catalytic activity and stability (Ebrahimi et al., 2024; Guo et al., 2022; Pan et al., 2018; Wang et al., 2021).

Metal-organic framework@metal-organic framework composites

These composites combine different MOFs to create materials with enhanced stability, improved catalytic activity, and tunable porosity. MOF@MOF composites can introduce new pores at the interfaces between different MOF phases, expanding their applications. These composites have been used in catalysis, where the combination of different MOF structures creates unique catalytic sites and boosts overall performance. MOF@MOF composites have also demonstrated improved performance in gas storage and separation, with their hierarchical pore structure improving gas selectivity and storage capacity. Additionally, they show potential in sensing applications, where different metal centers or organic linkers provide materials with increased sensitivity and selectivity toward specific analytes. In drug delivery, MOF@ MOF composites offer the potential to integrate various drug loading and release properties into a single material, enabling more controlled and sustained release profiles (Ding et al., 2019; Jin et al., 2022; Mohanty et al., 2024; Panchariya et al., 2018; Wang et al., 2020).

Perspectives on metal-organic framework composites

MOF composites have emerged as highly versatile and functional materials, attracting significant interest across numerous fields. These hybrid systems are synthesized by coordinating metal ions with organic linkers, yielding a highly porous architecture with an extensive surface area. This structural design allows MOF composites to accommodate functional entities either within their internal cavities or on their surface, making them excellent candidates for a wide array of applications, including supercapacitors and biomedical technologies. The range of potential applications spans energy storage, environmental remediation, catalysis, and gas sensing, where their tunable chemical and physical properties enable precise optimization for targeted functions.

One of the most promising applications of MOF composites lies in the field of supercapacitors. Supercapacitors demand materials capable of rapidly storing and releasing substantial energy, and MOF composites offer distinct advantages due to their large surface area and porous architecture, which promote faster charge-discharge cycles. By incorporating conductive nanoparticles, such as graphene or carbon nanotubes, into the MOF framework, their electrical conductivity can be significantly improved, thereby increasing the overall efficiency of energy storage systems. These composites exhibit superior energy and power density compared to traditional supercapacitor materials, positioning them as crucial components in the advancement of next-generation energy storage technologies.

In the field of carbon capture, MOF composites hold great promise due to their tunable pore size and high surface area. The ability to selectively adsorb CO_2 while excluding other gases makes them highly effective for carbon capture from industrial emissions. By functionalizing MOFs with amine groups, their affinity for CO_2 can be further enhanced, allowing for more efficient capture even in the presence of moisture, a condition that often impairs the performance of other materials. This moisture tolerance sets MOF composites apart from traditional carbon capture technologies and highlights their potential for large-scale environmental applications.

In fuel cells and solar cells, MOF composites are being explored as both catalysts and supporting materials. Their porous nature facilitates the movement of gases and liquids, which is critical for the efficiency of these devices. In fuel cells, MOF composites can act as electrocatalysts for the oxygen reduction reaction, a key process in energy conversion. Their ability to incorporate transition metals such as platinum, palladium, or nickel into their structure allows for improved catalytic activity and stability. Similarly, in solar cells, MOF composites can be used to improve light absorption and charge transport, making them integral components for next-generation photovoltaic devices.

EMI shielding is another area where MOF composites are making significant strides. With the increasing reliance on electronic devices, EMI shielding is essential to prevent interference between different electronic components. Traditional shielding materials, such as metals, tend to be heavy and bulky. MOF composites, by contrast, offer a lightweight and tunable alternative. By integrating conductive polymers or nanoparticles into MOFs, their shielding effectiveness can be significantly enhanced, offering protection across a broad frequency range without the drawbacks of traditional materials.

Water splitting, a process used to produce hydrogen fuel from water, also benefits from MOF composites. The high surface area of MOFs allows for greater interaction between water molecules and the catalytic sites, improving the efficiency of the reaction. When combined with metal nanoparticles, MOF composites exhibit enhanced catalytic activity, making them ideal for use in electrochemical cells designed for water splitting. The ability to fine-tune the pore size and surface chemistry of MOFs allows researchers to optimize them for specific types of water splitting, such as photoelectrochemical or electrocatalytic water splitting.

MOF composites are also being explored in food applications, particularly in food preservation and packaging. The porous structure of MOFs allows them to adsorb gases such as oxygen, which can lead to spoilage, thereby extending the shelf life of perishable goods. Furthermore, MOFs can be incorporated into food packaging materials to create active packaging that responds to environmental conditions, such as humidity or temperature, by releasing preservatives or scavenging harmful gases. This dynamic approach to food preservation is a promising area of research, with the potential to revolutionize food safety and shelf-life extension.

In polymer composites, MOFs are being incorporated to enhance the fire safety of materials. Polymers are widely used in construction and consumer products, but their flammability is a major safety concern. MOF composites can act as flame retardants by releasing gases such as CO_2 when exposed to high temperatures, thereby suppressing the combustion process. Additionally, the high thermal stability of MOF composites ensures that they do not degrade or lose their effectiveness at elevated temperatures, making them suitable for use in a variety of fire safety applications.

The adsorption of volatile organic compounds (VOCs) is another important application of MOF composites. VOCs are harmful pollutants emitted by various industrial processes, and their removal from the air is crucial for maintaining air quality. MOF composites, with their large surface area and tunable pore sizes, are highly effective at capturing VOCs from the air. Furthermore, by incorporating functional groups that enhance the interaction between the MOF and specific VOCs, the selectivity of these composites can be improved, making them ideal for use in air purification systems.

In antibacterial and biomedical applications, MOF composites are showing significant potential. Their high surface area allows for the loading of antibacterial agents, which can be released in a controlled manner to combat infections. Additionally, by incorporating biocompatible materials such as zinc or iron into the MOF structure, their toxicity can be minimized, making them suitable for use in medical devices and implants. In drug delivery, MOF composites offer a means to encapsulate drugs within their porous structure, protecting them from degradation and allowing for controlled release at the target site.

Electrocatalysts for conversion represent another area of significant interest. The conversion of CO_2 into useful chemicals or fuels is a key strategy for mitigating climate change, and MOF composites have shown promise as electrocatalysts for this process. By incorporating metal nanoparticles into the MOF structure, the catalytic activity for CO_2 reduction can be enhanced, leading to more efficient conversion of CO_2 into valuable products such as methanol or formic acid. This approach not only helps to reduce greenhouse gas emissions but also provides a pathway for producing renewable fuels.

In the catalytic desulfurization and denitrogenation of fuels, MOF composites have shown great promise due to their high surface area and tunable active sites. These composites can effectively remove sulfur and nitrogen compounds from fuels, which is critical for meeting environmental regulations and reducing pollution from combustion engines. The ability to functionalize MOFs with specific catalytic sites allows for selective removal of these contaminants, making them more efficient than traditional desulfurization and denitrogenation methods.

Environmental remediation is another area where MOF composites are making a significant impact. Their high adsorption capacity makes them ideal for removing pollutants from air and water. For instance, MOF composites can be used to remove heavy metals, organic pollutants, and even radioactive materials from contaminated water. Their reusability and stability under harsh environmental conditions make them an attractive option for large-scale environmental cleanup efforts.

Sensing applications, particularly gas sensing, benefit greatly from the use of MOF composites. The high surface area and tunable pore sizes allow for the detection of trace amounts of gases, making them highly sensitive sensors. By functionalizing the surface of MOFs with specific groups that interact with target gases, the selectivity of these sensors can be improved, allowing for the detection of specific gases in complex environments. This makes MOF composites ideal for use in industrial safety, environmental monitoring, and even medical diagnostics.

In the realm of energy storage, MOF composites offer a unique solution due to their high surface area and ability to incorporate conductive materials.

By integrating materials such as graphene or carbon nanotubes into the MOF structure, their electrical conductivity can be improved, making them more effective in batteries and supercapacitors. Additionally, the porous nature of MOFs allows for the efficient storage and release of ions, improving the performance of these energy storage devices.

The immobilization of enzymes within MOF composites represents another promising application. Enzymes are highly efficient biological catalysts, but their use in industrial processes is often limited by their instability and the difficulty of recovering them after the reaction. By encapsulating enzymes within the porous structure of MOFs, their stability can be enhanced, and their catalytic activity can be maintained over longer periods. Furthermore, the porous nature of MOFs allows for the easy diffusion of substrates to the active sites, improving the efficiency of the catalytic process.

Overall, the perspectives on MOF composites indicate a bright future for these materials across a wide range of applications. From energy storage to environmental remediation, the versatility and tunability of MOF composites make them ideal candidates for addressing some of the most pressing challenges of the 21st century. As research continues to advance, it is likely that MOF composites will become an integral part of many industrial and environmental processes, offering solutions that are both efficient and sustainable.

Challenges in the application of metal-organic framework composites

MOF composites have gained recognition as highly promising materials for a range of applications, including catalysis, gas storage and separation, drug delivery, and environmental remediation. However, several challenges hinder their broader implementation in practical settings. This analysis explores the primary obstacles encountered in the application of MOF composites and the current strategies to overcome them.

One major challenge is the stability of MOF composites under various operational conditions. Although many MOFs exhibit excellent thermal and chemical stability in controlled environments, their performance often deteriorates under real-world conditions (Zhang et al., 2014). For example, certain MOFs are sensitive to moisture and degrade when exposed to humid environments, which poses a significant problem for gas storage and separation, as water vapor can greatly reduce adsorption capacity and selectivity. To mitigate this issue, researchers have developed methods to improve the stability of MOF composites. One approach is to incorporate hydrophobic components into the MOF structure. Another strategy is to use high-valent metal ions or azolate-based ligands to create more robust MOF structures.

Scalability in the synthesis of MOF composites is another significant challenge. Many MOF composites are produced using laboratory-scale methods that are not easily scalable for industrial production. These processes often involve complex procedures, expensive precursors, and extended reaction times, making large-scale production economically unfeasible. Additionally, maintaining consistent quality and reproducibility across different batches of MOF composites is difficult. To address these issues, researchers are exploring alternative synthesis methods and continuous flow processes, which may enable large-scale production with more consistent quality.

Mechanical stability also presents a challenge, particularly for applications requiring high pressure or mechanical stress. Many MOFs are brittle and may collapse under pressure, leading to a loss of porosity and functionality. This is especially relevant in gas storage and separation, where high pressures are commonly used. To enhance mechanical stability, researchers have incorporated polymers and other supportive materials. For example, Matzger's group developed MOF-5-polystyrene composites that exhibited improved mechanical stability compared to pure MOF-5 - (Gamage et al., 2016). However, balancing mechanical stability with other desirable properties, such as high surface area and porosity, remains a challenge.

Integrating MOF composites into existing industrial processes is another significant obstacle. Many industrial applications require materials capable of withstanding harsh conditions, such as high temperatures, pressures, and corrosive environments. While some MOF composites have shown promise in laboratory settings, their performance in industrial environments often falls short. Researchers are now working on developing MOF composites specifically designed for industrial applications, focusing on improving stability, performance, cost-effectiveness, ease of handling, and compatibility with existing equipment.

Long-term stability and recyclability are also critical for the practical use of MOF composites, particularly in catalysis and environmental remediation. While many MOF composites show excellent initial performance, they often degrade or deactivate over time due to leaching of active components, structural collapse, or poisoning of catalytic sites. To improve long-term stability, researchers are developing core-shell structures with protective layers and using more durable linkers and metal nodes. Efforts are also being made to investigate regeneration methods to extend the lifespan and economic viability of MOF composites.

Environmental impact and safety concerns are becoming increasingly important as MOF composites move toward real-world applications. Some MOFs contain toxic metal ions or organic linkers, which could pose environmental or health risks if released. Additionally, the synthesis of MOF composites often involves hazardous solvents and generates considerable waste. To address these issues, researchers are working on greener synthesis methods and investigating the use of biocompatible and environmentally friendly components.

The cost of MOF composites remains a major barrier to widespread adoption. Many MOF composites rely on expensive precursors and complex synthesis methods, making them economically uncompetitive with existing materials. Reducing the cost of MOF composites while maintaining their desirable properties is essential for commercial viability. Researchers are exploring cheaper, more abundant precursors and more efficient synthesis methods. Efforts are also focused on improving the performance of MOF composites, which could allow for smaller quantities of material to achieve desired results.

In conclusion, MOF composites hold great potential for a variety of applications, but overcoming challenges related to stability under real-world conditions, scalability, mechanical properties, integration into industrial processes, long-term stability, environmental safety, and cost is essential for their broader adoption. Ongoing research is making significant progress in addressing these issues, and MOF composites are expected to play an increasingly important role in technological advancements in the near future.

Future directions

The future of MOF composites lies in overcoming existing challenges through advanced synthesis techniques and innovative material designs. Progress in computational modeling and machine learning is anticipated to accelerate the discovery of novel MOF structures and composites with tailored properties for specific applications. Additionally, integrating MOFs with cutting-edge materials, such as two-dimensional materials and nanostructures, could result in multifunctional composites capable of addressing multiple challenges simultaneously. In summary, while MOF composites face numerous challenges, their potential for innovation is significant. Ongoing research and development in this field could lead to breakthroughs that enhance the performance and applicability of MOF composites across a wide range of industries, contributing to the development of more sustainable and efficient technologies.

Conclusion

MOF composites have emerged as promising materials with great potential across various fields, such as catalysis, energy storage, gas separation, environmental remediation, biomedical applications, fire safety, water splitting, electromagnetic shielding, food safety, adsorption of volatile organic compounds, sensors, CO_2 conversion, solar cells, fuel cells, and supercapacitors. Their ability to combine the inherent properties of MOFs with additional functional elements results in materials with enhanced performance and versatility. However, fully realizing the potential of MOF composites requires addressing challenges such as ensuring stability under different operational conditions, scaling up synthesis processes, and integrating them into existing technologies. Research efforts are focused on improving stability through advanced material design and synthesis, developing scalable and cost-effective production techniques, and enhancing mechanical properties to withstand real-world applications. Addressing environmental and safety concerns is also critical for the widespread adoption of MOF composites. As research advances, MOF composites are expected to play an increasingly vital role in technological innovation, offering sustainable solutions to some of the world's most pressing challenges.

References

Afzal, S., Ur Rehman, A., Najam, T., Hossain, I., Abdelmotaleb, M. A. I., Riaz, S., Karim, M. D. R., Shah, S. S. A., & Nazir, M. A. (2024). Recent advances of MXene@MOF composites for catalytic water splitting and wastewater treatment approaches. *Chemosphere, 364*. Article 143194. https://doi.org/10.1016/j.chemosphere.2024.143194.

Brahmi, C., Benltifa, C., Vaulot, C., Michelin, L., Dumur, F., Millange, F., Frigoli, M., Airoudj, A., Morlet-Savary, F., Bousselmi, L., & Lalevée, J. (2021). New hybrid MOF/polymer composites for the photodegradation of organic dyes. *European Polymer Journal, 154*. Article 110560.

Chen, L., Zhan, W., Fang, H., Cao, Z., Yuan, C., Xie, Z., Kuang, Q., & Zheng, L. (2017). Selective catalytic performances of noble metal nanoparticle@MOF composites: The concomitant effect of aperture size and structural flexibility of MOF. *Chemistry – A European Journal, 23*(47), 11397–11403.

Ding, M., Cai, X., & Jiang, H. L. (2019). Improving MOF stability: Approaches and applications. *Chemical Science, 10*(44), 10209–10230.

Duan, H., Zhao, Z., Lu, J., Hu, W., Zhang, Y., Li, S., Zhang, M., Zhu, R., & Pang, H. (2021). When conductive MOFs meet MnO_2: High electrochemical energy storage performance in an aqueous asymmetric supercapacitor. *ACS Applied Materials & Interfaces, 13*(27), 33083–33090.

Ebrahimi, A., Krivosudský, L., Cherevan, A., & Eder, D. (2024). Polyoxometalate-based porphyrinic metal-organic frameworks as heterogeneous catalysts. *Coordination Chemistry Reviews, 508*. Article 215764.

Gamage, N. D., McDonald, K. A., & Matzger, A. J. (2016). MOF-5-polystyrene: Direct production from monomer, improved hydrolytic stability, and unique guest adsorption. *Angewandte Chemie International Edition, 55*(39), 12099–12103.

Giliopoulos, D., Zamboulis, D., Giannakoudakis, D., Bikiaris, D., & Triantafyllidis, K. (2020). Polymer/metal organic framework (MOF) nanocomposites for biomedical applications. *Molecules, 25*(1), 185.

Guntern, Y. T., Pankhurst, J. R., Vávra, J., Mensi, M., Mantella, V., Schouwink, P., & Buonsanti, R. (2019). Nanocrystal/metal-organic framework hybrids as electrocatalytic platforms for CO_2 conversion. *Angewandte Chemie International Edition, 58*(41), 12632–12639.

Guo, H., Chen, Y., Wu, N., Peng, L., Yang, F., Pan, Z., Liu, B., Zhang, H., Li, C., & Yang, W. (2022). A novel CDs coated polyoxometalate/metal-organic framework composite for supercapacitors. *Journal of Alloys and Compounds, 921*. Article 165730.

Jin, Y., Mi, X., Qian, J., Ma, N., & Dai, W. (2022). Modular construction of an MIL-101(Fe)@MIL-100(Fe) dual-compartment nanoreactor and its boosted photocatalytic activity toward tetracycline. *ACS Applied Materials & Interfaces, 14*(42), 48285–48295.

Li, Z., Yu, R., Huang, J., Shi, Y., Zhang, D., Zhong, X., Wang, Y., Wu, Y., & Li, Y. (2015). Platinum–nickel frame within metal-organic framework fabricated in situ for hydrogen enrichment and molecular sieving. *Nature Communications, 6*. Article 8248.

Liu, H., Chang, L., Bai, C., Chen, L., Luque, R., & Li, Y. (2016). Controllable encapsulation of "clean" metal clusters within mofs through kinetic modulation: towards advanced heterogeneous nanocatalysts. *Angewandte Chemie International Edition, 55*(16), 5019–5023.

Liu, X. W., Sun, T. J., Hu, J. L., & Wang, S. D. (2016). Composites of metal-organic frameworks and carbon-based materials: Preparations, functionalities and applications. *Journal of Materials Chemistry A, 4*(11), 3584–3616.

Liu, Y., Huang, D., Cheng, M., Liu, Z., Lai, C., Zhang, C., & Liang, Q. (2020). Metal sulfide/MOF-based composites as visible-light-driven photocatalysts for enhanced hydrogen production from water splitting. *Coordination chemistry reviews, 409*. Article 213220.

Liu, Y., Tang, C., Cheng, M., Chen, M., Chen, S., Lei, L., Chen, Y., Yi, H., Fu, Y., & Li, L. (2021). Polyoxometalate@Metal-organic framework composites as effective photocatalysts. *ACS Catalysis, 11*(21), 13374–13396.

Meskher, H. (2023). A critical review about metal organic framework-based composites: Potential applications and future perspective. *Journal of Composites and Compounds, 5*(14), 25–37.

Mohanty, B., Kumari, S., Yadav, P., Kanoo, P., & Chakraborty, A. (2024). Metal-organic frameworks (MOFs) and MOF composites based biosensors. *Coordination Chemistry Reviews, 519*. Article 216102.

Nabipour, H., Aliakbari, F., Volkening, K., Strong, M. J., & Rohani, S. (2023). New metal-organic framework coated sodium alginate for the delivery of curcumin as a sustainable drug delivery and cancer therapy system. *International Journal of Biological Macromolecules, 242*. Article 128875.

Nabipour, H., Aliakbari, F., Volkening, K., Strong, M. J., & Rohani, S. (2024). The development of a bio-based metal-organic framework coated with carboxymethyl cellulose with the ability to deliver curcumin with anticancer properties. *Materials Today Chemistry, 37*. Article 101976.

Nabipour, H., & Rohani, S. (2023). Zirconium metal organic framework/aloe vera carrier loaded with naproxen as a versatile platform for drug delivery. *Chemistry Papers, 77*(6), 3461–3470.

Nabipour, H., Rohani, S., Batool, S., & Yusuff, A. S. (2023). An overview of the use of water-stable metal-organic frameworks in the removal of cadmium ion. *Journal of Environmental Chemical Engineering, 11*. Article 109131.

Pan, Y., Sun, K., Liu, S., Cao, X., Wu, K., Cheong, W. C., Chen, Z., Wang, Y., Li, Y., Liu, Y., Wang, D., Chen, C., & Li, Y. (2018). Core-shell ZIF-8@ZIF-67-derived CoP nanoparticle-embedded N-doped carbon nanotube hollow polyhedron for efficient overall water splitting. *Journal of the American Chemical Society, 140*(8), 2610–2618.

Panchariya, D. K., Rai, R. K., Kumar, E. A., & Singh, S. K. (2018). Core–shell zeolitic imidazolate frameworks for enhanced hydrogen storage. *ACS Omega, 3*(1), 167–175.

Parsapour, F., Moradi, M., Safarifard, V., & Sojdeh, S. (2024). Polymer/MOF composites for metal-ion batteries: A mini review. *Journal of Energy Storage, 82*. Article 110487.

Pramanik, B., Sahoo, R., & Das, M. C. (2023). pH-stable MOFs: Design principles and applications. *Coordination Chemistry Reviews, 493*. 215301.

Qiao, C., Sun, L., Zhang, S., Liu, P., Chang, L., Zhou, C., Wei, Q., Chen, S., & Gao, S. (2017). Pore-size-tuned host–guest interactions in Co-MOFs via in situ microcalorimetry: Adsorption and magnetism. *Journal of Materials Chemistry C, 5*(5), 1064–1073.

Shen, Z., Peng, Y., Li, X., Li, N., Xu, H., Li, W., Guo, X., & Pang, H. (2024). Design principle and synthetic strategy for metal-organic framework composites. *Composites Communications, 48*. Article 101933.

Tshuma, P., Makhubela, B. C. E., Bingwa, N., & Mehlana, G. (2020). Palladium(II) immobilized on metal-organic frameworks for catalytic conversion of carbon dioxide to formate. *Inorganic Chemistry, 59*(10), 6717–6728.

Wang, H., Li, T., Li, J., Tong, W., & Gao, C. (2019). One-pot synthesis of poly(ethylene glycol) modified zeolitic imidazolate framework-8 nanoparticles: Size control, surface modification and drug encapsulation. *Colloids and Surfaces A: Physicochemical and Engineering Aspects, 568*, 224–230.

Wang, J., Yue, X., Xie, Z., Abudula, A., & Guan, G. (2021). MOFs-derived transition metal sulfide composites for advanced sodium ion batteries. *Energy Storage Materials, 41*, 404–426.

Wang, X. G., Xu, L., Li, M. J., & Zhang, X. Z. (2020). Construction of flexible-on-rigid hybrid-phase metal-organic frameworks for controllable multi-drug delivery. *Angewandte Chemie International Edition, 59*(32), 18078–18086.

Wang, Z. H., Wang, X. F., Tan, Z., & Song, X. Z. (2021). Polyoxometalate/metal-organic framework hybrids and their derivatives for hydrogen and oxygen evolution electrocatalysis. *Materials Today Energy, 19*. Article 100618.

Xu, M., Sun, C., Zhao, X., Jiang, H., Wang, H., & Huo, P. (2022). Fabricated hierarchical CdS/Ni-MOF heterostructure for promoting photocatalytic reduction of CO2. *Applied Surface Science, 576*. Article 151792.

Yang, Q., Yao, F., Zhong, Y., Chen, F., Shu, X., Sun, J., He, L., Wu, B., Hou, K., Wang, D., & Li, X. (2019). Metal-organic framework supported palladium nanoparticles: Applications and mechanisms. *Particle & Particle Systems Characterization, 36*(6), 1800557.

Yu, J., Mu, C., Yan, B., Qin, X., Shen, C., Xue, H., & Pang, H. (2017). Nanoparticle/MOF composites: Preparations and applications. *Materials Horizons, 4*(4), 557–569.

Zhai, X., Li, S., Wang, Y., Cao, S., Sun, W., Liu, M., Mao, G., Cao, B., & Wang, H. (2022). A magnet-renewable electroanalysis strategy for hydrogen sulfide in aquaculture freshwater using magnetic silver metal-organic frameworks. *Analytica Chimica Acta, 1195*. Article 339450.

Zhan, W.-W., Kuang, Q., Zhou, J.-Z., Kong, X.-J., Xie, Z.-X., & Zheng, L.-S. (2013). Semiconductor@metal-organic framework core–shell heterostructures: A case of ZnO@ZIF-8 nanorods with selective photoelectrochemical response. *Journal of the American Chemical Society, 135*(5), 1926–1933.

Zhang, W., Hu, Y., Ge, J., Jiang, H. L., & Yu, S. H. (2014). A facile and general coating approach to moisture/water-resistant metal-organic frameworks with intact porosity. *Journal of the American Chemical Society, 136*(48), 16978–16981.

Zhang, X., Qiao, J., Jiang, Y., Wang, F., Tian, X., Wang, Z., Wu, L., Liu, W., & Liu, J. (2021). Carbon-based MOF derivatives: Emerging efficient electromagnetic wave absorption agents. *Nano-Micro Letters, 13*, 135.

Zheng, S., Li, Q., Xue, H., Pang, H., & Xu, Q. (2020). A highly alkaline-stable metal oxide@metal-organic framework composite for high-performance electrochemical energy storage. *National Science Review, 7*(2), 305–314.

Index

Note: Page numbers followed by "*f*" and "*t*" refer to figures and tables, respectively.

A

Absorption-dominant shielding system, 251–252, 270
Acetone, 339–341
Acetonitrile, 27–29
Acetylcholinesterase (AChE), 82
Acetylene (C_2H_2), 140
Acetylsalicylic acid (ASA), 625
Activated carbon (AC), 73–74, 77–78, 112–113, 151–152, 341–342, 351
  composite, 315
Activation method, 38
  of metal-organic frameworks, 37–42
Active food contact material, 319–320
Adaptive water harvesting (AWH), 878
Additive manufacturing, 143–144
Adsorbed natural gas (ANG), 835–836
Adsorbents, 621–622
Adsorption, 355–356, 610–611, 828
  efficiency, 615
  isotherm, 615
  mechanism, 615–618
    influential factors on adsorption, 616–618
  in MOF composites, 354–355
  of pollutants, 621–626
    adsorptive removal of organic dye by MOF composites, 621–624
    adsorptive removal of pharmaceutical compounds by MOF composites, 624–626
  of volatile organic compounds by MOF, 343
Adsorption-based cooler (ADC), 878
Adsorptive removal
  of heavy metals by MOF composites, 626–631
  of organic dye by MOF composites, 621–624
  of pharmaceutical compounds by MOF composites, 624–626

Advanced design strategies for enhanced performance, 103–104
Alcohol dehydrogenase, 784–785
Alcoholysis, 791–792
Aldehydes (RCHO), 337–338
Alkaline fuel cell (AFC), 179–180
Alkenyl succinic anhydride (ASA), 881
Alq$_3$ (tris(8-hydroxyquinoline)), 316
Aluminum-based porous metal-organic framework (Al-PMOF), 455
Aluminum composite, 316–317
Aluminum hypophosphite (AHP), 718–719, 731–732
Ambient-pressure composite (AP composite), 826
Aminated graphite oxide (AGO), 154
Amine
  amine-based absorbents, 135–136
  amine-functionalized cellulose-based MOF composite, 159
  amine-functionalized MIL-101, 145
  amine-functionalized MOFs, 104, 144–151
  functionalities, 158–159
2-amino-1,4-benzenedicarboxylic acid (BDC), 880
4-aminoantipyrine, 659
Amino-functionalized MCF, 167
Amino-functionalized ZIF-8 (ZIF-8-NH$_2$), 161–162
Ammonia (NH$_3$), 586–587, 712–713, 749–751
Ammonium polyphosphate (APP), 710
Amorphous silicon (a-Si), 212
Anionic buffers, 779
Anthropogenic CO$_2$ emissions, 135
Antibacterial agents, 373–374, 923
Antibacterial application, MOF composites in, 387–398
Antibacterial MOFs
  challenges of making, 398–399
  fundamentals of, 380–381

931

Antibacterial properties, MOF composites classification based on, 380–386
Antibacterials, 381
Antibiotics, 790–791
Antimicrobial nanocarriers, 320
Antimicrobial peptides (AMPs), 373–374
Antioxidant, 322–324
Aptasensor (Apt), 75
Aquatic ecosystems, 610–611
Aquatic environments, 626–627
Aquatic oil spills, 639–640
Aramid nanofibers (ANFs), 269–270
Argon (Ar), 43–44
Artificial intelligence (AI), 375
Asia-Pacific region, 830
Asodalite, 163
Aspergillus
  A. phoenicis, 321
  A. saitoi, 321
Atmospheric water generators (AWGs), 875
Atmospheric water-harvesting, 875–878
Atomically dispersed metal sites (ADMSs), 106
Atomic layer deposition (ALD), 472–474
Auger electrons, 789–790
2,2'-azino-bis (3-ethylbenzothiazoline-6-sulfonic acid) (ABTS), 484–485

B

Bacterial cellulose (BC), 161–162
Bacterial diseases, 373
Bacterial infections, 790–791
Bactericidal mechanism, 383–386
Bacteriostatic agents, 381
  mechanisms of, 381–383
Bacteriostatic antibiotics, 383–384
Bacteriostatic antimicrobials, 381
Bamboo powder cellulose (BM), 159
Barrett-Joyner-Halenda (BJH) method, 64
Batteries, 841–845
Benzene-1,2,4-tricarboxylic acid (BTC), 839–840
Benzene-1,3,5-tribenzoate (BTB), 343
Benzene-1,4-dicarboxylate (BDC), 84–85
Benzene-1,4-dicarboxylic acid (BDC), 343, 663, 840
Benzene dicarboxylic acid (BDC), 780–781
1,4-benzenedicarboxylic acid (BDC), 571–572
1,3,5-benzenetricarboxylate (H$_3$BTC), 225, 863
1,3,5-benzenetricarboxylic acid (BTC), 343
Benzimidazolate (bzim), 184–185
Benzyl alcohol, 460

Benzylamines, 460–461
Bienzymes, 788
Bilayer zwitterionic nanochannels (BZN), 843
Bimetallic metal-organic frameworks (BMOFs), 106–107, 844–845
Biobased-functionalized MOFs (Bio-FUN-MOFs), 730
Biocatalytic activity, 774
Biogenic volatile organic compounds (*b*VOCs), 341
Bioimaging, 397
  metal-organic framework-based nanocomposites for, 873–875
Biomedical applications
  bulk engineering of MOFs composites for, 377–379
  surface engineering of MOFs for, 377
Biomedical engineering, 423
  application of MOFs in, 413–416
  application of MOF composites in, 416–424
Biosensing, 69
  materials, 424
Biosensor, 787–788, 845–863
  point-of-care testing sensors, 854–855
  smartphone sensors, 848–854
  turn-on processes, 847–848
  wearable diagnostic sensors, 857–863
Biphenyl-4,4-dicarboxylate (BPDC), 780–781
Bis(1H-1,2,3-triazolo [4,5-b],[40,50-i])dibenzo [1,4]dioxin)(BTDD), 829
Bismuth oxybromide (BiOBr), 620–621
Bisphenol A (BPA), 468–469, 788–789
Boron dipyrromethene (BODIPY), 662–663
Boron nitride nanosheets (BNNS), 168
Boron trifluoride (BF), 845
"Bottle around ship" method, 917–918
Bottom-up method, 554–555
Bovine hemoglobin (BHb), 659
Branched poly-(ethylenimine)-capped carbon quantum dots (BPEI-CQDs), 72
Brilliant cresyl blue (BCB), 661
Brønsted acidic ionic liquid, 590
Brunauer–Emmett–Teller (BET) method, 64, 97–98, 140, 747
Bucky-papers (BP), 72
Buffers, effect of, 778–779
(1-butyl-3-methylimidazolium bis (trifluoromethyl-sulfonyl) imide) (BMIM) (Tf$_2$N), 163–164
1-butyl-3-methylimidazolium tetrafluoroborate ((BMIM)(BF$_4$)), 163

Index   933

C
Cadmium (Cd), 609–610
Cadmium telluride (CdTe), 212
Calcium fumarate (CaFu), 324–325
CALF-20, 139
Calgary Frameworks (CALF), 807
Cambridge Structural Database (CSD), 9
Cambridge University KRICT (CUK), 13–15
Cancer therapy, MOF-enzyme composite for, 789–790
Cantilever sensors, 685–686
Capacitive sensors, 693
Capillary electrophoresis (CE), 322
Carbamazepine (CB), 76–77, 448–449
Carbon, 73, 106, 192–193
  allotropes, 835–836
  based composites, 119–122
  carbon capture technologies, 135–136
  carbon-based materials, 112–113, 284, 448–449, 760–761
    composites electrode, 112–113
    metal-organic framework@carbon nanotubes composites, 113
    metal-organic framework@graphene nanocomposites, 113
  carbon-based metal composites, 113
  carbon-based MOF composites, 918–919
  composites, 151–162, 155f
  layers, 761
  materials, 110, 688, 915–916
Carbonaceous materials, 265
Carbon capture, utilization, and storage (CCUS), 135
Carbon cloth (CC), 74–75, 841–842, 854–855
Carbon dioxide (CO_2), 135, 140, 277, 539–540, 583
  capture of, 822–826
  MOF composites for, 137–140
    shaped MOF composites, 137–140, 138f
  reduction, 456–457
Carbon dioxide reduction reaction (CO_2RR), 455, 539
  electrocatalyst materials for, 544–545
  electrocatalytic mechanism of, 542–543
  mechanistic study of, 545–547
    MOF composites as electrocatalysts, 546–547
  MOF composites for, 541–542
Carbon dots (CDs), 443–445, 633–634, 666
Carbon fiber (CF), 71, 443–445
Carbon hollow fibers (CHF), 152–153
Carbonization, 551

Carbon/MOF composites, 442–449
  catalytic performances of, 443–448
  electrocatalysis, 443–448
  photocatalysis, 448–449
  preparation methods of, 443
Carbon monoxide (CO), 277, 451, 542–543, 708–709, 749–751
Carbon monoxide-trimesic acid (Co-TMA), 778–779
Carbon nanofibers (CNFs), 182–183, 264, 761
Carbon nanotubes (CNTs), 62, 71, 108, 113, 151, 181, 258–259, 264, 278, 299–301, 343–345, 351, 443–445, 541–542, 578–579, 617, 658, 751–752, 783, 831, 921
  composites, 299–301
  mats, 72
Carbon quantum dots (CQDs), 62, 71, 314
Carboxylate, 613
  frameworks, 613
  groups, 779
Carboxyl groups (-COOH), 351, 684
Carboxylic zinc phthalocyanine, 398
Carboxymethylated cotton (CM Cotton), 351
Carboxymethyl chitosan (CMCS), 880
Carboxymethyl filter paper (CMFP), 395
Catalase enzyme, 784–785
Catalysis, 837–840, 918–919
  MOF composites for, 442–449
Catalysts, 441–442, 540–541, 635, 840
Catalytic dyes, 621–622
Catalytic efficiency, 499–502
Catalytic oxidation systems, 341–342
Catechin, 322–324
Cationic dyes, 621–622
Cellulose, 151, 318–319
Cellulose acetate (CA), 70, 162
Cellulose aerogel (CA), 627
Cellulose carboxymethyl cellulose (CMC), 349–350
Cellulose-modified MOFs, 158–159
Cellulose nanocrystal (CNC), 319
(2,2,6,6-tetramethylpiperidine-1-oxyl radical) cellulose nanofibril (TOCNF), 160–161
Cerium (Ce), 467–468
Cerium phenylphosphonate (CeHPP), 734–735
Charge coupled device (CCD), 848–853
Chemical activation, 41–42
Chemical adsorption, 615
Chemical and compositional analysis, 64–65
Chemical mechanisms (CM), 680–681
Chemical oxygen demand (COD), 70

934   Index

Chemical purification, MOF composites, 822–829
  capture
    of CO_2, 822–826
    of NO_x, 828
    of SO_2, 826–827
    of volatile organic compounds, 828–829
  petrochemicals purification, 829
Chemical reactions, 311
Chemical stability of metal-organic frameworks, 21–23
Chemical vapor deposition (CVD), 78
Chemical warfare agents (CWA), 472–474
Chemodynamic therapy (CDT), 789–790
Chemo-resistive gas sensors, 745–746
Chitosan (CS), 338–339, 382
Chitosan polyethylene oxide (CS-PEO), 880
Chlorides, 810–811
Chlorpyrifos (CHPS), 71
Chlortetracycline (CTC), 694–695
Cholesterol oxidase (ChO$_x$), 788
Christian-Albrechts-University (CAU), 13–15
Chromium MOF (Cr-MOF), 229
Chromogenic compounds, 659
Chronic kidney disease (CKD), 854
Ciprofloxacin (CIP), 479–480, 636–637
Claisen-Schmidt reaction, 840
Closed system, 29
Coating methods, 686
Cobalt (Co), 218, 380
Cobalt-based MOFs (CoMOFs), 312–313
Cobalt-doped ZIF-8 (Co/ZIF-8), 680
Cobalt oxide (Co_3O_4), 113
Cobalt sulfide (CoS), 301–304
Cocatalysts, 448–449
Colloidal crystals (CCs), 672–673
Combination therapy, 422
Composites, 284
  characterization of MOFS and, 63–65
  membrane, 201
  MOFs, 253–258, 284–304
  of molecular species, 318
  with porous materials, 291–299
  sensor, 670
  structural and functional regulation of MOFs and, 61–62
Compound annual growth rate (CAGR), 804
Computational modeling, 926
Computed tomography (CT), 423
Conducting polymers (CPs), 68–69
Conduction band (CB), 217, 282, 448–449, 580–581, 618–619

Conductive polymers (CPs), 111, 116, 259
  composites, 115–116, 117f
  composites electrode, 111–112
  metal-organic frameworks@poly(3,4-ethylenedioxythiophene), 112
  metal-organic frameworks@ polyaniline, 112
  metal-organic frameworks@ polypyrrole, 112
  MOF composites with intrinsically, 259–261
Conductive substrates, MOF on, 552–553
Conductor-like screening model for realistic solvents (COSMO-RS), 164
Cone calorimetry tests (CCT), 714–715
Congo red (CR), 633–634
Continuous flow reactors, 811–813
Conventional imidazolyl ILs, 165
Conventional synthesis techniques for MOF, 102–103
Conventional wastewater treatment methods, 626–627
Coordination bonds, 21
Coordination chemistry, 252
Coordinatively unsaturated sites (CUSs), 24, 594–595
Copper (Cu), 218, 380, 609–610
Copper-based metal-organic frameworks (Cu-based MOFs), 395, 462–463
Copper/dipicolinic acid-MOF/oxidized pectin/CS (Cu/DPA-MOF/OP/CS), 385
Copper indium gallium selenide (CIGS), 212
Copper mesh (CM), 630–631
Copper oxide (CuO), 195–198
Copper-trimesic acid (Cu-TMA), 778–779
Core-shell (CS), 484–485
  nanocrystals, 831
CoS_2, 222
Counter electrode (CE), 211–212, 221
  MOF composites as, 221–222
Covalent-organic frameworks (COFs), 1, 109–110, 295–296, 689
  composite, 295–296
Crotonaldehyde, 79
Crude oil spills, 639–640
Cryogenic distillation, 829
Crystal-glass composites (CGCs), 917–918
Crystal growth, 102
Crystalline silicon (c-Si), 212
Crystallizing MOFs around presynthesized polymers, 68
Crystallographic characteristics, 747
Cutting-edge hybrid materials, 918

Index 935

Cutting-edge materials, 926
Cyclohexane, 499–502
Cyclopentanol (CPL), 483

D
De-doped polyaniline (DPANI), 479–480
Degradation
  of pollutants, 631–639
  of heavy metals by MOF composites, 637–639
  of organic dye by MOF composites, 631–635
  of pharmaceutical compounds by MOF composites, 635–637
Density functional theory (DFT), 63, 545, 586–587, 619–620
Deoxyepinephrine (DXEP), 696
Department of Energy (DOE), 830
Desorption, 833
Desulfurization process, 584
Dibenzothiophene (DBT), 468
Dibenzothiophene dioxide (DBTO$_2$), 579–580
Dielectric sensor
  based on MOF composites, 673–684
  fiber-optic interferometric sensors based on MOF composites, 675–680
  localized surface-plasmon resonance sensor based MOF composites, 683–684
  surface-enhanced Raman scattering based in MOF composites, 680–683
Diesel consumption, 568
Differential pulse voltammetry (DPV), 693–694, 697
Diffuse reflectance spectroscopy (DRS), 633–634
Diffusion method, 30–31, 102
9,10-dihydro-9-oxa-10-phosphaphenanthrene-10-oxide (DOPO), 722
2,5-dihydroxyterephthalate (DHTP), 485–486
4,6-dimethyldibenzothiophene (4,6-DMDBT), 578
4,6-dimethyldibenzothiophene sulfone (4,6-DMDBTO), 578
Dimethylformamide (DMF), 784–785
Dimethyl sulfoxide (DMSO), 27–29
1,3-dimethylurea (DMU), 695
Dimethyl yellow (DY), 624
Diphenylalanine (DPA), 869–873
Dipyrrometheneboron difluoride (BODIPY), 873–875
Direct ethanol fuel cell (DEFC), 195

Direct methanol fuel cell (DMFC), 179–180, 185–186
Dispersive solid-phase extraction (DSPE), 324
Divinylbenzene (DVB), 165–166
Dopamine (DA), 69, 666–667
Double-shelled hybrid nanocages (DSNC), 486
Double-solvent method (DSM), 917–918
Doxorubicin (DOX), 75, 873
Dresden University of Technology (DUT), 13–15
Drug
  adsorption, 625–626
  delivery, 69, 863–873, 923
Drug delivery systems (DDS), 413–415
Dry-gel conversion method (DGC method), 813–817
Dual solvent method (DSM), 470
Dye-sensitized solar cells (DSSCs), 211–212, 217
  MOF composites in, 217–228
Dynamic sorption methods, 356–357

E
EcoFuel Asia Tour project, 835–836
Effective absorption bandwidth (EAB), 259
Electrical conductivity, 453–455, 747–748
  of metal-organic frameworks, 25
Electrical sensor based on metal-organic framework composites, 688–697
Electric double-layer capacitors (EDLCs), 99, 841
Electric double-layer theory, 99
Electrocatalysis, 486
Electrocatalyst(s), 284, 441–442, 923
  materials for carbon dioxide reduction reaction, 544–545
  MOF
    composites as electrocatalysts, 546–547
    composites as electrocatalyst support materials, 192–198
    and MOF composites as precursors for, 181–189, 183f
Electrocatalytic carbon dioxide reduction reaction, recent advances on MOF composites for, 547–556
Electrocatalytic materials, 552
Electrocatalytic mechanism of carbon dioxide reduction reaction, 542–543
Electrocatalytic reactions, 486
Electrocatalytic water splitting, 279–282, 280f, 281t
Electrochemical biosensors, 788

Electrochemical carbon dioxide reduction reaction, 539–541
Electrochemical CO reduction (ECR), 545
Electrochemical energy storage devices, 97
Electrochemical impedance spectroscopy (EIS), 224, 632–633, 689
Electrochemical method, 32, 102–103, 917–918
Electrochemical reduction, 455
Electrochemical sensing, 853–854
Electrochemical stability of MOFs, 24
Electrochemical surface area (ECSA), 195–198
Electrochemical synthesis, 102–103, 375–376
Electrode, 24, 551–552
  MOFs composites as electrode materials, 115–122
Electrolytes, 198
  functionalization of MOF composites in electrolyte of fuel cell, 199–201
  ion diffusion, 114–115
  MOF composites in electrolyte of fuel cell, 198–199
Electromagnetic interference (EMI), 251–252
  shielding, 253–254, 918–919, 922
Electromagnetic mechanisms (EM), 680–681
Electromagnetic wave (EMW), 251–252, 270, 918–919
Electron extraction layer (EEL), 236–237
Electronic noses (E-noses), 745–746
Electron paramagnetic resonance (EPR), 635–636
Electrons, 458, 580–581
Electron transport layer (ETL), 211–212, 230
  MOF composites in, 230–231
Electrospinning technique, 818
Electrospun nanofibrous membranes, 628–629
Encapsulated enzymes, 793
Encapsulated gold nanoparticles, 683
Encapsulated luminescent modules (ELMs), 671–672
Encapsulated nanoparticles, 592
Encapsulated phosphomolybdic acid, 578
Energy dispersive spectrometer (EDS), 167–168
Energy-dispersive X-ray spectroscopy (EDX spectroscopy), 42, 45, 64, 470
Energy storage, 920
  devices, 98–99
Energy transfer, 668–669
Enterococcus faecalis, 880
Environmental application, metal-organic framework-enzyme composites for, 792
Environmental contamination, 631–632
Environmental remediation, 820–822, 923
Environment, enzyme-immobilized metal-organic framework composites for, 788–789
Enzyme(s), 317, 774, 780–781, 784–785, 924
  application of enzyme-immobilized MOF composites, 787–792
  biosensors, 787–788
  enzyme-immobilized MOF composites for environment, 788–789
  MOF enzyme composites for cancer therapy, 789–790
  MOF-enzyme composites for environmental application, 792
  MOF-enzyme composites for wound healing-related applications, 790–791
  MOF–enzyme composites in food and industrial applications, 791–792
  composite, 317–318
  conformation, 776–777
  enzyme-immobilized MOF composites preparation, 775–781
  effect of buffers, 778–779
  effect of ligands, 779–780
  effect of MOF hydrophobicity in enzyme immobilization, 776–778
  effect of metals, 780–781
  immobilizers and stabilizers for enzymes and active compounds, 322–324
  improvement of immobilized enzyme stability in MOFs composite, 784–785
  MOF composite for improving enzyme immobilization, 781–784
  recent fabrication and application of enzyme-immobilized MOF composites, 785–787
Epoxy resin (EP), 712
  MOF composites applied in fire retardancy of, 713–717
Erbium-based MOF (Er-MOF), 301–304
Eriochrome Black T (EBT), 77–78
Escherichia coli, 321, 380, 880
Esterification, 791–792
Ethanol (EtOH), 542–543, 752–753, 784–785
  ethanol-exchanged samples, 39–40
Ethylbenzene adsorption, 828–829
Ethylene (C_2H_4), 140, 542–543
Ethylene-diamine-tetraacetic acid (EDTA), 869–873
Ethylene vinyl acetate (EVA), 712, 730–731

Index   937

2-ethylimidazolate (eim), 184–185
Eukaryotic cell microenvironment, 782
Europium (Eu), 668
Expandable graphite (EG), 727–728
Ex-situ synthesis techniques, 73
  direct mixing and mechanical process, 73
  self-assembly technique, 73
Extracellular matrix (ESM), 422–423
Extraction-coupled catalytic oxidative desulfurization (ECODS), 589

F

Fabrication
  of enzyme-immobilized MOF composites, 785–787
  methods, 481–483
  process, 262–263
Fabry-Perot interferometer (FPI), 675–677
Faradaic efficiency (FE), 455–456
Faraday efficiency (FE), 544
Fenton reaction, 182
Fiber, 349–351
  fiber-optic interferometric sensors based on MOF composites, 675–680
  optic sensors, 675–677
Field emission scanning electron microscopy (FESEM), 44, 63, 635–636, 747
Fill factor (FF), 213–214
Fire retardancy
  MOF composites applied in
    of epoxy resin, 713–717
    of polylactic acid, 722–725
    of polymers, 730–735
    of polystyrene, 728
    of polyurea, 729–730
    of polyurethane elastomer, 727–728
    of polyurethanes, 717–722
    of unsaturated polyester resin, 725–727
Flame retardancy, mechanism of, 709–711
Flame retardants, 709–711
  mechanism of MOF composites, 712–713
  MOF composites as flame retardants, 711–712
  polymers, 709
Flexible polyurethane foam (FPUF), 712
Flexible polyurethane sponge, 631
Flexible printed circuit board (FPCB), 857–862
Flexible symmetrical supercapacitors (FSSCs), 111
Fluorescein isothiocyanate (FITC), 670
Fluorescence imaging, 413–415
Fluorine doped tin oxide (FTO), 221, 553
Food
  applications of MOF composites in, 319–325
    active food contact material, 319–320
    antimicrobial nanocarriers, 320
    cleaning, 324–325
    immobilizers and stabilizers for enzymes and active compounds, 322–324
    material sensors, 321–322
    nanoreactors, 321
    packaging substance nanofillers, 321
    regulated discharge of nano-systems for active compounds, 320
  MOF composites in food applications, 311–319
    activated carbon composite, 315
    aluminum composite, 316–317
    composites of molecular species, 318
    enzyme composite, 317–318
    graphene oxide composite, 312–313
    heterostructures/hybrid composites, 317
    metal-organic framework cellulose composites, 318–319
    nanoparticle composite, 311–312
    organic polymer composite, 313–314
    polyoxometalate composite, 314
    quantum dot composite, 314
    silica composite, 313
    thin film on substrate, 315–316
  MOF–enzyme composite in, 791–792
  nanoreactors, 321
  packaging, 878–884
  technology, 881–884
  spoilage, 878–880
Formamidinium iodide (FAI), 233
Forster resonance energy transfer (FRET), 670
Fosfomycin (FOS), 382
Fossil fuels, 179, 341, 539, 567
Fourier-transform infrared spectroscopy (FT-IR spectroscopy), 42, 44, 65, 148–149, 470, 545–546, 747
Freeze-drying activation, 40
Fuel cell electric vehicles (FCEVs), 830
Fuel cells, 179, 192, 921–922
  functionalization of MOF composites in electrolyte of fuel cell, 199–201
  MOF composites in, 180–181
  in electrolyte of, 198–199
Fuel oils
  MOF composites-based porous ILs for ODS of, 589–591
  MOF composites-derived catalysts for ODN, 595–597

for ODS, 585–589
MOF composites for ODN of, 591–597
  using H_2O_2 as oxidant, 592–595
  using O_2 as oxidant, 595
MOF composites for ODS of, 571–591
MOF composites for photo-catalytic oxidative desulfurization of, 580–584
Functional groups, 18, 23–24, 105

G

Gadolinium-based MOF (Gd-MOF), 313–314
Gallium arsenide (GaAs), 212
γ-valerolactone (GVL), 504–508
Gas chromatography (GC), 322, 745–746
Gaseous hydrogen, 832
Gases storage, 830–837
  hydrogen storage, 830–833
  natural gas storage, 833–837
Gasoline consumption, 568
Gas phase, 707–708
  adsorption, 617–618
  approach, 79
  carbon monoxide, 451
  diffusion, 30–31
  infiltration, 549
  mechanism, 709–710
Gas sensing
  measurements, 752
  MOF-based
    carbon nanotube composites for gas sensing, 751–754
    graphene/graphene derivatives composites for, 748–751
    metal and metal oxide composites for gas sensing, 755–760
    MOF-on-MOF composites for gas sensing, 760–761
Gelatin methacryloyl (GelMA), 873
Gels, 102
  diffusion, 30–31
Gene therapy, 422
Glassy carbon electrode (GCE), 72, 694–695
Global energy
  demand, 97
  demands, 277
Glucose (Glu), 689
Glucose oxidase (GO$_x$), 678, 776–777, 782–783
Glutathione (GSH), 790–791
Gold (Au), 179–180, 290–291
Gold nanoparticles (AuNPs), 385, 694–695
Gold nanorods (AuNRs), 854–855

Gold superparticles (GSPs), 681
Grafting polymers, 67
  onto MOF ligands post synthesis, 67
Graphene, 6–7, 108, 264, 285–286, 552–553, 719, 749, 921
  composites, 285–286
  and derivatives MOF composites, 345–348
Graphene aerogel (GA), 317
Graphene nanoribbons (GNR), 120–122
Graphene oxide (GO), 71, 73–74, 107–108, 151, 154, 181, 285–286, 312–313, 347–348, 448–449, 783–785
  composite, 312–313
Graphene oxide-IL@MOF ternary composite, 157
Graphene oxide-nickel (GO-Ni), 679–680
Graphite and derivatives MOF composites, 345–348
Graphite oxide (GO), 343–345, 823
Graphitic carbon nitride (g-C$_3$N$_4$), 725–726
Green Science Alliance company, 840
Green solvents, 430

H

Half-wave potential (HWP), 182–184
Haloferax volcanii, 784–785
Hard and soft acids and bases theory (HSAB theory), 21–22, 747
Heavy metals, 609–610
  adsorptive removal by MOF composites, 626–631
  degradation by MOF composites, 637–639
Hemeprotein-MOF composites (H-MOFs composites), 659
Hemoglobin, 789–790
Henry's law constant, 356
Heteroatoms, 107–108
Heterocyclic compounds, 224–225
Heterogeneous catalysts, 837
Heterogeneous photocatalysis, 620
Heterostructures, 317
Hexadecyltrimethoxysilane (HDTM), 631
Hexagonal boron nitride (h-BN), 714–715
Heyrovsky reaction, 279–282
Hierarchical carbon monolith (HCM), 152
Hierarchical cluster analysis (HCA), 670
High-angle annular dark-field scanning transmission electron microscopy (HAADF-STEM), 167–168
High-entropy zeolitic imidazolate framework (HE-ZIF), 579–580
Highest occupied molecular orbital (HOMO), 222–223, 586–587

High internal phase emulsions (HIPEs), 817–818
High-performance liquid chromatography (HPLC), 322
High-resolution transmission electron microscope (HRTEM), 63–64, 747
Hole extraction layer, 236–237
Hole transport layer (HTL), 211–212, 231
  MOF composites in, 231–232
Hollow fibers, 152–153
Hollow polymeric fibers, 152–153
Hong Kong University of Science and Technology (HKUST), 13–15
Horseradish peroxidase (HRP), 777–778
Host-guest interactions, 658–659
Human interference, 745–746
Human serum albumin (HSA), 670
Hybrid composites, 317
Hybrid material, 286–288, 297–298
Hybrid nanorods sensor, 755–758
Hybrid Ni-MOF@Fe-MOF catalyst, 298–299
Hybrid sensor, 755–758
Hybrid systems, 921
Hydrodenitrogenation (HDN), 568
Hydrodesulfurization (HDS), 568
Hydrofluoric acid (HF), 84–85
Hydrogen, 286, 820–822, 830
  hydrogen-driven fuel cells, 180–181
  production, 277–278, 457
  storage, 830–833
    materials, 833
Hydrogenation reactions, 459–460
Hydrogen-bonded organic frameworks (HOFs), 1
Hydrogen evolution reaction (HER), 277–279, 443–445, 540–541, 920
  electrocatalysis, 445–447
Hydrogen oxidation reactions (HOR), 179–180
Hydrogen peroxide (H_2O_2), 382
  MOF composites
    for ODN of fuel oils using H_2O_2 as oxidant, 592–595
    for ODS with H_2O_2 as oxidant, 571–580
Hydrogen sulfide (H_2S), 343
Hydrolysis, 791–792
Hydrophilic carbon nanomaterials, 788
Hydrophilic solvents, 917–918
Hydrophobic interactions, 776
Hydrophobic mesoporous ionic copolymer, 165–166
Hydrophobic solvents, 917–918
Hydrophobic ZIF-8, 151–152
Hydrothermal, 26–27
  method, 26–27, 102, 215–216
  process, 120–122, 215–216
  synthesis, 26–27, 755–758
Hydroxides, 117–119
  composites, 117–119
(1-(2-hydroxyethyl)-3-methylimidazolium dicyanamide) (HEMIM)(DCA), 164
Hydroxyl groups, 581–582
Hydroxyl radicals (•OH), 382
5-(hydroxymethyl) furfural (HMF), 504–508
Hydroxypropyl cellulose (HPC), 139–140

I

Imidazolate (Im), 184–185
Imidazole, 12–13
Iminodiacetic acid (IDA), 869–873
Immediately dangerous to life or health (IDLH), 339–341
Immunosensor, 689
Indium (In), 231
Indium-based MOFs, 229, 231
Indocyanine green (ICG), 880
Indoor air quality (IAQ), 745–746
Inductively coupled plasma mass spectroscopy (ICP-MS), 45
Inductively coupled plasma optical emission spectroscopy (ICP-OES), 470
Inductively coupled plasma spectroscopy (ICP spectroscopy), 42, 45
Industrial applications
  MOF composites, 820–884
  MOF–enzyme composite in, 791–792
Industrial gases, 830
Industrial hydrogen production, 277
Industrial waste, 609–610
Infectious bacteria, 373
In-house extended growth method (IEGM), 486–487
Inorganic nanoparticles composites, 114
Inorganic sub networks dimensions, classification of MOFs-based on, 5
In situ dip coating (ISDC), 143
In situ encapsulation, 775–776
In situ growth technique, 83–84
In situ layer-by-layer assembly technique, 72
In situ polymerization, 65–66
In situ reduction route (ISRR), 470
In situ synthesis, 84–85
  techniques, 72
  layer-by-layer assembly, 72
  one-pot synthesis approach, 72
  seeded growth method, 72

Instituto de Tecnología Química Metal-Organic Framework (ITQMOF), 13–15
Interfacial layers (ILs), 211–212
  MOF composites as, 232–236
Interfacial polymerization (IP), 70
Internal extended growth method (IEGM), 485
Internal scattering, 254–255
International Union of Pure and Applied Chemistry (IUPAC), 1
Iodide, 217
Ion-and liquid-assisted grinding (ILAG), 103
Ion exchange, 610–611
Ionic liquid (IL), 29–30, 157, 162, 589, 687, 715–717
  MOF and ILs composites, 162–166
Ionic liquid molecular layers (ILMLs), 165
Ion mobility spectroscopy, 745–746
Ionothermal method, 29–30
Iridium (Ir), 277–278, 290–291
Iron (Fe), 380
  Fe-based metal-organic frameworks, 396
  Fe-based MOF, 182–183
  Fe-polyoxometalates, 221
Iron-1,3,5-benzenetricarboxylate (Fe-BTC), 778–779
Iron oxide NPs, 385
Isopropanol, 752
Isopropyl alcohol (IPA), 504–508, 753
Isoreticular metal-organic frameworks (IRMOFs), 9, 11*f*, 283, 611–612
IRMOF-3, 343

J

Jilin University China (JUC), 13–15

K

K-nearest neighbor (KNN), 662–663

L

Lab-on-chip system, 786–787
Lactobacillus helveticus, 341
Langmuir-Blodgett method (LB method), 316, 818–820
Langmuir isotherm model, 625
Lanthanide MOFs (Ln-MOFs), 25
Lateral flow assay (LFA), 322
Layer-by-layer (LBL), 143, 315–316
  assembly, 72, 719
  method, 677
Layered double hydroxides (LDHs), 62, 631, 720, 777–778
Lead (Pb), 609–610
Lead-based MOF-525, 233
Leaf-like zeolitic imidazolate framework (ZIF-L), 159–160
Lewis acids, 711–712
Lewis basic sites (LBSs), 136–137
Life cycle assessment (LCA), 809–811
Ligands
  effect of, 779–780
  impact of ligand engineering, 283–284
Limiting oxygen index (LOI), 712
Limit of detection (LOD), 321–322
Limits of quantification (LOQ), 321–322
Linear detection range (LDR), 321–322
Linear discriminant analysis (LDA), 662–663
Lipases, 780–781, 791–792
Liquid-assisted grinding (LAG), 103, 834–835
Liquid-exfoliation graphene (LEG), 120–122
Liquid impregnation method, 79
Liquid phase, 617–618
  diffusion, 30–31
Liquid phase epitaxial growth (LPE growth), 686–687
Liquid-state DSSCs, 225
Lithium bis(trifluoromethanesulfonyl)imide (Li-TFSI), 231
Localized surface plasmon (LSP), 683–684
Localized surface-plasmon resonance spectroscopy (LSPR spectroscopy), 458, 658
  sensor based MOF composites, 683–684
Long-period grating (LPG), 678
Lowest detection limit (LOD), 70–71
Lowest unoccupied molecular orbital (LUMO), 222–223, 586–587
Luminescence sensors, 847
  based on MOF composites, 665–673
Luminescent compounds, 666
Luminescent donor ligands, 673
Luminescent MOFs (LMOFs), 665–666
Lysozyme (lys), 780–781

M

Machine learning (ML), 375, 926
Mach-Zehnder interferometer (MZI), 675–677
Macroporous metal-organic frameworks, 8
Magnesium (Mg), 380
Magnesium-based MOF composite (Mg-MOF), 225
Magnetic nanocomposite, 628–629
Magnetic nanoparticles, 252, 311–312, 873–875
Magnetic properties of metal-organic frameworks, 25
Magnetic resonance imaging (MRI), 413–415

Index 941

Magnetic separation, 465–466
Magnetic solid-phase extraction (MSPE), 324, 348–349
Magnetism, 25
Magnetite nanoparticles, 808–809
Manganese (Mn), 380
Manganese oxide (MnO$_x$), 113
Manganese oxide nanoparticles (Mn-NPs), 593
Mass spectroscopy, 745–746
Material Institute Lavoisier-101 (MIL-101), 138–139, 145, 182–184
Materials from University of Tehran (MUT), 13–16
Materials Institute Lavoisier (MIL), 13–15, 311–312
Materials institute Lavoisier-101(Cr)/GO composites (MIL-101@GO), 345–346
Matrix solid-phase dispersion extraction (MSPDE), 324
Mechanical sensors based on MOF composites, 684–688
Mechanochemical method, 33–34, 103
Mechanochemical synthesis, 103
Meerwein–Ponndorf–Verley (MPV) reactions, 504–508
Melamine-based flame retardants, 709–710
Melamine cyanurate (MCA), 731–732
Melamine foam (MF), 673, 840
Melamine polyphosphate, 709–710
Membranes, 827
3-mercaptophenylboronic acid, 787–788
Mercury (Hg), 81
Mesocellular foam (MCF), 167
Mesoporous (mp), 219–220
    MOFs, 8
Mesoporous carbon (OMC), 343–345
Mesoporous poly(ionic liquids) (MPIL), 165–166
Mesoporous silica, 775–776
Metal(s), 106
    clusters, 9, 424
    effect of, 780–781
    metal-based electrocatalysts, 540–541
    metal-based material@MOF composites, 919
    metal compounds, 106
    metal elements, 106–107
    nodes, 283
    particles, 915–916
Metal-azolate frameworks (MAFs), 9, 12
    MAF-7, 776
Metal-biomolecule frameworks (MBioFs), 9, 13–15

Metal/carbon hybrids (M/C hybrids), 541–542
Metal hydroxides, 760–761
Metal ions, 611–612
Metallic nanoparticles, 548
Metal-linker bonds, 22
Metal nanoparticles (MNPs), 62, 278, 288–290, 373–374, 439–440
    composite, 288–291
Metal nanoparticles/MOF composites, 449–450
    catalytic performances of, 451–462
    common catalytic performances, 451–462
    electrocatalysis, 452–456
    photocatalysis, 456–462
    preparation methods of, 450
Metal nitrates, 810–811
Metal-organic framework composites, 167–168, 214–216, 297–304
    adsorptive removal
        of heavy metals by, 626–631
        oil water separation by, 629–631
        of organic dye by, 621–624
        of pharmaceutical compounds by, 624–626
    amine-functionalized, 144–151
    annual global CO$_2$ emission, 136f
    applications, 918–921
    carbon-based, 918–919
    metal-based material@MOF composites, 919
    metal-organic framework@MOF composites, 920–921
    metal oxide/MOF composites, 919
    metal sulfide/MOF composites, 919
    of MOF composites in food systems, 319–325
    in photovoltaic devices, 216–237
    polymer@MOF composites, 920
    polyoxometalate@MOF composites, 920
    applied in fire retardancy
        of epoxy resin, 713–717
        of polylactic acid, 722–725
        of polymers, 730–735
        of polystyrene, 728
        of polyurea, 729–730
        of polyurethane elastomer, 727–728
        of polyurethanes, 717–722
        of unsaturated polyester resin, 725–727
    background, 567–570
    and carbon composites, 151–162
    carbon nanotubes composites, 299–301
    for catalysis, 442–449

Index

carbon/MOF composites, 442–449
challenges
 in application of, 924–926
 and current limitations, 735–736
 and future prospects, 270–271
characterization of, 747–761
 MOF-based carbon nanotube composites for gas sensing, 751–754
 MOF-based graphene/graphene derivatives composites for gas sensing, 748–751
 MOF-based metal and metal oxide composites for gas sensing, 755–760
 MOF-on-MOF composites for gas sensing, 760–761
composite, 284–304
 with conductive materials, 284–291
 with graphene, 285–286
 with metal nanoparticles, 288–291
 with MXenes, 286–288
 with porous materials, 291–299
degradation
 of heavy metals by, 637–639
 of oil spills by, 639–641
 of organic dye by, 631–635
 of pharmaceutical compounds by, 635–637
 in dye-sensitized solar cells, 217–228
 as counter electrode, 221–222
 in organic solar cells, 236–237
 in perovskite solar cells, 228–236
 in RE, 224–228
 as sensitizer dye, 222–224
 as working electrode, 217–221
electrical sensor based on, 688–697
as electrocatalysts, 546–547
as electrocatalyst support materials, 192–198
as electrode materials, 115–122
 carbon based composites, 119–122
 conductive polymers composites, 115–116, 117f
 metal oxide and hydroxide composites, 117–119
in electrolyte of fuel cell, 198–199
in electron transport layer, 230–231
enhancement of immobilized enzyme activity using, 781–784
 MOF carbon composite-enzyme, 783
 MOF carbon MXene composites-enzyme, 783–784

MOF/metal oxide or MOF/metal nanoparticles composites-enzyme, 781–783
MOF polymer composite-enzyme, 781
fiber-optic interferometric sensors based on, 675–680
flame retardant mechanism of, 712–713
 as flame retardants, 711–712
flame retardants and mechanism of flame retardancy, 709–711
 gas phase mechanism, 709–710
 solid phase mechanism, 710–711
functionalization in electrolyte of fuel cell, 199–201
fundamentals of water splitting, 279–282
future directions, 926
global map for, 805f
global patent application trends in MOF, 805f
GO, Co-MOF, and MOF@rGO properties, 748f
in hole transport layer, 231–232
improvement of immobilized enzyme stability in, 784–785
for improving enzyme immobilization, 781–784
industrial applications, 820–884
 atmospheric water-harvesting, 875–878
 batteries and supercapacitors, 841–845
 catalysis, 837–840
 chemical purification, 822–829
 drug delivery, 863–873
 food packaging, 878–884
 gases storage, 830–837
 metal-organic framework-based nanocomposites for bioimaging, 873–875
 MOF composites as green and sustainable catalyst, 838t
 sensor and biosensor, 845–863
and ILs composites, 162–166
localized surface-plasmon resonance sensor based, 683–684
mechanical sensors based on, 684–688
merit of MOFs for water splitting, 282–284
MOF composites, 284–304
 for CO_2 capture, 137–140
 metal sulfides composites, 301–304
for ODN of fuel oils, 591–597
 using H_2O_2 as oxidant, 592–595
 MOF composites-derived catalysts for ODN of fuel oils, 595–597
 using O_2 as oxidant, 595

Index 943

for ODS of fuel oils, 571–591
  based porous ILs for ODS of fuel oils, 589–591
  derived catalysts for ODS of fuel oils, 585–589
  for ODS with H_2O_2 as oxidant, 571–580
  for ODS with O_2 as oxidant, 574–580
  for photo-catalytic oxidative desulfurization of fuel oils, 580–584
perspectives, 761–762, 921–924
preparation of, 917–918
prominent industrial methods for synthesis, 809–820
recent fabrication and application of enzyme-immobilized, 785–787
schematic diagram of combustion mechanism of polymer materials, 708f
and shielding mechanisms, 253–258, 256f
structural MOF-monolith composites, 141–144, 144f
surface-enhanced Raman scattering based in, 680–683
synthesis of, 215–216
trends in research on MOF composites over recent years, 916f
for volatile organic compounds removal, 343–351
  fiber and polymer MOF composites, 349–351
  graphene, graphite, and derivatives MOF composites, 345–348
  metal, semimetal, and salt MOF composites, 348–349
  ordered mesoporous carbon, carbon nanotube, activated carbon, and, 351
Metal-organic framework-derived carbon nanotubes on carbon fibers (MDCNT@CF), 268–269
Metal-organic framework/metal-organic framework composites, 481–486
  catalytic performances of, 483–486
  common catalytic performance, 483–486
  preparation methods of, 481–486
Metal-organic framework-monolith composite adsorbent (MOF-MCA), 141–143
Metal-organic framework-polymer mixed-matrix membranes (MMMs), 818–820
Metal-organic frameworks (MOFs), 1–3, 2f, 97–98, 111–113, 136–137, 180, 211,
252, 255, 278, 293–299, 311, 319, 338–339, 374, 411–412, 439, 540–541, 569–570, 611–615, 657–658, 711–712, 773–774, 803–804, 915
activation of, 37–42
  chemical activation, 41–42
  freeze-drying activation, 40
  heating activation, 38
  microwave activation, 40
  photothermal activation, 41
  solvent-exchange activation, 38–39
  supercritical carbon dioxide activation, 39–40
adsorption
  in MOF composites, 354–355
  of volatile organic compounds by, 343
advanced design strategies for enhanced performance, 103–104
  pore window and pore size optimization, 103–104
in antibacterial application, 387–398
  copper-based metal-organic frameworks, 395
  fundamentals, 387–388
  iron-based MOFs, 396
  silver-based MOFs, 388–395
  zeolitic imidazolate framework-based MOFs, 396–397
  zirconium-based MOFs, 397–398
application
  of MOF composites in biomedical engineering, 416–424
  of MOFs and MOF composites in supercapacitor, 114–122
  of MOFs in biomedical engineering, 413–416
based vaprochromic sensor, 660f
bulk engineering of MOF composites for biomedical applications, 377–379
for carbon dioxide reduction reaction, 541–542
carboxylate frameworks, 613
cellulose composites, 318–319
characterization of, 42–45, 63–65
  chemical and compositional analysis, 64–65
  energy dispersive X-ray spectroscopy, 45
  Fourier-transform infrared spectroscopy, 44
  inductively coupled plasma spectroscopy, 45

944 Index

nuclear magnetic resonance
   spectroscopy, 45
physisorption isotherm, 43–44
scanning electron microscopy, 44
solid-state nuclear magnetic resonance
   spectroscopy, 45
structural and morphological
   characterization, 63–64
surface area and porosity
   characterization, 64
thermogravimetric analysis, 44
transmission electron microscopy, 45
X-ray diffraction, 42–43
chemical stability, 21
classification based on antibacterial
   properties, 380–386
   bactericidal mechanism, 383–386
   fundamentals of antibacterial MOFs
      and, 380–381
   mechanisms of bacteriostatic agents,
      381–383
classification of MOF based on connection
   dimensions, 3–5
   0D metal-organic frameworks, 3
   1D metal-organic frameworks, 3
   2D metal-organic frameworks, 4
   3D metal-organic frameworks, 4–5
classification of MOFs-based on
   inorganic sub networks
   dimensions, 5
classification of MOFs-based on
   morphological dimensions, 5–7
classification of MOFs-based on pore
   size, 7–8
   macroporous MOFs, 8
   mesoporous MOFs, 8
   microporous MOFs, 8
composites, 229–230, 258–259, 422
   as interface layer, 232–236
composites recycle, 358
conventional synthesis techniques for,
   102–103
   diffusion method, 102
   electrochemical method, 102–103
   hydro/solvo thermal method, 102
   mechanochemical method, 103
   microwave method, 102
   sonochemistry method, 103
degradation, 98
effect of MOF hydrophobicity in enzyme
   immobilization, 776–778
emergence and properties of, 61
in fuel cells, 180–181

functionalization of, 412–413
fundamental of, 101–104, 101f
generic features of, 375–376
material composites, 113–114
   metal-organic frameworks/carbon-
      based metal composites, 113
   metal-organic frameworks/inorganic
      nanoparticles composites, 114
   mMOF@graphene nanocomposites, 113
MOF-based catalysts, 283–284
MOF-based composite materials, 110–114
   rationale for combining MOFs with
      materials, 110–111
   types of MOF composites for
      supercapacitor, 111–114
MOF-based composites, 252, 255–256
MOF-based optical sensors, 658–684
   dielectric sensor based on MOF
      composites, 673–684
   luminescence sensors-based on MOF
      composites, 665–673
   vapochromism sensors based on,
      658–665, 664f
MOF-based shielding composites, 270
MOF@carbon composites, 71–73
   applications, 73–78
   ex situ synthesis techniques, 73
   in situ synthesis techniques, 72
MOF@carbon nanotubes composites, 113
MOF-coated monolithic materials, 143
and MOF composites as precursors for
   electrocatalysts, 181–189, 183f
MOF–enzyme composites
   for cancer therapy, 789–790
   for environmental application, 792
   in food and industrial applications,
      791–792
   for wound healing-related applications,
      790–791
MOF@metal nanoparticles, 78–79
   applications, 80–82
   liquid impregnation method, 79
   solid grinding technique, 79
   solvent free gas phase loading
      technique, 78–79
   synthesis, 78
MOF@MOF composites, 167–168
MOF@MXene applications, 85
MOF@MXene composites, 82–85
MOF-patterned thin films, 315
MOF@poly(3,4-
   ethylenedioxythiophene), 112
MOF@polyaniline, 112

MOF-polymer composites, 313–314
MOF@polymer composites for
  applications, 68–71
MOF@polypyrrole, 112
nanoporous carbons derived from, 613–615
polymer composites, 65–68
  adding preformed MOFs to preformed
    polymers, 67–68
  crystallizing MOFs around
    presynthesized polymers, 68
  grafting polymers onto MOF ligands
    post synthesis, 67
  integrating prefabricated polymer
    ligands, 66–67
  in situ polymerization, 65–66
postsynthetic modification in, 104–105
presynthetic modification in, 105–110
properties of, 17–26
  customizable structure of, 19
  flexibility of, 19–21
  porosity of, 17–18
  tunable pore size of, 18, 19f
recent advances on MOF composites for
  electrocatalytic carbon dioxide
  reduction reaction, 547–556
  on conductive substrates, 552–553
  MOF-polymer/polyoxometalate
    composites, 554–556
  MXene composites, 556
  on porous materials, 553–554
  pristine metal-organic framework,
    547–548
  as supports, 548–552
scopes and objectives, 542
shielding composites, 258–266
  composites with intrinsically
    conductive polymers, 259–261
  composites with MXenes, 261–263
  metal-organic framework-based
    carbonaceous composites,
    264–266
stability of, 21–24
  chemical properties of, 24
  chemical stability of, 21–23
  electrical conductivity of, 25
  electrochemical stability of, 24
  host-guest interactions of, 26
  magnetic properties of, 25
  mechanical stability of, 23–24
  open metal sites in, 24
  optical properties of, 25
  proton conductivity of, 26
  thermal stability of, 23

structural and functional regulation of
  MOFs and composites, 61–62
structures based on configuration, 9–13
  isoreticular metal-organic frameworks,
    9, 11f
  metal-azolate frameworks, 12
  metal-biomolecule frameworks, 13
  porous coordination networks, 9
  zeolite-like metal-organic framework,
    11, 12f
  zeolitic imidazolate frameworks, 11
structures based on university
  names, 13–16
  Materials from University of
    Tehran, 15–16
  Materials Institute Lavoisier, 15
  metal-organic frameworks structure-
    based on MOFs network, 16
  university of Oslo, 15
structures of, 9–16
surface engineering of MOFs for
  biomedical applications, 377
synthetic processes of, 26–37
  diffusion method, 30–31
  electrochemical method, 32
  hydrothermal method, 26–27
  ionothermal method, 29–30
  mechanochemical method, 33–34
  microfluidics method, 34–35
  microwave method, 33
  postsynthetic modification method, 37
  reflux method, 29
  room temperature synthesis method, 30
  slow evaporation method, 31–32
  solvothermal method, 27–29
  sonochemical method, 33
  spray-drying method, 35
  template method, 35–36
topology, 105–106
types of, 3–8
zeolitic imidazolate frameworks, 612–613
Metal-organic framework/silica
  composites, 498
  catalytic performances of, 499–502
    common catalytic performances,
      499–502
    photocatalysis, 502
  preparation methods of, 498
Metal-organic frameworks-polymer
  composites, 65
Metal-organic gels (MOGs), 225
Metal oxide, 117–119, 760–761, 915–916, 919
  composites electrode, 113

946 Index

gas sensors, 755
Metal oxide nanoparticles (MONPs), 439–440, 548
Metal oxide nanoparticles/MOF composites, 462–463
   catalytic performance of, 464–470
   common catalytic performances, 464–465
   photocatalysis, 466–470
   preparation methods of, 463
Metal phosphides, 760–761
Metal-polydopamine framework (MPF), 322
Metal salts, 26–27, 215
Metal sites, 24
Metal sulfides, 760–761, 919
   composites, 301–304
Methane (CH_4), 542–543, 820–822
Methanol (CH_3OH), 542–543
Methanol (MeOH), 339–341, 459, 752–753
Methanol oxidation reaction (MOR), 195
2-methoxyterephthalic acid (MTA), 668
Methylammonium chloride (MACl), 233
Methyl ammonium iodide (MAI), 233
Methyl ammonium lead iodide (MAPbI), 233
Methylcyclopropene, 884
Methylene blue (MB), 70, 479–480, 617, 633–634
2-methylimidazolate (mim), 184–185
2-methylimidazole (2-MIM), 68, 678, 755–758, 863
Methyl orange (MO), 621–622
Methyl red (MR), 624, 661
Michelson interferometer (MI), 675–677
Microbial fuel cell (MFC), 179–180
Microcantilever (MCL), 684–685
Microcystin-leucine arginine (MC-LR), 639
Microemulsion method, 917–918
Microfluidics
   flow synthesis method, 786–787
   method, 34–35
   reactors, 813
   strategy, 813–817
Microperoxidase-8 (MP8), 779–780
Micropores, 64
Microporous materials, 831
Microporous metal-organic frameworks, 8
Microsolid-phase extraction (μ-SPE), 324
Microstructure, 422–423
Microwave (MW), 102, 375–376, 811
   absorption, 263
   activation, 40
   heating, 40

irradiation, 813–817
   method, 33, 102
   MW-assisted heating, 111
Mid-infrared (MIR), 873–875
Millifluidic reactors, 813
Minimum reflection loss (RL_{min}), 259
Molecularly imprinted polymers (MIPs), 314, 686
Molecular sieve (MS), 451–452
Molecular species, composites of, 318
Molten carbonate fuel cell (MCFC), 179–180
Molybdenum disulfide (MoS_2), 301–304
Molybdenum trioxide, 576
Monoclonal antibodies, 697
Monoclonal antibodies-loaded MPF (MPFmAb), 322
Monoethanolamine (MEA), 135–136
Monomers, 65–66
Morphological dimensions
   classification of MOFs-based on, 5–7
   one-dimensional MOFs with 1D morphology, 5–6
   three-dimensional MOFs with 3D morphology, 7
   two-dimensional MOFs with 2D morphology, 6–7, 7f
Multicationic oligomers (MCOs), 843–844
Multiwalled carbon nanotubes (MWCNTs), 153–154, 185–186, 223, 299–301, 447–448, 622–623, 753–754, 824–825
Multiwalled carbon nanotubes@MIL-53(Cr), 351
MXene@Co-MOF composite, 263
MXenes, 82–83, 258–259
   composites, 286–288, 556
   metal–organic frameworks composites with, 261–263

N

Nanocomposites, 380–381
Nanohybrid material, 749
Nano metal-organic frameworks (NMOFs), 873–875
Nano-metal oxides, 576
Nanoparticles (NPs), 2–3, 109, 373–374, 412–413, 440–441, 658, 808–809
   composite, 311–312
Nanoporous carbons, 551, 613–614
   derived from MOFs, 613–615
Nanoprobes, 873–875
Nanoscale, 157
Nanosheets, 722
Nanostructures, 97–98, 926

Index **947**

Nano-systems for active compounds, regulated discharge of, 320
Nanotubes (NTs), 69
Natural AMPs, 373–374
Natural antibacterial agents, 373–374
Natural gas storage, 833–837
N-doped carbon nanotubes (N-CNTs), 268–269
Near-infrared (nIR), 873–875
Near infrared radiation (NIR), 69, 783–784
N-ethylmorpholine (NEM), 472–474
Neutral red (NR), 624
Nickel (Ni), 218, 285–286
Nickel-based MOF (Ni-MOF), 728–729
Nickel bimetallic electrocatalyst, 195
Nickel foam (NF), 579–580, 689
Nickel-iron (Ni/Fe), 285–286
Nickel-MOF with rGO (Ni-MOF/rGO), 120–122
Nickel oxide (NiO), 113, 195–198
Nickel-palladium nanoparticles (NiPd NPs), 782–783
Nickel sulfide (NiS), 301–304
Nickel-trimesic acid (Ni-TMA), 778–779
Nile red (NR), 672–673
Nitroaromatics (NACs), 671–672
Nitrofurantoin (NFT), 71, 673
Nitrofurazone (NFZ), 71, 673
Nitrogen (N), 43–44
 compounds, 568, 591
 N-based flame retardants, 710–711
 N-doped carbon materials, 186–187
Nitrogen-carbon (NC), 186–187
Nitrogen-containing compounds (NCCs), 591
Nitrogen-containing microporous carbon (NC), 152
Nitrogen-doped porous carbon (NPC), 613–614
Nitrogen oxides (NO$_x$), 708–709, 828
 capture of, 828
Nitrogen reduction reaction (NRR), 455–456
3-nitropropionic acid (3-NPA), 673
N-methyl-2-pyrrolidone (NMP), 27–29
N,N-diethylformamide (DEF), 27–29
N,N-dimethylacetamide (DMA), 27–29
N,N-dimethylformamide (DMF), 27–29
Nonenzymatic electrochemical sensor, 693–694, 857–862
Nonheteroatom catalysts, 185–186
Non-PGM catalysts, 193
Nonplatinum group metals, 193
Northeast Normal University (NENU), 343
Northwestern University (NU), 13–15
Nottingham (NOTT), 13–15
N-rich porous carbon (NPC), 232
Nuclear magnetic resonance spectroscopy (NMR spectroscopy), 42, 45, 148–149, 745–746

O

Oil spills by MOF composites, degradation of, 639–641
Oil-water separation, 629–630
 by MOF composites, 629–631
Oleylamine (OLA), 680
One-dimension (1D), 3
 CNFs, 762
 configurations, 103–104
 materials, 658
 MOFs, 3
 one-dimensional MOFs with 1D morphology, 5–6
One-pot synthesis approach, 72
Open circuit voltage (OCV), 183–184, 213–214
Open coordination sites (OCS), 24
Open metal sites, 147–148
Open metal sites (OMSs), 24, 136–137
 in MOFs, 24
O-phenylenediamine (o-PDA), 484–485
Optical fiber-MOF (OF-MOF), 673–675
Optical fibers, 673–675
Ordered mesoporous carbon, 351
Ordered mesoporous nonactivated carbon (OMC), 152
Organic compounds, 35–36
Organic contaminants, 637–638
Organic contaminations, 613–614
Organic dyes, 789
 adsorptive removal of organic dye by MOF composites, 621–624
 degradation of organic dye by MOF composites, 631–635
Organic-inorganic hybrid structures, 2–3
Organic ligands, 26–27, 215, 283–284, 398–399, 411–412, 736, 779
Organic pollutants, 618
Organic polymers, 137
 composite, 313–314
Organic solar cells (OSCs), 211–212, 236
 MOF composites in, 236–237, 236*f*
Organic solvents, 216
Organophosphorus pesticides (OPPs), 82, 324
Overpotential, 279
Oxidant-to-sulfur molar ratio (O/S molar ratio), 581–582

Oxidation reaction, 460, 490–492, 596, 786–787, 840
Oxidative denitrogenation (ODN), 569
  MOF composites-derived catalysts for ODN of fuel oils, 595–597
  MOF composites for ODN of fuel oils, 591–597
    using H_2O_2 as oxidant, 592–595
    using O_2 as oxidant, 595
Oxidative desulfurization (ODS), 490–492, 569
  MOF composites-based porous ILs for ODS of fuel oils, 589–591
  MOF composites-derived catalysts for ODS of fuel oils, 585–589
  MOF composites for ODS of fuel oils, 571–591
    with H_2O_2 as oxidant, 571–580
    with O_2 as oxidant, 574–580
Oxygen (O_2), 286, 574, 922
  clusters, 314–315
  MOF composites
    for ODN of fuel oils using O_2 as oxidant, 595
    for ODS with O_2 as oxidant, 574–580
Oxygen evolution reaction (OER), 109, 277–278, 443–445, 579–580, 920
  electrocatalysis, 447–448, 453–455
  process, 279–282
Oxygen reduction reaction (ORR), 109, 179–180, 186–187, 443–445
  electrocatalysis, 448
(2,2,6,6 tetramethylpiperidin-1-yl)oxyl (TEMPO), 484–485
Oxytetracycline (OTC), 689
Ozone (O_3), 337–338

P
Palladium (Pd), 180–181, 290–291
Patulin (PAT), 324–325
Peak CO production rate (PCO production rate), 712
Peak heat release rate (PHRR), 712
Peak specific extinction area (PSEA), 734–735
Pellet-based systems, 150–151
Pentaerythritol (PER), 710
Perfluorosulfonic acid (PFSA), 199
Perovskite solar cells (PSCs), 211–212
  device, 228
  MOF composites in, 228–236
    electron transport layer, 230–231
    hole transport layer, 231–232
    interface layer, 232–236
    PL, 229–230
Petrochemicals purification, 829
Pharmaceutical compounds
  adsorptive removal of, 624–626
  degradation of, 635–637
Pharmaceutical pollutants, 625
Pharmaceutical products, 609–610
1,4-phenylenediacetic acid, 843
Phomopsis, 341
Phosphate, 779
Phosphate-buffered saline (PBS), 69, 778, 863–869
Phosphazine (PZN), 715
Phosphonitrilic chloride trimer (PCT), 728
Phosphoric acid fuel cell (PAFC), 179–180
Phosphorous-based flame retardants, 710–711
Phosphorus (P), 731–732
Photoanode, 211–212, 217
Photocatalysis, 486–487, 610–611, 618, 621–622, 919
Photocatalysis-extraction oxidation desulfurization system (PEODS), 468
Photocatalysts, 441–442
Photocatalytic activity, 462
Photocatalytic bacteriostatic agents, 381–382
Photocatalytic coupling reaction, 460–461
Photocatalytic degradation
  factors influencing, 620–621
  mechanism, 618–620
Photocatalytic efficiency, 508
Photocatalytic hydrogen production, 457
Photocatalytic oxidative desulfurization (PODS), 580–581
  MOF composites for PODS of fuel oils, 580–584
Photocatalytic reduction, 637–638
Photocatalytic system, 620
Photocatalytic water splitting, 282
Photochemical reduction, 549
Photodegradation process, 508–510, 631–632
Photodynamic therapy (PDT), 397, 880
Photo electrons, 789–790
Photo-excited electrons, 508
Photoluminescence spectroscopy (PL spectroscopy), 71, 229–230, 466–467, 632–633
Photonic stop band (PSB), 672–673
Photoreduction route method (PR method), 462
Photothermal activation, 41, 869–873
Photothermal effect, 41
Photothermal therapy (PTT), 75, 397
Photovoltaic devices (PV devices), 211

Index 949

application of MOF composites in, 216–237
Photovoltaic technology, 212–214
  fundamentals of, 213–214, 214f
P-hydroxybenzoic acid (PHBA), 732–733
Physical adsorption, 615
Physisorption isotherm, 42–44
Phytic acid-arginine-cobalt shell, 730
Piperazine pyrophosphate (PAPP), 723–725, 730–731
Plasmonic nanoparticles (PNPs), 680–681
Platinum (Pt), 179–181, 217, 445–447
Platinum bimetallic electrocatalyst, 195
Platinum group metal (PGM), 181
Pohang University of Science and Technology (POST), 13–15
Point-of-care testing (POCT), 854–855
Pollutants, 638–639
  adsorption of, 621–626
  degradation of, 631–639
Pollution, 610–611
Poly(2-hydroxyethyl methacrylate) (PHEMA), 668
Poly(2,5-benzimidazole), 200–201
Poly(3,4-dihydroxy-L-phenylalanine) (PDopa), 624
Poly(3,4-ethylenedioxythiophene) (PEDOT), 111, 221, 692
Poly(4-styrene sulfonate) (PSS), 198–199
Poly (acrylic acid) (PAA), 818–820
Poly(arylene ether ketone sulfone), 201
Poly(benzimidazole), 198–199
Poly(ether-ether-ketone), 198–199
Poly(L-lactic acid) (PLLA), 723–725
Poly(methyl methacrylate) (PMMA), 380, 818
Poly(styrene sulfonate) (PSS), 692
Poly(styrene-ethylene-butylene-styrene), 198–199
Poly(vinyl alcohol), 198–199
Poly(vinylidene fluoride), 198–199
(Poly (3,4-ethylene dioxythiophene)) (PEDOT), 316
Polyacrylates (PA), 349
Polyacrylic acid (PAA), 781
Polyacrylonitrile (PAN), 149–150, 380, 628–629
Polyamic acid (PAA), 267
Polyamide (PA), 70
Polyamide 6 (PA6), 712
Polyaniline (PANI), 68–69, 99–100, 186–187, 259, 349, 686
Polybutylene succinate (PBS), 712
Polycarbonate (PC), 712

Polycyclic aromatic hydrocarbons (PAHs), 324, 681–683, 708–709
Polydimethyl-siloxane (PDMS), 23, 385, 675–677
Polydopamine (PDA), 151–152, 188, 479–480, 720–721, 786–787
Polyether sulfone (PES), 70, 140
Polyethylene dioxythiophene (PEDOT), 68–69
Polyethylene glycol (PEG), 350–351
Polyethyleneimine (PEI), 145, 627, 786–787
Polyethyleneimine-embedded ZIF-8 composite (PEI@ZIF-8), 146–147
Polyethyleneimine ethoxylated (PEIE), 237
Polyethylene terephthalate (PET), 810–811
Poly(2-diisopropylamino) ethyl methacrylate (PDPA), 869–873
Polyhedral oligomeric silsesquioxane (POSS), 715
Polyimide (PI), 818
Polylactic acid (PLA), 380, 712
  MOF composites applied in fire retardancy of, 722–725
Poly-L-glutamic acid (PLGA), 781
Polymer(s), 111, 255, 354–355, 385, 688, 707
  composites, 922
  MOF composites, 349–351
  MOF composites applied in fire retardancy of, 730–735
  polymer-assisted immobilized enzyme, 787–788
Polymer/MOF composites, 471
  catalytic performances of, 472–480
  common catalytic performances, 472–480
  photocatalysis, 480
  preparation methods of, 471
Polymer@metal-organic framework composites, 920
Polymyxins, 384
Poly ortho aminophenol (POAP), 841–842
Polyoxometalate metalloporphyrin organic frameworks (PMOFs), 556
Polyoxometalate/MOF composites, 488–489
  catalytic performances of, 489–498
  common catalytic performances, 489–493
  photocatalysis, 496–498
  preparation methods of, 489
Polyoxometalate@metal-organic framework composites, 920
Polyoxometalates (POMs), 2–3, 62, 278, 293–294, 439–440, 547, 572–574, 576–577, 818–820, 915–916, 920

composite, 293–294, 314, 554–556
Polyoxometalates-based metal-organic frameworks (POMOFs), 231, 293–294
Polyphenol, 779–780
Poly(3,4-ethylenedioxythiophene) polystyrene sulfonate (PEDOT:PSS), 261
Polypropylene (PP), 472–474
Polypyrrole (PPy), 68–69, 111, 183–184, 259
Polypyrrole nanoparticles (PPY-NPs), 733–734
Polysiloxanes@Cu-MOF (PS@Cu-MOF), 385
Polystyrene (PS), 712, 818
  MOF composites applied in fire retardancy of, 728
Polystyrene-co-acrylic acid (PSA), 662–663
Polyurea (PUA), 712
  MOF composites applied in fire retardancy of, 729–730
Polyurethane elastomer (PUE), 712
  MOF composites applied in fire retardancy of, 727–728
Polyurethane foams (PUFs), 828–829
Polyurethanes, MOF composites applied in fire retardancy of, 717–722
Polyvinyl alcohol (PVA), 663
Polyvinyl butyral (PVB), 823–824
Polyvinylidene fluoride (PVDF), 70, 824–825
Polyvinylpyrrolidone (PVP), 68, 79–80, 787–788, 818
Poor electrical conductivity, 110
Pore structure, 108–109
Porosity
  of metal-organic frameworks, 17–18
  surface area and porosity characterization, 64
Porous building block, 270–271
Porous carbon (PC), 71, 152
Porous coordination networks (PCNs), 9, 807
Porous coordination polymers (PCPs), 2, 374, 499–502, 613–614
Porous crystalline materials, 829
Porous graphene (PG), 265–266
Porous hybrid materials, 641–643
Porous liquids (PLs), 589–590
Porous macroscopic MOF-based building blocks, 271
Porous materials, 1, 97–98, 214, 356–357, 398, 613–614, 834
  composites with, 291–299
    MOF/covalent-organic framework composite, 295–296
    MOF/metal-organic framework composites, 297–299

MOF/polyoxometalates composites, 293–294
MOF on
Mofs, 2–3, 2f, 553–554
  activation of, 37–42
  characterization of, 42–45
  properties of, 17–26
  structures of, 9–16
  synthetic processes of, 26–37
  types of, 3–8
Porous MOF-based shielding constructs, 266–270
Porous proton-conducting materials, 26
Porous silica nanoparticle (PSN), 662–663
Porous structures, 282–283
Postsynthesis method, 503–504
Postsynthesis modification (PSM), 728
Postsynthetic methods, 17–18
Postsynthetic modification (PsSM), 24, 104
  in metal-organic frameworks, 104–105
  method, 37, 747
Powder X-ray diffraction (PXRD), 42–43, 167–168
Power conversion efficiency (PCE), 212–213
Power generation, 179
Precious metals, 179–180
Predesigning, 17–18
Prefabricated polymer ligands, integrating, 66–67
Pressure-swing adsorption (PSA), 135–136, 833
Presynthesized metals, 79–80
Presynthesized polymers, crystallizing metal-organic frameworks around, 68
Presynthetic modification (PrSM), 105
  in MOFs, 105–110
  tailoring MOFs for specific supercapacitor applications, 106–110
  topology, 105–106
Primary building units (PBUs), 101
Principal component analysis (PCA), 670, 749–751
Pristine graphene, 346
Pristine metal-organic framework, 547–548
Proenzymes, 791–792
Prominent industrial methods for MOF composites synthesis, 809–820
Proteins, 776
Proton
  conductivity, 26
  proton conductivity of MOFs, 26

Index   951

Proton exchange membrane fuel cell (PEMFC), 179–180
Prussian blue (PB), 119–120
Pseudo capacitors (PCs), 99
Pseudo first-order (PFO), 622–623
Pseudo second-order (PSO), 622–623
Pure metal-organic frameworks electrode for supercapacitor, 114–115
Pyrazolate, 613
Pyrazole, 12–13
Pyrolysin, 780–781

Q
Quantum dots (QDs), 2–3, 314, 658, 873–875
  composite, 314
Quantum yield (QY), 485–486
Quartz crystal microbalance (QCM), 684–685, 753
Quasisolid-state (QSS), 225
Quinoline yellow (QY), 462

R
Raman spectroscopy, 65, 545–546
Rate-determining step (RDS), 279–282, 543
Ratiometric fluorescence sensors, 666
Ratiometric sensors, 665–666
Reactive orange (RO16), 70
Reactive oxygen species (ROS), 380
Reactive red (RR120), 70
Reduced graphene oxide (rGO), 71, 73–74, 99–100, 181, 218, 264, 312–313, 346, 448–449, 719, 747–748
Reduced graphene oxide aerogel (rGA), 823
Reflux method, 29
Refractive index (RI), 673–675
Relative humidity (RH), 146–147
Relative standard deviations (RSDs), 321–322
RE, metal-organic framework composites in, 224–228
Reticular chemistry, 61
Retinol palmitate, 322–324
Reverse osmosis, 610–611
Reversible hydrogen electrode (RHE), 182, 277–278, 455
Rhodamine B (RhB), 76–77, 666
Rhodium (Rh), 290–291
Rigid polyurethane foam (RPUF), 712
Robust enzymes, 776–777
Room temperature synthesis method, 30
Ruthenium (Ru), 222–223, 277–278

S
Sacrificial templating method, 775–776

Sagnac interferometer (SI), 675–677
Salicylic acid (SA), 625
Salt metal-organic framework composites, 348–349
Scalability, 924–925
Scanning electron microscopy (SEM), 42, 44, 63, 140, 167–168
Secondary building unit (SBU), 9, 61, 105, 673, 804
Secondary metal elements, 106–107
Secondary organic aerosols (SOA), 337–338
Seeded growth method, 72
Self-sustaining combustion process, 707–708
Semiconductor, 618–619
  industry, 845
  materials, 213–214
  metal oxides, 688
Semimetal-organic framework composites, 348–349
Sensing applications, 923
Sensor, 318, 657, 845–863
  point-of-care testing sensors, 854–855
  smartphone sensors, 848–854
  turn-on processes, 847–848
  wearable diagnostic sensors, 857–863
Seoul National University (SNU), 13–15
Sepiolite (SEP), 718–719
Shielding effectiveness (SE), 257
Short side chain (SSC), 199
Signal-to-noise ratio (S/N ratio), 690–692
Silica composite, 313
Silver (Ag), 380, 548
  Ag-based metal-organic frameworks, 388–395
  Ag-based MOFs, 114–115
  nanoparticles, 455
Single atom catalysis (SACs), 547–548, 837
Single-crystal Ordered Macropore (SOM), 479–480
Single crystal X-ray diffraction (SCXRD), 42
Single-layer graphene (SLG), 854–855
Single-mode coreless single-mode (SCS), 679
Single-mode fiber (SMF), 677
Single-site heterogeneous catalysts (SSHCs), 837
Single-stranded DNA (ssDNA), 67–68
Single-walled CNTs (SWCNTs), 299–301
Slow evaporation
  method, 31–32
  process, 31–32
Smartphone
  sensors, 848–854

952   Index

smartphone-based spectroscopic
  approaches, 853–854
Smart sensors, 745–746
Smoke factor (SF), 712
Smoke production rate (SPR), 712
Sodium alginate monomers, 818–820
Sodium hydroxide (NaOH), 813–817
Sodium hypochlorite (NaClO), 593
Solar cells, 212, 921–922
  technology, 211
Solar-photocatalysis, 635
Sol-gel method, 216
Solid grinding technique, 79
Solid oxide fuel cell (SOFC), 179–180
Solid phase mechanism, 710–711
Solid-state batteries (SSBs), 843–844
Solid-state electrolytes (SSEs), 843
Solid-state nuclear magnetic resonance
  spectroscopy (SS-NMR spectroscopy),
  42, 45
Solid-state resistive devices, 755
Solvent, 810–811
  exchange activation, 38–39
  free gas phase loading technique, 78–79
  free mechanochemical method, 346–347
Solvent-assisted ligand incorporation (SALI),
  104–105, 828
Solvo thermal method, 27–29, 102, 105, 116,
  216, 635–636, 641
Solvothermal synthesis, 412
Sonochemical method, 33
Sonochemistry method, 103
Sono crystallization technique, 625–626
Sorbent-assisted water harvesting systems, 875
Sorption-based atmospheric water
  generators, 875
Space-time-yield (STY), 809
Specific shielding effectiveness (SSE), 257
Spherical polyacrylonitrile beads, 157–158
Spray-drying method, 35, 813–817
Stable materials, 916–917
Staphylococcus aureus, 341, 380, 854
Strontium metal-organic framework (Sr-
  MOF), 630
Strontium MOF (Sr-MOF), 312
Sulfonated Co-based MOF (sCo-MOF),
  200–201
Sulfonated membrane using polysulfone
  (sPSF), 199–200
Sulfonated poly(ether ether ketone)
  (SPEEK), 382
Sulfonated polyimide, 198–199
Sulfonic groups (–SO$_3$H), 199–200

Sulfoxide, 576–577
Sulfur compounds, 591
Sulfur dioxide (SO$_2$), 583, 749–751, 826–827
  capture of SO$_2$, 826–827
Supercapacitor (SC), 68–69, 99, 841–845, 921
  applications of MOFs and MOF composites
    in, 114–122
  MOFs composites as electrode
    materials, 115–122
  pure MOFs electrode for, 114–115
  tailoring MOFs for specific supercapacitor
    applications, 106–110
  design of active sites, 106–108
  interface engineering, 108
  morphologies, 109–110
  pore structure, 108–109
  types of MOF composites for, 111–114
  MOFs/carbon-based materials
    composites electrode, 112–113
  MOFs/conductive polymers composites
    electrode, 111–112
  MOFs/material composites, 113–114
  MOFs/metal oxides composites
    electrode, 113
Supercritical carbon dioxide (ScCO$_2$), 37–40
Supercritical CO$_2$ (ScCO$_2$), 813–817
Superoxide radicals (O$_2^-$), 382
Supply chain management, 878–880
Surface acoustic wave (SAW) sensors,
  684–685
Surface-anchored, crystalline, and oriented
  metal-organic framework multilayers
  (SURMOFs), 315–316
Surface-enhanced Raman scattering (SERS),
  81–82, 322, 658
  based in MOF composites, 680–683
Surface plasmon resonance (SPR), 322, 462,
  680–681
Swift heavy ions (SHIs), 692

T

Tag, 37
Tannic acid, 779–780
Tarbiat Modares University (TMU), 13–15
Target molecules, 81–82
Techno-economic analysis, 809
Temperature-programmed desorption (TPD),
  155–156
Temperature swing adsorption (TSA),
  135–136, 833
Template method, 35–36
Template protection sacrifice method (TPS
  method), 465–466

Index 953

Template synthesis technique, 79–80
Terephthalic acid, 843, 863
Tertbutyl decarbonate (Boc), 149–150
Tert-butylpyridine (tBP), 224–225
Tetracycline (TC), 75, 322, 396, 579–580, 635–636, 670
Tetraethylenepentamine, 145–146
2,2',7,7'-Tetrakis(N,N-di(4-metoksifenil) amino)-9,9'-spirobifluoren, 231
Tetrakis (4-carboxyphenyl) porphyrin (TCPP), 221, 789–790
3,3',5,5'-tetramethylbenzidine (TMB), 659
Tetra-tertbutylpyridine (TBP), 231
Tetrathiafulvalene (TTF), 689
Tetrazole (Httz), 12–13
Theoretical density functional theory calculations, 291
Thermal stability, 717–718
  of MOFs, 23
Thermogravimetric analysis (TGA), 38, 42, 44, 747
Thermoplastic polyurethane (TPU), 712
Thin film
  solar cells, 216
  on substrate, 315–316
Thin-film composite membrane (TCM), 70
Thioflavin T (ThT), 670
Third-generation solar cells, 212
Three-dimension (3D), 3
  carbon/MOF composites, 443–445
  ceramics, 783–784
  conductive network, 265
  configurations, 103–104
  LDH, 777–778
  materials, 265–266, 658
  MOFs, 4–5
  three-dimensional MOFs with, 7
Three-dimensional carbon nanostructures (3DCNS), 268–269
Three-dimensional graphane-based network (3DGN), 218
Titanium (Ti), 277–278
Titanium oxide, 919
  nanoparticles, 782
Titanium (II) oxide, 219–220
Toluene, 27–29
Top-down method, 554–555
Total heat release (THR), 712
Total organic carbon (TOC), 76–77, 635–636
Total shielding effectiveness (SE_T), 257
Total smoke release (TSR), 712
Total suspended solids (TSS), 70
Traditional catalysts, 318

Traditional inorganic flame retardants, 711–712
Traditional shielding materials, 922
Trametes versicolor, 341
Transglutaminase 2 (TGM2), 854
Transition metal (TM), 544
Transition metal hydroxide, 118
Transition metal oxides, 117–118
Transmission electron microscopy (TEM), 42, 45, 63–64, 164, 470
1,2,3-triazole, 12–13
1,2,4-triazole (Htz), 12–13
Triethanolamine (TEA), 135–136
Triethylamine (TEA), 460–461, 813–817
Triethylenediamine (TEDA), 659
Triethylene tetramine lactate, 164–165
Triethyl phosphate (TEP), 722–723
Triiodide, 217
Tris(2-aminoethyl) amine (TAEA), 104–105
Tris (1,10-phenanthroline) iron (II) perchlorate (TPIBP), 184–185, 192
Tungsten trioxide, 576
Turn-on processes, 847–848
Turnover frequency (TOF), 299–301, 472–474, 596
Turnover numbers (TONs), 293–294
Twin-screw extrusion (TSE), 813–817
Two-dimension (2D), 3
  configurations, 103–104
  hierarchical structures, 630–631
  materials, 286–288, 658, 926
  MOF nanosheets, 6–7
  MOFs, 4, 232
  nonmetallic semiconductor g-C_3N_4, 552–553
  ratiometric luminescence approach, 671–672
  structure, 781
  two-dimensional MOFs with, 6–7, 7f

U

Ultra-capacitors, 841
Ultrasonic separation, 813–817
Ultrasound-assisted technique, 33
Ultraviolet-Vis spectroscopy, 635–636
University of Oslo (UiO), 13–15
University of Oslo-66, 155
University of Oslo-67, 347
Unsaturated polyester resin (UPR), 712
  MOF composites applied in fire retardancy of, 725–727
Upconversion nanoparticles (UCNPs), 873–875

Urea coating, 189
Uric acid (UA), 689
UV-visible light, 640

V

Valence band (VB), 217–218, 282, 448–449, 580–581, 618–619
Vapochromism sensors based on metal-organic frameworks, 658–665
Vector network analyzer (VNA), 256–257
Vinyl resin, 732–733
Viscose fibers (VF), 80–81
Vitamin A palmitate, 322–324
Volatile organic compounds (VOCs), 324, 337–338, 657, 745–746, 828–829, 922–923
  adsorption in MOF composites, 354–355
  capture of, 828–829
  definitions, 339
  disadvantages and possible costs, 357–358
  less noticed source of, 341
  MOF composites
    for volatile organic compounds removal, 343–351
    recycle, 358
    parameter of volatile organic compounds adsorption, 355–357
    dynamic sorption methods, 356–357
    thermodynamic parameters, 355–356
  strategies for reduction and elimination of, 341–342
  suitable MOFs for volatile organic compounds removal, 342–343
    adsorption of volatile organic compounds by MOFs, 343
  variety and hazardous of, 339–341
Volatile organic solvents (VOSs), 671–672
Volmer reaction, 279
Volumetric hydrogen capacity, 833

W

Wafer-based solar cells, 212
Wastewater treatment, 609–611
  current state of wastewater treatment, 610–611
  industrial waste and water pollution, 609–610
Water (H$_2$O), 445–447
  pollution, 609–610
  treatment, 617–618
  vapor, 875
Water splitting, 277–278, 282–284, 299–301, 922
  fundamentals of, 279–282
  electrocatalytic water splitting, 279–282, 280f
  photocatalytic water splitting, 282
  merit of MOFs for
    challenges and strategies water-splitting over MOF-based electrocatalysts, 284
    extremely high porosity combined with facile access to active sites, 282–283
    impact of ligand engineering, 283–284
    optimizing metal nodes' performance, 283
Water-stable metal-organic frameworks (WMOFs), 638
Wearable diagnostic sensors, 857–863
Wearable sensors, 857
Wettability, 629–630
White-light-emitting colors (WLE colors), 671–672
Working electrode (WE), 217
  metal-organic framework composites as, 217–221
World Intellectual Property Organization (WIPO), 61
Wound healing-related applications, MOF-enzyme composites for, 790–791

X

X-ray diffraction (XRD), 42–43, 63, 466–467, 635–636, 747
  powder X-ray diffraction, 43
  single-crystal X-ray diffraction, 42
X-ray photoelectron spectroscopy (XPS), 64–65, 164, 466–467, 545–546

Z

Zeolite-based MOFs, 758
Zeolite imidazolate framework-8 (ZIF-8), 218
Zeolite imidazole framework-8 (ZIF-8), 107–108, 163, 182, 185–186, 192
  MOF structure, 232
Zeolite-like metal-organic framework (ZMOFs), 9, 11, 12f
Zeolite/metal-organic framework composites, 502–503
  catalytic performances of, 504–510
    common catalytic performances, 504–508
    photocatalysis, 508–510
  preparation methods of, 503–504

Zeolitic imidazolate framework-8 (ZIF-8), 380, 715
　nanostructures, 788–789
　NPs, 396–397
Zeolitic imidazolate framework-67 (ZIF-67), 322, 857–862
Zeolitic imidazolate framework-based metal-organic frameworks, 396–397
Zeolitic imidazolate framework/graphene oxide (ZIF-67@GO), 312–313
Zeolitic imidazolate frameworks (ZIFs), 9, 11, 114–115, 574, 612–613
Zeolitic imidazole framework-8 (ZIF-8), 343
Zero-dimension (0D), 3
　configurations, 103–104
　metal-organic frameworks, 3
　particles, 62
Zinc (Zn), 231, 380
　Zn-based metal-organic framework, 639
　Zn-ox-mtz, 139–140
Zinc-based benzene dicarboxylate (Zn-BDC), 749–751
Zinc-based MOF, 229, 231
Zinc oxide (ZnO), 580–581, 919
Zinc oxide nanowires (ZnO NWs), 755
Zirconium (Zr), 380
　Zr-based metal-organic frameworks, 397–398
　Zr-based porphyrin MOF, 229
Zirconium-based MOF (Zr-MOF), 312–313, 472–474, 571

Made in Canada by LoginPOD
Powered by Publishers' Graphics
loginpod.ca